# 2008 13th International Power Electronics and Motion Control Conference

Poznan, Poland
1-3 September 2008

Pages 1555-2070

**IEEE Catalog Number:** CFP0834A-PRT
**ISBN 13:** 978-1-4244-1741-4

**Copyright © 2008 by The Institute of Electrical and Electronics Engineers, Inc.**
**All Rights Reserved**

*Copyright and Reprint Permissions:* Abstracting is permitted with credit to the source. Libraries are permitted to photocopy beyond the limit of U.S. copyright law for private use of patrons those articles in this volume that carry a code at the bottom of the first page, provided the per-copy fee indicated in the code is paid through Copyright Clearance Center, 222 Rosewood Drive, Danvers, MA 01923.

For other copying, reprint or republications permission, write to IEEE Copyrights Manager, IEEE Operations Center, 445 Hoes Lane, Piscataway, New Jersey USA 08854.  All rights reserved.

IEEE Catalog Number:     CFP0834A-PRT

ISBN 13:     978-1-4244-1741-4

ISSN:     2007906910

**Additional Copies of This Publication Are Available from:**

IEEE Service Center
445 Hoes Lane
Piscataway, NJ 08854
Phone:     (800) 678-IEEE
     (732) 981-1393
Fax:     (732) 981-9667
E-mail:     customer-service@ieee.org

# 2008 13th International Power Electronics and Motion Control Conference

Poznan, Poland
1-3 September 2008

IEEE Catalog Number: CFP0834A-POD
ISBN: 978-1-42441-741-4

# Table of Contents

**Electric Drive System for Automatic Guided Vehicles Using Contact-free Energy Transmission** ...... 1
*Marcel Jufer*

**State-of-the-Art High Power Density and High Efficiency DC-DC Chopper Circuits for HEV and FCEV Applications** ...... 7
*Atsuo Kawamura, Martin Pavlovsky, Yukinori Tsuruta*

**Current-Based Condition Monitoring of Electrical Machines in Safety Critical Applications** ...... 21
*Thomas G. Habetler*

**The Essence of Three-Phase AC/AC Converter Systems** ...... 27
*J. W. Kolar, T. Friedli, F. Krismer, S. D. Round*

**An Analysis on Turn-off Behaviour of 1.2kV NPT-CIGBT under Clamped Inductive Load Switching** ...... 43
*S.T. Kong, L.Ngwendson, M. Sweet, E.M. Sankara Narayanan*

**Turn-off behaviour of high voltage NPT- and FS-IGBT** ...... 48
*Hans-Guenter Eckel, Karl Fleisch*

**Exact Circuit Power Loss Design Method for High Power Density Converters Utilizing Si-IGBT/SiC-Diode Hybrid Pairs** ...... 54
*Kazuto Takao, Hiromichi Ohashi*

**A Forward Converter with a Monolithic Cascode Device: Design and Experimental Investigation** ...... 61
*F. Chimento, S. Musumeci, A. Raciti, L. Abbatelli, S. Buonomo, R. Scollo*

**Switching and conducting performance of SiC-JFET and ESBT against MOSFET and IGBT** ...... 69
*André Knop, W. Toke Franke, Friedrich W. Fuchs*

**In-Service Life Consumption Estimation in Power Modules** ...... 76
*Mahera Musallam, C Mark Johnson, Chunyan Yin, Hua Lu, Chris Bailey*

**Measurement Of Temperature Sensitive Parameter Characteristics Of Semiconductor Silicon And Silicon-Carbide Power Devices** ...... 84
*Mietek Nowak, Jacek Rabkowski, Roman Barlik*

**Unsymmetrical Gate Voltage Drive for High Power 1200V IGBT4 Modules Based on Coreless Transformer Technology Driver** ...... 88
*Piotr Luniewski, Uwe Jansen*

**A Novel RESURFed Double Gates IGBT with Superior Performance** ...... 97
*Dongming Wu, Kaihang Li, Lingling Yang*

**An Empiric Approach to Establishing MOSFET Failure Rate Induced by Single-Event Burnout** ...... 102
*Jeroen van Duivenbode, Bart Smet*

**Comparative Study on Paralleled vs. Scaled Dc-dc Converters in High Voltage Gain Applications** ...... 108
*Pawel Klimczak, Stig Munk-Nielsen*

**A Low-Loss Dc-Dc Converter For A Renewable Energy Converter** ...... 114
*David S. Thompson, Otu A. Eno*

**A Single Active Edge-Resonant Snubber Cell-assisted ZCS Half-Bridge DC-DC Converter with Constant Frequency Asymmetrical PWM Scheme** ...... 119
*Tomokazu Mishima, Mutsuo Nakaoka, Eiji Hiraki*

**A New Approach to High Efficiency in Isolated Boost Converters for High-Power Low-Voltage Fuel Cell Applications** ...... 127
*Morten Nymand, Michael A.E. Andersen*

**New Modulation Strategy with Low Switching Frequency and Minimum Baseband Distortion** ...... 132
*N. E. Ruger, O. Schnick, W. Mathis, A. Mertens*

# Table of Contents

**A Bit-Stream Based PWM Technique for Variable Frequency Sinewave Generation** ...... 139
*N. D. Patel, U. K. Madawala*

**Control Strategies of the Quasi-Resonant DC-Link Inverter** ...... 144
*Slawomir Mandrek, Piotr J. Chrzan*

**Consideration for Input Current-Ripple of Pulselink DC-AC Converter for Fuel Cells** ...... 148
*Kentaro Fukushima, Tamotsu Ninomiya, Masahito Shoyama, Isami Norigoe, Yosuke Harada, Kenta Tsukakoshi*

**New Practical Approach to Input Current Shaping in AC-DC Power Converters** ...... 154
*Kuno Janson, Viktor Bolgov, Lauri Kütt, Ants Kallaste, Heigo Mölder*

**LLCC-PWM Inverter for Driving High-Power Piezoelectric Actuators** ...... 159
*Rongyuan Li, Norbert Fröhleke, Joachim Böcker*

**Modelling and Analysis of a Matrix-Reactance Frequency Converter Based on Buck-Boost Topology by DQ0 Transformation** ...... 165
*Pawel Szczeniak, Zbigniew Fedyczak, Marius Klytta*

**A Modular AC/DC Rectifier Based on Cascaded H-bridge Rectifier** ...... 173
*H. Iman-Eini, Sh. Farhangi, JL. Schanen*

**Low Loss Soft Switching Boost Converter** ...... 181
*So-Ri Park, Sang-Hoon Park, Chung-Yuen Won, Yong-Chae Jung*

**Methods for Experimental Assessment of Component Losses to Validate the Converter Loss Model** ...... 187
*Yi Wang, Sjoerd de Haan, Jan Abraham Ferreira*

**Modified multistage semiconductor-Fitch generator topology with magnetic compression** ...... 195
*Stanislaw Kalisiak, Marcin Holub*

**Modeling and Measuring Results of a Shunt Current Source Active Power Filter with Series Capacitor** ...... 201
*P. Parkatti, M. Salo, H. Tuusa*

**A Multi-Drive System Based on a Two-stage Matrix Converter** ...... 207
*Dinesh Kumar, Patrick W Wheeler, Jon C Clare, Lee Empringham*

**Characteristics of the Single Active Bridge Converter with Voltage Doubler** ...... 213
*Andreas Averberg, Axel Mertens*

**Analysis of Capacitor Dividers for Multilevel Inverter** ...... 221
*Oleg Sivkov, Jiri Pavelka*

**Space Vector Modulation for a Capacitor Clamped Multi-level Matrix Converter** ...... 229
*Xu Lie, Jon C. Clare, Patrick W. Wheeler, Lee Empringham*

**New Family of Matrix-Reactance Frequency Converters Based on Unipolar PWM AC Matrix-Reactance Choppers** ...... 236
*Zbigniew Fedyczak, Pawel Szczesniak, Igor Korotyeyev*

**Consideration of Conduction Losses for the Series Resonant Converter by Means of a Simple Extension to the SPA Approach** ...... 244
*Alexander Bucher, Thomas Duerbaum, Daniel Kuebrich, Markus Schmid*

**Validation and Comparison of different PWM Converter Small Signal Models** ...... 250
*Alexander Bucher, Markus Schmid, Lukas Bendkowski, Thomas Duerbaum*

**Dynamic Behaviour of a Series - Connected Multilevel Converter with Interleaved Switching** ...... 256
*C. Fahrni, A. Rufer*

**Simple Analysis of a Flying Capacitor Converter Voltage Balance Dynamics for DC Modulation** ...... 260
*A. Ruderman, B. Reznikov, M. Margaliot*

**Simulation of Simplified Seven Level Multilevel Converter Circuit** ...... 268
*Gerardo Ceglia, Víctor Guzmán, Carlos Sánchez, Fernando Ibáñez, Julio Walter, María Giménez*

# Table of Contents

**SEPP High-Frequency Inverter Incorporating an Auxiliary Switch and Its Performance Evaluation** ........................ 275
*H.Ogiwara,Y.Fujita, R.Urabe, M.Itoi, T.Sugai, M. kuwata, M.Nakaoka*

**Multiphase coupled converter models dedicated to transient response and output voltage regulation studies** ............... 281
*Nadia Bouhalli, Marc Cousineau, Emmanuel Sarraute, Thierry Meynard*

**A 13.56 MHz Current-output-type Inverter Utilizing An Immittance Conversion Element** ........................ 288
*Yosei Sakamoto, Keiji Wada, Toshihisa Shimizu*

**Voltage Fed Zero-Voltage Zero-Current Switching PWM DC-DC Converter** ........................ 295
*Jaroslav Dudrik, Vladimír Ru1scin*

**PWM Spectrum Evaluation and Over-Modulation Phenomena in a Three-Phase Inverters - Analytical Approach** ........................ 301
*Miro Milanovic*

**Experimental Study of a Matrix Converter Excited Doubly-Fed Induction Machine in Generation and Motoring** ........................ 307
*Ivan Shapoval, Jon Clare, Eduard Chekhet*

**Effect of Type and Interconnection of DG Units in the Fault Current Level of Distribution Networks** ........... 313
*H.R. Baghaee, M. Mirsalim, M. J. Sanjari, G.B. Gharehpetian*

**An Isolated Full-Bridge DC/DC Converter..with Bidirectional Communication Capability** ............... 320
*Lon-Kou Chang, Ru-Shiuan Yang*

**Efficiency and Power Losses in PM BLDC Motor with Variable Bridge/half-bridge Structure Electronic Commutator** ........................ 326
*K. Krykowski, A. Bodora*

**Analysis of a device for converting a unipolar input voltage into two symmetric bidirectional output voltages with a magnetically coupled coil** ........................ 331
*Felix. A. Himmelstoss, Wilhelm Kraeftner*

**Invariant Modulation Strategy for Two-stage Direct Power Converter** ........................ 337
*Radiy Bekbudov*

**Experimental Study of A Multicell ac/ac Converter Balancing Circuit** ........................ 345
*Robert Stala, Andrzej Mondzik*

**A Comparison and Optimum Design of Reluctance-Controlled Classical Load-Resonant Converters** ........ 350
*Stefan V. Mollov, Michael P. Theodoridis*

**Capacitor Clamped Multilevel Matrix Converter Controlled with Venturini Method** ........................ 357
*Janina Rzasa*

**Reliability Consideration for a High Power Zero-Voltage-Switching Flyback Power Supply** ............... 365
*Arash Rahnamaee, Jafar Milimonfared, Kaveh Malekian, Mohammad Abroushan*

**The Traction Drive Topology Using the Matrix Converter with Middle-Frequency Transformer** ............... 372
*Martin Pittermann, Pavel Drábek, Marek Cédl*

**Analysis of Multipulse Rectifiers with Modulation in DC Circuit in Vector Space Approach** ................ 377
*Andrzej KAPLON and Jaroslaw ROLEK*

**High Efficiency Soft Switching Boost Converter for Photovoltaic System** ........................ 383
*Gil-Ro Cha, Sang-Hoon Park, Chung-Yuen Won, Yong-Chae Jung, Sang-Hoon Song*

**A Power Converter For Fault Tolerant Machine Development In Aerospace Applications** ........................ 388
*Liliana de Lillo, Patrick Wheeler, Lee Empringham, Chris Gerada, XiaoyanHuang*

**Optimal Bus Capacitance Design for System Stability in On-Board Distributed Power Architecture** ........... 393
*Seiya Abe, Masahiko Hirokawa, Masahito Shoyama, Tamotsu Ninomiya*

# Table of Contents

**Steady State Analysis of Hysteretic Control Buck Converters** ....................................................... 400
*L.K. Wong, T.K. Man*

**A Novel Control Method for IGBT Current Source Rectifier** ........................................................ 405
*Longcheng Tan, Yaohua Li, Ping Wang, Congwei Liu, Zixin Li, Yonggang Chen, Wei Xu*

**A procedure to optimize the inductor design in boost PFC applications** .......................................... 409
*Florent Liffran*

**Electric Vehicle Drive Inverters Simulation Considering Parasitic Parameters** ................................ 417
*Wen Huiqing, Liu Jun, Zhang Xuhui, Wen Xuhui*

**DC-DC Converters with FPGA Control for Photovoltaic System** .................................................. 422
*Jan Leuchter, Pavel Bauer, Vladimir Rerucha, Petr Bojda*

**Control of a Converter with Superconductive Energy Storage Inductor** ......................................... 428
*Rozanov Yurie Konstantinovich, Lepanov Michail Gennadevich, Kiselev Michail Gennadevich*

**FPGA-based Controllers for Switching Converters** .................................................................... 432
*Karel Jezernik*

**Gamesa DAC converter: the way for REE grid code certification** ................................................. 437
*Itziar Martinez, Daniel Navarro*

**Flatness-Based Voltage-Oriented Control of Three-Phase PWM Rectifiers** ................................... 444
*J. Dannehl, F.W. Fuchs*

**Control of a single phase H-Bridge multilevel inverter for grid-connected PV applications** ............. 451
*Elena Villanueva, Pablo Correa, Jose Rodriguez*

**Switching and Voltage Controls for a Flyback Switch-Mode Rectifier** .......................................... 456
*Yuan-Chih Chang, Chang-Ming Liaw*

**Method Of Designing ZVS Boost Converter** ............................................................................ 463
*Miroslaw Luft, Elzbieta Szychta, Leszek Szychta*

**A New DC-DC Converter with Multi Output: Topology and Control Strategies** ............................. 468
*Arash A Boora, Firuz Zare, Gerard Ledwich, Arindam Ghosh*

**Maximum Frequency for Hysteretic Control COT Buck Converters** ............................................. 475
*L.K. Wong, T.K. Man*

**Current Control Method Based on Hysteresis Control Suitable for Single Phase Active Filter with LC Output Filter** .............................................................................................................. 479
*Yukinori Kobayashi, Hirohito Funato*

**Optimal Slope Compensation for step load in peak current controlled dc-dc Buck Converter** .......... 485
*Susovon Samanta, Pradipta Patra, Siddhartha Mukhopadhyay, Amit Patra"*

**Performances of a PLL Based Digital Filter for double-conversion UPS** ...................................... 490
*Armando Bellini, Stefano Bifaretti*

**10A 12V 1 chip digitally-controlled DC/DC converter IC with high resolution and high frequency DPWM**............. 498
*Kazutoshi Nakamura\*, Toshiyuki Naka\*, Yuki Kamata\*, Toyoki Taguchi, Takaaki Shimizu, Yoshiko Ikeda, Akio Nakagawa, Dragan Maksimovic*

**Modelling and Modulation of Voltage Source Converter** ............................................................ 504
*Grzegorz Radomski*

**Sliding Mode Control of DC/DC Multiphase Power Converters** .................................................. 512
*Vadim Utkin*

**A New Digital Control Method for High Performance 400 Hz Ground Power Unit** ........................ 515
*Zixin Li, Ping Wang, Haibin Zhu, Yaohua Li, Longcheng Tan, Yonggang Chen, Fanqiang Gao*

# Table of Contents

**Single-phase 50-kW 16.7-Hz Four-Quadrant Line-Side Converter for Railway Traction Application** ....... 521
*C. Heising, R. Bartelt, V. Staudt, A. Steimel*

**Technique to Improve IGBT Converter Efficiency and Transient Response** ................................................ 528
*Robert W. Turner, Simon Walton, Richard Duke*

**The control of voltage converter rectifiers** ......................................................................................................... 536
*Krzysztof Szubert*

**Load Voltage Regulation and Line Loss Minimization of Loop Distribution Systems Using UPFC** ............ 542
*Mahmoud A. Sayed, Takaharu Takeshita*

**Control of Traction Single-Phase Current-Source Active Rectifier under Distorted Power Supply Voltage** ............... 550
*Jan Michalík, Jan Molnár, Zdenck Peroutka*

**Simulation Model Of Neural Network Based Synchronous Generator Excitation Control** ............................ 556
*Damir Sumina, Neven Bulic, Gorislav Erceg*

**Predictive Current Control of a 7-level AC-DC back-to-back Converter for Universal and Flexible Power
Management System** ............................................................................................................................................... 561
*Stefano Bifaretti, Pericle Zanchetta, Florin Iov, Jon C. Clare*

**Predictive Stator Current Control For Three-Level Voltage-Source Inverters With Output LC-Filters** ..... 569
*Tomasz Laczynski, Axel Mertens*

**Research on Dimming Control Method of Electronic Ballast for the Automotive HID Headlight** ............... 576
*P. Dong, K.W.E.Cheng, S.L.Ho*

**Control Method for a Three-Port Interface Converter Using an Indirect Matrix Converter with an Active
Snubber Circuit** ..................................................................................................................................................... 581
*Koji Kato, Jun-ichi Itoh*

**Precise Digital Control Method with Multi-rate deadbeat control for Single Phase Utility Interactive
Inverter with FPGA based Hardware Controller** ................................................................................................ 589
*Kenta Hayashi, Tomoki Yokoyama*

**A Digital Current Controller for Zero-Current Transition Bidirectional Converter** ...................................... 595
*Nobuyuki Kasa, Takahiko Iida*

**Control Method for a Single Phase Arbitrary Waveform-output Inverter** ....................................................... 600
*Satoshi Taniguchi, Keiji Wada, Toshihisa Shimizu*

**Elimination of Harmonics in Multilevel Inverters with Non-Equal DC Sources Using PSO** ....................... 606
*A. K. Al-Othman, Tamer H. Abdelhamid*

**Improved PFC Circuit Having Ladder Type Filter with Only Passive Devices** ............................................... 614
*Kenji Ando, Keiju Matsui, Nobuhito Takeuchi, Masaru Hasegawa*

**Fuel Cell Current Ripple Minimization using a bi-Buck Power Interface** ....................................................... 621
*Nicu Bizon, Marian Raducu, Mihai Oproescu*

**Power Control Strategy of Parallel Inverter Interfaced DG Units** .................................................................... 629
*H.R. Baghaee, M.Mirsalim, M. J. Sanjari, G.B. Gharehpetian*

**Implementation of Nonlinear power flow controllers to control a VSC** .......................................................... 637
*Nelson L. Díaz, Fabián H. Barbosa, Cesar L. Trujillo*

**Harmonic Distortion Reduction Technique for Uninterruptible Power Supplies with DC Voltage Boost
Technique** ................................................................................................................................................................ 643
*Juei Lung Shyu*

**Energy-based Modulation Error Control for High-Power Drives with Output LC-Filters and
Synchronous Optimal Pulse Width Modulation** ................................................................................................. 649
*Tomasz Laczynski, Timur Werner, Axel Mertens*

*vii*

# Table of Contents

**Voltage Harmonic Control of Z-source Inverter for UPS Applications** ........................... 657
*Arkadiusz Kulka, Tore Undeland*

**A Method of Optimal Control for Switched-Mode Power Converters** ........................... 663
*Anatoly Bekishev, Albert Iskhakov, Leonid Klyachko, Vladimir Pospelov, Sergey Skovpen*

**Experiment results with modified Hybrid PWM method for three phase induction motor** ........................... 669
*Daniel Lewandowski, Grzegorz Lisowski*

**Optimized Design of a Delay line based Analog to Digital Converter for Digital Power Management Applications** ........................... 674
*Mukti Barai, Sabyasachi Sengupta, Jayanta Biswas*

**Overmodulation Region of Multi-Phase Inverters** ........................... 682
*S. Halasz*

**Optimal Control of Induction Motor Using High Performance Frequency Converter** ........................... 690
*Jerkovic Vedrana, Spoljaric Zeljko, Valter Zdravko*

**Power Electronic Converter for the Reluctance Pump Drive** ........................... 695
*B. J. Szymanski, K. Kompa, N. Michalke, H. Kuß, U. Schuffenhauer*

**A Predictive Control Scheme for Current Source Rectifiers** ........................... 699
*Pablo Correa, Jose Rodriguez*

**Analysis and Design of New Switching Table for Direct Power Control of Three-Phase PWM Rectifier** ........................... 703
*Abdelouahab Bouafia, Jean-Paul Gaubert, Fateh Krim*

**Improvement of the performance for DC-DC Converter** ........................... 710
*X..She, Yun She*

**A Drive System With High-Speed Single-Phase Supplied Three-Phase Induction Motor** ........................... 714
*T. Binkowski, M. Grad, M. Latka, W. Malska, D. Sobczynski*

**A Pulse Width Modulation Technique for a Multilevel Converter in High Voltage High Frequency Applications** ........................... 718
*Jafar Adabi, Hamid Soltani, Firuz Zare*

**Bidirectional Positive Buck-Boost Converter** ........................... 723
*Arash A Boora, Firuz Zare, Gerard Ledwich, Arindam Ghosh*

**Control system of power electronics current modulator utilized in diode rectifier with sinusoidal source current** ........................... 728
*Michal Gwózdz, Michal Krystkowiak*

**Design and control of a half-bridge converter to drive piezoelectric actuators** ........................... 731
*Oriol Gomis-Bellmunt, Josep Rafecas-Sabate, Daniel Montesinos-Miracle, Josep-Maria Fernandez-Mola, Joan Bergas-Jane*

**Online Diagnosis of PEM Fuel Cell** ........................... 734
*Abdellah Narjiss, Daniel Depernet, Denis Candusso, Frederic Gustin, Daniel Hissel*

**Application of Kalman filters to the control of independent power electronic voltage sources** ........................... 740
*Ryszard Porada, Lukasz Nyczkowski*

**Verification of the load sharing characteristics in Autonomous Decentralized UPS system using FPGA based Hardware Controller** ........................... 744
*Nobuaki Doi, Tsuyoshi Saito, Tomoki Yokoyama*

**Fault Current Reduction in Distribution Systems with Distributed Generation Units by a New Dual Functional Series Compensator** ........................... 750
*H.R. Baghaee, M. Mirsalim, M. J. Sanjari, G.B. Gharehpetian*

**Dynamic Simulation of PM Motor Drive System based on Reluctance Network Analysis** ........................... 758
*Kenji Nakamura, Osamu Ichinokura*

# Table of Contents

**Performance Improvement of Direct Torque Controlled Interior Permanent Magnet Synchronous Motor Drive by Considering Magnetic Saturation** ............................................................763
*Behrooz Majidi, Jafar Milimonfared, Kaveh Malekian*

**Condition Monitoring for Mechanical Faults in Fully Integrated Servo Drive Systems** ...................769
*Jesus Arellano-Padilla, Mark Sumner, Chris Gerada*

**Feed-forward Compensation of Load and Parameter Variations of Electric Drive** ........................776
*Alon Kuperman, Yoram Horen, Saad Tapuchi, Uri Suissa*

**Thermal Effect of Short-Circuit Current in Low Power Induction Motors** ....................................782
*Leo.s Beran*

**Generalized Model for a Class of Switched Reluctance Motors** .....................................................787
*Constantin Pavlitov, Yassen Gorbounov, Radoslav Rusinov, Alexandar Alexandrov, Kliment Hadjov, Dimitar Dontchev*

**Neural Network based Fault Detection of PMSM Stator Winding Short under Load Fluctuation** ...........793
*J. Quiroga, D.A. Cartes, C.S. Edrington, Li Liu*

**Review of Electrical Machine in Downhole Applications and the Advantages** ..............................799
*Anyuan Chen, Ravindra. B. Ummaneni, Robert Nilssen, Arne Nysveen*

**Broken Rotor Bar Impact on the Closed Loop and Sensorless Control of Induction Machine** .............804
*Piotr Kotodziejek, Elzbieta Bogalecka*

**Coupled Magnetic Circuit Method and Permeance Network Method Modeling of Stator Faults in Induction Machines** .............................................................................................................810
*Amin Mahyob, Mohamed Y. Ould Elmoctar, Pascal Reghem, Georges Barakat*

**Explosion Protected Electrical Drives - Risk Assessment and Technical Diagnostics** ..................818
*Ivica Gavranic, Drago Ban, Damirarko Zarko*

**The effect of subharmonics on induction machine heating** ...........................................................826
*Piotr Gnacinski, Marcin Peplinski, Mariusz Szweda*

**Influence of Saturation Effects in a Transverse Flux Machine** ......................................................830
*M. Siatkowski, B. Orlik*

**A Model of Semiconductor Converter-Fed Asynchronous Machines Taking into Account Energy Losses and Thermal Processes** ...........................................................................................837
*M. Pronin, O. Shonin, Y. Koskin, A. Vorontsov, P. Kalatchikov*

**Use of an AC Self-excited Switched Reluctance Generator as a Battery Charger** .........................845
*Abelardo Martínez, Estanislao Oyarbide, Javier Vicuña, Francisco Perez, Eduardo Laloya, Bonifacio Martín-del-Brío, Tomás Pollán, Beatriz Sánchez, Juan Lladó*

**Direct Thrust Controlled Linear Induction Motor Including End Effect** .......................................850
*Berrin Susluoglu, Vedat M. Karsli*

**Analysis of Short-Circuit Forces at the Top of the Low Voltage U-Type and I-Type Winding in a Power Transformer** ..................................................................................................................855
*Leonardo Strac, Franjo Kelemen, Damir Zarko, Josipa Mokrovica*

**Anisotropy Comparison of Reluctance and PM synchronous Machines for Low Speed Position Sensorless Applications** ..........................................................................................................859
*H.W. de Kock, M.J. Kamper, R.M. Kennel*

**Analysis of VSI-DTC Fed 6-phase Synchronous Machines** .........................................................867
*Ibrahim Abuishmais, Waqas M. Arshad, Sami Kanerva*

**Optimal Rotor Flux Shape for Multi-phase Permanent Magnet Synchronous Motors** ...................874
*Roberto Zanasi, Federica Grossi*

# Table of Contents

**Modelling of Electrical Machines Using the Modelica Bond-Graph Library** ....................................... 880
*Mieczyslaw Ronkowski*

**Induction Motor Parameters Identification using Genetic Algorithms for Varying Flux Levels**................ 887
*Konstantinos Kampisios, Pericle Zanchetta, Chris Gerada, Andrew Trentin, Omar Jasim*

**Study of the sudden symmetrical short-circuit using the mathematical models of the synchronous machine and the numerical methods**.................................................................................................................. 893
*Petropol Serb Gabriela, Petropol Serb Ion, Campeanu Aurel, Sonia Degeratu, Anca Petrisor*

**Analytical Method of Calculation of the Current and Torque of a Reluctance Stepper Motor Using Fourier Complex Series**.................................................................................................................................. 899
*Pavel Zaskalicky, Maria Zaskalicka*

**Bearing Damage Analysis by Calculation of Capacitive Coupling between Inner and Outer Races of a Ball Bearing**................................................................................................................................................... 903
*Jafar Adabi, Firuz Zare, Gerard Ledwich, Arindam Ghosh, Robert D.Lorenz*

**The Model of the Squirrel Cage AC Motor including Rotor Slot Harmonics**.......................................... 908
*Eleonora Darie, Costin Cepisca, Emanuel Darie*

**Identification of mathematical model induction motor's parameters with using evolutionary algorithm and multiple criteria of quality**.................................................................................................................. 912
*Hudy Wiktor, Jaracz Kazimierz*

**Simulation Study on Control of Ultrahigh Speed Drives in Waste Energy Recovery Systems**................ 916
*Péter Stumpf, Miklós G. Simon, Rafael K. Járdán, István Nagy*

**Adaptive Back EMF Parameter Adjustment of Simplified Vector Control for Position Sensorless Permanent Magnet Synchronous Motors**............................................................................................... 924
*Kiyoshi Sakamoto, Yoshitaka Iwaji, Daigo Kaneko, Toshihiro Takeuchi, Tsunehiro Endo, Atsuo Kawamura*

**Identification and Control of Precision XY Stages with Active Vibration Suppression System** ............. 932
*Mayumi Nitta, Seiji Hashimoto*

**Sensitivity of the Currents Input-Output Decoupling Vector Control of the DFIM versus Current Sensors Fault**....................................................................................................................................................... 938
*Meriem Abdellatif, Maria Pietrzak-David, Ilhem Slama-Belkhodja*

**Extended Back EMF model for PM synchronous machines with different inductances in d- and q-axis** ..................... 945
*Andreas Eilenberger, Manfred Schroedl*

**Gait generation of a two-legged robot by using adaptive network based fuzzy logic control** ................ 949
*Umit Onen, Mete Kalyoncu, Mustafa Tinkir, Fatihm. Botsali*

**Walking robot HEXOR® II - a versatile platform for engineering education**........................................ 956
*M. Sajkowski, T. Stenzel, B. Grzesik*

**Motion Control of Steel Sheet Shears with Rocking Knife Mechanism**............................................... 961
*Jan Fetyko, Frantisek Durovsky, Viliam Fedak*

**Intelligent Adaptive Control and Monitoring Of Band Sawing** ......................................................... 967
*Ilhan Asiltürk, Ali Ünüvar*

**Hierarchical adaptive network based fuzzy logic controller design for a single flexible link robot manipulator**.............................................................................................................................................. 974
*Mete Kalyoncu, Mustafa Tinkir*

**Digital Controlled High Speed Synchronous Motor** ........................................................................ 982
*Zdenk Cerovský, Jaroslav Novák, Martin Novák, Marek Cambál*

**Analysis of combustion engine - electric Linear generator set operation** ........................................... 988
*Jirí Pavelka*

*x*

# Table of Contents

**Closed Loop Control of AC Drive with LC Filter**......994
*Jaroslaw Guzinski*

**Sensorless IPMSM based drive for reciprocating compressor**......1002
*Anton Dianov, Kim Young-Kwan, Lee Sang-Joon, Lee Sang-Taek, Yoon Tae-Ho*

**Controlling system of electrodynamic drive**......1009
*Josef Cernohorský*

**Expert System for Electric Drive Design**......1017
*Juhan Laugis, Valery Vodovozo*

**Improvement of Moving Characteristics of Cableless Micro-actuator and Consideration of Reversible Motion**......1020
*Hiroyuki Yaguchi, Kazumi Ishikawa, Toshihiro Zamma, Koichi Funayama*

**Sensorless Control of AC Machines using High-Frequency Excitation**......1024
*Heiko Zatocil*

**Adaptive PF Speed Control of SRAM Drives**......1033
*Laszlo Szamel*

**A Very Simple Fuzzy Control System for Inverter Fed Synhronous Motor**......1040
*Pawel Fabijanski, Ryszard Lagoda*

**Distributed control system of DC servomotors for six legged walking robot**......1044
*D. Belter, K. Walas, A. Kasinski*

**Optimization of Starting Process of the Frequency Controlled Induction Motor**......1050
*I.Ya. Braslavsky, A.V. Kostylev, D.P. Stepanyuk*

**3-Axes Satellite Attitude Control Based on Biased Angular Momentum**......1054
*Azam Ghaedi, Mohammad Ali Nekoui*

**Modelling and simulation of a signal injection self-sensored drive**......1058
*Alen Poljugan, Mark Sumner, Chris Gerada, Qiang Gao*

**Robust PI Cascade Control for a Multi-Mass System Optimized by Evolutionary Algorithms**......1064
*M. Joost, K. Zielinski, B. Orlik, R. Laur*

**Permanent Magnet Synchronous Servo-Drive with State Position Controller**......1071
*Lech M. Grzesiak, Tomasz Tarczewski, Slawomir Mandra*

**Closed-Loop Control of Virtual FPGA-Coded Permanent Magnet Synchronous Motor Drives using a Rapidly Prototyped Controller**......1077
*Christian Dufour, Vincent Lapointe, Jean Bélanger, Simon Abourida*

**Speed Sensorless Nonlinear Control Of Induction Motor In The Field Weakening Region**......1084
*MiroslawWlas, Haithem Abu-Rub, Joachim Holtz*

**Comparison of Dynamic Performances of Speed Control System Containing Time - Minimal Speed Controller with Control System Containing PI Speed Controller**......1090
*Andrzej Andrzejewski, Marian Roch Dubowski*

**Optimisation of Real-Time Complex Path Generation in Constrained Intelligent Motion Applications Based on IPM Motor Drives**......1097
*Silverio Bolognani, Roberto Petrella, Fabio Stefanutti, Piero Stocco*

**PMSM Sliding Mode Observer for Speed and Position Estimation Using Modified Back EMF**......1105
*Ilioudis Vasilios C., Margaris Nikolaos I.*

**Optimal Control of Electrical Drives with Induction Motors for Variable Torques**......1111
*Corneliu Botan, Marcel Ratoi, Vasile Horga*

# Table of Contents

**An Optimal Control for Saturated Interior Permanent Magnet Linear Synchronous Motors Incorporating Field Weakening** .................................................................................................................... 1117
*Mohammad Abroshan, Jafar Milimonfared, Kaveh Malekian, Arash Rahnamaee*

**Improved Direct Torque Control for Induction Machine Drives using Fuzzy Logic and Particle Swarm Optimization** ........................................................................................................................ 1123
*Mohammad Mehdi Rezaei, Mojtaba Mirsalim, Kaveh Malekian*

**Design and Implementation of High Performance Full-Digital Spindle Drives** ........................... 1128
*Liu Yang, Zhao Jin*

**Semi hierarchical adaptive network based fuzzy logic controller design for a multi-straight-line path tracing flexible robot manipulator with rotating-prismatic joint** ................................................ 1132
*Mete Kalyoncu, Mustafa Tinkir*

**Control System with the Set Point Observation** ........................................................................... 1140
*Algirdas Baskys, Vitoldas Gobis, Valerijus Zlosnikas*

**Electropneumatic Servo System with Adaptive Force Controller** .................................................. 1144
*Arunas Grigaitis, Vilius Antanas Gele~evicius*

**New fault tolerant DTC control for induction machine drives** ...................................................... 1149
*A.Ben Abdelghani Bennani, M. Ghodbane Cherif, I. Slama Belkhodja*

**Stability Analysis of the Natural Field Orientation Controlled Induction Machine Drive** ........... 1155
*G. Mirzaeva, A. Rojas*

**Control of SR motor EV by instantaneous torque control using flux based commutation and phase torque distribution technique** ........................................................................................................... 1163
*Ayumu Nishimiya, Hiroki Goto, Hai-Jiao Guo, Osamu Ichinokura*

**Simulation of IPM Motor by Nonlinear Magnetic Circuit Model for Comparing Direct Torque Control with Current Vector Control** ........................................................................................................ 1168
*Hiroki Goto, Kensuke Kimura, Hai-Jiao Guo, Osamu Ichinokura*

**A Simplified Model for Induction Machines with Faults to Aid the Development of Fault Tolerant Drives** ............. 1173
*O. Jasim, C. Gerada, M. Sumner, J. Arellano-Padilla*

**About the Experimental Results of an Electric Driving System Based on Asynchronous Motor and PWM Converter** .................................................................................................................................. 1181
*Petre-Marian Nicolae, Dan-Gabriel Stanescu, Ioana-Gabriela Sîrbu*

**Real-World Force Feedback Control for Mobile-Hapto** ................................................................ 1187
*Wataru Yamanouchi, Yuki Yokokura, Seiichiro Katsura, Kiyoshi Ohishi*

**The new numerical integration routine applied in sensorless drives** ............................................ 1193
*Arkadiusz Gardecki, Krystyna Macek-Kaminska*

**Application of Fuzzy Logic Techniques To Robust Speed Control of PMSM** ............................... 1198
*Tomasz Pajchrowski, Krzysztof Zawirski*

**Optimal control of current commutation of high speed SRM drive** ............................................. 1204
*Jan Deskur, Tomasz Pajchrowski, Krzysztof Zawirski*

**Comparison Between Direct Torque Control and Vector Control of a Permanent Magnet Synchronous Motor Drive** ..................................................................................................................................... 1209
*Rafa Souad, Houcine Zeroug*

**Detection and self-tuning compensation of periodic disturbances by the control of DC motor** ....... 1215
*Michael Ruderman, Frank Hoffmann, Johannes Krettek, Torsten Bertram*

**A Linear Switched Reluctance Motor Based Position Tracking System** ....................................... 1221
*S. W. Zhao, N. C. Cheung, Y. Lu, W. C. Gan, Z. G. Sun*

# Table of Contents

**Mobile Robot Navigation with Obstacle Avoidance Capability** ........................................................ 1225
*Anca Sorana Popa, Mircea Popa, Ioan Silea*

**Requirements for Power Electronics in Solid Oxide Fuel Cell System** ........................................... 1233
*T. Riipinen, V. Väisänen, M. Kuisma, L. Seppä, P. Mustonen, P. Silventoinen*

**Power Supply for a IGBT-Driver with High Insulation Voltage based on a Printed Planar Transformers** .............. 1239
*Günter Schmitt, Wolf Kusserow, Ralph Kennel*

**Variable Motor Operating Point by Integration of Power Electronic Device into Rotor** ........................... 1243
*Adrian Tulbure, Hans-Peter Beck, Mircea Risteiu*

**Magnetic Material Comparisons for High-Current Gapped and Gapless Foil Wound Inductors in High Frequency DC-DC Converters** ........................................................................................... 1249
*Marek S. Rylko, Brendan J. Lyons, Kevin J. Hartnett, John G. Hayes, Michael G. Egan*

**Feasibility Study of Half- and Full-Bridge Isolated DC/DC Converters in High-Voltage High-Power Applications** ...................................................................................................... 1257
*Dmitri Vinnikov, Tanel Jalakas, Mikhail Egorov*

**Evaluation of Different Loss Calculation Methods for High-voltage IGBT-s Under Small Load Conditions** ........... 1263
*T. Jalakas, D. Vinnikov, J. Laugis*

**Control of Power Supply Unit for Military Vehicles Based on Four-Leg Three-Phase VSI with Proportional-Resonant Controllers** ................................................................................... 1268
*Tomál Glasberger, Zdenek Peroutka*

**Optimal Design of a Half Wave Cockroft-Walton Voltage Multiplier with Different Capacitances per Stage** ....................................................................................................... 1274
*Ioannis C. Kobougias, Emmanuel C. Tatakis*

**Calculation of Leakage Inductance of Core-Type Transformers for Power Electronic Circuits** .................... 1280
*Reinhard Doebbelin, Marcel Benecke, Andreas Lindemann*

**Enhanced Current Pulsation Smoothing Parallel Active Filter for Single Stage Grid-connected AC-PV Modules** ...................................................................................................... 1287
*A.C. Kyritsis, N.P. Papanikolaou, E.C. Tatakis*

**Outline of the Design of a Cascaded H-bridge Medium Voltage STATCOM** ........................................ 1293
*R.E. Betz, B.J. Cook, T.J. Summers, R. Fisher, A. Bastiani, S. Shao, P. Stepien, K. Willis*

**Investigation of High Frequency Effects on Layered Coils** ..................................................... 1301
*Georgios S. Dimitrakakis, Emmanuel C. Tatakis*

**Soft Switching PWM Inverter for Induction Heating Applied to Heating of Ferromagnetic Metal** .................. 1309
*Sachio Kubota, Muneo Sato, Fumio Ito, Yoshihiro Shimaoka, Kunihiro Nishioka*

**Corona Treatment System with Resonant Inverter - Selected Proprieties** ......................................... 1316
*Mucko Jan*

**Power supply unit for an electric discharge machine** ............................................................ 1321
*Wojciech Mysinski*

**High Power, High Voltage, High Frequency Transformer / Rectifier for HV Industrial Applications** .............. 1326
*T. Filchev, D. Cook, P. Wheeler, A. Van den Bossche, J. Clare, V. Valchev*

**Small Power Laboratory Model and High Power Prototype of the Four-Level VSI** .................................. 1332
*Ryszard Michal Strzelecki, Pawel Szczepankowski, Andrzej Kasprowicz, Genady Stepanovic Zinoviev, Krzysztof Zymmer, Zbigniew Zakrzewski*

**AC Voltage Regulator Using PWM Technique and magnetic flux distribution** ...................................... 1337
*A.M. Dabroom*

**Minimum Reactive Power Filter Design for High Power Converters** ............................................... 1345
*Alex-Sander Amavel Luiz, Braz Jesus Cardoso Filho*

# Table of Contents

**Injection of a carrier with higher than the PWM frequency for sensorless position detection in PM synchronous motors** ...... 1353
*Roberto Leidhold, Peter Mutschler*

**Parallel Fixed Point FPGA Implementation of Sensorless Induction Motor Torque Control** ...... 1359
*Jacek D. Lis, Czeslaw T. Kowalski*

**Design of an FPGA-Based Real-Time Simulator for Electrical System** ...... 1365
*I. Bahri, M-W. Naouar, E. Monmasson, I. Slama-Belkhodja, L.Charaabi*

**A New, Ultra-low-cost Power Quality and Energy Measurement Technology** ...... 1371
*Alex McEachern, Andreas Eberhard*

**Rotor Time Constant Adaptation Using Radial Basis Function Network** ...... 1375
*Pavel Brandltetter, Ondfej Skuta*

**Application of Speed and Load Torque Observers in High Speed Train** ...... 1382
*Jaroslaw Guzinski, Marc Diguet, Zbigniew Krzeminski, Arka diusz Lewicki, Haithem Abu-Rub*

**Position Estimator including Saturation and Iron Losses for Encoder Fault Detection of Doubly-Fed Induction Machine** ...... 1390
*Kai Rothenhagen, Friedrich W. Fuchs*

**Wide Range Low Noise Current Sensor** ...... 1398
*F. Richter, C. Sourkounis*

**Transducerless Speed Control with Initial Position Detection for Low Cost PMSM Drives** ...... 1402
*Roman Filka, Peter Balazovic, Branislav Dobrucky*

**Study About the Possibility of Electrodes Motion Control in the EAF Based on Adaptive Impedance Control** ...... 1409
*Manuela Panoiu, Caius Panoiu, Sorin Deaconu*

**Asynchronous machine stator resistance estimation using integrated PWM modulator and sampler unit as FPGA application** ...... 1416
*Dag Samuelsen, Waldemar Sulkowski*

**Development of Monitoring System for Series HEV Bus with Touch Panel** ...... 1421
*Tae-Won Chun, Quang-Vinh Tran, Uk-Don Choi, Heung-Gun Kim*

**A Development System for Testing Integrated Circuits Used for Power and Energy Measurements** ...... 1426
*Vladimir Cuk, Aleksandar Nikolic, Aleksandar Zigic*

**State and parameter estimation in a hydraulic system - moving horizon approach** ...... 1432
*Jerzy Baranowski, Andrzej Tutaj*

**Technologies of Current Sensors Suitable for Hot High Density Power Electronics** ...... 1440
*Filip Grecki, Grzegorz Iwanski, Wlodzimierz Koczara, Jozef Lastowiecki*

**Nonlinear dynamical feedback for motion control of magnetic levitation system** ...... 1446
*Jerzy Baranowski, Pawel Platek*

**Speed and position estimation of SRM** ...... 1454
*Konrad Urbanski, Krzysztof Zawirski*

**Potential of Digital Gate Units in High Power Appliations** ...... 1458
*Harald Kuhn, Thies Koneke, Axel Mertens*

**Disturbance Currents of Inverters** ...... 1465
*Petr Vrana, Jiri Javurek*

**Improvement of the Energy Recovery of Traction Electrical Drives using Supercapacitors** ...... 1469
*Diego Iannuzzi*

# Table of Contents

**A Multi-Core PC-based Simulator for the Hardware-In-the-Loop Testing of Modern Train and Ship Traction Systems**.................................................................................................................1475
*Christian Dufour, Guillaume Dumur, Jean-Nicolas Paquin, Jean Bélanger*

**Energy Saving Control of Tram Motors Taking Light Signalling and City Disturbances into Account**....................1481
*Stanislaw Rawicki*

**Characterization and Improved Control of a Brushless DC Drive with In-Wheel Motor**.................................1491
*Manuele Bertoluzzo, Giuseppe Buja, Alessandro Pavoni*

**Supply of Electric Vehicles via Magnetically Coupled Air Coils**............................................................1497
*Slawomir Judek, Krzysztof Karwowski*

**Sliding-Mode Approach to Control Design for Induction Motor Drive fed by a Three-Level Voltage-Source Inverter**....................................................................................................................1505
*Sergey Ryvkin, Richard Schmidt-Obermoeller, Andreas Steimel*

**Analysis and configuration of supercapacitor based energy storage system on-board light rail vehicles**................1512
*R. Barrero, X. Tackoen, J. Van Mierlo*

**Design of High Power Electronic Building Block based on Parallel of IGBTs for Electric Vehicle**......................1518
*Wen Huiqing, Liu Jun, Zhang Xuhui, Wen Xuhui*

**Stability Analysis on the DC Power Distribution System of More Electric Aircraft**..................................1523
*H. Zhang, C. Saudemont, B. Robyns, N. Huttin, R. Meuret*

**Design Considerations for Control of Traction Drive with Permanent Magnet Synchronous Machine**....................1529
*Zden..k Peroutka, Karel Zeman*

**Control of Primary Voltage Source Active Rectifiers for Traction Converter with Medium-Frequency Transformer**.....................................................................................................................1535
*Vojtech Blahník, Zdenek Peroutka, Jan Molnár, Jan Michalík*

**Energy management strategy for Coupling Supercapacitors and Batteries with DC-DC converters for hybrid vehicle applications**................................................................................................1542
*M.B. Camara, F. Gustin, H. Gualous, A. Berthon*

**Dual-Source Fed Multiphase Traction System with Standard and Non-Standard Control Regimes Based on Synchronized PWM**...............................................................................................1548
*Valentin Oleschuk, Marian P. Kazmierkowski*

**Analysis of a H-NPC topology for an AC Traction Front-End Converter**....................................................1555
*I. Etxeberria-Otadui, A. Lopez-de-Heredia, J. San-Sebastian, H. Gaztañaga, U. Viscarret, M. Caballero*

**Hybrid - type system of power supply for a trolleybus with an asynchronous motor**..................................1562
*Zygmunt Gizinski, Marcin Gasiewski, Ireneusz Mascibrodzki, Michal Zych, Krzysztof Zymmer, Marcin Zulawnik*

**Control of rotor flux in AC tram drive during sudden braking operation**..................................................1568
*Andrzej Debowski, Piotr Chudzik*

**A New Novel Power Electronic Circuit to Reduce Stray Current and Rail Potential in DC Railway**.....................1575
*Reza Fotouhi, Siamak Farshad*

**Slip Control Upgrades for Light-Rail Electric Traction Drives**................................................................1581
*Madis Lehtla, Hardi Hõimoja*

**Practical Aspects on the Improved DC Driving System Used in Electric Urban Traction**.............................1585
*Petre Marian Nicolae, Ioana-Gabriela Sîrbu, Ileana-Diana Nicolae, Lucian Mandache*

**The study of using the traction drive topology with the middle-frequency transformer**.............................1593
*Martin Pittermann, Pavel Drábek, Marek Cédl, Jiří Foft*

**Control of a Linear Switched Reluctance Motor as a Propulsion System for Autonomous Railway Vehicles**..........1598
*L. Kolomeitsev, D. Kraynov, S. Pakhomin, F. Rednov, E. Kallenbach, V. Kireev, T. Schneider, J. Böcker*

*xv*

# Table of Contents

**Motion Copying System Based on Real-World Haptics in Variable Speed**.....................................................**1604**
*Yuki Yokokura, Seiichiro Katsura, Kiyoshi Ohishi*

**Adaptive Fuzzy Control of magnetically suspended Rotary Table** .................................................**1610**
*Thomas Schallschmidt, Denis Draganov, Frank Palis*

**Wideband Force Sensing for Haptic Energy Transmission Utilizing FPGA**...................................**1614**
*Seiichiro Katsura, Masaki Kondo, Kiyoshi Ohishi*

**On the development of BLDC motor control run-up algorithms for aerospace application** ...................**1620**
*Vladimir Hubik, Martin Sveda, Vladislav Singule*

**Rotor Levitation by Active Magnetic Bearing Using Digital State Controller** ................................**1625**
*Chip Rinaldi Sabirin, Andreas Binder*

**Dynamical Torque-Speed-Curve Adaption To Damp Load Peaks Occuring In Drive Trains Of Shredding Plants**...................................................................................................................................................**1633**
*Constantinos Sourkounis*

**Traction vehicle distributed control computer system architecture with auto reconfiguration features and extended DMA support** ...............................................................................................................**1638**
*Jiri Zdenek*

**Analysis and Position Control of a Linear Switched Reluctance Actuator Based on Sliding Mode Control** ..............**1646**
*António Espírito Santo, Maria R. A. Calado, Carlos M. P. Cabrita*

**Development and Control for a Reaction Wheel System Driven by Permanent Magnet Synchronous Motor** ...............................................................................................................................................**1652**
*Ming-Chang Chou, Chang-Ming Liaw, Sywe-Bin Chien, Fa-Hwa Shieh, Jih-Run Tsai, Hao-Chi Chang*

**Nonlinear control design for magnetic bearings via automatic differentiation**....................................**1660**
*Stefan Palis, Mario Stamann, Thomas Schallschmidt*

**Design of Energy Harvesting Generator Base on Rapid Prototyping Parts** .......................................**1665**
*Zdenek Hadas, Jan Zouhar, Vladislav Singule, Cestmir Ondrusek*

**Control of Bouc-Wen hysteretic systems: Application to a piezoelectric actuator** ...............................**1670**
*Oriol Gomis-Bellmunt, Faycal Ikhouane, Daniel Montesinos-Miracle*

**Electric drive for carding machine draft device**...............................................................................**1676**
*Martin Diblík*

**Two-level and Multilevel Converters for Wind Energy Systems: A Comparative Study** ......................**1682**
*R. Melício, V. M. F. Mendes, J. P. S. Catalão*

**A Stand-alone Photovoltaic Supercapacitor Battery Hybrid Energy Storage System** ..........................**1688**
*M.E. Glavin, Paul K.W. Chan, S. Armstrong, W.G Hurley*

**Integrated contactless power transmission systems with high positioning flexibility** ...........................**1696**
*Daniel Kürschner, Christian Rathge*

**A Transformerless Interface Converter for a Distributed Generation System**......................................**1704**
*Tzung-Lin Lee, Zong-Jie Chen*

**A Comprehensive Analysis and Comparison Between Multilevel Space-Vector Modulation and Multilevel Carrier-Based PWM** .................................................................................................................**1710**
*Constantinos Sourkounis, Ahmad Al-Diab*

**Identification of Electrical Parameters in a Power Network Using Genetic Algorithms and Transient Measurements** ......................................................................................................................**1716**
*Wei. Dong, Pericle Zanchetta, David W.P. Thomas*

**On Acoustic Noise Reduction Procedure for Inverter-Fed Induction Machines** ..................................**1722**
*Weiss Helmut, Zaucher Peter, Xiao Jian*

# Table of Contents

**Cascaded Doubly Fed Induction Generator for Mini and Micro Power Plants Connected to Grid** .......... 1729
*Marek Adamowicz, Ryszard Strzelecki*

**Contactless power transmission with new secondary converter topology** ........................................ 1734
*Matthias Dockhorn, Daniel Kürschner, Rudolf Mecke*

**Modeling Approach of a Generator with Non-linear Load in Embedded Electrical Network** .................. 1740
*Nicolas Amelon, Mourad Ait-Ahmed, Mohamed-Fouad Benkhoris*

**Optimal Use of the 14 V Alternator in 42 V Automotive Supply Systems** ...................................... 1748
*Vasile Comnac, Mihai Cernat, Adrian Mailat*

**New Dual Channel Quasi Resonant DC-DC Converter Topologies for Distributed Energy Utilization** .......... 1755
*J. Hamar, I. Nagy, P. Stumpf, H. Ohsaki, E. Masada*

**Output Filtering of the Customer-end Inverter in a Low-Voltage DC Distribution Network** .................. 1763
*Pasi Peltoniemi, Pasi Nuutinen, Pasi Salonen, Markku Niemelä, Juha Pyrhönen*

**Power Flow Control through a Multi-Level H-Bridge based Power Converter for Universal and Flexible
Power Management in Future Electrical Grids** ........................................................................ 1771
*Stefano Bifaretti, Pericle Zanchetta, Yue Fan, Florin Iov, Jon Clare*

**Energy Storage Systems The Flywheel Energy Storage** .......................................................... 1779
*Tomasz Siostrzonek, Stanislaw Piróg, Marcin Baszynski*

**Analysis of Wide Area Integration of Dispersed Wind Farms Using Multiple VSC-HVDC Links** .............. 1784
*S. González-Hernández, E. Moreno-Goytia, O. Anaya-Lara*

**Generator Selection for Offshore Oscillating Water Column Wave Energy Converters** ...................... 1790
*D.L. O' Sullivan, A.W. Lewis*

**A Novel Approach To Photovoltaic Powered Water Pumping Design** ............................................ 1798
*Michael James Case, Ernest Edward Denny*

**Direct Controls in Voltage-Source Converters - Generalizations and Deep Study** .......................... 1803
*Karoly Veszpremi, Istvan Schmidt*

**Multipolar double fed induction wind generator with a single phase secondary winding** .................. 1811
*Leonids Ribickis, Guntis Dilevs, Nikolajs Levins, Vladislavs Pugachevs*

**The measurement on the solar cells in Liberec city** ........................................................ 1815
*Jiri Kubin*

**Rotor Turn-to-Turn Faults of doubly-fed Induction Generators in Wind Energy Plants - Modelling,
Simulation and Detection** .......................................................................................... 1819
*Vincenz Dinkhauser, Friedrich W. Fuchs*

**Static and Dynamic Response of a Photovoltaic Characteristics Simulator** ................................ 1827
*Anastasios Ch. Nanakos, Emmanuel C. Tatakis*

**Modeling and Optimal Sizing of Hybrid Renewable Energy System** .......................................... 1834
*Rachid Belfkira, Cristian Nichita, Pascal Reghem, Georges Barakat*

**Photovoltaic System MPPTracker Investigation and Implementation using DSP engine and Buck- Boost
DC-DC converter** .................................................................................................... 1840
*Dimosthenis Peftitsis, Georgios Adamidis, Panagiotis Bakas, Anastasios Balouktsis*

**Multi Objective Distributed Generation Planning Using NSGA-II** .......................................... 1847
*Muhammad Ahmadi, Ashkan Yousefi, Alireza Soroudi, Mehdi Ehsan*

**Testing of the Grid-connected Photovoltaic Systems Using FPGA-based Real-Time Model** .................. 1852
*Robert Stala*

*xvii*

# Table of Contents

**Output Maximization Using Direct Torque Control for Sensorless Variable Wind Generation System Employing IPMSG**.................................................................................................................................1859
*Yukinori Inoue, Shigeo Morimoto, Masayuki Sanada*

**Improving Connection and Disconnection of a Small Scale Distributed Generator Using Solid-State Controller** ....................................................................................................................................................1866
*M.M.R. Ahmed*

**Research control of electric systems in wind generator systems**........................................................1872
*Stefan Winternheimer, Artem Kolesnikov, Evgeny Glushkin, Alexander Bukatov*

**Stand-alone Photovoltaic Generation System with Combined Storage using lead Battery and EDLC** .....1877
*Hiroaki Nakayama, Eiji Hiraki, Toshihiko Tanaka, Noriaki Koda, Nobuo Takahashi, Shuji Noda*

**Active Filter Action of Inverter Exciting Induction Generator for Wind Power Generation** .............1884
*Noriyuki Kimura, Tomoyuki Hamada, Katsunori Taniguchi, Toshimitsu Morizane*

**The Operation of Power Electronic Converters in Photovoltaic Drive Systems**................................1890
*Marek Niechaj*

**Experimental results of a hybrid wind/hydro power system connected to isolated loads** ..................1896
*Mehdi Nasser, Stefan Breban, Vincent Courtecuisse, Arnaud Vergnol, Benoît Robyns, Mircea M. Radulescu*

**Grid Connection of Multi-Megawatt Clean Wave Energy Power Plant under Weak Grid Condition**.......1904
*Kai Rothenhagen, Marek Jasinski, Marian P. Kazmierkowski*

**Improved sizing method of storage units for hybrid wind-diesel powered system** .............................1911
*A.M. Tankari, B. Dakyo, C. Nichita*

**A Research Platform for a Smart-Blade Wind Generation System** ......................................................1918
*J. Davey, Udaya K. Madawala, R. Sharma*

**Soft Switching Multi-Phase Boost Converter for Photovoltaic System**...............................................1924
*Joo-Hyuk Lee, Jae-Hyung Kim, Chung-Yuen Won, Su-Jin Jang, Yong-Chae Jung*

**Soft Switching Boost Converter for Photovoltaic Power Generation System** .....................................1929
*Doo-Yong Jung, Young-Hyok Ji, Jae-Hyung Kim, Chung-Yuen Won, Yong-Chae Jung*

**Optimisation Of Wind Power Pmsm To Grid Conversion System** .......................................................1934
*Ince Kayhan, Weiss Helmut*

**Analysis of Wind Farm and Multilevel Converter Interactions in Medium Voltage Networks Under Steady-State and Transient Conditions** ...................................................................................................1941
*J. Sosa-Ruiz, E. Moreno-Goytia, O. Anaya-Lara*

**A Simple, Low Cost Design Using Current Feedback to Improve the Efficiency of a MPPT-PV System for Isolated Locations** ....................................................................................................................................1947
*Herman Fernández, Abelardo Martínez, Víctor Guzmán, María Isabel Gímenez*

**A Single-Phase Active Power Filter Based in a Two Stages Grid-Connected PV System** ....................1951
*Kleber C.A. De Souza, Denizar C. Martins*

**Wide Bandwidth Power Flow Control Algorithm of the Grid Connected VSI under Unbalanced Grid Voltages**.................................................................................................................................................1957
*Zoran Ivanovic, Marko Vekic, Stevan Grabic, Evgenije Adzic, Vladimir Katic*

**The use of Switched Reluctance Generator in wind energy applications** ............................................1963
*Eleonora Darie, Costin Cepisca, Emanuel Darie*

**Active Line Shaping of a Single Phase Rectifier using the Switching Function Technique** ................1967
*Christos Marouchos*

**Control of Reactive Power in Double-Fed Machine Based Wind Park** ................................................1975
*Elzbieta Bogalecka, Michal Kosmecki*

# Table of Contents

**A Novel Hybrid Modulation Method for Cascaded H-bridge Active Power Filter** .................... 1981
*Yonggang Chen, Ping Wang, Yaohua Li, Zixin Li, Longcheng Tan*

**Apparent Power Ratio of the Shunt Active Power Filter** ......................................... 1987
*A. Kouzou, B.S Khaldi, S. Saadi, M.O. Mahmoudi, M.S. Boucherit*

**Shunt Active Power Filter with Improved Dynamic Performance** ............................... 1995
*Krzysztof Piotr Sozanski*

**The Research on the Active Power Filter Based on the Cascaded H-bridge Converter** ........... 2000
*Yonggang Chen, Junling Chen, Ping Wang, Yaohua Li, Longcheng Tan, Zixin Li, Wei Xu*

**E-laboratory in the Field of Electrical Drives** ................................................. 2005
*H.Hõimoja, A.Rosin, T.Möller, M. Müür*

**Laboratory Setup for Studying Ultracapacitors in Industrial Applications** .................... 2011
*I. Roasto, D. Vinnikov, T. Lehtla*

**Synchronous machine direct axis parameters estimation module from an iterative strategy** ..... 2015
*Emile Mouni, Slim Tnani, Gérard Champenois*

**Determination of the Characteristic Life Time of Paper-insulated MV-Cables based on a Partial
Discharge and tan(..) Diagnosis** ............................................................... 2022
*I. Mladenovic, Ch. Weindl*

**Elimination of Increased Excitation of Common- Mode Oscillations in Electrical Drive Systems with
Active Front End and Long Motor Cables** ...................................................... 2028
*Thomas Weidinger*

**Internal Short Circuit in a Tooth Wound PMSM with Stranded Conductors** .................... 2037
*Damien Birolleau, Christian Chillet, Laurent Albert*

**Implementation of a Virtual Laboratory for Low Power Electrical Drives** ..................... 2043
*Gh. BALUTA, V. HORGA, C. LAZAR*

**DQ-Transformation Approach for Modelling and Stability Analysis of AC-DC Power System with
Controlled PWM Rectifier and Constant Power Loads** ........................................ 2049
*K-N Areerak, S.V. Bozhko, G.M. Asher, D.W.P. Thomas*

**Genetic Identification of Parameters the Sandwich Piezoelectric Ceramic Transducers for Ultrasonic
Systems** ...................................................................................... 2055
*Pawel Fabijanski, Ryszard Lagoda*

**The Impact of Higher-Order Harmonics on Tripping of Residual Current Devices** .............. 2059
*Stanislaw Czapp*

**Estimation of the Untapped Regenerative Braking Energy in Urban Electric Transportation Network** .... 2066
*Leonards Latkovskis, Linards Grigans*

**Performance Evaluation of Electric Power Steering with IPM Motor and Drive System** ......... 2071
*Hamidreza Akhondi, Jafar Milimonfared, Kaveh Malekian*

**Optimal Control: Load Frequency Control of a Large Power System** ........................... 2076
*Sílvio José Pinto Simões Mariano\*, Luís António Fialho Marcelino Ferreira*

**LCL-Load Modular Converter For Induction Heating** ......................................... 2082
*Maciej A. Dzieniakowski, Jan Fabianowski, Robert Ibach*

**On-line PID Controller Tuning Using Genetic Algorithm and DSP PC Board** ................... 2087
*Pawel Fabijanski, Ryszard Lagoda*

**Regulation Properties of Pumping Station Control System In The Highest Efficiency Range** ..... 2091
*Szychta Leszek*

*xix*

# Table of Contents

**Inner Gas Pressure Measurement Based Life-span Estimation of Electrolytic Capacitors**......................................2096
A. Riz, D. Fodor, O. Klug, Z. Karaffy

**Robust Control Methodologies for Optical Micro Electro Mechanical System - New approaches and Comparison** ..................................................................................................................................................2102
Alireza Izadbakhsh, S.M.R. Rafiei

**Modeling a Buck-Based Switching Amplifier for Sinusoid Wide Band Tracking by Using a Nonlinear Time Varying Map** ........................................................................................................................................2108
A. El Aroudi, E. Alarcón, E. Rodriguez, R. Leyva

**Single Inductor Multiple Outputs Interleaved Converters Operating in CCM**......................................2115
Luis Benadero, Vanessa Moreno-Font, Abdelali El Aroudi, Roberto Giral

**Control of a two-cell dc/dc converter in presence of saturating duty cycle** .......................................2120
Moez Feki, Abdelali El Aroudi, Bruno Gerard Michel Robert, Nabil Derbel

**Bifurcations and Chaotic Dynamics in a Linear Switched Reluctance Motor** .....................................2126
M.R. De Castro, B.G.M. Robert, C. Goeldel

**Modular Architecture for Decentralized Hybrid Power Systems** ........................................................2134
E. Ortjohann, M. Lingemann, O.Omari, A. Schmelter, N. Hmasic, A. Mohd, W. Sinsukthavorn, D. Morton

**Design of a power management system for an active PV station including various storage technologies** ...................2142
Di Lu, Tao Zhou, Hicham Fakham, Bruno Francois

**Energy Management and Power Flow of Decoupled Generation System for Power Conditioning of Renewable Energy Sources** ......................................................................................................................2150
Wlodzimierz Koczara, Zdzislaw Chlodnicki, Nazar Al-Khayat, Neil L.Brown

**Inversion Based Control of a Diesel Fed Low Temperature Fuel Cell System** ....................................2156
Daniela Chrenko, Marie-Cecile Pera, Daniel Hissel

**Power Management in an Autonomous Adjustable Speed Large Power Diesel Gensets** .......................2164
Grzegorz Iwanski, Wlodzimierz Koczara

**Cost evaluation of Generator-set with Energy Storage for 4Q-load** ..................................................2170
Freek J.F.Baalbergen, Pavol Bauer

**Integrating renewable energy sources and storage into isolated diesel generator supplied electric power systems** .............................................................................................................................................2178
Chad Abbey, Jonathan Robinson, Géza Joós

**Performance comparison of different wind generator based hybrid systems** .....................................2184
Vincent Courtecuisse, Benoit Robyns, Marc Petit, Bruno Francois, Jacques Deuse

**First Approach for a Fault Tolerant Power Converter Interface for Multi-Stack PEM Fuel Cell Generator in Transportation Systems** ...........................................................................................................2192
Alexandre De Bernardinis, Gérard Coquery

**Development of Electrical System for Hybrid Vehicles Using the Free-swinging Piston Engine and Oscillating Rotating Generator** .................................................................................................................2200
Sigitas Kudarauskas

**Power flow control in different time scales for a wind/hydrogen/super-capacitors based active hybrid power system** ....................................................................................................................................2205
ZHOU Tao, LU Di, FAKHAM Hicham, FRANCOIS Bruno

**Neuro-Fuzzy Adaptive Control of the IM Drive with Elastic Coupling** .............................................2211
Teresa Orlowska-Kowalska, Krzysztof Szabat, Mateusz Dybkowski

**Control of Flexible Drive with PMSM employing Forced Dynamics**.................................................2219
Vittek Ján, Bris Peter, Makys Pavol, Stulrajter Marek, Vavrus Vladimír

# Table of Contents

**The problems of high dynamic drive control under circumstances of elastic transmission**..........2227
*Jan Deskur, Roman Muszynski*

**Protective Predictive Control of Electrical Drives with Elastic Transmission** .............................2235
*Mario Vasak, Nedjeljko Peric*

**Low-Cost High-Performance Predictive Control of Drive Systems with Elastic Coupling** ...........2241
*Marcin Cychowski, Kieran Delaney, Krzysztof Szabat*

**Development of an Expert System for Identification, Commissioning and Monitoring of Drives** ............2248
*Mario Pacas, Sebastian Villwock*

**Control of Axial Flux Permanent Magnet Motor by the PIPCRM Method at Standstill and at Low Speed** .............2254
*Janusz Wisniewski, Wlodzimierz Koczara*

**Zero Speed Position Estimation of a Matrix Converter Fed AC PM Machine using PWM Excitation** .....................2261
*Q. Gao, G. M. Asher, M. Sumner*

**Sensorless Direct Torque and Flux Control of an IPM Synchronous Motor at Low Speed and Standstill**................2269
*Gilbert Foo, S. Sayeef, M.F. Rahman*

**Sensorless Control of PM Synchronous Motors Using a Predictive Current Controller with Integrated INFORM and EMF Evaluation**..................2275
*Manfred Schrödl, Christian Simetzberger*

**Torque Sensorless Control of Induction Motor** ..................2283
*Karel Jezernik, Miran Rodic*

**Application of the induction motor torque - observer to the control of turbo - machines** ....................2289
*Andrzej Debowski, Daniel Lewandowski*

**Observer of induction motor speed based on exact disturbance model**..................2294
*Zbigniew Krzeminski*

**Experimental Performance Evaluation for Low Speed and Regenerating Operation of Sensor-less Vector Control System of Induction Motor Using Observer Gain Tuning**....................2300
*Kazuhiro Ohyama, Greg Asher, Mark Sumner*

**Application of the Stator Current-based MRAS Speed Estimator in the Sensorless Induction Motor Drive**............2306
*Mateusz Dybkowski, Teresa Orlowska-Kowalska*

**State and Parameter Estimation in Induction Motors using Sliding Modes** ...................2312
*Sachit Rao, Martin Buss, Vadim Utkin*

**Torque Transient Alleviation in Fixed Speed Wind Generators by Indirect Torque Control with STATCOM**..................2318
*Marta Molinas, Jon Are Suul and Tore Undeland*

**Flicker Study on Variable Speed Wind Turbines with Permanent Magnet Synchronous Generator** ........................2325
*Weihao Hu, Zhe Chen, Yue Wang, Zhaoan Wang*

**Power Output Characteristics Analysis of Wind Energy Converter Control Methods** ......................2331
*Bingchang Ni, Constantinos Sourkounis*

**A Cooperative Control Method for Output Power Smoothing and Hydrogen Production by Using Variable Speed Wind Generator**..................2337
*Rion Takahashi, Hirotaka Kinoshita, Toshiaki Murata, Junji Tamura Masatoshi Sugimasa, Akiyoshi Komura, Motoo Futami, Masaya Ichinose, Kazumasa Ide*

**A new interconnecting method for wind turbine/generators in a wind farm and basic characteristics of the integrated system** ..................2343
*Shoji Nishikata, Fujio Tatsuta*

**Educational aspects of mechatronic control course design for collaborative remote laboratory** ...............2349
*Andreja Rojko, Darko Hercog, Karel Jezernik*

*xxi*

# Table of Contents

**PEMCWebLab - Distance and Virtual Laboratories in Electrical Engineering: Development and Trends**............2354
*Pavol Bauer, Viliam Fedák, Otto Rompelman*

**Integrated multimedia educational program of a DC servo system for distant learning**............2360
*Gabor Sziebig, Istvan Nagy, Rafael Kalman Jardan, Peter Korondi*

**Electromechanical Actuators WEB-lab**............2368
*Dusan Maga, Jan Sitar, Juraj Dudak, Rene Hartansky, Peter Siroky, Jan Halgos, Pavol Bauer*

**Power Quality and Active Filters as Web-Controlled Experiment in the frame of PEMC WebLab**............2371
*Volker Staudt, Andreas Steimel, Pavol Bauer, Vítezslav Hájek*

**Distant learning of Pulse Width Modulation Techniques for Voltage Source Converters**............2378
*Bartlomiej Kamiski, Dariusz Sobczuk*

**Modern design optimisation exploiting field simulation**............2383
*Jan K. Sykulski*

**Transmission-Line Modelling of Wave Propagation Effects in Machine Windings**............2385
*Herbert De Gersem, Olaf Henze, Thomas Weiland, Andreas Binder*

**An efficient field-circuit coupling method by a dynamic lumped parameter reduction of the FE model**............2393
*F. Henrotte, E. Lange, K. Hameyer*

**Coupled field-circuit-mechanical model of an electromagnetic actuator operating in error actuated control system**............2400
*Lech Nowak*

**Simulation and Investigation of Magnetorheological Fluid Brake**............2406
*Wieslaw Lyskawinski, Wojciech Szelag, Cezary Jedryczka*

**Field and Field-Circuit Description of Electrical Machines**............2412
*Andrzej Demenko, Kay Hameyer*

**Interaction between Thermal Impedance and Parasitics in Power Sections**............2420
*Stefan Forster, Andreas Lindemann*

**Discussion of Internal and External High Frequency Common Mode Noise Current on a Chopper Circuit**............2428
*Tetsuya Mitani, Keiji Wada, Toshihisa Shimizu, Hiromichi Ohashi*

**A Novel Digital Control Method for DC-DC Converter**............2434
*Fujio Kurokawa, Masashi Okamatsu, Yuichi Sumida, Yasuhiro Mimura, Masahiro Sasaki*

**A Novel Single/Three-phase Matrix Converter For High Power Integration**............2439
*Makoto Saito*

**An Effective Design Method for High Power Density Converters**............2445
*Yusuke Hayashi, Kazuto Takao, Toshihisa Shimizu, Hiromichi Ohashi*

**Power Devices in Polish National Silicon Carbide Program**............2452
*Mariusz Sochacki, Andrzej Kubiak, Zbigniew Lisik, Jan Szmidt*

**SiC Power Semiconductor Devices for new Applications in Power Electronics**............2457
*Dominique Planson, Dominique Tournier, Pascal Bevilacqua, Nicolas Dheilly, Herve Morel, Christophe Raynaud, Mihai Lazar, Dominique Bergogne, Bruno Allard, Jean-Pierre Chante*

**Silicon carbide Schottky diodes and MOSFETs: solutions to performance problems**............2464
*Owen J. Guy, Michal Lodzinski, Ambroise Castaing, P. M. Igic, Amador Perez-Tomas, Michael R. Jennings, Philip A. Mawby*

**Characterization of the Static and Dynamic Behavior of a SiC BJT**............2472
*M.M.R. Ahmed, N-A.Parker-Allotey, P.A. Mawby, Muhammed Nawaz, Carina Zaring*

**An active network control method using distributed energy resources in microgrids**............2478
*Takayuki Tanabe, Yoshinobu Ueda, Toshihisa Funabashi, Shigeo Numata, Kimio Morino, Eisuke Shimoda*

*xxii*

# Table of Contents

**Energy Management in Solar Photovoltaic Plants based on ESS** ..................................................... 2481
*M. Lafoz, L. García-Tabarés, M. Blanco*

**A Method of Three-Phase Balancing in Microgrid by Photovoltaic Generation Systems** ........................... 2487
*Masahide Hojo, Yuta Iwase, Toshihisa Funabashi, Yoshinobu Ueda*

**Development of HILS(Hardware In-Loop Simulation) System for MMS(Microgrid Management System) by using RTDS** ................................................................................................................ 2492
*Jin-Hong Jeon, Jong-Yul Kim, Seul-Ki Kim, ong-Bo Ahn, JuneHo Park*

**Power Quality Analysis of Jeju Island Power System with Wind Farm and HVDC System** ........................ 2498
*Jae-Hong Kim, Eel-Hwan Kim, Se-Ho Kim, Jaeho Choi, Gil-Soo Jang, Seung-Ho Song*

**A New Control Method for Power Turbine Generators Using an Accurate Ship Plant System Model** ............ 2504
*Nobumasa Matsui, Fujio Kurokawa, Keiichi Shiraishi*

**Voltage profile support in distribution networks - influence of the network R/X ratio** ............................. 2510
*B. Bla~ic, I. Papic*

**Modeling, Simulation and Analysis of Conducted Common-Mode EMI in Matrix Converters for Wind Turbine Generators** ............................................................................................................. 2516
*S. Zhang, K.J. Tseng*

**Design of Frequency Shift Acceleration Contol for Anti-islanding of an Inverter-based DG** ..................... 2524
*Seul-Ki Kim, Jin-Hong Jeon, Heung-Kwan Choi, Jonng-Bo Ahn*

**Integrated Power Converter for Photovoltaic and Fuel Cell Systems in Home** .................................. 2530
*Yasuyuki Nishida, Shinichiro Sumiyoshi, Hideki Omori*

**A Comparison of Position Control Structures for Ironless Linear Synchronous Motor** ........................... 2538
*Martin Hrasko, Pavol Makys, Marek Franko, Jozef Kuchta*

**A Comparison of Sliding Mode Approaches to a Nanometre Position Control Application** ....................... 2543
*Paul Andreas Stadler, Stephen James Dodds*

**Sliding Mode Control of PMSM Drives Subject to Torsion Oscillations in the Mechanical Load** ............... 2551
*Stephen J. Dodds, Jan Vittek*

**Sliding Mode Vector Control of PMSM Drives with Minimum Energy Position Following** ....................... 2559
*Stephen J. Dodds*

# Analysis of a H-NPC topology for an AC Traction Front-End Converter

I. Etxeberria-Otadui, A. Lopez-de-Heredia, J. San-Sebastian, H. Gaztañaga, U. Viscarret, M. Caballero

IKERLAN –IK4 Technological Research Centre, Control Engineering and Power Electronics, Mondragón (Spain)
*ietxeberrria@ikerlan.es*, *alopezheredia@ikerlan.es*, *jsansebastian@ikerlan.es*

*Abstract*— **In this paper a H-NPC topology is analyzed as the front-end converter at an AC traction application. These converters permit to achieve higher voltage levels and better current quality spectra than conventional single-phase H-bridge converters. The objective here is to analyze the suitability of main modulation strategies for traction applications, and to study the design requirements of current and voltage control loops for this kind of applications. Presented control and modulation loops have been tested in simulation and have been verified experimentally in a prototype test bench, showing the appropriateness of the proposed converter topology and control structures.**

*Keywords*— **Modulation strategy, Multilevel converters, Converter control.**

## I. INTRODUCTION

The main task of a traction front-end converter consists in rectifying the AC voltage (that comes from the AC catenary) in order to feed traction motors through their inverters. A single phase transformer, which also plays the role of catenary connection filter, is used to adapt catenary voltage levels to the front-end converter input voltage.

The main design requirements for AC traction front-end converters are:

- Limited harmonic generation: to fulfill catenary requirements.
- Sufficient dynamics: to follow traction system load variations (accelerations and decelerations).

In a context of limited switching frequency, due to the relatively high transferred power and the necessity to limit power losses, two are the main solutions used for increasing input current harmonic quality and power: parallel association of several converters or using multilevel topologies.

Classically the first solution has been used, paralleling H-bridge converters and using interleaved modulation [1]. They allow sharing current between semiconductors and permit to reduce switching harmonics. Here, the input transformer must contain as much secondary windings as H-bridges are used (see top of figure 1).

More recently, H-NPC converters (see bottom of figure 1) have also been introduced [2]. In this case, if the same semiconductors are used, converter input and output voltage may be doubled keeping the current constant (in terms of magnitude and primary current harmonic

quality). The result is a more compact converter (less modular) and fewer secondary windings (of the same current but with higher voltage), using some additional semiconductors with respect to the previous case (neutral clamping diodes) and a more complex modulation strategy. If required, in this case also some converters can be connected in parallel in order to increase total power capability and to reduce primary current harmonics by interleaved modulation.

The choice between these two solutions depends as much on technical requirements as on economic ones: volume, component prices, DC link rated voltage, catenary requirements, etc.

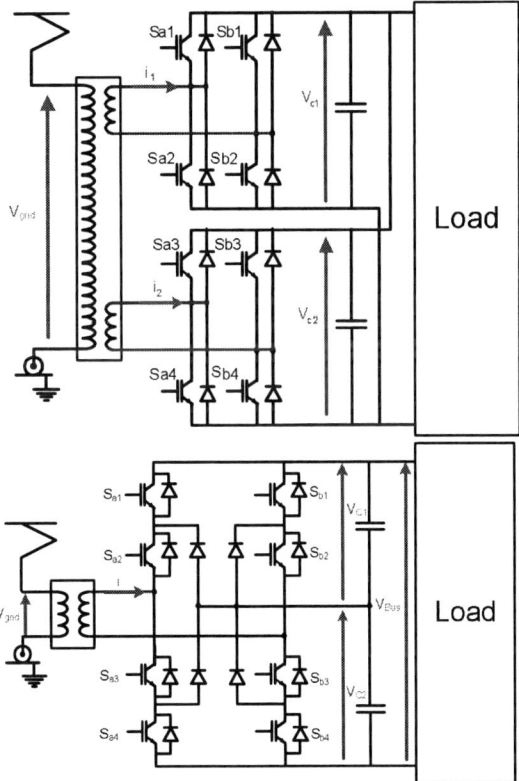

Fig. 1. Paralleling H-bridge converters (top) and multilevel H-NPC topology (bottom).

In this paper the modulation and control of a H-NPC bridge converter are studied considering an AC traction application. The objective of the paper is to analyze the suitability of main modulation strategies for traction applications, and to investigate on the design requirements

978-1-4244-1741-4/08/$25.00 ©2008 IEEE   1555

of current and voltage control loops. Presented control and modulation loops have been validated in simulation and experimentally.

## II. H-NPC CONVERTER MODULATION STRATEGIES

As in any multilevel converter (single or three-phase), there are two main concerns from the modulation viewpoint:

- ☐ To obtain the required fundamental output voltage (minimizing harmonics or losses).
- ☐ To balance DC bus capacitor voltages.

In order to fulfill the former condition, the correct output voltage amplitude must be applied during each interval. For the latter condition, the correct choice between equal amplitude switching (redundant) states must be done, to balance the power flow between DC link capacitors. In a single phase H-NPC converter there are a total of 9 switching states available (see table I) as well as 5 output voltage levels ($\pm V_{bus}$, $\pm V_{bus}/2$ and 0) and 4 voltage sections (I-IV) between them (see figure 2):

TABLE I.
AVAILABLE SWITCHING STATES IN A H-NPC CONVERTER

| $V_{ab}$ | $S_{a1}$ | $S_{a2}$ | $S_{a3}$ | $S_{a4}$ | $S_{b1}$ | $S_{b2}$ | $S_{b3}$ | $S_{b4}$ | State |
|---|---|---|---|---|---|---|---|---|---|
| 0 | 0 | 0 | 1 | 1 | 0 | 0 | 1 | 1 | 1 |
| $-V_{bus}$ | 0 | 0 | 1 | 1 | 0 | 1 | 1 | 0 | 2 |
| $-2V_{bus}$ | 0 | 0 | 1 | 1 | 1 | 1 | 0 | 0 | 3 |
| $-V_{bus}$ | 0 | 1 | 1 | 0 | 1 | 1 | 0 | 0 | 4 |
| 0 | 0 | 1 | 1 | 0 | 0 | 1 | 1 | 0 | 5 |
| $V_{bus}$ | 0 | 1 | 1 | 0 | 0 | 0 | 1 | 1 | 6 |
| $2V_{bus}$ | 1 | 1 | 0 | 0 | 0 | 0 | 1 | 1 | 7 |
| $V_{bus}$ | 1 | 1 | 0 | 0 | 0 | 1 | 1 | 0 | 8 |
| 0 | 1 | 1 | 0 | 0 | 1 | 1 | 0 | 0 | 9 |

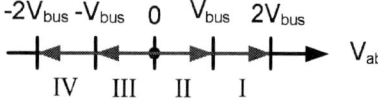

Fig. 2. Defined voltage sections (I-IV).

An important difference with respect to 3 phase converters is that here, in all 4 voltage sections and for any voltage to be synthesized, there are always available redundant switching states with opposite effect on DC bus voltage, and therefore balance is guaranteed for any modulation index and output power factor as long as the time during which redundant vectors are applied (to obtain the desired output voltage) is enough to balance DC bus capacitors.

Various modulation strategies have been proposed in the literature for H-NPC converters [3], [4]. Here the objective is to analyze their suitability for a traction application. In the same way as in the 3 phase case, there are 2 main families of modulation strategies: those applying 2 switching states per period (2SC) and those using 3 (3SC). In both cases a triangular modulation signal is used as in a conventional PWM.

### A. 2SC Modulation

In this modulation strategy two vectors are used symmetrically in order to minimize semiconductor switching frequency [3].

$$V_{ab} = V_1 \cdot t_1 + V_2 \cdot t_2 \qquad (1)$$

$V_1$ is the nearest redundant vector, $V_2$ is the nearest non-redundant vector and $t_1$ and $t_2$ is the time during each vector is applied.

Considering the particular case where the output voltage vector is located in section I, in 2SC modulation the nearest redundant vector is 6 or 8 and the nearest non-redundant vector is 7 (see figure 3).

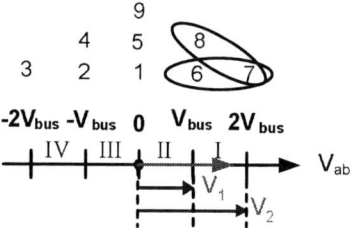

Fig. 3. 2SC Modulation in section I.

Thus, only one redundant state is applied per period depending on the output voltage ($V_{ab}$) sign and the voltage unbalance ($\Delta V$) and the current ($I_{ab}$) signs product, obtaining an output apparent frequency equal to the modulation triangular signal.

$$V_1 = f\{sign(V_{ab}), sign(\Delta V) * sign(I_{ab})\} \qquad (2)$$

In balanced conditions, redundant vectors are applied alternatively and switchings are homogeneously distributed between converter branches.

### B. 3SC Modulation

In this case, 3 states are applied per period, using both available redundant states per period [4].

$$V_{ab} = V_{1a} \cdot t_{1a} + V_2 \cdot t_2 + V_{1b} \cdot t_{1b} \qquad (3)$$

$V_{1a}$ and $V_{1b}$ are the nearest redundant vectors, $V_2$ is the nearest non-redundant vector and $t_{1a}$, $t_{1b}$ and $t_2$ is the time during each vector is applied.

Considering the particular case where the output voltage vector is located in section I, in 3SC modulation the nearest redundant vectors are 6 and 8 and the nearest non-redundant vector is 7 (see figure 4).

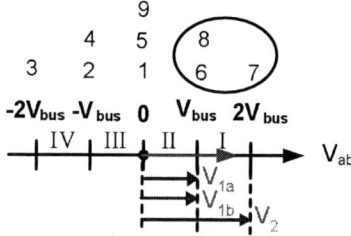

Fig. 4. 3SC Modulation in section I.

In this way, the apparent frequency is twice the triangular frequency, maintaining permanently the branch frequency equal to the triangular one.

In this modulation technique different methods can be identified depending on the way the application time is shared between redundant vectors ($t_{1a}$ and $t_{1b}$). Often a proportional (constant or adaptive) gain is considered. However, there is also the possibility of defining a region next to the zero voltage, where instead of the zero vector, redundant vectors are applied in order to accelerate the voltage balancing process. In this study, a proportional adaptive gain ($k$) has been considered, which is proportional to the voltage unbalance and the current sign.

$$\Delta = k \cdot \Delta V \cdot sign(I_{ab}) \qquad (4)$$

The algorithm used in this modulation is very similar to the one used in 2SC, but is more complex due to the fact that a third vector is applied at each period.

### C. Comparison

Therefore, with the 2SC modulation, the frequency of the triangular must be doubled in order to obtain the same apparent output frequency of the 3SC modulation, but with the advantage of a 50% smaller sample time, minimizing consequently the overall equivalent delay of the system (see figure 5).

In the case of a traction application, large unbalances between capacitor voltages can not be expected. These capacitors are charged in a controlled way and therefore there is no reason for a significant initial unbalance between their voltages.

Concerning continuous operation, a machine side NPC three phase converter, could provoke in some particular conditions unbalances between DC bus capacitors [5]. However, they could be easily removed with the support of the NPC front end converter.

Consequently, any of the presented strategies could be suitable for this application and the most robust and simple one should be chosen. In this paper, the 2SC strategy has been chosen due to its simplicity and minor sample time.

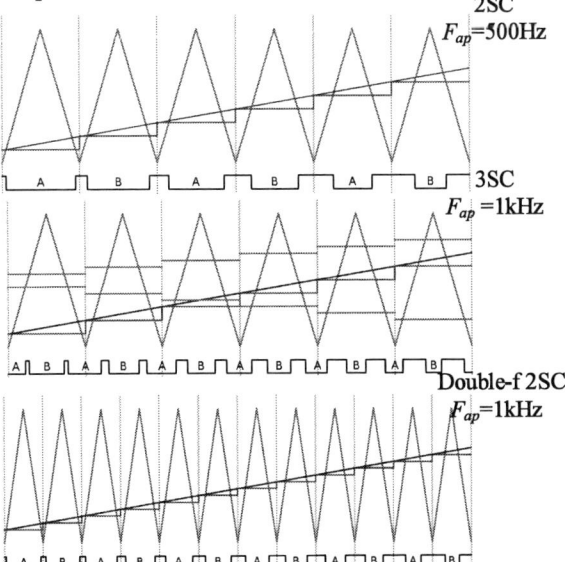

Fig. 5.   2SC, 3SC and double-frequency 2SC modulation strategies.

## III. CURRENT AND VOLTAGE CONTROL

The main role of the front-end converter is to keep constant the DC bus voltage, adapting the amount of exported/imported power to the requirements of the traction system with sufficient dynamics (in order to avoid a significant voltage error that could disturb traction motor operation and guarantee the integrity of the DC bus link components).

Therefore, the total control structure is composed of a voltage and a current loop: the voltage loop determines the current which is necessary to exchange with the catenary in order to keep the DC bus voltage constant (see figure 6). Here the objective is to analyze the design of these control loops considering the requirements of a traction application.

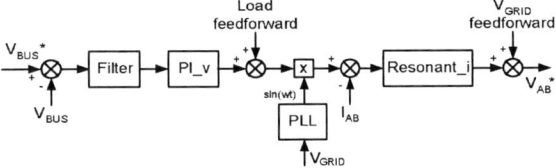

Fig. 6.   Voltage and current control loops.

The low switching frequency (some hundreds of Hz) imposed by the high power involved in traction applications (hundreds of kWs per converter) may harden the dynamics of the control system, due to stability and closed loop resonance problems provoked by relatively high implementation delays [6].

The proposed current control loop is based here on a resonant controller, which introduces a resonant term, i. e. a double imaginary pole, adjusted to the fundamental catenary voltage frequency in order to obtain an infinite open loop gain at this frequency.

Thus, this controller is able to eliminate the steady state error of selected alternative signals. It is especially adapted for single-phase applications, obtaining similar performances to three phase synchronous frame controllers but operating in a static frame [7].

$$C_R = k_p + \frac{2 \cdot k_r \cdot s}{s^2 + \omega_0^2} \qquad (5)$$

Concerning the tuning method, a frequency response approach has been used. The objective is to guarantee a sufficient phase margin for the system, taking into consideration PWM implementation and computational delays and avoiding close loop high resonances (see figure 7) [8].

Concerning the DC voltage loop, the main drawback for obtaining high dynamics is the necessity to use a low pass filter to remove the 100Hz component from the DC voltage measurement (derived from the single phase 50Hz power exchange). The conventional approach is to use low cutting frequency filters, obtaining good steady-state performances but with a poor transient behavior. The proposal here is to add a load feedforward, permitting to significantly improve system transient performances. The DC voltage loop is based on a conventional PI controller. In this case also the bandwidth of the voltage loop must be limited to obtain a sufficient global phase margin (>40°) and to avoid close loop high resonances (see figure 8).

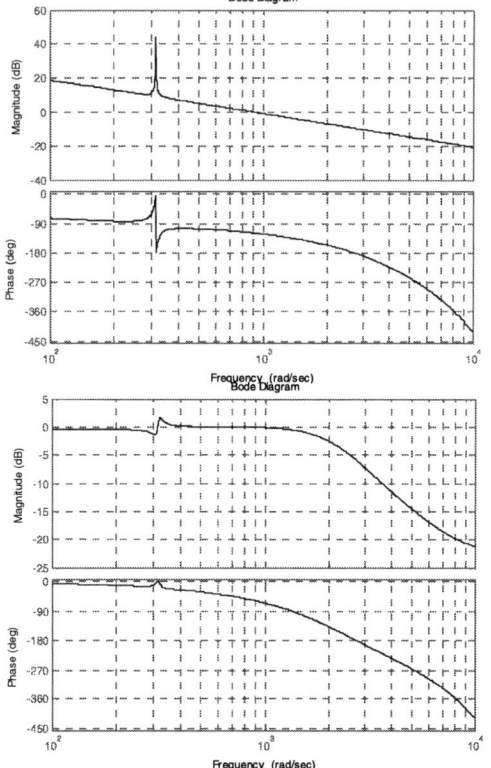

Fig. 7. Current control open-loop (top) and closed-loop (bottom) frequency responses.

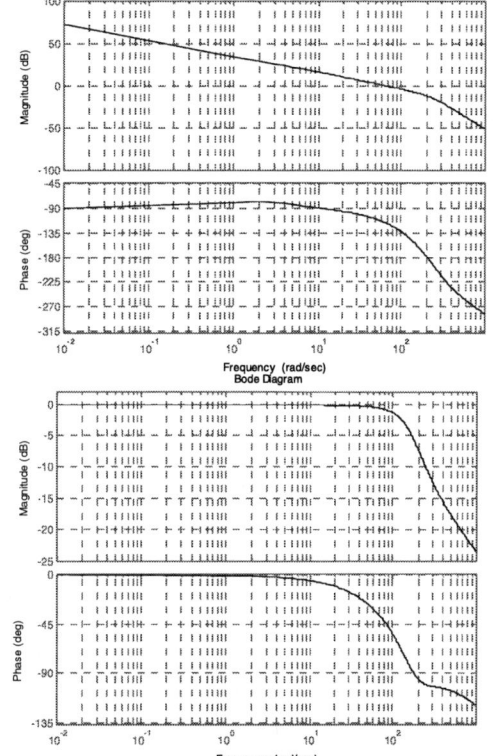

Fig. 8. Voltage control open-loop (top) and closed-loop (bottom) frequency responses.

## IV. CASE STUDY

Previously presented control and modulation loops have been applied to a traction application front-end H-NPC converter and tested in simulation based on a case study. The main characteristics of the system are presented in table II.

TABLE II.
MAIN CHARACTERISTICS OF THE STUDIED SYSTEM

| Symbol | Magnitude | Value |
|--------|-----------|-------|
| $F_{sw}$ | Switching frequency | 500Hz |
| $F_s$ | Sampling frequency | 1000Hz |
| $U_{dc}$ | DC bus voltage | 1500V |
| E | Grid voltage | 800V |

Two different types of analysis have been carried out. Firstly, in the steady-state response the voltage balancing capability has been analyzed in no load and full load conditions (300kW). Afterwards, the voltage control transient response has been evaluated with and without load feedforward.

### A. Voltage balancing capability

First of all, the voltage balancing capability of the 2SC modulation has been tested in simulation in no load conditions (only losses are compensated), considering an initial voltage unbalance between capacitors of 200V. As it can be seen in figure 9 both voltage buses are correctly balanced after a short transient.

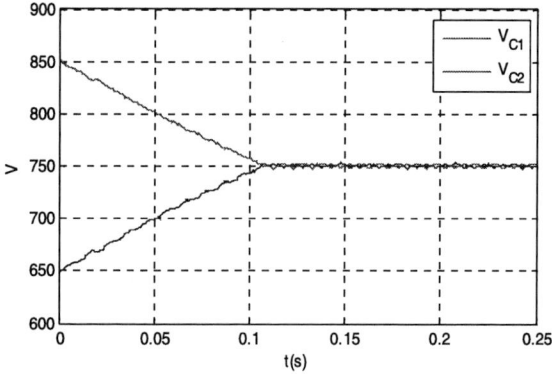

Fig. 9. Initial unbalance in no load conditions. Simulation results.

Secondly, the voltage balancing capability has been analyzed in full load conditions, considering a permanent power unbalance of 10% between the loads connected to both DC capacitors (starting at t = 0.3s). As it can be seen in figure 10, the selected 2SC modulation strategy is also able to correctly minimize load-provoked unbalances.

Figure 11 presents the output voltage (five levels) and the exchanged catenary current. Thanks to the correct balanced of the DC capacitors, the output voltage is not affected by load unbalances. Concerning catenary current, in case of unbalanced load, the fundamental is slightly increased due to the current flow that appears through the neutral point. However, in case of high currents and small unbalances (like this one), this small increment is negligible.

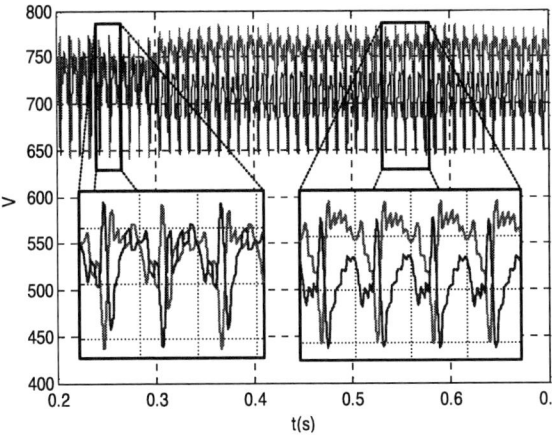

Fig. 10. DC voltage buses in a provoked load unbalance at t=0.3s in full load conditions. Simulation results.

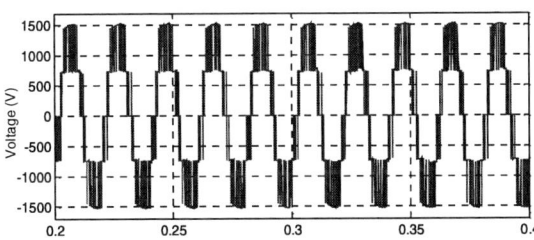

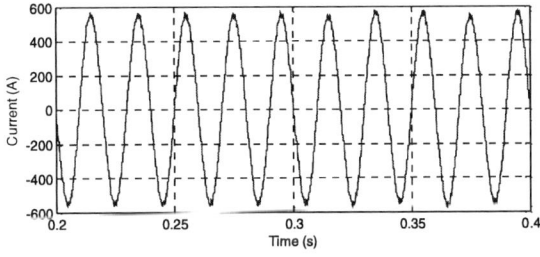

Fig. 11. Output voltage and grid current in a provoked load unbalance at t=0.3s in full load conditions. Simulation results.

### B. DC voltage transient response

During transient simulations, chosen analysis conditions are equivalent to a torque ramp requiring an increase of the absorbed power from zero to 100% in 300ms.

Figure 12 shows the DC bus voltage and the current responses obtained in simulation without and with the proposed load feedforward of the DC bus voltage loop. As it can be seen, the feedforward improves considerably the transient performance of the voltage control loop.

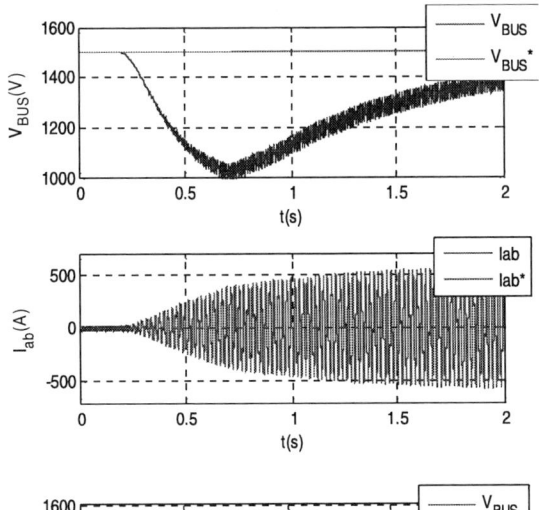

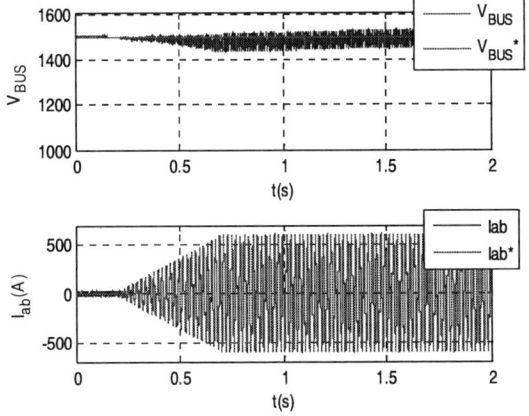

Fig. 12. Total DC bus voltage and input current responses without (top) and with (bottom) load feedforward. Simulation results.

## V. EXPERIMENTAL RESULTS

This simulation analysis has been confirmed experimentally using a 150kW prototype test-bench located in the laboratories of IKERLAN-IK4 (see figure 13). The H-NPC test-bench is composed of 1700V/800A Eupec IGBTs and clamping diodes. The experimental tests have been carried out considering the same main characteristics as in simulation (see table II).

As in the simulated case study, both voltage balancing capability and DC voltage transient response have been analyzed in no load and load conditions. Nevertheless, in this case the converter has been tested in reduced power conditions ($\approx$ 40kW). Experimental tests in load conditions have been carried out using balanced ($R_1=R_2$) and unbalanced ($R_1 \neq R_2$) resistive loads (see figure 14).

Fig. 13. IKERLAN-IK4 H-NPC converter test bench.

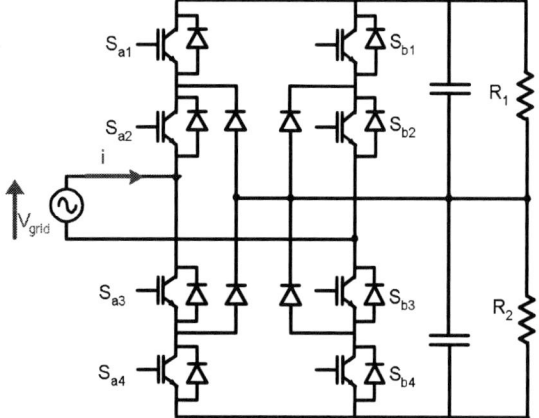

Fig. 14. H-NPC Converter.

## A. Voltage balancing capability

First of all, the voltage balancing capability of the 2SC modulation has been tested experimentally in no load conditions, considering as in simulation, an initial voltage unbalance between capacitors of 200V. As it can be seen in figure 15, both voltage buses are correctly balanced.

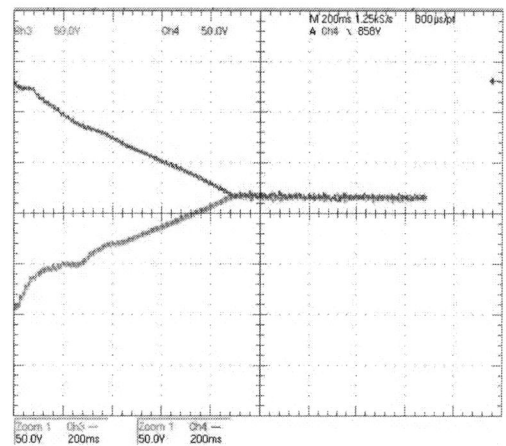

Fig. 15. Initial unbalance in no load conditions. Experimental results.

Secondly, the voltage balancing capability has been analyzed in load conditions. Figure 16 presents the five level output voltage and the grid current in case of balanced, $R_1=R_2=30\Omega$, (top) and unbalanced, $R_1=20$ and $R_2=40\Omega$, (bottom) loads. As in the simulation study, in both cases the output voltage is almost equal (as both DC capacitors are correctly balanced).

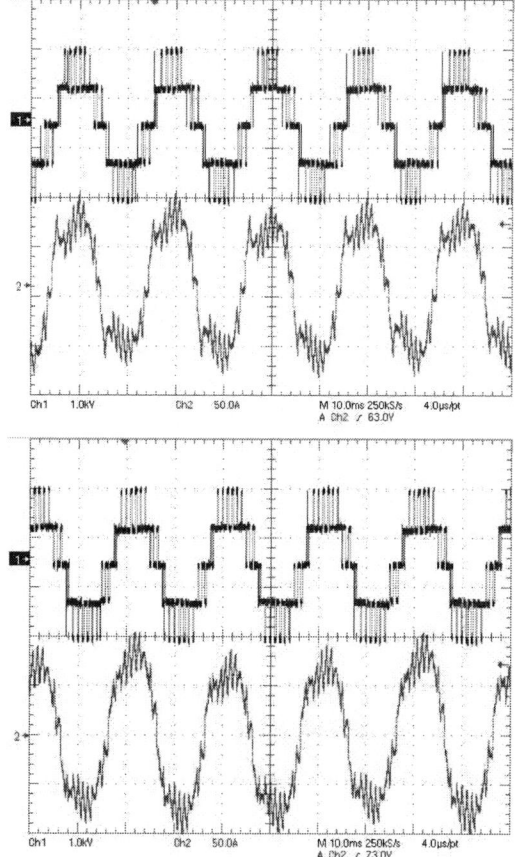

Fig. 16. Five-level output voltage and grid current in case of balanced (top) and unbalanced (bottom) loads. Experimental results.

However, in case of unbalanced loads, the grid current form is slightly different due to the current that flows through the neutral point (which increases the fundamental current and improves the harmonic distortion). It must be said that in both cases grid current is much more distorted than in the case study. This is due to the fact that the experimental tests have been carried out with a considerably smaller power (and current).

Figure 17 proves that in both cases correct voltage balance is achieved and that in case of unbalanced loads, this balancing is not exactly half of the DC bus.

## B. DC voltage transient response

Figure 18 shows the DC voltage buses and the grid current transient responses without (top) and with (bottom) the load feedforward. As it can be seen, without the load feedforward it takes longer to achieve the nominal voltage value while with the load feedforward the voltage response is almost instantaneous.

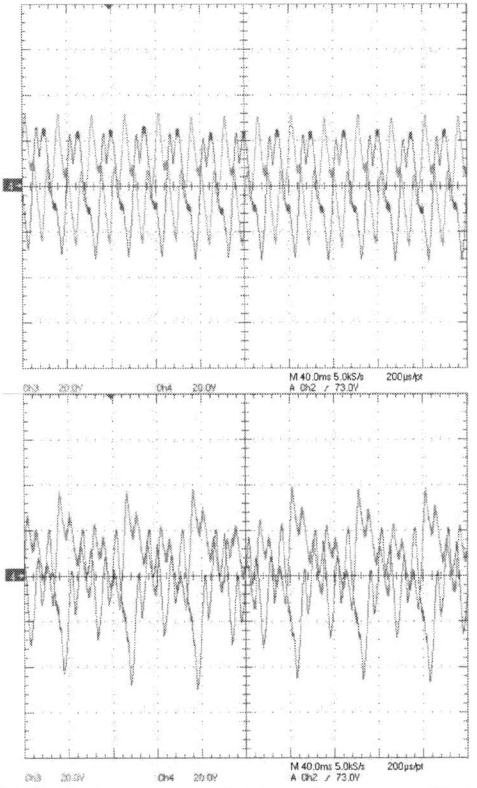

Fig. 17. Steady-state responses of DC bus voltages in case of balanced (top) and unbalanced (bottom) loads. Experimental results.

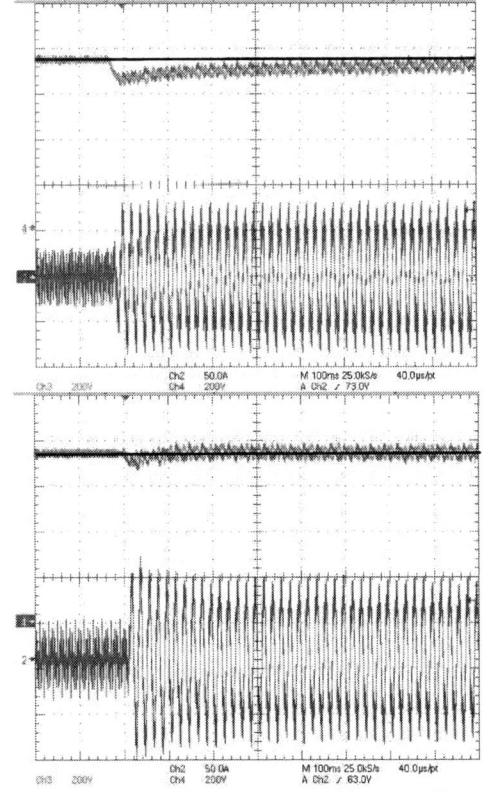

Fig. 18. Load step of DC bus voltages and grid current without (top) and with (bottom) load feedforward. Experimental results.

## VI. CONCLUSIONS

In this paper a H-NPC topology has been analyzed as the front-end converter of an AC traction application. This kind of converters permits to achieve higher voltage levels and better current quality spectra than conventional single-phase H-bridge converters.

The suitability of the 2SC modulation strategy has been evaluated for traction applications, as well as the design requirements of current and voltage control loops.

Presented control and modulation loops have been tested in simulation and experimentally. In the steady-state response, the correct voltage balancing capability has been verified in no load and load conditions. Concerning the transient response, the advantage of using a load feedforward on the voltage control loop has been proved.

## REFERENCES

[1] H-J. Ryoo, J-S. Kim, G-H. Rim, Y-J. Kim, M-H Woo, C-Y. Won. "Unit Power Factor Operation of Parallel Operated AC to DC PWM Converter for High Power Traction Application". PESC'01.

[2] G-W. Chang, H-W. Lin, S-K. Chen. "Modeling Characteristics of Harmonic Currents Generated by High-Speed Railway Traction Drive Converters". IEEE Transactions on Power Delivery, vol. 19, N 2, April 2004.

[3] J.-I. Leon, R. Portillo, L. G. Franquelo, S. Vazquez, J. M. Carrasco, E. Domínguez. "New Space Vector Modulation Technique for single-Phase Multilevel Converters". ISIE'07, Vigo (Spain), June 2007.

[4] J. Salaet. "Contributions to the use of rotating frame control and space vector modulation for multilevel diode-clamped single-phase AC-DC power converters". PhD Thesis of the Polytechnic University of Catalonia (Spain), July 2006.

[5] J. Pou i Fèlix. "Modulation and Control of Three-Phase PWM Multilevel Converters", PhD. Thesis of the Polytechnic University of Catalonia (Spain), November 2002.

[6] A. Lopez-de-Heredia, I. Etxeberria-Otadui, S. Aurtenetxea, G. Abad, M.A. Rodriguez, S. Bacha. "Application of Multiple integrator based controllers for low switching frequency Multilevel NPC Power Active Filters: limitations and improved structures". EPE, Aalborg (Denmark), September 2007.

[7] Y. Sato, T. Ishizuka, K. Nezu and T. Kataoka, "A New Control Strategy for Voltage Type PWM Rectifiers to Realize Zero Steady State Error in Input Current," *IEEE Trans. on Industrial Applications*, vol. 34, n° 3, p. 480-486, May-June 1998.

[8] I. Etxeberria-Otadui, A. Lopez-de-Heredia, H. Gaztañaga, S. Bacha and R. Reyero." A Single Synchronous Frame Hybrid (SSFH) Multi-Frequency Controller for Power Active Filters". IEEE Trans. on Industrial Electronics, October 2006.

# Hybrid – type system of power supply for a trolleybus with an asynchronous motor

Zygmunt Giziński, Marcin Gąsiewski, Ireneusz Maścibrodzki,
Michał Zych, Krzysztof Zymmer, Marcin Żuławnik,
The Electrotechnical Institute, Warsaw, Poland e-mail: npm@iel.waw.pl, nte@iel.waw.pl

*Abstract –* The paper presents results of research work on a prototype trolleybus with asynchronous motor drive system and with energy storage operated by the Urban Transportation Company in Lublin. There are shown operation test results with a capacitor type energy storage and with an additional accumulator battery and without the battery. Basing on the operation test results, the advantages are shown, resulting from the introduction of this technical solution, such as energy saving and the possibilities of travelling over sections of routes without power supply from the traction network.

*Keywords*: municipal transport, trolleybus, energy storage system.

## 1. INTRODUCTION

The Electrotechnical Institute has been for over two years conducting research work on application of capacitor – type energy storage in electric traction vehicles. Since 2006 the work has been conducted in the framework of a special research project COST/260/2006 ordered by the Ministry of Science and Higher Education. Application of capacitor – type energy storage in electric traction vehicles, especially in urban transport, offers measurable energy and operational advantages:
- Reduction in the power taken from the traction network by $20 \div 40\%$ and reduction in energy consumption by about $20 \div 25\%$;
- Significant shortening of the peak power taken from traction substations thus reduction in amounts paid for peak power;
- Accumulation of trolleybus braking energy;
- Travelling a part of the route without power supply from the traction network and travelling in the depot without necessity of having a traction network there and also in urban traffic to travel over sections with damaged network.

Fig. 1. The trolleybus tested in the streets of Lublin

## 2. TROLLEYBUS SYSTEM WITH CAPACITOR TYPE ENERGY STORAGE

In 2006 the Electrotechnical Institute developed and carried out an inverter-type drive system for a trolleybus of 180 kW, bought from Maxwell Company and tested a capacitor-type energy storage of 0.7 kWh/750 V and designed additional converters for controlling the processes of charging the storages with energy and loading them. Fig. 1 shows a diagram of the electric system of the vehicle.

The supplied equipment was assembled in the model trolleybus. The trolleybus with the storage type power supply system was tested in common by the user and the Electrotechnical Institute in June and July 2007 r. in normal operation conditions.

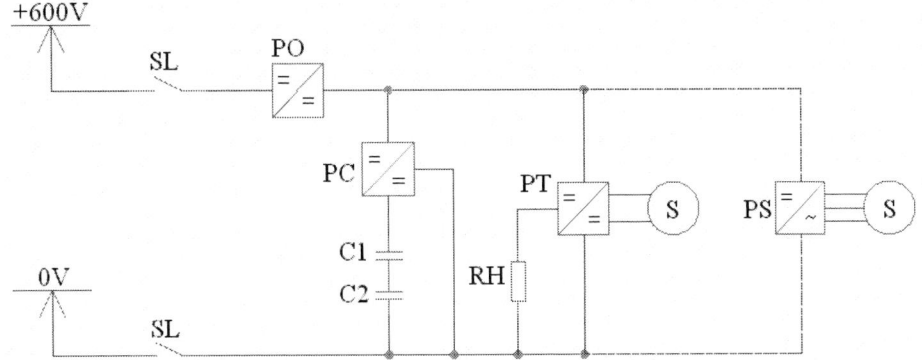

**Fig. 2. Block diagram of the power circuit of the trolleybus with capacitor – type energy storage.**

| | |
|---|---|
| **SL** | - line contactors |
| **PC** | - capacitor changing converter |
| **C1 – C2** | - two sets of 390 V capacitors connected in series |
| **PO** | - converter for returning energy to the traction network |
| **PT** | - transistor converter for DC motors drive |
| **PS** | - inverter for AC drive |
| **RH** | - braking resistor |
| **S** | - traction motor |

The capacitor-type storage C1-C2 is charged with energy from the traction network by the transistor-type converter PC, which controls the charging current and capacitors voltage.

During the vehicle start-up period the current is taken from the capacitors until their voltage drops to the value of the network voltage.

The approximate parameters of the system recorded while the vehicle is traveling, supplied with power from the capacitor-type storage, are shown in Fig. 2:
- mean speed: 23 km/h,
- energy taken from the network: 0.69 kWh,
- energy taken from the capacitor during start-up: 0.25 kWh,
- energy returned to the capacitor during braking: 0.20 kWh,
- mean energy consumption: 1.53 kWh/km.

For comparison, the corresponding data for a vehicle without the capacitor-type energy storage are:
- energy taken from the network: 0.88 kWh,
- mean energy consumption: 2.0 kWh/km.

During braking the energy is stored in capacitors up to the moment when they reach their maximum voltage values of 750 ÷ 800 V. If the vehicle speed at the start of braking is over 50 km/h then the excess of energy is received by the traction network (the PO converter) or lost in braking resistors.

Tests of travelling without network over the depot and along the street, using the capacitor storage drive proved that it was possible to cover 400 m. This makes it possible to travel in the depot without network as well as to cover a section with damaged network.

## 3. TROLLEYBUS WITH HYBRID – TYPE POWER SUPPLY ACCUMULATOR BATTERIES AND CAPACITOR TYPE ENERGY STORAGE.

The vehicle range without supply from traction network can be increased by adding an accumulator battery from which the capacitor – type storage is charged by means of a voltage increasing converter, Fig. 4.

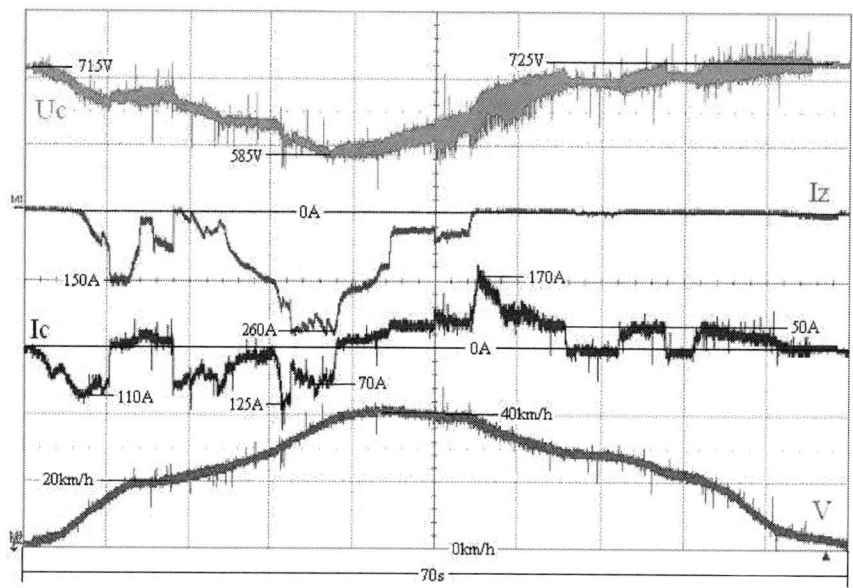

**Fig. 3. Travel of the trolleybus on traction network (ab. 450 m) from the capacitor-type storage (trolleybus loaded with 8 Mg)**
Uc –voltage of the capacitor type storage
**Ic – current of the capacitor type storage**
**Iz – current of the traction network**
**V – speed of the trolleybus**

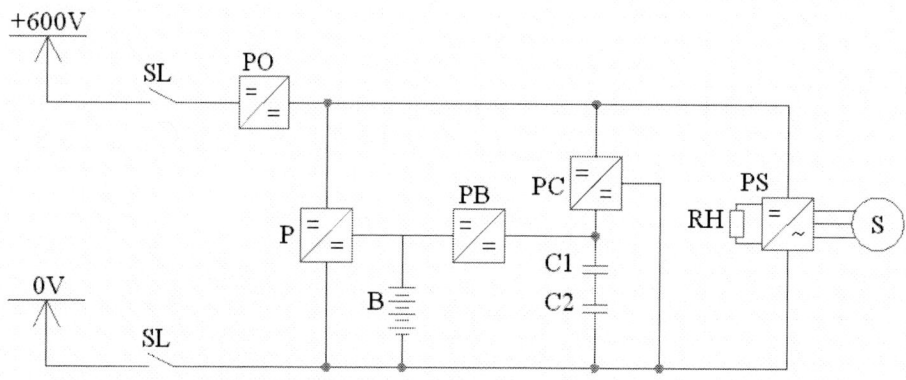

**Fig. 4. Block diagram of the main circuit of the trolleybus with hybrid supply.**
**SL - line contactors**

| | |
|---|---|
| **PC** | **- capacitor changing converter** |
| **C1** | **– C2 - two sets of 390 V capacitors connected in series** |
| **PO** | **- converter for returning energy to the traction network** |
| **P** | **- converter for charging the accumulator battery** |
| **B** | **- accumulator battery of 7 kWh** |
| **PB** | **- converter for charging capacitors from the battery B** |
| **PS** | **- inverter for AC drive** |
| **RH** | **- braking resistor** |
| **S** | **- traction motor** |

Such a solution was used in the electric system of the trolleybus discussed, by installing accumulators of 7 kWh. This allowed to increase the vehicle range (without supply from traction network) up to ab. 2 km. With such a solution applied during a travel of a trolleybus loaded with 8 Mg over a distance of 1600 m with three stops the capacitor voltage dropped from 700 V to 420 V.

The fully loaded trolleybus travelled a maximum distance of 4000 m, with discharging the accumulator battery by 50% (approx.) of the rated capacity. Exemplary diagrams of the vehicle velocity, capacitor-type storage voltage and

current, and accumulator battery current are shown for a distance of 350 m in Fig. 4:
- The initial voltage value of the capacitor-type storage is 670 V;
- The lowest voltage value of the capacitor-type storage is 430 V. It was obtained at V = 32 km/h;

- During braking the capacitor was charged up to the voltage of 640 V;
- After charging from the accumulator battery at the stop (approx. 8 s.) the capacitor voltage is aprox 680 V, which means that the supply system is ready for travelling the next distance.

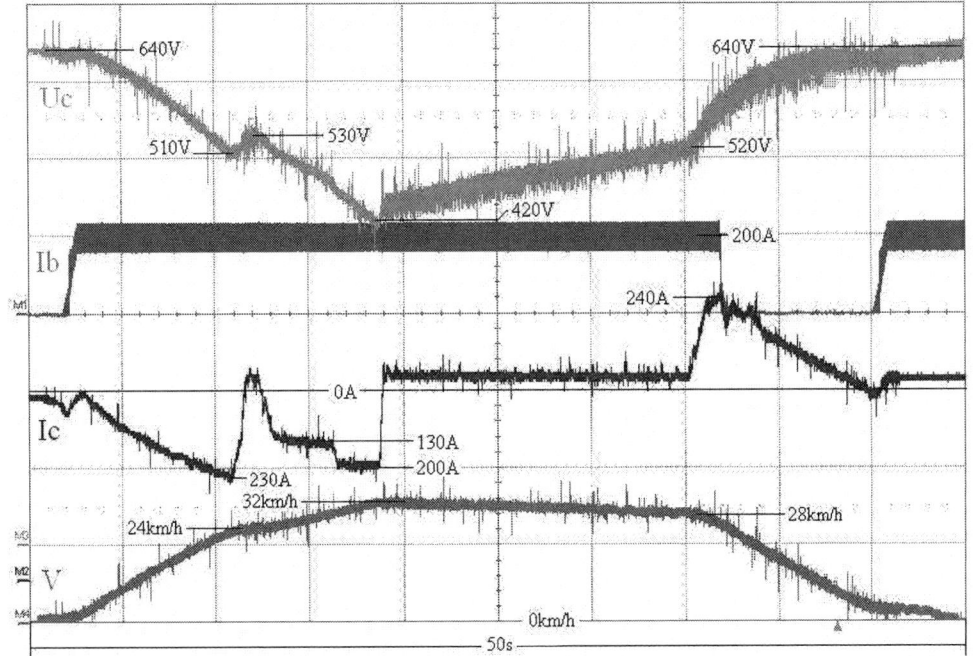

**Fig. 5. Travelling of the trolleybus without traction network (about 350 m) with power supply from the capacitor-type storage and accumulator battery (trolleybus loaded with 8 Mg)**

| Uc | –voltage of the capacitor – type storage |
| Ic | – current of the capacitor – type storage |
| Ib | – current of the accumulator battery |
| V | – trolleybus velocity |

Operational tests of the prototype trolleybus showed that for sections of $1200 \div 2000$ m without traction network planned by the UTC Lublin direction, the energy of the accumulator battery was chosen correctly. Discharging for those conditions does not exceed $20 \div 30\%$ of the rated capacity. For charging the battery back again to its rated capacity the length of the travel should be $10 \div 12$ km ($1.5 \div 2.0$ km without network, $8 \div 10$ km with power supply from the traction network). Life time of the battery used (with annual operation distance 60 000 km) will amount to – $2.5 \div 3.5$ years according to the producer's data.

### 4. VELOCITY CONTROL SYSTEM

The trolleybus is driven by an asynchronous motor with a power of 165 kW, rated torque 1185 Nm, and critical torque 3725 Nm and rated frequency 60 Hz with a power electronic velocity control system in the form of a microprocessor controlled transistorized voltage inverter. The motor operates in the range of angular velocity $0 \div 1200$ rpm with a constant torque while between 1200 and 3100 rpm with constant power. The motor is equipped on the shaft with a sensor detecting the velocity and the rotor position.

The velocity control system of the trolleybus must meet the following requirements:
- sufficient driving torque at zero velocity ensuring smooth start of a fully loaded vehicle, start on a slope with a possibility of overcoming a road obstacle – e.g. a kerbstone;
- operational velocity up to 60 km/h and a maximum velocity of 75 km/h;
- acceleration at start-up amounting to 1.3 m/sec$^2$ and deceleration at braking of 1.6 m/sec$^2$;
- smooth start and braking and smooth transition from motor operation to generator operation ensuring passenger's comfort;
- ensuring smooth starting uphill in case of rolling back downhill caused by loosening mechanical brakes when departing from a stop;
- ensuring smooth and reliable operation of the drive system at changing variable voltages in the traction network within $(400 \div 750)$ V;
- ensuring recovery of the energy of braking to the capacitor – type storage.

The velocity control system was assembled on the trolleybus and tested in urban traffic in Lublin. The test rides were carried out with 8 Mg load and without load at different operation conditions like starting, braking, rapid transitions from motor to generator operation and vice-versa. The tests were carried out using power supply from traction network and from the capacitor – type energy storage at the same time. Recorded variations of motor phase current, voltage across the inverter filter capacitor, and vehicle speed for different operation states (travelling forwards, braking, coasting) at power supply from traction network and from the capacitor – type storage at the same time are shown in Fig. 6. The corresponding recordings for "travelling forward", coasting at power supply only from the capacitor – type storage are shown in Fig. 7.

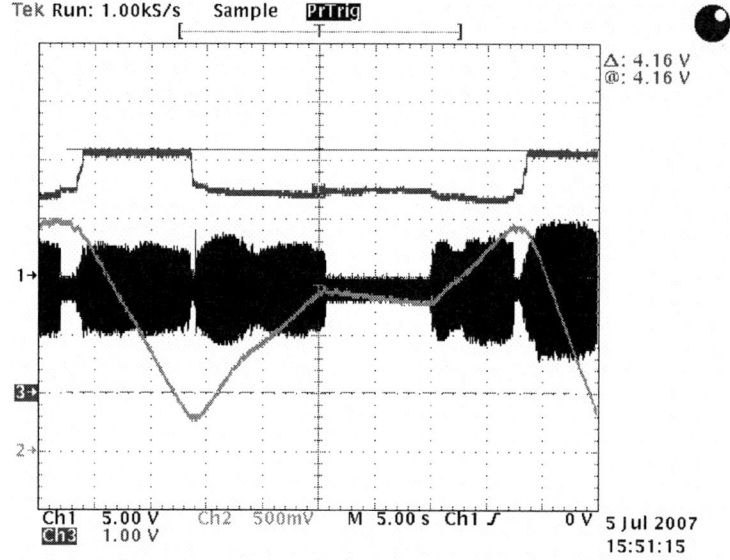

**Fig. 6. Trolleybus rides in urban traffic (starts, coasting and braking) at supply from traction network and the capacitor – type storage:**
**1 – motor phase current (400 A/div.), 2 – trolleybus speed (14 (km/h)/div), 3 – voltage across inverter filter capacitor (200 V/div).**

The operational tests showed that the power electronic speed control system of the bus meets the operational requirements posed by the Urban Transport Company in Lublin in the scope of technological parameters (acceleration at start and breaking, vehicle speed) as well as the passengers comfort related to smooth changes in operation states. The drive system also cooperates satisfactory with the capacitor – type energy storage where the breaking energy is returned and then used at the subsequent starts.

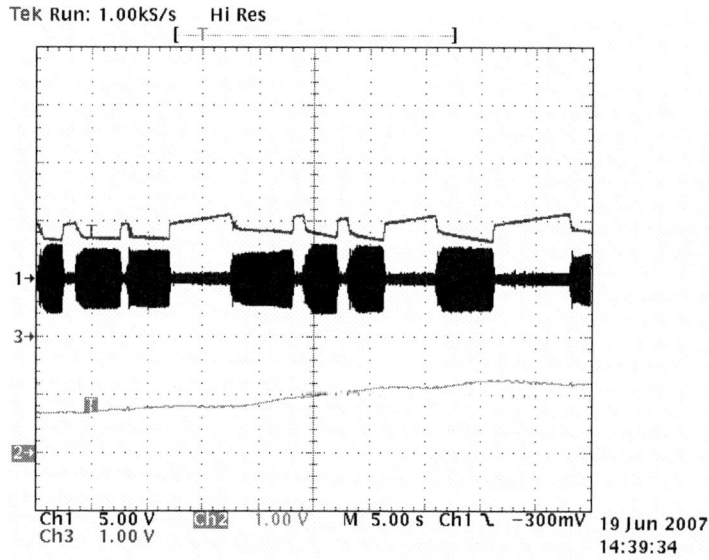

**Fig. 7. Travel recording of the trolleybus in the urban traffic (starts and coasting) without the traction network with, power supply from the capacitor – type storage: 1 – phase current of the motor (400 A/div.), 2 –trolleybus speed ( 28(km/h)/div),**
**3 – voltage across the inverter filter capacitor (200 V/div).**

## 5. CONCLUSIONS:

1) Operational tests of the trolleybus with energy storage have fully confirmed the advantages resulting from its hybrid – type power supply.

2) The capacitor – type energy storage of 0.7 kWh permits travelling without power supply from traction network both around the depot as well as along the route – e.g. in case of damaged traction network, to leave the street crossing. The energy stored in the capacitors is enough to travel $200 \div 400$ m.

3) The advantages of the trolleybus with capacitor type energy storage, confirmed by operation tests, are as follows:

   a) reduction the starting time and starting current taken from the network;

   b) storage of braking energy $0.1 \div 0.3$ kWh, practically the energy of braking at a velocity of $40 \div 50$ km/h;

   c) reduction of the mean power taken from the traction substation by $20 \div 30\%$ approx.

4) Testing the hybrid – type supply system with an additional accumulator battery of 7 kWh of a trolleybus loaded by (8 Mg) along the planned route and three replacement routes showed the possibility of travelling 2000 m with the accumulator battery discharged by $20 \div 30\%$ of its rated capacity ensuring $2 \div 3$ years durability of the battery. Such a system creates practical possibilities to introduce trolleybus traction into the historical parts of towns without any necessity of installing a traction network.

5) A trolleybus equipped with a drive with an asynchronous motor and an power electronics system of speed control with microprocessor control, was subjected to operational tests in the area of the town of Lublin. The tests showed that the speed control system satisfies the requirements in the range of operation parameters (the required driving torque at zero velocity, acceleration at starting and braking, operation and maximum velocity, smooth transition from motor operation to generator operation and vice versa). The system ensures co-operation with the capacitor – type energy storage in the aspect of storing the braking energy and using it at the vehicle start.

## REFERENCES:

[1] Giziński Z., Żuławnik M., Gąsiewski M., Zych M.: *Hybrid power supply system for a trolleybus*. TTS Technology of rail transport nr 9 2007 r.

[2] Giziński Z.: *Storage-type power supply system for a trolleybus*. Nowa Elektrotechnika nr 7-8 2007 r.

[3] Giziński Z.: *Capacitor-type energy store for electric traction vehicles*. Conference proceedings of the of the Power Supply Commission Nałęczów 2006 r.

[4] Żuławnik M.: *Hybrid-type power supply for a trolleybus*. Conference proceedings of the of the Power Supply Commission Nałęczów 2006 r.

[5] Giziński Z., Żuławnik M.: *Hybrid-type power supply system for a trolleybus*. International Conference Modern Electric Traction MET 2007 r.

[6] Juda Z., *Buforowe wtórne źródło energii z superkondensatorami w zastosowaniu do napędu elektrycznego pojazdów*, Journal of KONES Internal Combustion Engines 2005, vol. 12, 3-4,str.141-148.

[7] Jari Keskinen, *Ultracapacitor –A tool for hybrid power*, VTT, Technical Research Centre of Finland 2007.

[8] Maher, *Ultracapacitors and the Hybrid Electric Vehicle*, White Paper, Maxwell Technologies.

[9] John M. Miller,Dragan Nebrigic, Michael Everett, *Ultracapacitor Distributed Model Equivalent Circuit for Power Electronic Circuit Simulation*, Maxwell Technologies 2006.

[10] S. Pay,Y. Baghzouz: *Effectiveness of Battery-Supercapacitor Combination in Electric Vehicles*, 2003 IEEE Bologna Power Tech Conferene.

[11] A.Rufer, D.Hotellier, P.Barrade: *A Supercapacitor-Based Energy-Storage Substation for Voltage-Compensation in Weak Transportation Networks*, 2003 IEEE Bologna Power Tech Conferene.

# Control of rotor flux in AC tram drive during sudden braking operation

Andrzej Dębowski, Piotr Chudzik

Technical University of Łódź / Institute of Automatic Control – Łódź, Poland,
e-mail: *andrzej.debowski@p.lodz.pl , piotr.chudzik@p.lodz.pl*

*Abstract* — **The paper focuses on the behavior of the rotor flux indirect control system for an inverter fed AC tram drive. The control strategy used in this drive is based on the original state stimulator concept and has been applied in several modernized trams and in a trolley-bus in Poland. In these vehicles some oscillations of the modulus of the rotor flux vector during sudden braking operation were observed. The phenomenon of oscillations was tested in laboratory stand with adequate physical model of real tram drive as well as in a computer simulation model. These research allowed to explain the nature of the oscillations and to find some way to improve the control quality.**

*Keywords* — **Control of drive, Induction motor, Traction application, Vector control.**

## I. INTRODUCTION

In the last years modern trams are often equipped with induction motors. These drives are very reliable, they can be working without need of constant repairs, even in extreme conditions. AC drive has also good dynamical properties, especially if they have been controlled using more sophisticated vector control methods. Several years ago the team from Technical University of Łódź ' together with ENIKA company from Łódź decided to build an own AC tram drive based on the indirect rotor flux control of the induction motor [2]. This drive was designed for modernized trams in polish cities (Fig. 1).

Fig. 1. Modernized tram in Łódź.

Since then 5 drives have been built for the trams currently used in Łódź and Elbląg. Also, in 2005 this drive was used for modernization of a trolley-bus in Lublin. The next trams with drives should appear on the Poznań

---

' This paper describes a part of a research work supported by the Polish Ministry of Science and Higher Education, grant No. R01 014 01.

streets in July 2008. The research conducted during building and testing of these drives revealed many different control problems. One of them was connected with sourcing of traction drives with use of grid lines, returning energy during dynamical breaking or keeping high drive efficiency [1].

One of the most important problems which were given careful analysis was ensuring stability of drive during braking, especially during sudden braking. Proper functionality of the drive in such difficult situation as sudden braking is very important because decides on passengers safety. Sudden, often very high change of the desired torque in the induction motor can appear with different speed of the vehicle. During braking slide can appear which can lead to sudden, vast change of the wheel speed. Drive must react in such situations very fast to regain grip. For this to happen, drive has to change torque sign and value very fast. Authors decided to carefully analyse transients appearing in the drive and determine control system structure which assures proper drive behaviour during rapid changes of torque.

## II. CONTROL ALGORITHM OF THE DRIVE

### A. Dynamical description of the drive

The proposed control strategy for tram AC drive is based on indirect torque and rotor flux control approach with use of appropriate forcing of the stator current and slip frequency. The dynamical description of the electromagnetic state of the motor, represented by the rotor flux vector, can be expressed in rotated reference frame related to the stator current vector shown in Fig. 2.

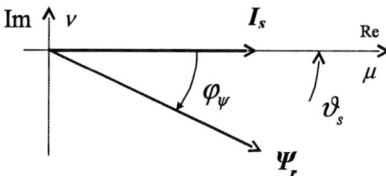

Fig. 2. Representation of rotor flux vector in rotating stator-current oriented coordinates system.

Selection of the AC motor for traction applications depends strongly on economical reasons. That means, that it is important to keep the dynamical performance of the vehicle as high as possible but the low price of the drive is also required. Usually the compromise in this choice is achieved thanks to the selection of the AC drive nominal parameters corresponding to the middle velocities of the vehicle. Unfortunately in this case, during the motor

---

978-1-4244-1741-4/08/$25.00 ©2008 IEEE     1568

operation with higher velocities we observe the lack of the voltage necessary for effective driving. Hence, because of the voltage limitation, for higher tram speed a control of the induction motor with the field weakening is necessary.

In proposed approach to the induction motor control problem, it is very convenient to introduce a decomposition of the motor dynamics description [1], [2], [4], which can be divided into two parts:
- stator-current producing part (stator voltage vector is a control input and stator current is an output),
- motor inner-torque producing part (where stator current vector and slip frequency are control inputs and inner-torque is an output).

The main part of the discussed induction motor control approach is the rotor flux dynamical model expressed in polar coordinates system

$$T_r \frac{d}{dt} \Psi_r^* = -\Psi_r^* + L_m I_s^* \cos \varphi_\psi^*$$
$$\frac{d}{dt} \varphi_\psi^* = -\frac{1}{T_r} L_m I_s^* \frac{1}{\Psi_r^*} \sin \varphi_\psi^* + \Omega_r^* \qquad (1)$$

where: $\Psi_r^*$ - rotor flux, $\varphi_\psi^*$ - torque angle, $\Omega_r^*$ -slip frequency, $I_s^*$ - stator current, $T_r = L_r / R_r$ - rotor time constant, $L_m$ - mutual inductance.

The desired motor torque value is

$$M^* = -\frac{3}{2} p \frac{L_m}{L_r} \Psi_r^* I_s^* \sin \varphi_\psi^* . \qquad (2)$$

It was assumed that the state of dynamical model (1) is controlled by two new control variables

$$T_r \frac{d}{dt} \Psi_r^* = S_\Psi , \qquad \frac{d}{dt} \varphi_\psi^* = S_\varphi . \qquad (3)$$

Equations (3) describe the dynamical system, being a kind of compensator (called by the authors the state stimulator), which should be followed by the motor. To achieve this effect is necessary to connect the model (3) with current control loop by coupling system, obtained by transformation of right sides of equations (1)

$$I_s^* = k_\psi \frac{\Psi_r^*}{L_m \cos \varphi_\psi^*} , \qquad \Omega_r^* = \Omega_{rm} - S_\varphi . \qquad (4)$$

where: $\Omega_{rm} = -k_\psi \frac{1}{T_r} tg \varphi_\psi^*$ - component of desired slip frequency, proportional to the torque,

$k_\psi = 1 + \frac{S_\psi}{\Psi_r^*}$ - correction coefficient.

The motor torque desired value takes now the form

$$M^* = \frac{3}{2} p \frac{(\Psi_r^*)^2}{R_r} \Omega_{rm} . \qquad (5)$$

In a solid coordinates system fixed to the stator windings, position of the current vector $I_s^*$ is described by the angle

$$\vartheta_s^* = \int_0^t (\Omega_r^* + \Omega) dt \qquad (6)$$

where: $\Omega$ - instantaneous value of motor shaft speed.

If the drive is working with constant excitation ($\Psi_r = const$), the component $\Omega_{rm}$ of desired slip frequency depends on torque angle directly. Although when a need of working with a changeable flux value the correction coefficient must be taken into account in control laws (4). So the desired slip frequency value which corresponds to the desired torque value is to calculate from

$$\Omega_{rzd} = \frac{2}{3} \frac{R_r}{p} \frac{M_{zd}}{(\Psi_{rzd})^2} . \qquad (7)$$

The given equations are leading to the block diagram of the nonlinear control algorithm shown in Fig. 3.

This system has a bit complicated structure but there is very important to control torque and flux very accurately during sudden braking because the voltage limitation at high speed of the motor require field weakening. Hence the desired value of rotor flux modulus can be expressed as a function of rotor speed and grid voltage (Fig. 4).

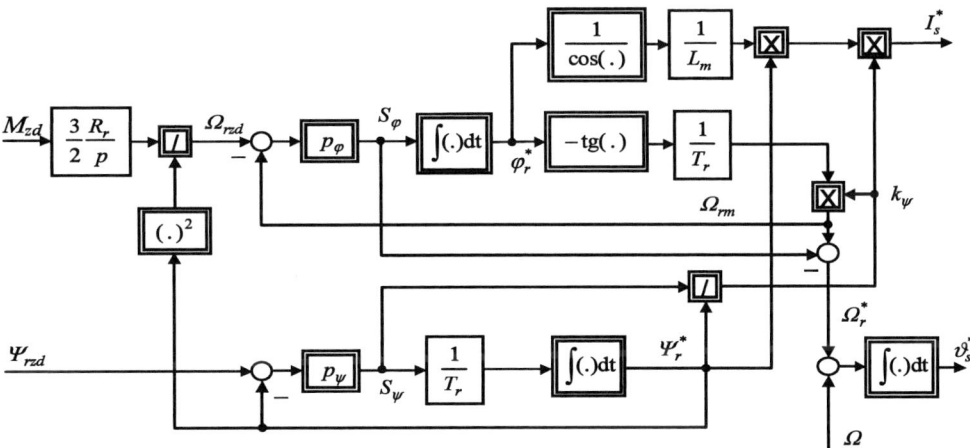

Fig. 3. Block diagram of the rotor flux stimulator being the base for IM indirect control algorithm.

$$\Psi_{rzd} = F(U_T, \Omega). \qquad (8)$$

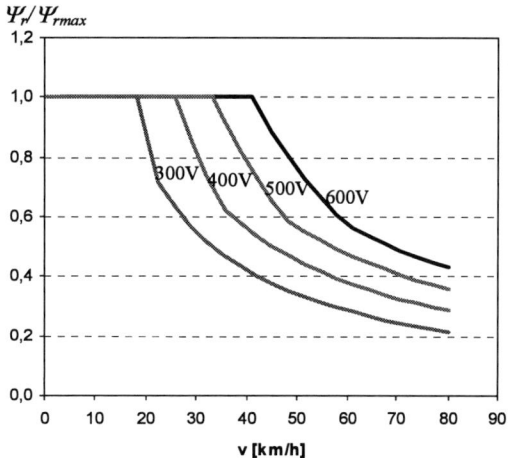

Fig. 4. Maximum value of rotor flux vector modulus as a function of grid voltage and tram velocity.

## B. Control unit

The vector control algorithm for steering the AC tram drive, mentioned above, is implemented in a control unit equipped with the DSP microprocessor. The procedures connected with realization of the indirect flux control algorithm and with measurement and converting all necessary analog physical signals to the digital form operate in microprocessor system in the real time. So, implementation of this part of chosen control algorithm consisting on calculating desired values of stator current vector and slip frequency to the drive control system had not posed a problem.

Much more difficult task for control unit is to supervise by microprocessor operating system the realization of the stator phase current transients in electric motor, corresponding to the calculated instantaneous values of the desired current vector. Compatibility of the real stator current vector in the motor with the calculated desired vector, also in dynamical states, plays a very important role to assure suitable dynamics of the drive and to keep the electromagnetic state of machine fully under control. In the described drive system this task is performed by a stator current vector controller steering the IGBT switches. This vector controller shapes the output voltage of the IGBT inverter using traditional pulse width modulation (PWM) or vector based pulse width modulation (VPWM), in such a way, that the current in stator windings of the motor are maximally similar to their desired values resulting from the desired current vector components worked out in flux indirect control subsystem (stimulator). The higher is the switching frequency so the better is desired stator current forced in the motor. A limitation for current control algorithm accuracy is the maximal switching frequency (specified by technical data of IGBT modulus) and actual value of the traction grid voltage limiting the maximal inverter output voltage.

## III. The Subject of Considerations

During the tests of the tram drive some unpleasant oscillations in stator current and motor torque transients

had been observed [2]. The stator currents are much simpler to observe then the torque on motor shaft or the flux vector inside the motor. Authors of this paper in earlier works [2], [4], and also other researchers [6], [7], have had supposed, that these oscillations are caused by non-perfect work of the stator current vector controller, because some of these oscillations in current transients could be significantly depressed by careful and proper tuning of this controller (Fig. 5).

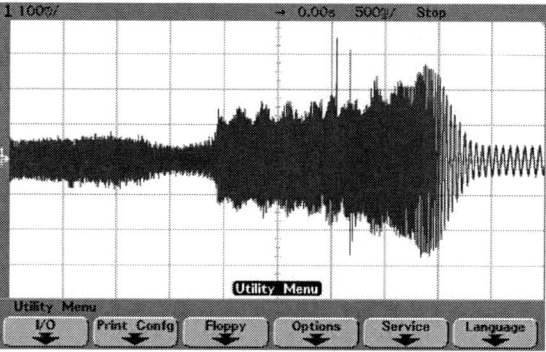

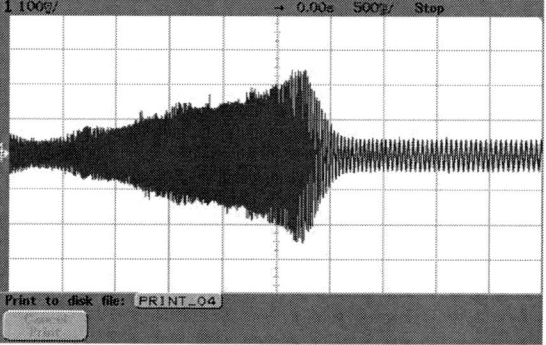

Fig. 5. Oscillations is stator current transients appearing by tram braking can be eliminated by accurate experimental tuning of the conventional current controller.

To make the controller tuning operation much easier, an improved structure of the current controller was proposed by authors [4]. This new structure with EMF compensation was provisionally checked by simulation of the laboratory model of the AC tram drive built in suitable scale with a smaller induction motor (of 5 kW). The results have shown, that the current oscillations still remain, even though the much more sophisticated current controller was applied.

The further investigations made by authors of this paper using the simulation model have shown, that current oscillations are the result not the reason of the oscillations observed in the motor. The authors supposed that the real reason of these oscillations are the disturbances appearing in flux vector replacement, especially during the sudden braking, and the oscillations observed in current and torque can be explained rather as a visible effect of this phenomenon (Fig. 6). Sudden braking of the tram introduces stepwise change in the sign of the desired torque command. In the considerate control system it means, that in stator-current related coordinates the rotor flux vector have to change stepwise its position in reference to to stator current vector.

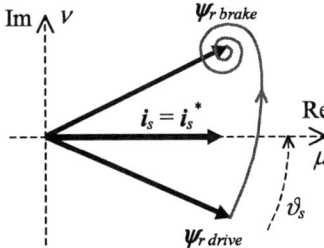

Fig. 6. Schematic vector diagram of the indirectly controlled motor showing the position change of rotor flux vector during the sudden braking of the drive, expressed in the desired stator-current vector related coordinates frame.

## IV. LABORATORY MODEL OF THE TRAM DRIVE

The laboratory stand with the AC tram drive built in smaller scale (shown in Fig. 6) makes possible the preliminary practical verifying of the simulation results and the testing of the drive behaviour in different operating conditions which are difficult to mathematical modelling.

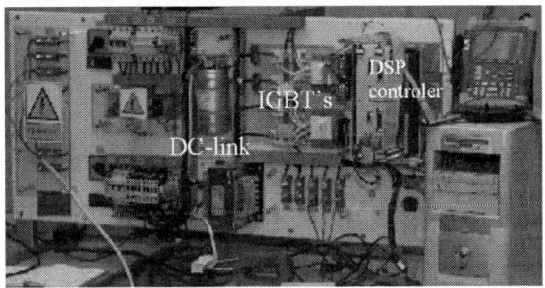

Fig. 7. Laboratory stand with the AC motor of 5 kW.

One of possible disturbances is wheel slipping during the braking process. The another one is a dynamical state when the wheel loses the contact with the rail for a while, when the tram is running across rail-track crossing. In this situation the wheel is supported itself on its flange for a while and then the speed of its rotation changed suddenly. It is very important to take such circumstances it into account by designing of the controllers in AC drives which will be applied in traction vehicles like trams or trolley-buses.

All of the phenomena, mentioned above, have an influence on the reliability of the drive and the comfort of the passenger's travel. The tested AC induction motor is coupled with additional DC motor. Thanks to the own active sourcing system of the DC motor connected to the DC generator in a Ward-Leonard system it was possible to arrange different AC motor-shaft load changes at the laboratory stand. The laboratory stand is also equipped with DC-link enclosing appropriate inductor and capacitor connected to the inverter, which enables generating of

most kinds of disturbances taking place in a real traction drive [3], [4], [5].

The most completely report about the investigations results performed at this laboratory stand, the authors intend to publish separately.

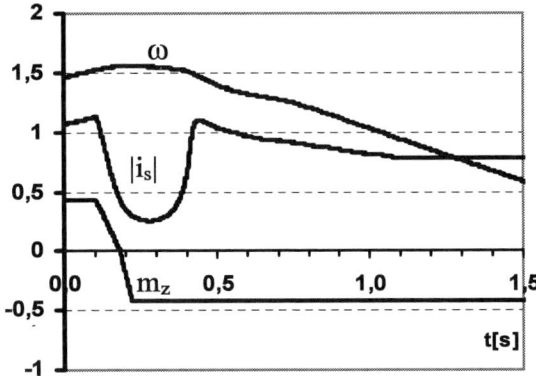

Fig. 8. Transients of desired modulus of stator current - $i_s$, desired torque – $m_z$ and real speed ω during sudden braking, registered in DSP controller memory.

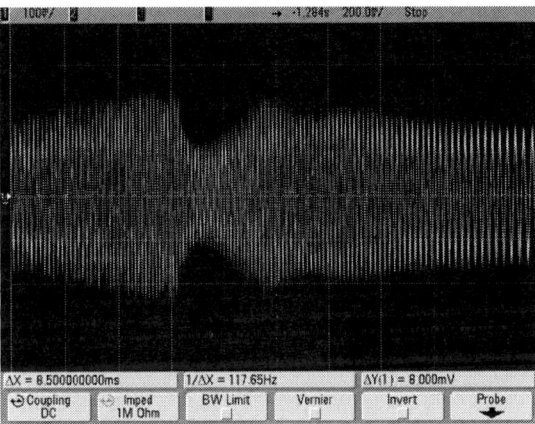

Fig. 9. Oscillogram related to desired changes defined in figure given above, showing that some oscillations in stator current transients still remain in spite of correct tuned current controller.

Results of tests arranged at the laboratory stand also with the correct tunned current controller confirmed, that stator current vector oscillations are still existing and that this phenomenon is evoking especially by high and sudden change of desired torque value (for example like these shown in Fig. 8). This effect is clearly visible by the high velocities of the drive.

## V. SIMULATION MODEL OF THE TRAM DRIVE

For the sake of complicated nature of current control problem and specific tram supplying conditions it was made a decision about constructing the tram drive simulation model, which has allowed for exact examination of the earlier proposed structure of current controller [4] and its properties in different operation condition. The simulation model was built using the PSIM programming package of the Powersys company. By constructing the model, some effort were made for accurate representation

1571

of the real drive system details. Special attention was turned for taking into account the temporal dependences between individual events occurring in control algorithm and power electronics elements. Model being under consideration is rather complex, because it contains all important drive components. Because the complete block scheme of the simulation model is to large, it will not be shown here and only some resulting transients are given in Fig. 10 and 11.

On these pictures are demonstrated transients corresponding to the exciting the flux in the standstill (from 0 to 1,3 sec.), driving the motor with nominal limitation of the torque (from 1,3 sec. to 2,5 sec.), braking

the motor with doubled nominal limitation of the torque (from 2,5 sec. to 3,1 sec.) and repeated driving the motor with nominal limitation of the torque (from 3,1 sec.).

Research made using this simulation model, which has parameters corresponding to the technical data of the laboratory AC drive, show that phenomenon of oscillation of current, torque and flux, observed at the physical laboratory stand, can be reproduced even if controller has the perfect structure described in [4]. The oscillations can be a bit reduced when the gain of flux modulus controller is set significantly low, i.e. when the demand on fidelity of reproducing the changes of the desired flux modulus value is decreased.

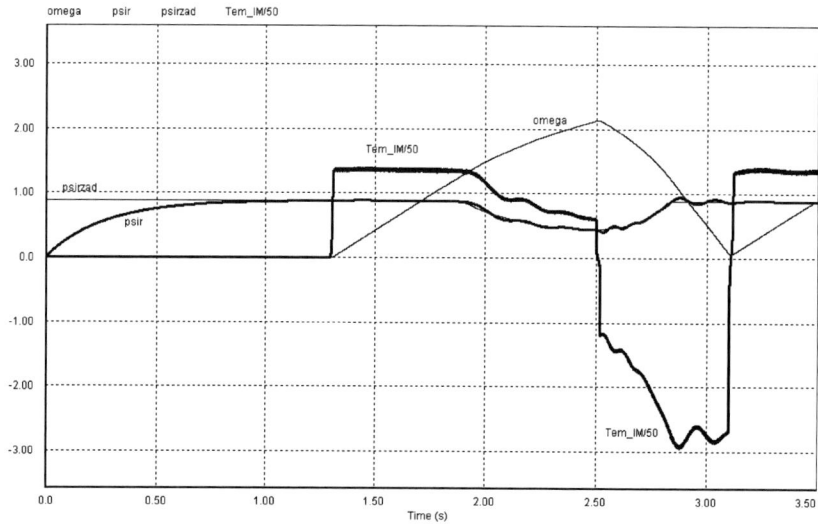

Fig. 10. Transients in full tram drive simulation model with stator current controller with compensation of the electromotive force (flux controller gain = 5).

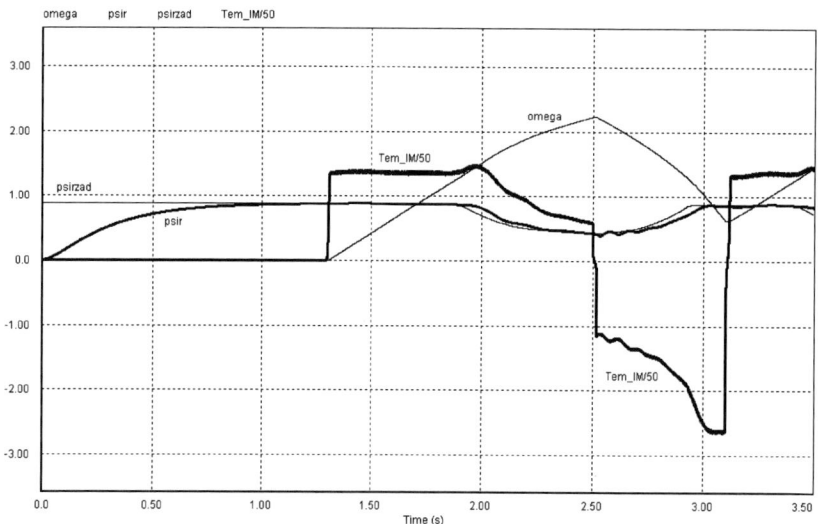

Fig. 11. Transients in full tram drive simulation model with stator current controller with compensation of the electromotive force (flux controller gain = 0.05).

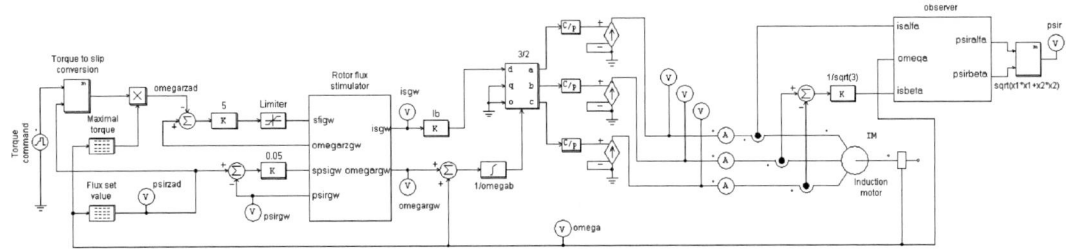

Fig. 12. Simulation scheme in PSIM of the drive with ideal current sourcing of the induction motor model

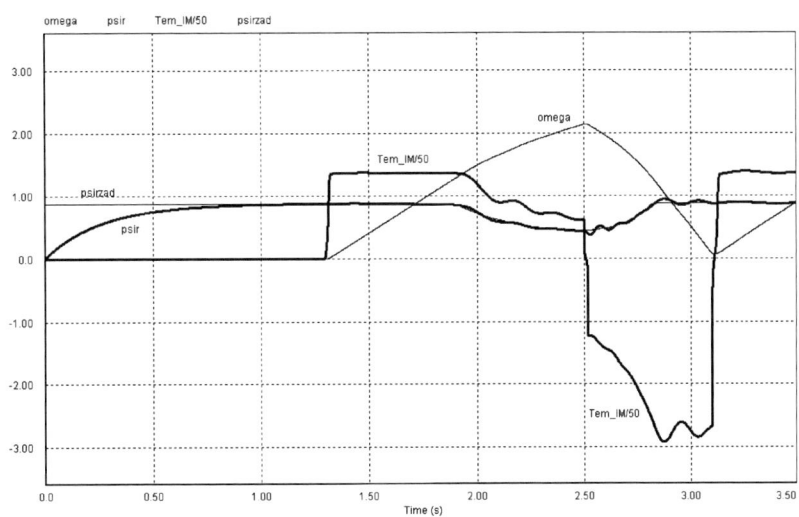

Fig. 13. Transients in tram drive simulation model with ideal current sourcing
of the motor model (flux controller gain = 5).

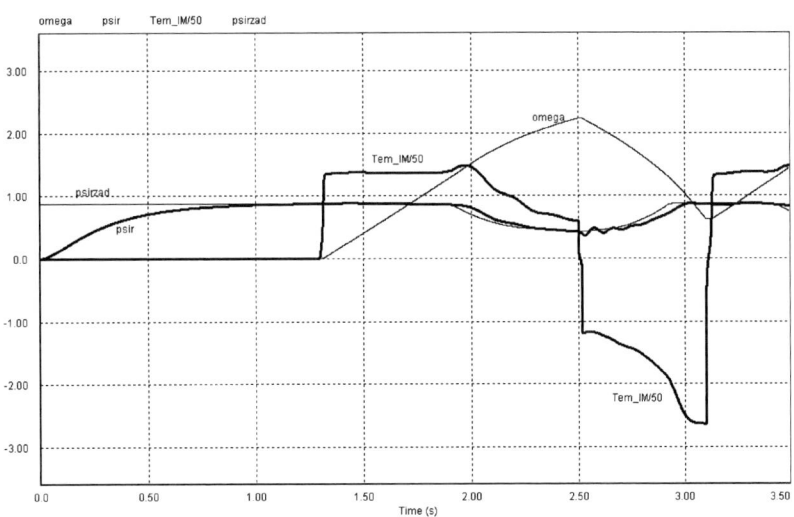

Fig. 14. Transients in tram drive simulation model with ideal current sourcing
of the motor model (flux controller gain = 0.05).

**1573**

Results of simulations shown in Fig. 10, 11, 13 nad 14 shown, that the supposition that the oscillation observed in vector controlled AC drives can be caused not only by improper work of the stator current control loop.

The theoretical explanation of this phenomenon can by easy given for indirectly controlled drive based on so called state stimulator. It is easy to see, that for exact control of the induction motor according to the model (1) using the control laws (4), not only the desired stator current vector $I_s^*$ should be forced perfectly.

It is fully understood, that the compensator (stimulator) dynamical model (3) with appropriate proportional controllers being an inertial system, which precedes the motor could not evoke any oscillation. Such an oscillation can appear only as an effect not proper cooperation between the indirect controller and the motor.

The tracking error between the rotor flux dynamical model (1) and appropriate part of the real motor dynamics description, expressed in rectangular coordinates system related to the desired stator current vector $I_s^*$, as it was more detailed shown in [4], has the form of equation

$$\frac{d\Delta \Psi_r(t)}{dt} = A(t)\Delta \Psi_r(t) , \qquad (9)$$

where flux tracking error

$$\Delta \Psi_r(t) = \Psi_r^*(t) - \Psi_r(t) , \qquad (10)$$

and the tracking equation matrix

$$A(t) = \begin{bmatrix} -\dfrac{R_r}{L_r} & \Omega_r^*(t) \\ -\Omega_r^*(t) & -\dfrac{R_r}{L_r} \end{bmatrix} . \qquad (11)$$

It is easy to check, that the eigenvalues of the matrix (11) are

$$\lambda_{1,2} = -\frac{R_r}{L_r} \pm j\Omega_r^*(t) . \qquad (12)$$

Because the eigenvalues of the matrix $A(t)$ have non-zero imaginary parts, the transients ot the flux tracking error components have always tendency to oscillations with frequency proportional to the motor loading (the higher is the motor load torque - the higher frequency have the flux oscillations). The negative real parts of eigenvalues show that the system has the own stability and both error components are always vanishing to zero.

The oscillations are appearing only when the tracking between the desired state in control system (stimulator) and the real state of the motor is in any way disturbed – i.e. when in any time instant the initial conditions of the stimulator and the motor are not the same. The sudden braking and the rapid change of the desired motor torque is such a disturbance. The expected salvation before this effect is not only to improve the quality of stator current control, what means the condition of very accurate forcing the stator current $I_s(t) \cong I_s^*(t)$, but also to ensure in the real control system the correct frequency synthesis, what means the similar condition for the slip frequency $\Omega_r(t) \cong \Omega_r^*(t)$.

## VI. RESUME

To avoid harmful oscillations in indirectly controlled AC drives, it is necessary to achieve compatibility of the real rotor flux vector in the motor with the calculated desired vector, not only in steady states, but also in dynamical changes of the working conditions of the motor. To ensure this, it would be advisable to construct not only the efficient current controller, but also to careful implement in microprocessor system all angle calculations connected with coordinates transformation.

## REFERENCES

[1] A. Dębowski, W. Błasinski, H. Mroczek: Indirect control system for induction motor based on state stimulator. Proc. of the Int. Conf. on Electrical Machines, ICEM'96, Vigo (Spain), 1996, pp. 248-253.

[2] P.Chudzik, A. Dębowski, G. Lisowski: Results of testing a vector controlled AC tram drive. Proc. of the Int. Conf. on Power Electronics and Intelligent Control for Energy Conservation, PELINCEC, Warsaw (Poland), 2005, pp. 161-1 - 161-6.

[3] A. Dębowski, P. Chudzik, G. Lisowski: State transitions in vector controlled AC tram drive. Proc. of the 12th Int. Power Electronics and Motion Control Conf., EPE-PEMC'2006, Portoroż (Slovenia), 2006, pp. 479-484.

[4] A. Dębowski, P. Łukasiak: Design of current controller in an AC drive using a state stimulator concept. Proc of the 12th European Conf. on Power Electronics and Applications, EPE'2007, Aalborg (Denmark), 2007, pp. P1 – P6.

[5] H. Mosskull: Controllability analysis of an inverter fed induction machine. Proc. of the 2004 American Control Conf., Boston, Massachusetts , 2004, pp.76-81

[6] Z. Peroutka, K. Zeman, J. Flajtingr: Active regenerative braking: braking of induction machine traction drive with maximum torque in high speeds. Proc.of the 12th Int. Power Electronics and Motion Control Conf., EPE-PEMC'2006, Portoroż (Slovenia), 2006, pp. 485 - 490.

[7] Z. Peroutka, K. Zeman: Robust field weakening algorithm for vector-controlled induction machine traction drives. Proc. of the 32nd Annual Conf. of the IEEE Industrial Electronics Society, IECON'06, Paris (France), 2006, pp. 856-861.

# A New Novel Power Electronic Circuit to Reduce Stray Current and Rail Potential in DC Railway

Reza Fotouhi
Railway Engineering Department
Iran University of science & technology
Tehran, Iran
Reza.fotouhi@gmail.com

Siamak Farshad
Railway Engineering Department
Iran University of science & technology
Tehran, Iran
Farshad@iust.ac.ir

*Abstract— The problem of stray current in DC railways has not been solved completely in spite of many years of research and investigation, specially where there is not good rail insulation to prevent corrosion risk to the infrastructure. A new novel circuit, based on power electronic semiconductors, which has the same advantage of a booster transformer in AC systems, is proposed in this paper for DC railways. This circuit forces the return current to flow through return current wire (RCW) instead of the rail or earth. As a result, the current of the contact wire (CW) and return current wire would be approximately the same. Then, not only has the stray current reduced but also the rail current and potential decreases as well.*

*Keywords—Traction application, insulation, rail vehicle*

## I. Introduction

In DC electrical railways, rails are used as the return current path. The series resistance of rails is usually about 40 to 80 milliohms per kilometer and there is a poor insulation from earth typically from 2 to 100 ohms per kilometer [1]. In the mass rail transit (MRT) systems the current required by the train will be up to thousands amperes [2, 3, 4]. Therefore, a significant voltage drop appears on the running rails - this voltage level could reach to 60-100 volts-[2, 5], as a result, the current leaks from the rail and flows into parallel circuits either directly through the soil or through buried conductors before returning onto the rail and the negative terminal of the DC rectifier. This current is called stray or leakage current [2].

When the stray current flows from a conductor to earth or from earth to a conductor, there is an electron to ion or ion to electron transfer [1]. This causes corrosion of metallic objects of infrastructure: stations, tunnels, bridges; and piping equipment and so on. The main solution to minimize the stray

current is insulating the rail from earth [6]; however, insulation of the track to reduce these currents may cause unacceptably high rail voltages and hence excessive step and touch potentials [7].

This paper introduces a new power electronic circuit which can decrease the stray current and also the rail potential by forcing the return current to flow through return current wire (RCW). Fig. 1 shows this circuit in a DC traction system.

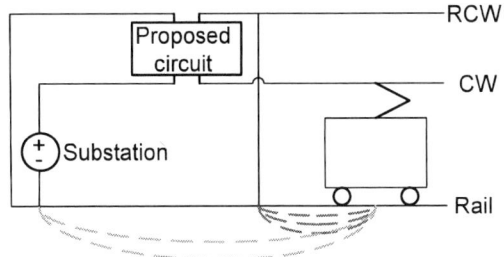

Fig. 1- DC Traction System with the proposed circuit

The development of high-power semiconductor switches such as GTO, IGBT and IGCT [8, 9] with high speed, low-loss operation allows implementation of this circuit.

## II. Operation and Analysis of Proposed Circuit

The proposed circuit shown in Fig. 2 consists of an inductor, a resistance and seven semiconductor switches. During different switching intervals the inductor is charged in the CW and will force the current to flow from RCW periodically. Changing the inductor's position from CW to RCW and vice versa needs some considerations:

- For avoiding voltage impulses on the load, loads current must not be interrupted during switching intervals. Switches $S_1$ and $S_2$ keep

978-1-4244-1741-4/08/$25.00 ©2008 IEEE

the current continuous in the intervals in which the loads current may be interrupted.

- For avoiding voltage impulses on the inductor, inductors current must not be interrupted during switching intervals. Switch $S_2$ keeps the current continuous in the intervals in which the inductors current may be interrupted.

- Commutation between the switches $S_{ai}$ and $S_{bi}$ causes instantaneous short circuit between CW and RCW which leads to voltage drop on the load. This kind of short circuit can be avoided by considering a dead time between switches $S_{ai}$ and $S_{bi}$.

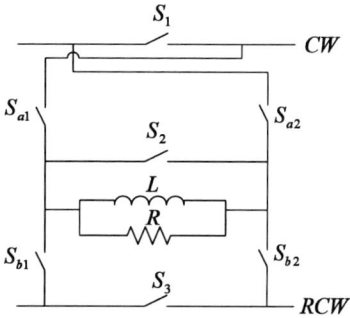

Fig. 2- The proposed circuit

To illustrate the function of the proposed circuit, operation can be divided in to eight operating intervals. The gate pulses and the equivalent circuits in different intervals of operation are shown in fig. 3 and 4.

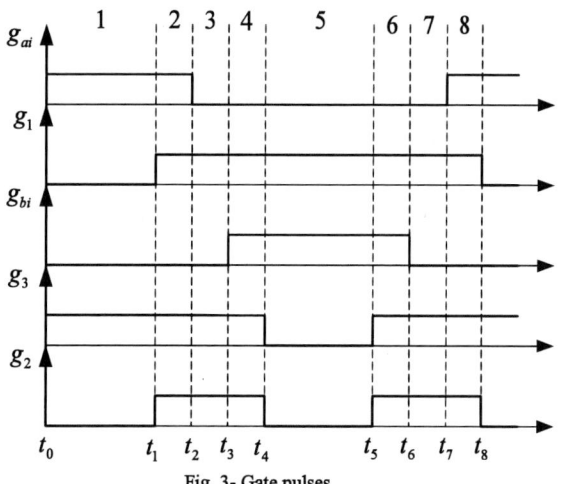

Fig. 3- Gate pulses

## A. Charging Interval

At the start of the switching period, it is assumed that the switches $s_{ai}$ (i=1, 2) and $s_3$ are on and other switches are off. As shown in fig. 4(a) the load current is flowing through the inductor, consequently, the inductor's current is equal to load's current in this interval. At the beginning of this interval there is a step in the inductor's current which leads to an impulse voltage on the inductor and the load. To avoid this impulse voltage a large parallel resistance R is applied.

In the charging interval ($t_0$-$t_1$) equations of inductor's current are:

$$v_L(t) = L\frac{di_L(t)}{dt} \qquad (I)$$

$$v_L(t) = R\big(I - i_L(t)\big) \qquad (II)$$

Where I is a constant value describing load's current during this interval.

$$i_L(t) = Ae^{-\frac{R}{L}t} + I$$

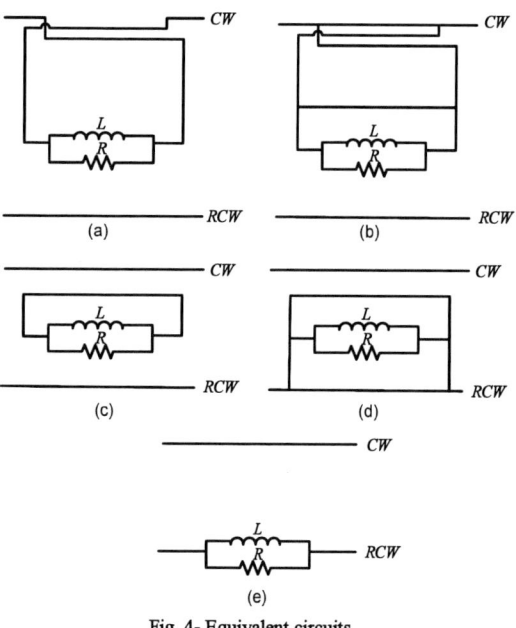

Fig. 4- Equivalent circuits

## B. A Period between Charging Interval and Forcing Interval

As the inductor's current reached to the CW's current at $t_1$ it should be serried with the RCW to force the return current flow through RCW. Changing the inductors position from the CW to RCW requires some considerations to avoid voltage impulses on the inductor and short circuits between CW and RCW. Note that as mentioned above during switching operations switches $S_{ai}$ and $S_{bi}$ should not be on simultaneously to avoid short circuit between CW and RCW. To prevent voltage impulses the inductor should not be open circuit in switching intervals and there should be no interruption in loads current. These following steps should be taken to avoid the aforementioned problems.

Interval $(t_1-t_2)$ [fig. 4(b)]: when the inductor's current reaches to the load current at $t_1$ the inductor is charged and the switches $S_1$ and $S_2$ should be gated. Duration of this interval is the turn on time of the switches $S_1$ and $S_2$ to avoid voltage impulses on the load.

Dead time interval $(t_2-t_3)$ [fig. 4(c)]: when the switches $S_1$ and $S_2$ are turned on at $t_2$, the switches $S_{ai}$ should be turned off. Duration of this interval is the turn off time of the switches $S_{ai}$ to prevent short circuit possibilities between CW and RCW.

Interval $(t_3-t_4)$ [fig. 4(d)]: when the switches $S_{ai}$ are turned off at $t_3$, the switches $S_{bi}$ (i=1, 2) should be gated. Duration of this interval is the turn on time of the switches $S_{bi}$. When the switches $S_{bi}$ turned on at $t_4$ $S_3$ should be turned off. This interval avoids voltage impulses on the load and the inductor.

## C. Forcing Interval

Forcing interval starts when the switch $S_3$ is turned off. As shown in fig. 4(e) the RCW's current flows through the inductor in this period, consequently the inductors current is almost equal to RCW's current. The inductor's current can't change suddenly due to its characteristics, thus, in this period the inductor forces the return current to path through the RCW instead of the rail and earth.

Considering $R_{eq}$ as equivalent resistance paralleled with the inductor, the time constant of the inductor's current would be:

$$\tau = \frac{L}{R_{eq}}$$

Duration of this interval should be less than one tenth of the time constant to prevent falling of inductor's current. Since RCW and rail are paralleled $R_{eq}$ is less than one ohm. As a result, time constant of this interval is about a hundred milliseconds. Therefore, if the duration of this interval is less than ten milliseconds the inductor's current would not change noticeably.

## D. A Period between Forcing Interval and Charging Interval

When the inductor's current reaches to its minimum value at $t_5$, it should be recharged. This period is similar to the period $(t_1-t_4)$ and is divided into three intervals. During these intervals the inductor's position changes from RCW to CW.

Interval $(t_5-t_6)$ [fig. 4(d)]: when the inductor's current reaches to its minimum value at $t_5$, the switches $S_2$ and $S_3$ should be gated. Duration of this interval is the turn on time of the switches $S_2$ and $S_3$. This interval prevents voltage impulses on the inductor and the load.

Dead time interval $(t_6-t_7)$ [fig. 4(c)]: when the switches $S_2$ and $S_3$ are turned on at $t_6$, the switches $S_{bi}$ should be turned off. Duration of this interval is the turn off time of the switches $S_{bi}$ to avoid short circuit possibilities between CW and RCW.

Interval $(t_7-t_8)$ [fig. 4(b)]: when the switches $S_{bi}$ are turned off at $t_7$, the switches $S_{ai}$ should be gated. Duration of this interval is the turn on time of the switches $S_{ai}$ to prevent voltage impulses on the load. When the switches $S_{ai}$ are turned on at $t_8$, the switch $S_1$ should be turned off.

## III. Modeling the Rail and Earth

Fig. 5 shows the arrangement of the train, the rails, CW, RCW and the substation. Moreover, distances between the proposed circuit with the traction substation and the joints between rail and

RCW. The train is represented as an ideal current source [10] and the substation is represented as an ideal voltage source [2].

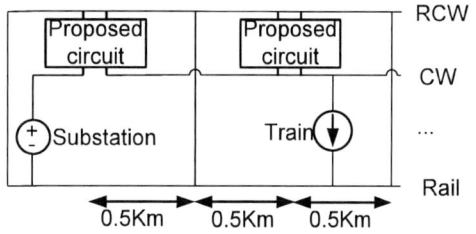

Fig. 5-Distances between train, substation and the proposed circuit

There are different ways to model the rail like finite cell modeling [10] and radial feed circuit model [5]. In this paper the principles of modeling power transmission lines are used which is more accurate than other methods [11].

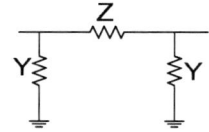

Fig. 6 Equivalent circuit of the rail

Rails have a finite longitude, or series, resistance $r_r$ and a poor insulation from earth $y_e$ [1]. Fig. 6 shows the equivalent circuit for a rail with the length of l, where Z and Y are [11]:

$$Z = R_r \frac{\sinh \gamma l}{\gamma l}$$

$$Y = \frac{Y_e}{2} \frac{\tanh(\gamma l / 2)}{\gamma l / 2}$$

;

$$\gamma = \sqrt{r_r y_e}$$

$$R_r = r_r l$$

$$Y_e = y_e l$$

### IV. Simulation Results

Analyzing the results of preliminary simulations indicates that symmetric switching has a better performance from the voltage drop and circuit efficiency point of view. What is meant by symmetric switching is that the charging interval is equal to the forcing interval. Determining optimal algorithm for minimizing the voltage drop require more precise investigations.

The proposed circuit is analyzed using MATLAB/SIMULINK software for its performance. Table 1 includes the simulation parameters taken for analysis. In this table $r_t$ is the longitude resistance of the rail (four parallel rails),$y_t$ is the rail to earth leakage admittance, $R_e$ is the earth resistance of the traction substation, $r_{RCW}$ and $r_{CW}$ are the longitude resistances of the CW and RCW and finally, R and L are the resistance and inductance of the proposed circuit. To simplify the analysis, all the components (switches, inductors and capacitors) are assumed ideal.

| $R_e$ | 0.5 Ω | |
|---|---|---|
| $r_t$ | 11 mΩ/km | |
| $y_t$ | 0.125 Ʊ/km | |
| $r_{RCW}$ | 35 mΩ/km | |
| $r_{CW}$ | 71.1 mΩ/km | |
| R | 100 Ω | |
| L | 100 mH | |
| $I_{load}$ | 500 A | |
| $V_{no\text{-}load}$ | 790 V | |
| $V_n$ | 750 V | |
| $t_{on}$ | 1μs | Switch parameters |
| $t_{off}$ | 3μs | |
| $t_{ch}$ | 495μs | Charging time |
| $t_f$ | 495μs | Forcing time |
| $f_s$ | 1kHz | Switching frequency |

Table 1- simulation parameters[13,9]

A railway system with the length of 2.5km and a traction substation in the beginning of it, is simulated using MATLAB/SIMULINK software. The distances between the proposed circuits and the connections between the rail & RCW are shown in fig. 5. Simulation results indicate that the best and the worst conditions happen close to the connections between the rail and RCW and at the middle of two connections (before and after the proposed circuit). Therefore, the best and the worst conditions of the stray current, the RCW's current and the CW's voltage in three conditions of with the proposed circuit, with the RCW-just a paralleled wire with the rail- and without the proposed circuit and the RCW, are shown in table 2, 3 & 4.

|  |  | $V_L$ | $I_{RCW}$ | $I_{stray}$ |
|---|---|---|---|---|
| 0.5km |  | 765.6V | 307.0A | 1.67A |
| 1km | before | 745.3V | 306.6A | 5.4A |
|  | after | 735.3V | 306.6A | 1.82A |
| 1.5km |  | 718V | 306.5A | 4.38A |
| 2km | before | 697.7V | 306.3A | 6.26A |
|  | after | 687.8V | 306.3A | 4.91A |
| 2.5km |  | 670.5V | 306.3A | 5.41A |

Table 2- simulation results with the proposed circuit

|  | $V_L$ | $I_{RCW}$ | $I_{stray}$ |
|---|---|---|---|
| 0.5km | 770.2V | 0A | 3.31A |
| 1km | 750V | 1.441A | 6.3A |
| 1.5km | 730.6V | 2.646A | 8.7A |
| 2km | 710.5V | 3.351A | 10.3A |
| 2.5km | 691V | 3.915A | 10.7A |

Table 3- simulation results with the RCW

|  | $V_L$ | $I_{stray}$ |
|---|---|---|
| 0.5km | 769V | 4.3A |
| 1km | 749.1V | 8.0A |
| 1.5km | 728.7V | 11.3A |
| 2km | 708.3V | 13.4A |
| 2.5km | 687V | 14.0A |

Table 4-simulation results without the proposed circuit & the RCW

The wave forms of loads voltage, stray and RCW's current for the case that train is at the 2km and before the proposed circuit are shown in figure 7 to 9.

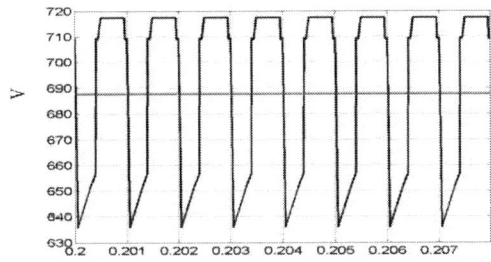

Fig.7 $V_l$ and it's mean value with the proposed circuit at the 2km and before the proposed circuit

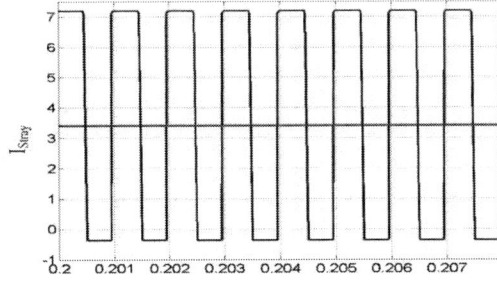

Fig.8 $I_{stray}$ and it's mean value with the proposed circuit at the 2km and before the proposed circuit

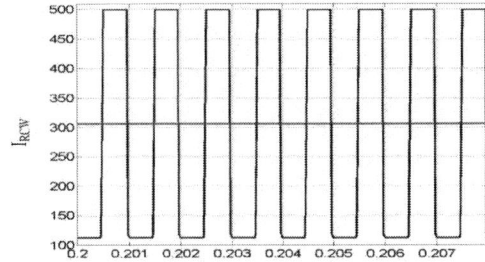

Fig.9 $I_{RCW}$ and its mean value with the proposed circuit at the 2km and before the proposed circuit

Comparing these results lead to:

1- Maximum leakage current decreases from 14A to 6.2A.

2- The proposed circuit forces 61.2 percent of the loads current to flow through the RCW. As a result, rail's current decreases 61.2 percent then rail's potential decreases noticeably as well.

3- The main disadvantage of the proposed circuit is to cause the voltage drop on CW to increase from 8% to 10.6%. As a result, the distance between substations would decrease and the energy dissipation would increase.

4- Due to moving loads and the track gradient profile, electrical railway load varies with time. This variation is not considered in these simulations for simplicity. While considering a variable load, the voltage drop increases on the proposed circuit. This occurs due to its inductive characteristics.

V. Conclusion

A new circuit for reducing stray current in DC traction systems is proposed. Operation of the proposed circuit is analyzed using MATLAB/SIMULINK software. Simulation results indicate 55% reduction in the stray current that leads to a significant reduction in the rail potential. The other advantages of the proposed circuit are: a simple control system which doesn't require any feedback or complicated processors like microcontrollers and DSPs; and the capability of installing it to the existing railway systems with no need to change fundamental parameters. However, the proposed circuit has some disadvantages like decreasing the system efficiency; and increasing voltage drop and

DC voltage ripple. In conclusion, this circuit is effective to decrease the stray current where there are electrified DC railways with poor rail to earth insulation.

## References

[1] Cotton, I., Charalambous, C., Aylott, P. and Ernst, P., "Stray Current Control in DC Mass Transit Systems", March 2005, IEEE Transactions on Vehicular Technology, vol. 54, NO. 2

[2] Liu, Y.C. and Chen, J. F., "Control Scheme for Reducing Rail Potential and Stray Current in MRT Systems", May 2005, IEE Proc.-Electr. Power Appl., Vol. 152, No. 3

[3] Yu, J.G.: 'The effects of earthing strategies on rail potential and stray currents in DC transit railways'. Proc. Int. Conf. on Developments in Mass Transit Systems, IEE, April 1998, pp. 303–309

[4] Lee, C.H., and Wang, H.M.: 'Effects of grounding schemes on rail potential and stray current in Taipei rail transit systems', IEE Proc. Electr. Power Appl., 2001, 148, (2), pp. 148–154

[5] Pham, K.D., and Thomas, R.S.: 'Analysis of stray current, track to earth potentials & substation negative grounding in DC traction electrification system'. IEEE/ASME Joint Conf., Toronto, Canada, April 2001, pp. 141–160

[6] Griffiths, K., "Stray Current Control an Application of Ohm's Law", The Institution of Engineering and Technology, IET, Michael Faraday House

[7] Hill, R.J. "Electric Railway Traction, Part 3 Traction Power Supplies", Power Eng. J., Dec 1994, pp. 275-286

[8] Pires, C. L., Nabeta, S.I. and Cardoso, J.R., "ICCG method applied to solve DC traction load flow including earthing models", Appl., 2007, IET Electr. Power, 1, (2), pp. 193–198

[9] "Short Form Catalogue High Power Semiconductors", 2007 edition, ABB, http://www.abb.com/semiconductors

[10] Yu, J. G. and Goodman C. J., "Modeling of Rail potential Rise and Leakage Current in DC Rail Transit Systems", IEE Colloq. Stray Current Effects DC Railway and Tramway, 1990, pp. 2/2/1-2/2/6

[11] Saadat, H., "Power System Analysis", second edition, 2002, McGraw-Hill Higher Education, US

[12] Dodds, R. "DC Traction Stray Current Control", 1999, IEE Railway industry group, Savoy place, London WC2A OBL, UK.

[13] Waterland, F. "Stray Direct Current: A View from the Main Line", 1999, IEE Railway industry group, Savoy place, London WC2A OBL, UK.

[14] Stillman, H. M., "IGCTs—Megawatt Power Switches for Medium Voltage Applications", 1997, ABB Review, No. 3, pp. 12–17

# Slip Control Upgrades for Light-Rail Electric Traction Drives

Madis Lehtla*, Hardi Hõimoja*

* Tallinn University of Technology/Dept.of Electrical Drives and Power Electronics, Estonia

*Abstract* — **Many obsolete drives such as a rheostat, contactor or camshaft-contact controlled DC motor drives have often poor dynamic properties in insufficient adhesion conditions. Light-rail vehicles running in many cities already have switched-mode traction drives with thyristor-based or IGBT-based converters that often lack high quality slip, antilock or creep control. Switched-mode converter based multi-motor drives could be partially upgraded to microcontroller-based control. This paper presents and analyses some approaches to be used for performance, efficiency and safety improvement.**

*Keywords* — **Rail vehicle, Traction application, System integration, Software for measurements, Real time processing, Multi axle drives, Motion control, Load sharing control, Driver assistant systems, Electrical Drive, Electric vehicle**

## I. INTRODUCTION

The aim of slip control in light-rail vehicles is to improve overall stability and as a result reduce overall braking and acceleration distance in real traffic conditions. These systems also help to reduce the load of mechanical braking systems and equalize wearing of wheels.

The systems consist of a microcontroller for signal processing and control. The microcontroller system is connected to speed feedback circuits, torque limitation circuits and reference signals. Microcontroller system on vehicles depends on the signals used in drive control system. Systems on older vehicles have bipolar analogue signals that have to be converted to levels acceptable to new modern microcontrollers. Some modern microcontrollers have analogue 0 to 2.5 V inputs. Analogue amplifiers affect signal delay and limits. Microcontroller card has analog amplifiers for signal conversion from 2.5 to 15 V level (Fig. 1).

Fig. 1. Sample microcontroller card with analog amplifiers

The compact DIP40 size microcontroller module has been used in reconstruction of the control system. Board includes supply voltage converters, error LED's, RS232 interface and optocouplers for isolation of some digital signals. Compact DIP40 size microcontroller module was used in control system (fig. 1). The control board designed is pin-compatible with the old analog control board. Thus rack backplane board changes are minimized and compatibility with old analog control boards is assured.

Traction and braking force depend on the interaction between the rail and the wheel. When traction or braking force is applied, it always causes some slip between the wheel and the rail. The rail-wheel adhesion (friction) coefficient (fig. 2) is mainly affected by such contaminants as water, ice, oil, trash.

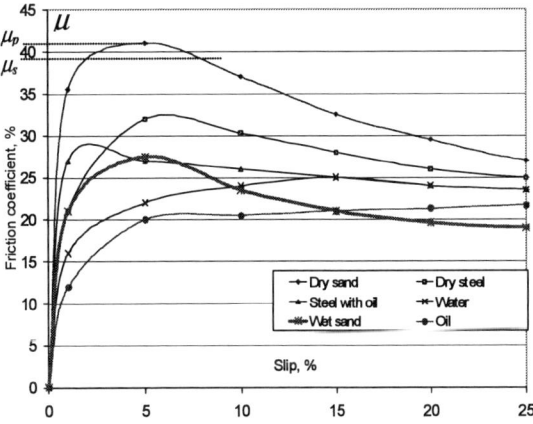

Fig. 2. Wheel friction on different slips

Most traction drive systems are convenient and efficient in normal conditions, but proper acceleration and braking in complicated conditions require too high driving skills. Electric drives have often limited operation speed range and stability issues at lower speeds (on blocking). Thus, wheel slip or spin should be avoided or limited using the control system. Modern traction control systems correct reference signals of a driver automatically in real-time and determine the peak adhesion. This enables better acceleration and braking.

## II. FEEDBACKS FOR SLIP OR SPIN DETECTION

Direct measurement of speeds requires speed sensors in all axles. This is a good solution, as the quality of feedback does not depend on the condition of a motor or another component of the drive system. Signals form high-resolution digital pulse sensors of different motorized-axles or wheels of multi-motor traction drive enable us to calculate velocity and acceleration differences between different axles without a significant

978-1-4244-1741-4/08/$25.00 ©2008 IEEE    1581

delay. This is simple on vehicles that have at least one non-motorized and non-braking freewheeling axle as an additional independent speed-signal source. The wheels that have maximal speed on braking and minimal speed on acceleration can be used as the closest to real speed on vehicles that do not have freewheeling axles. The speed difference method [1] uses slip velocity $v_s$:

$$v_s = \omega_r \cdot r - v_{ref}, \qquad (1)$$

where: $\omega_r$ – angular speed, $r$ – radius of wheel, $v_{ref}$ – vehicle speed estimated from the minimum of the angular wheel velocities. Larger amount of wheels used for minimum velocity detection leads to the higher accuracy of reference speed.

Fig. 3. Gearing wheel and digital pulse sensors integrated into tram gearing in Tallinn

Speed signals can be measured via sensors (Fig. 3) or estimated from motor models using the measured current and voltage values. There are two main possibilities for speed measurement for gear-integrated pulse sensor – counting of pulses in time interval or measuring time interval between pulses. The measurement of time interval between pulses allows smaller delay that is important for control. Angular speed of motor shaft using measurement system with 16-bit timer can be calculated accordingly:

$$\omega = \frac{2\pi}{60} \cdot \frac{\left(2^{16}-1\right)}{n_t} \cdot k_{RPM}, \qquad (2)$$

where $n_t$ is discrete time interval measured using timer and $k_{RPM}$ transfer gain for conversion of measured result to unit of rotates per minute (rpm). Transfer coeficient between motor shaft speed in rotates per minute (rpm) and discrete time interval of 16-bit timer is calculated using formula:

$$k_{RPM} = \frac{60 \cdot n}{z \cdot \left(2^{16}-1\right) \cdot t_{CLK} \cdot 2^{Tx}}, \qquad (3)$$

where $t_{CLK}$ - period of timer clock signal (that is $t_{CLK} = 50$ns on clock frequency 20 MHz), $z$ – amount of pulses per turn on wheel axle (amount of teeth on gear wheel), $Tx$ – divisor of clock frequency and $n$ – transfer ratio of gear.

The delay of the result depends on the speed. The time between the pulses is longer at low speeds, thus the delay is longer. The measurement range of the timer is chosen according to the minimum permitted delay, the lowest measured speed value and clock frequency is chosen according to the required accuracy of speed measurement.

An indirect comparison of speeds often uses speed values estimated from the models of separately controlled motors. This can be applied for speed comparison between separately supplied under-carriages (bogies). With separately controlled AC motors, speed estimation via motor current differences [2] is applicable. Instead of model-based methods, slip can also be detected via direct comparison of electrical parameters.

Obsolete drive systems with direct current motors do not facilitate separate control of axles and often have voltage difference relay systems to compare motor voltages and thus motor back-electromotive forces.

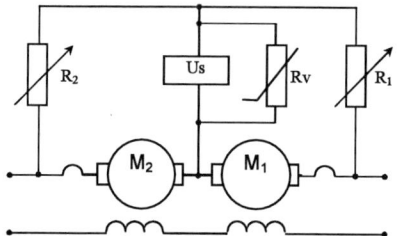

Fig.4. Circuit diagram of a simple DC traction motor based slip detection system

A bridge circuit (Fig. 4) is often used to compare back-electromotive forces when multiple series-connected motors are on the same bogie (wheel-set type Bo or M2). This is not applicable to systems with mono-motor bogies (wheelset type B or M1). A parallel-connected metal-oxide varistor Rv increases the delay and reduces the influence of high-frequency disturbances caused by chopper supply and commutators.

## III. CONTROL METHODS AND CONTROLLERS

Under slip control operation in light-rail vehicles, different operation ranges can be distinguished: stable traction without significant slip, large slip, reduction of torque for slip to be reduced, retrieval of the output torque to an estimated limit and retrieval of output torque to the set-point (according to pedal or lever position). Ramp-functions are used for smooth retrieval of the output torque. Also, control-input reference ramp should be reset to equalize torques on different axles. Instead of limitation functions such as ramps, controllers are needed for continuous control of output torques. The methods used for slip, antilock and re-adhesion control are shown in Fig. 5.

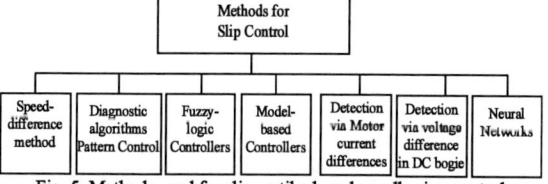

Fig. 5. Methods used for slip, antilock and re-adhesion control

The following methods can be used to control the traction force according to slip-conditions [3]: Neural Networks to estimate the parameters that cannot be measured on-line [4], Diagnostic Algorithms [5] such as Pattern Control Method [6], for example, thresholds (of slip) that will trigger the control process when exceeded, Detection via Motor Current Differences of AC drives [2] or using Voltage Differences on DC multi-motor bogies,

Model-Based Controllers [7], Hybrid re-adhesion control method [6], Steepest Gradient Method [8], Fuzzy Logic Based Slip Control [9][10], etc. In addition, other methods exist [11] that are more complex for application in real-time slip and spin detection such as vehicle dynamic behavior analysis, surface reflections observation the front of the vehicle via optical sensors, wheel sound analysis via acoustic sensors, and mechanical strain analysis using sensors in the wheels.

The most advanced methods include multiple control methods (an example [6]), such as binary rule-based controllers (fuzzy-logic, etc.) together with classical proportional and integral controllers for different operation ranges, such as micro-slip, macro-slip, creep etc. Some examples of control rules included are:

1. Minimal speed rule for drive system stability at low speeds
2. Speed difference threshold rule (to enable speed difference for dynamics)
3. Rules for macro-slip detection (separate for axles)

Controllers can include:
1. acceleration difference regulator
2. speed difference regulator (PI with dead zone)
3. reference signal ramp-up from the macro-slip level (common and separate for axles)
4. or hybrid systems that contain several of these.

The advanced system uses continuous control in combination with logic rules (Fig. 5). The operation of antilock and creep control described in Fig. 6 is based on speed measurements on each motorized axle for comparison of speeds of non-motorized freewheeling axle. Measurements with developed real-time control system were carried out on lubricated track with 4-axle articulated tramcar (Fig. 6).

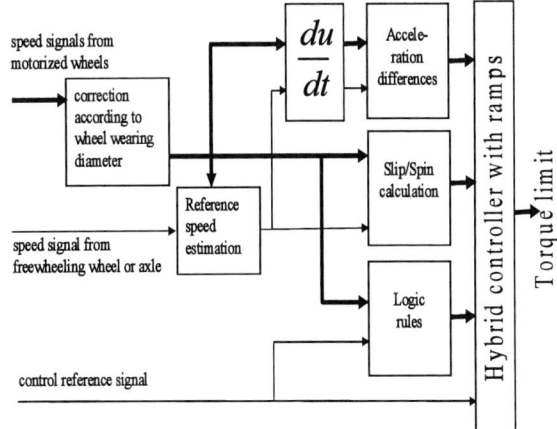

Fig. 6. Control structure for advanced slip and spin control

Acceleration values was calculated from the measured speed values and filtered using a low-pass filter. The filter delay should be optimal. The control uses logical rules (relational operators) which are used for speed comparison on different axles with threshold levels. Different threshold levels are foreseen for control of braking and acceleration, because the speed of accelerating axles is higher than the speed of the freewheeling axle and the speed of the braking axles is lower than that of freewheeling axle due to slip effects.

The controller includes ramp functions of the reference signal for smooth application of torque when slip has been reduced to a normal range. Stable (creep range) micro-slip control uses customized (dead-zone and anti-windup) PI regulators of acceleration and velocity difference signals. The sign of acceleration difference varies on acceleration and braking. Speeds on both controlled axles and uncontrolled freewheeling axle with controller output signals are shown in figure 7

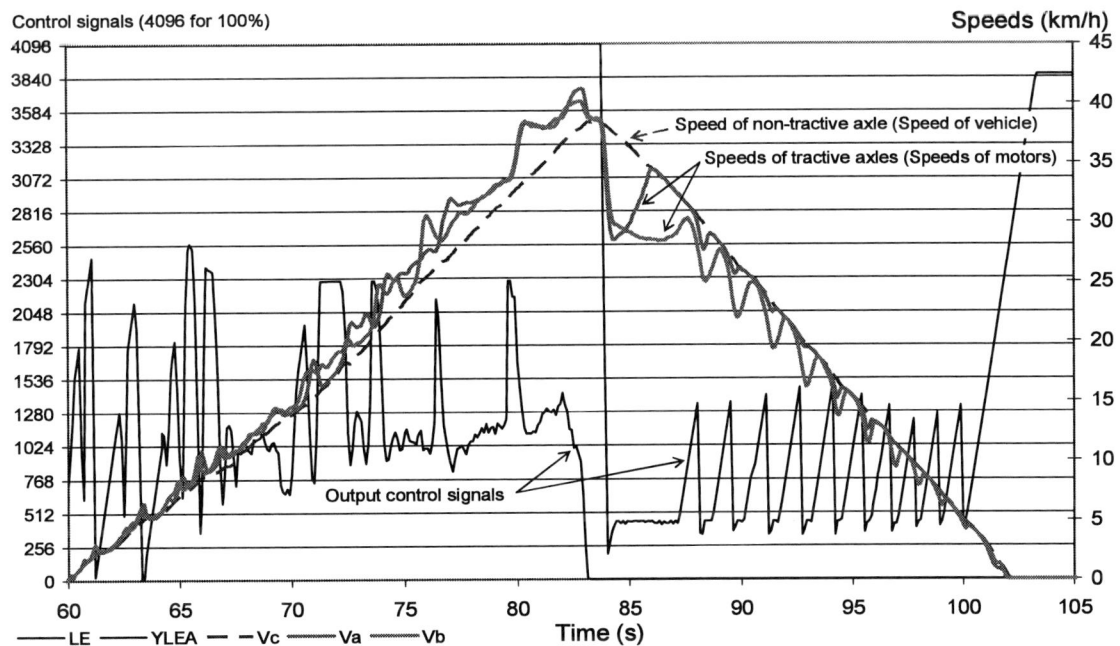

Fig. 7. Test cycle for tuning of the control system on articulated tramcar

. Traction converter allows separate control of axles on acceleration but control of axis on braking is bound up. Complete blocking of axles should be avoided for stable excitation of DC traction motors and for saving wheel surface and disks of mechanical brakes. Acceleration is shown up to 40 km/h and braking is following until stop.

Some ABS algorithms for road vehicles incl. passenger cars [12] can also be adapted and applied to light-rail vehicles. Different mechanical and electro-mechanical anti-slip systems and anti-blocking brakes are used in light-rail vehicles.

The following additional brake systems are used on trams:
1. separate mechanical brakes or track-braking magnets in the case of blocked wheels
2. electrically controlled sand-systems for the improvement of adhesion during braking
3. mechanical stop and park-brakes

## CONCLUSIONS

Switched-mode converter based multi-motor drives could be upgraded to the microcontroller-based slip and antilock control. The slip/spin signals enable rail-wheel adhesion conditions to be indicated on dashboard. In addition to regenerative braking control, these signals can be used together with sand systems and rail-braking magnets or other mechanical brake systems.

High traction and braking forces can be achieved due to very high forces in wheel-rail contact point, but even the molecular amount of contaminants significantly decreases adhesion. These effects are very random and cannot be determined or precisely modeled, but each traction drive system should operate in these conditions. Thus, tuning of anti-slip and creep control systems requires various tests on vehicles that cannot be completely replaced with computer simulations. The microcontroller system includes serial communication interface and has features in software that enable data logging to notebook computer hard-disk. This improves the tuning process and does not require additional instruments. Reconstructed system improves acceleration quality and stability of regenerative braking. Software-based checking and filtering of sensor and control reference signals improves safety and control quality.

## ACKNOWLEDGMENT

The authors thank the Estonian Ministry of Education and Research for the support through the target oriented project SF0140009s08 Energy saving and sustainable electrical power engineering. The autohrs thank also Dr.Eng. Jüri Joller and company *Energiatehnika OÜ* for contributing the experimental data about trams.

## REFERENCES

[1] Yasuoka, I., Henmi, T., Nakazawa, Y., Aoyama, I. *Improvement of re-adhesion for commuter trains with vector control traction inverter.* IEEE, Proceedings of the Power Conversion Conference, Vol. 1, Nagaoka, 1997, pp. 51-56.

[2] Watanabe, T., Yamashita, M. *Basic study of anti-slip control without speed sensor for multiple motor drive of electric railway vehicles.* IEEE, Proceedings of the Power Conversion Conference, Vol. 3, 2002, pp. 1026-1032.

[3] Frylmark, D., Johnsson, S. *Automatic Slip Control for Railway Vehicles* Master's thesis performed in Vehicular Systems, LiTH-ISY-EX-3366-2003, 6-th February, 2003

[4] Gadjár, T. Rudas, I., Suda, Y. *Neural network based estimation of friction coefficient of wheel and rail.* IEEE International Conference on Intelligent Engineering Systems, 1997, pp. 315-318.

[5] Gustafson, F. *Monitoring tire-road friction using the wheel slip.* IEEE Control Systems Magazine, Vol. 18 (4), 1998, pp. 42-49.

[6] Park, D-Y., Kim, M-S., Hwang, D-H., Kim, Y-J., Lee, J-H. *Hybrid re-adhesion control method for traction system of high-speed railway.* ICEMS 2001: Fifth International Conference on Electrical Machines and Systems, Vol. 2, 2001, pp. 739-742.

[7] Ohishi, K., Ogawa, Y., Miyashita, I., Yasukawa, S. *Adhesion control of electric motor coach based on force control using disturbance observer.* IEEE Proceedings. 6-th International Workshop on Advanced Motion Control, Nagoya, 2000, pp. 323-328.

[8] Ohishi, K., Miyashita, I., Nakano, K., Yasukawa, S. *Anti-slip control of electric motor coach based on disturbance observer.* IEEE Proceedings. 5-th International Workshop on Advanced Motion Control, Coimbra, 1998, pp. 580-585.

[9] Palm, R., Storjohann, K. Torque optimization for a locomotive using fuzzy logic. ACM symposium on Applied Computing, 1994, pp. 105-109.

[10] Hill, R.J., de la Vassière, J.-F. *A fuzzy wheel-rail adhesion model for rail traction.* EPE'97: European Conference on Power Electronics and Applications, Vol. 3. Norway: Trondheim, 1997, pp. 3416-3421.

[11] Gustafsson, F. *Slip-based estimation of tire - road friction.* Technical Report LiTH-ISY-R-1755, Sweden: Linköping, Department of Electrical Engineering, Linköpings Universitet, 1995, ftp.control.isy.liu.se/pub/Reports/1995/1730.ps.Z,

[12] Day, T.D., Roberts, S.G. *A Simulation Model for Vehicle Braking Systems Fitted with ABS.* SAE Technical Paper Series, 0559, Reprinted From: Accident Reconstruction 2002, SP-1666, Society of Automotive Engineers, 2002, 19 p.

[13] Daigle, J.L.; *Traction vehicle/wheel slip and slide control*, patent US6499815, 2002

# Practical Aspects on the Improved DC Driving System Used in Electric Urban Traction

Petre - Marian Nicolae[*], Ioana-Gabriela Sîrbu[*], Ileana-Diana Nicolae[†] and Lucian Mandache[*]

[*] University of Craiova, Electrical Engineering Faculty, Craiova, Romania, e-mail: *pnicolae@elth.ucv.ro*
[†] University of Craiova, Faculty of Automatics, Computer Science and Electronics, Craiova, Romania

*Abstract*—The paper presents some practical considerations and experimental results obtained by the exploitation of a tram that has an electric driving system with an improved technological solution. One presents the technological solution using chopper that substitutes the initial accelerators of the tram. Some aspects concerning the operation in running and respectively in braking regimes in the new configuration (including the regenerative braking regime) are discussed. Some considerations with respect to the electromagnetic compatibility problems and modalities to improve it are made. The recordings of the current and voltage waveforms, during the normal running stage, when the tram climbs a ramp and respectively goes down on a slope are presented and explained. The experimental determinations revealed that the system was designed correctly, in order to support all the operational conditions that could appear in its exploitation.

*Keywords*—electrical drive, power converters for EV, pulse width modulation (PWM), efficiency, safety.

## I. INTRODUCTION

In many industrial applications, it is required to convert a constant d.c. voltage of a supply source into a variable d.c. voltage. A chopper can perform this task [1], [2], [3], [4].

Choppers are widely used for traction motor control in electric vehicles, trolley-buses, trams etc. They provide smooth acceleration control, high efficiency and fast dynamic response [5]. Choppers can be used in regenerative braking of d.c. motors to return energy back to the d.c. voltage line; thus one obtains energy savings for transportation systems with frequent stops [6].

Due to this important advantage, in the modernization process of the trams an improved technological solution with chopper should be considered .

In some Romanian cities (including Craiova), tram vehicles were imported from Germany. They are endowed with traction motors of 600 Vd.c. (two d.c. series motors at rated voltage of 300 V d.c. each), but are supplied from the 825 V d.c. network. To fix this mismatch, a first solution was the introduction of a resistance in series with the d.c. motors in the rotor circuit. This technical solution presents a series of drawbacks, among them being a low energetic efficiency.

In order to get a compatibility of the Romanian urban travelers transportation with its European counterpart, one must consider the environment requirements, the energetic

This work was supported by ANCS through the 4th Program – Partnerships and under the Contract 71-145/2007.

consumptions savings and the adaptation of consumption to the travelers flux.

Recent electromagnetic interference (EMI) and electromagnetic compatibility (EMC) regulations such as FCC, VDE, CISPR, and EN have become tighter than ever. The modern power electronics is an engineering discipline that deals with the conversion of electrical power using fast semiconductor devices. These semiconductor devices are utilized as switches where the main designer's concern consists in reducing the power losses. Switching operation generates signals with significant $du/dt$ and $di/dt$ and, consequently, wide bandwidths of disturbances. Because the power electronics equipment is usually connected to the supplying lines, those wide-band signals are traveling through them and pollute the electromagnetic environment with unwanted interference.

The EMC directive CISPR 25 emphasized an increasing awareness of EMC issues that highlight the conducted and radiated EMI problem of switching power converters in automotive environment [7]. Exact prediction of EMI could help designer to satisfy EMI regulations cheaply and face with EMI problems easily before making the final hardware. Active components, such as switches and diodes, are the main noise sources in switched mode power supplies. Especially, the parasitic ringing voltage and current of the switches affect EMI pattern severely [8].

Therefore a solution was adopted for the modernization of existing tram vehicles through major changes in the power circuit [9], [10]. These changes focus both on the removing of series resistances and on the substitution of starting rheostats of classic driving by variable voltage regulators (choppers) realized with IGBT [11], [12], that provide the current control in the regimes "RUNNING" and "BRAKING". In the same time, one attempts to consider a great part of the problems of electromagnetic compatibility that could appear and to solve them [8].

## II. PRESENTATION OF THE NEW SOLUTION

Usually the trams have two motor bogies. Their driving was made initially through one or two accelerators, depending on the tram type and structure. Because the conventional accelerators (with servomotors and relay switches) are no longer necessary, available places are created. They can host the electric driving system with chopper.

On trams with two accelerators one can substitute them by two braking choppers (one for each bogie), each one requiring an intermediate circuit. The solution is too expensive.

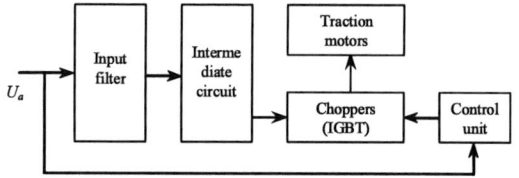

Fig. 1. Operational structure of the equipment.

On trams equipped with a single accelerator (e.g. the KT4D tram discussed here), one can use a single braking transistor and respectively a single intermediate circuit, because it does not operate in a continuous way. Instead of the accelerator, our solution involves an electric drive with static converter.

The electronic equipment with chopper for traction motors' driving has the following operational structure (Fig.1): input filter; recovering diode; filter for intermediate circuit; choppers with IGBT; electronic control unit with microprocessor.

The advantage of the technical solution that uses a single resistive braking chopper consists in the fact that the braking chopper module is not provided with a continuous driving. Thus one gets space saving and a decrease of material costs. Moreover, the system schematic is simpler. The braking chopper switches into the conduction regime only when the braking voltage is greater then the imposed maximum voltage. It is designed to get the energy from motors (when they operate as generators) during the braking regime.

At the presented solution, due to the advantages mentioned above, a single control structure was used. It contains two choppers for running/braking (Fig. 2), a chopper for resistive braking (Fig. 3) and an intermediate circuit (Fig. 4).

The choppers for running/braking from Fig. 2 are electrically linked to the other elements from the general schema as follows:

1 – the link with the resistive braking elements;

2 and 6 – the links with the additional resistances corresponding to the bogie A and respectively to the bogie B;

3 and 7 – the links with the motors' inductances from the bogie A and respectively from the bogie B;

4,5 and 8,9 - the links with the motors' excitations from the bogie A and respectively from the bogie B.

The IGBTs T1 and T2 from Fig. 2 realize the modification of the d.c. quantities values at the same d.c. power, changing the source mean voltage. Through the relative duration of the voltage pulse one gets the motors' angular velocity variation at constant excitation flow. When the running treadle is pressed, the running schema is accomplished (through a contactors schema) and the choppers transform the input voltage received from the contact line (750…825) V d.c. into a voltage used to supply the motors' group from the bogies (almost 600 V d.c.), this way controlling their currents and torques.

Each chopper serves one bogie, so an independent operation becomes possible and the tram might return to the depot with a single operational bogie in the case of erroneous operation.

After the running regime selection (backward/forward), the driver presses the running treadle and the schema corresponding to the selected regime is automatically selected.

The information received by the microcontroller from the control part (not depicted by Fig.2) are processed and therefore the signal "START" is issued for the running choppers, the current through the motors raising with a predetermined slope toward the prescribed value from the running treadle.

With respect to the treadle position the supplying line voltage is progressively transferred toward the d.c. motors that develop the corresponding torque and the tram is accelerated. Through the contacts not depicted by fig. 2, one can get a braking schema and if the running treadle is released, an accelerated regime is automatically established, with a braking at minimum current.

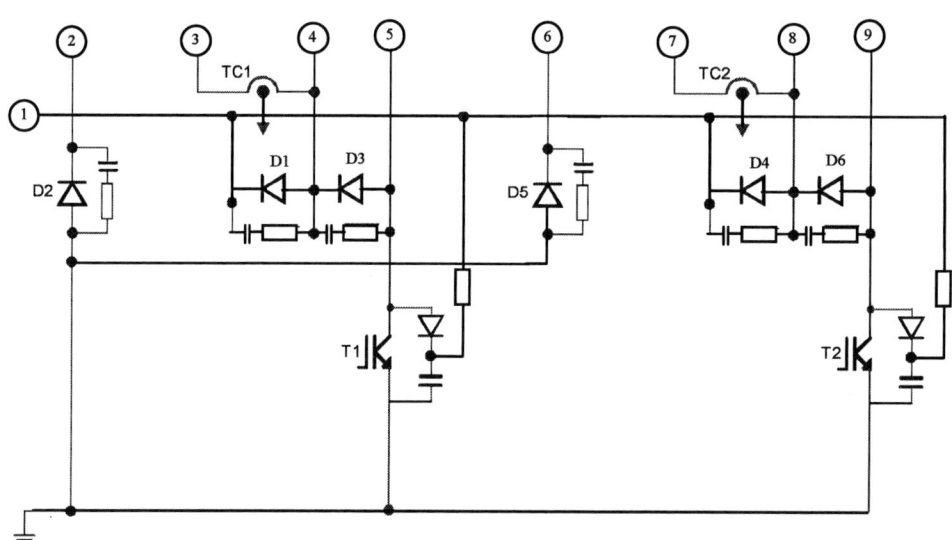

Fig. 2. Chopper's electric schema for running/braking

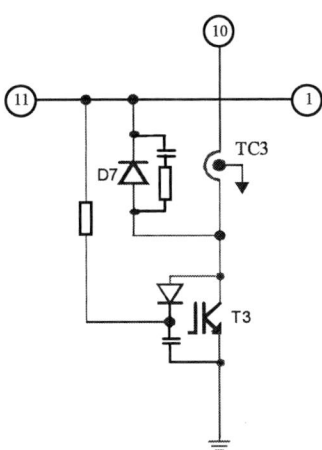

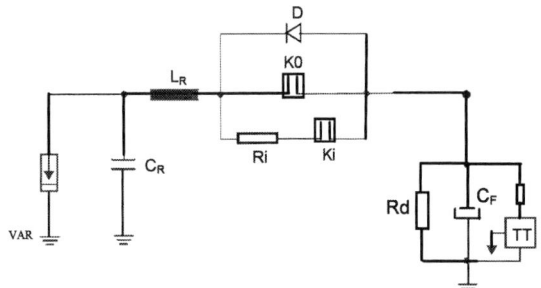

Fig. 3. Chopper electric schema for resistive braking

Fig. 4. Electric schema of the line filter from the intermediate circuit

When the braking treadle is pressed, the braking current raises toward the prescribed value and the motors that operate as generators provide energy toward the contact line, energy that might be used by other trams connected to the same line.

If no consumer is connected to the supplying network and therefore the regenerative braking is not possible, the energy stored by motor is dissipated over a resistance as heat. In this case the current that should have had to be oriented toward the supplying line is reoriented by the T3 IGBT transistor (Fig. 3) toward the resistance. In Fig. 3 one can see the electric links of the transistor T3 to the other elements from the general schema:

1 – the link with the running/braking choppers;
10 – the link with the braking resistance;
11 – the link with the intermediate circuit.

At the force circuit input a radio-frequency filter is placed ($L_R$, $C_R$) in order to reduce the perturbations level from the electrical driving systems to the supplying voltage (Fig. 4). EMI noise generation involves both the typical converter operation and the parasitic stray components. According to their generation mechanisms, two different kinds of noise are distinguished: Common Mode (CM) noise and Differential Mode (DM) noise.

CM noise is mainly due to high $du/dt$ and high frequency oscillations produced by the interactions of the rapidly switching semiconductor devices with stray inductance. It does not occur between converter's power lines, but between the power lines and ground. Since CM noise is strongly linked to voltage switching, and DM noise to current switching, CM and DM noise generations can take place simultaneously [13], [14].

Serrao et al. [15] suggest a modality to reduce the conducted emissions through the designing of a proper EMI filter by separating the noise into its common mode and differential mode components. Their experimental results underlined the importance of considering CM and DM noises separately in order to minimize the filter components count and size.

Since this capacitor is typically an electrolytic one, it is effective for the reduction of DM noise up to the switching frequency. DM noise is mainly caused by the

pulsating converter current and is attenuated by the input filter [16]. Concerning the CM noise, it is limited due to the position of the equipment on the electric vehicle and by using a shield.

The input filter ($L_R$, $C_R$) will diminish the current harmonics induced in the contact line with respect to the currents flowing through chopper. The designing was done so as to reduce up to an acceptable level the influence of the switches from chopper over the supplying network.

With respect to the common mode noise reducing, the line filter from the intermediate circuit ($L_R$, $C_F$, $K_0$, $K_i$, $R_i$) from Fig. 4 reduces up to acceptable limits the supplying network pulses for the running regime and takes over the instantaneous energy of the electric motor during the braking regime. Moreover, it provides pulses whose magnitudes do not influence the operation of the regulators from the control schema.

The contactor $K_i$ used for charging through resistor from the intermediate circuit is connected when the control voltage occurs and is automatically disconnected when the contactor $K_0$ is closed or when the braking is activated. $K_0$ is a contactor used to set the charging resistance into the short-circuit state and it is automatically connected when the voltage from the intermediate circuit overlaps the voltage minimum level $U_1$ and it is automatically disconnected when the voltage falls under the value $U_1$ or when the braking is activated.

III. EQUIPMENT OPERATION PRINCIPLE

The traction motor can be considered as a RL consumer supplied by a d.c. voltage of a variable value provided by the chopper. The supply voltage mean value can be adjusted through the chopper duty cycle. The chopper operating principle is based on the width variation of the pulses, method that is known as "control by pulse width modulation (PWM)". Figure 5 depicts a principle schema.

The chopper's operating principle can be divided as follows:

1) the chopper is in the running regime (Fig. 5 (a)):

- if it is set ON, the current flux flows from the source toward the consumer (traction motor);

- if it is set OFF, the current flux flows through the diode D.

2) the chopper is in the braking regime (Fig. 5 (b)):

- if it is set ON, the current flux flows through the diode D;

- if it is set OFF, the current flux flows from the traction motor toward the source.

1587

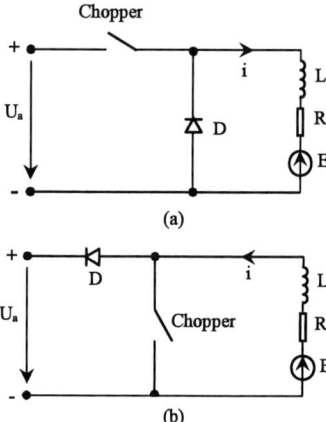

Fig. 5. Scheme of principle – (a) running regime, (b) braking regime.

In the running regime, when the chopper is ON, the load current $i_1$ is given by the expression [17], [18]:

$$i_1(t) = I_1 \cdot e^{-t \cdot R/L} + \frac{U_a - E}{R} \cdot \left(1 - e^{-t \cdot R/L}\right), \qquad (1)$$

where $U_a$ represents the line supplying voltage, $E$ represents the total electromotive force induced by motors, $R$ represents the equivalent resistance of the series motors (mainly due to the stator and rotor resistances), $L$ is the equivalent inductance of series motors (mainly due to the stator and rotor inductances) and $I_1 = i_1|_{t=0}$. For the sake of simplicity (to understand better the operating principle), one can assume as constant the values of $L$ and $R$.

When the chopper is OFF, the load current $i_2$ can be determined using the relation:

$$i_2(t) = I_2 \cdot e^{-t \cdot R/L} - \frac{E}{R} \cdot \left(1 - e^{-t \cdot R/L}\right), \qquad (2)$$

where $I_2$ is the value of the current at the end of the operating time ($t = t_1$). At the end of the OFF case, for the steady state conditions, the load current becomes again $I_1$.

In the braking regime, when the chopper is ON, the load current $i_3$ is determined with the relation [19]:

$$i_3(t) = I_3 \cdot e^{-t \cdot R/L} + \frac{E}{R} \cdot \left(1 - e^{-t \cdot R/L}\right), \qquad (3)$$

where $I_3 = i_3|_{t=0}$.

When the chopper is OFF, the load current $i_4$ can be determined with the relation:

$$i_4(t) = I_4 \cdot e^{-t \cdot R/L} + \frac{E - U_a}{R} \cdot \left(1 - e^{-t \cdot R/L}\right), \qquad (4)$$

where $I_4$ is the value of the current at the moment $t = t_1$. At the end of the OFF case, for the steady state conditions, the load current becomes again $I_3$.

The peak-to-peak value variation of the current [20], for both cases (running and braking regimes) is:

$$\Delta I = \frac{U_a}{R} \cdot \frac{1 - e^{-\alpha T \cdot R/L} + e^{-T \cdot R/L} - e^{-(1-\alpha) T \cdot R/L}}{1 - e^{-T \cdot R/L}}, \qquad (5)$$

where $T$ represents the pulse period and $\alpha = t_1/T$ is the duty cycle of the chopper.

Both the running and braking regimes are accomplished through the modification of the supplying voltage and implicitly of the current that is to be prescribed. When the network allows it, the braking will be a regenerative one, as it was presented previously. Otherwise it will be a rheostatic braking.

In *regenerative braking*, the motor acts as a generator and the kinetic energy of the motor is returned back to the supply. One assumes that the armature of the series motor is rotating due to the motor's inertia.

The average voltage drop across the chopper is:

$$U_{ch} = (1 - \alpha) U_a. \qquad (6)$$

If $I_a$ is the average armature current, the regenerated power can be found from:

$$P_g = I_a U_a (1 - \alpha). \qquad (7)$$

The voltage generated by the series motor acting as a generator is [17]:

$$E = K_m I_a \omega = U_{ch} + R I_a = (1 - \alpha) U_a + R I_a, \qquad (8)$$

where $K_m$ is the motor constant, $\omega$ is the motor speed and $R$ is the series motor total resistance.

The equivalent load resistance of the motor acting as a generator is:

$$R_{eq} = \frac{U_a}{I_a}(1 - \alpha) + R \qquad (9)$$

By varying the duty cycle $\alpha$, the equivalent load resistance seen by the motor can be varied from $R$ to $U_a/I_a + R$, and the regenerative power can be controlled.

Thus, from the equations (8) and (9), the minimum and the maximum braking speeds of the series motor can be obtained:

$$\begin{aligned} \omega_{min} &= \frac{R}{K_m} \\ \omega_{max} &= \frac{U_a}{K_m I_a} + \frac{R}{K_m} \end{aligned} \qquad (10)$$

The regenerative braking would be effective only if the motor speed $\omega$ is between the two speed limits. At any speed less than $\omega_{min}$, an alternative braking type would be required.

In the *rheostatic braking* (also known as dynamic braking), the energy is dissipated in a rheostat, that is an undesirable feature. Sometimes, this energy is used to heat the tram.

The average current of the braking resistor is:

$$I_b = I_a (1 - \alpha) \qquad (11)$$

and the average voltage across the braking resistor $R_b$ is:

$$U_b = R_b I_a (1 - \alpha) \qquad (12)$$

The equivalent load resistance of the series generator is:

$$R_{eq} = R_b (1 - \alpha) + R \qquad (13)$$

The power dissipated in the braking resistor is calculated as:

1588

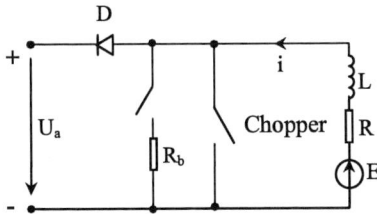

Fig. 6. Combined braking scheme

$$P_b = I_a^2 R_b (1-\alpha) \qquad (14)$$

Depending on $\alpha$, the load resistance can vary from $R$ to $R + R_b$ and the braking power can be controlled. The braking resistance determines the maximum voltage rating of the chopper.

Because the supply is partly receptive, a combined regenerative and rheostatic brake control is used (Fig. 6).

During regenerative braking, the line voltage is continuously tested. If it exceeds a certain preset value (20% above the line voltage), the regenerative braking is substituted by the rheostatic braking. It allows an almost instantaneous transfer from regenerative to rheostatic braking, even for a moment.

The operating principle is achieved by means of a microcontroller control, using the PWM technique, through the modification of the disconnection duration, maintaining a constant frequency of 2 kHz.

## IV. EXPERIMENTAL RESULTS

After the realization of numerical simulations of the operating regimes of the force and control parts, the driving system was designed for the new configuration that includes two choppers for running/braking and another one for the rheostatic braking. A KT4D tram was used for the experimental model implementation.

During the experimental tests one performed recordings using the scope of the load current and voltage waveforms for various operating regimes (running, regenerative braking and mechanical brake).

The test equipment consisted in a HAMEG scope (Fig. 7) and a laptop for the recording and analysis of the oscillograms (Fig.8). The oscillograms were recorded choosing the scope time basis for an optimum visualization, namely 1 s/div.

Fig.7. Recording of the waveforms using scope

Fig. 8. Transfer and waveforms analysis using a laptop

The current waveform is recorded using the first scope channel (waveform CH I) and the voltage waveform is recorded using the second channel (waveform CH II).

### A. Recording of the first chopper current and voltage waveforms, for the first running stage

On the current waveform (Fig. 9, waveform CH I), when the running treadle is pressed, one notices the current raise from 0 to 100 A along a time interval of 0.4 s, the raising slope having a value of 250 A/s. The current is kept constantly to 100 A (this value corresponds to the first running stage) for 6 seconds. On the current oscillogram one notices short duration pulses, generated by multiple phenomena: scope noise, vibrations, treadle unequal pressing, a.o. The driver releases the treadle and consequently the current falls from 100 A to 0; the transistor from the chopper enters in the blocking stage, and after 1,2 seconds the d.c. motors enter in the motor brake regime. After this stage, when no treadle is pressed, the driver progressively presses the brake treadle, the currents for each braking stage having the values 80A, 130 A and 160 A, as one can notice from the oscillogram.

The voltage oscillogram (waveform CH II) reveals that both during the running and braking the voltage is kept almost constant to a value of 820 V for 6.5 seconds. In this interval frequent voltage ripples can be noticed. When

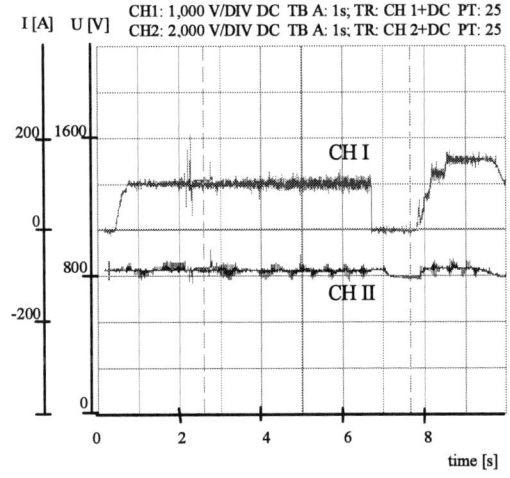

Fig. 9. Current (CH I) and voltage (CH II) corresponding to chopper 1 for the first running stage

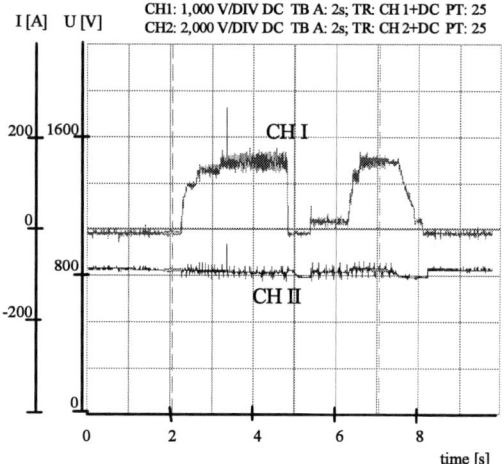

CH1: 1,000 V/DIV DC TB A: 2s; TR: CH 1+DC PT: 25
CH2: 2,000 V/DIV DC TB A: 2s; TR: CH 2+DC PT: 25

Fig. 10. Current (CH I) and voltage (CH II) corresponding to chopper 1 for the 3-rd running stage

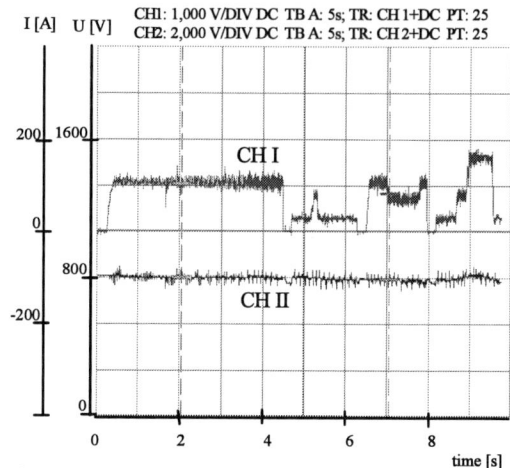

CH1: 1,000 V/DIV DC TB A: 5s; TR: CH 1+DC PT: 25
CH2: 2,000 V/DIV DC TB A: 5s; TR: CH 2+DC PT: 25

Fig. 11. Current (CH I) and voltage (CH II) corresponding to chopper 1 during climbing

the driver releases the running treadle, after 0.3 s, the voltage falls to 800 V with a slope of 67 V/s and the voltage keeps this value that is equal to the supplying voltage value.

When the braking treadle is pressed the motor enters in a regime of d.c. generator when one realizes braking with recovering, the voltage raises to a value of 880 V and keeps this value for 1.5 s. At the end of this interval the tram stops and the voltage comes back to its initial value of 800 V.

*B. Recording of the second chopper current and voltage waveforms, for the third running stage*

On the current waveform (Fig. 10) one notices that when the running treadle is pressed on the first stage the current raises to the value of 100 A for 0.6 s.

Then one passes to the second running stage for 1.6 s. and the current keeps constant to a value of 130 A. For 3.6s the third running stage is reached and the current becomes 160 A. When the acceleration treadle is released, the current drops to 0. The process continues and the tram enters in a regime with motor brake. One can see in the oscillogram all the 3 brake stages.

*C. Recording of the first chopper current and voltage waveforms, during a ramp climbing*

Experimental tests were made for the cases when the tram is climbing a ramp and when it is going down on a slope. These tests are necessary because in these situations additional resistance forces operate over the tram. The value of this force depends on the slope angle. Fig. 11 presents the waveforms of the current and of the voltage during the climbing.

When the motor is supplied, the current increases to 100 A; it is kept to this value for 40 seconds. Then the tram passes consecutively in regimes of running, braking and running. During the next acceleration, the current reaches again the value of 100 A for 3 seconds. Then the tram passes to the first braking stage for 4 seconds; afterward the driver presses again the acceleration treadle, the current raise to the same value of 100 A.

During the braking tests, the current reaches consecutively the values of 80 A for 2 seconds (the first braking stage), of 130 A for 0.6 seconds (the second braking stage) and of 160 A for 3 seconds (the 3-rd braking stage). At the end the driver releases the braking treadle and the current falls, due to the discharging of the filter capacitor.

On the voltage oscillogram one can notice that initially the supply voltage is equal to 810 V. When the acceleration process begins, the voltage falls to 800 V due to other consumers that were connected to the same line. During the running and braking regimes, the voltage remains constant to a value of 800 V. When the driver releases the braking treadle, the voltage falls to 780 V and then it returns to the initial value of 800 V when the current becomes zero.

*D. Recording of the first chopper current and voltage waveforms, when going down on a slope*

When going down on the slope, the braking treadle is pressed and the current has the value of 10 A due to motor's remaining energy (fig. 12).

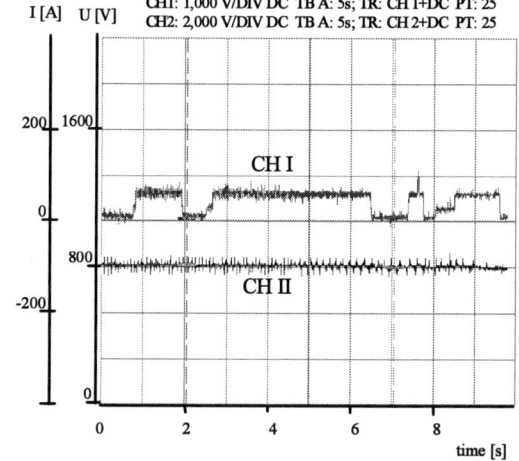

CH1: 1,000 V/DIV DC TB A: 5s; TR: CH 1+DC PT: 25
CH2: 2,000 V/DIV DC TB A: 5s; TR: CH 2+DC PT: 25

Fig. 12. Current (CH I) and voltage (CH II) corresponding to chopper 1 when the tram goes down on a slope

Then the tram is passing in the running mode and in braking regime for 20 seconds. Afterward the driver releases the braking treadle and the discharging of the capacitor can be noticed. In the next 5 seconds, the current is maintained to the value of 10 A, followed by an interval of 2 seconds when the current becomes 80 A (braking regime). At the end, the braking treadle is released.

### E. Additional tests

A series of tests accomplished on the test stands and on the tram were conceived in order to verify:

- the regenerative braking efficiency;

- the voltage limitation with the braking transistor during the regenerative braking for various braking stages;

- the current variation during the running and braking regimes depending on the type of the pressed treadle;

- the protection against over-currents;

- the current regulator efficiency for the running regime;

- the variation limits of the voltage from the intermediate circuit (CH II) at the acceleration current variations (CH I).

This paper presents only recordings for the first chopper, because similar results were obtained for the second one (both on stand and respectively on tram).

## V. CONCLUSIONS

From the analysis of the obtained results one concluded that the electric driving system with the improved technological solution works in accordance with the initial estimations.

No problems were noticed in the power circuit during the operation tests.

The control unit responds very well to different operating conditions. Moreover the new equipment eliminates the shocks corresponding to start, stop and speed regulation regimes, directly influencing the travelers comfort.

The input filter and the line filter for the intermediate circuit significantly reduced the electromagnetic interferences.

For the improvement of control and the reducing of the electromagnetic interferences between the electronic components, an optical fiber circuit is used in the control unit.

Now the modernized vehicle is under exploitation. Its monitoring continues in order to define exactly the reliability indicators and the maintenance program. One could notice savings of 50% of the overhauling and repairs costs and an energy consumption of 47.6% lower that in the case of the same tram in the classical structure.

The driving system was extended with a diagnosis and monitoring system, placed on the tram.

## ACKNOWLEDGMENT

The authors thank to their partners from S.C. INDAELTRAC S.A. and RAT Craiova, who helped them to develop and implement this technological solution.

## REFERENCES

[1] C.M. Ong, *Dynamic Simulation of Electric Machinery using Matlab / Simulink*, Prentice Hall, New Jersey, 1998

[2] M. Belloni, E. Bonizzoni and F. Maloberti, "A voltage-to-pulse converter for very high frequency dc-dc converters", *Proceedings of International Symposium on Power Electronics, Electrical Drives, Automation and Motion, SPEEDAM 2008*, June 11-13, 2008, Ischia, Italy, pp.789-791

[3] T.S. Lee, K.S. Tzeng and M.S. Chong, "A passivity-based controller design for three-phase active rectifiers without dynamic feedback", *Proceedings of IPEC 2005 - The International Power Electronics Conference*, Toki Messe, Niigata, Japan, April 4-8, 2005, Session Paper S69, pp.9-12

[4] C.Y. Chan, "A nonlinear control for dc-dc power converters", *IEEE Trans. On Power Electronics*, vol.22, no.1, 2007, pp.216-222

[5] S. Abourida, C. Dufour and J. Belanger, "Real-time and hardware-in-the-loop simulation of electric drives and power electronics: process, problems and solutions", *Proceedings of IPEC 2005 - The International Power Electronics Conference*, Toki Messe, Niigata, Japan, April 4-8, 2005, Session Paper S59, pp.8-13

[6] J. Luo, "Evaluating the relationship between transport service supply and urban fringe development", *Proceedings of ICSSSM '05. 2005 International Conference on Services Systems and Services Management*, Chongqing, China, June 13-15, 2005. Vol. 2, pp.1451-1456

[7] EN 55025 Document, "Radio disturbance characteristics for protection of receivers used on board vehicles, boats, and on devices — Limits and methods of measurements", 2004

[8] W. Chuchra, W. Zajac, "The problems of electromagnetic compatibility interferences generated by a trolley-bus and tram", *Proceedings of EMC Europe 2006 - The International Symposium on Electromagnetic Compatibility*, Barcelona, Spain, September 4-8, 2006, vol. 2, pp. 788-791

[9] P.M. Nicolae, I.D. Nicolae, L. Mandache and I.G. Sîrbu, "Improved technological solution for urban transportation system by tram", *Proceedings of the International Symposium on Power Electronics, Electrical Drives, Automation and Motion - SPEEDAM*, Taormina - ITALY, 23-26 May, 2006

[10] I.D. Nicolae, P.M. Nicolae, L. Mandache and V.D. Vitan, "Designing, simulation, implementation and testing of automatic current regulators from an electric driving system with d.c. voltage variator used in electric urban traction", *Proceedings of IPEC 2005 - The International Power Electronics Conference*, Toki Messe, Niigata, Japan, April 4-8, 2005, Session Paper S61, pp.9-15.

[11] R. Erbe and J. Reichard, "Drive technology alternatives", *Proceedings of IEEE International Electric Machines and Drives Conference*, Seattle, Washington, USA, 1999, p.710-712.

[12] A. Morar, "D.C. drive system with the insulated gate bipolar transistors", *Proceedings of AQTR - IEEE International Conference on Automation, Quality and Testing, Robotics*, Cluj-Napoca, Romania, May 25-28, 2006, vol. 1, pp. 265-270

[13] F. Mihalic and D. Kos, "Randomized PWM for conductive EMI reduction in DC-DC choppers", *HAIT Journal of Science and Engineering B*, vol 2, 2005, pp.594-608

[14] F. Mihalic, D. Kos, "Reduced conductive EMI in switched-mode DC-DC power converters without EMI filters: PWM versus randomized PWM", *IEEE Trans. On Power Electronics*, vol.21, no.6, 2006, pp.1783-1794

[15] V.Serrao et al., "Common and differential mode EMI filters for power electronics", *Proceedings of International Symposium on Power Electronics, Electrical Drives, Automation and Motion, SPEEDAM 2008*, June 11-13, 2008, Ischia, Italy, pp.918-923

[16] B. Choi, "Analysis of input filter interactions in switching power converters", *IEEE Trans. On Power Electronics*, vol.22, no.2, 2007, pp.452-460

[17] M.H. Rashid, *Power electronics - Circuits, Devices and Applications*, Prentice-Hall International, Inc., New Jersey, 1988.

[18] F. Ionescu et al., *Power Convertors (Convertisseurs statiques de puissance)*, Ed. Tehnica, Bucharest, Romania, 1995.

[19] J.R. Wells, P.L. Chapman and P.T. Krein. "Applications of ripple correlation control of electric machinery", *Electric Machines and Drives Conference*, 2003, IEEE International, vol.3, pp.1498-1503

[20] S. Sugimoto et al., "Current ripple control switching regulator with switching frequency constant control", *Proceedings of IPEC 2005 - The International Power Electronics Conference*, Toki Messe, Niigata, Japan, April 4-8, 2005, Session Paper S7, pp.9-15

# The study of using the traction drive topology with the middle-frequency transformer

Martin Pittermann, Pavel Drábek, Marek Cédl, Jiří Fořt

Department of Electromechanics and Power Electronics, University of West Bohemia, Plzen, Czech Republic
pitterma@kev.zcu.cz, drabek@kev.zcu.cz, alvist@kev.zcu.cz, fort@kev.zcu.cz

*Abstract*— **This paper presents research motivated by industrial demand for special traction drive topology devoted to minimization of traction transformer weight of traction vehicles. The main attention has been given to the special traction drive topology for AC power systems: input high voltage trolley converter (single phase) – middle frequency transformer (single phase) – output converter (single phase voltage-source active rectifier + three phase voltage-source inverter) - traction motor. This configuration can minimize traction transformer weight against topology with classical 50Hz traction transformer. Several variants of innovative topologies of the traction vehicles fulfill the weight reduce requirements have been presented.**

*Keywords*— **Locomotive, Traction drive, AC power system, Multilevel converters, Matrix converter.**

## I. INTRODUCTION

At the beginning we can start describing configuration of the classical traction drive topology with the normal 50Hz (and especially 16,7 Hz for Germany and Austria) transformer situated at the input part of the AC traction vehicle as shown in Fig.1. Classical traction transformer is situated in the input part of the train vehicles and transformer outputs are connected to the traction converters for regulation the traction drives and auxiliary drives. Adjusting of the high level input trolley voltage to the applicable level and isolating character as well are the main aims of the traction transformer.

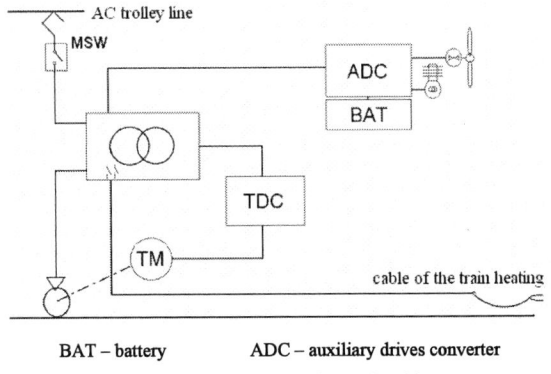

BAT – battery      ADC – auxiliary drives converter

MSW – main switch      TDC – main traction drive converter

TM – traction motor

Fig. 1: The topology of the classical AC train vehicle with 50Hz traction transformer at the input part.

The Czech railways dispose of two supply systems: DC supply system 3kV and AC supply system 25kV/50Hz. Fig.2 shows Czech's classical AC train vehicle of 3MW total power at supply system 25 kV / 50Hz. However our neighbor's railways (Germany and Austria) use different AC system: 15 kV / 16,7 Hz. This supply system has three times lower frequency which leads to the decreasing total loco power to one third. As shown in Fig. 2 presented loco drive with nominal power 3MW can use only 1 MW power in supply system 15 kV / 16,7 Hz.

Power (50Hz): 3 MW
Power (16,7 HZ): 1 MW

Fig. 2: The Czech railways AC loco train for abroad connection to Germany and Austria (15 kV / 16,7 Hz)

According to the demands of the Czech Ministry of Transport and Communications to create train corridor BERLIN (Germany) to Vienna (Austria) through the Prague and Brno (Czech Republic) it is necessary to calculate with both AC systems at new loco train development or purchasing for the Czech railways. This planned corridor unfortunately crosses three supply systems: AC system 15 kV/16,7Hz in Germany and Austria, DC system 3 kV in Czech Republic and finally AC system 25 kV/50Hz in Czech Republic.

Therefore the Czech railways also manage lots of multi-system loco drives (topology of the multi-system train is shown in Fig.3). The configuration is very similar to the classical AC loco train (Fig.1). Traction transformer is situated in the input part of the train vehicles and transformer outputs are connected to the traction converters for regulation the traction drives and auxiliary drives. Further there is connection of DC supply system which is directly connected to the DC link of TDC – main traction drive converter and supply VSI – voltage-source inverter.

978-1-4244-1741-4/08/$25.00 ©2008 IEEE      1593

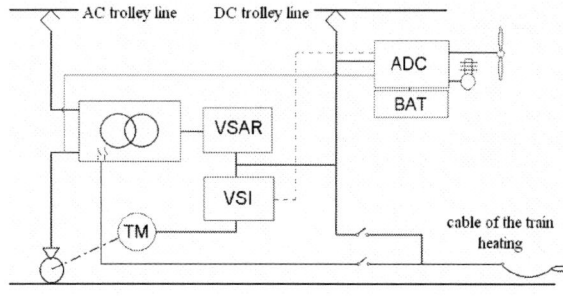

BAT – battery             ADC – auxiliary drives converter

VSAR – voltage-source active rectifier   VSI – voltage-source inverter

TM – traction motor

Fig. 3: The topology of the multi-system train vehicle with 50Hz

traction transformer at the input part.

The high weight of the traction transformer worked at the frequency 50 Hz (and especially 16,7 Hz) leads to limitation of the maximal value of installed power in the traction vehicle. For example in the standard 4-axle locomotive we can install maximal power 6 – 7 MW according to the maximal permitted track load (approximately 21-22,5 tons per one axle). Therefore in several applications (especially at the traction vehicles for high speeds) it is necessary to realise special topology of

the traction drive which enables to achieve higher power (you will see in following chapters). Mainly special vehicles – e.g. electrical motor units – need this special traction topology more than locomotives.

## II. TOPOLOGY WITH MIDDLE-FREQUENCY TRANSFORMER

Today we can see several low power applications using converter topology with middle-frequency transformer (e.g. UPS, switching supply sources, welding machines, etc.). According to the growing of the power electronic area (mainly the price decreasing of semiconductor components) we can look forward to the future using this idea for high power devices (e.g. traction application) as well. Basic topology of the electric drive with middle-frequency transformer is in Fig.4:

A) Input converter (HVC in Fig.4) regulates input line voltage to the appropriate waveform for the middle frequency transformer (e.g. AC course with high frequency)

B) Middle frequency transformer (MFT) galvanic insulates input and output and adjusts output voltage level

C) Output traction converter modifies middle frequency course from transformer to suitable waveform for traction drive supply

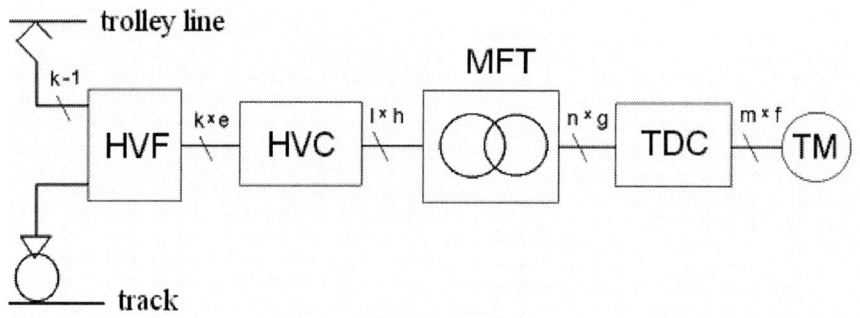

HVF – high voltage input filter      HVC – high voltage trolley line converter      MFT – middle frequency transformer

TDC – main traction drive converter      TM – traction motor

Fig. 4: Topology of the electric drive with middle-frequency transformer

Basically the configuration of the input (HVC) and output (TDC) train converter can be realised arbitrarily. In case of input and output AC voltage converters HVC and TDC can be design as direct (matrix converter [5] or similar topology [2,3]) or indirect frequency converter (voltage-source active rectifier + voltage source inverter [4]).

In the Fig.4 several variants of number of phases change (letters k,l,m,n) and possibility of realisation of several potential level systems (letters e,f,g,h) are symbolically indicated. The most synoptical situation is single phase trolley line (it means k=l=n=2 in the Fig.2) and 3-phase supplying of the traction motor (m=3 in the Fig.2). For lower voltage levels the single potential system is sufficient (e=f=g=h=1). However real traction applications present system which consists of several potential levels (see Fig.5B a 5D – mainly e=h and f=g).

## III. INPUT HIGH VOLTAGE CONVERTER REALIZATION

For designing high voltage converter it is necessary use high voltage semiconductor components (this should be idea in the future – e.g. SiC based material) or special converter topology (following paragraphs) when input high voltage is spread on several active switches in series connection. Design of high voltage converter (25kV) leads to high number of semiconductor devices or whole semiconductor structures. For example using of the IGBT devices with 6,5 kV presents series connection of the 13 transistors at the trolley line system 25 kV. This number of devices has negative influence on the price of the drive topology and also on the potential decrease of reliability in comparison with classical solution.

There are several variants to realize high voltage converter (it means that input voltage is higher than operating voltage of each semiconductor component – Fig.5):

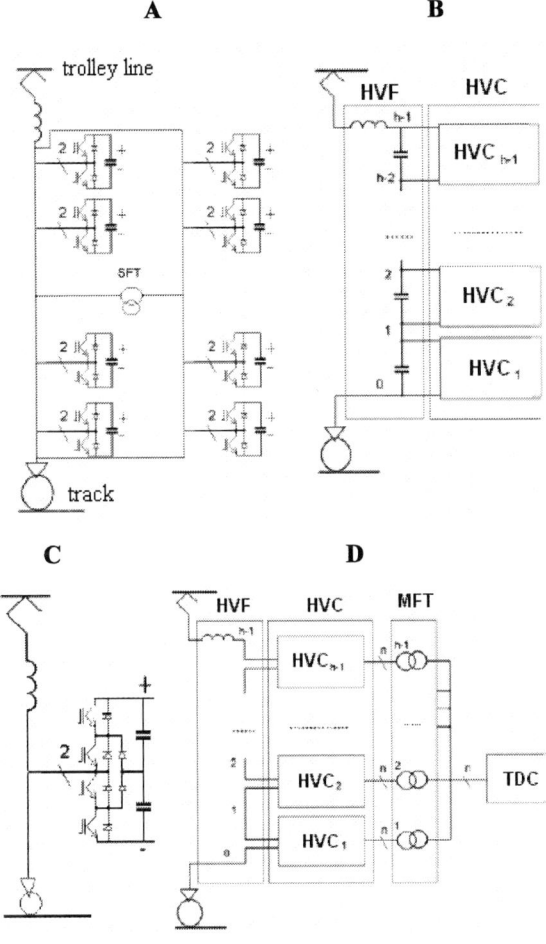

Fig. 5: Realisation of the input high voltage converter – the topology of the electric drive with middle-frequency transformer

A) Series connection of semiconductor components – series connection is very problematic and for realization it is necessary to guarantee voltage uniform spread – in steady state and transient state as well (e.g. using of special commutation circuits to correct balance of each layer).

B) Special converter topology called – Modular Multi Level Converter [2] (see Fig.5 A). This converter is created by few arms and each arm should be making as a series connection of single phase active rectifiers. So there is not series connection of component but series connection of converter arms.

C) Converter topology with input voltage dividing at the converter entrance – e.g. high voltage multi level converter is connected through serial connection input capacitors (Fig.5 B). There is very important to guarantee voltage uniform spread

at these capacitors [3] or using of special type of converter – for example matrix converter.

D) Converter disposition very similar to the multi level converters – e.g. converter input part designed as a multi level converter with clamping diodes [Fig.5 C].

E) Converter topology coming out of galvanic separate input converters connected to high voltage cascade in the input and connected to the different primary windings of middle frequency transformer in the output (see Fig.5 D and [4]).

Now we compare individual variants (we will mentioned variant A – modular-multi level converter with 4 arms, B – single phase matrix converter and D – indirect frequency converter consist of 1f VSAR + 1f VSI) by used number of switching components. For comparison we will think over just one level of input high voltage converter. By this simplification we have to calculate the following minimal number of IGBTs (+FWDs):

| Variant | Number of IGBT switches |
|---------|-------------------------|
| A       | 4 x 4 = 16              |
| B       | 4 x 2 = 8               |
| D       | 4 + 4 = 8               |

It is evident that variant A contains the most IGBTs of all. Single phase variant B and D includes the same number of IGBTs compared to the three phase variant (B – 18 and D - 12). The other point of view should be control algorithm difficulty (evidently D is simple) or the cost of the passive components (here D is second, because of high capacitor in the DC bus line and input reactor for VSAR). Upon these facts we decided for variant B – single matrix converter.

IV. SINGLE PHASE MATRIX CONVERTER

The single phase matrix converter is the one possibility of input high voltage converter according to the Fig. 5B. Matrix converter can be understood as an alternative today standard indirect frequency converter (Active voltage rectifier – DC voltage bus – Voltage source inverter) for AC drives supplying (more [5,6]).

Fig. 6 shows scheme of the traction drive with single phase matrix converter as an input traction converter which supplies middle-frequency transformer. The circuit consists of input high voltage filter (reactor $L_F$ and capacitor $C_F$) connected to the input of the matrix converter which supplies middle-frequency transformer. The output of the transformer connects single phase active voltage rectifier, three phase voltage source inverter supplying AC motor is added through the DC bus line. Filter with capacitor is situated in the input of the matrix converter and the inductive load (winding of the transformer) is connected at its output. These facts have to be taken into account at the control of matrix converter – it cannot short circuit input terminals and disconnect output terminals at the same time.

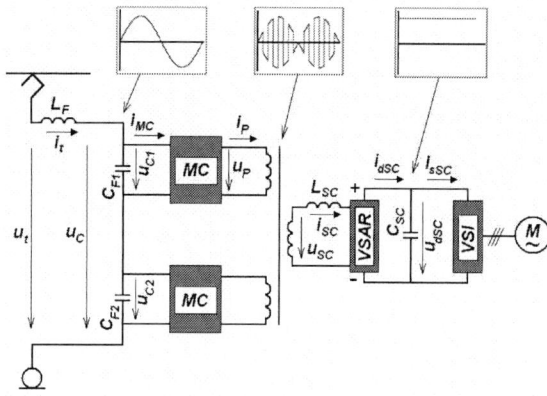

Fig.6: Topology with single phase matrix converter as a primary high voltage traction converter.

For control of the matrix converter following methods have been considered [8]:

- **PWM**
- **Frequency control**
- **Inserting of NULL vector**
- **Two valued control**
- **Control by secondary active voltage rectifier**
- **Inserted commutations**

Detailed scheme of 1f matrix converter is shown in Fig.7. Switching individual branches proceed by sequential crossing within switching states. More details about converter commutation you can find in [8]. To control matrix converter following stabile switching states will be used:

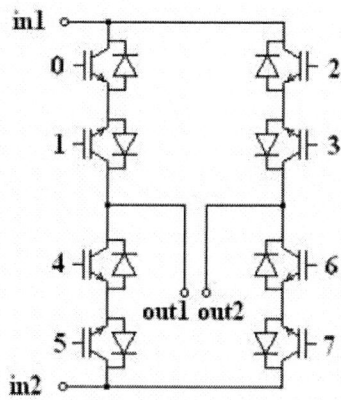

**0167** – Input of matrix converter is directly connected to the output (state **"1"**)

**2345** - Input of matrix converter is reversely connected to the output (state **"-1"**)

**0123 (or 4567)** – Input of the matrix converter is disconnect and the output is short circuit (so-called NULL vector) (state **"0"**)

Fig.7: Detailed scheme of 1f matrix converter with individual transistors

From control methods mentioned above the controlling by inserting NULL vector will be consider.

## V. CONTROL OF MATRIX CONVERTER INSERTING NULL VECTOR

At this control method the output active voltage rectifier takes square current waveform from middle-frequency transformer (two-value control) and primary high voltage converter (single phase matrix converter) ensures sinusoidal phase current taken from trolley line by inserting appropriate NULL vector to the control algorithm. At the same time the matrix converter controls phase shift between trolley line voltage and current by inserting NULL vector. Detailed block diagram with synoptical charts of mentioned control method is presented in Fig.8 and Fig. 9.

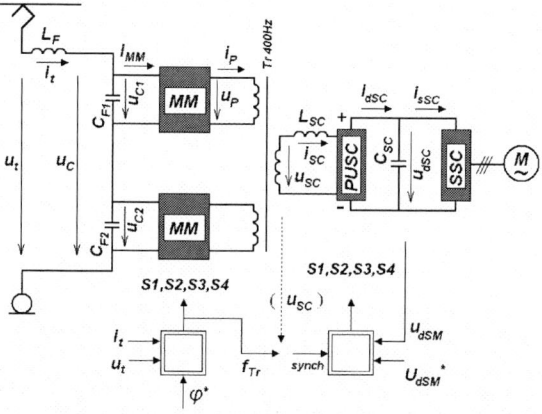

MC - 1f matrix converter – primary converter
VSAR - secondary voltage-source active rectifier,
VSI - secondary voltage-source inverter
$L_F$ - reactor of input filter, $C_F$ - capacitor of input filter
$L_{SC}$ - clamping choke of the secondary converter
$C_{SC}$ – capacitor of DC bus link

Fig.8: Power circuit of the drive with control circuit for control of the matrix converter inserting NULL vector

The proposed control method uses switching frequency 400Hz according to possibilities of high power semiconductor devices. Fig.10 and Fig.11 show appropriate simulation results of middle frequency topology. In the upper-hand picture you can see variables of primary matrix converter – voltage and current of the trolley line, output current and voltage of the matrix converter flow to the primary winding of the middle frequency transformer. In the lower-hand picture you can see variables of secondary active rectifier – input current and voltage (output variables of the transformer), hysteresis band of the input current, output voltage in the DC bus, average switching frequency (due to two-value control).

1596

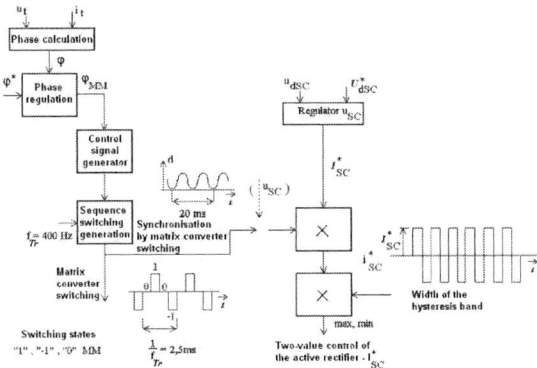

Fig.9: Block control scheme of the matrix converter inserting NULL vector and two-valued control of secondary active voltage rectifier

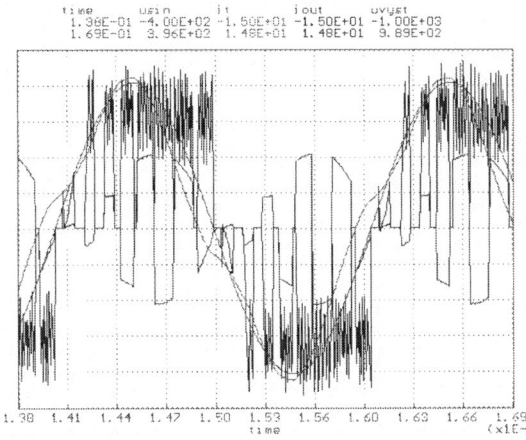

Fig.10: Simulation results of primary matrix converter (input and output variables) for following parameters: Ut=230V, P=3kW, LF=40mH, CF=50uF, LSC=1mH, CSC=4mF

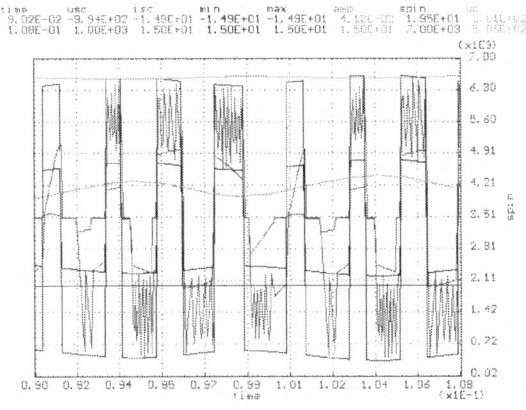

Fig.11: Simulation results of secondary active rectifier (input and output variables) for following parameters: Ut=230V, P=3kW, LF=40mH, CF=50uF, LSC=1mH, CSC=4mF

## VI. THE OTHER VARIANTS ENABLED WEIGHT REDUCTION

Except mentioned variants of the traction drive with middle-frequency transformer (Fig.4) it is necessary to think of other variants enabled weight reduction of the traction equipment. We can appoint for example using of the untraditional types of the traction motor (e.g. SMPM [1] or SRM [8]), also it is possible use traction transformer with more suitable parameters (e.g. superconductive traction transformer – [7]), respectively development of special variants of the drive topology absolutely without traction transformer. Some of these variants eliminate disadvantages of high power installed drives of the trains, eventually disadvantages of the high voltage converters as well.

## VII. CONCLUSION

Several variants of innovative topologies of the traction vehicles fulfil the weight reduce requirements have been discussed. The main attention has been given to the traction drive topology: input high voltage trolley converter – middle frequency transformer – traction motor converter. At the mentioned variant a lot of technical and economical problems can be supposed which will be necessary to solve. Concretely it is possible to suppose extensive problems especially at the realisation of the high voltage converter. However in the case of the successful solve of this ambitious project we can suppose using achieved knowledge in other parts of the electrical engineering as well (e.g. area of Power engineering).

### ACKNOWLEDGMENT

This research work has been made within research project of Ministry of Industry and Business of the Czech Republic No. MPO FT-TA2/035.

### REFERENCES

[1] V.Buba, *The control of the traction drive with permanent magnet synchronous motor*. PhD thesis. UWB. Plzen. 2005.

[2] M.Glinka and R.Marquardt, *A New Single Phase AC/AC – Multilevel Converter for Traction Vehicles Operating on AC Line Voltage*. In: Proc. EPE'03, Toulouse 2003.

[3] G. Kalvelage, P.Dubin, and T.Lequeu, *Reduction of Mass and Volume of On-Board Multi-Input Voltage Converters Using SPARC Topology*. In: Proc. EPE'03, Toulouse 2003.

[4] M.Victor, *Energieumwandlung auf AC – Triebfahrzeugen mit mittelfrequentransformator*. Fahrzeugtechnik eb 103 (2005) Heft 11, p. 505-510.

[5] P.Wheeler, J.Rodriguez J. Clare and L. Empringham: *Matrix Converters: A Technology Review*, IEEE Transactions on Industrial Electrinics, Vol. 49, No. 2, April, 2002.

[6] Domenico Casadei, Jon Clare, Lee Empringham, Giovanni Serra, Angelo Tani, Andrew Trentin, Patrick Wheeler, Luca Zarri, "Large-Signal Model for the Stability Analysis of Matrix Converters," IEEE Trans. on Industrial Electronics, vol. 54, no. 2, pp. 939-950, Apr. 2007.

[7] M.Krasl, R. Vlk and J. Rybá  *Loses and the cooling system of the superconductive traction transformer*. In: Electric drives. Plzen. 2005.

[8] M. Pittermann, *Application of matrix converters in traction – basic study*. Research report UWB. Plzen 2006.

[9] M. Pitterman, P. Drabek, M. Cedl, J. Fort, *The Study of the Traction Topology with the Middle-Frequency Transformer*. In: Proc. ISIE'08, Cambridge 2008.

# Control of a Linear Switched Reluctance Motor as a Propulsion System for Autonomous Railway Vehicles

L. Kolomeitsev[1], D. Kraynov[1], S. Pakhomin[1], F. Rednov[1], E. Kallenbach[2], V. Kireev[2], T. Schneider[3], J. Böcker[3]

[1] Emetron Ltd., Novotcherkassk, Russian Federation, e-mail: *rednov.emet@mail.ru*
[2] Steinbeis Transferzentrum Mechatronik, Ilmenau, Germany, e-mail: *eberhard.kallenbach@stz-mtr.de*
[3] University of Paderborn, Inst. of Power Electronics and Electrical Drives, Paderborn, Germany, e-mail: *boecker@lea.upb.de*

*Abstract*— A linear switched reluctance drive with large airgap was developed and manufactured for a test track of autonomous railway vehicles (RailCab). Control structure and current shaping algorithms are presented providing smooth transitions between the motoring and dynamic braking operation modes at low switching losses. In both modes low force and torque ripples are maintained with reasonable efficiency. The control is tested on the rotational variant of the motor. Details of the test bench and measurements of phase currents, force or torque, respectively, and efficiency are presented.

*Keywords*—linear and rotational switched reluctance drives, propulsion, control, dynamic braking, low ripple.

## I. INTRODUCTION

The RailCab project was founded at the University of Paderborn in 1998. The focus of this project is to investigate a novel railway system with individually operating autonomous vehicles called RailCabs. These RailCabs can drive in convoy without the mechanical couplings.

For real life validation of convoy operation, a test track in a scale of 1:2.5 was built at the University of Paderborn [1][2][3]. For building convoys, the propulsion and braking system must be independent of the weather conditions. For this reason a linear doubly-fed motor (LDFM) was chosen [1]. The LDFM offers the opportunity to transmit energy contactless into the vehicle. Therefore a catenary or a feeder rail is not necessary for energy supply of the vehicle. But the additional copper windings in the active stator increase the construction costs. That is the reason why two further types of linear electric motors have been considered: The linear induction motor (LIM) and the linear switched reluctance motor (LSRM). Switched reluctance motors generally offer a very simple and robust design. Thus, they are very suitable for highly reliable and fault-tolerant applications [4][5]. That is the reason why the LSRM was discussed in [6] as an alternative propulsion concept for RailCabs.

An important issue with the switched reluctance motors is the highly nonlinear magnetisation characteristic. If this issue is not addressed properly in motor design and control, the nonlinearity will considerably increase the force ripple and noise of switched reluctance motors. As a matter of fact, the motor geometry, winding parameters, and the control should be thoroughly designed using precise dynamic models describing the nonlinear force generation with respect to the position and current and taking into account the phase voltage and current limitations of the power converter as well.

To enable relative motion between the vehicles of a convoy, the LSRM needs active translators below the undercarriage whereas the stator located between the rails is the passive part. Therefore a catenary or a feeder rail is necessary for energy supply of the vehicle.

To compare the proposed novel LSRM [6] with the existing LDFM [1], the LSRM was designed within the same geometrical spacing, i.e. with a scale of 1:2.5 of a targeted system. The specifications of the test track which was built for real life validation of the RailCab concept are:

- Maximum slope of 5.3%
- RailCab tare weight of 1.2t (incl. motor weight)
- Maximum speed of 10 m/s
- Mechanical airgap of 12 mm

In order to meet the acceleration specification of 1 m/s² the new LSRM should provide at least a force of 1500 N. Furthermore, the propulsion system has to provide not only a smooth acceleration and cruising at constant speed, but also regenerative braking and reverse driving.

The paper is organized as follows: Section II describes the structure and the parameters of a six-phase LSRM for the RailCab test track. The results of transient finite element analyses are presented in Section III. The control for the presented LSRM in order to meet the force specifications is covered in Section IV. Details of the prototyped models and the experimental results for the torque and efficiency are presented in Section V and followed by conclusions in Section VI.

## II. SR MOTOR STRUCTURE AND PARAMETERS

To study the LSRM characteristics experimentally, two motor prototypes with drive control have been developed [6]. The first prototype (linear SRM with a structure shown in Fig. 1) is intended for static measurements of the propulsion force characteristics and of the normal force. The normal force (attraction force between stator and translator) is an important issue of the linear motor, because it implies additional load for the track and the bogie.

Because it is expensive to set up test facilities to run linear motors at constant speed, a second prototype was built as a rotational SRM. The construction of this motor,

---

This work was supported by the German Federal State of North Rhine-Westphalia, the University of Paderborn, STZ Mechatronik Ilmenau, Germany, and the Federal Agency of Science and Innovations, Russian Federation, Grant No.02.516.11.6100

in particular pole number, pole pitch, airgap, etc. are done as close as possible to that of the linear motor.

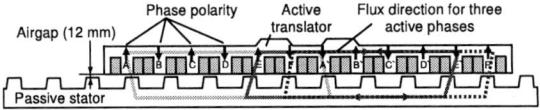

Fig. 1. Structure of a linear SRM.

### A. Design Procedure and Results

In the design phase, a combined approach of finite element analysis (FEA) together with an equivalent magnetic network model of the transient electromagnetic processes in switched reluctance motors with power converters is used [6]. The magnetic model is discussed in detail in [7]. Using the simplified model, power, efficiency as well as the DC-link and phase currents can be predicted with an error less than 3...5%. The overall computation time is much shorter compared to the complete transient finite element analysis of the motor described below in Section III. Due to these advantages, the geometry optimization and control synthesis tasks can be solved more effectively with this simplified model.

The main parameters of the linear and rotational motors determined during the design phase are presented in the Table I and Table II, respectively.

TABLE I
TECHNICAL DATA OF THE LINEAR SRM

| Supply voltage (RMS) | 400 V |
| --- | --- |
| Rated speed | 10 m/s |
| Rated power | 15 kW |
| Maximal speed | 20 m/s |
| Speed range (constant force) | 0...10 m/s |
| Speed range (15 kW constant power) | 10...20 m/s |
| Stator height | 103.5 mm |
| Stator core length | 1525 mm |
| Airgap | 12 mm |
| Stacking length | 140 mm |
| Active phase resistance at 20°C | 0.27 Ω |
| Propulsion force : normal force ratio | 1:4 |
| Efficiency at the speed 10 (20) m/s | 76 (84) % |
| Rated phase current (amplitude) | 70 A |
| Rated phase current (RMS) | 32.3 A |
| Current density in the conductors | 4.42 A/mm$^2$ |
| Copper weight | 83 kg |
| Turns per phase | 2×280 |

In contrast to the rotational motor, the linear SRM with an open magnetic circuit has some particular yoke cross-sections that have to conduct more magnetic flux than others. Depending on the direction of current commutation (clockwise or counterclockwise) the magnetic flux in these yoke segments can be by a factor of 1.5...2.5 higher in comparison to the weaker excited yoke segments. From the analysis of some variants of winding diagrams, a preferred combination of the magnetic pole directions has been found which minimizes the magnetic flux in the stator yoke (Fig. 1). In this manner, only two yoke sections conduct a magnetic flux that is by a factor

of 1.5...1.6 higher than the average flux of other segments.

TABLE II
TECHNICAL DATA OF THE ROTATIONAL SRM

| Supply voltage (RMS) | 400 V |
| --- | --- |
| Rated speed | 406 rpm |
| Rated power | 15 kW |
| Maximal speed | 812 rpm |
| Speed range (constant torque) | 0...406 rpm |
| Speed range (15 kW constant power) | 406...812rpm |
| Outer diameter of the stator | 698 mm |
| Inner diameter of the stator | 469.7 mm |
| Airgap | 12 mm |
| Stacking length | 140 mm |
| Active phase resistance at 20°C | 0.27 Ω |
| Efficiency at the speed 406 (812) rpm | 75 (83) % |
| Rated phase current (amplitude) | 70 A |
| Rated phase current (RMS) | 32.3 A |
| Current density in the conductors | 4.42 A/mm$^2$ |
| Copper weight | 83 kg |
| Turns per phase | 2×280 |

To maintain an acceptable level of flux density, the yoke segments have been enlarged as shown in Fig. 1 and Fig. 3.

### III. FINITE ELEMENT ANALYSIS

To validate the initial motor design and to evaluate the force ripple characteristics of the linear and rotational SRM, complete finite element analyses of both rotational and linear motors have been performed using *Maxwell 2D* from *ANSOFT*. Some of the results for the static FEA have been presented in [6]. The dynamic field distribution at rated speed and torque for the rotational and the linear SRM are shown in Fig. 2 and Fig. 3 respectively.

Due to the large airgap of 12 mm the windings should provide a sufficient magnetomotive force (MMF) to overcome high airgap reluctance and to produce sufficient magnetic flux with a reasonable efficiency. The peak value of the MMF is of 19600 A at rated speed of 406 rpm and rated current of 70 A. Therefore, the copper loss is dominant in the overall power balance (the iron loss is only 5% of the total loss).

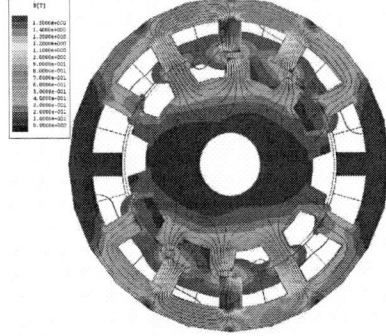

Fig. 2. Magnetic flux density and flux lines plot for the rotational SRM at rated speed of 406 rpm and torque of about 350 Nm.

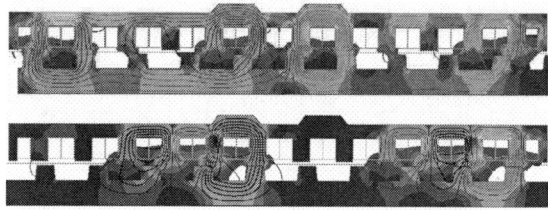

Fig. 3. Magnetic flux density distribution and flux line plots for the linear SRM at rated speed of 10 m/s and force of about 1560N.

The efficiency of the motor found by FEA is about 75% at 10 m/s (406 rpm) and approx. 82% at 20 m/s (812 rpm).

The results found by transient FEA at rated speed of 10 m/s (Fig. 4) and at 20 m/s (Fig. 5) also confirm the initial design relation of the tangential to the normal force $F_x:F_y = 1:4$ defined with an SRM design model. Because of a higher back-EMF level only phase currents of 52 A can be adjusted at 812 rpm.

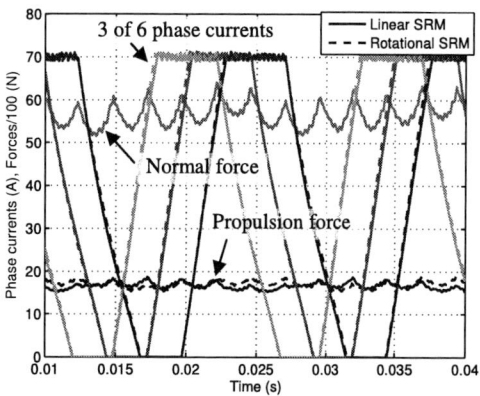

Fig. 4. Current shapes for three of six phases, propulsion force $F_x$ and normal force $F_y$ of linear SRM; phase currents and torque $M/r$ of the rotational SRM: FEA simulation at MMF of 19.6 kA, i.e. rated phase current amplitude of 70 A and rated speed of 10 m/s (406 rpm).

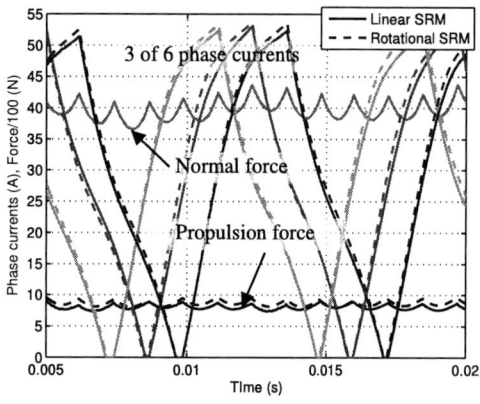

Fig. 5. Current shapes for three of six phases, propulsion force $F_x$ and normal force $F_y$ of linear SRM; phase currents and torque $M/r$ of the rotational SRM: FEAsimulation at a phase current amplitude of 53 A and speed of 20 m/s (812 rpm).

The diagrams in Fig. 4 and Fig. 5 also confirm that the characteristics of the linear and of the rotational motor are

very close by. Hence, it is assumed that the optimal control algorithm for the rotational SRM is also adequate for the LSRM.

## IV. CONTROL

The switched reluctance drive structure is presented in Fig. 6. Only three of six phases are shown for means of clarity. The motor phase windings are magnetized by a power converter using six IGBT modules. Front end converter is a three-phase bridge rectifier with a DC bus capacitor. Each phase module consists of an asymmetrical single phase half-bridge. A single break chopper and resistor are used for dumping the extra energy during braking operation.

The IGBT gate drivers comprise an integrated detection of the collector-emitter voltage desaturation and fault status feedback. Communication with the control system is done via fiber optics (Fig. 6).

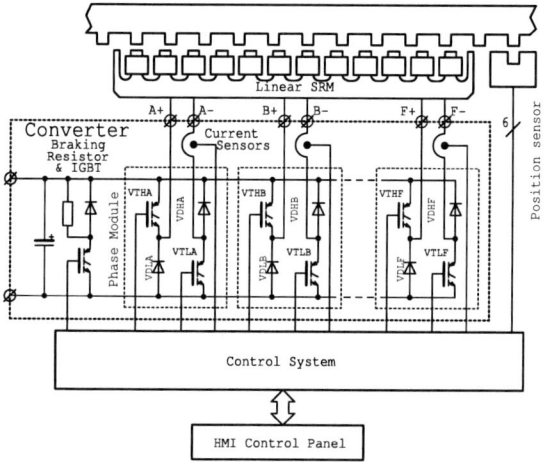

Fig. 6. SR drive structure

The control system is implemented using a 16-bit Infineon XC161 microcontroller. The block diagram of the control system is presented in Fig. 7. The firing signals are generated by the microcontroller and two CPLDs (XILINX XC95144). Each CPLD is comprised of eight 36V18 function blocks, providing 3,200 usable gates with propagation delays of 7.5 ns.

The control system is provided with a human-machine interface (HMI) with LCD and keyboard panel for communication with the operator.

The implementation of a two-quadrant phase current control scheme for an SRM drive implies the development of specialized control pulse sequences for the converter IGBTs (Fig. 6) depending on the required operating mode (motoring or dynamic braking). The objectives of the control algorithm are to maximize the output power and efficiency. In motor mode, a phase current has to be switched on during the positive slope of the phase inductance profile. For the highest power output, the current conduction angle should be nearly 180° (electrically) [8][9].

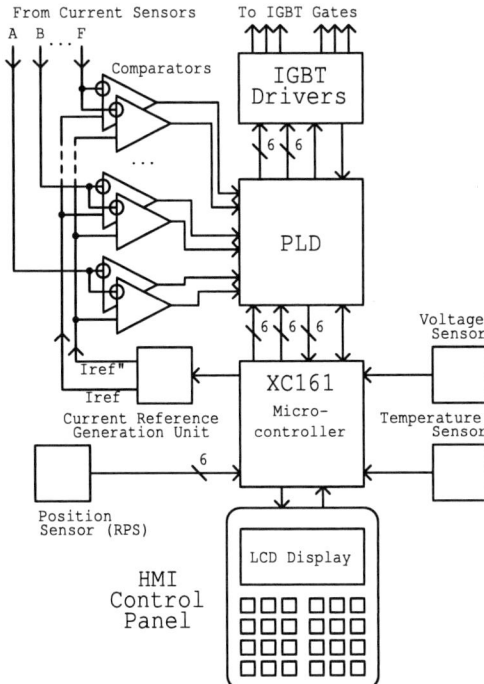

Fig. 7. Block diagram of the control system

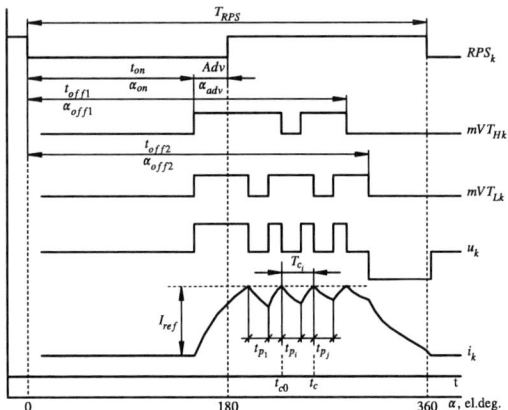

Fig. 8. Variable parameters of the control pulses.

For the generator mode or braking, the current conduction interval should coincide with the negative slope of the inductance profile. A smooth transition between the motor and generator modes is of great importance for a robust drive system operation [10].

The current shape, amplitude, advance angle and the pulse duration are controlled. The control algorithm has to adapt the control signals with respect to the reference input $I_{ref}$ as shown in Fig. 8. The parameter set is calculated in real time using a mathematical model of the motor implemented in the control algorithm. The vector of the reference parameters includes absolute value and direction of the speed and the torque (force). Additional control objectives are to maintain maximum efficiency with minimal torque ripple and acoustic noise.

The symbols in Fig. 8 are explained below:

- $RPS_k$: rotor position sensor signal for phase $k$ ($k = 1..6$);

- $mVT_{Lk}$, $mVT_{Hk}$: firing signals for low and high side IGBTs of the respective phase module;

- $I_{ref}$ : current reference value;

- $i_k$ , $u_k$: measured current and voltage of a SRM phase.

The period of the rotor position measuring system (RPS) is used as a base value ($T_{RPS}$) which is inversely proportional to the instantaneous drive speed. The calculated parameters are:

- $\alpha_{on}$: on-angle of the IGBTs,

- $\alpha_{off1}$, $\alpha_{off2}$: off-angles of the high side and low side IGBTs,

- $\alpha_{adv}$: advance angle which is calculated as a linear function of the motor speed.

The algorithm converts the calculated angles into the commutation times $t_{on}$, $t_{off1}$, $t_{off2}$ of the IGBTs for each motor phase.

The regenerative braking mode is implemented through a change of a switching sequence and shift of a current conducting interval by 180° (electrically) in comparison to the motoring mode. The switching angles $\alpha_{on}$, $\alpha_{off1}$, $\alpha_{off2}$ are varied in such a manner that the phase currents are excited at negative slope of the inductance profile with respect to the rotor position.

The control algorithm includes control loops for the phase currents, the speed and the DC link voltage. Traditionally the SRM phase currents are limited with a hysteresis controller [11]. The usual implementation known from the literature requires two comparators and results in a variable commutation frequency with a relatively wide spectrum. In contrast to the standard hysteresis controller, the implemented current limiting algorithm [12] generates an almost constant frequency (see Fig. 8), similar to the peak current mode control.

For the motoring mode, this method requires just one comparator per phase. The positive comparator input is connected with the current reference unit output ($I_{ref}$), the negative one with the phase current sensor. So, if the current exceeds the $I_{ref}$ value, the comparator output becomes active. The control system generates a command to switch the high and low side IGBTs of the active phase module in turn. The phase voltage becomes zero and the phase current decreases slowly. Using alternating transistor switching, the converter losses are better balanced between the high side and low side IGBTs in comparison to the well known soft switching technique with choppering of a single (e.g. only high side) transistor and another transistor being continuously closed.

With the presented algorithm, the current ripple frequency is not exactly equal to the $f_{c\,ref}$, but can be maintained closely to this value. At high rotation speed the phase current does not reach the $I_{ref}$ value due to a higher back EMF voltage, which implies the disappearance of the 8 kHz commutation current ripples.

In order to implement the safe transition to dynamic braking mode, the current feedback scheme has an additional comparator, which has a higher switching level ($I_{ref}$"). The value of this level is approximately 3.4 A higher than for the first comparator. This is less than 5% of the rated phase current amplitude which is reasonable for this application. The selected level difference makes the converter more robust in the environment with strong electromagnetic interference (EMI) due to high current switching. In the dynamic braking mode the back EMF is positive. After only one (high or low side) of the two IGBTs per phase is switched off, the phase current continues to increase. When it reaches the upper reference value $I_{ref}$", the upper comparator output becomes active and the control system generates a command to switch off both IGBTs. The phase voltage becomes negative and the phase current starts to decrease.

The speed controller is implemented as a proportional-integral (PI) controller. In the motor mode the speed controller adjusts the $I_{ref}$ value within the limits set by the external analog signal. The speed controller task is executed at each RPS event. Therefore, the speed signal is updated with a higher frequency and the speed dependent advance angle is changed more smoothly as opposed to a control algorithm presented in [9]. Additionally, in contrast to a method described in [9], our control implementation does not need any "mode detector" neither in current control loop nor in the speed control loop since the switching between motor and braking modes is done automatically using two switching levels. Therefore, we suppose that the presented method is more simple and reliable.

The control loop for the DC link voltage is primarily implemented to protect the power electronics. If the DC link voltage rises above a certain critical level (e.g. while braking), the voltage regulator reduces the $I_{ref}$ value in order to limit an energy flow from the converter to protect the DC link capacitors and the braking circuitry. The control system has also an additional hysteresis voltage controller, which controls the braking IGBT. Fig. 9 depicts a smooth transition from motor to regenerative braking mode and back to the motor mode.

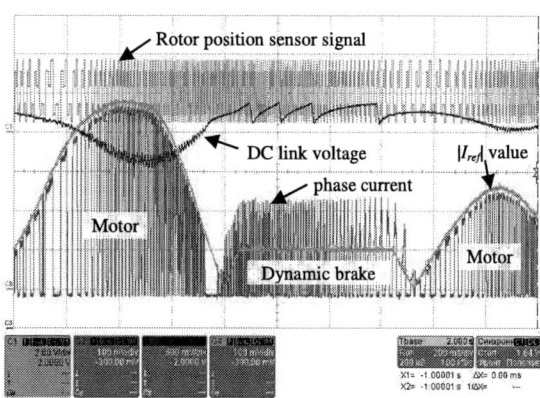

Fig. 9. Smooth transition between motor and braking modes.

The proportional and integral coefficients of both speed and voltage controllers are set low for a better visualization of the transitions. The operation of the brake voltage hysteresis controller can be recognized by the saw-tooth pulses of the DC link voltage caused by the switching of the braking resistor.

## V. TEST BED AND EXPERIMENTAL RESULTS

To evaluate the analytical results, prototypes of both SRM types were built up. Although the LSRM is of particular interest, it is only possible to measure static

forces with the linear test bed. The test bed of the equivalent rotational motor is used to evaluate the dynamic characteristics of the SRM and will be primarily discussed in the following.

The rotational SRM is designed and built to validate the control strategy and to investigate the dynamic characteristics of a six-phase SRM. The SRM is coupled to a DC motor to simulate the load. A torque sensor is mounted on the coupling shaft. The arrangement of test bed is shown in following Fig. 10.

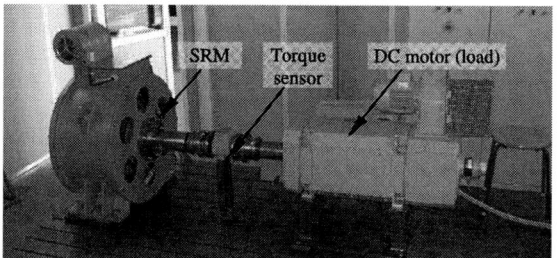

Fig. 10. Rotational SRM with DC motor and torque sensor

As measurement signals, three of six phase currents, the DC link current and the DC link voltage were acquired in each operating point. Torque and current signals from measurement and simulation are shown in Fig. 11. The comparison of simulations and measurements shows that the calculated torque can be achieved, which proves the combined approach of FEA together with the equivalent SRM magnetic network model used in the design phase. Furthermore it is possible to validate the proper functionality of the control algorithm by comparing the simulated and measured current shapes.

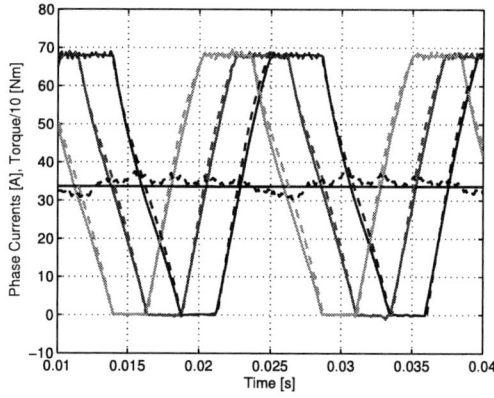

Fig. 11. Current shapes for three of six phases and torque at 406 rpm (10 m/s) and current amplitude of 68A.
——— measurement (filtered); ------ transient FEM simulation

To validate the characteristics of the rotational SRM in different operation points the average torque was measured with different currents and rotational speed (Fig. 12). This was realized by limiting the current with the analog interface of the microcontroller.

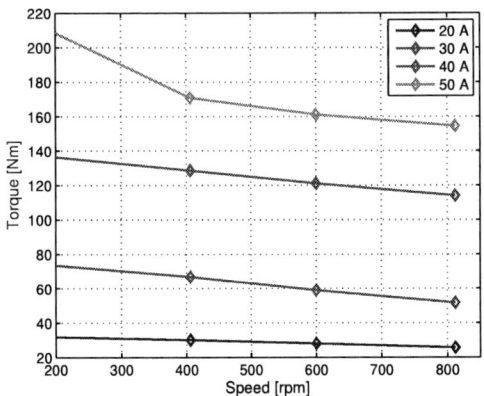

Fig. 12. Average torque characteristics for different phase currents

The mechanical power (Fig. 13) is the product of the measured torque and the rotational speed. The electrical power is calculated by the product of the measured DC link current and voltage. The efficiency shown in Fig. 14 is defined as the mechanical power divided by the electrical power.

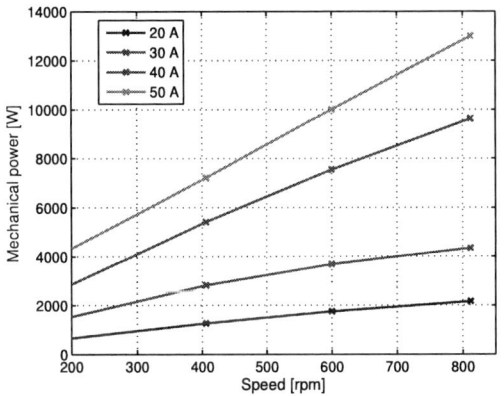

Fig. 13. Shaft power vs. speed for different phase currents

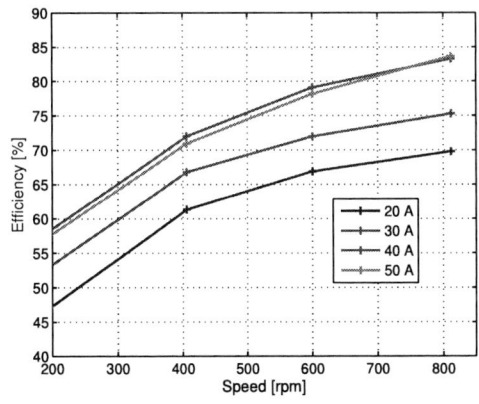

Fig. 14. SR-drive efficiency for different phase currents

## VI. CONCLUSIONS

The paper presents a study on model-based development and control of linear and rotational switched-reluctance motors. The models are validated with static and transient finite element analysis and measurements. Theoretical and practical results show good coincidence.

The optimization of the magnetic circuit geometry and winding parameters with respect to the control limitations (maximal voltage and current), provides a good performance of the overall drive system.

The presented control scheme provides low force ripple and smooth transition between the motor and dynamic brake mode.

The initial design tasks have been solved, therefore the linear SR motor can be supposed as a good candidate for use in an autonomous railway system.

## REFERENCES

[1] Henke, M.; Grotstollen, H.: *Modelling and Control of a Longstator-Linearmotor for a Mechatronic Railway Carriage.* Proc. of the 1st IFAC Symp. on Mechatronic Systems. Sept. 2000, Darmstadt, Germany, pp. 353-357.

[2] Yang, B.; Grotstollen, H.: *Control Strategy for a Novel Combined Operation of Long Stator and Short Stator Linear Drive System.* Proc. of the 10th European Conference on Power Electronis and Applications (EPE 2003), Sept., 2003, Toulouse, France.

[3] Henke, C.; Fröhleke, N.; Böcker, J.: *Advanced Convoy Control Strategy for Autonomously Driven Railway Vehicles.* Proc. of IEEE Conf. on Intelligent Transportation Systems, Sept. 2006, Toronto, Kanada.

[4] Schramm, A.; Gerling, D.: *Researches on the Suitability of Switched Reluctance Machines and Permanent Magnet Machines for Specific Aerospace Applications Demanding Fault Tolerance.* Int. Symp. on Power Electronics, Electrical Drives, Automation and Motion (SPEEDAM), Taormina, Sicily, Italy, 2006.

[5] Miller, T.J.E.: Optimal Design of Switched Reluctance Motors. - IEEE Transactions on Industrial Electronics, Vol. 49, No. 1, February 2002.

[6] Kolomeitsev, L.; Kraynov, D.; Pakhomin, S.; Rednov, F.; Kallenbach, E.; Kıreev, V.; Schneider, T.; Böcker, J.: *Linear Switched Reluctance Motor as a High Efficiency Propulsion System for Railway Vehicles.* Proc. Int. Symp. on Power Electronics, Electrical Drives, Automation and Motion (SPEEDAM), Ischia, Italy, 2008, pp. 155-160.

[7] Kolomeitsev, L.; Pahomin, S.; Krainov, D. et al.: *Mathematical Model for Calculation of the Electromagnetic Processes in a Multi-Phase Switched Reluctance Motor* (in Russian). Izv. vuzov, Electromechanika, No. 1, 1998, pp. 49-53.

[8] Electronic control of switched reluctance machines: edited by T.J.E. Miller, Oxford, U.K.: Newnes, 2001.

[9] Kraynov, D.V.; Duvakin, A.V.; Kolomeytsev, V.L. et al.: *Control method of a switched reluctance motor* (in Russian). Patent of Russian Federation RU 2260243, IPC H02P 8/12, 6/00. - 2003136649/09; Priority 17.12.2003; Published by 10.09.2005, Bulletin № 25.

[10] Mademlis, C.; Kioskeridis, I.: *Smooth Transition between Optimal Control Modes in Switched Reluctance Motoring and Generating Operation.* Int. Conf. on Power Systems Transients (IPST'07), Lyon, France, June, 2007.

[11] Gallegos-Lopez, G.; Walters, J. and Rajashekara, K.: Switched Reluctance Machine Control Strategies for Automotive Applications. SAE 2001 World Congress, March 5-8, 2001.

[12] Kraynov, D.V.; Duvakin, A.V.; Kolomeytsev, V.L. et al.: *Method of current shape synthesis in the phase winding of a switched reluctance motor* (in Russian). Patent of Russian Federation RU 2229768, IPC 7H02P 8/12. - 2002101418/09; Priority 11.01.2002; Published by 27.05.2004, Bulletin № 15.

# Motion Copying System Based on Real-World Haptics in Variable Speed

Yuki Yokokura*, Seiichiro Katsura†, Kiyoshi Ohishi‡

*Dept. of Electrical Engineering, Nagaoka University of Technology, Nagaoka, Niigata, Japan,
e-mail: *yuki@stn.nagaokaut.ac.jp*
†Dept. of System Design Engineering, Keio University, Kohoku, Yokohama, Japan,
e-mail: *katsura@sd.keio.ac.jp*
‡Dept. of Electrical Engineering, Nagaoka University of Technology, Nagaoka, Niigata, Japan,
e-mail: *ohishi@vos.nagaokaut.ac.jp*

*Abstract*—The paper proposes a novel motion copying system considering variable reproduction velocity. The motion copying system consists of motion saving system and a motion loading system. The motion saving system preserves a motion of a human operator to a motion data memory. The motion loading system operates according to saved position and force in the motion data memory, and reproduces saved motion. The reproduced position and force corresponds to stored position and force. The proposed motion copying system in this paper is able to vary the reproduction velocity. By the experiment, the paper confirmed that a reproduction velocity can be freely changed, when the motion loading system reproduces motion.

*Keywords*—Real-World Haptics, Motion Control, Disturbance Observer, Modal Decomposition, Bilateral Control

## I. INTRODUCTION

An information processing technologies has been rapidly grown for several decades. Preservation engineering and reproduction engineering of information are developed using such technologies. Acoustic information or visual information is recorded to tape cassettes, compact discs, digital versatile discs, flash memories and so on. This stored information to media is reproduced in the real-world at any time. The player of the acoustic and visual information is able to fast-forward and/or slow down. In other words, it makes a video of rapid movement of an object slow motion video. Also, such information can be processed using various methods. On the contrary, preservation, reproduction and processing engineering of a haptic information has not been enough researched and developed. A research using a visual information acquires a motion of a human operator [1]–[3]. However, these methods are not able to detect force between the human operator and an environment. The preservation of both position information and force information is very important, in order to save the arbitrary motion. Therefore, these methods are not enough to realize preservation and reproduction of the haptic information. A control method considering both the force information and the position information has been developed [4], [5]. Though, the viewpoint of preservation and reproduction of the haptic information is nothing in the researches. Additionally, a research based on virtual environment has been proposed [6], however it is not able to treat the haptic information

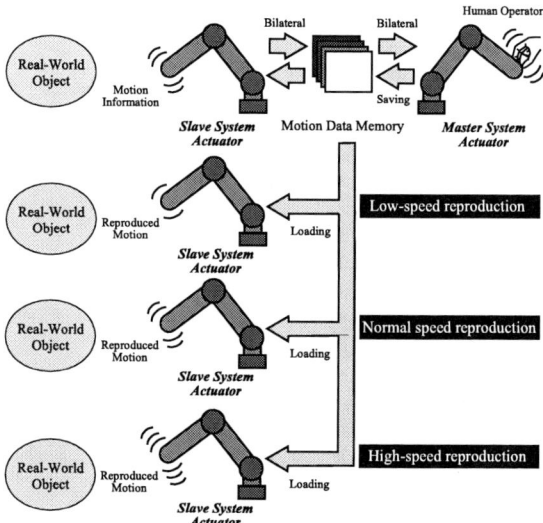

Fig. 1. Motion copying system considering variable reproduction velocity.

in the real-world. In addition, a method of reproduction according to recorded reaction force information has been developed [7].

A motion copying system has been already proposed [8], however the conventional motion copying system does not consider in regard to variable reproduction velocity. This paper proposes a novel motion copying system, which is considered variable reproduction velocity. Fig. 1 shows the expanded motion copying system. The motion copying system consists of a motion loading system and a motion saving system using quarry matrix [9]. The motion saving system is operated by bilateral controller with disturbance observer [11]. The motion saving system measures position and reaction force of a master motor by a linear encoder and reaction force observer [12], and saves it to a motion data memory.

On the other hand, the motion loading system is operated by a virtual bilateral controller. The virtual bilateral controller dose not has master motor in the real-world.

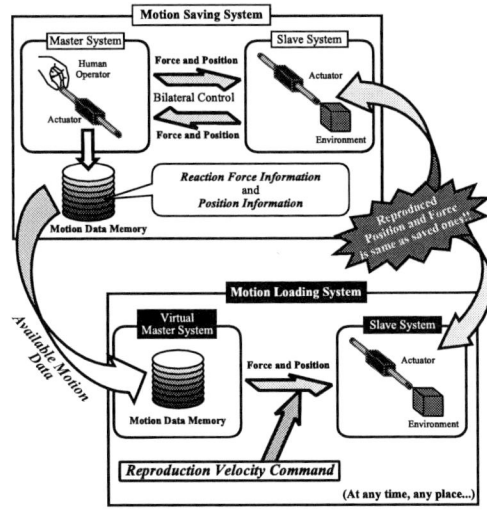

Fig. 2. Conceptual diagram of motion copying system in variable reproduction velocity.

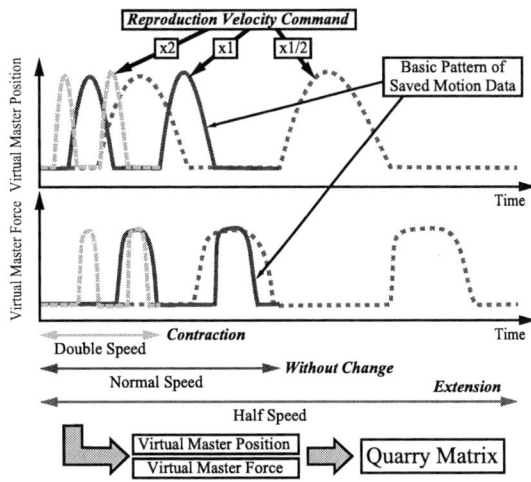

Fig. 3. Processing of force and position information in a motion data memory.

The motion loading system reproduces the motion of the human operator according to the stored motion data in the motion data memory. The reproduced position and force by the motion loading system corresponds to the stored motion data. Therefore, the motion copying system realizes reproduction of the motion. Also, a reproduction velocity can be freely changed, when the motion loading system reproduces the motion. Thus, this system shifts the saved motion of the human operator to low speed or high speed. In other words, slow motion or rapid motion of a object in the real-world is obtained by the proposed method as with a sound or video player.

This paper is organized as follows. A concept of the motion copying system considering variable reproduction velocity is described in Section II. In Section III, a motion control for the motion saving system is shown. Section IV explains the motion loading system considering variable reproduction velocity. Section V shows the experimental results of the proposed method. The last section summarizes this paper.

## II. CONCEPT OF MOTION COPYING SYSTEM IN VARIABLE SPEED

The proposed method in this paper utilizes the motion copying system to save and reproduce the motion of the human operator. Fig. 2 shows conceptual diagram of the motion copying system in variable speed. The motion copying system consists of two systems. First system is the motion saving system; other system is the motion loading system. This section explains the motion saving system and the motion loading system considering variable reproduction velocity.

### A. Motion Saving System

A master system and a slave system constitute the motion saving system. The human operator impresses

force and moves position of an actuator in the master side, at the same time, an actuator in the slave side synchronizes it. In short, the human operator is able to grasp haptic information of an environment in the master side. The motion saving system stores the impressed force and position of the actuator to a motion data memory, when the human operator operates the master system. The motion data in the motion data memory is used to reproduce the motion of the human operator using the motion loading system. Fundamentally, the utilized motion saving system in this paper is same as the conventional motion saving system. Therefore, the stored motion data by the conventional motion copying system is available to the proposed system.

### B. Motion Loading System considering Variable Reproduction Velocity

The motion loading system is constructed by a bilateral controller. However, the motion loading system does not utilize the actuator in the master side, because the motion data memory is used instead of the actuator in the master system. The motion data memory outputs stored the force and position information, which operates the slave system. The reproduced force and position in the slave system correspond to saved ones. Therefore, the actuator in the slave system is operated according to the saved motion of the human operator. The critical difference between the proposed method and the conventional motion copying system is the reproduction velocity can be freely changed. Fig. 3 shows the method of variable reproduction velocity. If a reproduction velocity command is set to normal speed, the virtual master system outputs force and position information to the quarry matrix without change. The detailed explanation of the quarry matrix is consequent section. In the case of the reproduction velocity is double speed, the force and position information in the motion data memory are contracted before output. On the con-

**1605**

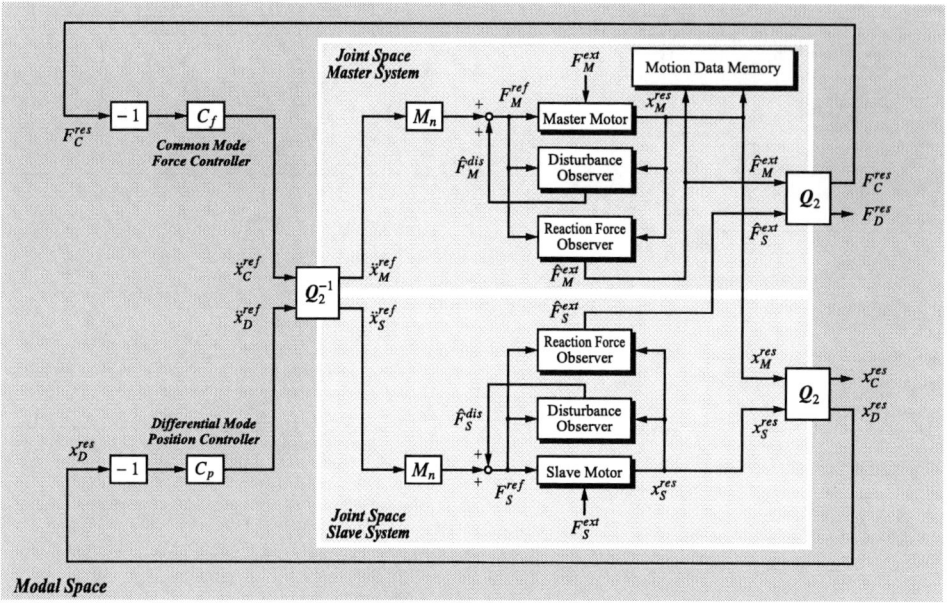

Fig. 4. Block diagram of motion saving system using bilateral control system.

trary, the motion loading system extends the force and position in the memory, when the velocity command is set to half speed. A zero-order hold, linear interpolation and spline interpolation are available in this processing. Finally, the proposed motion loading system is able to vary the reproduction velocity of the stored motion.

## III. MOTION SAVING SYSTEM USING BILATERAL CONTROLLER

Fig. 4 shows block diagram of the motion saving system. The system preserves the motion of the human operator. The master and slave system constitute the motion saving system. A force controller and a position controller control the force of the common mode and the position of the differential mode between respective systems. $F$, $\hat{F}$, $x$ and $\ddot{x}$ are force, estimated force, position and acceleration. The subscripts $M$, $S$, $C$ and $D$ of the paramerters are master system, slave system, common mode and differential mode, respectively. The superscripts $res$, $ref$, $dis$ and $ext$ show response, reference, disturbance and external, respectively. $M_n$ is mass of actuator components.

The quarry matrix is defined in (1) [9], [10]

$$Q_2 = \frac{1}{2} \begin{bmatrix} 1 & 1 \\ 1 & -1 \end{bmatrix}. \tag{1}$$

The expressed quarry matrix in (1) is used to control the force of the common mode and the position of the differential mode, independently. The quarry matrix $Q_2$ decompose the reaction force of the master side actuator $F_M^{ext}$ and the slave side actuator $F_S^{ext}$ to the each mode $F_C^{res}$, $F_D^{res}$

$$\begin{bmatrix} F_C^{res} \\ F_D^{res} \end{bmatrix} = Q_2 \begin{bmatrix} \hat{F}_M^{ext} \\ \hat{F}_S^{ext} \end{bmatrix} \tag{2}$$

In the case of treating the position of the common mode $x_C^{res}$ and differential mode $x_M^{res}$, the calculation method is same

$$\begin{bmatrix} x_C^{res} \\ x_D^{res} \end{bmatrix} = Q_2 \begin{bmatrix} x_M^{res} \\ x_S^{res} \end{bmatrix}. \tag{3}$$

where, $x_M^{res}$ is the position response of the master side actuator, $x_S^{res}$ is the slave side actuator position response. A force controller of the common mode $C_f$ calculates a acceleration reference of the common mode $\ddot{x}_C^{ref}$

$$\ddot{x}_C^{ref} = -C_f F_C^{res}. \tag{4}$$

An acceleration reference of differential mode $\ddot{x}_D^{ref}$ calculated using a position controller of differential mode $C_p$

$$\ddot{x}_D^{ref} = -C_p x_D^{res}. \tag{5}$$

In this paper, the force controller $C_f$ and the position controller $C_p$ are determined by

$$C_f = K_f \tag{6}$$

$$C_p = K_p + \frac{s\, g_{pd}}{s + g_{pd}} K_d. \tag{7}$$

where, $K_f$ is force servoing gain, $K_p$ and $K_d$ is proportional gain and differential gain, $g_{pd}$ is pole of a pseudo derivative. In short, the position controller is PD controller. Also, the disturbance observer estimates and compensates the disturbance force so as to realize acceleration control [11]. The control system in Fig. 4 does not use the force response of differential mode $F_D^{res}$ and the position response of common mode $x_C^{res}$. In addition, the 2nd-order inverse quarry matrix $Q_2^{-1}$ transforms to the acceleration reference in the joint space

**1606**

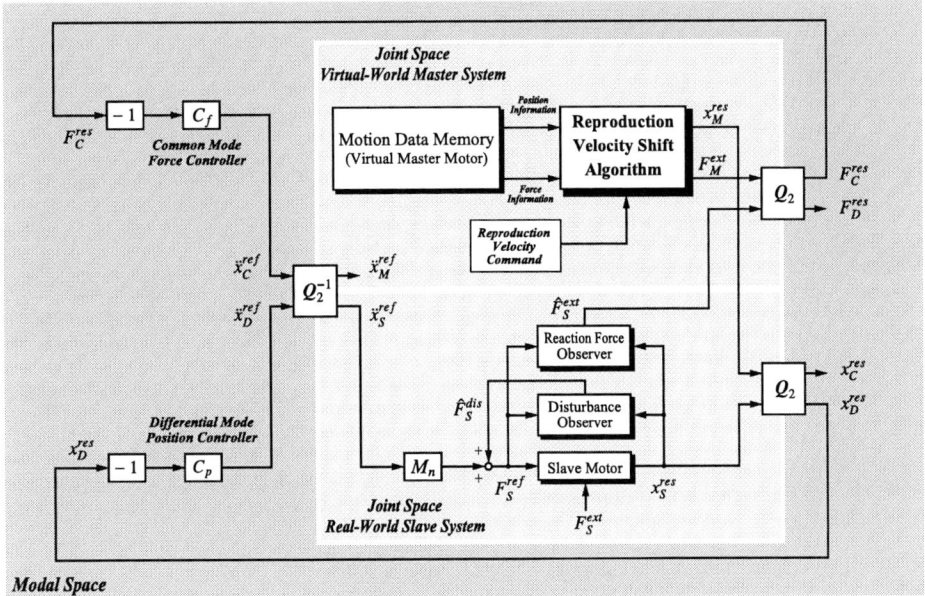

Fig. 5. Block diagram of motion loading system considering variable reproduction velocity.

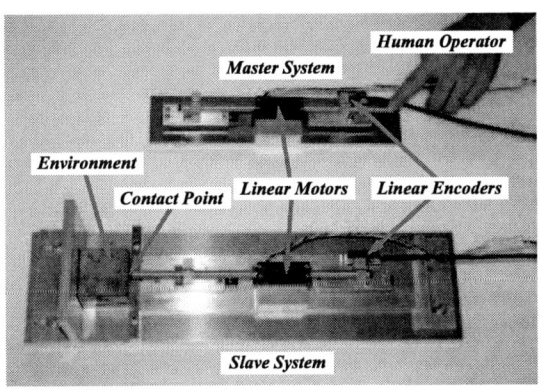

Fig. 6. Experimental device of motion saving system.

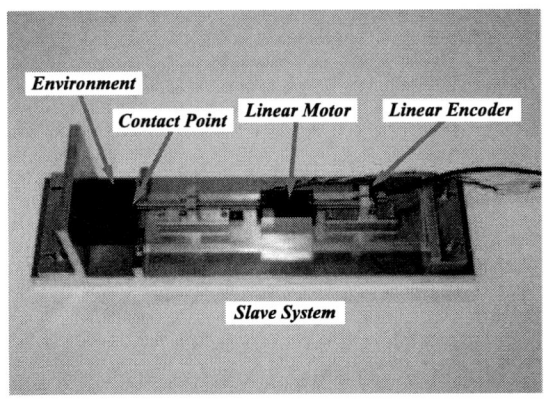

Fig. 7. Experimental device of motion loading system in variable speed.

$\ddot{x}_M^{ref}$, $\ddot{x}_S^{ref}$ using the calculated acceleration reference in the modal space by the controller

$$
\begin{bmatrix} \ddot{x}_M^{ref} \\ \ddot{x}_S^{ref} \end{bmatrix} = Q_2^{-1} \begin{bmatrix} \ddot{x}_C^{ref} \\ \ddot{x}_D^{ref} \end{bmatrix}
$$

$$
= \begin{bmatrix} 1 & 1 \\ 1 & -1 \end{bmatrix} \begin{bmatrix} \ddot{x}_C^{ref} \\ \ddot{x}_D^{ref} \end{bmatrix} \tag{8}
$$

The acceleration reference of respective systems $\ddot{x}_C^{ref}$, $\ddot{x}_D^{ref}$ drives the master side actuator and the slave side actuator. The presented method in this section achieves the bilateral control system, which is able to convey the haptic information of the environment in the slave side to the human operator in the master side. The motion saving system stores the reaction force $F_M^{ext}$ and the position $x_M^{res}$ in the master side to the motion data memory during operating the bilateral controller. More specifically, the system preserves the motion of the human operator. This saved motion data is utilized for the motion loading system.

## IV. MOTION LOADING SYSTEM CONSIDERING VARIABLE REPRODUCTION VELOCITY

Fig. 5 shows block diagram of the motion loading system considering variable reproduction velocity. The calculation method of parameters in the block diagram is same as previous section. However, the critical difference between the motion saving system and the motion load system is the master system. The actuator in the master side is unused; alternatively, the system uses the motion data memory. The motion data memory outputs the stored reaction force and position information to the quarry matrix. In the motion loading system, the actuator in the slave side reproduces the motion according to the

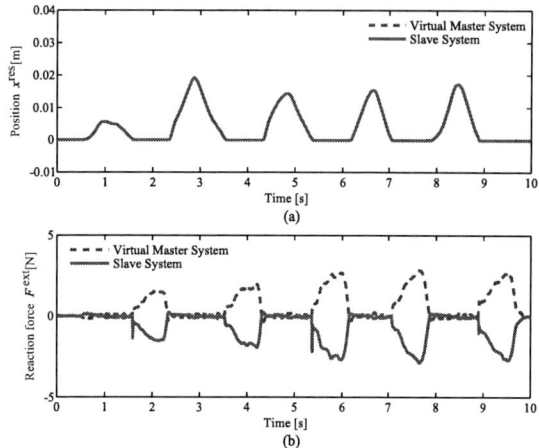

Fig. 8. Saved motion data using motion saving system.
(a) Saved position  (b) Saved force

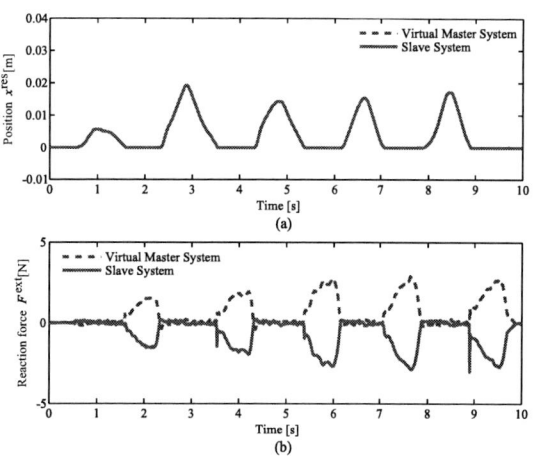

Fig. 9. Loaded motion data using motion loading system.
(Reproduction velocity is normal) (a) Loaded position (b) Loaded force

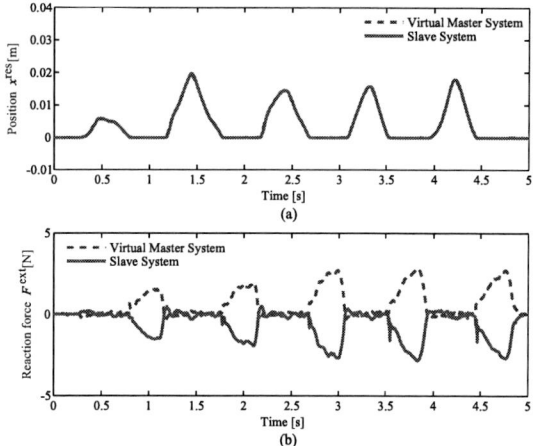

Fig. 10. Loaded motion data using motion loading system.
(Reproduction velocity is twice) (a) Loaded position (b) Loaded force

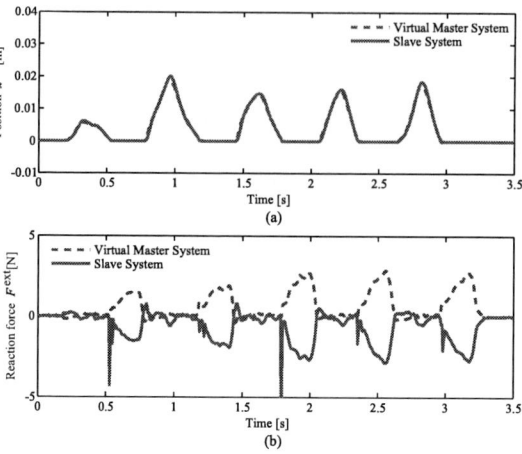

Fig. 11. Loaded motion data using motion loading system.
(Reproduction velocity is third-time) (a) Loaded position (b) Loaded force

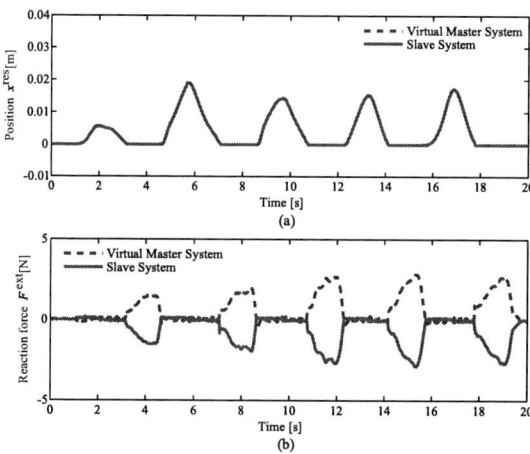

Fig. 12. Loaded motion data using motion loading system.
(Reproduction velocity is one-half) (a) Loaded position (b) Loaded force

information. A reproduction velocity of the virtual master motor depends on a reproduction velocity command. The reproduction velocity command can be freely changed, and is set to low speed or normal speed or high speed. If the reproduction velocity command is set to low speed, number of sample of saved motion data is lower than in case of reproduction motion. Accordingly, a interval of sample of saved motion data should be interpolated. A zero-order hold interpolates the interval of sample in this paper. On the contrary, several sampling data is ignored on condition that the reproduction velocity command is set to high speed. Thus, the motion loading system shifts saved motion of human operator to low speed or high speed.

## V. EXPERIMENT

### A. Experimental Setup

Figs. 6 and 7 show the experimental device of respective systems. The device of the motion saving system consists of two linear motor and two linear encoder. On

TABLE I
SETUP PARAMETERS

| $T_s$ | Period of control system | 100 | $\mu s$ |
|---|---|---|---|
| $K_{tn}$ | Force constant | 3.33 | N/A |
| $M_n$ | Mass of nominal | 0.18 | kg |
| $K_f$ | Gain of force controller | 15 | |
| $K_p$ | Proportional gain of the PD controller | 10000 | |
| $K_d$ | Differential gain of PD controller | 200 | |
| $g_{pd}$ | Pole of pseudo derivative for the PD controller | 10000 | rad/s |
| $g_{dis}$ | Pole of the disturbance observer | 1000 | rad/s |
| $g_{reac}$ | Pole of the reaction force observer | 1000 | rad/s |

the contrary, a linear motor and a linear encoder are used in the motion loading system as previously noted. A tip of the slave motor is able to contact to the environment. The slave motor of the motion saving system is moved according to a master motor. The human operator moves the master motor, and grasps the environment of the slave motor side. In this paper, the environment is an iron block. TABLE I shows the values of each parameter in this experiment.

*B. Experimental Results*

Fig. 8 shows saved motion data by the motion saving system in the experiment. Fig. 8(a) and Fig. 8(b) show the position and force, respectively. The experiment of impact motion to the iron block are carried out. The motion saving system using bilateral controller contrives law of action and reaction.

The experimental result of the motion loading system in the case of normal speed is shown in Fig. 9. The reproduced position and force by the motion loading system corresponds to the saved motion, which is shown in Fig. 8. Therefore, the motion loading system realizes reproduction of the motion of the human operator. Additionally, Figs. 10 and 11 show that the motion loading system reproduces the motion. In this case, the reproduction velocity is set to double or triple. Even then, these results confirm that the position and force are reproduced corresponding to stored motion data. Fig. 12 shows experimental results of the motion loading system in cases where the reproduction velocity command is one-half. The motion loading system normally operates even where reproduction velocity is low.

In this way, the motion saving system saves the motion of the human operator, and the motion loading system reproduces the motion. Moreover, the motion loading system is able to reproduce the saved motion even if the reproduction velocity is changed.

## VI. CONCLUSIONS

The paper proposes a novel motion copying system considering variable reproduction velocity. The motion copying system is constructed by the motion saving system and the motion loading system. The bilateral controller using the quarry matrix controls the motion saving system. The motion saving system observes position and force of the master motor, and saves the motion of the

human operator. The motion loading system is operated by the virtual master-slave controller according to the saved motion data. The reproduced position and force of the slave motor in the motion loading system corresponds to saved position and force. This paper confirms that the motion copying system normally operates. Additionally, the reproduction velocity can be freely changed when the motion loading system reproduces the motion. Thus, the motion copying system considering variable reproduction velocity is realized. The motion copying technology will be useful for industrial applications, medical and welfare human assistance and so on.

## ACKNOWLEDGMENT

This work was supported in part by Industrial Technology Research Grant Program 07C46047a in 2007 from New Energy and Industrial Technology Development Organization (NEDO) of Japan.

## REFERENCES

[1] Y. Kuniyoshi, M. Inaba, H. Inoue : "Learning by Watching: Extracting Reusable Task Knowledge from Visual Observation of Human Performance," *IEEE Transactions on Robotics and Automation*, Vol. 10, No. 6, pp. 799–822, Dec. 1994.

[2] K. Ogawara, J. Takamatsu, H. Kimura, K. Ikeuchi : "Extraction of Essential Interactions Through Multiple Observations of Human Demonstrations," *IEEE Transactions on Industrial Electronics*, Vol. 50, No. 4, pp. 667–675, Aug. 2003.

[3] C. Chang, R. Ansari, A. Khokhar : "Density propagation for tracking initialization with multiple cues [human motion visual tracking]," *IEEE International Conference, ICASSP '04*, Vol. 3, pp. 629, May. 2004.

[4] K. Ohishi, M. Miyazaki, M. Fujita : "Hybrid Control of Force and Position without Force Sensor," *Proceedings of the International Conference of the IEEE Industrial Electronics Society, IECON'92*, Vol. 2, pp. 670–675, Nov. 1992.

[5] T. Yoshikawa, K. Harada, A. Matsumoto : "Hybrid Position/Force Control of Flexible-Macro/Rigid-Micro Manipulator Systems,"

[6] J. Lloyd, J. Beis, D. Pai, D. Dowe : "Programming Contact Tasks using a Reality-based Virtual Environment Integrated with Vision," *IEEE Transactions on Robotics and Automation*, Vol. 15, No. 3, pp. 423–434, Jun. 1999. *IEEE Transactions on Robotics and Automation*, Vol. 12, No. 4, pp. 663–640, Aug. 1996.

[7] T. Shimono, S. Katsura, K. Ohnishi : "Bilateral Motion Control for Reproduction of Real World Force Sensation based on the Environmental Model," *IEEJ Transactions on Industry Applications*, Vol. 126, No. 8, pp. 1059–1068, 2006.

[8] Y. Yokokura, S. Katsura, K. Ohnishi : "Motion Copying System Based on Real-World Haptics," *The 10th IEEE International Workshop on Advanced Motion Control AMC '08-TRENTO*, pp. 613–618, Mar. 2008.

[9] S. Katsura, K. Ohnishi : "Advanced Motion Control for Wheelchair in Unknown Environment," *2006 IEEE International Conference on Systems, Man, and Cybernetics, SMC '06-TAIPEI*, pp. 4926–4931, Oct. 2006.

[10] Seiichiro Katsura, Kiyoshi Ohishi : "Modal System Design of Multi-Robot Systems by Interaction Mode Control," *IEEE Transactions on Industrial Electronics*, Vol. 54, No. 3, pp. 1537–1546, Jun. 2007.

[11] K. Ohnishi, M. Shibata, T. Murakami : "Motion Control for Advanced Mechatronics," *IEEE/ASME Transactions on Mechatronics*, Vol. 1, No. 1, pp. 56–67, Mar. 1996.

[12] T. Murakami, F. Yu, K. Ohnishi : "Torque Sensorless Control in Multidegree-of-freedom Manipulator," *IEEE Transactions on Industrial Electronics*, Vol. 40, No. 2, pp. 259–265, Apr. 1993.

# Adaptive Fuzzy Control of magnetically suspended Rotary Table

Thomas Schallschmidt, Denis Draganov and Frank Palis

Otto von Guericke University Magdeburg, Magdeburg, Germany, e-mail: *thomas.schallschmidt@ovgu.de*
Otto von Guericke University Magdeburg, Magdeburg, Germany, e-mail: *denis.draganov@ovgu.de*
Otto von Guericke University Magdeburg, Magdeburg, Germany, e-mail: *frank.palis@ovgu.de*

*Abstract*— the paper deals with an axial magnetic bearing system for a rotary table. The table is supported by 16 current controlled magnets. Special emphasis is laid on control system design. Here, a neuro-fuzzy based approach is analyzed and experimentally investigated. High performance of observer based cascade structure control systems in connection with an Adaptive Neural Fuzzy Inference System is shown.

*Keywords*— **Magnetic bearings, Machine tool drives, Adaptive control, Neuro-Fuzzy systems**

## I. INTRODUCTION

Active magnetic suspension systems are used in a wide range of technical applications. Their advantages are among others negligible friction, controllability and eco friendliness. The developed magnetic suspension system is meant to be utilized in a rotary table of machine tools. It will be shown that it is qualified by high position precision and dynamic performance. These features allow also vibration damping in a wide scale of frequencies. As known, magnetic bearing systems are characterized by highly non-linear behavior due to their variable inductance when the air gap is changing and the quadratic dependency between the current and the levitation force $F(i, \delta)$. However, the resulting magnetic resistance can be kept independent of the air gap length when using hybrid magnets in differential arrangement. The highly non-linear current-force-relationship can be linearized and linear control structures can be utilized. This approach was shown in [1],[2]. As known, this control strategy covers only a small range of working points where the conditions of linearization are true. Better results in a wide range of operation parameters can be obtained using non-linear and adaptive control approaches as shown in [3]. Here a non-linear control law is designed that guaranties high stability, satisfying dynamics, robustness and accuracy in the whole working area.

As known, intelligent techniques like neural nets and fuzzy systems offer good possibilities to implement effective adaptive control approaches. To this purpose, different methods with different net architectures can be used e. g. inverse model control, predictive control or model reference adaptive control. But despite of the great number of works carried out in the last decade they did not become widely accepted in practical applications. This shortage in acceptance is due to several reasons. Among them can be mentioned the need of training processes during start-up operation, incompatibility in structure with conventional control architectures like cascade control strategies or stability problems. The presented paper deals with simulation and practical investigations of a magnetic bearing system in traditional cascade structure where the inner velocity controller is designed using ANFIS (Adaptive Neural Fuzzy Inference System). To overcome the mentioned disadvantages the conventional cascade structure has been retained and the controller has been pre-designed using rapid prototyping.

## II. SYSTEM DESCRIPTION

The investigated magnetic suspension system of a rotary table consists of 12 levitation and 4 centering hybrid magnets. Their arrangement is shown in figure 1. In the inner part the 4 pairs of centering magnets are

Fig 1: magnet arrangement

represented. The outer part is composed by 12 pairs of levitation magnets. Due to the differential arrangement of the magnet pairs both positive and negative forces can be obtained and the magnetic resistance of the overall air gap can be kept constant, i. e the inductance of the installed current loop remains constant, too. This fact is of great importance when designing the current control loop.

Figure 2 presents the general view of the rotary table. The table is driven by a standard

Fig 2: genaral view of the rotary table

position controlled synchronous torque motor and possesses the following main construction parameters:

- Diameter of the table: 2.00 m,
- height: 0.5 m,
- rotor mass: 2700 kg,
- maximum load capacity: 7000 kg,
- rotation speed: 60 min$^{-1}$.

The construction of the rotary table represents a mechanically over-determined system. Consequently, the redundant DOF must be eliminated using appropriate co-ordinate transformations to pass from the 12 coordinates of the levitation system to determined co-ordinate system with 3 DOF and from 4 co-ordinates of the centring system to 2 DOF. The transition from 16 DOF co-ordinate System to the 5 DOF co-ordinate system is realized via multiplication with the matrices $[A^5_{16\ position}]$ and $[A^{16}_{5\ position}]$.

The force transformation is realized by using the matrices $[A^5_{16\ force}]$ and $[A^{16}_{5\ force}]$.

$[A^{16}_{5\ position}]$ provide a basis for all matrices. It contain a reference to the geometry of the rotary table.

$$
A^{16}_{position\ 5} =
\begin{bmatrix}
0 & 0 & 1 & 0 & -r_{TM} \\
0 & 0 & 1 & \sin(\pi/6)r_{TM} & -\cos(\pi/6)r_{TM} \\
0 & 0 & 1 & \cos(\pi/6)r_{TM} & -\sin(\pi/6)r_{TM} \\
0 & 0 & 1 & r_{TM} & 0 \\
0 & 0 & 1 & \cos(\pi/6)r_{TM} & \sin(\pi/6)r_{TM} \\
0 & 0 & 1 & \sin(\pi/6)r_{TM} & \cos(\pi/6)r_{TM} \\
0 & 0 & 1 & 0 & r_{TM} \\
0 & 0 & 1 & -\sin(\pi/6)r_{TM} & \cos(\pi/6)r_{TM} \\
0 & 0 & 1 & -\cos(\pi/6)r_{TM} & \sin(\pi/6)r_{TM} \\
0 & 0 & 1 & -r_{TM} & 0 \\
0 & 0 & 1 & -\cos(\pi/6)r_{TM} & -\sin(\pi/6)r_{TM} \\
0 & 0 & 1 & -\sin(\pi/6)r_{TM} & -\cos(\pi/6)r_{TM} \\
1 & 0 & 0 & 0 & h_{ZM} \\
\cos(\pi/6) & \cos(\pi/6) & 0 & -\cos(\pi/6)h_{ZM} & \cos(\pi/6)h_{ZM} \\
0 & 1 & 0 & -h_{ZM} & 0 \\
-\cos(\pi/6) & \cos(\pi/6) & 0 & -\cos(\pi/6)h_{ZM} & \cos(\pi/6)h_{ZM}
\end{bmatrix}
$$

Figure 4 illustrates the 5 DOF co-ordinate system., 3 axis (x, y, z) and the rotations φx and φy.

The overall control system in cascade structure with inner velocity and current control loop is shown in figure 3.

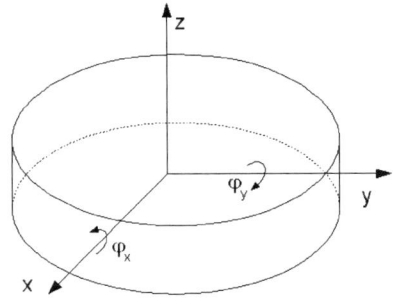

Fig 4: 5 DOF co ordinate system

Here, a state observer is used to filter the measured position signal and to determine velocity. The velocity controller is designed as Adaptive Neural Fuzzy Inference System *(ANFIS)* to cope with the non-linear characteristic of the levitation system $F(i, \delta)$ (fig. 5).

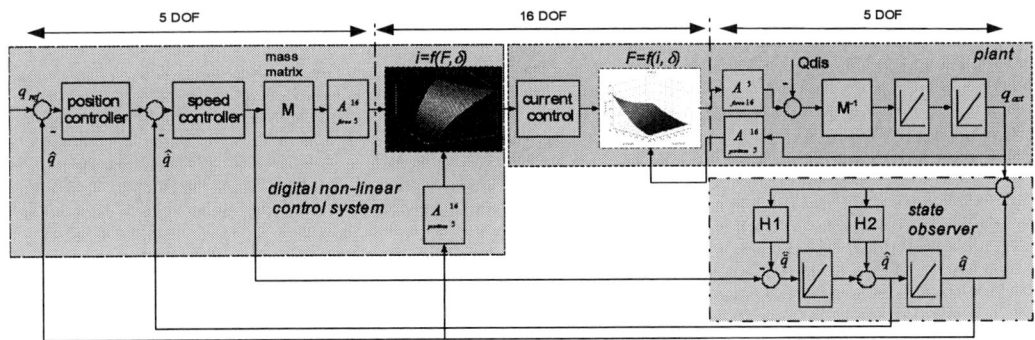

Fig 3: overall control structure

The inverse characteristic $i = f (F, \delta)$ has been pre-designed using measured value of the steady state characteristic of the levitation system (rapid prototyping).

## III. DESIGN OF THE ANFIS STRUCTURE

ANFIS is a training routine for Sugeno-fuzzy inference systems. (Adaptive Neural Fuzzy Inference System)

This routine uses a hybrid learning method to identify parameter of the Sugeno-Fuzzy-System. It is a combination of the least-squares method and the back propagation gradient method.

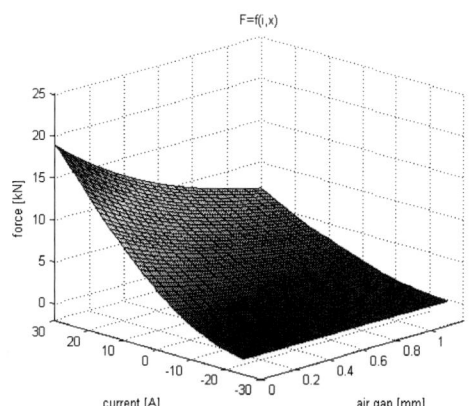

Fig 5: force characteristic of one magnet

The input layer is structured like a typically fuzzy system.

For the 2 inputs $(F, \delta)$ the Gaussian membership function was used. Each of them gets 7 membership functions.

The output membership functions are linear. So the output of a Sugeno rule is:

$$i_i = \alpha_i F + \beta_i \delta$$

and it is weighted by the firing strength $w_i$ of the rule.

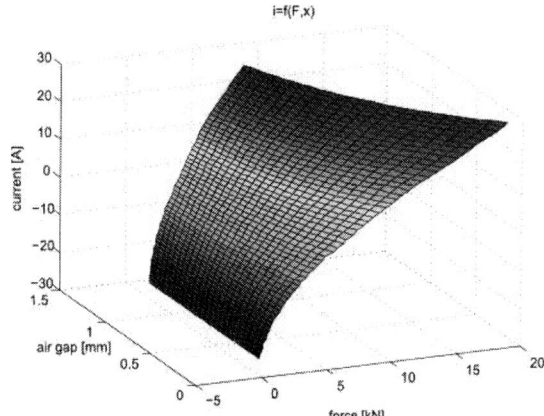

Fig 6: inverse force characteristic

The final output of the system is the weighted average of all rule outputs, computed as

$$i = \frac{\sum_{i=1}^{N} w_i\, i_i}{\sum_{i=1}^{N} w_i}$$

The non-linear force characteristic of each magnet was investigated before the assembly of the rotary table starts (fig. 5). So a training-data set [I δ F] can be calculated and the ANFIS Structure can be trained off-line to model the function $i = f (F, \delta)$ (fig. 6).

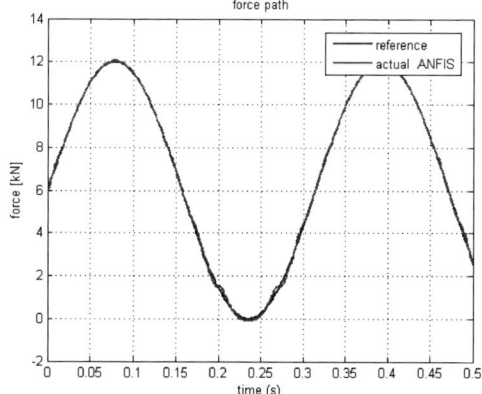

Fig 7: tracking performance of harmonic reference force

Knowing the current $i$ that is necessary to create a given force $F$ when having a given air gap $\delta$.

Here, the output of the velocity controller represents the force reference value that is converted into the current reference value via *ANFIS*. The current in turn creates a force that corresponds to the force reference value.

Fig.7 shows the actual force of the magnet by using the output current value of the ANFIS - Structure in comparison with the actual force input of the ANFIS – Structure. $F_{act} = (x, i = f(F_{ref}, x))$

It shows that the non-linear characteristic is nearly linearized. Due to the high dynamics of the current loop the magnetic bearing works like a linear system.

This system can optimised with the principles of cascade structure. The presented control scheme is characterised by simplicity, easy controller design and by very convenient startup operation. Moreover, rapid prototyping of the ANFIS does not need high accuracy because cascade structure systems are renowned for being relatively insensible against parameter variations and for possessing high stability.

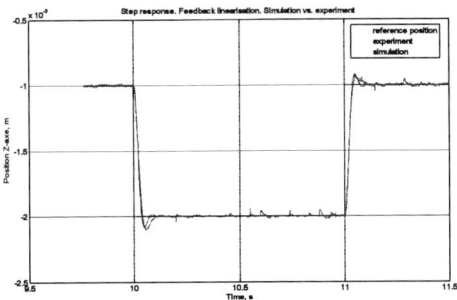

Fig 8: step response

## IV. MODEL BASED FEEDFORWARD CONTROL STRUCTURE

The presented approach starts from the supposition that a part of the plant can be modeled. Here we can build the magnet characteristic $F = f(i, \delta)$ via ANFIS. This function can be inverted to find $i = f(F, \delta)$. Hence, for a given air gap we can pre calculate the required current to create a desired force or, knowing the mass matrix $M$ to create desired acceleration. This signal is applied to current control and the reference speed to the speed controller. Consequently, the response to set point changes will be mainly governed by the dynamics of the rapid current control loop. Moreover, due to fade the inner speed loop is linerized it can be optimized the obtain high performance disturbance reaction. Fig.8 shows the comparison of simulation and experimental results of the step respond of the investigated levitation system.

## V. CONCLUSION

The investigations have shown that the non-linear characteristic of the magnetic system can be linearized. Due to the high dynamics of the current loop the system works in the whole operation range like a linear system that can be optimised in full accordance with the principles of cascade structure systems. The range of working points have been expanded.

The ANFIS input layer is structured like a Fuzzy system so it can be easily pre-designed by experience i.e. by covering the technically interesting input range $(F, \delta)$ with corresponding membership functions. To improve system performance on line training routines may be used for perfecting the ANFIS model.

## REFERENCES

[1] T. Schallschmidt, D. Draganov, F. Palis, *Design and experimental investigation of a magnetically suspended rotary table*, 12 th European Conference on Power Electronics and applications, 12. – 15. Sept. 2007 , Aalborg (Denmark)

[2] K.-D. Tieste, *Mehrgrößenregelung und Paramteridentifikation einer Liear- Magnetführung*, Fortschrittberichte der VDI Zeitschriften, Reihe 8 Nr. 656, Düsseldorf 1997

[3] S. Palis, M. Stamman, T. Schallschmidt: *Nonlinear adaptive control of magnetic bearing*, 12 th European Conference on Power Electronics and applications, 12. – 15. Sept. 2007 , Aalborg (Denmark)

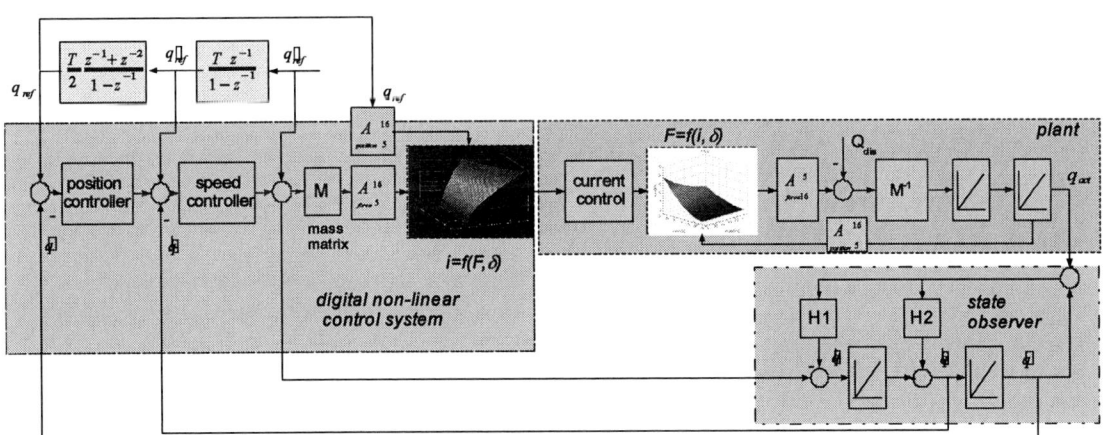

Fig 9: feedforward control structure

# Wideband Force Sensing for Haptic Energy Transmission Utilizing FPGA

Seiichiro Katsura *, Masaki Kondo **, and Kiyoshi Ohishi **

* Department of System Design Engineering, Keio University, Yokohama, Japan
** Department of Electrical Engineering, Nagaoka University of Technology, Nagaoka, Niigata, Japan
katsura@sd.keio.ac.jp; musuku@stn.nagaokaut.ac.jp; ohishi@vos.nagaokaut.ac.jp

*Abstract*— Haptic and/or tactile information has been paid attention as the third multimedia information. An electric motor with the disturbance observer is a good candidate to achieve haptic ability because it is possible to match actuation and sensation. Furthermore, it is possible to convert haptic energy into electric energy. The paper discusses necessity for the framework of haptic energy transmission. Especially, the paper proposes both theory and implementation method. To enlarge the bandwidth of a disturbance observer, the paper proposes super wideband force sensing using a field-programmable gate array (FPGA). Since an FPGA is implemented in it, the control performance is superior to human's sensation. The experimental results show viability of the proposed method.

## I. INTRODUCTION

Industrial technologies have been developed as extension devices of personal sensations. For example, artificial acquisition and reproduction of auditory and visual sensations are basic technologies of communication engineering. Auditory information is obtained by a microphone, and a speaker reproduces it by artificial means. A video camera and a television make it possible to transmit visual sensation by broadcasting. Such a multimedia technology connects a human with another human at the remote site. Furthermore, it is possible to save, process, and analyze the information by developments of digital signal processing technology.

Haptic and/or tactile information has been paid attention as the third multimedia information. There are some different features between haptic information and audio-visual information. One of the most important features is that haptic information is subject to the Newton's law of action and reaction in the real world. In other words, the haptic information is a kind of bilateral environmental information. It means that the reproduction of the haptic information does not permit the time delay. Furthermore, to reproduce vivid sensation, the information should have wide bandwidth. Thus a realization of a system which acquires, transmits and reproduces the haptic information has been demanded for new communication devices.

It is possible to extend human's sensations to the remote environment by bilateral teleoperation. There is much research about bilateral teleoperation in order to attain force feedback from the remote environment [1]–[10]. One of the major objectives in designing bilateral control systems is achieving transparency, which is defined as a correspondence of positions and forces between master and slave [2], or a match between the impedance per-

ceived by the operator and the environmental impedance [3]. Time delay compensation of bilateral teleoperation is an important problem to achieve transparency [1], [5], [6]. Not only time delay but also frequency range of force signal is important for achievement of vivid sensation [7], [11], [12]. Especially, sensor-less force sensing based on a disturbance observer [13]–[14] brings wide-band controller which satisfies above requirements [8].

An electric motor with the disturbance observer is a good candidate to achieve haptic ability because it is possible to match actuation and sensation. Furthermore, it is possible to convert haptic energy into electric energy. Once such haptic energy is converted into digital information, it becomes easy to be acquired, transmitted and reproduced. In other words, the future human support space should be composed of electric motors which are connected by haptic network.

Conventional power electronics technologies for motor drives have paid attention to improve power factor. They are not always suitable for such haptic energy transmission. For realization of future haptic network, novel theory and implementation methods should be developed. The paper discusses necessity for the framework of haptic energy transmission. Especially, the paper proposes both theory and implementation method utilizing a field-programmable gate array (FPGA).

Since a disturbance observer obtains the haptic information by the second-order derivative of a position response, its bandwidth is limited due to the derivative noise [15]. To solve such a problem, this paper proposes a novel structure of a disturbance observer. In the proposed method, the position sensor information and the acceleration sensor information is integrated to cover from DC to higher band range, and position-acceleration integrated disturbance observer (PAIDO) is introduced [16]. Furthermore, to enlarge the bandwidth of PAIDO, the paper proposes super wideband force sensing utilizing FPGA. Since an FPGA is implemented, the bandwidth of force sensing is able to outstrip the one of human's tactile sensation.

This paper is organized as follows. The following section shows the framework of haptic energy transmission. Wideband force sensing based on PAIDO is introduced in Section III. In Section IV, the implementation based on FPGA is described. he experimental results are shown in Section V. At the last section, this paper is summarized.

978-1-4244-1741-4/08/$25.00 ©2008 IEEE

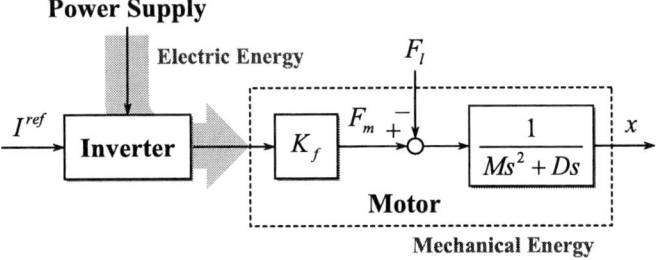

Fig. 1. Power electronics technology for motor drives considering improvement of power factor.

## II. HAPTIC ENERGY TRANSMISSION

### A. Energy Conversion by Electric Motor

An electric motor converts the electric energy to mechanical energy

$$F_m = K_f I^{ref} \tag{1}$$

$$M\ddot{x} = F_m - F_l - D\dot{x} \tag{2}$$

where

$F_m$ : generated force;
$K_f$ : force constant;
$I^{ref}$ : current reference;
$M$ : mass of motor;
$D$ : friction coefficient;
$x$ : position;
$F_l$ : load force.

Since it is possible for an electric motor to convert the electro-mechanical energy bi-directionally, it is the basis for industry applications.

Power electronics technology for motor drives supports their realization considering energy conservation. In other words, the design consideration is to improve power factor. The design procedure of such a power electronics technology for motor drives is shown in Fig. 1.

Although the human input is regarded as the load force form the motor drive point of view, it should be suppressed by controller fundamentally. Thus they are not always suitable for realization of haptic energy transmission.

### B. Force Feedback Control Based on Identification of Load Force

Force feedback control is the key technology for future haptic communication engineering. A touching motion is inherently bilateral, since an action is always accompanied by a reaction. Furthermore, recognition of the contact environment is attained only after touching it. This means that the artificial realization of touching needs very fast controller to keep the time-delay as small as possible. Thus, conventional model-based environmental recognition is not always suitable for future applications such as human support. The environmental recognition in the real world should be based on action, and it is necessary for the future communication to have force feedback ability.

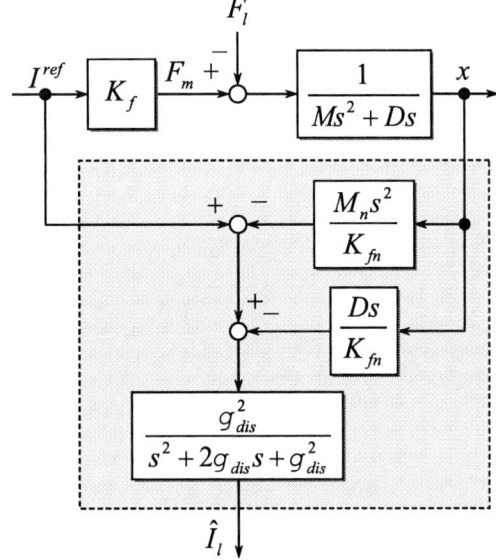

Fig. 2. Haptic energy conversion based on disturbance observer.

A disturbance observer [13], [14] is a good candidate for attainment of force feedback control. A disturbance observer identifies the total mechanical load and parameter changes.

The total disturbance force $F^{dis}$ is represented as (3)

$$\begin{aligned} F^{dis} &= (M - M_n)\ddot{x} + (K_{fn} - K_f)I^{ref} \\ &\quad + D\dot{x} + F_l \end{aligned} \tag{3}$$

where $M_n$ and $K_{fn}$ denote the nominal mass and the nominal force coefficient, respectively. In (3), the first term of the right side is the force due to the self-mass variation. The second term is force ripples due to the variation of the force constant. The third term is the viscous friction.

The equivalent current of the estimated disturbance force $\hat{I}^{dis}$ is obtained from the position $x$ and the current reference $I^{ref}$ as (4)

$$\hat{I}^{dis} = \frac{g_{dis}^2}{s^2 + 2g_{dis}s + g_{dis}^2}\left(I^{ref} - \frac{M_n s^2}{K_{fn}}x\right) \tag{4}$$

To suppress the derivative noise in calculating the equivalent current of the disturbance force, the second-order

**1615**

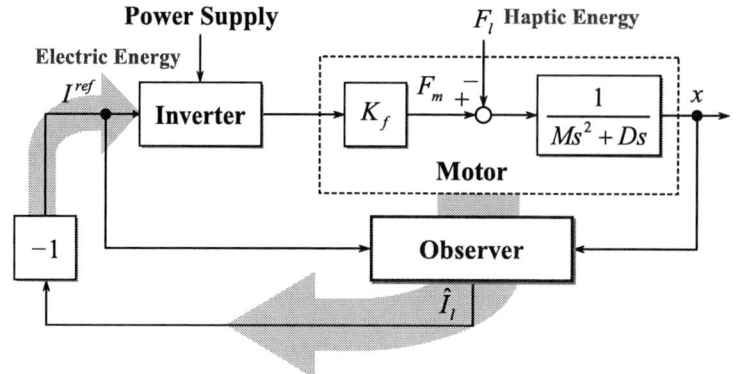

Fig. 3. Force feedback control based on haptic energy conversion.

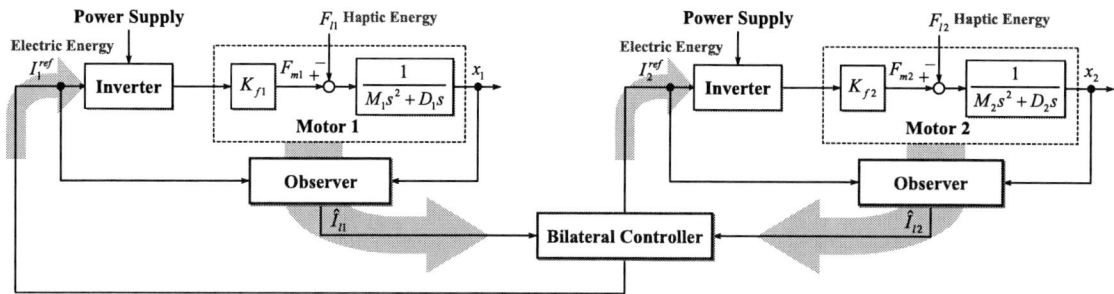

Fig. 4. Bilateral force feedback control for haptic energy transmission.

low-pass filter is inserted. The bandwidth of the low-pass filter is determined by $g_{dis}$.

In the actual application, the estimated disturbance force is effective for not only the disturbance compensation but also the parameter identification. The output of the disturbance observer is only the friction effect under the constant angular velocity motion. This feature makes it possible to identify the friction effect in the mechanical system. The load force effect is also identified by using the estimated disturbance [13]. Here it is assumed that the friction effects are known beforehand by the above identification process. By implementing the accelerated motion, the system parameters $K_{fn}$ and $M_n$ are adjusted in the observer design so that they are close to the actual values, respectively. As a result, the disturbance observer estimates only the equivalent current of the load force as (5)

$$\hat{I}_l = \frac{g_{dis}^2}{s^2 + 2g_{dis}s + g_{dis}^2} \left( I^{ref} - \frac{M_n s^2}{K_{fn}} x - D\dot{x} \right) \quad (5)$$

The identification procedure of the equivalent current of the load force is shown in Fig. 2.

Once load force is estimated as the current dimension, the force feedback control is attained by (6)

$$I^{ref} = -\hat{I}_l. \quad (6)$$

The design procedure of the force feedback control based on haptic energy conversion is shown in Fig. 3.

### C. Bilateral Force Feedback Control for Haptic Energy Transmission

In order to support an action of a human, future robots should recognize the real environment and transmit effective environmental information to a human through the human sensations. Thus, they should act as the physical agents between human and human and/or human and real environment.

Since tactile and/or haptic sensation is subject to the "law of action and reaction" in real world, a twin-motor system is required to transmit such bilateral information of action and reaction. Artificial realization of the "law of action and reaction" by bilateral force feedback is shown in Fig. 4. As shown in Fig. 4, such a framework of haptic energy transmission will be the basis for realization of haptic network.

### III. HAPTIC ENERGY CONVERSION BASED ON PAIDO

Transparency is one of the indices of the haptic energy transmission. Transmission performance of the environmental information to human depends on the transparency. There are two kinds of transparency: spatial transparency and temporal transparency. Higher spatial transparency transmits much environmental modes. Spatial transparency depends on the number of degree of freedom and sensing points. On the contrary, higher temporal transparency transmits wide frequency bandwidth. To improve the temporal transparency, the paper introduces a haptic energy conversion based on PAIDO.

**1616**

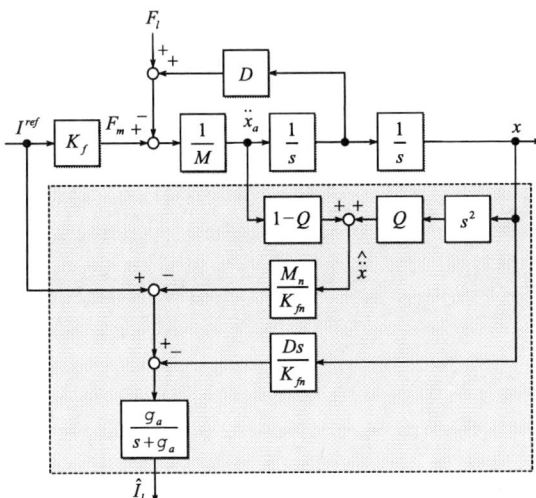

Fig. 5. Block diagram of PAIDO.

Fig. 6. FPGA board.

The primacy of a disturbance observer over a force sensor is considered as follows; estimation accuracy of the impact force, high-resolution force sensing, high signal/noise ratio, and so on. However, since the conventional disturbance observer uses the acceleration information by the second-order derivative of a position response, its bandwidth is limited due to the derivative noise [15].

This paper introduces wideband estimation of acceleration information based on PAIDO. PAIDO uses an acceleration sensor for enlargement of its bandwidth. Generally, the bandwidth of an acceleration sensor is covered from 1 Hz to more than 10 kHz. Namely, an acceleration sensor cannot obtain DC range. To cover DC range, the acceleration information by a position sensor is integrated to obtain wideband acceleration $\hat{\ddot{x}}$

$$\hat{\ddot{x}} = Qs^2x + (1 - Q)\ddot{x}_a \quad (7)$$

where $\ddot{x}_a$ denotes the acceleration response by an acceleration sensor. $Q$ means the second-order low-pass filter.

By using the estimated acceleration information, it is possible to attain wideband disturbance estimation. The disturbance force is estimated as follows;

$$\hat{I}_l = \frac{g_a}{s + g_a}\left(I^{ref} - \frac{M_n}{K_{tn}}\hat{\ddot{x}} - D\dot{x}\right) \quad (8)$$

where $g_a$ denotes the cut-off frequency of PAIDO. If the bandwidth of an acceleration sensor is beyond the one of current control loop, the disturbance force is perfectly estimated

$$\hat{I}_l \approx \frac{1}{K_{fn}}F_l. \quad (9)$$

The block diagram of PAIDO is summarized in Fig. 5. PAIDO is able to reduce the derivative noise of the conventional disturbance observer. As a result, the bandwidth of PAIDO is much wider than the conventional one.

## IV. FPGA IMPLEMENTATION OF FORCE SENSING FUNCTION

An FPGA has been widely used to implement a digital system because of its programmability. An FPGA is a type of large scale integration (LSI) where a user can design its internal logic. It has features of hardware, and it is effective in speeding up the motion control. Moreover, an FPGA is suitable for an experimental system, because it is easy to reprogram its internal logic. From these reasons, the FPGA is beginning to be focused on in the motion-control area [10], [17], [18]. When the motion controller is implemented on an FPGA, it operates faster than that implemented in a personal computer with a real-time operating system.

In this paper, the force sensing function based on PAIDO is implemented by FPGA. It is possible to raise the performance on time resolution and bandwidth by shortening the sampling time. The FPGA used in this paper is shown in Fig. 6. Stratix EP1S25 made by Altera is implemented in the FPGA. The clock is 100 MHz.

The block diagram of force sensing utilizing FPGA is shown in 7. The FPGA implements disturbance observer module, counter module, A/D-converter controller module, D/A-converter controller module, and DIO-converter controller module.

## V. EXPERIMENT

The experimental system is shown in Fig. 8. In the experiment, the force sensing function is installed in the linear motor 1. The linear motor 1 is controlled by the position regulator. The position regulator is designed as follows,

$$I_1^{ref} = \frac{M_n}{K_{fn}}\left\{(x_1^{ref} - x_1)K_p + (\dot{x}_1^{ref} - \dot{x}_1)K_v\right\} \quad (10)$$

where $I_1^{ref}$, $x_1^{ref}$, and $\dot{x}_1^{ref}$ are current, position, and velocity references of the linear motor 1, respectively. $x_1$ and $\dot{x}_1$ denote the position and velocity responses of the linear motor 1. $K_p$ and $K_v$ are the position and velocity control gains, respectively.

To evaluate the performance of the proposed method, another linear motor 2 is used to give physical force

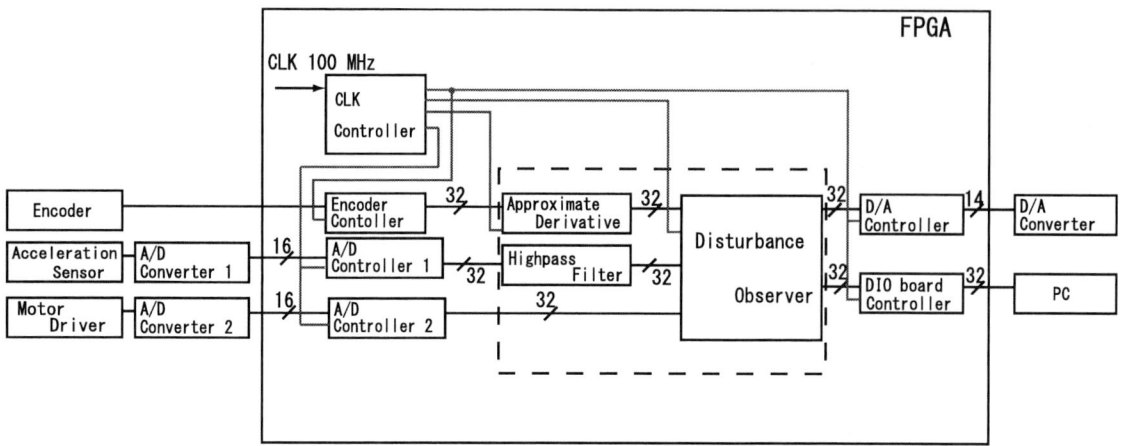

Fig. 7.   Block diagram of force sensing utilizing FPGA.

inputs. The linear motor 2 is controlled by the force servoing.

$$I_2^{ref} = \frac{M_n}{K_{fn}}(F_2^{ref} - \hat{F}_{l2})C_f \qquad (11)$$

where $I_2^{ref}$ and $F_2^{ref}$ are current and force references of the linear motor 2, respectively. $C_f$ is the force control gain. $\hat{F}_{l2}$ is the estimated external force by the disturbance observer.

The force reference is given stepwise as 1 N. The motion control of each linear motor is implemented by RT–Linux. Only force sensing function is implemented by the FPGA.

The experimental results of force sensing are shown in Fig. 9. In the experiment, the sampling time is set to 100 $\mu$s. The value is the minimum for the conventional RT–Linux implementation. The bandwidth of the force sensing is also limited to 12560 rad/s.

On the contrary, in the proposed FPGA implementation, the minimum sampling time is 2 $\mu$s. The experimental results of force sensing are shown in Fig. 10. The bandwidth of the force sensing is raised to 62800 rad/s. In Fig. 10, the dotted line by Linux is the observed value utilizing RT–Linux where the sampling time is 100 $\mu$s.

It is possible to convert haptic force to current dimension in high efficiency. As a result, since the force sensing ability is superior to human's sensation, it is useful for haptic energy transmission.

## VI. CONCLUSIONS

The paper proposes a novel framework of haptic energy transmission. First, the paper shows that the haptic energy conversion is possible by using the disturbance observer. Enlargement of bandwidth of the disturbance observer is the key issue for high-efficient conversion. Second, to enlarge the bandwidth of a disturbance observer, both a position encoder and an acceleration sensor are implemented to compose PAIDO. Finally, the paper shows the implementation method utilizing an FPGA. The bandwidth is raised to 62800 rad/s and its time resolution

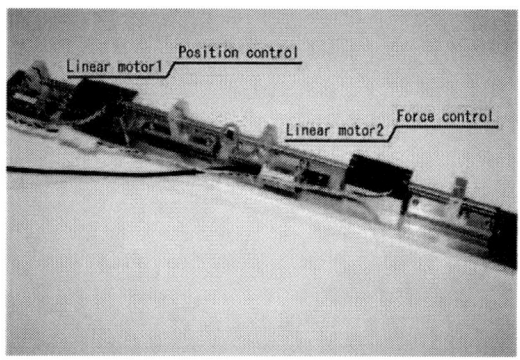

Fig. 8.   Experimental system.

is 2 $\mu$s in the experimental force sensing device. Since the control performance is superior to human's sensation, it will be the fundamental technique for realization of future human support.

## ACKNOWLEDGMENTS

This research was partially supported by the Ministry of Internal Affairs and Communications, Strategic Information and Communications R&D Promotion Programme (SCOPE), 072104001, 2007.

## REFERENCES

[1] R. Anderson, M. Spong : "Bilateral Control of Teleoperators with Time Delay," *IEEE Transactions on Automatic Control*, Vol. 34, No. 5, pp. 494–501, May, 1989.

[2] Y. Yokokohji, T. Yoshikawa : "Bilateral Control of Master-Slave Manipulators for Ideal Kinesthetic Coupling-Formulation and Experiment," *IEEE Transactions on Robotics and Automation*, Vol. 10, No. 5, pp. 605–620, October, 1994.

[3] D. A. Lawrence : "Stability and Transparency in Bilateral Teleoperation," *IEEE Transactions on Robotics and Automation*, Vol. 9, No. 5, pp. 624–637, October, 1993.

[4] K. Kosuge, T. Itoh, T. Fukuda, M. Otsuka : "Telemanipulation System Based on Task-Oriented Virtual Tool," *Proceedings of the 1995 IEEE International Conference on Robotics and Automation, ICRA'95*, Vol. 2, pp. 351–356, August, 1995.

[5] R. Oboe : "Web-Interfaced, Force-Reflecting Teleoperation Systems," *IEEE Transactions on Industrial Electronics*, Vol. 48, No. 6, pp. 1257–1265, December, 2001.

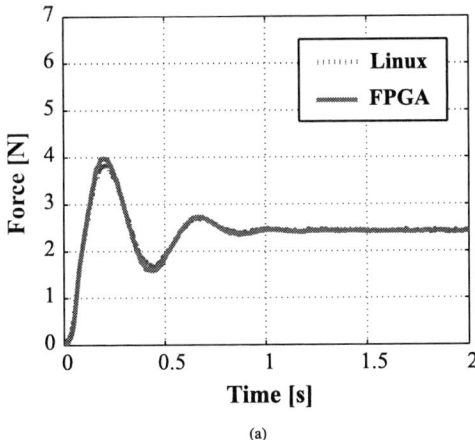 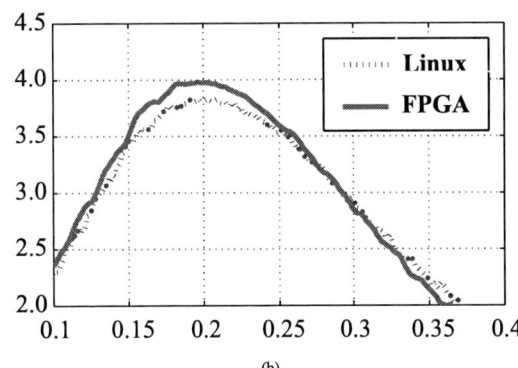

(a)　　　　　　　　　　　　　(b)

Fig. 9.  Experimental results of force sensing. (Sampling time is 100 $\mu s$, $g_a = 12560$ rad/s) (a) Estimated force responses. (b) Expanded figure of (a).

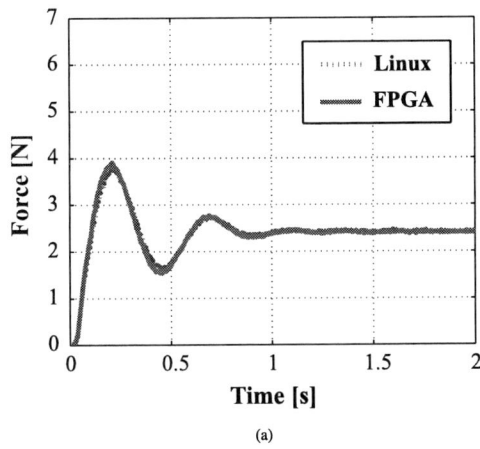 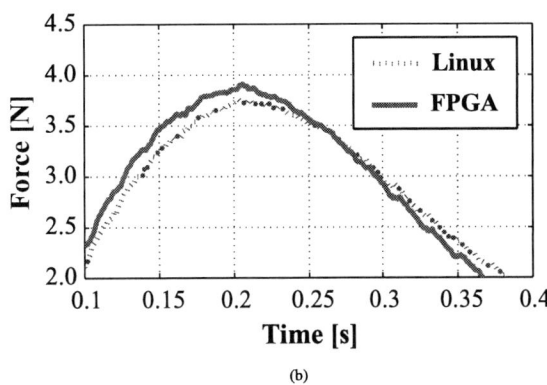

(a)　　　　　　　　　　　　　(b)

Fig. 10.  Experimental results of force sensing. (Sampling time is 2 $\mu s$, $g_a = 62800$ rad/s) (a) Estimated force responses. (b) Expanded figure of (a).

[6]　K. Hashtrudi-Zaad, S. E. Salcudean : "Transparency in Time-Delayed Systems and the Effect of Local Force Feedback for Transparent Teleoperation," *IEEE Transactions on Robotics and Automation*, Vol. 18, No. 1, pp. 108–114, February, 2002.

[7]　Y. Matsumoto, S. Katsura, K. Ohnishi : "An Analysis and Design of Bilateral Control Based on Disturbance Observer," *Proceedings of the IEEE International Conference on Industrial Technology, ICIT'03-MARIBOR*, pp. 802–807, December, 2003.

[8]　S. Katsura, W. Iida, K. Ohnishi : "Medical Mechatronics – An Application to Haptic Forceps –," *IFAC Annual Reviews in Control*, Vol. 29, No. 2, pp. 237–245, November, 2005.

[9]　T. Tsuji, K. Natori, H. Nishi, K. Ohnishi : "Controller Design Method of Bilateral Control System," *European Power Electronics and Drives Journal*, Vol. 16, No. 2, pp. 22–28, May, 2006.

[10]　E. Ishii, H. Nishi, K. Ohnishi : "Improvement of Performances in Bilateral Teleoperation by Using FPGA," *IEEE Transactions on Industrial Electronics*, Vol. 54, No. 5, pp. 1876–1884, August, 2007.

[11]　S. Katsura, Y. Matsumoto, K. Ohnishi : "Analysis and Experimental Validation of Force Bandwidth for Force Control," *IEEE Transactions on Industrial Electronics*, Vol. 53, No. 3, pp. 922–928, June, 2006.

[12]　S. Katsura, Y. Matsumoto, K. Ohnishi : "Modeling of Force Sensing and Validation of Disturbance Observer for Force Control," *IEEE Transactions on Industrial Electronics*, Vol. 54, No. 1, pp. 530–538, February, 2007.

[13]　T. Murakami, F. Yu, K. Ohnishi : "Torque Sensorless Control in Multidegree-of-freedom Manipulator," *IEEE Transactions on Industrial Electronics*, Vol. 40, No. 2, pp. 259–265, April, 1993.

[14]　K. Ohnishi, M. Shibata, T. Murakami : "Motion Control for Advanced Mechatronics," *IEEE/ASME Transactions on Mechatronics*, Vol. 1, No. 1, pp. 56–67, March, 1996.

[15]　M. Bertoluzzo, G. S. Buja, E. Stampacchia : "Performance Analysis of a High-Bandwidth Torque Disturbance Compensator," *IEEE/ASME Transactions on Mechatronics*, Vol. 9, No. 4, pp. 653–660, December, 2004.

[16]　K. Irie, S. Katsura, K. Ohishi : "Wideband Motion Control by Acceleration Disturbance Observer," *12th International Power Electronics and Motion Control Conference, EPE-PEMC'06-PORTOROZ*, pp. 361–366, August, 2006.

[17]　Z. Salcic, J. Cao, S. K. Nguang : "A Floating-Point FPGA Based Selftuning Regulator," *IEEE Transactions on Industrial Electronics*, Vol. 53, No. 2, pp. 693–704, April, 2006.

[18]　K. Sridharan, T. K. Priya : "The Design of a Hardware Accelerator for Real-Time Complete Visibility Graph Construction and Efficient FPGA Implementation," *IEEE Transactions on Industrial Electronics*, Vol. 52, No. 4, pp. 1185–1187, August, 2005.

# On the development of BLDC motor control run-up algorithms for aerospace application

Vladimir Hubik*, Martin Sveda†, Vladislav Singule‡

*Faculty of Mechanical Engineering, BUT, Brno, Czech Republic, e-mail: *vhubik@nbox.cz*
†UNIS a.s., Department of Mechatronic systems, Brno, Czech Republic, e-mail: *msveda@unis.cz*
‡Faculty of Mechanical Engineering, BUT, Brno, Czech Republic, e-mail: *singule@fme.vutbr.cz*

*Abstract*—Recent developments in aviation systems are leading to implementation of Power Optimized Aircraft by means of FBW and PBW technologies. Effort is being aimed at replacement of existing hydraulic actuators by intelligent EHA/EMA actuators. A fundamental element that drives the EHA/EMA actuator is the BLDC motor. In critical applications, it is necessary to ensure correct start-up of the motor. This article deals with the analysis and simulation of the start-up phase of the BLDC motor and summarises first results obtained during development of control algorithms.

*Keywords*—Actuator, aerospace, brushless drive, electrical drive, intelligent drive, mechatronics, modelling, motion control, sensorless control.

## I. INTRODUCTION

Recent developments in aviation systems are leading to implementation of Power Optimized Aircraft (POA) by means of Fly-by-Wire (FBW) and Powered-by-Wire (PBW) technologies. The aim is to replace existing actuators, which are driven from central hydraulic system, with intelligent Electro-Mechanical (Figure 1) and Electro-Hydraulic (EHA) actuators. These new types of actuators are capable of providing required energy on demand.

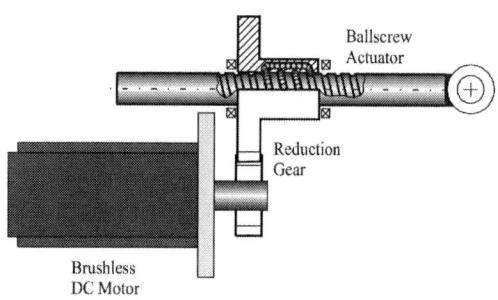

Fig. 1: An example of EMA actuator  [8]

Implementation of any new technology in the aerospace industry makes strict demands on the safety and reliability of on-board equipment. Before application onboard the aircraft, many simulations and tests have to be performed to prove the reliability and usability of any newly developed equipment.

A fundamental element that drives the EHA/EMA actuator is the DC motor. Nowadays, brushless DC motors (BLDC) are widely used, mainly because of their better characteristics and performance. BLDC motors can be controlled in sensor or sensor-less mode. The advantage of sensor-less BLDC motor control is that the sensing part can be omitted, and thus overall costs can be considerably reduced. The disadvantages of sensor-less control are higher requirements for control algorithms and more complicated electronics.

An analysis of input parameters has been performed and a BLDC motor Simulink model has been designed. Performing simulation and tests on the BLDC motor model dramatically speeds up development of new types of actuators. The BLDC motor model is based on BLDC parameter analysis so the Simulink model matches the real motor as closely as possible.

## II. BLDC MOTOR ARCHITECTURE

Brushless Direct Current (BLDC) [9] motors are one of the motor types rapidly gaining popularity. BLDC motors are used in industries such as Appliances, Automotive, Aerospace, Medical, and with their architecture are suitable for any safety critical applications.

The brushless DC motor is a synchronous electric motor that from a modelling perspective looks exactly like a DC motor, having a linear relationship between current and torque, voltage and rpm. It is an electronically controlled commutation system, instead of a mechanical commutation system (ie. brushes).

In the BLDC motor, the electromagnets do not move. The permanent magnets rotate and the armature remains static. This gets around the problem of how to transfer current to a moving armature. In order to do this, the brush-system/commutator assembly is replaced by an intellignet electronic controller. The controller performs the same power distribution found in the brushed DC motor.

BLDC motors have many advantages over brushed DC motors and induction motors. A few of these are :

- Better speed versus torque characteristics
- High dynamic response
- High efficiency and reliability
- Long operating life (no brush erosion)
- Noiseless operation
- Higher speed ranges
- Reduction of electromagnetic interference (EMI)

In addition, the ration of torque delivered to the size of the motor is higher, making it useful in applications where space and weight are critical factors - especially in aerospace. To reduce overall cost of actuating devices it is normally used sensor-less variation of the BLDC motor.

The operation of an actuator is safety critical and thus the actuator requires a reliable control algorithm that ensures safe start-up and operation of the BLDC motor that drives the actuator. In this article, we will discuss in detail the mathematical model of BLDC motor and will be shown the start-up scourses.

## III. MATHEMATICAL MODEL OF BLDC MOTOR

The mathematical model of the BLDC motor, created in the MATLAB/Simulink environment, consists of several independent blocks, which describe its real behaviour. For easier orientation it can be separated into two parts - *electrical* and *mechanical*, shown in the Figure 2. For this reason it is possible to apply the model to any control circuit or complex system.

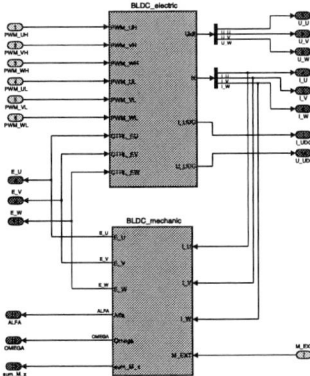

Fig. 2: Mathematical model of BLDC motor - electrical and mechanical part.

The mathematical model is designed to implement as many as possible of the parameters supplied by the BLDC motor's manufacturer. The aim is to have a model that reliably matches a real BLDC motor. Input parametrs are set in special M-file, in numbers about 20 values.

The electrical part models the internal conditions and wiring of the direct current motor. It provides fundamental electrical parameters at its outputs such as currents and voltages on individual windings. The component's design parameters and appropriate driving signals from superior blocks are required as the input parameters.

The differential equations 1, 2, 3 describe electrical behavior of direct current motor and are solved by numerical methods in Simulink environment. Individual phase voltages are evaluated and could be used for sensorless detection of actual rotor position.

$$u_U = R_U \cdot i_U + L_U \cdot \frac{\partial i_U}{\partial t} + u_{bemfU} \qquad (1)$$

$$u_V = R_V \cdot i_V + L_V \cdot \frac{\partial i_V}{\partial t} + u_{bemfV} \qquad (2)$$

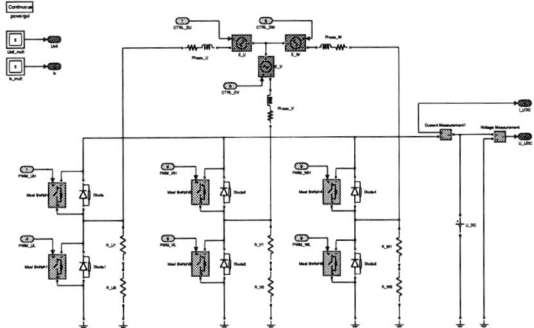

Fig. 3: Electrical part of the model with power stage.

$$u_W = R_W \cdot i_W + L_W \cdot \frac{\partial i_W}{\partial t} + u_{bemfW} \qquad (3)$$

where...

$\frac{\partial i_U}{\partial t}$, $\frac{\partial i_V}{\partial t}$, $\frac{\partial i_W}{\partial t}$ – time derivation of winding current,
$R_U$, $R_V$, $R_W$ – resistance of winding $[\Omega]$,
$L_U$, $L_V$, $L_W$ – inductance of winding $[H]$,
$u_{bemfU}$, $u_{bemfV}$, $u_{bemfW}$ – back electro-motive voltages $[V]$,
$u_U$, $u_V$, $u_W$ – phase voltages $[V]$.

The electrical part of the model is created from parts of the SimPower integrated environment. It can be thought of as an electrical schematic of a motor, Figure 3. Even if Simulink libraries have the better switching components, such as power MOSFET, IGBT or bipolar transistors, it was used only basic ideal switching device with inverse diode. That will dramatically decrease computation time to real value.

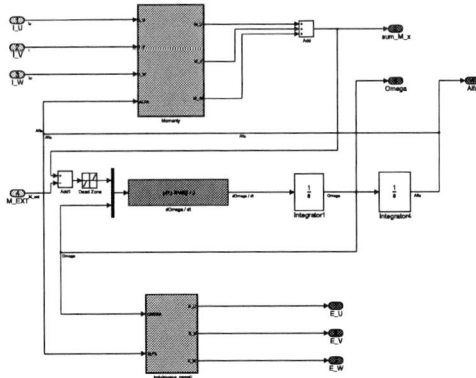

Fig. 4: Mechanical part of the model.

$$\frac{\partial \omega}{\partial t} = \frac{M_U + M_V + M_W - M_{EXT} - B \cdot \omega}{J} \qquad (4)$$

The mechanical part Figure 4 of the model is based on equation of motion 4 and simulates the interaction between mechanical and electrical values of the system.

Actual kinetic torques of individual windings are evaluated according to input currents, equations 5, 6, 7 from the electrical part, Figure 5. The model also takes into account attenuation forces arising from friction during rotation or from appropriate external braking torques on the rotor shaft:

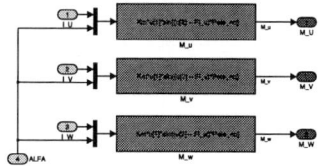

Fig. 5: Evaluation of the actual kinetic torques.

$$M_U = K_m \cdot i_u \cdot \sin\left((\alpha - F_U) \cdot POL_{num}\right) \quad (5)$$

$$M_V = K_m \cdot i_v \cdot \sin\left((\alpha - F_V) \cdot POL_{num}\right) \quad (6)$$

$$M_W = K_m \cdot i_w \cdot \sin\left((\alpha - F_W) \cdot POL_{num}\right) \quad (7)$$

where...
$\frac{\partial \omega}{\partial t}$ – time derivation of angular speed,
$M_U$, $M_V$, $M_W$ – torque of windings $[N \cdot m]$,
$M_{EXT}$ – external breaking torque $[N \cdot m]$,
$B$ – friction coeficient $\left[\frac{N \cdot m}{rad/s}\right]$,
$\omega$ – angular speed $\left[\frac{rad}{s}\right]$,
$J$ – inertia torque $[kg \cdot m^2]$,
$K_m$ – mechanical constant of motor $\left[\frac{N \cdot m}{A}\right]$,
$i_u$, $i_v$, $i_w$ – current of windings $[A]$,
$\alpha$ – actual angel of rotor $[rad]$,
$F_U$, $F_V$, $F_W$ – mutual angel of phases $[rad]$,
$POL_{num}$ – number of pole pairs.

An important output value is the amplitude of induced voltages in particular windings. This is used for detection of the correct moment for commutation during sensorless control conditions. These back EMF voltages are evaluated in subsystem, that is depicted in the Figure 6. The appropriate equations are mentioned below 8, 9, 10. Integration of the equation of motion is possible to enable evaluation of instantaneous rotor position and angular velocity. This part of model is created in the Simulink environment by means of basic integrated blocks, Figure 4.

$$u_{bemfU} = -K_e \cdot \omega \cdot \sin\left((\alpha - F_U) \cdot POL_{num}\right) \quad (8)$$

$$u_{bemfV} = -K_e \cdot \omega \cdot \sin\left((\alpha - F_V) \cdot POL_{num}\right) \quad (9)$$

$$u_{bemfW} = -K_e \cdot \omega \cdot \sin\left((\alpha - F_W) \cdot POL_{num}\right) \quad (10)$$

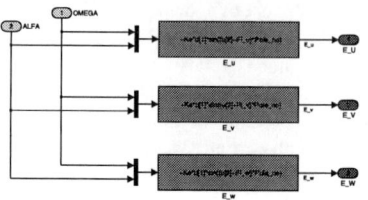

Fig. 6: Evaluation of the actual back electro-motive force (BEMF).

where...

$K_e$ – electrical constant of motor $\left[\frac{rad}{V}\right]$,
$u_{bemfU}$, $u_{bemfV}$, $u_{bemfW}$ – back electro-motive voltages $[V]$.

## IV. START-UP PHASE SIMULATION

The operation of an actuator is safety critical and thus the actuator requires a reliable control algorithm that ensures safe start-up and operation of the BLDC motor that drives the actuator.

BLDC motor start up in sensor-less mode runs, in principle, in two stages. In the first stage the motor is run up in so-called frequency mode, where rotational speed is increased in open control loop. During frequency mode run-up, the back electro-motive force (BEMF) is measured on individual windings. Closed loop control is turned on after reliable detection of zero-crossing.

The main issue in frequency run-up mode is the selection of optimal commutation speed. At the beginning of run-up, the rotor position is unknown and so the magnetic field could lead the rotor position, thereby reducing the smoothness of the motor run-up. In addition, reliable zero-crossing detection is possible only above a certain speed. Measurement of BEMF is almost impossible at low rpm because of its low value and the effect of interference. Thus detection of zero-crossing at low rpm speed is unreliable.

We have developed a new method of BLDC motor run-up that provides a smoother and more reliable BLDC motor run-up with possibility of zero-cross detection at low speed.

The results of the simulation are shown in Figures 7. Traditional methods of motor run-up are published in many publications. The disadvantage of these methods is irregular run-up caused by unknown rotor position during running up. Time behaviour of rotor position, including rotor positioning, is shown in the Figure 7a. Figure 7b shows motor run-up in 3-D projection where the regular run-up is depicted much more clearly.

From simulations it is obvious that the presented new run-up algorithm provides much better results. Start-up is smoother and detection of rotor position is also possible at low rpm, Figure 7. Figure 7 depicts the first 200 ms of motor start-up. BLDC motor made something between four or five revolutions.

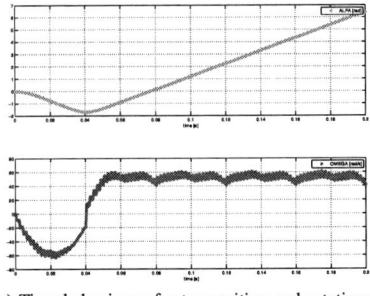

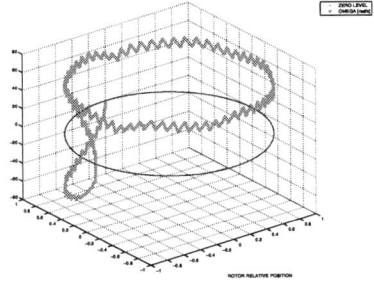

(a) Time behaviour of rotor position and rotational speed

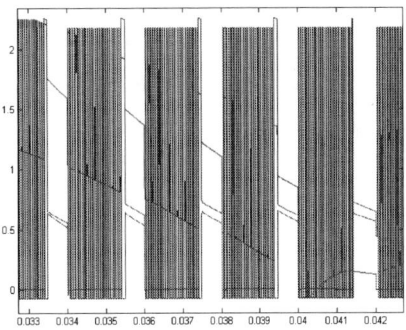

Fig. 8: Principle of BEMF sensing on a low speed.

(b) Speed dependency of position vector at BLDC motor run up

Fig. 7: Innovated sensor-less BLDC motor run-up algorithm

## V. BEMF SENSING ON A LOW SPEED

Electrical commutation in the first runnig stage is normally realized by clasical PWM signal that drives a MOSFET power stage. As we mentioned before, it is open-loop control without any position feedback. The aim is to sense back EMF voltage as soon as possible to determine actual position.

Figure 8 describes the principle how to sense a position of the rotor on a low speed up to 1000 rpm. Normal PWM cycle is on-off modulated by auxiliary signal with a lower frequency. For example, for the 20 kHz main PWM signal we can use 1 kHz modulation signal. In the empty windows it is possible to clearly sense generated back EMF voltages from all particular windings.

Depends on type of driven BLDC motor, course of this BEMFs could have trapezoidal or sinusoidal behave. From the level of induced voltages it is possible to determine actual rotor position for the next commutation cycle. The advantage of this methode is quality of measured signals. All windings are switched off and the courses are without any noise or interferences. Disadvantage could be little bit lower starting power due to interruption of the main PWM signal.

## VI. CONCLUSION

Intelligent EHA/EMA actuators and smart actuation systems are more and more promising technologies for future power optimised aircraft. Although their benefits in energy consumption, weight savings, easy assembly procedures and maintenance are obvious, there still remain some features must be investigate.

The project involves the development of control algorithms and electronics to improve run up performance of the BLDC motor in EHA/EMA actuating devices for safety critical applications. The most important feature of the device is that closed loop control is possible much sooner in the start up phase than was previously possible.

The simulation results will be tested on real equipment in the next stage. If these tests verify all theoretical assumptions, the new methode of BLDC starting will be implemented in a new generation of EHA/EMA actuators.

## VII. ACKNOWLEDGMENT

This project was supported by project No. MSM 0021630518 "Simulation modeling of mechatronics systems" solved on the Faculty of Mechanical Engineering, Institute of Production Machines, Systems and Robotics, Brno Technical University. Published results were acquired using the subsidization of the European Union, FP6 research project No.:30 888, "CESAR - Cost Effective Small Aircraft". Hardware and software development and physical modules were realized and project was supported by UNIS a.s. as well.

## REFERENCES

[1] G. EASON, B. NOBLE, and I. N. SNEDDON, "On certain integrals of Lipschitz-Hankel type involving products of Bessel functions," *Phil. Trans. Roy. Soc. London*, vol. A247, pp. 529–551, April 1955.

[2] J. CLERK MAXWELL, *A Treatise on Electricity and Magnetism*, 3rd ed., vol. 2. Oxford: Clarendon, 1892, pp.68–73.

[3] I. S. JACOBS and C. P. BEAN, "Fine particles, thin films and exchange anisotropy," in *Magnetism*, vol. III, G. T. Rado and H. Suhl, Eds. New York: Academic, 1963, pp. 271–350.

[4] K. ELISSA, "Title of paper if known," unpublished.

[5] R. NICOLE, "Title of paper with only first word capitalized", *J. Name Stand. Abbrev.*, in press.

[6] Y. YOROZU, M. HIRANO, K. OKA, and Y. TAGAWA, "Electron spectroscopy studies on magneto-optical media and plastic substrate interface," *IEEE Transl. J. Magn. Japan*, vol. 2, pp. 740–741, August 1987 [Digests 9th Annual Conf. Magnetics Japan, p. 301, 1982].

[7] M. YOUNG, *The Technical Writer's Handbook*. Mill Valley, CA: University Science, 1989.

[8] COCHOY, O., CARL, U., B., THIELECKE, F. Integration and control of electromechanical and electrohydraulic actuators in a hybrid primary flight control architecture. *Recent Advances in Aerospace Actuation Systems and Components 2007*. Insa Toulouse: June 13–15, 2007. ISBN: 978-2-87649-053-6.

[9] COMPTER, J. C. *Electrical drives for precision engineering designs.* specAmotor, www.specamotor.com: Eindhoven, 2007.

[10] LEONHART, W. *Control of Electrical Drives.* Third Edition. Springer: Berlin, 2001. ISBN 3-540-41280-2.

[11] GREISSNER, C., CARL, U. Control of an Electro-Hydrostatic Actuation System for the Nose Landing Gear of an "All Electric Aircraft". *Recent Advances in Aerospace Actuation Systems and Components 2004.* Insa Toulouse: November 24 – 26, 2004. ISBN: 2-87649-047-1.

[12] TODESCHI, M. A380 flight control actuation lessons learned on EHAs design. *Recent Advances in Aerospace Actuation Systems and Components 2007.* Insa Toulouse: June 13 – 15, 2007. ISBN: 978-2-87649-053-6.

[13] MITAL, C., JAKOVLJEVIC, M. Distributed embedded computing for advanced control systems based on time-triggered protocol (TTP) and architecture (TTA). *Recent Advances in Aerospace Actuation Systems and Components 2007.* Insa Toulouse: June 13 – 15, 2007. ISBN: 978-2-87649-053-6.

[14] BOTTEN, S., L., WHITLEY, C., R., KING, A., D. Flight Control Actuation Technology for Next-Generation All-Electric Aircraft. *Technology Review Journal – Millenium Issue.* Fall/Winter 2000.

# Rotor Levitation by Active Magnetic Bearing Using Digital State Controller

Chip Rinaldi Sabirin, Andreas Binder

Darmstadt University of Technology/Institute for Electrical Energy Conversion, Darmstadt, Germany

e-mail: *csabirin@ew.tu-darmstadt.de, abinder@ew.tu-darmstadt.de*

*Abstract*—**Active magnetic bearings (AMB) find extended application in high-speed rotating electrical machines due to the contact-free operation, which reduces the overall machine wear. Previous research on the control system for the AMB dealt mostly with PID controller, with its advantage of simple design. Improvements using faster control access to the plant can be realized by state-space control. Additionally, the magnetic bearing as a MIMO (multiple-input multiple-output) system can be better described by the state-space equations. A digital state controller is applied to levitate a rotor of a 40-kW 40000-rpm PM synchronous machine. Simulation and measurement results are given for zero speed operation at an AMB test-rig.**

*Keywords*—**Magnetic bearings, DSP, optimal control, test bench.**

## I. INTRODUCTION

Active magnetic bearings (AMB) find extended application in high-speed rotating electrical machines due to the contact-free operation, which reduces the overall machine wear, compared to mechanical ball or roller bearings. Mechanical bearings must stand the high centrifugal stress in the high-speed operation, which is avoided by using active magnetic bearings.

Numerous research works dealt with the design of digital control for AMB. Most of the investigations were considering PD [1], [4], [7] or PID controllers [1], [2], [5]. Typical behavior of a PID controller can be seen after the rotor lift-off that the rotor movement has an overshoot, before it comes to the steady-state at centered position. Main advantage of the PID controller is its simple design and availability as commercial control hardware. The user "just" needs to tune the proportional, derivative and integration terms [3]. There are some methods in calculating these parameters of the PID controllers [3]. In [2], the pole placement method was used due to its direct influence of the design to the closed-loop stability of AMB.

State controllers can give faster response to a control demand for a system output because of the direct access of the state controller algorithm to the inner states of the system. Hence, the state controller is able to avoid the overshoot in a controller step response and allows a faster reaction to any position deviation from the reference value [3]. Research work on state-controller design and application for AMB systems were published in [4], [6], [9], [10].

## A. Principles of Force Generation

Fig. 1 shows an iron C-core used as an electromagnet to levitate an iron I-core with a magnetic force that is generated due to the existence of an air gap with a nominal distance of $s_0$ between the C- and I-core. An electrical current $i$ through the coils with a winding number $N$ generates a magnetic flux, which will be flowing through the paths comprising of the iron C-core, both air gaps and the iron I-core. The magnetic flux linkage has generally a nonlinear relation to the electrical current $i$ due to the nonlinear magnetic characteristic of the magnetic material with the variable $\mu_r$ as the relative permeability value of the iron cores. If the path through both magnetic C- and I-core has a length, which is small enough compared to the product of $(2 \cdot (s_0 - x) \cdot \mu_r)$, then we may neglect the iron reluctance. In this case, a linear system can be assumed, considering the small displacement of the I-core (i.e. $x \ll s_0$) in the unsaturated condition of the magnetic cores [8].

The magnetic flux density relating to the flux in the air gap is $B_0$, the total cross-section area of both the right and left air gap equals $(2 \cdot A)$. Based on the physical principle of magnetic energy stored in the air gap and the mathematical principle of virtual displacement of the air gap $\delta x$, the magnetic force magnitude generated in both air gaps can be calculated by

$$F = \frac{B_0}{2 \cdot \mu_0} \cdot 2 \cdot A \qquad . \qquad (1a)$$

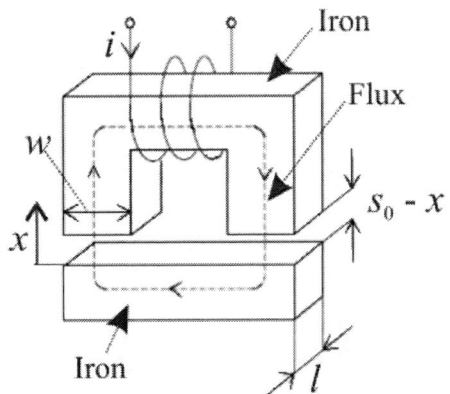

Fig. 1. A C-core as an electromagnet suspends a magnetic I-core. The air gap distance nominal value is $s_0$ with the displacement variable $x$, which results in varying air gap magnitude $s = s_0 - x$

Substitution of the magnetic variable $B_0$ results in an equation of magnetic force which depends on electrical parameters (i.e. electrical current $i$) and mechanical parameters (i.e. $N$, $A$, $s = s_0 - x$) [12]

$$F = \frac{N^2 \cdot i^2 \cdot \mu_0 \cdot A}{4 \cdot s^2} \quad . \quad (1b)$$

Thus, the force is proportional to the square values of current and inversed air gap magnitude:

$$F \sim \frac{i^2}{s^2} \quad (2)$$

If the I-core moves into the positive $x$-direction, then it is attracted to the C-core due to the magnetic force. Hence the air gap length decreases; this causes the magnetic force to increase, yielding an unstable mechanism of magnetic force (Fig. 1). This unstable mechanism will be confirmed by the pole positions of the active magnetic bearing system in the right half of the $z$-plane (Fig. 5).

The technical application in high-speed rotating electrical machines to suspend the rotor by a magnetic force is possible by using two active radial magnetic bearings (Fig. 3) and additionally one axial magnetic bearing.

### B. Hi-Speed Magnetic Levitated PM Synchronous Drive

There is a technical trend for the electrical drives towards increasing speed. This holds true for industrial drives, e.g. servo drives, industrial pumps, as well as for other applications, e.g. compressors. The main reason is to reduce the volume and the weight at a constant power level $P$ due to the equation $P = 2\pi \cdot n \cdot M$, where $n$ is the rotational speed and $M$ is the shaft torque. The use of the high-speed electrical drives allows a direct drive application. Omitting the gear increases the overall system efficiency and reliability.

TABLE I.
PARAMETER VALUES OF THE ELECTRICAL DRIVE

| | |
|---|---|
| Rated power $P_N$ | 40 kW |
| Rated rotational speed $n_N$ | 40000 rpm |
| Rated torque $M_N$ | 9.54 Nm |
| Pole number | 4 |
| Air gap | 1 mm |
| Stator inner bore diameter $d_{si}$ | 90 mm |
| Stator outer diameter $d_{sa}$ | 150 mm |
| Magnet height $h_m$ | 5 mm |
| Bandage thickness $h_b$ | 5.4 mm |
| Voltage-source inverter drive | Simodrive 611 |
| Rotational speed encoder | Incremental encoder sin/cos 1Vpp |
| Maximum inverter output frequency | 1400 Hz |
| Drive control method | PI-controller |

The permanent-magnet (PM) synchronous machine is a promising solution for the realization of the high-speed electrical drive. Compared to the induction machines, the stator of the permanent-magnet machines is identical. There are no excitation power losses in the rotor of the PM synchronous machine due to the application of the permanent-magnets as the excitation source on the rotor. Because of the absence of a needed magnetizing current, the air gap of the PM synchronous machines is generally designed with a higher value compared to the air gap in the induction machines. Hence, the air friction losses can also be considered to have a smaller value compared to the induction machines.

The PM synchronous machine (Fig. 2) used in our test-bench (TABLE I) is built with the surface-mounted permanent-magnets NdFeB. The permanent-magnets are glued onto the rotor surface and fixed by a carbon-fiber bandage, where its inner surface is cast with a glass-fiber compound to avoid local spots with higher stresses due to the permanent-magnet edges.

Fig. 2. The magnetic levitated PM synchronous machine

The rotor of our PM synchronous machine was tested for its stiffness up to a rotational speed of 44000 rpm successfully, which confirmed the strength of the rotor with the permanent-magnets and the bandage on it.

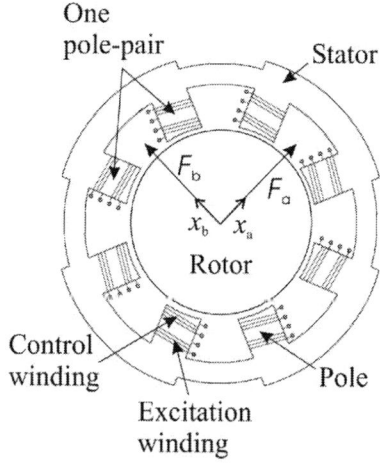

Fig. 3. Basic geometry of one radial active magnetic bearing with differential windings configuration

The rotor of the PM synchronous machine is levitated by the two radial active magnetic bearings. Their specification data are given in the TABLE II. The electromagnets are built as symmetrical 8-poles radial flux stators in differential winding mode, whereas the levitated part is the ferromagnetic shaft of the electrical machine. In each pole, there are two coils: 1) the excitation coil with number of turns $N_0$ and a constant current $i_0$, 2) the control excitation coil with the number of turns $N_c$ and a varying current $i_c$. The differential winding configuration allows a linearization of the relation in (2) and results in the linear equation of the magnetic force $F_a$ or $F_b$ respectively, generated by radial active magnetic bearing (Fig. 3).

$$F = k_i \cdot i_c + k_x \cdot x_{\text{Rotor}}, \qquad (3)$$

with: a) force-current factor ($\alpha = \pi/8$)

$$k_i = \mu_0 \cdot A \cdot \cos\alpha \cdot N_0 \cdot N_c \cdot \frac{i_0}{s_0^2}, \qquad (4)$$

b) force-displacement factor

$$k_s = \mu_0 \cdot A \cdot \cos\alpha \cdot N_0^2 \cdot \frac{i_0^2}{s_0^3}, \qquad (5)$$

$A$ is the cross-section area of one pole in each pole-pair. Depending on the chosen degree of freedom, the force magnitudes $F_a$ or $F_b$ are calculated by (3) at a detected rotor displacement $x_{\text{Rotor}}$ ($x_a$ or $x_b$) from the zero position and a supplied current, $i_0$ and $i_c$.

TABLE II.
PARAMETERS OF THE ACTIVE MAGNETIC BEARING IN THE PM SYNCHRONOUS MACHINE

| | |
|---|---|
| Rotor mass | 13.5 kg |
| Rotor moment of inertia of z-axis | $1.0265 * 10^{-5}$ kg.m$^2$ |
| Force of one magnetic bearing at rotor zero position at $i_0 = 6$ A and $i_c = 15$ A | 1526 N |
| Force-displacement factor $k_s$ | $2.02 * 10^6$ N/m |
| Force-current factor $k_i$ | 71.94 N/A |
| Rated current $i_0$ | 6 A |
| Rated current $i_c$ | 15 A |
| $N_0$ | 90 |
| $N_c$ | 36 |
| Pole cross-section area $A$ | 680 mm$^2$ |
| Nominal air gap $s_0$ | 0.4 mm |
| Pole axial length $l_{\text{pole}}$ | 40 mm |
| Nominal air gap of the auxiliary bearing $s_{0,\text{aux}}$ | 0.2 mm |

C. *Motion Dynamics of A Magnetic Levitated Rotor*

The stiff rotor (Fig. 3) is suspended by two radial active magnetic bearings (AMB), each with distance sensors (sensor A and sensor B) for detecting the rotor position (Fig. 4). The rotor may be described as a rigid body or as a flexible body, depending on the relation between the rated speed and the rotor's first bending natural frequency $f_{b1}$. At rated speed below approximately 70% of the first bending natural frequency, it is sufficient to assume the

rotor as a rigid body [14]. In our case, $f_{b1} = 1555.55$ Hz and $n_{\max} = 667$ s$^{-1}$, so we choose the rigid body model.

The rotor motion is calculated in the two planes of both radial active magnetic bearings (AMB). We define the motion vector [2]:

$$z_{\text{AMB}} = [x_A, x_B, y_A, y_B]^T, \qquad (6)$$

and the motion equation

$$M_{\text{AMB}} \cdot \ddot{z}_{\text{AMB}} + G_{\text{AMB}} \cdot \dot{z}_{\text{AMB}} = f_{\text{AMB}}, \qquad (7)$$

with $f_{\text{AMB}} = [F_{x_A}, F_{x_B}, F_{y_A}, F_{y_B}]^T$ as the force vector, which corresponds to the motion vector $z_{\text{AMB}}$. Each component of the force vector for each degree of freedom is calculated by (3). The matrix $M_{\text{AMB}}$ (4x4) represents the mass matrix depending on the rotor mass and rotor moments of inertia around the x-axis and y-axis direction (Fig. 4). The gyroscopic matrix $G_{\text{AMB}}$ (4x4) couples all four degrees of freedom $x_A$, $x_B$, $y_A$ and $y_B$. The gyroscopic effect depends on the rotational speed and on the rotor moment of inertia around the z-axis. The active magnetic bearing discussed in this paper is designed and built in a test-rig with calculated parameters given in TABLE III.

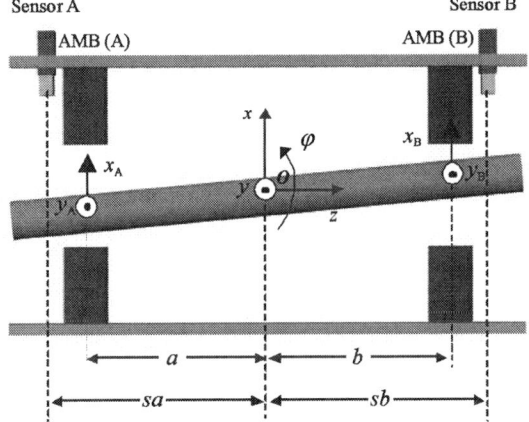

Fig. 4. Dynamic model of the levitated rotor

TABLE III.
PARAMETERS OF THE "TEST" ACTIVE MAGNETIC BEARING SYSTEM

| | |
|---|---|
| Rotor mass | 13.5 kg |
| Rotor moment of inertia of z-axis | $1.0265 * 10^{-5}$ kg.m$^2$ |
| Force of one magnetic bearing at rotor zero position at $i_0 = 6$ A and $i_c = 15$ A | 230 N |
| Force-displacement factor $k_s$ | $2.5 * 10^5$ N/m |
| Force-current factor $k_i$ | 10.9 N/A |
| Rated current $i_0$ | 6 A |
| Rated current $i_c$ | 15 A |
| $N_0$ | 46 |
| $N_c$ | 18 |
| Pole cross-section area $A$ | 680 mm$^2$ |
| Nominal air gap $s_0$ | 0.6 mm |
| Pole axial length $l_{\text{pole}}$ | 40 mm |
| Nominal air gap of the auxiliary bearing $s_{0,\text{aux}}$ | 0.2 mm |

## II. PRINCIPLES OF STATE CONTROLLERS

### A. Theoretical Background of State-Space Method

The idea of state-space comes from the state-variable method of describing differential equations. Ordinary differential equations of dynamic systems can be transformed into state-variable form, which is called the normal form of the differential equations. With the state-space method provided, the ordinary differential equations do not have to be linear [13].

The rotor position (potential energy) and velocity (kinetic energy) are selected as state variables. These states are related to the inputs and outputs. This is the main difference to the system description as a transfer function, which relates only the input to the output and does not show the internal behavior [13].

The state-space formulation fits better to computer-aided design of controllers with its mathematical calculation in matrices $F$, $G$, $H$ and $J$ (see (8)), directing faster to the set-point values for the plant to be controlled.

$$x(t) = F \cdot x(t) + G \cdot u(t)$$
$$y(t) = C \cdot x(t) + D \cdot u(t) \qquad , \qquad (8)$$

$x(k)$: state-variable vector
$u(k)$: input vector
$y(k)$: output vector
$F$: system matrix
$G$: input matrix
$C$: output matrix
$D$: direct transmission matrix

### B. State-Space Description of AMB

The basic motion differential equation of the active magnetic bearing in (7) is transformed first to its state-space version. The result is a state-space equation, which refers to the planes of both radial active magnetic bearings. The position and velocity of the suspended rotor are the state-variables. The position is the parameter, which has to be controlled in the AMB system. The velocity as the first time derivative of the position gives the corresponding kinetic energy, whereas the potential energy is given by the position. Due to four degrees of freedom, we have totally 8 state-variables.

$$x(t) = [x_A, x_B, y_A, y_B, x_A, x_B, y_A, y_B]^T, \qquad (9)$$

and four inputs in form of electrical control currents $i_c$ in the planes of A and B for the coils of the $x$- and $y$-axis direction of the magnetic force.

$$u(t) = [i_{xA}, i_{xB}, i_{yA}, i_{yB}]^T \qquad (10)$$

For digital control, a discrete formulation of (8) is needed. In (15) we have to define the system matrix $A$, which has the most influence on the open-loop and closed-loop behavior of the dynamic system. Having the position and velocity as state-variables, we define first the system matrix of (8) as

$$F = \begin{bmatrix} 0 & I \\ M_{AMB}^{-1} \cdot K_s & M_{AMB}^{-1} \cdot G_{AMB} \end{bmatrix} \qquad (11)$$

This system matrix $F$ (8x8) contains the zero matrix $0$ (4x4) and the unity matrix $I$ (4x4) in the first matrix row. $K_s$ is a diagonal matrix (4x4), where the value of each diagonal component equals to $k_s$ given in (5). The input matrix relates the inputs to the state-variables. It is defined as

$$G = \begin{bmatrix} 0 \\ M_{AMB}^{-1} \cdot K_i \end{bmatrix} \qquad (12)$$

Similar to $K_s$, $K_i$ is a diagonal matrix (4x4), where the values of each diagonal component equal to $k_i$, given in (4). The output of the state-space description is given by the positions in $x$- and $y$-direction in the A and B plane.

$$y(t) = [x_A, x_B, y_A, y_B]^T \qquad (13)$$

Hence, the output matrix is

$$H = \begin{bmatrix} I & 0 \end{bmatrix} \qquad . \qquad (14)$$

Again, $I$ is the unity matrix (4x4) and $0$ is the zero matrix (4x4). The direct transmission matrix $D$ has a value of zero, hence $D = 0$ (zero matrix).

For the controller design purpose, the mentioned state-space equations have to be transformed into a discrete system for discrete time steps. The Digital Signal Processor (DSP) used for the experiments in this paper works with zero-order sample-and-hold AD-converters. The sampling time, which equals to the control-loop-period, is determined mainly by the rotor behavior at high speed. The machine rated speed of $n = 40000$ rpm equals to a rotor mechanical rotational period of 1.5 ms. Any rotor in reality will suffer from mechanical imbalance, although the balancing process according to the standard ISO 1940 is performed. This "rest" imbalance results in an oscillating force with $f = n$ ("imbalance" force) during the rotational motion of the rotor, which has to be compensated. Hence, the sampling frequency $f_s$ of the compensator (controller) must be sufficiently higher than the mechanical frequency in order to obtain high-angle-resolution compensator $f_s > 10 \cdot n$. Hence, the sampling time will be configured with a value of $T = 53.33$ μs, so the discrete time steps for the chosen $z$-transform are given as $T = 53.33$ μs, giving a ratio of $f_s/n \approx 28$.

### C. Discrete State-Space Description

The discrete state-space formulation for a system with the state vector $x$ and the output vector $y$ is defined by (15) [3], where $k$ as a natural number is the counter of the time steps $T$.

$$x(k+1) = A \cdot x(k) + B \cdot u(k)$$
$$y(k) = C \cdot x(k) + D \cdot u(k) \qquad (15)$$

The transformation rules are defined at a given sampling time $T$ from the continuous to the discrete system in (16). The transformation algorithms are already available in the commercial computer-aided design tool MATLAB®.

$$A = \exp(F \cdot T)$$
$$B = \int_0^T \exp(F \cdot \eta) \cdot d\eta \cdot G \qquad (16)$$

We show the inherent instability of the active magnetic bearing by analyzing its state-space description. The eigenvalues of the system matrix $A$ in (15) provide us the positions of the poles of the AMB in the complex $z$-plane. The map of the zeros and the poles of the matrix $A$ of the AMB in the complex $z$-plane is shown in Fig. 5. Each two poles in the complex $z$-plane have the same value. This is the reason why only four pole positions are visible in Fig. 5. We find four of the eight poles of any open-loop AMB outside the unity circle ($|z_{pole}| > 1$), which confirms the inherent instability of the active magnetic bearing.

The calculation of feedback controller gain $K(t)$ for this system is possible by application of the pole placement method. An alternative solution is a systematic method to determine the desired pole locations, which is realized by an optimal method that searches for an optimal point between control effort and the response speed. It guarantees a stable system at the same time. Thus, this optimal method looks for a minimum of a cost function $J$ of weighted system inputs and weighted state-variables, like shown in (17). However, we may not have illusions that a true "optimal" solution is being achieved. The reason is that the weighting values as the parameters in the cost function $J$ are generally arbitrarily selected [3]. After choosing some starting values, iterations must be performed to obtain the "best" values of the feedback gain $K(t)$. Hence, the design is at best only partially optimal.

$$J = \frac{1}{2} \cdot \sum_{k=0}^{N} \left[ x^T(k) \cdot Q_1 \cdot x(k) + u^T(k) \cdot Q_2 \cdot u(k) \right] \quad (17)$$

The weighting matrix $Q_2$ represents, how big the control effort should be, and the weighting matrix $Q_1$ determines, how fast the response will be. Both matrices must be non-negative definite in order to keep both terms in the cost function $J$ to be non-negative. This can be done by picking the weighting matrices to be diagonal with all diagonal elements positive or zero. This optimal method uses the method of solving the **discrete** *Riccati*-equation and results in an "optimal" time-variant feedback gain $K(t)$. This method requires in practical systems a time-consuming calculation of the hardware, thus it is not really applicable. A steady-state (time-invariant) "optimal" control, which means a constant feedback gain $K$, can be calculated by solving the **algebraic** *Riccati*-equation, which generates a steady-state solution [3], which was finally done.

For the selection of the weighting values of $Q_1$ and $Q_2$, there is a recipe, which is called **Bryson's rules** to define "useable" starting values [13]. The main idea is to select the components in the weighting matrices $Q_1$ and $Q_2$ so that a fixed percentage change of each component of the

state-variables and of the input makes an equal contribution to the cost function. We may not forget that after choosing the starting values, iterations in computer-aided design tool must be performed to look for the best trade-off between the control effort and the response speed under the condition that the controlled system remains stable.

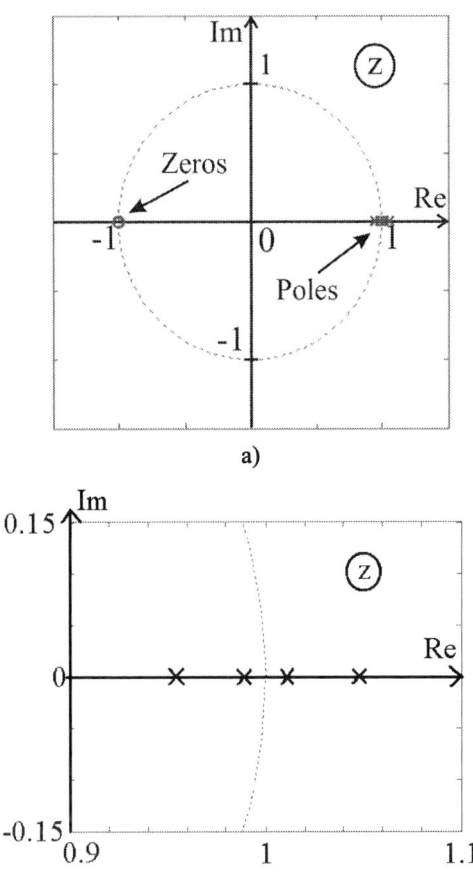

Fig. 5. Open-loop map of zeros and poles of the active magnetic bearing system in the complex $z$-plane: a) Poles and zeros of the active magnetic bearing; b) Zoomed position of the system poles

### D. Observer and Reference Input

The feedback gain $K$ was calculated by the optimal design method, where the goal was to drive all state-variables to zero. No consideration was included of how to obtain a good transient response to reference inputs, which are generally not always zero. In our case, the AMB inputs are defined by the electrical currents through the coils. The AMB outputs are the rotor positions. There can be only reference inputs $r$ as reference position values. A reference input filter $G_{filt}$ is necessary to transform the reference position values into the input values, which are the electrical currents through the coils. The calculation for the reference input filter is given in (18).

$$G_{\text{filt}} = \left( -C \cdot (A - (B \cdot K) - I)^{-1} \cdot B \right)^{-1} \quad (18)$$

Together with any given reference input vector $r$, a digital state-controlled dynamic plant will have a control structure as shown in Fig. 6. The input vector $u(k)$ in (19) is calculated as the sum of the feedback gain output and the reference input filter output.

$$u(k) = -K \cdot x(k) + G_{\text{filt}} \cdot r(k) \quad (19)$$

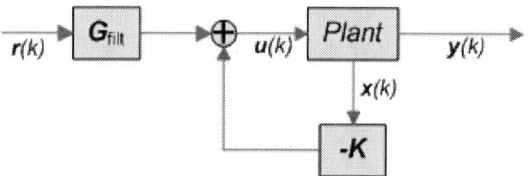

Fig. 6. Mathematical closed-loop structure of a state-controlled AMB [3].

In our AMB system, two important aspects have to be considered:

1) There are only position sensors available. The rotor velocity as the other state-variable can be only obtained by the time derivation of the measured position, as no velocity is measured. The time derivation of signals is sensitive to external noise. Hence, an alternative solution is to use an observer to estimate the velocity without having the disturbance of noise.

2) In practice, the external disturbance influences any system behavior. Depending on the disturbance magnitude, a disturbance compensation is necessary which will be integrated into the controller to keep the stability of the controlled system. Often, it is not possible to measure directly the external disturbance without a sophisticated configuration of instruments. A possible method to determine the disturbance magnitude is to define the disturbance as a state-variable (which means the previous state-vector $x(k)$ is now augmented to be a new state-vector $x^*(k)$) and estimate its magnitude by an observer.

This condition gives us two options of how to provide the augmented state vector $x^*(k)$ as input for the feedback controller matrix $K$, which now also contains the disturbance compensator:

a) Use of full state estimations: All the states (position, speed, disturbance force) used for the feedback control are provided by estimation, utilizing an observer.

b) Use of reduced state estimation: The position values are provided directly by the position sensors, while the speed and disturbance force will be estimated by an observer.

This paper shows the application of option b) with the reduced state estimation (Fig. 7). The state-vector $x_{\text{Pos}}(k)$ contains the measured positions as the measured state-variables. The other state-vector $x_{\text{Speed,Dist.}}(k)$ is calculated in real-time by the observer to obtain estimated values of speed and external disturbance. The control and

estimation (observer) algorithm is implemented on a DSP (Digital Signal Processor) TMS320F2812 of *Texas Instruments*. The coefficients of the feedback controller matrix $K$ were calculated with the computer-aided design tool MATLAB/Simulink® by solving the *Riccati*-equation of the L(inear)-Q(uadratic)-R(egulator) method. The observer matrix $L$ was designed by the method of pole placement for a MIMO system, taking into account that a sufficient higher response signal of the observer is necessary, by placing the observer's poles more left in the complex *Laplace* $s$-plane than the system's open-loop poles. After that, the poles were transformed for the digital control into their equivalent values in the complex $z$-plane of the $z$-transform, where the observer's coefficients were calculated.

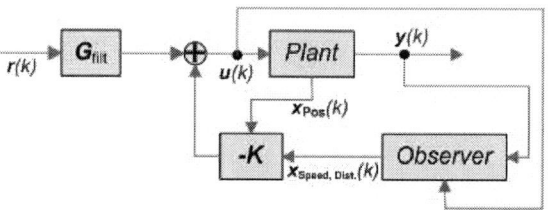

Fig. 7. Mathematical closed-loop structure of a state-controlled AMB augmented with the observer [3].

## III. EXPERIMENTS AND CALCULATION RESULTS

Fig. 8. Active-magnetic-bearing test-rig

First of all, the inherently unstable active magnetic bearing must be stabilized by the digital controller, which means that the discrete closed-loop poles of the active magnetic bearing must be moved within the unity circle in the complex $z$-plane to achieve a magnitude less than unity [3].

The distance of the pole locations from the $z$-plane origin of the open-loop AMB system are bigger than unity, confirming the instability of the open-loop system. These calculated eight open-loop poles of the AMB system are forming four double poles (see Fig. 5):

$z_{p,open} =$ [1.048, 1.011, 0.989, 0.954, 0.954, 1.048, 0.989, 1.011]$^T$

By designing the closed-loop system with the calculated controller matrices of Chapter II, the poles of the transfer functions are shifted to have now magnitudes less than unity:

$z_{p,closed} =$ [0.844, 0.844, 0.993, 0.993, 0.987, 0.987, 0.984, 0.984]$^T$

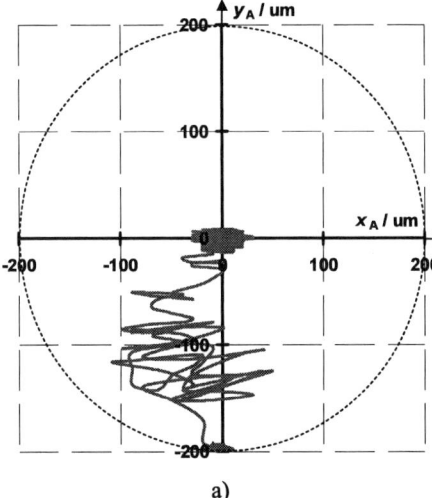

a)

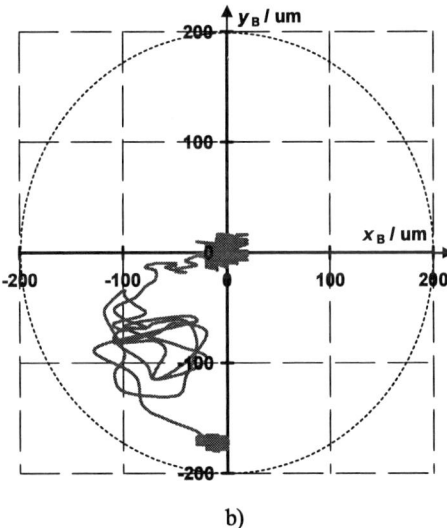

b)

Fig. 9. Measured lift-off of the rotor, rotor orbit in the *x-y* plane
Rotor centre orbits: a) A-side of the rotor;     b) B-side of the rotor
(Air gap in the auxiliary bearing: 200 µm)

The results in this Chapter are obtained for the AMB test-rig of Fig. 8, where the rotor is at standstill. In Fig. 9 and Fig. 10, the rotor lift-off from the retainer (auxiliary) bearing into the centre position in the stator bore is shown.

These motions in the axial *x-y* plane are seen from the drive-end side of the rotor. Typically for the state controller due to its capacity to control the system states is the absence of overshoot during the take-off. Unsymmetrical movements during the "take-off" are caused by the unsymmetrical initial position of the rotor on the auxiliary bearing. In practice, due to the non-ideal manufacturing of the machine casing we will have some deviations from the ideal AMB geometry. A typical tolerance of 10 µm in the AMB air gap must be considered. Therefore, the rotor will not be able to be positioned perfectly in the centre of the auxiliary bearing.

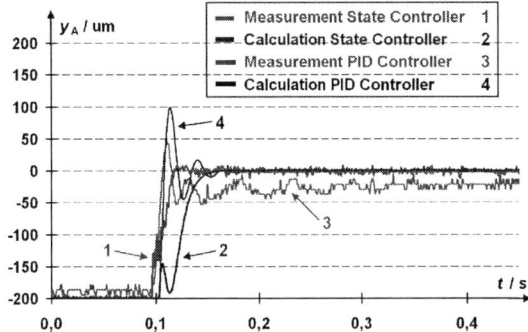

Fig. 10. Time behavior of $y_A$- axis of the rotor during the lift-off. Comparison of the state controller with a PID controller.

Fig. 10 shows the calculated (simulated) and the measured rotor motion as a time function: a) for the use of the state controller, and b) for the use of a decentralized PID controller.

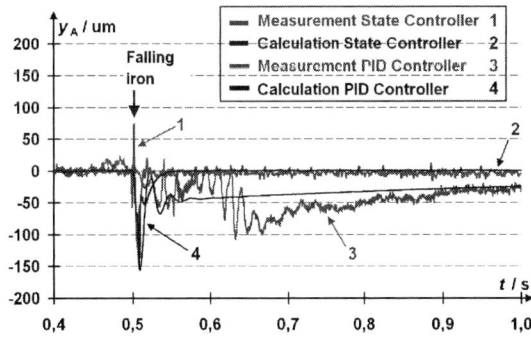

Fig. 11. Vertical position response of the AMB-controlled rotor ($y_A$– axis) during the compensation of an external disturbance due to a vertical force impact of 136 N on the rotor shaft midway between the A- and B- bearing. Comparison of the state controller with a PID controller.

Fig. 11 and Fig. 12 show the behavior of the rotor positions during an external disturbance. This external disturbance is realized by a falling iron block of 4.075 kg (falling distance: 2 cm) onto the rotor shaft in *y*-direction, which yields the disturbance. The impact of the iron block on the rotor shaft can be assumed to last approximately for 5 ms, as can be seen by the position change in Fig. 11. This duration of the impact yields a dynamic force of about 96 N [11], which is added to the static weight of the iron of 40 N. Hence, the total disturbance force during the impact is 136 N. The AMB with a nominal radial force of 230 N per bearing is controlled fast enough to keep the deviation of the rotor position from the centre within the clearing of the auxiliary bearing of 200 μm. Obviously in Fig. 10 and Fig. 11, the better dynamic performance of the state-controller over the PID-controller is visible.

## IV. CONCLUSIONS

Both by calculation and measurement the performance of two radial AMB with state-space digital control at a test-rig with a non-rotating rotor of 13.5 kg has been demonstrated. The radial AMB with 230 N nominal force (steady state at 77°C winding temperature, natural cooling) are for comparison also controlled with PID, showing the better dynamic response behavior of the state-space controller. In a next step the control will be applied to a magnetically suspended rotor of a permanent-magnet machine for 40 kW, 40000 rpm.

## REFERENCES

[1] R.D. Williams, F.J. Keith, P.E. Allaire, *"Digital Control of Active Magnetic Bearings"*, IEEE Trans. on Industrial Electronics, Vol. 37 No. 1, February 1990, pp. 19-27.

[2] C.R. Sabirin, A. Binder, D.D. Popa, A.Craciunescu, *"Modeling and Digital Control of An Active Magnetic Bearing System"*, Rev. Roum. Sci. Techn. – Electrotechn. Et Energ., 52, 2, 2007, Bucarest, pp. 157-181.

[3] G.F. Franklin, J.D. Powell, M. Workman, *"Digital Control of Dynamic Systems"*, 3rd Edition, Pearson Education, Inc., New Jersey, 2002.

[4] H. Kim, H.C. Kim, „*Modeling and Control of A Magnetic Bearing System for The Magnetically Suspended Centrifugal Blood Pump"*, The International Journal of Artificial Organs, Vol. 23, no.10, 2000, pp. 47-51.

[5] M. Spirig, J. Schmied, P. Jenckel, U. Kanne, „*Three Practical Examples of Magnetic Bearing Control Design Using A Modern Tool"*, Journal of Engineering for Gas Turbines and Power, Vol. 124, October 2002, pp. 1025 – 1031.

[6] C. Jungkunz, A. Binder, „*Entwurf und Bau des Prototypmagnetlagers SIMAB (Siemens Magnetic Bearing)"* 3. Zittauer Workshop Magnetlagertechnik, 1997, Zittau, pp. 21 – 31.

[7] T. P. Dever, G.V. Brown, R.H. Jansen, *"Estimator Based Controller for High Speed Flywheel Magnetic Bearing System"*, NASA TM-2002-211795, National Aeronautic and Space Administration, August 2002, Washington D.C.

[8] G. Schweitzer, A. Traxler, H. Bleuler, „*Active Magnetic Bearings – Basics, Properties and Application of Active Magnetic Bearings"*, vdf Hochschulverlag, Zürich, 1994.

[9] A. Baral, „*Entwicklung, Zustandsidentifikation und Regelung eines magnetisch gelagerten Hochgeschwindigkeitsantriebs"*, Dissertation an der Universität Kassel, Fortschrittberichte VDI, Reihe 8, Nr. 649, VDI Verlag, Düsseldorf, 1997.

[10] H. Bleuler, „*Decentralized Control of Magnetic Rotor Bearing Systems"*, Ph.D. Thesis, Prom. ETH 7573, Zürich, 1984.

[11] R.C. Hibbeler, *"Engineering Mechanics – Dynamics"*, 10th Edition, Pearson Education Inc., New Jersey, 2004.

[12] A. Chiba, T. Fukao, O. Ichikawa, M. Oshima, M. Takemoto, D.G. Dorrell, *"Magnetic Bearings and Bearingless Drives"*, 1st Edition, Elsevier, Oxford, 2005.

[13] G.F. Franklin, J.D. Powell, A. Emami-Maeini, *"Feedback Control of Dynamic Systems"*, 5th Edition, Pearson – Prentice Hall, New Jersey, 2006.

[14] R. Gasch, R. Nordmann, H. Pfützner, *"Rotordynamik"*, 2nd Edition, Springer Verlag, Berlin Heidelberg, 2002.

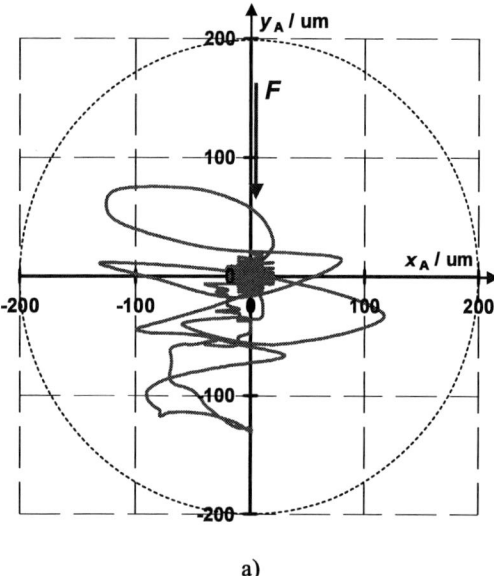

a)

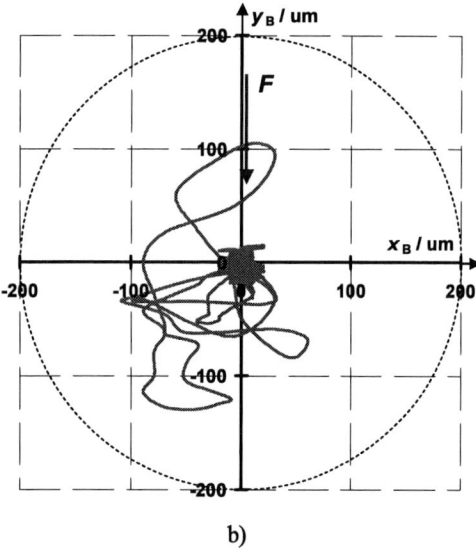

b)

Fig. 12. Measured rotor motion during a disturbance of a falling load onto the rotor axial middle position
(impact force *F* = 136 N)
Rotor centre orbits: a) A-side of the rotor;    b) B-side of the rotor

# DYNAMICAL TORQUE-SPEED-CURVE ADAPTION TO DAMP LOAD PEAKS OCCURING IN DRIVE TRAINS OF SHREDDING PLANTS

Constantinos Sourkounis
Research Group for Power Systems Technology
Faculty for Electrical Engineering and Information Sciences
Ruhr-University Bochum
Bochum, Germany
Phone: +49 234 32 25776
FAX: +49 234 32 14597
sourkounis@eele.rub.de

**Abstract** – Comminuting processes, which provide the required mechanical power for the process cause stochastical loads with high amplitudes in the electromechanical drive train. In the long run, peak loads cause premature material fatigue of the drive components alongside high thermal loads in the asynchronous machine.

A new drive train with a novel non-linear speed control was developed based on the operating performance of commercial comminuting units.

The non-linear speed control realises a rotational speed-flexible operation so that the rotating masses' kinetic energy is used to smooth the load peaks from the process.. In this way, it was possible to reduce the cumulative load in the drive train achieving high process performance.

*Keywords: active damping, asynchronous motor, control of drive, electrical drive, mechatronics, variable speed drive*

## I. INTRODUCTION

In comminuting of materials, the quality of a process, which is defined by its dwell time and the size of the produced scrap metal remnants, depends on the average circumferential velocity of hammers and eventually the mean rate of revolvements. Dynamical variations of the rotational speed around its optimum operational value do not affect the grade of the process noticeably [1]. Comminution plants can, therefore, be operated by drives with flexible rotational speed, so that load peaks from the process chamber are dampened.

In usual drive systems operating at constant speed, load peaks originating in the drive train are transmitted towards the main supply and cause early material fatigue in the mechanical and thermal overload in the electrical parts of the system. In addition, peak currents resulting in considerable voltage variations arise in a common drive system, where an asynchronous machine is directly connected to the supply.

Since active damping of load peaks requires only a narrow range around the optimum rotation with dy-

namic adjustment, classical speed variable drive systems, i.e. fully inverter fed electrical machines (e.g. asynchronous machine) are not preferably applied in this case. The issue is rather a robust yet simple layout of a drive with limited speed variability.

## II. LAYOUT OF A DRIVE WITH DYNAMICAL ADJUSTMENT OF THE TORQUE-SPEED-CHARACTERISTICS

A fully inverter fed asynchronous machine meets, as a rule, all demands resting on the necessarily speed variable operation. What concerns is that the full power fed to the process must pass the dc-link converter. Since this device must also address the overload problem, which is frequently encountered in comminution processes, technical effort can rise disproportionately high, depending on the required grade of performance. An applicable concept of a drive, where the actuator (converter) needs withstand only part of the power inherent to the plan, is shown in Fig.1. The power range of the actuator expands in dependence of the size of the speed variation where dynamical adjustment shall be effected for the purpose of damping load peaks. The concept for this drive is about an asynchronous machine with dynamic adjustment of the torque-speed-characteristics.

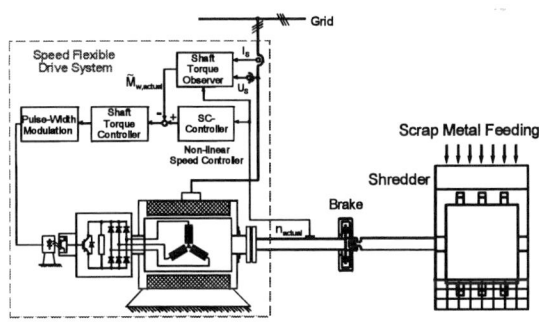

**Figure 1:** Drive train with adapting torque-speed characteristic

---

978-1-4244-1741-4/08/$25.00 ©2008 IEEE

An additional resistor is integrated into the rotor circuit using a power electronic device. The size of the resistor is adjusted as needed by pulse width modulation of the switching interval of the IGBT connected in parallel to the resistor.

$$R_{z,eff} = \left(1 - \frac{T_{on}}{T}\right) \cdot R_z = d \cdot R_z \qquad (1)$$

In accordance, since

$$s_k = \frac{R'_{2,tot}}{X_\sigma} \qquad (2)$$

with $R'_{2,tot} = R'_2 + \dfrac{\pi^2}{18} \cdot R'_{z,eff}$

the steady-state torque-speed-characteristics of the asynchronous machine change (cf. Fig.2). The effort for the actuator is reduced to a controller less B6-type AC/DC-converter and an IGBT switch (cf. Fig.1). The complete actuator and the additional resistor $R_z$ are mechanically mounted to the rotor and thus revolving with it. This way, slip-rings are not necessary. This is very important for the reliability and availability of the drive, in particular, where there is an effect of dust from the communition process on the wear of the contact strip of electrical machines.

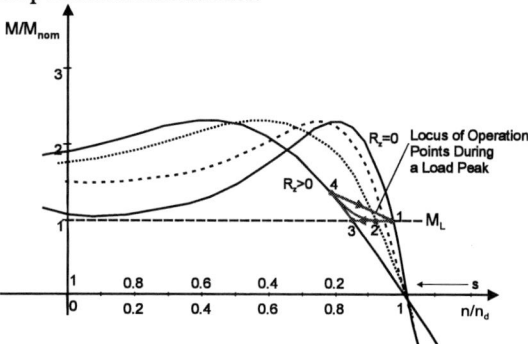

**Figure 2:** Torque-speed-characterisitc of asynchronous machine with variation of the additional rotor circuit resistor

### III. THE DYNAMICS OF THE DRIVE AND THE CONTROLLER STRUCTURE

By means of a cascaded controller structure, the dynamic operation was optimized so that, on the one hand, peak loads from the process chamber could be damponed by dynamical speed variation and, on the other, by the inner control circuit a shaft torque control is realized, so that torque oscillations, which were induced by the load peaks, could be suppressed.

The control of the torque and the superimposed non-linear speed control were based on a simplified model of the drive train with two body-masses [6] (cf. Fig.3).

$$\begin{pmatrix} \dot{n}_M \\ \dot{m}_w \\ \dot{n}_L \end{pmatrix} = \begin{pmatrix} 0 & -\dfrac{1}{T_M} & 0 \\ \dfrac{T_M}{(1+v)T_{ef}^2} & 0 & \dfrac{T_M}{(1+v)T_{ef}^2} \\ 0 & \dfrac{1}{T_R} & 0 \end{pmatrix} \begin{pmatrix} n_M \\ m_w \\ n_L \end{pmatrix} + \begin{pmatrix} \dfrac{1}{T_M} & 0 \\ 0 & 0 \\ 0 & -\dfrac{1}{T_R} \end{pmatrix} \begin{pmatrix} m_i \\ m_L \end{pmatrix} \qquad (3)$$

The shaft torque transfer function, $m_w$, characterises the load in the mechanical drive train. It is obtained after carrying out a Laplace-transformation and normalising with the air gap torque, $m_i$, and the load torque, $m_L$, as an input to:

$$m_w(s) = \frac{1}{v+1} \cdot \frac{1}{s^2 T_{ef}^2 + 1} \cdot \left(m_i(s) + v \cdot m_L(s)\right) \qquad (4)$$

with $T_{ef} = \sqrt{\dfrac{T_C T_M T_R}{T_M + T_R}}$ and $v = \dfrac{T_M}{T_R}$

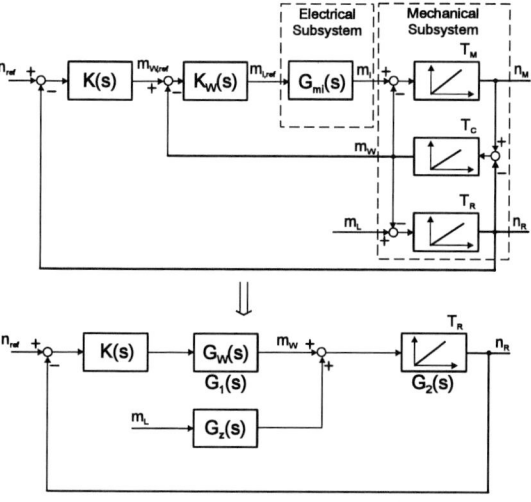

**Figure 3:** Functional block diagram of the speed-flexible electromechanical drive train

Since, supposing sufficient short power of the main supply, the electrical part of the drive train is determined by the performance of the asynchronous machine, the controller design was based on an extended average model of the asynchronous machine. Starting with the voltage equations for a squirrel cage asynchronous machine positioned within a stator voltage vector oriented coordinate system, the average model of the asynchronous machine with additional rotor circuit resistor was computed under consideration of the effective external resistor.

$$\underline{U}_1 = \underline{I}_1 R_1 + \frac{d\underline{\psi}_1}{dt} + j\omega_s \underline{\psi}_1 \qquad (5)$$

$$0 = \underline{I}_2 \left(R_2 + \frac{\pi^2}{18} d \cdot R_z\right) + \frac{d\underline{\psi}_2}{dt} + j\omega_2 \underline{\psi}_2$$

$$\underline{\psi}_1 = L_1 \underline{I}_1 + L_m \underline{I}_2$$

$$\underline{\psi}_2 = L_m \underline{I}_1 + L_2 \underline{I}_2$$

For steady-state operation of the asynchronous machine, provided that the additional resistor is short-circuited ($R_{zw}=0$) in nominal operation, it is valid:

$$\begin{pmatrix} 0 \\ 0 \\ 0 \\ 0 \end{pmatrix} = \begin{pmatrix} -\dfrac{R_1}{\sigma L_1} & \omega_s & \dfrac{R_1 L_m}{\sigma L_1 L_2} & 0 \\ -\omega_s & \dfrac{-R_1}{\sigma L_1} & 0 & \dfrac{R_1 L_m}{\sigma L_1 L_2} \\ \dfrac{L_m R_2}{\sigma L_1 L_2} & 0 & \dfrac{-R_2}{\sigma L_2} & \omega_2 \\ 0 & \dfrac{L_m R_2}{\sigma L_1 L_2} & -\omega_2 & \dfrac{-R_2}{\sigma L_2} \end{pmatrix} \cdot \begin{pmatrix} \psi_{1A} \\ \psi_{1B} \\ \psi_{2A} \\ \psi_{2B} \end{pmatrix} + \begin{pmatrix} 1 \\ 0 \\ 0 \\ 0 \end{pmatrix} \cdot U_1 \quad (6)$$

These dynamics can be described by the following state-space equations:

$$\begin{pmatrix} \dot{\psi}_{1A} \\ \dot{\psi}_{1B} \\ \dot{\psi}_{2A} \\ \dot{\psi}_{2B} \end{pmatrix} = \begin{pmatrix} \dfrac{-R_1}{\sigma L_1} & \omega_s & \dfrac{R_1 L_m}{\sigma L_1 L_2} & 0 \\ -\omega_s & \dfrac{-R_1}{\sigma L_1} & 0 & \dfrac{R_1 L_m}{\sigma L_1 L_2} \\ \dfrac{L_m R_2}{\sigma L_1 L_2} & 0 & \dfrac{-R_2}{\sigma L_2} & \omega_s \\ 0 & \dfrac{L_m R_2}{\sigma L_1 L_2} & -\omega_s & \dfrac{-R_2}{\sigma L_2} \end{pmatrix} \begin{pmatrix} \tilde{\psi}_{1A} \\ \tilde{\psi}_{1B} \\ \tilde{\psi}_{2A} \\ \tilde{\psi}_{2B} \end{pmatrix} \quad (7)$$

$$+ \begin{pmatrix} 0 & 0 & 0 & 0 \\ 0 & 0 & 0 & 0 \\ \dfrac{L_m \dfrac{\pi^2}{18} R_s \psi_{1AN}}{\sigma L_1 L_2} & 0 & -\dfrac{\pi}{18} R_s \psi_{2AN} & 0 \\ & & & \\ 0 & \dfrac{L_m \dfrac{\pi^2}{18} R_2 \psi_{1BN}}{\sigma L_1 L_2} & 0 & -\dfrac{\pi^2}{18} R_2 \psi_{2BN} \\ & & & \sigma L_2 \end{pmatrix} \cdot d$$

The transfer function, which is used for designing the controller, relates the air-gap torque $M_i$ to the PWM factor d. Assuming a very small stator resistor ($R_1 \ll$ ) it is formulated by the following:

$$\tilde{M}_i(s) = \frac{3}{2} Z_p \frac{L_m}{\sigma L_1 L_2} \cdot (\psi_{1B} \cdot \tilde{\psi}_{2A}) \quad (8)$$

with

$$\psi_{1B} = -\frac{U_{1A}}{\omega_s} = -\frac{U_1}{\omega_s}$$

$$\tilde{\psi}_{2A}(s) = \frac{\omega_2}{\omega_s} \cdot \frac{\pi^2}{18} \cdot \frac{L_m R_2}{\sigma L_2} \cdot U_1 \frac{\left(\dfrac{R_2}{\sigma L_2}\right)^2 - \omega_2^2 + \dfrac{R_2}{\sigma L_2}s}{s^2 + 2\dfrac{R_2}{\sigma L_2}s + \left(\dfrac{R_2}{\sigma L_2}\right)^2 + \omega_2^2} \cdot d$$

The speed flexible drive system is presented in detail by Fig.1. The mechanical power required for the comminution process is supplied by the asynchronous machine with dynamically adjustable torque-speed-characteristics. The asynchronous machine is directly grid connected. Following the dynamical adjustment of the characteristics resulting from the change of the additional rotor circuit resistor, the rotation speed can temporarily decrease in the presence of peak loads, so that part of the kinetic energy of the revolving masses is used to smooth the load peaks. The necessary controller structure displays a cascade of two control circuits. The control of the drive shaft torque has superficially the task of damping those torque oscillations induced in the

drive train by peak loads with a minimal correcting variable requirement. The control of the drive shaft torque was achieved with a single-loop PID controller. The shaft torque controller is designed according to the determined transfer function in equations (4) and (8). The actual drive shaft torque is computed by an observer. Process control and speed control respectively are based on the principles of S-curve control [1], [4]. The non-linear S-curve controller, by virtue of its general transfer function,

$$K_R = \frac{1}{(n_{ref} - n_{act})} + a\left(n_{ref} - n_{act}\right)^{2b} \quad (9)$$

$$\Rightarrow m_{w,ref} = 1 + a\left(n_{ref} - n_{act}\right)^{2b+1}$$

with a = 500 and b = 1 allows variations in the range of:

$$0,9\, n_N \leq n \leq n_0$$

This way, the kinetic energy of the revolving masses is used to smooth the peak loads from the process chamber.

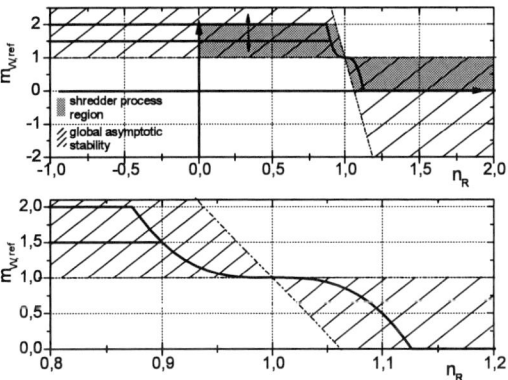

**Figure 4:** SC-Controller characteristic curve

## IV. COMPARATIVE TESTS

The drive system of the asynchronous machine with dynamically adjustable characteristics was examined with the help of simulation and a test plant especially designed for the purpose. In the first step, the examination was based on the synthetic load input function. This was derived from the knowledge gained in field tests. The load input function has a pulse shape evoked by the blow of the hammers on the scrap metal. The synthetic load input function can be reproduced easily so that the drives under consideration can be tested at exactly the same rate as the load input function.

In the present application, the time-diagram of the drive shaft torque is used for the comparison of different drives. The time-diagrams of the drive shaft torque from the experiments (cf. Fig.5, Fig.6) and, likewise, those simulated demonstrate that the drive system of the asynchronous machine with dynamic adapting of the torque-speed characteristics (DynAK) displays meaningful damping of both peak loads and oscillations. Torque peaks in the drive train were reduced to 45% of the nominal value versus 100% in the operation of the existing drive system with direct grid coupling. The amplitude of the load induced oscillations in the mechanical drive train is diminished by DynAK to 20% of the nominal torque by means of the underlying torque control mechanism, compared to 70% in the existing drive system. Furthermore, both drive systems are tested on a model shredder. Since many side effects of the comminution process cannot be modeled, testing the new drive concept constitutes a necessary step towards its technical application on a broad scale. Therefore, tests with different loads have been carried out. The load, caused by the process, was adjusted by the feeding of the scrap material having only an impact on the average load. The resulting time-diagram of the load is subject to a stochastical characteristic.

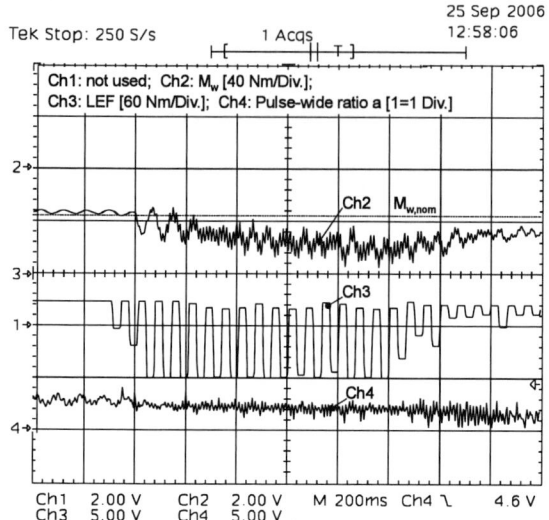

**Figure 6:** Record of the shaft torque during the action of the synthetic load input function from speed flexible drive system

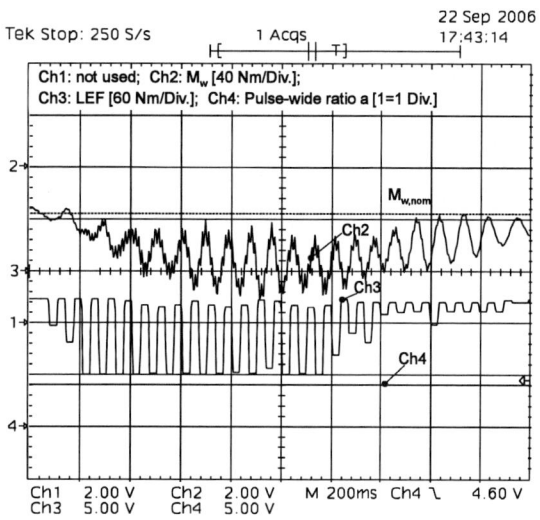

**Figure 5:** Record of the shaft torque during the action of the synthetic load input function from a drive system with squirrel cage machine

For the assessment of the considered drive systems, the distribution of load incidence in the drive train has proved useful. By means of speed variation and the underlying control mechanism working on the drive shaft torque, peak loads could be limited to 1.5-fold the nominal value, using minimum correcting value requirement. Overall, a marked shift of the load distribution towards lower amplitudes can be distinguished (cf. Fig. 7). The control operating range required for the speed flexible drive amounts to 12% of the nominal driving power, keeping the technical expenditure at a comparatively low level.

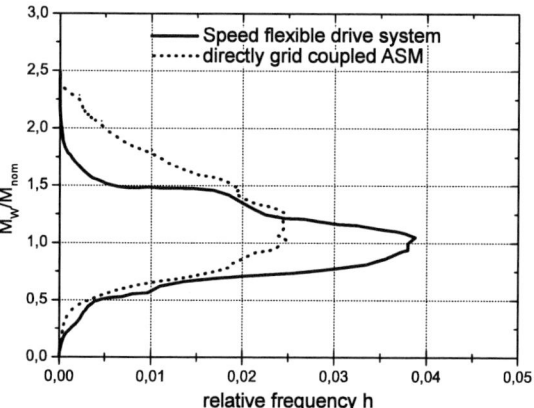

**Figure 7:** Frequency distribution of the shaft torque

The concept of the asynchronous machine with dynamical adjustment of the torque-speed-characteristics is, therefore, a meaningful approach to speed control resulting in minimum loads on comminution plants even at high power operation (e.g. some MW). The results obtained here for the example of a comminution process also point at advantages recommending their use in other process plants with similar load collectives.

## SYMBOLS

d:    duty factor (PWM factor)

$G_{mi}(s)$: Transfer function of the torque defining current

$G_W(s)$: Transfer function of the control path's response to setpoint changes

$G_z(s)$: Preliminary filter of the disturbance function

$I_1$:    stator current

$K(s)$:   transfer function of speed controller

$K_w(s)$: transfer function of shaft torque controller
$L_1$: stator inductance
$L_2$: stator inductance
$L_m$: main inductance
n: rotational speed
$n_0$: no-load speed
$R_z$: additional dc-rotor-circuit resistor
$R_{z,ef}$: effective additional dc-rotor-circuit resistor
$R_1$: stator circuit resistor
$R_2'$: rotor-circuit resistor transformed to stator
s: Laplacian operator
$s_k$: slip at breakdown torque
T: puls period
$T_C$: mechanical time constant for the spring stiffness
$T_{ef}$: time constant of the natural frequency of mechanical subsystem
$T_M$: mechanical time constant of ASM
$T_{on}$: puls duration
$T_R$: mechanical time constant of the shredder rotor
$U_1$: stator voltage
v: the mass inertia ratio
$x_\sigma$: leakage reactance
$z_P$: pole pair

$\Psi_1$: stator flux
$\Psi_2$: rotor flux
$\omega_1$: angular speed of stator flux
$\omega_2$: angular speed of rotor flux

Indices:
A: value component on direct axis
act: actual value
B: value component on quadrature axis
N: nominal
M: motor value
R: shredder rotor value
ref: reference value
tot: total
$\underline{U}$: vector
1: stator value
2: rotor value
′: transformed to stator circuit
~: alternating component

### REFERENCES

[1] Sourkounis C.: Drehzahlelastische Antriebe unter stochastischen Belastungen, Habilitationsschrift, TU Clausthal 2004 Papierflieger, Clausthal 2004; ISBN 3-89720-737-0

[2] Kirchner, J.; Schubert, G.: Shredder-Verbesserung des Prozesses, Berichte zu Ergebnissen aus dem Sonderforschungsbereich 180, Technische Universität Clausthal, 1996

[3] Fantoni, Isabelle; Lozano, Rogelio: Non-Linear Control for Underactuated Mechanical Systems, Springer, Berlin 2002, ISBN: 978-1-85233-423-9

[4] Sourkounis, C.: Verfahren und Regeleinrichtung zum Regeln eines Antriebssystems unter stochastischen Belastungen, Patentanmeldung: DE 10 2004 061 436 A1, Deutsches Patent- und Markenamt

[5] Quang, Ng. Ph.: Schnelle Drehmomenteinprägung in Drehstromstellantrieben, TU Dresden, Dissertation 1991

[6] Klöckner, J.:Berechnung des Schwingungsverhaltens von gedämpften, schwach nichtlinearen Schwingerdaten unter Berücksichtigung der Reduktion der Freiheitsgrade, Dissertation, TU Berlin, 1979

[7] Föllinger, O.: Nichtlineare Regelung, Bd. 1 und 2, Oldenbourg-Verlag; München 1998

[8] K. Sugiura and Y. Hori: Vibration suppression in 2- and 3-mass system based on the feedback of imperfect derivative of the estimated Torsional torque, IEEE Trans. Ind. Electron., vol. 43, pp. 56-64, Feb. 1996.

[9] Isidori, Alberto: Nonlinear Control Systems, Springer, Berlin 1995, ISBN 3-540-19916-0

[10] Dierk Schröder et al.: Intelligent Observer and Control Design for Nonlinear Systems, Springer, Berlin 2000, ISBN 3-540-63639-0

# Traction vehicle distributed control computer system architecture with auto reconfiguration features and extended DMA support

Jiri Zdenek

Czech Technical University in Prague, Faculty of Electrical Engineering, Prague,
Czech Republic, e-mail: *zdenek@fel.cvut.cz*

*Abstract*—The system design principles and hardware and software architecture of the distributed network control computer (DNCC) of a traction vehicle are presented. The system design is based on such criteria as functionality, reliability and maintainability. The number of computer network node types is minimized. Network node scans its physical position inside the network and automatically reconfigures itself to the required function using correct software to configure FPGA based hardware parts and to run correct control algorithms on the host processors. To get these properties special emphasis is placed on correct and optimal system function partition among distributed computer nodes to minimize the overall communication overhead and to minimize the hardware parts of the system and to move maximum of functions into the software. As an example of such design task the DNCC of an electric locomotive is presented. The vehicle DNCC is organized as a local computer network with master-slave node access method with the dual serial bus. The network communication services are uniformly used in all nodes to support reliable transfer of the process data. The communication overhead is minimized using exclusively DMA transfer method. DMA support is utilized to be the time information captured precisely (averaging ADCs) too. User task in the DNCC nodes are scheduled by the preemptive real time operation system with dynamic planning or by the simple static executive in the nodes with extra high requirements on the fast interrupt response.

*Keywords*—Traction application, Locomotive, Design, Software, Real time processing, Data transmission.

## I. INTRODUCTION

A digital control system of a traction vehicle is a typical example of a distributed local area network computer consisting of many independent programmable nodes communicating through an industrial computer network using messages. The distributed control computer (DNCC) of an electric locomotive can be used to demonstrate common design issues and methodology of distributed computer system design and development of software architecture. An electric locomotive as complex of electric drives, pneumatic brakes, security devices, engine driver stands and many other parts can be seen from point of view of hardware and software design as a mid-range design problem. The system architecture of computer controlled systems has many modifications and depends on the application size, required speed, available

design time, budget size, experience of software and designer, tight cooperation with hardware designer and many other factors. In a simple drive system (locomotive is not the case) with sufficient number of the hardware interrupt resources the interrupt-only (no-background) or the interrupt-background system is often used [1]. In more complex systems the selected architecture may depend on the number of application tasks required. If the number of the user tasks is from tenths to hundreds (say max 255) the preemptive Real Time Operating System (RTOS) [2], [3] may be used as a reliable basic layer to schedule and execute application tasks and to support user inter-task and/or inter-node communication including synchronization [4], [5]. The overhead of RTOS may be unacceptable in the lowest control levels where speed requirements are high so instead of RTOS simple static executive is often used. As the number of tasks increases a RTOS overhead increases rapidly and the total throughput and time response of a RTOS can be unsatisfactory. In such design cases using of the co-routines (cooperating routines) may be a good solution [6], [7] with low overhead even if we use excessive amount of the application tasks [8], [9] but the system integrity may be more easily disturbed by an incorrect programmer action in comparison with standard RTOS.

## II. SYSTEM SPECIFICATION REQUIREMENTS

The specifications of the distributed control computer features of traction vehicle before control system structure design stage starts were as follows (in short):

- Control of six main drives,
- Control of five slave converters (fans, comps.),
- Control of pneumatic brakes,
- Control of all logical devices,
- Control of engineer stands,
- Run-time diagnostics,
- Post-mortem diagnostics,
- High reliability,
- Fault tolerant system design,
- TMR system where necessary,
- Easily maintainable system,
- Reduced computer unit types,
- Common uniform system software,
- Uniform hardware node core.

## III. SYSTEM STRUCTURE PARTITION

The design keystone is correct distributed system functional partition to set up no node processor throughput or communication bottleneck. First of all it is necessary to collect all system functional requirements (this task is often very difficult) and to analyze them very carefully. To define control computer system partition we can use following methodology.

To define DNCC system partition we have to select from pool of contractor system requirements group of all user tasks that are to be assigned to suitable DNCC nodes. Further we have to collect together the user tasks by application of criterial (threshold) functions that can run together in one node of the DNCC. Finally for the defined groups of tasks we have to design suitable hardware and decide which system software is to be used for scheduling the user tasks. Selection of groups is iterative process which we have to continue until no user task is unassigned

Let us define following symbols:

$pjrq_i$ – project function requirements,
$ThUT_i$ – user task selection threshold function,
$pjSW$ – project software,
$ss_i$ – system software support, RTOS etc.,
$ut_i$ – user task,
$utg_i$ – user task group,
$utgXX_i$ – user task group selected by $XX$ criterion,
$ThFR_i$ – task function requirement threshold func.,
$ThCM_i$ – inter-task communication threshold function,
$ThTP_i$ – CPU throughput threshold function,
$hwNode_i$ – hardware network node with CPU.

Then we have:

$$ut_j = ThUT_j(\sum_i pjrq_i) \tag{1}$$

$$pjSW = \sum_i ut_i + \sum_j ss_j \tag{2}$$

Further we define application function requirements (3), inter-task data flow rate (4) and each CPU necessary required (or estimated) throughput (5).

$$utgFR_j = \sum_i ThFR_j(ut_i) \tag{3}$$

$$utgCM_j = \sum_i ThCM_j(ut_i) \tag{4}$$

$$utgTP_j = \sum_i ThTP_j(ut_i) \tag{5}$$

Finally we get group of tasks $utg_j$ (6) to that proper hardware $hwNode_j$ (7) will be assigned.

$$utg_j = utgFR_j \cap utgCM_j \cap utgTP_j \tag{6}$$

$$hwNode_j \leftarrow utg_j \tag{7}$$

Another important system structure design issue is to enable to be control system easily scalable and reconfigurable with minimal number of unit types. This feature is supported by automatic identification of computer node position inside local area network and by extensive usage of a field programmable gate arrays inside special hardware blocks. The presented system main final parameters have been defined as follows (Fig. 1):

Generally:

- Three level distributed hierarchical system,
- Maximum functions assigned to software,
- Serial communication channels
- Baude-rate (unformatted) 1(2) Mbps,
- Spare communication channels,
- Master-slave bus access,
- Uniform node hardware core,
- Automatic node position identification,
- Inter-node precise PLL time synchronization,
- Fault tolerant system design,
- TMR system where necessary,
- Common uniform system software,
- Common uniform communication software,
- Simple user friendly application APIs,
- Remote source code debugging available,
- Special hw blocks reconfigurable (FPGA),
- FLASH file storage system,
- Execution from RAM,
- Preemptive RTOS where possible,
- Simple fast static executive when necessary,
- Extended run-time monitoring available,
- Extended post-mort diagnostics available,
- Minimized number of unit types,
- DMA support for precise time measurement,
- Extensive DMA support for communication,
- Parallel local mono-master expansion bus,
- Computer unit core full galvanic insulation.

Inter-node communication:

- Common RTOS communication module,
- Uniform communication APIs in all nodes,
- Message transfer method (mail system),
- Message transfer fully buffered in RTOS,
- 1 Mbps communication speed (unformatted),
- Min 128 kbps (16 kBps) user data throughput,
- Two message transfer priority levels,
- No communication coprocessor,
- DMA message transfer only,
- Low host processor overhead (< 6 %),
- CPU loading on interrupt is very critical,
- CPU loading on background is uncritical,
- Message cycle time max 2 ms,
- 2 master transmit queues,

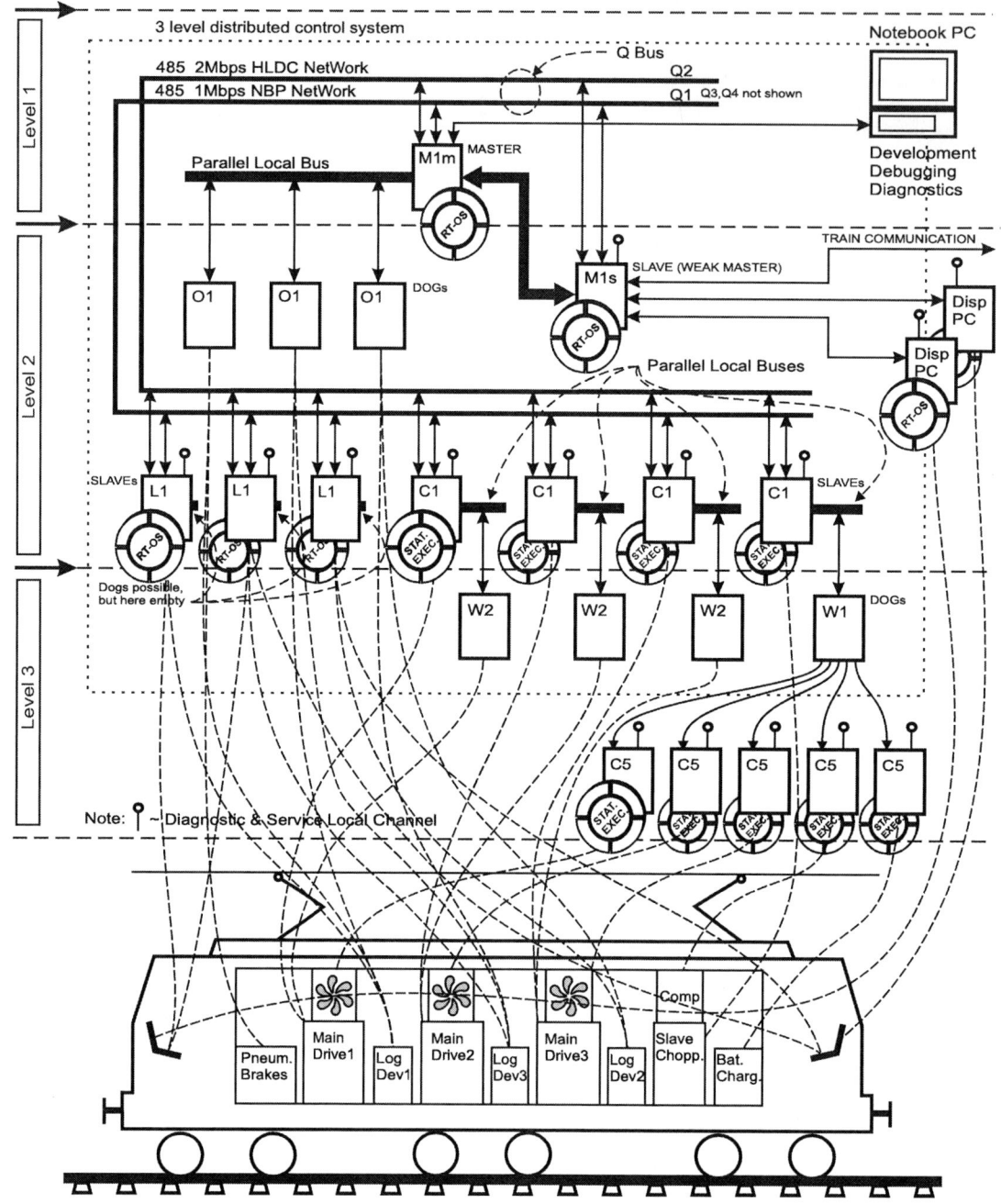

Fig. 1. Distributed network control computer – system structural view

- 15 master receiving queues,
- One slave receiving queue,
- One slave transmitting queue,
- Queue full/empty flags in all cases,
- Run-time error log available.

## IV. DNCC LAN SYSTEM ARCHITECTURE

DNCC of the electric locomotive (type 93E) (Fig. 1) is three level hierarchical system with serial communication channel between nodes. The nodes have local parallel busses to expand functionality of nodes easily. Levels one and two communicate through the Q-bus. Backbone communication channel of Q-bus is Q1 NBP protocol 1 Mbps (unformatted) serial bus with substantial transfer support of DMA controllers in both ends of transmission path. Q2 bus is spare and is in service when Q1 faults. Q3 bus supports mutual precise synchronization of control computer of the bogie converters. The precise synchronization is necessary to minimize the trolley current ripple of main drives.

## V. DNCC NODE CORE ARCHITECTURE

The each DNCC node type has the uniform host computer core (Fig. 2) including network communication section and a different specific hardware unit based on the target node function. The uniform hardware core facilitates implementation of system software including communication services. Fast multi-channel DMA controller, vectored interrupt controller and timers are integrated inside the core. Q1 bus utilizes fast (1 Mbps) NBP UARTs on both communication sides. UARTs have no hardware frame address detection so the address detection has to be implemented in software. The efficient multichannel DMA controller organizes virtual dual-port RAM area between CPU and UART [11]. The DMA controller cycle is not shadow type (no CPU bus cycle delay) it uses cycle stealing but it is very fast.

Let us define following symbols:

*tIRtst*    - interrupt service transfer time,

*tIRxxx*    - interrupt service xxx source overhead,

*tIRhpd*    - higher priority service max delay,

*tIRrql*    - interrupt request max time latency,

*tIRsct*    - save context time,

*tIRmdt*    - process and move data time,

*tIRrsct*   - restore context time,

*tIRirt*    - interrupt return time,

*tDMAxxx* - DMA service source overhead,

*tDMArql* - DMA request max time latency,

*tDMArdt* - read source data time,

*tDMAwr* - write data to destination time,

*tDMAtst* - DMA transfer time,

Then length of interrupt service to move data is:

$$tIRtst = \sum_i tIRxxx_i = \qquad (8)$$
$$tIRhpd + tIRrql + tIRsct + tIRmdt + tIRrct + tIRirt$$

and length of DMA service to move data is:

$$tDMAtst = \sum_i tDMAxxx_i = \qquad (9)$$
$$tDMArql + tDMArdt + tDMAwrt$$

For the used mid-range core CPU, relation between *tIRtst* (8) time and *tDMAtst (9)* time is:

$$tDMAtst \leq 70 \cdot tIRtst \qquad (10)$$

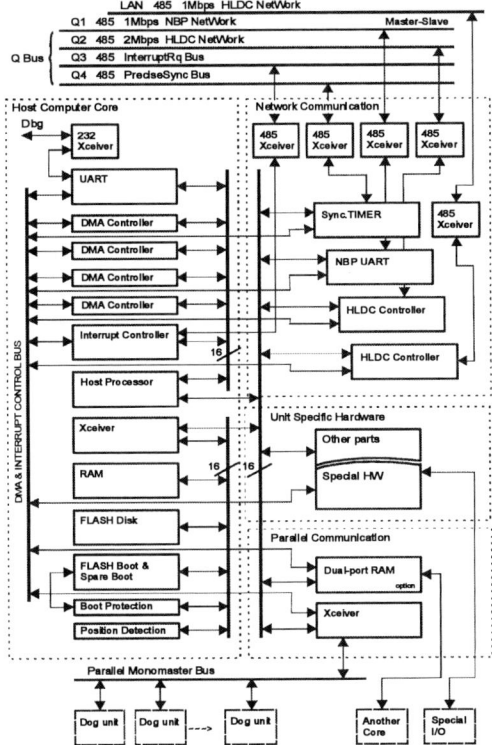

Fig. 2. Node uniform core structure

In the presented application and the communication speed 1 Mbps the received data byte has to be read every 10 μs. Time *tIRtst* ≈ 40 μs i.e. the interrupt system can not be used to process the incoming data. Time *tDMAtst* ≈ 600 ns and the received data are processed easily with unimportant influence on overall node throughput.

## VI. COMMUNICATION PROGRAMMER MODEL

The programmer system model of the communication service (mail service) (Fig. 3) is of the master/slave type and utilizes message queues on both sides. Main application important features of the mail service are as follows:

- No communication coprocessor,
- Network communication processed by host CPU,
- Message transfer method,
- Application messages fully buffered,
- Two levels of message priority,
- Up-to 512 kbps user data throughput i.e. 64 kBps,
- 1 Mbps communication speed (unformatted),
- Low application processor overhead – 3 % approx.,
- Message length configurable at initialization stage,
- Configurable message length - 4, 8, 16, 32 bytes,
- Mail system reconfigurable at run-time,
- Master-slave control, deterministic bus access,
- Up/down dual message cycle,

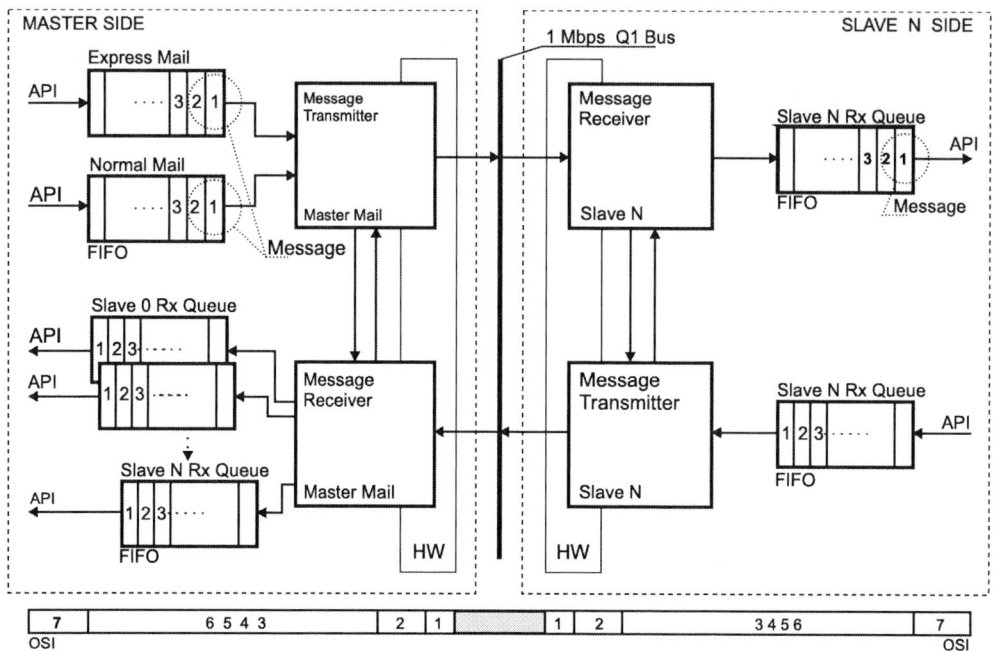

Fig. 3. Q1 bus 1 Mbps mail programmer model

- Message cycle time 1 ms,
- Max 15 slave nodes.

## VII. MAIL SOFTWARE ARCHITECTURE

The network communication utilizes two types of the data transfer cycles – Network Management Transfer (NMT) cycle and Process Data Transfer (PDT) cycle. The software design utilizes following resources of the node core: Master side – two DMA channels, NBP UART, timer, one interrupt level, and background. Slave side – two DMA channels, NBP UART, timer, four interrupt levels, background. The structure of PDT cycle, data flow and control actions are explained in Fig. 7 and Fig. 8. The details of NMT cycle timing see [11]. PDT cycle timing and mutual coordination of interrupts, DMAs, DMAs interrupts and background are depicted in Fig. 4.

The records of Q1 bus traffic from the logic analyzer show the frame header response time of the addressed slave ($T_{rt}$ = 30 µs) (Fig. 6) and one complete bus double message cycle (Fig. 7) (user data rate v ≈ 64 kBytes per sec, communication protocol overhead ≈ 49 %).

To hold transferred consistent and to make communication services robust the following design means are used. Bound dual NMT and PDT cycles are used. Address/control subframe from master is followed by user data subframe transmitted at first from slave and finally the user data subframe is sent from master. This method unburdens master from necessity to know precisely the slave interrupt latency which is highly dependent on the user interrupt disable discipline and which is difficult to predict. All user data messages are fully buffered. A fault of any slave node has no influence on communication with the other slaves. The reserves are

defined in safety time gaps. Correct PDT cycle transfer counters are used in each slave and in master side are processed independently for each slave. Tight transfer time supervision is used. Timeouts are checked in both communicating sides. Message block checksum is used in both transfer directions. A master quiet flag is used in slaves to be technology controlled by slaves set up to safe state in time. Synchronization by messages is supported by slave_x transmit queue empty flag. A current error log is available on the user request in both master/slave sides.

## VIII. CONCLUSION

The presented distributed network control computer with PDT cycle of 1 ms operates in 6 MW/3 kV$_{dc}$ electric locomotive. The network operation shows good behaviour and runs reliable with no apparent problems.

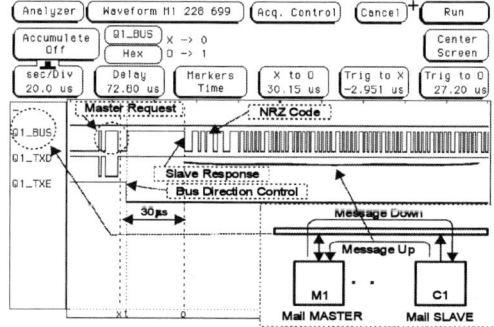

Fig. 6. Q1 bus 1 Mbps frame header slave response

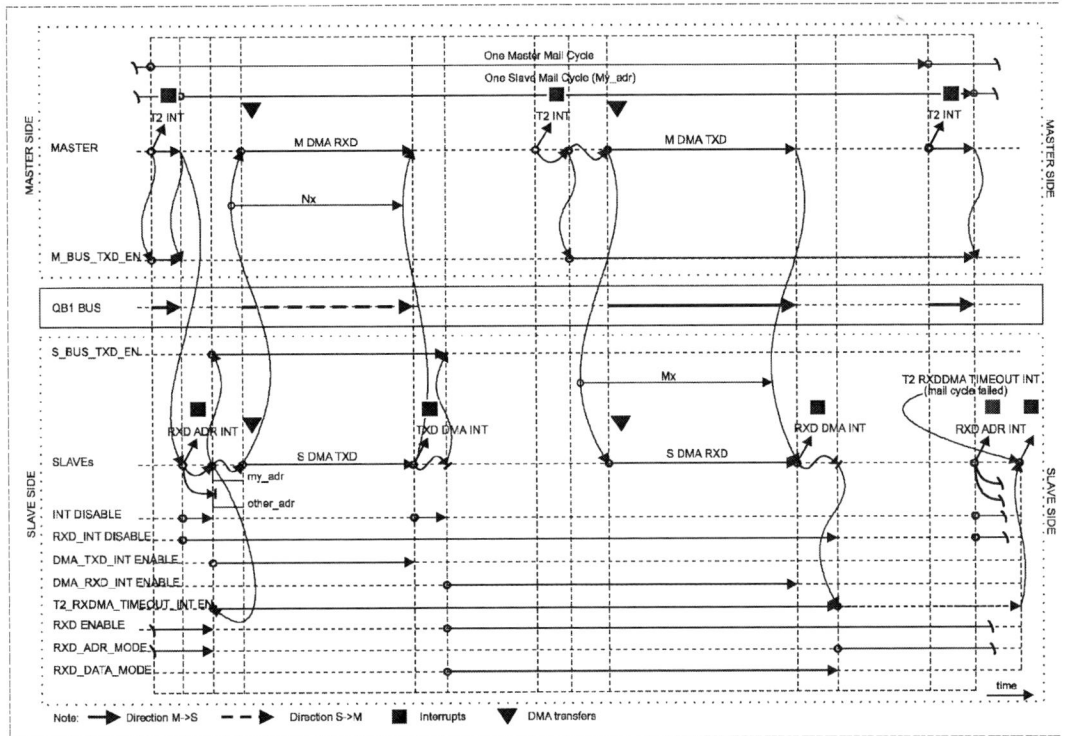

Fig. 4. Q1 bus 1 Mbps mail – PDT timing and master/slave cooperation

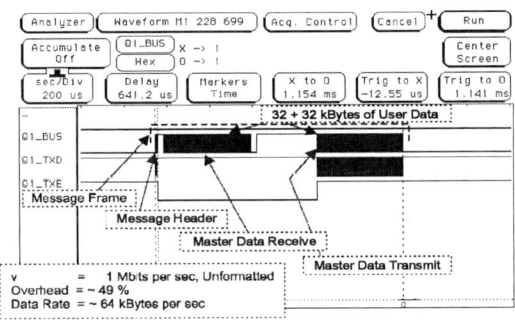

Fig. 5. Q1 bus 1 Mbps one PDT up/down message cycle

### ACKNOWLEDGMENT

The research was partly supported by the research program No. MSM6840770015 of the CTU in Prague sponsored by the MEYS of the Czech Republic.

### REFERENCES

[1] J. Lettl, S. Fligl, "Matrix Converter Control System" in Proceedings of International Symposium PIERS2005, Aug. 2005, Hangzou, China, pp. 395-398.

[2] H. Kopetz, *Real Time Systems: Scheduling, Design Principles for Distributed Embedded Applications*, Kluwer, 2003, ISBN-07923989947.

[3] A. M. K. Cheng, *Real-Time Systems: Scheduling, Analysis and Verification*, John Wiley, 2002. ISBN-0471184063.

[4] J. Zdenek, "Efficient DMA Based Local Computer Network Communication Services for Traction Vehicle." in Proceedings. of International Conference. SPRTS2005, October, 2005, Bologna, pp. 44-51.

[5] J. Zdenek, "Maximization Throughput of the Distributed Control Computer of Power System Using DMA and Optimal Partition", WSEAS Transactions on Power Systems, Issue 5, Vol. 1, May 2006, pp. 947-952

[6] J. Zdenek, "Efficient Scheduler-Dispatcher Software Architecture of the Space Power Facility Distributed Control Computer", in Proceedings of International Conference EPE2007, September 2007, Aalborg, CD ROM.

[7] J. Zdenek, "Control Electronics of Scientific Space Technological Facility." in Proceedings of 11th International Conference EPE-PEMC2004, September, 2004, Riga, CD ROM.

[8] J. Zdenek, L. Koucky, P. Mnuk, "Node for Drive Control of Scientific Equipment for High Temperature Material Processing in Space Station", in Proceedings of International Conference EDPE2003, September 2003, High Tatras, Slovak Republic, pp. 201-206.

[9] J. Zdenek, L. Koucky, P. Mnuk, "Network Node for Temperature Control of Scientific Equipment for High Temperature Material Processing in Space Station", in Proceedings of IFAC ws. PDS2003, January 2003, Ostrava, pp. 291-294

[10] J. Zdenek, "System Design and Software Architecture of Traction Vehicle Control Computer." in Proceedings of 12th International. Conf. EPE-PEMC2006, August. 2006, Portoroz, pp. 1205-1210.

[11] J. Zdenek, "Minimizing of the Communication Overhead in Distributed Control Computer of Power System Using Correct Partition and DMA", in Proceedings of IASME/WSEAS International Conference Energy and Environment 2006, May 2006, Chalkida, pp. 1790-1795.

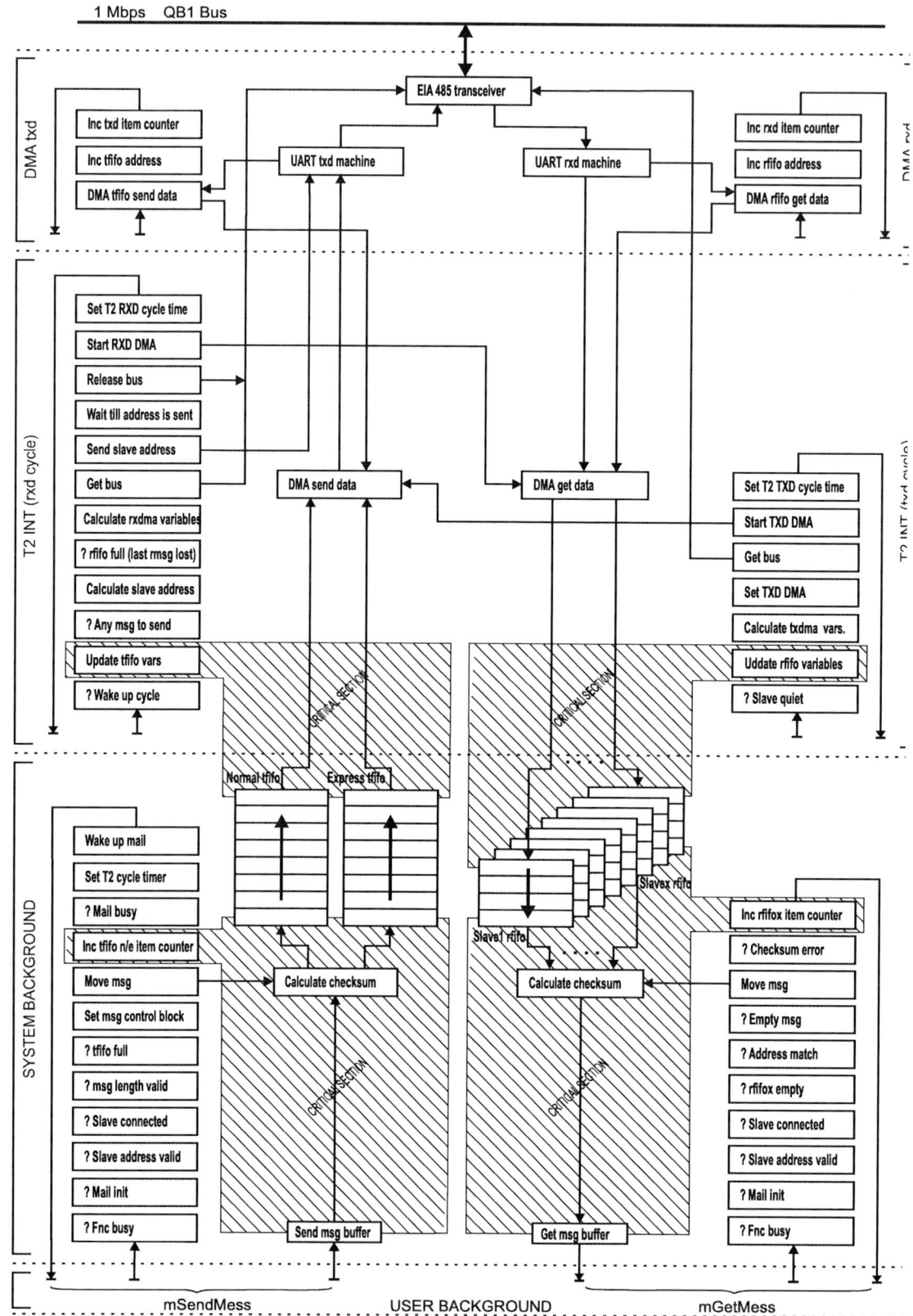

Fig. 7. Q1 bus 1 Mbps master mail – NMT architecture

1 Mbps    QB1 Bus

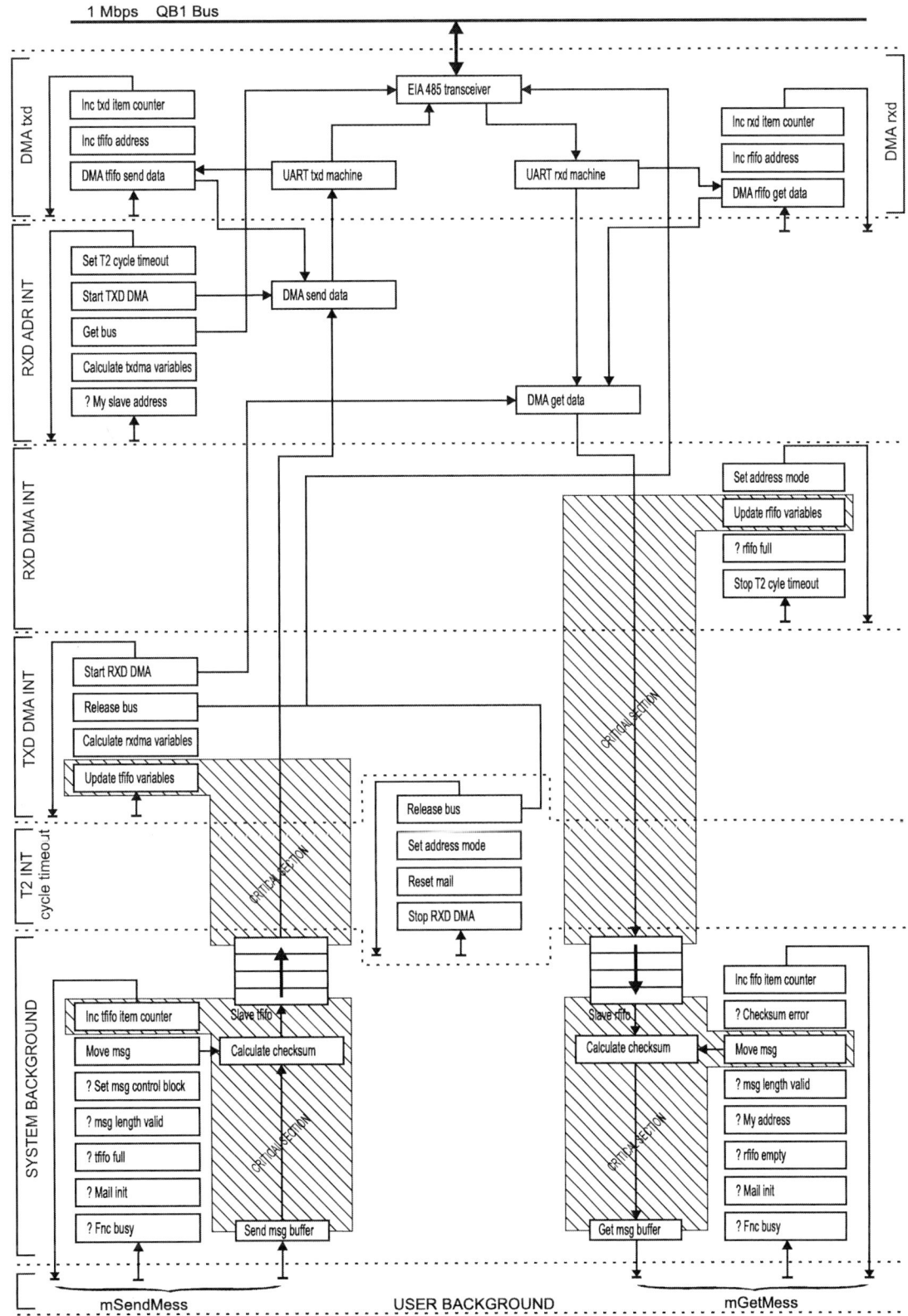

Fig. 8. Q1 bus 1 Mbps slave mail – NMT architecture

# Analysis and Position Control of a Linear Switched Reluctance Actuator Based on Sliding Mode Control

António Espírito Santo\*, Maria R. A. Calado[†], Carlos M. P. Cabrita\*\*

\*University of Beira Interior\Department of Electromechanical Engineering, Covilhã, Portugal, email: aes@ubi.pt
[†]University of Beira Interior\Department of Electromechanical Engineering, Covilhã, Portugal, email: rc@ubi.pt
\*\*University of Beira Interior\Department of Electromechanical Engineering, Covilhã, Portugal, email: cabrita@ubi.pt

*Abstract* — **This paper presents the finite elements analysis and the control of a linear switched reluctance actuator (LSRA). The energy conversion process is described, and based on a finite elements analysis, both traction and attraction maps are derived. This information is useful to understand the device working principle and its performance for different excitation and position conditions. The proposed control strategy, based on sliding mode control techniques, and the electronics used to proper drive the actuator are described and implemented with the TMS320F2812 eZdsp Start Kit. Control implementation takes advantage of the microprocessor internal peripherals and resources. Finally, experimental results are presented and discussed for the validation of proposed control strategy.**

*Keywords* – **Actuator, Control of Drive, Switched reluctance drive.**

## I. – INTRODUCTION TO LSRA

Linear Switched Reluctance Actuators (LSRA) exhibit high force/weight ratio, without permanent magnets, and a very low structural complexity, leading to low fabrication cost and high operation reliability, turning it suitable for precision robotics application [1],[2],[3]. The major problem concerns to develop and implement an adequate control strategy that provides the required actuator operation. A linear bidirectional actuator motion is obtained if all of the three phases are turned on and off sequentially. An electromagnetic device can convert electrical energy into mechanical energy, or vice-versa. This process is made through the device magnetic field.

The design of electromechanical devices requires the prediction of developed force, often derived from field solutions obtained through numerical analysis [4],[5],[6] based on different approaches, such as the classical virtual work; the Maxwell stress tensor; and the Coulomb's virtual work.

Energy is a state function for a conservative system. If losses are ignored, balance energy can be written as in (1), where $W_e$ is the system input energy, $W_{fe}$ the field storage, and $f_{em}$ the mechanical force that produces the mechanical work $dW_{em}$, when differential displacement $dx$ occurs. Expression (1) also shows that system energy, in a lossless device, depends on flux linkage $\lambda$ and position $x$.

$$dW_e = dW_{fe} + f_{em}dx \Leftrightarrow dW_{fe}(\lambda, x) = id\lambda - f_{em}dx. \quad (1)$$

A different energy entity, defined as co-energy $W_{fe}'$, without a real physical meaning, can be expressed by (2).

$$W_{fe}'(i, x) = i\lambda - W_{fe} \quad (2)$$

Thus, as can be seen, if a change in the linkage flux occurs, the system energy will also change. This variation can be promoted by means of a variation in excitation, a mechanical displacement, or both. The coupling field can be understood as a reservoir of energy, that receives it from the input system, in this case the electrical system, and delivers it to the output system, in this case the mechanical system.

After mathematical manipulation of (2), considering (1), (3) is obtained. This expression shows that co-energy $W_{fe}'$ depends on current $i$ and position $x$.

$$dW_{fe}'(i, x) = \frac{\partial W_{fe}'(i, x)}{\partial i}di + \frac{\partial W_{fe}'(i, x)}{\partial x}dx. \quad (3)$$

Because $i$ and $x$ are independent variables, the coefficients in equation (3) are given by (4).

$$\begin{cases} \lambda = \dfrac{\partial W_{fe}'(i, x)}{\partial i} \Rightarrow L = \dfrac{\lambda}{i} \, . \\[3mm] f = \dfrac{\partial W_{fe}'(i, x)}{\partial x} \end{cases} \quad (4)$$

The physical structure of the developed actuator is presented in Fig. 1, with dimensions listed in Table I. Each phase coil has 1100 turns with 10 Ω.

The here under control actuator presents a longitudinal configuration. A magnetic yoke and respective coil constitute each actuator phase, being magnetically independent from the others. This characteristic leads to two major advantages: the flux is completely independent for each phase. This is an important aspect because allows the simultaneous activation of two or more phases without loss of performance due to saturation; and the increasing in the number of phases is possible, increasing also the motion smoothness.

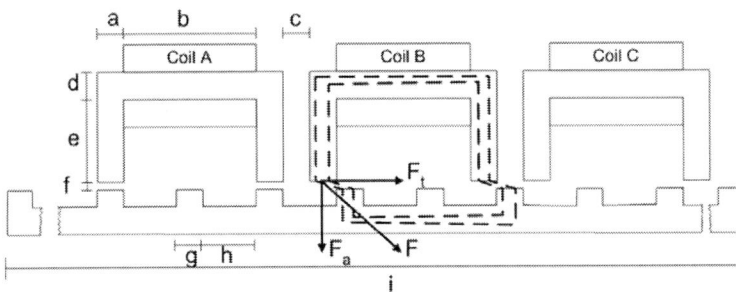

Fig. 1. LSRA physical dimensions

TABLE I.

LSRA PHYSICAL DIMENSIONS [MM]

| | |
|---|---|
| Yoke pole width (a) | 10 |
| Coil length (b) | 50 |
| Space between phases (c) | 10 |
| Yoke Thickness (d) | 10 |
| Yoke pole depth (e) | 30 |
| Air gap length (f) | 0.66 |
| Stator pole width (g) | 10 |
| Stator slot width (h) | 20 |
| LSRA length (i) | 2000 |
| LSRA stack width | 50 |

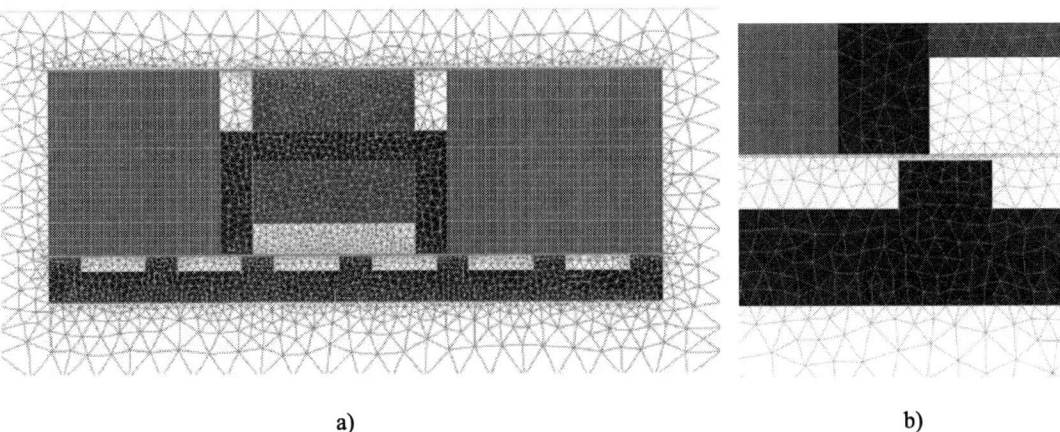

a)                                            b)

Fig. 2. Finite elements model of a single-phase actuator: a) global view and b) polar region detail

## II. – FINITE ELEMENT ANALISYS

A Finite Elements Model (FEM) of a single-phase actuator was constructed using FLUX2D [7]. FEM construction starts with geometric model definition, where each specific region is defined through points and lines.

The FEM constructed for the analysis of a single-phase actuator can be observed in Fig. 2. One of the finite element software features is the translation displacement possibility, allowing the longitudinal displacement of one, or a group, of regions, without the need of redefining the geometry of the model and respective finite element mesh. Two regions (magenta) are defined between the group of regions that must be moved in the longitudinal direction during static simulations. These regions are the actuator's primary (blue); the phase coil that carries current in the positive (inner region) and negative (outer region) direction (both in red), and respective surrounding air (white). Translation function demands also the definition of two narrow air-gap regions (yellow), located in both sides of the translations regions, and used for the displacement of the regions defined between them.

All regions use triangular elements in the finite elements mesh creation. The exceptions are the two displacement regions, that use quadrangular elements. Mesh creation must also observe that only a single layer

of elements can be established in the translations air-gap regions.

Primary and secondary regions are associated to materials with the magnetic characteristics identical to those of the ST-37 steel. All other regions inherit the vacuum magnetic characteristics. Inside the coil regions, a current source is defined with a positive value for the region that carries the current in the positive direction and an identical negative value for the region that carries the current in the negative direction. Dirichlet conditions are formulated in the model boundary, imposing a null flux across it.

Several simulations were performed for a set of primary positions and currents. Obtained phase traction ($F_t$) and attraction ($F_a$) maps are presented in Fig. 3 e Fig. 4, respectively.

As can be observed, when actuator poles are aligned with the secondary teeth traction force is zero. This same situation occurs also at non-aligned positions. From the analysis of the attraction map is possible to conclude that maximum attraction force is obtained at the aligned position. This force can be minimized introducing a different polar shape as described in [8].

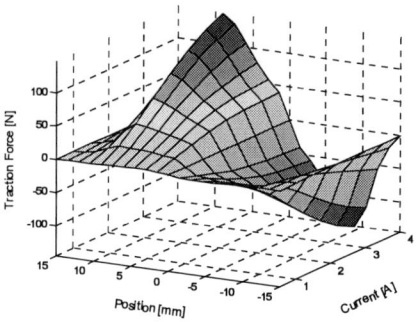

Fig. 3. Map of traction force as a function of position and current

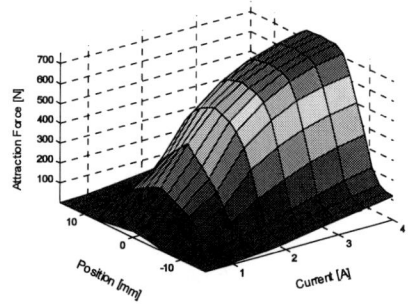

Fig. 4. Map of attraction force as a function of position and current

## III. – SLIDING MODE CONTROLLER STRUTURE

The LSRA electromagnetic model to n = {1,2,3} phases is described by (5), where $R_n$ is the coil's resistance, $v_n$ the supplied voltage and $i_n$ the phase current. Mathematical expression (6) describes actuator mechanical behavior where $a$ is the device acceleration and $M$ the mass. Actuator non-linearity is taken into account because both inductance $L(i,x)$ and traction force $F(i,x)$ change not only with position $x$, but also with current $i$.

$$
\begin{aligned}
0 &= R_n i_n(t) + \frac{dL(i_n,x)i_n(t)}{dt} - v_n \\
&= \frac{di_n}{dt}\left[L(i_n,x) + i_n(t)\frac{\delta L(i_n,x)}{\delta i}\right] \\
&\quad + \frac{dx}{dt}\left[\frac{\delta L(i_n,x)}{\delta x}\right] \\
&\quad + R_n i_n(t) - v_n
\end{aligned}
\tag{5}
$$

$$
a = \frac{F(x,t)}{M} \tag{6}
$$

The state space is defined as:

$$
\begin{cases}
\dfrac{dx}{dt} = y \\[2mm]
\dfrac{dy}{dt} = \dfrac{F}{M} \\[2mm]
\dfrac{di_n}{dt} = \dfrac{y\beta_n + R_n i_n - v_n}{\alpha_n}
\end{cases}
\tag{7}
$$

where $\alpha_n$ and $\beta_n$ stands for:

$$
\begin{cases}
\alpha_n = L(i_n,x) + i_n\dfrac{\delta L(i_n,x)}{\delta i} \\[3mm]
\beta_n = i_n\dfrac{\delta L(i_n,x)}{\delta x}
\end{cases}
\tag{8}
$$

Variable Structure Systems (VSS) belong to a particular case of automatic control systems [9],[10]. Intentional commutation is introduced between two different control actions, in one or more channels of control inputs.

$$
\dot{X} = A(X,t) + B(X,t)u . \tag{9}
$$

The possible type of motion of VSS in the state space is manifested by the appearance of the sliding mode regime. This kind of control was successively applied to the rotational switching reluctance motors. To achieve sliding motion regime, a switching surface is defined by $s(X,t) = 0$, the control structure $u(X,t)$ stated in (10) changes from one structure to another. When $s(X,t) > 0$ the variable structure control is changed in order to decrease $s(X,t)$. The same kind of action is taken when $s(X,t) < 0$. The control main goal is to keep the system state space sliding in the surface $S(X,t) = 0$. During this sliding motion, the system behaves like a reduced-order system, being insensitive to disturbances and parameters changes. Control structure can be expressed by:

$$
u(X,t) = \begin{cases}
u^+(X,t) & \Leftarrow \quad s(X) > 0 \\
u^-(X,t) & \Leftarrow \quad s(X) < 0
\end{cases} . \tag{10}
$$

The previously explained concept is used to develop the LSRA position structure control. At a specific moment, it is assumed that traction force can be developed in both directions with correct actuator phase choice. After turning off, an actuator phase still has the ability to produce traction force. This situation occurs because current phase do not goes down instantaneously, but diminishes by the free wheel diodes path, with a time constant that depends on phase inductance and resistance. This behaviour is responsible for the introduction of a delay in controller response, contributing to increase the oscillations (chattering) around the sliding surface.

Actuator's movement can be expressed as an VSS with two possible control actions. A control action produces a traction force in the left direction ($F_l$), while the other control action produces a traction force in right direction ($F_r$).

$$
\begin{cases}
\dot{x} = y \\[2mm]
\dot{y} = \dfrac{u}{M}
\end{cases} , \tag{11}
$$

The control law $u(t)$ is established as

$$u(t) = \begin{cases} F_r & \Leftarrow & s(e,\dot{e}) > 0 \\ F_l & \Leftarrow & s(e,\dot{e}) < 0 \end{cases}, \qquad (12)$$

The commutation function $s(e,\dot{e})$ depends on position error $e$ and derivative of the position error $\dot{e}$, and is defined by

$$s(e,\dot{e}) = me + \dot{e}, \qquad (13)$$

where $m$ is a positive constant, experimentally obtained.

The controller selects from the lookup Table II the phase that will provide the desired control action, in order to maintain $s(e,\dot{e}) = 0$. The aligned position for phase A is taken as the reference.

TABLE II.– RELATIVE ACTUATOR POSITION [MM]

| Traction Force | [0,10[ | [10,20[ | [20,30[ |
|---|---|---|---|
| Left direction ($F_l$) | Phase A | Phase B | Phase C |
| Right direction ($F_r$) | Phase B | Phase C | Phase A |

## IV. - DRIVING ELECTRONIC

Power converter topology is shown in Fig. 5. It can be divided in three main blocks: the Microprocessor Control System; the Regulation Electronics, and the Power Electronics. The TMS320F2812 firmware drive, for each actuator phase, the Regulation Electronics signal lines (PWM, $T_1$, $T_2$). The PWM signal is used to control the wave shape of the phase current in conjunction with the hysteretic controller. Signals $T_1$ and $T_2$ are used to establish the moments when to turn on/off the power transistors $Q_1$ and $Q_2$, respectively.

A comparator circuit is used to implement the Hysteresis Controller. The choice of this option is justified by simplicity reasons, and because allows power transistor operation in saturation state when conducting. This way, power transistor losses are reduced. As precaution RCD snubber circuits are associated to the power transistors in order to protect them.

## V. – EXPERIMENTAL RESULTS

Based on the theoretical obtained results, an experimental setup was built (Fig. 6). This setup possesses a mobile platform that supports the primary of the actuator. The tests to the evaluation of static traction forces were performed leaning the primary platform to a second one, that can be blocked and is equipped with a load cell. Data collected with this experimental setup allowed to verify FEM analysis results.

The TMS320F2812 eZdsp Start Kit was used to implement and apply the proposed control strategy, based on sliding mode control, to the LSRA [11]. Event Manager EVA was used to generate the PWM signal dedicated to each phase current. These analogue signals were acquired by the on-chip ADC. Actuator position is the information feedback to TMS320F2812 obtained from an incremental encoder, being the velocity and position derived from the QEP unit.

The sliding mode controller establishes the switching strategy, used to turn on and off the LSRA phases. Using Microcontroller GPIO, each phase signal lines $T_1$ and $T_2$ are properly switched. Regulation Electronics, based in the current signal reference, switches $Q_2$ to maintain the desired current profile. The most important software blocks are represented at Fig. 7.

Experimental results obtained for several small displacements from a starting position corresponding to the alignment of phase-C poles with secondary teeth, can be observed in Fig. 8. Chattering introduced by the sliding mode control can be observed at steady state.

The phase portrait trajectory along the sliding line to the origin can be observed in Fig. 9, for a specific case corresponding to the 25 [mm] displacement.

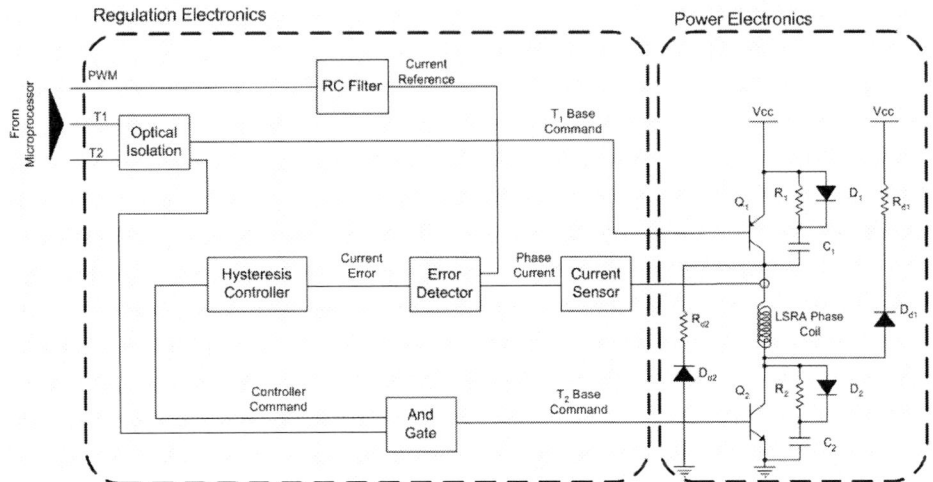

Fig. 5. Power converter general topology for one phase

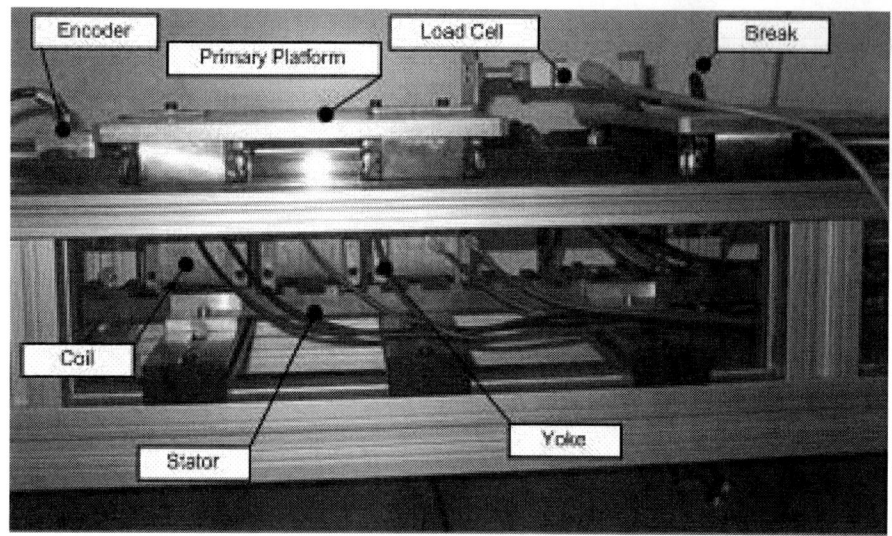

Fig. 6. Actuator experimental setup

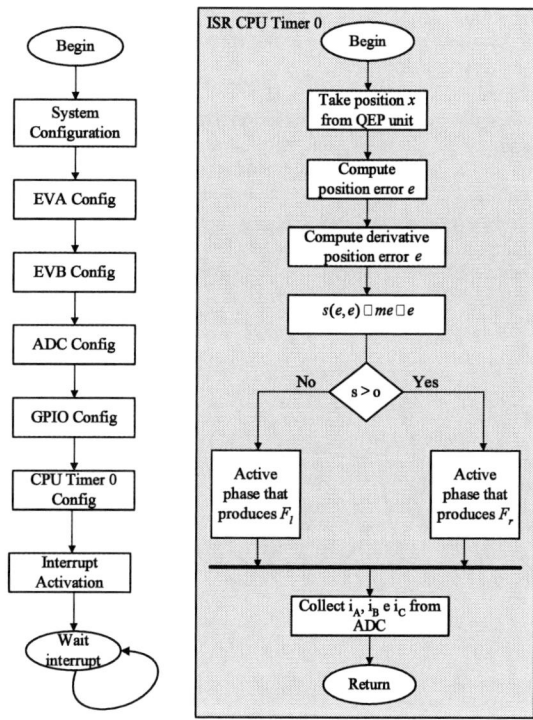

Fig. 7. TMS320F2812 eZdsp Start Kit code flow

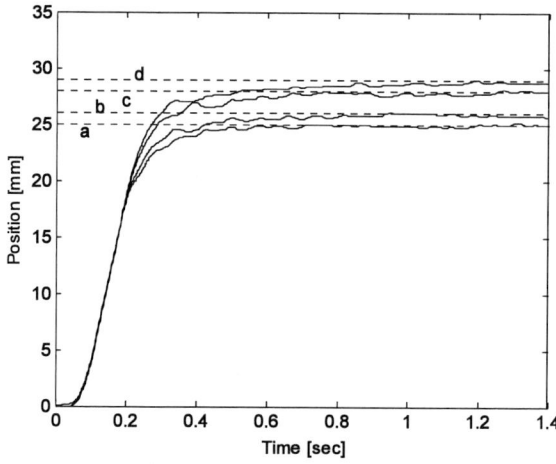

Fig. 8. Actuator position for small displacements: a) 25 [mm],
d) 26 [mm], c) 28 [mm], d) 29 [mm]

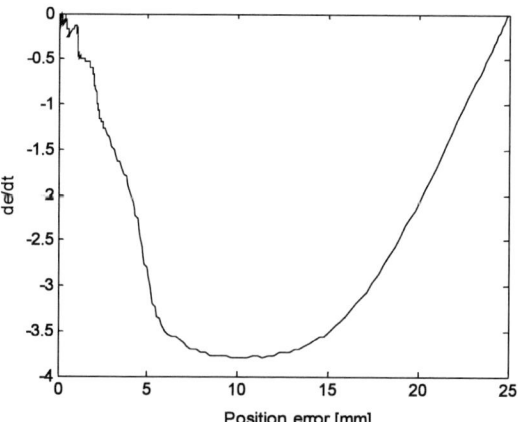

Fig. 9. Phase portrait for a 25 [mm] displacement

## VI – CONCLUSIONS

The finite elements analysis allowed to understand the working principle of a three phase Linear Switched Reluctance Actuator developed for robotics applications. Using this tool, traction and attraction map was obtained. This information allows to characterize the actuator behaviour.

Based on the obtained results a prototype was constructed with correspondent power and regulation electronics.

The establishment of proper position control for a high performance device was achieved through a proposed strategy based on sliding mode control. That task was performed by implementing the developed methodology on a TMS320F2812 eZdsp Start Kit, taking advantage from their built-in peripherals. Experimental results allowed to conclude that actuator can realize displacements with 1 [mm] resolution.

### REFERENCES

[1] J. Corda, E. Skopljak, "Linear Switched Reluctance Actuator," in *Proceedings of the Sixth International Conference on Electrical Machines and Drives*, pp.535-539, September 1993.

[2] Chuen Gan, Norbert C. Cheung, Li Qiu, "Position Control of Linear Switched Reluctance," *IEEE Transactions on Industry Applications*, vol. 39, no. 5, pp. 1350-1362, September/October 2003.

[3] A. Espírito Santo, M. R. A. Calado, C. M. P. Cabrita, "Variable Reluctance Linear Actuator Dynamics Analysis Based on Co-energy Maps for Control Optimization," in *proceedings of Linear Drives for Industry Application*, 2005.

[4] S. McFee, D. A. Lowther, "Towards Accurate and Consistent Force Calculation in Finite Elements Based Computational Magnetostatics," *IEEE Transactions on Magnetics*, vol. 23, no. 5, pp. 3771-3773, November 1987.

[5] W. Muller, "Comparison of Different Methods of Force Calculation," *IEEE Transactions on Magnetics*, vol. 26, issue 2, pp. 1058-1061, 1990.

[6] Coulomb J. L., "A methodology for the determination of global electromechanical quantities for a finite element analysis and its application to the evaluation of magnetic forces, torques and stiffness," *IEEE Transactions on Magnetics*, vol. 19, issue 6, pp. 2514-2519, 1993.

[7] Flux2D Tutorial (Version 7.40), *Cedrat*, August, 1999.

[8] A. Espírito Santo, M. R. A. Calado, C. M. P. Cabrita, "Influence of Pole Shape in Linear Switch Reluctance Actuator Performance," in *proceeding of Linear Drives for Industry Application*, 2007.

[9] K. David Young, Vadim I. Utkin, Umit Ozguner, "A Control Engineer's Guide to Sliding Mode Control," *IEEE Transactions on Control Systems*, vol. 7, pp. 328, May 1999.

[10] John Y. Hung, Weibing Gao, James C. Hung, "Variable Structure Control: A Survey," *IEEE Transactions on Industrial Electronics*, Vol. 40, No 1, pp.2-22, February 1993.

[11] TMS320C2000 Teaching ROM; Literature n°: SSQC011.

# Development and Control for a Reaction Wheel System Driven by Permanent Magnet Synchronous Motor

Ming-Chang Chou [1], Chang-Ming Liaw [1], Sywe-Bin Chien [2], Fa-Hwa Shieh [2], Jih-Run Tsai [3], Hao-Chi Chang [3]

[1] National Tsing Hua University/Electrical Engineering, Hsinchu, Taiwan, ROC., e-mail: d947902@oz.nthu.edu.tw
[2] Chung-Shan Institute of Science & Technology, Taoyuan, Taiwan, ROC., e-mail: sywebin@gmail.com
[3] National SPace Organization, Hsinchu, Taiwan, ROC., e-mail: chang297@nspo.org.tw

*Abstract*—This paper presents the robust current and torque controls for a satellite reaction wheel driven by permanent magnet synchronous motor (PMSM). First, the motor nominal key parameters are estimated, and the DSP-based motor drive is established. Then a winding current control scheme is proposed, wherein a traditional two-degrees-of-freedom controller (2DOFC) is augmented by an internal model feedback controller or a robust tracking error cancellation controller. Good sine-wave current tracking and back electromotive force (EMF) rejection performances are obtained. The similar robust control approach is also applied to perform the motor observed torque control with quick and close tracking response.

*Keywords*—Permanent magnet motor, synchronous motor, variable speed drive, brushless drive, robust control, modeling, PWM.

## I. INTRODUCTION

Reaction wheel is a key component used to control satellite attitude [1-4]. By applying suitable control torque to the motor shaft, the wheel reaction momentum can change the satellite axial position accurately. The permanent magnet synchronous motor (PMSM) is perhaps the most suitable actuator in this application owing to its many advantages, particularly for the interior PMSM (IPMSM) [5]. However, for yielding satisfactory driving performance, many affairs should be properly treated: (i) motor modeling and parameter estimation [6]; (ii) winding current tracking control [7,8]; (iii) sophisticated speed control [9]; and (iv) proper tuning of key parameters [7]. For a sine PMSM drive, the close sinewave varying frequency current tracking control under the existence of sinewave back-EMF and parameter variations is not a trivial issue. Internal model feedback control with resonant terms is known as an effective approach to handle the sinewave command tracking and disturbance rejection control problems [10-13]. The applications of internal model principle in the fixed-frequency sinewave current tracking controls of active filters can be found in [10,11]. In [12], the torque ripples of an AC permanent-magnet motor are modeled as a sinusoidal function with a frequency depending on the motor speed. And a gain scheduled robust 2DOF speed regulator based on internal model principle is developed to eliminate the torque and speed ripples and achieve the desired tracking response. The internal model principle has also been extended to perform non-sinusoidal PMSM current tracking control

[13]. However, the fast and precise sinewave winding current tracking control with changing frequency for a PMSM drive is still not easy to achieve via internal model control only. Hence in this research, the internal model control is augmented by command feedforward control to improve its dynamic response behavior.

On the other hand, there were also many researches concerning direct torque control (DTC) based on space vector PWM for PMSMs [14,15]. The concept of DTC for a PMSM is similar to those of induction motor. However some problems exist dedicatedly for PMSM, in particular, the existence of higher current and ripple torque. This problem is mainly due to the low control flexibility of zero voltage vectors.

In this paper, a DSP-based PMSM driven satellite reaction wheel is first established, and the key motor parameters are estimated. Then the sophisticated controls for PMSM driven reaction wheel to yield better current and torque dynamic responses are presented. In the inner current loop, the basic current 2DOFC consisting of a PI feedback controller and a command feedforward controller is augmented with an internal model controller or a robust tracking error cancellation controller. Very fast and close sinusoidal current waveform command tracking with good sinewave back-EMF rejection capability is obtained. As to outer torque loop, a feedback controller and a robust tracking error cancellation controller are employed to obtain fast observed torque tracking control performance.

## II. SYSTEM CONFIGURATION AND PROBLEM STATEMENT

### A. System Configuration

The developed PMSM driven reaction wheel system is shown in Fig. 1. It mainly consists of a photovoltaic cell set DC source with $V_{dc} = 22 \sim 38V$ (a DC source is used as a substitute); an inverter established using off-the-shelf intelligent power module (IR: IRAMS10UP60, 600V/10A (rms)); a PMSM (Kollmogen motor RBE 01511-C), which is rated as three-phase Y-connected, 12 poles, 131W, 5400rpm; a reaction wheel (J = 0.0217 $kg \cdot m^2$, speed $\leq 2000 rpm$, torque $\leq \pm 0.2 Nm$); and a DSP control board using TMS320F2812 manufactured by Texas Instruments. Hence the wheel drive system is operated in the speed range of $0 \leq \omega_r \leq 2000 rpm$ ($0 \leq f_1 \leq 200 Hz$, $f_1$ = inverter output frequency). And in making the proposed

978-1-4244-1741-4/08/$25.00 ©2008 IEEE

control scheme design, the nominal condition is chosen at $(V_{dc}=30V, 1000rpm, f_1=100Hz)$. The sampling rates are chosen to be $15kHz$ for both current and torque loops. All digital control algorithms are found using redesign approach with bilinear model transformation method.

### B. Problem Statement

As generally recognized, the close winding current tracking control for a sinewave type PMSM is a challenging issue. The affecting factors include: varied DC-link voltage, variations of motor parameters, sinusoidally varying current command, and the existence of speed-dependent sinewave back-EMF disturbance, etc. To overcome this difficulty, the simple robust current and torque controls for a PMSM driven reaction wheel are studied in this paper. In the proposed current controlled PWM scheme, the conventional 2DOFC is augmented with an internal feedback controller or a robust tracking error cancellation controller. Both transient and static current tracking characteristics under varying operating conditions and the effects of back-EMF can be obtained. As to the proposed robust torque control scheme, a torque estimate is obtained via a simple observer. And the PI feedback controller is added by a simple robust controller to yield quick and precise torque tracking control.

### III. MOTOR MODELING AND PARAMETERS ESTIMATION

#### A. Governing Model Equaions

The a-phase winding voltage equation of a PMSM can be written as [16]:

$$v_{as}=R_s i_{as}+\frac{d}{dt}(L_{asas}i_{as}+L_{asbs}i_{bs}+L_{ascs}i_{cs}+\lambda_m^{'r}\sin\theta_r) \qquad (1)$$

$$L_{asas}=L_{ls}+L_A+L_B\cos 2\theta_r \qquad (2)$$

$$L_{asbs}=-\frac{1}{2}L_A+L_B\cos 2(\theta_r-\frac{\pi}{3}) \qquad (3)$$

$$L_{ascs}=-\frac{1}{2}L_A+L_B\cos 2(\theta_r+\frac{\pi}{3}) \qquad (4)$$

where $v_{as}$= winding terminal voltage, $R_s$= resistance, $\lambda_m^{'r}=$

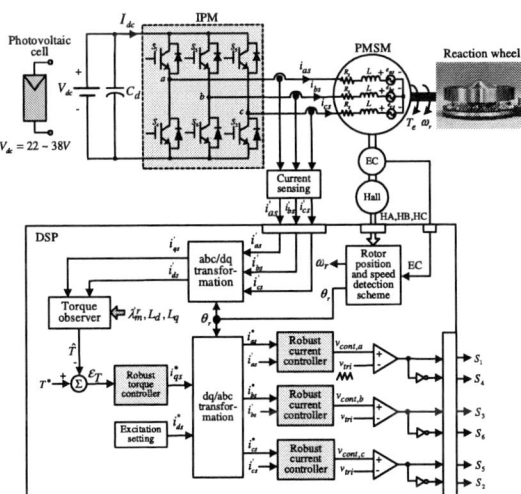

Fig. 1. The established PMSM driven reaction wheel system.

amplitude of flux linkage established by rotor permanent magnets, $\theta_r$ = rotor angular position, $L_{ls}$ =leakage inductance, and the expressions of $L_A$ and $L_B$ can be found in [16]. And the developed torque of a PMSM can be expressed as:

$$\begin{aligned}T&=\frac{3}{2}\frac{P}{2}(\lambda_m^{'r}i_{qs}+(L_d-L_q)i_{ds}i_{qs})\\&=\frac{3}{2}\frac{P}{2}[\lambda_m^{'r}i_{as}\cos\beta+\frac{L_q-L_d}{2}i_{as}^2\sin 2\beta]\end{aligned} \qquad (5)$$

where the q-axis inductance $L_q$ and d-axis inductance $L_d$ are defined as:

$$L_q=L_{ls}+\frac{3}{2}(L_A+L_B), \quad L_d=L_{ls}+\frac{3}{2}(L_A-L_B) \qquad (6)$$

In (5), the variable $\beta$ denotes the shift angle between a-phase current peak and the q-axis, which is kept in phase with the back-EMF and is regarded as a reference in making vector control of a PMSM.

### B. Model Parameter Estimation

The parameters estimated in this paper include: (a) PM flux linkage $\lambda_m^{'r}$ and speed constant $k_t$, where the rms value of the back-EMF is expressed as $E_d=k_t\omega_r$, $\omega_r=d\theta_r/dt$; (b) Equivalent circuit parameters, from the measured line-to-line winding inductance $L_{ab}(\theta_r)$ at $f=100Hz$, $L_d$ and $L_q$ are found as:

$$L_q=1/2(L_{ab,\max})=328.415\,\mu H \qquad (7)$$

$$L_d=1/2(L_{ab,\min})=214.635\,\mu H \qquad (8)$$

The average per-phase winding resistance at $f=100Hz$ is also found to be $\bar{R}_s=0.60625\Omega$.

### IV. CONTROL SCHEME

The proposed per-phase current and torque control schemes are shown in Figs. 2(a) to 2(d). The simple and practical control methods are developed to both loops to enhance their tracking control closeness and robustness against variations of source voltage, operating conditions and system parameters.

### A. Current Control Scheme

For simplicity, the ramp-comparison current-controlled PWM (RC-CCPWM) scheme is adopted in this PMSM drive. Generally, the current-controlled PWM scheme of a three-phase motor can be performed in abc-domain or dq-domain. The latter is easier to handle owing to the manipulation of DC quantities. However, the excellent and robust winding current tracking performance is still difficult to achieve. And moreover, the more frame transformation is required. As to the current control in abc-domain, it is more direct in the winding current PWM switching control action. Hence it is adopted in the established PMSM drive.

(a) Dynamic Modeling

Supposed that the PMSM armature windings are excited by three-phase sinusoidal balanced currents, then from (1)-(4) one can rearrange and yield the following simplified a-phase winding voltage equation:

$$v_{as} = R_s i_{as} + [L_{ls} + \frac{3}{2}(L_A + L_B)]\frac{d}{dt} i_{as} + \frac{d}{dt}(\lambda_m^{r} \sin\theta_r) \qquad (9)$$
$$\overset{\Delta}{=} R_s i_{as} + L\frac{d}{dt} i_{as} + e_{as}$$

where $L = L_q$ and $e_d = e_{as}$ denotes the phase back-EMF:

$$e_d = e_{as} = \omega_r \lambda_m^{r} \cos\theta_r \overset{\Delta}{=} \hat{E}_d \cos\theta_r \qquad (10)$$

According to (9), the per-phase motor dynamic model for current control loop is depicted in Fig. 2(a), where the subscript a is neglected, $K_{PWM}$ = PWM scheme transfer ratio and $K_s$ = current sensing factor. In the developed PMSM drive, $K_{PWM}$ =15.0V/V (at $V_{dc} = 30V$) and $K_s$ =1.0V/A is set.

For the sinewave PMSM drive, the current command possesses sinusoidal waveform with its magnitude and frequency being determined by the outer control loop. And moreover, the sinusoidal back-EMF disturbance always exits. Hence the satisfactory sinusoidal command tracking and sinusoidal disturbance rejection controls are difficult and challenging issues.

**(b) Classical Current Two-degrees-of-freedom Control (2DOFC) Scheme**

The proposed per-phase current 2DOFC and motor model are shown in Fig. 2(a), which consists of a PI feedback controller $G_{ib}(s)$ and a command feedforward controller $G_{if}(s)$. The functions and limitations of this traditional 2DOFC are introduced below.

**(1) PI feedback controller**

For the control scheme shown in Fig. 2(a), with only the feedback controller $G_{ib}(s)$ the closed-loop tracking and disturbance to output transfer functions can be derived from Fig. 2(a):

$$H_{dr}(s) \overset{\Delta}{=} \frac{i_s'}{i_s^*}\bigg|_{e_d=0} = \frac{G_{ib} K_{PWM} G_P K_s}{1 + G_{ib} K_{PWM} G_P K_s} \qquad (11)$$

robust tracking error cancellation control scheme; (d) robust torque control scheme.

$$H_{dd}(s) \overset{\Delta}{=} \frac{i_s'}{e_d}\bigg|_{i_s^*=0} = \frac{-G_P K_s}{1 + G_{ib} K_{PWM} G_P K_s} \qquad (12)$$

The traditional current feedback controller $G_{ib}(s)$ is set as the following PI-type:

$$G_{ib}(s) = \frac{K_{Pi} s + K_{Ii}}{s} \qquad (13)$$

It can be derived to find that: (i) $|H_{dr}| \neq 1$ and $\angle\theta_{H_{dr}} \neq 0$ for $\omega > 0$, thus the steady–state tracking error always exists for the sinusoidal reference current by PI feedback control; (ii) $|H_{dd}| \neq 0$ for $\omega > 0$, i.e., the asymptotical steady-state rejection for sinusoidal back-EMF also fails to achieve.

**(2) Command feedforward controller**

The command feedforward controller $G_{if}$ shown in Fig. 2(a) can assist the feedback controller to yield improved winding current tracking control performance. The design spirits of this controller are commented as follows: (i) Without back-EMF disturbance, the perfect command tracking control can be achieved if the controller $G_{if}$ can be set as the inverse of the actual motor model; (ii) The fact of (i) will be lost under the presence of sine-wave back-EMF for a PMSM. Fortunately, it can be proved that the ideal tracking current control under the effects of back-EMF can still be achieved if the parameters of $G_{if}$ are properly over-tuned. However, the successful tuning control is not a trivial task and worth studying.

In the proposed current control scheme, the simple PI-type feedback controller and the feedforward controller with nominal inverse motor model are first employed. Then the 2DOFC is augmented with an internal model feedback controller or a simple robust tracking error cancellation control scheme to yield improved sinusoidal reference current tracking and back-EMF disturbance rejection control performances.

**(c) The Proposed Control Schemes**

**(1) 2DOFC with internal model feedback controller**

The issues in achieving perfect sinusoidal current tracking and back-EMF rejection controls by feedback control are:

(i) For achieving the asymptotical tracking of sinewave reference current and the asymptotical steady-state rejection of sinewave back-EMF signal in a PMSM, the current loop gain transfer function should contain the same unstable poles as those of reference current signal ($s = \pm j\omega_r$ here), and the closed-loop transfer function must be stable.

(ii) For a PMSM drive, its reference current and back-EMF possess the same frequency. Hence the inclusion of ($s^2 + \omega_r^2$) in the denominator of the feedback controller $G_{ib}(s)$ can meet the requirements of (i). However, the adaptation of the term ($s^2 + \omega_r^2$) faithfully in accordance with the varying motor speed is critical issue in practical digital control realization.

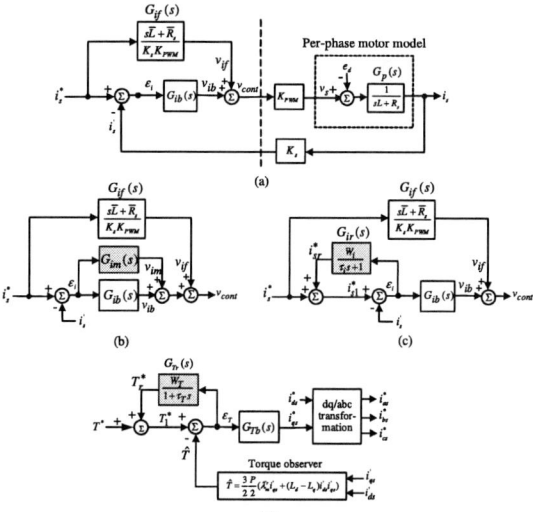

Fig. 2. The proposed control schemes: (a) per-phase current 2DOFC; (b) 2DOFC with internal model feedback controller; (c) 2DOFC with

As introduced previously, the internal model principle has been used in many occasions [10-13] to yield improved sinusoidal command tracking and disturbance rejection control performances. The proposed current internal model 2DOFC scheme for PMSM is shown in Fig. 2(b), which consists of a classical 2DOFC ( $G_{ib}(s) + G_{if}(s)$ ) and an augmented internal model feedback controller $G_{im}(s)$ . Through applying the proposed control structure with properly determined parameters, good transient and steady-state control characteristics can be achieve.

According to the above comments and the existing research [10], the internal model feedback controller $G_{im}(s)$ can either be sine type of $\omega_r/(s^2 + \omega_r^2)$ or the cosine type of $s/(s^2 + \omega_r^2)$ . Since the latter provides better phase margin characteristics, it is adopted here. Thus the resulted feedback and feedforward controllers are:

$$G_{ib}^{'}(s) \overset{\Delta}{=} G_{im}(s) + G_{ib}(s) = \frac{K_{Mi}s}{s^2 + \omega_r^2} + \frac{K_{Pi}s + K_{Ii}}{s} \quad (14)$$

$$G_{if}(s) = \frac{s\overline{L} + \overline{R}_s}{K_s K_{PWM}} \overset{\Delta}{=} as + b, \ a = \frac{\overline{L}}{K_s K_{PWM}}, \ b = \frac{\overline{R}_s}{K_s K_{PWM}} \quad (15)$$

where $\omega = \omega_r = \omega_e =$ inverter applying fundamental angular frequency, which is equal to the rotor angular speed $\omega_r$ for a PMSM.

From Figs. 2(a) and 2(b) one can derive the loop gain $LG(s)$ :

$$LG(s) = \frac{\begin{aligned} & K_{PWM} K_s K_{Pi}s^3 + K_{PWM} K_s(K_{Ii} + K_{Mi})s^2 \\ & + \omega_r^2 K_{Pi} K_{PWM} K_s s + \omega_r^2 K_{Ii} K_{PWM} K_s \end{aligned}}{Ls^4 + R_s s^3 + \omega_r^2 L s^2 + \omega_r^2 R_s s} \quad (16)$$

And the following transfer functions from controlled phase current to its command and disturbance, and those from tracking error to current command:

● With $G_{if}(s)$

$$H_{dr}(s) = \frac{\Delta i_s^{'}}{i_s^*}\bigg|_{e_d=0} = \frac{\begin{aligned} & \overline{L}s^4 + (K_{Pi} K_{PWM} K_s + \overline{R}_s)s^3 + [(K_{Ii} + K_{Mi})K_{PWM} K_s + \omega_r^2 \overline{L}]s^2 \\ & + \omega_r^2(K_{Pi} K_{PWM} K_s + \overline{R}_s)s + \omega_r^2 K_{Ii} K_{PWM} K_s] \end{aligned}}{\Delta(s)} \quad (17)$$

$$H_{dd}(s) = \frac{\Delta i_s^{'}}{e_d}\bigg|_{i_s^*=0} = \frac{-K_s s(s^2 + \omega_r^2)}{\Delta(s)} \quad (18)$$

$$H_{de}(s) = \frac{\varepsilon_i}{i_s^*}\bigg|_{e_d=0} = \frac{(L - \overline{L})s^4 + (R_s - \overline{R}_s)s^3 + \omega_r^2(L - \overline{L})s^2 + \omega_r^2(R_s - \overline{R}_s)s}{\Delta(s)} \quad (19)$$

where

$$\Delta(s) = Ls^4 + (K_{Pi} K_{PWM} K_s + R_s)s^3 + [(K_{Ii} + K_{Mi})K_{PWM} K_s + \omega_r^2 L]s^2 \quad (20)$$
$$+ \omega_r^2(K_{Pi} K_{PWM} K_s + R_s)s + \omega_r^2 K_{Ii} K_{PWM} K_s$$

● Without $G_{if}(s)$

$$H_{dr}(s) = \frac{\Delta i_s^{'}}{i_s^*}\bigg|_{e_d=0} = \frac{K_{PWM} K_s[K_{Pi}s^3 + (K_{Ii} + K_{Mi})s^2 + \omega_r^2 K_{Pi}s + \omega_r^2 K_{Ii}]}{\Delta(s)} \quad (21)$$

$$H_{dd}(s) = \frac{\Delta i_s^{'}}{e_d}\bigg|_{i_s^*=0} = \frac{-K_s s(s^2 + \omega_r^2)}{\Delta(s)} \quad (22)$$

$$H_{de}(s) = \frac{\varepsilon_i}{i_s^*}\bigg|_{e_d=0} = \frac{Ls^4 + R_s s^3 + \omega_r^2 L s^2 + \omega_r^2 R_s s}{\Delta(s)} \quad (23)$$

*Performance analysis:*

From (17) to (23), one can observe the following facts for the augmentation of $G_{if}(s)$ :

(i) Back-EMF rejection control

The $H_{dd}(s)$ in (18) and (22) are identical, i.e., the back-EMF rejection characteristics are not relevant to $G_{if}(s)$ . Since $|H_{dd}(s = j\omega_r)| = 0$ , the perfect fundamental back-EMF rejection is obtained by $G_{ib}^{'}(s)$ . The effect of $n - th$ order harmonic back-EMF on the current waveform can be found by $|H_{dd}(s = jn\omega_r)|$ from (18).

(ii) Command tracking control

From (19) and (23) one can find $|H_{de}(0)| = 0$ and $|H_{de}(j\omega_r)| = 0$ without and with $G_{if}(s)$ . Thus the perfect current tracking responses for DC and sinewave with $\omega = \omega_r$ are obtained by $G_{ib}^{'}(s)$ . However, for the transient dynamic frequency components with $\omega \gg \omega_r$ one can derive to yield the following expressions:

$$H_{de}(s) \approx \frac{\Delta L}{L} + \frac{1}{L}[\Delta R_s - (K_{Pi} K_{PWM} K_s + R_s)\frac{\Delta L}{L}]s^{-1} \approx \frac{\Delta L}{\overline{L}} + \frac{\Delta R_s}{\overline{L}}s^{-1}, \text{ with } G_{if}(s) \quad (24)$$

$$H_{de}(s) \approx 1 - \frac{1}{L}[K_{Pi} K_{PWM} K_s]s^{-1}, \text{ without } G_{if}(s) \quad (25)$$

where $L \overset{\Delta}{=} \overline{L} + \Delta L$ , $R_s \overset{\Delta}{=} \overline{R}_s + \Delta R_s$ , $\Delta L \ll L$ , $\Delta R_s \ll R_s$ are assumed. Hence form (24) and (25) one can find that the transient command tracking responses are greatly improved via applying $G_{if}(s)$ . It is worth noting form (24) that the parameter variations of $\Delta L$ and $\Delta R_s$ can be estimated from $\varepsilon_i$ and $i_s^*$ using the relationship of (24), and they can be employed for updating the parameters set in $G_{if}(s)$ to yield more precise transient tracking response.

(iii) The effects of internal model feedback controller parameters

From (17) and (19) one can find that with the precisely set value of $\omega_r$ in $G_{im}(s)$ , the parameter $K_{Mi}$ affects only the dynamic behavior around $\omega = \omega_r$ , and which can be determined from the loop gain $LG(s)$ expressed in (16). Generally speaking [10,11], the larger value of $K_{Mi}$ will lead to the larger magnitude of $|LG(j\omega)|$ around $\omega = \omega_r$ , and thus to yield more robust control with closer current command tracking around $\omega = \omega_r$ . However on the contrary, the phase margin will be slightly sacrificed.

(2) 2DOFC with robust tracking error cancellation control scheme

In the proposed robust current control scheme $G_{ir}(s)$ shown in Fig. 2(c), the conventional 2DOFC ( $G_{ib}(s) + G_{if}(s)$ ) is augmented with a simple robust current error cancellation control scheme. The tracking error $\varepsilon_i = i_s^* - i_s^{'}$ is regarded as a disturbance, and a robust compensating command $i_{sr}^* = W_i(s)\varepsilon_i$ is generated to reduce the tracking error ( $i_s^* - i_s^{'}$ ) as far as possible, where $W_i(s)$ denotes a weighting function

$$W_i(s) = \frac{W_i}{1 + \tau_i s}, \ \tau_i = \frac{1}{2\pi f_{ci}}, \ 0 \le W_i < 1 \quad (26)$$

with $W_i =$ weighting factor, $f_{ci} =$ cut-off frequency of low-pass filter. While the factor $W_i$ is used to determine the extent of robust error cancellation, the low-pass process $1/(1+\tau_i s)$ is arranged to reduce the effects of high-frequency contaminated noise on the closed-loop control behavior.

Through careful derivation from Fig. 2(c), one can find the output current $i_s'$ and modified current command $i_{s1}^*$ after applying the robust error cancellation control:

$$i_s' = i_s^* - (1 - W_i(s))\varepsilon_i \tag{27}$$

$$i_{s1}^* = \frac{1}{1 - W_i(s)} i_s^* - \frac{W_i(s)}{1 - W_i(s)} i_s', \quad i_{s1n}' \approx -\frac{W_i(s)}{1 - W_i(s)} n \tag{28}$$

where $n$ denotes the high-frequency noise contaminated in $i_s'$.

From (27), (28) and Fig. 2(c) some facts can be deduced: (i) the tracking error $(i_s^* - i_s') = (1 - W_i(s))\varepsilon_i$ is greatly reduced, although $\varepsilon_i$ will be larger than those without robust error cancellation control. The commands $i_s^*$, $i_{s1}^*$ and the resulted controls $v_{fc}$, $v_{fb}$ will be automatically adjusted to let $i_s'$ follows $i_s^*$ very closely. (ii) Although the closer tracking can be obtained by closing $W_i$ approaching to 1 and $\tau_i$ approaching to 0, the effects of high-frequency noise should be considered. This fact can be aware from (28). Generally speaking, for a PMSM drive, the general guideline for choosing $f_{ci}$ will be $\bar{f}_1 < f_{ci} < f_s$, where $\bar{f}_1$ = motor maximum excitation frequency, $f_{tri}$ = inverter switching frequency.

(3) Design results of the proposed control schemes

(i) PI feedback controller $G_{ib}(s)$

Refer to the control system shown in Fig. 3(a), the parameters of $G_{ib}(s)$ in the RC-CCPWM scheme are first determined. For the established motor drive, the following system parameters at maximum operation speed are used for making the controller design: $I_m^* = 3A$, $\omega_r = 2\pi \times 200$ (at 2000rpm), $\hat{E}_d = 10.78V$ (at 2000rpm), $L = L_q = 328.42\mu H$, $R_s = 0.60625\Omega$, $V_{dc} = 38V$, $K_s = 1$, $\hat{v}_{tri} = 1$, $f_{tri} = 15kHz$. From the large-signal stability criterion for a ramp-comparison current controlled PWM scheme, the upper limit of the proportional gain is found as:

$$K_{Pi} \le 0.5446 \tag{29}$$

Accordingly, $K_{Pi} = 0.39$ is first chosen. Then the simulated frequency responses of the loop-gain under $K_{Pi} = 0.39$ and different values of $K_{Ii}$ are plotted (not shown here). From which $K_{Ii} = 28.61$ is chosen to let the loop-gain crossover frequency $f_{cs} = 2.8$ kHz, which satisfies the general guideline $\bar{f}_1 < f_{cs} < 0.5 f_{tri}$ ( $\bar{f}_1 = 200Hz$, $f_{tri} = 15kHz$). Hence finally, the current feedback controller is determined as:

$$G_{ib}(s) = 0.39 + \frac{28.61}{s} \tag{30}$$

(ii) Current command feedforward controller $G_{if}(s)$

It is known that the accurate plant model parameters are difficult to obtain and hence the parameters of the command feedforward controller. Thus the command feedforward controller is set as the inverse of the nominal plant model. Accordingly from nominal motor model parameters given in Sec. III, one can obtain:

$$G_{if}(s) = \frac{s\bar{L} + \bar{R}_s}{K_s K_{PWM}} = 0.0000219s + 0.0404 \tag{31}$$

And the imperfect control will be automatically compensated by the proposed robust error cancellation control. To solve the unrealizable problem, the pure differentiator $s$ is replaced by a low-pass process $s/(1+\tau_d s)$, $\tau_d = 0.000159s$.

(iii) Internal model feedback controller $G_{im}(s)$

Till now, only the parameters in $G_{im}(s)$ are needed to be determined, which are made as followed:

- The angular speed $\omega_r$ should be accurately updated in accordance with the PMSM varying-frequency driving control.
- The parameter $K_{Mi}$: From the root locus for the closed-loop characteristic equation $\Delta(s)$ of (20), one can find that the location of poles is less affected by the reasonable value of $K_{Mi}$. In reality, $K_{Mi}$ affects only the dynamic behavior around $\omega = \omega_r$. Hence the frequency responses of the loop gain $|LG(j\omega)|$ with three values of $K_{Mi}$=10, 100, 1000 under 2000 rpm ( $f = f_r = 200Hz$ ) and ( $L = \bar{L}$, $R_s = \bar{R}_s$ ) are simulated (not shown here). From which $K_{Mi}$=100 is chosen taking the compromised considerations of robustness and phase margin.

(iv) Robust current controller

Taking the compromised considerations in control performance and effects of high-frequency noises, $W_i = 0.977$ and $\tau_i = 79.6 \times 10^{-6} s$ (corresponding to the cut-off frequency of $f_{ci} = 2kHz$ ) are chosen here.

### B. Torque Control Scheme

(a) Torque Observer

According to the analytic motor modeling made in Sec. II, the PMSM observed torque is obtained from (5). The estimated $\lambda_m^r$, the d-and q-axis inductances, and the real-time sensed currents $i_{ds}$ and $i_{qs}$ are employed for making the computation.

(b) Control Scheme

Basically, the torque control loop possesses the similar dynamic response speed requirement to those of current loops. Having obtained high performance current control behavior via the proposed approach, the torque control scheme can be much simpler. The proposed torque control scheme consists of a PI feedback controller and a simple robust torque error cancellation control scheme.

(1) Feedback controller

The torque feedback controller is also chosen to be the PI type:

$$G_{Tb}(s) = K_{PT} + \frac{K_{IT}}{s} \tag{32}$$

(2) Robust torque tracking error cancellation control scheme

A compensating torque command $T_r^* = W_T(s)\varepsilon_T$ is also generated to assist the feedback control in the further reduction of torque command tracking error. The robust control weighting function is set as:

$$W_T(s) = \frac{W_T}{1 + \tau_T s}, \quad 0 \le W_T < 1 \tag{33}$$

The methodology of this torque robust control scheme is similar to those presented above.

### (c) Numerical Design of the Proposed Control Scheme

The feedback controller and the robust control weighting function are respectively chosen as:

$$G_{Tb}(s) = 0.894 + \frac{1396}{s}, \quad W_T(s) = \frac{0.8}{1 + 79.6 \times 10^{-6} s} \tag{34}$$

## V. EXPERIMENTAL RESULTS

### A. Current Tracking Characteristics

In making the experimental current tracking control performance evaluation, the current commands $i_{qs}^* = 2A$ and $i_{ds}^* = 0$ are set to let the motor be accelerated, and the currents are recorded at a particular speed. Let $V_{dc} = 30V$ and only the PI current feedback controller $G_{ib}(s)$ be actuated, Fig. 3 shows the measured a-phase motor current $i_{as}$ and its command $i_{as}^*$ at $\omega_r = 1000rpm$. Large tracking error is observed from the result. Now the designed command feedforward controller $G_{if}(s)$ listed in (31) is added, the measured results at the same conditions of Fig. 3 shown in Fig. 4(a) indicate that although some improvement is yielded, there still exists significant tracking error. The key reasons, which have been analyzed previously, are (i) the inaccurate inverse plant model being employed; and (ii) the major error caused by sinewave back-EMF effect. Now the controller $G_{if}(s)$ is over-tuned to be $G_{if}(s) = 0.0000766s + 0.3342$ to yield the excellent tracking shown in Fig. 4(b). However, using this $G_{if}(s)$, the results at $\omega_r = 1500rpm$ plotted in Fig. 4(c) show that the tracking error exists. Hence, as the operating condition change occurs, the command feedforward controller can not yield good tracking control without properly tuning its parameters. This task can be overcome by using the proposed two approaches.

Let $V_{dc} = 30V$, $i_{qs}^* = 2A$, the measured phase-a currents $i_{as}$ and their commands $i_{as}^*$ at $\omega_r = 1000rpm$ and $1500rpm$ by $(G_{ib} + G_{im})$ are shown in Figs. 5(a) and 5(b). Very close current waveform tracking at steady-state are observed. The measured results at the same cases by $(G_{ib} + G_{im} + G_{if})$ are plotted in Figs. 6(a) and 6(b). No further improvement is observed by the addition of $G_{if}(s)$ for the already very small tracking error. Now without using $G_{im}(s)$, the conventional PI feedback controller and feedforward controller is augmented with the simple robust controller, the measured results using $(G_{ib} + G_{if} + G_{ir})$ shown in Figs 7(a) and 7(b) indicate that the good sinewave current command tracking control can also be obtained by the proposed robust control without using the internal model feedback controller $G_{im}(s)$.

For comparatively evaluating the transient waveform tracking performance, the measured $(i_{as}', i_{as}^*)$, $\varepsilon_{ia}$, $v_{conta}$ at $V_{dc} = 30V$ due to a step torque command change of $T^* = 0.05Nm$ to $0.15Nm$ under $\omega_r = 1500rpm$ by $(G_{ib} + G_{im})$, $(G_{ib} + G_{im} + G_{if})$ and $(G_{ib} + G_{if} + G_{ir})$ are compared in Figs.

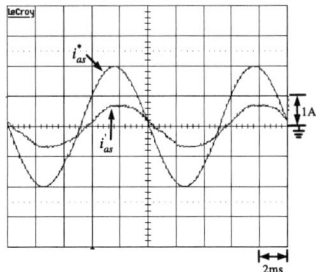

Fig. 3. Measured phase-a current and its command at ( $V_{dc} = 30V$ , $i_{qs}^* = 2A$ , 1000rpm) by $G_{ib}(s)$ only.

8(a) to 8(c). From the results one can be aware that: (i) by the augmentation of $G_{if}(s)$ to ( $G_{ib} + G_{im}$ ), a great reduction of tracking error is achieved; and (ii) the robust control scheme $G_{ir}(s)$ has the equivalent control action to $G_{im}(s)$. However, the implementation of the former is much easier.

### B. Torque Tracking Characteristics

In making torque control evaluation, the inner current loop is controlled using ( $G_{ib} + G_{if} + G_{ir}$ ) and the torque control scheme designed in (34) is employed. Let the torque command be the programmed profiles shown in Figs. 9(a) and 9(b), the measured torques, their commands, and speeds at $V_{dc} = 30V$ and $V_{dc} = 40V$ are plotted in Figs. 9(a) and 9(b). The results show that very smooth and close torque tracking control with low ripple component is obtained by the developed PMSM driven reaction wheel at $V_{dc} = 40V$ . The inaccurate torque tracking phenomenon with slightly larger ripple at higher speed is observed at $V_{dc} = 30V$ . This fact is due to the insufficient DC-link voltage to counter the back-EMF at larger speed. As well known, the commutation instant advanced shift is an effective means to achieve some extent of field weakening. Hence this issue is also worth of further studying.

## VI. CONCLUSIONS

This paper has presented the establishment of a DSP-based PMSM driven satellite reaction wheel system, and its robust current and torque controls. For performing the RC-CCPWM scheme design and the motor torque estimation, the dynamic modeling is made and the PMSM key parameters are estimated. In the current control scheme, the conventional 2DOFC is improved via the augmentation of an internal model feedback controller or a simple robust tracking error cancellation controller. Comparative study for the various control structures in sinewave current tracking control characteristics under the effects of sinewave PMSM back-EMF has been made. The improved transient and static current tracking responses by the proposed control schemes have been verified experimentally. As to torque control, the robust torque tracking error cancellation control scheme is also designed to yield very fast and close observed torque tracking response.

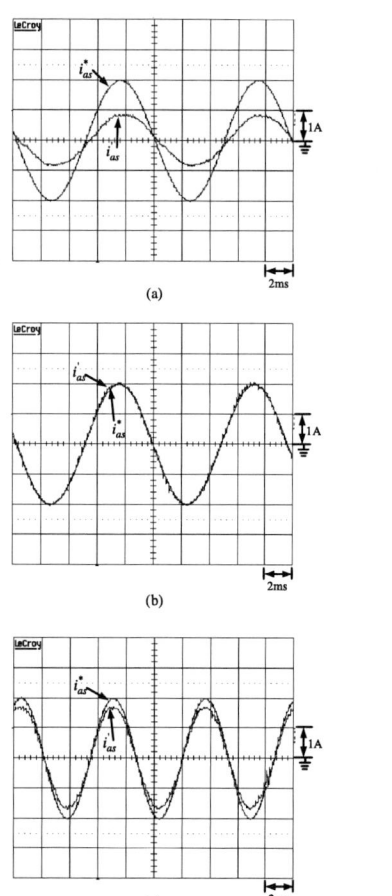

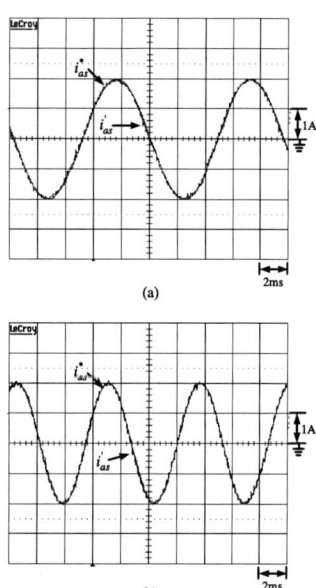

Fig. 4. Measured phase-a currents and their commands at $V_{dc}=30V$, $i_{qs}^*=2A$ by: (a) $G_{ib}(s)$ and nominal $G_{if}(s)$, 1000rpm; (b) $G_{ib}(s)$ and tuned $G_{if}(s)$, 1000rpm; (c) $G_{ib}(s)$ and tuned $G_{if}(s)$, 1500rpm.

Fig. 5. Measured phase-a currents and their commands by $G_{ib}(s)+G_{im}(s)$ at: (a) ($V_{dc}=30V$, $i_{qs}^*=2A$, 1000rpm); (b) ($V_{dc}=30V$, $i_{qs}^*=2A$, 1500rpm).

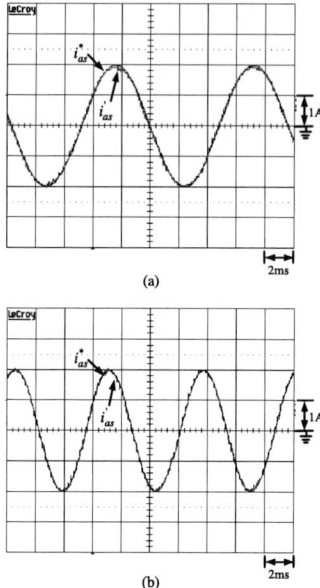

Fig. 6. Measured phase-a currents and their commands by $G_{ib}(s)+G_{im}(s)+G_{if}(s)$ at: (a) ($V_{dc}=30V$, $i_{qs}^*=2A$, 1000rpm); (b) ($V_{dc}=30V$, $i_{qs}^*=2A$, 1500rpm).

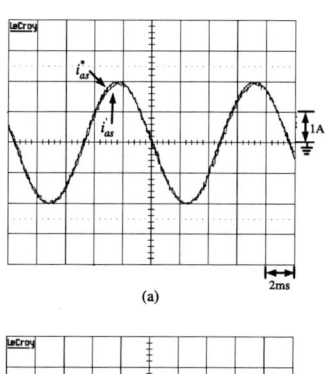

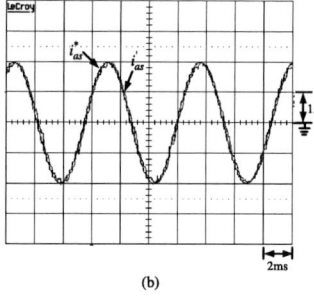

Fig. 7. Measured phase-a currents and their commands by $G_{ib}(s)+G_{if}(s)+G_{ir}(s)$ at: (a) ($V_{dc}=30V$, $i_{qs}^*=2A$, 1000rpm); (b) ($V_{dc}=30V$, $i_{qs}^*=2A$, 1500rpm).

## REFERENCES

[1] Y. Zhang, Y. Postrekhin, Ki Bui Ma and W. K. Chu, "Reaction wheel with HTS bearings for mini-satellite attitude control," Supercod. Sci. Technol., vol. 15, pp. 823-825, 2002.

[2] J. Zhou and K. J. Tseng, "A disk-type bearingless motor for use as satellite momentum- reaction wheel," in pesc 02. 2002 IEEE 33rd Annu. Power Electron. Specialists Conf., Jun. 2002, vol. 4, pp. 1971-1975.

[3] E. Lee, "A micro HTS renewable energy/attitude control system for micro/nano satellites," IEEE Trans. Appl. Supercond., vol. 13, no. 2, pp. 2263-2266, Jun. 2003.

[4] S. J. Dodds, and J. Vittek, "Spacecraft attitude control using an induction motor actuated reaction wheel with sensorless forced dynamic drive," in IEE Colloq. All Elect. Aircraft Dig. 1998/260, Jun. 1998, pp. 9/1-9/7.

[5] E. D. Ganev, "High-Performance electric drives for aerospace more electric architectures part I- electric machines," IEEE Power Engineering Society General Meeting, pp. 1-8, June 2007.

[6] M. Kondo, "Parameter measurements for permanent magnet synchronous machines," IEEJ Trans. Elect. and Electron. Eng., vol. 2, no. 2, pp. vii-viii, Feb. 2007.

[7] C. C. Liaw, C. M. Liaw, H. C. Cheng, Y. C. Chang and C. M. Huang, "Robust current control and commutation tuning for an IPMSM drive," in APEC '03 18th Annu. IEEE Appl. Power Electron. Conf. and Expo., Feb. 2003, vol. 2, pp. 1045-1051.

[8] H. C. Chen, M. S. Huang, C. M. Liaw, Y. C. Chang, P. Y. Yu and J. M. Huang, "Robust current control for brushless DC motor," in IEE Proc. Electric Power Appl., Nov. 2000, vol. 147, no. 6, pp. 503-512.

[9] T. S. Radwan and M. M. Gouda, "Intelligent speed control of permanent magnet synchronous motor drive based-on neuro-fuzzy approach," in PEDS 2005 Int. Conf. Power Electron. and Drives Syst., Jan. 2006, vol. 1, pp. 602-606.

[10] S. Fukuda and T. Yoda, "A novel current tracking method for active filters based on a sinusoidal internal model," IEEE Trans. Ind. Appl., vol. 37, no. 3, pp. 888-895, May/Jun. 2001.

[11] S. Fukuda and R. Imamura, "Application of a sinusoidal internal model to current control of three-phase utility-interface converters," IEEE Trans. Ind. Electron., vol. 52, no. 2, pp. 420-426, April 2005.

[12] W. C. Gan and L. Qiu, "Torque and velocity ripple elimination of AC permanent magnet motor control systems using internal model principle," IEEE/ASME Trans. Mechatronics., vol. 9, no. 2, pp. 436-447, Jun. 2004.

[13] P. Degobert, G. Remy, J Zeng, P. J. Barre and J. P. Hautier, "High performance control of the permanent magnet synchronous motor using self-tuning resonant controllers," in Proc. Thirty-Eighth Southeastern Symp. Sys. Theory, Mar. 2006, pp. 382-386.

[14] G. S. Buja and M. P. Kazmierkowski, "Direct torque control of PWM inverter-fed AC motors– a survey," IEEE Trans. Ind. Electron., vol. 51, no. 4, pp. 744-757, Aug. 2004.

[15] D. Ocen, L. Romeral, J. A. Ortega, J. Cusido and A. Garcia, "Discrete space vector modulation applied on a PMSM motor," in 12th Int. Power Electron. and Motion Control Conf., Aug. 2006, pp. 320-325.

[16] P. C. Krause, O. Wasynczuk and S. D. Sudhoff, Analysis of Electric Machinery and Drive Systems, 2nd ed. New York: Wiley-IEEE, 2002.

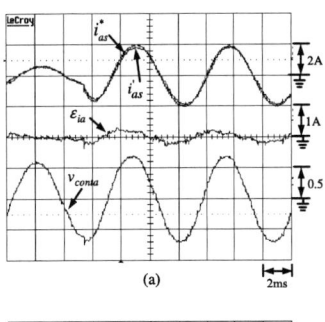

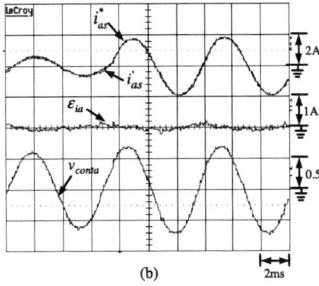

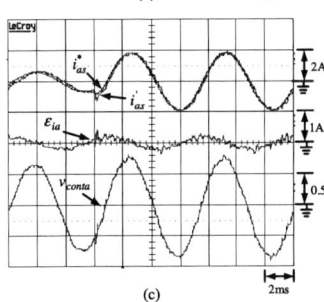

Fig. 8. Transient response under torque command change from $0.05Nm$ to $0.15Nm$ at 1500rpm, $V_{dc}=30V$ : (a) $G_{ib}(s)+G_{im}(s)$ ; (b) $G_{ib}(s)+G_{im}(s)+G_{if}(s)$ ; (c) $G_{ib}(s)+G_{if}(s)+G_{ir}(s)$ .

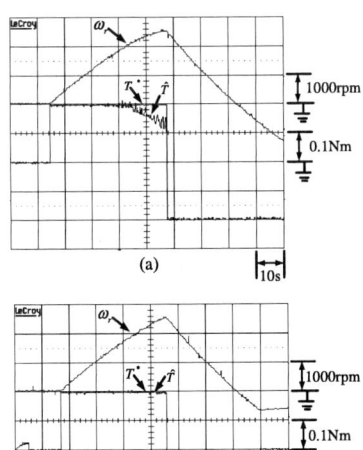

Fig. 9. Measured torque, its command and speed due to programmed torque command change with $G_{ib}(s)+G_{if}(s)+G_{ir}(s)$ : (a) ( $V_{dc}=30V$ ); (b) ( $V_{dc}=40V$ ).

**1659**

# Nonlinear control design for magnetic bearings via automatic differentiation

Stefan Palis *, Mario Stamann [†], Thomas Schallschmidt [†]

*Institute for Automation/Faculty for Electrical Engineering and Information Technology,
Magdeburg, Germany, e-mail: *stefan.palis@ovgu.de*
[†]Institute of Electric Power Systems/Faculty for Electrical Engineering and Information Technology,
Magdeburg, Germany, e-mail: *mario.stamann@ovgu.de*

*Abstract*—The presented paper shows how nonlinear control theory can be applied to magnetic bearings with permanent magnetic force. Because of the difficult expression for the magnetic force symbolic computation of Lie derivatives becomes unhandy. Therefore the Lie derivatives will be calculated applying automatic differentiation.

*Keywords*—Magnetic bearings; Nonlinear control; Automatic differentiation

## I. INTRODUCTION

Active magnetic bearings are getting more and more important in various applications. In contrast to conventional bearings the movable parts are not supported by mechanical contact or a fluid, but by magnetic forces. Therefore they are free of mechanical wear and need less maintenance. However, due to the attractive force between the magnet and the movable part the equilibrium is unstable and has to be stabilized by control. Although electromagnets provide a nonlinear behavior between current and magnetic force in industrial applications predominantly linear control laws (PID and state feedback) are used [9], [8]. The application of nonlinear controllers has been investigated [6], [7] for pure electromagnets with promising result.

Fig. 1. Prototype of a industrial magnetic bearing

For big industrial applications however hybrid magnets are used, providing a higher magnetic force. Unfortunately the magnetic force equation (1) for the hybrid magnets in the differential current topology (3) is much more difficult then for a pure electromagnet (2).

$$F_{M,hyb} = a \left[ \frac{(i + H_0)^2}{(k_2 - 2x)^2} - \frac{(-i + H_0)^2}{(k_1 + 2x)^2} \right] \quad (1)$$

$$F_{M,elec} = a \frac{i^2}{(k_2 - 2x)^2} \quad (2)$$

## II. MAGNETIC BEARING

In the following for illustrative reasons only the model equations for a one degree of freedom magnetic bearing consisting of two hybrid magnets in differential topology are used. The current is supplied via a separate current control loop, which characteristics have approximated by a $PT_2$ transfer function (fig. 2).

$$i = \frac{\omega_0^2}{s^2 + 2D\omega_0 s + \omega_0^2} u \quad (3)$$

| $\omega_0$ | $D$ |
|---|---|
| $8200s^{-1}$ | 0.45 |

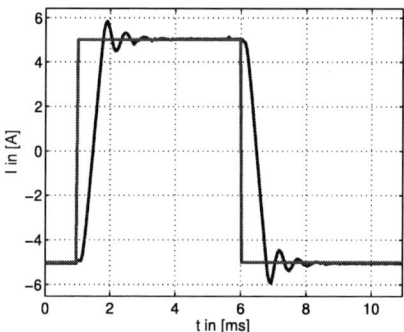

Fig. 2. Step response of the current control loop

The overall system equations for a one degree of freedom magnetic levitation system with two hybrid magnets are as follows:

$$\begin{pmatrix} \dot{x}_1 \\ \dot{x}_2 \\ \dot{x}_3 \\ \dot{x}_4 \end{pmatrix} = \begin{pmatrix} x_2 \\ \frac{1}{m} F_{M,hyb}(x_3, x_1) - g \\ x_4 \\ -\frac{2d}{T} x_4 - \frac{1}{T^2} x_3 + \frac{K}{T^2} u \end{pmatrix}, \quad (4)$$

$$F_{M,hyb} = a \left[ \frac{(x_3 + H_0)^2}{(k_2 - 2x_1)^2} - \frac{(-x_3 + H_0)^2}{(k_1 + 2x_1)^2} \right]. \quad (5)$$

The following four parameters $k_1$, $k_2$, $a$ and $H_0$ have been identified for one hybrid magnet in differential configuration.

978-1-4244-1741-4/08/$25.00 ©2008 IEEE

| $a$ | $H_0$ | $k_1$ | $k_2$ |
|---|---|---|---|
| $0.028 \frac{kNmm^2}{A^2}$ | $57.28A$ | $2.42mm$ | $5.42mm$ |

Thereby the hybrid magnets configuration is realized as shown in figure 3. Here the magnetic forces $F_o$ and $F_u$ are produced by the upper and lower hybrid magnet. By superposition of the fields of the permanent magnet and the inductor an attractive force is generated. Because of the contrary winding of the upper and lower electromagnet the upper force $F_o$ is strengthened, while the lower force $F_u$ is weakened, when applying a positive current. And vice versa, when a negative current is supplied. Therefore both positive and negative forces are applicable.

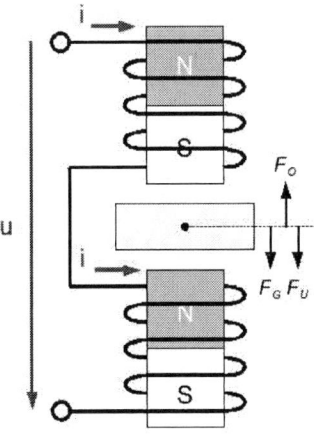

Fig. 3. Differential configuration of two hybrid magnets

Figure 4 shows the resulting force characteristic of two hybrid magnets in differential configuration.

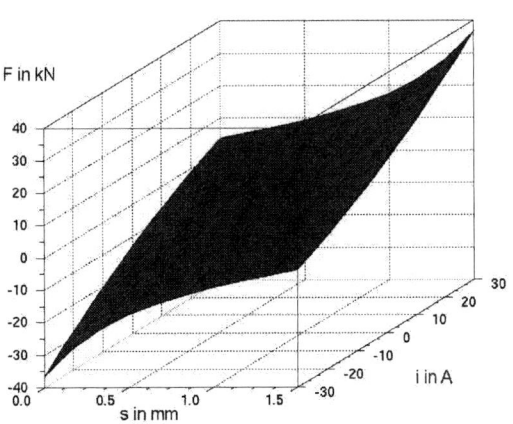

Fig. 4. Force characteristic of hybrid magnets in differential configuration

## III. EXACT LINEARIZATION VIA FEEDBACK

In the following nonlinear control theory, i.e. feedback linearization theory [5], [10], is applied to the magnetic bearing. Here a transformation is constructed by repeated differentiating the output $y = h(x)$ along the system trajectory as long as the control input $u$ appears, i.e

$$L_g y^{(j)} = L_g \left( L_f^{j-1} h(x) \right) \neq 0. \tag{6}$$

In the case that

$$L_g \left( L_f^{r-1} h(x) \right) = 0 \tag{7}$$

the system is said to have a relative degree of $r$. If the relative degree $r$ is equal to the system order $n$ the system is said to have full relative degree.

$$y^{(j)} = L_f^j h(x) + \underbrace{L_g \left( L_f^{j-1} h(x) \right)}_{=0} u = L_f^j h(x) \tag{8}$$

$$y^{(r)} = L_f^r h(x) + \underbrace{L_g \left( L_f^{r-1} h(x) \right)}_{\neq 0} u \tag{9}$$

For the time derivatives $y, \dot{y}, \ldots, y^{(r-1)}$ new state-variables $z_1, \ldots, z_r$ are introduced.

$$z = \begin{pmatrix} z_1 \\ z_2 \\ \vdots \\ z_r \end{pmatrix} = \begin{pmatrix} y \\ \dot{y} \\ \vdots \\ y^{(r-1)} \end{pmatrix} = \underbrace{\begin{pmatrix} h(x) \\ L_f h(x) \\ \vdots \\ L_f^{r-1} h(x) \end{pmatrix}}_{=\Phi(x)}. \tag{10}$$

Here $\Phi(x)$ is the state-transformation $x \to z$, which transforms the original system into the controllable canonical form (11), where $f_n(z)$ and $g_n(z)$ are derived transforming equation (9).

$$\begin{pmatrix} \dot{z}_1 \\ \dot{z}_2 \\ \vdots \\ \dot{z}_n \end{pmatrix} = \begin{pmatrix} z_1 \\ z_2 \\ \vdots \\ z_n + f_n(z) + g_n(z)u \end{pmatrix} \tag{11}$$

For a system in Brunovsky normal form (11) a simple nonlinear control law is:

$$u = -\frac{f_n(z) + v(z)}{g_n(z)} \tag{12}$$

compensating all nonlinearities and stabilizing the remaining chain of integrators through a linear state feedback $v(z)$. In terms of Lie derivatives the nonlinear control law reads:

$$u = \frac{v(z) - L_f^n h(x)}{L_g L_f^{n-1} h(x)}. \tag{13}$$

It can be shown that the system (4) is flat with respect to the output $y = x_1$. Therefore the relative degree $r$ is equal to the order of the system $n = 4$, meaning that the output $y$ has to be differentiated four times.

$$\dot{y} = x_2, \tag{14}$$

$$\ddot{y} = \frac{1}{m}F_{M,hyb} - g, \tag{15}$$

$$y^{(3)} = \frac{1}{m}\left(\frac{\partial F_{M,hyb}}{\partial x_1}x_2 + \frac{\partial F_{M,hyb}}{\partial x_3}x_4\right) \tag{16}$$

Because of the increased complexity of the magnetic force equation $F_{M,hyb}$ (1) compared to $F_{M,elec}$ (2), a symbolic derivation of $y^{(3)}$, $y^{(4)}$ and $u$ becomes error-prone and doesn't give any physical insights. To overcome this problem the Lie derivatives defining the control law (13) will be derived using automatic differentiation.

## IV. AUTOMATIC DIFFERENTIATION

Automatic differentiation is based on the application of elementary differentiation rules to elementary functions forming a function $f(x)$. In doing so the intermediate results are immediately evaluated, i.e. the result is not given in form of a symbolic expression but as a real number. Therefore in contrast to symbolic differentiation in automatic differentiation derivatives of a function $f(x)$ are evaluated only in a certain point $x$ [1], [2]. There are basically two modes in automatic differentiation: forward mode and reverse mode.

To derive the Lie derivatives $L_f^n h(x)$ and $L_f^{n-1}L_g h(x)$ Taylor series of order $n$ have to be calculated. Supposing the following autonomous system:

$$\dot{x} = f(x) \tag{17}$$

the solution $x(t)$ can be described by the following truncated Taylor series:

$$x(t) = x_0 + x_1 t + x_2 t^2 + \ldots + x_d t^d + \mathcal{O}(t^d). \tag{18}$$

Defining a function $z(t) = \dot{x}(t)$ the associated Taylor series reads:

$$z(t) = z_0 + z_1 t + z_2 t^2 + \ldots + z_d t^d + \mathcal{O}(t^d). \tag{19}$$

Hereby Taylor series coefficients $x_j$ and $z_j$ are as follows:

$$x_j = \frac{1}{j!}\frac{d^j x(t)}{dt^j}\bigg|_{t=0}, \tag{20}$$

$$z_j = \frac{1}{j!}\frac{d^j z(t)}{dt^j}\bigg|_{t=0} = \frac{1}{j!}\frac{d^j f(x)}{dt^j}\bigg|_{t=0} \tag{21}$$

Here the Taylor series coefficient $z_j$ can be calculated applying the forward mode of automatic differentiation. Since $x(t)$ is a solution of the initial value problem $\dot{x} = z$ the Taylor series coefficients $z_j$ have to satisfy:

$$z_j = \frac{1}{j!}\frac{d^j f(x)}{dt^j}\bigg|_{t=0} = \frac{1}{j!}\frac{d^j \dot{x}}{dt^j}\bigg|_{t=0}, \tag{22}$$

$$= \frac{1}{j!}\frac{d^{j+1}x}{dt^{j+1}}\bigg|_{t=0} = (j+1)x_{j+1}. \tag{23}$$

For known Taylor series coefficient $z_j$ the Taylor series coefficient $x_{j+1}$ can therefore be calculated applying equation (24).

$$x_{j+1} = \frac{z_j}{j+1}. \tag{24}$$

Using equation (21) and (24) the Taylor series coefficients $x_0, \ldots, x_j$ and $z_0, \ldots, z_j$ can be determined recursively, starting with $x_0$.

For the simple output equation $y = h(x) = x_1$ the Lie derivatives $L_f^j h(x) = y^{(j)} = x_1^{(j)}$ form a Taylor series for $x_1$:

$$x_1(t) = \sum_{j=0}^{n} x_j t^j = \sum_{j=0}^{n}\frac{1}{j!}L_f^j h(x)|_{t=0}t^j. \tag{25}$$

Therefore the Lie derivative $L_f^j h(x)$ can be calculated directly from the associated Taylor series coefficient $x_j$ [4].

$$L_f^j h(x) = j!x_j \tag{26}$$

Since the systems equation (4) are non autonomous with affine input:

$$\dot{x} = f(x) + g(x)u \tag{27}$$

mixed Lie derivatives $L_g L_f^{j-1}h(x)$ have to be calculated. A simple possibility is to calculate the mixed Lie derivative $L_g L_f^{j-1}h(x)$ from the Lie derivatives $L_f^j h(x)$ and $L_{f+g}^j h(x)$ [4].

$$L_g L_f^{j-1} = L_{f+g}^j h(x) - L_f^j h(x) \tag{28}$$

Let $\hat{y}$ denote the output of the system $\dot{x} = f(x) + g(x)$ and $y$ the output of $\dot{x} = f(x)$. Then the associated Lie derivates $L_f^i h(x)$ and $L_{f+g}^i h(x)$ are equal if $i$ is smaller than the relative degree $r$.

$$L_{f+g}^j h(x) = i!\hat{y}_i = i!y_i = L_f^j h(x) \text{ for } 0 \le i < r \tag{29}$$

Here $\hat{y}_i$ and $y_i$ denote associated the Taylor series coefficients. For $i = r$ one gets:

$$L_g L_f^{r-1} = L_{f+g}^r h(x) - L_f^r h(x) = r!\left(\hat{y}_r - y_r\right). \tag{30}$$

Hence applying (26) and (30) the Lie derivatives in the control law (13) can be derived. The overall control therefore reads:

$$u = \frac{v - r!x_r}{r!\left(\hat{y}_r - y_r\right)}. \tag{31}$$

## V. IMPLEMENTION USING ADOL-C

ADOL-C [3] is a C++ library facilating the evaluation of derivatives of C and C++ functions by automatic differentiation. Using operator overloading only minor changes in the program, which has to be differentiated are required. Therefore the so called active variable is defined representing the independent variables of a function. In the original program all independent variables have to be redefined as active variables.

```
void Lie_f(short int tape, double* z,
                          double* dz)
{
    adoublev x(4);
    adoublev dx(4);

    trace_on(tape);
    x<<=z;
    dx[0] = x[1];
    dx[1] = a/m*FM_hyb(x[0],x[2])-g;
```

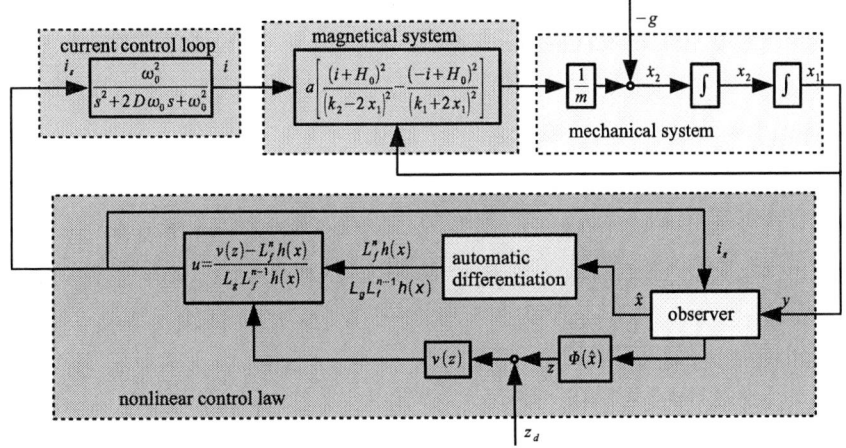

Fig. 5. Control scheme

```
    dx[2] = x[3];
    dx[3] = -2*d*x[3] - w02*x[2];
    dx>>=dz;
    trace_off ();
}

    short int tape_f = 0;
    Lie_f(tape_f,x,&dx);

    double **P;
    P = myalloc2 (N,D+1);
    P[0][0] = x[0];
    P[1][0] = x[1];
    P[2][0] = x[2];
    P[3][0] = x[3];
    forode(tape_f,N,D,P);
```

For simulation and a later real time code generation the automatic differentiation has been implemented in a MATLAB C++ S-function.

## VI. RESULTS

The above presented control law will now be used to stabilize the one degree of freedom magnetic bearing with two hybrid magnetic actuators in differential topology4. Applying the nonlinear feedback linerization all nonlinearities are compensated resulting in an open chain of four integrators. For them a linear state-feedback is designed solving the Riccati equation associated to the lqr problem.

$$PA + A^T P - PBR^{-1}B^T P + Q = 0 \qquad (32)$$

The state feedback is:

$$v(z) = -R^{-1}BPz. \qquad (33)$$

Applying the linear feedback law (33) in combination with the feedback linearizing control law (13), where the Lie derivatives are calculated through automatic differentiation, results in the following step responses (fig. 6), (fig. 7). The control scheme is shown in figure 5

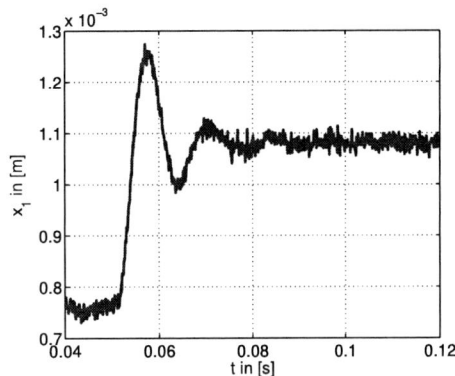

Fig. 6. Plot of position $x_1$

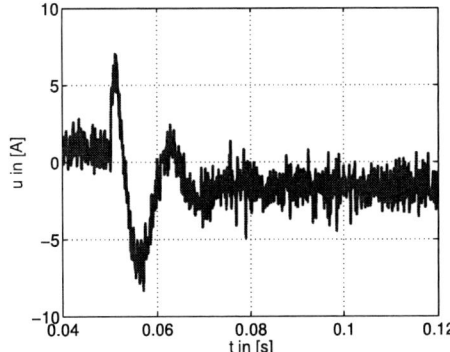

Fig. 7. Plot of reference current $u$

## VII. CONCLUSION

We have considered the design of a nonlinear control law for magnetic bearings with permanent magnetic force.

It has been shown, that calculating Lie derivatives for hybrid magnetic actuator systems with an additional current control loop, would result in long equations, when done symoblically. Therefore the application of automatic differentiation has been suggested. First promising results have been presented.

## REFERENCES

[1] A. Griewank. *Evaluation Derivatives, Principles and Techniques of Algorithmic Differentiation.* Frontiers in Appl. Math. 19, Phil., 2000.

[2] A. Griewank and J. Utke and A. Walther. *Evaluation higher derivate tensors by forward propagation of univariate Taylor series.* Mathematics of Computation 69:1117-1130, 2000.

[3] A. Walther and A. Griewank and O. Vogel. *ADOL-C: Automatic Differentiation Using Operator Overloading in C/C++.* Proc. Appl. Math. Mech. 2:41-44, 2003.

[4] K. Röbenack. *Automatic differentiation and nonlinear controller design by exact linearization.* Future Generation Computer Systems 21:1372-1379, 2005.

[5] Hassan K. Khalil. *Nonlinear systems.* Prentice-Hall, 1996.

[6] S. Palis.*Modified optimal control of magnetic levitation system.* APM 2007.

[7] S. Palis and M. Stamann. and T. Schallschmidt. *Nonlinear adaptive control of magnetic bearings.* EPE 2007.

[8] Thomas Schallschmidt. *Regelungstechnische Optimierung eines Magnetlagers.* Diplomarbeit, 2003.

[9] G. Schweitzer and A. Traxler. *Magnetlager - Grundlagen, Eigenschaften und Anwendungen berührungsfreier, elektromagnetischer Lager.* Springer, 1996.

[10] Qiang Lu and Yuanzhang Sun and Shengei Mei. *Nonlinear control systems and power system dynamics.* Kluwer academic publishers, 2001.

# Design of Energy Harvesting Generator Base on Rapid Prototyping Parts

Zdenek Hadas*, Jan Zouhar*, Vladislav Singule*, Cestmir Ondrusek[†]

* Faculty of Mechanical Engineering, Brno University of Technology, Brno, Czech Republic,
e-mail: *hadas@fme.vutbr.cz, zouhar@fme.vutbr.cz, singule@fme.vutbr.cz*
[†] Faculty of Electrical Engineering and Communication, Brno University of Technology, Brno, Czech Republic,
e-mail: *ondrusek@feec.vutbr.cz*

*Abstract*—This paper deals with an alternative design of an electromagnetic energy harvesting generator for supplying wireless sensors with energy. The developed device is complex mechatronics system which generates an electrical power from an ambient mechanical vibration by use of a suitable construction of electromagnetic generator. The developed design of generator has immobile parts base on rapid prototyping parts from ABS plastic material. It is suitable for product of device and it provides lightweight device with sufficient durability. As this device is excited by ambient mechanical vibration, it harvests electrical energy due to Faraday's law.

*Keywords*—Mechatronics, Design, Rapid Prototyping, Alternative energy, Resonant converter, Vibration.

## I. INTRODUCTION

This paper deals with an alternative design of an electromagnetic energy harvesting generator for supplying wireless sensors. The developed energy harvesting device is interest in low power wireless sensor which operates in environment excited by a mechanical vibration. If the expect power consumption of the wireless sensor is several mW then the ambient energy can be suitable source of the electrical energy. This generator can feed wireless sensor without the use of primary battery. The generator extends the lifetime of the wireless sensor that can be mounted without any problems inside engineering's constructions or can be placed inside embedded structures.

The developed generator is complex mechatronic system which contains a mechanical resonance oscillator and an electromechanical energy converter which are mutually affected. This complex system is optimally developed on the base of excited vibration, maximal volume and required output voltage and power.

The new design of this energy harvester is based on rapid prototyping methods of fast production of real parts, which can be used for development of some mechatronics systems [2]. The rapid prototyping is the process of creating physical objects from computer models generated using CAD program that a prototype machine can build a 3D prototype out of metal, plastic or resin. The rapid prototyping is used to quick and easy produce prototype parts. This fact is very interesting and useful for development of several mechatronic systems as this energy harvester. The complicated geometry of models and prototype parts can be tested during development cycle of this mechatronic system.

## II. STATE OF ART – ENERGY HARVESTING FROM VIBRATION

An energy harvesting device generates electric energy from its surroundings using some energy conversion method [1]. The different physical principles of harvesters for electrical energy from ambient vibration are mentioned.

The generally vibration energy harvester is shown in Fig. 1. The resonance mechanism is depicted as an oscillation mass $m$ suspended on a springy element with the known stiffness $k$. The tuning up of the generator depends on the mass and the stiffness ratio. The relatively movement is effected a mechanical damping $b_m$ provided by the resonance mechanism construction. The damper $b_e$ depicts the energy converter [3]. The harvesting of energy dissipates mechanical energy of an oscillating resonance mechanism.

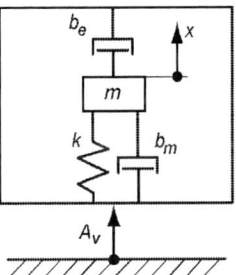

Fig. 1. Schematic diagram of energy harvester

The output power [4] is proportional of the frequency of vibration and it depends on an acceleration of excited vibration $Av$. The construction of the excited resonance mechanism provides a movement of the oscillating mass. The excited displacement $x$ of moving mass depends on the overall mechanical damping and the overall mechanical damping significantly affects the output power. Due to excited movement the energy converter generates the output power (1). The maximal output power corresponds to a maximal displacement and the construction of the energy transducer.

$$P_e = \frac{m^2}{2} \cdot \frac{b_e}{\left(b_e + b_m\right)^2} \cdot A_v^2 \qquad (1)$$

The principle behind energy harvesting from vibration is the excited relative movement of a moving mass or the

978-1-4244-1741-4/08/$25.00 ©2008 IEEE

mechanical deformation of some structure inside the energy harvesting device [1]. This relative movement can be converted to electrical energy by these methods:

## A. ELECTROSTATIC GENERATOR

This electrostatic generator relies on the changing capacitance of vibration dependant capacitors. A variable capacitor is initially charged and its plates are separated due to exciting of vibration then the mechanical energy is transformed into electrical energy. The most attractive feature of this method is its IC compatible nature, given that MEMS variable capacitors are fabricated through relatively mature silicon micro-machining techniques.

## B. PIEZOELECTRIC GENERATOR

This piezo-generator converts mechanical energy to electrical by strain of a piezoelectric material. The strain of the piezoelectric material causes charge separation across the device, producing an electric field and consequently a voltage drop proportional to the stress applied. The oscillating system is typically a cantilever beam structure with proof mass. The voltage produced varies with time and strain, effectively producing an irregular AC signal. Piezo-generator produces relatively higher voltage than the electromagnetic system but inner impedance is very high for low frequency.

## C. ELECTROMAGNETIC GENERATOR

This electromagnetic generator uses a magnetic field to convert mechanical energy to electrical. A permanent magnet attached to the oscillating mass moves through a stationary coil. The magnet moves through the coil causes varying magnetic flux through coil and an electromotive voltage is induced according to Faraday's law. The induced voltage depends on velocity and length of coil and it is usually small and must therefore be increased to viably source energy. Methods to increase the induced voltage include using a transformer or increasing the number of turns of the coil, and increasing the permanent magnetic field. However, each method is limited by the size constraints.

## III. DEVELOPED ELECTROMAGNETIC GENERATOR

This generator (vibration power generator) is based on an electro-magnetic energy converter and a unique resonance oscillator combination [5]. The design and parameters of oscillator (resonance mechanism) is tuned up to frequency and level of the ambient vibration. The exciting of the ambient vibration provides relative movement inside electromagnetic energy converter due to tuning up of the mechanism. The electromagnetic converter consists of a moveable magnetic circuit and a fixed coil. The magnetic circuit with permanent magnet is moved against the fixed coil and an electromotive voltage is induced due to Faraday's law. The relative movement inside generator depends on a quality of the resonance mechanism (overall mechanical damping of system) and level of vibration affects the relative movement which provides harvested power. The main problem of this device is mechanical damping inside resonance mechanism which affects movement. The design of generator is adapted to provide sufficient sensitivity of the energy harvesting generator (optimal relative movement of magnetic circuit against coil). Due to unknown mechanical damping of this system the development of

generator goes in several development cycles with aim to find optimal design of generator [8].

The generating of electrical energy causes feedback forces in resonance mechanism and affects relative movement inside generator too. Consequently the energy harvesting generator presents complex mechatronics system which includes mechanical system, electromagnetic system and feedbacks between systems. The connected electrical load, wireless sensor or another fed device affects behaviour of whole generator too and it provides additional subsystem of this energy harvesting device. Each parameter of this mechatronics system affects its construction and output power. The harvested power (1) depends on quality of the mechanism (overall mechanical damping) and the weight of moving mass, detailed power analysis is published in monograph [4].

The construction of the energy harvesting generator is designed on the base of the optimal size and the required harvested power with using of simulation modelling. The optimal parameters of the whole design have to be tailored exactly to level of vibration for generating of maximal electrical energy. The appropriate vibration power generator can produce the required power for wireless sensing. The design of this electromagnetic energy harvesting generator consists of:

- **Resonance mechanism** provides relative movement of a magnetic circuit against the coil.
- **Magnetic circuit** provides magnetic flux through coil. The most powerful permanent magnets FeNdB are used in the magnetic circuit.
- **Fixed Coil** is placed inside the moving magnetic circuit between permanent magnets.
- **Power management** is used for rectifying and stabilization of generated alternate voltage to the required value. The power management determines connected electric load. The power management can accumulate generated energy for non-excited time of the generator.

## A. Design of Energy Harvesting Generator

The design of this device is based on vibration power generator described in publication [5]. Parameters of the generator are tuned up to the stable resonance frequency (around 17 Hz) of the ambient mechanical vibration. The generator is designed to generating electrical energy with output power around 10 mW and minimal stable output DC voltage 3 V.

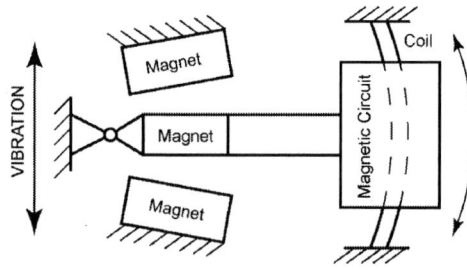

Fig. 2. Schematic diagram of generator design

The schematic diagram of generator design is shown in Fig. 2. The design of resonance mechanism is based on stiffness of repelled permanent magnets and moving arm. The set of permanent magnets determines only resonance frequency of moving arm and it does not participate to induce of voltage inside coil. The arm with moving magnetic circuit is moved against fixed air coil during exciting of vibration and this movement induces voltage inside coil. All parameters of generator are adjusted to required output power and voltage using simulation modelling of this mechatronics system published in [6]. The design of whole generator is in accordance with tuned up parameters and the developed generator is shown in Fig. 3. The frame of generator was manufactured from aluminium parts and mounted together by screws. The lab testing of generator [7] showed potential of this device as independent source of energy for wireless application.

Fig. 3. Aluminium frame of generator

The measurement of generator resonance characteristic is shown in Fig. 4. Due to non-linear behaviour of generator the resonance frequency is shifted in range 16.9 – 17.3 Hz, it depends on level of excited vibration and connected load which affect deflection and consecutively maximal amplitude of resonance frequency of generator. The generator harvests maximal power 26 mW with output voltage 9 V DC on resistance load 3 kΩ during exciting by acceleration of vibration 0.5 g and higher. The dimensions of aluminium generator are 50x40x40 mm and weight is 135 grams.

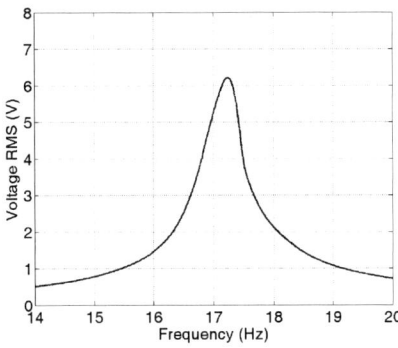

Fig. 4. Resonance characteristic of generator; vibration 0.1 g

The main problem of generator design is manufacturing of shape difficult frame and its weight. The traditional cutting production is very expensive and time consuming.

The mounting of generator by screws is not suitable for long time operation in vibrating environment too.

Therefore the unique way of modern manufacturing method was used. The design of frame and all immovable parts are based on rapid prototyping parts from plastic material. It provides lightweight energy harvesting generator with sufficient durability of the construction. The next advantage is very easy and quick manufacturing of these plastic parts. The very shape difficult parts are created in CAD software and the rapid prototyping method produces these parts with sufficient accuracy and durability during several hours. It can be very useful for many industrial application including mechatronics systems.

## IV. RAPID PROTOTYPING METHODS

Rapid prototyping substitutes classic production cycle of components. The rapid prototyping methods of fast production of real parts can be very useful for development of some mechatronic systems, robotics and other applications [2].

The basic principle of this technology is gradual forming of a layer with constant thickness from a material. The prototype component can be, according to the applied technology, manufactured from thermoplastic material (ABS), photopolymer, wax, sintered powder, paper etc. Additive manufacturing is extremely versatile because of its ability to create almost any geometry even parts that could not be commercially created with processes such as machine cutting or injection moulding.

### A. Fused Deposition Modelling

Dimension BST 3D printer, which works at the FDM (Fused Deposition Modelling) principle, is located at Faculty of Mechanical Engineering. The FDM is a rapid prototyping method and it produces ABS plastic prototype models, which have high strength and durability. ABS (acrylonitril butadien styrene – $(C_8H_8 \cdot C_4H_6 \cdot C_3H_3N)_n$) is a strong, durable production-grade thermoplastic used across many industries. The main typical properties are shown in Tab. 1. The input material is wire at the cartridge which is entered to the printing head where ABS is melted and it is pressed through the nozzle to the basic pad. The nozzle moves to produce a profile of the part then pad moves down and the next layer is built on top until the entire prototype model is fully built. A support construction from easy breakable ABS is used for the overhanging places. The maximal working space of this printer is 203x203x300 mm.

TABLE I.
MAIN PROPERTIES OF ABS

| Properties | Value |
|---|---|
| Tensile Strength | 22 MPa |
| Flexural Strength | 41 MPa |
| Heat Deflection Temperature @ 66 psi | 90°C |
| Dielectric Strength | 32 kV/mm |
| Dielectric Constant @ 60 Mhz | 2.4 |
| Rockwell hardness | R105 |
| Specific gravity | 1.05 g/cm³ |

ABS rapid prototype parts can be milled, drilled, tapped, painted and even plated with nickel, chrome or copper. The main properties of this plastic material are shown in Tab. 1.

The structure and requirements on total accuracy of final rapid prototyping parts have to be taken into account during rapid prototyping printing. The mechanical properties of model depend on direction of ABS layers and sandwich structure of printed model during prototyping.

### B. Tests of ABS Material

The tensile and bending tests of normalize specimen EN ISO 527(178) shown different material properties for individual direction of ABS layers. The specimen with horizontal layers has tensile strength 23 MPa and with vertical layers has 28 MPa during the tensile test. The bending tests show the similar results, the flexure strength of specimen with horizontal layers is 43 MPa and the 46 MPa for specimen with vertical layers.

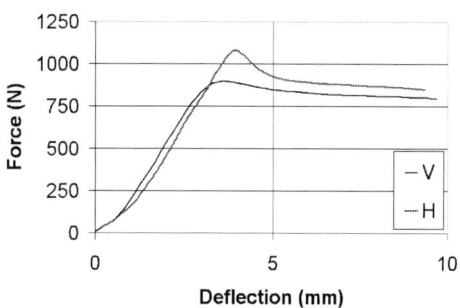

Fig. 5. Tensile test for vertical and horizontal layers

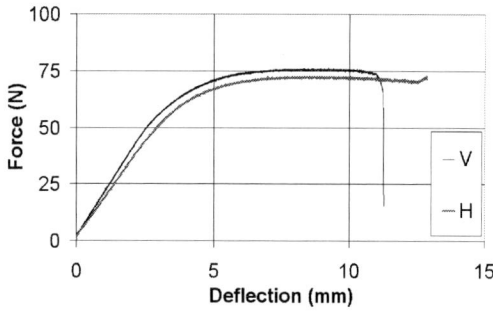

Fig. 6. Bending test for vertical and horizontal layers

The normalize specimen for tests has cross-section 4x10 mm and testing length of specimen is 60 mm. The results of tensile and bending tests for vertical and horizontal direction of ABS layers are shown in Fig. 5 and Fig. 6.

### C. Modification and Finishing of ABS Parts

Further, an influence of surface finishing of ABS models was investigated. Usually the rapid prototyping parts are processed by mastic, varnish or paint. A matter on the base of ethylmethylketone or toluene is used for improvement of material properties as porosity or absorbability. The matter on the base of hardening impregnation applied in vacuum chamber is used for another improvement of surface properties. The test showed that these modifications of surface do not affect tensile and flexure strength of rapid prototyping part. The advantage of these surface finishing is improvement of surface properties and strength increasing of part details.

The indisputable advantage of this rapid prototyping method is possibility of final machining and cutting of rapid prototyping parts. ABS parts can be milled, turned, and grinded or the thread cutting is applied in produced holes. This material has specific requirements for cutting due to very low melt temperature and sandwich structure. The suitable material of cutting tools is high speed steel or sintered carbide but the tool must be sharp-edged and smooth and the cutting operation has usually lower cutting speed with the smallest possible contact surface due to temperature generation by friction.

Fig. 7. Rapid prototyping parts of developed generator

The tensile and bending tests of ABS plastic specimen show that the rapid prototyping method of fused deposition modelling provides useful apparatus for manufacturing of very shape difficult plastic parts during development cycle of any prototypes in mechatronics and other branch.

## V. PLASTIC PRODUCT OF GENERATOR

The all immovable plastic parts, Fig. 7, were printed using rapid prototyping method from ABS plastic. The plastic parts are created directly from CAD model using FDM rapid prototyping method and subsequently the connecting surfaces of plastic parts were adapted to correct stick together. The parts were stick using epoxy glue and the assembly of these shape difficult parts is very simple and the final frame is durable and light for this application.

Fig. 8. Resonance mechanism of generator

The Fig. 8 shows resonance mechanism with moving arm placed inside plastic frame. The frame of air coil was produced from plastic too. The air coil was wound around plastic frame by enamelled cooper wire; it is shown in Fig. 9. The air coil was placed inside moving magnetic circuit and stick together with frame.

Fig. 9. Air coil with plastic frame

The final generator was placed inside housing which is produced by rapid prototyping from ABS plastic too. This housing includes generator, power management circuit and capacitor. The electronics as power management and storage circuit with capacitors were development in Brno University of Technology, Faculty of Electrical Engineering and Communication. The housing design was adjusted to location of these electronics and the volume of the whole energy harvesting device is 80x60x60 mm, Fig. 10.

Fig. 10. Plastic housing of generator with power management and capacitors

## VI. CONCLUSIONS

The developed energy harvesting generator is capable of generating useful electric power for feeding of concrete wireless sensor. The generator harvests the maximum output power around 26 mW with DC voltage around 9 V but the output power and voltage depend on level of vibration and the connected electrical load. The generator operates during exciting by vibration with frequency around 17 Hz and the vibration level in range of 0.1 – 1 g.

The rapid prototyping method of fused deposition modelling provides useful apparatus for manufacturing of very shape difficult plastic parts of generator product. It is suitable for product of this device and it provides lightweight with sufficient durability.

In real wireless application the power management circuit is connected to the generator and it rectifies and stables the required value of output voltage and accumulates the excess harvested electrical energy for non-excited time. The whole energy harvesting device with power management harvests the output power 3.3 mW with stable output voltage 3 V during excitation of sufficient vibration. The excess of electrical energy is storage in capacitors and this energy is used during transmission time of wireless sensor.

This duty time of wireless sensor consumes peak of current which is provided from capacitors because this dissipating of energy affects harvesting of generator (feedback of electromagnetic damping force affects relative movement inside generator). The level of vibration, it means harvested energy, determines real duty time of wireless sensor operation. The energy harvesting device was tested in lab with 5% duty time of wireless sensor and from level of vibration 0.3 g this device provides enough power for wireless sensing. Thus, the generator can be used as an unlimited source of electrical energy for wireless sensors working in such environment.

The energy harvesting generator has a great potential as an inexhaustible source of the electrical energy in future. This device can provide sufficient electrical energy for wireless sensing in several applications, which operates in vibration environment. Limits for using of generator are usually the operating temperature range, size, weight and sufficient vibration level.

### ACKNOWLEDGMENT

Published results were acquired using the subsidization of the Ministry of Education, Youth and Sports of the Czech Republic, research plan MSM 0021630518 "Simulation modelling of mechatronics systems".

### REFERENCES

[1] S. Roundy, J. M. Rabaey, P. K. Wright: *Energy Scavenging for Wireless Sensor Networks: With Special Focus on Vibrations*, Kluwer Academic Publishers, Boston MA, 2003.

[2] I. Ebert-Uphoff, C.M. Gosselin, D.W. Rosen, T. Laliberte, (2005) Rapid Prototyping for Robotics, book chapter in "*Cutting Edge Robotics*", Pro Literatur Verlag, Mammendorf, Germany, pp. 17-46.

[3] Williams, C. B., Yates, R. B: Analysis of a micro-electric generator for microsystems, Sensors and Actuators, A: Physical, A52(1), 1996, pp. 8-11.

[4] Hadas, Z., Singule, V., Ondrusek, C.: Optimal Design of Vibration Generator Function Product and Verification of Model, *Simulation Modelling of Mechatronic Systems II*, Brno University of Technology, Faculty of Mechanical Engineering, Brno, 2006.

[5] Hadas, Z., Kluge, M., Singule, V., Ondrušek, C.: Electromagnetic Vibration Power Generator, *6th IEEE International Symposium on Diagnostics for Electric Machines, Power Electronics and Drivers*, Cracow, 2007, pp. 451-455.

[6] Hadas, Z., Singule, V., Ondrusek, C., Kluge, M.: Simulation of Vibration Power Generator, *Recent Advances in Mechatronics*, Springer Berlin Heidelberg New York, 2007, pp. 350-354.

[7] Hadas, Z.; Ondrusek, C.; Singule, V.; Kluge, M. Vibration Power Generator for Aeronautics Applications. In *Proceedings of the 10th anniversary international conference of the europian society for precision engineering and nanotechnology Volume I.*, Zurich, 2008. pp. 46 - 50.

[8] Hadas, Z., Singule, V., Ondrusek, C.: Optimal Design of Vibration Power Generator for Low Frequency, *Journal Solid State Phenomena*, in print.

# Control of Bouc-Wen hysteretic systems: Application to a piezoelectric actuator

Oriol Gomis-Bellmunt*, Fayçal Ikhouane†, Daniel Montesinos-Miracle*

*Centre d'Innovació Tecnològica en Convertidors Estàtics i Accionaments (CITCEA-UPC),
Departament d'Enginyeria Elèctrica, Universitat Politècnica de Catalunya.
ETS d'Enginyeria Industrial de Barcelona, Av. Diagonal, 647, Pl. 2. 08028 Barcelona, Spain
Tel: +34 934016727, Fax: +34 934017433, e-mail: *oriol.gomis@upc.edu*
†Departament de Matemàtica Aplicada III
Escola Universitària d'Enginyeria Tècnia Industrial de Barcelona
Universitat Politècnica de Catalunya
Comte d'Urgell, 187, 08036 Barcelona, Spain

*Abstract*—**This paper deals with the control of hysteretic systems. The hysteretic nonlinearity is represented in this paper using the Bouc-Wen model and a time-varying PID controller is designed for micropositionning purpose. The performance of the controller is tested using numerical simulations of experimental data obtained from a piezoelectric actuator.**

*Index Terms*—**Piezoelectric actuator, Bouc-Wen model, Hysteresys.**

## I. INTRODUCTION

To describe the behavior of hysteretic processes several mathematical models have been proposed [1]: the Duhem model [2] uses the property that a hysteretic system's otput changes its character when the input changes direction; the Ishlinskii hysteresis operator has been proposed as a model for plasticity-elasticity [3]; the Preisach model has been used for the modeling of electromagnetic hysteresis [4]; the Bouc-Wen model has been used to model wood joints and structural systems [5]. A survey of the mathematical models for hysteresis may be found in [6]. These models have been applied to describe the behavior of piezoelectric actuators: Prandtl-Ishlinskii in [7], Preisach in [8]–[10] and Bouc-Wen in [11]–[13]. An energy based model has been employed in [14].

In this work we represent a piezoelectric actuator by the Bouc-Wen model for smooth hysteresis [15]. This model consists in a first-order nonlinear state equation, and an output equation where the input and state signals appear linearly. This model has received an increasing interest due to its ability to capture in an analytical form a range of shapes of hysteretic cycles which match the behavior of a wide class of hysteretical systems [16]. In particular, it has been used to model piezoelectric elements [11], magnetorheological dampers [17], [18] and wood joints [5]. The models, derived from experiments, have been used either to predict the behavior of the physical hysteretic element [17] or for control purposes as in [19]–[21].

Since the piezoelectric device is represented in this work using the Bouc-Wen model, the results of [21] are used and improved for the control of the piezoelectric element. In [21], a second-order mechanical system that includes a Bouc-Wen hysteresis is considered for control purposes. The control objective is to guarantee the global boundedness of all the closed loop signals, and the regulation of both the displacement and the velocity of the device to zero. This objective is achieved using a simple PID controller. However, the main drawback of this controller is that the equilibrium point of the closed loop system is not robust vis-à-vis perturbations which is undesirable in practice. The main advantage of the proposed control law over other existing control schemes, is that it is simple to implement in an industrial context.

## II. BACKGROUND RESULTS

### A. PID control of a Bouc-Wen hysteresis

We consider the second order mechanical system described by:

$$m\ddot{x} + c\dot{x} + \Phi(x)(t) = u(t) \qquad (1)$$

with initial conditions $x(0)$, $\dot{x}(0)$ and excited by a control input force $u(t)$. The output restoring force $\Phi$ is assumed to be described by the normalized Bouc-Wen model [22]:

$$\Phi(x)(t) = \kappa_x x(t) + \kappa_w w(t), \qquad (2)$$

$$\dot{x}(t) = \rho \left( \dot{x}(t) - \sigma \left| \dot{x}(t) \right| \left| w(t) \right|^{n-1} w(t) \right. \\ \left. + (\sigma - 1) \dot{x}(t) \left| w(t) \right|^n \right) \quad (3)$$

with an initial condition $w(0)$. The parameters $n \geq 1$, $\rho > 0$, $\sigma \geq 1/2$, $\kappa_x > 0$, $\kappa_w > 0$, $m > 0$ and $c \geq 0$ are unknown. The range of the parameters corresponds to the Class I Bouc-Wen model which is stable, asymptotically dissipative and thermodynamically consistent [22]. The displacement $x(t)$ and velocity $\dot{x}(t)$ are available through measurements, but the signal $w(t)$ is not. Let $y_r(t)$ be a (known) smooth and bounded reference signal whose (known) smooth and bounded derivatives are such that $\lim_{t \to \infty} y_r(t) = \lim_{t \to \infty} \dot{y}_r(t) = \lim_{t \to \infty} \ddot{y}_r(t) = \lim_{t \to \infty} y_r^{(3)}(t) = 0$ exponentially. This means that there

978-1-4244-1741-4/08/$25.00 ©2008 IEEE

exist some constants $a > 0$ and $b > 0$ such that $\left|y_r^{(i)}(t)\right| \leq ae^{-bt}$ for $t \geq 0$ and $i = 0, 1, 2, 3$.

The control objective is to globally asymptotically regulate the displacement $x(t)$ and velocity $\dot{x}(t)$ to the reference signals $y_r(t)$ and $\dot{y}_r(t)$ preserving the global boundedness of all the closed loop signals; that is $x(t)$, $\dot{x}(t)$, $w(t)$ and $u(t)$.

We assume the following:

*Assumption 1:* The unknown parameters lie in known intervals. That is we have $m \in [m_{\min}, m_{\max}]$ with $m_{\min} > 0$, $c \in [0, c_{\max}]$, $\kappa_x \in (0, \kappa_{x_{\max}}]$, $\kappa_w \in (0, \kappa_{w_{\max}}]$, $\sigma \in \left[\frac{1}{2}, \sigma_{\max}\right]$, $\rho \in (0, \rho_{\max}]$.

Note that the unknown structure parameter $n \geq 1$ is not required to lie in a known interval.

The problem of controlling the system (1)-(3) has been treated in [21], where it is demonstrated that a PID control insures that the displacement and velocity errors tend to zero. Introduce the variables:

$$x_1(t) = x(t) - y_r(t)$$
$$x_2(t) = \dot{x}(t) - \dot{y}_r(t)$$
$$x_0(t) = \int_0^t x_1(\tau)d\tau \qquad (4)$$

and choose as a control law the PID controller:

$$u(t) = -k_0 x_0(t) - k_1 x_1(t) - k_2 x_2(t) \qquad (5)$$

where the $k_i$'s are design parameters. Then we have [21]:

*Theorem 1:* Consider the closed loop formed by the system (1)-(3) and the control law (5). Define the following constants:

$$k_{2_{\min}} = \sqrt{2m_{\max}\left(\sigma_{\max}\rho_{\max}\kappa_{w_{\max}} + \kappa_{x_{\max}} + k_1\right)} \quad (6)$$

$$e_1 = \frac{(c_{\max} + k_2)^3}{m_{\min}^2}$$

$$e_2 = \frac{k_1^2}{m_{\max}^2}\left(k_2^2 - k_{2_{\min}}^2\right)$$

$$k_{0_{\max}} = \min\left(\frac{k_1 k_2}{m_{\max}}, -e_1 + \sqrt{e_1^2 + e_2}\right) \quad (7)$$

and choose the design gains $k_0$, $k_1$ and $k_2$ in the following way: take any positive value for $k_1$; then choose $k_2$ such that $k_2 > k_{2_{\min}}$; finally take $0 < k_0 < k_{0_{\max}}$. In this case we have the following:

1) All the closed loop signals $x_0$, $x_1$, $x_2$, $w$ and the control $u$ are globally bounded.
2) $\lim\limits_{t \to \infty} x(t) = 0$ and $\lim\limits_{t \to \infty} \dot{x}(t) = 0$.

### B. Forced limit cycle description of a Bouc-Wen hysteresis

In this section, we consider the system composed only of the two equations (2) and (3) where the input is $x(t)$ and the output is $\Phi(x)(t)$. We consider in this section that the displacement signal is periodic so that the output $\Phi(x)(t)$ is also asymptotically periodic [22]. The objective of this section is to characterize analytically the asymptotic limit cycle.

*1) Class of inputs:* In this section, we consider that the input signal $x(t)$ is wave $T$-periodic [22]. This means that it is continuous on the time interval $[0, +\infty)$ and periodic of period $T > 0$. Furthermore there exists a scalar $0 < T^+ < T$ such that the signal $x$ is $C^1$ on both intervals $(0, T^+)$ and $(T^+, T)$ with $\dot{x}(\tau) = \dfrac{dx(\tau)}{d\tau} > 0$ for $\tau \in (0, T^+)$ and $\dot{x}(\tau) < 0$ for $\tau \in (T^+, T)$. We denote $X_{\min} = x(0)$ and $X_{\max} = x(T^+) > X_{\min}$ the minimal and maximal values of the input signal respectively (see Figure 1). Periodic sine and triangular signals are also wave periodic.

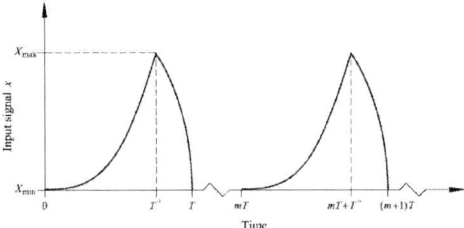

Fig. 1.  Example of a $T$-wave periodic signal.

*2) Analytic description:* Define the following functions:

$$\varphi_{\sigma,n}^-(\mu) = \int_0^\mu \frac{du}{1 + \sigma|u|^{n-1}u + (\sigma - 1)|u|^n} \quad (8)$$

$$\varphi_{\sigma,n}^+(\mu) = \int_0^\mu \frac{du}{1 - \sigma|u|^{n-1}u + (\sigma - 1)|u|^n} \quad (9)$$

$$\varphi_{\sigma,n}(\mu) = \varphi_{\sigma,n}^+(\mu) + \varphi_{\sigma,n}^-(\mu) \quad (10)$$

for any scalar $\mu \in (-1, 1)$. It has been shown in [22] that the functions $\varphi_{\sigma,n}^-(\cdot)$, $\varphi_{\sigma,n}^+(\cdot)$ and $\varphi_{\sigma,n}(\cdot)$ are strictly increasing on the interval $(-1, 1)$ so that they are bijective. Their inverses are denoted $\psi_{\sigma,n}^-(\cdot)$, $\psi_{\sigma,n}^+(\cdot)$ and $\psi_{\sigma,n}(\cdot)$ respectively. The limit cycle for the Bouc-Wen model is described by the following [22]:

*Theorem 2:* Let $x(t)$ be a wave $T$-periodic input signal. Define the functions $\omega_m$ and $F_m$ for any positive integer $m$ as follows

$$\omega_m(\tau) = w(mT + \tau) \quad \text{for} \quad \tau \in [0, T] \quad (11)$$

$$\phi_m(\tau) = \kappa_x x(\tau) + \kappa_w \omega_m(\tau) \quad \text{for} \quad \tau \in [0, T] \quad (12)$$

Then the sequence of functions $\{\phi_m\}_{m \geq 1}$ (resp. $\{\omega_m\}_{m \geq 1}$) converges uniformly on the interval $[0, T]$ to a continuous function $\bar{\Phi}$ (resp. $\bar{w}$) defined as

$$\bar{\Phi}(\tau) = \kappa_x x(\tau) + \kappa_w \bar{w}(\tau) \quad \text{for} \quad \tau \in [0, T] \quad (13)$$

$$\bar{w}(\tau) = \psi_{\sigma,n}^+\left(\varphi_{\sigma,n}^+\left[-\psi_{\sigma,n}\left(\rho\left(X_{\max} - X_{\min}\right)\right)\right]\right.$$
$$\left. + \rho\left(x(\tau) - X_{\min}\right)\right) \text{ for } \tau \in [0, T^+] \quad (14)$$

$$\bar{w}(\tau) = -\psi_{\sigma,n}^+\left(\varphi_{\sigma,n}^+\left[-\psi_{\sigma,n}\left(\rho\left(X_{\max} - X_{\min}\right)\right)\right]\right.$$
$$\left. - \rho\left(x(\tau) - X_{\max}\right)\right) \text{ for } \tau \in [T^+, T] \quad (15)$$

Furthermore we have for all $\tau \in [0, T]$

$$-1 < -\psi_{\sigma,n}\left(\rho\left(X_{\max} - X_{\min}\right)\right) \leq \bar{w}(\tau)$$
$$\leq \psi_{\sigma,n}\left(\rho\left(X_{\max} - X_{\min}\right)\right) < 1$$

the lower and upper bounds of $\bar{w}(\tau)$ being attained at $\tau = 0$ and $\tau = T^+$ respectively.

This result means that the output restoring force goes asymptotically to a periodic function. The transient behavior is captured by equations (11)-(12) while the steady-state is captured by equations (13)-(15). The loading part of the limit cycle is described by equations (13) and (14), while the unloading part is described by equations (13) and (15). Loosely speaking, the functions $\bar{\Phi}$ and $\bar{w}$ denote the steady-state responses of the functions $\Phi$ and $w$ respectively.

## III. SYSTEM MODELING

The system model is given by [19]:

$$m'\ddot{x}(t) + c'\dot{x}(t) + \kappa_a x(t) = \kappa_b \Phi(u)(t) \quad (16)$$

where $\kappa_a$ and $\kappa_b$ are elastic constants, $m'$ and $c'$ are the equivalent mass and damping coefficient of the piezoelectric actuator, $x(t)$ its relative position with respect to the sensor, and $\kappa_b \Phi(u)(t)$ is the force produced by the actuator. The term $\Phi(u)(t)$ is assumed to follow a Bouc-Wen equation so that the actuator may be represented by:

$$m'\ddot{y}(t) + c'\dot{y}(t) + \kappa_a \left(y(t) - y_0\right) = \kappa_b \kappa'_x u(t) + \kappa_b \kappa'_w w(t), \quad (17)$$

$$\dot{w}(t) = \rho\left(\dot{y}(t) - \sigma\,|\dot{y}(t)|\,|w(t)|^{n-1}\,w(t) + (\sigma - 1)\,\dot{y}(t)|w(t)|^n\right) \quad (18)$$

where $\kappa'_w$ and $\kappa'_x$ are constant gains. The nonlinear term $w(t)$ takes into account the effect of hysteresis.

Defining:

$$m = \frac{m'}{\kappa_a \kappa'_x}\ ,\ c = \frac{c'}{\kappa_a \kappa'_x}\ ,\ \kappa_x = \frac{\kappa_a}{\kappa_b k'_x}\ ,\ \kappa_w = -\frac{\kappa'_w}{\kappa'_x} \quad (19)$$

it can be seen that the actuator follows equations (1)-(3).

## IV. CONTROL LAWS

This section introduces three control laws for the piezoelectric device, which are based on the linear controller of Section II. These controllers are tested by means of numerical simulations.

### A. Control objective

The control objective is to insure the boundedness of all the closed loop signals, along with the regulation of the displacement and velocity of the piezoelectric actuator to zero. Furthermore, in steady-state, the control output has to have a unique value so that the closed loop system has a unique equilibrium point.

### B. PID Control

In this section we consider the closed loop formed by the system (1)-(3) along with the control law (5). The closed loop is then described by the equations:

$$\dot{x}_0 = x_1, \quad (20)$$

$$\dot{x}_1 = x_2, \quad (21)$$

$$\dot{x}_2 = m^{-1}\left(-\left(c + k_2\right)x_2 - \left(\kappa_x + k_1\right)x_1 - k_0 x_0 \right. \\ \left. -\kappa_w w - m\ddot{y}_r - c\dot{y}_r - \kappa_x y_r\right), \quad (22)$$

$$\dot{w} = \rho\left(x_2 + \dot{y}_r - \sigma|x_2 + \dot{y}_r|\,|w|^{n-1}w \right. \\ \left. +(\sigma - 1)\left(x_2 + \dot{y}_r\right)|w|^n\right). \quad (23)$$

In order to determine the PID constants $k_0$, $k_1$ and $k_2$, we need to have known bounds on the unknown parameters (Assumption 1). We use the following bounds:

- $m_{\min} = 3.98 \times 10^{-3}$ V s$^2$ m$^{-1}$
- $m_{\max} = 6.63 \times 10^{-3}$ V s$^2$ m$^{-1}$
- $c_{\max} = 13.43$ V s m$^{-1}$
- $k_{x\max} = 10.8 \times 10^6$ V m$^{-1}$
- $\sigma_{\max} = 0.9212$
- $\rho_{\max} = 9.510 \times 10^4$ m$^{-1}$
- $k_{w\max} = 48.74$ V

The PID controller parameters are determined using Theorem 1. The first design parameter to be chosen is $k_1 = 5 \times 10^6$ so that we get $k_{2min} = 567.6$. We choose $k_2 = 580$ so that we obtain $k_{0max} = 1.16 \times 10^9$. Finally we take $k_0 = 1 \times 10^9$.

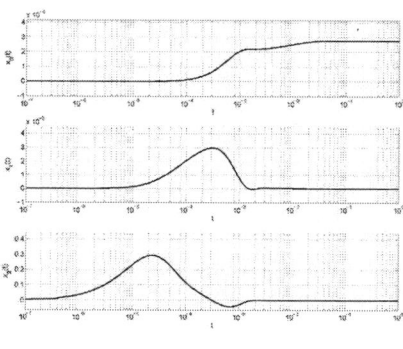
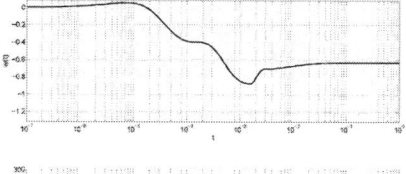
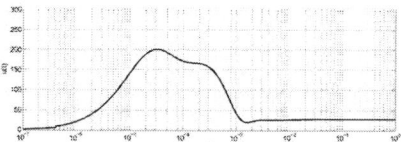

Fig. 2. Closed loop signals relative to the control law of Section IV-B.

Figure 2 gives the behavior of the closed loop signals with $m = 5.3 \times 10^{-3}$ V·s$^2$m$^{-1}$ and $c = 13$ V·s·m$^{-1}$. The initial conditions are $x_0(0) = 0$ m·s, $x_1(0) = 20 \times 10^{-6}$ m, $x_2(0) = 0.2$ m/s and $w(0) = 0$. For the reference signal, we choose $y_r$ as the output of the second order linear system $\dfrac{\omega_0^2}{s^2 + 2\xi\omega_0 s + \omega_0^2}$ with $\xi = 0.7$, $\omega_0 = 2\pi \times 500$ rad/s and zero input; that is, the linear system is driven only by the non-zero initial conditions $y_r(0) = x(0)$ and $\dot{y}_r(0) = \dot{x}(0)$. It can be seen that the outputs $x_1$ and $x_2$ are regulated to zero. Note that, although the control signal $u$ is zero for negative times, its asymptotic value is different from zero. This fact can be explained as follows. Taking $y_r = 0$ in equations (22)-(23), it can be seen that the four states system (20)-(23) has an infinite number of equilibrium points. These equilibria are defined by $\{x_1 = 0, x_2 = 0, k_0 x_0(\infty) = \kappa_w w(\infty) = u(\infty)\}$. It is not necessary that $x_0(\infty) = 0$ so that the control value may be nonzero asymptotically (see Figure 2). In practice, this behavior is undesirable as it implies that the actuator applies a control action at equilibrium, which means an unnecessary loss of energy. Another inconvenient of this behavior is the modification of the equilibrium point of the system.

*C. PID plus a sinusoidal component*

The previous section has pointed out to the possible modification of the equilibrium point of the system under the action of a PID controller. Since this behavior is not acceptable in practice, a modification of the controller is proposed in this section to reduce this effect. The reason for having a control which is not zero asymptotically is that $u(\infty) = \kappa_w w(\infty)$ where $w(\infty)$ is not necessarily zero. To solve this problem, the idea would be to force the hysteretic term to go to zero asymptotically, inducing the control to go to zero. Consider that the system (1)-(3) is in open loop and choose for $u(t)$ a wave periodic input signal (see Section II-B1). Numerical simulations show that the obtained displacement signal $x(t)$ is also wave periodic. On the other hand, we know from Theorem 2 that, if the signal $x(t)$ is wave periodic, then the hysteretic output $w(\cdot)$ is also wave periodic and that it belongs asymptotically to the interval $[-\psi_{\sigma,n}(\rho(X_{\max} - X_{\min})), \psi_{\sigma,n}(\rho(X_{\max} - X_{\min}))]$. On the other hand, it can be shown that, for fixed values of the parameters $\sigma$ and $n$, the function $\psi_{\sigma,n}(\mu)$ is increasing with its argument $\mu$. This implies that the interval $[-\psi_{\sigma,n}(\rho(X_{\max} - X_{\min})), \psi_{\sigma,n}(\rho(X_{\max} - X_{\min}))]$ can be made as small as desired if the quantity $X_{\max} - X_{\min}$ can be reduced arbitrarily. Numerical simulations suggest that if the amplitude of the wave periodic voltage input $u(t)$ is decreased, then the amplitude of the corresponding displacement signal is also decreased.

These remarks suggest the following control law for the system (1)-(3)

$$u(t) = -k_0 x_0(t) - k_1 x_1(t) - k_2 x_2(t) - A\sin(2\pi f t) \tag{24}$$

where $A$ and $f$ are positive design constants, and $k_0$, $k_1$, $k_2$ are computed using Theorem 1. The closed loop

behavior is given in Figures 3 and 4 with the values of $k_0$, $k_1$, $k_2$ that have been determined in the previous section, and for different values of the parameters $A$ and $f$. The initial states are $x_0(0) = 0$ m·s, $x_1(0) = 20 \cdot 10^{-6}$ m, $x_2(0) = 0.2$ m/s and $w(0) = 0$. The reference signal is chosen as in Section IV-B. As noticed before, the steady-state response of the closed loop is periodic, and it can be seen that the amplitude of the closed loop signals $x(t)$, $\dot{x}(t)$ and $u(t)$ decreases as $A$ decreases. The amplitude of the steady-state closed loop signals is independent of the frequency $f$. This frequency influences the settling time: the transient response of the system has a shorter duration for higher frequencies $f$.

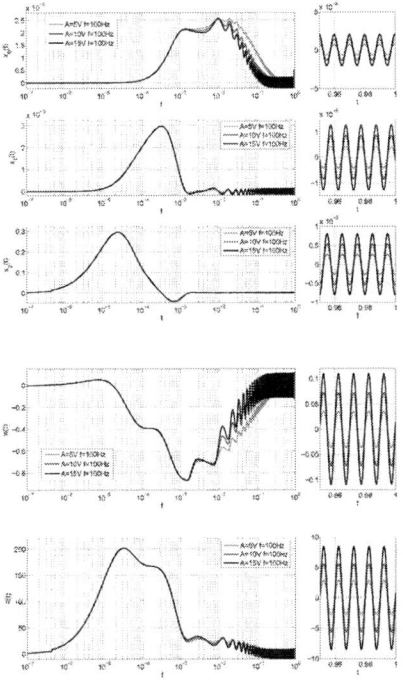

Fig. 3. Closed loop signals relative the control law of Section IV-C. The figures in the right are a zoom in the indicated region of the figures in the left.

As a conclusion, adding a term $A\sin(2\pi f t)$ to the PID controller makes the closed loop set point oscillating around zero. The amplitude of the oscillations can be made as small as desired by reducing the design parameter $A$.

*D. PID plus a sinusoidal component with a time varying amplitude*

The previous section has studied the behavior of a PID plus a sinusoidal component in the control law. It has been noticed that the set point of the closed loop steady-state systems oscillates around zero. As oscillations are also undesirable in practice, the control law has to be modified in order to eliminate them. Notice that the amplitude of the oscillations decreases with the amplitude

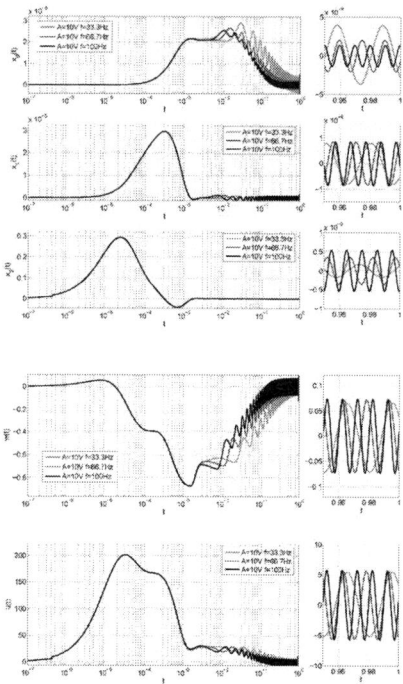

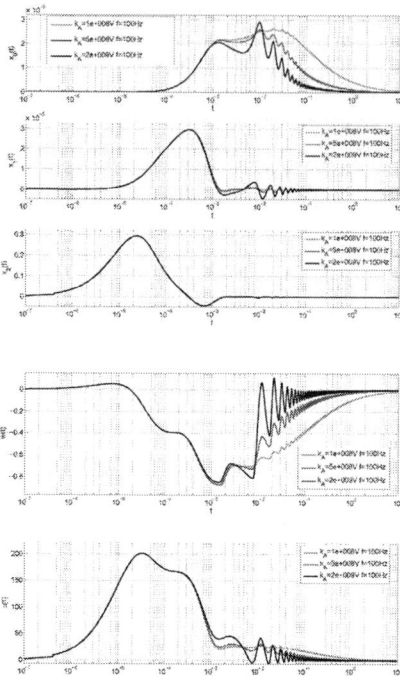

Fig. 4. Closed loop signals relative the control law of Section IV-C. The figures in the right are a zoom in the indicated region of the figures in the left.

Fig. 5. Closed loop signals applying the contol law introduced in Section IV-D with $f = 100$ Hz and different $k_A$ values.

of the sinusoidal component of the control law. This fact suggests to use for this component a time-varying amplitude that tends to decrease as the control law goes to zero. Since $u(\infty) = k_0 x_0(\infty)$ for the PID case, we choose as control law the expression:

$$u(t) = -k_0 x_0(t) - k_1 x_1(t) - k_2 x_2(t) - k_A x_0(t) \sin(2\pi f t) \quad (25)$$

where $k_A$ is a constant gain.

This control law has been tested using numerical simulations. The initial conditions are $x_0(0) = 0$ m·s, $x_1(0) = 20 \times 10^{-6}$ m, $x_2(0) = 0.2$ m/s and $w(0) = 0$. The reference signal is chosen as in Section IV-B.

The frequency $f$ is taken to be 100 Hz as this value makes the settling time shorter without harming the overall response (see the previous section). Three values of $k_A$ are chosen to study the effect of this parameter. The results of the closed loop simulations are given in Figure 5. It can be seen that the closed loop signals $x_1$ and $x_2$ converge to zero and that larger values of $k_A$ lead to a shorter settling time. Furthermore, the control value is the same before and after the perturbation so that the equilibrium point of the closed loop remains unchanged.

## V. Conclusion

This paper has presented a new control law for hysteretic systems, applying it to a piezoelectric actuator. The actuator has been represented using the Bouc-Wen

model for hysteresis. Three control laws have been tested numerically for the position regulation of the piezoelectric device. It has been observed that a PID with a time-varying component insures that the displacement and velocity of the actuator go to zero asymptotically, while maintaining the same equilibrium point for the closed loop system.

## Acknowledgment

Supported by CICYT through grants DPI2005-08668-C03-03 and DPI2005-08668-C03-01. The authors acknowledge the support of the Institute of Composite Structures and Adaptive Systems (German Aerospace Center) in Braunschweig, Germany, where the piezoelectric actuator has been assembled. The second author acknowledges the support of the Spanish Ministry of Education and Science through the "Ramón y Cajal" program.

## References

[1] A. Visintin, *Differential Models of Hysteresis.* Springer-Verlag, 1994.
[2] P. Duhem, "Die dauernden aenderungen und die thermodynamik," *Zeitschrift für Physikalische Chemie,* vol. 22, pp. 543–589, 1897.
[3] M. Krasnosel'skii and A. Pokrvskii, *Systems with Hysteresis.* Nauka, Moscow: Springer-Verlag, 1983.
[4] F. Preisach., "Über die magnetische nachwirkung," *Zeitschrift für Physik,* vol. 94, pp. 277–300, Feb 1935.
[5] G. Foliente, "Hysteresis modeling of wood joints and structural systems," *ASCE Journal of Structural Engineering,* vol. 121, no. 6, pp. 1013–1022, 1995.

[6] J. Macki, P. Nistri, and P. Zecca, "Mathematical models for hysteresis," *SIAM Review*, vol. 35, no. 1, pp. 94–123, 1993.

[7] C. Ru and L. Sun, "Hysteresis and creep compensation for piezoelectric actuator in open-loop operation," *Sensors and Actuators A: Physical*, vol. 122, no. 1, pp. 124–130, July 2005.

[8] S. A. Turik, L. A. Reznitchenko, A. N. Rybjanets, S. I. Dudkina, A. V. Turik, and A. A. Yesis, "Preisach model and simulation of the converse piezoelectric coefficient in ferroelectric ceramics," *Journal Of Applied Physics*, vol. 97, no. 64102, pp. 1–4, 2005.

[9] P. Ge and M. Jouaneh, "Modeling hysteresis in piezoceramic actuators," *Precision Engineering*, vol. 17, pp. 211–221, 1995.

[10] P. Ge, P. Ge, and M. Jouaneh, "Tracking control of a piezoceramic actuator," *IEEE Trans. Contr. Syst. Technol.*, vol. 4, no. 3, pp. 209–216, 1996.

[11] T. S. Low and W. Guo, "Modeling of a three-layer piezoelectric bimorph beam with hysteresis," *Journal of Microelectromechanical Systems*, vol. 4, no. 4, pp. 230–237, Dec 1995.

[12] F. Ikhouane, O. Gomis-Bellmunt, and P. Castell-Vilanova, "A new identification method for hysteretic systems: Theory and experiments," in *ECC2007 European Control Conference 2007*, Kos, Greece, July 2007.

[13] O. Gomis-Bellmunt, F. Ikhouane, P. Castell-Vilanova, and J. Bergas-Jané, "Modeling and validation of a piezoelectric actuator," *Electrical Engineering*, vol. on-line, pp. 1–10, 2006.

[14] R. Smith, *Smart Material Systems: Model Development*. Philadelphia: SIAM, 2005.

[15] Y. K. Wen, "Method of random vibration of hysteretic systems," *Journal of Engineering Mechanics Division, ASCE*, vol. 102(2), pp. 249–263, 1976.

[16] A. W. Smyth, S. F. Masri, E. B. Kosmatopoulos, A. G. Chassiakos, and T. K. Caughey, "Development of adaptive modeling techniques for non-linear hysteretic systems," *International Journal of Non-Linear Mechanics*, vol. 37, pp. 1435–1451, 2002.

[17] B. F. Spencer, S. J. Dyke, M. K. Sain, and J. D. Carlson, "Phenomenological model for magnetorheological dampers," *Journal of Engineering Mechanics ASCE*, vol. 123, pp. 230–238, 1997.

[18] S. B. Choi, S. K. Lee, and Y. P. Park, "A hysteresis model for the field-dependent damping force of a magnetorheological damper," *J. Sound. Vibr.*, vol. 245, no. 2, pp. 375–383, 2001.

[19] B. M. Chen, T. H. Lee, C. C. Hang, Y. Guo, and S. Weerasooriya, "An $h_\infty$ almost disturbance decoupling robust controller design for a piezoelectric bimorph actuator with hysteresis," *IEEE Trans. Contr. Syst. Technol.*, vol. 7, no. 2, pp. 160–174, 1999.

[20] F. Ikhouane, V. Mañosa, and J. Rodellar, "Adaptive control of a hysteretic structural system," *Automatica*, vol. 41, no. 2, pp. 225–231, 2005.

[21] F. Ikhouane and J. Rodellar, "A linear controller for hysteretic systems," *IEEE Transactions on Automatic Control*, vol. 51, no. 2, pp. 340–344, 2006.

[22] ——, "On the hysteretic Bouc-Wen model. part I: Forced limit cycle characterization," *Nonlinear Dynamics*, vol. 42, no. 1, pp. 63–78, 2005.

# Electric drive for carding machine draft device

Martin Diblík*

* Technical university of Liberec, Faculty of Mechatronics and Interdisciplinary Engineering Studies, Hálkova 6,
Liberec, Czech Republic, e-mail: *martin.diblik@tul.cz*

*Abstract*—Drafting of textile sliver is very important technology step in textile production. The draft device is equipped with speed-controlled electric drives and its high-dynamic response is necessary to obtain high-quality sliver. The article is focused on techniques and procedures that improve dynamic responses of controlled electric drives with synchronous machines with permanent magnets in rotor (PMSM). Significant place is dedicated to study current-feedforward control structure characteristics. Next part describes PMSM servodrive application to control the speed of carding machine draft device. According to the simulation, above-mentioned method of dynamics improvement leads into quality increase of output sliver in consequence.

*Keywords*—Permanent magnet motor, Electrical Drive, Highly dynamic drive, Simulation, Industrial application.

## I. INTRODUCTION

Draft process is one of the most important stages of the textile production chain. Low irregularity of drafted sliver is essential if high quality yarn is to be produced. An uncontrolled sequence of subsequent drafting and doubling operation was sufficient to achieve an acceptable sliver irregularity at older low-speed production machines. Modern draft devices can control the draft value in accordance with actual input sliver irregularity. The draft device must be driven by high-dynamics servodrive to eliminate short-time irregularity. To use high-dynamics drive leads to increase the energy consumption of machine but the energy requirements are lower in comparison with old machines.

The article deals with techniques and procedures that improve the dynamic responses of controlled electric drives. To achieve the high accuracy of drive speed, we have to focus on drives with synchronous machines with permanent magnets in rotor (PMSM). The main attention is focused to modification of standard control structures, especially the influence of current feedforward to the speed control loop bandwidth.

## II. MATHEMATIC MODELS OF SYNCHRONOUS MACHINE

Suitable mathematic model is basic precondition to investigate the properties and behaviour of PMSM by simulation. Published books and papers describe several ways of PMSM model synthesis. So called DQ-model is used very often for the simulation, its description can be found in [1], [2]. This model uses transformation of stator coordinates (axis are marked as $\alpha$, $\beta$) into the rotor coordinate system (marked as $d$, $q$). This system rotates with same angular velocity $\omega$ as the stator electromagnetic field.

In some cases the model is completed with the real machine losses term. If it is necessary to include this model into complex simulation structure, the simplified single-phase PMSM model is more suitable. Such model will be introduced below.

Fig. 1 shows the substitution electric scheme of PMSM, where $R_s$ stands for single phase winding resistance, $L_s$ means the one phase inductance. Permanent magnets in rotor agitate the magnetic field with $\Phi_F$ flux that induces the voltage $U_E$.

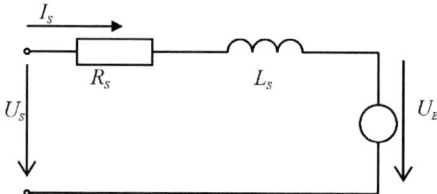

Fig. 1. Substitute electric diagram of permanent magnet synchronous machine.

### A. DQ-model of PMSM

This model can be derived from the model of general synchronous machine if we supose some simplification:

- The power voltage and induced voltage is harmonic.
- Parameters R and L of stator winding are same for all three phases and constant.
- Magnetic circuit of staror is linear.
- The iron losses are neglected.

The voltage equation is valid for the scheme at Fig.1, where superscript $S$ indicates the stator coordinate system quantities.

$$\mathbf{U}_1^S = R_a \cdot \mathbf{I}_1^S + \frac{d\mathbf{\Psi}_1^S}{dt} \tag{1}$$

Coupled magnetic flux can be defined as:

$$\mathbf{\Psi}_1^S = \mathbf{\Phi}_F^S \cdot e^{j\vartheta} + L_1 \cdot \mathbf{I}_1^S \tag{2}$$

Permanent magnet flux $\Phi_F$ is transformed into stator coordinates. The transformation procedure is described in [1]. Both equations can be converted into rotor coordinate system DQ (3), (4) and decomposed into orthogonal d- and q- componets.

$$\mathbf{U}_1^R = R_a \cdot \mathbf{I}_1^R + \frac{d\mathbf{\Psi}_1^R}{dt} + j\omega \mathbf{\Psi}_1^R \tag{3}$$

$$\mathbf{\Psi}_1^R = \mathbf{\Phi}_F^R + L_1 \cdot \mathbf{I}_1^R \tag{4}$$

978-1-4244-1741-4/08/$25.00 ©2008 IEEE

Motor electromagnetic torque can be derived from equation, where $p_p$ means the number of machine pole pairs.

$$M_E = \frac{3}{2} p_p \cdot \text{Im}\{\mathbf{\Psi}_1^* \cdot \mathbf{I}_1\} = \frac{3}{2} p_p [\Phi_F + (L_d - L_q)i_d]i_q \quad (5)$$

The motion equation (6) is the final part of PMSM mathematical model. The $\omega_m$ means the angular speed of machine and the symbol $M_L$ indicates the mechanical load torque of machine.

$$J \frac{d\omega_m}{dt} = M_E - M_L \quad (6)$$

The conversion between mechanical and electrical angular velocity is determined by number of pole pairs according to equation

$$\omega = p_p \cdot \omega_m \quad (7)$$

### B. Simple model of PMSM

The simple model of PMSM is often necessary for primary consideration or if it is required to include the PMSM model into complex control structures. Such simple model can be derived from common mathematical description of direct current electromotor excited by permanent magnets. We can suppose the same substitute electric scheme introduced in Fig. 1.

The voltage equation can be expressed as:

$$R_a \cdot i + L_s \frac{di}{dt} + U_E = U, \quad (8)$$

where induced voltage $U_E$ and machine torque $M_E$ is:

$$U_E = K_E \cdot \omega \quad (9)$$

$$M_E = K_M \cdot i \quad (10)$$

The motion equation has the same form as (6). Voltage constant $K_E$ and torque constant $K_M$ are parameters of particular simulated machine and have to be measured experimentally [3]. The simulation structure of simple model shows Fig. 2.

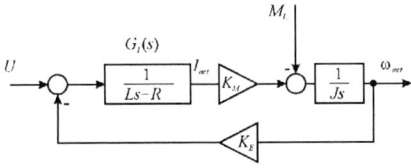

Fig.2. Simulation scheme of simple PMPS model.

The transfer (11) substitutes the dynamics of RL circuit.

$$G_I(s) = \frac{1}{L \cdot s + R} \quad (11)$$

### C. Parameters of real PMSM

It is always necessary to verify theoretical result with an aid of measurement at real machine or device.

The parameters of simulation models described above were filled with values that were measured experimentally. All measurements are described in [3].

For verification was used real Siemens synchronous servomotor 1FT6062-6AF71. The servomotor simulation parameters are summarized in next tables.

TABLE I.
PARAMETERS OF REAL SYNCHRONOUS SERVOMOTOR FOR D,Q-MODEL

| | | |
|---|---|---|
| $R_a$ [Ω] | 2,717 | Single phase stator winding resistance (lead cables included) |
| $L_d$ [H] | 0,0190 | Inductance of stator winding in d-axis direction |
| $L_q$ [H] | 0,0201 | Inductance of stator winding in q-axis direction |
| $\Phi_F$ [Wb] | 0,51334 | Coupled magnetic flux of rotor |
| $J$ [kg.m²] | 0,85.10⁻³ | Inertia of rotor |
| $p_P$ [-] | 3 | Machine pole-pairs |

TABLE II.
PARAMETERS OF REAL SYNCHRONOUS SERVOMOTOR FOR SIMPLE MODEL

| | | |
|---|---|---|
| $R_a$ [Ω] | 2,717 | Single phase stator winding resistance (lead cables included) |
| $L_S$ [H] | 0,0127 | Inductance of stator winding. |
| $K_E$ [V/rad.s⁻¹] | 0,726 | Voltage constants |
| $K_M$ [Nm.A⁻¹] | 0,725 | Torque constants |
| $J$ [kg.m²] | 0,85.10⁻³ | Inertia of rotor |
| $p_P$ [-] | 3 | Machine pole-pairs |

## III. COMMON REGULATION STRUCTURES OF ELECTRIC DRIVES

Mathematic model of any electric drive can be considered as dynamic system from the control engineering view. There is used the cascade layout of closed loops to control the motor behaviour. The basic loop is the current control loop (Fig.3). It has to speed up and optimize the current control process. The control element $R_I(s)$ is controller, mostly with common PI structure. The controller can be described with transfer function:

$$R_I(s) = K_{PI} \frac{T_{NI} \cdot s + 1}{T_{NI} \cdot s} \quad (12)$$

Electric controlled drives are mainly used as speed drives, the main regulated value is speed of drive $\omega_m(t)$. The speed loop must be closed in case of PMSM servodrives to achieve high accuracy and dynamic response. The speed controller $R_\omega(s)$ or $R_n(s)$ has PI structure with transfer function:

$$R_n(s) = K_P \frac{T_N \cdot s + 1}{T_N \cdot s} \quad (13)$$

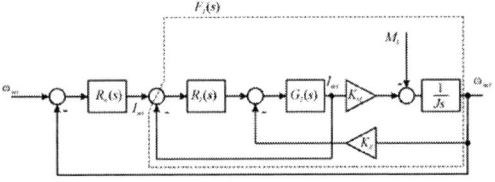

Fig.3. Simulation scheme of speed and current control loop with simple PMPS model

The Fig. 3 contains the speed and current control loop. The transfer function between current setpoint $i_{set}(t)$ and the actual revolutions $\omega_{act}(t)$ can be substitute as $F_I(s)$ transfer (14), where $R_I(s)$ is the current controller and $J$, $K_M$ and $K_E$ are parameters of servomotor.

$$F_I(s) = \frac{R_I G_I K_m \dfrac{1}{J \cdot s}}{1 + \dfrac{1}{J \cdot s} \cdot G_I K_e + G_I K_m K_e} \tag{14}$$

## IV. VERIFICATION OF SERVODRIVE SIMULATION MODEL

The necessary next step in our quest of dynamics improvement is verification of servodrive simulation model. This model contains PMSM block and two feedback control loops, as described above. Model was compared with real servodrive based on the Siemens Masterdrive MotionControl (MDMC) unit with 1FT6062-6AF71 servomotor.

Because the current controller values of real drive cannot be noticeably modified (and manufacturer does not recomend this changes), the original current controller setting is supposed. The controller setting in simulation was obtained by optimisation algorithm that compared step responses of real and model current control loops.

The same optimisation algorithm found suitable setting of speed controller in simulation model. Step responses of real drive and simulation model are shown at Fig. 4.

This setting of controllers was basic precondition for next steps.

## V. CURRENT FEEDFORWARD

The use of feedforward control structure allows eliminating the control deviation when the regulated quantity setpoint is time-variable. The current feedforward is used in case of speed servodrive. The size of this additive signal is deduced from the speed-setpoint first derivation and is added to the current setpoint. Control structure with this modification is shown at Fig. 5.

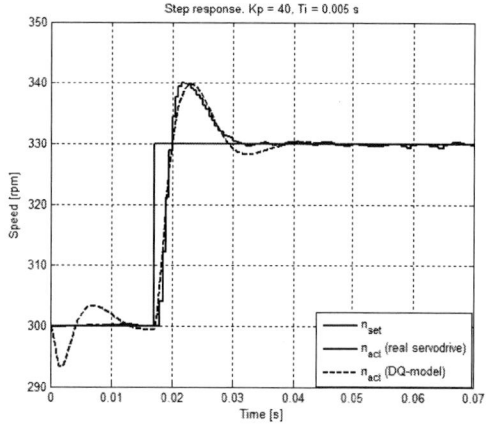

Fig. 4. Comparison of step responses of real servodrive (green) and simulation model (red).

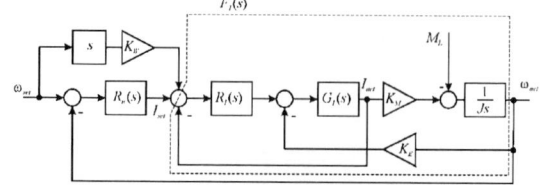

Fig. 5. Speed control loop with current feedforward, the PMSM model is simplified.

The transfer function of speed control loop including the current feedforward $K_W$ is:

$$F_n(s) = \frac{R_n F_I + s \cdot K_W F_I}{1 + R_n F_I} \tag{15}$$

If we want to investigate the influence of current feedforward on the speed control loop behaviour analytically, it is necessary to do some simplification of $F_I(s)$ transfer. In first step we can suppose that $G_I(s)$ and $R_I(s)$ terms are equal to 1. After decomposition of $F_I(s)$ in (12) we can write the transfer function as:

$$F_n(s) = \frac{K_W K_M T_N s^2 + K_P K_M T_N s + K_P K_M}{J T_N s^2 + K_P K_M T_N s + K_P K_M} \tag{16}$$

The desired dynamics of speed control loop can be described in frequency domain by condition:

$$\left| F_n(j\omega) \right| = 1 \tag{17}$$

If we compare the numerator and denominator of transfer function (13), the desired value of $K_W$ gain is then:

$$K_W = \frac{J}{K_M} = 0{,}00078 \, \text{Am}^{-1}\text{s}^2 \tag{18}$$

The numerical value of this gain is based on parameters of real servomotor 1FT6062-6AF71 mentioned in chapter II.

General solution of $K_W$ is more complicated when partial transfers $G_I(s)$ and $R_I(s)$ are not equal to 1. Transfer function of speed control loop $F_n(s)$ is then

$$F_n(s) = \frac{b_3 s^3 + b_2 s^2 + b_1 s + b_0}{a_4 s^4 + a_3 s^3 + a_2 s^2 + a_1 s + a_0} \tag{19}$$

where coefficients are follow:

$$\begin{aligned}
b_3 &= K_w T_N T_I K_M K_E \\
b_2 &= T_N K_M K_E (K_w + K_N T_I) \\
b_1 &= K_N K_M K_E (T_I + T_N) \\
b_0 &= K_N K_M K_E
\end{aligned}$$

$$\begin{aligned}
a_4 &= JLT_I T_N \\
a_3 &= JT_I T_N (R + K_I) \\
a_2 &= T_N (JK_I + K_M K_E T_I + K_M K_E K_N T_I) \\
a_1 &= K_M K_E (K_N T_N + T_I) \\
a_0 &= K_M K_E
\end{aligned} \tag{20}$$

The condition (17) is not achievable because of the $F_n(s)$ shape, thus it is necessary to verify the influence of $K_W$ gain by simulation. We test DQ-model and simple model to compare the results. Both models are used with current and speed control loop with current feedforward branch (as described in Fig. 5).

Step responses of both models are compared in Fig. 6. It is obvious that condition (17) cannot be fulfilled because the calculation of $K_W$ was made with simplifying conditions. The responses obtain considerable derivative aspect that indicates $K_W$ value too high.

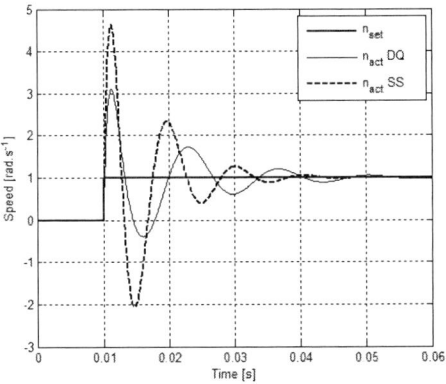

Fig. 6. Step response of speed loop with DQ-model and simple model of MPSM with analytically calculated value of current feedforward gain $K_W$.

To find out optimal value of current feedforward gain $K_W$ we have to compare the amplitude frequency characteristics of speed control loop. Our aim is to achieve the high loop bandwidth $f_{BW}$ and acceptable amplitude peak $L_{max}$ in this characteristic. In other words we should approximate the frequency characteristics to condition (17).

The simulation results (based on DQ-model of MPSM) are at Fig. 7. In first step the suitable $K_W$ value was found to obtain stable control process, then the repetitive simulations were made with $K_W$ in range 0 až $2.10^{-4}$.

The optimal bandwidth is marked by red curve ($K_W = 0,9.10^{-4}$). Blue curve shows the frequency characteristics for $K_W = 0$. It is obvious that the speed control loop is tuned very sharply and is not usable without proper current feedforward branch.

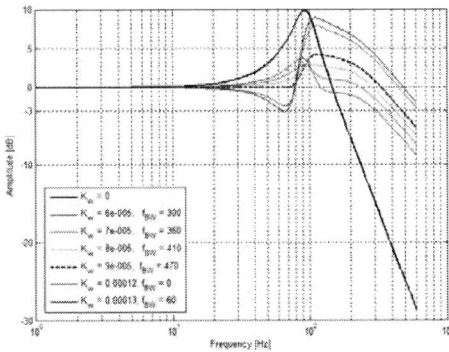

Fig. 7. Frequency characteristics of speed loop with DQ-model with different value of current feedforward gain $K_W$.

## VI. DRAFT DEVICE OF CARDING MACHINE

Middle phase of textile material processing means the transformation of raw material (e.g. preprocessed cotton) into the longtitudal form of sliver. This process is realized into the carding machine. In order to create next product form (yarn) with high quality, the constant density $T$ of sliver is required.

Quality of logtitudal form of textile products is defined as mass irregularity (see [4]) and for comparison is defined the quadratic mass irregularity $CV$ as (21), where $m_i$ is weight of sliver segment and dashed $m$ represents the weight of whole sliver. Value of $CV$ varies with length of tested sliver segments and can be graphically expressed as spectrogram (see Fig. 14).

$$CV = \frac{10^2}{\overline{m}} \sqrt{\frac{1}{p-1} \sum_{i=1}^{P} \left(m_i - \overline{m}\right)^2} \qquad (21)$$

The overall construction of carding machine has significant influence on the irregularity of yarn, but certain irregularity always lasts at carding machine output and must be eliminated by drafting process.

The draft mechanism (drafter) is formed by two or three pairs of glossy metal rollers (see Fig. 9). The value of draft $P$ (22) is defined as ratio between output $v_2$ and input $v_1$ circumferential velocity of neighbouring rollers or by ratio of input and output sliver density.

$$P = \frac{v_2}{v_1} = \frac{T_1}{T_2} \qquad (22)$$

If the speed of last pair of rollers is controlled, the total draft value can be variable to eliminate the input sliver irregularity. The actual value of input density is measured by tenzometric sensor (so called Trumpet). The servodrive for last pair of rollers has to have high dynamics (high frequency bandwidth) to eliminate mass irregularity at short distances (approx. $10^{-2}$ m) because the output production speed of carding machine is more then 200 m.min$^{-1}$.

The proper control of the last pair of rollers is the basic precondition for high quality of output product. This control used to be relatively less important in the past. The acceptable sliver irregularity was achieved by a sequence of subsequent drafting and doubling operations when low production machines were used.

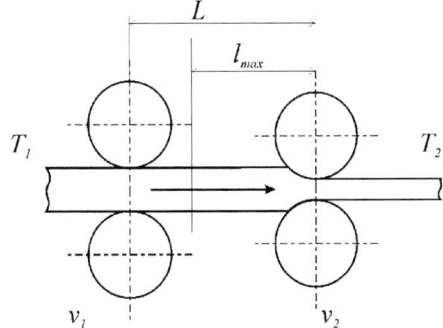

Fig. 8. Principal scheme of draft device.

Fig. 9. Real double-zone drafter implementation.

### A. Draft process model

Draft device can be simplified as only single zone for simulation purposes (see Fig.8). The arrangement may look similar to metal wire drafting, but there is an important difference between these two processes. Unlike metal wire, textile sliver is composed of many discrete fibres. These fibres do not change their length during the process (the Hook's law does not apply here), only their position respect to each other is changed. The lengths of individual fibres are random variables varying between minimum and maximum values $l_{max}$. The distance between pairs of rollers $L$ must be greater than $l_{max}$. If it is not the case, the different velocity of rollers breaks the individual fibres and degrades the final product.

The innovative approach to describe draft process was introduced in [5]. This model is nonlinear with transport delays, composed of two transfer functions $G_d(s)$ and $G_u(s)$ and can be aproximated by the transfer functions:

$$G_{dM}(s) = \frac{1}{P} \cdot \frac{0,0083s+1}{8,68 \cdot 10^{-5}s^2 + 4,08 \cdot 10^{-3}s+1} e^{-0,18s} \quad (22)$$

$$G_{uM}(s) = \frac{-4,59}{(0,008s+1)^2} \quad (23)$$

The aproximation is described in [6].

The simulated drafter has following parameters: input speed of sliver $v_1 = 0,49$ m.s$^2$, draft value $P = 1,7$ and distance between rollers is $L = 0,091$ m. The block expression of model is shown at Fig. 10. To avoid confusion between "time" ($t$) and "sliver density" ($T$) the sliver density will be marked as $d$ in the next articles.

The model is in difference form, where transfer $G_d(s)$ describes the change of output sliver density $\Delta d_2$ when the input sliver density changes $\Delta d_1$. The influence of output roller velocity change $\Delta v_2$ is described by transfer $G_u(s)$.

It is very complicated to verify the effect of servodrive-limited dynamics on real draft process quality because it is expensive and time consumpting experiment.

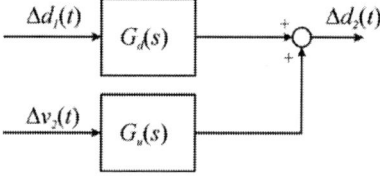

Fig. 10. Block expression of draft device model and ICM control structure.

That's why the simulation experiment is necessary. Because original transfers of error $G_d(s)$ and actuating $G_u(s)$ quantities have transport delays in internal feedbacks, the common design methods of control are hardly usable.

The internal model control structure (ICM) is suitable and the transfers of control elements were designed in [5]. The ICM principle goes out from the transfer inversion, thats why it is necessary to separate the noninvertable parts (mainly with transport delays) of transfer function.

The transfer function of plant can be written as

$$G(s) = G_D(s) \cdot G_0(s), \quad (24)$$

where $G_D(s)$ contains input transport delays and root coefficients corresponding to zeros of original transfer $G(s)$. The control element $R^*(s)$ is designed with an aid of decomposition as

$$R^*(s) = \frac{1}{G_0(s)} F(s), \quad (25)$$

and contains suitable low-pass filter $F(s)$ to achieve feasibility of $R^*(s)$. Simple control structure is shown in the Fig 11. Further theroretical background of ICM are e.g. in [7].

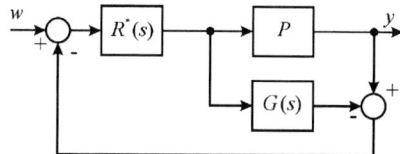

Fig. 11. ICM control structure for general plant $P$ control.

The control structure of draft process is more complicated, as can be seen in Fig. 12. Real draft process is simulated by original transfer functions $G_d(s)$ and $G_u(s)$, the internal model is represented by aproximated transfers (22) and (23).

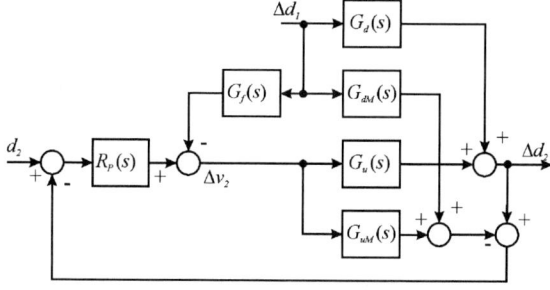

Fig. 12. Block expression of draft device model and ICM control structure.

Blocks $R_p(s)$ and $G_f(s)$ are regulator and filter designed according to [5] and [6]. Since $G_{uM}(s)$ is invertible, transfer functions of controller and feedforward compensator are:

$$R_p(s) = G_{uM}^{-1}(s) \frac{1}{(\lambda_1 s+1)^2} \quad (26)$$

$$G_f(s) = G_{uM}^{-1}(s) G_{dM}(s) \frac{1}{(\lambda_2 s+1)} \quad (27)$$

The values $\lambda_1$ and $\lambda_2$ should be selected to achieve an appropriate robustness and performance of regulation. In this example $\lambda_1 = 2$ ms and $\lambda_2 = 2.5$ ms is suitable.

### B. Draft process quality and servodrive performance

In paper [5] is supposed, that servodrive of draft device has got ideal transfer function $F_\omega(s) = 1$. This condition is not valid in reality therefore the quality of output sliver will be afected – the $CV$ value will be higher at short distances.

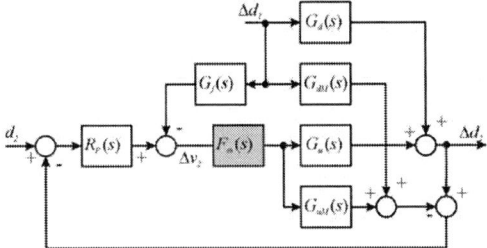

Fig. 13. Draft device model with ICM control structure and servodrive transfer function block.

The transfer of servodrive $F_\omega(s)$ should be inserted before the $G_u(s)$ and $G_{uM}(s)$ as sugested at Fig. 13.

Figure 14 shows the mass irregularity spectrogram of input sliver $CV_{in}$. The average value of sliver density is about $T_{1avg} = 6{,}8$ ktex. It is evident that short segments of sliver have high irregularity; the $CV$ value starts at 80 %.

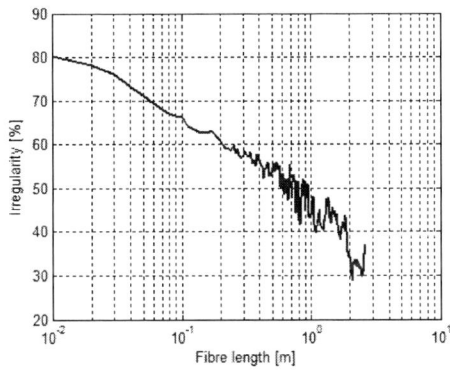

Fig. 14. Input sliver mass irregularity spectrogram

The transfer $F_\omega(s)$ contains the DQ-model of PMSM with current feedforward signal. The first simulation was realised without current feedforward (gain $K_w = 0$). The ICM control structure eliminates the irregularity of sliver markedly; the CV value decreases at 2 % (Fig. 15, solid curve)

Second simulation experiment reveals the influence of servodrive current feedforward. Irregularity of short segments falls at value 1,86 % (Fig. 15, dashed curve).

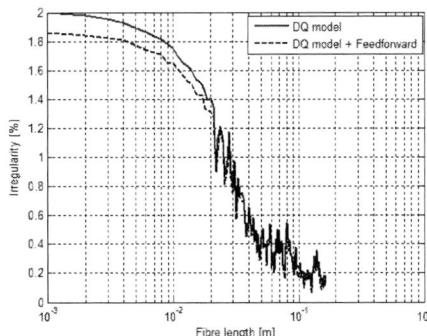

Fig. 15. Comparison of sliver mass irregularity spectrogram after draft process – the influence of servodrive current feedforward control.

The sliver quality improvement will be less significant at real draft device. The reason lays in higher production speed of these machines mainly and in differences between model of draft process and reality. Higher production speeds leads to transport delay reduction between density sensor and output pairs of rollers which means less time to accelerate or decelerate the output rollers.

## VII. CONCLUSION

The paper is applied to investigation how the limited dynamics of servodrive subsystem influence the overall quality of draft process. It has been shown that optimal current feedforward gain can be expressed analytically only for simple model of servodrive. For more complex DQ-model must be feedforward gain determined by simulation. The simulation of complex draft process model shows that higher dynamics of servodrive can improve the output quality of textile product.

### ACKNOWLEDGMENT

This work was supported by Ministry of Education of Czech Republic under project Research Center Textile II (1M0553).

### REFERENCES

[1] J. Pavelka, Z. Čeřovský, J. Javůrek, *Elektrické pohony*. 2. vydání, Praha: Vydavatelství ČVUT, 2001. 221 s. ISBN 80-01-02314-1.

[2] L. Kule a kol., *Technika elektrických pohonů*. 1. vydání, Praha: SNTL – Nakladatelství technické literatury, 1983. 580 s.

[3] M. Diblík, *Elektrické pohony pro dynamicky náročné aplikace*, dissertation thesis, TU Liberec, Fakulta mechatroniky, 2006.

[4] P. Ursíny, *Teorie předení 1.díl*. 2. vydání opravené, Liberec: VŠST v Liberci, 1987. 240 s.

[5] J. Hlava, *Modelling and Control of Cotton Sliver Drafting Process*, WSEAS Transactions on Circuits and Systems, Vol.2, No.4, pp.808-813, 2003, ISSN 1109-2734.

[6] J. Hlava, *Time Delay Systems Applications in Textile Industry - Modelling of Sliver Drafting Process*, Time Delay Systems 2003 - A Proceedings volume from the 4th IFAC workshop, Elsevier: Oxford, pp. 287-292.

[7] Zítek P., *Simulace dynamických systémů*. 1. vydání, Praha: SNTL - Nakladatelství technické literatury, 1990. 420 s. ISBN 80-03-00330-X.

# Two-level and Multilevel Converters for Wind Energy Systems: A Comparative Study

R. Melício[*], V. M. F. Mendes[†] and J. P. S. Catalão[*]

[*] University of Beira Interior, Covilha, Portugal, e-mail: *ruimelicio@gmail.com; catalao@ubi.pt*
[†] Instituto Superior de Engenharia de Lisboa, Lisbon, Portugal, e-mail: *vfmendes@isel.pt*

*Abstract*—This paper is concerned with modeling and simulation of a wind energy system with different topologies for the power converters, namely a two-level converter and a multilevel converter. Pulse modulation by space vector modulation associated with sliding mode is used for controlling the converters, and power factor control is introduced at the output of the converters. Finally, a case study is presented, showing the electric behavior of the power and the current at the output of the converters and the direct voltage at the multilevel converter.

*Keywords*—Multilevel converters, sliding mode control, wind energy.

## I. INTRODUCTION

Electricity restructuring has offered additional flexibility at both levels of generation and consumption. Also, since restructuring has strike the power system sector, developments in distributed generation technologies opened new perspectives for generating companies [1], in order to consider their energy supply portfolio with adequacy and advantage. Adequacy and advantage due to a better generation mix, concerning not only the traditional economic perspective, but also politic developments with strong social impact in power systems, imposing the internalization of costs formerly externalized.

Distributed generation technology is said to offer a clean emission-free energy source with fast ramp capability, and it goes on penetrating more and more power systems. Distributed generation technology includes, for instances, arrays of solar photovoltaic panels, wind farms, hydroelectric, biomass and tidal power plants. Among distributed generation technology, wind farms are the most commonly viewed on power systems, even envisaged as competing with the traditional fossil-fuelled thermal power plants in the near future.

The European Commission concerned with the climate change, due to the emission of greenhouse gases, put forward a set of proposals to create a new Energy Policy for Europe, cutting its own $CO_2$ emissions by at least 20% by 2020 and 50% until 2050, increasing the share of renewable energy sources in the overall generation mix.

Hence, it is expected that wind energy will turn out to be an important part of the future Energy Policy for Europe. In Portugal, the total installed wind power capacity reached 2037 MW in September 2007, and continues growing.

The increasing share of wind in power generation will change considerably the dynamic behavior of the power system [2], and may lead to a reduction of power system frequency regulation capabilities [3]. In addition, network operators have to ensure that consumer power quality is not compromised [4]. Hence, new technical challenges emerge due to the increase in wind power penetration: dynamic stability and power quality. These challenges imply research of more realistic physical models for wind energy systems.

Power electronic converters have been developed for integrating wind power with the electric grid. The use of power electronic converters allows for variable speed operation of the wind turbine, and enhanced power extraction. In variable speed operation, a control method designed to extract maximum power from the turbine and provide constant grid voltage and frequency is required [5].

This paper is concerned with modeling and simulation of a wind energy system with different topologies for the power converters, namely a two-level converter and a multilevel converter. Pulse modulation by space vector modulation associated with sliding mode is used for controlling the converters, and power factor control is introduced at the output of the converters. Finally, a case study is presented, showing the electric behavior for the two types of converters.

## II. MODELING

### A. Turbine and Electric Machine

The mechanical power of the turbine is given by

$$P_m = \frac{1}{2}\rho A u^3 c_p, \qquad (1)$$

where $P_m$ is the power extracted from the airflow, $\rho$ is the air density, $A$ is the area covered by the rotor, $u$ is the wind speed upstream of the rotor, and $c_p$ is the performance coefficient or power coefficient.

The power coefficient is a function of the pitch angle of rotor blades $\theta$, and of the tip speed ratio $\lambda$, which is the ratio between blade tip speed and wind speed upstream of the rotor. The computation of the power coefficient requires knowledge of issues like: blade element theory and blade geometry. These issues are considered in this paper using the numerical approximation developed in [6]. Hence, in this paper the power coefficient is given by

$$c_p = 0.73 \, \lambda_i \, e^{-\frac{18.4}{\lambda_{ii}}}, \qquad (2)$$

where $\lambda_i$ and $\lambda_{ii}$ are respectively given by

$$\lambda_i = \frac{151}{\lambda_{ii}} - 0.58\,\theta - 0.002\,\theta^{2.14} - 13.2\,, \qquad (3)$$

$$\lambda_{ii} = \frac{1}{\dfrac{1}{(\lambda - 0.02\,\theta)} - \dfrac{0.003}{(\theta^3 + 1)}}\,. \qquad (4)$$

The maximum power coefficient is given for a null pitch angle and is equal to

$$c_{p\,\max} = 0.4412\,, \qquad (5)$$

where the optimum tip speed ratio is equal to

$$\lambda_{opt} = 7.057\,. \qquad (6)$$

The mechanical power extracted from the wind is modeled by (1) to (4).

The equations for modeling rotor motion are given by

$$\frac{d\omega_m}{dt} = \frac{1}{J_m}(T_m - T_{dm} - T_{am} - T_{elas})\,, \qquad (7)$$

$$\frac{d\omega_e}{dt} = \frac{1}{J_e}(T_{elas} - T_{de} - T_{ae} - T_e)\,, \qquad (8)$$

where $\omega_m$ is the rotor speed of turbine, $J_m$ is the turbine moment of inertia, $T_m$ is the mechanical torque, $T_{dm}$ is the resistant torque in the turbine bearing, $T_{am}$ is the resistant torque in the hub and blades due to the viscosity of the airflow, $T_{elas}$ is the torque of the torsional stiffness, $\omega_e$ is the rotor speed of the electric machine, $J_e$ is the electric machine moment of inertia, $T_{de}$ is the resistant torque in the electric machine bearing, $T_{ae}$ is the resistant torque due to the viscosity of the airflow in the electric machine, and $T_e$ is the electric torque.

The equations for modeling a permanent magnetic synchronous machine, PMSM, can be easily found in the literature; using the motor machine convention, the following set of equations [7] is considered

$$\frac{di_d}{dt} = \frac{1}{L_d}(u_d + p\omega_e\,L_q\,i_q - R_d\,i_d)\,, \qquad (9)$$

$$\frac{di_q}{dt} = \frac{1}{L_q}[u_q - p\omega_e(L_d\,i_d + M\,i_f) - R_q\,i_q]\,, \qquad (10)$$

where $i_f$ is the equivalent rotor current, $M$ is the mutual inductance, $p$ is the number of pairs of poles; and where in $dq$ axes, $i_d$ and $i_q$ are the stator currents, $L_d$ and $L_q$ are the stator inductances, $R_d$ and $R_q$ are the stator resistances, $u_d$ and $u_q$ are the stator voltages. A unity power factor is imposed to the electric machine in order to minimize power losses, implying a null reactive electric power, $Q_e = 0$. The electric power $P_e$ is given by

$$P_e = [u_d \quad u_q \quad u_f][i_d \quad i_q \quad i_f]^T\,. \qquad (11)$$

The output power $P$ and $Q$ injected in the electric network in $\alpha\beta$ axes [8] is given by

$$\begin{bmatrix} P \\ Q \end{bmatrix} = \begin{bmatrix} u_\alpha & u_\beta \\ -u_\beta & u_\alpha \end{bmatrix} \begin{bmatrix} i_\alpha \\ i_\beta \end{bmatrix}\,, \qquad (12)$$

where in $\alpha\beta$ axes, $i_\alpha$ and $i_\beta$ are the phase currents, $u_\alpha$ and $u_\beta$ are the phase voltages. The apparent output power [8] is

$$S = (P^2 + Q^2 + H^2)^{1/2}\,, \qquad (13)$$

where $H$ is the harmonic power.

### B. Two-level Converter

The two-level converter is an AC/DC/AC converter, with six unidirectional commanded IGBT's $S_{ik}$ used as a rectifier, and with the same number of unidirectional commanded IGBT's used as an inverter.

The rectifier is connected between an electric machine and a capacity bank. The inverter is connected between this capacity bank and a filter, which in turn is connected to an electric network. In this paper, a three-phase symmetrical circuit in series models the electric network. The groups of two IGBT's linked to the same phase constitute a leg $k$ of the converter. The configuration of the wind energy system with two-level converter that will be simulated is shown in Fig. 1.

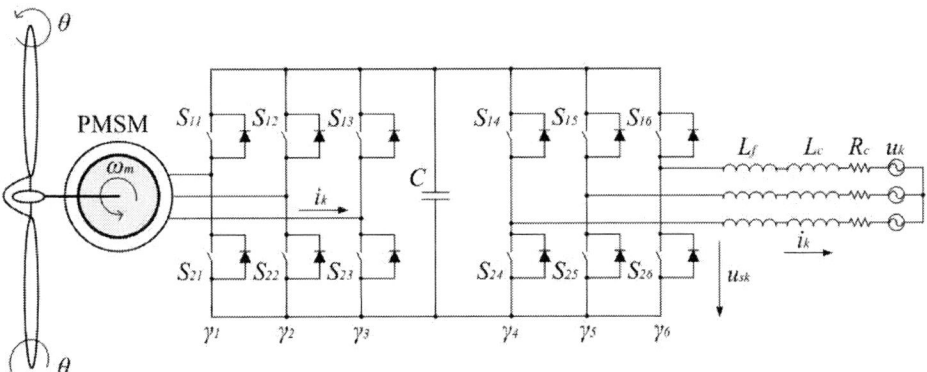

Fig. 1. Wind energy system with two-level converter.

1683

For the two-level converter modeling it is assumed that: 1) The IGBT's are ideal and unidirectional, and they will never be subject to inverse voltages, being this situation guaranteed by the arrangement of connection in anti-parallel diodes; 2) The diodes are ideal: in conduction it is null the voltage between its terminals and in blockade it is null the current that passes through it; 3) The voltage in the exit of the rectifier should always be $v_{dc} > 0$ ; 4) Each leg $k$ of the converter should always have one IGBT on conduction state.

For the switching function of each IGBT, the switching variable $\gamma_k$ is used to identify the state of the IGBT $i$ in the leg $k$ of the converter. The index $i$ with $i \in \{1,2\}$ identifies the IGBT. The index $k$ with $k \in \{1,2,3\}$ identifies the leg for the rectifier and $k \in \{4,5,6\}$ identifies the leg for the inverter. The two valid conditions [9] for the switching variable of each leg $k$ are as follows

$$\gamma_k = \begin{cases} 1, & S_{ik} = 1 \\ 0, & S_{ik} = 1 \end{cases} \text{ for } i \in \{1,2\} \text{ and } k \in \{1,...,6\}. \quad (14)$$

The topological restriction for the leg $k$ is given by

$$\sum_{i=1}^{2} S_{ik} = 1 \qquad k \in \{1,...,6\}. \quad (15)$$

Hence, each switching variable depends on the conduction and blockade states of the IGBT's. The phase currents injected in the electric network are modeled by the state equation

$$\frac{di_k}{dt} = \frac{1}{(L_c + L_f)}(u_k - R_c i_k - u_{sk}). \quad (16)$$

The output continuous voltage of the rectifier is modeled by the state equation

$$\frac{dv_{dc}}{dt} = \frac{1}{C}(\sum_{k=1}^{3} \gamma_k i_k - \sum_{k=4}^{6} \gamma_k i_k). \quad (17)$$

The two-level converter is modeled by (14) to (17).

### C. Multilevel Converter

The multilevel converter is an AC/DC/AC converter, with twelve unidirectional commanded IGBT's $S_{ik}$ used

as a rectifier, and with the same number of unidirectional commanded IGBT's used as an inverter.

The rectifier is connected between an electric machine and a capacity bank. The inverter is connected between this capacity bank and a filter, which in turn is connected to an electric network. In this paper, a three-phase symmetrical circuit in series models the electric network. The groups of four IGBT's linked to the same phase constitute a leg $k$ of the converter. The configuration of the wind energy system with multilevel converter that will be simulated is shown in Fig. 2.

For the multilevel converter modeling it is also assumed that: 1) The IGBT's are ideal and unidirectional, and they will never be subject to inverse voltages, being this situation guaranteed by the arrangement of connection in anti-parallel diodes; 2) The diodes are ideal: in conduction it is null the voltage between its terminals and in blockade it is null the current that passes through it; 3) The voltage in the exit of the rectifier should always be $v_{dc} > 0$ ; 4) Each leg $k$ of the converter should always have two IGBT's on conduction state.

For the switching function of each IGBT, the switching variable $\gamma_k$ is used to identify the state of the IGBT $i$ in the leg $k$ of the converter. The index $i$ with $i \in \{1,2,3,4\}$ identifies the IGBT. The index $k$ with $k \in \{1,2,3\}$ identifies the leg for the rectifier and $k \in \{4,5,6\}$ identifies the leg for the inverter. The three valid conditions [9] for the switching variable of each leg $k$ are as follows

$$\gamma_k = \begin{cases} 1, & (S_{1k} \text{ and } S_{2k}) = 1 \\ 0, & (S_{2k} \text{ and } S_{3k}) = 1 \qquad k \in \{1,...,6\}. \quad (18) \\ -1, & (S_{3k} \text{ and } S_{4k}) = 1 \end{cases}$$

The topological restriction for the leg $k$ is given by

$$(S_{1k}.S_{2k}) + (S_{2k}.S_{3k}) + (S_{3k}.S_{4k}) = 1 \quad k \in \{1,...,6\}. \quad (19)$$

With the two upper IGBT's in each leg $k$ ( $S_{1k}$ and $S_{2k}$ ) of the converters it is associated a switching variable $\Gamma_{1k}$ and also for the two lower IGBT's ( $S_{3k}$ and $S_{4k}$ ) it is associated a variable $\Gamma_{2k}$ , respectively given by

$$\Gamma_{1k} = \frac{\gamma_k(1+\gamma_k)}{2}; \ \Gamma_{2k} = \frac{\gamma_k(1-\gamma_k)}{2} \quad k \in \{1,...,6\}. \quad (20)$$

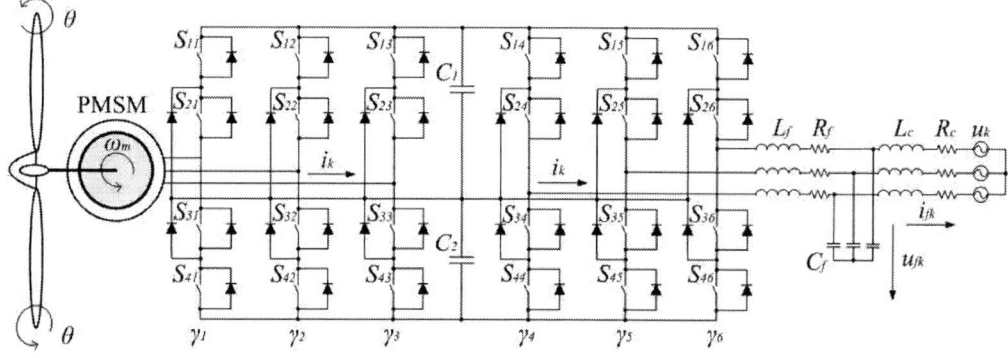

Fig. 2. Wind energy system with multilevel converter.

1684

Hence, each switching variable depends on the conduction and blockade states of the IGBT's.

The phase currents injected in the electric network are modeled by the state equation

$$\frac{di_{fk}}{dt} = \frac{1}{L_c}(u_{fk} - R_c i_{fk} - u_k) \qquad k = \{4,5,6\}. \quad (21)$$

The output continuous voltage of the rectifier is the sum of the voltages in the capacity banks $C_1$ and $C_2$, modeled by the state equation

$$\frac{dv_{dc}}{dt} = \frac{1}{C_1}(\sum_{k=1}^{3}\Gamma_{1k}i_k - \sum_{k=4}^{6}\Gamma_{1k}i_k) +$$
$$+ \frac{1}{C_2}(\sum_{k=1}^{3}\Gamma_{2k}i_k - \sum_{k=4}^{6}\Gamma_{2k}i_k), \quad (22)$$

where (18) to (22) model the neutral point clamped multilevel converter for high voltage and high power applications [9]-[11].

### III. CONTROL METHOD

Pulse modulation by space vector modulation associated with sliding mode is used for controlling the converters. The controllers used in the converters are PI controllers.

The output voltage vectors in the $\alpha\beta$ axes for the two-level converter are shown in the Fig. 3.

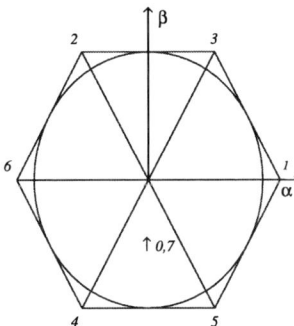

Fig. 3. Output voltage vectors for the two-level converter.

The output voltage vectors in the $\alpha\beta$ axes for the multilevel converter are shown in the Fig. 4.

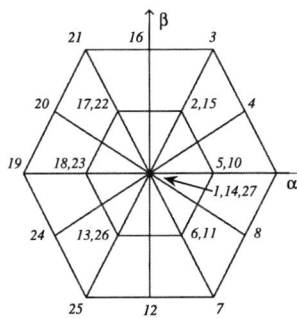

Fig. 4. Output voltage vectors for the multilevel converter.

The converters are variable structure, because of the on/off switching of their IGBT's. Hence, using the sliding mode control is due to be of fundamental importance for controlling the converters, guaranteeing the choice of the most appropriate space vectors.

The power semiconductors present physical limitations that have to be considered during design phase and during simulation. Particularly, they cannot switch at infinite frequency. Also, for a finite value of the switching frequency, an error $e_{\alpha\beta}$ will exist between the reference value and the control value. In order to guarantee that the system slides along the sliding surface $S(e_{\alpha\beta},t)$, it has been proven that it is necessary to ensure the state trajectory near the surfaces verifies the stability conditions [9], given by

$$S(e_{\alpha\beta},t)\,\frac{dS(e_{\alpha\beta},t)}{dt} < 0. \quad (23)$$

As power semiconductors can switch only at finite frequency, in simulation practice a small error $\varepsilon > 0$ for $S(e_{\alpha\beta},t)$ is allowed. Consequently, a switching strategy is considered, given by

$$-\varepsilon < S(e_{\alpha\beta},t) < +\varepsilon. \quad (24)$$

A practical implementation of the switching strategy considered in (24) could be accomplished by using hysteresis comparators at the simulation level.

### IV. SIMULATION

The mathematical models of the two-level and multilevel converters were implemented with Matlab/Simulink block system coding, because it is not yet available enough facilities in SimPowerSystem application for simulation of the model proposed in this paper.

The data used is for a wind energy system of 900 kW, a ramp wind speed upstream of the rotor with a speed between 4.5 and 25 m/s and a time horizon of 3.5 s.

Fig. 5 shows the behavior for the mechanical power of the turbine and the electric power of the electric machine. Also, it shows the difference between the two powers, i.e., the accelerating power.

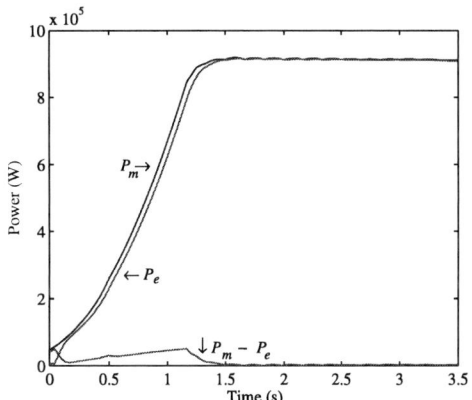

Fig. 5. Mechanical power and electric power.

**1685**

The output power performance for the two-level converter is shown in Fig. 6, while the current injected in the electric network is shown in Fig. 7.

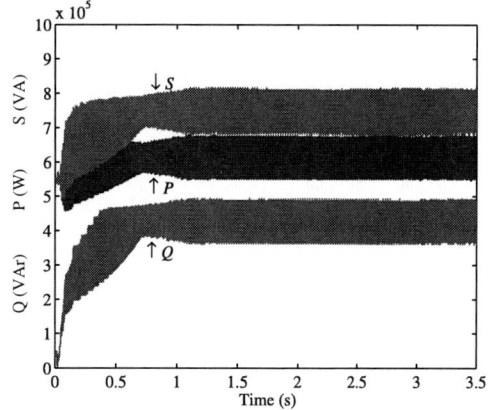

Fig. 6. Output power for the two-level converter.

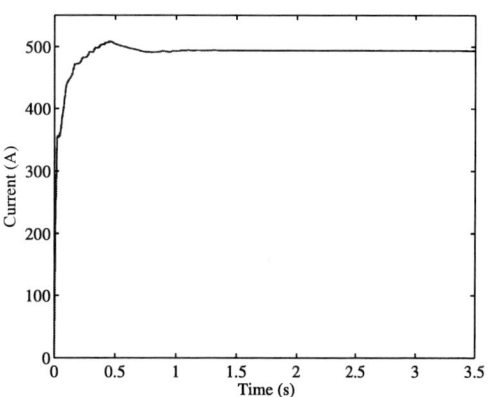

Fig. 7. Output current for the two-level converter.

The output power performance for the multilevel converter is shown in Fig. 8, while the current injected in the electric network is shown in Fig. 9.

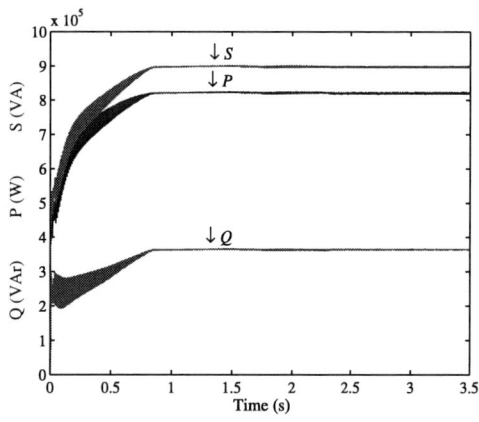

Fig. 8. Output power for the multilevel converter.

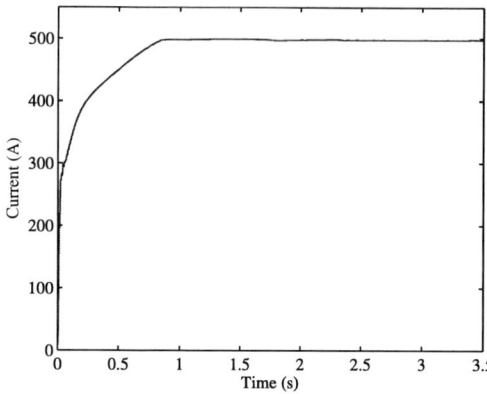

Fig. 9. Output current for the multilevel converter.

The output voltage $v_{dc}$, the voltages $v_{C1}$ and $v_{C2}$ for the multilevel converter are shown in Fig. 10.

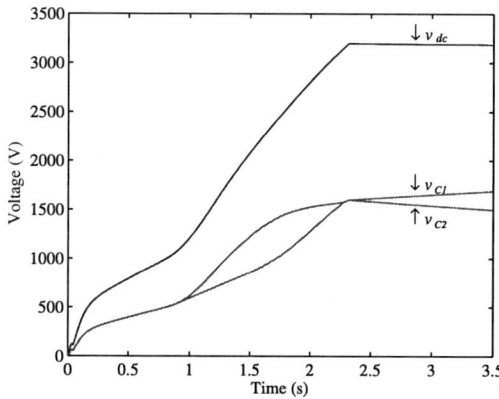

Fig. 10. Output voltages for the multilevel converter.

Hence, it is concluded that the multilevel converter presents a better behavior for the electric power and the electric current relatively to the two-level converter.

## V. CONCLUSIONS

The increased wind power penetration leads to new technical challenges, dynamic stability and power quality, implying research of more realistic physical models for modeling the wind energy systems behavior. This paper presents a more realistic modeling, considering a more accurate dynamic of the wind turbine, rotor, generator, converter and filter connecting the system to the electric network. A case study using the proposed modeling is presented for two power converter topologies: two-level and multilevel converters. Pulse modulation by space vector modulation associated with sliding mode is used for controlling the converters, and power factor control is used at the output of the converters. The results show that the multilevel converter has an enhanced behavior comparatively to the two-level converter. Although more complex, this modulation strategy presents more accurate results, which are necessary for facing the new technical challenges.

## REFERENCES

[1] J. A. Peças Lopes, N. Hatziargyriou, J. Mutale, P. Djapic, and N. Jenkins, "Integrating distributed generation into electric power systems: A review of drivers, challenges and opportunities," *Electr. Power Syst. Res.*, vol. 77, no. 9, pp. 1189–1203, July 2007.

[2] I. Erlich, J. Kretschmann, J. Fortmann, S. Mueller-Engelhardt, and H. Wrede, "Modeling of wind turbines based on doubly-fed induction generators for power system stability studies," *IEEE Trans. Power Syst.*, vol. 3, no. 3, pp. 909–919, August 2007.

[3] R. G. de Almeida and J. A. Peças Lopes, "Participation of doubly fed induction wind generators in system frequency regulation," *IEEE Trans. Power Syst.*, vol. 22, no. 3, pp. 944–950, August 2007.

[4] J. M. Carrasco, L. G. Franquelo, J. T. Bialasiewicz, E. Galvan, R. C. P. Guisado, A. M. Prats, J. I. Leon, and N. Moreno-Alfonso, "Power-electronic systems for the grid integration of renewable energy sources: A survey," *IEEE Trans. Ind. Electron.*, vol. 53, no. 4, pp. 1002–1016, August 2006.

[5] J. A. Baroudi, V. Dinavahi, and A. M. Knight, "A review of power converter topologies for wind generators," *Renewable Energy*, vol. 32, no. 14, pp. 2369–2385, November 2007.

[6] J. G. Slootweg, H. Polinder, and W. L. Kling, "Representing wind turbine electrical generating systems in fundamental frequency simulations," *IEEE Trans. Energy Convers.*, vol. 18, no. 4, pp. 516–524, December 2003.

[7] C. -M. Ong, *Dynamic Simulation of Electric Machinery*. New Jersey: Prentice Hall, 1998.

[8] E. H. Watanabe, R. M. Stephan, and M. Aredes, "New concepts of instantaneous active and reactive powers in electrical systems with generic loads," *IEEE Trans. Power Deliv.*, vol. 8, no. 2, pp. 697–703, April 1993.

[9] J. Barros and J. F. Silva, "Optimal predictive control of three-phase NPC multilevel inverter: comparison to robust sliding mode controller," in *Proc. IEEE Power Electronics Specialists Conf.*, 2007, pp. 2061–2067.

[10] J. Eloy-Garcia, S. Arnaltes, and J. L. Rodriguez-Amenedo, "Extended direct power control for multilevel inverters including DC link middle point voltage control," *IET Electr. Power Appl.*, vol. 1, no. 4, pp. 571–580, July 2007.

[11] E. J. Bueno, S. Cobreces, F. J. Rodriguez, F. Espinosa, M. Alonso, R. Alcaraz, "Calculation of the DC-bus capacitors of the back-to-back NPC converters," in *Proc. 12th International Power Electronics and Motion Control Conf.*, 2006, pp. 137–142.

# A Stand-alone Photovoltaic Supercapacitor Battery Hybrid Energy Storage System

M.E. Glavin, Paul K.W. Chan, S. Armstrong, and W.G Hurley, *IEEE Fellow*

Power Electronics Research Centre
National University of Ireland Galway, Galway, Ireland.
E-mail: margaret.glavin@nuigalway.ie

*Abstract*—Most of the stand-alone photovoltaic (PV) systems require an energy storage buffer to supply continuous energy to the load when there is inadequate solar irradiation. Typically, Valve Regulated Lead Acid (VRLA) batteries are utilized for this application. However, supplying a large burst of current, such as motor startup, from the battery degrades battery plates, resulting in destruction of the battery. An alterative way of supplying large bursts of current is to combine VRLA batteries and supercapacitors to form a hybrid storage system, where the battery can supply continuous energy and the supercapacitor can supply the instant power to the load. In this paper, the role of the supercapacitor in a PV Energy Control Unit (ECU) is investigated by using Matlab/Simulink models. The ECU monitors and optimizes the power flow from the PV to the battery-supercapacitor hybrid and the load. Three different load conditions are studied, including a peak current load, pulsating current load and a constant current load. The simulation results show that the hybrid storage system can achieve higher specific power than the battery storage system.

*Index Terms*—Photovoltaic, Lead acid battery, Energy Control Unit (ECU), Supercapacitor.

## I. INTRODUCTION

The world is approaching peak oil and the ability to produce high quality, inexpensive, and economically extractable oil on demand is diminishing. Peak oil and the environmental impact of fossil fuel utilization, has encouraged a growth in the area of renewable energies such as wind and solar power.

In remote areas stand-alone photovoltaic systems are most common. A typical stand-alone system Fig. 1(a) incorporates a photovoltaic panel, regulator, energy storage system, and load [1]. Generally the most common storage technology employed is the VRLA battery because of its low cost and wide availability. Photovoltaic panels are not an ideal source for battery charging; the output is unreliable and heavily dependent on weather conditions, therefore an optimum charge/ discharge cycle cannot be guaranteed, resulting in a low battery state of charge (SOC). Low battery SOC leads to sulphation and stratification, both of which shorten battery life [2, 3].

Certain load applications require high current for a period of time e.g. motor starting applications; the starting current requirement can be 6-10 times the normal operating current of the motor. Normally the peak current requirements are satisfied by the VRLA battery. VRLA batteries in this situation are large in order to deal with the high current being removed from the battery. The peak current demand might only need to be met for a few

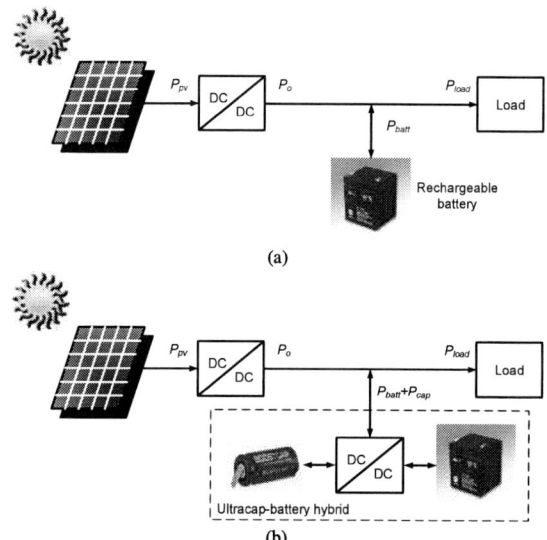

Figure 1. Block diagram of (a) conventional and (b) proposed photovoltaic system

seconds at a particular time. Sizing the battery around this can prove costly; in photovoltaic systems the batteries are replaced typically every 3-5 years depending on the application.

By utilizing a battery supercapacitor hybrid energy storage system as shown in Fig. 1(b) the battery size can be reduced and a higher SOC can be maintained. The supercapacitor has a greater power density than the battery, which allows the supercapacitor to provide more power over a short period of time. Conversely, the battery has a much higher energy density when compared to a supercapacitor allowing the battery to store more energy and release it over a long period of time. In Table 1 the battery and the supercapacitor are compared under various headings [4-6]. In the hybrid system the peak power requirements of the load are supplied by the supercapacitor and the VRLA battery supplies the lower continuous power requirements [7-10].

The proposed Energy Control Unit (ECU) aims to optimize the battery supercapacitor hybrid storage system to reduce the size of the battery and extend the life of the battery by avoiding deep discharge through high currents. The ECU monitors the battery, supercapacitor and photovoltaic panel current, voltage and temperature in addition to the load power requirements. The ECU estimates the battery and supercapacitor SOC, optimizes the energy from the photovoltaic panel and controls the flow of energy throughout the system.

978-1-4244-1741-4/08/$25.00 ©2008 IEEE

TABLE I. BATTERY VERSUS SUPERCAPACITOR PERFORMANCE [6]

| | Lead Acid Battery | Supercapacitor |
|---|---|---|
| Specific Energy Density (Wh/kg) | 10-100 | 1 – 10 |
| Specific Power Density (W/kg) | <1000 | <10,000 |
| Cycle Life | 1,000 | > 500,000 |
| Charge/Discharge Efficiency | 70 – 85% | 85 - 98% |
| Fast Charge Time | 1 – 5h | 0.3 – 30 sec |
| Discharge Time | 0.3 – 3h | 0.3 – 30s |

Matlab/Simulink is used for the design and optimization of the system. This paper outlines the models of the various components. The proposed VRLA battery supercapacitor hybrid storage model is described and simulations are presented comparing the proposed system with conventional battery storage under three different load types; a peak current load, pulsating current load, and a constant current load.

## II. MATLAB/SIMULINK MODELS

### A. Photovoltaic Model

A simple photovoltaic cell equivalent circuit model is shown in Fig. 2 [11]. The model consists of a current source $I_{ph}$ (represents cell photocurrent), a series resistance $R_s$ (the internal resistance of each cell) and a diode. The net output current of the photovoltaic cell is the differences between the photocurrent $I_{ph}$ and the diode current $I_D$ as described by the following equation,

$$I_s = I_{ph} - I_D = I_{ph} - I_o \left( e^{\left( \frac{q(V_s + I_s R_s)}{mkT} \right)} - 1 \right) \quad (1)$$

where $m$ is the ideality factor of the diode, $k$ is Boltsmann's constant, $T$ is the absolute temperature of the cell, $q$ is electron charge, $V_s$ is the voltage applied across the cell, and $I_o$ is the dark saturation current.

The cells are connected in series and parallel to form a PV module. The model simulates a BP solar BP 350 50W photovoltaic panel in Simulink. There are 36 cells in series and 2 parallel branches. In the model, the ideality factor, m, is equal to 2.0077 where it achieves the maximum power point at $V_s = 17.5V$ and $I_s = 2.9A$ at $T = 25°C$. Fig. 3 and Fig. 4 illustrate the simulated I-V and P-V characteristics of the photovoltaic panel under various temperature conditions respectively.

### B. Battery Model

Batteries are the main storage technology used in PV systems. The battery model is used to analyze the effects of different charge rates, state of charge (SOC), and state of health (SOH) of the battery. The optimum battery size for a particular application can be obtained by performing various test scenarios. Simulations are used to compare different storage technologies without the need for expensive test beds.

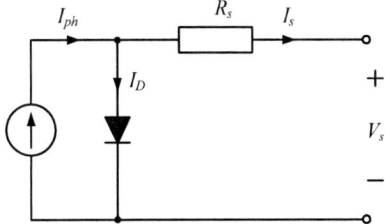

Figure 2. Photovoltaic cell model

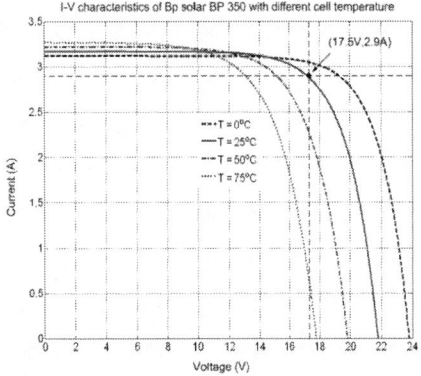

Figure 3. I-V characteristics of BP 350 photovoltaic module

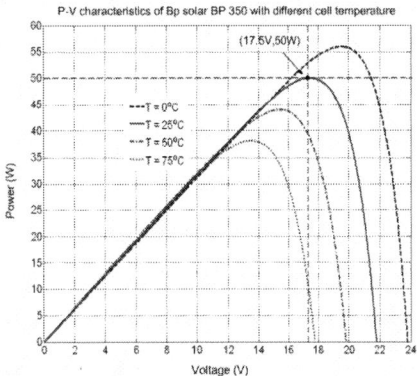

Figure 4. P-V characteristics of BP 350 photovoltaic module

A simple equivalent circuit battery model is shown in Fig. 5. The battery model takes into account the battery state of charge (SOC) and deep of charge (DOC). The battery's usable capacity decrease with increasing discharge current, the battery DOC measures the fraction of the battery's charge to usable capacity. The model includes an open circuit battery voltage $E_{oc}$, internal resistance $R_0$ and two $RC$ parallel branches [12-14]. The model equations are shown (2) - (6).

$$E_{oc} = E_0 - K_e (1 - SOC) \quad (2)$$

$$R_1 = R_{10} e(-K_1 (1 - SOC)) \quad (3)$$

$$R_2 = \frac{R_{20}}{DOC} \quad (4)$$

$$SOC = 1 - \frac{1}{C_n} \int i_{batt} d\tau \qquad (5)$$

$$DOC = 1 - \frac{1}{C(i_{avg})} \int i_{batt} d\tau \qquad (6)$$

where: $SOC$ is the state of charge of the battery
$DOC$ is the deep of charge of the battery
$C_n$ is the battery capacity
$C(i_{avg})$ is the current-dependent battery capacity (obtained in datasheet)
$E_0$ is the open circuit voltage when the battery is fully charge
$K_e$ is a constant
$R_{10}$ is the $1^{st}$ $RC$ branch constant in $\Omega$
$\tau_1$ is the $1^{st}$ $RC$ branch time constant in sec
$K_1$ is a constant
$R_{20}$ is the $2^{nd}$ $RC$ branch constant in $\Omega$
$\tau_2$ is the $2^{nd}$ $RC$ branch time constant in sec

Fig. 6 shows the simulated discharge characteristics curves for a Yuasa Np18-12 lead acid battery for various C-rates. From testing the Yuasa Np18-12 battery used has $E_0 = 12.85$, $K_e = 1.7$, $R_0 = 0.12\Omega$ for charging and $0.057\Omega$ for discharging, $R_{10} = 0.16\Omega$ for charging and $0.02\Omega$ for discharging, $K_1 = 7$, and $R_{20} = 0.0055\Omega$ for both charging and discharging. Fig. 7 and Fig. 8 show the 0.2C pulse charge and 0.2C pulse discharge of the battery respectively.

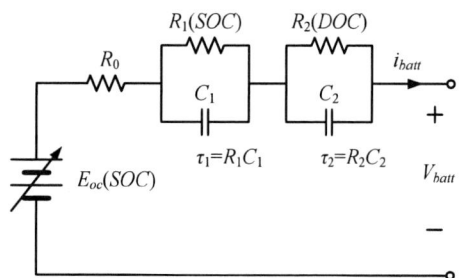

Figure 5.   Battery model

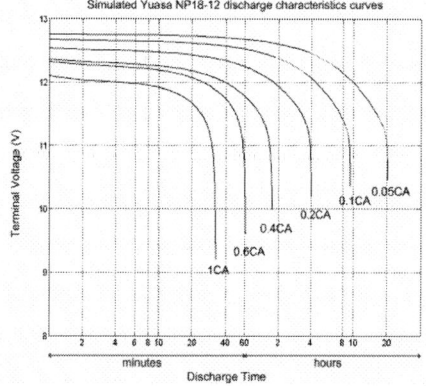

Figure 6.   Battery discharge characteristics

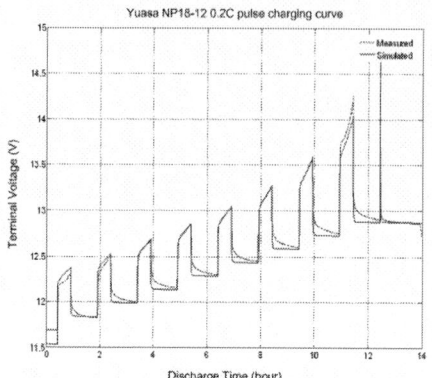

Figure 7.   0.2 pulse charge of Yuasa NP18-12 battery

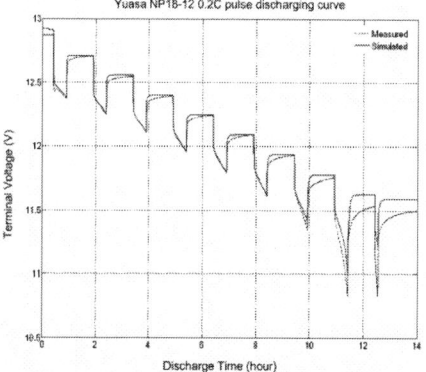

Figure 8.   0.2C pulse discharge of Yuasa NP18-12 battery

### C.  Supercapacitor Model

Fig. 9 shows the classical equivalent circuit model for the supercapacitor [15]. The model consists of three components, the capacitance, the equivalent series resistance (ESR), and the equivalent parallel resistance (EPR). The ESR is a loss term that models the internal heating in the capacitor and is most important during charging and discharging. The EPR models the current leakage effect and will impact the long term energy storage performance of the supercapacitor and $C$ is the capacitance. Equations (7)-(9) describe the ESR, EPR and terminal voltage of the supercapacitor.

$$ESR = \frac{\Delta V}{\Delta i} \qquad (7)$$

$$EPR = \frac{-(t_2 - t_1)}{\ln\left(\frac{V_2}{V_1}\right) C} \qquad (8)$$

$$v_c = ESR \cdot i_c + \frac{1}{C} \int \left(i_c - \frac{e_c}{EPR}\right) d\tau + V_{c\_init} \qquad (9)$$

where: $V_1$ is the initial self-discharge voltage at $t_1$
$V_2$ is the finial self-discharge voltage at $t_2$
$C$ is the rated capacitance

$\Delta V$ change in voltage at turn on of load
$\Delta I$ change in current at turn on of load
$V_{c\_init}$ is the initial capacitor voltage
$i_c$ is the capacitor current

The function of the voltage-dependent capacitor $C$ can be obtained with curve fitting from the charging/discharging measurements. The model is verified with Nesscap 2.7V/600F supercapacitor. Fig. 10 shows the 10A charging, rest and 5A discharging of the model with an ESR of 1mΩ and an EPR of 258Ω.

## III. BATTERY STORAGE SYSTEM

### A. Photovoltic Battery Storage Model

The most common setup for standalone photovoltaic systems, shown in Fig. 1(a), consists of a photovoltaic panel, converter, load, and battery storage. The energy produced from the photovoltaic panel is stored in the rechargeable battery to supply the load requirements when discrepancies arise between available and required energy. Deep discharge batteries are designed to be discharged down to as much as 80% depth of discharge (DOD) repeatedly and have thicker plates then car batteries making them the preferable choice for PV storage. Generally the battery is sized to enable it to supply power to the load for a period of 2-3 days, resulting in a large battery pack that will need to be replaced every few years.

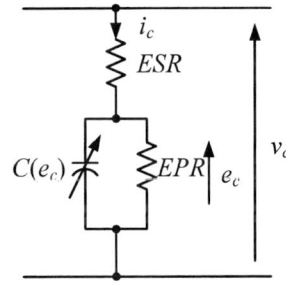

Figure 9. Supercapacitor equivalent circuit model

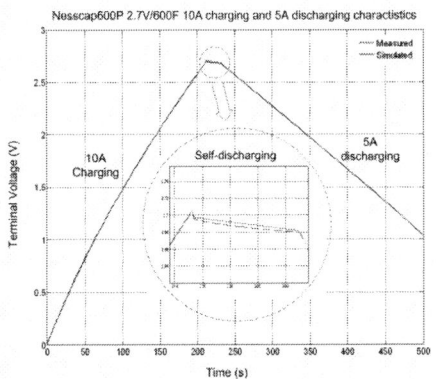

Figure 10. Supercapacitor charge/discharge characteristics

### B. Battery Management System (BMS)

The Battery Management System (BMS) controls the flow of energy from the photovoltaic panel to the battery and load. The BMS is responsible for calculating the battery SOC, varying the DC-DC converter duty cycle, and implementing the charging algorithm. The BMS is based on SOC estimation. The battery charging/ discharging is dependent on both the battery SOC and the load requirements as described by Table II.

The DC-DC converter implements Maximum Power Point Tracking (MPPT), charges the battery, and delivers energy to the load. Sensors and measurement circuits are responsible for measuring the voltages and currents of the solar panel, battery, and load along with the solar panel and battery temperature. This information is used by the control algorithm to enhance the performance of the system, making the best use of the available energy to maintain the battery at a high SOC but also ensuring that the load demand is met at all times.

## IV. HYBRID STORAGE SYSTEM

### A. Photovoltic Hybrid Storage Model

The proposed Hybrid storage model consists of a VRLA battery bank and a supercapacitor battery bank as shown in Fig. 1(b). The hybrid system adopts the advantages of both technologies, high power density from the supercapacitor and high energy density from the battery. The supercapacitor supplies the high peak power requirements and the battery bank supplies the low power requirements, resulting in a reduction in the battery pack size.

TABLE II.     BATTERY MANAGEMENT CONDITIONS

| Condition | Action |
|---|---|
| PV Power = Load<br>Battery SOC High | PV supplies load<br>No battery charging |
| PV Power = Load<br>Battery SOC Low | PV supplies load<br>No battery charging |
| PV Power > Load<br>Battery SOC High | PV supplies load<br>No battery charging |
| PV Power > Load<br>Battery SOC Low | PV supplies load<br>PV charges battery |
| PV Power < Load<br>Battery SOC High | PV supplies load<br>Battery supplies load |
| PV Power < Load<br>Battery SOC Low | PV supplies load<br>Battery supplies load until minimum SOC is reached then shut down load |
| No PV Power<br>Battery SOC High | Battery supplies load |
| No PV Power<br>Battery SOC Low | Shut down load |

## B. Proposed Energy Control Unit (ECU)

The Energy Control Unit (ECU) controls the complete photovoltaic system. The ECU is responsible for charging the battery/supercapacitor hybrid and supplying power to the load according to the conditions outlined in Table III.

The power available from the photovoltaic panel is used to supply load power, with excess energy being used for battery and supercapacitor charging. The ECU implements MPPT capturing the maximum power available from the panel. Various sensors are utilized throughout the system to measure the voltage and current of the battery, supercapacitor and panel along with the power requirement of the load. These observations enable intelligent decisions to be made about how to best utilize the available energy in order to avoid situations where the load must be shut down due to low battery and supercapacitor SOC under conditions of inadequate solar irradiation.

## V. SYSTEM LOAD COMPARISON

The battery management system (BMS) was compared to the proposed hybrid energy control unit (ECU) under different load profiles as outlined below. The solar irradiation profile utilized for the simulations is shown in Fig. 11.

### A. Peak Power Load

Fig. 12 shows a peak current load application that has been used to analyses the benefits of the supercapacitor. Examples of peak load applications are motor starting applications were the starting current maybe 6-10 times the continuous operating current of the motor. The profile of Fig. 12 has an initial current of 8.33A and a continuous current of 1.375A with the load operating for 45mins every hour throughout the day. Fig. 13 shows the battery SOC with BMS, battery SOC with ECU and supercapacitor SOC with ECU. In the Hybrid system the battery supplies a continuous current of 0.8A, a discharge rate of 0.05C, this current supplies power to the load and also recharges the ultracapacitor. A 12V 1200F supercapacitor supplies the remaining load current. The hybrid system results in the battery being maintained at a higher SOC.

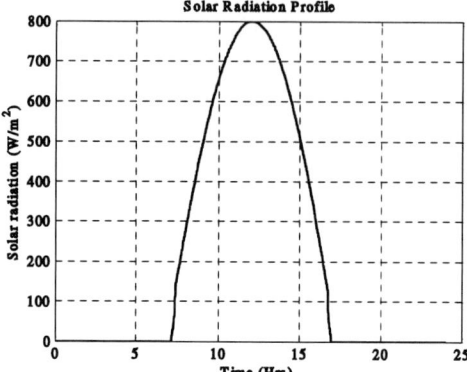

Figure 11. Solar radiation profile

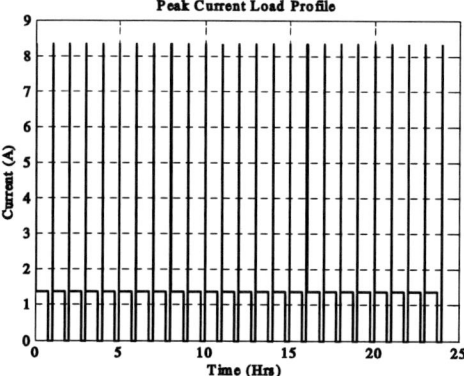

Figure 12. Peak current load profile

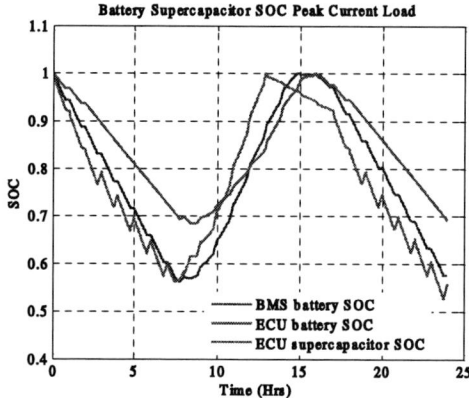

Figure 13. Battery supercapacitor SOC for peak current load

### B. Pulsating Load

The second load profile used in the analysis is a pulsating current load. A typical application is the transmitting system. In the simulation, the supercapacitors in the hybrid system deliver the pulse power while the battery supplies the remaining constant current.

TABLE III.    HYBRID SYSTEM CONDITIONS AND ACTIONS

|  | Photovoltaic Power | Battery SOC | Supercapacitor SOC |
|---|---|---|---|
| Supply Load | >0 | High | High |
| Charge Battery | >Load | Low | Low/High |
| Charge Supercapacitor | > Load | High | Low |
| Shutdown | None | Low | Low |

Fig. 14 shows the profile of the pulsating current load. The load operates for 200s out of 250s. The load has a low continuous current of 0.42A and a high pulse current of 2.08A with a duty cycle of 0.5 and a period of 20s, the load operating over 24 hrs.

Fig. 15 shows the battery SOC in BMS, battery SOC in ECU and supercapacitor SOC with ECU. The Hybrid system battery supplies a continuous current of 0.8A (0.05C) with the remaining current being supplied by the supercapacitor. The simulation results show that the hybrid system allowed the battery to be maintained at a higher SOC.

## C. Constant Power Load

A constant current load of 1.04A (0.06C of the battery) is simulated. The load was analyzed in both BMS and ECU. In the simulation, the battery current is limited at 1.04A. Without pulse current in the load profile, all the current is supplied from the battery in the hybrid system. Fig. 16 shows the SOC in both systems. In the simulation, the hybrid system has a lower SOC then battery system because the battery needs to charge the supercapacitors due to self discharge.

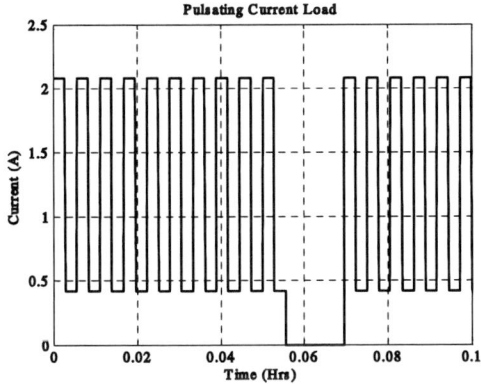

Figure 14. Pulse current load profile

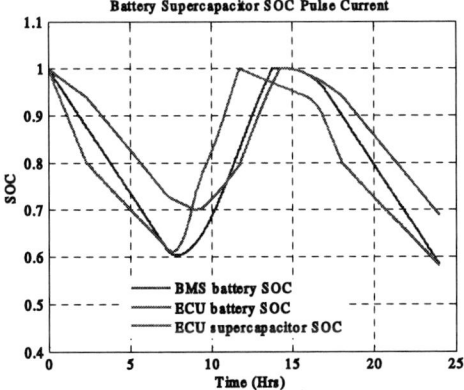

Figure 15. Battery supercapacitor SOC for pulsating load

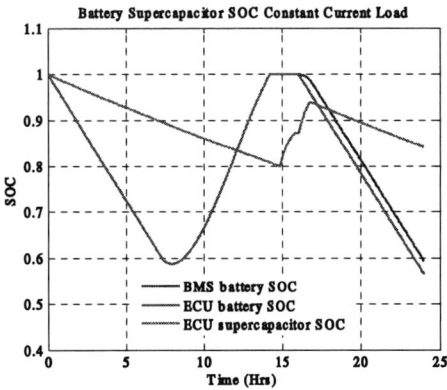

Figure 16. Battery supercapacitor SOC for constant current load

## VI. OPTIMIZATION

Photovoltaic are unreliable energy sources that are heavily dependent on weather conditions. The power output of the PV panel increases with increasing irradiation but decreases with increasing temperature, operating at its most efficient at high irradiation and low temperature. The power output from a BP 350 50W solar panel for a average day in June (best conditions) and December (worst conditions) is illustrated in Fig. 17, data obtained from [19] for Newcastle, England; which could be typical for a cloudy climate in northern Europe.

To ensure that the load requirements can be met throughout the year, photovoltaic systems are sized for worst case conditions, from Fig. 17 sizing is performed according to December figures. Other considerations are

➢ The allowable dept of discharge for VRLA batteries is 80%.

➢ The days of Autonomy, which refers to the number of days a battery system will provide a given load without being recharged by the photovoltaic array or other source is typically 3 to 5 days.

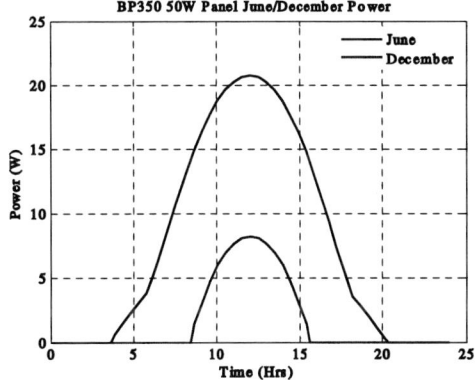

Figure 17. BP350W June and December average power

**1693**

Energy audits are performed to obtain information about the load profile that needs to be supplied from the PV system. Many appliances require higher starting power compared to operating power as outlined in Table IV [20]. The proposed ECU supplies this starting power from the supercapacitor. Fig. 18 shows the domestic profile obtained from a flat in Newcastle, England for a week in April. The average power consumption was recorded over 5 minute time intervals throughout 2005[21].

From Fig. 18 various spikes in power can be observed throughout the day. Spikes of approximately 9 times the continuous power requirements are observed, this would result in a large current being removed from the battery reducing the battery SOC. To ensure adequate power is available the battery pack size is increased to supply the large current. The supercapacitor can complement the PV panel and the battery to supply the high power requirements, allowing for a smaller battery pack.

The domestic load profile on Monday shown in Fig. 18 was scaled down and simulated with both the BMS and the ECU models to observe the benefits of the supercapacitor in a domestic application. Fig. 19 shows the solar radiation for a typical April day in Newcastle. The output current from the MPPT is shown in Fig. 21.

Fig. 22 shows the SOC of the Np18-12 Yuasa lead acid battery and a 12V 1200F supercapacitor. The battery in the hybrid system supplied continuous current of 0.5A (0.03C) with the supercapacitor supplying the remainder. It is observed from Fig. 22 that the hybrid system maintains the battery at a greater SOC with the final SOC for the hybrid system being 72% while the BMS has a battery SOC of 50%.

The hybrid system allows for an increase in battery SOC as outlined in Table V with the addition of a 12V 1200F supercapacitor pack. Supercapacitors are currently expensive components, but with the advances in technology and the increasing growth in the market the cost is being reduced.

VRLA batteries are generally large in photovoltaic systems and although there is an increase in SOC with the addition of the supercapacitor bank it will prove costly. Additional optimization is required to find the optimum balance between the VRLA battery and the supercapacitor bank. Other battery and fuel cell technology can be investigated to see if they have greater benefit.

TABLE IV.   APPLIANCE STARTING AND CONTINUOUS POWER

| Appliance | Starting Power | Continuous Power | Ratio |
|---|---|---|---|
| Clothes Washer | 5,042 | 225 | 22.4 |
| Well Pump 1/2hp | 1950 | 150 | 13 |
| Clothes Dryer | 4208 | 334 | 12.6 |
| Fridge/Freezer | 2700 | 600 | 4.5 |
| Freezer | 2100 | 800 | 2.6 |
| Vacuum cleaner | 2012 | 818 | 2.5 |

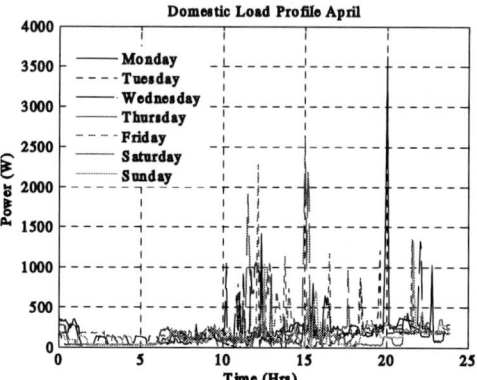

Figure 18. Domestic load profile for week in April

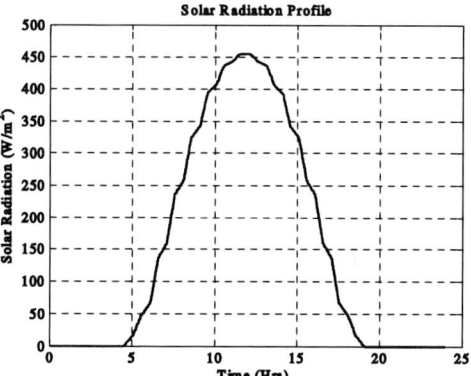

Figure 19. Solar radiation for April

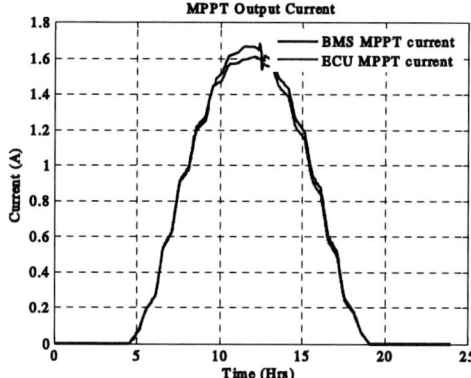

Figure 20. MPPT output current

VII.   CONCLUSIONS

An Energy Control Unit (ECU) for a photovoltaic battery supercapacitor hybrid system has been developed. Simulations have been performed to compare the ECU to the standard photovoltaic battery storage system under different load conditions; a peak current load, pulsating current load, constant current load, and a domestic profile. Maximum Power Point Tracking (MPPT) allows the maximum power to be gained from the photovoltaic panel and the proposed ECU is responsible for calculating the battery and supercapacitor SOC. The ECU controls the system based of the available power, battery/ supercapacitor SOC and the required load power. From the simulations performed the addition of a supercapacitor bank will increase the battery SOC for peak and pulse current loads.

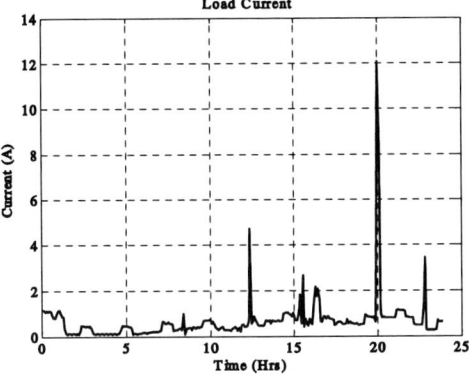

Figure 21. Load current

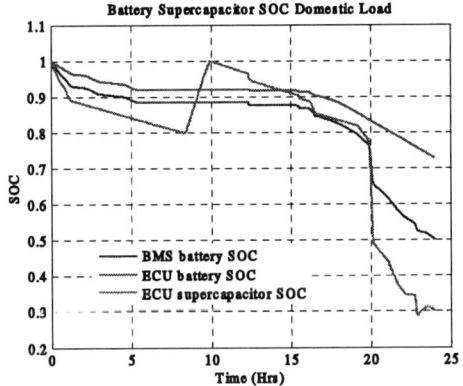

Figure 22. Battery supercapacitor SOC

TABLE V. BATTERY SOC COMPARISON

| LOAD | ECU Battery SOC | BMS Battery SOC | Increase in SOC |
|---|---|---|---|
| Peak Load | 69% | 57% | 12% |
| Pulse Load | 69% | 58% | 11% |
| Constant Load | 56% | 59% | (3%) |
| Domestic Load | 72% | 50% | 22% |

ACKNOWLEDGMENT

This project was supported by Enterprise Ireland under the Commercialisation Fund in Technology Development (CFTD).

REFERENCES

[1] S. Duryea, S. Islam, W. Lawrance, "A Battery Management System for Stand-Alone Photovoltaic Energy Systems", IEEE Industrial Applications Magazine, Vol.7, Issue 3, May-June, 2001, pp.67-72.

[2] J.P. Dunlop, B.N. Farhi, "Recommendations for Maximizing Battery Life in Photovoltaic Systems: A Review of Lessons Learned", Proceedings of Forum 2001 Solar Energy: The Power to Choose, Washington DC, April 21-25, 2001.

[3] T. Hund, "Capacity Loss in PV Batteries and Recovery Procedures", Photovoltaic System Applications Department, Sandia National Laboratories.

[4] A. Burke, "Ultracapacitors: Why, How, and Where is the Technology", Journal of Power Sources, Vol.91, pp.37-50, 2000.

[5] B.E. Conway, "Electrochemical Supercapacitors: Scientific Fundamentals and Technological Applications", Kluwer Academic Press/ Plenum Publishers, New York, 1999.

[6] http://www.ewh.ieee.org/r6/scv/pses/ieee_scv_pses_jan05.pdf

[7] K. Akiyama, Y. Nozaki, M. Kudo, T. Yachi, "NiMH Batteries and EDLC's Hybrid Standalone Photovoltaic Power System for Digital Access Equipment", 22nd International Telecommunications Energy Conference, 10-14 September 2000, pp.387-393.

[8] L. Gao, R.A. Dougal, S. Liu, "Power Enhancement of an Actively Controlled Battery Ultracapacitor Hybrid", IEEE Transactions on Power Electronics, vol. 20, no. 1, pp. 236-243, January 2005.

[9] R.A. Dougal, S. Liu, R.E. White, "Power and Life Extension of Battery Ultracapacitor Hybrids", IEEE Transactions on Component and Packaging Technologies, vol. 25, no. 1, pp. 120-131, March 2002.

[10] S. Liu, R.A. Dougal, E.V. Solodovnik, "Design of Autonomous Photovoltaic Relay Station", IEEE Proc-Gener. Transm. Distrib., vol. 152, no.6, pp.745-754, November 2005.

[11] G. Walker, "Evaluating MPPT converter Topologies using a Matlab PV Model", Proceedings of the Australasian University Power Engineering Conference, Brisbane, 2000.

[12] M. Ceraolo, "New Dynamic Models of Lead Acid Batteries" IEEE Transactions on Power Systems, vol. 15, no. 4, pp. 1184-1190, November 2000.

[13] S. Barsali, M. Ceraolo, "Dynamical Models of Lead Acid Batteries: Implementation Issues", IEEE Transactions on Energy Conversion, vol.17, no. 1, pp.16-23, March 2002.

[14] R.A. Jackey, "A Simple Effective Lead Acid Battery Modelling Process for Electrical Systems Component Selection", Mathworks INC..

[15] R.L. Spyker, R.M. Nelms, "Classical Equivalent Circuit Parameters for a Double Layer Capacitor", IEEE Tranactions in aerospace and electronic systems, vol. 36, no. 3, pp. 829-836, July 2000.

[16] Armstrong S., Glavin M.E., Hurley W.G., "Comparison of Battery Charging Algorithms for Stand Alone Photovoltaic Systems", Proceedings 39th Power Electronic Specialists Conference, Rhodes Greece, 15-19 June 2008.

[17] M. Coleman, W.G. Hurley, C.K. Lee, "An Improved Battery Characterisation Method using a Two Pulse Load Test", IEEE Trans. on Energy Conversion, vol. 23, no. 2, pp. 708-713, June 2008.

[18] Coleman M., Chi Kwan Lee, Chunbo Zhu, Hurley W.G., "State of Charge Determination from EMF Voltage Estimation: Using Impedance, Terminal Voltage, and Current for Lead Acid and Lithium-ion Batteries", vol. 54, no. 5, pp. 2550-2557, October 2007.

[19] http://re.jrc.ec.europa.eu

[20] R.L. Hammand, S. Everingham, "Stationary Batteries in Cycling Photovoltaic Applications", www.battcon.com/PapersFinal2003/HammondPaperFINAL2003.pdf.

[21] Dr. Ian Knight, N. Kreutzer, "Three European Domestic Electrical Consumption Profiles – July 2006".

# Integrated contactless power transmission systems with high positioning flexibility

Daniel Kürschner*, Christian Rathge*

* Institut für Automation und Kommunikation e.V., Magdeburg, Germany, e-mail: *daniel.kuerschner@ifak.eu*

*Abstract*— Contactless power transmission provides the vision for the future development of energy distribution in industrial, business or consumer electronics or where a constructive conditional air gap must be bridged. By using the inductive technology, sliding contacts, trailing cables or slip rings can be eliminated and the safety and reliability of the energy supply can be improved. To design the transmission system, the paper presents a possibility to design the magnetic system by using analytical and numerical analyses. Based on special FEA models it is shown in the paper how to analyse and minimise the power loss of coil configurations. Furthermore it is shown how to reach a higher positioning flexibility of two coils of a resonant transmission system by optimising the coil parameters and the compensation technique. The paper presents two solutions of contactless transmission systems in the power range of 60 W and 1 kW and by means of an experimental set up the theoretical investigations are demonstrated by measurements.

*Keywords*— Wireless power transmission, High power density systems, High frequency power converter, EMC/EMI, Flux model, Magnetic device, Resonant-mode power supply

## I. INTRODUCTION

To allow contactless energy transmission in the power range of several kilowatts, the overall efficiency of the magnetic system and the power electronics should be more than 90%. This can be reached by using resonant switching operation at 100kHz or above and by using HF-litz wire and ferrite components for the coils [1]. Important aspects of contactless transmission systems are reaching higher transferable power and at the same time smaller installation size and weight. This requires the analysis of the power loss of the overall system. However there are a lot of other technical demands like reaching higher positioning tolerance of both coils to each other, improving the operation security and minimising the magnetic leakage field near the air gap of the transmission system. A positioning tolerance means the displacement of both coils to each other out of their normal position. Examples for applications with high transferred electric power and the further condition of large positioning tolerances are power supplies for consumers on bad balanced shafts (e.g. cardan shafts or stranding machines), consumer electronics or battery charging systems for sensor modules, wafer processing equipment or medical devices. Positioning tolerances are mostly afflicted with unwanted electrical effects such as voltage rises or phase shifts. To prevent these effects, a controlling or stabilisation of the output voltage is necessary. This mostly either requires an additional contactless data-transfer to commit the measured value to the control unit on the primary side or needs the determination of the secondary load voltage from the primary quantities [2]. However when controlling, there are critical aspects that increase the complexity and the costs of the overall system like the dynamic transmission behaviour of the transformer, the implementation of control routines or the heat management.

## II. DESCRIBING THE RESONANT TRANSMISSION SYSTEM

In the same way as conventional transformers, transformers with air gap can be described by the T-equivalent circuit (Fig. 1). To determine the coupling of the primary and the secondary coil, i.e. the parameters $L_h$, $L_{1\sigma}$, $L_{2\sigma}$, the finite elements analysis (FEA) can be used. Fig. 2 shows the vector potential of a gyratory magnetic system (3D FEA) at an air gap of $\delta = 10$ mm and at a coil displacement of $v = 10$ mm (pot core ferrites, $d = 70$ mm). The primary coil is fed by a constant current density $J$ in $\varphi$-direction. After solving the symmetric FEA model you get the magnetic flux in the primary and secondary winding space (1) and then you can determine the main and leakage inductances of the T-equivalent circuit (2). In this way the influence of air gap ($\delta$), ferrite geometry and the horizontal displacement of both coils on the coupling of the magnetic system can be investigated.

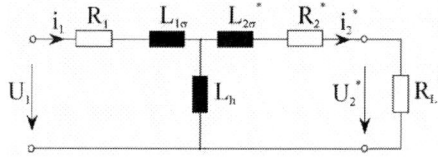

Fig. 1. T-equivalent circuit of the air gapped transformer (no core loss)

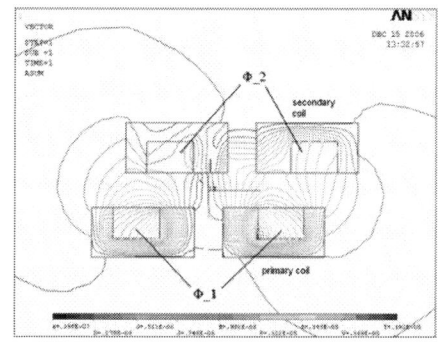

Fig. 2. Vector potential lines *A* at coil displacement (P core 70)

978-1-4244-1741-4/08/$25.00 ©2008 IEEE

$$\Phi = \frac{1}{N} \cdot \sum_{i=1}^{N} \int_{0}^{2\pi} A_{\varphi,i}(r) \cdot r_i \cdot d\varphi \qquad (1)$$

$$L_{ho} = \frac{\Phi_2}{i_1}; \; L_{1\sigma 0} = L_{2\sigma 0} = \frac{\Phi_1 - \Phi_2}{i_1} \qquad (2)$$

Contactless transmission systems are characterised by a small main and large leakage inductances. The transfer of appreciable electric power requires the compensation of the leakage inductances by resonance capacitors which can be placed in series or parallel to the primary and secondary coil. As a result, the primary feeding converter works in a resonant switching mode which enables higher transmission frequencies ($f$>100kHz). After tuning the transmission system in resonance, the step response of its transfer function gives you information about the electrical transfer behaviour. Fig. 3 (a) shows the equivalent circuit for a s-p-compensation and Fig. 3 (b) shows the transfer behaviour at $u_1$=150V, $L_{ho}$=65nH, $L_{1\sigma 0}$=81nH, $L_{2\sigma 0}$=100nH. When changing the frequency of $u_1$ there is one resonance peak at lower and two resonance peaks at higher load resistors $R_L$. This characteristic behaviour is a critical problem in the assembling process, where mechanical tolerances and tolerances of the compensation capacitors or ferrite components have to be considered. To be independent from detuning the resonance, the flat range that occurs here at 10 .. 60 Ω can be shifted along the $R_L$ - axis by changing the absolute windings of the coils without changing the winding ratio (Fig 3, c). In this option the peaks at high values of $R_L$ are eliminated and the peaks at very high values of $R_L$ can be prevented by providing a minimum resistor with very little power loss.

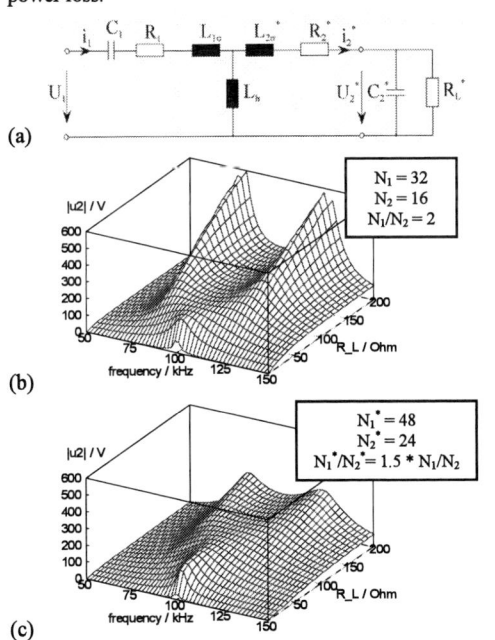

Fig. 3. Equivalent circuit (a) and the qualitative electrical behaviour of $u_2$ (b) at s-p-resonance, optimised windings (c)

## III. POWER LOSS ANALYSES

Depending on the power level of the transmission system, the loss of the magnetic system mostly amounts to more than 50% of the overall loss ($P_{l,total}$). The most important is the i²R loss of the copper wire, the hysteresis loss of the ferrite core and the eddy current loss in the conductive material of the housing.

$$P_{l,total} = P_{l,copper} + P_{l,core,eddy} + P_{l,core,hys} + P_{l,hosing,eddy} \qquad (3)$$

The power loss is a function of several parameters like

- Field frequency
- Air gap (-> coil currents and flux density)
- Coil geometry
- Wire thickness (if no litz wire, -> skin-depth)
- Count of strands (if litz wire)
- Permeability ($\mu_r$) of the ferrite elements
- Conductivity ($\sigma$) and geometry of the housing

### A. Determining the copper loss

The copper i²R loss include the power loss caused by the ohmic D.C. resistance and the resistance caused by the skin and proximity effect. Due to the non homogenous magnetic field and the use of ferrite cores and litz wire, the power loss can most simply be determined by the FEA. Therefore ANSYS and the free FEA tool *femm* [4] were used. To minimise the A.C. resistance of the wire, it must consist of many twisted strands so that the skin effect is minimised and the current density is constant over the copper cross section area. Fig. 4 shows the simulation ($f$ = 500kHz) of different litz wires (strand

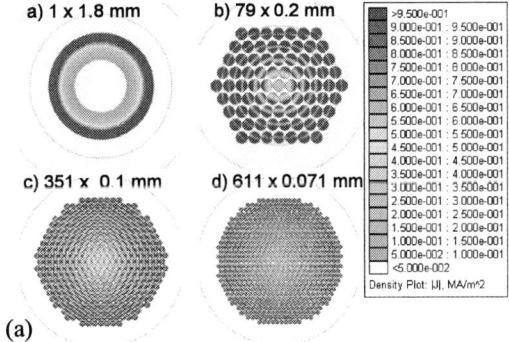

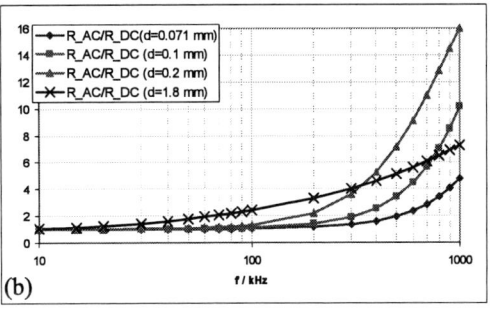

Fig. 4. Current density J of different litz wires at 500kHz (a), factor $R_{AC}$ / $R_{DC}$ (b)

diameters: 0.2, 0.1, 0.071 mm) and a single wire ($d = 1.8$ mm). The copper cross section area of all wires is constant $A_{Co} \approx 2.5$ mm². Compared to the single wire, in the litz wire the current density is constant over the whole area, however the current density in every single strand is not constant because of the proximity effect. This effect causes an increasing resistance when increasing the frequency. Fig. 4 (b) shows the factor $R_{A.C.}/R_{D.C.}$ as a function of the frequency. At 100kHz there is nearly the same A.C. resistance of all litz wires (b, c, d) but at higher frequencies the resistance increases rapidly caused by the proximity effect in the strands. Measurements with a RLC-Meter show that the resistance of litz wire with fewer count of strands can be even higher as the single wire resistance at high frequencies.

To design the magnetic system, the absolute winding resistances ($R_1$, $R_2$) must be known and they can be determined by the FEA, too. Therefore the copper cross section area can be modelled as shown in Fig. 5 (a). Because the absolute resistance depends on the absolute count of windings and their position inside the winding space, possibly an iterative determination is necessary. Furthermore because of the proximity loss every single strand of the wire has to be considered which causes an expensive model with a lot of nodes. To reduce the simulation effort, other models for twisted litz wire are needed. Fig. 5 (b) shows the simulation results of a magnetic system with p core halves ($d = 47$ mm) with 9 wires of 611 x 0.071 mm litz wire ($I_{abs} = 9 * 2$ A) which are conform to the measurement.

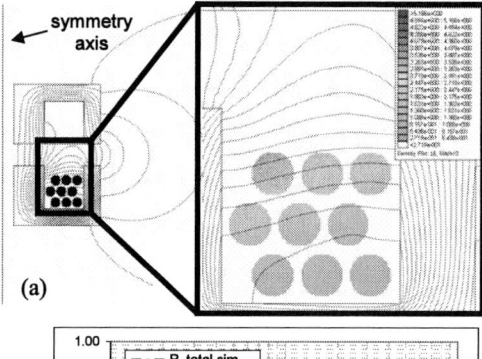

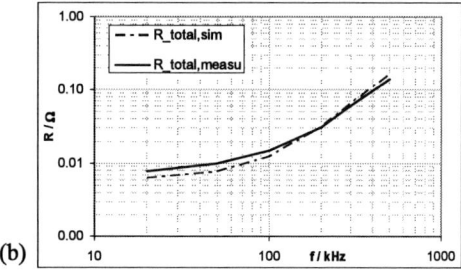

Fig. 5. Current density of litz wire in p cores (d = 47 mm) (a), comparison of the A.C. resistances with the measurement (b)

### B. Determining the core and eddy current loss

After determining the A.C. winding resistances, at one adjusted output power, the primary and secondary windings ($N_1$, $N_2$), the input and output currents of both coils and their phase shift ($i_1$, $i_2$, $\varphi_{i1i2}$) can be determined

with the T-equivalent circuit of Fig. 3. For now the core loss and the loss in the housing are neglected. With the core volume $V$ and the field frequency $f$, the hysteresis loss is given by the physical work of the hysteresis loop

$$P_{1,core,hys} = W_{hys} Vf = \oint_H BdH \cdot Vf . \qquad (4)$$

In another FEA-simulation, the hysteresis loss can be determined with a non-linear "effective" B-H curve without hysteresis [5]. Therefore firstly the "effective" B-H curve is used to evaluate a non-hysteretic permeability

$$\mu_{eff,max} = \frac{B_{eff,min}}{H_{min}} = \frac{1}{\mu_0} \cdot \lim_{H \to 0} \frac{B_{eff}}{H} . \qquad (5)$$

Secondly by using this effective permeability and a complex-valued B-H curve, the hysteresis loss can be calculated with the hysteresis angle $\phi_{h,max}$ which you can determine with the data sheet [5][6].

$$P_{1,core,hys} = \frac{V_{core} \cdot \pi f}{\mu_0 \mu_{eff,max}} \cdot |B|^2 \cdot \phi_{h,max} \qquad (6)$$

With the current density $J$ in the copper area you can determine the eddy current loss by integrating the i²R loss due to currents flowing in the $\varphi$-direction.

The power loss of the transmission system with p cores (Fig. 5, a) is investigated to find the optimal air gap of both coils to each other. In Fig. 6 the corresponding FEA-model is shown. The influences of outer conductive materials on the transmission system is minimised by using an aluminium housing ($f = 100$kHz, $N_1 = N_2 = 10$).

When enlarging the air gap between both coils the main inductance will decrease and the needed magneto motive force ($N*i$) in the coils will increase. By considering the transfer behaviour of the resonant transformer, the power transmission is set to $P_2 = 60$ W and the windings and the feeding currents ($N_1*i_1$, $N_2*i_2$) of both coils and their phase shift $\varphi_{i1i2}$ are determined (Fig. 7, a) and used as input parameters for the FEA-simulation. With the data sheet information of power loss of the materials SiFerrit N22 and N27, the hysteresis angle can be assumed with the typical value of $\phi_{h,max} = 20°$ [7].

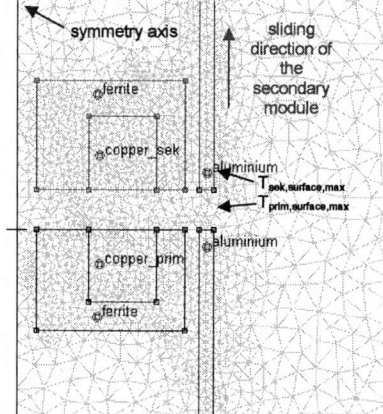

Fig. 6. 2D-FEA model for simulation of the power loss (p core 47)

Fig. 7 (b) shows the simulation results. At very small air gaps the hysteresis loss dominates caused by the very high flux density in the ferrite cores. At higher air gaps the copper loss and the eddy current loss in the housing dominate because of the increasing current density in the coils. The eddy current loss in the ferrite core can be neglected. There is a minimum of the power loss of the magnetic system near an air gap of $d = 4.5$ mm.

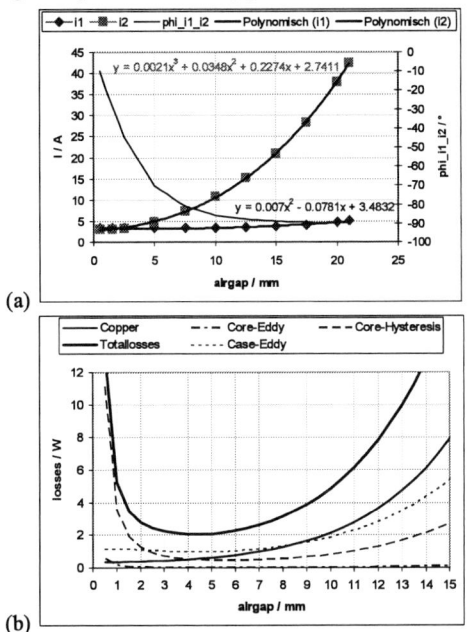

(a)

(b)

Fig. 7. FEA EM simulation, input parameters: $i_1(\delta)$, $i_2(\delta)$, $\varphi_{i1i2}(\delta)$ (a), simulation output: $P_l(\delta)$ (b)

After evaluating the power loss, it can be used as input parameter for a steady state heat flow simulation. On the surface of the housing the boundary condition "convection" is used. The heat transfer coefficient is assumed with $h = 10$ W/(m²/K) to emulate the measured thermal resistance of the housing of $R_{th} = 5°$ K/W. The ambient temperature is assumed with $T_{amb} = 20°$C. The simulation result of the example is shown by the absolute temperature ($T$) at $\delta = 4.5$ mm (Fig. 8).

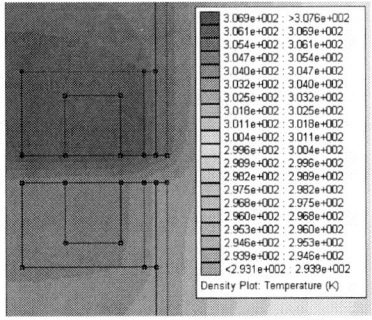

Fig. 8. Temperature (T) of the magnetic system (FEA, heatflow)

Caused by the higher value of $N_2{*}i_2$ the temperature of the secondary coil (top) is higher and caused by the

potting compound around the coils the heat is transferred to the housing. The surface temperature rise is only about $\Delta T = 13$ K (transferred electric power $P_2 = 60$ W). Fig. 9 shows the absolute maximum temperature rise on the surface of the housing as a function of the air gap (see Fig. 6). Similar to the power loss there is a minimum of temperature rise at approximately $\delta = 3 .. 5$ mm.

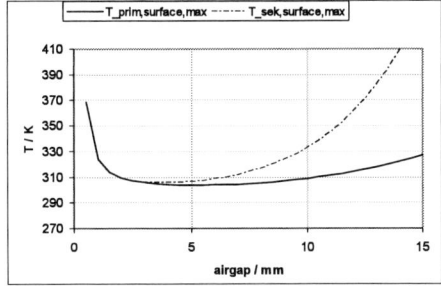

Fig. 9. FEA heat flow simulation, output: $T_{surface,max}(\delta)$

## C. Role of the eddy current loss in the housing

A great problem of resonant inductive transmission systems is the influence of conductive materials in the near of the coils on the magnetic flux lines because of eddy currents. They affect the inductive coupling and the T-parameters and often effort a re-adjusting of the feeding frequency. However, by using conductive housings, the influence of external conductive materials and the amount of the outer magnetic flux density can be minimised. Fig. 10 shows the eddy current analysis at $f = 100$kHz and the influence of an aluminium housing around the coils on the vector potential. On the left side, the material property air is used for the housing and the vector potential lines are not affected. On the right there are induced eddy currents in the housing material near the air gap region.

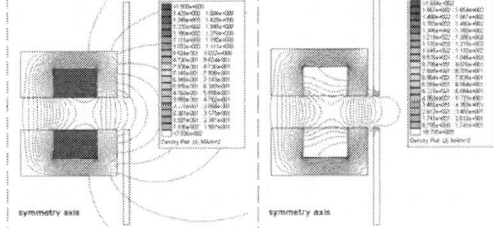

Fig. 10. FEA EM ($f = 100$kHz, $N{*}i{=}200$ A): left: air; right: aluminium

As shown in Fig. 7 the power loss in the housing is in the similar range as the copper loss. Laminating the housing could be a basic solution, however this is not recommendable to keep the grade of ingress protection. An important parameter to minimise the eddy current loss is the distance between the magnetic system and the housing. When increasing the diameter of the housing (Fig. 11, a) the main inductance will rise and the needed $N{*}i$ in the coils decreases. Fig. 11 (b) shows the power loss of the magnetic system from Fig. 6 (p core halves, $d = 47$ mm, $f = 100$kHz) as a function of the inner diameter of the housing around the coils. The transferred power is $P_2 = 60$ W $=$ const. and the air gap is 4.5 mm. The copper and the hysteresis loss are in a similar order of magnitude. Like shown in the diagram, the eddy current loss of the housing is very extensive at small distances to the coil and fall rapidly when enlarging the distance.

To investigate the power loss of a transmission system in the power range of 1 kW, a magnetic system with p core ferrites (d = 70 mm) was considered. The power loss as a function of the air gap shows a qualitative similar behaviour, however because of the higher absolute power loss, the use of a conductive housing for shielding is only usable with larger diameters of the housing, at smaller electric power or at smaller operating frequencies.

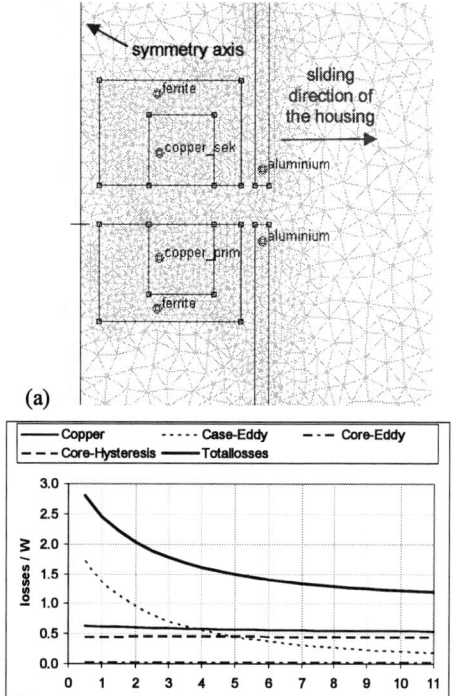

(a)

(b)

Fig. 11. FEA-model (a) and power loss as a function of the housing-coil distance (b)

## IV. COIL POSITIONING FLEXIBILITY

Another important aspect of inductive power transmission systems is the positioning flexibility of both coils to each other. For controlling the output voltage, the above-shown application includes a step up converter which causes additional power loss and decreases the overall efficiency. In the power range of several kilowatts this option is not recommended and other solutions to get a constant output voltage may be favourable.

It has been shown that changing the geometry of the coils and the ferrite elements will affect the electrical behaviour at a displacement of both coils and limit the positioning tolerance [3]. Rotary transmission systems can by designed with different types of ferrite elements. In Fig. 12 the p core halves are characterised by an embedded winding space (yellow/inner region) whereas system (b) consists of flat ferrite discs without a winding space. Due to the homogenous main field within the air gap region of the upper system, the inductive coupling is better and the magnetic leakage field can be minimised.

Another effect is that the condition of higher position tolerances (several centimetres) requires larger air gaps. For every axis-symmetric coil geometry there is the possibility to determine an equivalent air gap to compare these assemblies at horizontal displacements. At this equivalent air gap the magnetic couplings are equal. To investigate the influence of coil displacements, a 3D-FEA simulation has to be used. Fig. 12 shows the meshed p core halves ($d = 70$ mm) at $\delta = 10$ and of flat ferrite discs ($d = 130$ mm) at the equivalent air gap $\delta = 30$ mm.

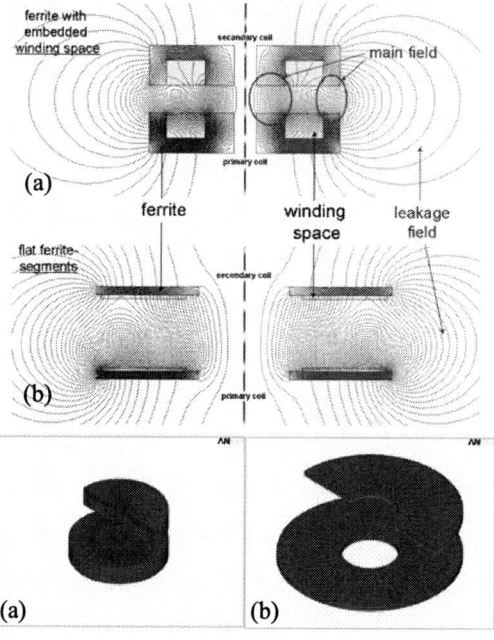

(a)

(b)

(a)

(b)

Fig. 12. Flux lines and FEA-models at p cores (a) and ferrite discs (b)

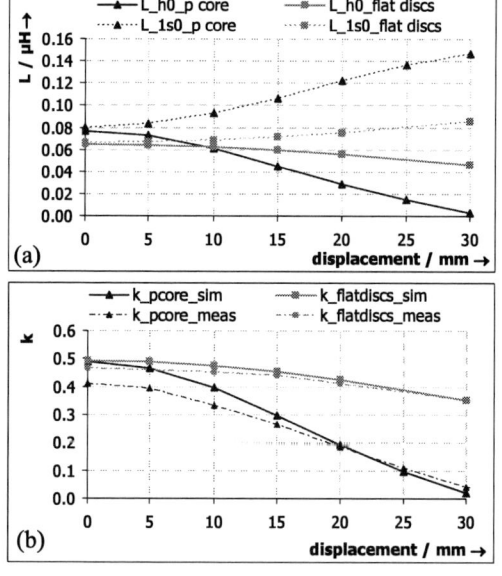

(a)

(b)

Fig. 13. Simulated T-parameters of p cores and ferrite discs (a) and simulated and measured coupling coefficients (b) at coil displacement

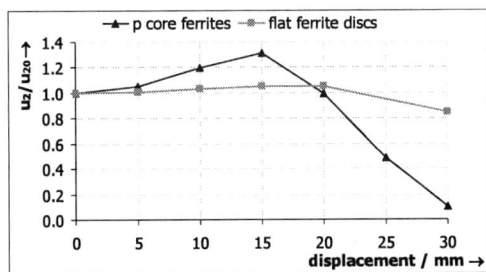

Fig. 14. Output voltage $u_2$ at p cores and ferrite discs ($R_L = 20\ \Omega$)

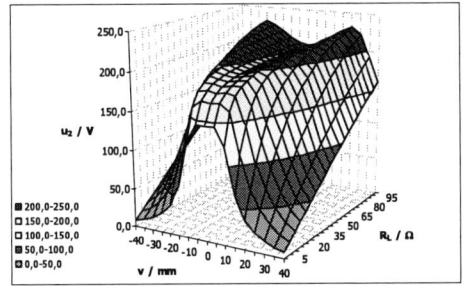

Fig. 15. Output voltage $u_2$ at coil displacement (flat discs, variable $R_L$)

Fig. 13 shows the simulated T-parameters at a horizontal coil displacement and the coupling coefficient $k$ which can be determined as follows:

$$k = \frac{M}{\sqrt{L_1 L_2}} = \frac{L_{h0}}{\sqrt{L_{10} L_{20}}} \qquad (7)$$

At identical coil geometries the coupling coefficient is

$$k = \frac{L_{h0}}{L_{10}} = \frac{L_{h0}}{(L_{h0} + L_{1\sigma0})} . \qquad (8)$$

With the use of flat ferrite discs the change of the main inductance and the coupling coefficient is more insensitive at coil displacement because the main flux will not predominate in a homogenous main field.

With the T-parameters the electrical behaviour can be determined. As shown in Fig. 3 the electrical behaviour of the entire system is very sensitive when the resonance circuit gets out of tune. Such an operation out of resonance can be caused by varying the inverter frequency but also by a displacement of both coils to each other or by varying the air gap. At such geometrical changes of the magnetic system, the reason for getting out of resonance is the change of the T-parameters ($L_{h0}$ and $L_{\sigma0}$). In any case of getting out of resonance there are mostly unwanted resonance effects of the input and output voltages or currents. Fig. 14 shows the normalised output voltage $u_2$ over a coil displacement of $v = 30$ mm. At the system with flat ferrite discs the voltage lies in the range of $u_2 = u_{2,nom}$ (+10, -20) %.

As a function of the load resistance there is a similar electrical behaviour at coil displacements like at frequency changes. Here also the resonance peaks at higher load resistors can be shifted along the $R_L$-axis as explained in Chapter II and shown in Fig. 3. Fig. 15 shows the transfer behaviour of the magnetic system with flat ferrite discs at well tuned coil parameters as a function of the coil displacement $v$ and the load resistance $R_L$. The changes of the output voltage $u_2$ are small in a defined range of displacement and over a large range of $R_L$.

Another important aspect at higher air gaps, higher transferred power and coils displacements are the requirements of the EMC standards. Therefore terms like radiation, conducted interference and magnetic field emission have to be observed. The most important issue for contactless inductive power transmission, especially for use in consumer electronics, is the mid frequency magnetic field and its effect on the biological environment. The permissible values of the magnetic flux density are prescribed in the electromagnetic compatibility regulations of the employer's liability insurance association [8] or in the ICNIRP guidelines [9]. Investigations and measurements have shown that the leakage field of magnetic systems with flat ferrite discs are quite greater than the field at the use of pot cores and that the limit values of the norm conditions can be kept in any case even in the power range up to 1 kW. However, at higher transferable electric power ($P_{RL} > 1$ kW) and for the use in consumer electronics, a compromise between the position flexibility and the observance of the outer magnetic leakage field has to be found [10].

## V. POWER LOSS MEASUREMENTS

To verify the power loss simulations and the electrical and thermal behaviour, a test module corresponding to Fig. 5 and Fig. 6 was built. With respect to the simulation results the air gap between the coils is set to $\delta = 5$ mm and the magnetic system and the power electronics are integrated in a concentric aluminium housing (Fig. 16) which lets the transmission system be insensitive against external conductive materials. The distance between the p ferrite cores and the aluminium housing is 3 mm to minimise the eddy current loss. The transmission system works as a plug substitution on the 24 V D.C. voltage level and it can be used for inductive energy and data supply of industrial sensor and actuator modules. The primary coil is fed by a full-bridge inverter and the leakage inductances are compensated by a series capacitor on the primary and a parallel capacitor on the secondary side. The secondary module contains a rectifier, a filter and an additional step up converter to control the output voltage when changing the air gap or the module position. In spite of the additional step converter and the aluminium housing, the overall efficiency is $\eta > 85\%$. The data transmission is realised with separated coils (Fig. 16, right) and the bit rate can be up to 115 kbit/s.

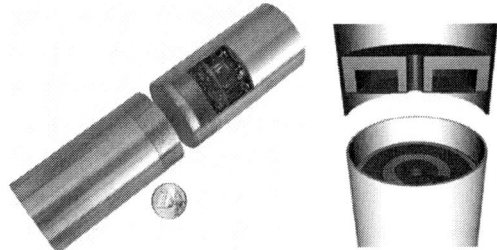

Fig. 16. Inductive energy and data transmission system, $P_2 = 60$ W, (left: prototype, right: schematic), _www.kontenda.de_

The complete module is optimised for small size and weight but not for eddy current loss in the housing. However, like the investigations have shown, an outer diameter of 55 mm and a length of 100 mm per module are the minimum required dimension to be able to dispose the power loss. Because of the high grade of ingress protection (IP 64) there is no possibility for active cooling. Fig. 17 (left) shows the infrared analysis of the housing of the magnetic system after reaching the final temperature at $t = 30$ min. The magnetic system was fed by auxiliary power electronics and the aluminium housing was prepared with white particle powder (right) to get correct measurement results (emissivity $\varepsilon = 0.9$). The absolute temperature rise of the housing is $\Delta T \approx 15$ K relative to the ambient temperature and conform to the simulation ($P_2 = 60$ W). Measurements of the complete transmission system show a temperature rise of about $\Delta T = 25$ K (housing surface) which can be explained by the additional loss of the power electronics.

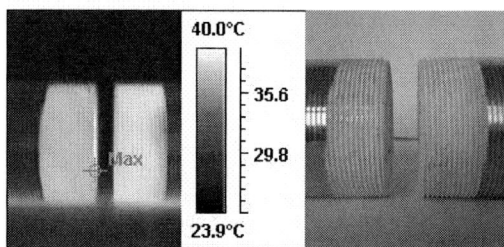

Fig. 17. Infrared analysis of the magnetic system

### VI. EXPERIMENTAL SET UP

To verify the simulation results for improving the coil positioning flexibilty, an experimental set up with flat ferrite discs at an air gap of $\delta = 30$ mm and with a secondary parallel compensation was realised. In comparison to the above-shown module, an optimised coil geometry should enable a greater positioning tolerance without the need of a voltage control. The primary coil is fed by a half bridge inverter at $f = 100$kHz and the primary leakage inductance is compensated by a series capacitor. The entire system in Fig. 18 shows the power electronic modules and the realised transmission system. It allows a transferable electric power of $P_{RL} = 1$ kW. To investigate the dependence from the load, a programmable electronic load was used.

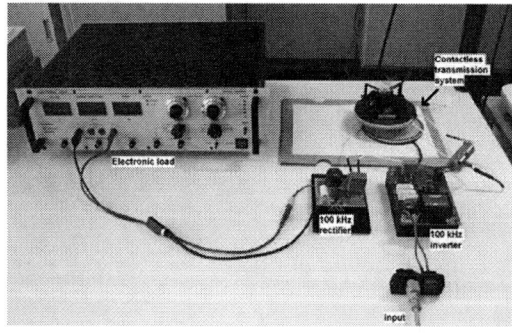

Fig. 18. Experimental set up for coil displacement

As shown in Fig. 19 the simulated and calculated voltage rise of $U_{RL}$ as well as the point of the maximum voltage are conform to the measurement. The small difference between the measured and simulated voltage is caused by non-ideal winding parameters of the coils. Over a horizontal coil displacement of $v = 30$ mm the output voltage lies inside the allowed range. To minimise the switching loss of the power electronics, a zero current switching operation (ZCS) is implemented (Fig. 20, a). At maximum displacement (Fig. 20, b) there is a phase lag between the primary current and voltage which let increases the switching loss. These additional switching loss and the over-voltage of the primary voltage $u_l$ over the IGBTs limit the maximum displacement.

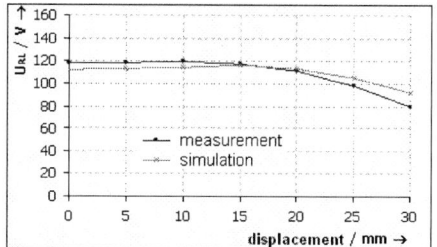

Fig. 19. Simulated and measured output voltage $u_{RL}$ at coil displacement

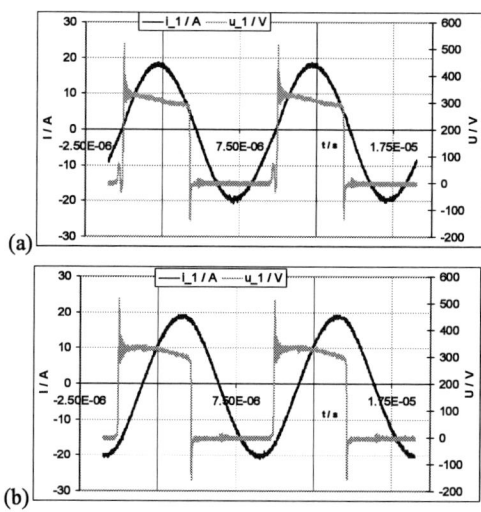

Fig. 20. Measured output voltage $u_1$ and current $i_1$ of the primary inverter at zero (a) and at 30 mm (b) coil displacement

Another example of power integration is shown in Fig. 21. It is geometrically similar to the presented system in Fig. 18 (ferrite discs), however, it enables the power transfer of about 1.5 kW through a tabletop. Because of the high power level, a combination of a PFC and a 100kHz full-bridge inverter on the primary side is necessary. On the secondary side a rectifier, a D.C. link and an additional 50 Hz inverter is integrated which generates the output voltage of 230 V A.C.. The air gap is 40 mm and the possible horizontal displacement is 40 mm. This "contactless power line coupler" enables the transmission of energy to any conventional plugged application. To identify and to detect the absolute position of the secondary device, the primary and secondary device contain an additional data transmission module. Therefore the data coils are realised with separated printed coils. It works inductively as well and it is geometrically optimised for a very low interference with the energy transmission. The identification module can be used for copy protection or to read special inverter settings after positioning the secondary device. With these parameters, a use in consumer electronics or even in kitchen equipment is imaginable.

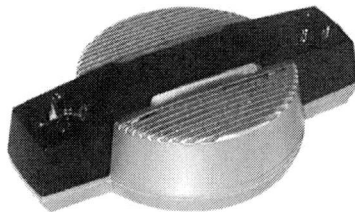

Fig. 21. Inductive energy transmission for home or business electronics

## VII. CONCLUSIONS

Inductive transmission technology can improve the energy supply of existing systems or open up new technical solutions wherever energy has to be transmitted and where a constructive conditional air gap must be bridged. The design process of contactless transmission systems can be divided into the design of the magnetic system and the power electronics.

To design a magnetic system in the paper very important aspects are analysed such as power loss of the magnetic system, improving the coil positioning flexibility and electromagnetic compatibility (EMC). Especially at higher transferable electric power the efficiency of the overall system can be improved by using higher transmission frequencies, ferrite elements for the coils and compensating the leakage inductances by resonance capacitors which allow the power electronics a resonant switching mode.

The power loss of the magnetic system can be reduced and the coil positioning flexibility can be improved by optimising its geometric parameters. Therefore it is necessary to be able to describe the behaviour of the resonant transmission system. The steady-state and the dynamic behaviour can be modelled by the T-equivalent circuit of the transformer and the T-parameters can be determined by the FEA.

To let the magnetic system be independent of external influences like magnetic fields or conductive materials, a conductive housing can be used. The evaluation of the core and copper loss of the magnetic assembly as well as the additional eddy current power loss can also be done by the FEA. In the paper favourable geometric parameters of the magnetic device and the housing are estimated and their influence on the overall loss and heat flow are simulated and verified by infrared analysis of a test system.

Increasing the positioning tolerance of the coils without the need of a voltage control can be reached by using flat ferrite discs, using a secondary parallel compensation and larger air gaps. Therefore special aspects like the behaviour of the T-parameters and electrical quantities at coil displacements are discussed and the simulation results are verified by measurements with an experimental set up.

## VIII. REFERENCES

[1] R. Mecke, C. Rathge, W. Fischer, B. Andonovski, "Analysis of inductive energy transmission systems with large air gap at high frequencies." *European Conference on Power Electronics and Applications*, 2003, Toulouse

[2] M. Bonke, „Modellbasierte Berechnung der Zustandsgrößen kontaktloser Übertragungsanordnungen", *Master's Thesis*, Hochschule Anhalt (FH), 2006, Köthen

[3] D. Kürschner, Ch. Rathge, E. Schulze: "Optimisation of contactless inductive transmission systems for high power applications", *PCIM 2007*, Nürnberg, 22.05.-24.05.2007

[4] D. C. Meeker, "Finite Element Method Magnetics Version 4.0.1", *http://femm.foster-miller.net*

[5] D. Meeker, A. Filatov, E. Maslen, "Effect of Magnetic Hysteresis on Rotational Losses in Heteropolar Magnetic Bearings", *IEEE TRANSACTIONS ON MAGNETICS VOL. 40 NO. 5*, 09/2004

[6] EPCOS AG, "Ferrites and accessories, Materials and General – Definitions", *Data Sheet*, September 2006

[7] R. L. Stoll, "The analysis of eddy currents", *Oxford University Press*, 1974

[8] Berufsgenossenschaft für Feinmechanik und Elektrotechnik BGFE (Hrsg.), „BGV B11", *Unfallverhütungsvorschrift Elektromagnetische Felder*, Köln, 1. Juni 2001

[9] International Commission on Non-Ionizing Radiation Protection, "Guidelines for limiting exposure to time-varying electric, magnetic, and electromagnetic fields (up to 300 GHz)", *Health Physics 74 (4)*: 494-522; 1998.

[10] D. Kürschner, Ch. Rathge, "Contactless energy transmission systems with improved coil positioning flexibility for high power applications", *Power Electronics Specialists Conference Rhodos, Greece*, 2008

## IX. NOMENCLATURE

| | |
|---|---|
| $\delta$ | air gap |
| $d$ | diameter |
| $N$ | count of windings |
| $v$ | horizontal coil displacement |
| $R$ | ohmic resistance |
| $R_{th}$ | thermal resistance |
| $L_{h0}$ | main inductance at one winding |
| $L_{\sigma 0}$ | leakage inductance at one winding |
| $k$ | coupling coefficient |
| $J$ | current density |
| $N*i$ | magnetomotive force (mmf) |
| $\varphi$ | phase angle (or angle of cylindrical coordinates) |
| $\sigma$ | conductivity |
| $P_2$ | output power ($P_{RL}$) |
| $P_l$ | power loss |
| $\eta$ | efficiency |
| $\phi_{h,max}$ | core hysteresis angle |
| $\mu$ | permeability |
| $\mu_{eff,max}$ | effective non- hysteretic permeability |
| $|B|$ | magnetic flux density |
| $B_{eff}$ | flux density of a non-linear effective B-H-curve without hysteresis |
| $H$ | magnetic field strength |
| $W_{hys}$ | work per hysteresis (B-H) loop |
| $A_{\varphi}$ | magnetic vector potential ($\varphi$-direction) |
| $\Phi$ | magnetic flux |
| $f$ | frequency |
| $T$ | temperature |

| | |
|---|---|
| EMC | electromagnetic compatibility |
| FEA | finite elements analysis |
| ZCS | zero current switching |
| IGBT | insulated gate bipolar transistor |

# A Transformerless Interface Converter for a Distributed Generation System

Tzung-Lin Lee*  Zong-Jie Chen**

* Department of Electrical Engineering, National Sun Yat-sen University, TAIWAN
** Department of Electrical Engineering, Chang Gung University, TAIWAN
Email: tzunglin.lee@gmail.com

*Abstract*—Integrating various interface converters of small-rated distributed generators into the power system has become a promising solution to rapidly increasing demand of premium electric power. The output voltage of distributed generators, such as PVs, is usually so low that a low-frequency coupling transformer is required to place between the interface inverter and the grid for proper power delivery. However, this transformer may impede practical installation due to its large weight and size. This paper proposes a transformerless interface converter for a distributed generation system. Instead of the low-frequency transformer, a coupling capacitor is installed between the inverter and the grid to sustain a part of the fundamental voltage. By dynamically adjusting inverter output voltage vector, the power can be converted between the inverter and the grid without any low-frequency transformer. Based on this scheme, the low-voltage generating source would be directly able to deliver available power into the high-voltage grid with unity power factor operation, or to supply reactive power for grid voltage regulation. The proposed method can also suppress power system harmonics due to its voltage control nature. Operation principles and design issues of the coupling capacitor are explained in detail, and computer simulation results are provided to validate the effectiveness of the proposed approach.

## KEYWORDS

Converter control, Renewable energy systems

## I. INTRODUCTION

Due to expansion of industries, the electric power consumption has been growing in an unprecedented pace. For delivering premium electric power in term of both reliability and quality, integrating interface converters of small-rated distributed generators, such as PVs or wind turbines, into the power system has become a critical issue in rescent years [1], [2], [3], [4], [5].

Most renewable-generated sources are variable with operating environment, and normally fabricated by stacked low-voltage generating cells, thus a power-processing interface converter is required for efficient and proper power delivery. Various converters and their control techniques for distributed generation have been presented previously [6], [7], [8], [9], [10], [11]. They demonstrate effective power flow control performances whether in the grid-connected or in the islanding operation. These methods usually need a low-frequency transformer or a voltage booster to raise voltage level for connecting to the power system. However, practical applications may be limited due to large weight of this transformer or

EMI of high voltage switching. Akagi et al. have presented a hybrid shunt active filter for suppressing power system harmonics, in which a LC passive filter replaces the low-frequency transformer for reducing converter size [12].

This paper proposes a transformerless interface converter for a distributed generation system. Instead of a low-frequency transformer, a coupling capacitor is installed between the generation converter and the grid. This capacitor can sustain a part of fundamental voltage, so the interface converter of the low-voltage generating source can deliver power into the high-voltage grid in a transformerless fashion. Based on this scheme, the power flow between the converter and the grid can be controlled by adjusting inverter output voltage vector in response to available power of the generator and requirement of grid voltage regulation. Therefore, the inverter can operate with unity power factor or support reactive power from a low-voltage source without any low-frequency transformer. This method also can damp power system harmonics and reduce switching EMI due to low-voltage inverter, which is the significant advantage of the proposed approach.

## II. OPERATION PRINCIPLES

A simplified one-line diagram of the proposed distributed generation system is shown in Fig. 1. The interface converter of each distributed generation unit (DGU), DGUx (x=1···N), transfers power from distributed generators, such as PVs, to the utility grid by connecting a coupling capacitor $C_x$. This capacitor is designed to support a part of fundamental voltage, so the inverter can deliver the real power or the reactive power from a low-voltage generation source to a high-voltage grid without any low-frequency coupling transformer. The control of each DGU consists of a voltage controller and a current controller. The voltage controller determines the inverter voltage command to delivery the generator available power and the reactive power for grid voltage regulation. The current controller sequentially follows the current command producing from the voltage controller to derive the switching pattern of the inverter. Because the inverter is controlled as a voltage source, the proposed approach also provides harmonic damping capability for the power system. Detail operational principles are described as follows.

### A. Voltage controller

The voltage controller calculates the real power and reactive power of the DGU, then determines the voltage command. The

978-1-4244-1741-4/08/$25.00 ©2008 IEEE

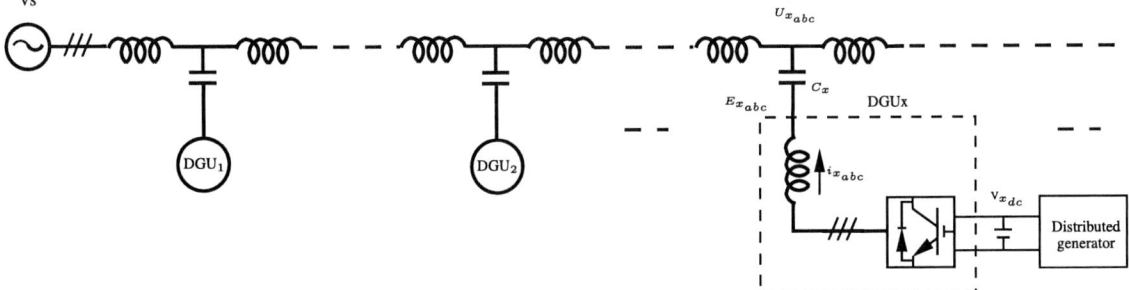

(a) One-line circuit diagram of the proposed distributed generation system.

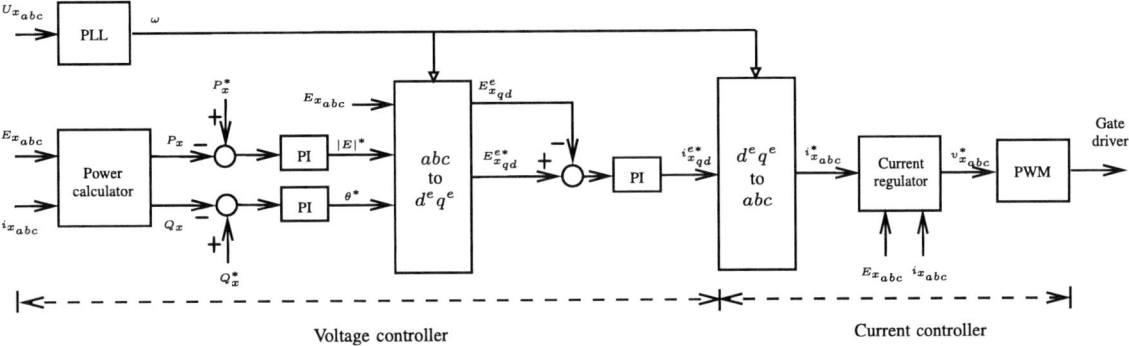

(b) Control block diagram of the proposed transformerless interface converter.

Fig. 1.   The proposed transformerless interface converter for a distributed generation system and the associated control.

real power $P_x$ and the reactive power $Q_x$ of the DGUx can be obtained by using the instantaneous reactive power theory [13].

$$\begin{bmatrix} P_x \\ Q_x \end{bmatrix} = \begin{bmatrix} E_{x_q}^s & E_{x_d}^s \\ -E_{x_d}^s & E_{x_q}^s \end{bmatrix} \cdot \begin{bmatrix} i_{x_q}^s \\ i_{x_d}^s \end{bmatrix} \qquad (1)$$

where $E_{x_{qd}}^s$ and $i_{x_{qd}}^s$ represent the voltage and the current of the DGUx in the stationary frame, respectively. The voltage command is based on the real power command $P_x^*$, which would be determined by operating status of the generator. In practice, $P_x^*$ may be variable with operating environment due to renewable-generated source, so it can be determined by the generator or by regulating dc voltage of the inverter, as shown in Fig. 2(a). On the other hand, the reactive power $Q_x^*$ would be obtained by a line voltage regulator to accomplish grid voltage regulation. As illustrated in Fig. 2(b), both $U_{x_{qd}}^e$ and $U_{x_{qd}}^{e*}$ are used to derive $Q_x^*$. After that, the voltage magnitude command $|E|^*$ and phase angle command $\theta^*$ of the inverter can be obtained by using two PI controllers, as shown in Fig. 1.

The output voltage regulation is implemented in the synchronous reference frame (SRF) [14]. By using a PI controller in the SRF, the voltage command $E_{x_{dq}}^{e*}$ and the measured voltage $E_{x_{dq}}^e$ can generate the current command $i_{x_{qd}}^{e*}$ of the inverter. The decoupling between the $d^e$ and $q^e$ axes components is also included. Based on this algorithm, the inverter current command is adjusted according to generator available power and grid voltage variation.

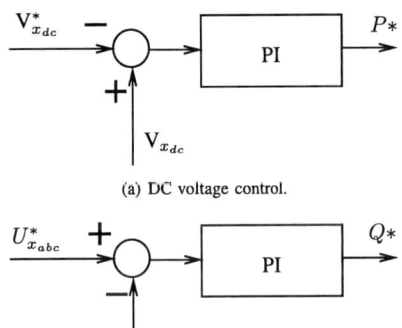

(a) DC voltage control.

(b) Grid voltage regulation control.

Fig. 2.   Real power and reactive power command.

### B. Current controller

Based on the current command $i_{x_{abc}}^*$, the measured current $i_{x_{abc}}$, and the measured voltage $E_{x_{abc}}$, the current regulator calculates the voltage command $v_{x_{abc}}^*$ as follows[15],

$$v_{x_{abc}}^* = E_{x_{abc}} - \frac{L_x}{\Delta T}(i_{x_{abc}}^* - i_{x_{abc}}) \qquad (2)$$

where $L_x$ is the output inductor of the inverter, and $\Delta T$ is the sampling period. Then, the space-vector PWM is employed

1705

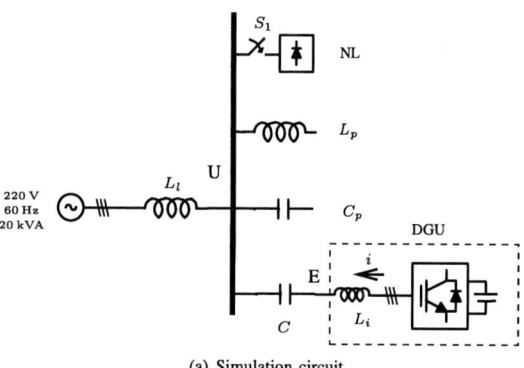

(a) Simulation circuit.

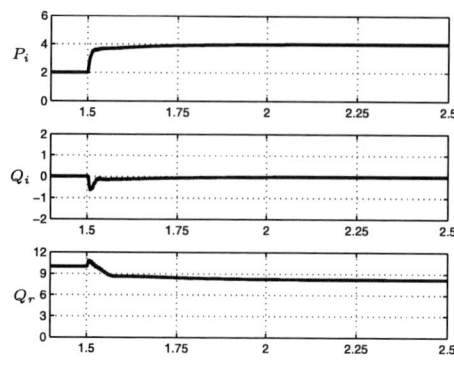

(b) Inverter real power $P$, inverter reactive power output $Q_i$, and reactive power input to the grid $Q_r$.

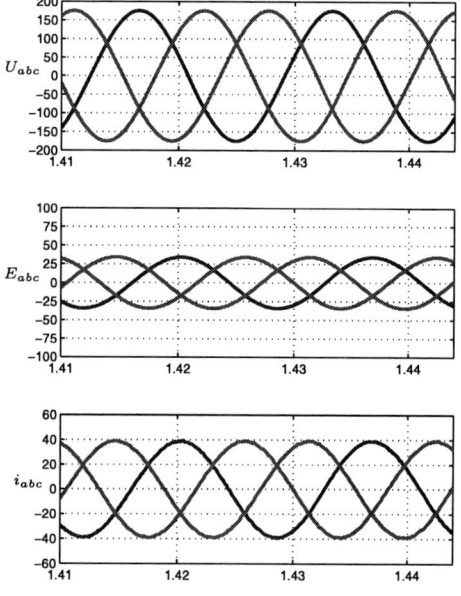

(c) Inverter voltages and currents if $P^* = 2.0kW$.

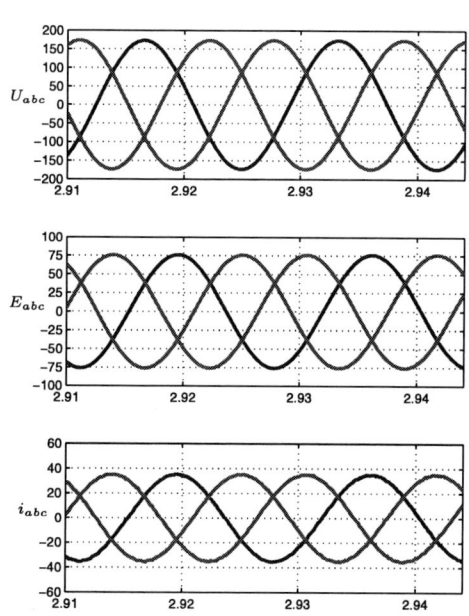

(d) Inverter voltages and currents if $P^* = 4.0kW$.

Fig. 3. Simulation circuit and results when the DGU is at unity opwer factor operation.

to synthesize the gating signals of inverter switches, so the DGU can produce the desired voltage and current with a high bandwidth.

## III. SIMULATION RESULTS

The proposed transformerless interface converter of distributed generation is simulated by using the alternative transient program (ATP) to demonstrate power delivery capability. In this section, unity power factor operation, grid voltage regulation, and harmonic filtering performance of the proposed converter are discussed. Fig. 3(a) shows a simplified one-line circuit model. Power system configuration and inverter parameters are given as follows:

- Power system: 220 V(line-to-line), 20 kVA, 60 Hz.
- $L_l$=1.0 mH(15.5 %) represents line inductance. It is larger than the nominal value for the sake of generating voltage sag and swell in testing the proposed voltage regulation control.
- Inverter coupling capacitor: $C = 600 \, \mu F$ (13.7 %).
- $C_p$=100 $\mu$F represents a power factor correction capacitor. $L_p$ are inductive loads. Its value is 8.0 mH.

- A nonlinear load NL is a diode rectifier with filter inductor, DC capacitor, and load resistor. It is rated at 1.5 kVA(30 %).
- The DGU is implemented with a conventional three-phase voltage source inverter. $L_i$ is 1.0 mH. The PWM frequency is 10 kHz.

### A. Unity power factor operation

In this test, $S_1$, $S_2$ are open. $Q^* = 0$ is set for unity power factor operation. Based on the real power command $P^*$ and the reactive power command $Q*$, the inverter would dynamically adjust it output voltage vector to make sure its voltage is exactly in phase with its current. Fig. 3(c) and Fig. 3(d) show the grid voltage $U_{abc}$, the inverter voltage $E_{abc}$, and the inverter current $i_{abc}$ for both $P* = 2kW$ and $P* = 4kW$. As illustrated, the inverter voltage is always maintained in phase with the inverter current, and the inverter voltage is raised in response to increase of output power, as expected in Section II. At steady state, $U_{abc}$=124 V, $E_{abc}$=34 V, and $i_{abc}$=28 A when $P^* = 2kW$, $U_{abc}$=122.5 V, $E_{abc}$=54 V, and $i_{abc}$=25 A when $P^* = 4kW$.

Fig. 3(b) shpws the response of $P_i$, $Q_i$, and $Q_r$, where $P_i$, $Q_i$, and $Q_r$ represent inverter output real power, inverter output reactive power, and the reactive power injecting into the grid from the coupling capacitor, respectively. Step change of $P^*$ occurs at 1.5s. Results illustrate the effective power tracking capability of the proposed approach. Inverter output power $P_i$ and $Q_i$ can be well controlled at the desired level. Because the reactive power $Q_r$ into the grid from the coupling capacitor is decreased, $U_{abc}$ would be slightly dropped after $P^*$ is raised.

### B. Grid voltage regulation

The voltage control of Fig. 2(b) is evaluated to verify capability of grid voltage regulation in the proposed method. After the voltage regulation control is engaged at t=1.5s, more reactive power is delivered into the grid, and then the grid voltage is restored to the nominal value 127 V as shown in Fig. 4(a). The reactive power output $Q_i$ of the DGU is 4.0 kvar, and the reactive power $Q_r$ is increased from 10 kvar to 13.9 kvar. The real power output $P_i$ is still maintained at the presetting level (2.0 kW). When reaching the steady state, $E_{abc}$=40 V, and $i_{abc}$=37 A.

If the inductive loading $L_p$ is reduced to 16.0 mH, grid voltage becomes 132 V without grid voltage control. After the voltage regulation control is enabled at t=1.5s, Fig. 4(b) shows the grid voltage would be reduced and maintained at 127 V. At the steady state, the DGU delivers $Q_i$=2.3 kvar, inverter voltage is 60 V, inverter current is 37 A, and $Q_r$ is decreased from 11 kvar to 6.0 kvar, respectively.

Therefore, the proposed interface converter can control the reactive power flow for grid voltage regulation in response to grid voltage sag or swell without degrading real power control.

### C. Nonlinear loading

When the nonlinear load is switched on ($S_1$ is closed), the grid voltage is severely distorted. As shown in Fig. 5(a),

voltage THD at U ($U_{THD}$) is at 5.1%. After the DGU is in operation, the DGU can deliver real power and also draw harmonic current. The grid voltage distortion is significantly reduced to 1.0%, as illustrated in Fig. 5(b). This feature allows the proposed inverter with harmonic damping capability for the utility grid.

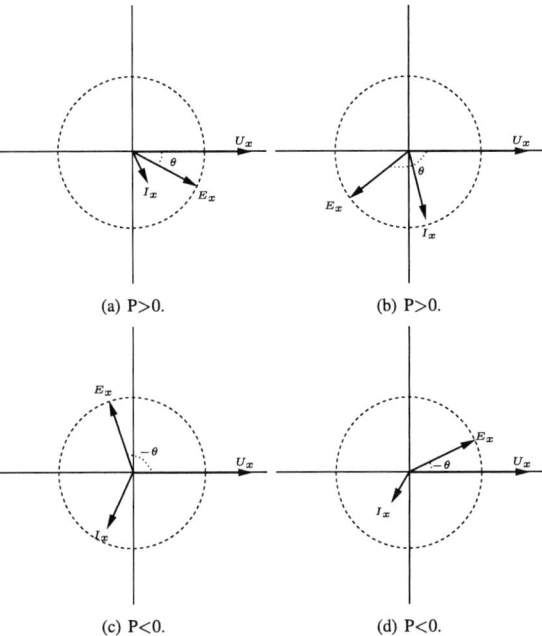

(a) P>0.   (b) P>0.

(c) P<0.   (d) P<0.

Fig. 6. Phasor diagrams of the grid voltage $U$, inverter voltage $E_x$, and inverter current $I_x$.

## IV. Discussion

In contrast to the conventional grid connected inverter, the proposed interface converter adopts a coupling capacitor to accomplish proper power delivery between the high-voltage grid and the low-voltage generating source. The inverter operation and the coupling capacitor must be further addressed.

### A. Inverter operation

Fig. 6 shows the phasor diagrams of the grid voltage $U_x$, the inverter voltage $E_x$, and inverter current $I_x$, respectively. $|E_x|$ is based on $P_x^*$ for controlling real power delivery, whereas $\theta$ is dependent on $Q_x^*$ for determining reactive power flow. As indicated in Fig. 6, $E_x$ moving in a concentric circle would vary the real power and the reactive power. If $E$ is located in the fourth or third quadrant, the inverter can output real power, and especially operate at unity power factor mode ($Q_x^*$=0), as shown in Fig. 6(a) and Fig. 6(b). When $E$ is in the first or second quadrant, Fig. 6(c) and Fig. 6(d) show the inverter is at charging mode (P<0). On the other hand, the inverter would be able to control the reactive power flow by varying inverter voltage angle, but the inverter branch (including coupling capacitor) always presents capacitive.

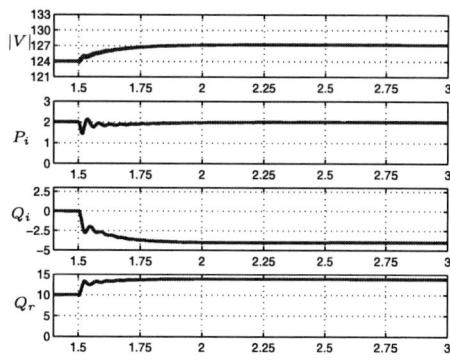
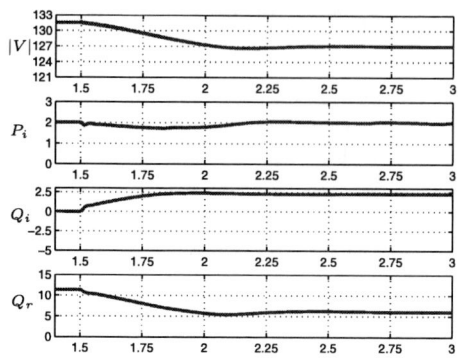

(a) Grid voltage and DGU output power in grid voltage sag.

(b) Grid voltage and DGU output power in grid voltage swell.

Fig. 4.    Simulations of grid voltage regulation.

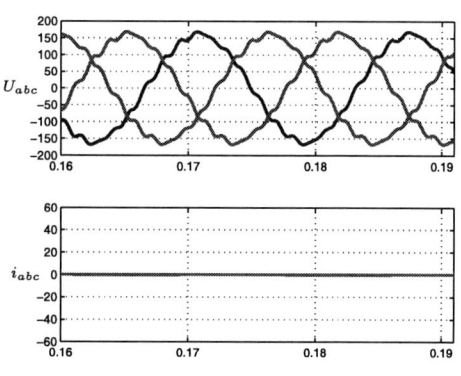
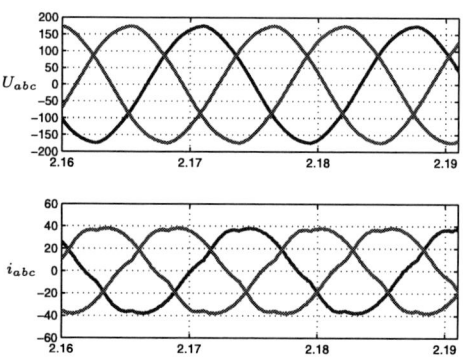

(a) Before the DGU is started, U$_{THD}$ is 5.1%.

(b) AFter the DGU is in operation, U$_{THD}$ drops to 1.0%.

Fig. 5.    Simulations of harmonic damping capability.

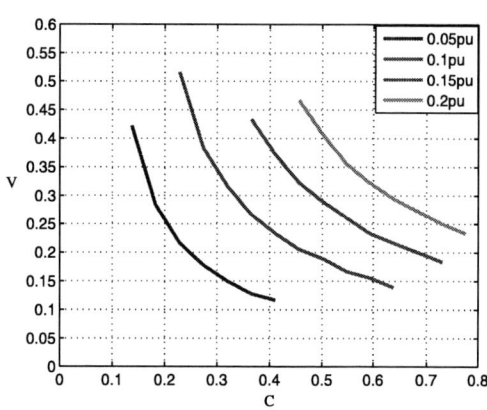

Fig. 7.    Required minimum dc bus voltage (V) to the coupliing capacitor (C) for various real power output ( 0.05, 0.1, 0.15, 0.2 p.u.).

### B. Coupling capacitor

In the proposed method, the inverter can deliver power to the grid from lower dc volatge source by installing a coupling capacitor between them. Clearly, how to determine the coupling capacitor is very critical for this application. Fig. 7 shows simulations of the required dc voltage to the coupliing capacitor for various real power output ( 0.05, 0.1, 0.15, 0.2 p.u.) if the inverter is operated in the linear region of sinusoidal PWM. The base voltage for dc bus voltage of the inverter is set as 360 V. Test results indicate the lower dc voltage is operated, the larger capacitor is required for the same real power output. The minimum capacitor can be determined to supply a rated real power for a fix dc voltage source. For example, if dc voltage source is 0.4 p.u., 0.5 p.u. coupling capacitor is suitable choice for supplying 0.2 p.u. power. On the other hand, if dc voltage source is only 0.25 p.u., 0.73 p.u. coupling capacitor must be used. Therefore, according to Fig. 7, a suitable capacitor would be determined to accomplish required power delivery to the grid from a specific low-voltage source.

## V. SUMMARY

A transformerless interface converter for a distributed generation system is proposed in this paper. By installing a coupling capacitor between the DGU and the grid, the available power can be directly delivered from the low-voltage source to the grid without any low-frequency transformer. The DGU can either operate as unity power factor mode or generate reactive power for supporting grid voltage by adjusting inverter output voltage vector. The proposed method also supplies active damping capability for power system harmonics. In addition, the operation of the DGU for bidirection delivery of both real power and reactive power and design issues of determining the coupling capacitor are discussed. Computer simulations are implemented to validate the effectiveness of the proposed approach.

## ACKNOWLEDGMENT

This research is funded by both National Science Council of TAIWAN under grant NSC 97-2218-E-182-004 and Chang Gung University of TAIWAN under grant UERPD-270121.

## REFERENCES

[1] R. Lasseter and P. Piagi, "Providing premium power through distributed resources," in *Proceedings of the 33rd Annual Hawaii International Conference on System Sciences*, 2002, pp. 1437–1445.

[2] R. Lasseter, "Microgrids," in *IEEE Power Engineering Society Winter Meeting*, 2002, pp. 305–308.

[3] C.-C. Hua, K.-A. Liao, and J.-R. Lin, "Parallel operation of inverters for distributed photovoltaic power supply system," in *IEEE 33rd Annual Power Electronics Specialists Conference*, 2002, pp. 1979–1983.

[4] Y. Li, D. M. Vilathgamuwa, and P. C. Loh, "Design, analysis, and real-time testing of a controller for multibus microgrid system," *IEEE Trans. Power Electron.*, vol. 19, no. 5, pp. 1195–1204, Sept. 2004.

[5] T.-L. Lee and P.-T. Cheng, "Design of a new cooperative harmonic filtering strategy for distributed generation interface converters in an islanding network," *IEEE Trans. Power Electron.*, vol. 42, no. 5, pp. 1301–1309, Sept. 2007.

[6] R. W. De Doncker and J. P. Lyons, "Control of three phase power supplies for ultra low THD," in *IEEE 6th Annual Applied Power Electronics Conference*, 1991, pp. 622–629.

[7] M. C. Chandorkar, D. M. Divan, and R. Adapa, "Control of parallel connected inverters in standalone AC supply systems," *IEEE Trans. Ind. Applicat.*, vol. 29, no. 1, pp. 136–143, Jan./Feb. 1993.

[8] S.-J. Chiang, C.-Y. Yen, and K.-T. Chang, "A multimodule parallelable series-connected PWM voltage regulator," *IEEE Trans. Ind. Electron.*, vol. 48, pp. 506–516, June 2001.

[9] T. Erika and D. G. Holmes, "Grid current regulation of a three-phase voltage source inverter with an LCL input filter," *IEEE Trans. Power Electron.*, vol. 18, no. 3, pp. 888–895, May 2003.

[10] M. N. Marwali and A. Keyhani, "Control of distributed generation systems–part I: Voltages and currents control," *IEEE Trans. Power Electron.*, vol. 19, no. 6, pp. 1541–1550, Nov. 2004.

[11] A. V. Timbus, R. Teodorescu, F. Blaabjerg, M. Liserre, and P. Rodriguez, "Linear and nonlinear control of distributed power generation systems," in *IEEE Industry Applications Conference 41st IAS Annual Meeting*, 2006, pp. 1015–1023.

[12] R. Inzunza and H. Akagi, "A 6.6-kV transformerless shunt hybrid active filter for installation on a power distribution system," *IEEE Trans. Power Electron.*, vol. 20, no. 4, pp. 893–900, Jul. 2005.

[13] H. Akagi, Y. Kanagawa, and A. Nabase, "Instantaneous reactive power compensator comprising switching devices without energy storage components," *IEEE Trans. Ind. Applicat.*, vol. IA-20, May/Jun. 1984.

[14] S. Bhattacharya, D. Divan, and B. Banerjee, "Synchronous frame harmonic isolator using active series filter," in *the 4th European Conference on Power Electronics and Applications*, 1991, pp. 30–35.

[15] T. G. Habetler, "A space vector-based rectifier regulator for AC/DC/AC converters," *IEEE Trans. Power Electron.*, vol. 8, no. 1, pp. 30–36, Jan. 1993.

# A Comprehensive Analysis and Comparison Between Multilevel Space-Vector Modulation and Multilevel Carrier-Based PWM

Constantinos Sourkounis, Ahmad Al-Diab

Ruhr-University Bochum / Research Group for Power Systems Technology, Bochum, Germany,
sourkounis@eele.rub.de, Ahmad.Aldiab@rub.de

*Abstract*—This paper presents an analysis and a comparison of Multilevel three-phase Inverters based PWM modulators. Two different PWM strategies are used to turn on and off the switching devices, Multilevel Space Vector Modulation and Multilevel Carrier Based Modulation. The performance characteristics regarding the harmonic contents, switching losses and conduction losses have been computed. These characteristics have been determined for a 10KWThree-level diode-clamped inverter supplying a 230V, 11A, 4% Inductive inner impedance of the grid through a simulation program to validate the proposed approaches.

Key Words: Space Vector Modulation (SVM), Carrier Based Modulation (CBPWM), Diode-Clamped Inverter, Total Harmonic Distortion (THD).

## I. INTRODUCTION

Inverters are expected to play an essential role in the field of power production, especially in the rural areas where over two billion people today have no access to electricity. As a result of the variety of inverters applications, a number of inverter topologies have been developed ranging from single-phase half-bridge to three-phase multilevel inverters.

Multilevel inverters are ideal for connecting renewable energy sources such as photovoltaic to the grid. They have been developed in such to overcome shortcomings in solid-state switching device ratings by using a series connected semiconductors devices to block the higher voltage levels involved [1]. The desired high ac voltage is synthesized from several of smaller dc voltages levels. For this reason, additional applications of multilevel inverters include such uses as medium voltage adjustable, speed motor drives, static VAR compensation [2].

The main advantages of this approach are summarized as follows:

*1) The semiconductors are wired in a series-type connection, which allows operation at higher voltages [3].*

*2) The voltage capacity of the existing devices can be increased many times without the complications of static and dynamic voltage sharing that occur in series-connected devices [4].*

*3) The smaller voltage steps lead to the production of higher power quality waveforms with low distortion harmonics without the use of transformers [5].*

*4) Spectral performance of multilevel waveforms is superior to that of their two level counterparts [5].*

*5) Multilevel waveforms naturally limit the problems of large voltage transients that occur due to the reflections on cables, which can damage the motor windings and cause other problems [1].*

*6) The use of a multilevel inverter to control the frequency, voltage output (including phase angle), and active and reactive power flow at a DC/AC interface provides significant opportunities in the control of distributed power systems [4].*

*7) Reduce the dv/dt stresses on the load and reduce the electromagnetic compatibility (EMC) concerns [1].*

The performances of multilevel inverters are generally based on the spectrum analysis of the generated output voltage or current. As the number of levels increases, the synthesized output waveform has more steps, which produces a staircase wave that approaches the desired waveform with a low harmonic distortion.

Several modulation strategies, differing in concept and performance, have been developed in order to achieve a variety of aims including: wide linear modulation range, less switching loss, less total harmonic distortion (THD) in the spectrum of switching waveform, easy implementation and less computation time [4]. With the development of microprocessors, space-vector modulation (SVM) has become one of the most important PWM methods for three-phase inverters due to its ability to reduce commutation losses and the harmonic contents of output voltage, as well as obtaining higher amplitude modulation indexes [5].

In this paper, an analysis and a comparison between Multilevel Carrier Based PWM (CBPWM) and Space Vector Modulation (SVM) algorithms for two and three level voltage source inverters will be introduced, analyzed. The main steps for implementing the algorithms will be simplified and analyzed. Simulation results will be shown to validate the proposed approach. The definition of Total Harmonic Distortion (THD) and the Switching Losses are used as a performance indexes.

## II. VOLTAGE SOURCE INVERTERS

Voltage source inverters will provide significant advantages: it can control the output frequency and voltage from energy sources and it can also control the active and reactive power flow from a utility connected to power source. This offer many benefits for power applications. In particular, uninterruptible power supplies (UPSs), static *VAR* compensation, harmonic compensation and active filters.

### A. Two-Level Three-Phase Inverter

Half bridge inverters can be extended to three-phase inverters, see Fig.1, in this case the three currents are balanced with 120° phase shift between them. There are eight switching states (vectors), Si = (S1, S3, S5), where the output voltage of the inverter is composed by one of these states.

The poor quality of the output current and voltage of a power system fed by a traditional two level inverter has led to find new topologies that can overcome the technical difficulties of these systems. Multilevel inverters are receiving increased attention recently, especially for use in high power applications [1], [2].

Multilevel inverter structures have been developed to overcome shortcomings in solid-state switching device ratings so that they can be applied to high-voltage electrical systems [6].

### B. Diode-Clamped Multi-Level Inverter

The most common used of multilevel inverters is the diode-clamped multilevel inverter. It provides a significant advantage it can be extended to any number of levels by increasing the number of capacitors. Early descriptions of this topology were limited to three-levels where two capacitors are connected across the dc bus resulting in one additional level. The additional level was the neutral point of the dc bus, so the terminology neutral point clamped (NPC) inverter was introduced. However, with an even number of voltage levels, the neutral point is not accessible, and the term multiple point clamped (MPC) is sometimes applied. Due to capacitor voltage balancing issues, the diode-clamped inverter implementation has been mostly limited to the three-level [3].

In general in *n-level* diode-clamped inverter there are (n − 1) adjacent transistors gated on for producing each switching state.

### C. Three-Level Diode-Clamped Inverter

This topology has twice as many transistors as well as added diodes compared to the two-level inverter as it shown in Fig.2, each of the switches must block only one half of the DC link voltage although the structure is more complex and the switching is straightforward.

Three level inverters are capable in producing three different levels of output voltage (+VDC, 0, -VDC). There are a total of 27 ($n^3 = 3^3$ where n is the number

of levels) possible switching combinations (vectors), Table. I presents the switching possibilities for phase-a.

## III. PULSE WIDTH MODULATION TECHNIQUES

The objective to achieve a variety of aims such as: wide linear modulation range; less switching loss; less total harmonic distortion (THD) in the spectrum of switching waveform, and easy implementation and less computation time [7] were the motivation for developing different modulation strategies with different concept and performance.

Carrier Based Pulse width modulation (CBPWM) technique has been extensively used, because it improves the harmonic spectrum of the inverter by moving the voltage harmonic components to higher frequencies [2]. With the development of microprocessors, space-vector modulation was introduced in the mid of 1980s as an alternative method for determining the switched pulse widths. It becomes the most important PWM methods for three-phase converters due to its ability to reduce commutation losses and the harmonic contents of output voltage, as well as obtaining higher amplitude modulation indexes [4].

### A. Carrier-Based Pulse Width Modulation

Carrier-based PWM (CBPWM) methods compare a reference waveform with a triangular or saw-tooth carrier at a higher frequency in order to generate the gating signals to switch on or off the switching device.

The operation of PWM can be divided into two modes [8]:

1) Linear Mode: in the linear mode, the peak of a modulation signal is less than or equal to the peak of the carrier signal. The maximum boundary for this mode when the value of the modulation index (m) reaches 1, (m=1), which gives the maximum peak amplitude of the fundamental output voltage as Vo(max)= Vs.

2) Nonlinear Mode: this operation is called *overmodulation*, where the modulation index is greater than 1, (m>1), leads to get a square-wave operation and

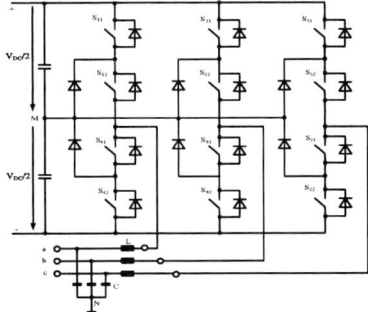

Fig. 2. Three-level diode-clamped inverter.

TABLE I.
THREE-LEVEL INVERTER OPERATION.

| Voltage Level | Switches | | | | Output Voltage |
|---|---|---|---|---|---|
| | $S_{11}$ | $S_{12}$ | $S_{41}$ | $S_{42}$ | |
| 2 | 1 | 1 | 0 | 0 | $+V_{DC}/2$ |
| 1 | 0 | 1 | 1 | 0 | 0 |
| 0 | 0 | 0 | 1 | 1 | $-V_{DC}/2$ |

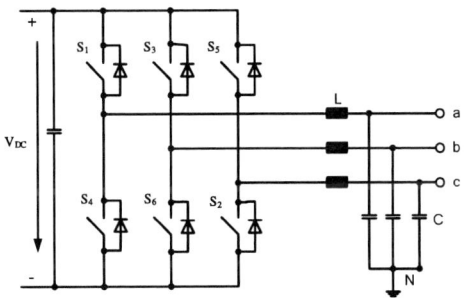

Fig. 1. Two-level three-phase inverter.

1711

adds more harmonics as compared to operation in the linear Mode. This operation is normally avoided in applications requiring low distortion for example in (UPSs) applications.

In general Carrier Based Modulation has high Harmonic distortion at a high modulation index and when the switching frequency for the switching devices is low, which is almost inevitable for high power applications.

### B. Multi-Level Carrier Based Pulse Width Modulation

For a n-level inverter, n-1 carrier signals with the same frequency and same peak to peak amplitude are disposed such that the bands they occupy are contiguous, see Fig.4.

As mentioned earlier, one of the important advantages of the proposed 3-level inverter is that it can be operated as a 2-level inverter in the lower output voltage range. This is accomplished by comparing the modulating sine wave with only one triangular carrier wave for the generation of PWM signals in the lower output voltage range and with two triangular carrier waves in the higher output voltage range [9].

The number of switchings per modulation cycle at each level of the inverter is dependent on the carrier frequency for that level and the duration of time that the reference waveform dwells within the level's corresponding time band.

If the carrier frequency for all of the levels is identical, the top and the bottom levels will have many more switchings than the intermediate levels.

### C. Space Vector Modulation

Space vector modulation is based on vector selection in the dq or in the αβ stationary reference frame, where the reference vector has a circulating path if a three phase set of voltages are required on the load [8]. SVM has proved to be one of the most popular and favourable PWM schemes due to its high dc link voltage utilization, low output distortion, and ability to minimize the switching and conduction losses. The algorithm for the SVM can be achieved through the following steps [9]:

1) Determining the switching combination and corresponding vectors.

2) Calculating the voltage drop related to each vector.

3) Identifying the position for each vector in the αβ-space vector diagram.

4) Identifying reference vector position.

5) Calculating duty cycles

6) Building a vector sequence.

7) Computing pulse pattern.

A coordinate transformation in the output space from the abc-space to the αβγ-space and it can be obtained using the following transformation:

$$\begin{bmatrix} V_\alpha \\ V_\beta \end{bmatrix} = \frac{2}{3} \cdot \begin{bmatrix} 1 & -\frac{1}{2} & -\frac{1}{2} \\ 0 & \frac{\sqrt{3}}{2} & -\frac{\sqrt{3}}{2} \\ 0 & 0 & 0 \end{bmatrix} \cdot \begin{bmatrix} V_a \\ V_b \\ V_c \end{bmatrix} \qquad (1)$$

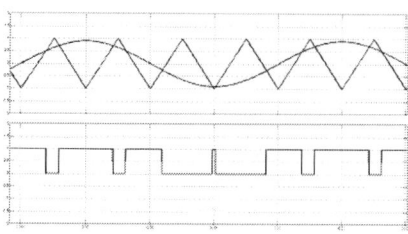

Fig. 3. Two-level carrier-based PWM generation.

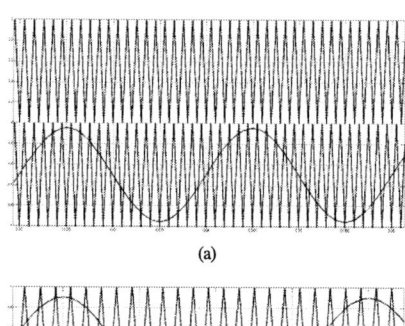

(a)

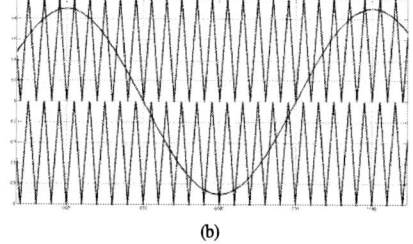

(b)

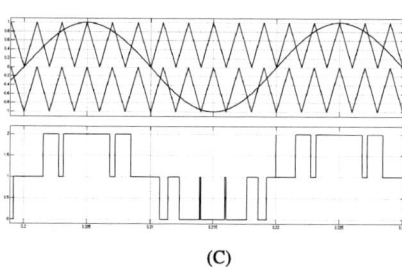

(C)

Fig. 4. Three-level carrier based PWM, (a) Inverter runs at the Two-Level Mode, (b) Inverter runs in the Three-Level Mode, (c) Three-level PWM signals generation.

In the case of three-level inverter, the space vector diagram for a three-level inverter is shown in Fig.5 with 27 switch combinations and 19 unique voltage vectors since some of the combinations produce the same voltage vector. These different combinations are based on the way of connecting the load to the DC bus.

From Fig.5, the Space vector diagram is composed into six sectors; each sector is divided into four regions, which is limited by three vectors, as is in Fig.6.

Assuming that the reference vector ($V_{ref}$) remains approximately constant during a modulation period, which is acceptable, if $Ts$ is much smaller than the line period ($T$), then (2) can be approximated as:

$$\underline{V}_{ref} = \underline{P}_a + \underline{P}_b + \underline{V}_3 \qquad (2)$$

In which $d_1$, $d_2$ and $d_3$ are the duty cycles of vectors $V_1$ to $V_2$ and to $V_3$, respectively, they must satisfy the following condition:

$$d_1 + d_2 + d_3 = 1 \qquad (3)$$

Duty cycles are defined by the projection of the reference vector after normalizing it by the $\alpha\beta$ transformation, the vectors $P_a$ and $P_b$ in Fig.7 are the projections from the reference vector $V_{ref}$ onto the segments that join the extreme of the $V_3$ to $V_1$ and to $V_2$, respectively, in the $\alpha$-$\beta$ plane.

The reference vector Vref can be expressed as follows:

$$\underline{V}_{ref} = \underline{P}_a + \underline{P}_b + \underline{V}_3 \qquad (4)$$

And implementing the lengths $I_1$ and $I_2$ of the vectors $V_1 V_3$ and $V_2 V_3$ respectively in equation (5) it will be as follow:

$$\underline{V}_{ref} = P_a \frac{\underline{V}_1 - \underline{V}_3}{I_1} + P_b \frac{\underline{V}_2 - \underline{V}_3}{I_2} + \underline{V}_3 \qquad (5)$$

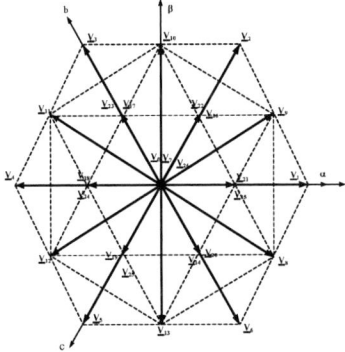

Fig. 5. Space Vector Diagram for Three-Level Diode-Clamped Inverter.

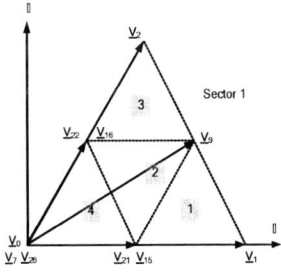

Fig. 6. Regions 1 to 4 in Sector 1 in for Three-level Diode-Clamped Inverter Space Vector Diagram.

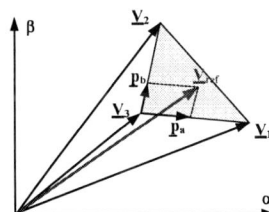

Fig. 7. Reference vector projection [2].

And simplifying equation (6) the reference vector can be expressed:

$$\underline{V}_{ref} = \frac{P_a}{I_1}\underline{V}_1 + \frac{P_b}{I_2}\underline{V}_2 + \left(1 - \frac{P_a}{I_1} - \frac{P_b}{I_2}\right)\underline{V}_3 \qquad (6)$$

From equation (7) the duty cycles of the vectors can be directly found as follows:

$$d_1 = \frac{P_a}{I_1}, \quad d_2 = \frac{P_b}{I_2} \quad \text{and} \quad d_3 = 1 - \frac{P_a}{I_1} - \frac{P_b}{I_2} \qquad (7)$$

To have triangle regions with unity lengths where $I_1$ and $I_2$ will be equal to 1 in this case the balanced SV-diagram is normalized and the calculations of the duty cycles will be simplified as follows:

$$d_1 = P_a, \ d_2 = P_b \quad \text{and} \quad d_3 = 1 - P_a - P_b \qquad (8)$$

*B. SYMMETRIC MODULATION*

In order to reduce the current ripple, switching vectors adjacent to the reference vector should be selected since the adjacent switching vectors produce non-conflicting voltage pulses (same voltage polarity) [8].

Symmetric modulation is characterized by using four vectors per modulation sequence. Dealing with another variable in the calculation of the duty cycles allows the equation of the NP current to be included in the equation system:

$$\underline{V}_{ref} = d_1\underline{V}_1 + d_2\underline{V}_2 + d_3\underline{V}_3 \qquad (9)$$

Symmetric modulation uses four switching vectors on the same region where the duty cycle is shared between both of them. If the reference vector in a three level inverter lies in Region 1 in the first sector, the sequence will be $V_{15}$-$V_1$-$V_9$-$V_{21}$, Fig.8. The best vector sequences in the first sector are shown in Table. II. The pulse sequence can be achieved by comparing the duty cycles with a carrier signal.

## IV. MULTI-LEVEL INVERTER LOSSES

The losses in a power-switching device constitute of conduction losses, off-state blocking losses, switching losses.

Since the leakage current during the, off state of the device is negligibly small, the power loss during the off

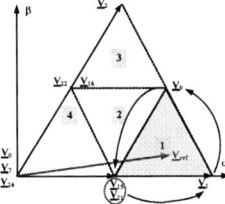

Fig. 8. Symmetric Modulation.

TABLE II.
SYMMETRIC MODULATION FOR SECTOR 1.

| Region | Sequence | Steps |
|---|---|---|
| 1 | $\underline{V}_{15}$-$\underline{V}_1$-$\underline{V}_9$-$\underline{V}_{21}$ // $\underline{V}_{21}$-$\underline{V}_9$-$\underline{V}_1$-$\underline{V}_{15}$ | 3//3 |
| 2 | $\underline{V}_{16}$-$\underline{V}_9$-$\underline{V}_{15}$-$\underline{V}_{22}$// $\underline{V}_{22}$-$\underline{V}_{15}$-$\underline{V}_9$-$\underline{V}_{16}$ | 3//3 |
| 3 | $\underline{V}_{16}$-$\underline{V}_9$-$\underline{V}_2$-$\underline{V}_{22}$ // $\underline{V}_{22}$-$\underline{V}_2$-$\underline{V}_9$-$\underline{V}_{16}$ | 3//3 |
| 4 | $\underline{V}_0$-$\underline{V}_{15}$-$\underline{V}_{16}$-$\underline{V}_7$ // $\underline{V}_7$-$\underline{V}_{16}$-$\underline{V}_{15}$-$\underline{V}_0$ | 3//3 |

state can be neglected. Thus the power loss of the device can be considered only as conduction loss and switching loss.

The Conduction losses are the losses that occur while the power device is in the on-state and conducts current. Therefore, it can be calculated in a straightforward manner as the product of the device current and the forward saturation voltage; and the blocking loss is the product of the blocking voltage and the leakage current [10] of the anti-parallel diode.

$$P_{cond} = I_c V_{ce}.D \qquad (10)$$

Where

Ic  Collector current

D  Duty Cycle

$V_{ce}$ Collector-emitter voltage drop

There is No general expression for the voltage and current during a switching transient. The datasheet parameters concerning the switching losses have to be used in this case. These parameters are referenced to a specific test circuit that simulates a clamped inductive load operated with a specific diode.

In [10] and [11] a brief explanation for switching losses in multilevel inverters was introduced and explained with approximated formulae for the losses in IGBTs. Where the on energy and the off energy are approximately estimated via:

$$E_{on} = \frac{V_{dc}.I_c}{2} t_{on} \qquad (11)$$

$$E_{off} = \frac{V_{dc}.I_c}{2} t_{off} \qquad (12)$$

Where

Ic  Collector current

$V_{dc}$ DC Input Voltage

$t_{on}$ Turn-on delay time

$t_{off}$ Turn-off delay time

In this study a SKW07N120 Infineon Fast IGBT was implemented on the inverters.

## V. SIMULATION RESULTS

As a testing case, a three-phase, three-level diode-clamped inverter is supplied by 700V DC-source with a 10 KHz switching frequency. A low-pass LC-filter, L=3mH and C=10μF is connected in series to the output terminals of the inverter, supplying a 3X 400V/230V, 4% Inductive Electrical Grid to evaluate the performance of the inverter based the proposed PWM techniques, operating at 10 kHz switching frequency. Total Harmonic Distortion (THD) is used as a performance index for the inverter at each load condition.

Fig.9 shows the output three-phase current respectively with a SVM modulator, while on Fig.10 the harmonics spectrums for the output current and for the voltage are shown, the THD for the voltage equals 0.25% and for the current is 2.85%.

The same for a Carrier Based Modulator on Fig.11 and Fig.12 the outputs and the harmonics spectrums are shown, where the THD for the voltage 0.36% and for the current is 4.53%.

## VI. CONCLUSIONS

This paper presents two different PWM algorithms for three-phase multilevel diode clamped inverters. The idea of this algorithm is to have a general procedure in order to implement SVM or CBPWM for diode-clamped inverters for any *n* voltage levels. The switching vectors, Space Vector diagrams and the boundary planes equations, as well as the calculations of the duty cycles and switching sequences are presented in details to implement SVM.

*n -1* carrier signals with the same frequency and same peak to peak amplitude are needed to implement CBPWM for *n-level* inverter. From the Simulation results it can be conclude that "As the number of levels increases, the synthesized output waveform has more steps, which produces a staircase wave that approaches the desired waveform with a low harmonic distortion. It can be concluded that Carrier Based Modulation has high Harmonic distortion and high switching and conduction losses".

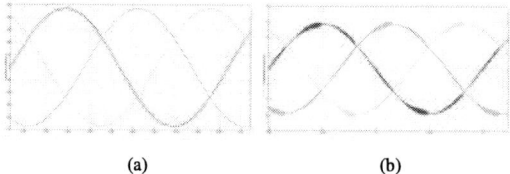

(a)                      (b)

Fig. 9. Three Phase Output for Three Level Inverter Based SVM Modulator: (a) Three Phase Voltage, (b) Three Phase Current.

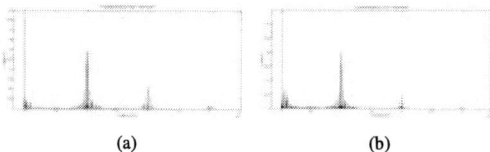

(a)                      (b)

Fig. 10. Total Harmonic Distortion: (a) Three Phase Voltage THD=0.25%,    (b) Three Phase Current THD=2.85%.

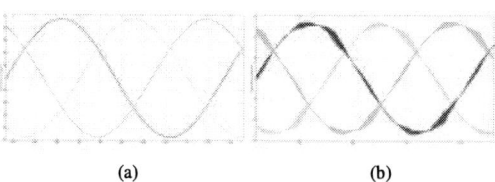

(a)                      (b)

Fig. 11. Three Phase Output for Three Level Inverter Based CBPWM Modulator: (a) Three Phase Voltage, (b) Three Phase Current.

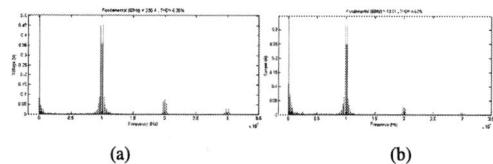

(a)                      (b)

Fig. 12. Total Harmonic Distortion : (a) Three Phase Voltage THD=0.36%,     (b) Three Phase Current THD=4.53%.

## REFERENCES

[1] B. McGrath, D.Holmes, T. Meynard "Reduced PWM Harmonic Distortion for Multilevel Inverters Operating Over a Wide Modulation Range," IEEE Transactions on power electronics, vol. 21, No. 4, July 2006.

[2] Celanovic, Nikola "Space Vector Modulation and Control of Multilevel Converter," Ph.D. Dissertation, Virginia Polytechnic Institute and State University, 2000

[3] Manuel A. Perales, M. M. Prats, Ramón Portillo, José L. Mora, José I. León, Leopoldo G. Franquelo, "Three-Dimensional Space Vector Modulation in abc Coordinates for Four-Leg Voltage Source Converters", IEEE power electronics leters, VOL. 1, NO. 4, December 2003.

[4] Ahmad Al-Diab, "Analysis for Space Vector Modulation In Multi-level/Multi-leg Inverters," Master dissertation, Westphalia University of Applied Sciences/Division Soest, 2006.

[5] Pou, Josep. "Modulation and control of Three-Phase PWM Multilevel Converters," Ph.D. Dissertation, Universtat Politecnica De Catalunya, 2002.

[6] Keliang Zhou, Danwei Wang, "Relationship Between Space-Vector Modulation and Three-Phase Carrier-Based PWM: A Comprehensive Analysis", IEEE Transaction on Industrial Electronics, VOL. 49, NO. 1, February 2002.

[7] V.T. Somasekhark, K. Gopakumar, M . R. Bauu, K. K. Mohapatraa, L. Umanand, "A PWM scheme for a 3-level inverter cascading two 2-level inverters", Centre for Electronics Design and Technology, Indian Institute of Science.

[8] Nikola Celanovic, "A Fast Space-Vector Modulation Algorithm for Multilevel Three-Phase Converters", IEEE International Conference on Industrial Electronics, Control and Instrumentation, MARCH/APRIL 2001.

[9] E. Ortjohann, A. Mohd, N. Hamsic, A. Al-Daib, "Three-Dimensional Space Vector Modulation Algorithm for Three-Leg Four-Wire Voltage Source Inverters", first International Conference on Power Engineering, Energy and Electrical Drives (POWERENG), April 2007.

[10] A.M.Massoud, S.J.Finney, B.W.Williams, "Conduction Loss Calculation for Multilevel Inverter: A Generalized Approach for Carrier Based PWM Technique", Department of computing and electrical engineering Heriot-Watt University, Riccarton Campus, Edinburgh, Scotland, UK.

[11] A.D. Rajapakse,, A.M. Gole, P. L. Wilson, "Approximate Loss Formulae for Estimation of IGBT Switching Losses through EMTP-type Simulations".

[12] L.M. Tolbert, F.Z. Peng. "*Multilevel converters as a utility interface for renewable energy systems*", Power Engineering Society Summer Meeting, 2000. IEEE.

[13] Keliang Zhou; Danwei Wang, "Relationship between Space-Vector Modulation and Three-Phase Carrier-Based PWM: A Comprehensive Analysis," IEEE Transactions on Industrial Electronics, 2002.

[14] H. Pinheiro, F. Botterón, C. Rech, L. Schuch, R. F. Camargo, H. L. Hey, H. A. Gründling, J. R. Pinheiro "Space Vector Modulation for Voltage-Source Inverters: A Unified Approach," IEEE 28th Annual Conference of the Industrial Electronics Society, 2002.

[15] R. Zhang, V. H. Prasad, D. Boroyevich, F. C. Lee, "Three-Dimensional Space Vector Modulation for Four-Leg Voltage-Source Converters" IEEE Transactions on power electronics, vol. 17, No. 3, May 2002.

[16] Ning-Yi Dai, Man-Chung Wong,Ying-Duo Han, "Application of a Three-level NPC Inverter as a Three-Phase, Four-Wire Power Quality Compensator by Generalized 3DSVM " IEEE Transactions on power electronics, Vol. 21, No.2, March 2006.

[17] D. Grahame Holmes, Thomas A. Lipo, "Pulse Width Modulation For Power Converters: Principles and Practice," IEEE Press& Wiley Interscience.

[18] J. Holtz. "Pulsewidth Modulation for Electronic Power conversion," Proc. of the IEEE, Vol. 82, No. 8, pp. 1194 – 1213, Aug 1994.

[19] Zhenxue Xu, "Advanced Semiconductor Device and Topology for High Power Current Source Converter", Ph.D. Dissertation, Virginia Polytechnic Institute and State University, 2003.

**Prof. Dr.-Ing.Constantinos Sourkounis** received his Dipl.-Ing and his Dr.-Ing degrees from the TU Clausthal in 1989 and 1994 respectively. After he received his Ph. D. at the Institute of Electrical Power Engineering at the TU Clauthal, he occupied the position of the chief engineer. In 2003 Dr.-Ing. Sourkounis received his habilitation. Since 2003 he is professor at the Ruhr-University Bochum, heading the Power System Technology Research Group. His main research areas are mechatronic drive systems, renewable energy sources, decentralized energy systems supplied by renewable energy sources, and energy supply systems for transportation systems.

**M.Sc.-Ing. Ahmad Al-Diab** received his B.Sc.-Ing degree in Mechatronics engineering from the Al-Balqa'a Applied University / Jordan in 2004. Since 2006 he has received his M.Sc. in electronic Systems and Engineering Management from South Westphalia University of Applied Sciences, Soest campus (Germany). Since 2007 he is a scientific assistant at the Power System Technology Research Group at the Ruhr-University Bochum. His main research area is the Grid Parameters Detection and Control for Inverters in Distributed Energy Supply Systems.

# Identification of Electrical Parameters in a Power Network Using Genetic Algorithms and Transient Measurements

Wei. Dong, Pericle Zanchetta and David W.P. Thomas

University of Nottingham, School of Electrical and Electronic Engineering, Nottingham, UK
Emails: eexwd@nottingham.ac.uk, pericle.zanchetta@nottingham.ac.uk, dave.thomas@nottingham.ac.uk

*Abstract*— The knowledge of the parameter values of a power network is very valuable for power system modelling, simulation, protection and control since it is a fundamental parameter for solving many problems such as minimizing the effect of voltage distortion, active filter control or relay setting. The electrical parameters in a power network are usually unknown or poorly quantified and can not be measured directly. A new effective and reliable method for power network parameters identification based on voltage transient measurements using genetic algorithm optimization is successfully demonstrated. Simulation tests performed using Matlab-Simulink show the effectiveness of the proposed identification strategy.

*Keywords*— System Identification, Genetic Algorithms, Power networks

## INTRODUCTION

Several methods have been proposed in literature to measure the parameters and the impedance of a power system. Usually these methods employ a disturbance injection on the power system and the voltage and current transients are measured. The harmonic impedance is then evaluated using signal processing techniques. In [1-3] high frequency voltage and current transients are created through a capacitor bank switching; the spectral analysis of the transients provides the frequency domain impedance characteristics. In these methods, the injected disturbances are uncontrolled. In [4], a technique for measuring the power system harmonic impedance using a power electronic converter to inject a voltage transient on to the network via an inductor is described. The injected transient in this case is fully controlled. It can be triggered at any instant, and employs a current control loop which can be used to limit the size of the peak of the current disturbance.

A novel parameter identification technique is presented in this paper. This technique also involves a power electronic converter to inject a small disturbance to a power network and the power system parameters are identified at the point of the injection. This is achieved by measuring the resulting transient voltage and optimising an equivalent circuit through minimisation of the simulated errors compared to the measured quantities with a genetic algorithm (GA).

In the past, many optimization methods have been used for system identification or structural identification. These methods are usually classified as classical and non-

classical methods [5]. Classical methods can be defined as the methods which are derived from traditional mathematical theories. The least square method is perhaps the simplest classical method for system identification. It estimates the unknown parameters of structural systems by minimizing the sum of squared errors between the predicted outputs and the measured outputs. In [6] the authors identified the stiffness and damping parameters of engineering structures.

During the optimization, there is a high possibility of converging to a local optimal point rather than the global optimum point when a classical method is used. Some classical methods even need a good initial guess of the unknown parameters to be successful.

Non-classical methods are mainly represented by heuristic optimization algorithms, among which Genetic Algorithms are the most renowned. They are less likely to prematurely converge to local optima solutions; they can be used for wider search problems and result in general more effective, but depending on computing power for the extensive and robust searches. GA belongs to the family of optimization strategies based on natural evolution theories together with three more groups of methods: evolutionary programming (EP), evolutionary strategies (ES), genetic programming (GP) [7]. The GA method was first introduced in the 1960s and aims to imitate animal evolution by natural selection. Together to what mentioned earlier, this method has another advantage over classical methods in optimization or identification. The GA is a self-start method and doesn't need the initial values of unknown parameters, but just some bounds. Many identification issues have been solved successfully by using the GA method. For example Authors in [5] have identified 52 unknown parameters of a large structural system with the help of a GA-based routine.

It has therefore been concluded that the GA approach was very suitable to this type of system identification. In the following, section 2 gives a general overview on GA, while section 3 explains the procedure for system identification. In section 4 the GA used is described with its implemented variants and section 5 presents the simulation results.

## GENETIC ALGORITHMS

The genetic algorithm represents an advanced numerical

978-1-4244-1741-4/08/$25.00 ©2008 IEEE          1716

search and optimisation technique based on the biological theory of evolution. The idea of survival of the fittest and an interbreeding population proposed by Charles Darwin in "On the Origin of Species by Means of Natural Selection" (1859) is employed in GAs to create an innovative search strategy. In the 60s, the genetic algorithms have been widely studied and implemented in many fields of the engineering sector. The concept of using genetic algorithm in engineering was first introduced by John Holland of the University of Michigan. His book 'Adaption in Natural and Artificial Systems (1975)' has provided a general framework for viewing all adaptive systems and showed how the evolutionary process can be applied to artificial systems [8].

The GA is initialized with little knowledge about a problem to be solved and a searching process is carried out in parallel for a large and complex defined search space.

The basic scheme of a typical GA can be summarized in the following points:

(1) In the beginning, an initial population is created with a group of individuals which are selected randomly.

(2) The individuals in the population are then evaluated and a fitness function value, defined on how they perform in a given task, is associated to every one of them.

(3) The genetic operators are used to create subsequent generations. Some individuals are selected from the population based on their fitness value. The higher the fitness value, the higher the chance of being selected.

(4) These individuals are coupled and reproduced to create one or more offspring; after that each offspring undergo crossover and mutation operations.

(5) The process will repeat until an optimum solution has been found or a certain termination criterion is met.

A flow chart of the generic GA routine is also shown in Fig 1.

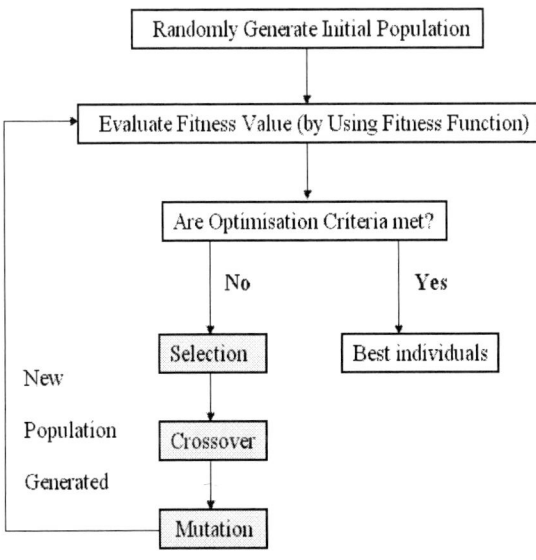

Fig. 1: The basic scheme of typical GA routine

## A. Population representation and initialization

Each individual in the population represents a potential solutions to a specified problem. The search space for the problem solution is defined in the population representation. Also a possible range for the parameters and variables needs to be defined. Each of the variables that compose an individual are known as chromosome. The chromosomes are commonly coded into a string to form the individual and the most commonly used representation for an individual is the binary alphabet. Other representations such as integer or real valued (used in this work) have been widely used in literature. For a better understanding of the basic idea of GA, a binary coded population has been used as an example in the following discussion. The real valued population is commonly used in the engineering fields when dealing with continuous time systems. At the initial step of the genetic algorithm, a specified number of individuals and generations are chosen. The initial set of the population is generated randomly. This is called the first generation pool.

## B. Fitness measure

Each individual in the population is assigned a fitness value, which determines how fit is the solution. The GA maintains a population of n individuals (possible solutions) with associated fitness values evaluated according to the problem nature and specification. The fitness value is then used to determine the fittest individuals that will propagate to subsequent generations. Consequently highly fit chromosomes are given more opportunities to be selected and reproduced than the worst performing solutions.

## C. Genetic operators

**Selection**

The purpose of selection is to choose individuals from the current population, using the fitness value assigned to the strings, to evolve in the group (mating pool) which will enable the formation of the new generation. As mentioned earlier an individual with high relative fitness value has a high probability of being selected. There are several types of selection operators such as the roulette wheel selection, the tournament selection and stochastic universal sampling. The latter has been used in this paper.

**Crossover**

Once a new population has been created and the best elements selected, an operator is needed to enable the population to evolve. This operator plays a very important role for the variety of the new generation. The crossover operator is used to exchange genetic elements between two parent chromosomes to create new offspring for the next generation. Strings are selected randomly in couples from the mating pool and the crossover operator is then applied.

When a GA is implemented using a real numbers representation, the two main kinds of crossover operators are the intermediate recombination and the line recombination. The intermediate recombination (used in this paper) is a method of producing new individuals around and between the values of the parents individuals. Offspring are produced according to the following rule:

$$Offspring\ 1 = Parent\ 1 \times \alpha\ (Parent\ 1 - Parent\ 2) \qquad (1)$$

Where $\alpha$ is a scaling factor chosen uniformly with random distribution over some interval, typically [-0.25, 1.25] and Parent 1 and Parent 2 are the parent chromosomes. Each variable in the offspring is the result of combining the variables in the parents according to the above expression with a new $\alpha$ chosen for each pair of parent chromosomes.

## Mutation

The mutation operator is used to maintain diversity from one generation of population to the next preventing the population of chromosomes become too similar among them. It is controlled by a mutation probability which is normally very low and indicates the frequency at which the mutation occurs; therefore the mutation operates on only few individual. For the real value mutation, the implementation rule is shown below:

$$New\ Chromosome = Old\ chromosome + s\Phi \qquad (2)$$

Where s is a positive constant and $\Phi$ is a random vector to produce a small perturbation on the old chromosome. The GA settings used in this work are shown in table 1.

Table 1: GA settings

| Selection operator: | Stochastic Universal Sampling (SUS) |
|---|---|
| Crossover operator: | Intermediate Recombination |
| Generation gap: | 0.9 |
| Crossover rate: | 0.7 |
| Mutation rate: | 0.2 |
| Number of individual: | 50 |
| Number of generation: | 30 |

Mutation rate is normally set with a small value from 0.1 to 0.3 in order to maintain diversity the generation individuals and in the meanwhile avoid sending the population into chaos. The experienced boundary for crossover rate is from 0.6 to 0.9. Individuals and generations numbers need to be set with a reasonable value for a compromise between accuracy of the results and simulation time, as big number of individuals and generations will cause the optimization to become too time consuming.

## D. Hybrid Strategy

The goal of using GA is to find the global optimal solution for the given problem. However, for structural identification processes, the use of GA alone may not be always effective because of the nature of the problem since many suboptimal solutions may be reached easily causing the effect called 'premature convergence'. Several variants to the basic GA routines have been

studied and applied in recent years to increase the diversity of the population and extend the search to a much wider space. For this particular application two variants have proved to be very effective in the optimization process and are described in the following together with their implementation.

## Migration scheme

The population is divided into a number of subpopulations. Each subpopulation is independent from the others and the optimization is being executed at the same time for all of them. A number of individuals are transferred from one subpopulation to another frequently at regular intervals during the routine run. As it is shown in Fig 2, according to the unrestricted migration strategy, subpopulation 1 accept new individuals from subpopulation 2 to 6 and in the meanwhile transferring new individuals to subpopulation 2 to 6. Any path between any two subpopulations is bidirectional.

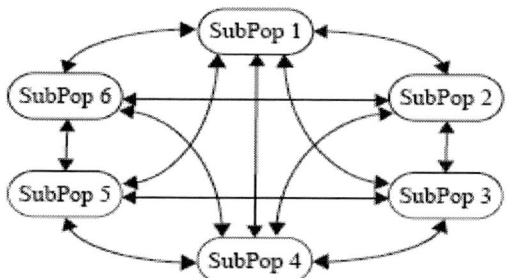

Fig. 2: the unrestricted migration strategy

A sufficient level of diversity in the whole population can be maintained in this way. Some parameters of the migration function have to be set, which are the number of subpopulations, the migration rate and migration generations. The settings used in this paper are shown in table 2.

Table 2: migration settings

| Number of subpopulation | 3 |
|---|---|
| Migration rate | 0.2 |
| Migration generation | 5 |

## Cataclysmic mutation

The GA routine is initialized with a specified number of generations. At the end of each generation, every individual will have its own fitness function value. It can happen that the GA converges after a few generations and it gets to a stage where the optimization starts producing more or less the same individuals with very small differences among them and the best solution does not improve significantly anymore. More diversity is therefore needed at this stage and this can be achieved by using an operator called cataclysmic mutation. The algorithm can be designed to check where the previous condition happens and in that case a completely new population is randomly built with the exception of the best previous solution which is sent forward in the new population. A higher mutation rate is also used instead of

1718

the initial mutation rate. An alternative used in this paper is to perform the cataclysmic mutation periodically after a fixed number of generations. This number is usually chosen based on experience and in this paper is fixed in 10 generations, as well as the new mutation rate which is fixed at 0.35.

## THE IDENTIFICATION METHOD

The basic principle of the identification method is simple. A current transient is injected onto the power network via a power converter (an active rectifier or an active power filter installed at the PCC) and a coupling inductor and the power system parameters are identified at the point of the injection. This is achieved by measuring the resulting transient voltage and optimising an equivalent circuit through minimisation of the simulated errors with the described genetic algorithm. The simulated network is constructed with the same structure as the experimental power network but with the parameters unknown. The parameter identification is performed by the GA which selects, at each iteration, a set of parameter values and evaluates the response of the simulated network. If the mismatch between experimental and simulated results is below a fixed satisfactory level, then we could say that the electrical parameters are identified. Otherwise, the optimization procedure will continue until the target condition is verified. A flow chart of the described procedure is shown in Fig 3. The fitness function used in the optimization which measure the mismatch between experimental and simulated results is the integral of the absolute error between the two responses.

In this paper a test of the proposed identification procedure is performed only via simulation by emulating the real power network with a known model. Fig 4 shows the schematics of the simulated circuit emulating the experimental laboratory set-up. It represents a real power system with 1 km transmission line, 800kW inductive load and 3.3kV power supply. The power factor is assumed to be 0.8. R1 represents the resistance of the 1 km transmission line; L1 is the inductance of the transmission line and the power supply. R2 and L2 are the values of load resistance and inductance. The values of these electrical parameters are shown in table 3.

In the identification process the values of the inductance of the power supply (L1), the capacitors bank (C1), the resistance and inductance of the load (R2, L2) have been supposed unknown and optimized.

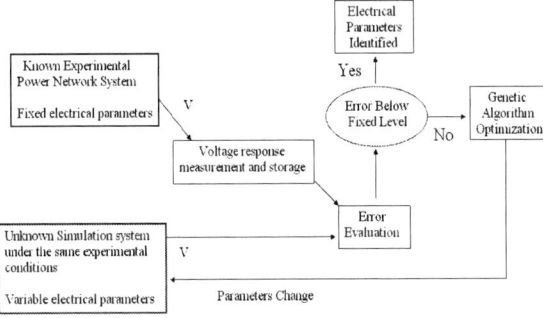

Fig. 3: The identification scheme

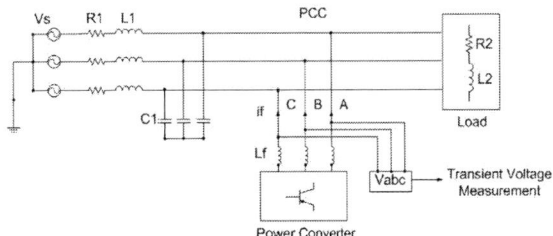

Fig. 4: Schematics of the simulated circuit emulating the experimental laboratory set-up

Table 3: Electrical parameters value

| Power supply : Vs | 3300V |
|---|---|
| Resistance of the power supply : R1 | 0.196Ω |
| Inductance of the power supply : L1 | 12.3482mH |
| Capacity bank : C1 | 0.1754mF |
| Load resistance : R2 | 8.712Ω |
| Load inductance : L2 | 20.8mH |

A controlled voltage source is connected to the network via a coupling inductor resulting in a current transient injection. The controlled voltage source is realized through a pulse width modulated (PWM) converter (for example an active filter or an active rectifier) with a transient signal added in the modulating waveform. The simplified scheme of the transient source and PWM generator is given in Fig 5.

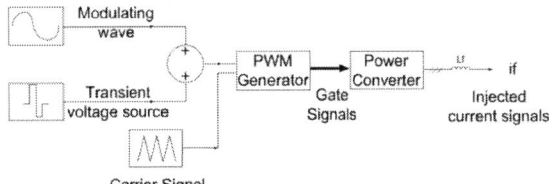

Fig. 5: The simplified scheme of the transient source and PWM generator

The dc-link capacitor of the inverter is approximated as a constant voltage source so it does not need control. A voltage matching control between the voltage at the point of common coupling and the PWM output voltage however is performed to avoid high fundamental current flowing into the coupling inductor and damage the inverter devices. The feedback signal of this control represents the modulating waveform, to which the transient signal is added. So if the inverter is used only for this transient injection, a very low fundamental current will flow before and after the transient.

Two different types of transient injections have been used in this paper in the modulating signal of the inverter, classified as short term injection (0.6ms) and medium term injection (10ms). The waveforms of these injections and the relative system voltage response, on which the fitness function is evaluated, are shown in Fig. 6. A with noise (1% of the power supply voltage rms value) has been taken into account during the simulation to emulate practical situations.

## THE OPTIMIZATION PROCEDURE USING GA

The GA algorithm has been written and implemented using the GA toolbox in MATLAB. The fundamental GA functions are written mostly in m-files, which can operate in Matlab environment interacting with Simulink simulation models.

The GA toolbox has basic built-in functions for creating initial population, reproduction function, genetic operator functions and termination criterion. These functions are commonly used for most of the problems. Other customized functions implementing the variants to the main Genetic Algorithm according to section II-D have been created and interfaced with the rest of the toolbox. Specified parameters need to be assigned to the functions in order to specify the optimization problem. Usually those parameters values are assigned at the first stage of the main program. Possible solutions for the string of the four unknown parameters of the simulated circuit that need to be identified (L1, C, R2, L2) represent the algorithm population. The search boundaries for these variables are set through experience and a rough calculation of reasonable values for the specific circuit: L1 [90e-4 to 15e-3], C [5e-7 to 5e-4], R2 [2.4 to 2400], L2 [0.01 to 0.04].

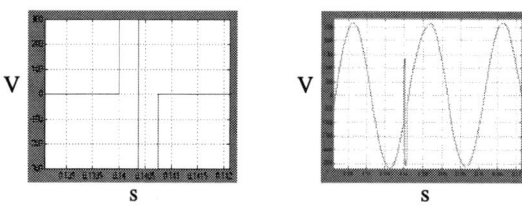

Fig 6(a): Short term injection and the voltage response

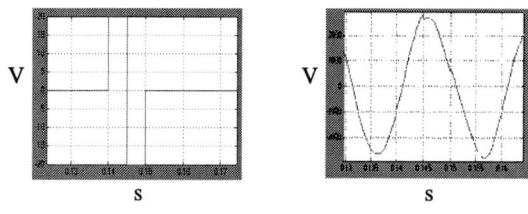

Fig. 6(b): Medium term injection and the voltage response

The population size is set at 50. Since the generation gap applied to the population is 0.9, then 45 out of 50 individuals are selected from the previous population. After the genetic operation has finished and created a new population, the counter is increased by one. The best individuals of the generation are selected and stored for further comparison.

The number of generations has been set at 50. Migration and Cataclysmic schemes have been implemented during the first 20 generations in order to improve the diversity of the population and to avoid the premature convergence. A number of individuals have been exchanged between 3 subpopulations every 5 generations and the population has been rebuilt with random individuals twice (every 10 generations). Then, a conventional GA has been run for the last 30 generations.

## GA OPTIMIZATION RESULTS

The identification results presented in table 4 and 5 show the estimated values of the 4 parameters (L1, C, R2 and L2) and the percentage error that affects the measurements. From these results it clearly appears that a very low estimation error has been achieved in both situations even if a quite extreme condition of 1% measurement noise has been taken into account.

## CONCLUSIONS

A new effective and reliable method for power network parameters identification based on voltage transient measurements using genetic algorithm optimization is successfully achieved in this work. This method has been tested in simulation using Matlab and Simulink. The estimated parameters values in the unknown system agree with the real ones with low error percentage The method has been proved efficient also in practical conditions when a PWM inverter is used as transient voltage generator and measurement noise is included.

The GA requires nothing more than a fitness assessment on transient signals and the boundaries of the parameters for a successful parameter identification. It is also a very flexible method since the program script can be updated or modified in order to improve the optimization performance. The identification method may provide improvements to several areas of power quality control. For example, when used within a stand alone piece of instrumentation it will be possible build up mesh equivalent networks of unknown power and distribution systems, which can then be used for load flow studies, protection or harmonic penetration prediction.

Table 4: optimization results with short term injection

| Parameter | L1 (mH) | C (mF) | R2 (Ω) | L2 (mH) |
|---|---|---|---|---|
| Initial values | 12.3482 | 0.1754 | 8.712 | 20.8 |
| Estimated values | 12.1 | 0.1779 | 8.6354 | 20.6 |
| Percentage error | 2% | 1.46% | 0.88% | 0.96% |

Table 5: optimization results with medium term injection

| Parameter | L1 (mH) | C (mF) | R2 (Ω) | L2 (mH) |
|---|---|---|---|---|
| Initial values | 12.3482 | 0.1754 | 8.712 | 20.8 |
| Estimated values | 12.1 | 0.17838 | 8.6187 | 20.5 |
| Percentage error | 2% | 1.7% | 1.1% | 1.4% |

## REFERENCES

[1] Agbabian MS, Masri SF, Miller RK, Caughey TK, "System identification approach to detection of structural changes," J Eng Mech, ASCE 1991; 117(2):370-90.

[2] Dumitrescu D. "Evolutionary computation". Boca Raton, FL: CRC Press; 2000.

[3] A. A. Girgis and R. B. McManis, "Frequency domain techniques for modeling distribution or transmission networks using capacitor switching induced transient," IEEE, Trans. Power Del., vol. 4, pp. 1882-1890, July 1989.

[4] C. G.. Koh, Y. F. Chen, C. Y. Liaw, "A hybrid computational strategy for identification of structural parameters," Computers and structures 81 (2003) 107-117.

[5] John R. Koza, "Genetic Programming: On the Programming of Computers by Means of Natural Selection," 1992.

[6] A. S. Morched and P. Kundur, "Identification and modeling of load characteristics at high frequencies," IEEE Trans. Power Syst., vol. PWRS-2, Feb. 1987.

[7] M. Nagpal, W. Xu, and J. Sawada, "Harmonic impedance measurement using three phase transients," IEEE Trans. Power Del., vol. 13, pp. 272-277, Jan 1998.

[8] Mark Sumner, Ben Palethorpe, David W. P. Thomas, Pericle Zanchetta, and Maria Carmela Di Piazza, "A technique for power supply harmonic impedance estimation using a controlled voltage disturbance," IEEE Trans. Power Electronics, vol. 17, NO.2, March 2002.

# On Acoustic Noise Reduction Procedure for Inverter-Fed Induction Machines

Weiss Helmut[1], Zaucher Peter[2], Xiao Jian[3]

[1] University of Leoben/Institute for Electrical Engineering, Leoben, Austria; e-mail: helmut.weiss@unileoben.ac.at
[2] University of Leoben/Institute for Designing Plastics and Composite Materials, Leoben, Austria;
e-mail: verbund@unileoben.ac.at
[3] Southwest Jiaotong University/School of Electrical Engineering, Chengdu Sichuan, China; e-mail: jian_x@126.com

*Abstract*— **Converter-fed variable speed drives are undesired sources of acoustic noise. Due to principle of operation magnetostriction and mechanical torque oscillations cause noise. For the prevailing variable speed realization through PMW inverter and induction machine mainly the machine with its outer components like cooling fins and ventilator hood can be the emitter of noise in audible area. Cooling fins practically shall stay metal material. Fan and fan hood favorably are from plastic material and could be made of composite material. Careful investigation of types of resonant bodies and their specific degree of freedom in mechanical resonant deformations help establish suitable designs for noise reduction. In addition, we shall take into consideration the advanced damping opportunities linked to composite material and the opportunity to define material stiffness parameters. This contribution provides a composite material based promising approach for efficient noise reduction.**

*Keywords* — **Variable speed drive, induction motor, converter machine interactions, acoustic noise**

## I. Acoustic Noise Generation with Converter-Fed Drives

Heavy acoustic noise is the standard phenomenon aligned with variable speed electrical drives. A drive block diagram with the basic links is displayed in Fig. 1.

The grid can supply linear loads as well as nonlinear loads. Nonlinear loads (like a converter) cause nonlinear currents and grid voltage deviations whose effects will again influence the induction machine. However, the main source of noise will be the converter-induction machine pair. At the variable frequency side, there are non-sinusoidal currents because of the PMW voltage waveform. The induction machine as electrical-to-mechanical power converter produces not only the desired torque at defined speed but also thermally dissipated power as well as acoustic poise.

For identifying origins of acoustic noise we have to look into the torque generation principle and the interactions inside the machine (Fig. 2).

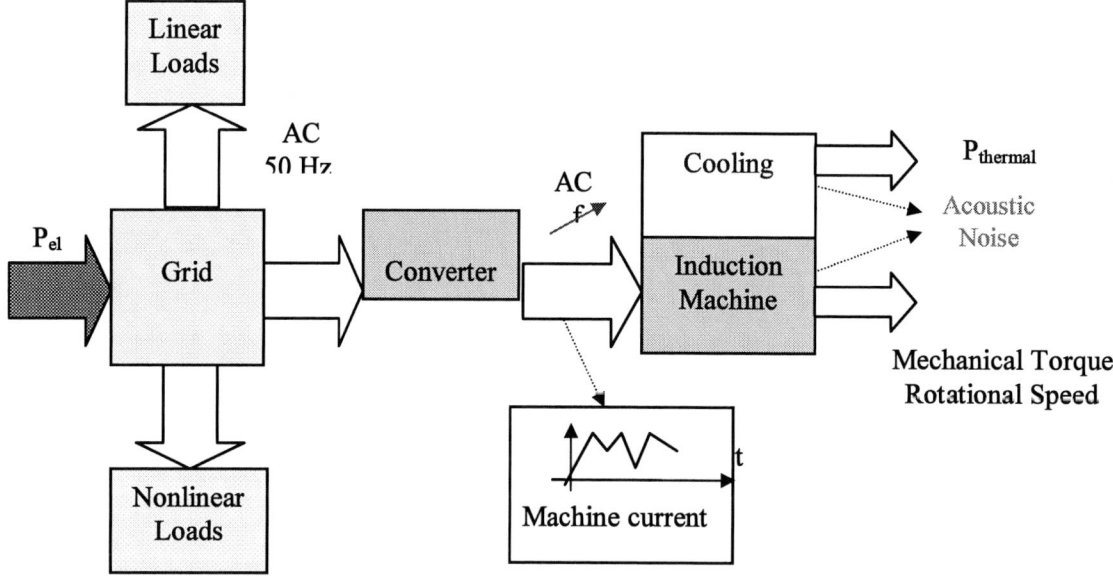

Fig. 1. Grid powered variable speed drive with interactions to acoustic noise

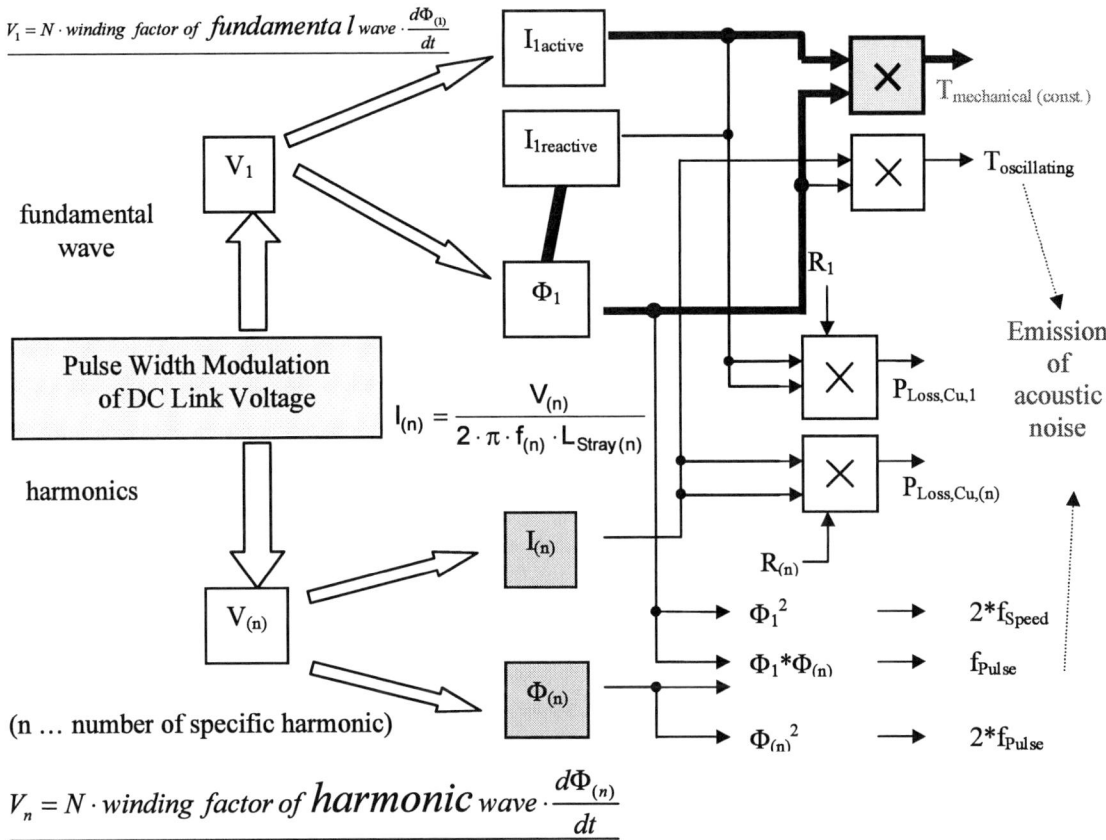

$$V_1 = N \cdot winding\ factor\ of\ \textbf{\textit{fundamental}}\ wave \cdot \frac{d\Phi_{(1)}}{dt}$$

$$I_{(n)} = \frac{V_{(n)}}{2 \cdot \pi \cdot f_{(n)} \cdot L_{Stray(n)}}$$

$$V_n = N \cdot winding\ factor\ of\ \textbf{\textit{harmonic}}\ wave \cdot \frac{d\Phi_{(n)}}{dt}$$

Fig. 2. Fundamental and harmonic wave effects for torque generation, losses and flux waves

We recognize that pulse width modulation of the DC-link voltage not only produces the desired fundamental wave optimized for present rotational speed but also harmonics. The usual output torque $T_{mechanical,(const.)}$ comes from force generation in a magnetic field ($\Phi_1$, linked with reactive current) by an active current ($I_{1active}$) both of fundamental wave (simplified approach). Unfortunately, all other outputs actually are troublesome. Also the fundamental wave (index 1) generates losses $P_{Loss,Cu,1}$ and acoustic noise by magnetostriction at $2*f_{Speed}$.

Fourier analysis of the inverter pulse pattern provides the harmonic voltages $V_{(n)}$. Each harmonic of the voltage is responsible for a certain harmonic current $I_{(n)}$ as well as for a certain flux harmonic $\Phi_{(n)}$. Really critical for oscillating torque $T_{oscillating}$ and losses $P_{Loss,Cu,(n)}$ are the low order harmonics as explained by estimation formula $I_{(n)}$. Concerning acoustic noise we have to take into consideration the sensitivity of the human ear with a maximum around 1 kHz. Therefore, pulse frequencies around 500 Hz or 1 kHz (early GTO inverters and transistor inverters) were troublesome origins of heavy acoustic noise. But also the IGBT inverters of today produce more acoustic noise than appreciated if operated in critical environments and in inappropriate manner.

Critical is the acoustic noise actually emitted into air. We have to pay attention to the main noise emitting structures at the machine and here for the dominant (resonant) frequencies. We identify the fan with air stream as a source and emitter but this is inevitable due to cooling

requirements and should exhibit not the critical (single frequency or few defined frequencies) tunes at certain frequencies but a wide-band signal which is not so annoying for the human ear.

## II. Acoustic Noise Emission Components

### A. Noise Origin and Propagation Identification

Standard electric machine look pretty much the same. We always see heat fins and a fan hood with a fan inside.

Fig. 3. Standard machine (20 kW range), with metal cooling fins and metal fan hood (in the rear)

Fig. 4. Standard machine (2 kW range), with metal cooling fins and polymer material fan hood

Dominating mechanical structures like heat fins (Fig. 3, 4) at the surface of the machine with the task of transferring thermal dissipation power (so called copper losses and iron losses of the induction machine) exhibit defined resonant frequencies where acoustic noise emission is optimized but in wrong way for us who wish to get a silent variable speed drive fed by power electronics that create many excitations over a wide frequency band.

The other element is the fan hood (Fig. 3, 4, 5) usually from metal sheet material for cost and robustness, sometimes a non-optimized cheap polymer material.

Fig. 5. Acoustic noise emitting elements at standard induction machine

Inside the fan hood we find the fan blades of metal or polymer material. Resonances of fan blade structure alone or fan blade / fan hood or fan hood alone have to be considered. There are many opportunities to obtain a resonance by exciting these elements capable of noise emission (cooling fins of variable length and fan hoods) at their resonance frequency through the general multi-frequency and variable-frequency generating PWM-fed drive motor. Just "hiding" resonances outside of 50 Hz and 60 Hz feeding as done with line-operated drives is not feasible and therefore no solution. But we have to analyze the resonances and provide damping.

### B. Noise Reduction Procedure

Summarizing the basics mentioned previously and opening up a way for designing noise emission minimized drives we propose to follow the procedure displayed in Fig. 6.

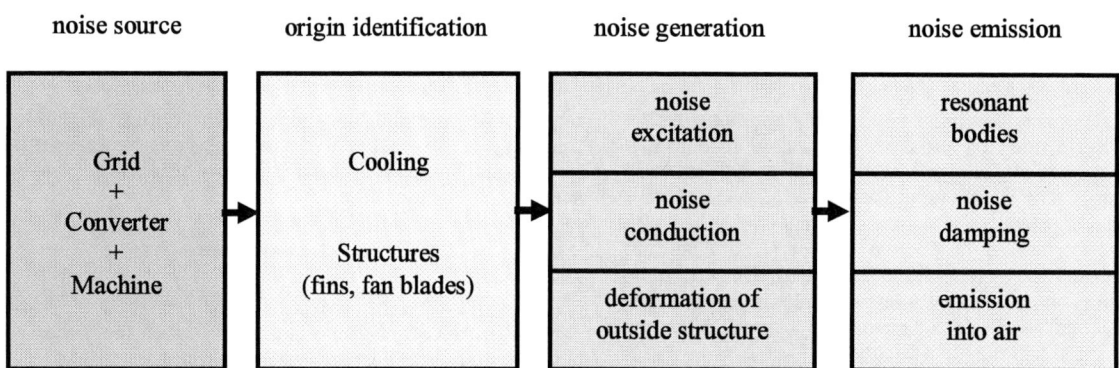

Fig. 6. Procedure for drive design with acoustic noise emission minimization.

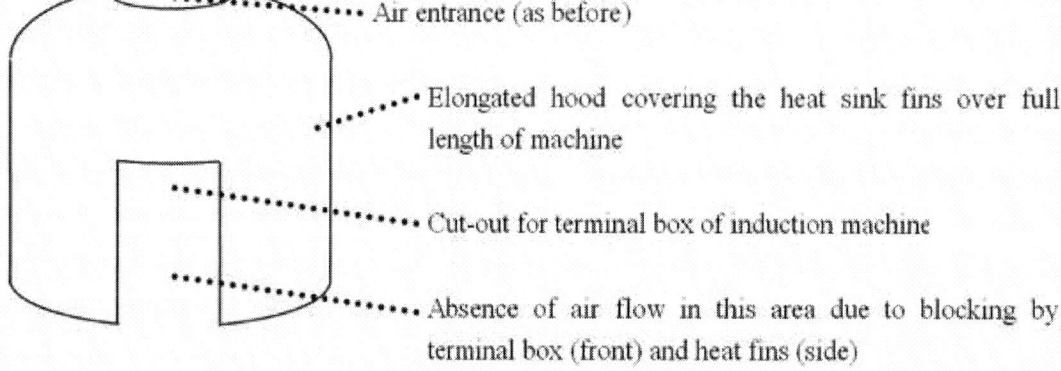

Fig. 7. Sketch of modified fan hood

We identify the converter as the main original initiator of emitted noise but consider also contributions from the machine and at latter extent from the grid. Then we identify the origin as the elements where acoustic noise comes from (cooling itself = air stream, and machine elements). Further on we define the noise propagation from excitation (inside machine, e.g.) over conduction (e.g. stator, iron yoke) to the mechanically deformed element outside that couples to air. In order to obtain the noise emission of the drive finally, these main elements are identified according to their resonant body properties, their opportunities to damp down internal excitation and finally to their actual emission behavior. The interrelations are described in Fig. 6.

### C. Modified Fan Hood

We propose a fan hood of modified outline (Fig. 7) and a fan hood made of composite material rather than standard metal sheet or simple polymer material. This fan hood shall cover the whole length of the machine and thus the whole length of the cooling fins.

This fan hood prevents direct noise radiation from fins into free air, it may damp fin movement at fin top and it provides a cleaner air stream even at the wheel side (opposite side of air inlet). Noise-optimized shapes of the hood in fan area can be realized very easily.

### III. CALCULATION METHOD FOR RESONANCES

Finite Element Analysis (FEA) is a well-proven method to obtain mechanical data of structure elements like resonances and damping/amplification at these frequencies. We have to use such a method to optimize the actual design chosen.

### IV. COMPOSITE PROPERTIES AND ADVANCED FEATURES

Before we can start a calculation or computation we definitely need a clear idea of properties of the composite material and the structure we want to use in our application. Composite material is not homogeneous but includes fibers in composite matrix arrangement. Various types of fibers and various types of arrangements and layouts can be used. Only with this knowledge as prerequisite we can continue our design effort.

Fig. 8, 9, 10, 11 explain basic differences between some types of composite material..

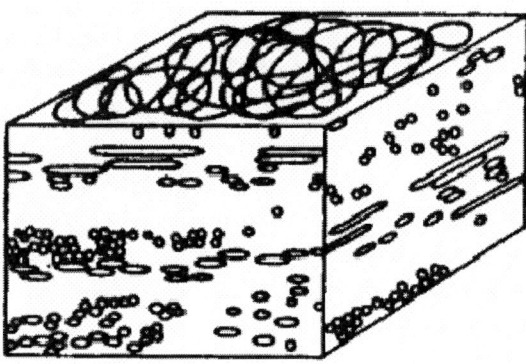

Fig. 9. Short / long fiber composites (2)

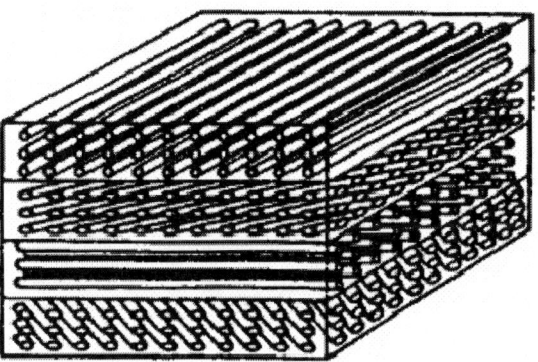

Fig. 10. Unidirectional fiber composites

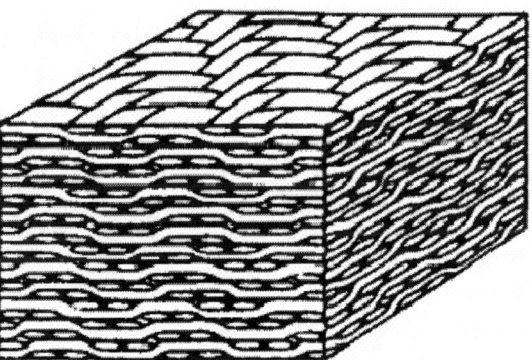

Fig. 11. Fiber fabric composites

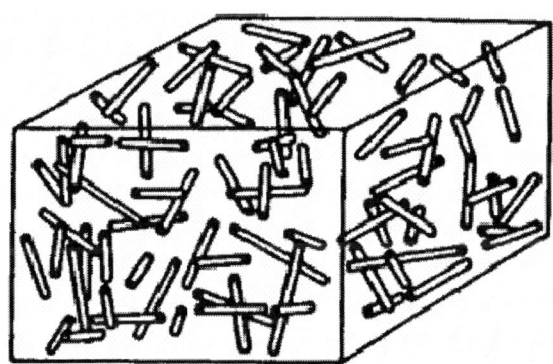

Fig. 8. Short / long fiber composites (1)

Fig. 12. Multi directional laminate consisting of UD layers (1)

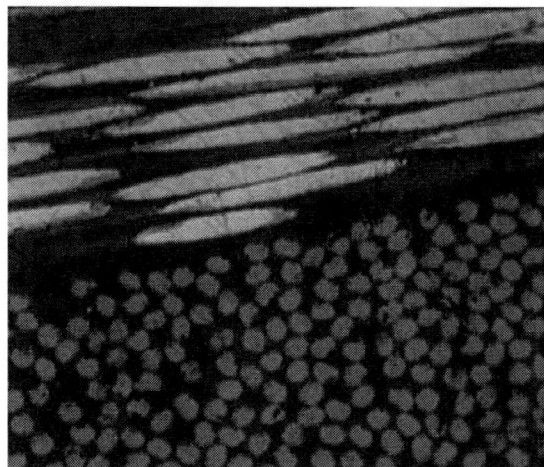

Fig. 13. Multi directional laminate consisting of UD layers (2)

Fig. 12, 13 provide information about internal structures of a typical multi-directional laminate with uni-directional (UD) layers.

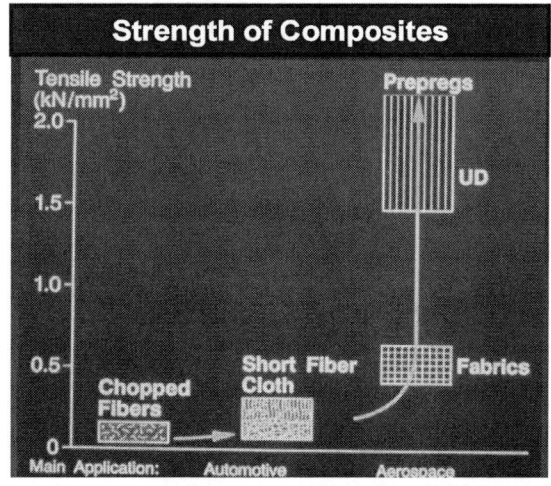

Fig. 14. Overview on tensile strength of composites

Fig. 14 displays application related stress values for certain types of composite materials.

### V. RESONANCES OF BAR AND PLATE STRUCTURES

A mass-spring-system is the basic approach to resonances. Depending on mass we calculate the acceleration force and depending on spring stiffness we get the counter force through elongation. The simple ideal mass-spring-system has just one resonance frequency. We have to consider all the resonances possible for certain structures. We identify simple structure elements like bars (Fig. 15) and plates (Fig. 16) and cylinders (Fig. 17). Resonance properties of these structures are presented here in a simplified way as follows (full details in [6]).

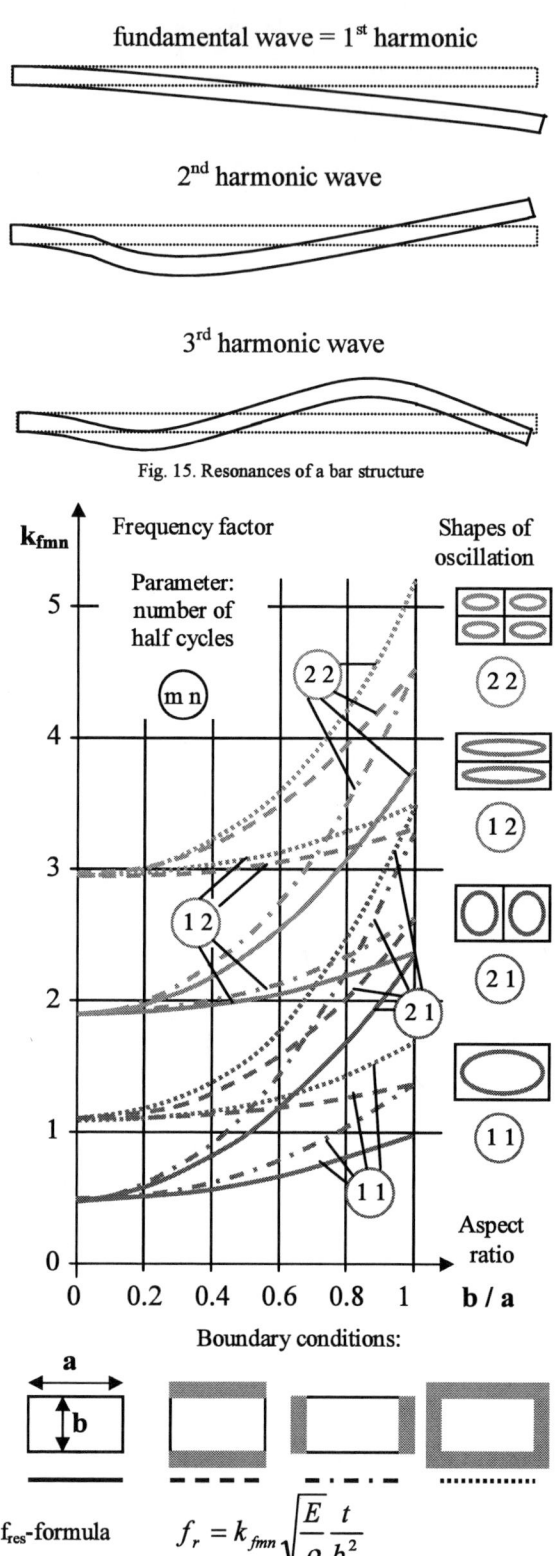

Fig. 15. Resonances of a bar structure

$f_{res}$-formula
$$f_r = k_{fmn} \sqrt{\frac{E}{\rho}} \frac{t}{b^2}$$

$E$ ... elasticity module        $t$ ... plate thickness
$\rho$ ... mass per volume of polymer material

Fig. 16. Resonances of a plate structure [6]

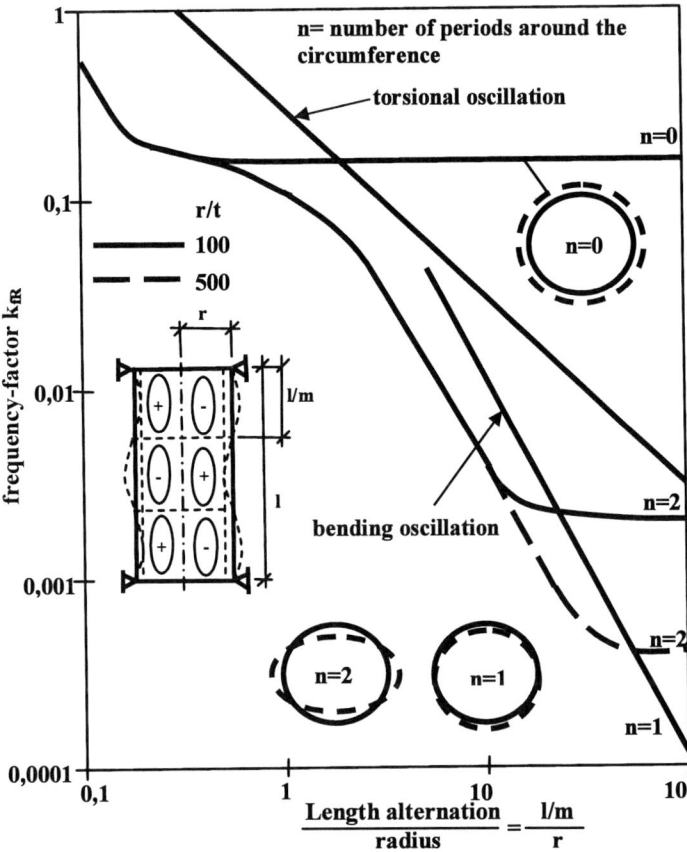

Fig. 16 explains calculation results for various resonance shapes of plate structures and the corresponding resonant frequency calculation depending on elasticity module, mass per volume, thickness and dimensions of the rectangular plate and finally boundary conditions in a graphical layout.

In fig. 17 we see resonance calculations for cylindrical bodies depending on material properties, layout and again oscillation shape.

$$f_r = k_{fR} \sqrt{\frac{E}{\rho}} \frac{1}{r}$$

We have to keep in mind that actual structures of machine parts are NOT as simple as these special cases where we can use closed form solutions [6]. However, a clear idea on relations helps us to get a reasonable design for a good start and an efficient process for optimization.

When we go into material details we have to care for the special and specific coupling effects of laminated polymer material (Fig. 18). This takes into account the various opportunities of the structural body to be flexible in several directions and modes.

Fig. 17. Resonances of a cylindrical structure [6]

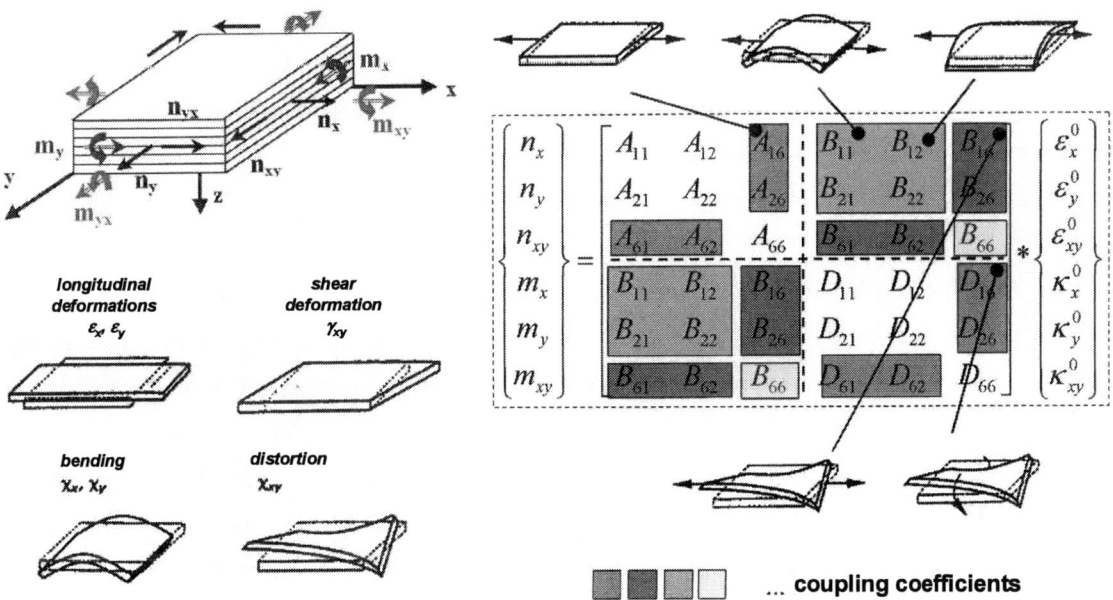

Fig. 18. Coupling effects in multi-layer composites [5]

After this pre-investigation and material selection we come to the FEA calculations on a pre-optimized design. Fig. 19 and 20 give examples for results and verification by test bench.

Fig. 19. FEA result in usual display (stress value = color) [5]

Fig. 20. Stress test bench [5]

## VI. Design Procedure

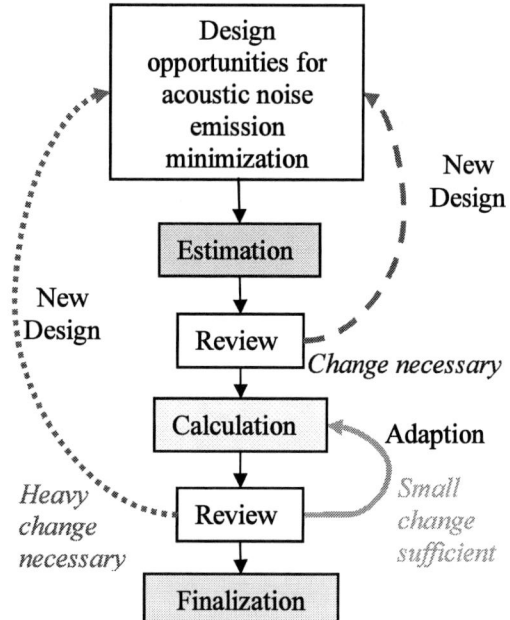

Fig. 21. Optimization process

Fig. 21 explains the design procedure. Certain design ideas are feasible with given equipment. Experience and estimations by formulas given in previous chapters give a first approach. Provided the reviewing is positive we continue by time consuming FEA to get the first optimization result. Again a review will decide if we reached the final result already but more often it will lead to adaptations or even completely new beginning design steps.

## VII. Conclusion

This contribution proposes an advanced procedure for reducing emitted acoustic noise from critical inverter-fed induction motor drives. After identifying the process of noise generation we analyze in detail the initial sources of this noise and the propagation way up to the emitting element. Cooling fins of the motor and fan hood are identified as critical main emitters of acoustic noise. The procedure does not influence the noise generation inside the machine as this would lead to an unfavorable new design of the machine that is operating well under electrical considerations. In order to meet acoustic noise limitation demands in a very cost efficient way with good results we suggest to run an interdisciplinary effort between the electrical engineering side (noise source, machine layout) and the mechanical engineering side (resonant bodies, mechanical structure analysis of plates and shells) and the polymer engineering side (composite material properties, design issues, calculation of resonances and degrees of freedom for deflection, cost efficient manufacturing of final design). The advantage of the procedure described lies in the ability to yield high rates of noise reduction with standard designed and realized electrical machines by simply replacing the existing fan hood by a composite polymer material fan hood of advanced design.

## References

[1] RADAN Ahmad, GHARAKHANI Arbi. Analytical Study of Affecting Characteristic of Voltage Vectors of a Three-level NPC Inverter on Torque and Flux of DTC controlled Drives. Proceedings on CD-ROM of the 12th European Conference on Power Electronics and Applications, EPE 2007, Aalborg, Denmark, paper no. 142

[2] TISBORN Guido, ORLIK Bernd. Optimal Utilization of Induction Motor Drives. Proceedings on CD-ROM of the 12th European Conference on Power Electronics and Applications, EPE 2007, Aalborg, Denmark, paper no. 545

[3] WEILAND Thomas, HENZE Olaf, ROCKS Alexander, DE GERSEM Herbert, BINDER Andreas, HINRICHSEN Volker. A network model for inverter-fed induction-motor drives, Proceedings on CD-ROM of the 12th European Conference on Power Electronics and Applications, EPE 2007, Aalborg, Denmark, paper no. 867

[4] HINKKANEN Marko, CEDERHOLM Mikaela, REPO Anna-Kaisa, LUOMI Jorma. Small-Signal Model for Saturated Induction Machines. Proceedings on CD-ROM of the 12th European Conference on Power Electronics and Applications, EPE 2007, Aalborg, Denmark, paper no. 92

[5] KARALL T. Funktionsgerechtes Gestalten von Kunststoffen am Beispiel der Faserverbunde. www.sfg.at/termine/docs/1921_Vortrag_Karall.pdf, (08-06-30)

[6] J. Wiedemann. Leichtbau 1: Elemente. Springer-Verlag, 2.Auflage, S.82-83.

# Cascaded Doubly Fed Induction Generator for Mini and Micro Power Plants Connected to Grid

Marek Adamowicz[*] and Ryszard Strzelecki[†]

[*] Gdynia Maritime University, Gdynia, Poland, e-mail: *madamowi@am.gdynia.pl*
[†] Gdynia Maritime University, Gdynia, Poland, e-mail: *rstrzele@am.gdynia.pl*

*Abstract—* **Mini hydropower and wind power plants have a promising future in Poland. Variable speed cascaded induction generators could be an attractive alternative to conventional double output induction generators habitually applied in mini hydropower stations. Cascaded induction generators require lower maintenance due the absence of slip rings and brushes. Unidirectional power converters can be employed in such power plants to provide simple power electronics, reduce of the maintenance and increase availability. Different topologies of rectifiers with a function of power factor correction could carry the slip power from the secondary side of the cascaded generator during the supersynchronous operation. The main benefit of these clean power rectifiers in comparison with diode rectifiers is the ability of canceling the torque harmonics of cascaded generator. The paper discuss which topology of clean power rectifiers could be applied as machine side rectifier in mini and micro power plants utilizing cascaded induction generator connected to grid.**

*Keywords—***Renewable energy systems, adjustable speed generation system.**

## I. INTRODUCTION

Small hydro and wind power plants have a promising future in Poland. Their potential has not yet been even half fully exploited. The small hydropower (SHP) development in Poland has followed a constant and an impressive upward trend and the SHP sector will continue to grow in the future [1]. The power rating of the SHP is less than few MW. SHP can be further subdivided into mini hydro, usually defined as < 500kW and micro hydro <100 kW [2]. Micro hydropower plants are applied especially in remote areas. Mini and micro hydropower plants (MHP) arouse a wide interest of non-industrial user's in Poland.

The electro-mechanical systems applied in a mini and micro hydropower plant should be robust and as simple as possible to reduce costs and decrease maintenance and to guarantee profitability of the investments. The majority of grid connected MHP plants is equipped with squirrel cage induction generators because of their robustness, mechanical simplicity and a low price. However doubly output induction generators are more useful for a local-use water power plants [3, 4] as they provide variable speed constant frequency operation. They can use a fractional power size converter in the rotor circuit as it handles only the slip power. Moreover application of a power electronic converter connected to the rotor can provide decoupled control of the active and reactive power flow of the doubly output induction generator. Unfortunately wound rotor induction generators require slip-rings to transfer power to and from the rotor circuit. These slip ring contacts need to be regularly maintained. Slip-ring maintenance intervals of few months are typically achieved.

To minimize the investment costs, small and mini hydropower plants are often equipped with low cost power electronic interfaces [4] which ensure variable speed operation and enable double output power generation only over the synchronous speed. A power factor correction using a capacitor bank is necessary in most of these applications according to the standard international grid connection specifications [2]. The capacitor bank is mostly connected in parallel with the generator to compensate the reactive power consumption. The basic system configuration of an adjustable speed compact MHP plant is shown in Fig. 1.

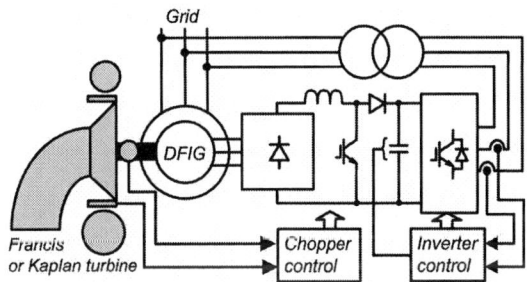

Fig. 1. Basic system configuration of the simple MHP.

The MHP system from Fig. 1 comprises Francis or Kaplan turbine connected to the shaft of a doubly fed induction generator (DFIG). The stator of the DFIG is directly connected to grid while the rotor side is connected to unidirectional power converter. The converter consists of a diode rectifier, smoothing reactor, step-up chopper and a grid side PWM inverter connected to grid by a step-up transformer. At partial load as well for higher pole numbers, the power factor is significantly lower than 1. The low frequency slip power is converted back to line frequency. A most of the slip power is returned to the power system through the inverter and isolation transformer. Except few advantages i.e. the low price, low voltage drops on the rectifier and system simplicity, the electric system from Fig.1 possess some significant drawbacks:

- maintenance problems with slip rings and brushes,

---

978-1-4244-1741-4/08/$25.00 ©2008 IEEE     1729

- lack of possibility of supplying the reactive power to the rotor,
- harmonic distortion and generator torque ripples caused by diode rectifier,
- low frequency pulsation of stator current,
- oversized transformer.

To eliminate the drawbacks of simple power electronic interface from Fig.1, the three phase PWM boost rectifier with bidirectional switches has been proposed in [5]. Three bidirectional switches have been added to diode rectifier. Additionally the single chopper from Fig.1 has been replaced by a double chopper with four switches. Control of the PWM rectifier adjusted to the slip frequency has been also proposed in [5].

To overcome maintenance problems with the slip rings, the cascaded doubly fed induction generator (CDFIG) have been proposed in [6]–[8] as an alternative to conventional doubly fed induction generators habitually installed in hydroelectric power plants. The rotor power of CDFIG is magnetically transferred to secondary stator terminals without the use of slip-rings. Application of four quadrant converter in the secondary stator circuit enables decoupled control of active and reactive power [9, 10]. However the power flow control of the CDFIG is much more complicated than for single conventional double fed induction generator and the reliability of control of the four quadrant converter connected to CDFIG is still low.

Configurations of small hydropower plants equipped with CDFIG and unidirectional converter have been proposed in [6]–[8]. For improving the performance of the power generation the secondary stator of the CDFIG has been modified and 12-pulse rectifier have been applied as it is shown in Fig.2.

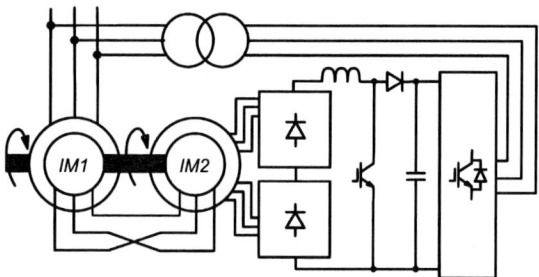

Fig. 2. SHP with cascaded generator with dual three phase windings of secondary stator [6]-[8].

Dual three phase windings of secondary stator relatively shifted by 30° have been connected through two diode rectifiers with a common step-up chopper and a grid side PWM inverter.

Using the topology from Fig.2, with the six-phase windings connected to 12-pulse rectifier, the mmf components of the harmonic numbers $|1 \pm 6m|$ (m: odd) can be canceled .

Several topologies of clean power rectifiers with low effects on the mains [11, 12] could be potentially applied in slip power recovery schemes with CDFIG without any modifications of the generator windings. Vienna rectifiers [11], [13] have recently been used in PM generator systems [14] and motor drive applications [15]. The unidirectional Vienna converter seems a good candidate for slip power recovery from cascaded generator as it has a good efficiency and small voltage drops which are smaller even than in PWM full-bridge [16]. Moreover some narrow-band reactive power control of Vienna rectifier may be also possible [13].

The paper describes two topologies of unidirectional converters with harmonic reduction and discuss the possibility of adopting them in mini and micro power plants utilizing cascaded induction generator connected to grid.

## II. CASCADED DOUBLY FED INDUCTION GENERATOR

The cascaded doubly fed induction generator analyzed in the paper consists of two identical 2.2kW 4-pole wound rotor induction machines IM1 and IM2 and is schematically shown in Fig. 3. Machine parameters are given in Table I.

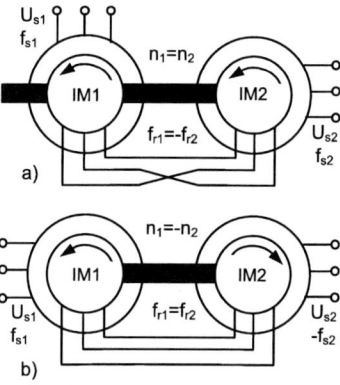

Fig. 3. Two equivalent realizations of cascaded doubly fed induction generator.

Rotors of both machines are electrically and mechanically coupled while both stators terminals are accessible. Natural speed $n_{nat}$ of CDFIG, which is the synchronous speed of the cascaded generator, can be obtained at zero frequency of the secondary stator $f_{s2}$ from

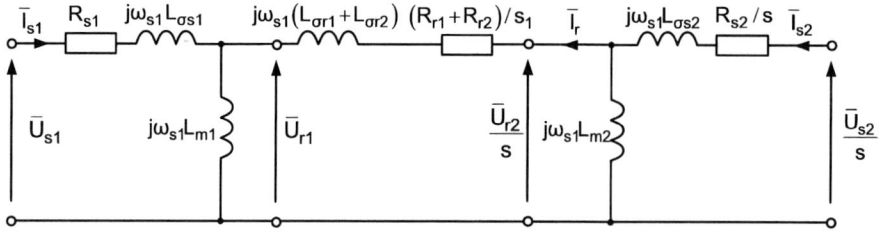

Fig. 4. Steady state equivalent circuit of the cascaded doubly fed induction generator

$$n = \frac{(f_{s1} + f_{s2}) \cdot 60}{p_1 + p_2}. \qquad (1)$$

For the CDFIG consisting of two identical induction machines ($p_1 = p_2$) the natural speed $n_{nat} = n_s/2$ is equal half of the synchronous speed of the single IM. For the parameters given in Table I the natural speed of the CDFIG is 750 rpm. More details of advanced modeling of the wound rotor induction machines can be found in [17].

TABLE I.
PARAMETERS OF UNIT INDUCTION MACHINE

| | | |
|---|---|---|
| Rated power | 2.2 | (kW) |
| Rated power Factor | 0.88 | (-) |
| Rated stator Voltage | 380 | (V) |
| Rated stator Current | 4.6 | (A) |
| Rated rotor Voltage | 72 | (V) |
| Rated rotor Current | 19 | (A) |
| Rated frequency | 50 Hz | (Hz) |
| Pole number | 4 | (-) |
| Stator resistance | 2.3 | (Ω) |
| Stator leakage inductance | 17.42 | (mH) |
| Rotor resistance | 4.68 | (Ω) |
| Rotor leakage inductance | 17.42 | (mH) |
| Mutual inductance | 469 | (mH) |

Steady state equivalent circuit of CDFIG is shown in Fig. 4. All phase quantities are referred to the first stator side. The slips $s$ and $s_1$ in Fig.4 can be obtained from

$$s_1 = \frac{f_{r1}}{f_{s1}} = \frac{f_{s1} - p_1 n_1 / 60}{f_{s1}} \qquad (2)$$

$$s_2 = \frac{f_{s2}}{f_{r1}} = \frac{f_{s2}}{f_{s2} - p_2 n_2 / 60} \qquad (3)$$

$$s = s_1 \cdot s_2 = -\frac{f_{s2}}{f_{s1}} \qquad (4)$$

where the signs of the secondary stator and rotor frequencies $f_{s2}$, $f_{r2}$ and sign of secondary shaft speed $n_2$ have to be chosen in accordance with configuration of the cascade described in Fig.3.

Fig. 5 shows the variation of secondary stator voltage $U_{s2}$ as a function of the shaft speed $n$.

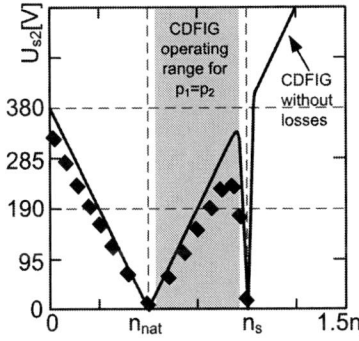

Fig. 5. Secondary stator voltage of the CDFIG versus shaft speed.

The secondary side power converter ratings depend on a speed range of the CDFIG. Cascaded system can be controlled in such a way that the slip power $P_{IM2}$ recovered through the stator of the second machine IM2 is a fraction of the maximum power $P_{IM1}$ of the first stator directly connected to grid, which can be obtained from

$$P_{IM2} = s \cdot P_{IM1} = \frac{n - n_{nat}}{n_{nat}} P_{IM1}. \qquad (5)$$

As it can be seen from Fig.5 at operating point with the shaft speed equal to twice natural speed of the CDFIG, the coupling of two machines in the cascade is lost. The secondary stator voltage and the generated power are then equal zero. The operating range of the CDFIG is therefore between natural speed of the cascade and a twice of its value.

### III. COMPARISON OF RECTIFIER TOPOLOGIES WITH LOW EFFECT ON THE CASCADED GENERATOR

The three-phase diode bridge rectifier used in the basic configuration of simple MHP from Fig.1 suffers from high total harmonic distortion of the secondary stator currents. Application of dual three phase windings of secondary stator relatively shifted by 30° and two diode bridges in Fig.2 [6]-[8] refers to multipulse methods of harmonic elimination. Shifted stator windings play similar role to the phase shifting transformers and the harmonics generated by one rectifier can be therefore canceled by harmonics produced by the second rectifier.

Reduction of harmonics of the secondary stator current is an important topic in CDFIG system as they also impact on the torque and stator currents of the first machine.

Two methods of harmonic reduction of secondary stator currents with using unidirectional rectifiers are presented in this section. First a three-phase diode bridge rectifier applying current injection [18] called Minnesota rectifier [11] is introduced. The second considered topology is a three-phase/switch/level PWM rectifier called Vienna rectifier [11] - [16].

#### A. Third-Harmonic Modulated (Minnesota) Rectifier

The system configuration of an adjustable speed MHP with CDFIG and Minnesota rectifier is shown in Fig.6.

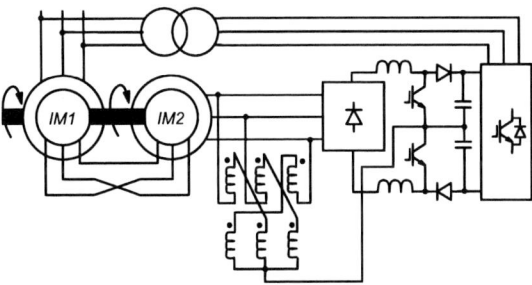

Fig. 6. MHP system with cascaded generator and Minnesota rectifier.

The Minnesota rectifier consist of two stages [18]. The first stage is the three phase diode rectifier similarly as in Fig.1. The second stage consists of two boost DC-DC converters. Undesirable diode bridge effects on the performance of cascaded generator are reduced by insertion of a zigzag autotransformer and by applying a modulation of DC current using two transistors with three times the secondary stator frequency. The sum of two third-harmonic currents is circulated through the generator side of the rectifier by a zigzag autotransformer [18]. The low-frequency harmonics in the IM2 stator current can be limited by controlling the magnitude and the phase of the third harmonic current. The benefit of this concept is that it is insensitive to stator voltage unbalances of the IM2. But it is required to synchronize the third harmonic modulation current to the phase stator voltage of the IM2.

The amplitude and the phase of a third harmonic current are adjusted in the control system and depend on the inductance of the source [19].

Simulations of the generator system with two 2.2kW machine CDFIG and Minnesota rectifier have been carried out using PSIM simulation software for machine parameters from Table 1. Voltage drops on diodes have been considered in simulation program. Only the possibility of improving secondary stator current shapes and reduction of torque ripples during the oversynchronous operation were investigated in simulations. Optimal power generation control has not been taken into account in this paper.

Simulation results of an adjustable speed MHP system with CDFIG and Minnesota rectifier for the slip value $s = -0.2$ and power generated from both generator sides are shown in Fig.7 and Fig.8. The switching frequency for Minnesota rectifier was 3.5kHz

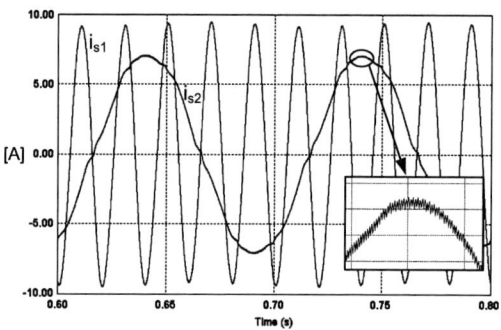

Fig. 7. Simulation results of MHP system with CDFIG and Minnesota rectifier (s = -0.2; 3.5kHz): stator currents of IM1 and IM2

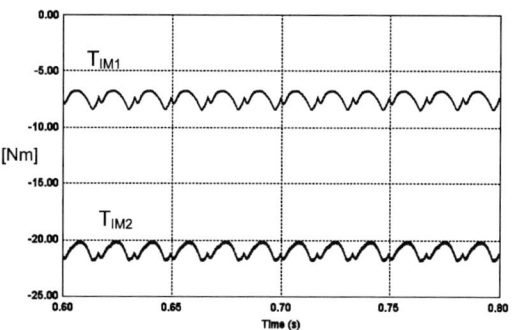

Fig. 8. Simulation results of MHP system with CDFIG and Minnesota rectifier (s = -0.2): torques of IM1 and IM2 .

## B. Three-Phase/Switch/Level PWM (Vienna) Rectifier

Overall scheme of an adjustable speed MHP with CDFIG and Vienna rectifier is shown in Fig.9.

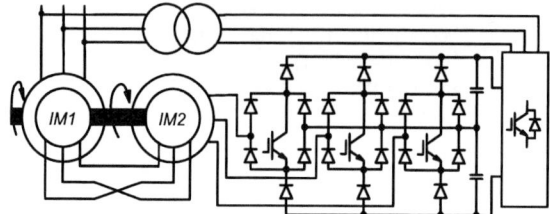

Fig. 9. MHP system with cascaded generator and Vienna rectifier .

The topology of Vienna converter is obtained by insertion of a turn-off semiconductors into each phase leg of a three-phase diode bridge [11]. The benefit of the Vienna rectifier is that it gives possibility of manipulation of phase shift of the CDFIG secondary stator current in the range of from 30° capacitive to 90° inductive. Pure sinusoidal shape of the both stators currents can be obtained. Due to nonlinear nature of the Vienna rectifier, advanced control methods have to be used. Moreover control restrictions i.e. reduced output voltage area, should be considered in control algorithm.

Simulation results of an adjustable speed MHP system with CDFIG and Vienna rectifier for the slip value $s = -0.2$ are shown in Fig.7 and Fig.8

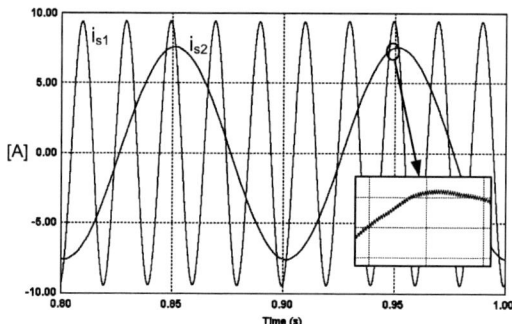

Fig. 10. Simulation results of MHP system with CDFIG and Vienna rectifier (s = -0.2; 5kHz): stator currents of IM1 and IM2

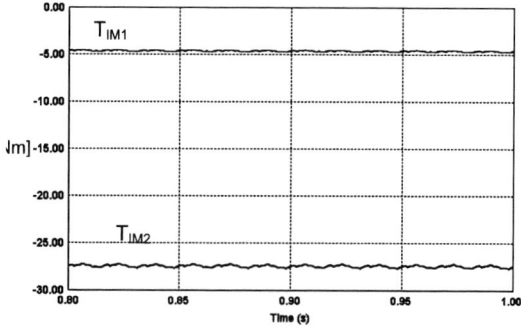

Fig. 11. Simulation results of MHP system with CDFIG and Vienna rectifier (s = -0.2): torques of IM1 and IM2 .

## IV. DISCUSSION

In this section two topologies of unidirectional rectifiers with low effects on the cascaded induction generator are shortly compared. Following [11] the comparison concerns

- number of components
- voltage and current stresses on the components (conductions and switching lossess)
- control complexity.

The third harmonic modulation of DC current used in Minnesota rectifier could not guarantee exact sinusoidal shape (Fig.7). The performance of Minnesota rectifier could be further improved by modifying the shape of the current fed back to the diode bridge inputs [11].

Similarly as for fulfilling requirements of high quality grid connected applications reported in [11] the Vienna rectifier with 5kHz switching frequency represents an ideal topology as unidirectional rectifier for MHP with CDFIG.

The Minnesota rectifier represents lower control complexity and lower number of discrete components. It consists of diode bridge and four discrete elements: two IGBT and two fast recovery diodes. Moreover three magnetic elements are required: the zigzag transformer and two inductors.

For mini and micro power plants the large number of discrete elements of Vienna rectifier can be replaced by integrated easy-to-mount rectifier modules for three phase power factor correction [20]. The IM2 stator inductance plays the role of input AC-filter. As it is relatively high in comparison to input inductances in typical grid connected applications of Vienna rectifier [11]-[16] it is easier to obtain purely sinusoidal currents of analyzed system (Fig.10). Another benefit achieved with Vienna rectifier is a good efficiency. For mini power plants the topology of Vienna rectifier allows the use of MOSFETs instead of IGBTs which ensures low current distortions and relatively low losses particularly at low loads.

## V. CONCLUSION

Two novel solutions for low cost mini and micro power plants has been presented in the paper. The CDFIG system and two rectifier topologies with low effects on the generator has been proposed to apply in mini and micro power plants. Proposed systems have the advantage of lower costs than bi-directional PWM converter as well lover voltage drops on the elements and characterize a good performance of the a slip power recovery.

The speed range of the adjustable speed micro power plant with CDFIG is restricted to supersynchronous operation. The grid side inverter has fractional ratings and can ensure only partial compensation of reactive power consumed by CDFIG. The power factor of the generator can be corrected using a capacitor bank.

## ACKNOWLEDGMENT

This work was financed by the Polish Ministry of Scientific Research and Information Technology (2005– 2008).

## REFERENCES

[1] Punys P., Pelikan B., "Review of small hydropower in the new Member States and Candidate Countries in the context of the enlarged European Union", *Renewable and Sustainable Energy Reviews*, vol. 11, pp. 1321-1360, 2007.

[2] "Status report on variable speed operation in small hydropower", *Energie*, European Communities, 2000.

[3] Kuge K., "Constant frequency and constant voltage control of induction generator", *Electrical Engineering in Japan*, vol. 112, pp. 89 – 94, Published Online: 22 Mar 2007

[4] Koichiro U., Koji S., Tomoki O., Mamoru N., Takashi Y., Kazumasa I., "Development of Controlled Speed Induction Generator System for Small Hydroelectric Power Plants", *Papers of Technical Meeting on Rotating Machinery, IEE Japan*, pp. 7-11, 2000.

[5] Hoshi N., Oguchi K., "A Novel Slip-Power Recovery System Using a PWM Boost Rectifier", *Electrical Engineering in Japan*, vol. 139, no. 2, pp. 52-60, 2002.

[6] Kato S., Hoshi N., Oguchi K., "A Low-Cost System of Variable-Speed Cascaded Induction Generators for Small-Scale Hydroelectricity", Thirty-Sixth IEEE IAS Annual Meeting Conference, vol. 2, pp.1419 - 1424, 30 Sep-4 Oct, 2001.

[7] Kato S., Hoshi N., Oguchi K., "Small scale hydro-power", *IEEE Industry Applications Magazine*, pp. 32-38, July/August 2003.

[8] Kato S., Hoshi N., Oguchi K., "Brushless Slip-Power Recovery System Simulation by Using Modified Nodal Analysis", *IEEJ Transactions Industrial Applications*, vol. 124, no. 12, pp. 1252 - 1260, 2004.

[9] Nicolas Patin N., Monmasson E., Louis J.P., " Analysis and control of a cascaded doubly-fed induction generator" in Proceedings of 31st Annual Conf. of IEEE Industraial Electronics Society IECON'05, 6 pp., 2005.

[10] Basic D., Zhu J.G., Boardman G., "Transient Performance Study of a Brushless Doubly Fed Twin Stator Induction Generator", *IEEE Trans. Energy Conv.*, vol. 18, pp. 400-408, September 2003.

[11] Kolar, J.W.; Ertl, H., "Status of the techniques of three-phase rectifier systems with low effects on the mains" Proc. of the 21st Int. Telecommunications Energy Conference INTELEC'99, 16 pp., 1999.

[12] Singh B., Singh B. N., Al-Haddad K., Chandra A., Pandey A., Kothari D. P. "A Review of Three-Phase Improved Power Quality AC–DC Converters ", *IEEE Transactions On Industrial Electronics*, vol. 51, pp. 641-660, June 2004.

[13] Todtermuschke K., Gensiort A., Rudolph J.,Webert J.,Guildnert H., "Flatness based control of the VIENNA-rectifier allowing for reactive power compensation", 37th IEEE Power Electronics Specialists Conference PESC '06, pp. 1-5, 2006.

[14] Sadowski T., "Einsatz des vienna- stromrichters als gleichrichter in windkraftanlagen", Dissertation, Technischen Universität Berlin, 2007.

[15] Lai R. at all., "A Systematic Evaluation of AC-Fed Converter Topologies for Light Weight Motor Drive Applications Using SiC Semiconductor Devices", IEEE International Electric Machines & Drives Conference IEMDC '07, vol. 2, pp. 1300 - 1305, 2007.

[16] Viitanen T., Tuusa H. "A Steady-State Power Loss Consideration of the 50kW VIENNA I and PWM Full-Bridge Three-phase Rectifiers", IEEE 33rd Annual Power Electronics Specialists Conference, vol. 2, pp. 915 - 920, 2002.

[17] Strzelecki R., Wilk A., Moson I., "Non-linear circuit model of a single doubly-fed induction machine formulated in natural axes for drive systems simulation purposes", XX Symposium Electromagnetic Phenomena in Nonlinear Circuits, Lille France, pp. 109 - 110, 2008.

[18] Naik R., Rastogi M., Mohan N., "Third-Harmonic Modulated Power Electronics Interface with Three-phase Utility to Provide a Regulated DC Output and to Minimize Line-Current Harmonics", *IEEE Trans. on Industry Applications*, vol. 31, no. 3, pp. 598-602, 1995.

[19] Mohan N., Rastogi M., Naik R., "Optimization of a novel dc-link current modulated interface with 3-phase utility systems to minimize line current harmonics", *IEEE PESC'92 Records*, pp. 162-167, 1992.

[20] VUM 85-05A Rectifier Module for Three Phase Power Factor Correction, www.ixyspower.com.

# Contactless power transmission with new secondary converter topology

Matthias Dockhorn[1], Daniel Kürschner[2] and Prof. Dr. Rudolf Mecke[1]

[1] Hochschule Harz, University of Applied Studies and Research, Wernigerode, Germany, e-mail: *rmecke@hs-harz.de*
[2] Institut f. Automation u. Kommunikation e.V., Magdeburg, Germany, e-mail: *daniel.kuerschner@ifak.eu*

*Abstract*—An innovative solution to slip rings and plug connectors is the contactless energy transmission system which is aligned to resonance. For an efficient energy transmission high frequencies (around 100 kHz) as well as an operation in resonance is useful. Some applications need a single-phase (50 Hz) supply A new energy conversion principle can be used to solve the problem. This paper describes a new converter topology and the development of the necessary control signals for the secondary and the primary circuit which provides a single-phase 50 Hz voltage on the secondary side without a need for a DC link. Moreover, practical tests and measurements are presented.

*Keywords*—AC/AC converter, Converter control, High frequency power converter, Magnetic device, Power transmission, Single phase system, Wireless power transmission.

## I. INTRODUCTION

Contactless inductive energy transmission technology has been developed within the last few years. Typical applications are assembly robots, machine tools, elevators, linear movable systems and non-contact battery chargers for electric vehicles. By means of the contactless transmission technology, conductor rails, sliding contacts, trailing cables or slip rings can be eliminated. As a result of abandoning any mechanical contact the safety and reliability of the energy supply can be improved and the maintenance can be reduced. Moreover the limits for speed and acceleration of movable consumers can be increased. Other advantages are: no wear and tear on the electrical contacts, no contact resistance, no spark formation (can be used in explosion-endangered environment) and no non-protected voltage-carrying contacts. Such an inductive power transmission system consists of a primary coil and the movable secondary pick up coil which contains the consumer.

## II. CONTACTLESS MAGNETIC SYSTEM

Investigations in the frequency range up to 200 kHz have shown that the steady-state as well as the dynamic behaviour of any magnetic assembly with an air gap can be described by an equivalent electric circuit, which contains the main inductance, leakage inductances and ohmic resistances of the windings. The main inductance and the leakage inductances of contactless systems mainly depend on the dimensions of the primary and secondary system, the existence of ferrite cores on the primary or secondary side and the air gap length. Contactless transmission systems are characterised by a small main inductance and large leakage inductances. The

inductances can be obtained by means of magnetic flux simulations (Fig. 1).

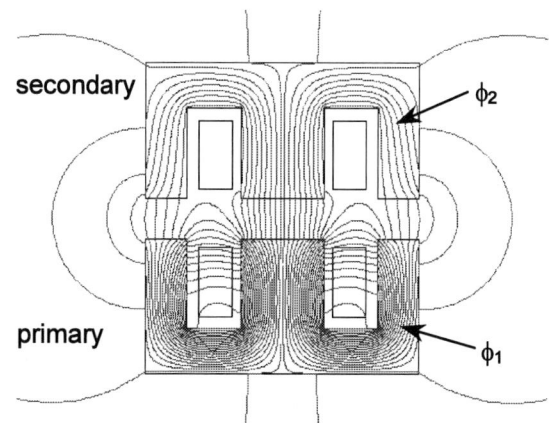

Fig. 1. Magnetic flux lines of a contactless energy transmission assembly

A fundamental problem of contactless energy transmission systems are the large primary and secondary leakage inductances. Consequently, the transfer of appreciable electric power requires the compensation of the leakage inductances by resonance capacitors on the primary and secondary coil. With a proper choice of these capacitors, the contactless system represents an ohmic load for the primary inverter. The primary current is nearly sinusoidal and the phase angle between primary voltage and current is zero. Therefore the switching events of the inverter can take place at zero current without auxillary commutation elements. This zero current switching (ZCS) is an important precondition for reaching higher transmission frequencies.

The transmission frequency is the most important electrical parameter for optimisation of contactless transmission systems. Fig. 2 shows the output power for different transmission frequencies as a function of air gap lengths, whereby the output power is related to the square of the primary magnetomotive force. The reachable output power strongly decreases with the air gap length nearly in the same way as the main inductance. On the other hand, by using higher transmission frequencies greater than 100 kHz, the transferable electric output power and the efficiency of contactless systems can be increased considerably. Moreover, high transmission frequencies lead to smaller filter elements and ferrite components.

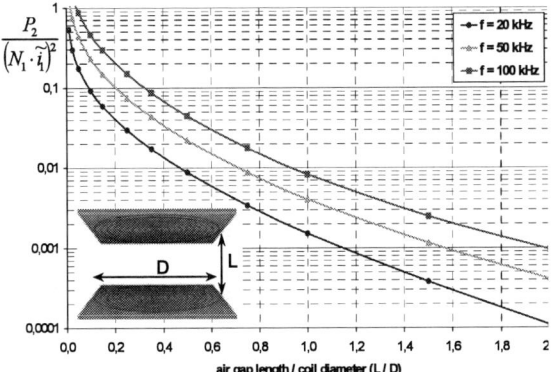

Fig. 2. Output power for different transmission frequencies as a function of air gap length

The following diagram shows the efficiency as function of the air gap length at different transmission frequencies. The efficiency is reduced with an increasing air gap due to the lower magnetic coupling. However, the efficiency of a magnetic assembly with a large air gap can be improved by using higher transmission frequencies in the range of 100 kHz. For smaller air gaps, the efficiency is greater than 95 % already at low transmission frequencies.

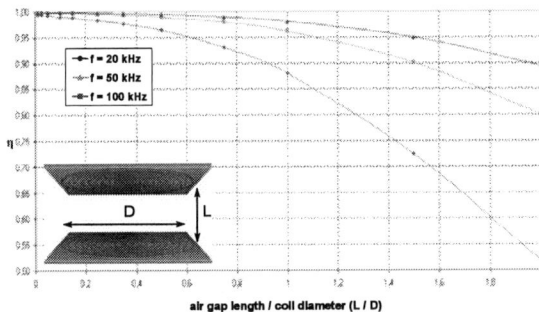

Fig. 3. Efficiency of contactless magnetic assembly

## III. CONVENTIONAL POWER ELECTRONIC SYSTEM

The conventional method (Fig. 4) commutates the 50 Hz alternating input voltage to DC voltage. The primary power inverter produces the 100 kHz transmission frequency. On the secondary side another converter module generates an AC output voltage (50 Hz) to provide a single-phase load. In this way the secondary side needs a lot of space particularly for the energy storage and the cooling elements.

Moreover, the different energy conversion steps produce unnecessarily semiconductor losses and the electrolytic capacitors limit the life time of the whole system. The idea is, to reduce the energy conversion steps. The primary inverter can additionally modulate the 50 Hz output voltage.

To reduce the size, weight and costs of the secondary converter system a new secondary converter topology without an energy storage was developed. To simplify the direct conversion concept on the secondary side an amplitude modulation on the primary side is necessary and has to be included in the control algorithm.

## IV. NEW POWER ELECTRONIC TOPOLOGY

### A. Basic Idea

The new power electronic topology is shown in Fig. 5. The general function is illustrated in Fig. 6. On the primary side the AC converter generates the amplitude modulated primary voltage. The new AC/AC converter at the secondary side only has to turn the pulses of the secondary voltage into the right direction and in this way it produces the output voltage as shown in Fig. 6. The output voltage consists of the fundamental 50 Hz sinusoidal wave and additionally 100 kHz harmonics, which can be filtered by passive elements. For this purpose, the secondary side needs a bidirectional direct AC/AC converter.

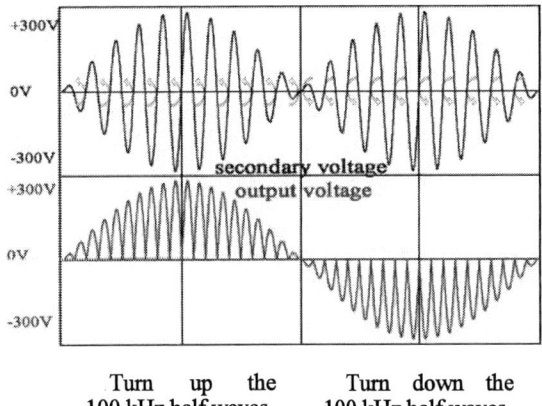

Turn up the 100 kHz half waves

Turn down the 100 kHz half waves

Fig. 6. Operation principle of the new power electronic topology

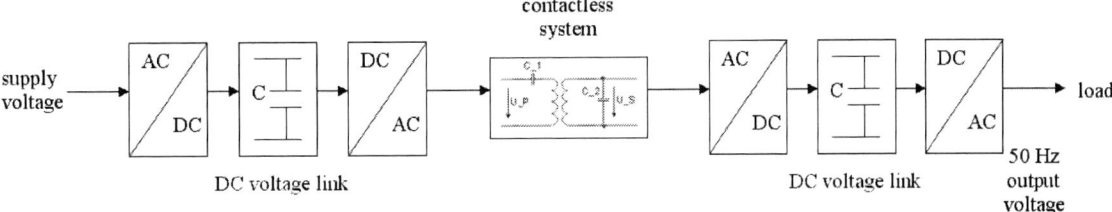

Fig. 4. Conventional power electronic system

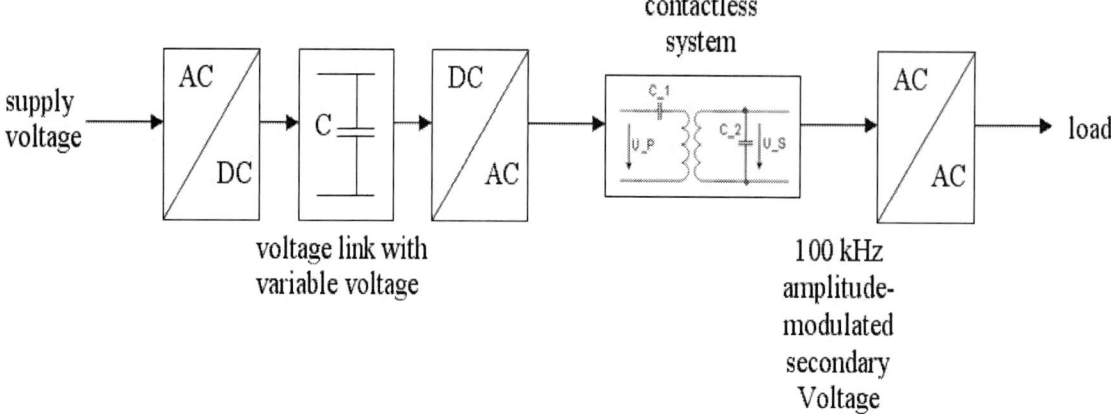

Fig. 5. New power electronic topology

### B. Control Strategy for the Primary Inverter

To generate the amplitude modulated primary voltage, the capacitor of the primary DC link has a small capacitance, so that the DC link voltage has sinusoidal half wave shape. This makes the 50 Hz modulation easier. The DC/AC primary inverter produces the 100 kHz transmission frequency. This is a simple method to realise an amplitude modulated secondary voltage. To generate the correct gate control signals, the phase-shift modulation (Fig. 7) is chosen to generate the 100 kHz transmission frequency with the possibility to change the rms-value of the primary voltage.

Because of the variable voltage on the primary side the amplitude-modulated secondary voltage can be realised very simple. But to be able to determine the correct primary capacitance, it is necessary to know the system's point of resonance operation. For this reason a reducing factor is provided to be able to change the phase-shift angle of the shift-control. Without using a variable voltage of the DC link it would be necessary to change the shifting-angle to small values. But for a small shifting-angle the DC/AC primary inverter has to switch a capacitive current which leads to high power losses.

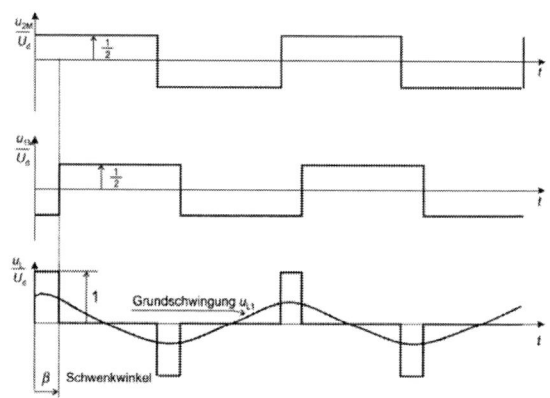

Fig. 7. Pulse pattern for phase-shift modulation

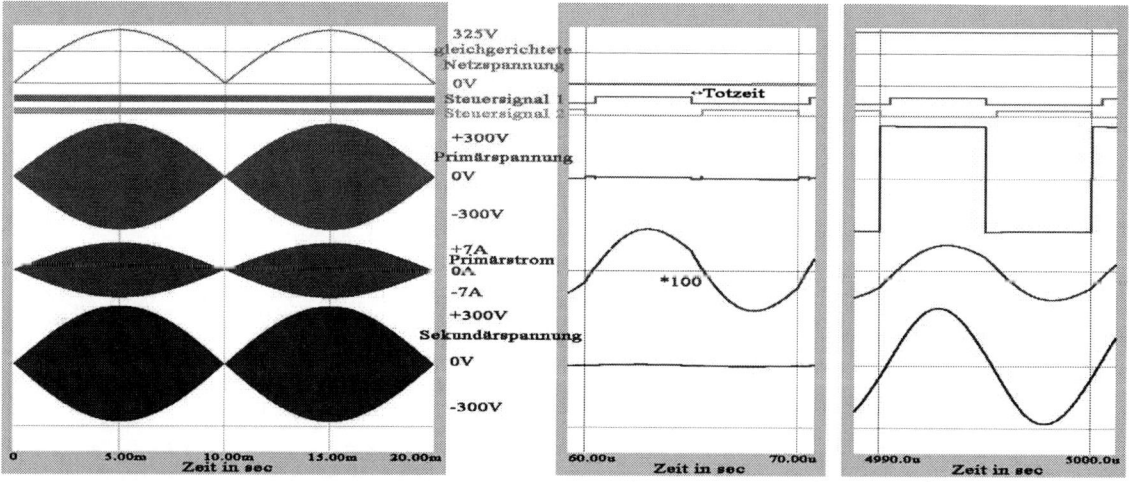

Fig. 8. Control strategy for the primary inverter

## C. Bidirectional Converter at the Secondary Side

The secondary AC/AC converter is shown in Fig. 9. It is build up with bidirectional switches and enables switching the secondary voltage to a defined direction. The right control algorithm leads to a correct current path even at ohmic-inductive loads. The control algorithm has to ensure, that the load current any time has a free-wheeling path. The primary inverter generates the 100 kHz transmission frequency and additionally modulates the 50 Hz output voltage. There is no secondary electrolytic capacitor in order to reduce the size and weight of the secondary side.

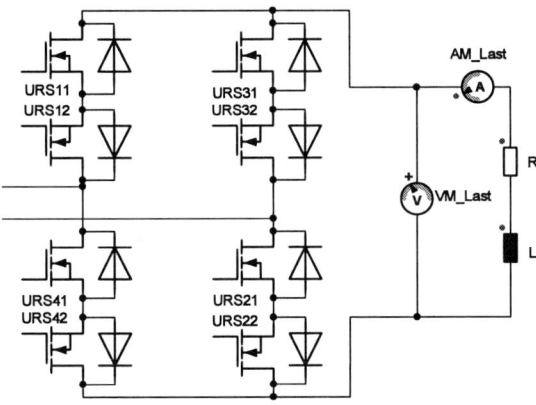

Fig. 9. Secondary AC/AC Converter

## D. Control Strategy for the Bidirectional Converter

The control signals for the secondary AC/AC converter are generated by measuring the secondary voltage and the direction of the load current. For one half wave of the 50 Hz load current, 4 switches are permanently turned on and the others have to toggle the 100 kHz secondary voltage to the desired direction. In Fig. 10 to 13 the secondary voltage in relation to the output voltage and output current in combination to the control signals are shown.

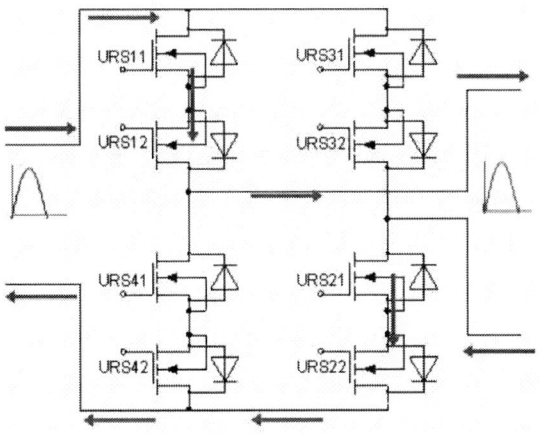

Fig. 10. Positive Input Voltage – Positive Output Voltage

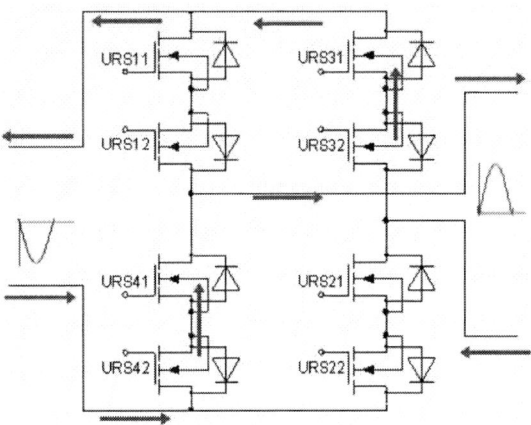

Fig. 11. Negative Input Voltage – Positive Output Voltage

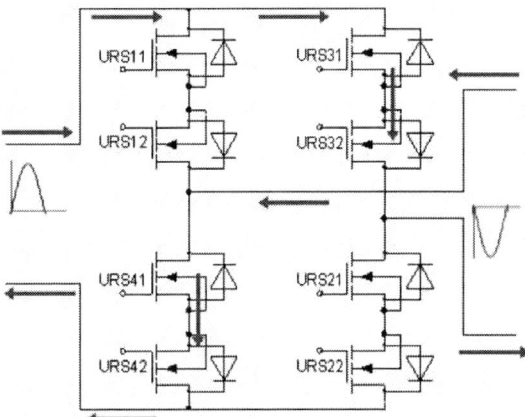

Fig. 12. Positive Input Voltage – Negative Output Voltage

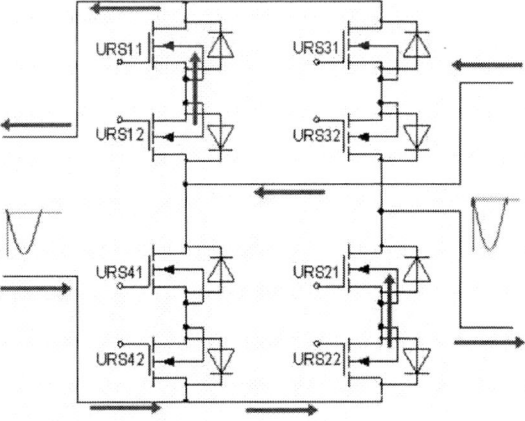

Fig. 13. Negative Input Voltage – Negative Output Voltage

The output voltage consists of the fundamental 50 Hz sinusoidal wave and additional 100 kHz harmonics. The output current (load current) is sinusoidal. The values are shown in Fig. 14.

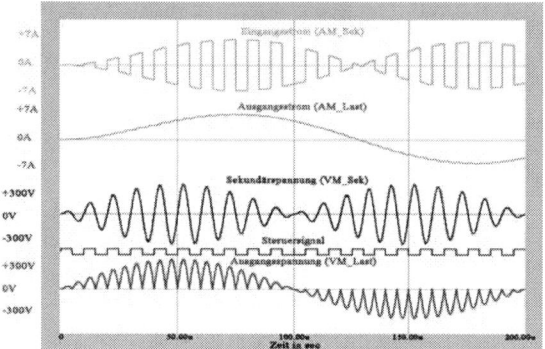

Fig. 14. Voltages and currents at ohmic-inductive Load

## V. EXPERIMENTAL RESULTS

### A. Test Equipment

This chapter describes the simulation of the whole system and it discusses the measurements as well. Fig. 15 shows the practical test configuration. The primary inverter is simulated by a programmable power amplifier. The contactless magnetic assembly with an air gap of 40 mm (Fig. 16) is adjusted to resonance. The secondary AC/AC converter is build up, as described before.

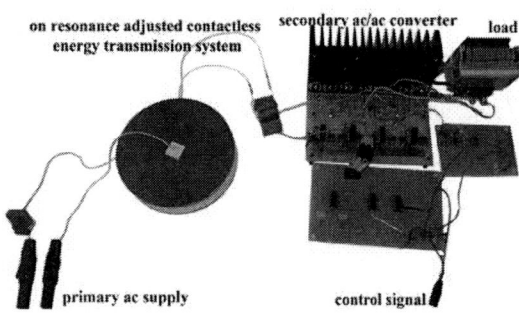

Fig. 15. Test Configuration

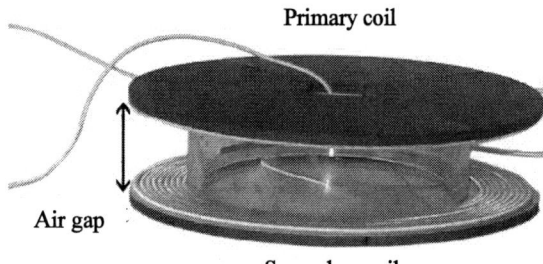

Fig. 16. Contactless Magnetic System

### B. Results at ohmic load

Fig. 17 shows the simulated signals of the amplitude modulated primary voltage and current, which leads to the amplitude-modulated secondary voltage which will be switched to the output voltage. The output voltage and the output current consist of the 50 Hz sinusoidal wave and additional 100 kHz harmonics. At this resolution the area of the 100 kHz transmission waveform is filled out.

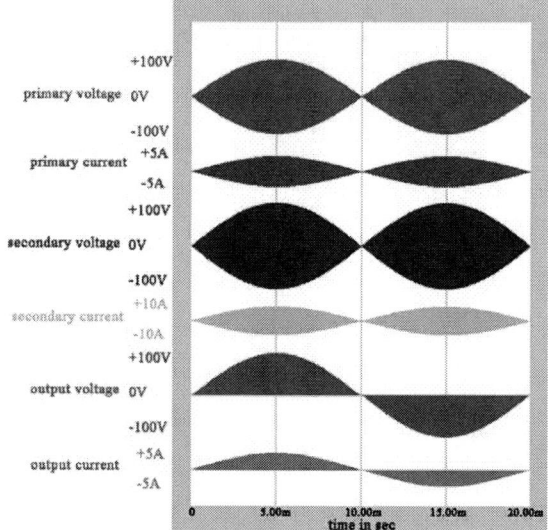

Fig. 17. Simulation results at ohmic load

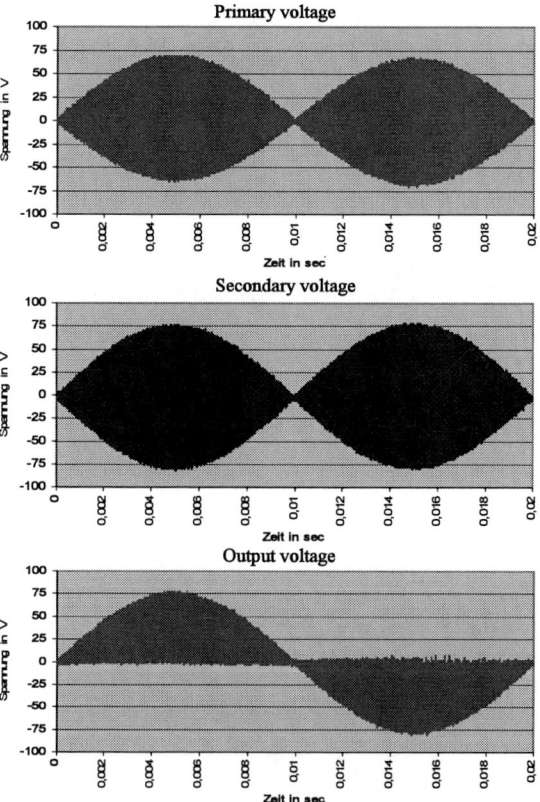

Fig. 18. Experimental results at ohmic load

## C. *Results at ohmic-inductive load*

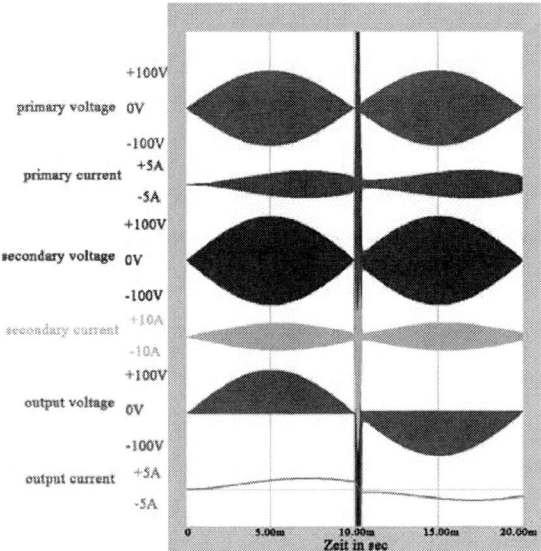

Fig. 19. Simulation results at ohmic-inductive load

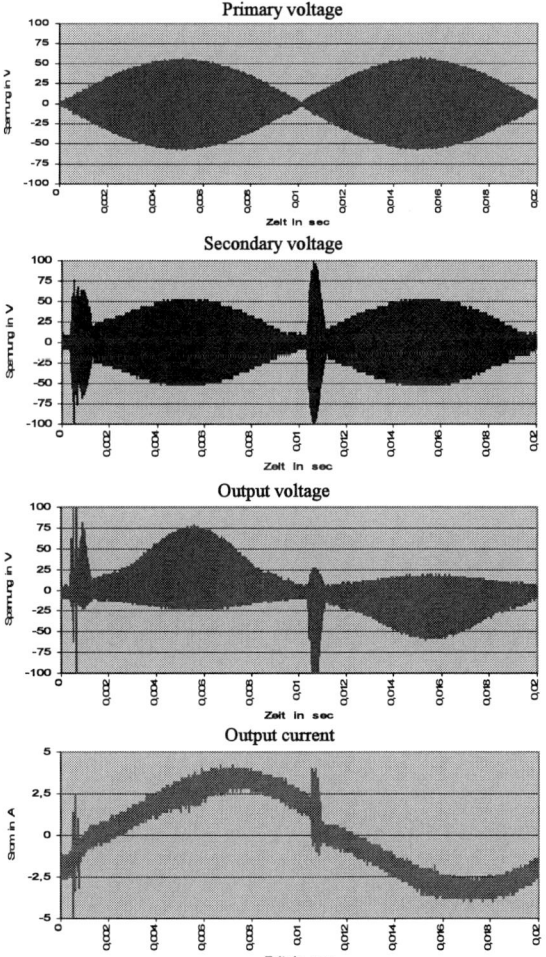

Fig. 20. Experimental results at ohmic-inductive load

The measurements show the correct behaviour of the secondary AC/AC converter which switches the secondary 100 kHz voltage halve-waves to a desired direction. The comparison of the simulated signals to the measurements shows a similar behaviour for the transmission system with the use of an ohmic load.

At ohmic-inductive load, oscillations and overvoltages occur at the secondary side. This results from the asymmetric topology of the magnetic assembly. By means of the series resonance capacitor at the primary side and a parallel capacitor at the secondary side it is only optimised for energy transmission from the primary to the secondary side. During reactive power flow from the secondary to the primary side the overvoltages occur. But this problem can be solved by optimizing the magnetic assembly for both energy directions.

## VI. CONCLUSIONS

A direct conversion of the high frequency secondary voltage of a contactless energy transmission system to the necessary 50 Hz output voltage with an AC/AC converter together with an amplitude modulated secondary voltage is possible. This reduces the number of energy conversion steps, allows a 4-quadrant operation, avoids the need for expensive voluminous electrolytic capacitors and reduces switching losses in power semiconductors. The converter function was simulated and tested at an experimental setup. For ohmic load conditions the theoretical investigations and simulation results so far are conform to the measurements. Further research will extend the application also for ohmic-inductive load conditions.

## REFERENCES

[1] Mecke, Rudolf: Hochfrequente kontaktlose Energieübertragung für Anwendungen aus dem Maschinenbau und dem Haushalts- und Bürobereich, Essen, Haus der Technik, Seminar „Berührungslose Energieübertragung", 17.05.2006

[2] Rathge, Christian; Mecke, Rudolf: Die Technologie der kontaktlosen induktiven Energie und Datenübertragung, 7. Magdeburger Maschinenbau-Tage Magdeburg, Universität Magdeburg, Tagungsband, S. 249-258, 11.-13.10.2005

[3] Mecke, Rudolf; Rathge, Christian: Kontaktlose Energie- und Datenübertragung im Maschinenbau - die Alternative zum Schleifring, 6. Magdeburger Maschinenbau-Tage Magdeburg, Universität Magdeburg, 24.-26.09.2003

[4] Kürschner, Daniel; Mecke Rudolf: Matrix converter with advanced control for contactless energy transmission., EPE 2007 - 12th European Conference on Power Electronics and Applications, Aalborg, Denmark, 02.-05.09.2007

[5] Wolfgang, Stephan: Leistungselektronik interaktiv - Aufgaben unter Simplorer und Mathcad; 1. Auflage, Leipzig, Fachbuch-verlag Leipzig, 2001

# Modeling Approach of a Generator with Non-linear Load in Embedded Electrical Network

Nicolas Amelon*, Mourad Aït-Ahmed[†], Mohamed-Fouad Benkhoris[‡]

*IREENA, 37 boulevard de l'université, 44602 Saint Nazaire, *nicolas.amelon@unit-nantes.fr*

[†] *mourad.ait-ahmed@unit-nantes.fr*

[‡] *mohamed-fouad.benkhoris@unit-nantes.fr*

*Abstract*—This paper deals with a modeling approach of an embarked electrical network. This network is composed of a synchronous generator feeding linear and non linear load. In order to simplify the identification of inaccessible part of the generator, a specific modeling approach of the machine is established. Thus, the rotor is modeled by generalized transfer functions. Its order is chosen to have the desired accuracy and short computation time. Transfer function parameters can be easily identified by SSFR *(StandStill Frequency Response)* identification method. The generated model takes into account the strong interaction between all network components and variable topology of the non linear load. Some time domain simulation results are given in the end of the paper and illustrate the efficiency of this modeling approach.

*Keywords*—Converter machine interaction, embarked network, synchronous machines, modeling, simulation

## I. INTRODUCTION

Embedded electrical network in transport are more and more present. In the planes, hydraulic and pneumatic actuators are progressively changed by electric actuators. In the navy, new ships are equipped of electric propulsion. These networks are generally complexes, with variable topology and strong interaction between generators and loads. Precise dynamical models and methodology to build the system state space model are required to the simulation. In embedded network, electrical generators are classically synchronous machines. Due to the constraint of these system (compactness, accessibility, temperatures, etc.), machine identification procedure can be complex, particularly for the damper windings parameters.

In this paper, a modeling approach of a simplified embarked network is proposed (fig. 1). This network is constituted of a synchronous machine feeding a linear load and a thyristor bridge. To simplify the model construction, it can be interesting to have a generalized formulation of the machine with easily identifiable parameters. In the proposed methodology, machine rotor will be represented by a two-port network on the direct axis and a dipole on the quadrature axis, as suggested in [4].

This approach is more flexible, well suited to StandStill Frequency Response *(SSFR)* tests and allows characterizing the rotor without determining the equivalent rotor circuit (fig. 2).

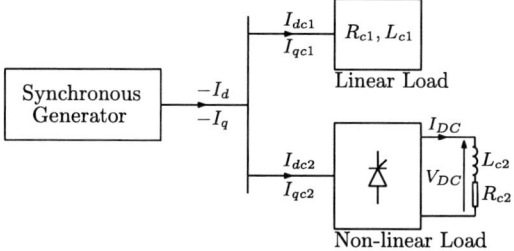

Fig. 1. Electrical Network Scheme

To compute the global model, the state equations of each system component are established in the dq frame of the synchronous machine. The rotor of the generator is described by transfer functions whose order depends on the desired accuracy. Standstill frequency response test are used to identify the machine model parameters ([8], [9], [1]). The topology variation of the non linear load is represented by multi state space model depending on the rectifier operating mode. Loads are both modeled in the machine reference frame, then machine and loads equations are combined to build a global state space model ([6], [2]) with Kirchhoff's laws.

The transformation of *abc* to *dq* variables in the rotor reference frame is defined by

$$x_{dq0} = P(\theta)x_{abc} \qquad (1)$$

where $\theta$ is the electric rotor position and

$$P(\theta) = \frac{2}{3}\begin{pmatrix} \cos\theta & \cos(\theta - \frac{2\pi}{3}) & \cos(\theta + \frac{2\pi}{3}) \\ \sin\theta & \sin(\theta - \frac{2\pi}{3}) & \sin(\theta + \frac{2\pi}{3}) \\ 1/2 & 1/2 & 1/2 \end{pmatrix}. \quad (2)$$

It will be assumed in all paper that the homopolar component of currents and voltages are negligible.

## II. STATE SPACE MODEL OF THE GENERATOR

The first stage of the global model construction is the machine characterization. In suggested methodology, stator remains modeled in conventional way. Non-accessible parts of the machine, principally damper windings, are in consideration.

978-1-4244-1741-4/08/$25.00 ©2008 IEEE

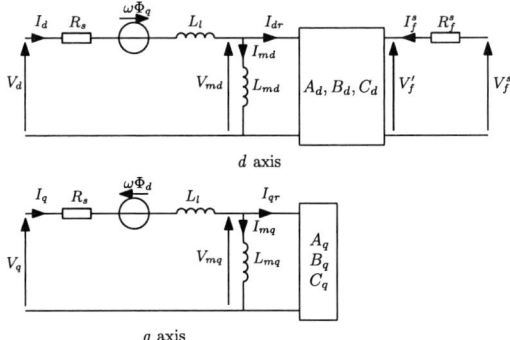

d axis

q axis

Fig. 2. Synchronous machine model

The order of the system, which depends on the desired precision, does not modify the equation set. Rotor operation is globally reproduced by the model and with the desired precision. In the following equations, matrix quantities are noted in bold fonts. $(I_d, I_q)$, $(I_{dr}, I_{qr})$, $(I_{md}, I_{mq})$ and $(\Phi_d, \Phi_q)$ respectively denote stator current, rotor current, magnetizing current and flux linkage. All rotor quantities are referenced to the stator. The stator flux linkage is separated into leakage flux and magnetizing flux as:

$$\Phi_d = L_l I_d + L_{md} I_{md} \quad (3)$$

$$\Phi_q = L_l I_q + L_{mq} I_{mq} \quad (4)$$

### A. Stator Equations

Stator modeling is classic. Fig. 2 gives:

$$\begin{pmatrix} V_d \\ V_q \end{pmatrix} = R_s \begin{pmatrix} I_d \\ I_q \end{pmatrix} + \omega \begin{pmatrix} 0 & -1 \\ 1 & 0 \end{pmatrix} \begin{pmatrix} \Phi_d \\ \Phi_q \end{pmatrix}$$
$$+ L_l \begin{pmatrix} \dot{I}_d \\ \dot{I}_q \end{pmatrix} + \begin{pmatrix} L_{md} & 0 \\ 0 & L_{mq} \end{pmatrix} \begin{pmatrix} \dot{I}_{md} \\ \dot{I}_{mq} \end{pmatrix} \quad (5)$$

### B. Rotor Equations

Rotor is represented by a sub state space. Matrix models are computed from admittance functions:

For $d$ axis: $I_{dr} = y_{11}(s) V_{md} + y_{12}(s) V_f' \quad (6)$

$$I_f^s = y_{12}(s) V_{md} + y_{22}(s) V_f' \quad (7)$$

For $q$ axis: $I_{qr} = Y_q(s) V_{mq} \quad (8)$

$y_{ij}(s)$ and $Y_q(s)$ are transfer function, with $s$ *Laplace* operator, of the following general form:

$$y(s) = \frac{a_0 + a_1 s + \cdots + a_{n_r-1} s^{n_r-1}}{b_0 + b_1 s + \cdots + b_{n_r-1} s^{n_r-1} + b_{n_r} s^{n_r}} \quad (9)$$

Standstill frequency response test are used to obtain machine parameters (particularly admittance function parameters). These tests are widely described in literature ([8], [9], [1]). It is possible to build transfer functions corresponding to each *SSFR* test from measured characteristics and by considering fig. 2 and (6) to (8). Order $n_r$ of these transfer functions are fixed, then a particle swarm optimization algorithm is used to determine machine parameters in agreement with measures. It is important to emphasize that precision of the model depends on $n_r$.

A bigger value of $n_r$ increases precision, but increases computation time too.

The $q$ axis state space model computation is straight forward. It can be easily deduced from equation (8) using the Matlab command `tf2ss`:

$$\frac{dX_q}{dt} = \mathbf{A_q} X_q + \mathbf{B_q} V_{mq} \quad (10)$$

$$I_{qr} = \mathbf{C_q} X_q \quad (11)$$

In order to establish the state space model for $d$ axis, $I_{dr}$ and $I_f^s$ are decomposed in two components:

$$I_{dr} = I_{dr1} + I_{dr2} \quad (12)$$

$$I_f^s = I_{f1}^s + I_{f2}^s \quad (13)$$

Where $I_{dr1} = y_{11} V_{md}$, $I_{dr2} = y_{12} V_f'$, $I_{f1}^s = y_{12} V_{md}$ and $I_{f2}^s = y_{22} V_f'$. Then, by using the Matlab command, the $d$ axis state space model can be built:

$$\frac{dX_{11}}{dt} = \mathbf{A_{11}} X_{11} + \mathbf{B_{11}} V_{md} \ ; \ I_{dr1} = \mathbf{C_{11}} X_{11} \quad (14)$$

$$\frac{dX_{12}}{dt} = \mathbf{A_{12}} X_{12} + \mathbf{B_{12}} V_f' \ ; \ I_{dr2} = \mathbf{C_{12}} X_{12} \quad (15)$$

$$\frac{dX_{21}}{dt} = \mathbf{A_{12}} X_{21} + \mathbf{B_{12}} V_{md} \ ; \ I_{f1}^s = \mathbf{C_{12}} X_{21} \quad (16)$$

$$\frac{dX_{22}}{dt} = \mathbf{A_{22}} X_{22} + \mathbf{B_{22}} V_f' \ ; \ I_{f2}^s = \mathbf{C_{22}} X_{22} \quad (17)$$

These matrices describe the $d$ axis state space model of the synchronous machine rotor. It can be represented as follow:

$$\frac{dX_d}{dt} = \mathbf{A_d} X_d + (\mathbf{B_{d1}} \quad \mathbf{B_{d2}}) \begin{pmatrix} V_{md} \\ V_f' \end{pmatrix} \quad (18)$$

$$\begin{pmatrix} I_{dr} \\ I_f^s \end{pmatrix} = \begin{pmatrix} \mathbf{C_{d1}} \\ \mathbf{C_{d2}} \end{pmatrix} X_d \quad (19)$$

With:

$$X_d = (X_{11} \ X_{12} \ X_{21} \ X_{22})^t \quad (20)$$

$$\mathbf{A_d} = \begin{pmatrix} \mathbf{A_{11}} & 0 & 0 & 0 \\ 0 & \mathbf{A_{12}} & 0 & 0 \\ 0 & 0 & \mathbf{A_{12}} & 0 \\ 0 & 0 & 0 & \mathbf{A_{22}} \end{pmatrix} \quad (21)$$

$$(\mathbf{B_{d1}} \quad \mathbf{B_{d2}}) = \begin{pmatrix} \mathbf{B_{11}} & 0 \\ 0 & \mathbf{B_{12}} \\ \mathbf{B_{12}} & 0 \\ 0 & \mathbf{B_{22}} \end{pmatrix} \quad (22)$$

$$\begin{pmatrix} \mathbf{C_{d1}} \\ \mathbf{C_{d2}} \end{pmatrix} = \begin{pmatrix} \mathbf{C_{11}} & \mathbf{C_{12}} & 0 & 0 \\ 0 & 0 & \mathbf{C_{12}} & \mathbf{C_{22}} \end{pmatrix} \quad (23)$$

The terminal voltage $V_f'$ of the two port network is immeasurable. Then, the state space model must be improved by introducing the terminal measurable voltage $V_f^s$ which is given by:

$$V_f' = V_f^s - R_f^s I_f^s \quad (24)$$

Finally, using (18), (19) and (24), the rotor $d$ axis state space model becomes:

$$\frac{dX_d}{dt} = (\mathbf{A_d} - R_f^s \mathbf{B_{d2}} \mathbf{C_{d2}}) X_d + (\mathbf{B_{d1}} \quad \mathbf{B_{d2}}) \begin{pmatrix} V_{md} \\ V_f^s \end{pmatrix} \quad (25)$$

It is important to emphasize that $X_{d,q}$ denote states vectors of each axis $(d, q)$. The order of each one depends on the chosen $n_r$.

1741

$$\mathbf{R}_{\text{machine}} = \begin{pmatrix} R_s & -\omega(L_l + L_{mq}) & 0 & \omega L_{mq}\mathbf{C_q} \\ -\omega(L_l + L_{md}) & R_s & -\omega L_{md}\mathbf{C_{d1}} & 0 \\ 0 & 0 & R_f\mathbf{B_{d2}C_{d2}} - \mathbf{A_d} & 0 \\ 0 & 0 & 0 & -\mathbf{A_q} \end{pmatrix} \qquad (26)$$

$$\mathbf{L}_{\text{machine}} = \begin{pmatrix} L_l + L_{md} & 0 & -L_{md}\mathbf{C_{d1}} & 0 \\ 0 & L_l + L_{mq} & 0 & -L_{mq}\mathbf{C_q} \\ -L_{md}\mathbf{B_{d1}} & 0 & L_{md}\mathbf{B_{d1}C_{d1}} + \mathcal{I} & 0 \\ 0 & -L_{mq}\mathbf{B_q} & 0 & L_{mq}\mathbf{B_qC_q} + \mathcal{I} \end{pmatrix} \qquad (27)$$

## C. Coupling Management

Finally, equations (5), (10) and (25) must be expressed in function of the selected state vector (for example $[I_d \ I_q \ [X_d]^t \ [X_q]^t]^t$) to eliminate linkage quantities. From (11) and (19), theses equations become:

$$V_d = R_s I_d - \omega(L_l + L_{mq})I_q + \omega L_{mq}\mathbf{C_q}X_q$$
$$+(L_l + L_{md})\frac{dI_d}{dt} - L_{md}\mathbf{C_{d1}}\frac{dX_d}{dt} \quad (28)$$

$$V_q = -\omega(L_l + L_{md})I_d + R_s I_q - \omega L_{md}\mathbf{C_{d1}}X_d$$
$$+(L_l + L_{mq})\frac{dI_q}{dt} - L_{mq}C_q\frac{dX_q}{dt} \quad (29)$$

$$\mathbf{B_{d2}}V_f^s = (R_f\mathbf{B_{d2}C_{d2}} - \mathbf{A_d})X_d - \mathbf{B_{d1}}L_{md}\frac{dI_d}{dt}$$
$$+(\mathbf{B_{d1}}L_{md}\mathbf{C_{d1}} + \mathcal{I})\frac{dX_d}{dt} \quad (30)$$

$$0 = -\mathbf{A_q}X_q - \mathbf{B_q}L_{mq}\frac{dI_q}{dt}$$
$$+(\mathbf{B_q}L_{mq}\mathbf{C_q} + \mathcal{I})\frac{dX_q}{dt} \quad (31)$$

with $\mathcal{I}$ identity matrix. The global state space model can be easily deduced from (28) to (31) and expressed as follow:

$$\begin{pmatrix} V_d \\ V_q \\ [\mathbf{B_{d2}}V_f^s] \\ [0] \end{pmatrix} = R_{\text{machine}} \begin{pmatrix} I_d \\ I_q \\ [X_d] \\ [X_q] \end{pmatrix}$$
$$+ L_{\text{machine}}\frac{d}{dt}\begin{pmatrix} I_d \\ I_q \\ [X_d] \\ [X_q] \end{pmatrix} \quad (32)$$

The expression of $R_{\text{machine}}$ and $L_{\text{machine}}$ are given in (26) and (27).

## III. STATE SPACE MODEL OF THE LOADS

### A. Linear Load

The model of a three-phase linear R-L load is simple. Its equation in the $dq$ reference frame is:

$$\begin{pmatrix} V_{dc1} \\ V_{qc1} \end{pmatrix} = \begin{pmatrix} R_{c1} & -\omega L_{c1} \\ \omega L_{c1} & R_{c1} \end{pmatrix} \begin{pmatrix} I_{dc1} \\ I_{qc1} \end{pmatrix}$$
$$+ \begin{pmatrix} L_{c1} & 0 \\ 0 & L_{c1} \end{pmatrix}\frac{d}{dt}\begin{pmatrix} I_{dc1} \\ I_{qc1} \end{pmatrix} \quad (33)$$

with $V_{dc1} \ V_{qc1}$ the terminal load voltage in the machine frame and $I_{dc1} \ I_{qc1}$ the current flowing through the load.

### B. Non-linear Load

The thyristors are supposed to be perfect switches and have two possible states: opened or closed. From the generator point of view, the load is variable. It will be represented by multi state space model depending on the rectifier operating mode. Then, it is assumed that two operating modes are possible:

- normal mode (or mode 2): two switches closed (fig. 3a);
- switching mode (or mode 3): three switches closed (fig. 3b).

The output voltage of the rectifier is always given by:

$$V_{DC} = R_{c2} \cdot I_{DC} + L_{c2} \cdot \frac{dI_{DC}}{dt} \quad (34)$$

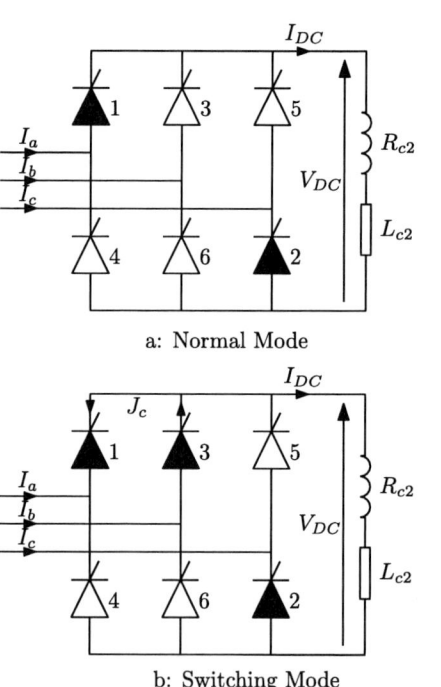

a: Normal Mode

b: Switching Mode

Fig. 3. Thyristor Bridge in conduction state (top) and switching state (bottom)

*1) Normal mode study:* In example of fig. 3a, the relations between inputs and outputs of the thyristor bridge are:

$$V_a = V_{DC} + V_c \qquad (35)$$
$$I_a = I_{DC} = -I_c \qquad (36)$$
$$I_b = 0 \qquad (37)$$

In the global state space model, all quantities are defined in the same Park reference frame. Thus, (35), (36) and (37) must be expressed with $V_{dqc2}$ and $I_{dqc2}$:

$$V_{DC} = \sqrt{3}V_{dc2}\sin\left(\theta + \frac{\pi}{3}\right)$$
$$\qquad -\sqrt{3}V_{qc2}\cos\left(\theta + \frac{\pi}{3}\right) \qquad (38)$$
$$I_{dc2} = \frac{2}{\sqrt{3}}I_{DC}\sin\left(\theta + \frac{\pi}{3}\right) \qquad (39)$$
$$I_{qc2} = -\frac{2}{\sqrt{3}}I_{DC}\cos\left(\theta + \frac{\pi}{3}\right) \qquad (40)$$

These equations can be generalized for each pair of conducing thyristors in normal mode. Let $\xi$ be a constant depending on the thyristor bridge topology. The different values of $\xi$ are given in the table I.

TABLE I

VALUES OF $\xi$ IN NORMAL MODE

| | Conducting Thyristors | | | | | |
|---|---|---|---|---|---|---|
| Mode 2 | 1,2 | 2,3 | 3,4 | 4,5 | 5,6 | 6,1 |
| $\xi$ | $\pi/3$ | 0 | $-\pi/3$ | $-2\pi/3$ | $\pi$ | $2\pi/3$ |

Thus, (38), (39) and (40) becomes:

$$V_{DC} = \sqrt{3}V_{dc2}\sin\left(\theta + \xi\right) - \sqrt{3}V_{qc2}\cos\left(\theta + \xi\right) \quad (41)$$
$$I_{dc2} = \frac{2}{\sqrt{3}}I_{DC}\sin\left(\theta + \xi\right) \qquad (42)$$
$$I_{qc2} = -\frac{2}{\sqrt{3}}I_{DC}\cos\left(\theta + \xi\right) \qquad (43)$$

These equations associate inputs and outputs of the thyristor bridge in normal mode for all pairs of conducing semiconductors.

*2) Commutation mode study:* In commutation mode, a new current must be introduced. Let $J_c$ be the current flowing through a switching semiconductor (corresponding to the current of the thyristor 3 in the example of fig. 3b). The commutation is over when $I_{DC} = J_c$. The thyristor bridge is then in normal mode and $J_c = 0$. Evolution of $I_{DC}$ and $J_c$ are given in fig. 4. Fig. 3b gives the following relation:

$$V_a = V_{DC} + V_c \qquad (44)$$
$$V_a = V_b \qquad (45)$$
$$I_a = I_{DC} - J_c \qquad (46)$$
$$I_b = J_c \qquad (47)$$
$$I_c = -I_{DC} \qquad (48)$$

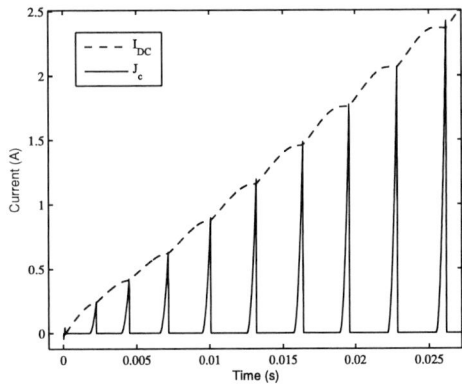

Fig. 4. Evolution of the rectified and commutation current

In the Park reference frame, (44) to (48) becomes

$$V_{DC} = \sqrt{3}V_d\sin\left(\theta + \frac{\pi}{3}\right) - \sqrt{3}V_q\cos\left(\theta + \frac{\pi}{3}\right) \quad (49)$$
$$0 = \sqrt{3}V_d\sin\left(\theta + \frac{2\pi}{3}\right) - \sqrt{3}V_q\cos\left(\theta + \frac{2\pi}{3}\right) \quad (50)$$
$$I_{dc2} = \frac{2}{\sqrt{3}}I_{DC}\sin\left(\theta + \frac{\pi}{3}\right) - \frac{2}{\sqrt{3}}J_c\sin\left(\theta + \frac{2\pi}{3}\right) \quad (51)$$
$$I_{qc2} = \frac{2}{\sqrt{3}}J_c\cos\left(\theta + \frac{2\pi}{3}\right) - \frac{2}{\sqrt{3}}I_{DC}\cos\left(\theta + \frac{\pi}{3}\right) \quad (52)$$

As in normal mode, a constant can be defined to generalize (49) to (52):

TABLE II

VALUES OF $\xi$ IN COMMUTATION MODE

| | Conducting Thyristors | | | | | |
|---|---|---|---|---|---|---|
| Mode 3 | 1,2,3 | 2,3,4 | 3,4,5 | 4,5,6 | 5,6,1 | 6,1,2 |
| $\xi$ | $\pi/3$ | 0 | $-\pi/3$ | $-2\pi/3$ | $\pi$ | $2\pi/3$ |

From table II and (49) to (52), it is possible to deduce a set of equations that describes the Thyristor bridge when three semiconductors conducting:

$$V_{DC} = \sqrt{3}V_d\sin\left(\theta + \xi\right) - \sqrt{3}V_q\cos\left(\theta + \xi\right) \quad (53)$$
$$0 = \sqrt{3}V_d\sin\left(\theta + \xi + \frac{\pi}{3}\right)$$
$$\qquad -\sqrt{3}V_q\cos\left(\theta + \xi + \frac{\pi}{3}\right) \qquad (54)$$
$$I_{dc2} = \frac{2}{\sqrt{3}}I_{DC}\sin\left(\theta + \xi\right)$$
$$\qquad -\frac{2}{\sqrt{3}}J_c\sin\left(\theta + \xi + \frac{\pi}{3}\right) \qquad (55)$$
$$I_{qc2} = -\frac{2}{\sqrt{3}}I_{DC}\cos\left(\theta + \xi\right)$$
$$\qquad +\frac{2}{\sqrt{3}}J_c\cos\left(\theta + \xi + \frac{\pi}{3}\right) \qquad (56)$$

Thus, the state equation of the non linear load in the machine reference frame is

$$
\sqrt{3}\left(\sin\left(\theta+\xi\right)-\cos\left(\theta+\xi\right)\right)\begin{pmatrix}V_{dc2}\\V_{qc2}\end{pmatrix} =
$$
$$
R_{c2}I_{DC}+L_{c2}\frac{dI_{DC}}{dt} \ . \tag{57}
$$

In commutation mode, the condition of (54) on the voltage must be considered. Expressions of the currents feeding the non linear load are required to build the final model. These equations depend on the rectifier working mode and must be expressed in the machine reference frame. In normal mode, (42) and (43) are used, while (55) and (56) are used in commutation mode.

## IV. STATE SPACE MODEL GENERATION PRINCIPLE

All required equations to modeling loads are now expressed in the synchronous machine's Park reference frame. Thus, Kirchhoff laws can be used to bind quantities of the system (fig. 1):

$$
\begin{aligned}
-I_d &= I_{dc1}+I_{dc2} & (58)\\
-I_q &= I_{qc1}+I_{qc2} & (59)\\
V_d &= V_{dc1} = V_{dc2} & (60)\\
V_q &= V_{qc1} = V_{qc2} & (61)
\end{aligned}
$$

Noting that

- the current $I_{dqc1}$ flowing through the linear load can be deduced from $I_{dq}$ and $I_{dqc2}$;
- the thyristor bridge input current is a function of $I_{DC}$, $J_c$ and $\theta$,

the selected state vector is:

$$
X = \begin{bmatrix}I_{DC} & I_d & I_q & [X_d]^t & [X_q]^t\end{bmatrix}^t
$$

in normal mode and

$$
X = \begin{bmatrix}I_{DC} & J_c & I_d & I_q & [X_d]^t & [X_q]^t\end{bmatrix}^t
$$

in commutation mode.

### A. Normal Mode Model determination

In order to construct the model describing the system when only two thyristors conduct, linkage quantities must be canceled.

The proposed procedure is:

- Replace $V_{DC}$ with expression (34) in (41). The state equation is written from (41) and (32):

$$
\begin{pmatrix}\sqrt{3}V_d\sin(\theta+\xi)-\sqrt{3}V_q\cos(\theta+\xi)\\V_d\\V_q\\\left[\mathbf{B_{d2}}V_f^s\right]\\[0]\end{pmatrix}
$$
$$
=\begin{pmatrix}R_{c2} & 0 & \cdots & 0\\0 & & &\\\vdots & & \mathbf{R}_{\text{machine}} &\\0 & & &\end{pmatrix}\begin{pmatrix}I_{DC}\\I_d\\I_q\\[X_d]\\[X_q]\end{pmatrix}
$$
$$
+\begin{pmatrix}L_{c2} & 0 & \cdots & 0\\0 & & &\\\vdots & & \mathbf{L}_{\text{machine}} &\\0 & & &\end{pmatrix}\frac{d}{dt}\begin{pmatrix}I_{DC}\\I_d\\I_q\\[X_d]\\[X_q]\end{pmatrix} \tag{62}
$$

- Express $I_{dqc1}$ in function of $I_{dq}$ and $I_{dqc2}$ with (58) and (59);
- Replace $I_{dqc2}$ with its expression in normal mode ((42) and (43));
- Inject the result of these two manipulation in (33) to compute the expression of $V_{dq}$ in function of the state variables. Linear load equation becomes

$$
\begin{pmatrix}V_d\\V_q\end{pmatrix} = \begin{pmatrix}R_{c1} & -\omega L_{c1}\\\omega L_{c1} & R_{c1}\end{pmatrix}\chi
$$
$$
+\begin{pmatrix}L_{c1} & 0\\0 & L_{c1}\end{pmatrix}\dot{\chi} \tag{63}
$$

with

$$
\chi = \begin{pmatrix}-I_d-\frac{2}{\sqrt{3}}I_{DC}\sin(\theta+\xi)\\-I_q+\frac{2}{\sqrt{3}}I_{DC}\cos(\theta+\xi)\end{pmatrix}; \tag{64}
$$

- Cancel $V_{dq}$ in (62) with (63) and (64);
- Rearrange (62) to obtain the global system equation in normal mode (or mode 2):

$$
\begin{pmatrix}0\\0\\0\\\left[\mathbf{B_{d2}}V_f^s\right]\\[0]\end{pmatrix} = \mathbf{R}_{\text{mode }2}(\xi,\theta)\begin{pmatrix}I_{DC}\\I_d\\I_q\\[X_d]\\[X_q]\end{pmatrix}
$$
$$
+\mathbf{L}_{\text{mode }2}(\xi,\theta)\frac{d}{dt}\begin{pmatrix}I_{DC}\\I_d\\I_q\\[X_d]\\[X_q]\end{pmatrix} \ . \tag{65}
$$

1744

### B. Commutation Mode Model Determination

The method to construct the state space model in commutation mode is similar to the method proposed in previous part:

- Replace $V_{DC}$ with expression (34) in (53). The state equation is written from (53), (54) and (32):

$$
\begin{pmatrix}
\sqrt{3}V_d \sin(\theta+\xi) - \sqrt{3}V_q \cos(\theta+\xi) \\
\sqrt{3}V_d \sin(\theta+\xi+\frac{\pi}{3}) - \sqrt{3}V_q \cos(\theta+\xi+\frac{\pi}{3}) \\
V_d \\
V_q \\
[\mathbf{B_{d2}}V_f^s] \\
[0]
\end{pmatrix}
$$

$$
= \begin{pmatrix}
R_{c2} & 0 & \cdots & 0 \\
0 & 0 & \cdots & 0 \\
\vdots & \vdots & & \mathbf{R}_{\text{machine}} \\
0 & 0 & &
\end{pmatrix}
\begin{pmatrix}
I_{DC} \\
J_c \\
I_d \\
I_q \\
[X_d] \\
[X_q]
\end{pmatrix}
$$

$$
+ \begin{pmatrix}
L_{c2} & 0 & \cdots & 0 \\
0 & 0 & \cdots & 0 \\
\vdots & \vdots & & \mathbf{L}_{\text{machine}} \\
0 & 0 & &
\end{pmatrix}
\frac{d}{dt}
\begin{pmatrix}
I_{DC} \\
J_c \\
I_d \\
I_q \\
[X_d] \\
[X_q]
\end{pmatrix} ; \quad (66)
$$

- Express $I_{dqc1}$ in function of $I_{dq}$ and $I_{dqc2}$ with (58) and (59) and replace $I_{dqc2}$ with its expression in commutation mode: (55) and (56);
- Use expression of $I_{dqc1}$ in function of state variables to deduce $V_{dq}$. The linear load equation is similar to (63), except for $\chi$:

$$
\chi = \begin{pmatrix}
-I_d - \frac{2}{\sqrt{3}}I_{DC}\sin(\theta+\xi) \\
\quad + \frac{2}{\sqrt{3}}J_c \sin(\theta+\xi+\frac{\pi}{3}) \\
-I_q + \frac{2}{\sqrt{3}}I_{DC}\cos(\theta+\xi) \\
\quad - \frac{2}{\sqrt{3}}J_c \cos(\theta+\xi+\frac{\pi}{3})
\end{pmatrix} ; \quad (67)
$$

- Inject the new expression of $V_{dq}$ in (66). After rearrangement and simplification, the equation of the system in commutation mode (or mode 3) can be deduced:

$$
\begin{pmatrix}
0 \\
0 \\
0 \\
0 \\
[\mathbf{B_{d2}}V_f^s] \\
[0]
\end{pmatrix}
= \mathbf{R}_{\text{mode }3}(\xi,\theta)
\begin{pmatrix}
I_{DC} \\
J_c \\
I_d \\
I_q \\
[X_d] \\
[X_q]
\end{pmatrix}
$$

$$
+ \mathbf{L}_{\text{mode }3}(\xi,\theta)\frac{d}{dt}
\begin{pmatrix}
I_{DC} \\
J_c \\
I_d \\
I_q \\
[X_d] \\
[X_q]
\end{pmatrix} . \quad (68)
$$

### V. SIMULATION AND RESULTS

To simulate the system, state space models are computed and switched in function of the static converter topology. This method is precise and suited to systems with limited number of switches. It can be interesting to have a methodology to systematically compute the state space model if many electronic switches are used. An algorithm to build a complex electrical system is proposed in [5], [7] or [10].

Simulation procedure is given in fig. 5. First, the simulation is initialized with machine and loads parameters, simulation time, initial state vector, speed and field voltage. From these parameters, a state space model is computed. Thyristors commutation conditions are tested during the simulation. If thyristor extinction is detected or activation is commanded, the program save the state variables and compute the state space model corresponding to the new topology. The mechanical equation is resolved during electrical model computation to determinate the rotor electrical position $\theta$.

Simulations of the model are presented in fig. 6. Angular speed of the rotor and field voltage are kept constant to respectively 314 rad.s$^{-1}$ and 10 V. Initial load parameters are:

- for the linear load: $R_{c1} = 50\Omega$ and $L_{c1} = 0.05$H;
- for the thyristor bridge load: $R_{c2} = 50\Omega$ and $L_{c2} = 0.05$H.

The simulation involves a sudden change in the $R_{c1}L_{c1}$ load. At $t = 0.4$s, linear load becomes $R_{c1} = 10\Omega$ and $L_{c1} = 0.01$H. Machine parameters are given in [3] and are reminded here for convenience:

$$
\begin{array}{llll}
R_s &=& 0.108 \ \Omega & L_{mq} &=& 8.75 \ \text{mH} \\
R_f^s &=& 15.3 \ \text{m}\Omega & L_l &=& 0.97 \ \text{mH} \\
L_{md} &=& 14.26 \ \text{mH} & &
\end{array}
$$

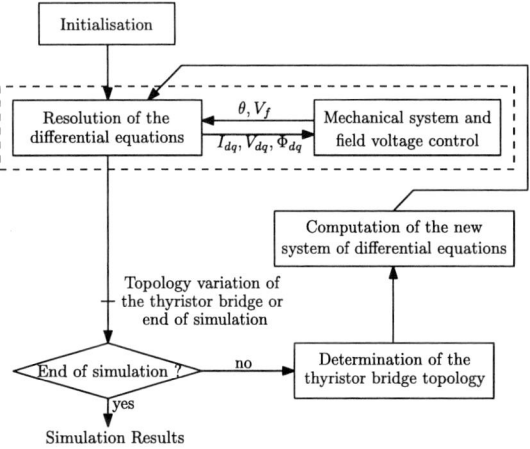

Fig. 5. Simulation principles of the system

$$y_{11} = 1239.6\frac{1 + 0.0182p}{p(1 + 0.0015p)}$$

$$y_{12} = -1239.6\frac{1 + 0.0128p}{p(1 + 0.0015p)}$$

$$y_{22} = 1239.6\frac{1 + 0.0092p}{p(1 + 0.0015p)}$$

$$Y_q = \frac{5.82}{1 + 0.0014p}$$

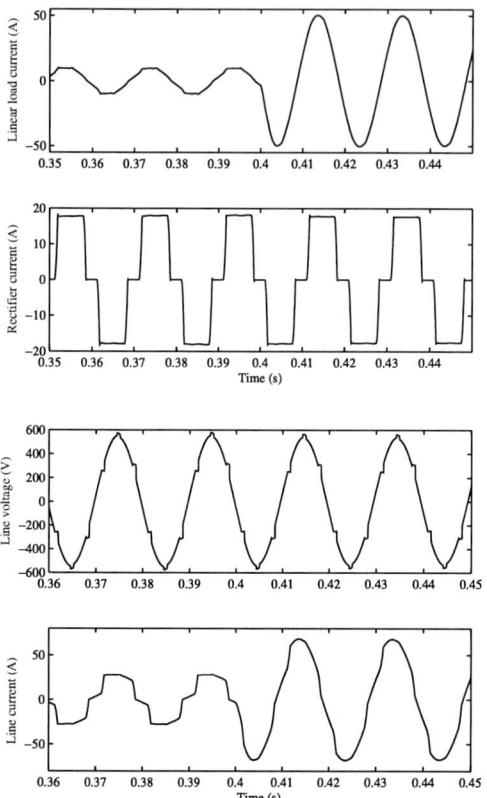

Fig. 7. Details of the simulated line voltage and current and load currents

Fig. 7 gives the details of the simulated currents in the generator and loads and the line voltage. The non linear load absorbs a non sinusoidal current that causes a line voltage deformation. This deformed voltage is applied to the linear load. Thus, this load consumes non-sinusoidal current. Moreover, the step of linear load causes a voltage drop that have an influence on the rectifier comportment.

It appears that all components of this system were in strong interaction. This justifies the necessity of using a modeling methodology that allows defining a global state space model.

## VI. CONCLUSION

In this paper, a dynamic modeling procedure of a simplified embedded network is proposed. This network consists of a synchronous machine, a linear load and a non-linear load. The synchronous generator model allows avoiding rotor structure determination problem. Generator's rotor is modeled using transfer functions expressed in state space model. Its parameters can be identified from *SSFR* tests.

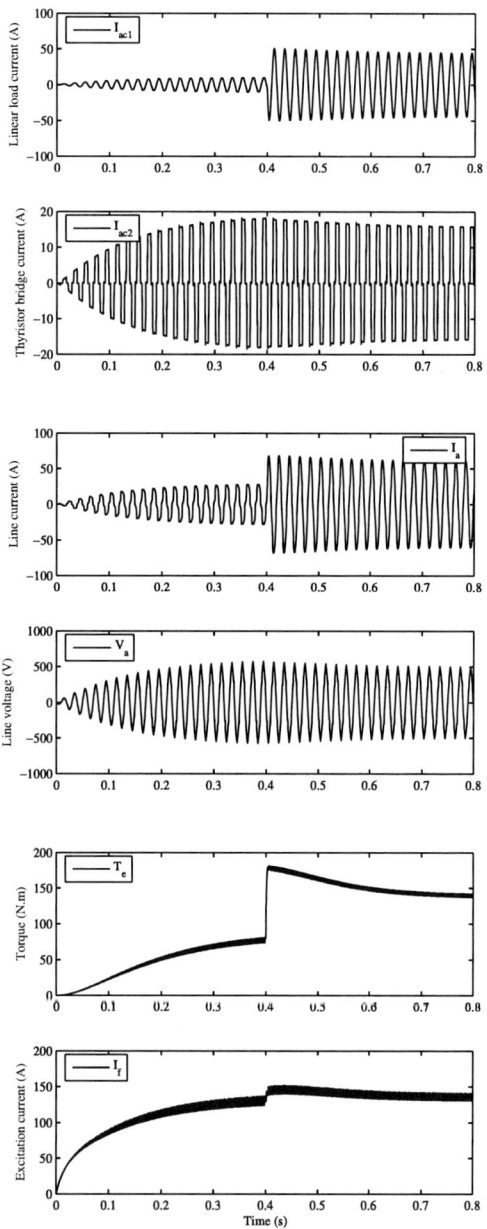

Fig. 6. Simulation results for a sudden linear load change

The choice of the rotor transfer functions order is very important. High order transfer functions can improve the precision of the synchronous machine model. Nevertheless, the number of parameters describing the rotor increases and the optimization procedure can become complex. Moreover, it is imperative to have very good experimental results. Thus, *SSFR* tests must be carried out several times to improve their precision. Lastly, a high order model increases simulation time. A compromise must be made between the model time computing and the model precision.

The linear and non-linear loads are modeled in the Park reference frame of the synchronous machine. Thus, simple relations deduced from Kirchhoff laws allow to simplify the model construction procedure. Equations are deduced from two particular states of the static converter (in normal mode and commutation mode) before being generalized. A state space model of the system is deduced for each converter configuration. The program checks the thyristor bridge topology and switch between state space models. This method is precise, well suited to systems with limited number of switches, and takes account of all interactions in the embedded network.

## REFERENCES

[1] Ieee guide: Test procedures for synchronous machines. Technical Report Std 115, The Institute of Electrical and Electronics Engineers, 1995.

[2] L. Abdeljalil, M-F. Benkhoris, and M. Aït Ahmed. Méthodologie de modélisation dynamique des réseaux électriques embarqués. In *RIGE*, 2006.

[3] Dionysios C. Aliprantis, Scott D. Sudhoff, and Brian T. Kuhn. Experimental characterization procedure for a synchronous machine model with saturation and arbitrary rotor network representation. *IEEE transactions on energy conversion*, 20:595–603, 2005.

[4] Dionysios C. Aliprantis, Scott D. Sudhoff, and Brian T. Kuhn. A synchronous machine model with saturation and arbitrary rotor network representation. *IEEE transactions on energy conversion*, 20:584–694, 2005.

[5] J.H. Allmeling and W.P. Hammer. Plecs - piece-wise linear electrical circuit simulation for simulink. In *International Conference on Power Electronics and Drive Systems*, 1999.

[6] M-F. Benkhoris and F. Terrien. Modélisation dans l'espace d'état des systèmes multiconvertisseurs. *RIGE*, 6(5-6):701–729, 2003.

[7] W.P. Hammer. *Dynamic Modeling of Line and Capacitor Commutated Converters for HVDC Power Transmission*. PhD thesis, Swiss Federal Institute of Technology, 2003.

[8] A. Keyhani and H. Tsai. Identification of high-order synchronous generator models from ssfr test data. *IEEE Transaction on Energy Conversion*, 9(3):593–603, 1994.

[9] P.A.E. Rusche, G.J. Brock, L.N. Hannett, and J.R. Willis. Test and simulation of network dynamic response using ssfr and rtdr derived synchronous machine models. *IEEE Transaction on Energy Conversion*, 5(1):145–155, 1990.

[10] O. Wasynczuk and Sudhoff S.D. Automated state model generation algorithm for power circuits and systems. *IEEE Transaction on Power Systems*, 11(4):1951–1956, November 1996.

# Optimal Use of the 14 V Alternator in 42 V Automotive Supply Systems

Vasile Comnac, *Member, IEEE*, Mihai Cernat, *Senior Member, IEEE*, and Adrian Mailat

"Transilvania" University of Brasov / Faculty of Electrical Engineering and Computer Science
Blvd. Eroilor nr. 29, 500036 Brasov, Romania
Email: comnac@unitbv.ro, m.cernat@unitbv.ro, mailat@vega.unitbv.ro

*Abstract*—This paper is aimed at the study of conventional 14 V automotive electrical supply systems as a basis for the development of a novel 42 V supply system capable to provide maximum power to both the vehicle's battery and the on-board consumers. By analyzing the operating characteristics of the 14 V Lundell alternator system, the conditions for maximum output power at various alternator driving speeds are determined. The modification of the conventional system by introducing a boost converter is proposed. The operating characteristics of the new system are analyzed by performing simulations using the MATLAB SIMULINK programming environment.

*Keywords*—Electrical machine, Automotive component, Automotive electronics, Automotive applications, Generation of Electrical Energy, Energy control unit..

## I. INTRODUCTION

The continuing demand for increased electrical power in automobiles as well as the emergence of new materials and techniques with widespread applications in electrical engineering, electronics and automation, allow for and also require the development and adoption of a novel type of automotive electrical power supply system, capable of increased power performances by operating at a higher output voltage.

After elaborate investigations, the 42 V system was proposed and is currently considered to be a future solution, provided that it is generally adopted by automotive and equipment manufacturers [1-6]. While the 42 V voltage level was established several years ago, the structure of the new system is still a fluid one, with little prospects of industrial standardization [7-9].

The problem consists in the currently used, low-voltage conventional, 14 V and 28 V systems. It is almost certain that the above-mentioned systems will be gradually abandoned to be replaced by the new 42 V system. Since the conventional low-voltage systems are found in most automobiles, literature refers more and more frequently to the dual-voltage electrical supply system.

In dual-voltage supply systems the storing function for the energy required by consumers, which must be supplied during engine standstill (lighting, alerts, clock, radio etc.), is separated from the vehicle's internal combustion engine start-up function. The low-power consumers which need to be operated during engine standstill are supplied from the 12 V battery via the 14 V sub-system. The electric starter is connected via the 42 V sub-system to the 36 V battery along with higher-power consumers and the additional new consumers, recently introduced into automobile construction. With respect to the required battery there are two possible options: dual systems employing two batteries of 12 V and 36 V respectively, or with a single 36 V battery. Irrespective of the structure utilised for energy storing, a number of different solutions, some of which will be only temporary, are currently examined to ensure functional integration of the 36 V battery along with the 42 V consumers in the conventional 14 V alternator system.

The paper proposes the development of a novel supply system for the vehicle's 42 V electrical consumers departing from the conventional alternator system with three-phase rectifier bridge (1180-14 V production alternator type) with its original windings, manufactured by "Electroprecizia" Săcele Inc.

The new supply system is obtained by adding a boost converter to the conventional alternator system in order to match the 36 V battery as well as the vehicle's 42 V consumers. The control strategy of the boost converter is achieved so as to provide maximum alternator output power for a given field current at any speed within its operating range. Using the simplified model of the alternator system presented in [10, 11], the paper approaches the specific problem area by deriving simple expressions, very useful in practical applications for calculating the main quantities (output power, alternator load current, battery charging current) of the new 42 V alternator system, depending on its parameters as well as the operating speed.

## II. MODELLING AND ANALYSIS OF THE CONVENTIONAL ALTERNATOR SYSTEM

We consider a conventional alternator system, whose simplified electrical model is represented in the diagram of Fig. 1. Its terminal voltage is regulated through field control. Dynamic performances are determined mainly by the resistances and inductances of the field and stator windings; of which the latter are usually relatively high in conventional alternators.

The field current of the synchronous machine is controlled using pulse width modulation (PWM) of the voltage applied to the field winding. The dynamic behaviour of the field winding current (its waveform over time) depends mainly on the field winding time constant, $\tau_f = L_f / R_f$, which is usually in excess of 100 ms. The stator winding is modelled with a three-phase circuit with its back-emfs $u_{ea}$, $u_{eb}$ and $u_{ec}$ forming a symmetrical system, connected in series with the synchronous inductances $L_s$ forming a balanced system. For this model, the stator winding resistances were neglected.

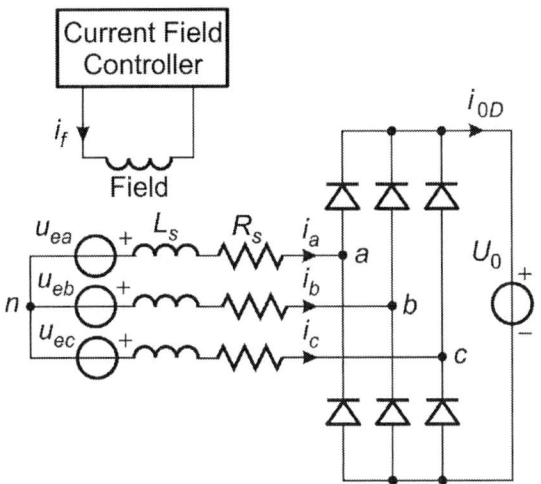

Fig. 1. Simplified electrical model of the conventional alternator system.

The field current of the synchronous machine is controlled using pulse width modulation (PWM) of the voltage applied to the field winding. The dynamic behaviour of the field winding current (its waveform over time) depends mainly on the field winding time constant, $\tau_f = L_f / R_f$, which is usually in excess of 100 ms. The stator winding is modelled with a three-phase circuit with its back-emfs $u_{ea}$, $u_{eb}$ and $u_{ec}$ forming a symmetrical system, connected in series with the synchronous inductances $L_s$ forming a balanced system. For this model, the stator winding resistances were neglected.

The electrical angular velocity is:

$$\omega = p\,\omega_m, \qquad (1)$$

where $\omega_m$ is the mechanical angular velocity of the rotor and $p$ the number of pole pairs of the alternator.

The magnitude of the back-emf induced in the stator windings depends on the electrical angular velocity and the field current:

$$U_e = k\,\omega I_f. \qquad (2)$$

An extremely useful synthetic expression, which will be employed in the development of the new 42 V alternator system, is its output power which can be determined using [12-13]:

$$P_o = \frac{3}{\pi} U_0 \frac{\sqrt{(U_e)^2 - \left(\frac{2}{\pi}U_0 + U_F\right)^2}}{\omega L_s}, \qquad (3)$$

where $U_0$ is the battery voltage, $U_e$ is the maximum value of the back-emf, $\omega$ is the angular velocity (puls-

ation) of the electrical quantities, $L_s$ is the synchronous inductance, $U_F$ is the forward voltage across the conducting diodes in the three-phase rectifier bridge.

Fig. 2 presents the output power characteristics of a conventional 14 V alternator versus output (battery) voltage for constant field current ($I_f = I_{fN} = 3.5$ A), at different rotor speeds. The alternator used in determining the curves of Fig. 2 has the following parameter values:

- $R_s = 0$ (The actual value of ~50 mΩ was neglected to simplify the relations);
- $L_s = 165$ µH (The actual value of ~140 µH was increased to compensate the neglected stator resistance);
- $k = 0.004$ V s/A.

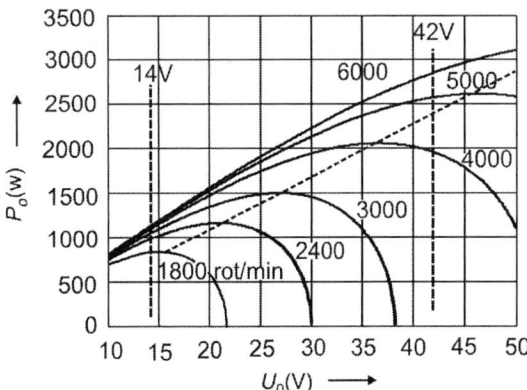

Fig. 2. Output power versus output voltage (for constant field current).

These curves show how the maximum power transfer is achieved in the presented case. As can be seen, for a given speed, the output power of the alternator varies considerably depending on the output voltage. It can be observed that for each speed there is a specific value of the output voltage corresponding to zero output current which will consequently result in zero output power. This voltage corresponds to the peak value of the line-to-line back-emf above which the diodes become reverse biased and no longer conduct.

The output power curves exhibit a maximum at a single value of the output voltage which is much lower than the peak value of the line-to-line back-emf. This aspect is specific to conventional Lundell machines with claw-pole rotor due to their large synchronous inductances which cause significant voltage drops of variable magnitude depending on load current and speed changes. Consequently, in Lundell alternators with diode rectifiers the magnitude of the output voltage is controlled mainly by the load. These machines usually need to produce stator back-emfs that are 4 to 6 times higher than the required output voltage of 14 V.

Therefore, a conventional alternator can be modelled with an open-circuit dc source in series with a large nonlinear, speed and current dependent impedance.

The curves of Fig. 2 illustrate how the power transfer occurs in the considered case. The above-mentioned impedance determines a large dc-side output impedance

**1749**

of the alternator system which calls for a high back-emf in order to ensure the rated output current. For this reason, the alternator is capable of delivering its maximum output power at constant speed only at a single output voltage.

On examining the alternator speeds at which the curves of Fig. 2 were obtained, it can be observed that these are covering the entire speed range of the alternator from 1800 rot/min, which corresponds to the engine's idling speed and up to 6000 rot/min which corresponds to the vehicle's cruising speed. At a given speed and constant output voltage, the output power can be reduced by decreasing the field current which will result in back-emf and output current decrease and finally will reduce the output power. For the alternator operating at 14 V, the available power is obtained at the intersection point of the output curves with the vertical corresponding to this voltage. The maximum output power (~ 820 W) of the conventional alternator is delivered very near to its rated 14 V design operating voltage. Considering the output power availability at higher speeds while maintaining the 14 V output voltage level, the alternator's output power will increase up to 1100 W at 6000 rot/min but will not reach the maximum power capability at higher speeds. The output power capability at 42 V is in excess of 2850 W at 6000 rot/min for the studied alternator.

Nevertheless, this conventional alternator with diode rectifier cannot be operated at 42 V because its output power will be zero at speeds ranging between 1800 rot/min and 3300 rot/min.

This paper proposes a relatively simple structural modification by including a boost converter in the alternator system in order to ensure maximum power capability of the machine.

### III. ADAPTING THE CONVENTIONAL ALTERNATOR TO THE 42 V SUPPLY SYSTEM

There are several possible implementations for the boost converter. In the following, we consider the basic diagram [1, 5] used for implementing a boost converter. It comprises a controlled switch (MOSFET) T, connected across the rectifier's output terminals and a diode D connected in series with the battery (Fig. 3). The new power unit consisting of diode rectifier and boost converter will be referred to as switched-mode-rectifier (SMR). The MOSFET is PWM-controlled with the duty ratio $d$. The rectifier terminal voltage consists of a train of pulses (blocks) with the average value depending on the output (battery) voltage $U_0$ and the duty ratio $d$, defined as the ratio between the length of the MOSFET's conduction interval $d \cdot T_p$ and the PWM cycle length $T_p$ (Fig. 4). The average voltage across the rectifier bridge terminals and the average current supplied to the consumers and the battery can be expressed as:

$$U_d = \widetilde{u}_d = (1-d)\, U_0 , \tag{4}$$

$$I_{0S} = \widetilde{i}_{0S} = (1-d) I_{dS} . \tag{5}$$

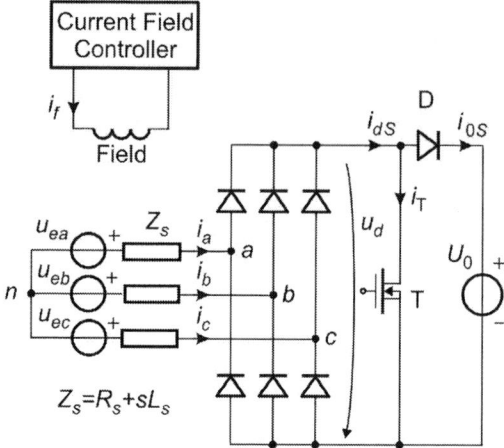

Fig. 3. Modified conventional alternator with switched-mode-rectifier (SMR).

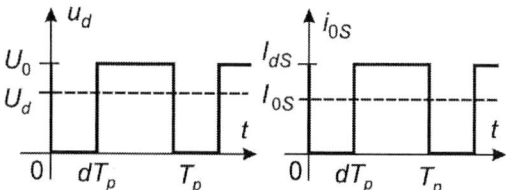

Fig. 4. Idealised waveforms of rectifier output voltage and output current.

For a more efficient use of the alternator in a 42 V system, the duty ratio $d$ is adopted so as to provide maximum power transfer to the battery at any operating speed

$$d_{\mathrm{M}}(\omega) = 1 - \frac{U_0}{U_{0\mathrm{M}}(\omega)} = 1 - \frac{42}{U_{0\mathrm{M}}(\omega)} \tag{6}$$

where $U_{0\mathrm{M}}(\omega)$ is the rectifier output voltage ensuring maximum power transfer from alternator to battery at a given angular speed.

The voltage $U_{0\mathrm{M}}(\omega)$ corresponding to the maximum output power is calculated by setting the derivative of (3) with respect to $U_0$ equal to zero.

$$\frac{\partial P_o}{\partial U_0} = \frac{3}{\pi \omega L_s} \left( \sqrt{\left(k\,\omega I_f\right)^2 - \frac{16}{\pi^2}\left(\frac{U_0}{2} + U_F\right)^2} - \right.$$

$$\left. - \frac{\frac{16}{\pi^2} U_0 \left(\frac{U_0}{2} + U_F\right)\frac{1}{2}}{\sqrt{\left(k\,\omega i_f\right)^2 - \frac{4}{\pi^2}\left(\frac{U_0}{2} + U_F\right)^2}} \right) = 0 \tag{7}$$

and the equivalent equation

$$U_{0\mathrm{M}}^2 + 3\,U_F\,U_{0\mathrm{M}} + 2\,U_F^2 - 2\left(\frac{\pi k\,\omega I_f}{4}\right)^2 = 0$$

which yields

$$U_{0\mathrm{M}} = \frac{-3\,U_F + \sqrt{8\left(\dfrac{\pi k\,\omega I_f}{4}\right)^2 + U_F^2}}{2} \qquad (8)$$

For practical purposes, the approximate relation is

$$U_{0\mathrm{M}} \cong \frac{\pi k\,\omega I_f}{2\sqrt{2}} - U_F \qquad (9)$$

By substituting (9) in (3), we obtain the expression of the alternator's maximum output power, given the speed and field current values:

$$P_{o\mathrm{M}} = \frac{3}{\pi} U_{0\mathrm{M}} \frac{\sqrt{\left(k\,\omega I_f\right)^2 - \dfrac{16}{\pi^2}\left(\dfrac{U_{0\mathrm{M}}}{2} + U_F\right)^2}}{\omega L_s} \qquad (10a)$$

$$P_{o\mathrm{M}} \cong \frac{6}{\pi^2}\frac{U_{0\mathrm{M}}^2}{\omega L_s} \qquad (10b)$$

$$P_{o\mathrm{M}} \cong \frac{6}{\pi^2 \omega L_s}\left(\frac{\pi k\,\omega I_f}{2\sqrt{2}} - U_F\right)^2 \qquad (11)$$

Thus, the average output current of the alternator at maximum output power is

$$I_d = \tilde{i}_d = \frac{P_{o\mathrm{M}}}{U_{0\mathrm{M}}} \cong \frac{6}{\pi^2}\frac{U_{0\mathrm{M}}}{\omega L_s}, \qquad (12)$$

and

$$I_d = \frac{6}{\pi^2 \omega L_s}\left(\frac{\pi k\,\omega I_f}{2\sqrt{2}} - U_F\right). \qquad (13)$$

The current delivered to the battery by the SMR which is proportional with the delivered power can be calculated as:

$$I_0 = \tilde{i}_0 = \frac{P_{o\mathrm{M}}}{U_{0N}} \cong \frac{6}{\pi^2}\frac{U_{0\mathrm{M}}^2}{42\,\omega L_s} \qquad (14a)$$

$$I_0 = \frac{1}{7\pi^2\,\omega L_s}\left(\frac{\pi k\,\omega I_f}{2\sqrt{2}} - U_F\right)^2 \qquad (14b)$$

Expressions (9), (11), (13) and (14) will be employed in a comparative analysis regarding the ratings of both the new 42 V alternator system and its 14 V conventional counterpart. Fig. 5 shows the alternator's output power versus its speed at two values of the field current ($I_f = 3.5\,\mathrm{A}$ and $2\,\mathrm{A}$). In the next figures, the quantities

bearing the additional subscript „D" describe the conventional diode bridge rectifier alternator while those employing the additional subscript „S" describe the alternator with a SMR-system.

The curves of Fig. 5 were obtained using relation (3) for $P_{oD}$ and expression (11) for $P_{oS}$, the output power in the SMR alternator system.

It should be noted that the extra output power is obtained for lesser machine losses since the currents $I_{dS}$, in the SMR system, are lower than the currents $I_{0D}$, in the stator windings of the conventional alternator, over a wide range of speeds (Fig. 6).

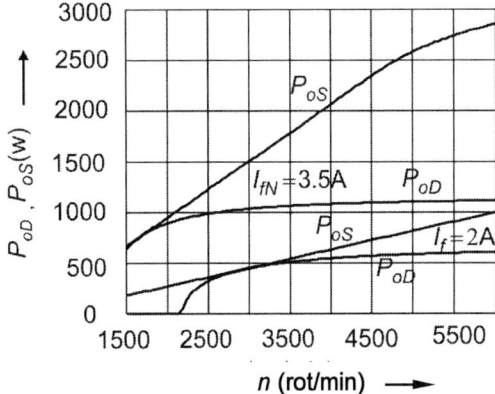

Fig. 5. Output power versus speed (at two field current values).

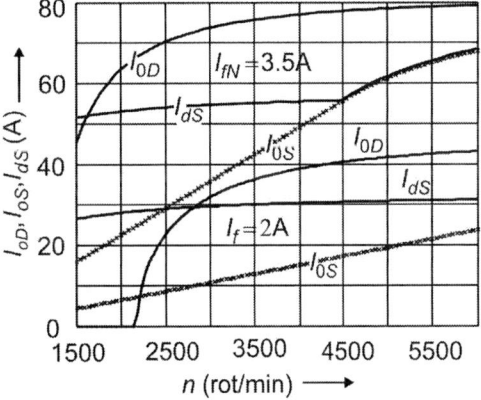

Fig. 6. Output currents of the 42 V and the conventional 14 V alternators, respectively.

The current $I_{0S}$, delivered by the SMR alternator to the battery, varies almost linearly with speed. This becomes obvious when neglecting the forward diode voltage, that is by setting $U_F = 0$ in (13) which yields:

$$I_0 = \tilde{i}_0 = \frac{k^2 I_f^2}{14\sqrt{2}\,L_s}\,\omega \qquad (15)$$

Strictly speaking, this relation is valid on the speed range where the voltage which corresponds with the maxi-

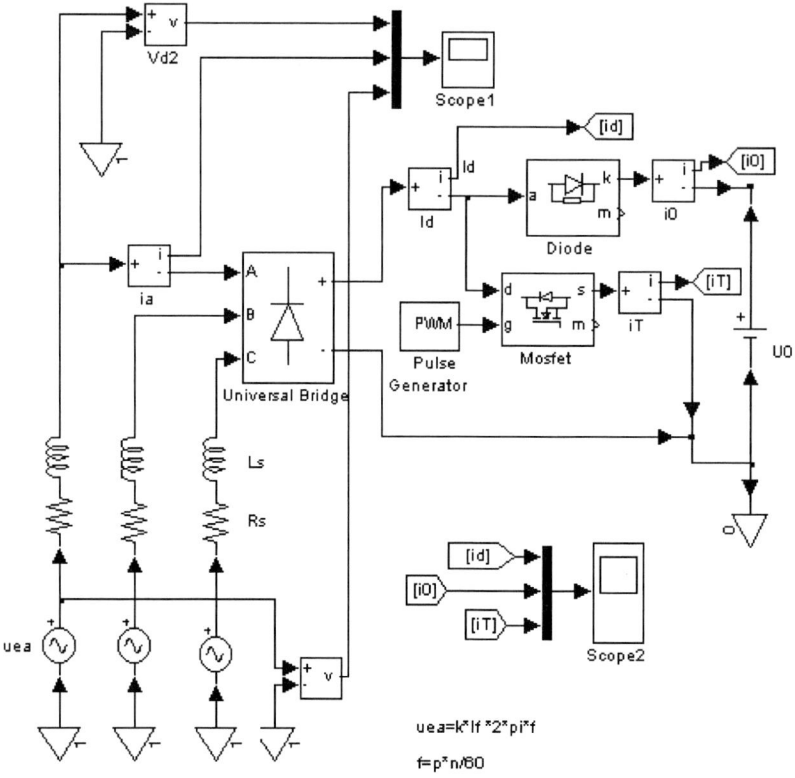

Fig. 7. SIMULINK model of the new 42 V alternator system.

mum output power is below the 42 V value, necessary for battery charging. This linear variation can be extrapolated with sufficient accuracy for the cases in which this condition is not satisfied (for example at speeds above 4500 rot/min as in Fig. 6 at $I_f = 3.5$ A ). Based on the above, it results that the process involving the alternator with SMR can be modelled with a current source, linearly depending, on the angular speed. We also note that the output current of the alternator is proportional with the square of the field current.

### IV. SIMULATION RESULTS

In order to verify and validate the above relations, several simulations were performed using the SIMULINK programming environment. The developed model (Fig. 7) uses components of the PSBs (POWER SYSTEM BLOCKSET) toolbox to simulate alternator with SMR from the diagram of Fig. 3. The alternator used in the model has the following specifications: $R_s = 0.05$ Ω, $L_s = 140$ µH, $k = 0.004$ V s/A , $p = 6$.

The simulation starts with the verification of the validity of relation (3) used in calculating the alternator's approximate output power depending on battery voltage. The simulations were performed by assuming that the MOSFET switch is maintained permanently blocked (in "off" state) and thus the alternator operates in its "conventional" mode with the battery voltage $U_0$ varying

between 10 V and 50 V. The dashed curves in Fig. 8 were obtained by cubical interpolation using the measurement points, marked with small circles, to achieve the simulations. During simulations, the alternator's output power is evaluated in terms of the continuous component of its current.

As can be observed from Fig. 8 there is good concordance between theory and simulation results which is also confirmed by the position of the output power maximum values that are used in the design of the control system of the new 42 V alternator.

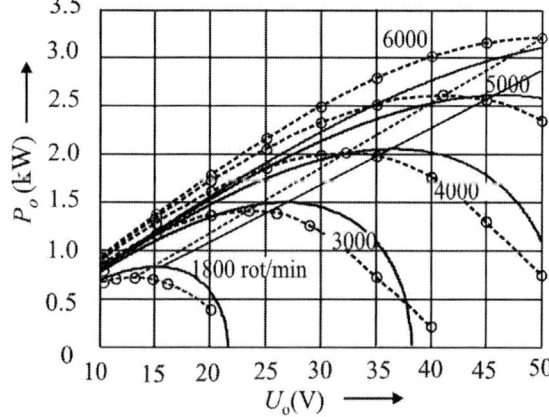

Fig. 8. Output power versus voltage. Solid and dashed curves illustrate calculation and simulation results respectively.

Finally, the behaviour of the alternator with SMR is analyzed. In this case, the MOSFET switch is controlled by a PWM voltage with a frequency of 20 kHz, having a variable "on-state" interval (depending on alternator speed and field current).

Fig. 9 and Fig. 10 illustrate the waveforms of the most significant electrical quantities of the SMR alternator system.

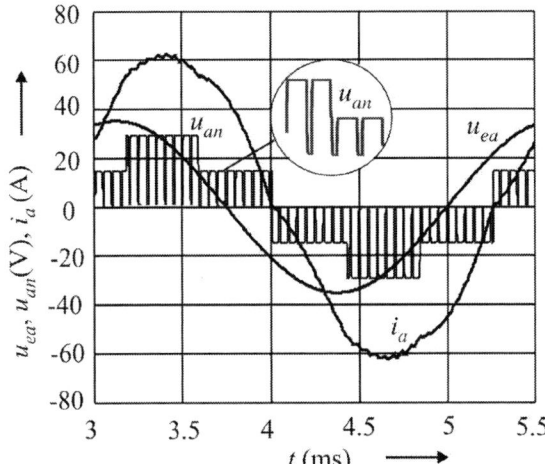

Fig. 9. Phase "$a$" alternator waveforms (back emf $u_{ea}$, voltage $u_{an}$ and phase current $i_a$).

The curves shown in Fig. 9 were obtained at a speed of 4000 rot/min and a duty ratio $d \cong 0.23$ for a battery terminal voltage of 42 V.

The current delivered to the battery consists of a pulse train of relative length $1-d \cong 0.77$ whose amplitude follows the waveform of the alternator's output current $i_d$, with its characteristic ripple, as resulting from Fig. 10.

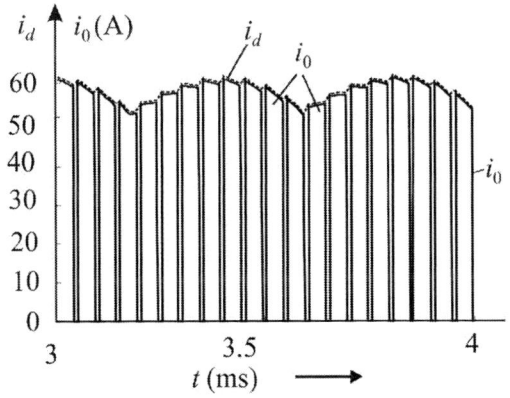

Fig. 10. Output current waveforms - alternator $i_d$, battery charging current $i_0$.

The validity of expression (11) allowing for approximate output power calculation, along with the validity of the expressions (13), (14) for the currents of the 42 V SMR alternator system was also verified.

The dashed curves of Fig. 11 were obtained using cubical interpolation at the measurement points which were represented by squares or small circles.

For simulation purposes, the output power of the alternator was evaluated using the continuous component of the conventional (diode-rectifier) alternator's output current.

The curves $P_{oS}$ represent the power of the 42 V SMR alternator system, whereas the curves labelled $P_{oD}$ describe the performances of the 14 V conventional alternator system.

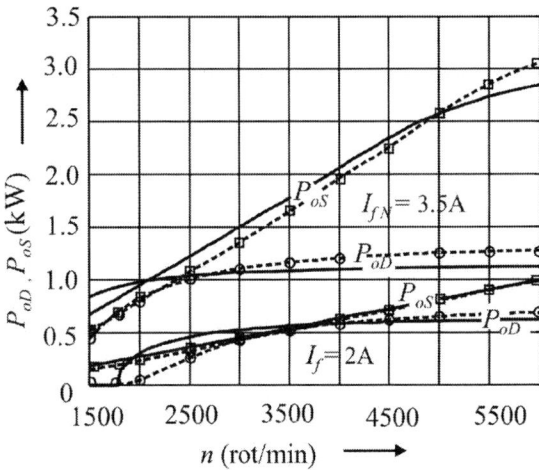

Fig. 11. Output power versus speed. The solid curves were calculated whereas the dashed curves were obtained through simulation.

Fig. 12 presents the average alternator output current $I_{dS}$ along with the average battery charging current, $I_{0S}$ versus alternator speed, at two field currents ($I_f = 3.5$ A and 2 A).

By examining the curves it can be observed that the current delivered to the battery by the SMR alternator system varies linearly with speed throughout its entire range.

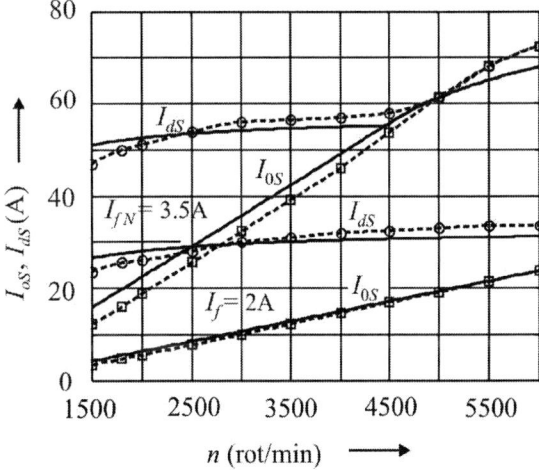

Fig. 12. The currents of the 42 V alternator system. Solid lines are calculated; dashed lines are simulations.

By examining the curves of Fig. 11 and Fig. 12, it can be observed that the experimental results obtained by simulation provide a sufficiently good approximation to validate the calculated results. Also, accuracy increases as the field current decreases.

## V. CONCLUSIONS

In this paper a new 42 V electrical supply system for automotive on-board consumers is developed, based on the 1180 conventional alternator system (with original, non-rewound stator) with three-phase diode rectifier bridge, manufactured at "Electroprecizia" S□cele Inc. and designed for 14 V consumers.

The new electrical supply system is obtained by adding a boost converter unit to the conventional system in order to match the latter with the 36 V battery and the 42 V vehicle consumers respectively. The boost converter is controlled so that for a given field current the output power of the alternator will be maxim at any operating speed.

Based on the simplified alternator model presented in [10, 11], in this paper are derived very simple, approximate relations, very useful for practical purposes in determining the principal quantities (output power, alternator output current or the battery charging current respectively) of the new 42 V alternator system depending on the main machine parameters and the operating speed.

Simulations performed using the SIMULINK programming environment show a good concordance between calculated results obtained by employing the approximate relations (11), (12) and (13) and the results obtained experimentally, by simulation.

## ACKNOWLEDGEMENT

The authors would like to acknowledge the support provided by the Romanian National Research Authority by financing the CEEX AMTRANS Project X2C33/2006.

## REFERENCES

[1] J.V. Hellmann and R.J. Sandel, "Dual/high voltage vehicle electrical systems," in *Proc. Future Transportation Technology Conf. Expo.*, Portland, OR, Aug. 1991.

[2] M.F. Matouka, "Design considerations for higher voltage automotive electrical systems," in *Proc. Future Transportation Technology Conf. Expo*, Portland, OR, Aug. 1991.

[3] J.G. Kassakian, H.C. Wolf, J.M. Miller, and C.J. Hurton, "Automotive electrical systems circa 2005," *IEEE Spectrum*, Aug. 1996, pp. 22-27.

[4] S. Muller and X. Pfab, "Considerations for implementing a dual voltage power network," in *Proc. IEEE-SAE International Conference on Transportation Electronics (Convergence)*, Deaborne, MI, Oct. 1998.

[5] F. Liang, J. Miller, X. Xu, "A vehicle electrical power generation system with improved output power and efficiency," *IEEE Trans. On Industry Applications*, Vol. 35, No. 6, November/ December 1999, pp. 1341-1346

[6] J.G. Kaliskan, "Automotive electrical systems – The power electronics market of the future," in *Proc. IEEE Applied Power Electronics Conference (APEC '00)*, Vol. 1, New Orleans, LA, Feb. 2000, pp. 3-9.

[7] D.J. Perreault, V. Caliskan, "A new design for automotive alternators," In *Proc. Int. Congress Transportation Electronics (Convergence '00)*, October 2000, pp. 583-594.

[8] D.J. Perreault and V. Caliskan, "Automotive power generation and control," *IEEE Trans. on Power Electron.* Vol. 19, No. 3, May 2004, pp. 618-630.

[9] J.M. Rivas, D.J. Perreault, and Th. Keim, "Performance improvement of alternators with switched-mode rectifiers," *IEEE Trans. on Energy Conversion*, Vol. 19, No. 3, September 2004, pp.561-568.

[10] V. Caliskan, D.J. Perreault, T.M. Jahns, and J.G. Kassakian, "Analysis of three-phase rectifiers with constant-voltage loads," *IEEE Trans. on Circuits and Systems – I: Fundamental Theory and Applications*, vol. 50, no. 9, Sept. 2003, pp. 1220-1226.

[11] M. Cernat, V. Comnac, A. Mailat, "Conventional electrical systems for automobiles. Part I: Modelling," *Proc.* 3rd *International Conference on Interdisciplinarity in Education (ICIE '07)*, March 2007, Athens, Greece, ISBN 978-960-89028-4-8, ISSN 1790-661X, pp. 426-431.

[12] J.G. Kassakian, M.F. Schlecht, and G.C. Verghese, *Principles of power electronics*. Reading, MA, Addison Wesley, 1991.

[13] N. Mohan, T.M. Undeland, and W.P. Robbins, *Power electronics: converters, applications, and design*, John Wiley & Sons, Inc., U.S.A., 1989.

[14] J. Vittek, S.J. Dodds, *Forced dynamics control of electric drives*, EDIS – Publishing Center of Zilina University, Slovakia, 2003, ISBN 80-8070-087-7.

# New Dual Channel Quasi Resonant DC-DC Converter Topologies for Distributed Energy Utilization

J. Hamar[1/2], I. Nagy[1], P. Stumpf[1], H. Ohsaki[3], E. Masada[4]

[1] Budapest University of Technology and Economics, Department of Automation and Applied Informatics
H-1111 Budapest, Budafoki ut 8. Hungary, European Union, Phone: +36 1 463-1165,
Fax: +36 1 463-3163, E-mail: *hamar@elektro.get.bme.hu, nagy@elektro.get.bme.hu*

[2] Invention and Research Center Services Co. Ltd., Budapest, Hungary

[3] The University of Tokyo, Tokyo, Japan

[4] Railway Technology Research Institute, Tokyo, Japan

*Abstract* — The paper is concerned with new resonant converter configurations, belonging to the previously published dual channel dc-dc converter family. On one hand, the new configurations solve a significant shortcoming of the original ones, namely, the output midpoint potential was floating in comparison with the input midpoint. This caused a high-frequency oscillation of the output midpoint, often significantly limiting the application fields. Common ground can be used at the input and output sides at the new configurations. In addition, two of the new topologies also include electrical isolation, performed by an isolating high-frequency transformer. This property is often necessary in order to fulfill the strict industrial requirements against the power converters.

*Keywords* — DC power supply, Distributed power, Converter circuit, Resonant converter, Simulation

## I. INTRODUCTION

The application of these dual channel DC-DC converters is recommended in the medium or higher power ranges, up to the several kW range. One of the main application fields is in the distributed electrical energy generation. Different sources can be applied at the input of the channels, and the power-flow can be well controlled between the two channels, maintaining the required constant symmetric or asymmetric output voltages even in case of an outage of any of the sources. As the combination of the input sources, different cases can emerge, e.g. photovoltaic panel as one input, battery as the other one etc. The converters can be applied as general power supply, dc power source for telecommunication equipments, computers, solar energy recovery system, etc. Their most favorable properties are the resonant operation and the small power loss due to soft switching at distinct operation points.

## II. CONFIGURATIONS & OPERATION

The configurations of the converter family and their basic operation were described in detail in [1] and [2]. Here the basic configurations (Fig. 1) and their operations are summarized for the general (asymmetrical) case. (Note

that the polarity of the output voltage is reversed in the buck & boost converter in Fig. 1b).

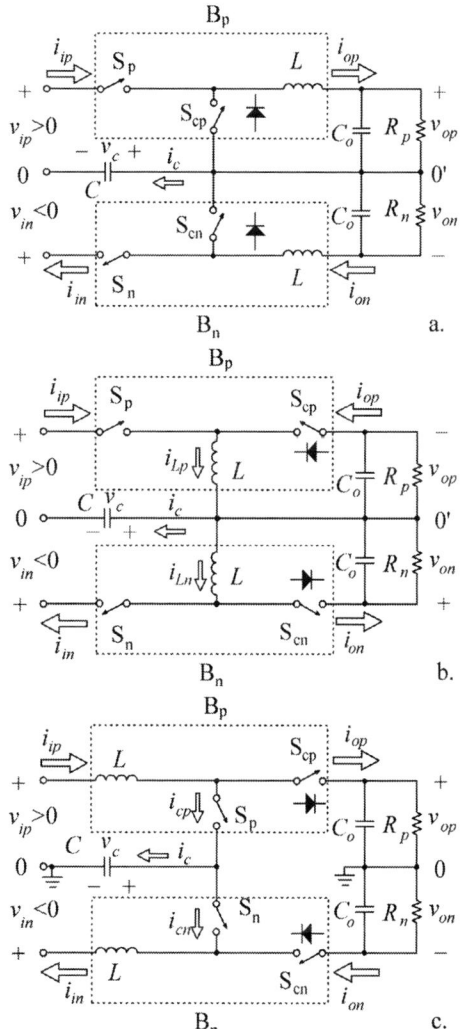

Fig. 1 Basic buck (a), buck & boost (b) and boost (c) configurations

Each configuration includes two channels the so-called positive **p** and negative **n** ones. The converters have two switches in each channel denoted by $S_p$ and $S_{cp}$ in channel **p** and by $S_n$ and $S_{cn}$ in channel **n** in Fig. 1. $S_p$ and $S_n$ are always controlled switches, the remaining two ones can be either controlled switches or diodes. The two channels are coupled by capacitor $C$ (Fig. 1). Two switches and inductor $L$ build up block $\mathbf{B_p}$ and $\mathbf{B_n}$. Connecting the three terminals of $\mathbf{B_p}$ and $\mathbf{B_n}$ to the rest of the circuit results in the buck, buck & boost or the boost configurations (Fig. 1).

At the description of the operation, ripple-free input and output voltages, discontinuous current conduction (DCM operation) in inductance $L$ and lossless components are assumed, furthermore the commutation intervals between $S_p$ and $S_{cp}$ ($S_n$ and $S_{cn}$) are neglected. The switch pairs $S_p$—$S_{cp}$ and similarly $S_n$—$S_{cn}$ are turned on and off alternately, that is, when $S_p$ is on $S_{cp}$ is off and vice versa. By turning switch $S_p$ on and $S_{cp}$ off, a sinusoidal current pulse $i_{1p}=i_c$ is developed from $\omega t=0$ to $\alpha_p$ (Fig. 2a, $\omega=1/\sqrt{LC}$) in circuit $S_p$, $L$, $V_{op}$, $C$ and $V_{ip}$ in the buck converter and in circuit $S_p$, $L$, $C$ and $V_{ip}$ in the buck & boost and in the boost converters.

It makes the voltage $v_c$ swing from $V_{cn}$ to $V_{cp}$ ($V_{cn}<0$) [see Fig. 2b]. Reaching $\omega t=\alpha_p$ switch $S_p$ is turned-off and $S_{cp}$ on and the capacitor $C$ is clamped on voltage level $v_c=V_{cp}$. The choke current commutates from $S_p$ to $S_{cp}$.

The choke current $i_{2p}=i_{op}$ decays like a ramp in circuit $S_{cp}$, $L$ and $V_{op}$ in the buck, and in the buck & boost converters and in circuit $L$, $S_{cp}$, $V_{op}$, $V_{ip}$ in the boost converter between $\alpha_p \leq \omega t \leq \alpha_{ep}$ and reaches zero at the extinction angle $\alpha_{ep}$. (At continuous conduction mode (CCM) the choke current does not reach zero before the initialization of the next period.) Similar process takes place in the next half period from $T_s/2$ in the negative channel of the converter.

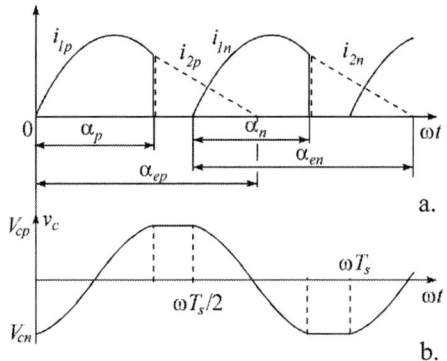

Fig. 2 Time functions of input and output currents and voltage $v_c$ (discontinuous operation)

$i_{ip,n}$ input and $i_{op,n}$ output currents can be composed from $i_{1p,n}$ and $i_{2p,n}$ for all three converters (Table 1).

Table 1 Composition of currents

|        | $i_{ip,n}$         | $i_{op,n}$          | $i_c$           |
|--------|--------------------|---------------------|-----------------|
| Buck   | $i_{1p,n}$         | $i_{1p,n}+i_{2p,n}$ | $i_{1p}-i_{1n}$ |
| B & B  | $i_{1p,n}$         | $i_{2p,n}$          | $i_{1p}-i_{1n}$ |
| Boost  | $i_{1p,n}+i_{2p,n}$| $i_{2p,n}$          | $i_{1p}-i_{1n}$ |

Switches $S_{cp}$ and $S_{cn}$ can also be replaced by diodes. However, in this case only the switching frequency ($f_s$) can freely be adjusted, the commutation from $S_p$ to $S_{cp}$ (and similarly from $S_n$ to $S_{cn}$) takes place when $v_c$ reaches $V_{ip}$ ($V_{in}$) in the buck, $V_{op}$ ($-V_{on}$) in the boost and $V_{ip}+V_{op}$ ($V_{in}-V_{on}$) in the buck & boost converters, respectively. In discontinuous conduction mode (DCM) the sinusoidal current in S starts from zero (zero current turn-on). Both in CCM and in DCM the current commutates from S to D practically under zero voltage across S and D when diodes are used in places of $S_{cp}$ and $S_{cn}$.

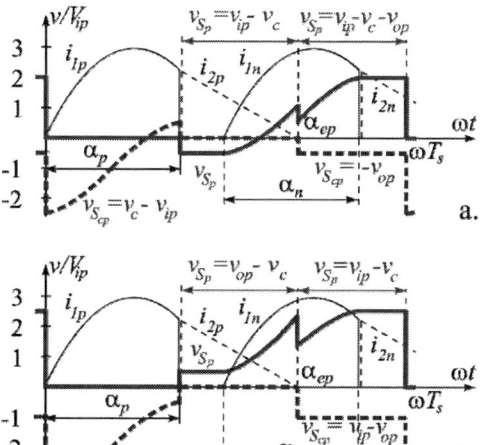

Fig. 3 Voltage and current waveforms of the switches in discontinuous current conduction mode.

Fig. 3 presents the voltage and current waveform of the two switches $S_p$ and $S_{cp}$ in discontinuous current conduction mode. Fig. 3a and Fig. 3b shows the waveforms for buck and for boost converter, respectively. The voltage waveforms are drawn by bold line the current waveforms are drawn by thin line.

It was assumed that the controlled switches are realized by the series connection of an ideal MOSFET and diode. Current can flow only in one direction through them. The two controlled switches are in complementary state, when one is on the other one is off and vica-versa. The only exception is the very short overlap interval in commutation.

## III. POWER FLOW

Regarding the current conduction in channel **p** and **n** four modes can be distinguished (Table 2). The positive ($x_1$) and negative ($x_2$) sequence symmetrical components or using other terminology common and differential mode components will be applied in the current paper beside the p and n channel variables or parameters.

Table 2 Possible operation modes

| Channels | Operation | | | |
|---|---|---|---|---|
| Positive (**p**) | DCM | DCM | CCM | CCM |
| Negative (**n**) | DCM | CCM | DCM | CCM |
| Mode | A | B | C | D |

They are defined as follows:

$$x_1 = x_c = (x_p + x_n)/2 \; ; \quad x_2 = x_d = (x_p - x_n)/2 \quad (1),(2)$$

$$x_p = x_1 + x_2 \; ; \qquad x_n = x_1 - x_2 \qquad (3),(4)$$

where $x_p$ and $x_n$ is the same kind of variables in positive and in negative side, respectively. The introduction of symmetrical components substitutes the asymmetrical operation by the superposition of two symmetrical modes [eq.(3), (4)]. Note that $v_{in} < 0$ and $v_{cn} < 0$ and in these cases $v_{i1}=(v_{ip}-v_{in})/2$; $v_{i2}=(v_{ip}+v_{in})/2$; $v_{c1}=(v_{cp}-v_{cn})/2$ and $v_{c2}=(v_{cp}+v_{cn})/2$ where $v_{cp}$ and $v_{cn}$ are peak values (Fig. 2b).

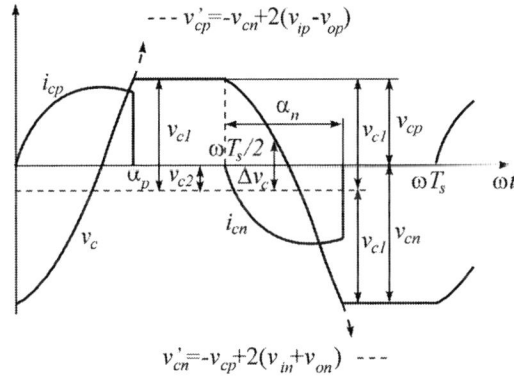

Fig. 4 Time functions in asymmetrical operation

Let us also introduce the following three binary variables expressing the identity of the corresponding configuration: $u_d$=1 if the converter is step-down (buck) type, otherwise $u_d$=0. $u_{ud}$=1 if the converter is step-up/down (buck & boost) type, otherwise $u_{ud}$=0. $u_u$=1 if the converter is step-up (boost) type, otherwise $u_u$=0. As an example the positive $v_{c1}$ and the negative $v_{c2}$ sequence component of the capacitor voltage defined by its peak values are shown in Fig. 4. The instantaneous value of $v_c$ is $v_c=\Delta v_c+v_{c2}$ (Fig. 2). When $v_c=v_{cp}$ (or $v_c=v_{cn}$) then $\Delta v_c=v_{c1}$ (or $\Delta v_c=-v_{c1}$). $v_{cp}'$ and $v_{cn}'$ would be the peak capacitor voltage without turning-on $S_{cp}$ and $S_{cn}$, respectively. Symmetrical components will not be

introduced for currents $i_i$, $i_o$ and $i_c$ (Table 1). Therefore notation $i_1$ and $i_2$ create no misunderstanding.

In symmetrical operation each input current pulse carries the energy $w_{ip}=w_{in}$ and the same amount of energy $w_{op}=w_{ip}$ and $w_{on}=w_{in}$ is drawn by the load in the output within one period in steady-state. There is no energy exchange between the two channels. The energy stored in the switched capacitor $C$ is the same at the beginning and at the end of each half-period. On the other hand, in asymmetrical operation (Fig. 4) the energy $w_c$ stored in $C$ is higher (smaller) at the beginning of the half period at $t=0$ than at the end of the half period at $t=T_s/2$. The capacitance $C$ plays an energy-balancing role [1].

## IV. NEW CONFIGURATIONS

Proceeding from the basic dual channel converters, new, improved converter topologies have been derived. First, revised configurations of buck as well as buck & boost converters will be discussed offering remedies for a disadvantageous feature of the original ones, without causing any alteration in the operation of the converters.

Two additional configurations have also been developed providing electrical isolation between input and output, originating them from the buck and from the boost converters. The viability of the proposed converters will be verified by simulations.

### A. No Floating Output Midpoint Potential

The potential of midpoint 0' in the output is floating compared to the potential of the midpoint 0 of the input voltages with the voltage of the switched capacitor (Fig. 1), which can be a drawback in the buck and in the buck & boost configurations.

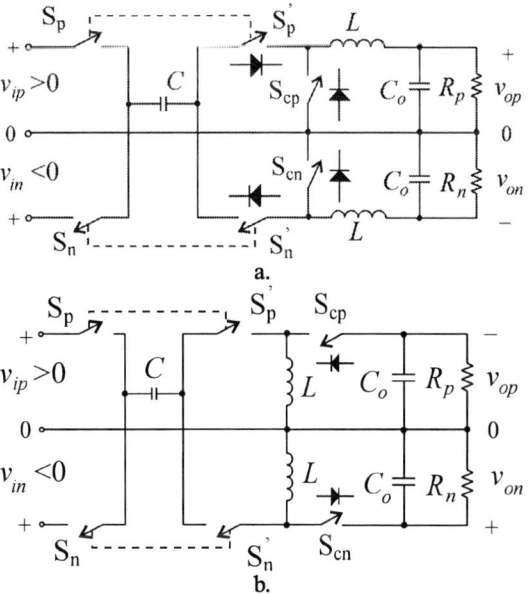

Fig. 5. Topologies for the buck (a) and the buck & boost (b) configurations

1757

The boost configurations do not suffer from this drawback because the two midpoints 0 and 0' are short-circuited.

To overcome this disadvantageous property, the revised buck and buck & boost configurations are shown in Fig. 5 a and in Fig. 5 b, respectively. The switches connected by dashed lines in the figure are turned on and off simultaneously. The clamping switches $S_{cp}$ and $S_{cn}$ can be replaced by diodes with similar effects as in the original configuration. In case of the buck converters $S'_p$ and $S'_n$ can also be replaced by diodes together with the clamping switches (Fig. 5 a). The operation of these converter is practically the same as the original one presented above. The relations coming from the analysis of the circuit in the Appendix can be used for the practical calculations.

The price paid for avoiding the floating output potential is the application of an additional series controlled switch (diode) in both channels.

### B. Isolated Converters

In order to isolate the input and output terminals from each other, two converter configurations are proposed. The two-channel resonant forward converter is derived from the buck configuration (Fig. 6 a). The isolated boost converter can be operated only by four-controlled-switches (Fig. 6 b).

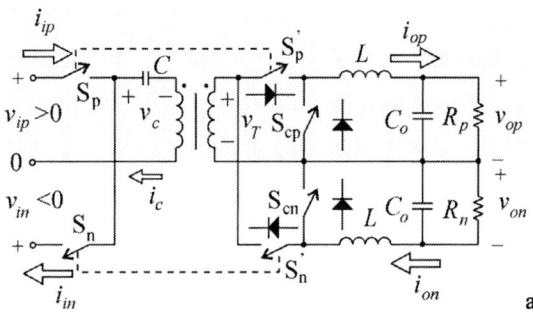

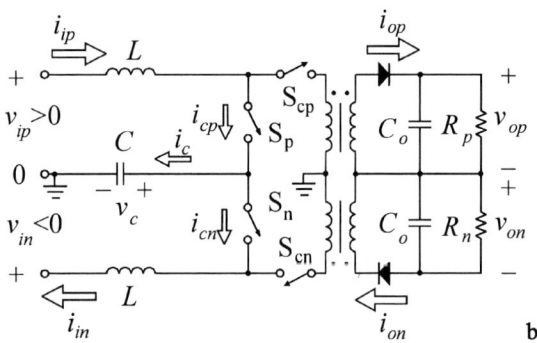

Fig. 6. Isolated configurations for the buck (a)
and for the boost (b) converters

The reason is that there is a direct path for the DC current through the chokes and the transformer windings, when diodes are applied in place of $S_{cp}$ and $S_{cn}$. Even in case of four-controlled-switch-operation the switches $S_{cp}$ and $S_{cn}$ must be turned off when they reach zero in DCM operation.

The steady state relations of the converters can be calculated on the basis of *Appendices*, taking into account the transformer's winding ratio $a=N_1/N_2$ between the primary and secondary side parameters. Note that the resonant frequency of the forward converter (Fig. 6 a) is changed to $1/a$ times of its original value.

### V. COMPUTER SIMULATIONS

In order to verify the theoretical results, computer simulations were carried out for each new topology presented, using PSpice program package. The aim of the simulations was to confirm the viability of the new topologies and to verify that the developed numerical relations hold in the investigated operating points. The calculation results are summarized in Table 3 and Table 4.

After executing the computer simulations, it could be concluded, that they were practically the same as the theoretical ones, that is, the results summarized in the tables correspond to the simulation results as well. The same can be confirmed by the time functions of capacitor voltage $v_c$ and choke currents $i_{Lp}$ and $i_{Ln}$ shown in Fig. 7 and Fig. 8 assigned to each row of the Tables.

Table 3. Calculation results
(isolated buck converter)

| Switching components: 6 Controlled Switches | | | | | | | | | |
|---|---|---|---|---|---|---|---|---|---|
| Fig. 7 | $a=$ $N_1/N_2$ | $f_s$ [kHz] | $R'$ [Ω] | $V_i$ [V] | $V_o$ [V] | $V_e$ [V] | $i_{L,peak}$ [A] | $\alpha$ [°] | $f'_r$ [kHz] |
| a. | 1.0 | 18.0 | 133 | 34.0 | 17.0 | 17.8 | 0.98 | 91 | 45.0 |
| b. | 2.0 | | 532 | | 12.4 | 9.5 | 0.53 | | 22.5 |
| c. | 1.0 | 45.0 | 133 | 34.0 | 17.0 | 7.1 | 0.62 | 66 | 45.0 |
| d. | 0.5 | | 33.3 | | 20.2 | 10.1 | 0.88 | | 90.0 |

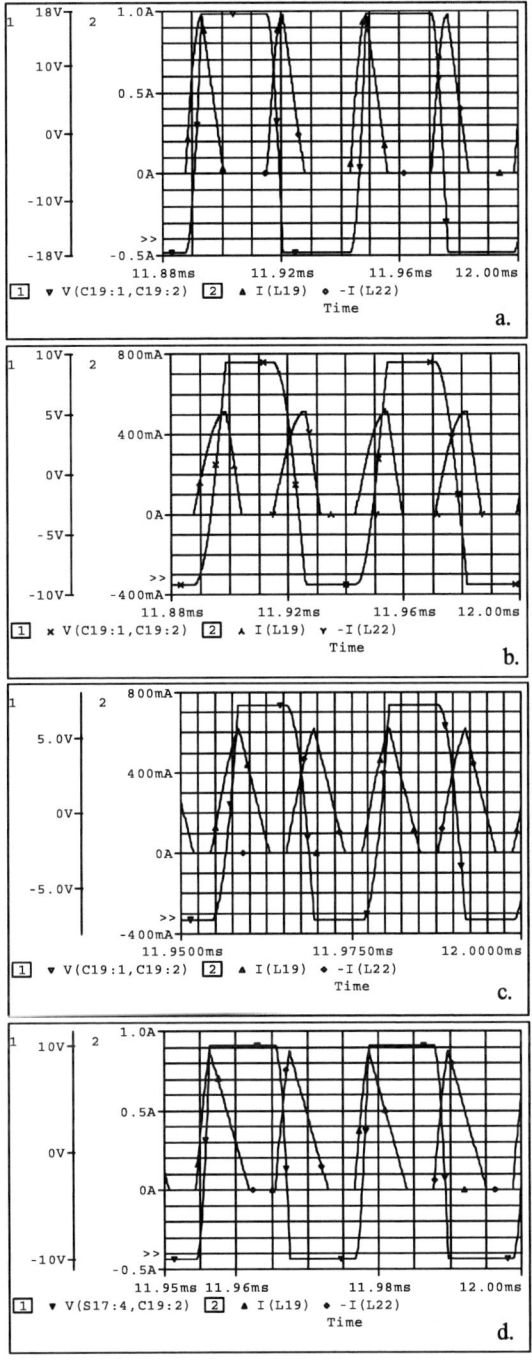

Fig. 7. Computer simulations (capacitor voltage and choke currents at isolated buck converter)

The variables (parameters) marked with prime indicate quantities reflected to the primary side of the transformer. The numerical results confirmed the identical operation of the original and revised converters. The basic circuit parameters commonly applied in each simulation were as follows: $L$=125μH, $C$=100nF and $C_o$=100μF.

Table 4. Calculation results
(isolated boost converter)

| Fig. 8 | $a = N_1/N_2$ | $f_s$ [kHz] | $R'$ [Ω] | $V_i$ [V] | $V_o$ [V] | $V_c$ [V] | $i_{L,peak}$ [A] | $\alpha' = \alpha$ [°] | $f_r' = f_r$ [kHz] |
|---|---|---|---|---|---|---|---|---|---|
| a. | 1.0 | 18.0 | 133 | 24.0 | 34.0 | 29.6 | 1.52 | 96 | 45.0 |
| b. | 2.0 | | 532 | | 25.4 | | | | |
| c. | 1.0 | 45.0 | 133 | | 34.0 | 11.8 | 0.95 | 70 | |
| d. | 0.5 | | 33.3 | | 61.3 | 28.3 | 1.62 | | |

The input voltage $v_i$, the load $R$ and the control variables $\alpha'$ and $f_s$ were kept constant in each simulation pairs (Table 3, row a-b, c-d) only the transformer winding ratio was changed to 2 or 0.5 in rows b and d. These changes caused however alterations in the output voltage, capacitor voltage, choke peak current and even in the resonant frequency.

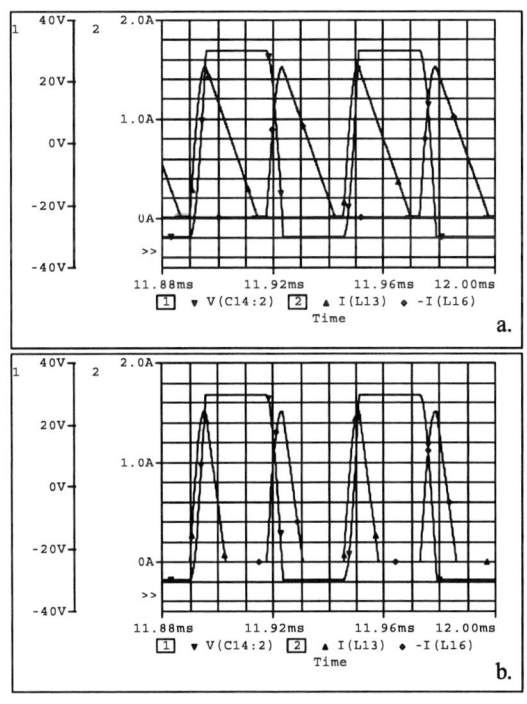

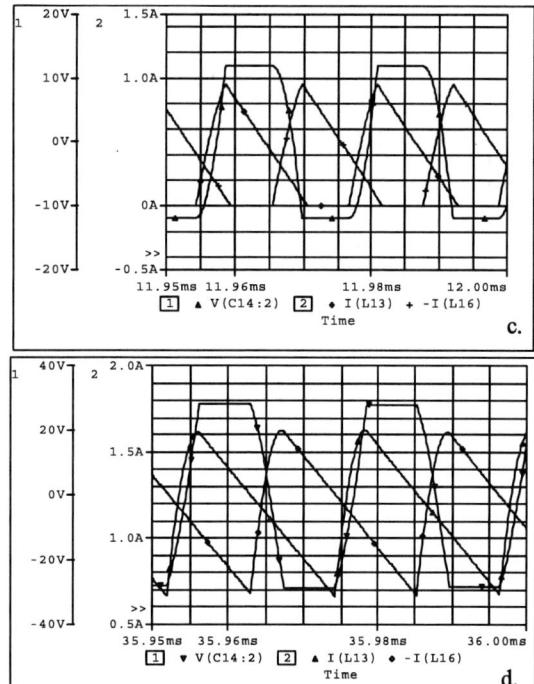

Fig. 8. Computer simulations (capacitor voltage and choke currents at isolated boost converter)

Rows b and d of Table 4 deviate from a and c by applying transformer winding ratio $a$=2 and $a$=0.5. The parameter values and control variables ($\alpha$=$\alpha'$ and $f_s$) were kept constant at each simulation pair now as well (Table 4 a-b, c-d). Adjusting $a$=0.5 in row d causes alteration in the current conduction mode (DCM→CCM). $\alpha$=const. involves $V_c$=const. and vice-versa in DCM independently of $a$. The transformer winding ratio does not modify the resonant frequency since components $L$ and $C$ situate on the same side of the isolation transformer.

## VI. CONCLUSIONS

New quasi-resonant dc-dc converter topologies have been presented. They overcome an adverse property of the original configurations, namely the floating output midpoint potential. In two cases they also provide electrical isolation between the input and output using high-frequency transformer. The price of the updated topologies is the increased number of switching devices, causing higher switching losses, however the barriers in their industrial application can significantly be reduced.

## ACKNOWLEDGEMENTS

The authors wish to thank the Hungarian Research Fund (OTKA TO46240, TO49640, FO49152, K62836), and the Control Research Group of the Hungarian Academy of Science (HAS) for their financial support and the support stemming from the cooperation between HAS and the Japanese Society for the Promotion of Science (JSPS), as well as the Janos Bolyai Research Fellowship of the HAS.

## APPENDICES

### Appendix A, Energy and Power Relations (CCM and DCM)

Using the relation between the capacitor voltage and current $i_c$=$Cdv_c/dt$ and taking into consideration that $\dfrac{1}{T_s}\displaystyle\int_0^{T_s} i_{op} dt = \dfrac{v_{op}}{R_p}$ the input energy pulse in the p channel delivered during one period is

$$w_{ip} = v_{ip}\int_0^{\alpha_p/\omega} i_c dt + u_u v_{ip}\int_0^{T_s} i_{op} dt = 2C v_{ip} v_{c1} + u_u \frac{v_{ip}v_{op}}{R_p f_s} \tag{a1}$$

The corresponding energy change in the switched capacitor C is

$$\Delta w_{cp} = C(v_{cp}^2 - v_{cn}^2)/2 = 2C v_{c1} v_{c2} \tag{a2}$$

and the output energy pulse is

$$w_{op} = w_{ip} - \Delta w_{cp} = 2C\cdot v_{c1}(v_{ip} - v_{c2})\left(\frac{v_{op}}{v_{op} - u_u v_{ip}}\right) \tag{a3}$$

where $\quad v_{op}/(R_p f_s) = w_{op}/v_{op} \tag{a4}$

is used. Introducing the dimensionless number

$$\delta = \frac{v_{i2} - v_{c2}}{v_{i1}} \tag{a5}$$

$$w_{op} = 2C\cdot v_{c1}v_{i1}(1+\delta)\left(\frac{v_{op}}{v_{op} - u_u v_{ip}}\right) \tag{a6}$$

$$w_{on} = 2C\cdot v_{c1}v_{i1}(1-\delta)\left(\frac{v_{on}}{v_{on} + u_u v_{in}}\right) \tag{a7}$$

Using eq. (a4) and (a6)

$$v_{op} = \frac{u_u}{2}v_{ip} + \sqrt{2R_p C f_s v_{c1} v_{i1}(1+\delta) + \frac{u_u}{4}v_{ip}^2} \tag{a8}$$

$$v_{on} = -\frac{u_u}{2}v_{in} + \sqrt{2R_n C f_s v_{c1} v_{i1}(1-\delta) + \frac{u_u}{4}v_{in}^2} \tag{a9}$$

In case of the buck and the buck & boost converters

$$\frac{w_{o2}}{w_{o1}} = \frac{p_{o2}}{p_{o1}} = \delta \tag{a10}$$

### Appendix B Derivation of $v_{op}$ and $v_{on}$ in Continuous Conduction Mode (CCM)

Assuming continuous conduction in the chokes the current time functions are

$$i_{Lpa}(\omega t) = I_{op}\cdot\cos\omega t + I_{kp}\sin\omega t \tag{b1}$$

in the interval $0 \le \omega t \le \alpha_p$, and

$$i_{Lpb}(\omega t) = I_{op}\cos\alpha_p + I_{kp}\sin\alpha_p - \frac{1}{Z}\big(v_{op} - u_u v_{ip}\big)\big(\omega t - \alpha_p\big) \tag{b2}$$

in the interval $\alpha_p \le \omega t \le \omega T_s$, where $I_{op} = i_{Lp}(0)$ and $Z = \sqrt{L/C}$ the characteristic impedance, furthermore

$$I_{kp} = (v_{ip} - u_d v_{op} - v_{cn})/Z \qquad (b3)$$

The choke current is the same in steady-state at $\omega t = 0$ and at $\omega t = \omega T_s$

$$i_{Lpa}(\omega t = 0) = i_{Lpb}(\omega t = \omega T_s) = I_{op} \qquad (b4)$$

The capacitor voltage change is the result of current $i_c = i_{Lpa}$ in interval $0 \le \omega t \le \alpha_p$

$$Z \int_0^{\alpha_p} i_{Lpa}\, \mathrm{d}(\omega t) = v_{cp} - v_{cn} = 2v_{c1} \qquad (b5)$$

Using the positive and negative sequence components of the variables $v_{i1}$, $v_{i2}$ and (a5) the output voltage is

$$v_{op} = \frac{u_u v_{ip}\sqrt{A_p}(\omega T_s - \alpha_p) + 2(1+\delta)v_{i1}}{\sqrt{A_p}(\omega T_s - \alpha_p) + 2u_d} \qquad (b6)$$

If $u_u=1 \rightarrow u_d=0$, if $u_u=0$ $u_d$ can be either one or zero, that is, (b6) can be rewritten in form

$$v_{op} = u_u v_{ip} + (1+\delta)\cdot \frac{2v_{i1}}{(\omega T_s - \alpha_p)\sqrt{A_p} + 2u_d} \qquad (b7)$$

and

$$v_{on} = -u_u v_{in} + (1-\delta)\cdot \frac{2v_{i1}}{(\omega T_s - \alpha_n)\sqrt{A_n} + 2u_d} \qquad (b8)$$

where

$$A_{p,n} = (1+\cos\alpha_{p,n})/(1-\cos\alpha_{p,n}) \qquad (b9)$$

## Appendix C $v_{c1}$ and $v_{c2}$ in Discontinuous Conduction Mode (DCM)

Assuming discontinuous choke currents $i_{Lp}(0)=I_{op}=0$. $v_{c1}$ and $v_{c2}$ are on the basis of (b1), (b5) and (b3) and their **n** channel equivalents

$$v_{c1} = 2\frac{v_{i1} - u_d v_{o1}}{A_p + A_n} = \frac{v_{i1} - u_d v_{o1}}{A_1} \qquad (c1)$$

and

$$v_{c2} = v_{i2} - u_d v_{o2} - \frac{A_2}{A_1}(v_{i1} - u_d v_{o1}) \qquad (c2)$$

## Appendix D Border between CCM and DCM

The extinction angle of choke current $i_{Lp}$ from the condition $i_{Lp}(\omega t=\alpha_{ep})=0$, and $I_{op}=0$ [see (b1) and (b2)] is

$$\alpha_{ep} = \alpha_p + \frac{2v_{c1}}{v_{op} - u_u v_{ip}}\sqrt{A_p} \qquad (d1)$$

The operation is just on the border between CCM and DCM when

$$\alpha_{ep} = \omega T_s \qquad (d2)$$

The operation is continuous when $\alpha_{ep} > \omega T_s$ and discontinuous when $\alpha_{ep} < \omega T_s$.

## REFERENCES

[1] J. Hamar, I. Nagy, "Control Features of Dual Channel DC-DC Converters", IEEE Transactions on Industrial Electronics, 2002, pp.1293-1305.

[2] J. Hamar, I. Nagy "Asymmetrical Operation of Dual Channel Resonant DC-DC Converters", IEEE Transactions on Power Electronics, 2001, pp.83-94.

[3] van Wyk JD, Lee FC, Boroyevich D "Power electronics technology: Present trends and future developments" Proceedings of the IEEE 89 (6): 799-802 Sp. Iss. SI JUN 2001

[4] Orabi M, Ninomiya T, " Analysis of PFC Converter Using Energy Balance Theory", IECON'03, IEEE-IES, 2003, Roanoke, Virginia, USA, pp. 544-549.

[5] B. Grezsik, Z. Kacymarzyk, "Model of Commutation for Power Electronic Converter", International Conference on Electrical Drives and Power Electronics, EDPE 2003, 24-26 September, High Tatras, Slovakia. Pp. 596-601.

[6] V. Oleschuk, F. Blaabjerg, "Novel Simplifying Approach for Analysis and Synthesis of Space Vector PWM Algorithms", EPE'2003, Toulouse, France, 2003, ISBN: 90-75815-07-7.

[7] Grzegorz Iwanski, Wlodzimierz Koczara, "Sensorless Direct Voltage Control of the Stand-Alone Slip-Ring Induction Generator", Trans. on Industrial Electronics, vol. 54, no. 2, pp. 1237-1239, Apr. 2007.

[8] Funato, H.; Kawamura, A.; Kamiyama, K., "Realization of negative inductance using variable active-passive reactance (VAPAR)", Power Electronics, IEEE Transactions on, Volume 12, Issue 4, Jul 1997 Page(s):589 - 596Digital Object Identifier 10.1109/63.602553

[9] L. Benadero, R. Giral, A. El Aroudi and J. Calvente, " Stability Analysis of Single Inductor Dual Switching Dc-Dc Converter" in Proceedings of the 8th International Conference on Modeling and Simulation of Electric Machines, Converters and Systems (ELECTRIMACS'05),(Hammamet, Tunisia), Apr. 17-20, 2005. CD Rom ISBN:2-921145-51-0.

[10] B. Robert, H. H. C. Iu, and M Feki, "Widening the Stability Range of a PWM Interver Using a Robust Chaos Control" in Proceedings of the 10th European Conference on Power Electronics and Applications (EPE2003), (Toulouse, France), Sept. 2-4, 2003. CD Rom ISBN:90-75815-07-7.

[11] M.Z. Youssef, H. Pinhiero, and P.K. Jain," A Sampled-Data Reduced Order Dynamic Model for a Self-Sustained Series-Parallel Resonant Converter", EPE (European Power Electronics and Drives) Journal, vol. 15, pp.5-15, Feb.2005.

[12] P.Bauer, J.Leuchter, V.Rerucha, O.Kurka: "Power Peaks of Mobile Electrical Power Sources with VSCF Technology", Proceedings of the EDPE International Conference on Electrical Drives and Power Electronics, Dubrovnik 2005, September 26-28, ISBN 953-6037-14-3-2

[13] Dudrík,J., Dzurko,P.," An Improved Soft-Switching Phase-Shifted PWM Full-Bridge DC-DC Converter" Proc. of the Int. Conf. on EPE-PEMC 2000, Vol. 2, 2000, Košice, pp.65-69.

[14] A. El Aroudi, M. Debbat, G. Olivar, J. Calvente, R. Giral, and L. Martínez-Salamero, "Stability Analysis and Bifurcations of Switching Regulators with PI and Sliding Mode Control" in Proceedings of the International Conference EPE-PEMC2002, (Cavtat-Dubrovnik, Croatia), Sept. 9-11, 2002.

[15] M. Meinhardt, G. Cramer, "Multi-String-Converter: The Next Step in Evolution of String-Converter Technology" EPE 2001, Graz

[16] Chan W. C. Y., Tse C. K., "Study of Bifurcations in Current-Programmed DC-DC Boost Converters: from Quasi-Periodicity to Period Doubling", IEEE Transactions on Circuits and Systems, I. 44 (12) pp. 1129-1142.

[17] E. Vidal-Idiarte, L. Martinez-Salamero, H. Valderrama-Blavi, F. Guinjoan, J. Maixe, "Analysis and Design of H∞ Control of Nonminimum Phase-Switching Converters", IEEE Trans. Circuits and Systems I, October 2003, Vol. 50, No. 10, pp. 1316-1323.

[18] C. K. Tse and M. di Bernardo, "Complex Behavior in Switching Power Converters", Proceedings of IEEE, Special Issue on Applications of Nonlinear Dynamics to Electronic and Information Engineering, vol. 90, May 2002, pp. 768-781.

[19] Yu. V. Kolokolov, S.L. Koschinsky, A. Hamzaoui," Comparative Study of the Dynamics and Overall Performance of Boost Converter with Conventional and Fuzzy Control in Application to PFC" , IEEE Power Electronics Specialist Conference PESC'04, June 20-25, 2004, Aachen, Germany, pp.2165-2171.

[20] Zd. Čerovskỳ, Vl. Pavelka "DC-DC Converter for Charging and Discharging Super-Capacitors Used in Electric Hybrid Cars", EPE'2003, Toulouse, France, 2003

[21] G. Buja, S. Castellan, "Active filter for high-power medium-voltage diode rectifiers", EPE'2003, Toulouse, France, 2003.

[22] Y. Nishida, "A New Buck-And-Boost DC-DC Converter (Tokusada Converter)", EPE'2003, Toulouse, France, 2003.

[23] M.Y.Ayad, S.Raël, B.Davat, "Hybrid Power Source Using Supercapacitors and Batteries, EPE'2003, Toulouse, France, 2003

[24] M. Pavlovsky, S.W.H. de Haan, J.A. Ferreira, "The ZVS, Quasi-ZCS Converters, a Family of Topologies with an Optimal Current Waveform for High Frequency Switching", EPE'2003, Toulouse, France, 2003.

# Output Filtering of the Customer-end Inverter in a Low-Voltage DC Distribution Network

Pasi Peltoniemi, Pasi Nuutinen, Pasi Salonen, Markku Niemelä, Juha Pyrhönen

Lappenranta University of Technology, Department of Electrical Engineering,

Lappeenranta, Finland

e-mail: *pasi.peltoniemi@lut.fi*

*Abstract*—**Passive filter topologies used at the output of the customer-end inverter in a low-voltage DC-distribution (LVDC) network are compared. The comparison of filter topologies is based on the requirements set for the voltage quality and issues such as the physical size of the filter, losses, the complexity of the needed control and the cost. It is concluded that most of the losses of LVDC system are coming from the customer-end filters. Therefore, to make the LVDC network as efficient as or even more efficient than the current three phase low voltage distribution network, filter topologies and their properties are evaluated. Because filters are used in distribution system, the cost and the losses have the governing roles. The study reveals the most suitable filter topology to be used in a LVDC distribution network, when using the specific properties presented.**

## I. INTRODUCTION

A low-voltage distribution network is traditionally based on a three-phase AC system. However, a growing demand on the renewable energy systems and other distributed energy production solutions present new kind of challenges to the power distribution systems and their structures. On the other hand, in some of the European countries the age of the power distribution systems is closing to a point where they need to be renovated. It means huge investments from the network holder. Therefore it should be considered what kind of present technical solutions could bring more reliability and secure better power quality in the future network with affordable cost. One possible solution to this problem could be that some parts of the present systems' low-voltage distribution would be replaced with a DC distribution system. Equipment to be used in the system are covered by the LVD 2006/95/EC[1], that is, equipments designed for use with a voltage rating between 50-1000 VAC or 75-1500 VDC. One of the possible structures of the LVDC distribution system is illustrated in Fig. 1. In the system at the customer-end the dc voltage is inverted with either a three-phase or a single-phase inverter to an ac voltage. When a residential power supply is considered, in most applications there is no need for three-phase supply. Therefore, only the single-phase customer-end system is considered here.

## II. SYSTEM REQUIREMENTS

The output voltage of the customer-end inverter has to fulfill the requirements set by EN50160[2]. The standard defines, for example, these three requirements for the phase voltage:

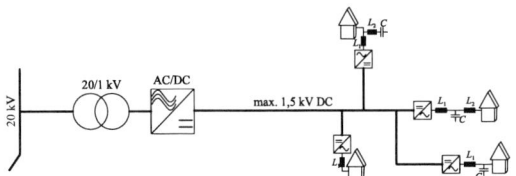

Fig. 1. A low-voltage DC distribution network representing the analysed filter topologies

- frequency deviation (50 Hz $\pm$ 1 % 99.5 % of the year),
- voltage level and fluctuation (230 V (rms) $\pm$ 10 %) and
- harmonic distortion (THD < 5 %).

In the case of dc distribution the level of the harmonic distortion of the customer's voltage is mostly determined by the customer inverter. With present day power electronic switches the switching frequency must remain moderate and therefore, to fulfill the requirement, the inverter output voltage is filtered with filters having expensive components. To design a filter that fulfills the requirement for the harmonic distortion in the nominal operating point is quite straightforward, but in a distribution system the quality of voltage should fulfill the requirements in every operating point, including very small partial loads.

To make the filters suitable for a power distribution system, several practical issues has to be solved. Especially the role of costs and losses are emphasized. That is, better efficiency is wanted with a reasonable cost. This makes the filter design challenging, since the efficiency of the system should be at least as good as the efficiency of the current system in every mode of operation.

While being cost effective and having low losses and good attenuation properties of harmonics, small physical dimensions and low audible noise are also preferred properties of the filter. Finally, from the implementation point of view the complexity of the needed control algorithm and number of needed measurements should be as simple and as few as possible.

The aim of this paper is not to present a new method to define the inductance and capacitance values for filters that have good attenuation capability of switching harmonics. It illustrates the effect of different options, such as modulation method and core material of the

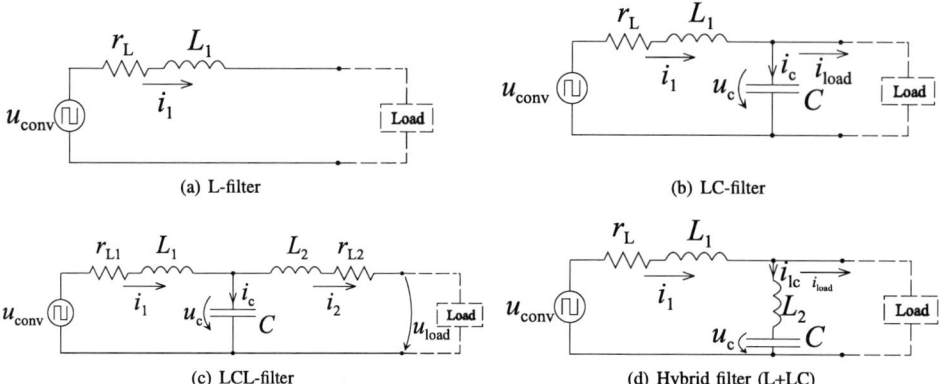

Fig. 2. The equivalent circuits of investigated filter topologies.

inductor, the losses, physical size and the investment cost of the filter topology. It also includes the discussion about the accuracy of the output voltage that depends on the measurements and the complexity of the control needed, which are important when simple implementation and low costs of control electronics are wanted.

### III. FILTER TOPOLOGIES AND SELECTION PROCEDURE

Nowadays, single-phase inverters are typically used in renewable energy solutions such as solar panels as the interface to the grid [3]. Furthermore they are used in uninterruptable power supply (UPS) equipments [4]. In these and many other single-phase inverter applications an LC filter (Fig.2b) is commonly used. Other typically used filter topologies include L- and LCL-filter (Fig.2a and Fig.2c). Especially the LCL filter has proven to be an attractive solution in three-phase applications [5]. In addition to previous topologies also a hybrid filter topology (L+LC) is considered (Fig.2d). In the hybrid filter the purpose of the LC-branch is to attenuate harmonics in the vicinity of the first group of switching frequency harmonics.

The properties of the filter can be affected with several factors. These factors include at least the switching frequency of the inverter $f_{sw}$, the modulation method of the inverter, the core material of the filter inductor and parameters of the passive filter components ($R,L,C$). Therefore, a selection procedure, to choose the most suitable filter topology, is made. The selection is given as step-by-step procedure

1) the switching frequency $f_{sw}$ and the modulation method are chosen,
2) inductor(s) inductance(s) and capacitor's capacitance are obtained from simulations, where single-phase full-bridge inverter is driven with chosen specifications,
3) the filter inductors are designed, different core materials are evaluated and the physical properties of the different inductors are approximated,
4) the filter losses are approximated,
5) the cost is approximated,

6) the needed measurements are selected and complexity control is discussed.

The comparison gives the most suitable filter topology to be used in a LVDC system, when the specific parameters, such as, modulation method and switching frequency, are used. The comparison does not give the global optimum solution and by varying the switching frequencies, modulation methods or core materials a different result may be reached. However, in a nowadays single-phase inverter applications the modulation methods, core materials and switching frequency values could be regarded as commonly used.

In the selection procedure the dc voltage of the dc link is set to 750 VDC and the rms value of the fundamental frequency of the customer voltage is 230 V. The nominal power of the customer inverter is 10 kVA that corresponds to 43.5 A of current. The filter inductors are assumed to have a toroidal core.

### IV. HARMONIC DISTORTION

The frequency spectrum of the inverter output voltage waveform includes, in addition to fundamental frequency, a number of frequencies due to a modulation process. The amount of harmonic distortion in the voltage waveform is typically measured using total harmonic distortion (THD) defined as

$$\text{THD} = \frac{\sqrt{\sum_{i=1}^{n} U_i^2}}{U_1}, \qquad (1)$$

where $n$ is the number of harmonic frequency components included in the calculation. In this paper all harmonic frequency components $f_n < 0.5 f_{sample}$ are included in a calculation of the THD ($f_{sample}$ is the sampling frequency).

To find out the frequency components of the output voltage some form of discrete fourier transform (DFT) is typically used. Especially the use of DFT makes the definition of THD vulnerable, since different results can be obtained when using different specifications. Therefore, the issues discussed in the following should be given a special attention to make the results comparable. The main focus in making the results comparable among

TABLE I

SIMULATED RESULTS, WHEN $P = 10$ kW, $U = 230$ V, $I = 43.5$ A, MINIMUM PULSE LENGTH $t_{min} = 2$ $\mu$s AND DEAD-TIME $t_d = 1$ $\mu$s.

| Mod. method | $f_{sw}$ | Filter parameters, when 8 $\mu$H capacitor is used | | | |
|---|---|---|---|---|---|
| | [kHz] | Filter Type | $L_1$ [$\mu$H] (pu) | $L_2$ [$\mu$H] (pu) | THD [%] |
| Bipolar | 7.5 | L | 3200 (0.193) | - | 4.93 |
| | | LC | 870 (0.052) | - | 4.94 |
| | | LCL | 390 (0.023) | 195 (0.012) | 4.94 |
| | | Hybrid | 1900 (0.113) | 56 (0.003) | 4.91 |
| | 10 | L | 2400 (0.142) | - | 4.97 |
| | | LC | 535 (0.032) | - | 4.94 |
| | | LCL | 264 (0.016) | 132 (0.008) | 4.95 |
| | | Hybrid | 1180 (0.070) | 32 (0.002) | 4.97 |
| Unipolar | 7.5 | L | 1450 (0.086) | - | 4.97 |
| | | LC | 550 (0.033) | - | 4.96 |
| | | LCL | 216 (0.013) | 107 (0.006) | 4.96 |
| | | Hybrid | 175 (0.010) | 14 (0.001) | 4.94 |
| | 10 | L | 1200 (0.071) | - | 4.89 |
| | | LC | 340 (0.020) | - | 4.99 |
| | | LCL | 186 (0.011) | 92 (0.005) | 4.95 |
| | | Hybrid | 108 (0.0064) | 8 (0.0005) | 4.98 |

different switching frequencies is the relation between the switching frequency and the sampling interval. More specifically, the relation between the frequency of the first group of switching harmonics $f_{sw,1st}$ and the sampling interval $f_s$, that is

$$r = \frac{f_{sw,1st}}{f_s}. \qquad (2)$$

This is due to a different location of the first group of harmonics in different modulation methods. Second, spectrums are comparable only when their frequency resolution is equal. The frequency resolution is a ratio of sampling frequency and the number of samples.

Now, the aim of simulations is to find out the values of the passive components of different filter topologies that fulfill the requirement set for voltage THD (<5 %). In these simulations two different modulation methods are used, namely, the bipolar and the unipolar modulation method specified in [6]. In the simulation model the dead-time $t_d$ is modeled and in simulations it is 1 $\mu$s. Modulation methods run at 7.5 and 10 kHz of switching frequency. In the bipolar modulation method the first group of switching harmonics is located at the switching frequency, therefore, 75 and 100 kHz of sampling frequencies are used. However, in the unipolar method the first group of switching harmonics is located at twice the switching frequency and therefore, 150 and 200 kHz of sampling frequencies are used. To obtain a frequency resolution of 5 Hz, ten cycles of fundamental frequency (50 Hz) is measured, that is, 0.2 s in steady-state with a nominal resistive load. In addition to the sampling time of the output voltage the calculation interval of the model should be proportional to the highest frequency that is investigated. In these simulations the sampling interval of the simulation model was one tenth of the sampling interval of the voltage waveform. In the simulations a capacitance of 8 $\mu$F is used to make the comparison between inductors, which are considered to make the differences between different filter types.

The results of the simulation study are given in Table I. It can be concluded from the table, that using 10 kHz as a switching frequency results in smaller inductance values. Also the use of unipolar modulation method leads to even smaller values of inductances, mainly because of the shifting the switching harmonics to higher frequencies. One very interesting observation is made from the table, that is, the hybrid filter produces the smallest inductance values at unipolar method while at bipolar method it produces almost the biggest values of $L_1$ inductances. The reason for this is that with the unipolar method the series LC circuit can be tuned to twice the switching frequency and also that the L filter's passband is higher than with the bipolar method that allows lower inductance values. It seems that the hybrid filter becomes an attractive solution when the unipolar modulation method is used.

Next, filters used with the unipolar modulation at 10 kHz switching frequency are designed to compare different core materials and properties of filters made of those specific materials.

## V. INDUCTOR DESIGN

The goal of the design is to make as small inductor cores as possible that do not saturate in normal operation. It means that filter cores are designed in such a way that the material maximum flux density $B_{max}$ is achieved with the maximum current value $I_{max}$ at nominal load. The maximum current value $I_{max}$ also includes the assumed current ripple coming from the modulation. Therefore, at nominal load the rms current equals 43.5 A, which corresponds to 61.5 A peak value and the ripple is assumed to be approximately 5 % of the peak value, so $I_{max}$ equals 65 A.

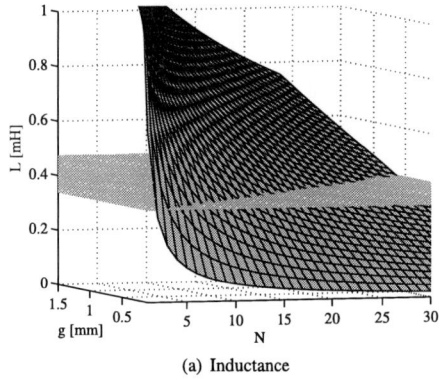

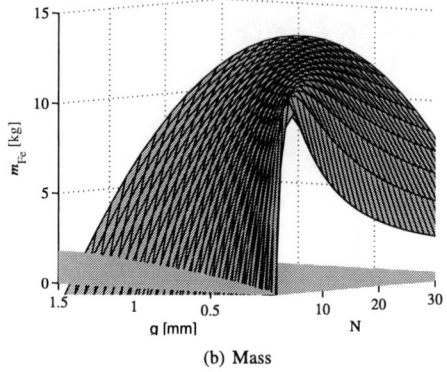

(a) Inductance    (b) Mass

Fig. 3. Surface plots of (a) the inductance and (b) the mass of the core. The solutions exists in grid points of the surface.

TABLE II
PROPERTIES OF CORE MATERIALS

| Material | $B_{max}$ [T] | $\mu_r$ | $\rho$ [kg/m$^3$] |
|---|---|---|---|
| SiFe | 1.5 | 7000 | 7650 |
| Amorphous | 0.82 | 3000 | 8270 |
| Ferrite | 0.4 | 5000 | 5280 |

As mentioned in Section III filters are designed to be as toroidal cored. The thermal designing of the core is not considered in this paper. For that reason cores may become even bigger when thermal issues are taken into account. The peak flux density $B_{max}$ is varied according to material used and it is chosen somewhat lower than the saturation flux density of the material $B_{sat}$. The materials considered here are silicon-iron (SiFe), amorphous metal and ferrite. Some of the properties of the materials used in the design are given in Table II. To control the peak flux density $B_{max}$ in the material, the airgap is used. In the design only single airgap is used, but in case of a bigger airgap length $g$, in practice, it can be divided into several parts.

*A. Design method*

When the inductance $L$ is known and the airgap length $g$ and the number of winding turns $N$ are selected, the cross-section area of the core $S_{Fe}$ can be solved as

$$S_{Fe} = \frac{Lg}{\mu_0 N^2}, \tag{3}$$

where $\mu_0$ is the permeability of free space. Now, the cross-section area of the airgap $S_g$ is determined as

$$S_g = S_{Fe} + 2\sqrt{S_{Fe}}g + g^2. \tag{4}$$

Because material properties (TableII) and $I_{max}$ are known, the length of the airgapped core $l_{Fe}$ can be solved as

$$l_{Fe} = \frac{\mu_0 \mu_r}{B_{max}} \left[ N I_{max} - B_{max} \frac{S_{Fe}g}{\mu_0 S_g} \right]. \tag{5}$$

The volume of the core $V_{Fe}$ can be calculated as

$$V_{Fe} = S_{Fe} l_{Fe} \tag{6}$$

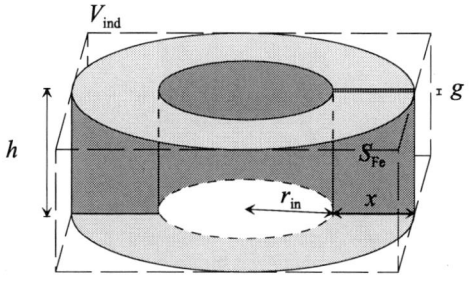

Fig. 4. Toroidal core of the filter with dimensions. Figure also illustrates the estimated volume of the inductor $V_{ind}$.

and the mass $m_{Fe}$ as

$$m_{Fe} = \rho V_{Fe}, \tag{7}$$

where $\rho$ is the density of the core material. By determining the minimum inner radius $r_{in,min}$, that is needed for winding arrangements, the other dimensions of the core cross-section can be solved as

$$x = 2\left(\frac{l_{Fe} + g}{2\pi} - r_{in,min}\right). \tag{8}$$

The height of the core is given as

$$h = \frac{S_{Fe}}{x}. \tag{9}$$

The dimensions of the cross section of the core $h$ and $x$ are illustrated in Fig. 4.

In order to obtain the optimum combination of the number of turns $N$ and the airgap length $g$, several different values of $N$ and $g$ are given to plot surfaces for the inductance $L$ and the mass of the core $m_{Fe}$. The surfaces of the 340 $\mu$H (LC-filter, unipolar, 10 kHz) inductor are illustrated in Fig. 3, where the grid points on the surface shows the solutions. It can be noticed that the grid represents a certain resolution. This resolution determines how accurately the inductance can be achieved. The inductance value $L$ is chosen as near as possible to the wanted value. The value of the core mass $m_{Fe}$ is determined according to point where inductance meets the

1766

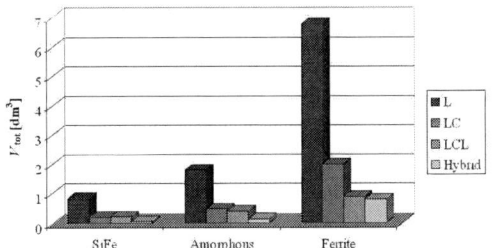

Fig. 5. The total volumes of the toroidal cores of different filters with different core materials.

requirement. The resolution in inductance values results in a wide variation of the core mass. The steep form of the core mass surface can be seen in Fig. 3b.

### B. Volume of the filter

The design method guarantees that the maximum flux density of the core $B_{\max}$ is the one chosen for the material (and given in TableII). Therefore, the volume of the core is in that sense the minimum. The total volume required by the inductor $V_{\mathrm{ind}}$ is estimated as a box. The box is illustrated in Fig. 4. The total volume of the filter is then calculated as the sum of volumes of its inductors. The volumes of filters are calculated with the filter topologies used at unipolar modulation at 10 kHz of switching frequency.

The total volumes of inductors with different materials are shown in Fig. 5. The chart shows that the minimum volumes are obtained with SiFe because it has the highest maximum flux density $B_{\max}$. It can be also noticed that L filter produces the highest volumes in spite of its minimum number of components. When the maximum flux density of the material is lower the difference between filters become noticeble and therefore, for example with ferrite, only LCL and hybrid filters produce reasonable choices when the sizes of the filters are considered.

## VI. INDUCTOR LOSSES

Inductor losses in the inverter application still remains an unsolved problem. Issues that make the modeling of inductor losses difficult are the rich frequency content of the output voltage and current and the unknown behaviour of materials at those frequencies. Therefore, typically the losses are approximated using either data provided by the material manufacturer or they are approximated using one of the various methods developed.

### A. SiFe cores

In this paper the *loss separation*-method (LSM) is used to approximate the losses of laminated SiFe core inductors. The method has been previously used especially to estimate the losses of laminated iron core inductors and electrical machines. In LSM the inductor core losses are separated into parts. The core losses are divided into

hysteresis $P_{\mathrm{hy}}$, eddy current $P_{\mathrm{ec}}$ and excess losses $P_{\mathrm{ex}}$. The total iron losses can be approximated as

$$P_{\mathrm{Fe}} = P_{\mathrm{hy}} + P_{\mathrm{ec}} + P_{\mathrm{ex}}. \qquad (10)$$

According to [7] when the losses are considered with a single frequency (sinusoidal), the formulation for the hysteresis losses $P_{\mathrm{hy}}$ could be expressed as

$$P_{\mathrm{hy}} = V_{\mathrm{Fe}} f \int_0^B H dB, \qquad (11)$$

and the eddy current losses $P_{\mathrm{ec}}$ as

$$P_{\mathrm{ec}} = \frac{\sigma d^2}{24} \omega^2 B_{\max}^2, \qquad (12)$$

where $\sigma$ is the conductivity of the core material and $d$ is the thickness of the steel plate. Finally, the excess losses $P_{\mathrm{ex}}$ can be expressed as

$$P_{\mathrm{ex}} = 8.76 \sqrt{\sigma G V_0 S} B_{\max}^{1.5} f^{1.5}, \qquad (13)$$

where $G$ and $V_0$ are parameters related to metallurgical properties of the material and $S$ is the cross section of the iron sheet.

However, when inverter supply is considered, formulation made for single frequency is not applicable nor sufficient and therefore loss approximations must be carried out using a different kind of an formulation. Here the formulation is transformed in such a way that it also includes the approximation of core losses that occur in the inductor at higher frequencies (i.e. at switching frequency and its harmonic multiples). The formulation is of the form

$$P_{\mathrm{Fe}} = \sum_n c_{n,\mathrm{hy}} P_{n,\mathrm{hy}} + \sum_n c_{n,\mathrm{ec}} P_{n,\mathrm{ec}} + \sum_n c_{n,\mathrm{ex}} P_{n,\mathrm{ex}}, \qquad (14)$$

where $c_n$'s are factors that describe the frequency-dependence of the amplitude at specific frequency, while being 1 at the fundamental frequency.

### B. Ferrite and amorphous cores

Next, the losses of the filters, whose inductor cores are of amorphous metal or ferrite, are calculated. In this paper the core losses are approximated using the data that manufacturer of the material has provided, because the behaviour of these materials is not well-known and therefore the LSM could not be used to approximate them.

With the ferrite material the core losses are approximated by using the equation given by the manufacturer, that is

$$P_{\mathrm{Fe}} = a f^c B_{\max}^d, \qquad (15)$$

where $a$, $c$ and $d$ are the parameters that the manufacturer has used for fitting the loss curve to the measured losses of the material. However, the given equation is not valid to calculate the losses of the PWM inverter, because it only takes a single frequency into account. Therefore, to get a better approximation for the core losses, (15) is modified

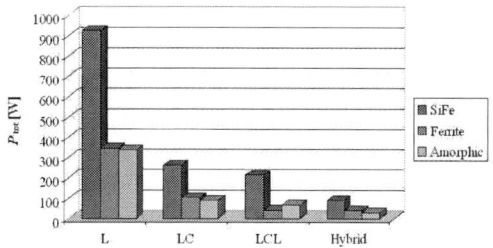

Fig. 6. The total calculated losses of different filters with different core materials at nominal load.

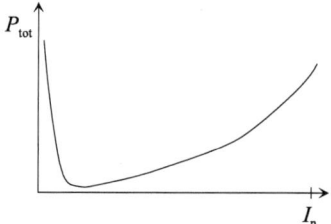

Fig. 7. Expected behaviour of losses of $L_1$ inductors in LC, LCL and hybrid filters as a function of load current. The figure is exaggerated for illustration purposes.

to include the core losses at switching frequency. Now the equation is expressed as

$$P_{\text{Fe}} = a f_{\text{fund}}^c \hat{B}_{@f_{\text{fund}}}^d + a f_{\text{sw}}^c \left( \hat{B}_{@f_{\text{sw}}}^d + 2a \hat{B}_{@2f_{\text{sw}}}^d \right), \tag{16}$$

where $\hat{B}$'s are the maximum flux densities calculated at the specific frequency. The equation returns the core losses as W/kg. The losses of the first inductor at nominal load with 10 kHz switching frequency are 10 W/kg and for the second inductor, that is, inductors in LCL- and hybrid filter is 8 W/kg.

The method to approximate the losses for amorphous metal is based on the same idea as with the ferrite material. The losses are given in charts by the manufacturer but they only consider losses at single frequency. By weighting the flux densities according to frequency spectra and summing up the losses at different frequencies the more accurate approximation is obtained. At nominal load with 10 kHz switching frequency the losses of amorphous metal equals 30 W/kg for the first inductor and 20 W/kg for the second inductor.

Finally, the total losses $P_{\text{tot}}$ also include copper losses $P_{\text{Cu}}$, which can be approximated simply using

$$P_{\text{Cu}} = I_{\text{rms}}^2 r_{\text{L}}, \tag{17}$$

where $I_{\text{rms}}$ is the rms phase current and $r_{\text{L}}$ is the resistance of the wire. In (17) the skin and proximity effects are not included.

The results of the study are shown in Fig. 6. From the figure it is concluded that the losses of filter with SiFe core become very high when compared with the losses of inductors made of other materials. It also reveals that the losses of hybrid filter are the smallest. However, differences between LCL and hybrid filter become quite small when using amorphous metal or ferrite. But even small differences may become important, when lifetime expenditures of filters are considered.

### C. Discussion about inductor losses

Typically the losses are given as a function of frequency or as a function of peak flux density $B_{\text{max}}$. In the case where the inductor is excited with a single frequency these charts provided by the manufacturer give a good approximation. The core losses approach zero as the flux density approaches zero. With the inductor that is used

in inverter application the behaviour of losses can be quite different. Let us consider, for example, the LC filter. When operating at nominal or partial loads the losses of the inductor behave in the same manner as with the inductor that is excited with a single frequency. However, the losses are increased, because of the presence of the switching harmonics (i.e. increased eddy current losses). When the system operates at no-load or with very small partial loads the core losses increase exponentially with laminated iron core inductors. The reason for this phenomenon is that the frequency of the inductor current could be considered to be changed from 50 Hz to a switching frequency. In no-load operation the polarization of magnetic domains of the material is changed according to switching frequency thereby increasing the losses. So the loss profile as a function of load current for the $L_1$ inductors in LC, LCL and hybrid filters reminds the one presented in Fig. 7. On the contrary, the behaviour of the losses in $L_2$ inductor of LCL filter could be considered to be very close to a behaviour of the losses of an inductor excited with a single frequency. Furthermore, the core losses with cores made of ferrite or amorphous metal could be assumed to remain almost constant as a function of load. The reason is that the core losses at low frequency with these materials are quite negligible and the only core losses occur at switching frequencies, of which amplitudes remain constant with the constant modulation index.

### VII. COST

The total cost of the filter includes investment and maintenance costs and also the costs caused by the losses. In order to make a comparison, the maintenance costs could be assumed equal among different filter types and the comparison is made according to investment costs and the costs caused by the losses. The investment costs consists of production, material and installation costs. Here only the core material costs are considered in addition to estimation of the costs caused by the losses.

The typical behaviour of the customer average energy usage during one year living in a single-family house in Finland is illustrated in Fig 8a. To approximate total average losses, the losses at the nominal operating point are weighted as

$$P_{\text{loss}} = P_{\text{tot}} * \left( \frac{r}{100} \right)^2, \tag{18}$$

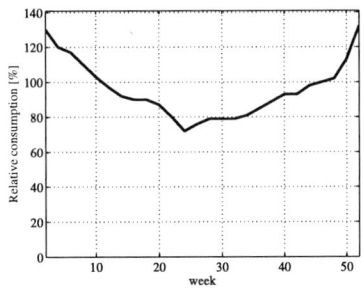
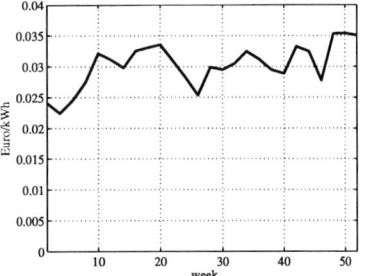

(a) Average consumption of energy     (b) Electric energy price variation at year 2006

Fig. 8. In (a) a typical behaviour of the customer average energy usage during one year, living in a single-family house in Finland is shown and in (b) the variation of the electricity price during the year 2006 [8] is illustrated .

where $P_{\text{tot}}$ is the losses at nominal load with specific filter and material and $r$ is the relative comsumption given in Fig. 8a. From (18) the average losses during the time period are obtained. The average losses must be multiplied by the time of the observation period $t_{\text{period}}$ to obtain electrical energy $E_{\text{loss}}$, that is

$$E_{\text{loss}} = \int_0^t P_{\text{loss}} dt \approx \sum_i P_{i,\text{loss}} t_{i,\text{period}}. \quad (19)$$

This electrical energy is then multiplied by the price of the single unit of electric energy as

$$\text{price} = E_{\text{loss}} \cdot \text{price per energy unit.} \quad (20)$$

The average cost of the losses with different filter types are shown in Fig.9a. The price profile used in the calculation is shown in Fig. 8b.

Next, the investment cost coming out of the core material is approximated. The cost is assumed to be linear with respect to weight of the core when cores with same current rating are considered. Then the material cost can be evaluated as

$$\text{Cost}_i = H_i * m_{\text{Fe},i}, \quad (21)$$

where $i$ is refers to the type of material. The prices per mass unit of different materials are approximated to be $H_{\text{SiFe}}$=15 €/kg, $H_{\text{fer}}$=65 €/kg and $H_{\text{Am}}$=85 €/kg for iron, ferrite and amorphous, respectively. The prices of the materials are approximated according to the price level of year 2008 and furthermore they are approximated to be somewhat higher not to present too optimistic results. The material costs of different filter types are shown in Fig.9b.

The filter topologies which have proven to be advantageous according to previous studies in the paper, are more expensive in material costs. For example, the core material costs of an SiFe hybrid filter equals 6 € and 33 € for amorphous material. The cost coming from the losses are 22 € and 7.3 €, respectively. Now, if the time of service for filter is 15 years in average, then the total lifetime costs for a single SiFe cored filter equals 338 €, where the same price for amorphous cored filter is 142 €. This means that with 10 customers using hybrid filter,

of which core is made of amorphous material instead of SiFe, 2000 € profit is made during the lifetime. In addition, todays electric energy price is higher than the one during the year 2006 and it makes the amorphous and ferrite materials even more attractive solutions from the economical point of view. However, it should be noticed that these values give an coarse approximation of prices and should not be taken as a reference values.

## VIII. VOLTAGE CONTROL AND IMPLEMENTATION

The filters will be used in LVDC distribution system, where the main focus will be on the quality of the output voltage. Therefore the control system is designed to maintain the quality of voltage inside the limits given by EN50160. To maintain the quality of the voltage, the accurate knowledge of the output voltage becomes important. It means that we must either measure, which is preferred, or to estimate the output voltage of the customer-end system. The estimation presents inaccuracy to control system and estimate's reliability can always be questioned. From this point of view the control system can become more complex with the filters whose output voltage is estimated. On the other hand, the more the filter has passive elements the higher is the degree of the filter and to ensure the stability of every filter state, sometimes the control algorithm may become more complex.

First, let us consider the L-filter, which is the simplest filter topology. To control the voltage either inductor current $i_1$ or its output voltage $u_{\text{load}}$ needs to be measured. If the current is measured then the voltage needs to be estimated. If the output voltage is measured there is no need for estimation, however, to make a fast response control, the measurement of the inductor current might also be needed.

With the LC filter the inductor current $i_1$ and the capacitor voltage $u_c$ are measured. Now, because there is no extra elements after the capacitor, the output voltage $u_{\text{load}}$ equals the capacitor voltage $u_c$. It means that the accuracy of the voltage is determined by the used measurement equipment. With LC filter the output voltage (i.e $u_c$) can be accurately controlled and typically both states are measured.

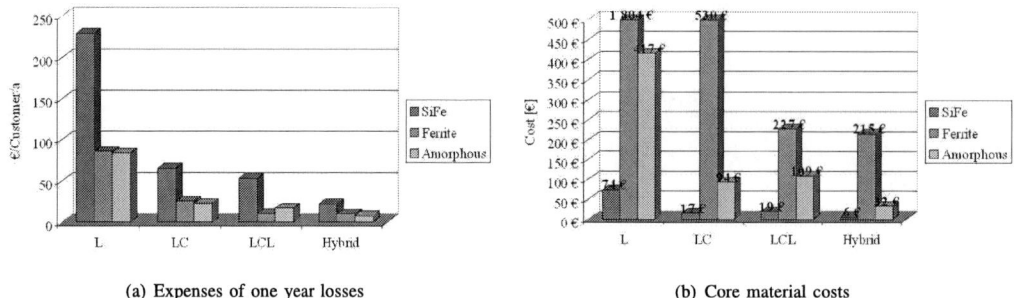

(a) Expenses of one year losses                    (b) Core material costs

Fig. 9.  Expenses of different filter types, when core material costs and losses are evaluated.

Now, let us consider the LCL-filter. Assume that the current of the inductor $i_1$, the capacitor voltage $u_c$ and also the current of the second inductor $i_2$ are measured. Even though inductor currents and capacitor voltage are measured the load voltage must be estimated. By measuring the output voltage the problem is solved, but then three measurements are needed. If we use two measurement and measure the inductor current $i_1$ and the capacitor voltage $u_c$, then the output voltage $u_{\mathrm{load}}$ has to be estimated. The drawback of the estimation is that it probably presents some extra inaccuracy to the voltage. To control the voltage as accurately as possible, the measurement of the output voltage is needed. So, it is concluded that by measuring the inductor current $i_1$ and the output voltage $u_{\mathrm{load}}$ the best voltage accuracy is obtained. To ensure the stable behaviour of the filter the capacitor voltage $u_c$ can be estimated. The estimation error of the capacitor voltage has no effect on output voltage, but it may have an effect on the control system performance.

The operation principle of the hybrid filter is different from other filters. First, let us assume that the inductor current $i_1$ and the capacitor voltage $u_c$ are measured. The capacitor voltage is not equal to output voltage, therefore, output voltage has to be estimated when this arrangement is used. Furthermore, the voltage of the LC branch capacitor voltage is probably not the best variable for control. Again, to maximize voltage accuracy the output voltage $u_{\mathrm{load}}$ should be measured. As a conclusion by measuring the inductor current $i_1$ and the output voltage $u_c$ the best performance is obtained.

It is concluded that for the voltage control, the output voltage can be either measured or estimated. However the estimation may present some extra inaccuracy to the voltage and measuring of the output voltage in the LVDC distribution application is needed. The measuring of the output voltage may lead to a need for estimating the states, which are also used in the control, with some filter types. The estimation always presents some amount of inaccuracy and may affect to the performance of the control system.

## IX. CONCLUSIONS

In LVDC distribution system a single-phase inverter is

supplying the customer. In order to fulfill the requirements set in standards for phase voltage, the output voltage of the inverter must be filtered. Four different filter topologies were studied. First, the filters parameters were determined using simulations. In simulations a capacitor's capacitance was given a fixed value, since the comparison between inductors was wanted. The comparison between filters' attenuation capability was carried out, using THD as a measure. Filters made of different core materials, matching specifications obtained from simulations, were designed and their volumes were compared. Approximation of losses of filters with different core materials were calculated at nominal load. The costs were determined and it was noticed that by choosing amorphous metal as a core material the average lifetime costs become lower than the costs with silicon-iron cored inductors although iron cored inductors core investment costs were lower than respective costs of amorphous metal cored inductor's. Finally, issues regarding filter implementation on a control system, control complexity and voltage accuracy were discussed. The study revealed that hybrid filter becomes the most attractive solution when losses and costs are considered.

## REFERENCES

[1] *Low voltage directive 2006/95/EC*, European Commission, Directive. [Online]. Available: http://ec.europa.eu/index\_en.htm[24. 5.2008]
[2] *Voltage characteristics of electricity supplied by public distribution systems*, CENELEC, Belgium, European Standard EN-50160, 1994.
[3] F. Blaabjerg, Z. Chen, and S. B. Kjaer, "Power electronics as efficient interface in dispersed power generation systems," *IEEE Transactions on Power Electronics*, vol. 19, no. 5, pp. 1184–1194.
[4] M. J. Ryan, W. E. Brumsickle, and R. D. Lorenz, "Control topology for single-phase UPS inverters," *IEEE Transactions on Industry Applications*, vol. 33, no. 2, pp. 493–501.
[5] M. Liserre, F. Blaabjerg, and A. Dell'Aquila, "Step-by-step design procedure for a grid-connected three-phase PWM voltage source converter," *Int. Journal of Electronics*, vol. 91, no. 8, pp. 455–460.
[6] N. Mohan, T. Undeland, and W. Robbins, *Power Electronics: Converters, Applications and Design.* Wiley, 2002.
[7] J. P. A. Bastos and N. Sadowski, *Electromagnetic Modeling by Finite Element Method.* CRC Press, 2003.
[8] "Electricity price statistics, In Finnish," Energiamarkki-navirasto. [Online]. Available: http://www.sahkonhinta.fi/summariesandgraphs[15.6.2008]

# Power Flow Control through a Multi-Level H-Bridge based Power Converter for Universal and Flexible Power Management in Future Electrical Grids

Stefano Bifaretti[*], Pericle Zanchetta[†], Yue Fan[†], Florin Iov[**], Jon Clare[†]

[*] University of Rome Tor Vergata/Department of Electronic Engineering, Rome, Italy, *bifaretti@ing.uniroma2.it*
[†] University of Nottingham/School of Electrical and Electronic Engineering, Nottingham, UK, *pericle.zanchetta@nottingham.ac.uk*
[**] Aalborg University/Institute of Energy Technology, Aalborg, Denmark, *fi@iet.aau,dk*

*Abstract*—The paper proposes a novel power conversion system for Universal and Flexible Power Management (UNIFLEX-PM) in Future Electricity Network. The structure is based on three AC-DC converters each one connected to a different grid, (representing the main grid and/or various distributed generation systems) on the AC side, and linked together at DC side by suitable DC isolation modules. Each port of the UNIFLEX-PM system employs a conversion structure based on a three-phase 7-level AC-DC cascaded converter. Effective and accurate power flow control is demonstrated through simulation in Matlab and Simulink environment on a simplified model based on a two-port structure and using a Stationery Reference Frame based control solution. Control of different Power flow profiles has been successfully tested in numerous network conditions such as voltage unbalance, frequency excursions and harmonic distortion.

*Keywords*—Distributed power, Converter control, Multilevel converters, Power management.

## I. INTRODUCTION

The present architecture of the electricity network is the result of technological and institutional development over many years, with most of the electricity generated in large power stations and transmitted, using a passive transmission line, through high voltage transmission systems. Power is then delivered to consumers via medium and low-voltage distribution system [1]. The power flow in this arrangement is only in one direction: from the central power stations to the consumers. Typically, such a layout for the electricity network leads to a national or regional monopoly of the supplier acting both at the transmission and distribution level.

Since most of the European countries have started to liberalize the electricity market, this monopoly will disappear. In order to enable the electricity market, multiple transmission system operators (TSOs) as well as distribution system operators (DSOs) will operate on the electricity network transparently and without discrimination under the governance of a regulator [1]. This scenario requires an increase penetration of the renewable energy resources (RES) and other distributed generation (DG) and an active role for DSOs in controlling the network stability, optimising central and distributed power inputs into the network, interconnection, etc.

Moreover, in order to reach this goal, the entire architecture of the electricity network must be redesigned on the basis of models, such as Micro-grids, an "Internet" model and Active Networks, derived by the information and communication technologies (ICT) that will transform the existing electrical grid into a smart one [2].

According to [1], active networks technically and economically may be the best way to facilitate DG initially in a deregulated market. Its architecture employs an increased number of power input nodes, as a result of DG, bi-directional energy flow is possible and new technologies are emerging that can enable the direct routing of electricity. New power electronics systems offer ways of controlling the routing of electricity and also provide flexible DG interfaces to the network.

Among the key technologies required to make these new network concepts a reality, Power Electronics is to play a major role as 100% of electricity produced by renewable energy sources has to be converted by power electronic equipments. Additionally, power flow control using power electronic converters is needed to ensure proper and secure work of the grid.

In such framework, the paper analyses a new power conversion structure to be employed for Universal and Flexible Power Management (UNIFLEX-PM) proposing a stationary reference frame control technique suitable to manage the energy exchanges between two different electricity networks and to obtain high quality power at both grids.

## II. CONVERSION STRUCTURE

The main objective of the Uniflex-PM system is to provide a flexible and modular power electronic interface able to connect different kind of sources and loads including MV electrical networks, RES and energy storage systems [3]. The foreseen structure of UNIFLEX-PM conversion system is shown in Fig. 1.

The conversion system is composed by three power converters, each one connected to a PCC. Port One and Port Two are used for MV electrical networks, while Port Three is used mainly to connect the conversion structure to an energy storage system and to a low voltage electricity network.

The conversion structure must be able to satisfy the following requirements:

- Bi-directional power flow operation in all ports with active and reactive power control capabilities;
- Compliance to most of European and International grid standards for DG connection in terms of injected harmonics, robustness to grid voltage distortions and excursions;
- Galvanic isolation among the ports;
- Modular architecture providing high reliability and easy maintenance.

A basic schematic block of the conversion structure, based on a multi-stage architecture, that can provide all the above mentioned capabilities, is illustrated in Fig. 2.

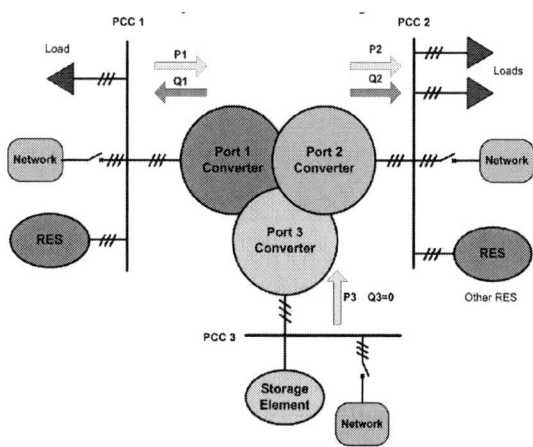

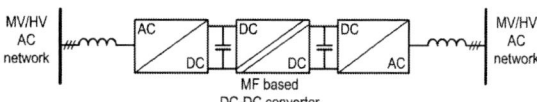

Fig. 1. General Structure of UNIFLEX model.

Fig. 2. Basic schematic block of the power conversion structure.

Such architecture consists of two AC/DC power conversion stages and a DC/DC conversion stage, based on Medium Frequency (MF) transformer to achieve the galvanic isolation between the AC terminals.

Different topologies of power converters can be considered for medium and high voltage grid applications [4]-[7]. However, due to the recent development in power semiconductor devices, particularly in the IGBT technology, there is an increasing interest in the last years, in multi-level power converters especially for medium to high-power, high-voltage [8]-[10]. Multi-level converters present different advantages such as the reduced harmonics content in the input and output voltage, reduced switching losses at the same harmonic performance as for a two-level converter. Among the different topologies, to assure also a modular architecture, a multi-level cascaded single phase H-bridge topology has been selected for the implementation of the UNIFLEX-PM system.

To simplify the control implementation, the basic concept will be demonstrated on this paper on a model based on two-port structure, as shown in Fig.3.

Each port of the UNIFLEX-PM system employs a conversion structure based on a three-phase 7-level AC-DC cascaded converter.

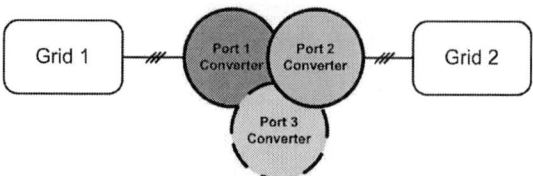

Fig. 3 – Block diagram of the reduced UNIFLEX-PM converter model

One phase of the two-port based model is shown in Fig. 4 where each block denoted as Ax, (x=1..3), represents a H-bridge converter. Since isolation modules on DC side have an independent control for each branch, an equivalent capacitor C is used in the model.

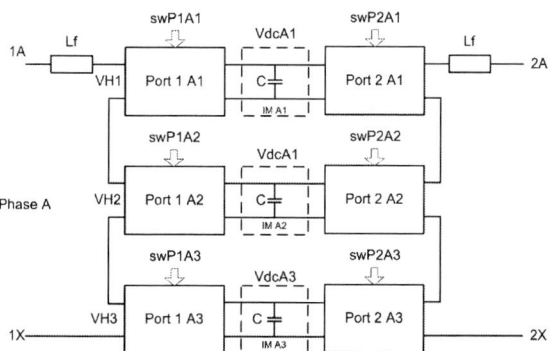

Fig. 4 - Equivalent model per phase for the UNIFLEX -PM model.

## III. CONTROL SYSTEM

Fig. 5 shows the overall control block diagram while Fig. 6 shows controller schemes for port 1 and port 2 respectively. A cascaded control structure is selected where the outer loop is the power/DC-link voltage loop while the inner loop is responsible for the current control.

Two *dq* frames in positive and negative coordinates are implemented to regulate the supply current taking unbalanced supply conditions into consideration [11]-[13]. Instead of a normal PI controller, a PI+Resonant controller is used to control the positive and negative *dq* currents in case of a distorted supply voltage [14].

The positive and negative signals are extracted via the delay signal cancellation method. Moreover, a PLL filter is used to obtain the phase angle that allows generating a rotating frame for voltages and currents synchronous with the grid.

In port 1, the DC-link voltage and reactive power are regulated generating the references for the current control. The reactive power and DC-link voltage are regulated by PI control. One PI controller is utilized to control the average value of the DC-link voltage. Different power control strategies can be applied depending on the chosen objective [15]. In this paper, to obtain sinusoidal and balanced grid currents, only the positive sequence current references are generated while the negative sequence current references are set to be zero.

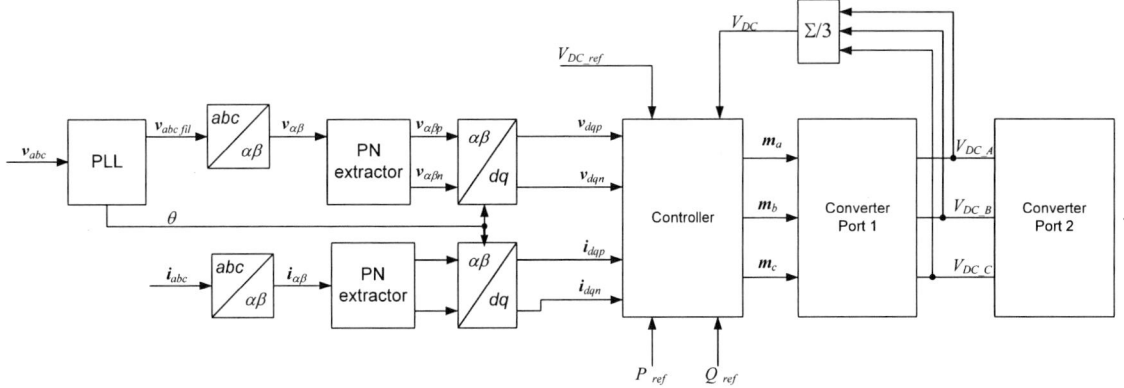

Fig. 5. Block diagram of stationary reference frame Control.

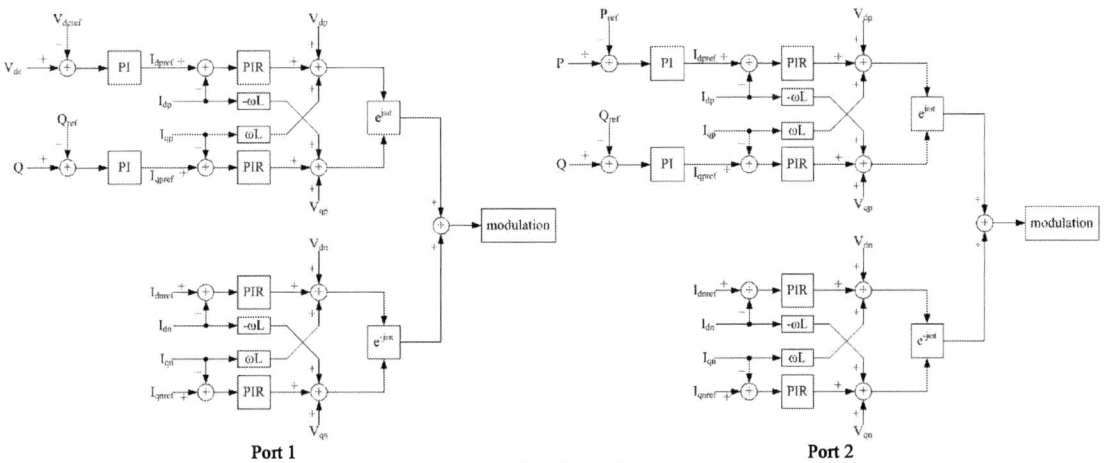

Fig. 6. Detailed Control Scheme.

In port 2, active and reactive powers are controlled. The active power control generates the current reference for the positive sequence d-axis current while the reactive power control generates the current reference for the positive sequence q-axis current. The negative sequence currents are regulated to be zero.

A horizontally phase-shifted multicarrier modulation is used for this system. The multilevel cascaded H-bridge converter (CHB) with $H$ bridge cells per phase leg has $m$ voltage levels:

$$m=2H+1 \qquad (1)$$

Hence, a number of ($m$-1) triangular carriers are necessary. These triangular carriers have the same frequency and the same amplitude. The phase shift $\varphi_{carrier}$ between any two adjacent carriers can be obtained by:

$$\varphi_{carrier} = \frac{2\pi}{m-1} \qquad (2)$$

This modulation strategy has as input three-phase reference signals for the phase voltages that are provided by the control strategy. In order to obtain the gate signals for the power switches these references are compared with the triangular carriers. The principle of generating these gate signals is illustrated in Fig. 7 for a seven level CHB converter. The carriers one, two and three are used

to generate gate signals for the upper switches in the left legs of the H-bridge cells. The inverted signal of theses carriers will produce the gate signals for the upper switches in the right leg. The output voltage of a H-bridge cell is switched between 0 and $V_{dc}$ during the positive cycle of the reference signal or between 0 and -$V_{dc}$ during the negative cycle respectively as shown in Fig. 8; in the same figure, an example of modulated 7-level voltage waveform produced by the converter is illustrated.

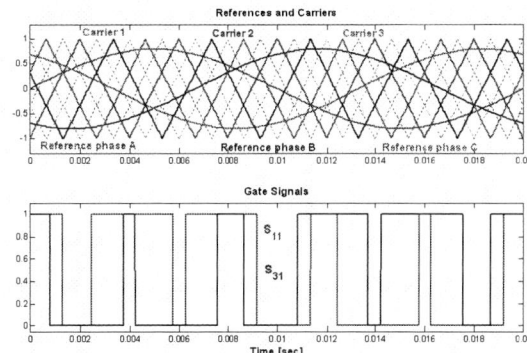

Fig. 7. Gate signal generation for a seven level CHB inverter

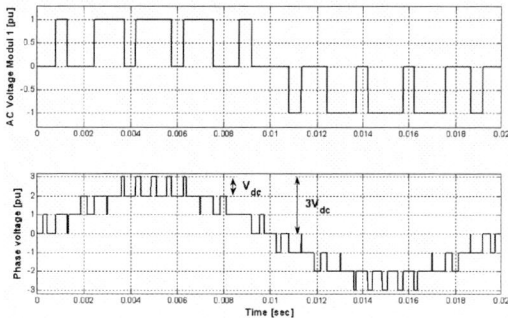

Fig. 8. AC voltages for a seven level CHB inverter

## IV. SIMULATION RESULTS

The proposed control strategy was evaluated under different power flow profiles, grid conditions and situations such as harmonic content, voltage excursions, voltage unbalances, frequency excursions and distorted grid voltages as shown in the following six cases study. The simulations, carried out in Matlab-Simulink, account the following parameters for the power conversion structure: rated power 300 kVA, rated line-to-line voltage 3.3 kV, DC-link voltage on each capacitor 1100 V, DC-link capacitor 6.2 mF, input filter inductance 16 mH. A bidirectional power flow is simulated using an average model of the UNIFLEX system in the first 4 cases; harmonics on the grid voltages are added in case E and finally a switching model of the converter including PWM modulation is used in case F.

### A. Power Flow

A bidirectional power flow is simulated with a reference profile as shown for port 1 in Fig. 9a and 9b blue line. This power profile corresponds to a leading/lagging operation at 0.9 power factor. Simulation results for active and reactive power are presented respectively in Fig. 9a and 9b green line. It is to note that the instantaneous active and reactive power traces the reference with good dynamics. Fig. 10 presents the three-phase input currents in port 1 while Fig 11 shows active and reactive power flow in port 2 (reference blue line).

### B. Voltage Unbalance

Further tests are included when an unbalance of 3% is considered in the grid voltages at port 1. This unbalance is defined as the ratio between the negative and the positive sequence of the voltage. Figs 12a and 12b show active and reactive power flow control on port 1 (reference blue line). As it can be noticed, the active power is well regulated while the effect of the grid voltage unbalance causes ripples on the reactive power which is not possible to eliminate in this case. Fig. 13a and Fig. 13b show, respectively, the three-phase input currents in Port 1 and the average DC-link voltage on Phase A. Note that the line currents are essentially sinusoidal and the DC-link voltage presents negligible variations.

### C. Frequency Excursion

The control strategy is evaluated for a ±6% frequency excursions profile in Port 1 as shown Fig. 14. The simulation results presented in Fig. 15a and 15b show active and reactive power flow control on port 1 (reference blue line). It can be seen that a short period of slight frequency excursions has a limited effect on the system performance.

### D. Voltage Excursion

Voltage excursions of 75% and 120% from the rated voltages in the PCC of the Port 1 is considered as shown in Fig. 16. It can be seen from the simulated waveforms in Figs. 17 and 18 that that the designed control system has strong disturbance rejection ability to the voltage excursion. Meanwhile the line current waveform is well controlled to be sinusoidal.

### E. Effects of Distorted Grid

Another interesting aspect is the performance evaluation of the converter control when the grid voltages are distorted. In this case, the analysis considers that the power modules deliver 90% active power into the PCC while the reactive power is set to zero.

The grid voltages are polluted with $5^{th}$, $7^{th}$, $11^{th}$ and $13^{th}$ harmonics according to the amplitudes furnished by standard EN 61000-2-4.

Fig. 19a and 19b show respectively the waveforms of the grid voltages and the grid currents in Port 1, while Table I summarises the voltage and current harmonic contents. It can be seen that the resonant controller can eliminate up to 75% of the $5^{th}$, $7^{th}$, $11^{th}$ and $13^{th}$ harmonics in the supply current. Due to bandwidth limitation of the controller, the current harmonics can not be ideally removed.

### F. Harmonic content

Harmonic study tests were performed to evaluate the current harmonics injected into the grid by the converter modulation. The analysis considers that the power modules deliver 100% active power into the PCC while the reactive power is set to zero.

For the current harmonics analysis a 1800 Hz switching frequency was imposed to the converter and undistorted grid voltages are considered. Fig. 20 shows the phase A current on Port1 under the mentioned operating conditions.

The complete harmonic contents and the THD are summarized in Fig. 21 that shows individual harmonics values for a phase current up to $50^{th}$ harmonic vs EN 61000-2-4 harmonic levels, represented in the figure respectively with a green bar and a red bar.

A few current harmonic levels that exceeds the standard values are positioned at harmonic orders between $27^{th}$ and $45^{th}$, i.e. around the switching frequency. For lower harmonics all the values are drastically reduced compared to the values required by standard EN 61000-2-4. Finally, a value of less than 4% is obtained for the current THD, which is lower than 8%, limit imposed by the considered standard.

## V. CONCLUSIONS

From the above simulation results, the effectiveness of the proposed PI controller-based method in dual synchronous frame is verified. It allows to accurately tracing active and reactive power reference under different conditions. The supply currents are essentially sinusoidal even under extreme unbalanced supply conditions. Voltages on the different DC links are well regulated with a reduced ripple.

The system performance relies on the angular information, so the PLL design is a crucial part. A reliable and fast PLL is expected to have deep effect on the controller bandwidth and reliability.

Furthermore the delay signal cancellation method, which allows positive and negative signals extraction, also plays an important role to improve the controller bandwidth.

### ACKNOWLEDGMENT

This research has been carried out within the project UNIFLEX-PM, EC Contract n°: 019794 (SES6) EUROPEAN COMMISSION DIRECTORATE J – ENERGY

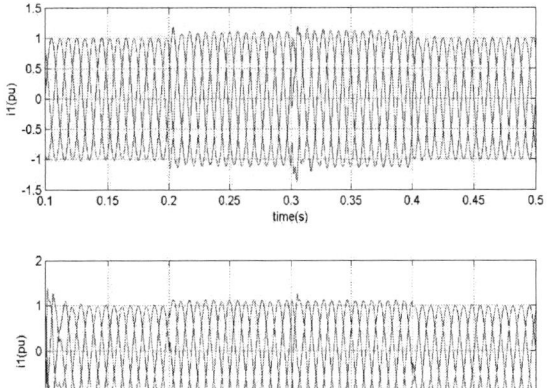

Fig. 10. Power flow study.

Input currents Port 1

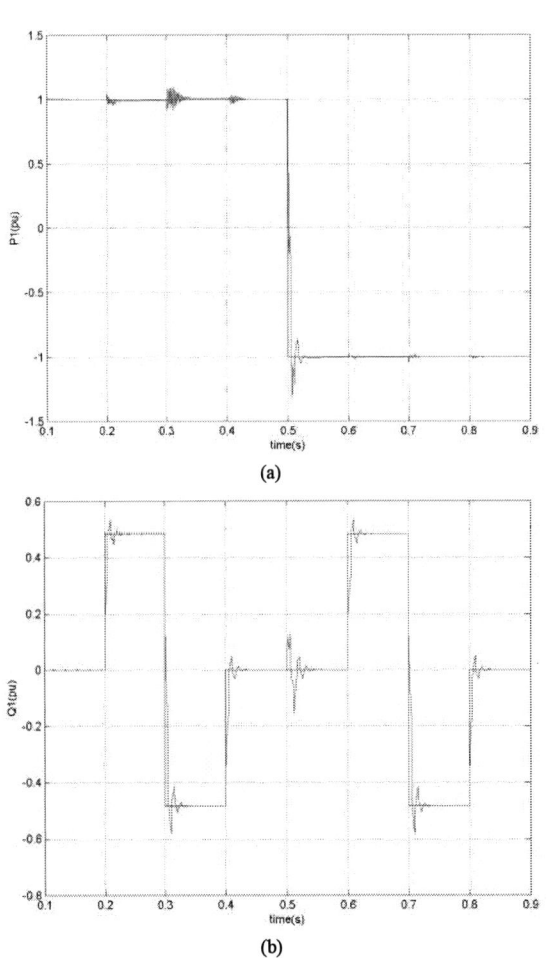

Fig. 9. Power flow study.

(a) Active power Port 1 and (b) Reactive power Port 1

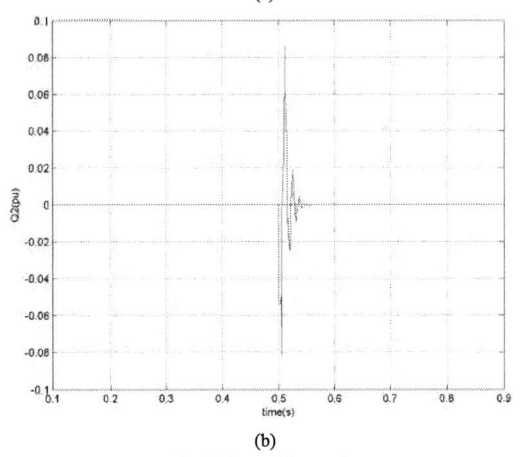

Fig. 11. Power flow study.

(a) Active and (b) Reactive power profiles Port 2

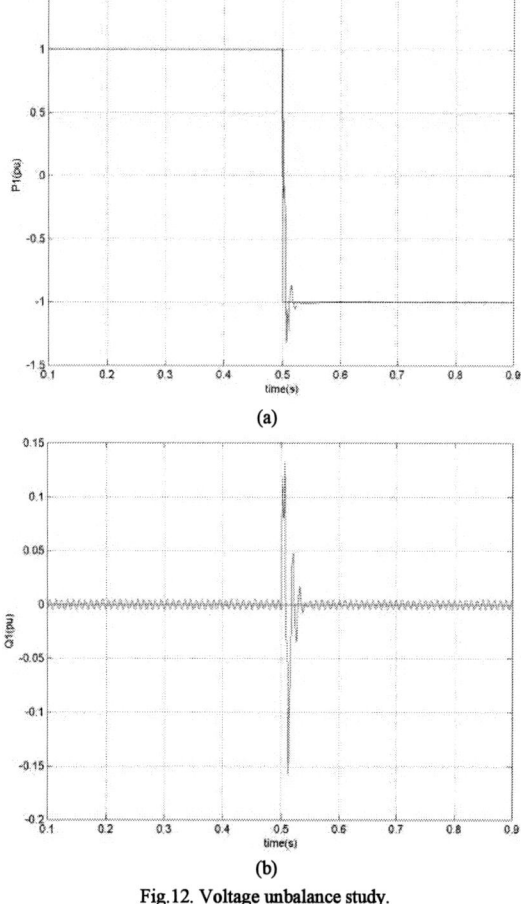

(a)

(b)

Fig.12. Voltage unbalance study.
Active (a) and Reactive (b) power Port 1

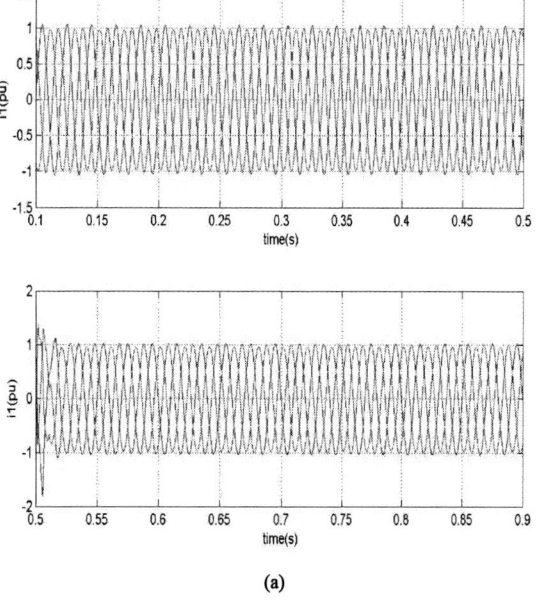

(a)

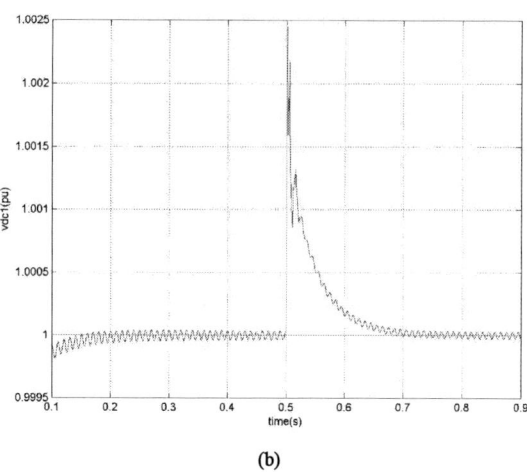

(a)

(b)

Fig. 13. Voltage unbalance study.

(a) Input currents in Port 1 (b) DC-link voltage on Phase A

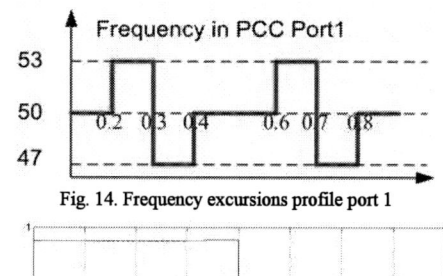

Fig. 14. Frequency excursions profile port 1

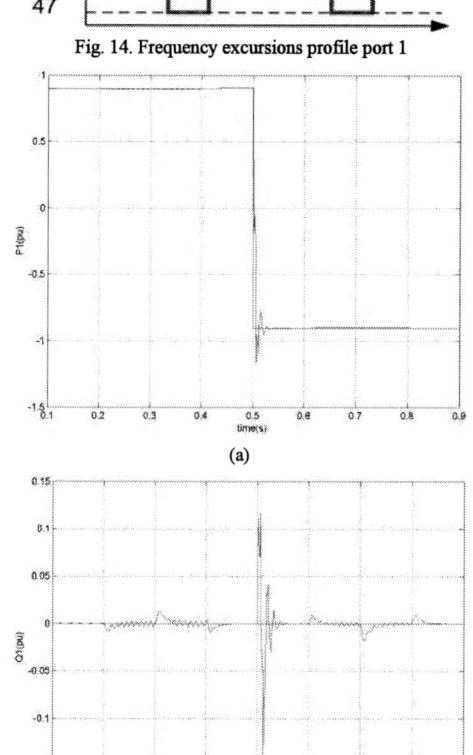

(a)

(b)

Fig.15. Frequency excursion study.
(a) Active and (b) Reactive power Port 1

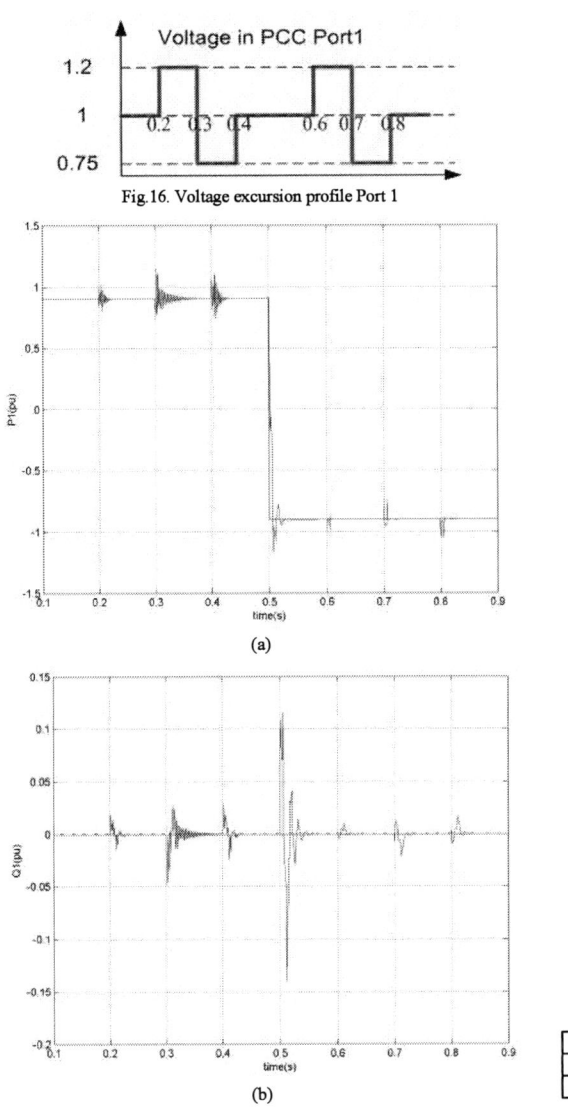

Fig.16. Voltage excursion profile Port 1

(a)

(b)

Fig.17. Voltage excursion study.
(a) Active and (b) Reactive power Port 1

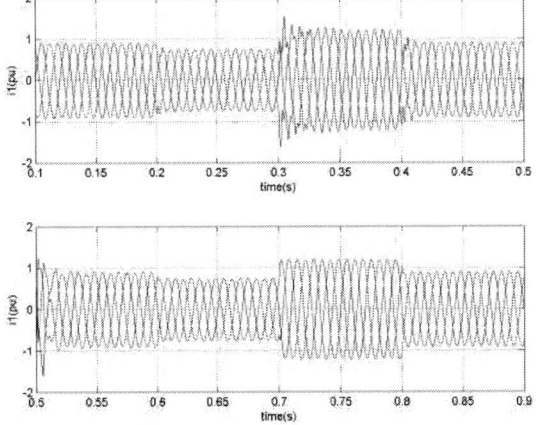

Fig. 18. Voltage excursion study. Input currents port 1

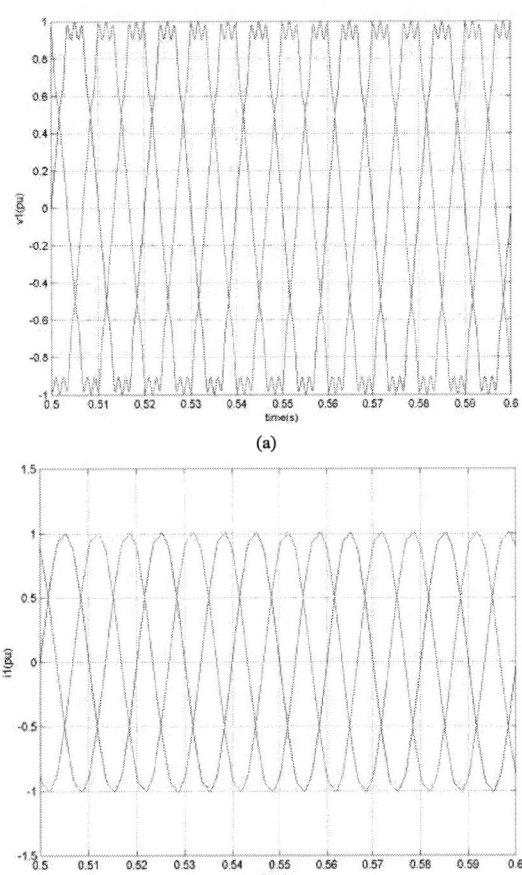

(a)

(b)

Fig. 19. Effects of distorted grid voltages in Port 1:
(a) grid voltages, (b) grid currents.

Table I. Harmonic content of grid voltages and currents.

|  | 5th(%) | 7th(%) | 11th(%) | 13th(%) |
|---|---|---|---|---|
| V1 | 6 | 5 | 3.5 | 3 |
| I1 | 0.64 | 0.62 | 0.63 | 0.64 |

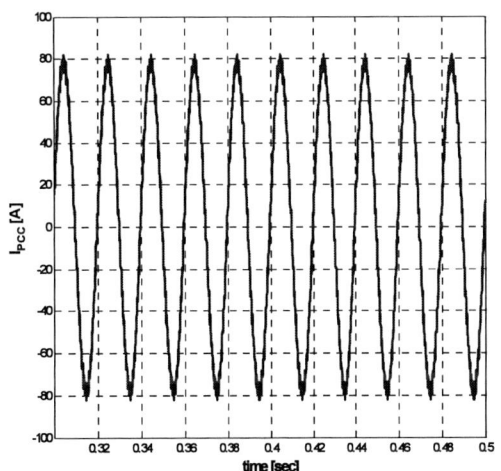

Fig.20. Phase A current in port 1 considering converter modulation

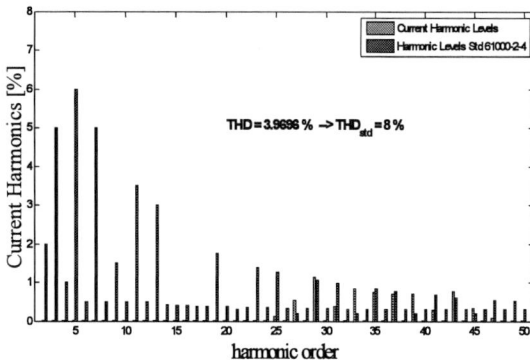

Fig. 21. Harmonic spectrum of Port 1 phase A current.

### REFERENCES

[1] European Commission – New ERA for electricity in Europe. Distributed Generation: Key Issues, Challenges and Proposed Solutions, EUR 20901, 2003, ISBN 92-894-6262-0.

[2] European Commission – Towards Smart Power Networks. Lessons learned from European research FP5 projects, EUR 21970, 2005, ISBN 92-79-00554-5.

[3] F. Iov, F. Blaabjerg, R. Bassett, J. Clare, A. Rufer, S. Savio, P. Biller, P. Taylor and B. Sneyers, "Advanced Power Converter for Universal and Flexible Power Management in Future Electricity Network," in *Proc. CIRED 2007,* Vienna, Austria, May 2007.

[4] M. Marchesoni and M. Mazzucchelli "Multilevel converters for high power ac drives: a review," in *Proc. of IEEE International Symposium on Industrial Electronics*, pp. 38-43, 1993.

[5] D. Soto and T.C. Green, "A comparison of high-power converter topologies for the implementation of FACTS controllers," *IEEE Trans. Ind. Electron.*, vol.49, no.5, pp. 1072-1080, Oct. 2002.

[6] Y. Cheng, C. Qian, M.L. Crow, S.Pekarek and S. Atcitty, "A comparison of diode-clamped and cascaded multilevel converters for STATCOM with energy storage," *IEEE Trans. Ind. Electron.*, vol.53, no.5, pp. 1512-1521, Oct. 2006.

[7] F. Blaabjerg and F. Iov, "Wind power – a power source now enabled by power electronics," in *Proc of 9th Brazilian Power Electronics Conference* COBEP 07, Blumenau, Santa Catarina, Brazil, ISBN 978-85-99195-02-4, October 2007.

[8] Bin Wu, *High-Power Converters and AC Drives*, IEEE Press, Wiley Interscience 2006, ISBN 10-0-471-73171-4.

[9] V.G. Agelidis, G.D. Demetriades and N. Flourentzou, "Recent Advances in High-Voltage Direct-Current Power Transmission Systems", in *Proc. IEEE Int. Conf. on Industrial Technology ICIT 2006, Dec. 2006.*

[10] A. Rufer, "Today's and Tomorrow's Meaning of Power Electronics within the grid interconnection," keynote paper presented at *The 12th European Conf. on Power Electronics, EPE 2007,* Aalborg, Denmark, September 2007.

[11] Long-Seok Song and Kwanghee Nam, "Dual current control scheme for PWM converter under unbalanced input voltage conditions," *IEEE Trans. Ind. Electron,* vol. 46, Issue 5, pp. 953 – 959, Oct. 1999.

[12] S. Yongsug and T.A Lipo, "Modeling and analysis of instantaneous active and reactive power for PWM AC/DC converter under generalized unbalanced network," *IEEE Trans. on Power Delivery*, vol. 21, Issue 3, pp. 1530 – 1540, July 2006.

[13] R. Teodorescu and F. Blaabjerg, "Flexible control of small wind turbines with grid failure detection operating in stand-alone or grid-connected mode," *IEEE Trans. Power Electron.*, vol. 19, no. 5, pp. 1323–1332, Sep. 2004.

[14] W. Lenwari, M. Sumner, P. Zanchetta and M. Culea, "A High Performance Harmonic Current Control for Shunt Active Filters Based on Resonant Compensators," in *Proc. of IECON 2006*, pp. 2109 – 2114, Nov. 2006.

[15] A. V. Timbus, P. Rodriguez, R. Teodorescu, M. Liserre and F. Blaabjerg, "Control Strategies for Distributed Power Generation Systems Operating on Faulty Grids," in *Proc. of IEEE International Symposium on Industrial Electronics*, Montreal, Canada, pp. 1601-1606, July 2006.

# Energy Storage Systems
# The Flywheel Energy Storage

Tomasz Siostrzonek, Stanisław Piróg and Marcin Baszyński
AGH University of Science and Technology
Faculty of Electrical Engineering, Automatics, Computer Science and Electronics
Department of Electrical Drive and Industrial Equipment
Krakow, Poland
tsios@agh.edu.pl, pirog@agh.edu.pl, mbaszyn@agh.edu.pl
phone number: +48 12 617 40 56
fax: +48 12 633 22 84

*Abstract*— Storage of energy is one of the main problem of contemporary technology. Currently used manners of the energy store are listed below:
- the magnetic accumulator – the energy is kept in the magnetic field of superconductive inductor,
- the accumulator with supercapacitors. The low voltage (1,6-2,5V) is the fault of this one,
- the accumulator with lead-acid or alkaline accumulator. The fault of this solution is very low charging and discharging efficiency,
- the electromechanical accumulator. Flywheels store energy mechanically in the form of kinetic energy.

In this article the flywheel energy storage will be described precisely and compared with other energystorage technologies.

*Keywords*—Battery management system (BMS), brushless drive, energy storage, flywheel system, supercapacitor.

## I. INTRODUCTION

Electrical energy in an AC system cannot be stored electrically. Energy can be stored as electromagnetical, electrochemical, kinetical or potential energy. Each energy storage technology usually includes a power conversion unit to convert energy from one form to another. The applications of an energy storage technology are characterized by two factors:
- the amount of energy that can be stored in the device,
- the rate at which energy can be transferred into or out of the storage device.

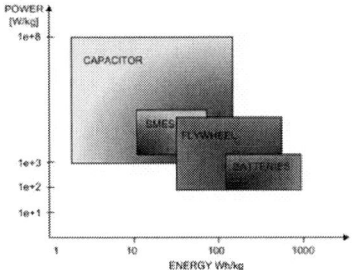

Fig.1 Specific power and specific energy ranges [1]

The energy ranges for near-to-midterm technology are shown in Fig.1. Integration of these possible energy storage technologies with flexible AC transmission systems (FACTS) and custom power device are among the possible power applications utilizing energy storage [1].

## II. ENERGY EFFICIENT OF SUPERCAPACITOR ENERGY STORAGE SYSTEM [13]

The double-layer capacitors (ultracapacitors, supercapacitors) are made of carbon, which have huge effective surface so the capacitance could attain several farad even thousands farad. The supercapacitors doesn't have electrochemical reaction and only have electric charges absorption and desorption when charge and discharge. It has many merits such as high charge/discharge current, less maintenance, long life and some other perfect performance. At the same time its small leakage current enables it has long time of energy storage and the efficiency could exceed 95%. China has produced 400V and 0,58F products [13].

Efficiency is a very important issue for supecapacitor. Supercapacitors are new electrochemical storage device based on the principle of the double-layer electrolyte capacity. Organic electrolyte and activated carbon electrodes represents the most promising and mature technologies for the success capacitors. Comparing with traditional capacitor, supercapacitor possesses the maximum energy density and ratio capacitance that could be the magnitude of farad.

Several varieties of advanced capacitors are in development, with several available commercially for low power applications. This capacitors have significant improvements in one or more of the following characteristics: higher permittivities, higher surface area, or higher voltage-withstand capabilities.

## III. SUPERCONDUCTING MAGNETIC ENERGY STORAGE (SMES)[1]

SMES system have attracted the attention of both electric utilities and high efficiency (a charge-discharge efficiency over 95%). The SMES unit in a device that stores energy in the magnetic field generated by the DC

978-1-4244-1741-4/08/$25.00 ©2008 IEEE     1779

current flowing through a superconducting coil. The inductively stored energy and rated power are commonly given specifications for this device. The device consists of a large superconducting coil at the cryogenic temperature. This temperature is maintained by a cryostat or dewar that constains helium or nitrogen liquid vessels. A power conversion/conditioning system connects the SMES unit to an AC power system and it is used to charge/discharge the coil.

The SMES system is costly, when compared with other energy storage. However, the integration of an SMES coil into existing FACTS devices eliminates the cost for the inverter unit, which is typically the largest portion of the cost for the entire system. In the fig.2 the SMES system is shown.

## IV. BATTERY ENERGY STORAGE SYSTEM [1]

Batteries are the storage which the energy is stored as electrochemically energy. There are the most cost-effective energy storage technologies. The batteries system consists of low voltage/power battery modules connected in parallel and series to achieve a desired

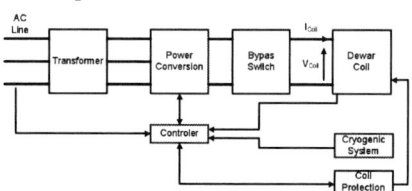

Fig.2 The SMES system

electrical characteristic. Key factors of batteries for storage applications include: high energy density, high energy capability, cycling capability, life span, initial cost.

Batteries store DC charge, so power conversion is required to interface a battery with an ac system. Battery energy storage system (BESS) have recently emerged as one of more promising near-term storage technologies for power applications such as area regulation, area protection, spinning reserve, power factor correction.

## V. FLYWHEEL ENERGY STORAGE SYSTEM

The "Flywheel Energy Storage" or "Flywheel batteries" describes a system, which consist of three parts (Fig.3):
- flywheel,
- motor/generator,
- power electronic system.

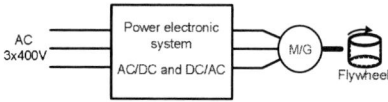

Fig.3 The Flywheel Energy Storage System

The electric energy from the source, store as a kinetic energy of rotation, and delivers it to the load (as electric energy). The machine (working as motor) spins up the flywheel and stores energy mechanically. When the machine works as generator, the electrical energy is delivered to the load. Modern high-speed flywheels differ from their forebears in being lighter and spinning much faster.

Today's system operate somewhere between atmospheric pressure (100 kPa) and 0,1 Pa, with the highest speed devices operating at the lowest pressure. Lower speed flywheels are larger, but can be operated in air with acceptable losses.

The parameters of the FES constructed by authors:
- supply voltage – 3x400V AC,
- motor current – 8A (rated curent),
- electrical power 2,9 kW (rated power of Brushless DC motor),
- rated speed 6000 rpm,
- the angular velocity of the barrel:

$$\omega = \frac{2 \cdot \pi \cdot n}{60} = 628\,[1/s] \qquad (1)$$

- the energy stored:

$$E_k = \frac{1}{2} \cdot J \cdot \omega^2 = 0,584\,[MJ] = 0,162[kWh] \qquad (2)$$

Three phase rectifier is a part of Power Electronic system for supply of brushless DC motor (BLDCM). In this device the three phase transistor converter which worked as a rectifier with sinusoidal source current was introduced. Additionally, bidirectional power flow (possibility of energy return from drive system to supply system) and DC voltage stabilization was possible. Most computations (the rotor position, actual speed, current error, hysteresis regulator) are carried out by FPGA. Output data, in the form of transistor's driving pulses, are fed back to the FPGA structure, where control logic was generated allowing for safe switching of transistors in the inverter's branch (Fig.4) .

The current regulation of BLDC motor can be worked out in the same way as for classic DC machine with separately excited, by means of PI controller. The feedback signal for PI controller is a signal, which is proportional to current wave ($I_d$) of DC source. This signal can be received:

a) directly from DC current sensor,
b) as signal proportional to the sum of module of load phase AC current (Fig. 4).

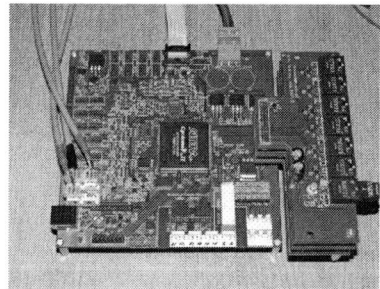

Fig.4 The controller with FPGA Device (Cyclone II)

## VI. THE RECTIFIER WITH SINUSOIDAL SOURCE CURRENT

A unity input power factor control of a three-

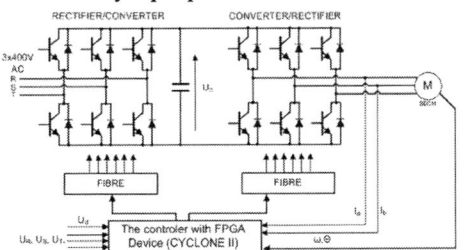

Fig.5 The scheme of FES with controlling board

phase step-up converter is feasible in the rotating co-ordinate frame because in this system the source frequency quantities are represented by constant values. The diagram of the rectifier connection to a supply network is shown in figure 6. Since $X_L \gg R$, the resistances of reactors are disregarded in the diagram.

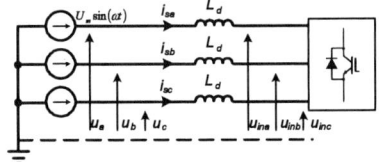

Fig.6 Diagram of the rectifier connection to a supply network

The following designations are used the diagram of figure 6: $i_{sn}$- phase currents, $u_{sn}$- the supply line phase-to-neutral voltages, $u_{inn}$- the converter output voltage (where n= a, b, c). The phase currents, according to the diagram, are described by equation (3).

$$u_{sn} - u_{inn} = L\frac{di_{sn}}{dt} \tag{3}$$

Converting the equation (3) into the rotating reference frame $dq$ we obtain equation (4).

$$\boldsymbol{u_{sdq}} - \boldsymbol{u_{indq}} = \Delta\boldsymbol{u_{dq}} = L_d\frac{di_{sdq}}{dt} + j\omega L_d\boldsymbol{i_{sdq}} \tag{4}$$

Decomposing the equation (3) into $dq$ components we obtain (4).

$$u_{ind} = u_{sd} - \Delta u_d = u_{sd} - \left(L_d\frac{di_{sd}}{dt} - \omega L_d i_{sq}\right) \tag{5}$$

$$u_{inq} = u_{sq} - \Delta u_q = u_{sq} - \left(L_d\frac{di_{sq}}{dt} + \omega L_d i_{sd}\right) \tag{6}$$

Equations (5) and (6) describe the converter input voltages. Inserting the required line current values into these equations we can determine the output voltage waveforms forcing the required current. The components $L_d(di_{sdq}/dt)$ represent the converter dynamic states (load switching or changes in the load parameters). Assuming the control system comprises only proportional terms we obtain from equations (5) and (6) relationships describing the control system (7) and (8).

$$u_{ind} = u_{sd} - (K_R\Delta i_{sd} - K_d\Delta i_{sq}) = u_{sd} - [K_R(i_{sdr} - i_{sd}) - K_d(i_{sqr} - i_{sq})] \tag{7}$$

$$u_{inq} = u_{sq} - (K_R\Delta i_{sq} + K_q\Delta i_{sd}) = u_{sq} - [K_R(i_{sqr} - i_{sq}) - K_q(i_{sdr} - i_{sd})] \tag{8}$$

Figure 7 shows block diagram of the control system and the power circuit. The following designations are used in the diagram: TP – switch-on delay units (blanking time), PI – proportional-integral controller, KS- sign comparator, SAW- triangle wave generator, $K_R$, $K_d$, $K_q$- proportional terms, ST- contactors, $R_a$, $R_b$, $R_c$ - resistors limiting the capacitor $C_F$ charging current, $\Sigma$- adder.

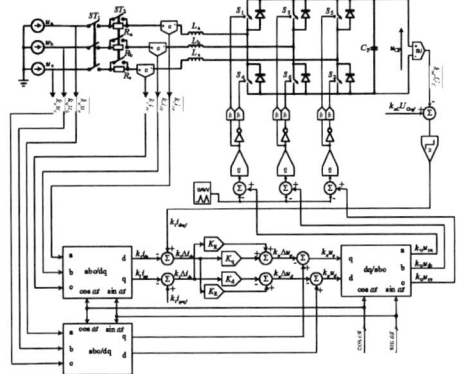

Fig.7 Block diagram of the control system and the power circuit

In the control circuit of the diagram 6 was applied transformation from the thee-phase system to the rotating co-ordinate system (abc→dq), described by equation (3).

$$\begin{bmatrix} v_d \\ v_q \end{bmatrix} = \begin{bmatrix} \cos\omega t & \sin\omega t \\ -\sin\omega t & \cos\omega t \end{bmatrix}\begin{bmatrix} v_a \\ \frac{1}{\sqrt{3}}(v_b - v_c) \end{bmatrix} \tag{9}$$

Where:

$$\begin{cases} v_a = V_m\cos\omega t \\ v_b = V_m\cos\left(\omega t - \frac{2}{3}\pi\right) \\ v_c = V_m\cos\left(\omega t + \frac{2}{3}\pi\right) \end{cases} \tag{10}$$

In order to determine the transformation abc→dq it is necessary to generate functions $\cos\omega t$ and $\sin\omega t$, as follows from equation (9), such that the function $\cos\omega t$ will correspond (i.e. be cophasal) to $v_a = V_m\cos\omega t$. In practical solutions various methods for generating the $\cos\omega t$ and $\sin\omega t$ functions are employed, e.g. synchronization with a single, selected phase (normally a) employing a single-phase PLL loop. The advantage of this method is an easy implementation in digital technique. Microprocessor systems employ an external, specialized device performing the functions of a phase-locked loop and connected with a microprocessor port dedicated for counting external events. Therefore the CPU workload due to generating the $\cos\omega t$ and $\sin\omega t$ functions is reduced to minimum. A drawback of this method is the generated function is related to only one phase of the synchronizing signal and the system does not control the other phases. In the event of a disturbance starting in phase c (a phase jump in the synchronizing voltage caused by switching a large active power load) the control system will respond with large delay. In order to protect the converter from effects of a phase jump the synchronization circuit should control all phases of the synchronizing voltage. Substituting equations (10)

1781

describing the three-phase synchronizing voltage into equation (9), the transformation abc→dq takes the form (11).

$$\begin{bmatrix} v_d \\ v_q \end{bmatrix} = \begin{bmatrix} V_m(\cos^2 \omega t + \sin^2 \omega t) \\ V_m(-\sin \omega t \cos \omega t + \sin \omega t \cos \omega t) \end{bmatrix} = \begin{bmatrix} V_m \\ 0 \end{bmatrix} \quad (11)$$

It follows from equation (11) that if the functions $\cos \omega t$ and $\sin \omega t$ are generated correctly ($\cos \omega t$ is cophasal with voltage in phase a), the component in axis d equals the amplitude of the synchronizing voltage whereas the component q is zero. This property of the abc→dq transformation is employed in the design of the three-phase synchronization circuit depicted in figure 3.

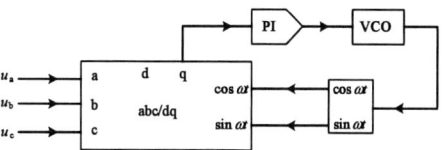

Fig.8 Block diagram of the synchronization circuit

The following designations are used in figure 8: PI-proportional-integral controller, VCO- voltage controlled square-wave generator. The PI controller input signal is the instantaneous value of the q-axis component of abc→dq transformation. The controller tunes the VCO oscillator, whose output signal controls the $\cos \omega t$ and $\sin \omega t$ generation circuit. The controller connected to the q-axis controls the PI controller error to zero (the value q-axis component equals zero) what means that, according to equation (4), the generated $\cos \omega t$ signal is cophasal with synchronizing voltage $u_a$. The control circuit shall attain the state in which the q-axis component value is zero. This condition (q= 0) is satisfied in two cases:

1.      The generated function $\cos \omega t$ is cophasal with the synchronizing voltage $u_a$. This case is described by equation (11). Figure 9 shows oscillograms of the synchronizing voltage ($u_a$, $u_b$, $u_c$) and the generated function $\cos \omega t$. The simulation w

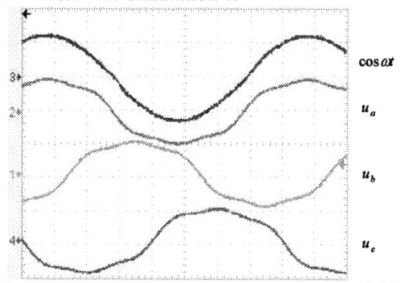

Fig. 9 The synchronizing voltage waveforms and the generated function $\cos \omega t$

aveforms are computed using a model of the system implemented in FPGA.

2.      The generated function $\cos \omega t$ is phase-shifted in with respect to the synchronizing voltage $u_a$ by $\pi$. The transformation abc→dq then takes the form described by equation (12).

$$\begin{bmatrix} v_d \\ v_q \end{bmatrix} = \begin{bmatrix} -\cos \omega t & -\sin \omega t \\ \sin \omega t & -\cos \omega t \end{bmatrix} \begin{bmatrix} v_a \\ \frac{1}{\sqrt{3}}(v_b - v_c) \end{bmatrix} \quad (12)$$

Substituting (10) to equation (12) yields (13).

$$\begin{bmatrix} v_d \\ v_q \end{bmatrix} = \begin{bmatrix} -V_m(\cos^2 \omega t + \sin^2 \omega t) \\ V_m(\sin \omega t \cos \omega t - \sin \omega t \cos \omega t) \end{bmatrix} = \begin{bmatrix} -V_m \\ 0 \end{bmatrix} \quad (13)$$

The phase shift of $\cos \omega t$ function (by $\pi$) with respect to the voltage $u_a$ results in erroneous power relationships in the converter operation and therefore is inadmissible. This case is illustrated in figure 10 which shows the simulation waveforms computed using the system model.

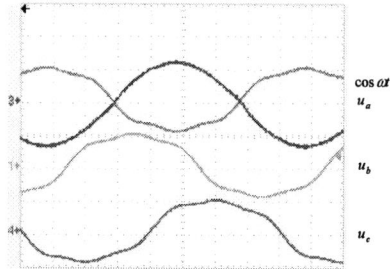

Fig.10 The synchronizing voltage waveforms and the generated function $\cos \omega t$

The operating point of the synchronizing circuit from figure 8 (case 1 or 2) is determined by the initial phase of synchronizing voltages at the instant of the converter start. In order to prevent a random selection of the operating point the synchronization circuit should be modified to enforce the system operation according to equation (11) (case1). Figure 11 shows two possible versions of the synchronizing circuit.

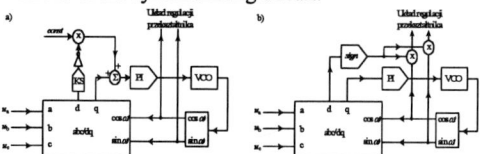

Fig.11 Two versions of the synchronizing circuit enforcing the correct generation of $\cos \omega t$ function

The designations are used in figure 11: const – a constant value, KS- sign comparator, sgn- sign of the signal ("1" if the signal is greater than zero, conversely "-1"), other elements according to figure 6. If the d-axis component value (figure 6a) is less than zero, the case described by equation (13), the constant value const is added to the instantaneous value in axis q (the PI controller input error) therefore tuning the VCO oscillator, even when the q-axis component value is zero. When the control circuit attains the state in which the value of the d-axis component is greater than zero only the q-axis component is applied to the PI controller input (the controller error). In the circuit from figure 11b no modification is made to the synchronization circuit. The functions $\cos \omega t$ and $\sin \omega t$ are multiplied by the d-axis component sign prior to being applied to the converter control circuit. When the system operates correctly, i.e. according to equation (9), the functions $\cos\omega t$ and $\sin\omega t$ are directly applied to the converter control circuit. When the function $\cos\omega t$ is phase-shifted in with respect to the synchronizing voltage $u_a$ by $\pi$ (equation 13) the multiplication of $\cos\omega t$ and $\sin\omega t$ functions by the sign of d-axis component ("-1") reverses their phases and the correct values of $\cos\omega t$ and $\sin\omega t$ functions are applied to the converter control circuit. The diagram of the synchronization circuit (the version in figure 11b), implemented in FPGA using the Quartus II v7.2 Altera package is depicted in figure 12. The functions $\cos\omega t$ and $\sin\omega t$ were discretized and stored in separate tables denoted in figure 12 lpm_rom0 and lpm_rom1, respectively. Other elements of the circuit have been developed using the VHDL language (Very high speed integrated circuit Hardware Description Language).

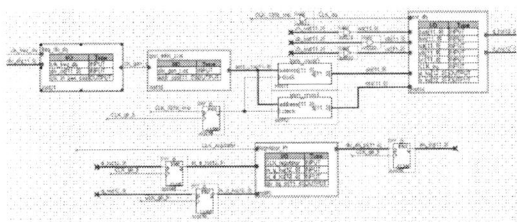

Fig.12 The diagram of the synchronization circuit (the version according to figure 11b), implemented in FPGA using the Quartus II v7.2 Altera software

Figure 13 illustrates the synchronization circuit (figures 11b and 12) operation, the time graphs are recorded in the laboratory setup using the SignalTap II Logic Analyzer tool (a part of the Quartus II package).

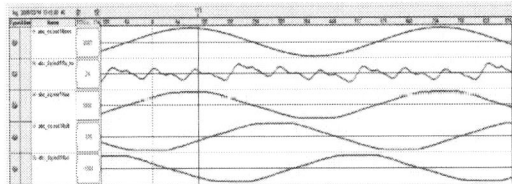

Fig.13 The time graphs of $\cos\omega t$ function, and the synchronizing voltages waveforms recorded in the experimental setup

Figure 13 shows the oscillograms of phase voltages $u_a$, $u_b$, $u_c$, the generated $\cos\omega t$ function, and q-axis component (the controller in figure 11 input error). In the presented synchronizing circuit implementation all quantities are represented by eleven-bit numbers plus a sign bit. Instantaneous values of these quantities can vary within the range $\pm2047$. According to equation (11) the values of q-axis component are equal zero if the synchronizing voltages are consistent with (10), i.e. are not distorted. Due to the power system voltage harmonic distortion with $1\pm6n$ harmonics (where $n= 0, 1, 2...$) the values recorded in axis q are different from zero. These values were varying over the range $\pm98$, what makes 4,78% of permissible range.

## VII. CONCLUSIONS

The storage of energy is characterised by the energy density in mass unit can be the parameters. In the Table 1 the maximum energy density in mass unit for different storages is compared. The flywheel storage has the higher value of energy in mass unit.

Table 1

|  | $W_{max}/m$ [kJ/kg] |
|---|---|
| Capacitor (polyethylene) | 0,02 |
| Air-core coil | 2,78 |
| Flywheel (steel) | 230 |
| Flywheel (composite) | 1600 |

## ACKNOWLEDGMENT

This work is supported by Polish Ministry of Science and Information Society Technologies under project PBZ-KBN-109/T10/2004 in period 2006-2009.

## REFERENCES

[1] P.F. Ribeiro, B.K. Johnson, M.L. Crow, A. Arsoy, Y. Liu: Energy Storage System for Advanced Power Applications. Proceedings of the IEEE, vol. 89, No.12, December 2001, p.:1744-1756.

[2] R. Hebner, J. Beno, A. Walls: Flywheels batteries come around again. IEEE Spectrum, April 2002, p.: 46-51.

[3] D. Boutacoff: Energy Storage. Emerging strategies for energy storage. IEEE Power Engineering Review, December 1989, p.: 3-6.

[4] C. Abbey, G. Joos: Supercapacitor energy storage for wind energy applications. IEEE Transactions on Industry Applications, vol. 43, No. 3, May/June 2007, p.: 769-776.

[5] J.P. Barton, D.G. Infield: Energy storage and its use with intermittent renewable energy. IEEE Transactions on Energy Conversion, vol. 19, No. 2, June 2004, p.: 441-448.

[6] J.N. Baker, A. Collinson: Electrical energy storage at the turn of the Millennium. Power Engineering Journal, June 1999, p.: 107-112.

[7] S. Ginter, G. Gisler, J. Hanks, D. Havenhill, W. Robinson, L. Spina: Spacecraft. Energy Storage Systems. IEEE AES Systems Magazine, May 1998, p.:27-32.

[8] W.R. Lachs, H. Tabatabaei-Yazdi: Energy storage in power systems. IEEE 1999. International Conference on Power Electronics and Drives Systems, PEDS '99, July 1999, Hong Kong, p.: 843-848.

[9] S.M. Lukic, S.G. Wirasingha, F. Rodriguez, J. Cao, A. Emadi: Power management of an Ultracapacitor/Battery hybrid energy storage system in an HEV. IEEE 2006.

[10] K. Murakami, M. Komori, H. Mitsuda: Flywheel Energy Storage System Rusing SMB and PMB. IEEE Transactions on Applied Superconductivity, vol. 17, No. 2, June 2007, p.: 2146-2149.

[11] M.D. Anderson, D.S. Carr: Battery energy storage technologies. Proceedings of the IEEE, vol. 81, No.3, march 1993, p.: 475-479.

[12] A Rufer., P. Barrade: A supercapacitor-based energy storage system for elevators wirh soft commutated interface. IEEE 2001.

[13] Y. Zhong, J. Zhang, G. Li, A. Liu: Research on energy efficiency of supercapacitor energy storage system. International Conference on Power System Technology, 2006.

[14] Y.Y. Yao, D.L. Zhang, D.G. Xu: A study of supercapacitor parameters and characteristic. International Conference on Power System Technology, 2006.

# Analysis of Wide Area Integration of Dispersed Wind Farms Using Multiple VSC-HVDC Links

S. González-Hernández[*], E. Moreno-Goytia [*], and O. Anaya-Lara [†]

[*] Instituto Tecnológico de Morelia/Postgraduate Studies in Electrical Engineering, Morelia, México, e-mail:
*servandogh@ieee.org, elmg@ieee.org*
[†] University of Strathclyde/Institute for Energy and Environment, Glasgow, United Kingdom, e-mail:
*olimpo.anaya-lara@eee.strath.ac.uk*

*Abstract-* Two multi-terminal HVDC topologies are investigated in this paper to connect large-scale wind farms to the grid: two single-input single-output HVDC corridors and a double–input single-output double HVDC corridor. The first topology has two DC links operating in parallel which implies using four converter stations. The second topology has two DC corridors both connected at the output end to a single HVDC dual-input station by means of a DC busbar, which implies using three converter stations. The first topology collects power generated from widely dispersed wind farm sites, whilst the second topology is intended to efficiently interconnect nearby wind farms. The simulations results presented in this paper are building blocks towards the development of platforms helpful in the design and testing of new control strategies for operating multiple interconnected HVDC corridors. The steady state and transient performance of each topology is clearly presented throughout various study cases. The modeling and simulation work has been carried out using Simulink©.

*Index Terms—* Wind Farms, Multiterminal High Voltage DC, Voltage Source Converter (VSC), DC corridor.

## I. INTRODUCTION

Hard-pressed by environmental concern but also favoured by public funds and incentives as tax relaxation, renewable energy sources are these days a viable alternative to satisfy the growing world demand on clean and sustainable power energy, mainly from wind and solar sources. However, in the case of wind energy, the sources are commonly located far from the distribution and consumption centres, which may imply significant electrical losses [1]. In addition, wind farms are geographically dispersed and are growing in number and installed capacity, then large-scale integration of wide-area interconnected wind farms may become a challenge in the near future [2].

HDVC links are constantly increasing their presence in electric networks all around the world. Classical HVDC links and back-to-back schemes are mature and well-established technologies, and the DC link concept is being seriously considered as an alternative solution for large-scale integration of dispersed wind power and its transmission and distribution to also dispersed customers. Furthermore, the use of VSC-HVDC based systems appears to be a better option to transport the increasing amounts of energy over long distances. In addition, as more VSC-HVDCs appear on play the need to interconnect them is a future scenario which seems to be a venturous complex task to focus on if the research on

the topic is not carried on. As tendency points towards large multi-branch DC transmission-distribution networks, the so called multifeed VSC HVDC and multi-HVDC based systems are nowadays the newest concept on the fore to face the challenge for transmission and distribution of power energy from regionally dispersed wind farms. It should also be mentioned that technological advancements such as IGBT-based VSC, multi-level VSC, control schemes and PWM strategies are all pushing forward the application of DC links at different voltage levels.

The investigation presented in this paper focuses on the analysis of two Multi-HVDC configurations for various wind farms interconnection configurations. The results provide further understanding about the complex nature of such type of system. Figure 1.a illustrates two Simple Input-Simple Output DC link configurations, SISOC-HVDC, and Figure 1.b shows a Double Input-Single Output DC link configuration, DISOC-HVDC.

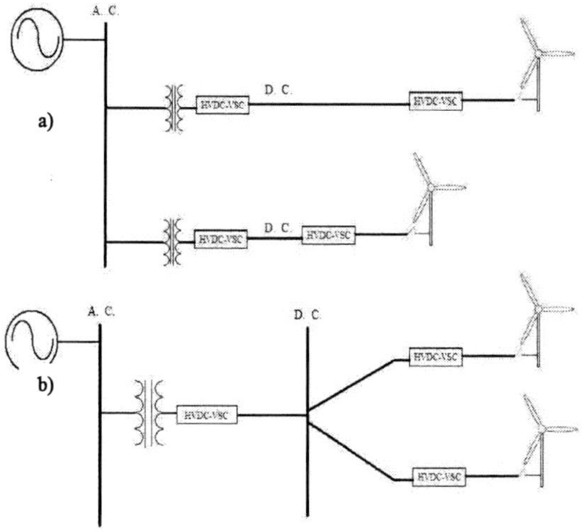

Fig. 1. HVDC-wind farm configurations for of study cases.

The operation performance of each topology is investigated under steady-state conditions, assessing technical aspects such as: AC-DC network integration effects, AC power variations and the power variations at the DC corridor ends due to different faults types. Also, the pros and cons of the two topologies under consideration are discussed.

## II. GENERAL SYSTEM DESCRIPTION

The operation performance of the SISOC-HVDC and DISOC-HVDC topologies are examined under steady-state and transient conditions. Besides the information provided by simulations on the operation of multiterminal DC links connected to wind farms, the models and study cases are built also looking to testing novel control strategies for multiple interconnected multiterminal HVDC corridors in research work to come.

The study cases for SISOC-HVDC utilize the topology 1 shown in Fig. 2. This multiterminal HVDC structure includes two SISOC-HVDC corridors both implemented using Simulink© built-in blocks and various 3 level VSC converters specially built by the authors for the study cases. Each wind farm for the study cases is assembled with a "cluster" of 4 groups having each 6x1.5MW DFIG wind turbines for a total of 36MW per cluster, an equivalent model is used to represent the behavior of 6 wind turbines by a single block, but it is possible to represent the behavior of more than 20 individual elements (DFIG) on a single block model [3]. Each wind farm is interconnected to the AC network trough a single HVDC VSC-Based link, or in other words, two single input-single output HVDC corridors are employed.

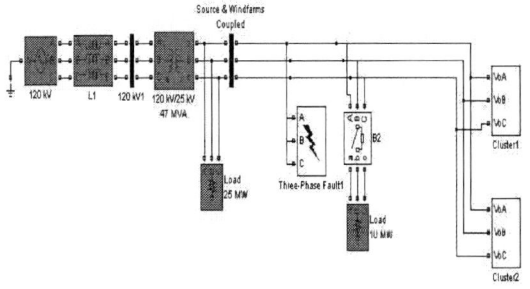

Fig. 2 Matlab/simulink Model Layout for SISOC-HVDC Topology 1

Each of the two SISOC-HVDC, enclosed in cluster 1 and cluster 2, systems is basically built by two three-level IGBT-based VSC-NPC stations and a 30km long DC cable. The VSC-NPC stations are open-loop controlled and for driving the IGBTs, a SPWM strategy is used. The DC link, -VSC-NPC stations plus DC cable-, altogether with ancillary equipment, -such as transformers, harmonics filters, coupling and smoothing inductors, as well as capacitors-, are shown in Fig 3. Each cluster in Fig. 2 represents a 36 MW wind farm. In these grounds, the total installed capacity for these study case is 72 MW. Both clusters are interconnected to the main AC network by means of an AC link. In this condition, the power generated by wind farm is first rectified by an open-loop VSC station operating as rectifier. The DC power is transmitted by the bipolar DC link through the DC cable and converted back to AC by a VSC IGBT-based operating as inverter and its ancillary equipment.

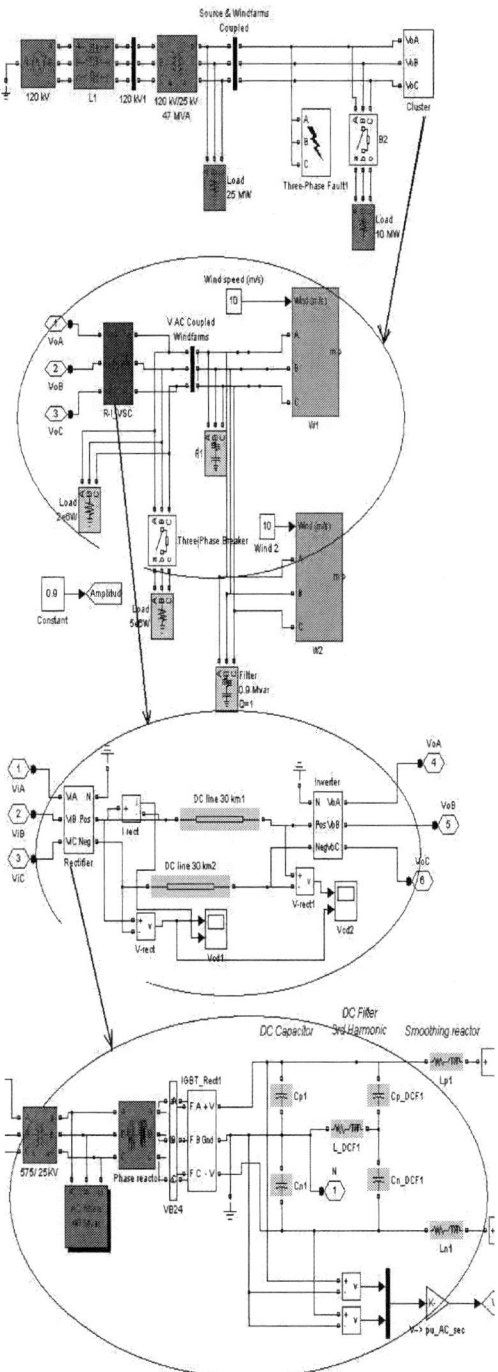

Fig. 3. Detailed Simulink Model used by topology 1

Fig. 4 illustrates the inner components of cluster 1 or wind farm 1, which include the equivalent model of 6x1.5MW DFIG wind turbines, coupled by an AC link the a VSC rectifier, DC transmission line and VSC Inverter. The same structure applies for cluster 2.

1785

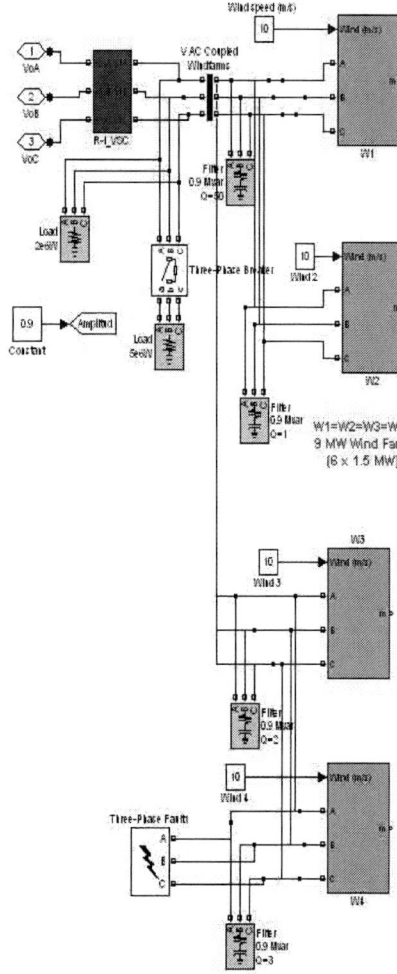

Fig. 4. Matlab/Simulink distributed area cluster1 wind farm model for topologies 1 and 2

### III. TOPOLOGY 1: STUDY CASES AND RESULTS

In this section a number of tests performed to the SISOC-HVDC, or topology 1, are presented. To study transient conditions, four different faults are applied to the DC link: a) phase A-B to ground fault at the wind farm side at $t$=3.5s for .01s; b) phase A-B to ground at the AC-DC interconnection node at $t$=4s for .01; c) +DC line-to-ground fault at inverter side at $t$=5s for .01 and finally d) +DC line-to-ground fault at rectifier side at $t$=6.5 s for .01sec.

#### A. Test Results

For SISOC-HVDC scheme, all tests have been performed applying the faults only at wind farm 1 side but the collateral effects on cluster 2 are also analyzed. Four relevant signals are monitored on the AC side and other four are monitored in the DC side.

For Cluster 1 the waveforms are presented in Fig. 5 as follows:

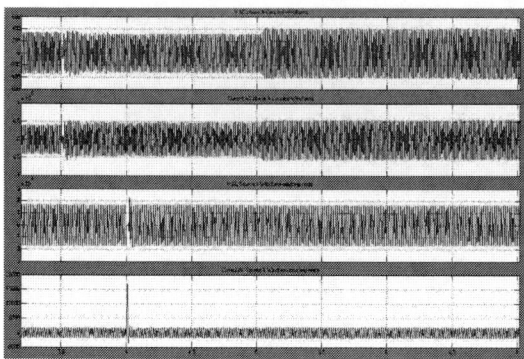

Fig. 5a System response Cluster1 topology 1, from up to down: a)windfarm Generated AC Voltage, b)windfarm generated AC current, c) AC voltage in coupling node of AC source and windfarm d)AC current in coupling node of AC source and windfarm

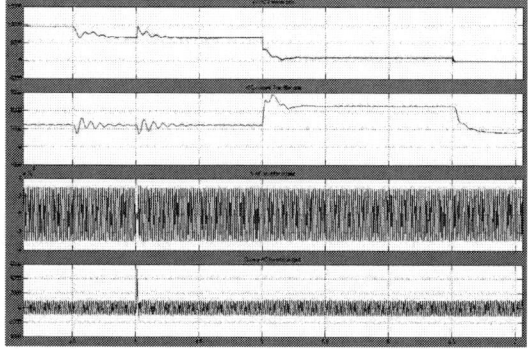

Fig. 5b. System response in <u>Cluster1</u> topology 1, from up to down: a) DC Link voltage b)DC Link current c) VSC output AC Voltage d) VSC output AC Current

Fig. 5 System Response <u>Cluster1</u> topology 1

As Fig. 5a and 5b show, at $t$=3.5s the line voltage and current collapses due the presence of the fault event. The voltage overshoot in the DC bus is a noticeable collateral effect at the DC side of the VSC station. However there is no evidence of significant perturbation over the AC system close to the VSC station, at the AC system- wind farm coupling point. Once the fault condition is cleared, the network returns to its pre-fault operating conditions showing the network steady-state stability and robustness. At $t$ =4 s the fault occurs on the AC side and it has a significant adverse impact all over the entire grid including the surrounding of the VSC-station closer to the AC network. Furthermore, at $t$=5s, a fault at the DC side, in the neighborhood of the VSC station, causes a heavy drop on the DC voltage but a current increase on the opposite VSC station (operating as rectifier). In this case, the voltage and current from windfarms are close to collapse and even when the fault condition is cleared the system does not return to the post-fault conditions. Finally, at $t$=6.5s a second fault in the DC line causes a total collapse dropping the DC line to 0.

For Cluster 2 the key waveforms are shown in Fig. 6.

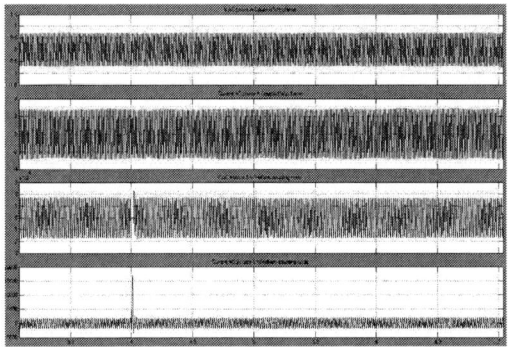

Fig. 6a. System response Cluster2 topology 1 from up to down: a) windfarm generated AC Voltage, b) windfarm generated AC current, c) AC voltage in coupling node of AC source and windfarm d) AC current in coupling node of AC source and windfarm

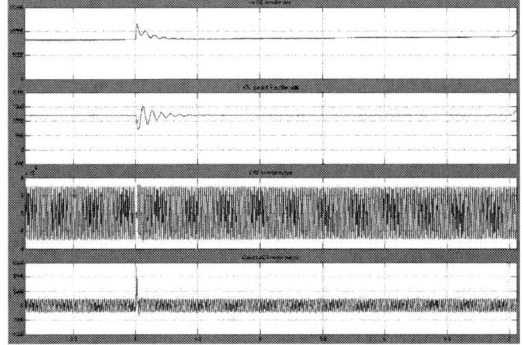

Fig. 6b. System response Cluster2 topology 1 from up to down: a) DC Link voltage b) DC Link current c) VSC output AC Voltage d) VSC output AC Current

Fig. 6 System Response Cluster2 topology 1

Fig. 6a and 6b illustrate the behavior of the AC grid-DC link-wind farm (cluster 2) scheme under the same transient events as those analyzed for cluster 1. The study cases for topology 2 are similar than those performed on topology 1. As first study case, at $t = 3.5$ s the fault conditions is applied on cluster 1, however no voltage nor current drop or wave distortion is noticed neither in the AC side of the grid nor the DC link and even no voltage overshoot in the DC link. Apparently, this fault provokes no adverse effects on cluster 2, which can be problematic conditions for a protection relay. This condition should be further investigated looking for collateral effects all over the interconnection. At $t = 4$ s, when the fault occurs on the AC side the effects on cluster 2 are similar to those showed in cluster 1, this is, the event affects the entire grid. At $t = 5$s, the fault at the VSC station side, closer to cluster 1, causes no voltage drop on the DC link, and the DC current for the VSC station, closer to the AC grid, shows no change, contrary to what happened in cluster 1. However, some voltage and current oscillations are evident in the AC side. Finally, at $t = 6.5$ s, a second DC line fault causes no significant effect over cluster 2 apart from the voltage oscillations already mentioned.

## IV. TOPOLOGY 2: STUDY CASES AND RESULTS

The second case under study is the named topology 2, DISOC-HVDC, which integrates two clusters linked by a DC bus as shown in Fig 7. Each cluster is equipped with 4 groups of 6x1.5MW DFIG-wind turbines for a total installed wind power capacity of 72 MW (as considered for topology 1). In this case a single 3 level IGBT-based inverter is used to drive power from two clusters linked by a DC Bus (represented by a pair of DC sources); in this case two rectifiers are needed. Fig. 7 shows the details of the interconnection of the wind farms to the DC and AC grid and the structure of the cluster.

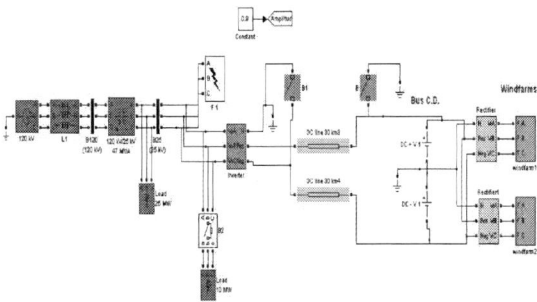

Figure 7 Simulink-based distributed windfarms model for topology 2

### A. Test Results for Topology 2

As has been mentioned for topology 1, only the four most representative signals from the AC side are selected and monitored to illustrate the effects on the topology of the changes and conditions applied to it. Also, for the DC link, four DC values are monitored. The behavior of the AC-DC- wind farm scheme is primarily analyzed applying four different types of faults: a) phase A-B fault to ground at the wind farm side at $t = 3.5$s for .01s; b) phase A-B to ground fault at AC interconnection point at $t = 4$s for .01s; c) +DC line-to-ground fault at the VSC station closer to the AC grid at $t = 5$s for .01 s and d) +DC line-to-ground fault at the VSC closer to the wind farms at $t = 6.5$ s for .01s. The next figures illustrate the topology 2 behavior.

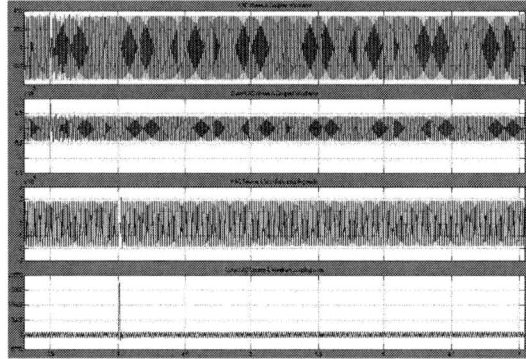

Fig. 8a System response Cluster1 topology 2 from up to down: a) windfarm Generated AC Voltage, b)windfarm generated AC current, c) AC voltage in coupling node of AC source and windfarm d)AC current in coupling node of AC source and windfarm

1787

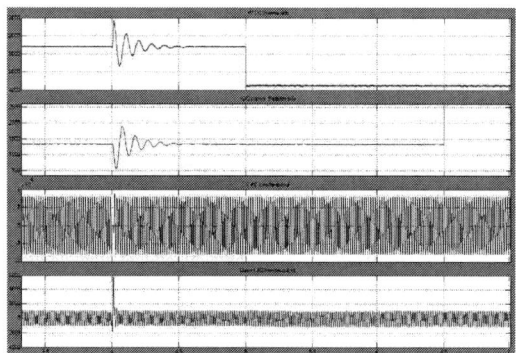

Fig. 8b System response Cluster1&2 topology **2** from up to down: a) DC Link voltage b) DC Link current c) VSC output AC Voltage d) VSC output AC Current

It is seen from Fig. 8 that at $t = 3.5s$ little changes are noticed in the voltage and current waveforms in AC grid. Only small fluctuations on the DC voltage, comparatively greater than the fluctuation of the DC current, are observed. At $t = 4s$ the voltage and current waveforms at DC and AC sides are significantly affected mainly at the VSC output connected to the AC side. For a fault occurring in the DC link, at $t = 5$ s, both voltage and current collapse after less than four cycles. It should be noticed that the post-fault voltage and current levels do not return to the nominal pre-fault values and remain close to zero. Worth to be mentioned is that the current and voltage deviations between $t = 5s$ and $t = 6s$ is a system response to steady-state load changes. For this particular case the DC side system response is the same for cluster 1 and 2 because they are linked by a single DC Bus to one single VSC inverter as shown in Fig. 7

For Cluster 2 the DC side response is exactly the same as for Cluster 1. The response for Cluster 2 AC side is shown in Fig. 9.

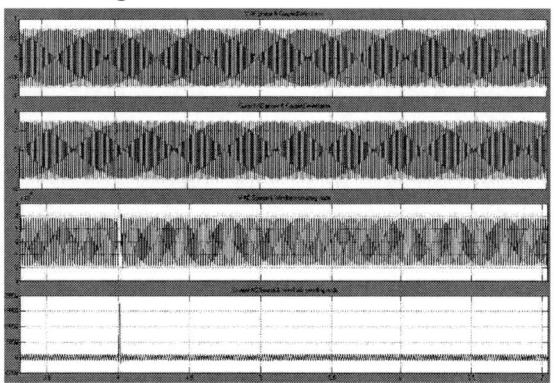

Fig. 9. System response cluster2 topology 2 from up to down: a)windfarm Generated AC Voltage, b)windfarm generated AC current, c) AC voltage in coupling node of AC source and windfarm d)AC current in coupling node of AC source and windfarm

It is possible to observe that it is almost the same response than that of Cluster 1, but it is clear that a fault in phases A & B at t=3.5 in cluster 1 does not affect the AC side coupling node on cluster 2 but phase A-B to

ground fault at AC interconnection point at $t = 4s$ for .01s clearly affects and generates and important current peak on the AC coupling node (for both Clusters 1 & 2).

As complement to de DC side system response for clusters 1&2 Fig. 10 shows the DC side current peak during a fault at t=6.5 s (+DC line-to-ground fault at the VSC closer to the wind farms for .01s. by means of switch "B" (see Fig. 7) which does not occur during +DC line-to-ground fault at the inverter side by means of switch "B1" (see Fig. 7).

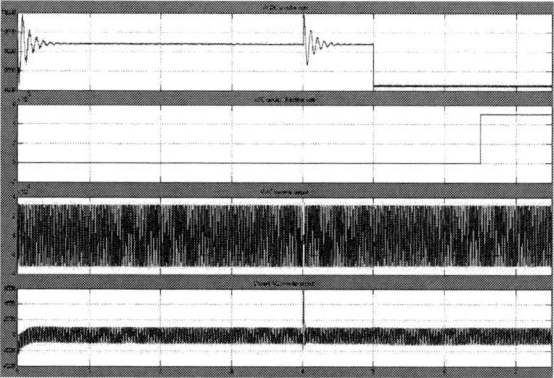

Fig. 10 System response Cluster1&2 topology 2 from up to down: a) DC Link voltage b) DC Link current c) VSC output AC Voltage d) VSC output AC Current

CONCLUSIONS

In all cases presented in this paper, the DFIG wind turbines models in use have shown a strong dependence between the transformer parameters and DFIG controller. A mismatch between the latter and the former makes it difficult to reach a stable steady state operation at nominal transformer power. In this work the network reaches steady state in approximately 3 s.

The DFIG model has various operation modes for self-regulation. In all simulations the control mode $Q=0$ for the DFIGs have been selected. This is evident because the almost constant DC link current and such regulation seems to help reduce the adverse effects of faults.

Furthermore, in the specific cases performed for topology 1, it can be clearly observed that the core of DFIG wind generators enclosed in cluster 2 gives a significant power support to the loads at the AC side when a fault occurs in cluster 1. After a comparative analysis it becomes evident that the use of topology 1 represents a more robust system than topology 2 system but this second one could be probably less expensive and less difficult to be built at full scale.

REFERENCES

[1] Stephan Meier, *Novel Voltage Source Converter based HVDC Transmisión System for Offshore Wind Farms*, Royal Institute of Technology, Department of Electrical Engineering Electrical Machines and Power Electronics, Stockholm 2005

[2] Sahi Towito, Mario Berman, Gil Yehuda and Raul Rabinivici "Distribution generation case study: electric wind farm doubly fed induction generators," *2006 IEEE 24TH Convention of Electrical and Electronics Engineers in Israel* , Nov. 2006, pp. 393-397

[3] L.M. Fernández, C. A.García, J.R. Saenz, F. Jurado "Reduced model of DFIGs wind farm using aggregation of wind turbines and equivalent wind," *Proc. of the 2006 IEEE Mediterranean Electrotechnical Conference MELECON 2006*, May 2006, pp. 881 – 884.

[4] D. Jovcic, *Member IEEE* "Interconnecting offshore wind farms using multiterminal VSC-based HVDC" proc. of *2006 IEEE Power Engineering Society General Meeting*, June 2006

## BIOGRAPHIES

 **Servando Gonzàlez Hernàndez** was born in Morelia Michoacan, Mèxico, on february 11, 1968. He obtained his BSc and MSc form the Instituto Tecnológico de San Luis Potosi, and the Instituto Tecnològico de Morelia, respectively.

His fields of interest included power electronics, HVDC and renewable power resources. He is now pursuing a PhD. degree at Instituto Tecnológico de Morelia on HVDC field.

# Generator Selection for Offshore Oscillating Water Column Wave Energy Converters

D.L. O' Sullivan [*] and A.W. Lewis [†]

[*] Hydraulics & Maritime Research Centre, University College, Cork, Ireland, e-mail: *Dara.OSullivan@ucc.ie*
[†] Hydraulics & Maritime Research Centre, University College, Cork, Ireland, e-mail: *T.Lewis@ucc.ie*

*Abstract*— In the field of wind energy, electrical generator solutions have converged on a small number of technologies, for specific technical and economic reasons. This paper investigates whether a similar rationale exists within the field of oscillating water column, wave energy converters. The predicted performance of the various generator options within a typical wave regime is validated by means of a time-domain MATLAB/Simulink model. The suitability or otherwise of brushed machines in the offshore marine environment is examined in detail. Each generator configuration is then modelled and the annual energy output for a typical wave climate assessed with consideration being given to the impact of speed control on the energy output.

*Keywords*— Wave energy, Renewable energy systems, AC machine, Adjustable speed generation system.

## I. Introduction

Renewable energy technology is steadily gaining importance in the world energy market, due to the limited nature of fossil fuel supplies. Wind power has been rapidly increasing its penetration levels in the past decade, and its technology is maturing to the point where the electrical system configuration is converging to a small number of options. Wave energy has historically struggled to break through to a commercial implementation, but recent years have begun to see a maturing of some technologies, with several commercial companies now conducting sea trials with quarter-scale prototypes and actively developing full-scale prototypes. Three such prototypes are depicted in Figure 1.

Figure 1: Prototype wave energy converters: (t) Ocean Energy, (b, l-to-r) Wavebob, Pelamis.

The focus of the wave energy developers to date has been primarily on sea performance and survival, as well as on the primary power take-off mechanisms, rather than on the electrical power take-off. This is necessary and understandable, as these are the system components that have the most significant influence on infrastructural cost and on overall energy efficiency. However, the technology has now reached the point where it is important to begin examining the different elements of the electrical system in more detail. This is vital from the point of view of system optimisation, both in terms of robustness and efficiency, both of which affect cost, and in terms of the direct infrastructural cost of the electrical system, as well as in the context of overall system control. In this regard, it is important to both learn from, and distinguish from where appropriate, the developments in wind energy electrical systems.

### A. Power Take-Off in Wave Energy Converters (WEC)

There are three primary forms of power take-off in wave energy converters:

- Rotary turbo-generators – typically driven by an oscillating airflow
- Hydraulic motor-generators – typically driven by a pressurised fluid
- Direct-drive linear generators – typically driven directly by sea motions

While direct-drive linear generators seem rather far from commercial realisation, hydraulic devices and air-driven devices -known as oscillating water columns (OWC)- have reached the pre-commercial stage, with OWC converters at the most advanced stage of development. Part of the reason for this is the fact that standard rotary generators can be deployed in these devices.

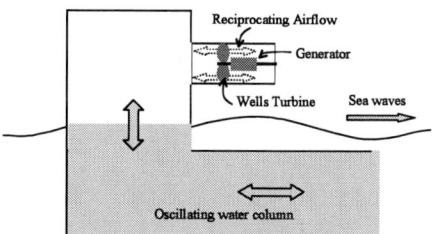

Figure 2: Floating OWC device (Bent Backward Duct Device)

Both brushed and brushless induction machines [1],[2],[3],[4] as well as permanent magnet machines [5],[6] have been considered for such devices. However, no clear consensus regarding the most suitable machine type exists. Hence, the work in this paper will focus on the generator machine selection for offshore, floating OWC-type converters, such as depicted in Figure 2.

## B. Electrical Machines in Renewable Energy Generation

Synchronous generators (SG) form the backbone of traditional fossil fuel power generation. They are cost effective at multi-MW power levels, rotate at constant speed, and can be easily controlled to provide both real and reactive power as required. However the characteristics of renewable energy sources are very different from fossil fuel supplies in that they are by and large, highly variable energy sources. For instance, in wind or wave power generation, a fixed speed generator will extract a much smaller fraction of the available power than a variable speed generator [7]. Likewise, under heavy gust or swell conditions, a fixed speed generator will experience severe shock loads on the generator shaft, whereas if the speed is allowed to increase, the inertia of the system will absorb some of the extra power input. This led to the adoption of asynchronous generators in wind turbines where the slip range was utilised to provide a small measure of speed variation. Extensions to the speed range were provided for by pole changing or rotor resistance variation [8] In recent years, the improvement in cost and performance levels of high power switching transistors has led to the adoption of fully variable speed controlled generators. These have typically taken the form of either gear-coupled Doubly Fed Induction Generators (DFIG) with power electronics control of the rotor voltage and frequency, or direct-coupled Synchronous Generators (SG) with power electronics control of the stator voltage and frequency, and either brushless field excitation or permanent magnet excitation (PMG).

On examination of the technology employed by the worlds leading wind turbine manufacturers, it can be concluded that the geared DFIG represents the state of the art in the world's leading wind energy manufacturers, with the adoption of the direct-coupled brushless SG/PMG representing a more recent trend, and in actual numbers, the DFIG still representing around 65% of the market share in wind turbines[8],[9]. There are clear reasons for the choice of these two generator technologies. In the case of the DFIG, system cost, size, and power efficiency is optimum for a variable speed system, since the power electronics must only be rated at 30-50% of the system rated power. In the case of the PMG, initial cost is significantly higher due to the high cost of the large permanent magnets, as well as the 100% rating of the power electronics. However, the dominant factor in this case is reliability and maintenance reduction. This is due to the lack of brushes in the system, as well as the direct coupling, which eliminates the gearbox with its associated maintenance. Furthermore, direct coupling enhances bearing lifetime due to the significantly reduced number of revolutions per year of the generator, as a result of the lower rotational speed.

What is clear from the evolution of wind energy to a stage of relative maturity is that there appears to be a convergence towards one or two electrical machine technologies, and that there is a clear technical and economic rationale for such. The purpose of this paper is to seek to identify whether a similar rationale can be identified in the case of wave energy conversion. While it is likely that practical experience will be a strong deciding factor in the coming years, it is also possible to learn from the evolution of wind technology.

## II. OWC WEC Generator Comparison

Gearing is not typically an issue in OWC WECs as the turbine can be designed to rotate at speeds consistent with standard off-line machines such as 750/1500/3000rpm. Hence, the machines to be considered in this study are all directly shaft-coupled to the turbine, and are listed as follows:

- Doubly Fed Induction Generator (DFIG)
- Squirrel Cage Induction Generator (SCIG)
- Permanent Magnet Synchronous Generator (PMG)
- Field Wound Synchronous Generator (SG)

These generator technologies are compared under the headings of:

- General suitability for offshore environment
- Energy efficiency over typical wave climate
- Grid connection
- Cost

In each case the capability to implement a wide speed range control, through power electronic frequency converters is assumed so that optimum aerodynamic damping can be provided to the varying air flow and the turbine efficiency can be maximised, as described for the development of wind turbine technology.

### A. Suitability to Offshore Environment

Suitability to sustained and reliable operation of the generator in the harsh environmental conditions of the offshore marine environment is clearly a significant and important requirement. This is examined in terms of the mechanical and environmental wear sustained by machines in such an environment. One important issue to be tackled is the feasibility of the use of brushed machines in such an environment, and the consequent maintenance requirements. Both the DFIG and the brushed version of the SG utilise brushes, so before considering these machines any further it is necessary to examine the suitability of brushed machines in general in offshore WECs.

### 1) Brush Operation

Brush wear in brushed machines is the result of mechanical friction and electrical erosion. Friction produces carbon dust, while the result of electrical erosion is the vaporization of carbon with little physical residue. In order to achieve a good coefficient of friction between the carbon brush and the slip ring, it is necessary to establish a good carbon composite film[10]. A good film layer can reduce the coefficient of friction to 10% of the original bare coefficient. In order to maintain a good working film, brushes should ideally operate close to the rated load current. Operation in over-current causes slip ring blackening and reduced brush life, whereas protracted light load operation results in film removal and increased brush friction and wear. The differing power profiles of typical wave and wind energy converters are depicted in Figure 3 for a time series of 12s. While the power input to the wind device is oscillatory, its variation around the average power level is significantly less severe than for the wave device, where the power input fluctuates to zero once every half-wave cycle. The high pulsating nature of wave energy power flows is thus clearly not well matched to the desired electrical operating point of a brush-slip ring arrangement.

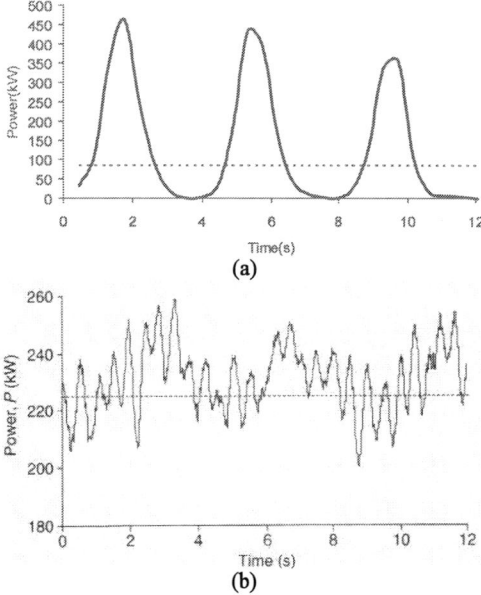

Figure 3: Power time series for (a) WEC (b) wind turbine over a 12s time slot.

It is evident from Figure 3 that wave energy converters require a high peak to average rating, and so will operate close to their peak or overload current ratings for a significant proportion of the time. This represents a further complication in the employment of brushed machines. In a recent publication sponsored by Vestas Wind Systems [20], it was discovered that under high current operation brush-slip ring systems can periodically enter film instability modes that are characterised by severe brush wear and high brush temperature, increasing the wear rate by approximately 40% more than the expected wear rate.

The other factors that inhibit good brush film formation are high humidity and the presence of chemical contaminants in the air. While these environmental factors are issues for offshore wind turbines also, humidity and air vapour control are being built into modern offshore wind turbines for these reasons [11]. The inclusion of such air quality control in the generator enclosure of OWC WEC is less feasible since the generator is located in an airflow path that is in direct contact with the sea.

The minimum brush life of a general purpose, slip ring machine is typically 3,500-8,750 hrs. The effective operating period of a WEC at a good site is around 5,000 hrs annually. Thus, in order to avoid costly outages, or even generator damage, the brushes ought to be changed at least twice annually, which corresponds with best practice in the wind industry.

### 2) Operation and Maintenance

The option of using the DFIG machine or the brushed SG machine requires the presence of brushes in the system which, as previously mentioned, must be maintained and replaced on a regular basis, typically twice a year. Brushless DFIGs [14] are still only at the concept stage and are a long way from commercial viability. Depending on the site specifics this is potentially a more serious issue for wave energy converters than for wind energy, even offshore wind. For instance, if on-board maintenance is considered safe up to a 1m swell, then this only allows a probable 7 days in the year for maintenance. If on-board maintenance is allowed up to a 1.5m swell, there will be on average about 55 days in the year when such

maintenance is possible. This problem became apparent for the first time in the Bockstigeen wind farm approximately 12km of the Swedish west coast [13]. Docking the maintenance boat proved to be extremely difficult even at a wave height of a little over 1m. Subsequent to this experience, alternative approaches are being explored in accessing offshore wind turbines, including submarine vehicle and diver access points, as well as helicopter pads located on the nacelle [12], [13]. Moreover, once access is possible, the actual maintenance procedure can proceed in a relatively stable and protected environment. Clearly the situation is even less straightforward for floating offshore WECs. Access is likely to only be by boat, and the working environment itself is not stable. These factors and the consensus of the industry and research community [21] appear to strongly support low-maintenance, long lifetime designs.

These considerations would appear to rule out the use of the DFIG (as well as the brushed SG) in offshore OWC WECs, despite its clear advantage in terms of size, cost and efficiency. In the medium term, there may be some hope for the DFIG if alternative brush technologies come to successful fruition. Metal fibre brushes [15] potentially offer significantly enhanced service lifetimes, and more robust operation in harsh environments. Initial calculations, based on wear rate figures supplied by Hipercon [15], appear to suggest brush lifetimes of up to 17 years. However, this technology is as yet unproven, and it is unclear if this level of performance is actually achievable.

### 3) Corrosive environment

All machine types will be protected to a high degree from the worst effects of the environment. However, given the location of the generator in an OWC turbine duct, and the normal requirement for cooling air to flow around the rotor, it is likely that the machine will experience airflow with high saline content. This will have the most detrimental effects on a PMG machine since *NdFeB*, which is the material of choice for high performance permanent magnet machines, is very sensitive to corrosion [22]. This fact represents a significant disadvantage for the PMG in this regard.

### 4) Mechanical Issues

Typical heave motions of the OWC device during a severe sea state are illustrated in Figure 4. Motions up to 6m in amplitude can occur in time periods of 3s. It is evident that such conditions apply severe mechanical stress on the system components, and bearings and couplings will have to be rated to absorb these shock loads

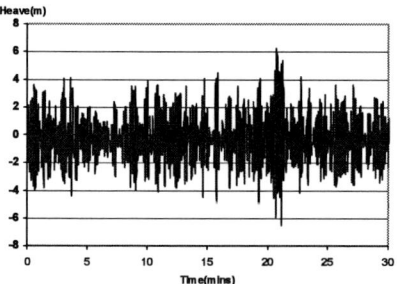

Figure 4: OWC heave motions

Generators with a higher power to weight ratio will have an advantage in this regard, as the shock loading and bearing ratings will not be as severe. In the power range of interest, the IG has a ratio of 4-4.5 kg/kW, as compared to 3-3.8kg/kW for the SG and the PMG [23]. Hence the IG has a weight penalty of 20-30%. The surface magnet PMG, however, has an additional problem in that the permanent magnets are brittle and can be easily cracked under mechanical shock unless precautions are taken.

### B. Energy Efficiency Over Typical Wave Climate

A typical yearly average wave power variation in kW/m, as well as its statistical distribution is illustrated in Figure 5. It is clear that there is a significant variation between the minimum and maximum average levels, and also that the lower average power levels are statistically more important.

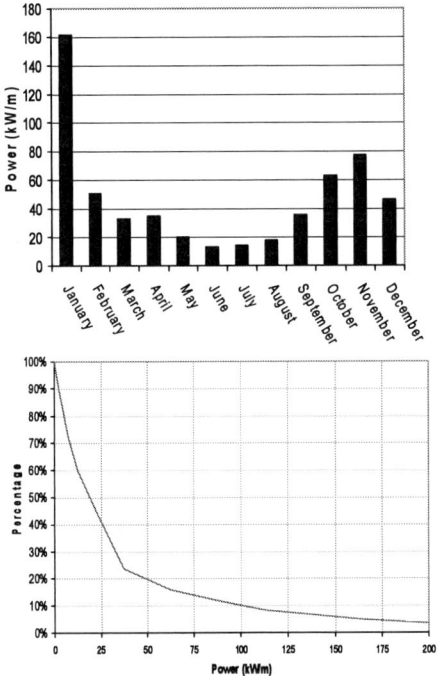

Figure 5. Typical annual wave climate power variation off the SW coast of Ireland

If the wave climate is examined in more detail it is evident that the variation is even more significant. Typical monthly data, as depicted in Figure 6 reveal that the peak power levels over a month (January) can be greater than ten times the average power for the same month. This large peak-to-average ratio is also apparent on a second by second time scale, as illustrated previously in Figure 3(a).

Each 'sea state' in a given wave climate is represented by a significant wave height, $H_S$, and peak wave period, $T_P$, which represents the peak of the statistical energy density spectrum of that sea state. The climate can then be described by a scatter diagram of the individual sea states, as illustrated in Figure 6. The performance of the WEC in the different sea states represented in the scatter diagram will have a significant filtering effect on the power curves. The majority of WECs are designed to be at their most efficient at the prevailing wave period, with a significant reduction in efficiency at wave periods removed from this. This efficiency data is typically obtained through numerical modelling of the WEC, or scale tank tests, or a combination of both [16].

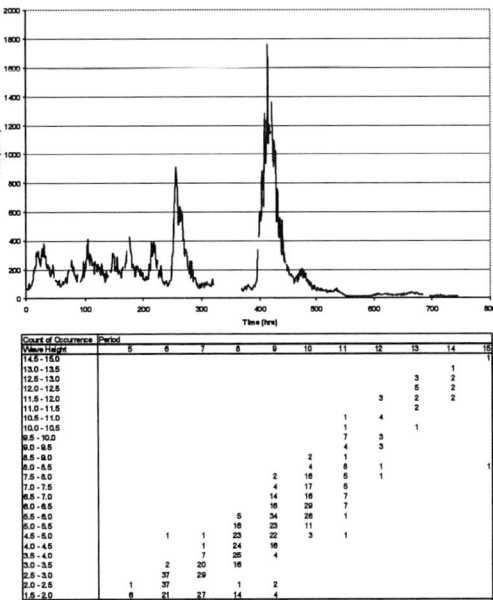

Figure 6: Typical monthly wave power variation (t to b): Power vs time format; wave height and period scatter diagram.

The efficiency of the OWC device under consideration is plotted in Figure 7, as a function of wave period. This efficiency curve is generated from model scale tests of the device in the wave basin in the Hydraulics & Maritime Research Centre (HMRC), shown in Figure 8. The input wave power is calculated as the incident wave power per metre of wave front multiplied by the device width. However, the device is capable of absorbing energy from waves beyond its own body width. Hence, given this definition of input power, it is possible for a wave energy device to have an efficiency value greater than unity, as illustrated in Figure 7. The output power in these calculations is the pneumatic power flow in the device air chamber.

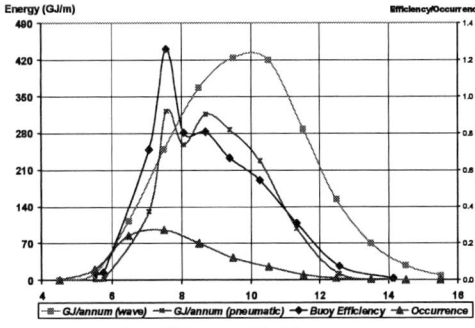

Figure 7: OWC device efficiency and incident wave energy per metre of device length

The annual incident wave energy per metre and the statistical occurrence of this energy as a function of wave period are also plotted. The product of these with the OWC efficiency gives an indication of the pneumatic output energy per annum, as a function of wave period. Clearly, the OWC response must be tuned to the prevailing wave periods in order to maximize energy production.

Figure 8: Model testing in the HMRC wave basin.

### 1) Efficiency Modelling

The conceptual layout of the system model is illustrated in Figure 9, and the Simulink model is depicted in Figure 10.

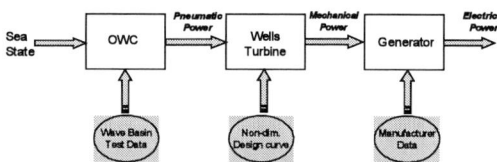

Figure 9: System model conceptual layout

Data has been collected from wave basin tests of the device over 15 representative sea states. From these, the 10 most statistically significant sea states are selected, and the pneumatic power data time series form the input to the remainder of the system model. The system is modelled mathematically in MATLAB, and also in a Simulink time domain simulation.

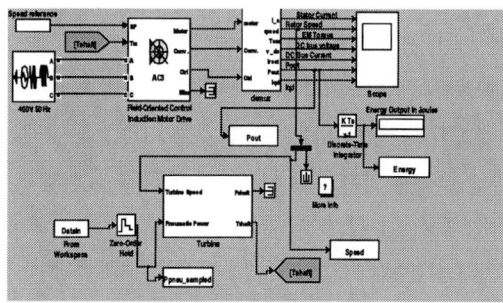

Figure 10: Simulink system model

If the wave climate is defined by a finite number of different sea states, $N=1,2..N_{max}$, then the shaft power $P_{G_N}$ applied to the generator during an individual sea state $N$, can be expressed as:

$$P_{G_N}(t) = P_{W_N}(t)\,\eta_{WEC_N}\,\eta_{turb}(t,\omega_m), \quad (1)$$

where $P_{W_N}(t)$ is the incident wave power time series for sea state $N$ and $\eta_{WEC_N}$ is the average WEC efficiency over that sea state. The turbine efficiency characteristic, $\eta_{turb}$, is a function of the applied pneumatic air flow rate and the turbine rotational speed $\omega_m$ and typically has a maximum value of 50-60% [18]. The efficiency characteristic for the Wells turbine[18] utilised in this study is plotted in Figure 11, as a function of flow coefficient, $\phi$, where $\phi$ is the ratio of airflow velocity across the turbine to the tip velocity of the turbine blades. It can be seen that at high flow coefficients the turbine stalls, and efficiency reduces

drastically. Also, at low flow coefficients, aerodynamic drag dominates.

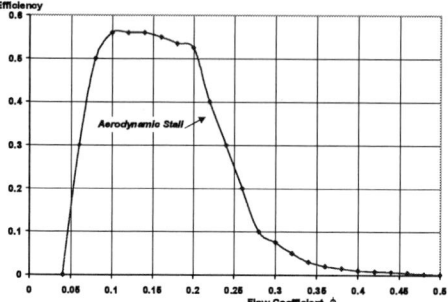

Figure 11: Turbine Efficiency Characteristic

The total energy output, $E_N$ of the generator for a given sea state $N$, over the time period $T$ of the sea state, can then be calculated as:

$$E_N = \int_0^T \eta_{gen}(t) P_{G_N}(t)\,dt, \quad (2)$$

where $\eta_{gen}$ is the generator efficiency (including the generator-side inverter). The energy output over a year is then calculated by summing the energy outputs over the individual sea states,

$$E_{1yr} = \sum_{N=1}^{N_{max}} p_N E_N \frac{T_{1yr}}{T} = \\ \sum_{N=1}^{N_{max}} p_N \left[ \int_0^T \eta_{gen}(t) P_{G_N}(t)\,dt \right]. \quad (3)$$

The generator efficiency is a function of load power (and also of turbine speed), and can be expressed as a function of the machine parameters.

$$\eta_{gen}(t) = \frac{P_{G_N}(t) - P_{MLoss}(t) - P_{InvLoss}(t)}{P_{G_N}(t)}, \quad (4)$$

where $P_{MLoss}$ is the power loss in the machine, and $P_{InvLoss}$ is the power loss in the generator-side inverter. The machine power losses for the various machine options are given in (5), in terms of available machine parameters. In these equation derivations, field oriented vector control is assumed in each case, with maximum torque per ampere production.

$$P_{MLoss}(t,\omega_{m_N}) = \begin{cases} \dfrac{8P_{G_N}^2(t)(R_s+R_r)L_r^2}{3\omega_{m_N}^2 p^2 \lambda_m^2 L_m^2} + \dfrac{3\lambda_m^2 R_s}{2L_m^2} + P_{CFW}(\omega_{m_N}) & \text{SCIG} \\[3mm] \dfrac{8P_{G_N}^2(t)R_s}{3\omega_{m_N}^2 p^2 \lambda_m^2} + I_F^2 R_F + P_{CFW}(\omega_{m_N}) & \text{SG} \\[3mm] \dfrac{8P_{G_N}^2(t)R_s}{3\omega_{m_N}^2 p^2 \Lambda_{md}^2} + P_{CFW}(\omega_{m_N}) & \text{PMG} \end{cases} \quad (5)$$

$$P_{InvLoss}(t) = \frac{P_{G_N}^2(t)}{3k_\omega^2 \omega_{m_N}^2} R_{inv} + \frac{P_{G_N}(t)}{\sqrt{3}k_\omega \omega_{m_N}} V_{Linv} \quad (6)$$

where $R_s, R_r, R_f, L_r, L_m, L_{md}$ are the usual machine equivalent circuit parameters which are either provided or can be extracted from manufacturer data, $p$ is the number of pole pairs, $\omega_{m_N}$ is the machine angular speed corresponding to the damping required for the sea state $N$,

$\lambda_m$ and $\Lambda_{md}$ are the rated flux linkage and permanent magnet flux linkages respectively, $k_\omega$ is the speed constant, and $R_{Inv}$ and $V_{LInv}$ are an equivalent series resistance and voltage drop corresponding to the fixed and current proportional parts of the switching and on-state IGBT losses of the power inverter,

The comparison of machine technology from the point of view of energy efficiency is evaluated by combination of the power loss equations for each machine technology with (3). The DFIG is not considered. It is without doubt the most efficient solution due to the significantly smaller rating of the power converters; however, from the discussion of Section A it is not considered a suitable option

### 2) Energy Efficiency Results

The average output electrical power for each modelled sea state is illustrated in Figure 11 and Figure 13 for the three remaining generator types for two different turbine designs. In Figure 12, the turbine is designed to stall at a relatively low volume flow rate, resulting in significant stall effects at high sea states, but in better performance at lower sea states, which can be statistically more significant. In Figure 13, the turbine stall flow level is set to a higher value. This allows more power to be absorbed from the higher sea states, but the turbine is less efficient at the lower sea states. Both of these designs utilize a fixed speed control.

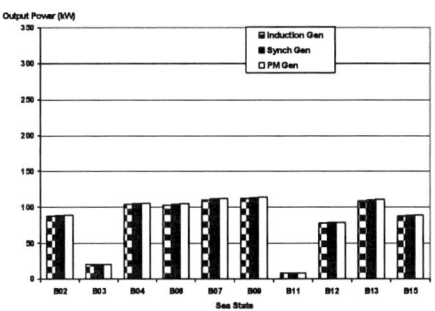

Figure 12: Output power at fixed speed and low turbine stall flow design

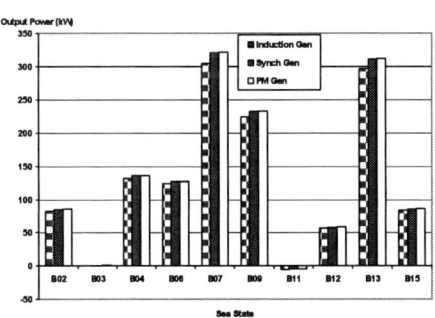

Figure 13: Output power at fixed speed and optimum turbine stall flow design

From these results, it is evident that the energy efficiency of the IG design is several percent less than the SG and PMG designs. These results are tabulated in Table 1.

### C. Impact of Speed Control on Energy Efficiency

The turbine speed is now controlled in the model, in order to maintain the turbine flow coefficient at the optimum value of 0.15, as seen from Figure 11. Maximum

and minimum limits are set on this speed variation to allow for the physical capabilities of the generator control. The difference in available mechanical shaft power from the turbine with variable speed control is plotted in Figure 14.

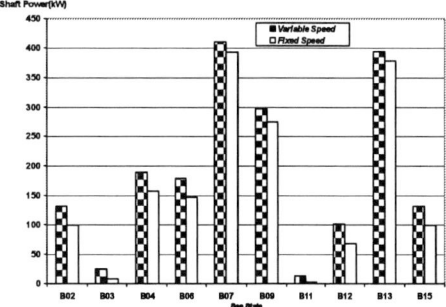

Figure 14: Shaft power comparison

The output power results are plotted in Figure 15, and the yearly energy yield is tabulated in Table 1.

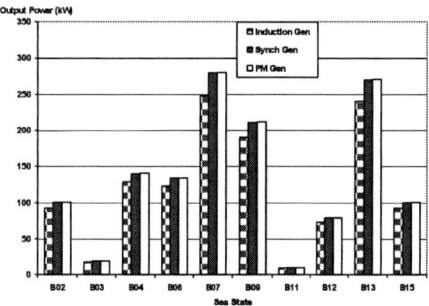

Figure 15: Output power with variable speed control

From Table 1, it is evident once again, that the PMG and SG significantly outperform the IG from an energy perspective.

Table 1: Energy efficiency results

| Machine | Yearly MWh Fixed Speed | | Yearly MWh Variable Speed | |
|---------|------|--------|------|----------|
| IG | 1153 | 1pu | 1089 | 1pu |
| SG | 1196 | 1.037pu | 1199 | 1.101pu |
| PMG | 1201 | 1.042pu | 1204 | 1.106pu |

An interesting result that emerges from this study is the marginal difference in MWh produced by the variable speed strategy compared to fixed speed operation. If the fixed speed output power is plotted beside the variable speed output power for the SG as illustrated in Figure 16, it can be seen that variable speed operation is more efficient at low power sea states, while fixed speed operation is more important at the higher power sea states. These results point to the need to examine more closely the optimum control strategy. The assumption that variable speed control may be the best approach under all conditions, and in all wave climates, may not in fact be true, and this issue needs to be addressed in significantly greater detail.

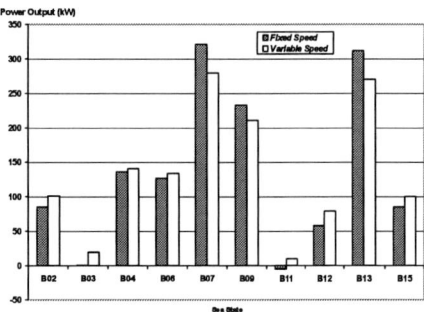

Figure 16: Output power – fixed vs. variable speed

### D. Grid Connection

With regard to grid connection, the DFIG is at a disadvantage to the PMG, SCIG and SG. The DFIG stator is directly connected to the grid, and so is much more influenced by grid faults and failures, and likewise draws its starting current and reactive current directly from the grid. The PMG, SCIG and SG all have full frequency converters in place between the machine stators and the grid, and so are more significantly decoupled from its influence. The presence of a full frequency converter with a capacitive dc bus also allows the potential to provide reactive power to the grid and thus participates in grid voltage support. Moreover a frequency converter will not feed a grid fault beyond the overcurrent rating of the converter. This can be a significant advantage in areas where the existing network equipment is operating close to the limit of its fault level capacity. It can also, however, be seen as a disadvantage in that frequency converters may not be able to provide enough fault current in the event of a fault occurrence to trip the system breakers, resulting in a fault remaining undetected.

### E. Cost

Once again, the DFIG is not considered in this section. From an initial cost point of view, there is little to distinguish between the IG and the SG. At power levels below about 800kW, the SG is slightly more expensive, but above this power level the IG loses this advantage. Both machines have cost levels within the range of €30-40/kW at the power levels of interest [23]. It is difficult to obtain precise cost data for high power PMGs, partly because they are not very widely manufactured at 100kW-1MW power levels. This is in part due to the high cost associated with, and difficulties in the production of, large volumes of rare earth permanent magnet material.

Figure 17: Neodymium (Nd) and Dysprosium (Dy) cost

The challenges facing PMG manufacturers are illustrated in Figure 17. Rare earth metals Neodymium and Dysprosium (which helps maintain the magnetization of the material in demanding environments) are rapidly rising in cost, driven by applications such as motors for electric power steering and hybrid electric vehicle drives [24].

## III. CONCLUSIONS

In this paper, generator options for offshore, floating OWC WECs are considered. Generator configurations in the technologically similar field of wind generation are considered, and distinctions made with the wave energy environment. Maintenance requirements are assessed for brushed machines, and the conclusion is reached that these are unsuitable due to the particular characteristics of wave energy converters in the offshore marine environment. The remaining brushless options are analysed and assessed for energy efficiency, grid connectivity and cost.

Based on the considerations outlined in this paper, a crude ranking table can be constructed for the generator options as depicted in Table 2.

Table 2: Generator Ranking

|      | Offshore Suitability | Energy Efficiency | Cost |
|------|------|------|------|
| IG   | 1 | 3 | 1 |
| SG   | 2 | 2 | 2 |
| PMG  | 3 | 1 | 3 |
| DFIG | 4 | N/A | N/A |

It appears that the SG and the IG are the strongest contenders for the generator technology of choice. The IG performs better in the ranking table shown above, however on balance it is felt that the brushless SG may be a better option overall due to its improved energy yield, weight, and controllability.

### ACKNOWLEDGEMENTS

This work was funded under the Charles Parsons Initiative under the auspices of the Dept. of Communications, Energy and Natural Resources of the Irish Government. The author wishes to acknowledge the technical assistance and input of Dr. R. Alcorn and S. Barrett of the Hydraulics & Maritime Research Centre

### REFERENCES

[1] Jayashankar, V.; Udayakumar, K.; Karthikeyan, B.; Manivannan, K.; Venkatraman, N.; Rangaprasad, S., "Maximizing power output from a wave energy plant," *Power Engineering Society Winter Meeting, 2000. IEEE* , vol.3, pp.1796-1801.

[2] Kiran, D.R.; Palani, A.; Muthukumar, S.; Jayashankar, V., "Steady Grid Power From Wave Energy," *IEEE Transactions on Energy Conversion*, , vol.22, no.2, pp.539-540, June 2007

[3] Yegna Narayanan, S.S.; Murthy, B.K.; Sridhara Rao, G., "Dynamic analysis of a grid-connected induction generator driven by a wave-energy turbine through hunting networks," *Power Electronics, Drives and Energy Systems for Industrial Growth, 1996.*, vol.1, pp.445-451.

[4] Marques, G.D., "Stability study of the slip power recovery generator applied to the sea wave energy extraction," *IEEE Power Electronics Specialists Conference, 1992.* ,vol.1, pp.732-738

[5] Chan, Tze-Fun; Lai, Loi Lei, "Permanent-Magnet Machines for Distributed Power Generation: A Review," *IEEE Power Engineering Society General Meeting, 2007.* pp.1-6, 24-28.

[6] Jasinski, M.; Malinowski, M.; Kazmierkowski, M.P.; Sorensen, H.C.; Friis-Madsen, E.; Swierczynski, D., "Control of AC/DC/AC Converter for Multi MW Wave Dragon Offshore Energy Conversion System," *IEEE International Symposium on Industrial Electronics, 2007.*, pp.2685-2690.

[7] Datta, R.; Ranganathan, V. T., "Variable-Speed Wind Power Generation Using a Doubly Fed Wound Rotor Induction Machine: A Comparison with Alternative Schemes," *IEEE Power Engineering Review*, vol.22, no.7, pp.52-52, July 2002

[8] Hansen, L.H.; Madsen, P.H.; Blaabjerg, F.; Christensen, H.C.; Lindhard, U.; Eskildsen, K., "Generators and power electronics technology for wind turbines," *The 27th Annual Conference of the IEEE Industrial Electronics Society, 2001.*, vol.3, pp.2000-2005.

[9] www.btm.dk/documents/pressrelease.pdf

[10] Hamilton, R.J., "DC motor brush life," *IEEE Transactions on Industry Applications,* , vol.36, no.6, pp. 1682-1687, Nov/Dec 2000

[11] *Elsam. Offshore Wind Farm. Horns Rev Annual Status Report for the Environmental Monitoring Programme, 2002,* http://www.offshore-wind.de/page/fileadmin/offshore/documents/Umweltmonitoring/HornsRev 2001 Annual Status Report for the Environmental Monitoring Programme.pdf

[12] *Offshore Wind Energy and Industrial Development in the Republic of Ireland*, Sustainable Energy Ireland, 2004.

[13] Hau, E., *Wind turbines : fundamentals, technologies, application, economics,* Springer; 2nd ed. (October 1, 2005)

[14] http://thefraserdomain.typepad.com/energy/2007/11/new-brushless-g.html

[15] http://www.hipercon-llc.com/

[16] Barrett, S.N., *Offshore Wave Energy Devices: Model Testing of the Backward Bent Duct Device (B2D2)*, M.Eng. Sc. Thesis, 2002.

[17] http://www.geindustrial.com/publibrary/checkout/Application%20and%20Technical%7CGEZ-6209%7CPDF

[18] Curran R.; Whittaker T.J.T.; Stewart T.P, "Aerodynamic Conversion of Ocean Power From Wave to Wire", *Energy Conversion and Management,* Vol. 39, No. 16, Nov 1998 , pp. 1919-1929(11)

[19] http://www.asahi.com/english/Herald-asahi/TKY200711220085.html

[20] Jensen, M.V.R.S., "Long-term high resolution wear studies of high current density electrical brushes," *Electrical Contacts, 2005. Proceedings of the Fifty-First IEEE Holm Conference on* , vol., no., pp. 304-311, 26-28 Sept. 2005

[21] http://www.wave-energy.net/index_files/documents/Workshop_CA-OEt.pdf

[22] Puranen, J., "Induction Motor Versus Permanent Magnet Synchronous Motor In Motion Control Applications: A Comparative Study", *D.Sc. Thesis, University of Lappeenranta, Finland, 2006.*

[23] Hansen, L.H. et al, "Generators and Power Electronics Technology for Wind Turbines", *IEEE Industrial Electronics Conference, 2001, pp. 2000-2005.*

[24] Campbell, P., *High Coercivity,* Magnetics Business & Technology, Feb/Mar 2008

# A Novel Approach To Photovoltaic Powered Water Pumping Design

Michael James Case*, Ernest Edward Denny[†]

*University of Johannesburg, Johannesburg, South Africa, e-mail: mcase@uj.ac.za
[†] Vaal University of Technology, Van der Bijl Park, South Africa, e-mail: denny1@mweb.co.za

*Abstract-* **Modelling a photovoltaic (PV) system is generally a complex, time-consuming and computationally intensive exercise. This paper presents a simplified approach to using component efficiency and de-rate factors to generate a design estimate for a stand-alone, true North orientated, direct-coupled, PV powered water pumping system.**

*Keywords-* **Alternative energy, Design, Photovoltaic, Solar Cell System, Industrial Application.**

## I. INTRODUCTION

Water pumping has a history dating back to the earliest civilizations and many methods have been developed over the years to realize this task with the minimum of effort. Various power sources including human energy, animal power, wind and hydro-power, electrical (grid) power and fossil fuel powered generator/pump pumping sets have been utilised to accomplish this. Photovoltaic-powered pumping systems are a relatively recent addition to this list, with the first systems being installed in the 1970s [1].

Photovoltaic water pumping applications are one of the most common uses of PV power throughout the world, with thousands of solar-powered water pumps installed both in industrialised and developing nations [2]. Owing to their inherent low maintenance and high reliability, they are also increasingly being used as a replacement for mechanical windmill pumps.

A true North fixed orientation, direct-coupled, PV-powered water pumping system typically consists of an energy source (sunlight), an energy conversion process (photovoltaic – light to direct current electricity), a transmission medium (wiring) and a power consumer (motor/pump set). These components combine to produce a basic process whereby sunlight is converted into electrical power and subsequently into mechanical energy, which can then be used to pump water according to site and equipment constraints. This concept is illustrated in block diagram form in Figure 1.

The dominant problems associated with this type of design (direct-coupled PV) are the non-linear supply of power and the resultant complexity in providing an optimised load match. The output from a PV array is a non-linear and time-dependant source of power that changes according to the variation in solar irradiance throughout a day, as well as the attenuating effects of PV cell temperature. This directly influences the performance characteristics of a DC motor, which is generally designed for, and operated from, a fixed voltage source.

In a PV Pumping System (PVPS), a pump driven by a DC motor performs optimally at a specific PV array configuration and pumping head profile [3]. PV array configuration is a function of the electrical load requirements, which in turn is a function of the motor/pump set and related balance-of-system (BOS) efficiencies. The motor/pump set is chosen on the basis of hydraulic load requirements, derived from the site water requirements and borehole/well parameters such as the aquifer water level and pumping head. Hydraulic efficiency is reduced by mechanical BOS factors such as piping friction and pressure losses over fitting *etc.* All these factors are interrelated and must be considered when doing a PVPS design.

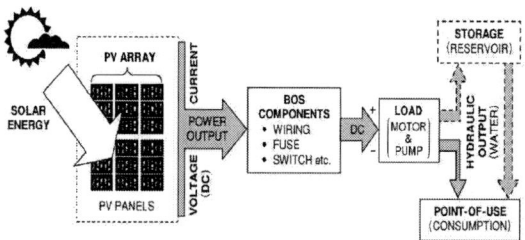

Figure 1. Block diagram of a direct-coupled Photovoltaic Water Pumping System

Knowing the efficiency of the individual components allows for a simplified approach to calculating system design parameters such as system size and output.

## II. DESIGN APPROACH

Assuming the local meteorology parameters are known, the process starts with a total daily water requirement ($Q_{Dr}$, in litres/day). Once this has been established, it is possible to calculate pump rate requirements ($Q_P$, in litres/hour).

The next requirement is to calculate the Total Dynamic Head (TDH) of the pumping installation. Typical data required include static level, draw-down and discharge head, together with installed pipe diameter, pipe length, and fittings used. Compensation for friction and other losses is then computed and combined with other data to determine the TDH for the design.

The next step is to determine compensation for friction losses caused by pipes and fittings. A PVPS piping installation consists of lengths of straight pipe, bends, elbows, tees, valves and various other flow impediments, each requiring consideration in the calculation of TDH.

978-1-4244-1741-4/08/$25.00 ©2008 IEEE

A number of empirical formulae have been developed to solve problems involving pressure drop and friction losses in pipes. The most commonly accepted approximations for water flow applications are the Darcy-Weisbach Formula and the Hazen-Williams Equation [4]. Standard practice is to use the less accurate but acceptable Hazen-Williams Equation [4]. The Hazen-Williams friction loss equation is considered valid for water at temperatures between four degrees and twenty-five degrees Celsius [5] and is therefore expected to function adequately for South African operating conditions. The equation delivers friction loss results in metres of water head loss per metre of installed pipe.

The Hazen-Williams formula provided (1) is a metric adaptation of formulae extracted from Technical Report TR-14 [4], supplemented by information from 'The Engineering Toolbox' [6] and Thomas [7].

$$H_{FL} = Q_{Pm}^{1.852} \cdot \left( \frac{67.03455}{C_{HW}^{1.852} \cdot d_i^{4.8655}} \right), \tag{1}$$

*where*

$H_{FL}$ = Friction head loss (metres per metre of installed pipe)

$Q_{Pm}$ = Volume flow rate (litres per minute; = $Q_p/60$)

$C_{HW}$ = Hazen-Williams Roughness Coefficient (dimensionless)

$d_i$ = Installed pipe inner diameter (millimetres)

Analysis of various commercial friction loss tables showed common use of the Hazen-Williams equation, with an element of conservatism added by incorporating a safety multiplier. The multiplier values varied greatly, so two broad ranges were defined and the average multiplier value for the ranges was obtained. These are 126 (large: most conservative), 6 (small: less conservative) and a value of 1 providing no safety multiplier.

TDH may now be calculated according to (2).

$$TDH = L_V + \lambda_{FF} + P_{comp}, \tag{2}$$

*where*

TDH = Total Dynamic Head (metres)

$L_V$ = Total Vertical Lift (metres)

$\lambda_{FF}$ = Total friction and fittings losses (equivalent pipe length, in metres)

$P_{comp}$ = Tank Pressurisation Compensation (metres of head)

At this point enough data are available to enable calculation of the hydraulic energy required for the design, which is done using (3).

$$E_{Hyd} = \frac{Q_{Dr} \cdot TDH}{366.972}. \tag{3}$$

*where*

$E_{Hyd}$ = Hydraulic energy required for the project (Wh/day)

$Q_{Dr}$ = Total daily water requirement (litres/day)

TDH = Total Dynamic Head (metres)

366.972 = Energy conversion factor (Constant derived from water density, force of gravity and time)

The next step in the calculation process is to select a suitable motor/pump and calculate its relative efficiency according to (4).

$$\eta_{Pump} = \frac{P_{out}}{P_{in}} = \frac{TDH \cdot Q_P \cdot 27.249 \times 10^{-4}}{P_{Pump}}, \tag{4}$$

*where*

$\eta_{Pump}$ = Pump efficiency fraction ($\times 100$ = percent)

TDH = Total Dynamic Head (metres)

$Q_P$ = Required Pump Rate (litres/hour)

$P_{Pump}$ = Rated Pump Power (Watts)

$27.249 \times 10^{-4}$ = Reciprocal of the 'Energy conversion factor' from (3)

This formula assumes an integrated motor/pump combination and calculates a 'wire-to-water' efficiency factor.

Enough data are now available to enable calculation of preliminary PV array energy ($E_{Array}$, in Wh/day) and system load ($S_{Load}$, in Ah/day), calculated using (5) and (6) respectively.

$$E_{Array} = \frac{E_{Hyd}}{\eta_{Pump}}, \tag{5}$$

$$S_{Load} = \frac{E_{Array}}{V_n}, \tag{6}$$

*where*

$\eta_{Pump}$ = Pump efficiency fraction

$V_n$ = Nominal system voltage (Volts generally in multiples of 6, starting at 12)

The next step is to estimate the wire losses for the design. This is done for a given voltage, cable length, diameter and ambient temperature and results in an estimated electrical current requirement for the design. The system load should now be de-rated by the wire loss factor, using (7), thus allowing calculation of an estimated design current according to (8).

$$S_{Load}' = \frac{S_{Load}}{F_{WL}}, \tag{7}$$

$$I_{DesP} = \frac{S_{Load}'}{H_t}, \tag{8}$$

*where*

$S_{Load}'$ = De-rated system load (Ah/day)

$F_{WL}$ = Wire loss factor (fraction)

$I_{DesP}$ = Preliminary design current (Ampere)

$H_t$ = Mean monthly irradiation per day (kWh/m²/day)

The next step in the design process is to select an appropriate PV module. Data required for module evaluation include material type, short circuit current ($I_{SC}$), open circuit voltage ($V_{OC}$), maximum power point current ($I_{MP}$), voltage ($V_{MP}$) and power ($P_{MP}$), nominal operating cell temperature (NOCT) and difference between name-plate power and datasheet warranted power (dP%). These data are commonly available from the manufacturer via the datasheet supplied with the PV module.

PV module selection is dependant on final design current requirements. In order to calculate a final design current, the preliminary current obtained in (8) must be modified by PV module and ancillary equipment efficiency factors. This modification factor is typically less than one and is therefore generally a de-rate factor. The 'Panel Adjustment Factor' accounts for various losses or enhancements incurred due to the PV array and ancillary equipment. These factors are summarised in and originate from cross-referenced data between the California Energy Commission [9], providing default values and PVWATTS System documentation [10], providing range values. Three exceptions are the temperature compensation algorithm, which is adapted from work by Evans [11] as well as the boost factor and name-plate versus warranted power compensation (dp%).

TABLE I. PV MODULE ADJUSTMENT FACTORS

| Component De-Rate Factors | Default Value (%) | Typical Range (%) |
|---|---|---|
| Difference between warranted module minimum power & nameplate power rating | (dP%) From database | 0 to 20 |
| Module Mismatch | 2 | 0.5 to 3.0 |
| Diodes & Connections | 0.5 | 0.3 to 1 |
| Soiling | 5.5 | 0.5 to 70 |
| Shading | 0 (= no shade) | 0 to 100 |
| Age | 0 (= new) | 0 to 30 |
| Temperature | From calculation | 87% to 92% of STC Power Rating |
| Component Boost Factors | Default Value | Typical Range |
| Cumulative Boost Factors | 0 | 0 to 100 |

Excluding Temperature and Boost, the component de-rate factors are processed according to (9) to obtain fractional values, which are then multiplied together to obtain a total de-rate fraction ($F_{d\_total}$) according to (10).

$$F_d = 1 - \frac{F_{CDR}}{100},$$
(9)

$$F_{d\_total} = F_{d1} \cdot F_{d2} \cdot F_{dN}...,$$
(10)

*where*

$F_d$ = Fractional de-rate adjustment factor
$F_{CDR}$ = Individual component de-rate factors

The fractional boost value is obtained using (11).

$$F_b = 1 + \frac{Cumulative\ Boost\ Factors}{100},$$
(11)

*where*

$F_b$ = Fractional boost adjustment factor (use if required; Default = 1)
Cumulative Boost Factors = Total efficiency increase gained from a Linear Current Booster or equivalent electronic/mechanical system performance enhancer (Percent).

An array can be characterised by its average efficiency ($\eta_P$), which is a function of average module temperature ($Tc$):

$$\eta_P = \eta_r \cdot \left[ 1 - \frac{\beta_p \cdot (T_c - T_r)}{100} \right],$$
(12)

*where*

$\eta_r$ is the PV module efficiency at reference temperature Tr (= 25 degrees Celsius), and $\beta_p$ is the temperature coefficient for module efficiency (in percent per degree Celsius).
Tc is related to the mean monthly ambient temperature Ta through use of the Evans formula [11], with Tc modified to cater for $\overline{Kt}$ = 0, as given in (13) and obtaining Cf from (14). $\eta_r$ and $\beta_p$ depend on the type of PV module considered. This information can be gained from manufacturer data sheets.

$$Tc = \left[ \left( 223.45 + 832\overline{Kt} \right) \cdot \frac{NOCT - 20}{800} \right] \cdot Cf + T_{amb}$$
(13)

*where*

Tc = PV cell operating temperature (degrees Celsius)
NOCT = Nominal operating cell temperature rating (degrees Celsius)
$\overline{Kt}$ = Monthly maximum clearness index
$T_{amb}$ = Ambient system operating temperature (degrees Celsius)

The PV array tilt compensation factor (Cf) is equal to a value of one when the PV array's tilt is set at optimal ($\theta_{OPT}$). If the actual tilt angle ($\theta$, in degrees), differs from the optimum, then Cf becomes:

$$Cf = 1 - 1.17 \times 10^{-4} \left( \theta_{OPT} - \theta \right)^2,$$
(14)

The temperature de-rate fraction ($F_{Td}$) value is obtained using (15) and (16).

$$F_{Td} = 1 - \frac{F_{TDR}}{100},$$
(15)

$$F_{TDR} = \frac{\eta_r - \eta_p}{\eta_r} \cdot 100,$$
(16)

{Equation (16) is the temperature de-rate factor (Percent)}

*where*

$\eta_r$ = PV module efficiency at reference temperature Tr
$\eta_p$ = Average PV array efficiency, from (12)
The energy delivered by the PV array ($E_P$) is then:

$$E_P = S_{PV} \eta_p \overline{Ht},$$
(17)

*where*

$\overline{Ht}$ is the monthly mean irradiation and $S_{PV}$ is the effective surface area (m²) of the array.

Array power is reduced by various array losses ($\lambda_p$, from mismatch, diodes and connections, *etc.*) and power conditioning losses ($\lambda_c$, from wire losses, connections, *etc.*):

$$E_A = E_P(1 - \lambda_P)(1 - \lambda_c), \qquad (18)$$

*where*
$E_A$ = Energy available to the load.

The total PV module current modification factor is now calculated according to (19).

$$F_{PVm} = F_{d\_total} \cdot F_b \cdot F_{Td}, \qquad (19)$$

*where*
$F_{PVm}$ = Total PV module adjustment factor

Final design current is calculated using the preliminary current obtained in (8), adjusted by the total PV module adjustment factor using (20).

$$I_{DES} = \frac{I_{DesP}}{F_{PVm}} \qquad (20)$$

*where*
$I_{DES}$ = Final design current (Ampere)
$I_{DesP}$ = Preliminary design current (Ampere)
$F_{PVm}$ = Total PV module adjustment factor

The PV modules needed to fulfil design requirements can now be calculated according to (21), (22) and (23). An assumption is made that all the modules to be used are of the same type and rating.

$$M_{PT} = \frac{I_{DES}}{I_{MP}} \qquad (21)$$

$$M_{ST} = \frac{V_n}{V_{MP}} \qquad (22)$$

$$M_{Total} = M_{PT} \cdot M_{ST} \qquad (23)$$

*where*
$M_{PT}$ = Total number of parallel module strings required in the PV Array
$I_{MP}$ = Peak module current (Ampere)
$I_{DES}$ = Final design current (Ampere)
$M_{ST}$ = Total number of series modules required per parallel string in the PV Array
$V_n$ = Nominal system voltage (volts)
$V_{MP}$ = Peak module voltage (volts)
$M_{Total}$ = Total number of modules required for the PV Array

The array power for the design is given by (24). The maximum power point voltage for the design can also be obtained and is defined as seventy-nine percent of the cumulative open circuit voltage of the system [12], calculated according to (25).

$$P_{Array} = (I_{MP} \cdot M_{PT}) \cdot (V_{MP} \cdot M_{ST}) \qquad (24)$$

$$V_{MPP} = 0.79 \cdot M_{ST} \cdot V_{OC} \qquad (25)$$

*where*
$P_{Array}$ = PV array power (watts)
$I_{MP}$ = Peak individual module current (Ampere)

$M_{ST}$ = Number of series connected PV modules per parallel string in the array
$V_{MP}$ = Peak individual module voltage (volts)
$M_{PT}$ = Total number of parallel module strings in the PV array
$V_{MPP}$ = Maximum power point voltage for the array (volts)
$V_{OC}$ = Open circuit voltage of the individual modules used in the array (volts)

The efficiency of the 'Final Design' can now be evaluated by back-calculating through the design process, using previously calculated design values and comparing component-based pumped water volume and pump rate with what was initially input and/or calculated as user requirements. Equations (26) and (27) are used for this purpose.

$$Q_{Dr\_design} = \frac{P_{Array} \cdot \eta_{Pump} \cdot \overline{H_t} \cdot F_{PVm} \cdot 366.972}{TDH} \qquad (26)$$

$$Q_{P\_design} = \frac{Q_{Dr\_design}}{\overline{H_t}} \qquad (27)$$

*where*
$Q_{Dr\_design}$ = 'Final Design' water pumped (litres/day)
$Q_{P\_design}$ = 'Final Design' pump rate (litres/hour)
$P_{Array}$ = PV array power (watts)
$\eta_{Pump}$ = 'Wire to water' pump/motor combination efficiency factor
$\overline{H_t}$ = Mean monthly irradiation per day (kWh/m²/day), *i.e.* ESH or PSH (Equivalent or Peak) sun hours per day.
TDH = Total dynamic head (metres)
$F_{PVm}$ = Total PV module adjustment factor

Ensure that the water source is able to deliver the maximum $Q_{Dr\_design}$ volume. If it is unable to deliver the design volume, then the design must be down sized until an equitable figure is achieved. Do not exceed the capacity of the water source, as this will cause the pump to run dry and could result in serious equipment damage.

## III. CONCLUSION

Iterative variation of PV module type/ratings and/or motor/pump set allows manual component-matching and system optimisation as far as available components and site conditions allow. Calculation and comparison allows for a 'Best Fit' analysis of the design based on equipment used versus PVPS efficiency (water pumped) and quantitative analysis (number of PV modules required).

The basic methodology can easily be extrapolated to cater for other design modes by including efficiency functions for battery, MPPT and inverter operation, *etc.* The methodology described in the text has been successfully applied in a software design aid described fully in [13]. Short-term trials of the software application have delivered results within three and a half percent of actual water pumped [14].

Meteorological and ancillary design data such as Clearness Index, Optimum Tilt Angle and Irradiation Values *etc.* for Southern Africa, as applied in this text, are available in tabular format or via a design aid software application [14], available from the Vaal University of Technology (VUT).

## ACKNOWLEDGMENT

This work was funded in part by a Research Award from the Vaal University of Technology.

## REFERENCES

[1] B van Campen, D Guidi & G Best, *Solar Photovoltaics for Sustainable Agriculture and Rural Development"*. Food and Agriculture Organisation of the United Nations (FAO), 2000

[2] G Leng, N Meloche, A Monarque, G Painchaud, D Thevenard, M Ross & P Hosette, *Clean Energy Project Analysis: RETScreen Engineering & Cases Textbook - Photovoltaic Project Analysis*. CANMET Energy Technology Centre, 2004

[3] JP Dunlop, Analysis and Design Optimisation of Photovoltaic Water Pumping Systems. *Proceedings of the 20th IEEE Photovoltaics Systems Conference*, Las Vegas, NV, pp. 1182-1185, 1988

[4] *Water Flow Characteristics of Thermoplastic Pipe*, Plastics Pipe Institute, Technical report: TR-14, 2000

[5] LMNO Engineering, Research & Software, Ltd. http://www.lmnoeng.com/hazenwilliams.htm [Accessed: 28/06/2006]

[6] The Engineering Toolbox: Hazen-Williams Equation - Calculating Friction Head Loss in Water Pipes. http://www.engineeringtoolbox.com/ hazenwilliams-water-d_797.html [Accessed: 22/02/2006]

[7] MG Thomas, *Water Pumping: The Solar Alternative*, Document: SAND87-0804, Sandia National Laboratory PV Systems Design Assistance Centre, 1996

[8] *Solar Water Pumping Applications Guide*, Kyocera Solar Inc., 2002

[9] California Energy Commission: *A Guide to Photovoltaic (PV) System Design and Installation*. Publication Number 500-01-020, 2001

[10] Renewable Resource Data Centre: *PVWATTS Version 2 - A Performance Calculator for Grid-connected PV Systems: Changing System Parameters*. http://rredc.nrel.gov/solar/codes_algs/ PVWATTS/system.html [Accessed: 31/05/2006]

[11] DL Evans, Simplified Method for Predicting Photovoltaic Array Output, *Solar Energy*, Vol. 27, No. 6, pp. 555-560, 1981

[12] MJ Case, MJ Joubert & TA Harrison, A Novel Photovoltaic Array Maximum Power Point Tracker. *Proceedings of 2002 EPE-PEMC Conference*, Dubrovnik, Croatia, pp. T5-005, 2002

[13] EE Denny & MJ Case, Development of a Computer-based Aid for General Use in the Design of Directly Coupled Stand-Alone Solar Powered Water Pumping Systems. *Proceedings of 16th South African Universities Power Engineering Conference*, Cape Town, pp. 7-13, 2007

[14] EE Denny, *Development of a Computer-aided Programme for General Use in the Design of Solar Powered Water Pumping Systems*. MTech. Dissertation, Vaal University of Technology, 2007

# Direct Controls in Voltage-Source Converters - Generalizations and Deep Study

Károly Veszprémi and István Schmidt

Budapest University of Technology and Economics/Department of Electric Power Engineering, Budapest, Hungary

e–mail: *veszpremi@vet.bme.hu; schmidt@vet.bme.hu*

*Abstract*—The well–known principle of the direct torque control applied mainly to squirrel–cage induction motor drives can be generalized and applied to other machines, quantities, converters and systems. These are: other AC machines (double–fed induction and permanent magnet synchronous machine), active and reactive power, direct power control of the line side converter, direct controls of both of the converters in a DC link inverter. The idea is the same for all of them and can be generalized. Looking deeper into the operation of the direct controls, some special effects can be found: different resulting error bands and switching frequency in different operation modes of the converter, large delay in reference tracking. These can badly affect the operation of the control: larger (not the expected) error, larger switching frequency, slower control. Methods are given against them: modified switching tables, increasing the number of sectors, selecting proper sector orientation.

*Keywords*— Direct power control, Direct torque and flux control, Voltage Source Converter (VSC), Windgenerator systems

## I. INTRODUCTION

The direct torque control (DTC) thanks to its well–known robustness and simplicity has found wide application to (mainly squirrel–cage) induction motor drives [1]. It implements vector control scheme with closed torque and flux loops without current controllers and PWM modulators to control the two–level voltage source inverter (VSI) supplying the machine. The converter switching states are appropriately selected by a switching table based on the instantaneous errors between the reference and feedback values of the controlled signals. It has become the counterpart of the field oriented control, since the algorithm is much simpler, no current loops, no coordinate transformations, no decoupling, no PWM modulators are required. But the switching frequency is variable and for accurate control the sampling frequency must be higher, requiring larger computing power and faster AD converters [1].

The basic idea can be generalized and applied to other machines, quantities, converter and system [6].

*1) Other machines:* As the development of the torque and the effect of the voltage on the flux are similar in any AC machine, the same principle can be used for all AC machines [4], [5].

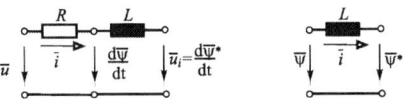

Fig.1. General equivalent circuits of the controlled system.

*2) Other quantities:* Every quantity which has close relation to the torque can be controlled in the same way as the torque. It can be the machine active power. Also, every quantity, which has close relation to the flux can be controlled in the same way as the flux. It can be the reactive power of the machine. This is the direct power control (DPC) [2], [3].

*3) Other converter:* The three–phase AC network connected to a line–side converter (LSC) can be modelled by an induced voltage (of the distant generator) and series inductance and resistance, which is the same as an AC machine model in Fig.1. (If it is a machine, the inductances are reduced to the inverter-supplied side). Virtual fluxes (of the distant generator) can be introduced. In this way the active and reactive power of the LSC can be controlled similarly to the torque and the flux respectively [2], [3].

*4) Other system:* All converters of a complete system, (e.g. a variable–speed generator in a wind power plant) can use direct controls. The machine–side converter (MSC) is controlled with DTC to implement speed control, providing maximum efficiency. The LSC is controlled with DPC, providing high dynamic control of the instantaneous active and reactive power fed to the lines with sinusoidal and symmetrical currents.

Looking deeper into the operation of the direct controls, some special effects can be found: different resulting error bands and switching frequency in different operation modes of the converter, large delay in reference tracking. These can badly affect the operation of the control: larger error (larger than expected), larger switching frequency, slower control. Methods are given against them: modified switching tables, increasing the number of sectors, selecting proper sector orientation. Quantitative parameters are introduced to compare the different switching table versions.

The proposed methods are validated and investigated by simulation. In the investigations per–unit system is used.

## II. DIRECT CONTROLS GENERALLY

Describing generally, the torque–like quantity is denoted by $c_I$ and the flux–like quantity by $c_{II}$. There are always two fluxes (see Fig.1. and Fig.2.): one of them is

---

This work was supported by the Hungarian N.Sc. Found (OTKA No. K75116) for which the authors express their sincere gratitude.

controlled by the direct control, it is $\overline{\psi}$. The other is constrained (e.g. by a constrained voltage or by the closed rotor circuit or by the excitation), it is $\overline{\psi}^*$, its angular speed is $\omega^*$ (in steady state $\omega^*$=const.). $c_I$ can be quickly controlled by the $\delta$ angle between the two fluxes. $c_{II}$ can be controlled by the $\psi$ magnitude of the controlled flux. The effect of the control values ($\delta$, $\psi$) on the controlled values ($c_I$, $c_{II}$) must be the same at any operating point.

The $\overline{\psi}$ flux vector can be quickly controlled by the two–level VSI, switching proper $\overline{u} = \overline{u}_k$ ($k$=1..7) voltage vector to the terminals: $d\overline{\psi}/dt = \overline{u} - \overline{i}R \approx \overline{u}$.

The voltage vector $\overline{u}_7 = 0$ can be developed in two ways: connecting all machine terminals to the positive bar (7P) or all to the negative bar (7N).

Proper voltage vectors can be selected for the desired flux vector modification (Fig.2.). The quickest $\psi$ or $\delta$ modification can be achieved by voltage vectors approximately in phase or perpendicular to the $\overline{\psi}$ flux respectively. Generally and most commonly, six sectors can be defined for the position of the $\overline{\psi}$ flux (Fig.2., $N$=1..6). Examining generally the sector $N$=$i$, the $\overline{u}_k$ voltage vectors are denoted according to Fig.3. ($k$ overflows at 6). Neglecting the resistances, the achievable flux derivatives are equal to the possible voltage vectors (see above). It can be proven geometrically, if the $\overline{\psi}$ flux vector is in the $i$th sector, its magnitude (and $c_{II}$) can be increased by the $\overline{u}_i$, $\overline{u}_{i+1}$, $\overline{u}_{i+5}$ and can be decreased by the $\overline{u}_{i+3}$, $\overline{u}_{i+2}$, $\overline{u}_{i+4}$ voltage vectors. Whereas the $\delta$ angle (and $c_I$) can be increased by the $\overline{u}_{i+1}$, $\overline{u}_{i+2}$ and can be decreased by the $\overline{u}_{i+4}$, $\overline{u}_{i+5}$ voltage vectors. The $\overline{u}_7 = 0$ voltage does not change the flux magnitude, but by stopping the $\overline{\psi}$ vector $\delta$ and $c_I$ are decreased if $\omega^*$>0 and increased if $\omega^*$<0. The positive direction of $\delta$ corresponds to positive $c_I$. The positive direction of $c_I$ corresponds to positive $\omega^*$ (convention). Fig.2. is drawn for these conditions for one case. The other case is if the relative position of $\overline{\psi}$ and $\overline{\psi}^*$ is opposite as in Fig.2. to provide positive $c_I$. In this case $\delta$ must be measured from $\overline{\psi}$ and the effect of the voltage vectors to increase or decrease $c_I$ is opposite as described above.

On this basis by hysteresis control $c_I$ and $c_{II}$ can be controlled directly and can be kept in the tolerance band around their reference values. One possible realization of the direct $c_I$ and $c_{II}$ control is presented generally in Fig.4.

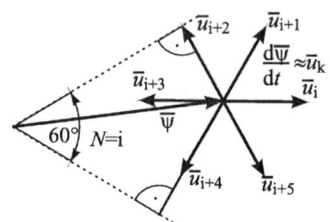

Fig.3. The ith flux sector.

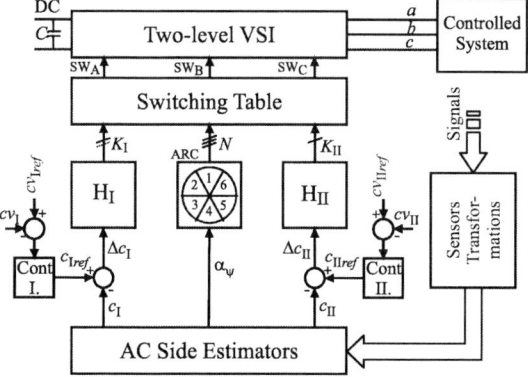

Fig.4. General block diagram of the direct control.

The hysteresis controls are $H_I$ and $H_{II}$. The number of the necessary levels of the hysteresis control depends on the required smoothness of the control and on the controlled quantity. $H_{II}$ is generally a two–level hysteresis control. $H_I$ is not so obvious. Generally, if $\omega^*$ is bidirectional $H_I$ is three–level, if unidirectional, under normal condition it can be two–level hysteresis control. If the $\overline{u}_7$ zero vectors are not used, it is always a two–level control. The later is good, if $c_I$ must be changed quickly, but the switching frequency is significantly increased in this case.

The voltage vector to be switched is determined by three quantities: $\Delta c_I$ error of $c_I$, $\Delta c_{II}$ error of $c_{II}$ and the $\alpha_\psi$ angle of $\overline{\psi}$. The feedback signals for the final controls are calculated by the AC Side Estimators, using the necessary sensed and transformed (to the used coordinate system) signals. This is always a proper model of the controlled system. If the terminal voltages of the PWM inverter are to be sensed their sensing can be saved, if the $U_{dc}$ dc link voltage and the switching states (SW$_A$, SW$_B$, SW$_C$) of the inverter are known.

The reference values of the finally controlled quantities ($c_{Iref}$, $c_{IIref}$) can be set by outer closed–loop controllers (Cont.I, Cont.II). or can be set to constant or to a calculated value. The $H_I$ and $H_{II}$ hysteresis controllers determine $K_I$ and $K_{II}$, forming one part of the address of the Switching Table. Generally, switching between two states, the larger value of $K$ (e.g. 1) corresponds to increasing, the smaller (e.g. 0) to decreasing $c$. The tolerances of $H_I$ and $H_{II}$ are $\pm\Delta C_I$ and $\pm\Delta C_{II}$ respectively. The other part of the address is determined by the output of the sector selection (ARC). The content of the Switching Table depends on the application [6].

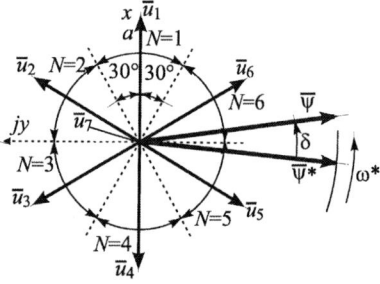

Fig.2. Voltage and flux vectors and the sectors of the controlled flux.

## III. DIRECT POWER CONTROLS

In this case the Controlled System is the network (lines), $c_I$ and $c_{II}$ are the active ($p_\ell$) and reactive ($q_\ell$) powers of the lines, $cv_I$ and Cont. I. are the dc link voltage and its closed-loop control, $cv_{II}$ is the reactive power acting as a constant reference. The quantities corresponding to the general descriptions are given in Table I.

**TABLE I.**
THE CORRESPONDING QUANTITIES IN DPC.

| General | DPC | Meaning |
|---|---|---|
| $a, b, c,$ | $La, Lb, Lc,$ | line phases and |
| $x, y$ | $Lx, Ly$ | coordinate system connected to line phase $a$ |
| $\bar{u}, \bar{\psi}$ | $\bar{u}_L, \bar{\psi}_L$ | line (inverter) voltage at the converter terminals and virtual flux from $\bar{u}_L$ |
| $\bar{u}_i$ | $\bar{u}_\ell$ | the induced voltage of the distant generator |
| $\alpha_\psi$ | $\alpha_{\psi L}$ | angle of $\bar{\psi}_L$ in $Lx$–$Ly$ ($\alpha$ in the following) |
| $\bar{\psi}^*, \omega^*$ | $\bar{\psi}_\ell,$ $\omega_{\psi_\ell} = \omega_\ell$ | virtual flux from $\bar{u}_\ell$ constrained by the network and its angular speed |
| $R, L$ | $R_\ell, L_\ell$ | line resistance and inductance |
| $\bar{i}$ | $-\bar{i}_\ell$ | line current (see Fig.1. and Fig.5.) |
| $c_I, c_{II}$ | $p_\ell, q_\ell$ | active and reactive power of the network |
| $cv_I$ Cont.I | $u_{dc}$ PI | DC link voltage closed–loop $u_{dc}$ control |
| $cv_{II}$ Cont.II | $q_\ell$ open–loop | reactive power of the network constant reactive power reference |
| $K_I, K_{II}$ | KP, KQ | |

The block scheme of the system, defining the positive direction of the quantities is presented in Fig.5. According to these positive directions, the vector diagram of the quantities is also drawn (Fig.6.). Comparing the orientation of the fluxes, it is opposite to Fig.2., the effect of the voltage vectors to increase or decrease $p_\ell$ is opposite as described generally (e.g. the $\bar{u}_7 = 0$ vector increases it).

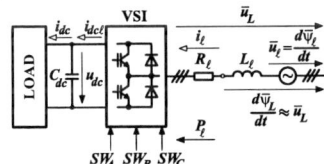

Fig.5. The block scheme of the controlled system.

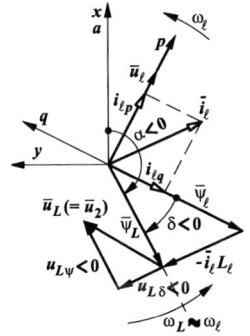

Fig.6. The vector diagram of the quantities.

As can be seen, consumptive positive directions are used:

$$p_\ell = U_\ell i_{\ell p} = -U_\ell \frac{\psi_L \sin\delta}{L_\ell} \approx -\frac{U_\ell \psi_L \delta}{L_\ell}, \quad (1)$$

$$q_\ell = U_\ell i_{\ell q} = U_\ell \frac{\psi_L \cos\delta - \Psi_\ell}{L_\ell} \approx \frac{U_\ell}{L_\ell}(\psi_L - \Psi_\ell). \quad (2)$$

According to these considerations the generally used six-sector switching table is presented in Table II., where for active power increase either zero vectors (version 6a, this is the default case) or active vectors (version 6b in brackets) can be used. The problems and phenomenon arising with the following special effects need the modification of this default switching table.

**TABLE II. .**
THE SWITCHING TABLE FOR DPC; VERSIONS: 6a, (6b)

| KQ | KP | N | | | | | |
|---|---|---|---|---|---|---|---|
| | | 1 | 2 | 3 | 4 | 5 | 6 |
| 1 | 0 | 2 | 3 | 4 | 5 | 6 | 1 |
| | 1 | 7P (6) | 7N (1) | 7P (2) | 7N (3) | 7P (4) | 7N (5) |
| 0 | 0 | 3 | 4 | 5 | 6 | 1 | 2 |
| | 1 | 7N (5) | 7P (6) | 7N (1) | 7P (2) | 7N (3) | 7P (4) |

## IV. SLOW POWER REFERENCE TRACKING

### A. The Problem

When the reference value of the power is changing slowly, the control operates well with the default 6a zero vector version. However a periodic slow-down of e.g. $q_\ell$ modification in the hysteresis band can be identified, near the sector borders. It can be identified in Fig.7. at the border of sector 6 and 1. The $\bar{u}_{i+1}$ and $\bar{u}_{i+2}$ vectors are used to increase and decrease the $q$ respectively by modifying the magnitude of $\bar{\psi}_L$. But according to Fig.3., these voltage vectors are perpendicular to the flux vector at the beginning and at the end of the sector respectively, having no effect on the flux magnitude: the flux magnitude decrease is slow at the end of the sector, the flux magnitude increase is slow at the beginning of the sector.

If the reference value of the power (in this case the reactive) is changing fast (e.g. for indirect current vector control, filtering harmonics or compensating unbalance [7]), the reference tracking for the reactive power will be poor. Fig.8. presents such a case, when the direct power

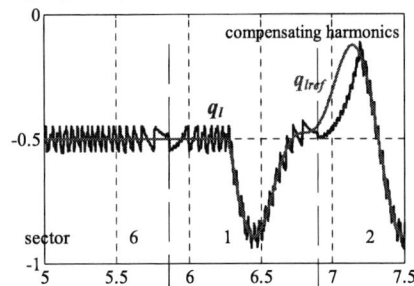

Fig.7. Reactive power and its reference vs. time to demonstrate slow-down of $q_\ell$ modification at sector border in version 6a (zero vectors).

**1805**

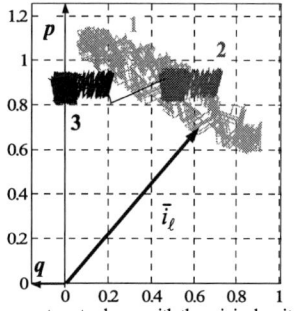

Fig.8. Line current vector locus with the original switching table (version 6a).

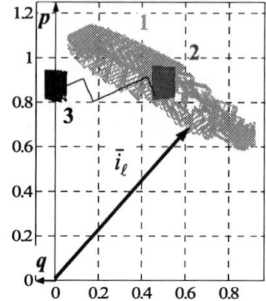

Fig.10. Line current vector locus with the modified switching table (version 6b).

control of the line-side converter of a wind turbine generator is used to compensate the harmonics and the reactive power demand of a thyristor converter controlled *dc* drive connected to the lines (1: existing 5th and 7th harmonics; 2: compensated harmonics; 3: compensated reactive power; compensations are done by injecting compensating currents to the network by the line-side converter). The periodicity of power pulsation caused by these harmonics coincides with the periodicity of the sector changing (Fig.7.), resulting in a periodic occurrence of the above described poor reference tracking in *q* direction. In the presented case, the firing angle of the thyristor converter is 30°, fast $q_{\ell ref}$ increase occurs at the beginning of the sector (Fig.7.). Similar effect can be expected, if fast reference decrease occurs at the end of the sector.

### B. The Solution

During *q* modification there are switching caused by the *p* control. The zero vectors, used for *p* increase do not change the flux magnitude. Using instead non-zero vectors ($\overline{u}_{i+4}$ and $\overline{u}_{i+5}$), the flux magnitude modifications are more effective, the *q* reference tracking is much better (Fig.9.). and the compensation is perfect (Fig.10.). This is the switching table version 6b. The side-effect of this switching table modification is the increasing switching frequency, caused by the double-switches (not the neighbour vector is switched) and by the faster controlled values' change in the hysteresis band (Fig.9.).

### C. Effectiveness of Voltage Vector Selection

To compare the dynamic properties of the different switching tables quantitatively, the effectiveness of the voltage vectors are investigated by the caused derivative of the modified quantity.

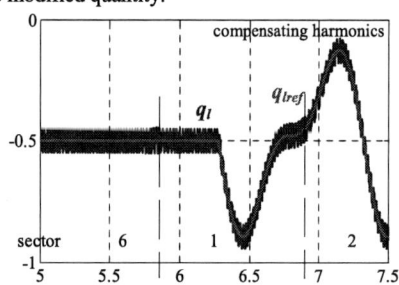

Fig.9. Reactive power and its reference vs. time to demonstrate perfect reference tracking in version 6b (active vectors).

These derivatives can be got using (1) and (2):

$$\frac{dp_\ell}{dt} = -\frac{U_\ell}{L_\ell}\frac{d}{dt}\left(\psi_L\delta\right) = -\frac{U_\ell}{L_\ell}\left(\frac{d\psi_L}{dt}\delta + \psi_L\frac{d\delta}{dt}\right), \quad (3)$$

$$\frac{dq_\ell}{dt} = \frac{U_\ell}{L_\ell}\frac{d\psi_L}{dt}. \quad (4)$$

In stator (*x-y*) coordinates:

$$\frac{d\overline{\psi}_L}{dt} = \frac{d}{dt}\left(\psi_L e^{j\alpha}\right) = \frac{d\psi_L}{dt}e^{j\alpha} + j\psi_L e^{j\alpha}\frac{d\alpha}{dt}, \quad (5)$$

$$\frac{d\overline{\psi}_L}{dt} \cong \overline{u}_L = \left(u_{L\psi} + ju_{L\delta}\right)e^{j\alpha}. \quad (6)$$

$u_{L\psi}$ and $u_{L\delta}$ are the inverter voltage vector components in $\overline{\psi}_L$ and perpendicular direction ($\psi_L$ and $\delta$ modification directions, see Fig.6.).

The angle modification is the following:

$$\frac{d\alpha}{dt} = \omega_\ell + \frac{d\delta}{dt}. \quad (7)$$

Comparing (5), (6) and using (7):

$$\frac{d\psi_L}{dt} = u_{L\psi}, \quad \frac{d\delta}{dt} = \frac{u_{L\delta}}{\psi_L} - \omega_\ell. \quad (8a,b)$$

Substituting to (3):

$$\frac{dp_\ell}{dt} = \frac{U_\ell}{L_\ell}\left(-u_{L\psi}\delta - u_{L\delta} + \psi_L\omega_\ell\right) \quad (9)$$

and (4):

$$\frac{dq_\ell}{dt} = \frac{U_\ell}{L_\ell}u_{L\psi}. \quad (10)$$

Because of δ, (9) is operating point dependent.

These derivatives are depending on the switching state of the inverter ($\overline{u}_L$). They can be calculated, considering the content of the switching table.

*1) The calculated and displayed quantities are:*

Power time derivatives dPmn, dQmn (m=KQ, n=KP) as the function of the flux vector position in the sector.

e.g.: dP01 is the derivative of $p_\ell$ at KQ=0 and KP=1.

Weighted derivatives: dPn, dQm (m=KQ, n=KP). The meaning and explaining of the weighting is the following:

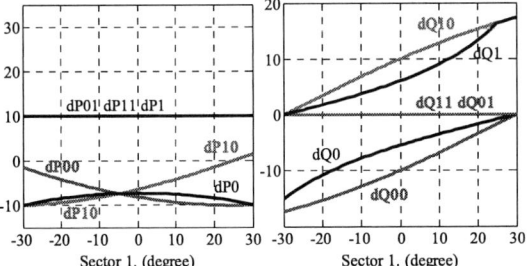

Fig.11. Derivatives in one sector for version 6a (zero vector).

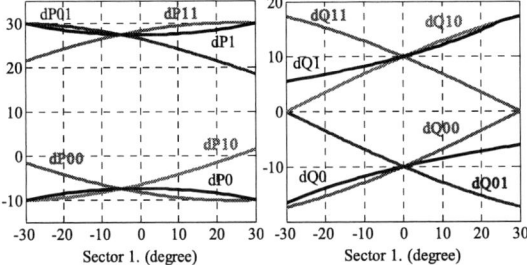

Fig.12. Derivatives in one sector for version 6b (active vector).

During the modification of the examined quantity between the two limits of the hysteresis band the other quantity also can cause switching state (voltage vector) modification, depending on whether its increase or decrease is necessary (two different voltage vectors are switched). The effectiveness of the examined quantity modification in the given direction is the mean value (in time) of the effectiveness of the two switching states caused by the other quantity (weighted mean). The base of averaging is the time interval of the switching state in the switching period, which is inversely proportional to the caused derivative of the corresponding other quantity.

To make it more clear, here is an example:

dQ1 is the weighted derivative of $q_\ell$ for its increase (m=KQ=1). During $q$ increase, the $p$ controller switches voltage vectors corresponding to KP=1 and 0. dQ1 should be combined from dQ10 and dQ11 with proper weighting. E.g. the weight of dQ10 can be given by the relative time duration of $p$ decrease (KP=0) in the switching cycle. It is inversely proportional to dP10. It can be proven that it is: dP11/(-dP10+dP11), the negative signs are for using absolute values.

Expressing generally:

$$dQm = dQm0 \frac{dPm1}{-dPm0 + dPm1} + dQm1 \frac{dPm0}{-dPm0 + dPm1},$$

(11a,b)

$$dPn = dP0n \frac{dQ1n}{-dQ0n + dQ1n} + dP1n \frac{dQ0n}{-dQ0n + dQ1n}.$$

*2) Remarks:*

The negative signs are for using absolute values. Can be that the sign of the derivative is not the expected: dPm0 can be positive caused by the effect of the positive-signed $\omega_\ell$ in the active power derivative (9). In this case the ratios of dQm0 and dQm1 are 1 and 0 respectively (no switching caused by $p_\ell$).

Can be, that the denominator of the ratios in (11) is zero (only if both derivatives there are zero, because of the absolute values). It can be only for zero vector versions (e.g.6a) to calculate dP1, because of the zero voltage vector selection for $p_\ell$ increase (dQ01=dQ11=0). In this case only one vector is switched to modify $p_\ell$ ($q_\ell$ does not cause switching). All the same which vector, considering the weighting, since all zero vectors cause the same $p$ derivative: dP01=dP11. Their any weighting results in the same value: dP1=dP01=dP11.

*3) Results:*

The calculated values for the 6 sector versions (6a and 6b) are presented in Fig.11. and Fig.12. The angles are measured from phase $a$ ($x$) axis. The data used for calculations are: $U_\ell$=1pu; $L_\ell$=0.1pu; $\delta\approx-5°$ (corresponding to the operating point of the compensation); $\omega_\ell$=1pu; $U_{dc}$=3pu.

The calculated values coincide well with the effect identified at the problem definition and can be used well to explain the reason of the problem and the solution:

The dQ1 value representing reactive power increase effectiveness is very weak at the beginning of the sector, if zero vector is used (Fig.11.). By using active vector (Fig.12.) it is improved significantly. (The same can be told about dQ0 at the end of the sector.) No problem with the active power in both cases.

*D. The Effect of Larger Sector Number*

Using the developed tool to express the effectiveness of a switching table version, other switching tables also can be examined and compared in this respect.

The effectiveness of controlled values' modification could be increased by using more sectors, namely twice more: 12 [2]. This expectancy is supported by the larger resolution of sectors (30°), resulting in more variants of voltage vector selections and larger freedom to select the orientation of the sectors.

Two versions of the 12-sector cases are examined, using two different orientations: version 12A (Fig.13.) and version 12B (Fig.14.).

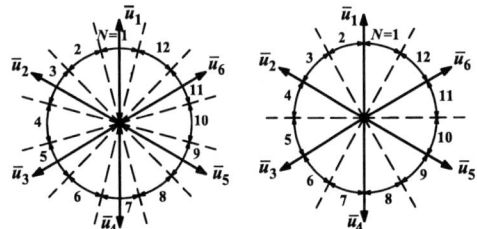

Fig.13. Sectors of version 12A.　　Fig.14. Sectors of version 12B.

TABLE III.
SWITCHING TABLE OF VERSION 12A: 12Aa, (12Ab)

| KQ | KP | N | | | | | | | | | | | |
|---|---|---|---|---|---|---|---|---|---|---|---|---|---|
| | | 1 | 2 | 3 | 4 | 5 | 6 | 7 | 8 | 9 | 10 | 11 | 12 |
| 1 | 0 | 2 | 2 | 3 | 3 | 4 | 4 | 5 | 5 | 6 | 6 | 1 | 1 |
| | 1 | 7P (6) | 7P (1) | 7N (1) | 7N (2) | 7P (2) | 7P (3) | 7N (3) | 7N (4) | 7P (4) | 7P (5) | 7N (5) | 7N (6) |
| 0 | 0 | 3 | 4 | 4 | 5 | 5 | 6 | 6 | 1 | 1 | 2 | 2 | 3 |
| | 1 | 7N (5) | 7P (5) | 7P (6) | 7N (6) | 7N (1) | 7P (1) | 7P (2) | 7N (2) | 7N (3) | 7P (3) | 7P (4) | 7N (4) |

## TABLE IV.
### SWITCHING TABLE OF VERSION 12B: 12Ba, (12Bb), {12Bc}, [12Bd]

| KQ | KP | N | | | | | | | | | | | |
|---|---|---|---|---|---|---|---|---|---|---|---|---|---|
| | | 1 | 2 | 3 | 4 | 5 | 6 | 7 | 8 | 9 | 10 | 11 | 12 |
| 1 | 0 | 2 | 2 | 3 | 3 | 4 | 4 | 5 | 5 | 6 | 6 | 1 | 1 |
| | 1 | 7P | 7P | 7N | 7N | 7P | 7P | 7N | 7N | 7P | 7P | 7N | 7N |
| | | (7P) | (7P) | (7N) | (7N) | (7P) | (7P) | (7N) | (7N) | (7P) | (7P) | (7N) | (7N) |
| | | {6} | {6} | {1} | {1} | {2} | {2} | {3} | {3} | {4} | {4} | {5} | {5} |
| | | [6] | [1] | [1] | [2] | [2] | [3] | [3] | [4] | [4] | [5] | [5] | [6] |
| 0 | 0 | 3 | 4 | 4 | 5 | 5 | 6 | 6 | 1 | 1 | 2 | 2 | 3 |
| | | (3) | (3) | (4) | (4) | (5) | (5) | (6) | (6) | (1) | (1) | (2) | (2) |
| | | {3} | {3} | {4} | {4} | {5} | {5} | {6} | {6} | {1} | {1} | {2} | {2} |
| | | [3] | [3] | [4] | [4] | [5] | [5] | [6] | [6] | [1] | [1] | [2] | [2] |
| | 1 | 7N | 7P | 7P | 7N | 7N | 7P | 7P | 7N | 7N | 7P | 7P | 7N |
| | | (7N) | (7N) | (7P) | (7P) | (7N) | (7N) | (7P) | (7P) | (7N) | (7N) | (7P) | (7P) |
| | | {5} | {5} | {6} | {6} | {1} | {1} | {2} | {2} | {3} | {3} | {4} | {4} |
| | | [4] | [5] | [5] | [6] | [6] | [1] | [1] | [2] | [2] | [3] | [3] | [4] |

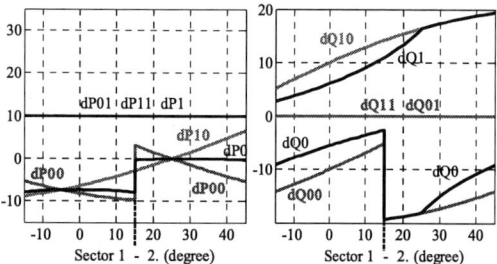

Fig.15. Derivatives in two sectors for version 12Aa.

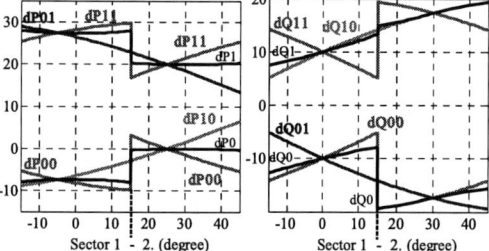

Fig.16. Derivatives in two sectors for version 12Ab.

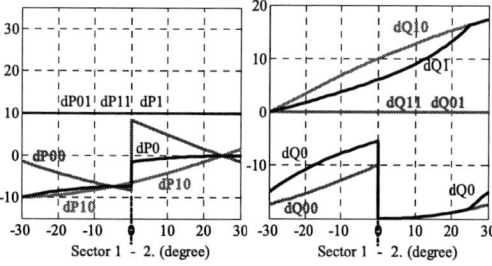

Fig.17. Derivatives in two sectors for version 12Ba.

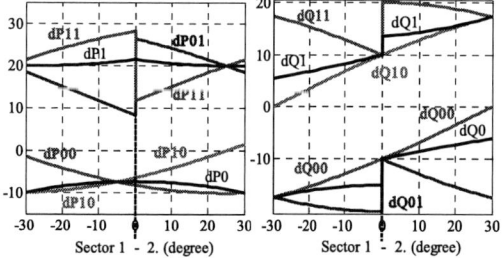

Fig.18. Derivatives in two sectors for version 12Bd.

In the same way, as for the 6-sector versions (Chapter II.), the possible voltage vector selections can be determined. The corresponding switching tables are presented in Table III. and IV.

Only two variant can be got for version 12A:

– Version 12Aa with zero vectors.

– Version 12Ab with active vectors.

Whereas four variants can be found for version 12B:

– Version 12Ba with zero vectors (the switching table is the same as in version 12Aa, only the orientation of the sectors is different).

– Version 12Bb with more effective vector selection at KQ=0, KP=0 for $p$ decrease.

– Version 12Bc with more effective vector selection at KP=1 for $p$ increase.

– Version 12Bd with more effective vector selection at KP=1 for $q$ modification.

The modifications are done in this listed order (from up to down in table rows, the modified vectors are bold and underlined), they are initiated by the identified poor reference tracking of the controlled powers. To find out the necessary switching table modification, the previously derived parameters characterizing the effectiveness of the voltage vector selection can be used very well. These are displayed in Fig.15-18. They are presented not for all cases, since as can be identified, the versions 12Bb and 12Bc are not new, they are the same as version 6a and 6b respectively.

The odd and even sectors behave differently. A new effect can be identified in the even sectors: the $p$ decrease is very weak in the original versions (Fig.15. Fig. 17.), the derivative can be even positive (dP00, dP10). It is caused by the effect of the rotation ($\omega_\ell$ in (9)). To demonstrate this weakness, again the same compensation process is presented, with firing angle 60° (firing angle of the converter, generating the harmonics), which has larger effect in $p$ direction than the previous 30°. The time functions and the current vector are presented in Fig.19. and Fig.20. respectively (version 12Ba).

The usefulness of the developed quantitative merit of effectiveness of a switching table can be proven with explaining the reason of this behaviour. The weighted derivative dP0 in Fig.17. shows clearly the reason: it is very low at sector 2, even lower at the end of the sector (zero). The weighting is also validated, since without weighting the behaviour at the beginning and at the end of the sector can not be explained.

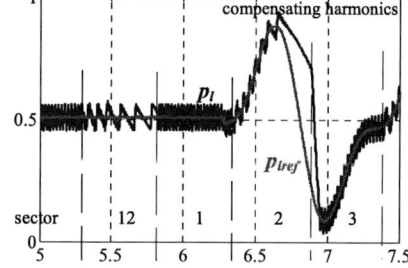

Fig.19. Active power and its reference vs. time to demonstrate slow-down of $p_\ell$ modification at even sectors, mainly at the end of the sector, in version 12Ba.

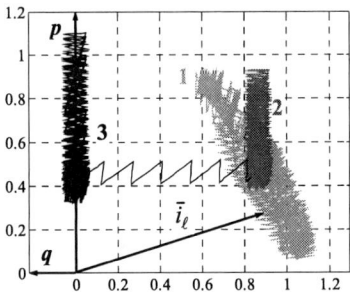

Fig.20. Line current vector locus with the original 12B sector switching table (version 12Ba).

*1) Version 12A:* The modification of the original switching table is done to improve the very weak dP0 in the even sector (Fig.15.). It is a pity that the only possible modification does not modify this parameter (Fig.16.), since it modifies $p$ increase, not decrease.

*2) Version 12B:* There are more possibilities at this version, but only one is different from the previously described ones: the version 12Bd. As can be identified in Fig.18., the dP0 parameter is improved significantly.

### E. Comparison of the Different Versions

To compare the examined versions quantitatively, our aim is to calculate one-value representation of the effectiveness of a given table for a given quantity in each modification direction:

In each examined case, the presented derivatives in the sectors are repeated with 60° periodicity, the values are calculated for this range. An average numerical value of the weighted derivatives can represent the effectiveness of the switching table for that value and that modification direction (Q0, Q1, P0, P1). These are given in rows "all" of Table V. Since the 12-sector versions have 30° partition, a more detailed information can be got if the average is calculated also for the 30° sectors separately. These are given in the same table in S1 and S2 rows. The improvements of the targeted parameters can be seen well (e.g. 12Ba P0=−0.5 → 12Bd P0=−8.2). But be careful with the average: e.g. the weakness of dQ1 at the beginning of the version 6a sector cannot be seen from Q1.

TABLE V.
NUMERIC DATA FOR THE EXAMINED VERSIONS.

| Swich. Table | Average weighted derivative | | | | Switching number | | |
|---|---|---|---|---|---|---|---|
| | Q0 | Q1 | P0 | P1 | Single | Double | Triple |
| **6a all** | -6.1 | 7.1 | -8.2 | 10 | 2369 | 0 | 48 |
| **(6b) all** | -10.4 | 10.6 | -8.2 | 28.2 | 5808 | 2005 | 0 |
| 12Aa all | -10.3 | 11.4 | -3.8 | 10 | 4395 | 1502 | 16 |
| 12Aa S1 | -5.6 | 6.3 | -7.5 | 10 | | | |
| 12Aa S2 | -14.9 | 16.4 | -0.1 | 10 | | | |
| (12Ab) all | -13.8 | 13.7 | -3.8 | 23.8 | 6243 | 2582 | 0 |
| (12Ab) S1 | -10.1 | 10.1 | -7.5 | 27.5 | | | |
| (12Ab) S2 | -17.4 | 17.2 | -0.1 | 20.1 | | | |
| 12Ba all | -14.2 | 7.1 | -4.3 | 10 | 4036 | 1361 | 22 |
| 12Ba S1 | -9.6 | 2.8 | -8.2 | 10 | | | |
| 12Ba S2 | -18.8 | 11.4 | -0.5 | 10 | | | |
| [12Bd] all | -11.9 | 11.3 | -8.2 | 20.5 | 6053 | 2148 | 0 |
| [12Bd] S1 | -15.8 | 7.6 | -8.2 | 20.5 | | | |
| [12Bd] S2 | -7.9 | 15.1 | -8.2 | 20.5 | | | |

Besides the weighted mean derivative, the switching numbers of the examined compensation process (the same for all) are also given in Table V. The "zero vector" versions (6a, 12Aa, 12Ba) have less switching number. Using "active vectors" instead, the switching numbers are increased (6b, 12Ab, 12Bd), including the double-switch number also.

Finally, there are two versions which can compete: 6b and 12Bd. They are almost identical in $p$ parameters. The version 12Bd is slightly better in $q$ parameters and it causes only a slightly larger switching frequency. If the reference tracking speed (good dynamic behaviour) is important, 12Bd should be selected.

## V. OPPOSITE BEHAVIOUR FOR OPPOSITE ACTIVE POWER DIRECTIONS

This phenomenon is important, if the active power is bidirectional.

### A. The Effect

The opposite behaviour is identified from the different width of the actual band of the active power changing and its different pulsation frequency (all presented figures are calculated using 2DP=0.1 hysteresis band and $u_{dcref}$=3pu):

– When the power is flowing towards the $dc$ link ( $p_\ell > 0$ , $i_{dc}$=0.5), the band is wider (0.118pu>2DP) and the pulsation frequency is smaller (~11.4kHz) (Fig.21).

– For $p_\ell < 0$ ( $i_{dc}$=−0.5) the band is narrower (0.087<2DP) and the pulsation frequency is higher (~15.2kHz) (Fig.22).

Its reason should be determined, since significant hysteresis band and switching frequency modification can occur, which must be considered during designing the system.

### B. The Reason

To determine the reason, the reference value of the active power ( $p_{\ell ref}$ ) is also drawn in the figures. As can be identified, it is changing differently in the two cases, comparing with the $p_\ell$ active power signal:

– For $p_\ell > 0$ , the sign of the $dp_{\ell ref}/dt$ is the same as the sign of the $dp_\ell/dt$ .

– For $p_\ell < 0$ , they have opposite signs.

The $p_{\ell ref}$ is the output of the $dc$ voltage controller. It is responding to the $u_{dc}$ voltage change, if it is increasing, the output (and $p_{\ell ref}$ ) is decreasing and vice versa. For the two cases ( $p_\ell < 0$ and $p_\ell > 0$ ) $u_{dc}$ responds oppositely to the voltage vector used for $p_\ell$ increase (in the presented Fig.21-22. figures, the version 6a was used, applying zero vectors for $p_\ell$ increase):

Zero voltage vector switching means, that no current is flowing to/from the capacitance from/to the network ( $i_{dc\ell} = 0$ ). But there is a current $i_{dc}$ to/from the load (the other side of the $dc$ link, Fig.5.):

– If $p_\ell > 0$ , $i_{dc}$>0 discharges the capacitance, $u_{dc}$ is decreasing, $p_{\ell ref}$ is increasing.

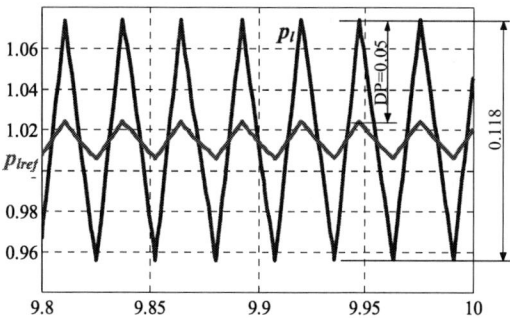

a) The active power and its reference (in black and white: the reference is the smaller signal).

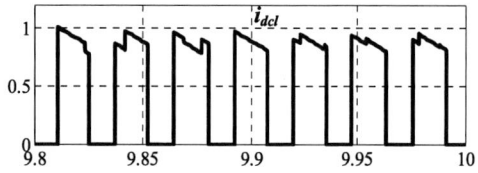

b) The *dc* current on the inverter (line) side.

Fig.21. Time functions for positive line power (version 6a).

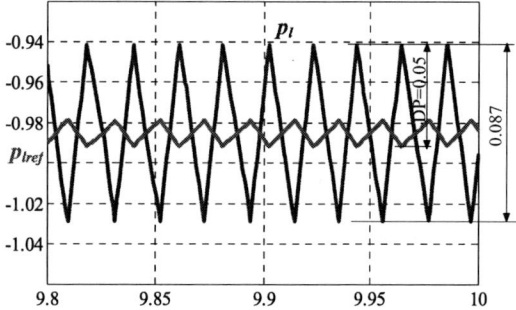

a) The active power and its reference (in black and white: the reference is the smaller signal).

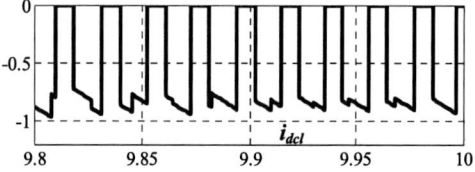

b) The *dc* current on the inverter (line) side.

Fig.22. Time functions for negative line power (version 6a).

– If $p_\ell < 0$, $i_{dc} < 0$ charges the capacitance, $u_{dc}$ is increasing, $p_{\ell ref}$ is decreasing.

For both cases, the $p_\ell$ decrease is done by active voltage vectors, resulting in an $i_{dc\ell}$ current $|i_{dc\ell}| > |i_{dc}|$ (Fig.21b, Fig.22b), modifying $u_{dc}$ and $p_{\ell ref}$ oppositely as for $p_\ell$ increase.

The different behaviour is the reason of the different $p_{\ell ref}$ behaviour. The hysteresis band of the active power controller is measured between $p_{\ell ref}$ and $p_\ell$:

– If they are changing in the same direction, the resultant $p_\ell$ changing band is wider, as identified (Fig.21a).

– If they are changing in opposite direction, the resultant $p_\ell$ changing band is narrower, as identified (Fig.22a).

The effect of the switching table modification on this phenomenon also can be investigated. The non-zero vector versions can be expected to be different, but they are not. They are using active vectors for $p_\ell$ increase instead of zero vector. The resulting non-zero $i_{dc\ell}$ current has the same effect on $u_{dc}$ and $p_{\ell ref}$ as the $i_{dc\ell} = 0$ (even higher, resulting in faster changes and larger switching frequency). The opposite behaviour remains the same.

This phenomenon exists at the direct torque control also, but there its effect is negligible. The reason is the following: in that case the corresponding control loop is a speed control, including a much higher time constant, caused by the inertia of the system.

## VI. CONCLUSIONS

One result of our investigations is the generalization of the direct controls. The basic idea can be generalized and applied to other machines, quantities, converter and system.

Looking deeper into the operation of the direct power controls, two special effects are found, explained and the possible improvements are developed:

– The dynamics of the power controls depend on the switching table. If good dynamic behaviour is important, it must be considered. Effective quantitative parameters are introduced to compare the different switching table versions, supporting the selection.

– Different behaviour is found for active power control at different power directions. The reason is found and the effect must be considered at system design.

## REFERENCES

[1] G.S. Buja and M.P. Kazmierkowski, "Direct torque control of PWM inverter–fed AC motors–a survey", *IEEE Trans. on Industrial Electronics*, Vol.51. No.4. pp744–757. August 2004.

[2] T. Noguchi, H. Tomiki, S. Kondo and I. Takahashi, "Direct power control of PMW converter without power–source voltage sensors", *IEEE Trans. on Industry Applications*, Vol.34. No.3. pp473–479. May/June 1998.

[3] M. Malinowski, M.P. Kazmierkowski, S. Hansen, F. Blaabjerg and G.D. Marques, "Virtual–flux–based direct power control of three–phase PWM rectifiers", *IEEE Trans. on Industry Applications*, Vol.37. No.4. pp1019–1026. July/August 2001.

[4] K. Gierlotka and M. Jelen, "Control of double–fed induction machine using DTC method", *in Proc. EDPE'03*. pp476–481. Sept. 2003. Slovakia.

[5] M.F. Rahman, L. Zhong and K.W. Lim, "A direct torque–controlled interior permanent magnet synchronous motor drive incorporating field weakening", *IEEE Trans. on Industry Applications*, Vol.34. No.6. pp1246–1253. Nov/Dec. 1998.

[6] I. Schmidt and K. Veszprémi, "Application of Direct Controls to Variable-Speed Wind Generators", *IEEE International Conference on Industrial Electronics and Control Applications, ICIECA 2005* CD-Rom paper, ISBN:0-7803-9420-8, Quito, Equador. November 2005.

[7] I. Schmidt and K. Veszprémi, "Additional Application Field of a Modern Wind Generator Even at No-Wind", *12th European Conference on Power Electronics and Applications, EPE'2007* Proceedings on CD-ROM. 10 pages. ISBN:9789075815108. 2007. September, Aalborg, Denmark

# Multipolar double fed induction wind generator with a single phase secondary winding

Leonids Ribickis*, Guntis Dilevs*, Nikolajs Levins[†], Vladislavs Pugachevs[†]

* Riga Technical University, Faculty of Power and Electrical Engineering, 1 Kalku Str., LV-1050, Riga, LATVIA, e-mail: *guntis.dilevs@gmail.com*

[†] Institute of Physical Power Engineering, 21 Aizkraukles Str., LV-1006, Riga, LATVIA, e-mail: *magneton@edi.lv*

*Abstract*—The project concerns the practical applications of multipolar induction machines as generators for small and medium wind turbines. Such a multipolar generator should be built in a way that the primary and secondary windings are placed on the stator, and the rotor is tooth-like without windings, with each tooth equivalent to one pole pair. The most reliable design is that with a single-phase secondary winding, which provides efficient control of the generator.

*Keywords*—IEEEtran, double fed, induction generator, paper, multipolar.

## I. INTRODUCTION

Nowadays the use of wind energy is developing in the direction of large and medium wind generators, which are able to work in parallel with the grid or independently. The most common solution for such systems is a direct-drive synchronous generator, which provides high efficiency and installed power factor.

A conventional induction generator possesses a number of advantages – such as high reliability and stability when working to the power grid [5]. However, it grooves/bores of stator 1. Accordingly rotor 2 is tooth-like and has no windings. Each tooth 3 of rotor 2 corresponds to one pole pair. The total number of teeth can be rather big, which does not lead to enlargement of winding coils of the stator and, therefore, of the magnetizing current. The secondary winding is single phased and is also placed in the stator's grooves. In the secondary circuit capacitors and electrical devices (load) $R_{L1}$ – $R_{L5}$ are arranged. The secondary winding plays a major role in magnetizing the system and reducing the losses of the electrical machine. The system requires neither brushes nor slip-rings.

Owing to the single phase secondary winding the number of coils in the stator is limited. In this case, at the number of pole pairs equal to 22, there are four coils in each phase. The primary windings, A-X, B-Y and C-Z, are situated in pole extensions 4. Each pole extension between two grooves can have up to five smaller teeth with a step $t_{zp}$, which is equal to the step $t_z$ for the teeth on the rotor.

also has some significant disadvantages – e.g. the low efficiency and power factor when the number of pole pairs is greater than/at the number of pole pairs exceeding four [6,8]. They are impossible to use as direct driven wind generators for small and medium wind turbines. Accordingly, the direct driven generators are based on synchronous machines with controlled or permanent magnet excitation systems [2,3].

In this project we overview the practical application of multipolar induction machines as generators for small and medium wind turbines. In the considered model all the windings are placed on the stator, and the rotor is tooth-like with no windings.

## II. SPECIAL FEATURES OF THE MULTIPOLAR INDUCTION GENERATOR WITH A SINGLE-PHASE SECONDARY WINDING

Figure 1 shows the circuit diagram of an induction generator whose primary (A – X, B – Y, C – Z) and secondary (a – x) windings are situated in the

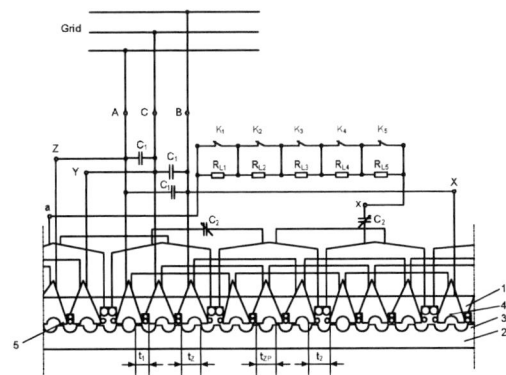

Fig. 1. The construction and circuit diagram of the induction generator with primary and secondary windings on the stator: A-X, B-Y, C-Z – primary windings; a-x – secondary windings; 1 – stator; 2 – rotor; 3 – rotor teeth; 4 – pole extension; 5 – teeth on the pole extension; $C_1$, $C_2$ – capacitors in the primary and secondary circuits; $R_{L1}$ – $R_{L5}$ – controlled secondary load resistance; $t_Z$, $t_{ZP}$, $t_1$, $t_2$ – stator and rotor tooth steps; $k_1$ – $k_2$ – NC contacts

The practical application of the single-phase secondary winding is that it simplifies the transfer of energy to the load. At rotation the rotor teeth change their location with respect to the stator pole extension – from teeth to grooves, which, in turn, causes a periodical

978-1-4244-1741-4/08/$25.00 ©2008 IEEE

change in the magnetic conductivity from minimum to maximum. The conductivity of the K-th pole extension can be described by a Fourier's periodic function series [9] as

$$\lambda_k = a_0 + a_1 \cos(Z_R\alpha - \varphi_k) + a_2 \cos 2(Z_R\alpha - \varphi_k) + \\ + a_3 \cos 3(Z_R\alpha - \varphi_k) + \ldots + a_v \cos v(Z_R\alpha - \varphi_k)$$
(1)

To avoid the total harmonics distortion (THD), the width of stator's teeth has to be $t_Z/3$, and the gap between the teeth should be calculated assuming the radius of $t_Z/3$ [9]. Accordingly, equation (1) can be rewritten as

$$\lambda_k = a_0 + a_1 \cos(Z_R\alpha - \varphi_k)$$
(2)

where

$a_0$ is the constant magnetic conductivity of the pole extension;
$a_1$ is the amplitude of first harmonics of a pole extension's magnetic conductivity;
$Z_R$ is the number of rotor teeth (pole pairs);
$\varphi_k$ is the phase angle.

The periodic change in magnetic conductivity causes a periodic change in the magnetic-flux linkage, which means the generation of energy transferred to the power grid and to the secondary load.

## III. THE BASIC EQUATIONS OF THE INDUCTION GENERATOR AND ITS OPERATING MODES

For creation of the required electromagnetic linka between windings it is necessary to have appropriate phase shifts between the processes going in pole extensions. For this purpose (see Fig.1) the pole extensions are divided into four groups, each of them with a coil of secondary winding a–x. The step between the mixed pole extensions belonging to one group equals $t_1 = 2t_Z/3$, and that between the mixed pole extensions belonging to mixed groups equals $t_2 = 7t_Z/6$. This provides a phase shift of the pole extensions of 240 degrees within the same group, and of 60 degrees for mixed group. The equations are similar to those for the conventional induction machine having only one secondary winding on the rotor.

According to the above mentioned and to Fig.1 it is possible to obtain equations for the magnetic flux linkage as

$$\Psi_A = w_{k1}\begin{pmatrix}(\lambda_1 + \lambda_7)(w_{k1}i_A + w_{k2}i_a) + \\ + (\lambda_4 + \lambda_{10})(w_{k1}i_A - w_{k2}i_a)\end{pmatrix} = \\ = 4a_0 w_{k1}^2 i_A + 4a_1 w_{k1} w_{k2} i_a \cos(Z_R\alpha)$$

$$\Psi_B = w_{k1}\begin{pmatrix}(\lambda_3 + \lambda_9)(w_{k1}i_B + w_{k2}i_a) + \\ + (\lambda_6 + \lambda_{12})(w_{k1}i_B - w_{k2}i_a)\end{pmatrix} = \\ = 4a_0 w_{k1}^2 i_B + 4a_1 w_{k1} w_{k2} i_a \cos(Z_R\alpha - 120^0)$$

$$\Psi_C = w_{k1}\begin{pmatrix}(\lambda_2 + \lambda_8)(w_{k1}i_C + w_{k2}i_a) + \\ (\lambda_5 + \lambda_{11})(w_{k1}i_C - w_{k2}i_a)\end{pmatrix} = \\ = 4a_0 w_{k1}^2 i_C + 4a_1 w_{k1} w_{k2} i_a \cos(Z_R\alpha - 240^0)$$

$$\Psi_a = w_{k2}\begin{pmatrix}\left(\sum_{k=i}^{12}\lambda_k\right)w_{k2}i_a + \\ + (\lambda_1 + \lambda_7 - \lambda_4\lambda_{10})w_{k1}i_A + \\ + (\lambda_3 + \lambda_9 - \lambda_6 - \lambda_{12})w_{k1}i_B + \\ + (\lambda_2 + \lambda_8 - \lambda_5 - \lambda_{11})w_{k1}i_C\end{pmatrix} = $$

$$= 12a_0 w_{k2}^2 i_a + 4a_1 w_{k1} w_{k2} i_A \cos(Z_R\alpha) + \\ + 4a_1 w_{k1} w_{k2} i_B \cos(Z_R\alpha - 120^0) + \\ + 4a_1 w_{k1} w_{k2} i_C \cos(Z_R\alpha - 240^0)$$
(3)

Next, it is possible to calculate the equations for the EMF and currents:

$$u_{AB} = i_A R_A + \frac{d\Psi_A}{dt}$$

$$u_{BC} = i_B R_B + \frac{d\Psi_B}{dt}$$

$$u_{CA} = i_C R_C + \frac{d\Psi_C}{dt}$$

$$i_{1A} = i_A + i_{C1A} - i_C - i_{C1C}$$

$$i_{1B} = i_B + i_{C1B} - i_A - i_{C1A}$$

$$i_{1C} = i_C + i_{C1C} - i_B - i_{C1B}$$

$$0 = i_a = (R_2 + R_{sl}) + 2U_{C2} + \frac{d\Psi_a}{dt}$$
(4)

where

$u_{AB}$, $u_{BC}$, $u_{CA}$ are the grid line voltages;
$i_{1A}$, $i_{1B}$, $i_{1C}$ are the grid currents;
$i_A$, $i_B$, $i_C$ are the phase currents of the generator;
$i_{C1A}$, $i_{C1B}$, $i_{C1C}$ are the capacitor $C_1$ currents/currents of cap.-*better*;
$R_1$, $R_2$ are the active resistance of primary and secondary windings;
$R_L$ is the active resistance of the secondary load;
$U_{C2}$ is the voltage drop on capacitor $C_1$.

With these equations it is possible to solve the most difficult tasks met in application of multipolar induction generators - e.g. analysis of transient process.

The project is based on the idea that during the rotation the secondary winding generates the EMF proportional to the slip $s = 1 - \dfrac{nZ_R}{60f}$ (where f is the frequency of primary current). If the absolute value of the slip increases, the power generated in the secondary part/by the secondary winding? also increases [10].

The advantages of the proposed design are as follows.

First, the generated power for the slip s= – (1..2) can be transferred through the primary and secondary windings? part through the load resistance $R_L$. There are no limits for raising the output power if the wind speed is increasing.

Second, the capacitors in the secondary winding increase the output power and help to magnetize the generator, with the power factor also increasing.

Third, the controlled secondary capacitors allow the limits of the generator's installed power to be extended. They also reduce the influence of the THD on the power grid.

## IV. THE RESULTS OF PRACTICAL RESEARCH

The results of the practical research are collected in Table 1. There has been also done a comparison to a double fed induction generator (DFIG) with secondary winding on the rotor.

TABLE I.
THE COMPARISON OF MEASURED VALUES AND PARAMETERS BETWEEN THE MULTIPOLAR INDUCTION GENERATORS

| No | Measured values and parameters | unit | DFIG with secondary windings on the stator | DFIG with a single phase secondary winding on the rotor (with slip rings) |
|---|---|---|---|---|
| 1 | Stator diameter, D | mm | 300 | 300 |
| 2 | Stator length, l | mm | 250 | 220 |
| 3 | Number of pole extensions | - | 12 | - |
| 4 | Number of teeth in one pole extension | - | 2 | - |
| 5 | Number of teeth in the rotor, $Z_R$ / pole pairs | - | 26 | 26 |
| 6 | Number of phases in the primary winding, $m_1$ | - | 3 | 3 |
| 7 | Number of phases in the secondary winding, $m_2$ | - | 1 | 1 |
| 8 | Primary voltage, $U_1$ | V | 380 | 380 |
| 9 | Secondary voltage, $U_2$ | V | 114 | 220 |
| 10 | Rotation frequency, n | min$^{-1}$ | 244 | 244 |
| 11 | Grid frequency, $f_1$ | Hz | 50 | 50 |
| 12 | Power transferred to the grid, $P_1$ | W | 1980 | 1200 |
| 13 | Power transferred to the secondary circuit, $P_2$ | W | 1870 | 1000 |
| 14 | Slip, s | - | -1,11 | -1 |
| 15 | Capacitors, $C_1/C_2$ | μF | 56/37 | - |
| 16 | Power factor, cosφ | - | 0,95 | 0,42 |
| 17 | Efficiency factor, η | - | 0,79 | 0,35 |
| 18 | THD level generated in the grid | % | < 5% | <4% |

The design of the proposed machine is showed in the Fig. 2.

Fig. 2. Multipolar double fed induction machine

## V. CONCLUSIONS

The research was targeted at extension of the practical application of directly driven induction machines in the wind turbines. For small powers (400 – 500 W) it is possible to use a generator with 8..10 pole pairs and a squirrel cage rotor. The conventional design cannot provide increasing output power with simultaneous decrease in the number of pole pairs. In the proposed low-speed induction generators all windings should be placed on the stator, with a tooth-like windingless rotor. To increase the output power of such a generator it is recommended to use a double fed system. The number of pole pairs in a generator can reach ≥50. The research was done for a 52-pole induction machine with three-phase primary and one-phase secondary windings, and with tooth-like unwound rotor. The results have shown that with such a machine a reliable power generation can be achieved.

## REFERENCES

[1] Daškova J., Kamoliņš E., Levins N. Multipolar directly driven wind generators. Latvian journal of Physics and Technical sciences, Nr.4, 2004.gads. pages 41 – 47. *(in Latvian)*

[2] The new generation of wind energy converters. Pitch Wind Systems AB, Sweden, 2001.

[3] Enercon grid properties and wind farm managemant, Enercon, Hamburg, 2002. 15 pages.

[4] Pugachev V., Levin N., Ribickis L., Manonov M., A multipolar inductor generator of annular design for windmills. Riga RTU Power and electrical engineerings 4, 10. 2003.gads, 10 – 15 pages.

[5] Nordex N54/1000 kW MK3, Technical Overview Svindblek 7323 Give Denmark, 1996, 4 pages.

[6]   Волдек А., Electrical machines. Энергия 1974, 640 pages. *(in Russian)*

[7]   Копылов И., Клоков Б., Морозкин В., Токарев Б. Design of electrical machines, Москва, ВШ. 2005. 768 pages. *(in Russian)*

[8]   Постников И. Design of electrical machines, ГИТЛ УССР, Киев 1952. 736 pages. *(in Russian)*

[9]   Левин Н., Серебряков А., Electrical machines and drives, 3.ed, Electrical drives based on the motors with static windings, Рига 1976, 192 pages.

[10]  Diļevs G., Levins N., Pugačevs V., Ribickis L. Double fed induction generator. Patents application P – 03 – 101, 20.09.2007, 8 pages.

# The measurement on the solar cells in Liberec city

## Jiri Kubin

Technical university Liberec/Institute of Mechatronics and Computer Engineering, Liberec, Czech republic, e-mail: jiri.kubin@tul.cz

*Abstract*— In the years 2005 – 2007 was realised the project „Pilot project of the power-produsing yield of the solar photovoltaic system in our climatic conditions for requirement of the supplying of the information and monitoring system of Liberec city". The project was solved at Technical University of Liberec (TUL) with the cooperation of High School for Electrotechnics and Mechanical Engineering in Liberec (SPŠSE). The project was financed by statutory city of Liberec. The paper deals with realisation of the project, measurements proceeding and some particular results.

*Keywords*—IEEEtran, Solar Cell System, Renewable energy systéme, Photovoltaic

## I. Introduction

The solar energy is one form of power with more and more higher significance. For example the installed output of solar power stations (in Czech Republic??) was 167 kW in 2005 and the total volume of produced electric energy grew up to 120,1 MWh. This value represents interannual growth aboout 55,4 %. It is expected, according to State Energy Inspection, that this groving will be continue till 2010. It is done mainly by the technology progress in this field, that makes photovoltaic technology more accesible and also by increasing amount of capital expenditure. In spite of it the share of this technology at total amount of produced energy will be still very small.

The Nordic World Ski Championship (Ski 2009) will take place in Liberec on March 2009. At this occasion the City of Liberec plans to use the solar energy as power supply for information panels. Liberec city is located at the foothills on the $50^{th}$ parallel of northern width and therefore the solar radiation intensity is not too high. Thus it is necessary to take all possible measures to increase the efficiency of conversion solar energy to electrical. The following of the Sun position by solar panel with the help of the turning gear represents one of the possibilities. The pilot project had the objective to create two measurement workplaces - the first workplace with turning solar panel at the High School for Electrotechnics and Mechanical Engineering and second workplace with static system at Technical University. The main effort led to longtime measurement of output power and analyse the efficiency of solar photovoltaic panels at both workplaces.

## II. The principle of conversion of solar energy to electrical energy

The solar cells are constructed as large-area semiconductor crystal with PN junction. It rises if the part of silicon crystal is subsidized like semiconductor of P type and neighbouring part like semiconductor of N type. In the place of junction the part of free electrons pass over N type area into the P type area and the part of free holes pass over P into N type area. The bound electric charge of ionized addition creates the areas of charge space, between them the electric field grow up. This field blocks the consequent flow of free carriers.

PN junction in the solar cells is oriented perpendiculary to the frontal area between front and back face of the panel. The electron – hole couple is generated when the photons impacts on the solar cell. Thereby photons give back theirs energy and are absorbed. The redundant energy of the photons is passed on the oscillations of the grid that leads to heating of the panel. The couples electron – hole are separated from each other by electric field of PN junction between area of the charge space. The holes are accelerated towards the direction of the electric field and electrons in the opposite direction. The separation of the electron – hole couple results in the issue of the electric voltage between the opposite poles of the solar cell. The direct current passes through the connected load, which is proportional to the area of the cell and the intensity of the incident radiation. If we use serial-parallel connection of cells we can obtain the photovoltaic panel with defined output voltage.

## III. Used panels and description of turning gear

Two identical solar panel of the type SI72 – 110 – 12 produced by Solartec company were bought in the terms of the project.

TABLE I.
PARAMETERS OF THE SOLLAR CELLS

| Un | 12 V |
|---|---|
| Pmax | 110 W |
| Umax | 17,4 V |
| Imax | 6,3 A |
| U0 | 21,6 V |
| rozměry (š x l x v) | 654 mm x 1310 mm x 40 mm |

The panels are made from monocrystalline silicon photovoltaic cells and their parameters are inserted in the table 1..

The first panel (Fig 1.) is static and is situated in the campus of the Technical university in Liberec on the pavilon A (southern face of the building).

Fig. 1. Magnetization as a function of applied field.

The second panel is situated on the roof of High school in Liberec city and can be oriented perpendiculary to the solar radiation. The swiweling mechanism is used for positioning of the panel. The rotation around the vertical and horizontal axes is realised by stepper motors with worm gears. The vertical axis of rotation is parallel to the rotation axis of the Earth. The High school lies at a lattitude of 50°46′37″ North. Vertical rotation axis of the solar panel has direction $\Psi$ = 90° - 50°46′23″ = 39°13′37″. The axis aims to the star Polaris – so called parallactic mounting. The turning of the panel around this vertical axis is active from the morning 6:00am till the evening 19:00 in the range of angle $\alpha$ = 195° with the speed 15° per hour. The movement is realized every twenty minutes. The acuracy of positioning is checked by means of the panel orientation to the South in the 12:00am. After sunset the panel rolls back to the east starting position.

The solar panel turns horizontally about ±23° from base position as well. The base position is determined in the time of the spring and autumn equinox. This tilting of the panel takes place according to next formula:

$$\varphi = \operatorname{arctg} [0{,}435 \sin (0{,}0172\ d)] \qquad (1)$$

Horizontal orientation of panel is realized manually every 14 days. The stepper motors are managed from control system situated in the switchboard in the classroom.

## IV. THE POSIBILITIES OF THE MEASUREMENT ON THE SOLAR PANELS

For the measurement of the main electric quantities were both solar panel workplaces equipped with the measurement cards. The workplace on the High school contains the measurement card PCA – 7228AS from Tedia company with eight analog inputs with 14-bits resolution. The isolating amplifier OPT – 181 was purchased to ensure the galvanic isolation of measured voltage. The card is supplied with program ScopeWin – 32 to set measuring, save the measured data and visualise them. The picture of the isolating amplifier with the connected load on the SPŠSE is it the figure 2.

Fig. 2 Connection of the load with isolating amplifier

The workplace on the Technical University contains the measuring card PCMCIA – 6036E. The problems with different voltage range and current measuring were necessary to solve too. The data saving and visualization software was designed on the Technical university of Liberc in LabView development tool. The measuring circuit is in the figure 4.

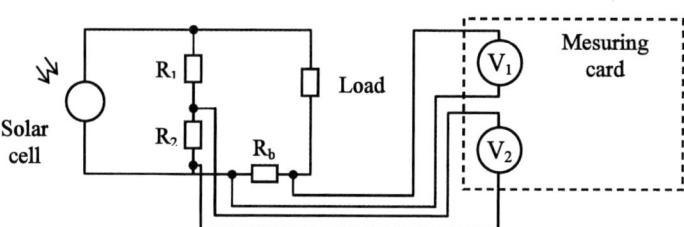

Fig. 3 Measuring circuit

Fig. 4 Measuring workplace of the solar cells on the TUL

Common 100W bulbs are used as load for both solar systems for first experiments. The resistance of bubl filament depends significantly on the electric current supplied by solar panel. Thats why the bulb is not suitable load if we want to compare the amount of electric energy produced by panels.

If we want to obtain maximal electric power from any electric source, the impedance of load should be complex conjugate value equally to sum of source impedance and line wiring impedance. Beacuse the solar panels are DC sources the reactance of circuit can be neglected. Solar panel is nonlinear source (figure 5) and its properties depend on the weather very strongly.

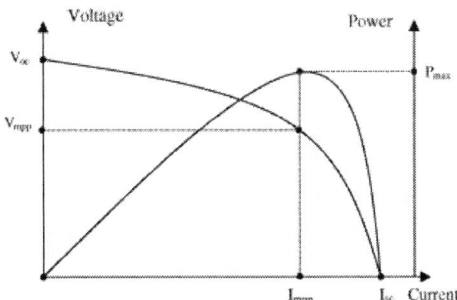

Fig. 5 Voltampere curve of the solar cells

The suitable load impedance is therefore very complicated to estimate. The programmable load was obtained to determine the right load value and to optimize the measurement circuit.

## V. PROCEEDING MEASUREMENT

The measurements are proceeding now every working day from morning to evening hours, on the SPŠSE at weekend too. Actual output voltage and current are automatically recorded with the measuring card every minute and the instantaneous output power is calculated. The curves of the actual power in the 15th week of year 2007 are in the figure 5.

We can observe the influence of positioning mechanism on this figure. This mechanism consumes the most of electric energy from the solar panel. We switched off this mechanism from 13.4.2007 till 16.4.2007 and measured the amount of produced energy. Because the period was very sunny, it is possible to compare the individual curves of the produced power in the different days. At 14th, 15th, 16th March was panel stable and so we can see the gradual rise of the power to the steady value. In the afternoon we can see slow fall of the produced power. The positioning mechanism was switched on at 16th March at 8 a.m. and the panel produced the energy about one hour longer in comparison with the previous days. The influence of panel orientation is not so striking in the overall values of produced energy, despite of the first sight on the figure 5. This is done by the whole exposure time of the panel with solar radiation. In any case it is seen, that the absence of the slewing reduces the amount of the produced energy. The influence of the horizontal slewing is much more evident, as could be seen from the power curves from both panels during the long time period.

## VI. CONCLUSION

Useful data for evaluation we have from 8th January 2007 (the second week) till the 17th week Although it is relatively short period, we can evaluate the influence of slewing on the whole amount of produced energy. If we focus on the maximal power in the monitored period, we find out, that the movable panel has reached over 85 W at the 4th March 2007 and the unmoveable panel over 65 W at the 18th March. If we watch the power curves after the whole weeks, it is obvious that the produced power of unmoveable panel gradually rise.

Large difference can be seen at sunny days. The swivelling panel has reached maximal produced power (about 70 W) in very short time after the sunset, but the power of static panel grew very slowly and reached a short-term maximum in the period between 12:00 am and 2:00 pm.

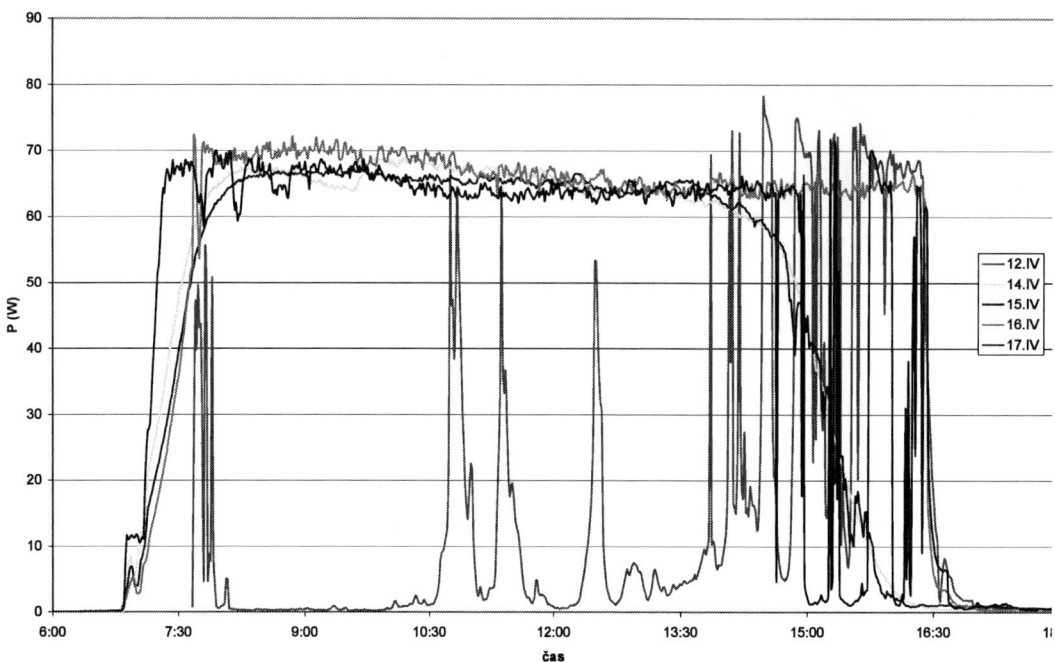

Fig. 5. Time curve of the output power during 12th – 17th March.

The differences in the amount of produced electrical energy from both panels are caused by swivelling equipment. But electrical energy is necessary for swivelling of the panel too. Most of the energy is consumed for a vertical swivelling during the day. Step motor input is approximately 120 W and the panel turns from morning position to evening position and at midnight comes back. The total time of motor operation is 1108,8 s. The electrical energy consumption in one day with vertical swivelling is approximately 34 Wh. In this value is not count in energy needed at horizontal swivelling, which takes place one in fourteen days and power consumption of stepper motor controller and microcontroller.

If we focus on the period of the Championship (18th February till 1st March) the situation is not very favourable. It was cloudy from 18.2 till 20.2 and the overall produced energy per one day was only about units of watts. At 21.2 it was slightly clouded sky and in the afternoon half covered sky. The panel at SPŠSE produced about 180 Wh after substraction of swivelling mechanism power consumption. This is about 2,5 more then at panel TUL. When it is half covered sky or cloudy the total energy per day is about 60W but rotary panel still generate two-times more energy then static one.

Definite conclusion about the possibility to depend on solar energy in the period of championship can be based on last year measurements only. Weather conditions for electric power production were not too convenient last year and it is not possible to forecast if the weather will be better or worse in next year. It is clear from the results that the sufficient amount of energy can be produced only in really sunny days. Therefore it is impossible to rely only on solar panels. They can be recommended as source of the electric power only for sunny days, in the case of cloudy weather the sun must be replaced with other source of energy.

### REFERENCE

[1] Libra, M; Poulek, V – Physical principle of photovoltaic conversion solar energy, Sv□lo 2005/01 str. 32-36

[2] www sides – http://www.i-ekis.cz

[3] Doležel, J – Report about filling of the indicative aim of the electric power producing from renewable source in the year 2005, www sides of ministry of industry and business

[4] Krieg, B: Electricity from the shine, HEL, Ostrava 1993

# Rotor Turn-to-Turn Faults of doubly-fed Induction Generators in Wind Energy Plants – Modelling, Simulation and Detection –

Vincenz Dinkhauser*, Friedrich W. Fuchs†
Institute of Power Electronics and Electrical Drives
Christian-Albrechts-University of Kiel
Kaiserstr. 2, 24143 Kiel Germany
* e-mail: vvd@tf.uni-kiel.de
† e-mail: fwf@tf.uni-kiel.de

*Abstract*— This work considers the seldomly treated case of rotor turn-to-turn fault of a doubly-fed induction generator. This fault can take place where the doubly-fed induction generator is a standard solution in wind power stations. For investigations an error-adoptable three phase machine model is derived and simulated. Furthermore the model is integrated in a wind energy plant model and simulated for a turn-to-turn fault in one rotor phase. Motor current signature analysis, wavelet transformation and the Luenberger observer are used for detection and are compared.

*Keywords*— Diagnostics, doubly-fed induction machine, Fault handling strategy, Wind energy, Simulation.

## I. INTRODUCTION

The doubly-fed induction machine is a widely spread system in the wind energy industry. Due to the smaller rotor-sided converter compared to stator side converter system, there are considerable cost benefits. However seldom faults – like a rotor turn-to-turn fault – with a small probability as other kinds of faults can take place. The rotor turn-to-turn fault can be a result different reasons as ageing, over-voltages and so on. So far, no publication about this fault has been found. Publications about turn-to-turn faults of permanent magnet synchronous machines and induction machines are found. But only few detailed models exist for the doubly-fed induction generator, which contain the internal fluxes at asymmetries of the system [1,2]. So a corresponding model will be derived.

The structure of the paper is as follows. In section II.A. a three-phase machine model will be developed and in II.B. it is compared to a standard d-q model in order to validate it. Then in section III. a rotor turn-to-turn fault is implemented and simulated. Afterwards in section IV. three algorithms of fault detection are described and validated for rotor turn-to-turn fault. Finally in section V. a conclusion will be given. References and an appendix complete the paper.

## II. THE THREE PHASE MACHINE MODEL

### A. Machine modeling

At first an induction machine model with smoothed air gap and concentrated windings has to be derived like in [3] and [4], which is based on the general matrix equation

for the voltages of induction machines (1). The three phase model has been chosen, due to the easy integration of asymmetrical faults.

$$\mathbf{u}(t) = \mathbf{R}\mathbf{i}(t) + \frac{d\psi(t)}{dt} \tag{1}$$

$$\mathbf{u}(t) = \begin{pmatrix} u_{SA}(t) & u_{SB}(t) & u_{SC}(t) & u'_{Ra}(t) & u'_{Rb}(t) & u'_{Rc}(t) \end{pmatrix}^{\mathrm{T}} \tag{2}$$

$$\mathbf{R} = \begin{pmatrix} R_S & 0 & 0 & 0 & 0 & 0 \\ 0 & R_S & 0 & 0 & 0 & 0 \\ 0 & 0 & R_S & 0 & 0 & 0 \\ 0 & 0 & 0 & R'_R & 0 & 0 \\ 0 & 0 & 0 & 0 & R'_R & 0 \\ 0 & 0 & 0 & 0 & 0 & R'_R \end{pmatrix} \tag{3}$$

$$\mathbf{i}(t) = \begin{pmatrix} i_{SA}(t) & i_{SB}(t) & i_{SC}(t) & i'_{Ra}(t) & i'_{Rb}(t) & i'_{Rc}(t) \end{pmatrix}^{\mathrm{T}} \tag{4}$$

$$\psi(t) = \begin{pmatrix} \psi_{SA}(t) & \psi_{SB}(t) & \psi_{SC}(t) & \psi_{Ra}(t) & \psi_{Rb}(t) & \psi_{Rc}(t) \end{pmatrix}^{\mathrm{T}} \tag{5}$$

$\mathbf{u}(t)$ - stator side referred time dependent phase voltages
$\mathbf{i}(t)$ - stator side referred time dependent phase currents
$\psi(t)$ - stator side referred time dependent phase fluxes
$R'_R$ - stator side referred rotor resistance
$R_S$ - stator side referred stator resistance

The relation between flux, inductivity and current is as follows:

$$\psi(t) = \mathbf{L}(t)\mathbf{i}(t) \tag{6}$$

$$\mathbf{L}(t) = \begin{pmatrix} \mathbf{L}_{SS} & \mathbf{L}_{RS}(t) \\ \mathbf{L}_{SR}(t) & \mathbf{L}_{RR} \end{pmatrix} \tag{7}$$

$$\mathbf{L}_{SS} = \begin{pmatrix} L_S & -\dfrac{L_{m,ph}}{2} & -\dfrac{L_{m,ph}}{2} \\ -\dfrac{L_{m,ph}}{2} & L_S & -\dfrac{L_{m,ph}}{2} \\ -\dfrac{L_{m,ph}}{2} & -\dfrac{L_{m,ph}}{2} & L_S \end{pmatrix} \tag{9}$$

978-1-4244-1741-4/08/$25.00 ©2008 IEEE

$$\mathbf{L}_{RR} = \begin{pmatrix} L'_R & -\dfrac{L_{m,ph}}{2} & -\dfrac{L_{m},ph}{2} \\ -\dfrac{L_{m,ph}}{2} & L'_R & -\dfrac{L_{m,ph}}{2} \\ -\dfrac{L_{m,ph}}{2} & -\dfrac{L_{m,ph}}{2} & L'_R \end{pmatrix} \qquad (8)$$

$$\mathbf{L}_{SR}(t) = L_{m,ph}\begin{pmatrix} b1 & b2 & b3 \\ b3 & b1 & b2 \\ b2 & b3 & b1 \end{pmatrix} = \mathbf{L}_{RS}^{\mathrm{T}}(t) \qquad (10)$$

$$\frac{\mathrm{d}\mathbf{L}(t)}{\mathrm{d}t} = \dot{\mathbf{L}}(t) = \begin{pmatrix} \mathbf{0} & \dot{\mathbf{L}}_{RS}(t) \\ \dot{\mathbf{L}}_{SR}(t) & \mathbf{0} \end{pmatrix} \qquad (11)$$

$$L_{S,ph} = L_{m,ph} + L_{S\sigma} \qquad (12)$$

$$L'_{R,ph} = L_{m,ph} + L'_{R\sigma} \qquad (13)$$

$L_{m,ph}$ - stator side referred phase mutual inductance
$L_{S,ph}$ - stator side referred phase stator inductance
$L'_{R,ph}$ - stator side referred phase rotor inductance
$L'_{R\sigma}$ - stator side referred rotor leakage inductance
$L_{S\sigma}$ - stator side referred stator leakage inductance
$b_1 = \cos(\theta_m)$

$b_2 = \cos\left(\theta_m - \dfrac{2\pi}{3}\right)$

$b_3 = \cos\left(\theta_m + \dfrac{2\pi}{3}\right)$

$\theta_m$ - electrical rotor angle between stator and rotor

So (1) can be combined with (6) to:

$$\mathbf{u}(t) = \mathbf{R}\mathbf{i}(t) + \mathbf{L}(t)\frac{\mathrm{d}\mathbf{i}(t)}{\mathrm{d}t} + \mathbf{i}(t)\frac{\mathrm{d}\mathbf{L}(t)}{\mathrm{d}t} \qquad (14)$$

The mechanical behavior is defined by (15)-(17), which is also described in [3].

$$T_{ag} = -\frac{1}{2} p\,\mathbf{i}(t)\frac{\mathrm{d}\mathbf{L}(\theta_m)}{\mathrm{d}\theta_m}\mathbf{i}(t)^T \qquad (15)$$

$$\omega_{mech} = \frac{1}{\Theta}\int T_{ag} - T_{load}\,\mathrm{d}t \qquad (16)$$

$$\theta_m = p\int \omega_{mech}\,\mathrm{d}t = p\theta_{mech} \qquad (17)$$

$T_{ag}$ - air gap torque
$\theta_{mech}$ - mechanical rotor position
$\omega_{mech}$ - mechanical rotor speed
$\Theta$ - inertia of the system
$p$ - number of pole-pairs

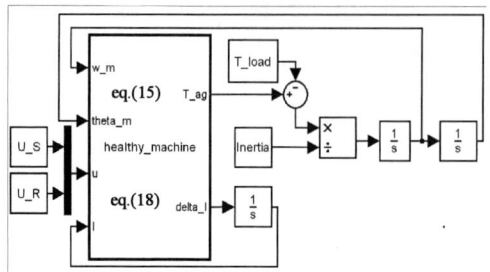

Fig.1:    Simulink® model of the three phase machine

Solving (14) for the derivate of $i(t)$ yields (18), which can be directly implemented in Matalb/Simulink for solving the equation.

$$\frac{\mathrm{d}\mathbf{i}(t)}{\mathrm{d}t} = \mathbf{L}(t)^{-1}\left(\mathbf{u}(t) - \mathbf{R}\mathbf{i}(t) + \mathbf{i}\frac{\mathrm{d}\mathbf{L}(t)}{\mathrm{d}t}\right) \qquad (18)$$

The resulting differential equation system of (18) has been programmed in an *Embedded Matalb Function* and solved in a numerical simulation by Matalb/Simulink. Fig. 1 shows the complete machine model. In the upper right part mechanical behavior with motor speed calculation behind the first integrator and rotor angle behind the second integrator is implemented. In the lower part electrical behavior is implemented with input voltages and behind the integrator machine currents. The calculation of the matrices, the derivate of the currents and the air gap torque is located in the block *healthy machine*.

### B.  Machine model validation

To ensure the validity of the used model, it has been compared with a d-q reference frame machine model set in synchronous reference frame [5]. Due to space vector theory the mutual inductance changes to

$$L_m = \frac{3}{2} L_{m,ph}, \qquad (19)$$

while the leakage inductances are the same.

The stator voltage was used as reference frame, so the electrical d-q reference frame machine model is described by (20) and (21).

$$\frac{d}{dt}\Psi_{Sd}^{U_s} = -\frac{R_s}{\sigma L_S}\Psi_{Sd}^{U_s} + \omega_k\Psi_{Sq}^{U_s} + \frac{R_S L_m}{\sigma L_S L_R}\Psi_{Rd}^{U_s} + U_{Sd}^{U_s}$$

$$\frac{d}{dt}\Psi_{Sq}^{U_s} = -\omega_k\Psi_{Sd}^{U_s} - \frac{R_s}{\sigma L_S}\Psi_{Sq}^{U_s} + \frac{R_S L_m}{\sigma L_S L_R}\Psi_{Sq}^{U_s}$$

$$\frac{d}{dt}\Psi_{Sd}^{U_s} = -\frac{R_R L_m}{\sigma L_S L_R}\Psi_{Sd}^{U_s} - \frac{R_R}{\sigma L_R}\Psi_{Rd}^{U_s} + \omega_R\Psi_{Rq}^{U_s} + U_{Rd}^{U_s}$$

$$\frac{d}{dt}\Psi_{Sq}^{U_s} = \frac{R_R L_m}{\sigma L_S L_R}\Psi_{Sq}^{U_s} - \omega_R\Psi_{Rd}^{U_s} + \frac{R_R}{\sigma L_R}\Psi_{Rq}^{U_s} + U_{Rq}^{U_s}$$

$$(20)$$

$$I_{Sd}^{U_s} = \frac{1}{\sigma L_S}\Psi_{Sd}^{U_s} - \frac{L_m}{\sigma L_S L_R}\Psi_{Rd}^{U_s}$$

$$I_{Sq}^{U_s} = \frac{1}{\sigma L_S}\Psi_{Sq}^{U_s} - \frac{L_m}{\sigma L_S L_R}\Psi_{Rq}^{U_s}$$

$$I_{Rd}^{U_s} = \frac{1}{\sigma L_R}\Psi_{Rd}^{U_s} - \frac{L_m}{\sigma L_S L_R}\Psi_{Sd}^{U_s}$$  (21)

$$I_{Sq}^{U_s} = \frac{1}{\sigma L_R}\Psi_{Rq}^{U_s} - \frac{L_m}{\sigma L_S L_R}\Psi_{Sq}^{U_s}$$

$\sigma$ - leakage factor

And the mechanical equations are:

$$T_{ag} = \frac{3}{2}p\frac{L_m}{L_S}\left(\Psi_{Sq}^{U_s}I_{Rd}^{U_s} - \Psi_{Sd}^{U_s}I_{Rq}^{U_s}\right)$$  (22)

$$\frac{\mathrm{d}}{\mathrm{d}t}\omega_m = T_{ag} - T_{load}$$  (23)

$$\omega_R = \omega_S - p\omega_{mech}$$  (24)

Start-up and a load change with shorted rotor phases were used to verify steady-state and transient behaviour of the model.

Both models are compared in fig. 2 and 3. There are shown the reference frame components of rotor and stator current of the three phase model, which have the highest amplitude, the speed of the machine represented by three phase model and the differences between three-phase and d-q reference frame model states. This composition was necessary, due to the small differences between both models. In fig. 2 the start-up and in fig. 3 a load change up to rated torque are shown. As already mentioned, both

figures show, that even at high values and changing rates differences are very small - lower than $2 \cdot 10^{-3}$ -. These differences can be caused by computation delay, by small asymmetries during magnetization or by the relative tolerances during numerical simulation, which were set to $10^{-8}$. So the models aren't identical, but a very good validity of the three-phase model can be stated.

The machine model was implemented in a wind energy plant model presented in [6]. This model consists of several sub models such as wind turbine, gearbox, generator, and converter as described in detail in [6]. The dynamic model of the drive train is reduced to a two mass system and the converter is modeled as continuous sinusoidal voltage source. The control is field oriented in stator voltage reference frame for speed and grid-side voltage for dc voltage and reactive power.

### III. FAULTED MACHINE MODEL

#### A. Fault Modeling

In case of a turn-to-turn shorting the shorted turns can be modeled as a new phase ([1], [2] and [7]). This new phase is represented by $u'_{Rf}(t)$, $i_{Rf}(t)$ in input and state vector and it is inserted in the fifth column and row of the inductance matrices of (14). At the shortcut turns is no external voltage drop so $u'_{Rf}(t)$ equals zero volts, but a current is induced by the machine flux. The changed vectors and matrices for the faulty machine are described in (25)-(31).

$\mu$      $= N_{sc}/N_{Ph}$     severity of the fault

$N_{Ph}$ - number of turns per phase

$N_{sc}$ - number of shorted turns

$$\mathbf{u}(t) = \left(u_{SA}(t) \; u_{SB}(t) \; u_{SC}(t) \; u'_{Ra}(t) \; u'_{Rf}(t) \; u'_{Rb}(t) \; u'_{Rc}(t)\right)^T$$  (25)

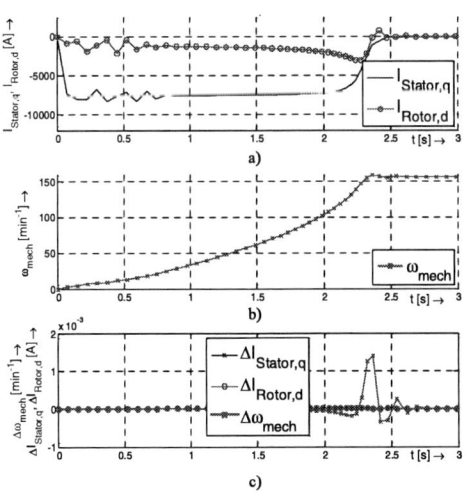

Fig.2: Machine behaviour of the three phase model at start-up with absolute current of stator-q and rotor-d component in a) and speed in b) and in c) the absolute differences of the mentioned states to the fourth order model

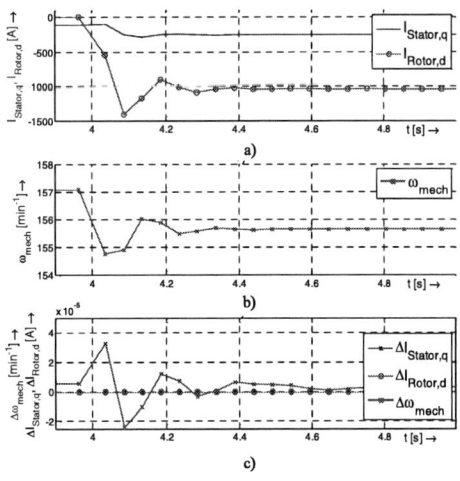

Fig.3: Machine behaviour of the three phase model with a load change at 4 sec. from no-load to rated torque with absolute current of stator-q and rotor-d component in a) and speed in b) and in c) the absolute differences of the mentioned states to the fourth order model

$$\mathbf{i}(t) = \begin{pmatrix} i_{sA}(t) & i_{sB}(t) & i_{SC}(t) & i'_{Ra}(t) & i'_{Rf}(t) & i'_{Rb}(t) & i'_{Rc}(t) \end{pmatrix}^{\mathrm{T}} \quad (26)$$

$$\mathbf{L}_{SRf}(t) = L_{m,ph} \begin{pmatrix} (1-\mu)b_1 & (1-\mu)b_2 & (1-\mu)b_3 \\ \mu b_1 & \mu b_2 & \mu b_3 \\ b_3 & b_1 & b_2 \\ b_2 & b_3 & b_1 \end{pmatrix} = \mathbf{L}_{RSf}{}^{T}(t) \quad (27)$$

$$\mathbf{L}_{RRf} = \begin{pmatrix} (1-\mu)^2 L'_{R,ph} & (1-\mu)\mu L_{m,ph} & -\dfrac{(1-\mu)L_{m,ph}}{2} & -\dfrac{(1-\mu)L_{m,ph}}{2} \\ (1-\mu)\mu L_{m,ph} & \mu^2 L'_R & -\dfrac{\mu L_{m,ph}}{2} & -\dfrac{\mu L_{m,ph}}{2} \\ -\dfrac{(1-\mu)L_{m,ph}}{2} & -\dfrac{\mu L_{m,ph}}{2} & L'_R & -\dfrac{L_{m,ph}}{2} \\ -\dfrac{(1-\mu)L_{m,ph}}{2} & -\dfrac{\mu L_{m,ph}}{2} & -\dfrac{L_{m,ph}}{2} & L'_R \end{pmatrix}$$

$$(28)$$

$$\mathbf{L}_f(t) = \begin{pmatrix} \mathbf{L}_{SS} & \mathbf{L}_{RSf}(t) \\ \mathbf{L}_{SRf}(t) & \mathbf{L}_{RRf} \end{pmatrix} \quad (29)$$

$$\dot{\mathbf{L}}_f(t) = \begin{pmatrix} \mathbf{0} & \dot{\mathbf{L}}_{RSf}(t) \\ \dot{\mathbf{L}}_{SRf}(t) & \mathbf{0} \end{pmatrix} \quad (30)$$

$$\mathbf{R}_f = \begin{pmatrix} R_S & 0 & 0 & 0 & 0 & 0 & 0 \\ 0 & R_S & 0 & 0 & 0 & 0 & 0 \\ 0 & 0 & R_S & 0 & 0 & 0 & 0 \\ 0 & 0 & 0 & (1-\mu)R'_R & 0 & 0 & 0 \\ 0 & 0 & 0 & 0 & \mu R'_R & 0 & 0 \\ 0 & 0 & 0 & 0 & 0 & R'_R & 0 \\ 0 & 0 & 0 & 0 & 0 & 0 & R'_R \end{pmatrix} \quad (31)$$

The dependency of the selfinductance in square on the number of turns can be extracted from [2] and is obvious, because selfinductances depend on the square of the number of windings. The linear relation of the mutual-inductances can also be extracted from [2]. Differences to [2] are explained by the simplification to a machine model with one winding per phase.

*B. Simulation Results*

A period of 8 seconds at a mechanical speed of 1274 min$^{-1}$ has been simulated. In this time at 2 seconds a turn-to-turn fault for 0.5 % of the turns of rotor phase 'a' is generated. Fig. 4(b) shows the high induced short cut current of the shorted turns. Additionally only small changes in the electrical values are visible in the figures 4(a) and 4(c).

Beside switch off [5] the operation at the synchronous speed is a possibility for protection against overheating and destroying. At this point of operation the induced voltage to the rotor is equal to zero referring to (32).

$$U_{i,sync} = \frac{\mathrm{d}\,\psi_{R,sync}}{\mathrm{d}t} = 0 \quad (32)$$

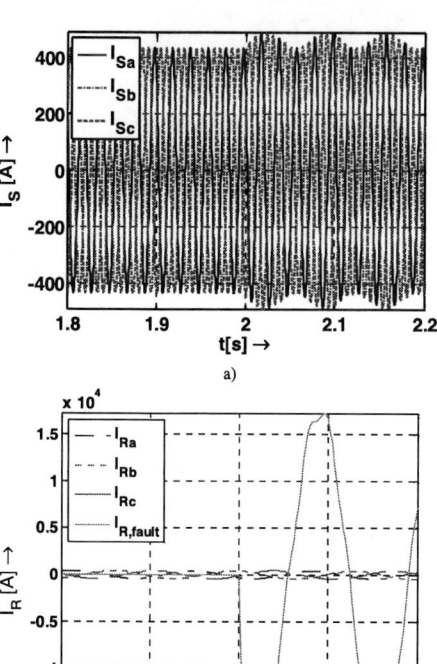

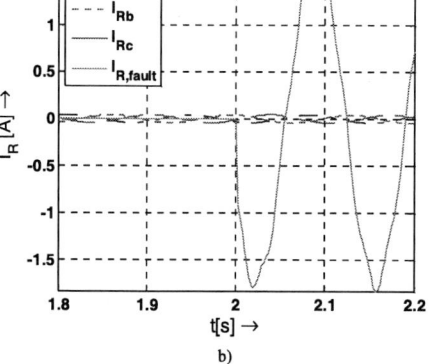

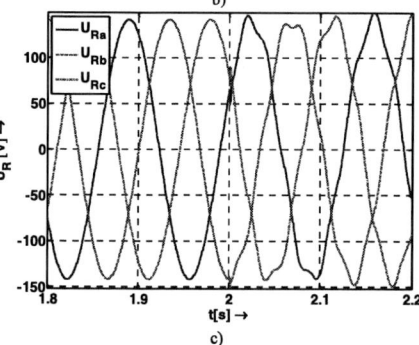

Fig.4: Machine currents (a),(b) and voltages (c) until 2 sec. without and after 2 sec. with shorted turns.

By this method a continued energy capture out of the wind is possible. In order to switch to the fault-tolerant operation in time, the fault has to be detected fast.

## IV. FAULT DETECTION ALGORITHMS

For fault-detection the algorithms of Motor Current Signature Analysis (MCSA) and discrete wavelet transformation and the state based Luenberger observer have been selected and will be compared in their reactions on the turn-to-turn fault.

*A. Motor Current Signature Analysis (MCSA)*

MCSA is the analysis of sidebands of the motor current spectrum. It uses the effects of machine faults and torque changes on the electrical flux [8]. So for example a broken rotor bar changes the flux path and amplitude, which is shown in [9]. Other faults like excentricities [10], bearing defects [11], gearbox faults [12] and turn-to-turn faults can be detected [13].

For the investigated rotor turn-to-turn fault sidebands relations between the fundamentals of stator and rotor current have to be considered, which is done in the appendix. The result is a fault frequency of:

$$f_{fault} = f_S - 2f_R. \tag{33}$$

This harmonic can be seen in the spectrum of the stator current shown in fig. 5, which increases slowly. The slow reaction is caused by the 2 sec transformation window, which is necessary for a frequency resolution of 0.5 Hz. The often mentioned sideband at $f_S+2f_R$ has not been observed in this simulation [5].

The problem of this method is the duration and the minimum resolution needed in time and frequency for separation of different fault frequencies. Another disadvantage is the operation close to synchronous speed, where the frequencies overlap. So even if the fault has been detected, a different method has to be used for a satisfying monitoring of the faulted phase.

### B. Wavelet transformation

Wavelet transformation avoids a high computation effort with high resolution in frequency and time. It uses windowing-functions – so called wavelets –, which contain a limited frequency band. Used as window on the original signal shown in (34) only the signal parts of the wavelet frequency band remain.

$$w(k) = \sum_{k_0}^{k_0+b} s(k)\psi_{a,b}(k) \tag{34}$$

$w(k)$    - wavelet transformed signal
$s(k)$    - measured signal
$\psi_{a,b}(k)$ - wavelet
$a$       - scale of the wavelet
$b$       - length of the wavelet

The frequency band of the used wavelet is dependent on the length. The shorter the wavelet is in time the higher

in frequency and wider the frequency band becomes. Length and scale are linked by the level of a wavelet. The level represents the power of two of the scale $a$ and the length $b$ divided by the power of two. This means, that the frequency resolution increases in square to the level, while the time resolution decreases in square to the level. To know where the frequency band is located, a so called pseudo frequency can be computed from the scale $a$, the sampling rate $\Delta t_{sample}$ and the center frequency $f_{center}$ of the wavelet (35).

$$f_{pseudo} = \frac{f_{center}}{a\Delta t_{sample}} \tag{35}$$

At this the center frequency equals the fundamental of the wavelet.

In case of the discrete wavelet transformation the original signal is split into detail coefficients $d_i$ and approximation coefficients $a_i$, where $i$ denotes the level of the wavelet. The detail coefficients are the high pass filtered (wavelet transformed) signal parts and the approximation coefficients represent the low pass filtered signal, where the low pass filter is called scaling function, which is adapted to the wavelet. This algorithm is started at level 1 and iteratively repeated, until the level of interest is reached. So the signal can be reconstructed by the following equation:

$$s(k) = a_n(k) + \sum_{i=1}^{n} d_i(k) \tag{36}$$

$n$   - maximum level of the used wavelets

Fig. 6 shows the results of the discrete wavelet transformation applied to the stator currents. It was used a discrete db6-wavelet up to level 8 with a sampling rate of 4 kHz. The bar on the right shows the brightness of the wavelets in relation to its amplitude. A change in relation to the fault is obvious at $d_7$ of the stator current.

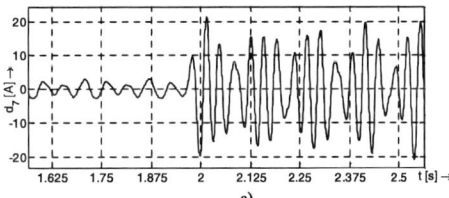

a)

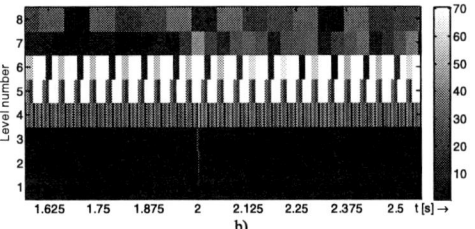

b)

Fig.6: Discrete wavelet transformed signal decomposition with detail coefficients of level 7 in a) and the top view of all detail coefficients from level 1 to 8 in b).

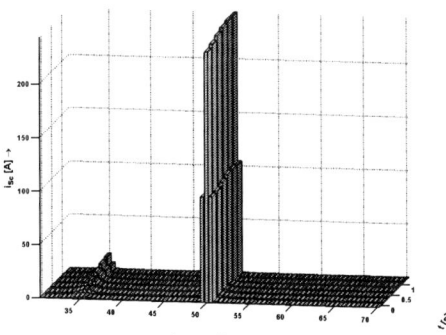

Fig.5: hamming filtered stator current spectrum for rotor turn-to-turn faults, with a fault severity of μ=0.5% and a mechanical speed of 1274 min⁻¹

The level related pseudo frequencies are shown in TABLE II. The reaction is faster than MCSA, caused by the shorter sample length. But it still needs a minimum number of samples and it has more problems of detection near synchronous speed, because the fault frequency changes into the wavelet of the fundamental, which already has a high amplitude.

### C. The Luenberger observer

The Luenberger observer is a state-based method for reconstruction of unmeasured states. It can also be used for faults [14]. It uses a reference model, which gets the same input values for the estimation of the system states. For the adaption of the model states to the system states a feedback of the differences between system and model -called residuals- with a feedback gain is used. Fig. 7 shows the structure of an observer.

In case of a Luenberger observer the gain is constant. It is computed from the chosen eigenvalues of the observer equation (14).

$$\dot{\tilde{x}} = \left(A - HC^{T}\right)\tilde{x} \qquad (37)$$

The real parts of these eigenvalues have to be more negative than the real parts of the system eigenvalues [14]. But they shouldn't become too negative for robustness against noisy input values. In the presented case the eigenvalues of TABLE III. in the appendix were generated by the Matlab® command 'place' and used for analysis.

The results in fig. 8 show, that the Luenberger observer with measurement of all the electrical states and the speed. It is visible, that the fault leads to an oscillation of the residuals within less than 3 µs. This reaction is faster than MCSA and as fast wavelet transformation, whereas sampling and computing of the wavelet needs more time. The residual amplitude is at 30 amps per reference

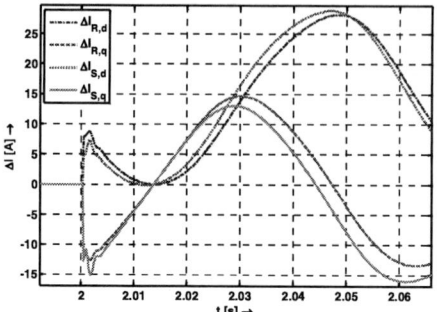

Fig.8: Observer resiudals in stator voltage reference frame d,q-coordinates

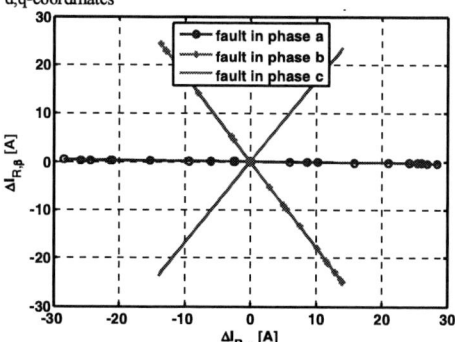

Fig.9: Observer residuals of the rotor current in α,β-coordinates

component. Which is very high compared to the fault-free residual amplitudes. This can be seen in Fig. 8 in the time before fault occurrence. So the residuals are a good indicator for this fault. Transforming into α-β-space phasor, it is also possible to locate the faulty phase, as shown in fig. 9.

## V. CONCLUSION

Modeling simulation and detection of rotor turn-to-turn faults of doubly-fed induction generators is essential. Due to increasing installation of these generators in wind power stations this will be needed and help to get failure tolerant generators. By this fault a very high current will be induced to the shorted turns, which will cause catastrophic faults. The turn-to-turn fault has been modeled and simulated. The behavior of the electrical values has been shown and fitting fault detection algorithms have been considered, whereby the Luenberger observer has the best performance for the detection of turn-to-turn faults.

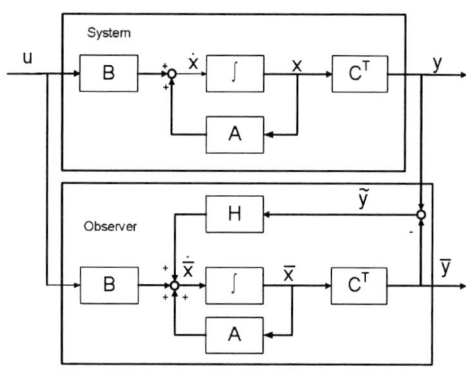

Fig.7:   Structure of an observer with known inputs
u           -input vector
y           - output vector
A           - system matrix
B           - input matrix
$C^{T}$        - output matrix
H           - observer feedback matrix
x           - system state vector
$\bar{x}$           - observer state vector
$\tilde{x} = x - \bar{x}$ - difference between system and model state vectors

## VI. APPENDIX

### TABLE I. STATOR-SIDE REFERRED MACHINE PARAMETERS

| stator frequency | $f_s$ | 50 Hz |
|---|---|---|
| rated power | $P_{rated}$ | 1.5 MW |
| rated Voltage | $U_S$ | 1000 V |
| rated torque | $T_{rated}$ | 9500 Nm |
| mutual inductance | $L_m$ | 27 mH |
| stator leakage inductance | $L_{S\sigma}$ | 280.1 µH |
| rotor leakage inductance | $L'_{R\sigma}$ | 117.7 µH |
| stator resistance | $R_S$ | 10.3 mΩ |
| rotor resistance | $R'_R$ | 8.28 mΩ |
| number of pole pairs | p | 2 |
| machine inertia | $\Theta_{machine}$ | 116 kg/m² |
| wind rotor inertia | $\Theta_{wind\_rotor}$ | $1.2 \cdot 10^6$ kg/m² |

### TABLE II. PSEUDO FREQUENCIES OF THE DETAIL COEFFICIENTS

| wavelet level | pseudo frequency [Hz] |
|---|---|
| 1 | 1.454 |
| 2 | 727 |
| 3 | 364 |
| 4 | 182 |
| 5 | 90.9 |
| 6 | 45.5 |
| 7 | 22.7 |
| 8 | 11.4 |

### TABLE III. EIGENVALUES OF MACHINE AND OBSERVER

| | Machine eigenvalues | Observer eigenvalues |
|---|---|---|
| $e_1$ | -25.93 + j313.31 | -150 + j85.533 |
| $e_2$ | -25.93 - j313.31 | -150 - j85.533i |
| $e_3$ | -20.96 + j313.31 | -100 + j85.533i |
| $e_4$ | -20.96 - j313.31 | -100 - j85.533i |

### A. Consideration of linkages between stator and rotor currents

The linkage between both currents can be shown by the flux. As example the flux of stator phase 'A' is used.

$$\psi_{SA}(t) = L_{Sf}i_{SA} - \frac{1}{2}L_{mf}i_{SB} - \frac{1}{2}L_{mf}i_{SC} + L_{mf}\left\{ i_{Ra}\cos(\theta) + i_{Rb}\cos\left(\theta + \frac{2\pi}{3}\right) + i_{Rc}\cos\left(\theta - \frac{2\pi}{3}\right)\right\}$$

The fundamental currents are defined by the following expressions:

$$i_{SA} = \hat{i}_S \cos(\omega_S t)$$

$$i_{SB} = \hat{i}_S \cos\left(\omega_S t - \frac{2\pi}{3}\right)$$

$$i_{SC} = \hat{i}_S \cos\left(\omega_S t + \frac{2\pi}{3}\right)$$

$$i_{Ra} = \hat{i}_R \cos(\omega_R t)$$

$$i_{Rb} = \hat{i}_R \cos\left(\omega_R t - \frac{2\pi}{3}\right)$$

$$i_{Rc} = \hat{i}_R \cos\left(\omega_R t + \frac{2\pi}{3}\right)$$

$$\omega_{m,el} = p\omega_m = \begin{cases} \omega_S - \omega_R & \omega_S > p\omega_m \\ \omega_S + \omega_R & \omega_S < p\omega_m \end{cases}$$

$\hat{i}_S$ = stator current amplitude
$\hat{i}_R$ = rotor current amplitude
$\omega_R$ = rotor voltage rotational speed
$\omega_S$ = stator voltage rotational speed

With use of the addition theorem:

$$\cos(x)\cos(y) = \frac{1}{2}\left(\cos(x-y) + \cos(x+y)\right)$$

The flux caused by the fundamentals of the current is as follos:

$$\psi_{SA}(t) = L_{Sf}\hat{i}_S \cos(\omega_S t) - \frac{1}{2}L_{mf}\hat{i}_S \cos\left(\omega_S t - \frac{2\pi}{3}\right)$$
$$- \frac{1}{2}L_{mf}\hat{i}_S \cos\left(\omega_S t + \frac{2\pi}{3}\right)$$
$$+ L_{mf}\hat{i}_R \cos(\omega_R t)\cos((\omega_S - \omega_R)t)$$
$$+ L_{mf}\hat{i}_R \cos\left(\omega_R t - \frac{2\pi}{3}\right)\cos\left((\omega_S - \omega_R)t + \frac{2\pi}{3}\right)$$
$$+ L_{mf}\hat{i}_R \cos\left(\omega_R t + \frac{2\pi}{3}\right)\cos\left((\omega_S - \omega_R)t - \frac{2\pi}{3}\right)$$

$$= L_{Sf}\hat{i}_S \cos(\omega_S t) - \frac{1}{2}L_{mf}\hat{i}_S \cos\left(\omega_S t - \frac{2\pi}{3}\right)$$
$$- \frac{1}{2}L_{mf}\hat{i}_S \cos\left(\omega_S t + \frac{2\pi}{3}\right) +$$
$$+ \frac{L_{mf}\hat{i}_R}{2}\left(\cos((\omega_S - 2\omega_R)t)(+\cos(\omega_S t))\right.$$
$$+ \left(\cos\left((\omega_S - 2\omega_R)t + \frac{4\pi}{3}\right) + \cos(\omega_S t)\right)$$
$$\left. + \left(\cos\left((\omega_S - 2\omega_R)t - \frac{4\pi}{3}\right) + \cos(\omega_S t)\right)\right)$$

So the fluxes, which are coupled to the rotor current, partly contain rotational speeds of $\omega_S - 2\omega_R$. In symmetrical systems these parts eliminate each other by their phase shift of $4\pi/3$. In asymmetrical systems these parts don't vanish. In supersynchronous operation, the order inverts and rotational speeds of $\omega_S + 2\omega_R$ arise.

## VII.  REFERENCES

[1] RM. Tallam, TG. Habetler, and R. Harley, "Transient model for induction machines with stator winding turn faults," in *Conference Record of the 2000 IEEE Industry Applications Conference. No.00CH37129*, 1 ed, 2000.

[2] QF. Lu, ZT. Cao, and E. Ritchie, "Model of stator inter-turn short circuit fault in doubly-fed induction generators for wind turbine,", 2004, pp. 932-937.

[3] P. Vas, "Parameter Estimation, Condition Monitoring and Diagnosis of Electrical Machines," Oxford: Calendron Press, 1993.

[4] LM. Popa, BB. Jensen, E. Ritchie, and I. Boldea, "Condition monitoring of wind generators," in *Conference Record of the 2003 IEEE Industry Applications Conference Cat. No.03CH37459*, 3 ed, 2003.

[5] D. Schröder, *Electrical drives -Control of drive systems-Elektrische Antriebe –Regelung von Antriebssystemen–*, 2. ed. Berlin: Springer Verlag, 2001, p. 424.

[6] R. Lohde, S. Jensen, FW. Fuchs, and A. Knóp, "*Analysis of Three Phase Grid Failure and Doubly Fed Induction Generator Ride-through using Crowbars*," Aalborg: EPE 2007 - 12th European Conference on Power Electronics and Applications , 2007.

[7] WT. Thomson, "On-line MCSA to diagnose shorted turns in low voltage stator windings of 3-phase induction motors prior to failure," in *IEMDC 2001. IEEE International Electric Machines and Drives Conference Cat. No.01EX485*, 2001, p. 891.

[8] SMJ. Fatemi, H. Henao, and GA. Capolino, "The effect of the mechanical behavior on the stray flux in an induction machine based electromechanical system,".

[9] F. Pedrayes, CH. Rojas, MF. Cabanas, MG. Melero, GA. Orcajo, and JM. Cano, "Application of a dynamic model based on a network of magnetically coupled reluctances to rotor fault diagnosis in induction motors,".

[10] RR. Schoen and TG. Habetler, "A new method of current-based condition monitoring in induction machines operating under arbitrary load conditions," *Electric Machines and Power Systems*, vol. 25, no. 2, pp. 141-152, /2.

[11] Wei Zhou, TG. Habetler, and RG. Harley, "Stator current-based bearing fault detection techniques: a general review,".

[12] AR. Mohanty and C. Kar, "Fault detection in a multistage gearbox by demodulation of motor current waveform," *IEEE Transactions on Industrial Electronics*, vol. 53, no. 4, pp. 1285-1297, June2006.

[13] J. Royo and FJ. Arcega, "Machine current signature analysis as a way for fault detection in squirrel cage wind generators," *IEEE Symposium on Diagnostics for Electric Machines Power Electronics and Drives (SDEMPED). IEEE. pp. 383-7. Piscataway.*

[14] H. Unbehauen, *Control techniques II -Regelungstechnik II-*, 8 ed. Braunschweig/Wiesbaden: Vieweg Verlag, 2000, pp. 81-94.

# Static and Dynamic Response of a Photovoltaic Characteristics Simulator

Anastasios Ch. Nanakos, Emmanuel C. Tatakis
University of Patras, Department of Electrical and Computer Engineering,
Laboratory of Electromechanical Energy Conversion, Rion - Patras, Greece,
e-mails: *tnanakos@ece.upatras.gr, e.c.tatakis@ece.upatras.gr*

*Abstract*— **In this paper a simulator of photovoltaic generators with programmable electrical characteristics is presented. The proposed system has the ability to generate the current-voltage curves of photovoltaic modules under any desirable insolation and temperature conditions. The system is also capable of integrating any maximum power point tracking algorithm under a unified control. The simulator's aim is the introduction of a faster, spherical and more effective approach in experimental investigation of photovoltaic systems, either in standalone or grid connected applications, independently from the atmospheric conditions. Towards this aim, the use of a DC power supply, controlled through a data acquisition card by appropriate algorithms, is proposed. These algorithms are implemented on a personal computer. Special effort was given in the development of a simplified user interface that monitors and controls the entire system offering effortless and faster conclusions.**

*Keywords*— **Photovoltaic, Solar Cell System, Renewable energy systems, Simulation, Virtual instrument.**

## I. INTRODUCTION

Nowadays, dominant factor of human evolution is the intense energy and environmental crisis. The conventional energy resources are consumed rapidly and their exclusive use for a lot of decades has caused serious repercussions to the world ecosystem. The need for energy is increasing but the environment is asking for a fundamental change in energy production. The wide use of clean energy sources forms a serious solution in order to fulfil both these two contradictory necessities [1-3]. The major interest on renewable energy is concentrated on solar and aeolian energy. The practical problem in the installation of wind generators in urban regions makes the production of energy via photovoltaic frames the only way for clean energy production in these regions.

Intense research effort is observed in the pursuit for most optimal and economic ways of the interconnection between photovoltaic generators and the low voltage grid. Each photovoltaic generator is characterized from a voltage-current curve with dominant factors the insolation level and the cell temperature. The form of these curves is presented in Fig. 1. Taking into consideration that the solar irradiation as well as the temperature varies during the day, the characteristic curve also alters. Every voltage-current curve is characterized by the maximum power point. This point for given irradiation and cell temperature is the maximum product of voltage and current, leading to the maximum generated power.

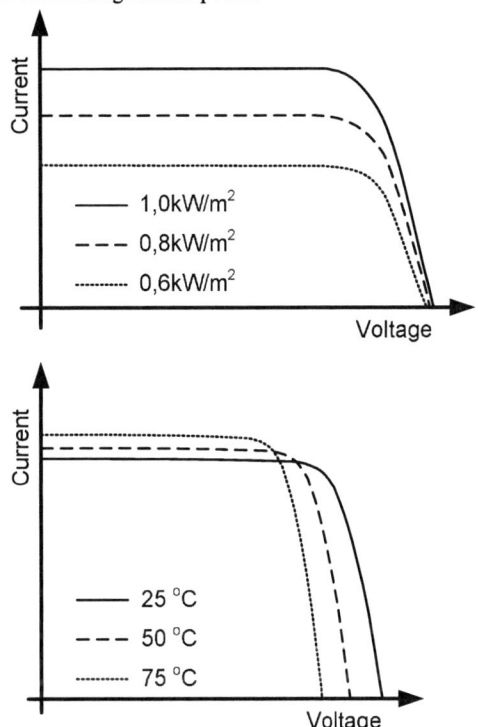

Fig. 1. V-I Characteristics of a P/V module for various insolation and temperature values.

For example in autonomous systems it may be required continuous current 12V or 24V for accumulator charging, or alternating current for e.g. asynchronous machines in pumping systems [4]. Another example of the photovoltaic application usage is their ability for direct interconnection to the low voltage grid [5].

In any case, the apparatus that converts the electric power, provided by the P/V system, should not only be able to adapt in different input and output characteristics but also it should absorb the maximum power that the photovoltaic generator can provide. So, the operation point of the converter is supposed to be able to: a) move along the characteristic curve of the photovoltaic frame, b) track down and stabilize to the maximum power point. Thus, it is possible the maximum exploitation of solar energy, satisfying one from the most essential reasons of

---

This work was funded by the European Social Fund (75%), the Greek Secretariat Research and Technology (25%) as well as ANCO S.A. and ENERGY SOLUTIONS S.A., within the framework of Measure 8.3 of the Operational Programme "Competitiveness" and the 3rd Community Support Programme (PENED 2003, 03EΔ400).

978-1-4244-1741-4/08/$25.00 ©2008 IEEE

the development of such systems. Towards this, rich bibliography exists on algorithms of maximum power point tracking (M.P.P.T.) that, at most times, are integrated in solar systems [6-8].

After the design and implementation of such solar systems, it is crucial the experimental verification and the validation of their proper operation and efficiency. Because the PV characteristics depend directly from the climatic conditions, the reliable investigation is laborious or many times impossible, leading to an incomplete trial. In order to perform decent experimental measurements the researcher must consume valuable time at the field waiting the repetition of conditions.

At this point the existence of a simulator for photovoltaic output characteristics seems to be irreplaceable. Using the simulator the user-researcher can program the output in order to approximate accurately the output characteristic curve of any desirable real photovoltaic generator. He can also, regulate the solar radiation and temperature of the frame changing the characteristic output voltage-current (V-I) curve.

Among related papers in international bibliography, different systems that assimilate the output characteristics of photovoltaic generators are presented. In all these systems we can distinguish four basic aspects of their operation. First of all is the theoretical model that reproduces the characteristic V-I curve. The dominant models are the parametric and the interpolation model. The parametric model [9,10] is very accurate but it is used when all the parameters are known. The interpolation model [11-16] needs only three specific points of the real curve to reproduce the whole curve. Secondly is the type of the control implementation. This control is responsible to force the hardware to act as a PV generator. In some papers the control is implemented by a microcontroller [12-14,16], in others by a personal computer [11,15] or a special real time digital simulator (RTDS) that combines hardware and software [9]. The third aspect that a simulator includes is the hardware that supplies the power as it is generated from the sun. In most papers [11,12,15,17] the researchers propose custom made buck or buck-boost converters in order to generate PV characteristic curves. In work [17] the simulator is based on an active power load without power consumption. On the other hand in paper [9], the basis is a RTDS flexible

real – time simulator. Power system networks are created on the RTDS by arranging electrical components from the model libraries. Analog signals can be interfaced between the RTDS and the external equipments via analog input ports for real time simulation.

The fourth and final part of a PV simulator consists of the user interface the extra features and the representation manner of the results and graphs.

Taking into consideration all these facts becomes obvious that the presence of such a simulator into the research procedure imports a powerful tool in the experimental analysis of a complete solar system. Thus, with the use of the simulator, it is possible for the researcher in the laboratory comfort, independently from the climatic conditions, to examine thoroughly with repeated similar experiments the efficiency of any converter in a wide spectrum of radiation and temperature. Using the proposed simulator is also possible the combined operation between the photovoltaic generator's simulator and the control of a built-in M.P.P.T. algorithm in order to check a complete solar system including the converter. Also the proposed system has the ability of integrating any fully parametric M.P.P.T. algorithm, for the verification of its proper operation. It is also feasible the reception of measurements, their graphical representation and the efficiency comparison between various maximum power point tracking M.P.P.T. algorithms. The proposed simulator can support a large number of PV modules [18] with maximum values for different panels up to 500W, 50V and 10A.

## II. THE PROPOSED PV SIMULATION SYSTEM

The implemented system is presented in Fig. 2 and concists of:

i.   A switch-type DC power supply (SM120-25D of Delta Elektronika [19])
ii.  A personal computer
iii. A data acquisition card (DAQ card, PCI-6024E of National Instruments [20])
iv.  An Inductor
v.   An appropriate software (Labview of National Instruments [20])

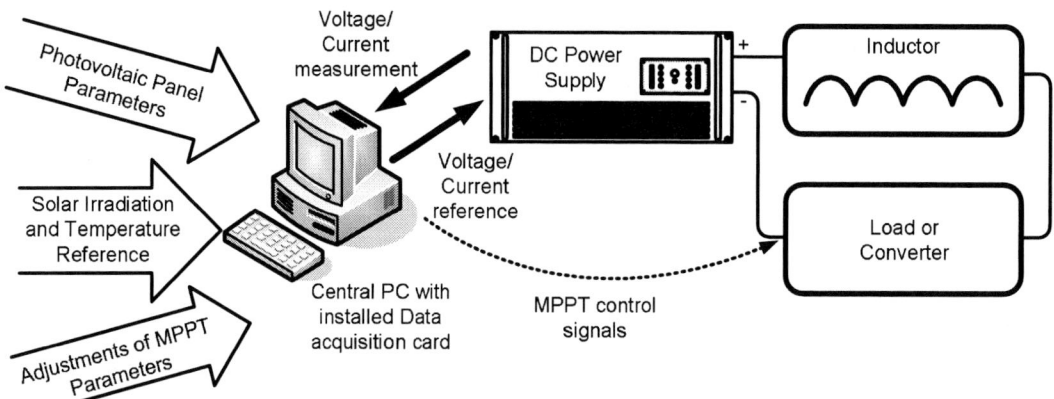

Fig. 2. General diagram of the photovoltaic generators simulator

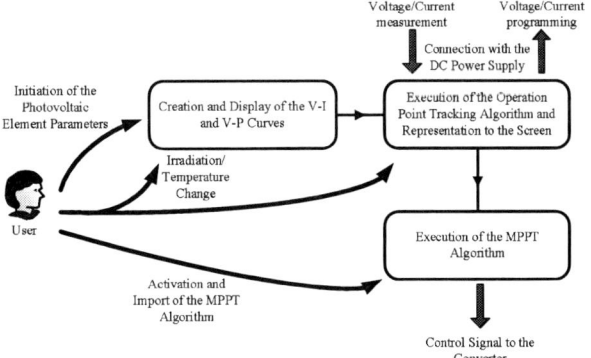

Fig. 3. Diagram of software structure.

The user starts the operation of the simulation system and provides the values of the essential parameters that describe the desirable under simulation photovoltaic generator. These parameters belong to the interpolation model of a photovoltaic generator. This model will be examined in the next paragraph. The general diagram of the software structure is presented in fig. 3.

Then, typical characteristic curves appear on the display for various insolation levels with the highlighted curve to be the active one. A built-in algorithm controls the power supply in order to force its output to approximate the values of the highlighted V-I curves. The voltage and current adjustment is accomplished through the DAQ card with the use of two analog signals that control the output of the power supply. Also the DAQ card is connected with the measurement transducers (output voltage and current) transferring the crucial feedback to the operation point tracking algorithm. Using the above layout and control, the software forces the power supply to behave as a photovoltaic generator.

Besides the implementation of the algorithm for tracking the operation point along the V-I curve, it is also feasible the activation of the second algorithm responsible for the maximum power point tracking (M.P.P.T). So the user can easily activate the built-in M.P.P.T. algorithm or integrate any other algorithm to the special made algorithm box of the program.

Finally, in order the power supply to behave as a current source a coil is connected in series with its output increasing the output impedance. Towards to fulfill all the above requirements a mathematical model must be presented that verifies the output characteristics of the PV generator.

### III. THE PV MATHEMATICAL MODEL

The most important issue in hardware simulation is the mathematical modeling of the PV generator. There are two methods in simulating the PV element.

The first [9,10] method is realized with the use of the parametric model and the assumption that all the parameters are known. The difficulties in using the parametric model are the need for the precise value for many natural parameters, the complexity in calculations and the implicit form of the function.

Therefore, the second method, based on the interpolation model [11,15], is used when the natural

parameters are unknown with a very high approximation to the real characteristic curve. In this case three points are necessary to calculate the curve:

1. The maximum power point ($V_{mp}$, $I_{mp}$).
2. The open circuit point ($V_{oc}$, 0).
3. The short circuit point (0, $I_{sc}$).

The parameters a, b are the temperature coefficients, given usually at the manufacturer's datasheet and they are different for each module. These coefficients customize the way the characteristic curve changes when the insolation or the temperature alters.

This method was adopted and used in this paper and is implemented by the Labview software.

### IV. SOFTWARE IMPLEMENTATION

The software implementation has multifunction role. It is responsible not only to provide a friendly user interface but also to execute the control algorithms.

First of all, the software has to satisfy the user communication demands. So the system provides parameter import tables, visual displays and graphs for output information.

In Fig. 4. is shown the system's control panel which is composed of:

a. Menu to import the special parameters that define the desirable PV generator. Requisite elements: (The maximum power point ($V_{mp}$, $I_{mp}$), the open circuit point ($V_{oc}$, 0), the short circuit point (0, $I_{sc}$)).

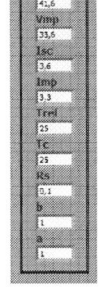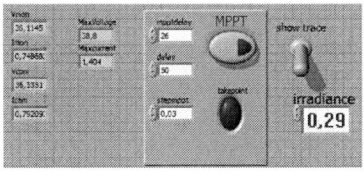

Fig.4. System control panel.

**1829**

b. Menu to import the desirable irradiation percentage (0 to 1 of 1000Watt/m$^2$).

c. Menu to import the temperature coefficients (a, b) taken from the datasheet.

d. Digital displays of the measured output voltage and current ($V_{mon}$, $I_{mon}$).

e. Digital displays of the interpolation model's calculated values ($V_{com}$, $I_{com}$).

f. Digital displays of the panel's maximum voltage and maximum current for the specific level of insolation.

g. Menu to import the MPPT algorithm's parameters and special control buttons.

Besides the control menus of the system, the developed software also includes representation displays of the characteristic curves V-P and V-I. In the same displays also appear, the current operation point and if it is selected its track over time along the curve as the load varies.

The software is responsible to control the whole system and executes two loos:

A. The operation point tracking algorithm responsible to control the power supply and force it to simulate the output characteristics of a PV generator.

B. The maximum power point tracking responsible to force the interconnected converter to absorb the maximum feasible solar energy.

### A. Operation Point Tracking Algorithm

The parametric model is the basis of the operation point algorithm. Using the Labview software the implemented loop routine ensures the fast and precise operation point tracking. The methodology presented in [12] requires extremely accurate voltage and current measurements in order the successive approximation points to converge to the real operation point. In addition, that methodology does not exclude possible V-I points located outside the characteristic curve. It is very possible among the successive V-I points some of them to have values greater than the edge values $I_{sc}$ and $V_{oc}$. For these reasons, in the proposed system is used a different approach in the operation point approximation procedure. The combined bisection methodology in regard to both the voltage and current axis is used. Due to the high efficiency of modern microcomputers the operation point is located fast and precisely. The structure of the implemented loop is presented in Fig. 5.

Fig. 6. shows an operation point tracking example when a resistor is connected at the output of the simulator. Starting the system the simulator's output stands in point (1) and the operation point tracking algorithm commences. With the use of the parametric model's equations [11,15] the measured values $V_{mon1} = 0$ and $I_{mon1} = 0$ are converted to the new values $V_{com1}$ and $I_{com1}$. This two values under the equations $V_{com1}$.

$$V_{prog1} = V_{mon1} + \frac{V_{com1} - V_{mon1}}{2}, \quad I_{prog1} = I_{mon1} + \frac{I_{com1} - I_{mon1}}{2} \text{ are}$$

converted to the new output values for the dc power supply. The next point (2) is the point in which is firstly met one of the conditions $V_{mon2} = V_{prog1} \quad I_{mon2} = I_{prog1}$.

In this particular point is the first one that is valid. The exact same procedure is executed to find the next point (3). The second condition is valid only for very small load resistances where current comes very close to the short circuit current.

The operation point tracking loop runs until the locking conditions $V_{mon} - V_{com} < V_{error}$ and $I_{mon} - I_{com} < I_{error}$ become valid. $V_{error}$ and $I_{error}$ are the maximum permissible errors that the current point is rated as the operation point. The maximum error, the distance between the real point and the operation point, is limited by the measurement and the power supply's programming accuracy.

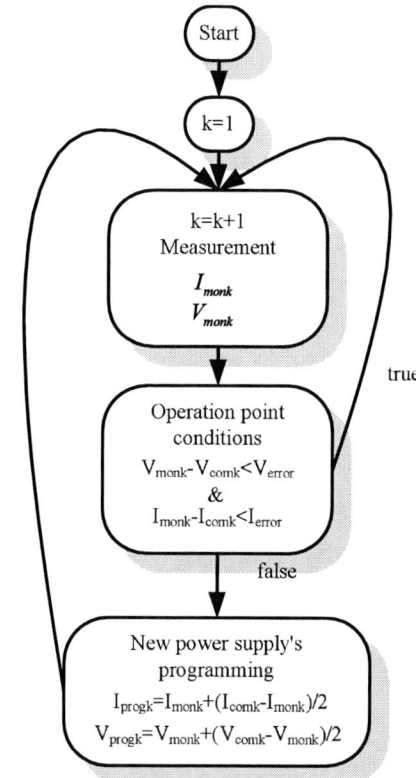

Fig. 5. The two algorithms' execution correlation.

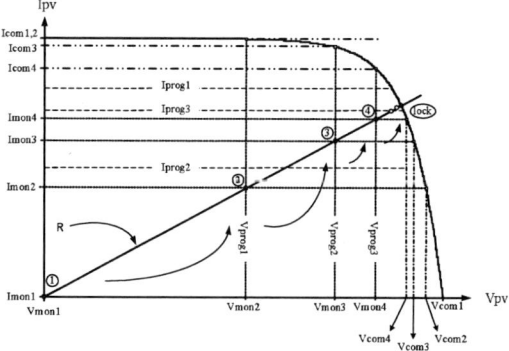

Fig. 6. Operation point tracking procedure of the proposed simulation system

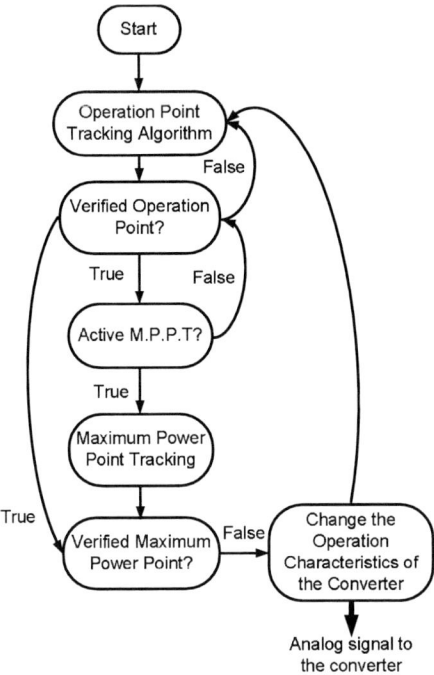

Fig. 7. Operation point tracking algorithm.

### B. Maximum Power Point Tracking Capability

Eminent place in the implementation of the proposed simulation system is the capability to integrate and control a large number of various M.P.P.T algorithms. The built-in algorithm [8] can be substituted by other and is feasible the manipulation of basic aspects of its operation as it is, the execution step length, the execution speed and the edge values.

The user researcher can add his own algorithm in a special area of the code and execute it in order to investigate its operation. The output data of the algorithm adjusts the operation of the interconnected converter by moving the operation point along the characteristic V-I curve until it coincides with the maximum power point. When this condition is fulfilled the loop locks and continues to execute in order to detect any possible change in the maximum power point location due to irradiance or temperature alternation. The combination between the execution of the M.P.P.T. and the execution of the operation point tracking is shown in Fig. 7.

## V. EXPERIMENTAL RESULTS

To verify the proper operation of the proposed simulator, various tests and experiments were conducted not only in static but also in dynamic conditions.

In order to investigate the static behaviour, two kinds of experimental procedures were realised. Firstly, only a variable resistor was connected to the system's output. It is worthy mentioning that all experiments were conducted under small load change between measurements. At these points the voltage and current had been stabilized at specific values. These values correspond with the intersection points between the PV and resistor

characteristics. In the figures 8-11 is shown the successful operation point tracking for two different resistor values. In Fig. 8 and 9 the resistor's value is small so the current is limited near the short circuit value. Fig. 10 and 11 show the operation point for bigger load resistance.

Moreover, the second test layout consists of the simulation system an interconnected buck converter. This test is aiming to verify the maximum power point tracking. The M.P.P.T. algorithm is integrated to software and controls the buck's operation parameters through an analog signal.

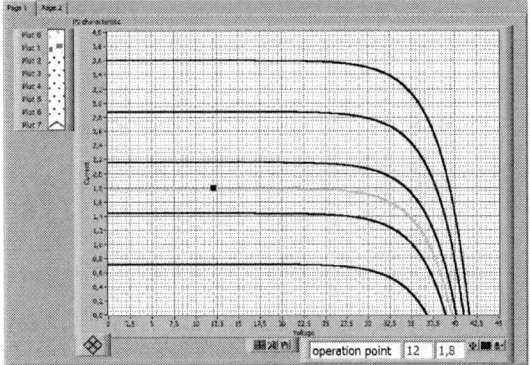

Fig. 8. Active output V-I characteristic (light colour) and operation point.

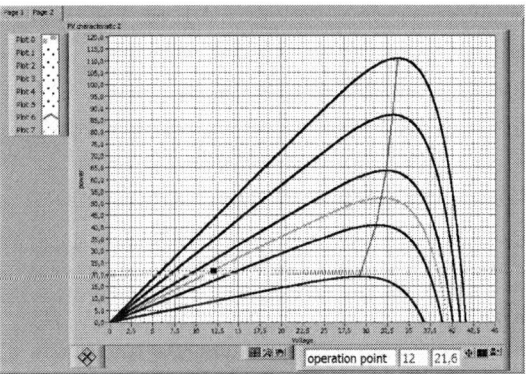

Fig. 9. Active output V-P characteristic (light colour) and operation point.

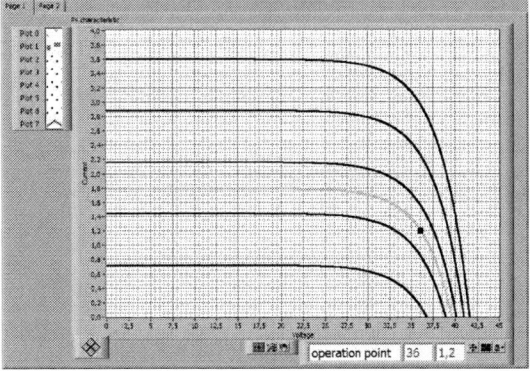

Fig. 10. Active output V-I characteristic (light colour) and operation point.

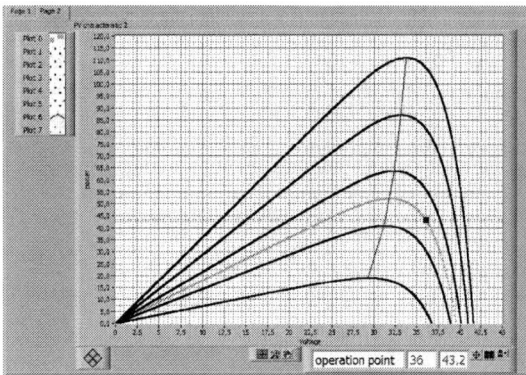

Fig. 11. Active output V-P characteristic (light colour) and operation point.

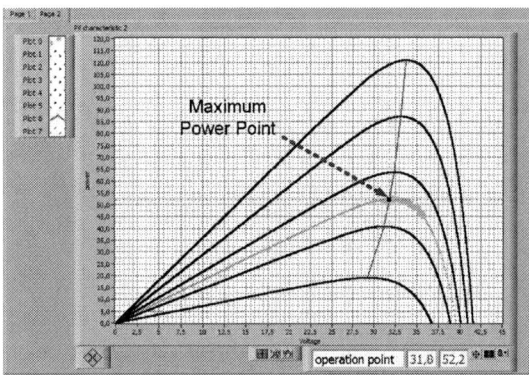

Fig. 12. Verified maximum power point tracking.

The changes in duty cycle are very small and infrequent so the system is able to stabilize on valid operation points between each M.P.P.T. algorithm's iteration. So, every new input in M.P.P.T. algorithm belongs to the PV characteristic curve. The come outs of this experiment are shown in Fig. 12.

Along the experimental procedure the simulator's dynamic behaviour was also tested. Because of the large output capacitors of the DC/DC power supply the response time is limited to 20ms. Besides this, the general use personal computer is not dedicated only to control the simulator but also must serve the operation system. So it slows down the whole system to 1msec maximum response time. In order to accelerate the system's execution the power supply should be replaced with a faster one and the control should be implemented by a dedicated microcontroller. These changes will improve the dynamic performance making the interconnection between the simulator and rapid changing load feasible.

## VI. CONCLUSION

In this paper, a laboratory simulator of a photovoltaic generator using a personal computer, a DAQ card and a power supply, was presented. This system was developed to examine the proper operation and investigate the efficiency of photovoltaic systems. In addition, it is feasible the activation of an M.P.P.T. algorithm, which in combination with the irradiation selection ability, can simulate a full photovoltaic system. So, not only the

operation of the PV panel but also the efficiency of the M.P.P.T. algorithm can be analyzed. With the new design and improvement that are under development at this time, aiming to improved dynamic behavior, the simulator will compose a complete tool for photovoltaic systems research.

## VII. REFERENCES

[1] Mitchell J.F.B, "The greenhouse Effect and Climate Change", February 1989, Internet Report, available: http://www.bom.gov.au/info/GrenhouseEffectAndClimateChange. pdf.

[2] Campbell C., Laherrere J., "The end of cheap oil", *Science American*, Vol. 278, March 1998, pp. 78-83

[3] Runci P.J., "Energy R&D in the European Union", *Pacific Northwest National Laboratory, U.S. Department of Energy*, Contract DE-AC06-76RLO 1830, May 1999, No PNNL-12218, Internet Report.

[4] Chen Kunlun, Zhao Zhengming, Yuan Liqiang, "Implementation of a stand-alone photovoltaic pumping system with maximum power point tracking", *5th International Conference on Electrical Machines and Systems (ICEMS 2001)*, Vol. 1, 18-20 August, 2001, pp. 612-615.

[5] Goldstein L., Mortensen J., Trickett D., "Grid-Connected Renewable - Electric Policies in the European Union", *National Renewable Energy Laboratory (NREL), U.S Department of Energy*, NREL/TP.620.26247, May 1999, Internet Report, available: http://www.nrel.gov/ docs/ fy99osti/26247.pdf.

[6] Salas V., Olías E., Barrado A., Lázaro A., "Review of the maximum power point tracking algorithms for stand-alone photovoltaic systems", *Solar Energy Materials & Solar Cells*, Vol. 90, No 11, pp. 1555-1578, 6 July 2006.

[7] Esram T., Chapman P.L., "Comparison of Photovoltaic Array Maximum Power Point Tracking Techniques", *IEEE Transaction on Energy Conversion*, Vol. 22, No 2, June 2007, pp. 439-449.

[8] Hussam Al-Atrash and Khalid Rustom, "Statistical Modeling of DSP-based Hill-Climbing MPPT Algorithms in Noisy Environments", *20th IEEE Applied Power Electronics Conference and Exposition (APEC'2005)*, Vol. 3, 6-10 March, 2005, pp. 1773-1777.

[9] Minwon Park, In-Keun Yu, "A Novel Real-Time Simulation Technique of Photovoltaic Generation Systems Using RTDS", *IEEE Transactions on Energy Conversion*, Vol. 19, No. 1, March 2004, pp. 164-169.

[10] N. Locci, F. Mocci, M. Tosi, "A programmable simulator of photovoltaic generators", *IEEE, 1986, PESC*, pp. 633-638.

[11] K. Khouzam, C. Ly, C. Khoon Koh, Poo Yong Ng, "Simulation and Real – Time Modelling of Space Photovoltaic Systems", *IEEE 1st World Conference on Photovoltaic Energy Conversion (WCPEC'1994)*, Vol. 2, Dec. 5-9, 1994, Hawaii, pp. 2038-2041.

[12] Byung-Hwan Jeong, Byung-Hee Kang, Gyu-Ha Choe, Heung-Geun Kim, "Electronic-Controlled Power Supply System for Solar Cell Characteristics", *Proceedings of the 4th IASTED Conference on European Power and Energy Systems (EuroPES'2004)*, June 28-30, 2004, Rhodes (Greece), paper on CD, Nr. 442-191.

[13] Easwarakhanthan T., Botin J., El-Slassi A. and Ravelet S. "Microcomputer-controlled simulator of a PV generator using a programmable voltage generator", *Solar Cells*, Vol. 17, No 2-3, pp. 383-390, 1986.

[14] Qingrong Zeng, Pinggang Song, Liuchen Chang, "A Photovoltaic Simulator Based on DC Chopper", *IEEE Conf. of Canadian Electrical and Computer Engineering (CCECE 2002)*, Vol. 1, 12-15 May, 2002, pp. 257-261.

[15] K.Khouzam, and K. Hoffman, "Real- Time Simulation of Photovoltaic Modules", *Solar Energy*, Vol. 56, No. 6, pp. 521-526, 1996.

[16] Jae-hyun Yoo, Jeok-seok Gho, Gyu-ha Choe, "Analysis and Control of PWM Converter with V-I Output Characteristics of Solar Cell", *IEEE Proc. of the International Symposium on Industrial Electronics (ISIE'2001)*, Vol. 2, 12-16 June, 2001, Pusan (Korea), pp. 1049-1054.

[17] Hiroshi Matsukawa, Koukichi Koshiishi, Hirotaka Koizumi, Kosuke Kurokawa, Masayasu Hamada, Liu Bo, "Dynamic evaluation of maximum power point tracking operation with PV array simulator", *International Photovoltaic Science and Engineering Conference (PVSEC 12)*, Vol. 75, Issues 3-4, February, 2003, Cheju Island (Korea), pp. 537-546.

[18] Kyritsis A.Ch., Kobougias J.C., Klimis D.S., Tatakis E.C., "Comparison between AC PV Modules Topologies for Decentralised Grid Connected Applications", CIGRE Symposium on Power Systems with dispersed generation, Athens (Greece), 13-16 April, 2005, paper on CD, Nr. 101.

[19] Delta Elektronika web site www.delta-elektronika.nl

[20] National Instruments web site www.ni.com

# Modeling and Optimal Sizing of Hybrid Renewable Energy System

Rachid Belfkira*, Cristian Nichita, Pascal Reghem, Georges Barakat[†]

GREAH, Groupe de Recherche en Electrotechnique et Automatique du Havre
University of Le Havre, 25 rue Philippe Lebon, BP 540
76058 Le Havre, France
Tel.: +33 / (0) – 232744331
Fax: +33 / (0) – 232744348
* e-mail: *rachid.belfkira@univ-lehavre.fr*
† e-mail: *georges.barakat@univ-lehavre.fr*

*Abstract—* **This paper presents a new methodology of sizing optimization of a stand-alone hybrid renewable energy system. The developed approach makes use of a deterministic algorithm to minimize the life cycle cost of the system while guaranteeing the availability of the energy. Firstly, the mathematical modeling of the principal elements of the hybrid wind/PV system is exposed showing the main sizing variables. Then, the deterministic algorithm is presented and implemented to minimize the objective function which is equal to the life cycle cost of the hybrid system and finally, the obtained results are exposed and discussed.**

*Keywords—* **Renewable energy systems, energy storage, power supply, modeling.**

## I. Introduction

Around two billion people world-wide do not have access to electricity services, of which the main share in rural areas in developing countries. Renewable energy resources are a favorable alternative for rural energy supply [1].

Renewable energy sources essentially have unpredictable random behaviors. However, some of them, like solar radiation and wind speed, have complementary profiles. Stand-alone hybrid power systems (Fig. 1) usually take advantage of this particular characteristic combining photovoltaic (PV) panels and wind turbines (WT).

Because of the intermittent solar irradiation and wind speed characteristics, which highly influence the resulting energy production, the major aspects in the design of PV and wind generator (WG) power generation systems are the reliable power supply of the consumer under varying atmospheric conditions and the corresponding total system cost. Then it is essential to select the number of PV modules, WGs and batteries, and their installation details such that power is uninterruptedly supplied to the load and simultaneously the minimum system cost is achieved [2].

The use of renewable energy technology to meet the energy demands has been steadily increasing over the years.

Several research tasks concerning the design and the sizing of the hybrid systems were carried out. In [3], based on the available hourly average data of wind speed,

insolation, and the power demand, the generation capacity is determined to best match the power demand by minimizing the difference between generation and load ($\Delta P$) over a 24-hour period. The objective function to be minimized is the sum of the annual cost of the capital over the life of the generating system and its annual maintenance cost. The iterative procedure is adopted for selecting the wind turbine size and the number of PV panels needed for a stand-alone system to meet a specific load.

An alternative methodology for the optimal sizing of stand-alone PV/WG systems has been proposed by Koutroulis et al. [2], which the purpose is to suggest, among a list of commercially available system devices, the optimal number and type of units ensuring that the 20-year round total system cost is minimized subject to the constraint that the load energy requirements are completely covered, resulting in zero load rejection. The 20-year round total system cost is equal to the sum of the respective components capital and maintenance costs. The decision variables included in the optimization process are the number and type of PV modules, WGs and battery chargers, the PV modules tilt angle, the installation height of the WGs and the battery type and nominal capacity. The minimization of the cost (objective) function is implemented employing a genetic algorithms approach.

In [4] the authors have developed the HOGA program (Hybrid Optimization by Genetic Algorithms) to calculate the optimal configuration of the hybrid PV-Diesel system. This optimal configuration is described very precisely: the number and the type of PV panels, the number and the type of batteries, the inverter power, the Diesel generator power, the optimal control strategy of the system with its parameters, the Total Net Present Value (cost of the investments plus the discounted present values of all future costs) of the system and finally, the number of running hours for the Diesel generator per year.

Chedid and Rahman have used linear programming techniques to determine the optimal sizes of the PV and WG power sources and the batteries by minimizing the system total cost function which consists of both initial cost and yearly operation and maintenance costs [5].

In this paper, the proposed optimization procedure is based on a dynamic evaluation of the wind and solar energetic potential based on statistical models of wind speed and solar radiation of the site of production. This

978-1-4244-1741-4/08/$25.00 ©2008 IEEE

dynamic evaluation of the energetic potential of the site permits the introduction of new constraints making the optimization procedure more flexible like the maximum acceptable time of energy unavailability and the minimum power level authorised regarding the power demand. Consequently, this approach results in a more realistic optimization.

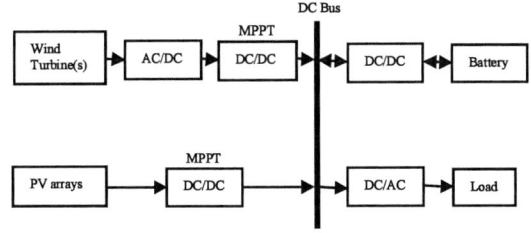

Fig.1. Block diagram of a hybrid WT/PV system

## II. HYBRID SYSTEM MODELING

### A. Wind Turbine Model

Using the wind speed at a reference height $h_r$ from the database, the velocity at a specific hub height for the location is estimated on an hourly basis throughout the specified period through the following expression [2]

$$v(t) = v_r(t)\left(\frac{h}{h_r}\right)^\gamma \qquad (1)$$

where:

$v$ is the wind speed at projected height $h$,

$v_r$ is wind speed at reference height $h_r$,

$\gamma$ is the power-law exponent (~1/7 for open land).

In function of this wind speed, the model used to calculate the output power, $P_{WT}(t)$ (W), generated by the wind turbine generator is as follows:

$$P_{WT}(t) = \begin{cases} av^3(t) - b.P_R & v_{ci} < v < v_r \\ P_R & v_r < v < v_{co} \\ 0 & otherwise \end{cases} \qquad (2)$$

where $a = P_r / (v_r^3 - v_{ci}^3)$, $b = v_{ci}^3 / (v_r^3 - v_{ci}^3)$, $P_r$ is the rated power, $v_{ci}$, $v_r$ and $v_{co}$ are respectively the cut-in, rated and cut-out wind speeds of the wind turbine.
Fig. 2 shows typical wind turbine characteristics.

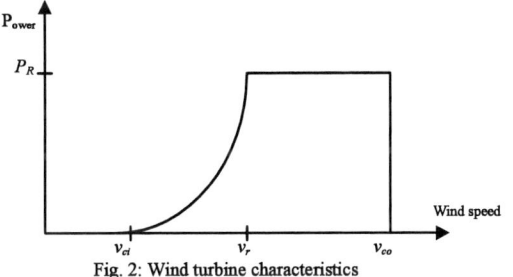

Fig. 2: Wind turbine characteristics

### B. PV Array Modeling

The output power from a PV panel can be calculated by an analytical model given by France Lasnier and Tony Gan Ang [6], which defines the current-voltage relationships based on the electrical characteristics of the PV panel. This model includes the effects of radiation level and panel temperature on the output power. With a maximum power point tracker (MPPT), the output power from a PV panel is given as:

$$\begin{cases} P_{PV} = V_{mpp} I_{mpp} \\ V_{mpp} = V_{mpp,ref} + \mu_{V,oc}\left(T_c - T_{c,ref}\right) \\ I_{mpp} = I_{mpp,ref} + I_{sc,ref}\left(G_T / G_{ref}\right) + \mu_{I,sc}\left(T_c - T_{c,ref}\right) \end{cases}$$

$$(3)$$

where $P_{PV}$ is the PV panel power (W) at the maximum power point at hour $t$, $V_{mpp}$ is the PV panel voltage at the maximum power point (V) at hour $t$, $V_{mpp,ref}$ is $V_{mpp}$ at reference operating conditions (V), $I_{mpp}$ is the PV panel current at the maximum power point (A) at hour $t$, $I_{mpp,ref}$ is $I_{mpp}$ at reference operating conditions (A), $I_{sc,ref}$ is the short circuit current at reference operating conditions (A), $E_T$ is the daily irradiance on a tilted surface (W/m²), $E_{ref}$ is the irradiance of 1000W/m² at reference operating conditions, $\mu_{V,oc}$ and $\mu_{I,sc}$ are the temperature coefficients for open circuit voltage (V/°C) and short circuit current (A/°C) respectively, $T_{c,ref}$ is the PV panel temperature of 25°C at reference operating conditions and $T_c(t)$ corresponds to the PV panel operating temperature (°C) at hour t and which can be expressed as follows [7]

$$T_c(t) = T_a(t) + \frac{NOCT - 20}{800}.E_T \qquad (4)$$

where $T_a(t)$ is the ambient temperature (°C) of the site under consideration at hour $t$ and $NOCT$ (Normal Operating Cell Temperature) is defined as the cell temperature when the PV panel operates under 800W/m² of solar irradiation and 20°C of ambient temperature, $NOCT$ is usually between 42°C and 46°C.

Most local observatories provide only solar irradiation data on a horizontal plane [8]. Thus, an estimate of the solar irradiation incident on any sloping surfaces, as analyzed by [9], is needed.

The PV panels are connected in series to form strings, where the number of panels to be connected in series $N_{PV,s}$ is determined by the selected DC bus voltage ($U_{Bus}$) as follows [10]

$$N_{PV,s} = \frac{U_{Bus}}{U_{PV,nom}} \qquad (5)$$

where $U_{PV,nom}$ is the nominal PV panel voltage. Then $N_{PV,s}$ is not subject to the optimization, whereas the number of parallel strings $N_{PV,p}$ is the design variable that needs optimization.

## C. Model of Battery

It is evident that the power generated by the hybrid system and the amount of energy stored are time dependent. So, the power input to the battery bank, is controlled by the equation

$$\Delta P(t) = P_{re}(t) - P_L(t) \qquad (6)$$

in which $P_{re}(t)$ is the total power produced by the renewable resources (PV panels and wind turbine(s)) at hour $t$ and $P_L(t) = P_{load}(t)/\eta_i$, where $P_{load}$ is the power demanded by the load.

For the charging process ($\Delta P(t)>0$) and discharging process ($\Delta P(t)<0$) of the battery bank, the state of charge ($SOC$) can be calculated from the following equation

$$SOC(t+1) = SOC(t) + \eta_{bat} \cdot \left( \frac{P_{re}(t) - P_L(t)}{U_{bus}} \right) \cdot \Delta t \qquad (7)$$

where $\eta_{bat}$ is equal to the round-trip efficiency in the charging process and is equal to the 100% in the discharging process [2], $U_{bus}$ is the DC bus voltage and $\Delta t$ is the hourly time step is set equal to 1hour.

For longevity of the battery bank, the maximum charging rate, $SOC_{max}$, is given as the upper limit, where $SOC_{max}$ is equal to the total nominal capacity of the battery bank, $C_n$, which is related to the total number of batteries, $N_{BAT}$, the number of batteries connected in series, $N_{BAT,s}$ and the nominal capacity of each battery, $C_B$ (Ah), as follows [11]

$$C_n = \frac{N_{BAT}}{N_{BAT,s}} \cdot C_B \qquad (8)$$

and the lower limit that the state of charge of the battery bank does not have exceeded at the time of discharging is $SOC_{min}$ which may be expressed as follows

$$SOC_{min} = (1 - DOD) \cdot SOC_{max} \qquad (9)$$

where DOD is the Depth of Discharge of battery.

The batteries are connected in series to give the desired nominal DC operating voltage and are connected in parallel to yield a desired Ah system storage capacity. Then, the number of batteries connected in series depends on the DC bus voltage and the nominal voltage of each individual battery $U_{Bat,nom}$ and it is calculated as follows

$$N_{BAT,s} = \frac{U_{Bus}}{U_{Bat,nom}} \qquad (10)$$

The number of batteries to be connected in series is therefore not subject to the optimization but is a straightforward calculation, whereas the number of parallel battery strings, each consisting of $N_{BAT,s}$ batteries connected in series, is a design variable that needs optimization.

## III. Deterministic Algorithm and Developed Methodology

The sizing optimization process consists in solving problems of the form [12]:

$$\begin{cases} \min_x f_i(x) & i \in [1,n] \\ h_k(x) = 0 & k \in [1,p] \\ g_j(x) >= 0 & j \in [1,q] \\ x_l \le x \le x_u \end{cases} \qquad (11)$$

where $f_i \in R^n$ are the objective functions and $h_k \in R^p$, $g_j \in R^q$ are respectively the equality and the inequality constraints.

One of the major steps of the optimization process consists in the minimization of the objective functions. The optimization algorithms are generally divided into two groups: deterministic and stochastic. Many researchers have recently proved that the DIRECT algorithm is an effective deterministic algorithm to find the global optimum of the problem (11).

Developed by Jones et al. [13] and acronym for DIviding RECTangles, the DIRECT algorithm is a deterministic global optimization technique that is used to find the minimum of a Lipschitz continuous function without knowing the Lipschitz constant. The objective function and constraints must be Lipschitz-continuous in the research space $\mathfrak{I}$, satisfying

$$|f(x_1) - f(x_2)| \le L \|x_1 - x_2\| \quad \forall \, x_1, x_2 \in \mathfrak{I} \qquad (12)$$

This assumption means that the rates-of-change of the objective function and constraints are bounded. Traditionally, when this assumption (12) is satisfied, the global optimization problem was solved by the Lipschitz optimization method, which had been considered as a practical and deterministic approach to many science and engineering problems for several decades.

DIRECT evolved from the one-dimensional Piyavskii-Shubert algorithm and was further extended from one dimension to multiple dimensions by adopting a center-sampling strategy. Its corresponding 1-D description contrasted with Piyavskii-Shubert's algorithm can be found in [14]. Here, only the multidimensional DIRECT algorithm, which is of more interest for our application, is described. DIRECT's behavior in multiple dimensions can be viewed as taking steps in potentially optimal directions within the entire design space. The potentially optimal directions are determined through evaluating the objective function at center points of the subdivided boxes. The multivariate DIRECT algorithm can be described by the following steps [14]

1) Normalize the search space $\mathfrak{I}$ to unit hypercube.
2) Sample the center point $c_1$ of the hypercube; Evaluate $f(c_1)$. Set $f_{min} = f(c_1)$, $m = 1$ (evaluation counter), and $t = 0$ (iteration counter).
3) Identify the set $S$ of potentially optimal boxes.
4) Select any box $j \in S$.
5) Divide the box $j$ as follows:
   a) Identify the set $I$ of dimensions with the maximum side length $\varepsilon$. Let $\delta$ equal one-third of this maximum side length ($\delta = 1/3\ \varepsilon$).
   b) Sample the function at the points $c \pm \delta e_i$, for all $i \in I$, where $c$ is the center of the box and $e_i$ is the $i^{th}$ unit vector.
   c) Divide the box $j$ containing $c$ into thirds along the dimensions in $I$, starting with the dimension with the lowest value of $w_i = min\{f(c+\delta e_i),\ f(c-\delta e_i)\}$, and continuing to the dimension with the highest $w_i$. Update $f_{min}$, $x_{min}$ and $m$.

**1836**

6) Set $S = S - \{j\}$. If $S = \varnothing$ GO TO Step 4.
7) Set $t = t + 1$. If iteration limit or evaluation limit has been reached, stop. Otherwise, GO TO Step 3.

This algorithm has been applied to optimize a hybrid wind/PV system. In the developed method, the DIRECT optimal sizing methodology outputs the optimum numbers and the types of the components of the hybrid wind/PV system, ensuring that the system total cost is minimized subject to the constraint that the load energy demand is completely covered.

The optimization procedure is achieved by minimizing the total cost function consisting of the sum of the individual system devices capital, the 20-year round maintenance costs and the installation costs

$$
\begin{aligned}
F_C &\left( N_{PV,p}, N_{WT}, N_{BAT,p} \right) \\
&= \sum_{i=1}^{n_{PV}} N_{PV}^i \cdot \left( C_{PV}^i + 20.M_{PV}^i + C_{I,PV}^i \right) \\
&+ \sum_{j=1}^{n_{WT}} N_{WT}^j \cdot \left( C_{WT}^j + 20.M_{WT}^j + C_{I,WT}^j \right. \\
&\qquad \left. + C_h^j + 20.C_{hm}^j + C_{I,h}^j \right) \\
&+ \sum_{k=1}^{n_{BAT}} N_{BAT}^k \cdot \left( C_{BAT}^k + C_{I,BAT}^k \right. \\
&\qquad + y_{BAT}^k \left( C_{BAT}^k + C_{I,BAT}^k \right) \\
&\qquad \left. + \left( 20 - y_{BAT}^k - 1 \right).M_{BAT}^k \right)
\end{aligned}
\tag{13}
$$

where $N_{PV,p}$, $N_{WT}$ and $N_{BAT,p}$ represent the sizing variables, where $N_{PV,p}$ is the total number of parallel PV strings, $N_{WT}$ is the total number of wind turbines and $N_{BAT,p}$ is the total number of parallel battery strings, $n_{PV}$, $n_{WT}$, $n_{BAT}$ are the total numbers of PV panel types, wind turbine types and battery units types, respectively, and $C_{PV}^i$, $C_{WT}^j$, $C_{BAT}^k$ are the corresponding capital costs (€), $M_{PV}^i$, $M_{WT}^j$, $M_{BAT}^k$ are the corresponding maintenance costs per year (€/year) and $C_{I,PV}^i$, $C_{I,WT}^j$ and $C_{I,BAT}^k$ are the corresponding installation costs (€). $C_h^j$ is the WT tower capital cost (€), $C_{hm}^j$ is the WT tower maintenance cost per year (€/year), $C_{I,h}^j$ is the WT tower installation cost (€) and $y_{BAT}^k$ is the expected number of battery replacements during the 20-year system operation, because of limited battery lifetime. The costs of converters and of other components are included in the installation cost.
$N_{PV}^i = N_{PV,p}^i \times N_{PV,s}^i$ is the total number of PV panels of type $i$, and $N_{BAT}^k = N_{BAT,p}^k \times N_{BAT,s}^k$ is the total number of batteries of type $k$ in the battery bank.

The minimization of the objective function is subject to the constraints that the power produced by the system is equal to the power demanded by the load and the state of charge of the battery bank is limited between $SOC_{min}$ and $SOC_{max}$ as follows

$$
\begin{cases}
P_P(t) = P_L(t) \\
SOC_{min} \leq SOC(t) \leq SOC_{max}
\end{cases}
\tag{14}
$$

where $P_P(t)$ is the power produced by the system and it is calculated as follows

$$
P_P(t) = P_{re}(t) - \Delta P(t)
\tag{15}
$$

where $P_{re}(t)$ is the power produced by the renewable resources as follows

$$
P_{re}(t) = \sum_{i=1}^{n_{PV}} N_{PV}^i . P_{PV}^i(t) + \sum_{j=1}^{n_{WT}} N_{WT}^j . P_{WT}^j(t)
\tag{16}
$$

and $\Delta P(t) > 0$ during the charging process of the battery and $\Delta P(t) < 0$ in the discharging process as calculated in the eq. (6).

Additional constraints to be imposed are

$$
\begin{aligned}
1 &\leq N_{PV,p}^i \leq N_{PV,p\,max}^i \\
1 &\leq N_{WT}^j \leq N_{WT\,max}^j \\
1 &\leq N_{BAT,p}^k \leq N_{BAT,p\,max}^k
\end{aligned}
\tag{17}
$$

where $N_{PV,pmax}^i$, $N_{WTmax}^j$ and $N_{BAT,pmax}^k$ were calculated according to the nominal power of PV panel, wind turbine and nominal capacity of battery, respectively, and the peak of the load demand.

## IV. OPTIMIZATION RESULTS AND DISCUSSION

The optimization methodology developed above was applied to sizing a hybrid energy system supplying a variable load. In fig. 3, the hourly power demand during a day is presented. This power reaches the maximum values between 13 h and 15 h and between 21 h and 23 h in the day; this is due to the utilization of the household electrical appliances in these periods.

For the site of Fecamp in the region of "Haute-Normandie", in France, where the hybrid energy system is assumed to be installed, a long-term data of wind speed and ambient temperature were recorded for every hour of the day during the period of six months, are used for the calculation of the power produced by the hybrid system and are plotted in fig. 4. The wind speed was measured at a 40 meters height which is considered as the reference height for the site (cf. eq. (1)).

In this example, two types of each component of the hybrid wind/PV system have been used. The specifications and the related capital, maintenance and installation costs of each component type, which are input to the optimal sizing procedure, are listed in Tables I-III. The maintenance cost of each unit per year and the installation cost of each component have been set at 1% and 10% respectively of the corresponding capital cost.

The serial connection numbers of the two types 1 and 2 of the PV arrays and of the batteries which are determined by the operating voltage of the system which is chosen to be equal to a standard value of 48 V, take respectively the values: $N_{PV,S}^1 = 2$, $N_{PV,S}^2 = 3$, $N_{BAT,S}^1 = 4$ and $N_{BAT,S}^2 = 4$. The expected battery lifetime has been set at 3 years with proper maintenance resulting in $y_{BAT}^k = 6$. Since the tower heights of wind turbines affect the results significantly, 30 meter high tower is chosen.

Using all these data and parameters, the minimization of the system total cost is achieved by selecting an appropriate system configuration. The optimal sizing results, consisting of both the device types and their number, are shown in table IV. From these results, one can deduce that the rate of penetration of the wind power

**1837**

is higher than that of the PV power; this is due to the highly speed of the wind of the site of Fecamp compared to the solar radiation.

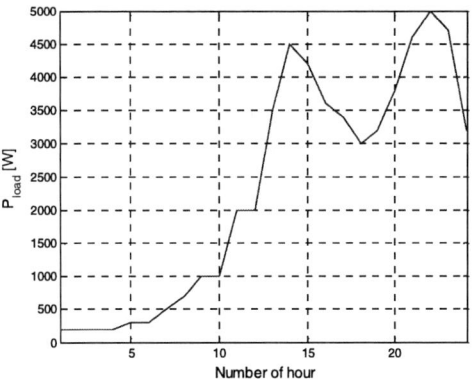

Fig. 3. Hourly demand power in a day

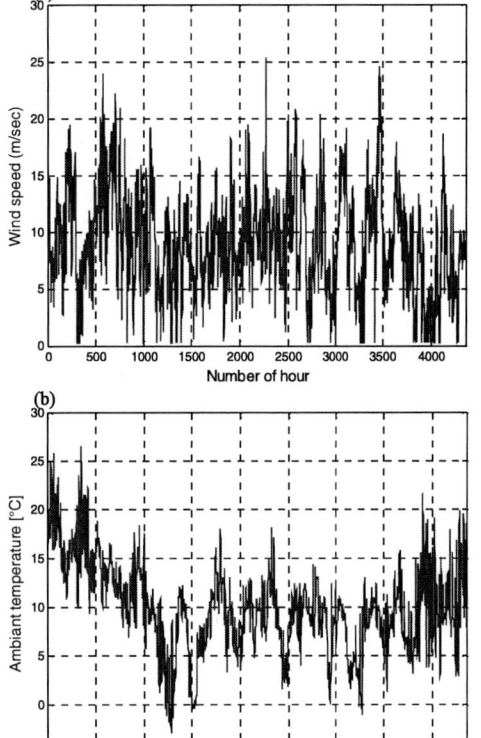

Fig. 4. Hourly mean values during a period of seven months of meteorological conditions: (a) wind speed and (b) ambient temperature

Fig. 5 presents the variation of the system total cost (fitness function) during the optimization procedure. It can be noted that a near optimal solution was derived during the early stages of the function evaluations.

Fig. 6 shows the state of charge (SOC) of the battery bank for the obtained configuration. One can deduce that

the SOC reaches the lower limit of discharge nearly between 300 h and 500 h and between 860 h and 2100 h, this is due to the low power produced by the renewable resources in these periods. Also, one can verify that the state of charge of the battery bank can never exceed the permissible maximum value, $SOC_{max}$ (100% of SOC) and can never be below the permissible minimum value, $SOC_{min}$ (20% of SOC).

TABLE I
PHOTOVOLTAIC PANELS SPECIFICATIONS

| Type | 1 | 2 |
|---|---|---|
| $V_{oc}$ (V) | 32.6 | 21 |
| $I_{sc}$ (A) | 7.87 | 7.22 |
| $V_{max}$ (V) | 25.9 | 17 |
| $I_{max}$ (A) | 6.95 | 6.47 |
| NCOT (°C) | 45.9 | 43 |
| Capital cost (€) | 603 | 519.14 |
| Installation cost (€) | 60.3 | 51.9 |
| Maintenance cost per year (€/year) | 6.03 | 5.19 |

TABLE II
WIND TURBINES SPECIFICATIONS

| Type | 1 | 2 |
|---|---|---|
| Power rating (W) | 10000 | 7500 |
| $v_r$ (m/s) | 13.8 | 13.8 |
| $v_{ci}$ (m/s) | 3.1 | 3.1 |
| $v_{co}$ (m/s) | 25 | 25 |
| Capital cost (€) | 20682 | 16978 |
| Installation cost (€) | 2068.2 | 1697.8 |
| Maintenance cost per year (€/year) | 206.82 | 169.78 |
| Tower capital cost (€) | 741 | 741 |
| Tower installation cost (€) | 7.41 | 7.41 |
| Tower maintenance cost per year (€/year) | 74.1 | 74.1 |

TABLE III
BATTERIES SPECIFICATIONS

| Type | 1 | 2 |
|---|---|---|
| Nominal capacity (Ah) | 100 | 230 |
| Voltage (V) | 12 | 12 |
| DOD (%) | 80 | 80 |
| Efficiency (%) | 80 | 80 |
| Capital cost (€) | 126 | 264 |
| Installation cost (€) | 12.6 | 26.4 |
| Maintenance cost per year (€/year) | 1.26 | 2.64 |

TABLE IV
OPTIMAL SIZING RESULTS

| Type | 1 | 2 |
|---|---|---|
| $N_{PV,p}$ | 1 | 0 |
| $N_{WT}$ | 1 | 0 |
| $N_{BAT,p}$ | 1 | 1 |
| Cost (€) | 41242 | |

1838

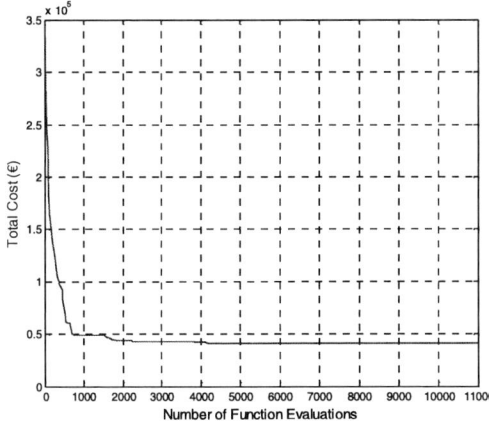

Fig. 5. The system total cost during the DIRECT optimization

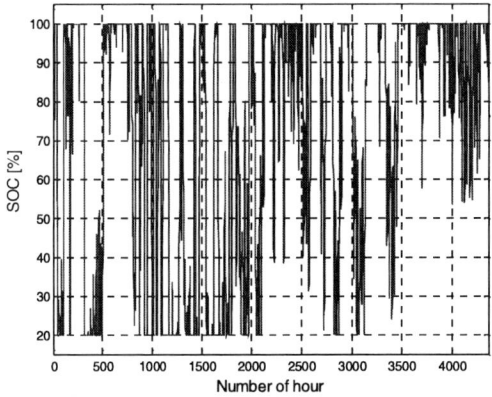

Fig. 6. Hourly variation of SOC of battery bank

## V. CONCLUSION

In this paper, a methodology of sizing a stand-alone hybrid wind/PV system using the DIRECT algorithm has been explained. This developed methodology is based on the use of the collection of six months data of the wind speed and the ambient temperature on one hand and on the other hand on the estimation of the solar irradiation for the site under consideration.

The optimum numbers of wind turbines, PV panels and batteries depend on the particular site, load profile and the specifications and the related cost of each component of the hybrid system.

## REFERENCES

[1] T. Gul, "Integrated analysis of hybrid systems for rural electrification in developing countries," *TRITA-LWR Master Thesis*, Stockholm 2004.

[2] E. Koutroulis, D. Kolokotsa, A. Potirakis, K. Kalaitzakis, "Methodology for optimal sizing of stand-alone photovoltaic/wind-generator systems using genetic algorithms," *Solar Energy*, 2006.

[3] W. D. Kellog, M. H. Nehir, G. Venkataramanan, V. Gerez, "Generation unit sizing and cost analysis for stand-alone wind, photovoltaic, and hybrid wind/PV systems," *IEEE Trans. on Energy Conversion*, vol. 13, no. 1, March 1998.

[4] R. Dufo-Lopez, J. L. Bernal-Agustin, "Design and control strategies of PV-Diesel systems using genetic algorithms," *Solar Energy*, vol. 79, pp. 33-46, 2005.

[5] R. Chedid, S. Rahman, "Unit sizing and control for hybrid wind-solar power systems," *IEEE Trans. on Energy Conversion*, vol. 12, no. 1, pp. 79-85, 1997.

[6] F. Lasnier, T. G. Ang, *Photovoltaic engineering handbook*. Bristol, England: A. Hilger, 1990.

[7] T. Markvar, *Solar Electricity*, 2nd ed, J. Wiley & Sons, 2000.

[8] H. Yang, L. Lu, W. Zhou, "A novel optimization sizing model for hybrid solar-wind power generation system," *Solar Energy*, vol. 81, pp 76-84, 2007.

[9] J. Bernard, *Energie solaire: calculs et optimisation*, Ellipses-Paris, 2004, pp. 53-93.

[10] G. Seeling-Hochmuth, "Optimisation of hybrid energy systems sizing and operation control," *Dissertation in Candidacy for the Degree of Dr.-Ing*, University of Kassel, October 1998.

[11] B. S. Borowy, Z. M. Salameh, "Methodology for optimally sizing the combination of a battery bank and PV array in a Wind/PV hybrid system," *IEEE Trans. on Energy Conversion*, vol. 11, no. 2, June 1996.

[12] J. Azzouzi, R. Belfkira, N. Abdel-Karim, G. Barakat, B. Dakyo "Design optimization of an axial flux PM synchronous machine: comparison between DIRECT method and GAs method," *EPE-PEMC*, August 30–September 1, 2006.

[13] D. R. Jones, C. C. Perttunen, B. E. Stuckman, "Lipschitzian optimization without the Lipschitz constant," *J. Optim. Theory Appl*, vol. 79, pp. 157–181, 1993.

[14] M. Björkman, K. Holmström, "Global optimization using the DIRECT algorithm in Matlab," *AMO - Advanced Modeling and Optimization*, vol. 1 no. 2, 1999.

# Photovoltaic System MPPTracker Investigation and Implementation using DSP engine and Buck-Boost DC-DC converter

Dimosthenis Peftitsis, Georgios Adamidis, Panagiotis Bakas, Anastasios Balouktsis

Democritus Univ. of Thrace/Dep. of Electr. and Computer Engineering, Xanthi, Greece, e-mail: dp2253@ee.duth.gr
Democritus Univ. of Thrace/Dep. of Electr. and Computer Engineering, Xanthi, Greece, e-mail: adamidis@ee.duth.gr

*Abstract*— In this paper, solar radiation simulation using clearness index $k_t$, hour of day $\omega_s$ and day of year n is studied. Daily distributions of solar radiation are presented for various clearness indexes $k_t$. Additionally, taking into account the studies about solar radiation, a photovoltaic array system and a DC/DC buck – boost converter are studied. Simulation of the whole system is presented focusing on DC/DC converter's control strategy so that the system operates in maximum power point (MPP) and converter's output voltage remains constant. Incremental conductance algorithm is used for maximum power point tracker (MPPT) implementation. A simplest method for controlling duty cycle D and photovoltaic array's voltage by using a new variable $d = \dfrac{D}{1-D}$ is proposed. Simulation results are shown and analyzed. For the implementation of this system, the dsPIC30F2010 Microchip's microcontroller was programmed to provide pulses to the semiconductor element of the DC-DC converter in order to track the Maximum Power Point (MPP) of the photovoltaic array.

*Keywords*—MPPT, Solar Cell System, Converter Control, DC/DC Converter, Incremental conductance algorithm, DSP- Microcontroller

## I. INTRODUCTION

Recent years, due to big demand of electrical energy, extended research in electricity production from solar energy using Photovoltaics (PV) has been done. Basic advantage of these energy sources is the abundance of solar radiation in nature and environmental friendly way of electricity production. Although the high cost of PV panels and their low efficiency, big PV parks have been installed in recent years around the world. Research, also has been done in increasing PV's efficiency and consequently maximizing output power. Furthermore, many papers have been published for PV's coupling with varied loads and PV's grid connection, using Power Electronics devices such as DC/DC converters and DC/AC single or three phase inverters [1], [2], [3], [4], [5]. For PV's output power maximization many algorithms have been developed and improved [13], [14], [15], such as Incremental Conductance, Perturbation and Observation (P&O), Parasitic Capacitance and Constant Voltage and Current algorithm. The most usual of these algorithms above is the Incremental Conductance one. According to this method maximum power points are detected by comparing for each step the derivative of conductance with the instant conductance. Derivative of conductance and instant one can be calculated by sensing instant PV voltage and current values and using previous.

## II. DISTRIBUTION OF SOLAR RADIATION

### A. Theoretical Study – Mathematical Model

Clearness index $k_t$, hour angle $\omega_s$ and air mass m constitute basic parameters for solar radiation study and simulation. Air mass, m, will not be taken into account in this paper analysis.

The clearness index probability density distribution can be constructed as a mixture [12] of two normal distributions, as shown below:

$$f(k_t) = \tau \cdot f_1(k_t) + (1-n) \cdot f_2(k_t) \qquad (1)$$

Where $f_1(k_t)$ and $f_2(k_t)$ are density functions of normal random variables as shown in (2) with σ1, σ2 and μ1, μ2 variations and mean values respectively.

$$f_i(k_t) = \left(2 \cdot \pi \cdot \sigma_i^2\right)^{-0.5} \cdot \exp\left(\frac{-0.5 \cdot (k_t - \mu_i)^2}{\sigma_i^2}\right), \quad i = 1,2 \qquad (2)$$

"Mixing Factor" τ is defined as:

$$\tau = \frac{\left(\overline{\mu_2} - \overline{k_t}\right)}{\left(\overline{\mu_2} - \mu_1\right)} \qquad (3)$$

Where $\mu_1$ and $\mu_2$ are the mean values of normal distributions $f_1$ and $f_2$ respectively and $\overline{k_t}$ is the mean value of clearness index. The range of τ values is 0<τ<1.

Clearness index $k_t$ is defined as the ratio of solar radiation G to extraterrestrial radiation $G_{ref}$ as shown in (4).

$$k_t = \frac{G}{G_{ref}} \qquad (4)$$

Equation (5) gives the extraterrestrial solar radiation:

$$G_{ref} = G_{SC} \cdot \left[ 1 + 0.033 \cdot \cos\left( \frac{360 \cdot n}{365} \right) \right] \cdot$$
$$\cdot (\cos\phi \cdot \cos\delta \cdot \cos\omega + \sin\phi \cdot \sin\delta) \qquad (5)$$

Where:

$G_{SC}$ = 1367 W/m$^2$ is the solar constant

$\omega$ is the hour angle

$\varphi$ is the latitude

n is the day of the year (1<n<365)

$\delta$ is given by (6)

$$\delta = 23.45 \cdot \sin\left( 360 \cdot \frac{284 + n}{365} \right) \qquad (6)$$

*B. Simulation Results for Solar Radiation*

Taking into account specific conditions for simulation ($\varphi$=41.08° and n=284, which refer to Xanthi, Greece and on October 11th respectively) and using Matlab/Simulink, figures 1 to 3 presents daily distribution of solar radiation for three different values of clearness index 0.4, 0.65 and 0.8 correspondingly during a day.

These three figures present an "extreme" case of solar radiation distribution. Thus, rapid changes are occurred in next figures.

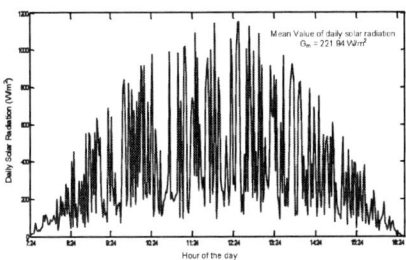

Fig. 1. Daily solar radiation for $k_t$=0.4.

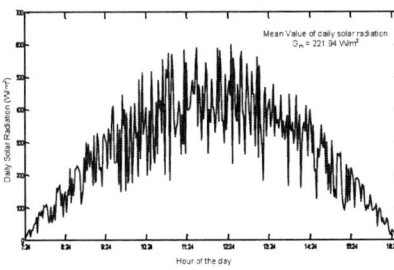

Fig. 2. Daily solar radiation for $k_t$=0.65.

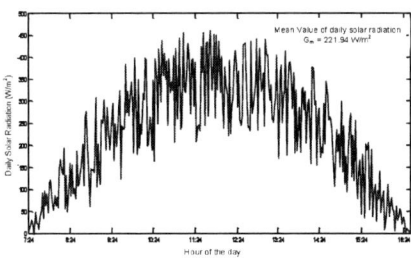

Fig. 3. Daily solar radiation for $k_t$=0.8.

## III. MATHEMATICAL MODEL OF PHOTOVOLTAIC

The simplest equivalent circuit of PV cells consists of a current source in parallel with a diode as shown in Fig. 4.

Photo current $I_{ph}$ is directly proportional to solar radiation G. Temperature T and photo current $I_{ph}$ have a linear relationship according to equation (7), where $I_{ph(Tref)}$ is the photo current which corresponds to reference temperature $T_{ref}$, and is equal to (8) $K_o$ is a constant given by (9) and T is the temperature. In equation (8) and (9) $G_{ref}$ is the nominal radiation given by PV's constructor $I_{SC}$ is the short circuit current.

$$I_{ph} = I_{ph(T_{ref})} \cdot \left( 1 + K_o \cdot \left( T - T_{ref} \right) \right) \qquad (7)$$

$$I_{ph(Tref)} = \frac{G}{G_{ref}} \cdot I_{SC(Tref)} \qquad (8)$$

$$K_o = \frac{I_{SC(T)} - I_{SC(Tref)}}{T - T_{ref}} \qquad (9)$$

Diode's current is given by (10), where V and I are PV's output voltage and current correspondingly, $I_o$ is saturation current of diode, $V_T$ thermal voltage of it and $R_S$ is in series resistance.

$$I_D = I_o \cdot \left[ \exp\left( \frac{V + I \cdot R_S}{V_T} \right) - 1 \right] \qquad (10)$$

Current $I_{SH}$ through shunt resistance $R_{SH}$ according to Ohm's law is equal to:

$$I_{SH} = \frac{V + I \cdot R_S}{R_{SH}} \qquad (11)$$

Taking into account equations (7) – (11) and applying Kirchhoff's current law, it results the I –V characteristic (12) for PV cell:

$$I = I_{ph} - I_o \cdot \left[ \exp\left( \frac{V + I \cdot R_S}{V_T} \right) - 1 \right] - \frac{V + I \cdot R_S}{R_{SH}}$$

(12)

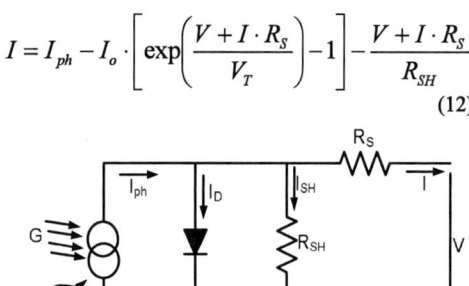

Fig. 4. Equivalent circuit of PV.

Analyzing equation (12) for s PV panels in series and p in parallel and ignoring current through shunt resistance, equation (13) gives the general I – V characteristic for PVs. Equation (14) gives the output power of PV cells.

$$I = p \cdot I_{ph} - p \cdot I_o \cdot \left[ \exp\left( \frac{V + I \cdot \left( \dfrac{s}{p} \right) \cdot R_S}{s \cdot V_T} \right) - 1 \right]$$

(13)

$$P = V \cdot I$$

(14)

## IV. PROPOSED CONTROL METHOD

Incremental Conductance algorithm is based on differentiation of PV power to its voltage and on condition of zero slope of P – V curve in maximum power point (MPP). Especially, differencing PV power and replacing power with equation (14), arises equation (15).

$$\frac{dP}{dV} = \frac{d(V \cdot I)}{dV} = I \cdot \frac{dV}{dV} + V \cdot \frac{dI}{dV} = I + V \cdot \frac{dI}{dV}$$

(15)

By using equation (15) and taking into account the basic condition of zero slope of P –V curve equation (16) is deduced.

$$\frac{dP}{dV} = 0 \Rightarrow I + V \cdot \frac{dI}{dV} = 0 \Rightarrow -\frac{I}{V} = \frac{dI}{dV}$$

(16)

Where –I/V represents the opposite of instant conductance of PV cells and dI/dV the incremental conductance. According to (16) these two quantities must be equal in MPP. Moreover, in the right of MPP is dI/dV<–I/V, thus a reduction in PV's voltage is essential to achieve MPP. Similarly, in the left of MPP is dI/dV>–

I/V, thus an increase in PV's voltage is essential to achieve MPP. These changes in PV's voltage may be done by coupling a DC/DC converter to PV and controlling properly its duty cycle, D. Most common used DC/DC converters in MPPT are the buck and boost, due to easy way of duty cycle control.

In this paper is presented a tracker for MPP using a buck – boost DC/DC converter (fig. 5) which is able to operate in a wide range of output voltages and different loads demands. Equation (17) gives the basic relationship between input and output voltage of this converter.

$$V_{out} = \frac{D}{1-D} \cdot V_{in}$$

(17)

As shown of (17), duty cycle is not support directly a linear relationship between $V_{out}$ and $V_{in}$. This means that a linear change in duty cycle D, will not imply for example a linear change of $V_{in}$ for keeping $V_{out}$ constant. Thus, an alternative way of PV's voltage control for achieving MPP is proposed. This new control scheme makes use of a new variable d, which is defined as shown in equation (18).

$$d = \frac{D}{1-D}$$

(18)

Figure 5 presents relationship between duty cycle D and new variable d for a constant change in D equal to $\Delta D = 0.01$. This relationship tends to be linear for duty cycles smaller than 0.5, but for values bigger than 0.5 is occurred a clearly non – linear one. By assuming that, controlling input voltage in DC/DC converter for keeping constant the output, by changing D, may drive the system to oscillations around maximum power point. Block diagram for control operation is shown in figure 6.

Flow chart in figure 7 shows the whole operation of Incremental Conductance algorithm. In each algorithm cycle, d is calculated using the instant value of duty cycle, D and after this d is changed properly. Finally, new value of D is calculated again using the new value of ratio d and switch is driven.

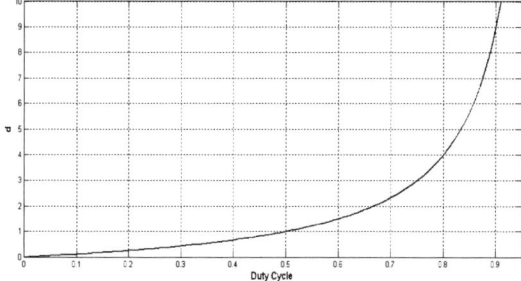

Fig. 5. Duty Cycle D versus proposed variable d.

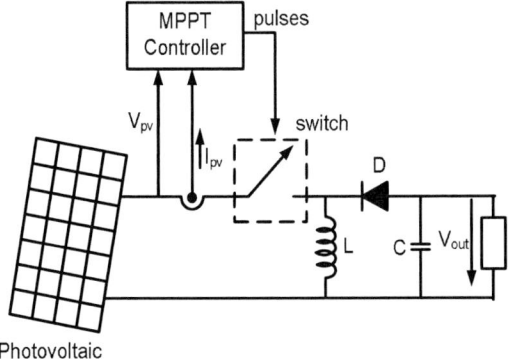

Fig. 6. Block diagram for the whole operation

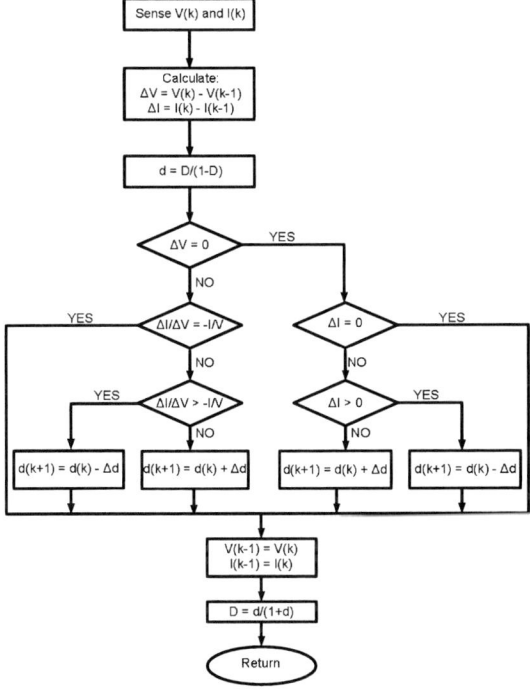

Fig. 7. Flow chart for proposed control method.

## V. SIMULATION RESULTS

Matlab/Simulink was used for whole system simulation. Before system implementation it is necessary to calculate the appropriate values of inductor and capacitor, by using equations (19) and (20), on condition of non – intermittent inductor current and capacitor voltage ripple vary in particular limits.

$$\frac{\Delta V_O}{V_O} = \frac{D \cdot T}{R \cdot C} = \frac{D}{R \cdot C \cdot f} \qquad (19)$$

$$L_{min} = \frac{R}{2 \cdot f} \cdot (1 - D)^2 \qquad (20)$$

Making use of these two equations, driving the switch with a frequency of 20 KHz and taking into account a percent of tolerance, these values were calculated as follows: L = 330μH and C = 220μF.

Simulation took place under a gradual change of solar radiation from 400 W/m2 to 500 W/m2, as shown in figure 8.

Making use three different values of d change, Δd (Δd = 0.0005, 0.001 and 0.01), and keeping the sampling time for A/D converters, which measure instant values of voltage and current, constant and equal to 0.5 ms, figures 9 to 11 presents MPPT's operation.

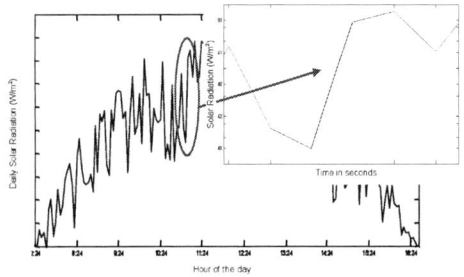

Fig. 8. Gradual change of solar radiation from 400 to 500 W/m², using for system simulation.

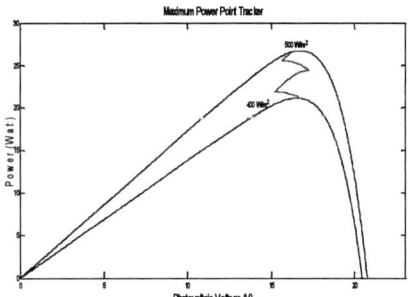

Fig. 9. MPPT's operation for Δd = 0.0005 for a gradual solar radiation change and sampling time 0.5 ms.

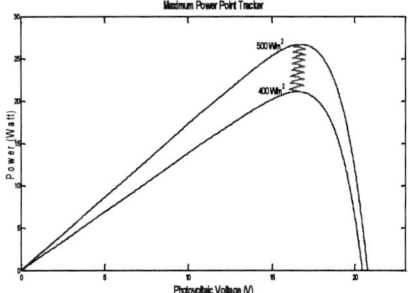

Fig. 10. MPPT's operation for Δd = 0.001 for a gradual solar radiation change and sampling time 0.5 ms.

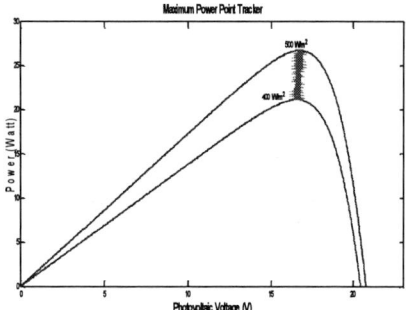

Fig. 11. MPPT's operation for Δd = 0.01 for a gradual solar radiation change and sampling time 0.5 ms.

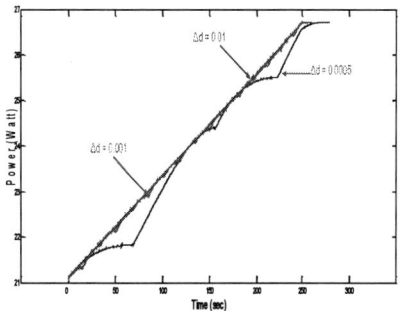

Fig. 12. Comparison between tracker's operations for different Δd variations.

Figure 12 shows a comparison between MPPT's operation for the three different variations, Δd.

From figure 12 is shown that the bigger Δd variation is used, the more linear operation of tracker occurs. Especially for very small values of Δd, such as Δd = 0.0005, extensive ripple is existed, which is undesirable.

Furthermore, simulation of the system has been done for different sampling times (T$_S$ = 0.5ms, 1ms and 2.5ms) for A/D converters with constant variation Δd. Δd is taken equal to 0.001, due to the fact that this value of Δd variation gives a small ripple as shown with green color in fig. 12. Figures 13 and 14 present simulation results for conditions quoted above. Curve for T$_S$ = 0.5ms and Δd = 0.001 is the same as fig. 10.

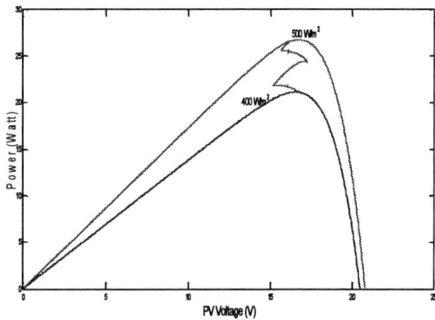

Fig. 13. MPPT's operation for T$_S$ = 1ms for a gradual solar radiation change as shown in fig. 8 and Δd = 0.001.

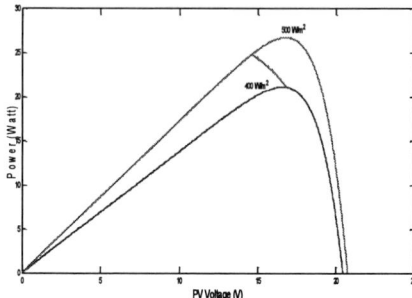

Fig. 14. MPPT's operation for T$_S$ = 2.5ms for a gradual solar radiation change as shown in fig. 8 and Δd = 0.001.

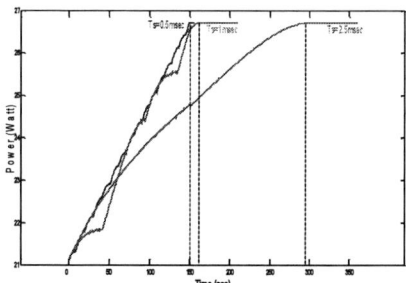

Fig. 15. MPPT's operation for T$_S$ = 2.5ms for a gradual solar radiation change as shown in fig. 8 and Δd = 0.001.

A comparison between three different sampling time conditions with constant variation Δd, is resulted figure 15. In this figure is shown the big difference in time for achieving MPP. Especially, the red line represents sampling time 2.5 ms and the time for achieving MPP is far bigger than the other two cases. Operation for achieving MPP for T$_S$ = 0.5 ms and T$_S$ = 1 ms (blue and green lines in fig. 15) needs about the same time, but operation for T$_S$ = 1 ms occurs some ripple.

## VI. EXPERIMENTAL RESULTS

The system that was used for the implementation of this project is presented in the Block Diagram of Fig. 16. The microcontroller that was used is Microchip's dsPIC30F2010 and was programmed to provide pulses to the semiconductor element (MOSFET – IR3805) of the Buck – Boost DC – DC converter in order to perform MPPT of the photovoltaic.

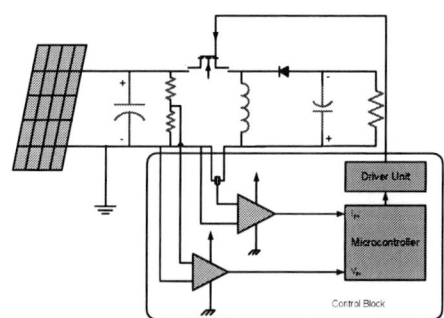

Fig. 16 : Block Diagram of the implemented circuit.

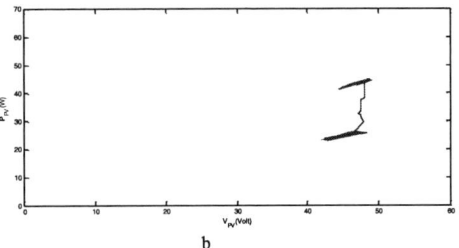

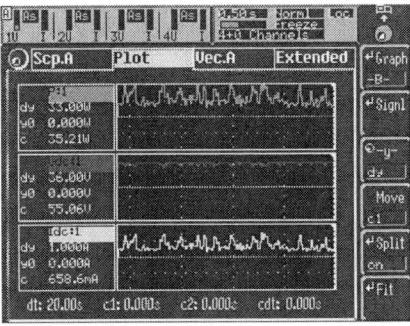

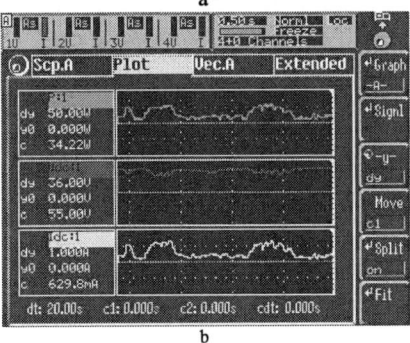

Fig. 17: Variation of power (green), voltage (red) and current in reference to time a) for step value of Δd=0,001, b) for step value of Δd=0

Fig. 18: a. Maximum power point Current-Voltage curve for radiation change

b. Maximum power point Power-Voltage curve for radiation change

## VII. CONCLUSION

This paper presents a new control method for MPPT using DC/DC buck boost converter, which allows more flexibility. Moreover, the proposed method implies a linear relationship between input and output voltages of converter. It is also, proved that according to a gradual change of solar radiation, as variation Δd is increased, the maximum power point approach is getting smoother. Additionally, a small sampling time for measuring voltage and current of PV is necessary for reducing MPP's achieving time. Generally, an appropriate combination between variation Δd and sampling time $T_S$ of A/D channels is essential for more efficient operation. This means that undesirable ripples during MPPT's operation will be eliminated and MPP will be achieved faster.

### ACKNOWLEDGMENT

The Project is co-funded by the European Social Fund and National Resources – (EPEAEK–II) ARHIMIDIS.

### REFERENCES

[1] Kazutaka Itako, Takeaki Mori, "A Single Sensor Type MPPT Control Method for PV Generation System", EPE 2007, *12th European Conference on Power Electronics and Application, Aalborg.*

[2] Sachin Jain, Vivek Agarwal, "A New Algorithm for Rapid Tracking of Approximate Maximum Power Point in Photovoltaic Systems", IEEE POWER ELECTRONICS LETTERS, VOL. 2, NO. 1, MARCH 2004.

[3] Masafumi Miyatake, Fuhito Toriumi, Tsugio Endo, Nobuhiko Fujii, "A Novel Maximum Point Tracker Controlling Several Converters Connnected to Photovoltaic Arrays with Particle Swarm Optimization Technique" EPE 2007, *12th European Conference on Power Electronics and Application, Aalborg.*

[4] D. P. Hohm and M. E. Ropp, "Comparative study of maximum power point tracking algorithms", *Progress in Photovoltaics: Research and Applications,* April 2003, 11: pp. 47-62.

[5] Eftichios Kountroulis, Kostas Kalaitzakis, Nicholas Voulgaris, "Development of a Microcontroller-Based, Photovoltaic Maximum Power Point Tracking Control System" *IEEE Transactions on Power Electronics,* Vol. 16, No. 1, January 2001.

[6] K. K. Tse, Billy M. T. Ho, Henry Shu-Hung Chung, and S. Y. Ron Hui, "A comparative study of maximum power point trackers for photovoltaic panels using switching frequency modulation scheme", *IEEE Transactions on Industrial Electronics,* Vol.51, No.2, April 2004, pp. 410-418.

[7] Eugene V. Solodivnik, Shengyi Liu, Roger A. Dougal, "Power Controller Design for Maximum Power Tracking in Solar Installations" *IEEE Transactions on Power Electronics,* Vol. 19,

The experimental measurements have been realized in a laboratory model system of the Power Electronics Lab. This model includes a photovoltaic panel with the characteristics of 64.5 Volt open-circuit voltage and 2.1 A short-circuit current for 1000 W/m² of solar radiation. The measurements have been realized outdoors under real conditions and with different step values Δd of d parameter.

In this paper the measurement waveforms for step values of Δd=0,001 and Δd=0,0005 are presented. In Figures 17.a and 17.b the waveforms of the panel's power, voltage and current, for step values of Δd=0,001 and Δd=0,0005 respectively, are presented. In these Figures the variation of power, voltage and current in respect to solar radiation and the oscillations due to step value Δd, are shown. For example, the oscillations in Figure 17.a, where the step value is Δd=0,001, are greater than the oscillations in Figure 17.b, where the step value is Δd=0,0005. The Power – Voltage and Current – Voltage maximum power point tracking curves for an increase in solar radiation, are presented in Figures 18.a and 18.b.

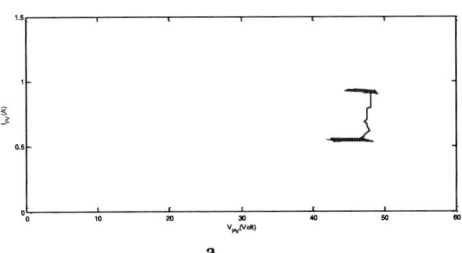

No. 5, September 2004.

[8] Trishan Esram and Patrick L. Chapman, "Comparison of Photovoltaic Array Maximum Power Point Tracking Techniques", *IEEE Transactions on Energy Conversion*, Vol.22, No.2, June 2007, pp. 439-449.

[9] M. Jurado, J.M. Caridad and V. Ruiz, "Statistical distribution of the clearness index with radiation data integrated over five minutes intervals", *Solar Energy*, Vol.55, No.6, 1995, pp. 469-473.

[10] Chee Wei Tam, Tim C. Green, Carlos A. Hernandez, "A Current Mode Controlled Maximum Power Point Tracking Converter for Building Integrated Photovoltaics" *EPE 2007, 12th European Conference on Power Electronics and Application*, Aalborg.

[11] K. K. Tse, Billy M. T. Ho, Henry Shu-Hung Chung, and S. Y. Ron Hui, "A comparative study of maximum power point trackers for photovoltaic panels using switching frequency modulation scheme", *IEEE Transactions on Industrial Electronics*, Vol.51, No.2, April 2004, pp. 410-418.

# Multi Objective Distributed Generation Planning Using NSGA-II

Muhammad Ahmadi*, Ashkan Yousefi †, Alireza Soroudi*, Mehdi Ehsan *

* Sharif University of Technology, Tehran, Iran, e-mail: *m_ahmadi@ee.sharif.edu*

† Tarbiat Modares University, Tehran, Iran

*Abstract*— With the increasing share of distributed generation in electrical energy supply, the proper planning procedure for using these units, has become more vital for power system planners. In this paper, the application of multi-objective optimization techniques for siting, sizing and determination of the proper technology to be used, has been investigated in such a way that, the load demand during the planning horizon is met and the technical and environmental constraints are satisfied.

Unlike the traditional optimization methods used in power system optimization, NSGA-II multi-objective optimization algorithm does not consider techno-economical and environmental parameters in the constraints of optimization problem. This method optimizes them simultaneously as multiple cost functions to find out non-dominated solutions set, named Pareto optimal front, instead of aiming to find single solution.

*Keywords*—Distributed generation, Multi-objective optimization, Pareto optimal front.

## I. INTRODUCTION

Most real-world optimization problems are multi-objective in nature, since they normally have several possibly conflicting objectives that must be satisfied simultaneously [1]. This matter is so clear in power system optimization problems, since system should be capable of supplying the demand at a minimum cost and important factors such as losses, reliability indices, voltage profile and environmental pollution factors should be put simultaneously at optimum levels.

In this paper, distributed generation planning has been considered using multi-objective optimization by NSGA-II algorithm. DG planning means determining how to supply loads of feeders at the presence of DGs, in order to minimize fixed and variable costs, and also satisfying different factors as mentioned earlier. So this paper presents a methodology for optimal DG feasibility study, siting and sizing in distribution system, in order to optimize simultaneously three objective functions, including total costs of power supply, electrical network losses and voltage profile.

By using the technique that is further demonstrated in section-II, eventually several decision making options can be achieved in a way that, none of them dominates another one theoretically, and decision maker can choose an option according to his/her unique opinion, problem environment and the weights considered for each objective depending on that special condition. Of course, the presence of several options and comparing them with each other, gives the decision-maker the chance to make the best decision.

Advantages of this kind of planning against pervious works can be summarized in two following notes:

1) Goal of nearly all planning problems is to minimize investment cost of project over the planning horizon [2-4], but technological and environmental parameters have been considered as constraints of problem. In [5] the main purpose is minimizing cost, but stability is known as optimization constraint. The methodology presented in [6] aims to optimize the allocation and sizing of DG in order to minimize the primary distribution network losses and to guarantee acceptable reliability level and voltage profile. Also in [7], with the same objective function, voltage profile has been kept constant in an acceptable level.

2) One of the most sensible advantages of this methodology is achieving to various optimum points, which gives flexible options to choose from, according to different decision-making policies. Since these constraints are not crisp and decision-making conditions are not crystal clear practically, the application of this feature is vital for reliable planning.

The rest of this article is organized as follows. Section II describes proposed algorithm for solving optimization problem, and then in section III, procedures of problem formulation and test grid will be introduced as the case study. Thereafter diagrams of Pareto front are depicted in section IV. Finally at the conclusions of the paper, results of proposed approach will be given and discussed.

## II. INTRODUCTION TO NSGA-II

In principle, multi-objective optimization is very different with the single-objective optimization. In single-objective, one attempts to obtain the best design or decision, which is usually the global optimum [8]. But in the case of multiple conflicting objectives, there may not exist one solution which is best with respect to all objectives. Therefore we normally look for "Trade-offs" rather than single solution, when dealing with multi-objective optimization. These solutions are known as Pareto-optimal solutions or non dominated solutions. Since none of solutions in the non dominated set is absolutely better than any other, any of them is an acceptable solution. The choice of one solution over the others requires problem knowledge and a number of problem related parameters. Thus, one solution chosen by a designer may not be acceptable to another designer or in a changed environment [9].

The first implementation of evolutionary multi-objective optimization approach was VEGA, which was introduced by Schaffer in 1984. To overcome the speciation problem of this algorithm, Deb introduced "NSGA" in 1994, which differed to the simple GA, only

978-1-4244-1741-4/08/$25.00 ©2008 IEEE

in the way in which selection operator was performed. Because of complexity computing, Deb purposed NSGA-II in 2002.

The NSGA-II procedure is also shown in Fig. 1 [10]. Before explaining the main loop of algorithm, two sorting technique used in this algorithm should be described. These sorting techniques are defined as follow:

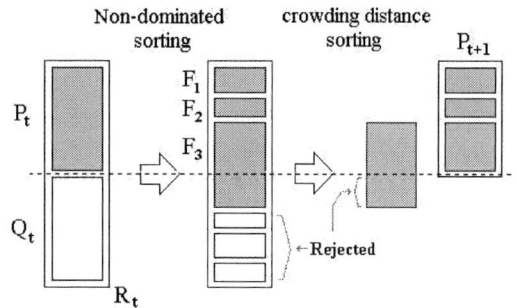

Fig. 1. NSGA-II Procedure.

1- In order to divide the population into separate fronts, fast non-dominated sorting approach is applied, in which for each solution two entities are calculated:

a) Domination count $n_p$, the number of solution which dominates the solution p.

b) Sp, a set of solutions that the solution 'p' dominates.

All the solutions in the first non-dominated front will have their domination count as zero, and they are eliminated to get the other fronts. It should be mentioned that, if 'm' is the number of objective function, dominance of chromosome 'x' against chromosome 'p' is defined as:

for $i = 1: m$

$fitness^i(x) \leq fitness^i(p)$

2- Thereafter for the purpose of maintaining sustainable diversity in population, chromosomes in each front should be sorted with crowded-comparison approach. For each chromosome 'i', the quantity of crowding distance '$i_{distance}$' is the average distance between two points on either side of this point along each objective. Therefore in comparison between two solutions, we prefer solution with lower rank and in the case of same ranks, the solution that is located in lesser crowded region (with larger quantity of $i_{distance}$ ) is chosen.

After implementation of these sorting procedures, half population is directly chosen as $P_{t+1}$ like the elitist method. Thereafter the new population $P_{t+1}$ is now used for tournament selection, crossover and mutation to create the other half of population so-called $Q_{t+1}$. With calculation the merit of total population, next iteration is commenced.

III. CASE STUDY

As illustrated furthermore, the goal of this study is finding the best combination of DG units in appropriate places of distribution network, in order to supply the demand of electricity during the planning horizon. Distribution system for testing our proposed methods has been depicted in Fig. 2.

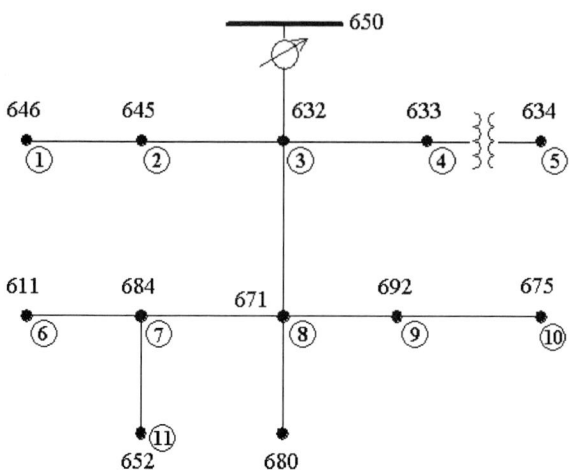

Fig.2. IEEE 13 node test feeder

The numbers 1 to 11 are to designate load points which have the possibility of DG installation. Since, in most practical problems, might not be possible to install DG in some nodes because of environmental problems, therefore Node-680 will not be counted as DG installation point. In Table I, some of available units for installation and their capacities are presented that we have considered them for present study.

TABLE I
DG TECHNOLOGY CONSIDERED IN THIS CASE STUDY

| Number | Type | Size(kw) |
|--------|------|----------|
| DG1 | Micro Turbine | 500 |
| DG2 | Combustion Turbine | 1000 |
| DG3 | Combustion Turbine | 2000 |
| DG4 | IC Engine | 1400 |
| DG5 | MC Fuel cell | 250 |
| DG6 | Photovoltaic | 500 |

The current study is based on some assumptions as follows:

a. DG units don't have capability of reactive power generation and only modeled as a negative active load at load points.

b. Planning period is considered 5 years, and it is assumed that DG units could be installed at the beginning of the base year.

c. Distribution system's loads are constant within a year and total load at the base year is $4.17^{MW} + j0.68^{MVAR}$, but annual growth rates are assumed 10%, 8%, 12% and 12% respectively.

By these assumptions, the structure of proposed chromosome of population in integer genetic algorithm is as shown in Fig. 3.

| | DG1 | ... | DG6 |
|---|---|---|---|
| Chromosome | (0 to 11) | .... | (0 to 11) |

Fig. 3. Configuration of proposed chromosome.

1848

Each gene in a chromosome is designated to one of DG1 to DG6 units and its numerical value is an integer between 0 and 11 that introduced bus number in which respective DG is installed. It should be noticed zero in a gene means that, respective DG unit is not used in system configuration.

By this structure, it is obvious that for evaluation each chromosome, first step is decoding the chromosome, and then, network configuration is changed accordingly. Thereafter by executing power flow for each year within planning horizon, each objective function is calculated.

To find the best configuration, we should define objective functions in an efficient manner. We should go in a way that, all of objectives getting approach their optimum value. Three objectives for this case study are considered as follows:

*A. Cost*

Network cost during planning period of time is consisted of:

*1)* Installation cost of DG units which are decided in chromosome to be used.

*2)* Variable costs are those which are dependent on the produced energy by used DG units and involve fuel cost, operation and maintenance costs.

*3)* Expended cost to purchase energy from transmission system. (The amount of this energy will be achieved after solving power flow).

$$
\begin{aligned}
Cost &= \sum_{used\,DG} IC_i \times \frac{T}{Life\,Time_i} \\
&+ \sum_{t=1}^{t=T} \frac{[P_{grid} \times C_{grid} + (O\&M + F) \times Cap] \times 8760}{(1 + d.r)^t}
\end{aligned}
$$

There are following definitions in above-mentioned phrase:

$IC_i$: Fixed cost that is incurred to purchase and install DG units, which respective values in the gene on chromosome is not zero.

$P_{grid}$: is the amount of supplied power from grid.

$C_{grid}$: The price of purchased energy from grid.

*Cap*: is the vector of capacity of used DGs in distribution system.

*d.r*: discount rate.

*T*: planning horizon.

$Life\,Time_i$: is the life time for i-th DG.

*B. Active and Reactive Losses*

Losses are calculated by algebra sum of losses in all lines. In this study to considering both active and reactive kind of power, loss is defined as below:

$$Loss = \sqrt{lossP^2 + lossQ^2}$$

$$lossP = \sum_{t=1}^{t=T} \sum_{l \in L} (P_{i \to j}^l + P_{j \to i}^l), \; lossQ = \sum_{t=1}^{t=T} \sum_{l \in L} (Q_{i \to j}^l + Q_{j \to i}^l)$$

*C. Voltage Profile*

It is desired usually that voltage magnitude in each bus should not exceed a certain allowable range. Otherwise, it may cause irreparable damages and side effects to consumers. In this case study allowable margin is defined in the range of [0.95 1.05] p.u, and voltage profile is the average of all deviations of voltage from this range in all buses and also in all over period of planning horizon.

## IV. SIMULATION RESULTS

NSGA-II evolutionary algorithm has been entered to an IEEE standard distribution system with 13 buses, in three different modes of choosing objective functions. In any mode that will be introduced later, parameters of this algorithm have been selected as following:

- Population size : 400 chromosome
- Iteration numbers: 200 times
- Mutation rate: 4 %

Also in every following mode, the best Pareto front is chosen through performing the algorithm for ten times.

*A. First mode*

It is supposed that all three mentioned objective functions are important for designer in this mode, so the problem is triple-objective optimization problem and it is apparent that, Pareto fronts are three-dimensional. The last Pareto optimal front is depicted in Fig. 4.

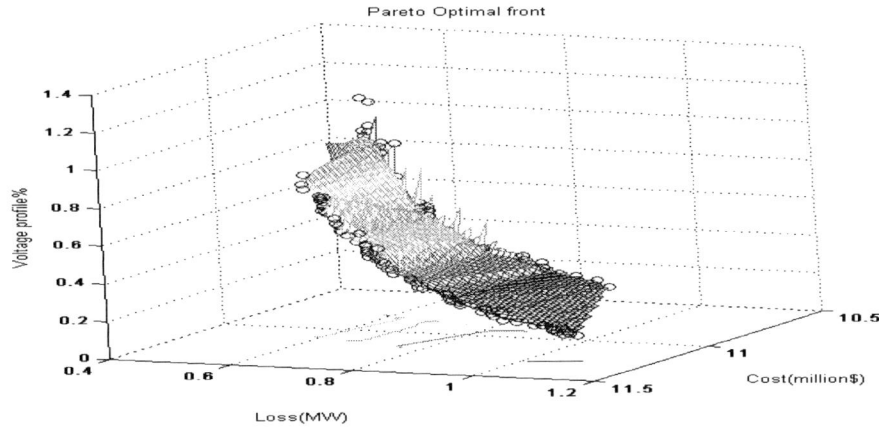

Fig. 4. Pareto optimal front in the first mode.

As it shown, the best front of the last generation has 112 members that none of them could dominate the others. In this case, if the decision maker capable to accept $0.7^{MW}$ of losses and 0.6% of voltage profile, optimal point could be chosen as Figs. 5a and 5b.

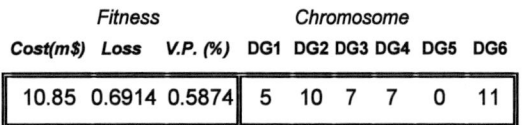

| Fitness | | | Chromosome | | | | | |
|---|---|---|---|---|---|---|---|---|
| Cost(m$) | Loss | V.P. (%) | DG1 | DG2 | DG3 | DG4 | DG5 | DG6 |
| 10.85 | 0.6914 | 0.5874 | 5 | 10 | 7 | 7 | 0 | 11 |

Fig.5-a. Decided chromosome in the first mode.

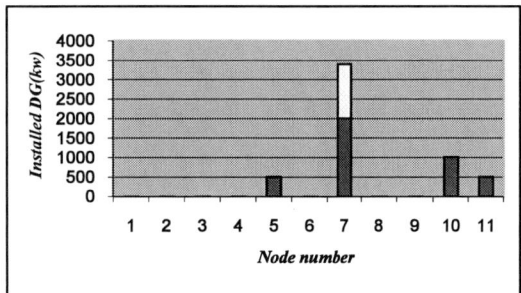

Fig.5-b. Installed capacity on each bus in the first mode.

### B. Second mode

The problem is solved again with taking into account voltage profile and cost as objective functions. It's obvious that, in this mode the Pareto optimal front should be a curve of voltage profile versus cost in two dimensional areas. The best set of solutions in final generation consists of 59 points of solution as shown in Fig. 6.

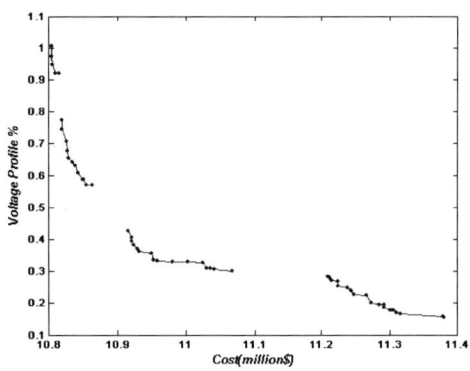

Fig. 6. Pareto optimal front in the second mode.

Now, if system designer involve in the previous circumstance of decision-making, according to Figs. 7a and 7b, for the same voltage profile, investment cost reduces for about 96000$. Because in this case, decision maker does not care about losses, that brings about different solution with different configuration (due to different objectives).

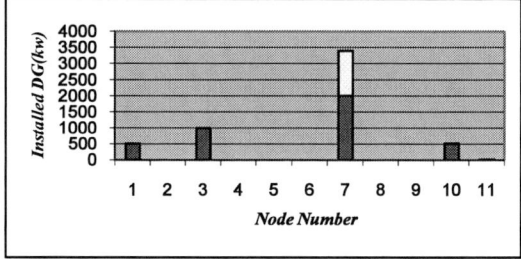

| Fitness | | Chromosome | | | | | |
|---|---|---|---|---|---|---|---|
| Cost(m$) | V.P. (%) | DG1 | DG2 | DG3 | DG4 | DG5 | DG6 |
| 10.741 | 0.5874 | 3 | 10 | 7 | 7 | 0 | 11 |

Fig.7-a. Decided chromosome in the second mode.

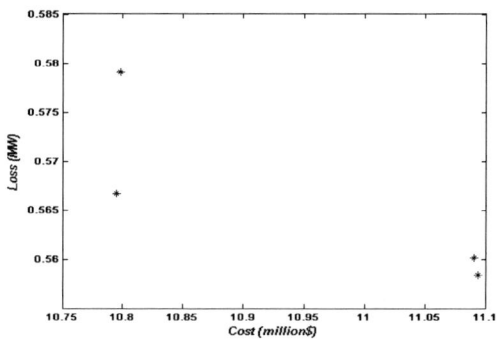

Fig.7-b. Installed capacity on each bus in the second mode.

### C. Third mode

The problem is solved once again with the target of minimizing investment cost and losses regardless of voltage profile. As shown in Fig. 8, in this mode, because of particular problem topology, only four points of solutions are obtained as Pareto front.

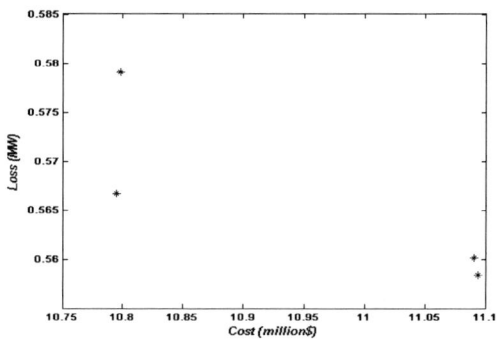

Fig. 8. Pareto optimal front in the third mode

According to Figs. 9-a and 9-b, in comparison to the first mode, by less investment cost we can achieve to less amount of losses in this mode. Because voltage profile is not important in this mode, so there is no need to investment for reducing it and investment is carried out only for loss reduction.

| Fitness | | Chromosome | | | | | |
|---|---|---|---|---|---|---|---|
| Cost(m$) | Loss | DG1 | DG2 | DG3 | DG4 | DG5 | DG6 |
| 10.795 | 0.5667 | 1 | 3 | 7 | 7 | 0 | 10 |

Fig.9-a. Decided chromosome in the third mode

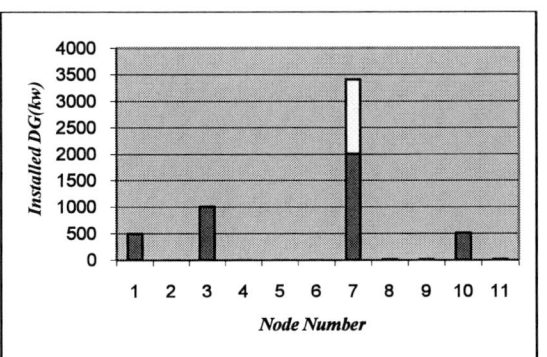

Fig.9-b. Installed capacity on each bus in the third mode

By considering all of the above modes, notice that large units are surely used in configuration due to having less generation cost, but fuel cell does not participate in generation, because of high investment cost. In other hand, if we consider environmental factors as objective function, having less pollution makes fuel cells a good option for electricity generation. Meanwhile majority percentage of generating capacities of DG units have been concentrated on bus no.7, because of its situation is much proper in order to deliver energy to other load points due to the grid topology.

## V. CONCLUSIONS

With considering different kind of objective functions for the test feeder and applying the algorithm, in any case Pareto optimal fronts are obtained in a way that any solution points of each front give us information about location, type and size of DG units. The important result is that, by comparing modes of problem, it is indicated that in multi-objective case, optimum of each objective is further than their optimum in single-objective mode. Because the objective function in multi objective are conflicting each others. As mentioned before, this technique aims to optimize technical and economical factors simultaneously, so obtained solutions are more realistic and practical and decision-maker has many points in allowable range of objective functions rather than only single crisp solution. This gives his/her the capability of decision-making in a flexible environment.

It should be noticed here that, any number of objective functions such as reliability, amount of $CO_2$ emission can be added easily in this planning method and according to importance and weights of each objective, the appropriate solution is chosen.

Number of things could be carried out along with developing this technique, are modeling load more actually in form of base, intermediate, and peak load. Indeed installation time can be extended during time horizon instead of installation at the beginning of planning period. Future works will be focused on considering reliability effects of DG units and also harmonic and protection problems associated with distribution networks at the presence of DGs.

## REFERENCES

[1] Maria João Alves, "A review of interactive methods for multi-objective integer and mixed-integer programming," European Journal of Operational Research, vol 180, Issue 1, pp. 99-115, 1 July 2007.

[2] Favuzza, S. Graditi, G. Ippolito, M.G. Sanseverino, "Optimal Electrical Distribution Systems Reinforcement Planning Using Gas Micro Turbines by Dynamic Ant Colony Search Algorithm," IEEE Trans. on power system, vol 22, no. 2, pp. 580-587, May 2007.

[3] Gareth P. Harrison, et al, "Distributed Generation Capacity Evaluation Using Combined Genetic Algorithm and OPF," International Journal of Emerging Electric Power Systems, vol. 8, Issue. 2, May 2007.

[4] Alireza Soroudi , Mehdi Ehsan, "A Life cycle Model for Optimal DG Placement on Distribution Networks to reduce active Losses and Investment Costs," International Conference on Genetic and Evolutionary Methods, Las Vegas, Nevada, USA (June 25-28, 2007).

[5] William Rosehart, Ed Nowiki, "Optimal placement of Distributed Generation", 14th PSCC, Sevilla, June2002.

[6] Carmen L.T. Borges and Djalma M.Falao, "Optimal distributed generation allocation for reliability, losses and voltage improvement", International Journal of Electrical Power and Energy Systems, vol 28, Issue 6, pp. 413-420, July 2006.

[7] Edwin Hasen et al, "Optimal placement and sizing of Distributed Generator units using Genetic Optimization Algorithms," Electrical Power Quality and Utilization Journal, vol XI, no.1, 2005.

[8] N. Srinivas and K. Deb., "Multi-objective optimization using non-dominated sorting in genetic algorithms," Evolutionary Computation, vol 2, no. 3, pp. 221–248, 1994.

[9] C.M. Fonesca and P.J. Fleming, "Genetic Algorithms for Multi-objective Optimization: formulation, discussion and generalization" , Proceeding of the fifth international conference on genetic algorithms, pp. 416-423, San Mateo, California 1993.

[10] Kalynmoy Deb et al, "A Fast and Elitist Multi objective Genetic Algorithm NSGA-II", IEEE Transactions on evolutionary computation, vol 6, no. 2,Page 182-197, April 2002.

[11] Dan Zhu, Robert P.Broadwater, "Impact of DG placement on Reliability and efficiency with Time-Varying loads", IEEE Trans. on power systems, vol 21, no. 1, February2006.

[12] Naresh Acharya, "An analytical approach for DG allocation in primary distribution network", International Journal of Electrical Power & Energy Systems, vol 28, Issue 10, Pages 669-678, December 2006.

# Testing of the Grid-connected Photovoltaic Systems Using FPGA-based Real-Time Model

Robert Stala

AGH University of Science and Technology/Department of Electrical Drive and Industrial Equipment,
Krakow, Poland, e-mail: stala@agh.edu.pl

*Abstract*—The paper describes FPGA-based real-time simulation method of photovoltaic systems. FPGA devices are widely used for power electronic systems control, but the same chip can have implemented real-time model of controlled system. Prototyping the control circuit before application in power system increases the safeness, and can reduce time and costs of implementation. FPGA-based prototyping of power systems enables research of the system sensitivity on parameters variations, or topology conceptions. The paper presents examples of designs of the FPGA-based models of PV array, PV grid-connected transformerless systems with single phase inverter and L and LCL filters, the dc/dc converter. The conception of FPGA-based realization of spectral analysis is also presented.

*Keywords*—DSP, generation of electrical energy, photovoltaic, real time simulation, renewable energy systems

## I. INTRODUCTION

FPGA (Field Programmable Gate Arrays) devices are widely used in power electronics control systems. Modern FPGAs are supported in logic resources but also in memory and hardware implemented DSP-type modules [3], [4] (Fig. 1). They can realize very precise and untypical PWM functions [2], and also regulators algorithms, or the protection and management functions. Single FPGA chip can deliver the entire control of complex power electronic systems (Fig. 2). The functional blocks work in parallel by using different parts of the FPGA structure.

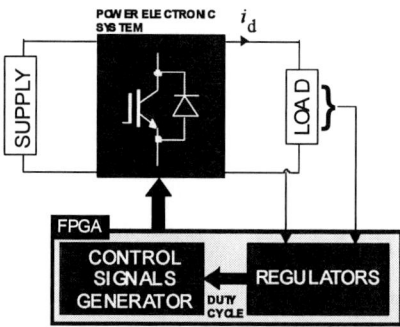

Fig. 2. The power electronic control system functions realized in different elements of FPGA chip

Paralleling of computations, very fast clocking (increased with use of internal PLLs) makes the FPGAs useful also in HF systems control. Programmability feature, and CAD-supported design enable shorting time to market by design verification using simulation and debugging tools. Then, during application of the control system in industrial systems, the most difficult, expensive, time-consuming and unsafe becomes the stage of verification of control and protection algorithms on physical process. The features of FPGAs can help to improve this process by enabling testing the control algorithms on real-time models of controlled systems (Fig. 3). Using selected FPGA device the model can be implemented in the same chip like control algorithms. This method increases the safe and decreases time to market of mechatronic systems implementation.

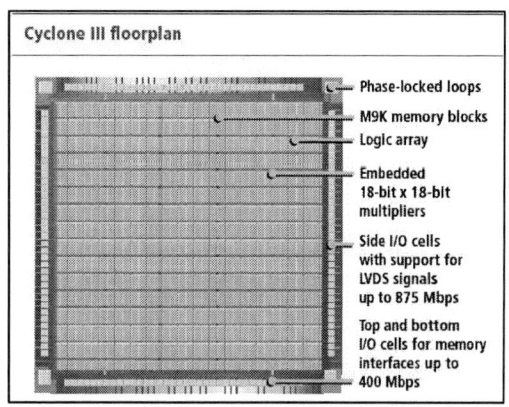

Fig. 1. Altera Cyclone III FPGA resources [4]

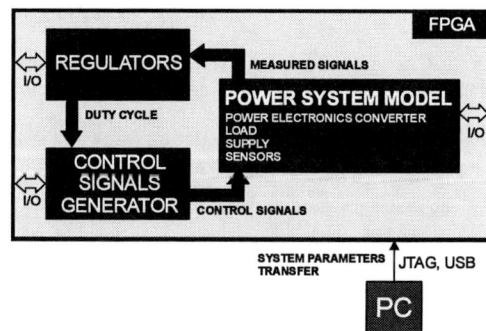

Fig. 3. FPGA (or external DSP) control verification in real-time model

Photovoltaic systems are very predisposed for the FPGA prototyping and testing, because of their complexity and specific behavior (e.g. PV source characteristic). Many

978-1-4244-1741-4/08/$25.00 ©2008 IEEE

algorithms of control must cooperate with each other. Besides correctness of the control system operation may be tested under emulated, real operating conditions (e.g. MPPT on PV featured source, grid voltage and impedance, islanding operation, etc.). FPGA-based prototype of the entire PV system may be adequate to the actual system components in details with possibility of making changes.

## II. THE FPGA-BASED MODEL COMPONENTS OF PV GRID-CONNECTED SYSTEMS

### A. PV Sources

Using FPGA, the families of P-I characteristics of different solar panels (Fig. 4) can be generated (by pre-definition in memory - LUT, calculation from mathematical model or approximation). Conceptions of PV source modeling and the model application are described in [5]-[10]. The PV source characteristics can be generated very detail but for some applications (e.g. preliminary testing of MPPT algorithms) the model may be simplified even to linear approximation. Power of the PV source is a function of 3 components: current, temperature and the irradiation.

$$p = f(I, T, E) \qquad (1)$$

For pre-definition of PV characteristic in internal FPGA memory, each point on $p(i)$ used area must have defined 3 coordinates ($I$, $T$, $E$). Even not large resolution of every components cause large amount of used memory. This solution can be implemented for generation P-I or I-U characteristics for discrete values of $T$ and $E$ and it can be very useful for detailed MPPT algorithms testing.

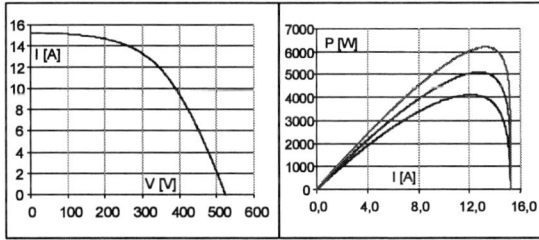

Fig. 4. Data samples of PV module characteristic for LUT generation in FPGA (data from ICAP/4 software for 0, 25 and 47 °C (curve of lowest $P_{max}$ and current) – series/parallel connection of poly-crystaline modules about max. open voltage 25.6V, max. MPP power 12.4W)

Many procedures of solar cell characteristic generation use a model of diode but this method is not convenient for FPGA implementation due to large resources requirements of the IC's. In [7] the mathematical description of maximum current vs. maximum voltage is presented. Such relation could be very important because enables significant simplification of FPGA-based simulation of solar cell behavior, e.g. for every tests with exception of MPPT it can be assumed that he PV array is operates in the MPP=$f(E, T)$. It enables testing the power electronic system in a wide range of environment conditions.

Very interesting procedure of simulation of PV panel behavior is presented in [5]. The method can be easily implemented for discrete simulation in FPGA. The PV panel is modeled using switched resistor that emulates its equivalent resistance (Fig. 5).

| | |
|---|---|
| (circuit: $I_{Req}$, $I_{PV}$, $I_{SC}$, $R_{eq}$) | - The output voltage: $V_{PV} = (I_{SC} + I_{PV}) \cdot R_{eq}$<br>- $I_{SC}$ – Short circuit current (a function of irradiation of the panel),<br>Model of $R_{eq}(I_{PV})$ enables modeling of the PV panel |
| (circuit: $R_1$, $R_2$) | In [5] the $R_{eq}$ is parallel connection of the resistors $R_2$, and the switched resistor $R_1$<br><br>$R_{eq} = \dfrac{R_1 \cdot R_2}{R_1 + R_2 \cdot D}$ |
| $V_{Mod} = K_0 + K_1 \cdot I_{PV} + K_2 \cdot (I_{SCmax} - I_{SC})$ | PWM control signal is a function of $I_{PV}$ and $I_{SC}$ |
| The value of the $K_0$, $K_1$, $K_2$ parameters are chosen on the basis of real PV panel parameters [5] | |

Fig. 5. Conception of PV panel simulation presented in [5]

In the FPGA calculation the switching behavior of $R_{eq}$ (Fig. 5) is not modeled, but the relations which determine $V_{Mod}$, $R_{eq}$ and $V_{PV}$ can be calculated directly. Tab. 1, Tab. 2 and Fig. 6 presents the set of parameters and exemplary results of FPGA-based model of PV panel, with simulated irradiation.

TABLE I.
DESIGN CONDITIONS - FPGA-BASED (ALTERA CYCLONE II EPM35) EMULATION OF PV SOURCE BEHAVIOR WITH USE OF METHOD PRESENTED IN FIG. 5.

| Input/Output signals resolution | 18bits |
|---|---|
| Accuracy of parameters and results | Fixed-point arithmetic using 9-bits left shifted |
| Signal Tap debug parameters | - Data Input Width: 21<br>- Sample Depth: 4096 |

TABLE II.
RESOURCES USAGE IN THE TEST OF FPGA-BASED (ALTERA CYCLONE II EPM35) PV SOURCE BEHAVIOR (FIG. 5, TAB. 1). QUARTUS II RESULTS.

| Resource | Usage |
|---|---|
| LC combinationals | 1335 (4%) |
| LC registers | 20 (<1%) |
| Embedded multiplier 9-bit elements | 8 (11%) |

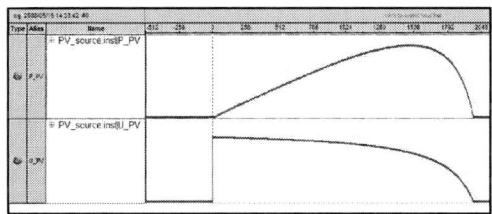

Fig. 6. FPGA-based simulation results of output voltage and power vs. the load current for parallel connected resistor and PWM resistor (Fig. 5). Results from Quartus II/SignalTap Logic Analyzer software

1853

## B. Sun Trajectory

For some applications a very detail behavior of solar source should be simulated. The actual position of the sun can be also considered in the FPGA model. In PV systems the $E(t)$ characteristic during the day is different in the case of stationary devices and the panels with sun tracking. Both type characteristic can be generated in the FPGA with consideration $E(t)$ in $p(I, T, E)$ characteristic.

## C. Energy Conversion System.

The PV systems can operate as grid connected or autonomic. In both cases there are many solutions for power conversion systems and control [13]. The model should be consisted of power converters, and filters (e.g. ac *LCL* filter). FPGA-based modeling of switch-mode converters is presented in [14] and [15].

## D. The Power Electronic Converters Topologies and Control

There are many topologies of power electronic converters applied in PV systems [13], [1], e.g. multistring, AC-Modules, transformer or transformerless, with high frequency or low frequency transformer, with half-bridge or full-bridge inverter, with or without dc/dc stage, and many untypical cases. The control strategy is also the field where many ideas are tested.

Fig. 7 presents the model of a single-phase grid connected PV system, and the steady state waveforms of the grid and inverter signals. In presented case the inverter operates in bipolar mode. It is one of the method of operation that decreases common mode currents due the constant value of common voltage [17]:

$$v_{cm} = (v_{AO} + v_{BO})/2 \qquad (2)$$

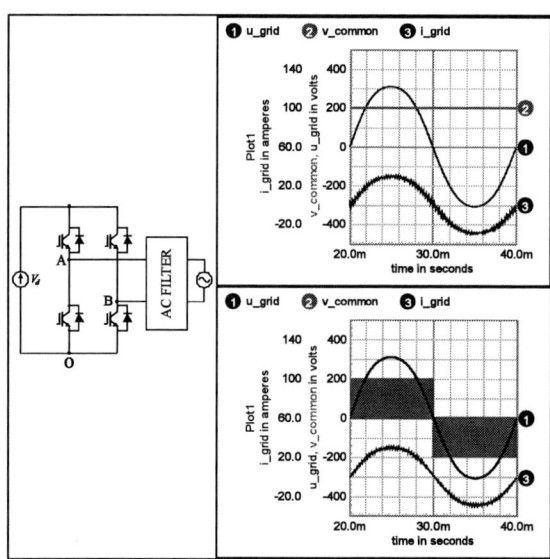

Fig. 7. Waveforms in the single-phase grid-connected system in case of bipolar and unipolar inverter operation: (1) the grid voltage, (2) common-mode voltage, (3) the grid current. Results of ICAP/4 Software.

In the grid connected systems it is also necessary to explore the interaction between different dc/dc converters

or inverters (e.g. in AC-Modules systems, presented in Fig. 8)) [13].

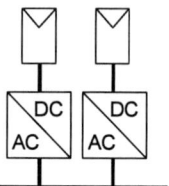

Fig. 8. AC-Module system [13]

## E. MPPT Algorithms

The FPGA-based real-time model of the PV system can be very functional tool for MPPT algorithms verification [7]. The MPPT algorithms can be tested in different solutions of power conversion systems, using dc/dc converters or directly by inverter control.

## F. AC Filters and Protective Functions

The FPGA prototyping of the PV grid-connected systems with ac filters (Fig. 9) gives possibility of spectral analysis of the grid current (e.g. for comparison to harmonics limitation) for different topologies and control strategies. The revision of ac filters can be found in [18] and [19].

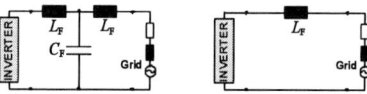

Fig. 9. Examples of AC filters solutions

The system behavior with any type of filter can be tested in FPGA with use their discrete differential equations description, or the frequency response (placed in the FPGA memory). Both methods can be used simultaneously. In the first method any transient state can be observed, and in the other method the grid current frequency characteristic is obtained.

In the method of direct grid current spectrum calculation the spectrum of the inverter voltage (difference of the inverter and the grid voltage) is used for calculation of the grid current spectrum. The same calculation procedure is realized for different filters by the frequency response characteristic replacement (Fig. 10).

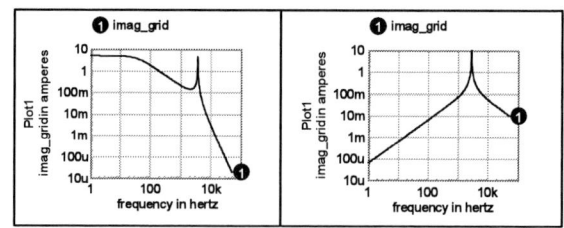

Fig. 10. The grid current frequency response characteristic of LCL filter (from the inverter side and from the grid side). Results of ICAP/4 Software.

For *L* filter (Fig. 8) the grid current calculation is realized by solving one (discrete) first-order differential equation:

1854

$L \cdot di_2/dt + R \cdot i_2 = u_{INV} - u_{GRID}$     (2)

Where: $L = L_F + L_{Grid}$; $R = R_F + R_{Grid}$

The example of results in such case is presented in Fig, 14, and Tab. 4.

For LCL filter application it is necessary to solve the set of equations describing $i_{inv}$, $i_F$, $u_F$ [11]:

$$\begin{bmatrix} \dfrac{du_{CF}}{dt} \\[2mm] \dfrac{di_{inv}}{dt} \\[2mm] \dfrac{di_G}{dt} \end{bmatrix} = \begin{bmatrix} 0 & \dfrac{1}{C_F} & \dfrac{-1}{C_F} \\[2mm] \dfrac{-1}{L_1} & \dfrac{-R_1}{L_1} & 0 \\[2mm] \dfrac{1}{L_2} & 0 & \dfrac{-R_2}{L_2} \end{bmatrix} \begin{bmatrix} u_{CF} \\[2mm] i_{inv} \\[2mm] i_G \end{bmatrix} + \begin{bmatrix} 0 \\[2mm] \dfrac{1}{L_1} \\[2mm] 0 \end{bmatrix} (u_{inv}) - \begin{bmatrix} 0 \\[2mm] 0 \\[2mm] \dfrac{1}{L_2} \end{bmatrix} u_{GRID}$$

(3)

The example of discrete model of the PV grid connected system with LCL AC filter is presented in Fig, 15, Fig. 16, Tab. 5 and Tab. 6.

*G. Anti-Islanding Algorithms*

One of the most useful application of the FPGA-based real-time models of PV systems is capability of verification of anti-islanding algorithms or other protection procedures. The procedure is safe and enables as well passive as active methods of grid testing realization. Many anti-islanding algorithms examples can be found in [20] and [21].

## III. REAL-TIME SPECTRAL ANALYSIS WITH USE OF WALSH-FOURIER TRANSFORMATION

For decreasing the FPGAs resources used for real-time spectral analysis and increasing the accuracy, the orthogonal Walsh system can be used. The Walsh functions system generation and Walsh transformation is described in [22]-[28]. The Walsh functions can be very simply generated from logic functions because they have two discrete values {-1, 1}. Then, there is no large amount of FPGA logic resources necessary for the base functions generation and there is no errors in their representation. The linear relation between the Walsh and the Fourier coefficients also exists [22].

Fig. 11 presents waveforms of the first four Rademacher functions and 10 Walsh functions. Rademacher functions are square wave {-1;1} functions about 50% duty cycle and frequency $f = 2^n f_1$ ($n = 0, 1, 2, \ldots$; $f_1$ - the lowest frequency of the Rademacher functions systems).

$r_n(t) = \text{sign}(\sin(2^n \pi t'))$; $r_n(t) \in (-1 \div 1)$; $t' = t/T_1$   (4)

Rademacher functions system is not complete, then can not be used for a function approximation by series, but from his functions the Walsh system can be generated. One of the method of the Walsh system generation is multiplication proper Rademacher functions (Fig. 11) [22]:

$$wal_n(t) = \prod_{i=1}^{k} \left[ r_i(t) \right]^{b_i} , \quad n = b_k 2^{k-1} + b_{k-1} 2^{k-2} + \ldots + 2b_2 + b_1; \quad (5)$$

$b_i \in \{0,1\}$, $wal_n(t) \in \{-1 \div 1\}$;

Representation of the Walsh function index ($n$) by $\mathbf{b} = \{b_1, b_2, \ldots, b_n\}$ using Gray code gives ordering of the functions similar to sinusoidal harmonics (frequencies and phase shift). There are also possible simplifications of the method of generation with use of symmetry and shifting [25].

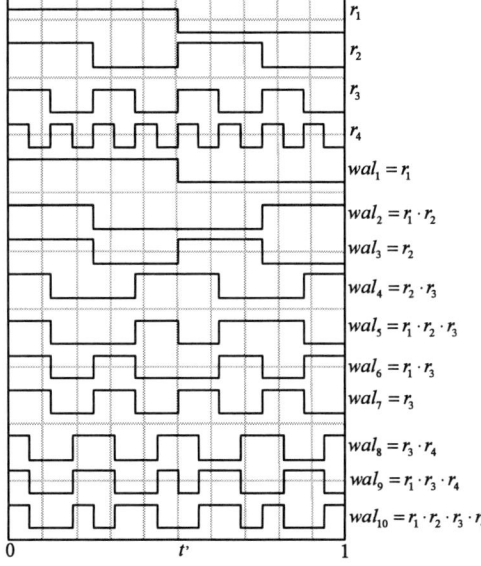

Fig. 11. The set of four Rademacher functions (r1 ÷ r4) and ten Walsh functions (wal1 ÷ wal10), and their relation. Index represented by Gray coded vector **b**

Walsh functions are orthonormal and can be used for a function approximation:

$$\int_0^T wal_n(t)dt = 0 \quad \text{dla } n = 1, 2, \ldots \quad (6)$$

$$x(t) = \sum_{k=0}^{\infty} W_k wal_k(t') , \quad k = 0, 1, 2, \ldots; \quad t' = t \cdot (1/T); \quad (7)$$

$$W_k = \frac{1}{T} \int_0^T x(t) wal_k(t') dt \quad (8)$$

RMS value of the signal can be calculated on the basis of Fourier or Walsh coefficients.

For impedance calculations (or other purposes) the conversion from Walsh harmonics to Fourier is possible [22]. Inserting the series of Walsh to Fourier coefficients calculations, the relation between the coefficients can be calculated, e.g.:

$$F_1 = \frac{\sqrt{2}}{T} \int_0^T \sum_{k=0}^{\infty} W_k wal_k(t') \sin(2\pi f t) dt , \quad k = 0, 1, 2, \ldots; \quad t' = t/T)$$

(9)

$$F_1 = \frac{\sqrt{2}}{T} \left( W_1 \int_0^T wal_1(t') \sin(2\pi f t) dt + W_2 \int_0^T wal_2(t') \sin(2\pi f t) dt + \ldots + W_n \int_0^T wal_n(t') \sin(2\pi f t) dt \right)$$

(10)

$$\frac{\sqrt{2}}{T}W_1\int_0^T wal_1(t')\sin(2\pi ft)dt = \frac{\sqrt{2}}{2\pi}W_1 4\int_0^{\frac{\pi}{2}} wal_1(t')\sin(\omega t)d\omega t = 0.9W_1$$

(11)

$$\frac{\sqrt{2}}{T}W_2\int_0^T wal_2(t')\sin(2\pi ft)dt = 0 \text{ , similarly for } n=3, 4, 6, \dots$$

(12)

$$F_1 = \frac{\sqrt{2}}{T}\int_0^T\sum_{k=0}^{\infty}W_k wal_k(t')\sin(2\pi ft)dt \text{ , } k=0, 1, 2, \dots; t'=t/T)$$

(13)

Finally:

$$F_1 = 0.9W_1 - 0.373W_5 - 0.074W_9 - 0.177W_{13} - 0.014W_{17} + \dots \quad (14)$$

$$F_2 = 0.9W_2 + 0.373W_6 - 0.074W_{10} + 0.178W_{14} - 0.015W_{18} + \dots (15)$$

The Walsh-Fourier analysis can be used for:
- spectral analysis (e.g. for grid impedance measurements)
- rms value measurement
- signals generation, e.g. the grid voltage about defined spectrum for the FPGA-based simulation

Signal analysis and measurements can be used in practical grid connected systems for the inverter to-public-network interface protection (voltage and frequency parameters) [16]

## IV. DESIGN EXAMPLES – FPGA-BASED MODELS OF PV SYSTEM COMPONENTS

### A. Grid-connected Single-Phase Inverter Control

Fig. 12 presents the example of the entire grid-connected transformerless PV system.

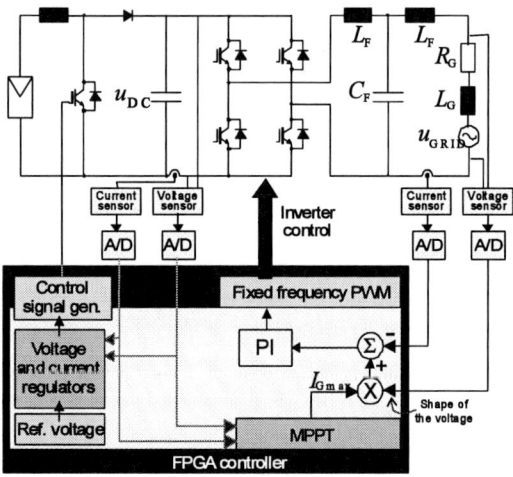

Fig. 12. Components of the grid-connected power system with PV array modeled in FPGA for control testing

The system is composed of PV source, dc/dc converter for DC link voltage stabilization, inverter for the line current shaping and regulation with MPPT, AC filter and grid, and also A/D converters.

Fig. 14 presents the results of FPGA-based simulation of fragment of the system responsible for grid current shaping (Fig. 13) and regulation. This section enables testing of the practical FPGA ac line current regulator (dynamics, operating range and precision, saturations etc.). Tab. 3 and Tab. 4 present design specification and the FPGA resources usage for this fragment of the project.

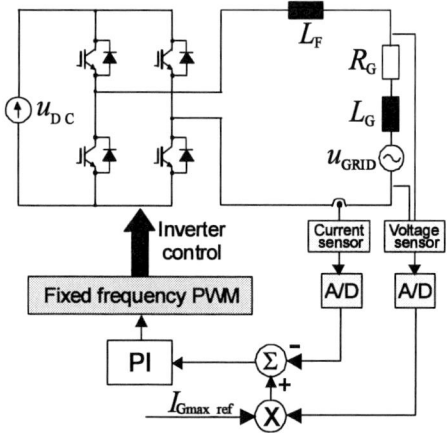

Fig. 13. The grid-connected inverter modeled in FPGA for ac-line current regulator testing in PV systems

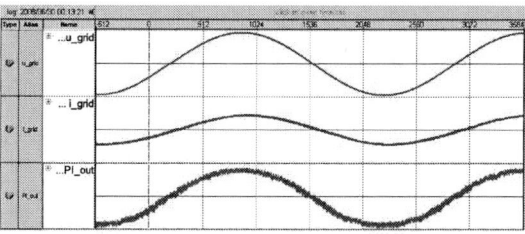

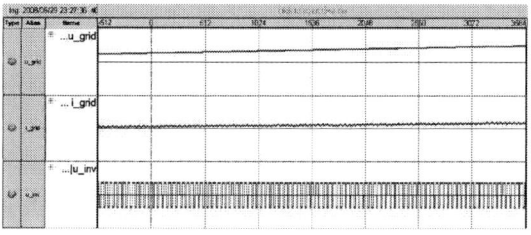

Fig. 14. Results of FPGA-based simulation of the grid-connected single-phase full bridge inverter (Fig. 13): the grid voltage and current, and the inverter voltage (bipolar modulation, $L$=1.5mH ac filter, operation in closed system). Results from QuartusII/SignalTap Logic Analyzer software.

### TABLE III.
DESIGN CONDITIONS - FPGA-BASED (ALTERA CYCLONE II EPM35) GRID CURRENT REGULATOR FOR PV SYSTEMS (WITH SINGLE-PHASE FULL BRIDGE INVERTER UNDER BIPOLAR OPERATION, $L$ AC FILTER, LUT GRID VOLTAGE GENERATOR).

| Inverter type | Full-bridge, bipolar operation |
|---|---|
| Regulator type | PI |
| Filter type | $L$ |
| Signals resolution | - Grid current, Grid voltage, inverter voltage, PI input/output, current error: 18 bits, <br> - PI parameters ($Kp$, and $Tp/Ti$): 9bits |
| Signal Tap debugging parameters | - Data Input Width: 54 <br> - Sample Depth 4096 |
| Grid voltage sampling | 4096 words of 18 bits |

### TABLE IV.
THE ENTIRE MODEL RESOURCES USAGE IN THE TEST OF FPGA-BASED (ALTERA CYCLONE II EPM35) GRID CURRENT REGULATOR (FIG. 13, TAB. 3). QUARTUSII RESULTS.

| Resource | Usage |
|---|---|
| LC combinationals | 633 |
| LC registers | 114 |
| Embedded multiplier 9-bit elements | 4 (6%) |

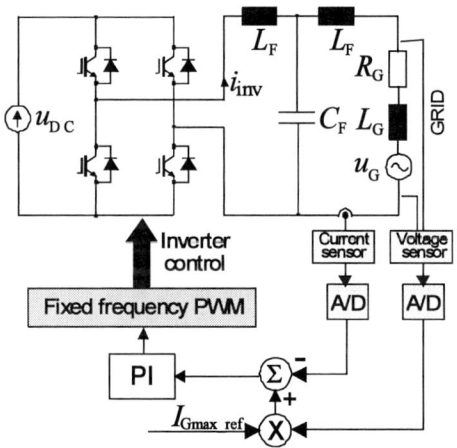

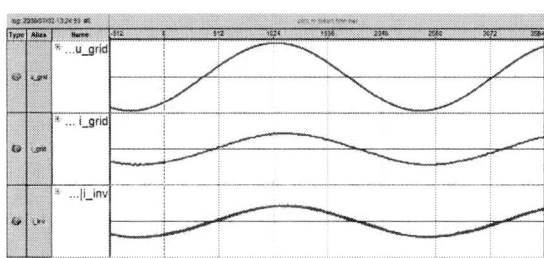

Fig. 15. The grid-connected inverter with LCL filter modeled in FPGA for ac-line current regulator testing in PV systems

Fig. 16. Results of FPGA-based simulation of the grid-connected single-phase full bridge inverter with LCL filter: the grid voltage and current, and the grid current (bipolar modulation, operation in closed system). Results from QuartusII/SignalTap Logic Analyzer software.

### TABLE V.
DESIGN CONDITIONS - FPGA-BASED (ALTERA CYCLONE II EPM35) GRID CURRENT REGULATOR FOR PV SYSTEMS (WITH SINGLE-PHASE FULL BRIDGE INVERTER UNDER BIPOLAR OPERATION, LCL AC FILTER, LUT GRID VOLTAGE GENERATOR).

| Inverter type | Full-bridge, bipolar operation |
|---|---|
| Regulator type | PI |
| Filter type | $LCL$ |
| Signals resolution | -Grid current, Grid voltage, inverter voltage, PI input/output, current error: 18 bits, <br> - PI parameters ($Kp$, and $Tp/Ti$): 9bits <br> - Filter parameters: 9 bits |
| Signal Tap debugging parameters | - Data Input Width: 54 <br> - Sample Depth 4096 |
| Grid voltage sampling | 4096 words of 18 bits |

### TABLE VI.
THE ENTIRE MODEL RESOURCES USAGE IN THE TEST OF FPGA-BASED (ALTERA CYCLONE II EPM35) GRID CURRENT REGULATOR (FIG. 15, TAB. 5). QUARTUSII RESULTS.

| Resource | Usage |
|---|---|
| LC combinationals | 1445 |
| LC registers | 333 |
| Embedded multiplier 9-bit elements | 22 (31%) |

Fig. 15, Fig. 16, Tab. 5 and Tab. 6 present the model conception and results of simulation the system with LCL AC filter. In this example the grid current is calculated using discrete differential equations (3).

### B.  DC/DC converter

In some PV installations (e.g. in grid-connected systems with low voltage PV panels) the DC/DC converters can be used.

### TABLE VII.
DESIGN CONDITIONS - FPGA-BASED (ALTERA CYCLONE II EPM35) MODEL OF DC/DC BOOST CONVERTER UNDER OPEN SYSTEM OPERATION

| Converter type | |
|---|---|
| Signals resolution | - Inductor, switch and diode current: 18 bits, <br> - Output/input voltage: 18 bits, <br> - 3.125MHz calculation frequency |
| Signal Tap debugging parameters | - Data Input Width: 73 <br> - Sample Depth 1024 |

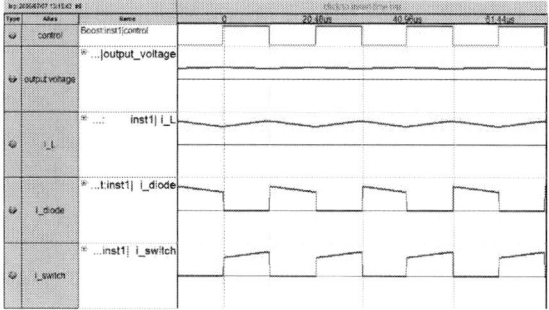

Fig. 14. Results of FPGA-based simulation of the DC/DC boost converter under open system operation: the control signal, the output voltage the inductor current, and the diode and switch currents. Results from QuartusII/SignalTap Logic Analyzer software.

TABLE VIII.
RESOURCES USAGE FOR FPGA-BASED MODEL (ALTERA CYCLONE II EPM35) OF DC/DC BOOST CONVERTER (TAB. 7). QUARTUSII RESULTS.

| Resource | Usage |
|---|---|
| LC combinationals | 279 |
| LC registers | 68 |
| Embedded multiplier 9-bit elements | 10 (14%) |

Tab. 7, Fig. 17 and Tab. 8 present scheme, design specifications and results of testing of FPGA-based model of DC/DC boost converter. This model is suitable for verification of the real system control.

## V. CONCLUSIONS

In the paper the FPGA-based real-time simulation method for verification of control of power electronic converters in grid-connected systems of energy generation from photovoltaic sources was demonstrated. Prototyping the control circuit before application in power system increases the safeness, and can reduce time and costs of implementation.

The design examples of FPGA-based real-time models of PV array, single-phase grid connected inverter with L and LCL filter, and also dc/dc converter was presented. For each described model the signals resolution and the usage of the most important FPGA's resources was presented.

The idea of using Walsh functions for real-time spectral analysis with use of FPGA was also presented.

### REFERENCES

[1] J.M. Carrasco, L.G. Franquelo, J.T. Bialasiewicz, E. Galvan, R.C. Portillo Guisado, M.A.M. Prats,. J.I. Leon, N. Moreno-Alfonso, *Power-Electronic Systems for the Grid Integration of Renewable Energy Sources: A Survey*. IEEE Trans. on Ind. El. Vol. 53, No. 4, August 2006.

[2] S. Pirog, M. Baszynski, J. Czekonski, S. Gasiorek, A. Mondzik, A. Penczek, R. Stala: *Multicell DC/DC Converter with DSP/CPLD Control. Practical Results*. 12h International Power Electronics and Motion Control Conference EPE-PEMC 2006, Portoroż Słowenia, 31 August - 2 September 2006, CD Proceedings.

[3] D. Maliniak: *Basics of FPGAs Design*, A Supplement to Electronic Design/Dec. 4, 2003, Sponsored by Mentor Graphics Corp.

[4] Altera Product Catalog 2007

[5] L.A.C. Lopes and A. -M. Lienhardt, *A Simplified Nonlinear Power Source For Simulating PV Panels*, Power Electronics Specialist Conference, 2003. PESC; 03. 2003 IEEE 34th Annual Volume 4, Issue , 15-19 June 2003 Page(s): 1729 – 1734.

[6] P. T. Krein, *Tricks of the Trade: A Simple Solar Cell Model*, IEEE Power Electronics Society Newsletter, April 2001

[7] M.A. Vitorino, L.V. Hartmann, A.M.N. Lima, M.B.R. Corrêa, *Using the model of the solar cell for determining the maximum power point of photovoltaic systems*, 12th European Conference on Power Electronics and Applications EPE'07, 2007–, CD Proceedings.

[8] O.M. Midtgard: *A simple photovoltaic simulator for testing of power electronics*, 12th European Conference on Power Electronics and Applications EPE'07, 2007–, CD Proceedings.

[9] U. Böke: *A simple model of photovoltaic module electric characteristics*, 12th European Conference on Power Electronics and Applications EPE'07, 2007–, CD Proceedings.

[10] Ch.W. Tan, T. C. Green and C. A. Hernandez-Aramburo. *A Current-Mode Controlled Maximum Power Point Tracking Converter for Building Integrated Photovoltaics*, 12th European Conference on Power Electronics and Applications EPE'07, 2007–, CD Proceedings.

[11] K. Masoud and Gerard Ledwich: *Grid Connection Via Third Order Filter: Near-Deadbeat Control*, AUPEC 2000.

[12] J.C. Alfonso-Gil, F.J.Gimeno-Sales, S. Seguí-Chilet, S. Orts, J. Calvo and Vte. Fuster: *New Optimization in Photovoltaic Installations with Energy Balance with the Three-Phase Utility*, IEEE ISIE 2005, June 20-23, 2005, Dubrovnik, Croatia

[13] F. Blaabjerg, F. Iov, R. Teodorescu, Z. Chen, *Power Electronics in Renewable Energy Systems*, 12h International Power Electronics and Motion Control Conference EPE-PEMC 2006, Portoroż Słowenia, 31 August-2 September 2006, CD Proceedings.

[14] R. Ruelland, G. Gateau, T.A. Meynard, J.M. Hapiot, *Design of FPGA-Based Emulator for Series Multicell Converters Using Co-Simulation Tools*, IEEE Trans. On Power. Electron, 18 (2003), n. 1, 455-463.

[15] R. Stala: *Analiza sterowania przeksztaltnika wielokomorkowego ac/ac na podstawie modelu zrealizowanego w ukladzie FPGA*. Przegląd Elektrotechniczny › 2007-10, p. 28, (Polish).

[16] B. Bletterie, R. Bruendlinger, Ch. Mayr: *Impact of power quality disturbances on PV inverters – Performance of integrated protective functions*, EPE 2007.

[17] R. González, J. López, P. Sanchis, E. Gubía, A. Ursúa, L. Marroyo: *High-Efficiency Transformerless Single-phase Photovoltaic Inverter*, EPE-PEMC 2006, Portoroż, Slowenia, CD Proceedings.

[18] S. Vasconcelos Araújo, A. Engler, B. Sahan, F. Luiz, M. Antunes: *LCL Filter design for grid-connected NPC inverters in offshore wind turbines*, The 7th International Conference on Power Electronics October 22-26, 2007 / EXCO, Daegu, Korea.

[19] M. Raou, M. T. Lamchich: *AVERAGE CURRENT MODE CONTROL OF A VOLTAGE SOURCE INVERTER CONNECTED TO THE GRID: APPLICATION TO DIFFERENT FILTER CELLS*, Journal of ELECTRICAL ENGINEERING, VOL. 55, NO. 3-4, 2004, 77-82.

[20] F. De Mango1, M. Liserre, A. Dell'Aquila and A. Pigazo: *Overview of Anti-Islanding Algorithms for PV Systems. Part I: Passive Methods*, EPE-PEMC 2006, Portoroz, Slowenia, CD Proceedings.

[21] F. De Mango1, M. Liserre, A. Dell'Aquila and A. Pigazo: *Overview of Anti-Islanding Algorithms for PV Systems. Part II: Active Methods*, EPE-PEMC 2006, Portoroz, Slowenia, CD Proceedings.

[22] R. Stala: *Realizacja FPGA detekcji czestotliwosci rezonansowej obwodu balansujacego przeksztaltnika wielokomorkowego ac/ac z wykorzystaniem funkcji Walsha*, Przeglad Elektrotechniczny (2008), (Polish)

[23] K. Choeisai: *Low Fourier Harmonics PWM Pattern Synthesis Method Based on Walsh Function*, International Conference on Advanced Intelligent Mechatronics (AIM 2003), p. 1172

[24] D. Jankovic, R. S. Stankovic, and R. Drechsler: *Decision Diagram Method for Calculation of Pruned Walsh Transform*, IEEE Trans. On Computers, Vol. 50, No. 2, (2001), p. 147

[25] B.J. Folkowski: *Generation Of Gray Code Ordered Walsh Functions By Symmetric And Shift Copies*, 0-7803-1254-6/93$03.080 (1993) IEEE, p.758.

[26] N.M. Blachman: *Spectral Analysis with Sinusoids and Walsh Functions*, IEEE Trans. On Aerospace And Electronic Systems Vol. Aes-7, No. 5 (1971), p.900.

[27] Ch.K. Yuen: *Function Approximation by Walsh Series*, IEEE Trans. on Computers, (1975), p. 590

[28] P. Porwik: *Widmowe modelowanie systemów cyfrowych o zadanych cechach*, Wyd. Uniwersytetu Śląskiego, Katowice (2000), (Polish).

# Output Maximization Using Direct Torque Control for Sensorless Variable Wind Generation System Employing IPMSG

Yukinori Inoue[*], Shigeo Morimoto[*], and Masayuki Sanada [*]

[*] Osaka Prefecture University, Sakai, Japan,

e-mail: *inoue@eis.osakafu-u.ac.jp, morimoto@eis.osakafu-u.ac.jp, sanada@eis.osakafu-u.ac.jp*

*Abstract*—A variable-speed wind power generation system, based on direct torque control (DTC) of an interior permanent magnet synchronous generator (IPMSG), is proposed herein. The proposed system can achieve the MPPT control without wind speed in addition to the speed and position sensorless control as well as the current control method. DTC has several advantages such as simple system configuration, easy control of flux-weakening, and it doesn't need a position sensor. In particular, the proposed control method can maintain the voltage and current at the rated value in the high speed operating region. Furthermore, as the armature resistance is the only necessary parameter, the system is insensitive to parameter variations. The performance of the proposed wind generation system was validated experimentally.

*Keywords*—Brushless drive, Direct torque and flux control, Generation of electrical energy, Permanent magnet motor, Renewable energy system, Sensorless drive, Wind generator system.

## I. INTRODUCTION

Wind power is one source of renewable energy which is already being extensively utilized. Generally, for wind generation, induction generators are employed. However, the use of permanent magnet synchronous generators (PMSGs) has also been investigated and established as being suitable for small generation systems such as distributed power generation systems. This is because, like permanent magnet synchronous motors (PMSMs), PMSGs have many advantages, such as high efficiency, maintenance-free operation and high-controllability. PMSGs, which are often utilized in high-performance variable-speed drives, can also be employed for variable-speed generation. In variable-speed wind power generation systems, maximum power point tracking (MPPT) control, which adjusts the generator speed to best suit the wind speed, has proved to be an important element in extracting maximum power from the wind. Many methods utilize wind speed information [1], [4]-[9], which requires an anemometer or other estimate of wind speed. Other estimation methods, based on the parameters of the wind turbine, have been reported in [1], [4]. Additionally, a neural-network based method has been reported in [5]. In contrast, methods not requiring wind speed have been investigated in [2], [10], [11]. In [2], fuzzy-logic based control is applied, whereas [10], [11] utilize the quadratic relationship between the optimal generator speed and the torque.

Generator control systems using PMSMs typically adopt a current control method in the rotating reference frame, which is synchronized with the N-pole of the rotor magnet. This requires the rotor position. In [4], the rotor position is estimated using the estimated flux for a sensorless drive. In [10], [11], speed and position estimation is based on estimation of an extended electromotive force. In contrast, other cases not employing position information and which use a diode rectifier were reported in [5], [9]. In these studies, generator control is achieved by control of a DC chopper or a line-side inverter.

Small wind power generation systems, those without mechanical sensors such anemometers or generator position or speed sensors, have cost and reliability advantages. This study examines a wind power generation system which employs neither a wind speed sensor nor a generator position sensor.

Herein, direct torque control (DTC) is applied to achieve optimal control of the wind power generation system. Several studies have examined DTC of permanent magnet synchronous motors (PMSMs) [12]-[15]. Generally, DTC has several advantages over current control methods. First, DTC does not require an accurate motor model or the associated parameters, except for the armature resistance. Second, a rotor position sensor is not required. Finally, it is easy to apply flux-weakening control since the control of the stator flux is easily available.

A variable-speed wind power generation system, using an interior permanent magnet synchronous generator (IPMSG) subject to DTC, is proposed herein. When applied to wind power generation systems, DTC has several advantages over conventional current control. First, DTC, which uses a reference torque as the controller input, is well suited to MPPT control, also applied herein. This is because MPPT control can be implemented by controlling the generator torque, without measuring the wind speed [10], [11]. Second, a method to maintain the terminal voltage at the limiting value is proposed. Controlling the terminal voltage of the generator in high speed operation is simple using flux-weakening control. Finally, a novel method is proposed for controlling the armature current which makes it easy to maintain the armature current at the rated value in high speed operation. Conventionally, a model-based control method is adopted; however deviations from the rated values occur [16]. The proposed method accomplishes current control using the reactive torque, which is calculated as the inner product of the flux

978-1-4244-1741-4/08/$25.00 ©2008 IEEE

and current. This does not require generator parameters such as the magnet flux or inductance.

This paper explains a control method for maximizing the generated power and also optimal vector control of the IPMSG based on DTC. Experimental results demonstrate the effectiveness of the proposed wind power generation system.

## II. WIND GENERATION SYSTEM

### A. System Overview

Fig. 1 illustrates the proposed DTC based wind power generation system. This is composed of a speed estimator, reference torque and reference flux calculator, and a DTC system. The estimated speed of the generator is used for MPPT control of the wind turbine in the absence of a wind speed measurement. The reference torque $T^*$ is calculated so as to maximize the generated power. The reference flux $|\Psi_s^*|$ is calculated so as to minimize losses and control the flux-weakening. The IPMSG, which is driven by the wind turbine, is connected to a DTC controlled PWM converter.

### B. Direct Torque Control

For DTC, the stator flux linkage vector $\hat{\Psi}_s$, in the $\alpha$-$\beta$ reference frame, is estimated based on the PWM converter's reference voltage vector $v_a^*$ and the armature current vector $i_a$. Fig. 2 shows a block diagram of the DTC system. Also, the stator flux linkage and the torque are estimated using (1)-(3) [12]-[14]:

$$\hat{\Psi}_s = \frac{1}{s + \omega_c}(v_a^* - R_a i_a) \tag{1}$$

$$\hat{\theta}_s = \arg(\hat{\Psi}_s) \tag{2}$$

$$\hat{T}_f = P_n(\hat{\Psi}_\alpha i_\beta - \hat{\Psi}_\beta i_\alpha) \tag{3}$$

where: $\hat{\theta}_s$ is the estimated stator flux position, $R_a$ is the armature resistance, $P_n$ is the number of pole pairs, $\hat{\Psi}_\alpha$ and $\hat{\Psi}_\beta$ are the $\alpha$- and $\beta$-axis components of the estimated flux vector $\hat{\Psi}_s$, $i_\alpha$ and $i_\beta$ are the $\alpha$- and $\beta$-axis components of the current vector $i_a$, and $\omega_c$ is the cut-off angular frequency.

Herein, the stator flux is estimated using a first-order low-pass filter instead of a pure integrator. This minimizes the effect of DC offset caused by the current transducer.

Basic DTC [12] uses hysteresis comparators and a switching table. Thus, compared to the current control method, it requires a short sampling period. To negate this, DTC utilizing a reference flux vector calculator [13], [14] (called RFVC DTC) is adopted. In RFVC DTC, the PWM signals can be generated using the PWM technique. This is consistent with the current control method.

In Fig. 2, the reference flux vector calculator generates the reference vector of the stator flux. This is calculated from the torque error and the amplitude and angle of the flux [13], [14]. The reference voltage vector calculator employs time subtraction of the flux vector.

### C. Generator Speed Estimation

The estimated position of the stator flux can be used to estimate the generator speed. This is because, in the steady-state, the velocity of the generator corresponds to the velocity of the stator flux. The estimated speed $\hat{\omega}_g$ is calculated using time subtraction of the estimated position of the stator flux obtained by (2) and passing it through a low-pass filter.

### D. Compensation of the Estimated Torque

Herein, the IPMSG model is assumed to include core losses [11]. Thus, an error occurs between the estimated torque, obtained from (1) and (3), and the actual generator torque. The error in the estimated torque due to core losses can be estimated by (4). Therefore the actual generator torque can be estimated using (5).

$$\hat{T}_i = \frac{P_n^2 |\hat{\Psi}_s|^2 \hat{\omega}_g}{R_c} \tag{4}$$

$$\hat{T} = \hat{T}_f - \hat{T}_i \tag{5}$$

where: $R_c$ is the equivalent core loss resistance and is a function of the generator speed.

## III. GENERATOR CONTROL METHOD

In order to maximize generated power, the references for the torque and the stator flux must be controlled optimally in accordance with the wind conditions. The

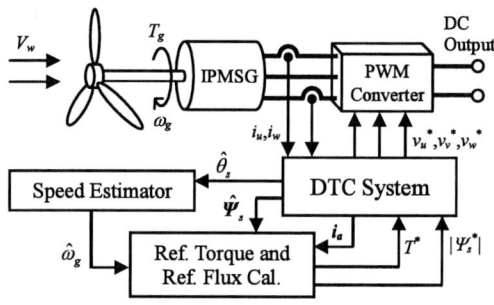

Fig. 1. DTC based wind power generation system.

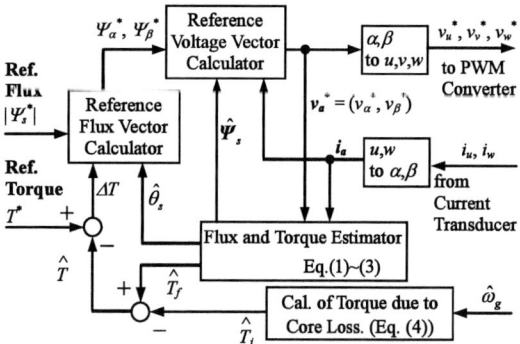

Fig. 2. Block diagram of the DTC with RFVC.

control method for the generator and PWM converter is described below.

### A. Reference Torque Calculation

The reference torque calculation method varies depending on the IPMSG armature current amplitude.

#### 1) Maximum Power Point Tracking Control

When the armature current $I_a$ is below the limiting current $I_{am}$, MPPT control is applied to maximize the mechanical input power. This is implemented in Mode I and II control. The optimum torque $T_{opt}$ is given by a quadratic function of the generator speed [10], [11]:

$$T_{opt} = K_{opt} \omega_g^2 \qquad (6)$$

where: $K_{opt}$ is a constant derived from the wind turbine characteristics.

The value calculated from (6) is used as the DTC reference torque.

#### 2) Control under Limiting Current Conditions

When the terminal voltage and armature current reach their limiting values, MPPT control can no longer be applied. Consequently, the reference torque must be controlled to maximize the generated output power within the limiting values of voltage and current. This is implemented in Mode III control. For this, the reference torque can be calculated from the reactive torque.

The torque equation normally used in DTC is given in (3) and involves the cross product of the flux and the current vectors. The value obtained from the inner product of the flux and the current vectors is defined as the reactive torque and can be estimated from:

$$\hat{T}_r = P_n(\hat{\psi}_\alpha i_\alpha + \hat{\psi}_\beta i_\beta) \qquad (7)$$

The reactive torque is also defined in [15] which proposes control for the DTC system based on the maximum torque per flux. However, though defined, the reactive torque is not actually used for control.

Neglecting core losses, the power in the IPMSG is given by:

$$(V_o I_a)^2 = (T_f \omega_g)^2 + (T_r \omega_g)^2 \qquad (8)$$

where: $V_o$ is the induced voltage, the term on the left corresponds to the apparent power, and the first and the second terms on the right correspond to the active and reactive components, respectively.

The relationship between the induced voltage and stator flux is:

$$V_o = P_n \omega_g |\psi_s| \qquad (9)$$

Substituting (9) into (8) yields the maximum torque $T_{max}$ for the case $I_a = I_{am}$:

$$T_{max} = \sqrt{(P_n |\psi_s| I_{am})^2 - T_r^2} \qquad (10)$$

Herein, because the estimated torque includes the torque due to core losses, the reference torque is:

$$T^* = -\hat{T}_i + \sqrt{(P_n |\hat{\psi}_s| I_{am})^2 - \hat{T}_r^2} \qquad (11)$$

Note: when the reference torque, given by (11), is applied to the system shown in Fig. 2, the error in the estimated torque due to core losses $\hat{T}_i$ cancels. Consequently, for Mode III control, it is unnecessary to estimate $\hat{T}_i$ and only the estimated flux and the detected current are required. This method may well also be effective for other kinds of generator, e.g. an induction generator. This is because (11) is independent of the generator parameters.

### B. Reference Flux Calculation

The reference flux calculation method varies depending on the inverter output voltage amplitude.

#### 1) Loss-minimization Control

When the IPMSG terminal voltage $V_a$ is below the limiting value $V_{am}$, restricted by the DC link voltage, loss-minimization control [10], [11] is applied in order to minimize the total losses in the IPMSG, i.e. the copper loss plus the core loss. This is implemented in Mode I control. The relationship between the stator flux linkage $|\psi_s|$ and the torque $T$ is derived from the IPMSG model and the torque equation in the d-q reference frame. This can be approximated to:

$$|\psi_s|_{MaxEff.} = K_{me3} T^3 + K_{me2} T^2 + K_{me1} T + K_{me0} \qquad (12)$$

where: $K_{me3}$, $K_{me2}$, $K_{me1}$ and $K_{me0}$ are the coefficients of the approximating polynomial and are determined by the IPMSG parameters.

#### 2) Control under Limiting Voltage Conditions (Flux-Weakening Control)

When the terminal voltage reaches the limiting value, flux-weakening control is applied, allowing the generation system to operate in the high speed region. This is implemented in Mode II and III control.

In steady state, the relationship between the $\alpha$- and $\beta$-axis voltage components is:

$$\begin{bmatrix} v_\alpha \\ v_\beta \end{bmatrix} = V_o \begin{bmatrix} \cos(\theta_s + \frac{\pi}{2}) \\ \sin(\theta_s + \frac{\pi}{2}) \end{bmatrix} - R_a \begin{bmatrix} i_\alpha \\ i_\beta \end{bmatrix} \qquad (13)$$

This can be rewritten in scalar form:

$$V_a^2 = (-V_a \sin\theta_s - R_a i_\alpha)^2 + (V_o \cos\theta_s - R_a i_\beta)^2 \qquad (14)$$

When (14) is solved for the variable $V_o$ and the relationship between induced voltage and stator flux is applied, the reference flux can be calculated:

$$|\psi_s|_{FW} = \frac{1}{\omega} \left\{ -R_a \cdot f_1 + \sqrt{(R_a \cdot f_1)^2 + V_{am}^2 - R_a^2 \cdot f_2} \right\} \qquad (15)$$

where: $f_1(i_\alpha, i_\beta, \theta_s) = -i_\alpha \sin\theta_s + i_\beta \cos\theta_s$,

$f_2(i_\alpha, i_\beta) = i_\alpha^2 + i_\beta^2$ and $\omega = P_n \omega_g$

The only generator parameter required in (15) is the armature resistance. The proposed system utilizes these estimated values instead of the position of the stator flux $\theta_s$ and the generator speed $\omega_g$.

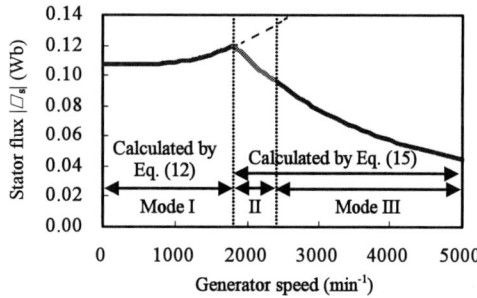

Fig. 3. Stator Flux Trajectory.

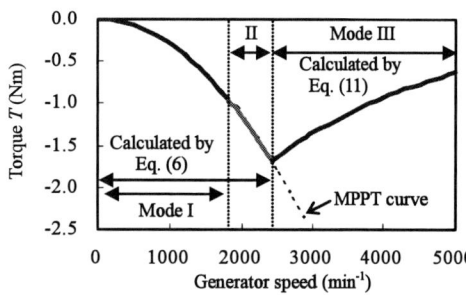

(a) Electromagnetic torque

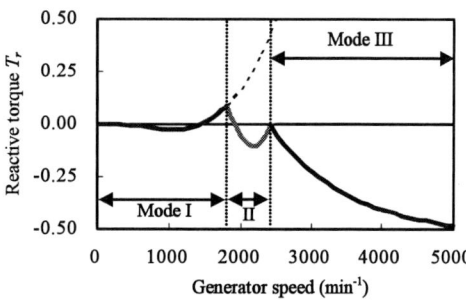

(b) Reactive torque
Fig. 4. Torque Trajectory.

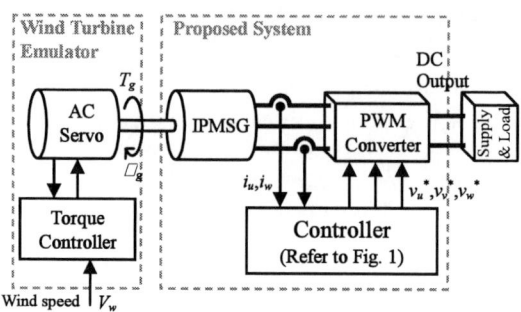

Fig. 5. Block diagram of experimental setup.

TABLE I. EXPERIMENTAL SYSTEM PARAMETERS

| Optimum coefficient: $K_{opt}$ | $2.64 \times 10^{-5}$ Nm / (rad/s)$^2$ |
|---|---|
| Number of pole pairs: $P_n$ | 2 |
| Magnet flux linkage: $\Box_a$ | 0.108 Wb |
| $d$-axis inductance: $L_d$ | 8.7 mH |
| $q$-axis inductance: $L_q$ | $28.3 - 0.657|i_q|$ mH |
| Armature resistance: $R_a$ | 0.64 Ω |
| Iron loss resistance: $R_c$ | $260 \times 40 / (40 + 260/\Box)$ Ω |
| DC link voltage | 61 V |
| Max. terminal voltage: $V_{am}$ | 43.1 V |
| Max. armature current: $I_{am}$ | 8.66 A |

TABLE II. COEFFICIENTS OF APPROXIMATING POLYNOMINAL

| Term | Value |
|---|---|
| $K_{me3}$ | 0.0103 |
| $K_{me2}$ | 0.0255 |
| $K_{me1}$ | 0.0029 |
| $K_{me0}$ | 0.1077 |

## IV. EXPERIMENTAL RESULTS

### A. Experimental Apparatus

The performance of the proposed wind power generation system was verified experimentally. Fig. 5 shows a schematic representation of the experimental apparatus. Instead of an actual wind turbine, an emulator based on an AC servomotor was employed. The AC output of the IPMSG was converted to DC by the PWM converter. The generated power is completely consumed by the electronic load and, consequently, the DC link voltage remains constant.

All of the controls were processed through a digital signal processor (Texas Instruments TMS320VC33). For DTC and speed estimation, the sampling period was 100 μs. For the reference torque calculation and the reference flux calculation, the sampling period was 5 ms.

Table I shows the parameters for the experimental apparatus. Also, Table II gives the coefficients of the approximating polynomial in (12). These were obtained by curve-fitting the characteristics shown in Figs. 3 and 4(a).

### C. Torque and Flux Trajectories

Fig. 3 shows the stator flux trajectory characteristics for the proposed system. In Mode I control, the stator flux increases as the generator speed increases. In contrast, in Mode II and III control (flux-weakening) the flux decreases.

Fig. 4 shows the torque trajectory characteristics. In Mode I and II control (MPPT control), the electromagnetic torque, shown in Fig. 4(a), is controlled using (6). In Mode III control, the torque is regulated to maintain the current at the rated value using (11). In contrast, the reactive torque, shown in Fig. 4(b), varies in a more complex fashion than the flux or the electromagnetic torque. In Mode III control, the reactive torque grows more negative. When operating as a generator, the phase difference between the induced voltage and the armature current becomes 180 degrees when the reactive torque is zero.

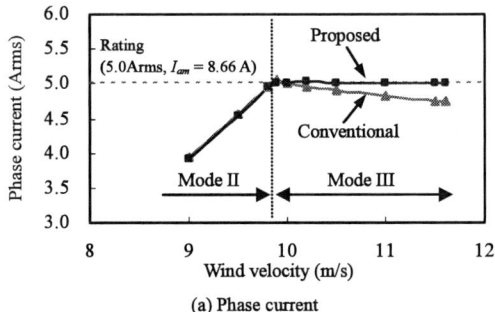

(a) Phase current

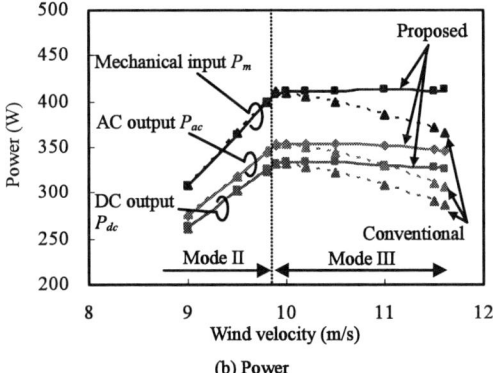

(b) Power

Fig. 6 Steady-state operation at constant voltage.

### B. Steady-state Characteristics in Mode III

In Mode III, where the generator speed becomes high, the only generator parameter required by the controller is the armature resistance. The generator is controlled independently of parameter variations in the magnet flux and the inductance. The experimental Mode III steady-state characteristics are shown in this section.

In the conventional method [16], the reference torque is calculated from an approximate function which, in turn, is obtained from an IPMSG model in the *d-q* reference frame. This method is complicated. For constant voltage and current, the relationship between generator speed and torque is:

$$T_{max} \square K_{tl3} \square_g^3 \square K_{tl2} \square_g^2 \square K_{tl1} \square_g \square K_{tl0} \qquad (16)$$

where: $K_{tl3}$, $K_{tl2}$, $K_{tl1}$ and $K_{tl0}$ are the coefficients of the approximating polynomial and are determined by the parameters of the IPMSG.

Fig. 6 details the steady-state characteristics for constant voltage. In Mode III, the controller maintains the current at the limiting value. Using the conventional method, the phase current reaches the limiting value (5Arms) at a wind speed of 10 m/s. Above this, the current decreases as the wind speed increases and the error becomes significant. This error appears to be due to parameter variations.

In contrast, using the proposed method, the phase current is constant and invariant with wind speed. This is because the stator flux and the reactive torque used in (11) are estimated according to the generator condition. Also,

measured parameters of magnet flux and inductance are not required. This minimizes sensitivity to parameter variations.

Fig. 6(b) shows the characteristics of the mechanical input and the AC output of the IPMSG, and the DC output of the PWM converter. Using the conventional method, as the wind speed increases both the mechanical input power and the DC output power reduce. However, using the proposed method, the generated power does not reduce because the controller can maintain the current at the rated value.

It has been confirmed that the torque control method defined in (11) has several advantages. These include insensitivity to parameter variations and simplicity of calculation.

### C. Dynamic Characteristic

The experimental results for the dynamic performance of the proposed wind generation system are shown in Fig. 7. Here, the wind speed ranges between 5 m/s and 11 m/s, shown in Fig. 7(a). Fig. 7(b) details the estimated speed of the generator and the error of the estimated speed. The generator speed is estimated accurately.

Fig. 7(c) shows the waveform of the stator flux which, in Mode I (low generator speed), is controlled using (12). In Modes II and III, where the generator speed increases and the terminal voltage reaches the limiting value, the flux is controlled by (15). Consequently, as the generator speed increases the flux decreases. Figs. 7(d) and (e) show the estimated torque and the reactive torque, respectively. With reference to Fig. 7(d), because the torque is controlled by (6) in Modes I and II, the generator torque ($\square \hat{T}$) increases with generator speed. In contrast, in Mode III the generator torque decreases to maintain the constant voltage and current. Fig. 7(e) shows that the characteristic of the reactive torque corresponds to the trajectory characteristic shown in Fig. 4(b).

Fig. 7(f) gives the armature current characteristics. Because the torque is controlled by (11), the current does not exceed the limiting value $I_{am}$ (= 8.66 A).

Fig. 7(g) shows the generated power at the DC side of the PWM converter. The power increases during Modes I and II. In Mode III, the power becomes nearly constant.

Fig. 8 details the trajectory of input power to the generator in order to verify the performance of the output maximization control. In Fig. 8, the power versus speed characteristics of the wind turbine and the optimum power curve are superimposed. Point A in Fig. 8 is the maximum output power and occurs at a wind speed of 5 m/s. The maximum output power changes as the wind speed increases and, because the torque is controlled by (6), the power trajectory follows the optimum power curve. When the wind speed reaches 10 m/s (Point B), the terminal voltage and the armature current reach their limiting values. Above this, MPPT control is not applied; instead the torque is controlled within the limiting voltage and current values. Consequently, from here, the power trajectory moves to Point C. The operating condition between Points B and C corresponds to Mode III in Fig. 7. This clearly demonstrates that the proposed system maximizes output without the need for sensors.

1863

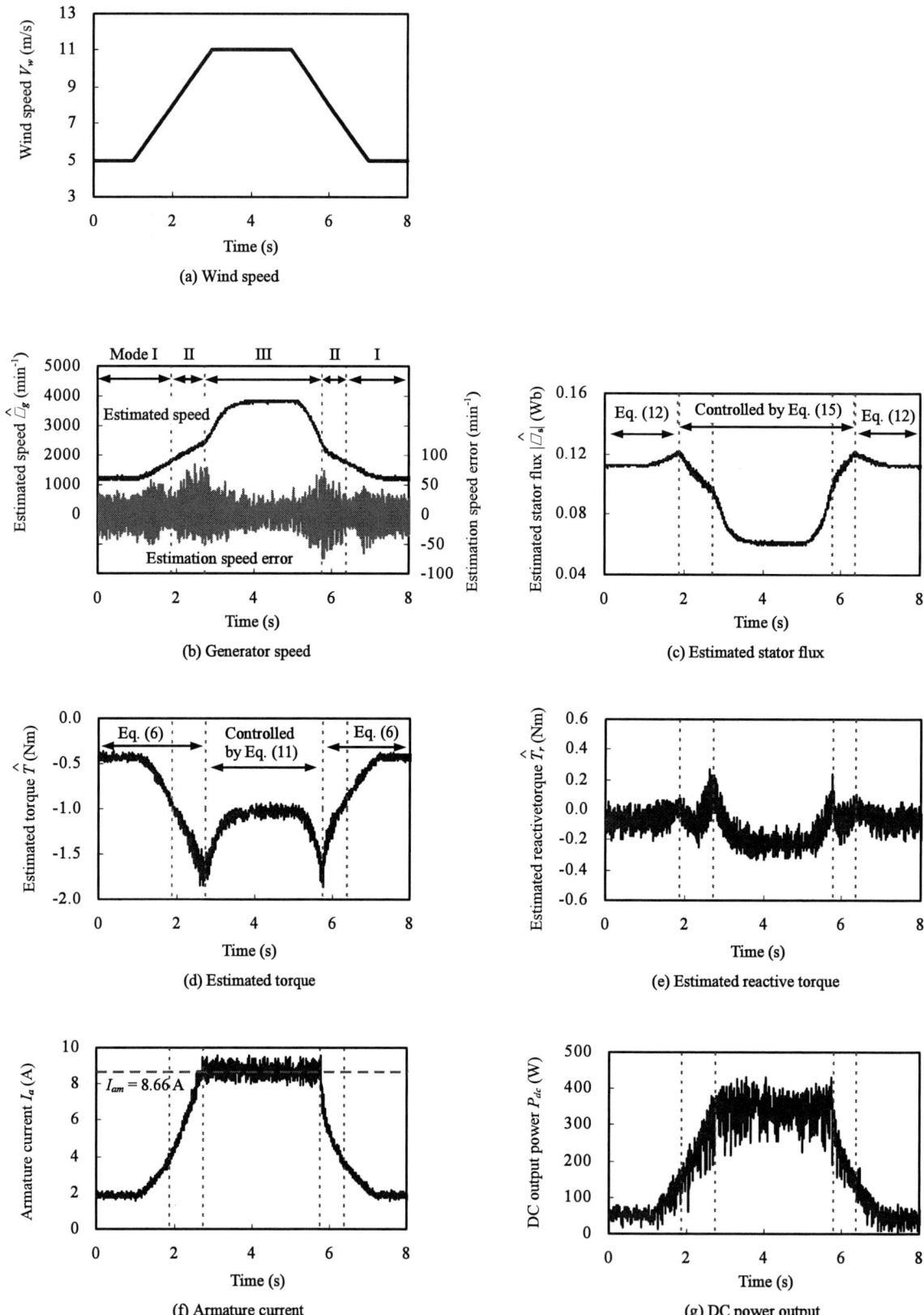

Fig. 7. Characteristics of wind power generation.

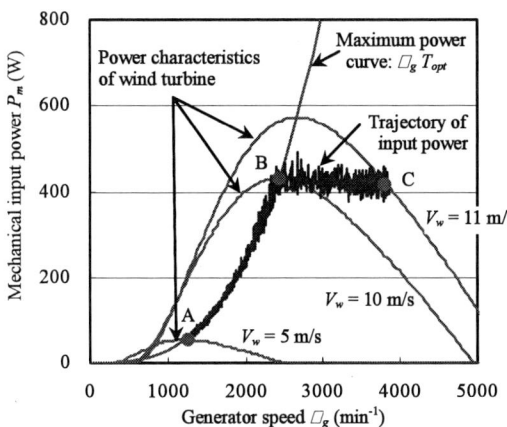

Fig. 8 Trajectory of generator input power.

## V. CONCLUSIONS

The wind generation system investigated herein uses an IPMSG and a PWM converter, both subject to DTC. It was confirmed that MPPT control of the wind turbine, in the absence of wind speed measurement, is possible in such a system. Additionally, control of the IPMSG was achieved without speed and position sensors. The maximum generated power was obtained within the limiting current and voltage of the IPMSG and converter.

## ACKNOWLEDGMENT

This work was partly supported by a Grant-in-Aid for Scientific Research from the Japan Society for the Promotion of Science (JSPS), No. (C) 18560281.

## REFERENCES

[1] S. Bhowmik, R. Spée, and J. H. R. Enslin, "Performance optimization for doubly fed wind power generation systems," *IEEE Transactions on Industry Applications*, vol. 35, no. 4, pp. 949-958, 1999.

[2] A. G. Abo-Khalil, D. Lee, and J. Seok, "Variable speed wind power generation system based on fuzzy logic control for maximum output power tracking," *2004 35th Annual IEEE Power Electronics Specialists Conference Record*, pp. 2039-2043, 2004.

[3] H. Li, and Z. Chen, "Optimal direct-drive permanent magnet wind generator systems for different rated wind speeds," *EPE2007 Conference Record*, 2007.

[4] T. Senjyu, S. Tamaki, N. Urasaki, K. Uezato, H. Higa, *et al.*, "Wind velocity and rotor position sensorless maximum power point tracking control for wind generation system," *2004 35th Annual IEEE Power Electronics Specialists Conference Record*, pp. 2023-2028, 2004.

[5] H. Li, K. L. Shi, and P. G. McLaren, "Neural-network-based sensorless maximum wind energy capture with compensated power coefficient," *IEEE Transactions on Industry Applications*, vol. 41, no. 6, pp. 1548-1556, 2005.

[6] W. Qiao, L. Qu, and R. G. Harley, "Control of IPM synchronous generator for maximum wind power generation considering magnetic saturation," *2007 IEEE IAS Annual Meeting Conference Record*, 2007.

[7] M. Chinchilla, S. Arnaltes, and J. C. Burgos, "Control of permanent-magnet generators applied to variable-speed wind-energy systems connected to the grid," *IEEE Transactions on Energy Conversion*, vol. 21, no. 1, pp.130-135, 2006.

[8] J. Dai, D. Xu, and B. Wu, "A novel control system for current source converter based variable speed PM wind power generators," *2007 Annual IEEE Power Electronics Specialists Conference Record*, pp. 1852-1857, 2007.

[9] K. Ohyama, S. Arinaga, and Y. Yamashita, "Modeling and simulation of variable speed wind generator system using boost converter of permanent magnet synchronous generator," *EPE2007 Conference Record*, 2007.

[10] S. Morimoto. M. Sanada, and Y. Takeda, "Sensorless optimum control of wind generation system with interior permanent magnet synchronous generator," *EPE2003 Conference Record*, 2003.

[11] S. Morimoto, H. Nakayama, M. Sanada, and Y. Takeda, "Sensorless output maximization control for variable-speed wind generation system using IPMSG," *IEEE Transactions on Industry Applications*, vol. 41, no. 1, pp. 60-67, 2005.

[12] M. F. Rahman, L. Zhong, and K. W. Lim, "A direct torque-controlled interior permanent magnet synchronous motor drive incorporating field weakening," *IEEE Transactions on Industry Applications*, vol. 34, no. 6, pp. 1246-1253, 1998.

[13] M. Fu, and L. Xu, "A sensorless direct torque control technique for permanent magnet synchronous motors," *1999 IEEE IAS Annual Meeting Conference Record*, vol. 1, pp. 159-164, 1999.

[14] L. Tang, L. Zhong, M. F. Rahman, Y. Hu, "A novel direct torque control for interior permanent-magnet synchronous machine drive with low ripple in torque and flux – a speed-sensorless approach," *IEEE Transactions on Industry Applications*, vol. 39, no. 6, pp. 1748-1756, 2003.

[15] J. Faiz, and S. H. Mohseni-Zonoozi, "A novel technique for estimation and control of stator flux of a salient-pole PMSM in DTC method based on MTPF," *IEEE Transactions on Industrial Electronics*, vol. 50, no. 2, pp. 262-271, 2003.

[16] Y. Inoue, S. Morimoto, and M. Sanada, "Sensorless output maximization control for variable wind generation system using IPMSG based on direct torque control," *Proceedings of the 2007 JIAS Conference*, vol. 1, pp. 403-406, 2007. (in Japanese)

# Improving Connection and Disconnection of a Small Scale Distributed Generator Using Solid-State Controller

M.M.R. Ahmed[*], member of IEEE

[*] Warwick University / Electrical Department, Coventry, UK, e-mail: *emohamed.ahmed@warwick.ac.uk*

*Abstract— Self excited induction generator is extensively used in renewable energy sources such as small & large wind farms, small hydroelectric power plant and recently in micro combined heat and power systems μCHP. When connecting this unit to the grid a high transient current will flow. There will be a large, very fast, magnetic inrush current similar to that which occurs when a transformer is energised. This paper presents a solid state controller using CoolMOS devices for improving connection and disconnection of the Induction generator to and from the grid. The behaviour of the grid with connection/disconnection of the induction generator (islanding) through solid state controller have been simulated. Experimental model for solid state controller has been built using CoolMOS devices. Both simulations and experimental tests based on a laboratory 1.1 kW induction generator driven by a dc motor are clearly compared and show that the proposed controller are reliable and cost effective.*

***Keywords—Induction generator, distributed generator, solid state switch, IGBT, CoolMOS, Islanding.***

## I. INTRODUCTION

Recently, self excited induction generator SEIG have received increased attention and they have been widely used in been broadly used renewable energy sources such as small & large wind farms, small hydroelectric power plant and recently in micro combined heat and power systems μCHP [1-7]. A μCHP power generation unit based induction generator is going to replace the householder boiler. The new unit will supply household heating but will provide electricity too, with excess power perhaps being sold to the local grid [8]. The SEIG is the most suitable candidate for this application due to its advantages over synchronous generator such as lower cost, simple construction, lower maintenance requirements due to their ruggedness, no need for dc supply for excitation, no need for synchronization equipment and better transient characteristics [9].

In grid-connected mode, the excitation current to the induction generator is supplied through the grid. A mechanical arrangement is used to maintain the rotor speed above the synchronous speed. In self-excitation mode, the excitation current is supplied by a local source. The simplest way of self-excitation is the use of fixed capacitors connected across induction generator (IG) terminals. The excitation capacitor value is chosen so as to regulate the generated voltage with changing speed and

load [10-12]. The excitation capacitance value can be changed by switching capacitor banks or using a thyristor-controlled reactor (TCR) for smooth variation [13].

The problem arises when connecting IG to the grid a high transient current will flow. There will be a large, very fast, magnetic inrush current similar to that which occurs when a transformer is energized which are undesirable particularly in the case of weak grids and can also cause severe torque pulsations and probably damage to the coupling [14].

There are few techniques have been found in the literature which are extensively used as a soft-starter in wind power generation system. A soft-starter based on semiconductor devices such as SCR, TRIAC and power resistor have been used for 15kW or more rated wind energy conversion system [15-17]. However further research and development is needed to be extended on soft-connection strategy for 1-5 kW μCHP system to provide the quality power to the grid and also to reduce the electrical and mechanical stresses on the entire system.

The aim of this paper is to develop an alternative technique using a solid-state controller, which will improve the connection and disconnection of small scale distributed generators to/from the low voltage grid and also it can limit and interrupt the fault currents. A simulation result has been carried to examine the performance of the proposed controller using MATLAB/ SIMULINK software Package. Experimental results obtained from a laboratory 1.1 kW induction generator driven by a dc motor are also performed. Both simulation and experimental results prove system practical effectiveness in term of fast response and high performance.

The layout of this paper is as follows. The construction and operation of the solid state controller is given in Section II, while hardware implementation of solid state controller of SEIG and self excitation of SEIG are presented in Sections III. The simulation and experimental results describing the connection / disconnection of SEIG to/from the grid are given in section IV. Conclusion is made in Section VI.

## II. CONSTRUCTION AND OPERATION OF SSC

Fig. 1 shows the construction of the SEIG based solid state controller. The SSC consists of three solid state switch and low power resistance through which the system is connected to the grid by S1. In addition gate drivers and controller to operate the SS controller. In order to limit the inrush current during the grid connection, a simple analog control logic has been used which makes it low cost and meaningful for small distributed generation system.

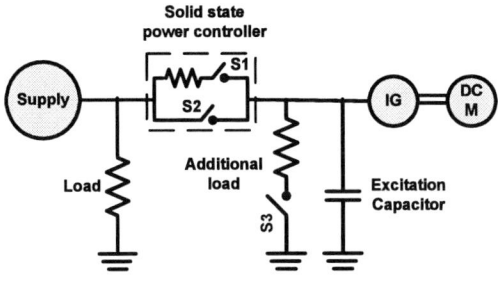

Fig. 1. Construction of SSC

The configuration of the Solid State Switch used in a solid state controller is shown in Fig. 2. It consists of a fast acting, bidirectional switch realized using power semiconductor devices [Super Junction transistors (CoolMOSs) and diodes as shown in the Fig. 2], a varistor (nonlinear resistor), and a snubber circuit, all connected in parallel. There are alternative ways to construct the switch [18]. The topology shown in Fig. 2 was chosen after considering the conduction and switching losses, transients associated with the reverse recovery current of the diodes and on state voltage drop across the switch.

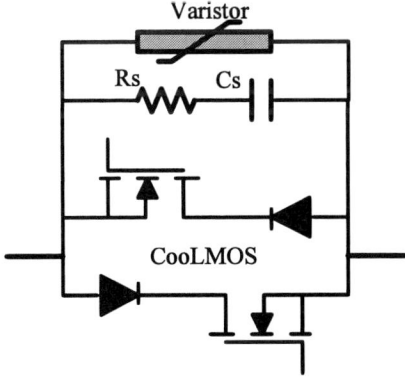

Fig. 2. Construction of the SSC

The principle operation of the solid state controller is as follows; first the SEIG is connected to the grid through a resistor by turning on S 1 for few ms then the connection is set up by S1 & S2 after a suitable time delay S 1 is open again and the connection is set up by S2. Timing for switching S 1 and S 2, and the suitable value of the power resistor are determined through both simulation and experimental investigation.

## III. HARDWARE IMPLEMENTATION

Fig. 3 shows the schematic diagram of the experimental hardware implementation of a solid state controller of SEIG. A separately excited DC machine has been used to drive the SEIG. In order to control the motor speed one variable dc power supply excites the field winding of the separately excited dc motor, while its armature current is maintained constant. The dc motor drives the single-phase induction generator above its synchronous speed. Through the solid state controller the induction generator is connected to the grid. In this work a normal circuit breaker has been used to connect the excitation capacitor across the IG terminals after it reach synchronous speed or slightly higher than it. The control part starts with sensing the line to line terminal voltages and the machine stator currents, these sensed voltage and current signals are fed to the control board to generate the adequate signals to the gate drive of the solid state controller switches by the comparison of the grid voltage and SEIG terminal voltage. Then the gate drive sends the required signals to the solid state controller based on the mode of operation.

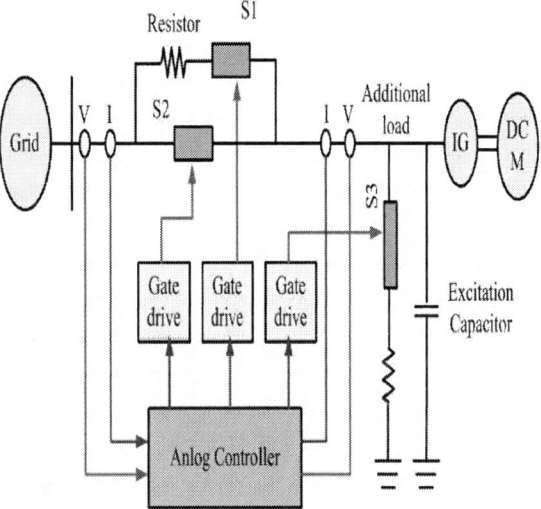

Fig. 3. Schematic diagram of the solid state controller

### A. Self excitation of the induction generator

The induction generator is driven at the speed of 3035 rpm and at time $t\_0$, one single-phase capacitor of value 40 mf are switched between IG terminals. Fig. 4 shows the build-up of phase voltage and speed during the self- excitation. It is observed that the transient build-up of terminal voltage continues until the magnetic circuit of the machine is saturated and it settles down to a steady state value, depending upon the prime-mover speed and the value of excitation capacitors. The relation between excitation capacitor values and generated voltage is shown in Fig. 5 once the SEIG connected to the grid it follows the grid voltage and frequency without need for synchronization equipment as shown in Fig. 6.

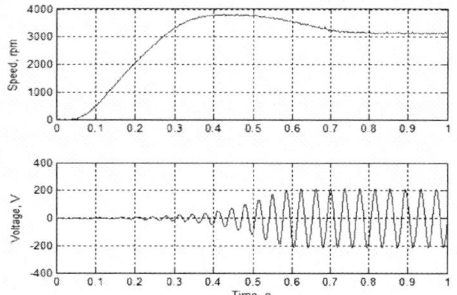

Fig. 4. Voltage build up across SEIG terminals

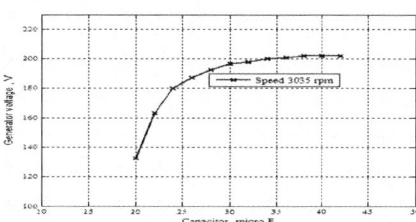

Fig. 5. generated voltage verses excitation capacitance.

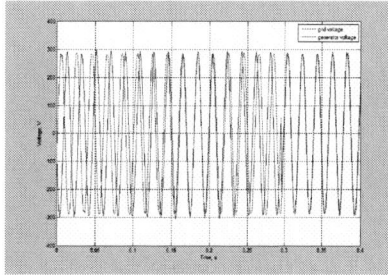

Fig. 6. Generator and grid voltage.

## IV. SIMULATION AND EXPERIMENTAL RESULTS

The proposed system in Fig. 3 was modelled and simulated in Matlab, using Simulink and Power System Blockset. In all simulations, the IG is excited using fixed capacitors bank of 40 mF. In order to test the influence of the solid state controller on the connection and disconnection of SEIG to/from the grid three Scenarios have been carried out using simulation and experimental work.

### A. First Scenario driect Connection of SEIG to the grid

Fig. 7 shows the circuit diagram of the test system under direct connection of SEIG to the grid through normal circuit breaker without any control. Figs. 8-10 show the simulation waveforms of generator current, voltage and Speed. Figs. 11-13 show the experimental waveforms of generator current, voltage and Speed. From these results it can be observed that connection of SEIG to the grid can led to high inrush current 75 A which may lead to serious damage to the reset of the system.

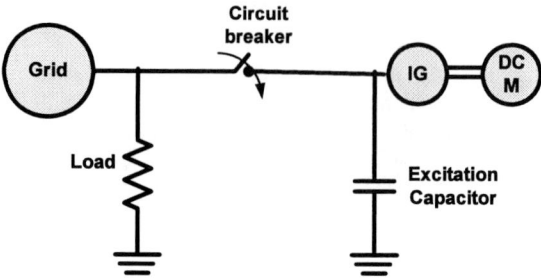

Fig.7. Test circuit

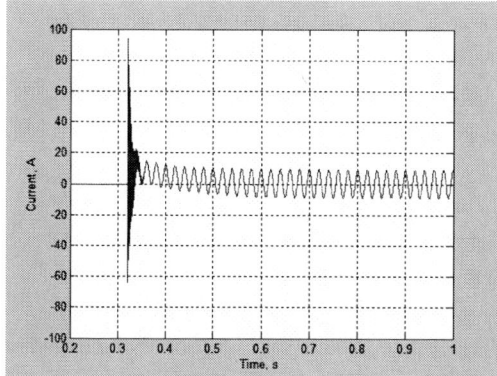

Fig. 8. Current waveform (simulation)

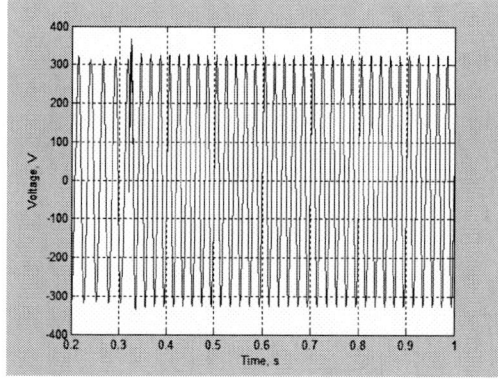

Fig. 9 Voltage waveform (simulation)

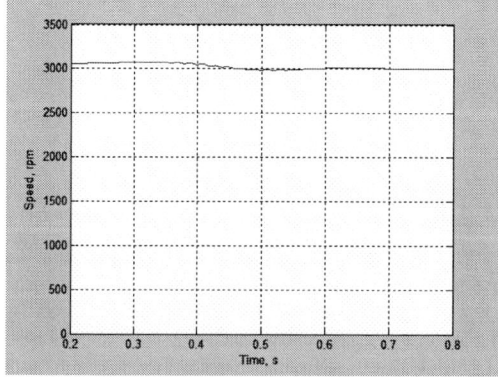

Fig. 10. Speed waveform (simulation)

1868

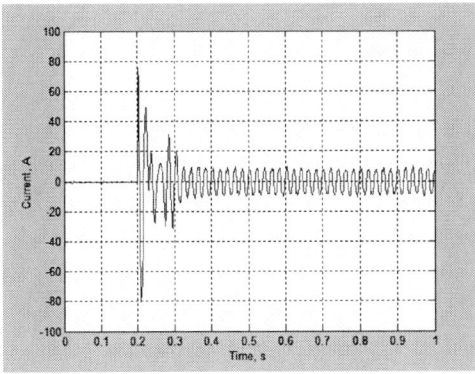

Fig. 11. Current waveform (experimental)

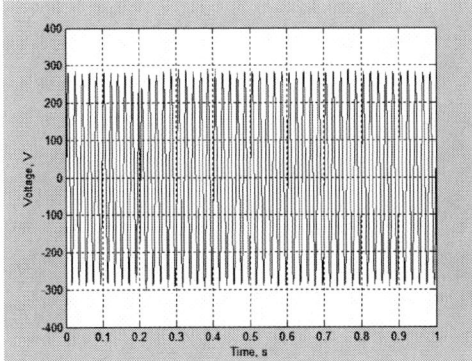

Fig. 12 Voltage waveform(experimental)

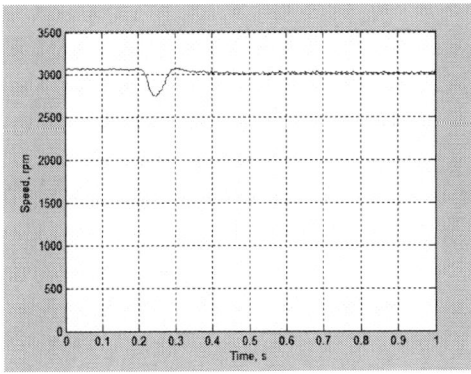

Fig. 13 Speed (experimental)

### B. Second Scenario Connection of SEIG to the grid via Solid State Controller

Figs. 14 & 15 show the circuit diagram of the test system during direct connection of SEIG to the grid through Solid state controller. First the generator connected to the grid through a low power resistor by closing S1 (Fig. 14) then after few ms S2 is closed after suitable time S1 is opan and the SEIG become connected to the grid via S2 (Fig. 15). Figs. 16-18 show the simulation waveforms of generator current, voltage and Speed. Figs. 19-21 show the experimental waveforms of generator current, voltage

and Speed. From these results it can be seen that the current is limited and after few ms is settle to its steady state value. Both values of resistor and delay time between switching S1 & S2 has been determined experimentally after running different tests. We found that 30 and delay time of 50 ms give better performance and limit the inrush current to 1.5 of the generator nominal current.

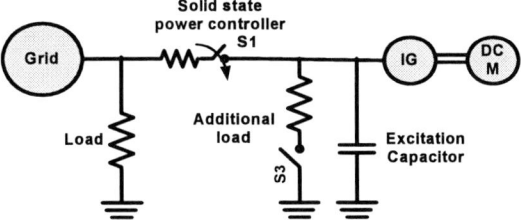

Fig. 14 Test system

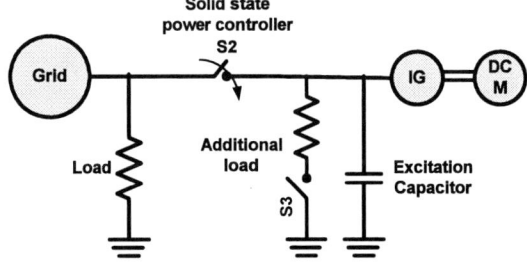

Fig. 15. Test system

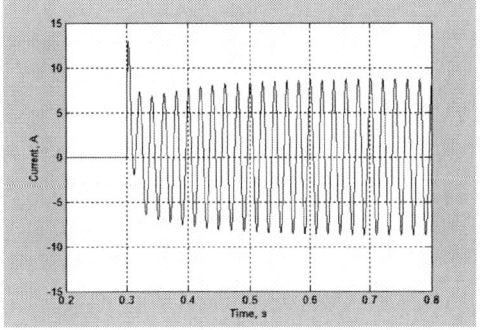

Fig. 16. Current waveform (simulation)

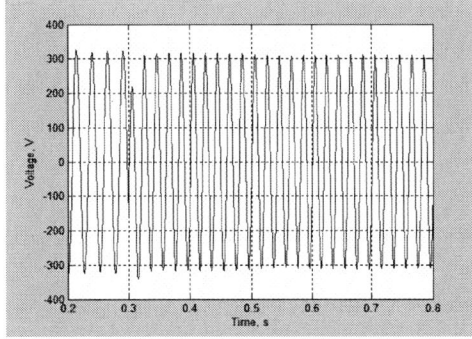

Fig. 17 Voltage waveform (simulation)

1869

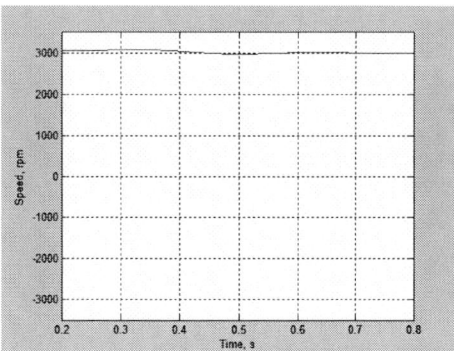

Fig. 18 Speed (simulation)

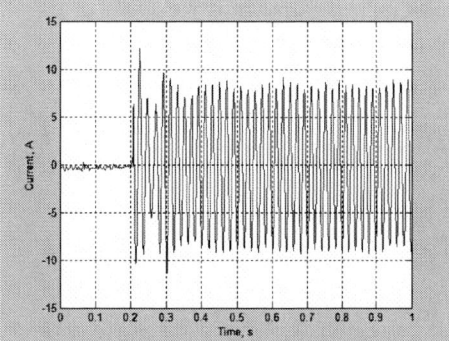

Fig. 19 Current waveform (experimental)

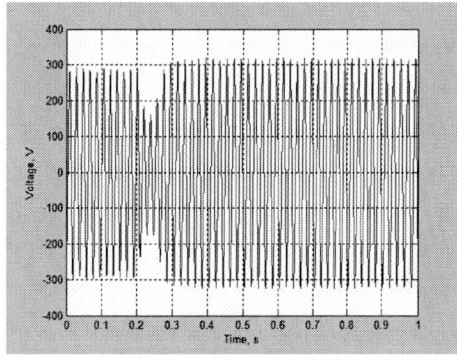

Fig. 20 Voltage waveform (experimental)

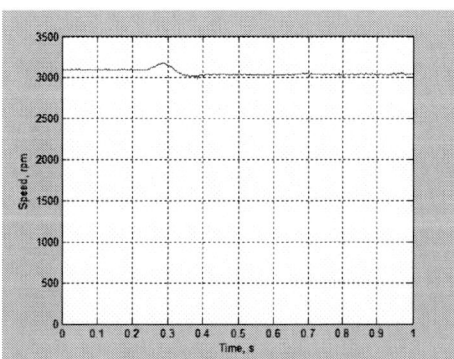

Fig. 21. Speed (experimental)

## C. Third Scenario Disconnection of SEIG from the grid via Solid State Controller

Fig. 22 shows the circuit diagram of the test system during disconnection of SEIG from the grid by using solid state controller. Figs. 23 shows the generator current feeding the grid (the SEIG fed the grid by 184 W during period from 0 to 0.2 s), then the generator has been disconnected from the grid at t = 0.2 s and reconnected to additional load at t = 0.38 s. Fig. 24 shows the SEIG voltage response after disconnection from the grid and during connection to the load with reduction of 36% of nominal voltage. A voltage & frequency regulator needs to be integrated to the solid state controller to prevent voltage sag during islanding operation. Fig. 25 shows the dc motor start to increases after the SEIG disconnected from the grid and the then back to the initial value 3065 rpm after SEIG connected to the grid [19-20]. The data of the tested system is given in appendix.

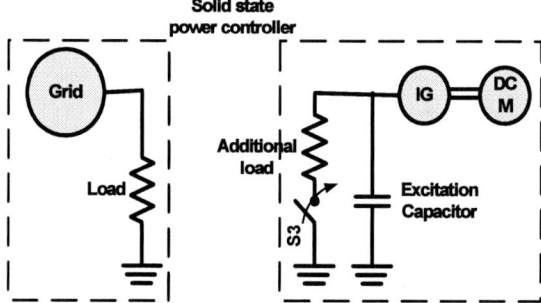

Fig. 22. Test system

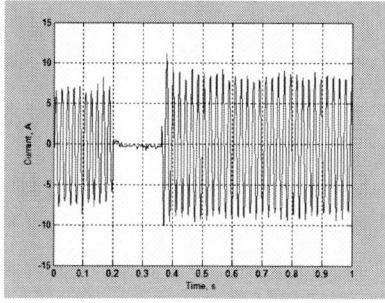

Fig. 23. Current waveform (experimental)

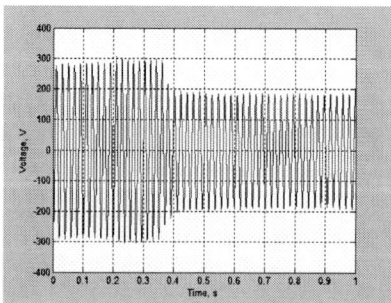

Fig. 24. Voltage waveform (experimental)

1870

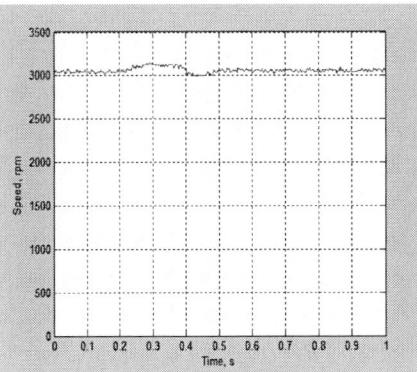

Fig. 25 Speed (experimental)

## V. CONCLUSIN

This paper has presented both simulated and experimental results of a 1.1 kW single phase self excited induction generator during connection/disconnection to/from the grid through a solid state controller.

Both simulation and experimental results has shown satisfactory operation of the proposed solid state controller of SEIG. This controller is capable to do the following functions:

- Limit the inrush current during connection of SEIG to the grid.

- Limit the induction generator voltage after disconnection of SEIG from the grid by connecting resistive load across the IG terminals.

- Limit and interrupt the fault current occurs at the grid side by inserting a series resistor between the IG and grid.

- Further work need to be done to integrate the voltage & frequency regulator to prevent voltage sag during islanding operation.

The proposed controller is expected to be used in both single and three phase system as high performance, reliable and cost effective interface device between SEIG and grid.

## APPENDIX

Rating of a single phase 1.1 Kw, 220V, 50 Hz, 2 pole induction machine is:
Ras=2.2Ω  Llqs=1.3mH, Rds=7.62Ω  Llds = 6.3 mH, Rqr = 1.3Ω Rdr = 3.5Ω, Llqr = 2.8mH, Lmq = 0.485 mH, J = 0.00875 Kgm2, C=40 ☐F. Current limiting Resistor = 30 Ω. CoolMOS IXKC 20N60C 600 V, 15 A RDS(on) max = 190 mΩ. DC motor 1.1 kW, 220 V.

## REFERENCE

[1] C. Grantham, D. Sutanto and B. Mismail, "Steady-state and transient analysis of self-excited induction generators", IEE Proc. B, Vol. 136, No. 2, pp. 61-68, March 1989.

[2] M. H. Salama, and P.G. Holmes, "Transient and steady-state load performance of stand-alone self-excited induction generator", IEE Proc. -Electr. Power Appl., Vol. 143, No. 1, pp. 50-58, January 1996.

[3] Murthy S. S., "A novel self-excited self-regulated single-phase induction generator Part 1: Basic system and theory," IEEE Transactions on Energy Conversion, Vol. 8, No. 3, pp. 377-382, Sept. 1993.

[4]. Murthy S. S., Rai H. C., Tandon A. K., "A novel self-excited self-regulated single-phase induction generator Part II:Experimental investigation," IEEE Transactions on Energy Conversion, Vol. 8, No. 3, pp. 383-88, Sept. 1993.

[5] Rahim Y. H. A., Alolah A. II and Al-Mudaiheem R. I., "Performance of single phase induction generator," IEEE Trans. on Energy Conversion, Vol. 8, No.3, pp. 389-395, Sept. 1993.

[6] Ojo O., "Performance of self excited, single phase induction generators with shunt, short-shunt and long shunt excitation connections," IEEE Trans. on Energy Conversion, Vol. 11, No. 3, pp. 477-482, Sept. 1996.

[7] L. B. Shilpakar, B. Singh and B. P. Singh, "Dynamic behavior of threephase self-excited induction generator for single-phase power generation", Electric Power Systems Research, Vol. 48, pp. 37-44, 1998.

[8] P. Breeze, " Power Generation Technologies", Book, Newnes, 2005, PP. 64-67.

[9] J.M Elder, J.T. Boys and J.L. Woodward, "Self-excited induction machine as a small low-cost generator", IEE Proc. C, Vol.131, No. 2, pp. 33-41, March 1984.

[10] L. Shridhar, B. Singh, C. S. Jha, B. P. Singh and S.M. Murthy, "Selection of capacitance for the self regulated short shunt self excited induction generator", IEEE Trans. on Energy Conversion, Vol.10, No.1, pp. 10-17, March 1995.

[11] N.H Malik and A.H. Al-Bahrani, "Influence of the terminal capacitor on the performance characteristics of a self excited induction generator",IEE Proc C., Vol. 137, No. 2, pp. 168-173, March 1990.

[12] Al Jabri, A. K. and Alolah, A. I., "Capacitance requirements for isolated self-excited induction generators", IEE Proc. B, Vol. 137, No. 3, pp. 154-159, May 1990.

[13] Al-Saffar M.A. Nho E. and Lipo T.A., "Controlled Shunt Capacitor Self-Excited Induction Generator", in Proc. 1998 IEEE Industry Applications Conference, Vol. 2, pp. 1486-1490.

[14] M. A. Ouhrouche, D. Xuan and R. Chane, "EMTP based simulation of a self-excited induction generator after its disconnection from the grid", IEEE Transactions on Energy Conversion, 1998, Vol. 13, No. 1, pp. 7-14.

[15] Torbjorn Thiringer, "Grid-friendly connecting of constantspeed wind turbines using external resistors", IEEE transactions on energy conversion,Vol. 17, No. 4, December 2002.

[16] F. Iov, Hansen, F. Blaabjerg, Remus Teodorescu, "Modeling of soft-starters for wind turbine applications", Power quality proceedings,May 2003, pp. 179-182.

[17] R. Ahslan, M.T. Iqbal and George K. I. Mann, Power Resistors Based Soft-starter for a Small Grid Connected Induction Generator Based Wind Turbine, Proceeding of The The Seventeenth Annual Newfoundland Electrical nd Computer Engineering Conference, 2007, pp. 1-5.

[18] C. Meyer, S. Schroder, and R. W. De Doncker, "Solid-state circuit breakers and current limiters for medium-voltage systems having distributed power systems," IEEE Trans. Power Electron., vol. 19, no. 5, pp. 1333–1340, Sep. 2004.

19] S. P. Singh, S. K. Jain and J. Sharma, "Voltagr regulation optimization of compensated self-excited induction generator with dynamic load",IEEE Trans. on Energy Conversion, Vol. 19, no. 4, pp. 724-732, 2004.

[20] B. Singh, S. S. Murthy and Sushma Gupta, "An Electronic Voltage and Frequency Controller for Single-Phase Self-Excited Induction Generators for Pico Hydro Applications", Proceedings of The IEEE International Conference on Power Electronics and Drive systems, 2005, pp. 240-245.

# Research control of electric systems in wind generator systems

Stefan Winternheimer[*], Artem Kolesnikov[†], Evgeny Glushkin[†], Alexander Bukatov[†]

[*] University of Applied Sciences of Saarland, Saarbruecken, Germany, e-mail: Stefan.winternheimer@htw-saarland.de
[†] Khakassky Technical Institute – branch of the Siberian Federal University, Abakan, Russia,
e-mail: kolesnikov_a_a@mail.ru

*Abstract*—**The problem of the operation's quality improvement of existing control systems of the wind-turbine electric plant is investigated in dynamic processes. The computer model of dynamic processes of considered system is developed for the decision of the given problem. On the basis of modern optimization methods and by means of computer simulation the optimization of existing control systems of the wind-turbine electric plant is executed. The new method of the model predictive control is offered for the researched electric system.**

*Keywords*—**windgenerator systems, control methods for electrical systems, optimal control, flicker, modelling.**

## I. INTRODUCTION

Nowadays there is great interest both in the countries of the European Union and in Russia to renewable energy sources. One of the basic and the most perspective renewable energy sources is application the wind-turbine electric plant [1]. However for the given type of energy sources there is a lot of problems which require the decision. One of these problems arises because of non-stationary of the disturbance acting on the wind-turbine electric plant, and is caused by impossibility of the classical subordinated control systems to consider these changes [2]. It leads to degradation of electrical energy in the wind-turbine electric plant in dynamic processes, in particular, to occurrence flicker. Modern methods of computer modeling in system Matlab and realized on its base optimization methods allow investigating accuracy of operation of existing classical subordinated control systems, to define their optimum parameters in view of non-stationary disturbance, to develop new methods model predictive control (MPC) of the wind-turbine electric plant.

The aim of the paper is decreasing the current changes caused by the network voltage flickers, by means of usage optimized parameters for classical subordinated control systems and development of model predictive control.

## II. DESCRIPTION OF ELECTRIC SYSTEM OF WIND-TURBINE ELECTRIC PLANT

In the paper is studied electric system (ES) of the wind-turbine electric plant the scheme of which presented by fig. 1. The paper will consider ES type of object dynamics in the form of state-space equations which describes dynamics of ES in the best way and most actually in practical applications:

$$\begin{cases} \dot{I}_d = \left( \dfrac{D}{L_d} - \dfrac{1}{L_d} \right) U_{zk} + \dfrac{U_d}{L_d}, \\ \dot{U}_{zk} = \left( \dfrac{1}{C_{zk}} - \dfrac{D}{C_{zk}} \right) I_d - \dfrac{I_{zk}}{C_{zk}}. \end{cases} \quad (1)$$

For research of operating of disturbing influence from the network, approximation of the proportional subordinated control (P-control) and the converter, the system (1) is added by following system equations:

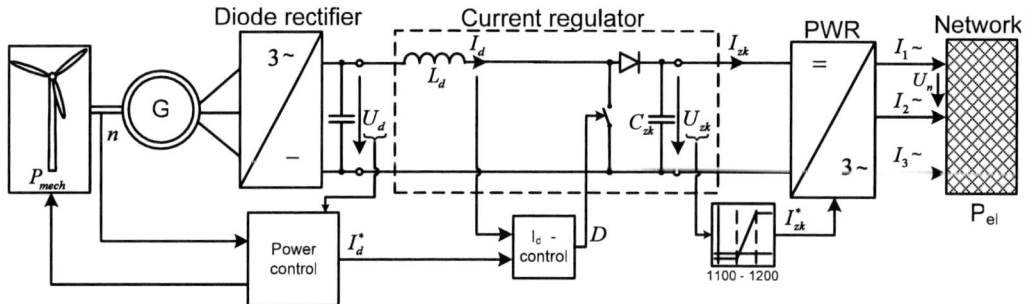

Fig. 1. Key diagram of the wind-turbine electric plant.

$$\begin{cases} \delta\dot{U}_{zk} = -\dfrac{1}{T_g}\delta U_{zk} + \dfrac{U_{zk} - U_{zkn}}{T_g}, \\[2mm] \dot{I}^*_{zk} = -\dfrac{1}{T_r}I^*_{zk} + \dfrac{K_r}{T_r}\delta U_{zk}, \\[2mm] \dot{I}_{zk} = -\dfrac{1}{T_n}I_{zk} + \dfrac{1}{T_n}I^*_{zk}, \end{cases} \quad (2)$$

where variables systems (1)-(2) are presented by fig. 1.

### III. PARAMETER OPTIMIZATION OF P-CONTROL OF ELECTRIC SYSTEM

The flowchart (fig. 2) on the basis of systems (1)-(2) allows realizing ES in Matlab (fig. 3), investigating operation of existing classical subordinated control (P-control) systems in dynamic modes in conditions of voltage fluctuation from the external electrical network $U_n$, signal of power control $I^*_d$, voltage on an output of the diode rectifier $U_d$, etc.

Practical work has shown that ES with P-control in conditions of operating disturbance from the load point of view function sometimes unsatisfactorily. It can result in undesirable effect e.g. the deviation of a voltage and a frequency, flicker occurrence. Using of the system Matlab and application of Nonlinear Control Design gives effective opportunity of optimization real P-control parameters in conditions of existence flicker [3].

For example by fig. 4 oscillograms $I_{zk}$, $I^*_{zk}$, $U_{zk}$, $U_n$ in conditions of flicker influence from an external network is presented at $I^*_d = \text{const}$ and $D = 0$ (with index «p»). Oscillograms show currents (b: $I_{zk}$ and $I^*_{zk}$) and voltage (c: $U_{zk}$) under the flicker influence with frequency 8 Hz on the disturbing value (a: $U_n$). Operation P-control with the optimized parameters (with index «o») yields more effective results and leads to decrease in amplitude $I_{zk}$ per 86.4 %. However optimization of P-control parameters requires constant updating of parameters and does not exclude completely existence flicker.

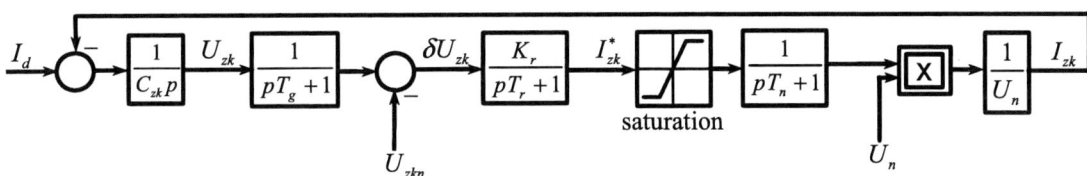

Fig. 2. Flowchart of ES with P-control in Matlab.

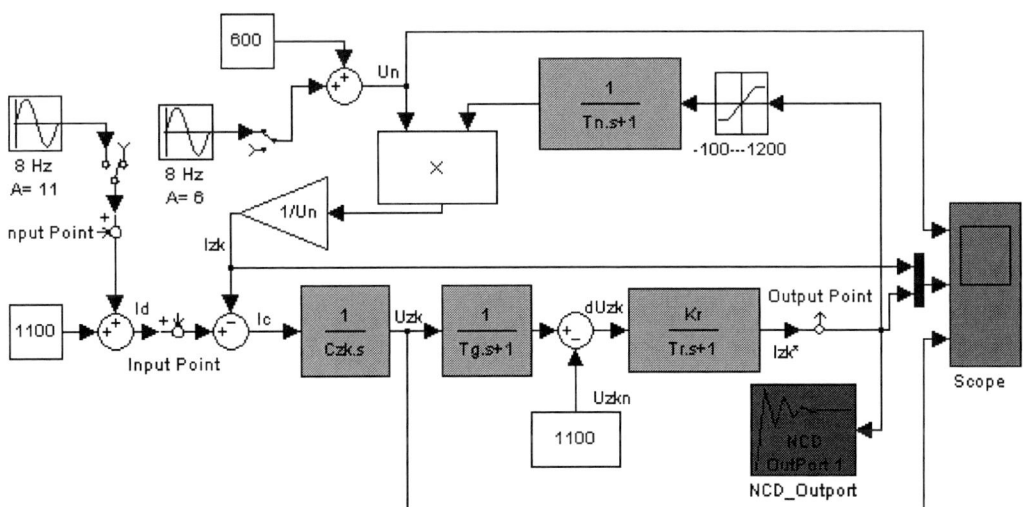

Fig. 3. S-model of ES with P-control in Matlab.

## IV. COMPUTER REALIZATION OF ELECTRIC SYSTEM WITH MODEL PREDICTIVE CONTROL

In that case, more effective and modern approach in control ES can be reached by application MPC the flowchart is presented on fig. 5 and the model of which is realized in Matlab and is presented on fig. 6 [4].

For the decision of MPC problem as usual we shall write down equations system of observable coordinates (1)-(2) in the following general kind:

$$y = Lx + Mu ,$$

where

$$L = \begin{pmatrix} A \\ A^2 + A \\ \dots \\ \sum_{i=1}^{P} A^i \end{pmatrix}, \quad M = \begin{pmatrix} B & 0 & \dots & 0 \\ AB + B & B & \dots & 0 \\ \dots & \dots & \dots & \dots \\ \sum_{i=0}^{P-1} A^i B & \sum_{i=0}^{P-2} A^i B & \dots & B \end{pmatrix},$$

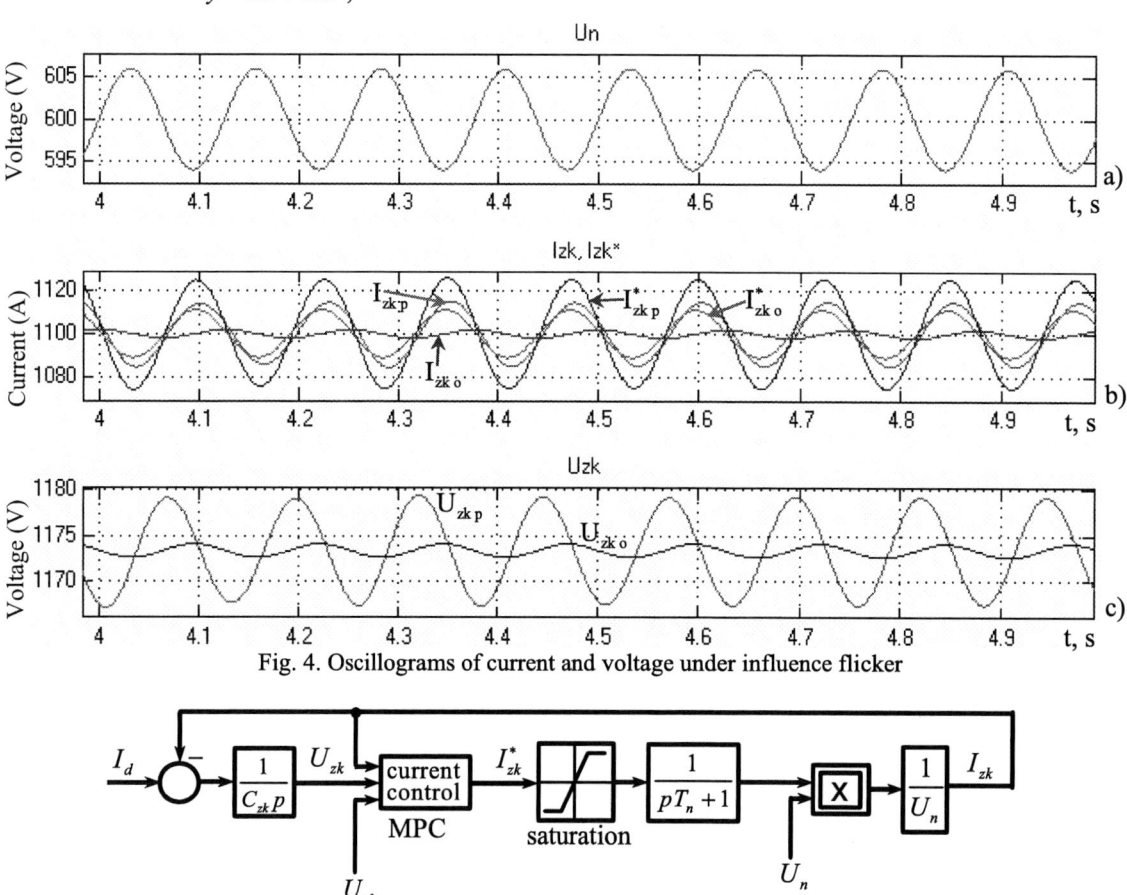

Fig. 4. Oscillograms of current and voltage under influence flicker

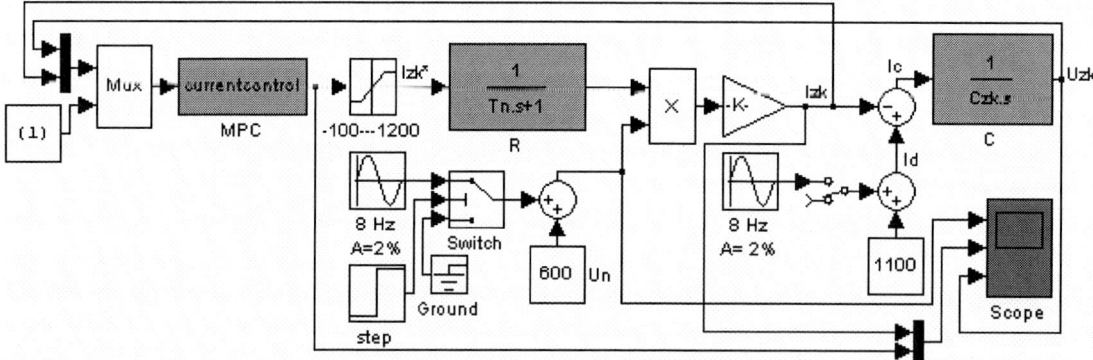

Fig. 5. Flowchart of ES with MPC in Matlab.

Fig. 6. S-model of ES with MPC in Matlab.

$A$, $B$ – matrixes of state-space coordinates and control, $P$ – prediction horizon.

Vector of the signal control change on $i$ step of control process is defined with the help of the formula:

$$\Delta u_i = u_i - u_{i-1}.$$

In the process of searching the optimized problems decision of the vector $\Delta u_i$ the functional on *i-interval* of time is minimized:

$$J_i = \min\left(\Delta u_i^T H \Delta u_i + 2f^T \Delta u_i + g\right), \quad (3)$$

where

$$H = M^T R M + Q, \; f = M^T R L p_k - M^T R r,$$

$$g = p_k^T L^T R L p_k + r^T R r - 2p_k^T L^T R r,$$

$Q$, $R$ – matrixes of weight coefficients,

$$p_k = \begin{pmatrix} x_k - x_{k-1} \\ C x_k \end{pmatrix} - \text{vector of initial conditions,}$$

$r$ – vector of desirable values of coordinates of the state of system. Thus, on each interval of control limitations should be fulfilled:

$$u_i^- \leq u_i \leq u_i^+,$$
$$\Delta u_i^- \leq \Delta u_i \leq \Delta u_i^+, \quad (4)$$

and limitations on state-space coordinates of system:

$$x_i^- \leq x_i \leq x_i^+, \quad (5)$$

where $u_i^-$, $u_i^+$, $\Delta u_i^-$, $\Delta u_i^+$, $x_i^-$, $x_i^+$ – set values of limitations.

Thus the formula

$$u_i = u_{i-1} + \Delta u_i$$

defines control which is optimum on a current step. The next step, according to the general conception of the MPC, repeats the process of calculations.

The question of minimization integrated square-law functional (3) with restrictions (4)-(5) is converted to the standard problem of the numerical analysis– to the problem of convex square-law programming which can be executed by the method of active limitations [5]. In this case the procedure of searching the problem (3) solution consists of the following steps:

1. **Checking up of the breaking conditions:** If in the actual point $\hat{u}_k$ the breaking conditions are fulfilled, calculations stop and $\Delta u_i = \hat{u}_k$ is solution of the problem (3).

2. **Choice of a logical branch:** It is being checked, whether it is necessary to continue optimization in the actual subspace or if it is meaningful to delete any limitations from the working list (4)-(5). If it is decided to save the working list, the transition to step N 3, – otherwise to step N 6 is carried out.

3. **Calculation of the searching direction:** The nonzero $(n - t_k)$-dimensional vector $p_z$ and the direction of search is calculated

$$p_k = Z_k p_z,$$

where $Z_k$ is a matrix made of basis vectors a subspace, orthogonal to the row of the matrixes of limitations and $t_k \times n$ is dimension of the matrix of limitations in the actual subspace.

4. **Calculation of step length:** As much as possible the admissible size of the step $\hat{u}_k$ lengthways $p_k$ is calculated. The positive step $\alpha_k$, such as

$$F\left(\hat{u}_k + \alpha_k p_k\right) < F\left(\hat{u}_k\right), \alpha_k \leq \bar{\alpha},$$

where $\bar{\alpha}$ – the set of limitation of step length – is defined. If $\alpha_k < \bar{\alpha}$, transition to step N 7, – otherwise to step N 5 is carried out.

5. **Introduction of limitations in the working list:** If $\alpha_k$ – step at which any limitation becomes equality – this limitation is added to the active set. Other attributes of the working list are updated and go to a step N 1.

6. **Output of limitations from the working list:** Limitation which should be deleted from the working list are selected and eliminated from the active set. Other attributes of the working list are updated and go to step N 1.

7. **Recalculation of approximation:** Assignment is fulfilled

$$\hat{u}_{k+1} = \hat{u}_k + \alpha_k d_k$$

and return to step N 1.

As an example on fig. 7 oscillograms $I_{zk}$, $I_{zk}^*$, $U_{zk}$, $U_n$ for ES with MPC in typical conditions similar for P-control. The oscillograms show currents (b: $I_{zk}$, $I_{zk}^*$) and voltage (c: $U_{zk}$) under the influence a flicker with frequency 8 Hz on the disturbing value (a: $U_n$). The criterion of optimization is the maximal reduction in amplitude of the current $I_{zk}$. Oscillograms show, that realization of the given control system provides decrease of amplitude $I_{zk}$ per 40.13 %.

**1875**

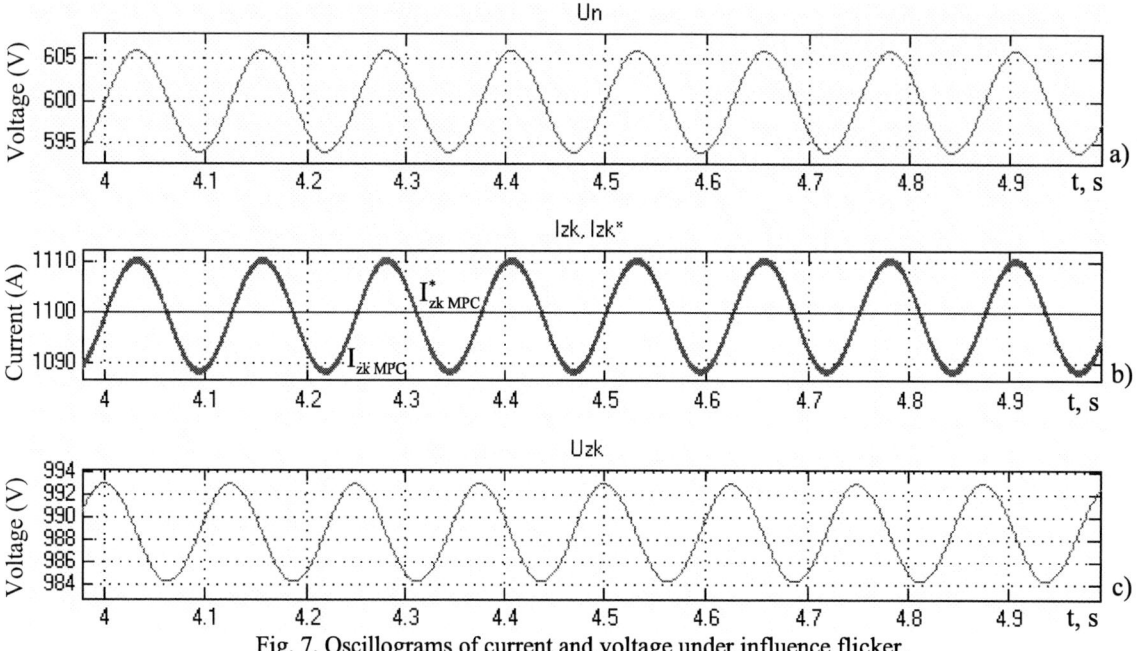

Fig. 7. Oscillograms of current and voltage under influence flicker.

### V. CONCLUSION

From results of research ES of the wind-turbine electric plant with classical subordinated control systems and MPC follows:

The usage of Matlab system for research of the considered ES allows to optimize parameters of now being in use regulating systems of the wind-turbine electric plant and to improve their operation in dynamic modes, for example in conditions of existence flicker.

The research of new MPC method shows the effective opportunity of its application in control ES of the wind-turbine electric plant. The given method provides qualitatively higher level of operation of considered systems in dynamic modes in conditions of existence of non-stationary disturbing influences.

### REFERENCES

[1]  Bianchi, Fernando D. Wind Turbine Control Systems: Principles, Modelling and Gain Scheduling Design / Fernando D. Bianchi, Hernan De Battista, Ricardo J. Mantz. – Springer, 2006. – 228 p.

[2]  Sterninson, L. D. Transients at Frequency/Power Control in Electric Power Systems / L. D. Sterninson. – Moskau: «Energy», 1975. – 216 p.

[3]  Klee, H. Simulation of Dynamic Systems with Matlab and Simulink / Harold Klee. – CRC, 2007. – 787 p.

[4]  Camacho, E. F. Model Predictive Control / E. F. Camacho, Springer-Verlag London Ltd., 1999. – 405 p.

[5]  Gill, P. E. Practical Optimization / P. E. Gill, W. Murray, M. Wright. – Moskau: Mir, 1985. – 509 p.

# Stand-alone Photovoltaic Generation System with Combined Storage using lead Battery and EDLC

Hiroaki Nakayama[*], Eiji Hiraki[*], Toshihiko Tanaka[*], Noriaki Koda[†], Nobuo Takahashi[†] and Shuji Noda[#]

[*]Graduated School of Yamaguchi University, Ube, Japan, e-mail: *nakayama@pe-news1.eee.yamaguchi-u.ac.jp*
[†]Matsue National College of Technology, Matsue, Japan,
[#] Shimane Institute for Industrial Technology, Matsue, Japan

*Abstract*—This research is mainly concerned with a stand-alone photovoltaic (PV) generation system for low power DC applications placed where the commercial AC grid connection is not supported. In this paper, basic and improved circuit topologies and the control schemes of stand-alone photovoltaic generation system are proposed. To prevent battery from deep discharge, Electric Double Layer Capacitor (EDLC) is incorporated into the proposed systems as combined energy storage device in parallel with conventional lead battery. The fundamentals of operating performances and effectiveness of proposed stand-alone PV system are illustrated and evaluated on the basis of a practical point of view.

*Keywords*—Stand alone photovoltaic system, DC-DC converters, combined power storage and EDLC.

## I. INTRODUCTION

Nowadays, the available application area and the installation of PV system are rapidly growing by a number of factors such as global warming, energy security, technology improvements and decreasing costs. In particular, stand-alone PV generation systems are attractive and indispensable electricity source such as the security camera devices, streetlights, electric signs and weather observation systems these are placed in remote or mountainous locations. These stand-alone PV generation systems usually consist of two converters; back or boost for maximum power point tracking (MPPT) of PV array and voltage regulation for DC-load. In addition, these power systems inevitably need batteries for storing energy during the fine day and releasing energy during the night or clouded condition. In systems employing rechargeable batteries, it is important to prevent the battery from the deep discharge, which may destroy or damage the cell. Usually, most of system turns off the circuit before the battery enters deep discharge. So, additional power source (DC energy storage device) is indispensable to keep continuous activities of the load. As for the small power application, EDLC may be suitable for this additional DC power source.

In this paper, two types of stand-alone photovoltaic generation system with combined storage using lead battery with EDLC for live-view camera system are introduced. One is step down DC-DC converter based basic circuit topology, and the other is newly developed soft switching converter based improved circuit topology. Simplified photovoltaic MPPT control method and

energy management procedure suitable for combined power storage are proposed and evaluated from experimental points of view.

## II. SYSTEM CONFIGURATION

Fig. 1 depicts the schematic diagram of the proposed stand-alone PV generation system. This system consists of the PV modules, two step down converters (SDC1 and SDC2) and combined storage that is composed of lead battery and parallel-connected EDLC. The electric power generated from the solar cells is MPPT-controlled and regulated by SDC1, and stored into the combined storage device. The stored power is regulated and supplied to DC-load (live-view camera system) by SDC2.

Capacity of the combined storage device and solar cells are mainly determined by charge/discharge efficiency of lead battery and EDLC, power conversion efficiency of power circuit, electric power consumption of the control circuit and DC load capacity as well as weather condition. In Japan, belonging to the "Cfa" of Köppen's climate classification, the probability in the fine day is 30%, the daylight length is less than 10 hours and typical intensity of sunlight is 50klx in winter. In order to drive the 5V-3.5W live-view camera system 24h continuously, more than 575Wh of battery capacity and 750F of EDLC capacitance are necessary. In this case, three modules of 13W/50klx PV panels are indispensable as shown in Fig. 2.

### A. Simplified MPPT and Step Down Operation

The PV modules in this system are ETFE film-based flexible type a-Si modules (Fuji electric systems Co. Ltd. FPV1024S). Maximum output power is 24W, nominal open-circuit voltage is 110V and maximum output current is 0.35A. Fig. 3 shows the measured P-V characteristics and MPP track of FPV1024S. From this

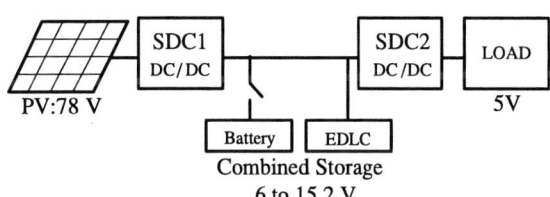

Fig. 1. Schematic diagram of stand-alone photovoltaic system.

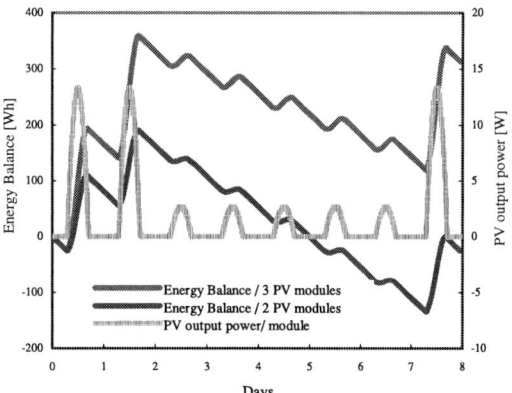

Fig. 2. Energy balance of solar cells and weather condition. Winter in Japan; maximum sun light 50 klx, daylight length 10 hours, 30% fine days in a week.

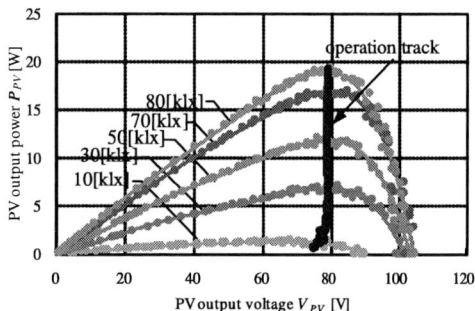

Fig. 3. P-V characteristic and MPP track of PV module (FPV1024S)

figure, MPP track can be achieved by tracing the output voltage 78 V line. Furthermore, 95 % of the maximum power can be guaranteed by holding the PV output voltage from 75V within 81V. Therefore, MPPT control can be realized by simple PI-control of the input voltage of SDC1 (= output voltage of PV modules) to 78V constantly.

*B. Operation Modes*

To prevent the lead battery in the combined storage device from deep discharge and over charge, three operation modes exist in this system. Fig. 4 shows the voltage area of the battery and EDLC.

1) MPPT mode (terminal voltage of combined storage; 10 to 15.2V): PV modules are MPPT controlled by SDC1. Lead battery and EDLC are able to charge by surplus power from PV modules.

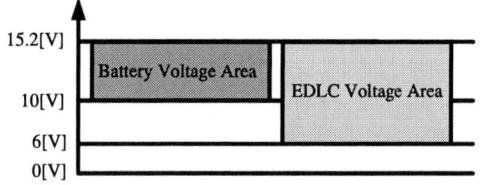

Fig. 4. Voltage area of lead battery and EDLC in combined storage.

2) Trickle charge mode (terminal voltage of combined storage; 15.2 to 12.5V with hysteresis): To prevent lead battery from over charge, PV power from SDC1 is inhibited. In this mode, lead battery and EDLC are able to charge and discharge.

3) EDLC mode (terminal voltage of lead battery; under 10V): to prevent lead battery from deep discharge, lead battery is separated off from the system. Electric power is supplied from EDLC to the DC-load through SDC2.

### III. BASIC CIRCUIT TOPOLOGY AND ITS EVALUATION

To evaluate a stand-alone photovoltaic system, a basic circuit was built and tested. Fig.5 indicate a basic circuit topology of the proposed system. PV modules are connected to SDC1 through LPF. Combined storage is composed of an EDLC and a lead battery. The lead battery can be separated off from the system by MOSFET switch $Q_3$. $R_{out}$ is connected in substitution for the live-view camera system. SDC1 and SDC2 are basic step down DC-DC converter. The total system is controlled by SH-tiny (SH7125) based CPU board. $Vpv$ is controlled at 78V and $V_{out}$ is regulated to 5V. Table I indicates the circuit parameters of the experimental circuit shown in Fig. 5. Switching frequency of SDC1 and SDC2 are 20kHz. Battery capacity and EDLC capacitance are set larger than calculated value mentioned above. Fig. 6 depicts the control circuit configuration of the photovoltaic system. MOSFET switch $Q_1$, $Q_2$ and $Q_3$ are driven by the photo

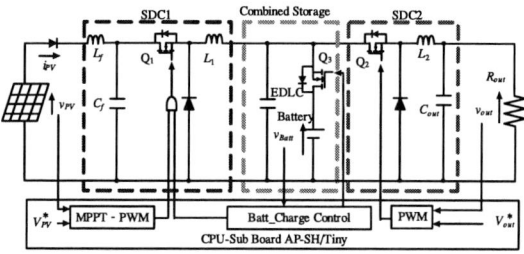

Fig. 5. Configuration of the basic circuit topology.

TABLE I.
EXPERIMENTAL BASIC CIRCUIT PARAMETERS

| Part | Item | Symbol | Value |
|---|---|---|---|
| Solar Cell | Maximum Power Point Voltage | $V_{pm}$ | 78 [V] |
| | Maximum Output Current | $I_{pm}$ | 0.9 [A] |
| | Maximum Output Power | $P_{max}$ | 72 [W] |
| LPF | Filter Inductance | $L_f$ | 75 [μH] |
| | Filter Capacitance | $C_f$ | 47 [μF] |
| SDC1 | SDC1 Switching Frequency | $f_{SW\_SDC1}$ | 20 [kHz] |
| | Output Inductance | $L_1$ | 1.3 [mH] |
| EDLC | EDLC Capacitance | $C_{EDLC}$ | 1100[F] |
| Battery | Lead Battery Capacity | $B$ | 50 [Ah] |
| SDC2 | SDC2 Switching Frequency | $f_{SW\_SDC2}$ | 20 [kHz] |
| | Output Inductance | $L_{out}$ | 230 [μH] |
| | Output Capacitance | $C_{out}$ | 150 [μF] |
| Load | Registance | $R_{out}$ | 7.14[orm] |
| | Input Voltage | $V_{out}$ | 5 [V] |

**1878**

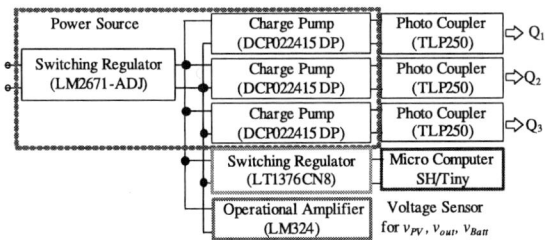

Fig.6. Block diagram of the control circuit.

driver IC TLP250. The input terminal of main switching regulator IC LM2671-AdJ is connected to the lead battery. Therefore, this experimental set-up is approved as stand-alone system. However, power consumption of this control circuit reaches 2.8W, which is 80% of the load capacity. To reduce the capacity of combined storage and PV modules, low power control circuit is necessary.

Fig. 7 shows the operation waveforms of experimental circuit of basic type. In this experiment, 35V-2200μF aluminum electrolytic capacitors are connected instead of lead battery and EDLC for brief period simulation. Fig. 7(a) indicates the mode transition between MPPT mode and trickle charge mode. In MPPT mode, PV output voltage is regulated to 78V constantly, and combined

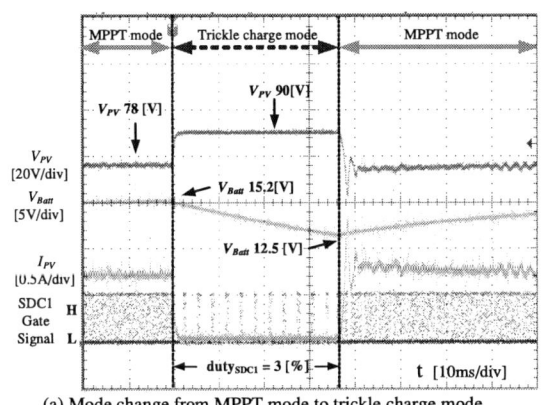

(a) Mode change from MPPT mode to trickle charge mode

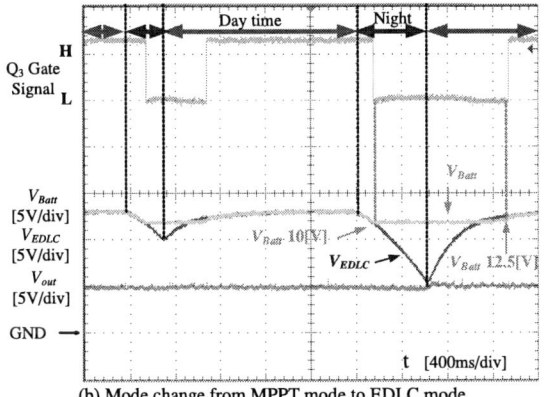

(b) Mode change from MPPT mode to EDLC mode

Fig. 7. Operation waveforms of experimental circuit

storage is charged in this condition. When sensed battery terminal voltage $V_{batt}$ reaches to 15.2V, on-duty of MOSFET switch $Q_1$ in SDC1 is limited to 3% (Trickle charge mode). In this mode, discharged power from combined storage is larger than PV output power, so $V_{batt}$ start to decease. By trickle charge mode, lead battery in the combined storage is prevented from over charge, which may involve explosion caused by the hydrogen gas generated in battery sells. Fig. 7(b) depicts the typical operation of combined storage. When $V_{batt}$ become lower than 10V by releasing stored energy at night or cloudy condition, battery is separated off from the system by MOSFET switch $Q_3$ to prevent battery from deep discharge. 10V of $V_{batt}$ is comparable to 70% of DOD (Depth of Discharge). Cycle life of the battery can be extended to more than 700 times by limiting DOD to 70%. This means that more than 10 years battery life may be guarantied. In this mode, EDLC provide the DC-load power through SDC2 (EDLC mode). From this figure, DC output voltage $V_{out}$ is regulated to 5V by SDC2, in spite of voltage variation of EDLC.

From these experimental results, it was proved that the basic circuit topology has enough functional performances. On the contrary, in order to reduce power capacities of the PV modules and the combined storage, some disadvantages were clarified including huge power loss in the control circuit mentioned above. One is low power conversion efficiency of SDC1 and SDC2 by the hard switching operation. The other one is delicate duty control in SDC1 caused by the significant step down ratio from input voltage 80V to output voltage 10V to 15.2V. In addition, the discharge termination voltage of EDLC was limited to higher value than output voltage $V_{out}$, because of the characteristics of SDC2. To solve these problems, the improved circuit topology is proposed in the following chapter.

## IV. IMPROVED SOFT-SWITCHING CIRCUIT TOPOLOGY OF THE STAND-ALONE PHOTOVOLTAIC SYSTEM

Fig. 8 shows the circuit configuration of improved topology of the stand-alone photovoltaic system. The constitutive architecture is same as the basic circuit topology as shown in Fig.5. The basic step down converters SDC1 and SDC2 are replaced by the coupled inductor assisted soft switching step down converter and the three-winding transformer assisted soft switching fly-back converter.

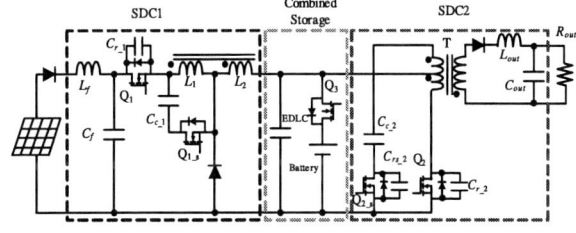

Fig. 8. Circuit configuration of improvement type.

## A. Operation principle of the new topology of SDC1; Coupled inductor assisted soft switching step down converter

A newly proposed coupled inductor assisted soft switching step down converter for improved SDC1 is illustrated in Fig. 9. In this figure, the combined storage, SDC2 and DC-load are replaced by $C_{EDLC}$ and $R_{out}$. $Q_{m1}$ ($S_{m1}/D_{m1}$) and $Q_{a1}$ ($S_{a1}/D_{a1}$) are main and auxiliary MOSFET switches. $C_{r1}$ is the resonant capacitor connected in parallel with $Q_{m1}$. $C_{C1}$ is the clamp capacitor. $S_{m1}$ and $S_{a1}$ are turned on and off alternately dead time $t_d$ as a. $S_{m1}$ and $S_{a1}$ can be operated under ZVS condition. As shown in (1), high voltage step down ratio can be settled by turn ratio (N1/N2) in addition to switching duty cycle D ($= T_{on}/T_{samp}$) of the main switch $S_{a1}$.

$$V_{EDLC} = V_{PV} \frac{D\frac{N_2}{N_1+N_2}}{D\frac{N_2}{N_1+N_2}+(1-D)} \tag{1}$$

In Fig. 10, typical voltage and current waveforms on this circuit topology are illustrated and equivalent circuits in each operation are represented in Fig. 11. The steady state operation of this circuit is described as follows;

**Mode0** ($t_0 - t_1$)
Current through $D_{m1}$ decreases to zero at $t_0$. $S_{m1}$ turns on under the conditions of ZVS and ZCS.

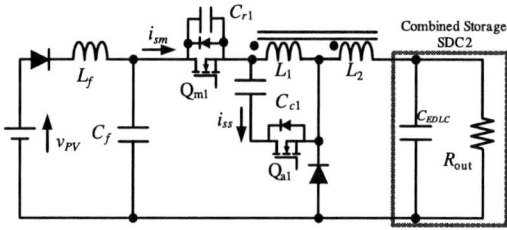

Fig. 9. Coupled inductor assisted soft switching step down converter for SDC1

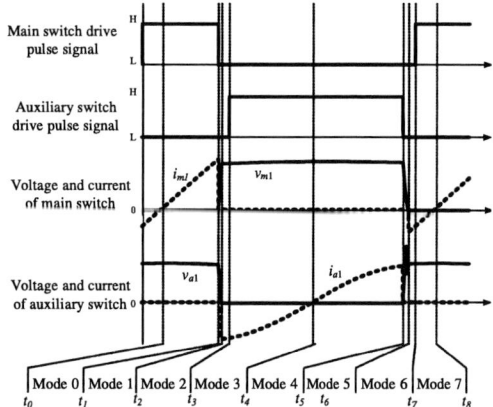

Fig. 10. Typical operation waveforms of switches $Q_{m1}$ and $Q_{a1}$ of newly proposed SDC1.

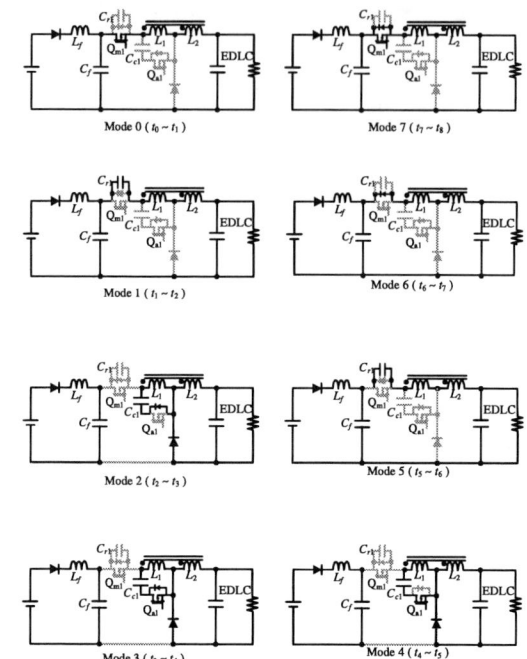

Fig. 11. Switching mode and equivalent circuits for each commutation stage of newly proposed SDC1.

**Mode1** ($t_1 - t_2$)
At time $t_1$, $S_{m1}$ is turned off under ZVS condition. Voltage across $C_{r1}$ increases linearly.

**Mode2** ($t_2 - t_3$)
Voltage across $C_{r1}$ reaches $v_{PV} + v_{L1}$; diode $D_{a1}$ start to conduct.

**Mode3** ($t_3 - t_4$)
While $D_{a1}$ is conducting, the gate signal is supplied to $S_{1a}$.

**Mode4** ($t_4 - t_5$)
Current through $D_{a1}$ decreases to zero. $S_{a1}$ turns on under conditions of ZVS and ZCS.

**Mode5** ($t_5 - t_6$)
At $t_5$, $S_{a1}$ is turned off under ZVS.

**Mode6** ($t_6 - t_7$)
At $t_6$, $C_{r1}$ voltage decreases zero; and diode $D_{m1}$ start to conduct.

**Mode7** ($t_7 - t_8$)
While $D_{m1}$ is conducting, the gate signal is supplied to $S_{m1}$.

## B. Operation principle of the new topology of SDC2; Three-winding transformer assisted soft switching fly-back converter

A three-winding transformer assisted soft switching fly-back converter for output voltage regulation (SDC2) is illustrated in Fig. 12. In this equivalent circuit, $Q_{m2}$ ($S_{m2}/D_{m2}$) and $Q_{a2}$ ($S_{a2}/D_{2a}$) correspond to the main and the auxiliary MOSFET switches. $C_{rm2}$ is the loss less snubber capacitor connected in parallel with $Q_{m2}$. $C_{C2}$ is the clamp capacitor. $S_{m2}$ and $S_{a2}$ are turned on and off alternately with dead time $t_d$. $S_{m2}$ and $S_{a2}$ can be operated

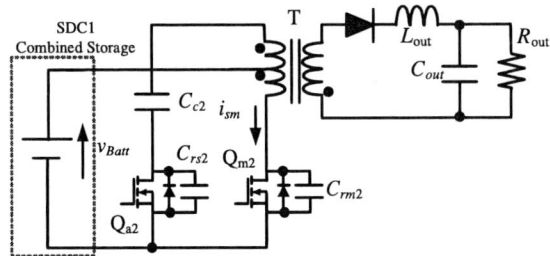

Fig. 12. Three-winding transformer assisted soft switching fly-back converter for SDC2.

under ZVS or ZV&ZCS condition. Typical voltage and current waveforms on this proposed converter circuit are illustrated in Fig 13. Furthermore, each operation stage of the converter is represented in Fig. 14.

Mode0 ($t_0$ - $t_1$)
　　Current through $D_{m2}$ decreases to zero at $t_0$. $S_{m2}$ turns on under ZV&ZCS conditions.
Mode1 ($t_1$ - $t_2$)
　　At $t_1$, $S_{m2}$ is turned off under ZVS condition. Voltage across $C_{rm2}$ increases linearly.
Mode2 ($t_2$ - $t_3$)
　　At $t_2$, $C_{rm2}$ voltage reaches $2V_{Batt}$; when $v_{cra2}$ becomes zero, diode $D_{2a}$ start to conduct.
Mode3 ($t_3$ - $t_4$)
　　While $D_{a2}$ is conducting, the gate signal is supplied to $S_{2a}$.
Mode4 ($t_4$ - $t_5$)
　　Current through $D_{a2}$ decreases to zero at $t_4$. $S_{a2}$ turns on under the conditions of ZV & ZCS.
Mode5 ($t_5$ - $t_6$)
　　At $t_5$, $S_{a2}$ is turned off under ZVS.
Mode6 ($t_6$ - $t_7$)
　　At $t_6$, $C_{rm2}$ voltage decreases zero, and diode $D_{m2}$ start to conduct.

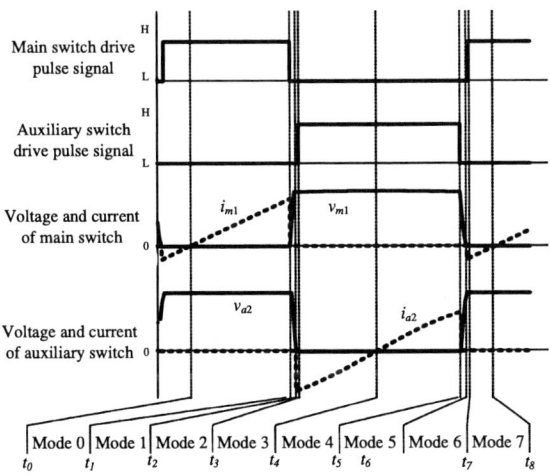

Fig. 13. Typical operation waveforms of switches Q2m and Q2aof newly proposed SDC2.

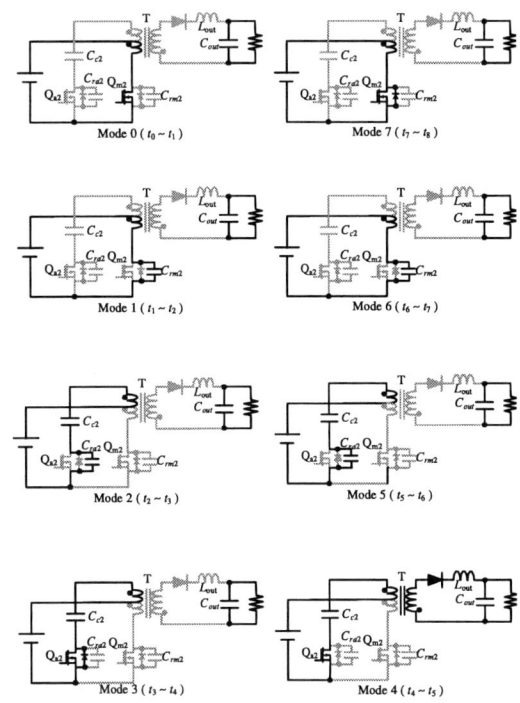

Fig. 14. Switching mode and equivalent circuits for each commutation stage of newly proposed SDC2.

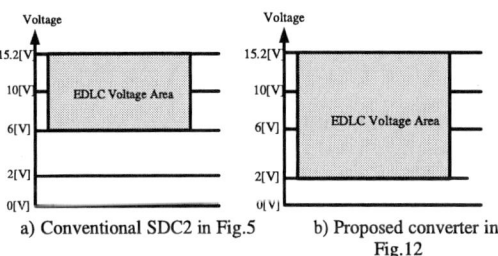

a) Conventional SDC2 in Fig.5　　b) Proposed converter in Fig.12

Fig. 15. Voltage area of EDLC in the Combined Storage.

Mode7 ($t_7$ - $t_8$)
　　While $D_{m2}$ is conducting, the gate signal is supplied to $S_{m2}$.

The available voltage range of EDLC in the combined storage is shown in Fig. 15. In comparison with the basic circuit topology shown in Fig. 5, available lower voltage limit can be expanded by back-boost operation of improved type of SDC2. As a result, demanded capacitance of EDLC in the combined storage can be reduced from 1100F to 750F.

## V. EXPERIMENTAL EVALUATION OF THE IMPROVED CONVERTERS FOR STAND-ALONE PHOTOVOLTAIC SYSTEM

Prototypes of the improved DC-DC converters shown in Fig. 9 and Fig. 12 have been built and experimentally tested to validate the basic circuit topology shown in Fig. 5. Table II summarizes the circuit parameters of the

TABLE II.
EXPERIMENTAL IMPROVED CIRCUIT PARAMETERS

| Part | Item | Symbol | Value |
|------|------|--------|-------|
| Solar Cell | Maximum Power Point Voltage | $V_{pm}$ | 80 [V] |
| | Maximum Output Current | $I_{pm}$ | 0.9 [A] |
| | Maximum Output Power | $P_{max}$ | 72 [W] |
| LPF | Filter Inductance | $L_f$ | 80 [μH] |
| | Filter Capacitance | $C_f$ | 20 [μF] |
| SDC1 | SDC1 Switching Frequency | $f_{SW\_SDC1}$ | 20 [kHz] |
| | Coupled Inductor Winding Turn Ratio | $N_1/N_2$ | 1/1 |
| | SDC1 Clamp Capacitor | $C_c1$ | 50 [μF] |
| | SDC1 Lossless Snubber Capacitor | $C_r1$ | 4.7 [nF] |
| EDLC | EDLC Capacitance | $C_{EDLC}$ | 750 [F] |
| Battery | Lead Battery Capacity | $B$ | 50 [Ah] |
| SDC2 | SDC2 Switching Frequency | $f_{SW\_SDC2}$ | 20 [kHz] |
| | Trans Winding Turn Ratio | $N_{1m}/N_{1a}/N_2$ | 2/1/2 |
| | SDC2 Clamp Capacitor | $C_c2$ | 30 [μF] |
| | SDC2 Lossless Snubber Capacitor ( Main Switch ) | $C_{rm}1$ | 4.7 [nF] |
| | SDC2 Lossless Snubber Capacitor ( Sub Switch ) | $C_{ra}1$ | 4.7 [nF] |
| | Output Inductance | $L_{out}$ | 1 [μH] |
| | Output Capacitance | $C_{out}$ | 150 [μF] |
| Load | Registance | $R_{out}$ | 7.14[orm] |
| | Input Voltage | $V_{out}$ | 5 [V] |

stand-alone photovoltaic system including improved converters. Compared to the basic system parameters shown in table I, capacitance of EDLC has been reduced by back-boost SDC2.

Experimental voltages and current waveforms of $Q_{m1}$, $Q_{a1}$ in the coupled inductor assisted soft switching step down converter for SDC1 are shown in Fig. 16 (a) and (b). As shown in these figures, both main and auxiliary switching power MOSFET have been operated with ZV & ZCS turn on, ZVS turn off. Experimental waveforms of the main switch $Q_{m2}$ and auxiliary switch $Q_{a2}$ in the three-winding transformer assisted soft switching fly-back converter for SDC2 are respectively illustrated in Fig. 17. From these experimental results, each MOSFET can be turned on with ZV & ZCS and turned off with ZVS, as well as SDC1. Voltage and current spikes are effectively dumped and reduced by slow $di/dt$ and $dv/dt$.

Fig.18 depicts power conversion efficiency of SDC1 (from PV modules to the combined storage). In case of more than 30W of PV module output power, power conversion efficiency of the improved converter is superior to the basic converter, actual efficiency can be kept to be more than 95% in fine or slightly cloudy weather condition. Lower efficiency at low power area of the improve converter may able to be improved by optimal design of the coupled inductor and resonant components shown in Fig. 9.

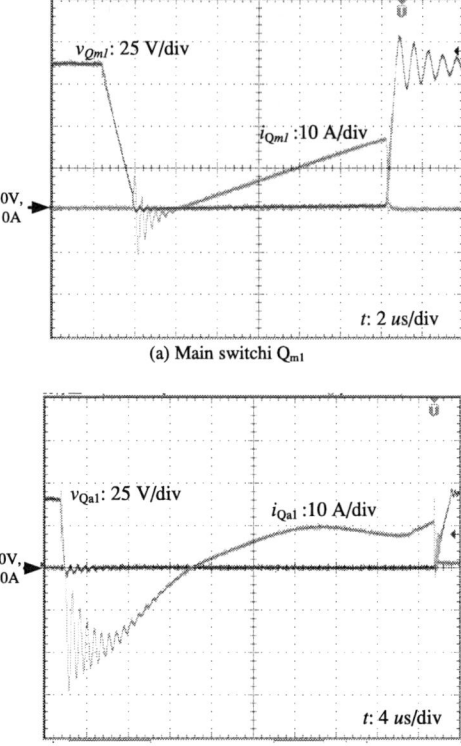

Fig.16. Experimental waveforms of MOSFET in improved SDC1.

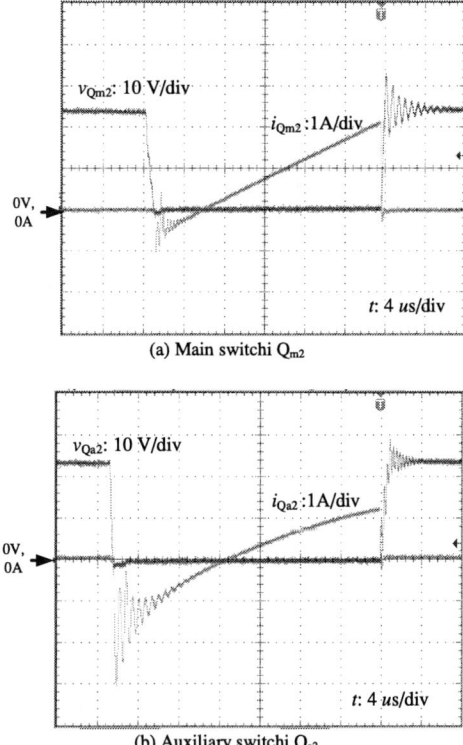

Fig.17. Experimental waveforms of MOSFET in improved SDC2.

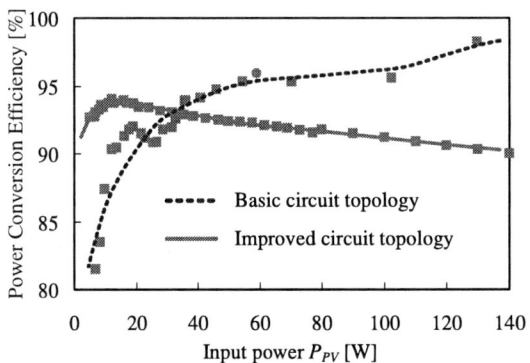

Fig.18. Power conversion efficiency from PV module to the combined storage

## VI. CONCLUDION

In this paper, basic and improved circuit topologies for stand-alone photovoltaic generation system with combined storage using lead battery and EDLC were introduced. The control principle of PV modules with simplified MPPT and combined storage were discussed and experimentally tested with the basic circuit topology. Furthermore, prototypes of the improved converter were tested to validate the basic circuit.

The prototype test results indicate that a high performance stand-alone photovoltaic system may be realized using introduced circuit designs and control scheme.

### REFERENCES

[1] Hussam Al-Atrash and Issa Batarseh, "Boost-Integrated Phase-Shift Full-Bridge Converter for Three-Port Interface", in proc. IEEE Power Electronics Specialist Conference, pp.2313-2321, USA, September, 2007.

[2] Hussam Al-Atrash, Justin Reese, and Issa Batarseh, "Tri-Modal Half-Bridge Converter for Three-Port Interface", in proc. IEEE Power Electronics Specialist Conference, pp.1702-1708, USA, September, 2007.

[3] Wissam MELHEM, Georges-William BAPTISTE, Philippe ERNEST, "EXPERIMENTALINVESTINGATION OF THE CURRENT SYNCHRONIZED CONTROL: APPLICATION TO HIGH VOLTAGE/HIGH POWER DC/DC CONVERTERS", The proceeding CD-ROM of 13th EPE Conference, September, 2007.

[4] S. Chandhaket, Y. Konishi, K. Ogura, E. Hiraki, and M. Nakaoka, "A Sinusoidal Pulse Width Modulated Inverter Using Three-Winding High-Frequency Flyback Trancformer for PV Power Conditioner", in proc. IEEE Power Electronics Specialist Conference, pp.1197-1201, Mexico, June, 2003.

[5] H. Terai, S. Sumiyoshi, T. Kitaizumi, H.Omori, K. Ogura, H. Iyomori, S. Chandhaket and M. Nakaoka, "Utility-Interactive Solar Photovoltaic Power Conditioner with Soft Switching Sinewave Modulated Inverter for Residential Applications", in proc. IEEE Power Electronics Specialist Conference, Australia, June, 2002.

[6] E. Hirai, M. Nakaoka, "Power Electronics in Latest Small – Scare High Frequency Switching Power Converters", IEEJ Trans. on Industory Applications, Vol. 125-D, pp. 955-963, 2005.

[7] C. Y. Inaba, Y. Konishi, M. Nakaoka, "Pulse Curent Regenerative Resonant Snubber – Assisted Two-Switch flyback ZVS PWM DC-DC Converter", IEEJ Trans. on Industory Applications, Vol. 124-D, pp. 255-261, 2004.

# Active Filter Action of Inverter Exciting Induction Generator for Wind Power Generation

Noriyuki Kimura[*], Tomoyuki Hamada[*], Katsunori Taniguchi[*] and Toshimitsu Morizane[*]

[*] Osaka Institute of Technology, e-mail: *kimura@ee.oit.ac.jp*

*Abstract*—This paper shows new induction generation system for wind power generation. This system has the voltage source converter (VSC) exciting induction generator and the PFC converter to transfer the real power. Rated power of the VSC is minimized by transferring all the real power into the PFC converter. However, the harmonic components in the current from the VSC has larger peak than expected for the fundamental component. Therefore, active filter action to eliminate these harmonic components in the current from the VSC is investigated in this paper. Simulation results show the reduction rate of 1/5 of harmonic components by the active filter action.

*Keywords*—Wind Power generation, Induction machine, PFC converter, voltage source converter, Active filter action.

## I. INTRODUCTION

Wind energy is most attractive resource of electrical power in the near future[1,2]. More reduction of the cost and the increments of the output is the important issue to promote the installation of the wind power generator. The induction generator is the most popular and the lowest cost machine for wind power generation. However it cannot generate power at lower speed than the synchronous speed. In Japan, the wind speed changes largely and rapidly. So, sometimes, the permanent magnet synchronous generator with frequency conversion system is installed. It is effective for maximize the output from the wind energy since it can change the rotating speed of the wind turbine depending on the wind power. However, the cost of the machine is much higher than the induction machine. Hence the authors have proposed to use the induction generator with voltage sourced converter (VSC) [3]. This paper shows how to reduce the output current of the inverter by using active filter action in the proposed system with the induction machine.

## II. PROPOSED WIND POWER GENERATION SYSTEM.

The induction generator system as shown in Fig.1 is the most popular and the lowest cost system for the wind power generation. However it cannot generate power at lower speed than the synchronous speed. In Japan, the wind speed changes largely and rapidly. So, sometimes,

the synchronous generator using permanent magnet with frequency conversion system as shown in Fig.2 is used. It is effective for maximize the output from the wind energy at any wind speed. However, the cost of the machine is much higher than the induction machine. Hence the authors have proposed to use the induction generator with voltage sourced converter (VSC) [3] as shown in Fig.3. Fig.4 and Fig.5 show the circuit of the PFC converter and the voltage sourced converter. This paper shows how to reduce the output current of the inverter by using the duty factor control of the PFC converter in the proposed system with the induction machine.

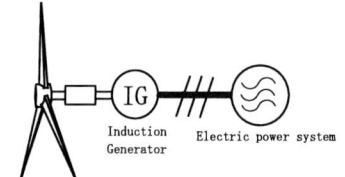

Fig. 1   Wind power generation system with induction machine

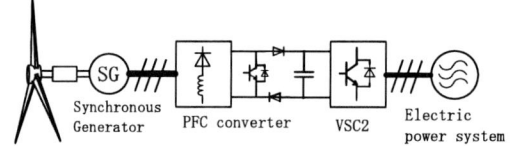

Fig. 2 Wind power generation system with synchronous machine and frequency conversion system

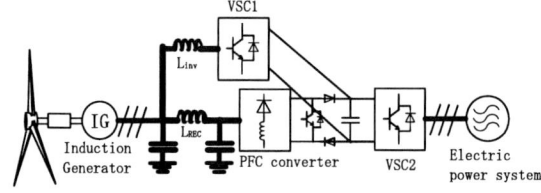

Fig. 3 Proposed wind power generation system

## III. CHARACTERISTICS OF PROPOSED SYSTEM

### A. Prevention of circulating current

Proposed wind power generation system has the PFC converter and VSC1 in parallel. Both ac circuit and dc circuit are common to the PFC converter and VSC1. Then, the circulating current can flow through these converters. An example of circulating current route is shown in Fig. 4.

978-1-4244-1741-4/08/$25.00 ©2008 IEEE          1884

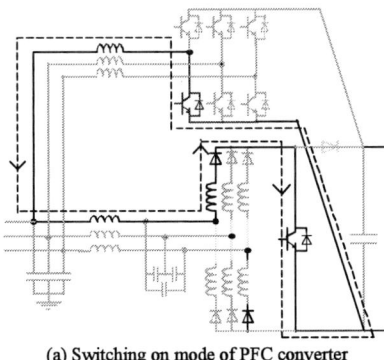

(a) Switching on mode of PFC converter

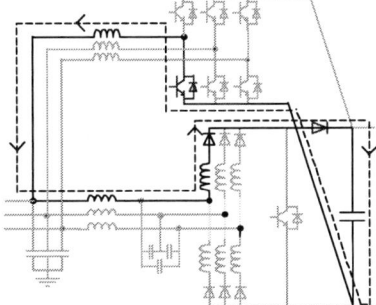

(b) Switching off mode of PFC converter

Fig. 4 Circulating current circuit of PFC converter and VSC1

Usually output dc current of the PFC converter becomes zero when the switch is on-state. However, since the PFC converter is connected in parallel with VSC1 at both ac and dc side, the dc current can flow through the negative output line of the PFC converter into the diode of the lower arm of VSC1. This current causes the offset both in the dc and ac currents. Simulation results are shown in Fig. 5. The shape of the positive line current and the negative line current are same. However, the dc offset appears in the negative line current, and the reverse current flows as indicated in Fig. 4(b). To prevent this dc offset in the output dc current of the PFC converter, additional diode at the negative line is installed as shown in Fig. 6.

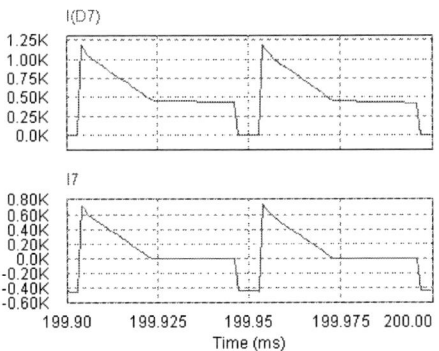

Fig. 5 DC output current waveforms of PFC converter
(upper : positive line current、 lower : negative line current)

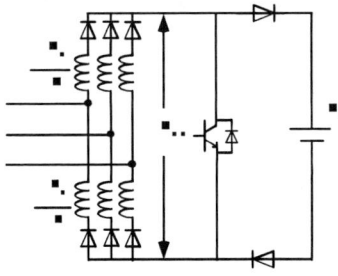

$V_{in}$ : Rectified voltage

Fig. 6 PFC converter circuit with additional diode

This additional diode prevents the circulating current when the switch is on-state. Simulation results of this configuration are shown in Fig. 7. The offset in the negative line has been eliminated.

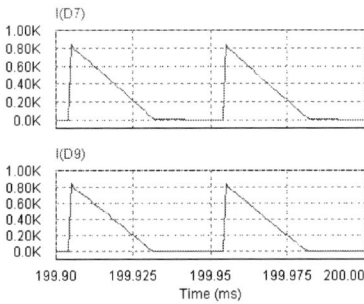

Fig. 7 DC output current waveforms of PFC converter with additional diode (upper : positive line、 lower : negative line)

### B. Relation between duty factor and DC side current of PFC converter

The PFC converter has to absorb all the real power from the induction generator to reduce the ac current of the VSC1. Duty factor of the PFC converter controls amount of the absorbed power. Simulation results and analytical results between the duty factor and the absorbed real power are shown in Fig. 8. It indicates the sudden change of the characteristics. The real power increases largely when the duty factor becomes larger than 0.23. We have investigated the cause of this change and concluded that it is the mode of the reactor current[5].

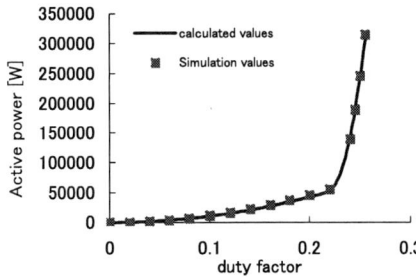

Fig. 8 Relation between duty factor and active power of PFC converter

1885

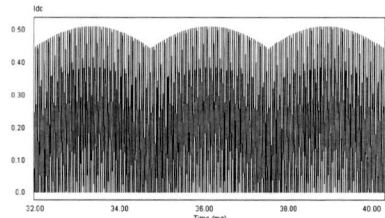

(a) Discontinuous mode

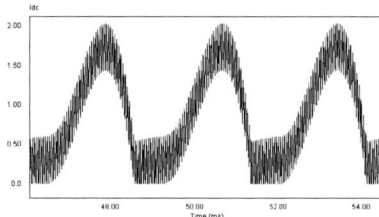

(b) The mixed mode

Fig. 9 DC reactor current of PFC converter

Fig. 9 shows the two modes of the dc side current of the PFC converter. Figure. 9(a) and 9(b) shows the current waveforms of the dc reactor of the PFC converter in the discontinuous mode and the mixed mode[5], respectively. The sudden change of the characteristics between duty factor and the PFC converter current is caused by the change of the current mode. Discontinuous mode appears in the following condition.

$$d_F \le 1 - \frac{V_{in}}{V_O} \qquad \text{...(1)}$$

The real power in the discontinuous mode is as follows.

$$P_{rec} = \frac{9V_{in}{}^2 V_o}{2\pi L(\pi V_o - 3V_{in})} d_F^2 \mathrm{T} \quad d_F^2 \le 1 - \frac{V_{in}}{V_O} \text{ ...(2)}$$

Where $T$ is the switching period of the PFC converter. To design the duty factor controller, eq. 2 is linearized as follows.

$$\Delta P_{rec} = \left[ \frac{9V_{in}{}^2 V_o}{2\pi L(\pi V_o - 3V_{in})} 2d_F \mathrm{T} \right] \Delta d_F \qquad \text{...(3)}$$

Fig. 10 shows the block diagram of the system with the duty factor controller. Fig. 11 shows the current waveforms from the induction generator and VSC1, and the FFT result of the VSC1 current.

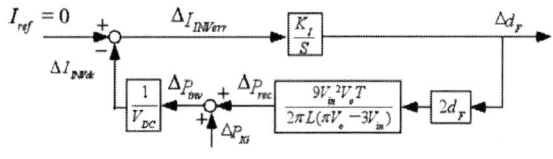

Fig. 10 Block diagram of duty factor control system.

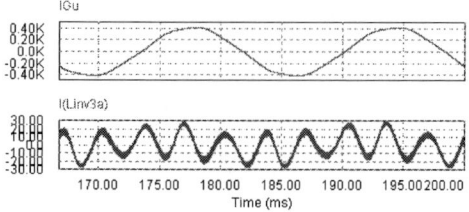

(a) Current waveform of induction generator and VSC1

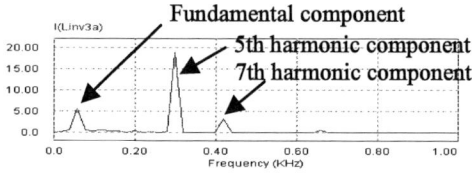

(b) FFT results of VSC1 current

Fig. 11 Simulation results of proposed wind power generation system.

The peak value of the VSC1 current is sufficiency small in comparison with the induction generator current. However, the peak of VSC1 current is not small enough since it has higher harmonic components. The VSC1 current has $5^{th}$ and $7^{th}$ harmonics of certain amount. These are caused by the distortion in the current of the PFC converter. Equivalent circuit for harmonic current analysis is shown in Fig. 12. The VSC1 is modeled by the voltage source of the fundamental component, while the PFC converter is modeled by the current sources of $5^{th}$ and $7^{th}$ harmonics. The induction generator is represented by the excitation reactance only. Then the $5^{th}$ and $7^{th}$ harmonic components appear in the currents of the VSC1 and the induction generator. As a result, the voltage at the capacitor includes the $5^{th}$ and $7^{th}$ harmonic components, $V_{hi}$.

The VSC1 current, $i_{inv}$, has the $5^{th}$ and $7^{th}$ harmonic components as follows, by using $V_{hi}$.

$$i_{inv} = -j\frac{V_{hi}}{\omega_{hi}L_{inv}} \qquad \text{...(4)}$$

If the VSC1 has the voltage output as same as $V_{hi}$, as shown in Fig. 14, then the $5^{th}$ and $7^{th}$ harmonic components in the VSC1 current can be eliminated as shown in the following equation.

$$i_{inv} = -j\frac{V_{hi} - V_{hi}}{\omega_{hi}L_{inv}} = 0 \qquad \text{...(5)}$$

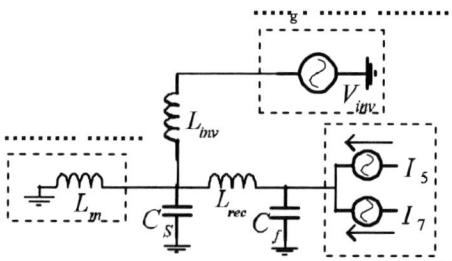

Fig. 12 Equivalent circuit from current source.

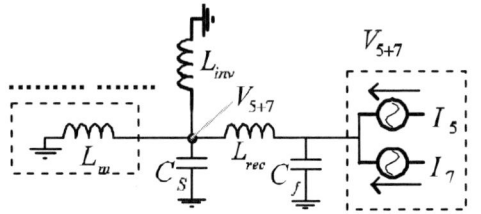

Fig. 13 Equivalent circuit for harmonic current source

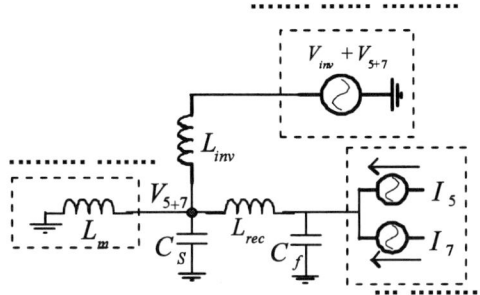

Fig. 14 Equivalent circuit from active filter operation

## IV. ACTIVE FILTER OPERATION OF VSC1

The $5^{th}$ and $7^{th}$ harmonic components in $V_{hi}$ are calculated as follows.

$$V_{hi} \square \square j \frac{x_{cs} x_{cf} x_{Linv} x_{Lg}}{x_{cs} x_{Linv} x_{ig} + (x_{cf} \square x_{Lrec})(x_{Lg} x_{cs} + x_{Linv} x_{cs} \square x_{Linv} x_{Lg})} I_{hi}$$

...(6)

If active filter action is ideally performed, eq.6 turns out as eq.7, since $x_{Linv}$ is eliminated from eq.6.

$$V'_{hi} \square \square j \frac{x_{cs} x_{cf} x_{Lg}}{x_{cs} x_{ig} + (x_{cf} \square x_{Lrec})(x_{cs} \square x_{Lg})} I'_{hi} \quad ...(7)$$

When the ac reactance of VSC1, $x_{Linv}$, is removed from eq.6, then the harmonics voltage has changed as eq.7.

Phasor diagrams are shown in Fig. 15. The ratio of voltage change are calculated and shown in Table I and Fig.16.

Simulation was used to investigate the performance of the active filter action. Table II and Table III show the system parameters at 50Hz. Capacitors for reactive power compensation is optimized at the frequency 50Hz.

We have designed two different controllers. Proposed wind power generation system can change the frequency from 40Hz to 60Hz. Output voltage of VSC1 is controlled in V/f=constant.

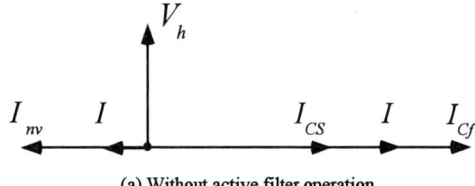

(a) Without active filter operation

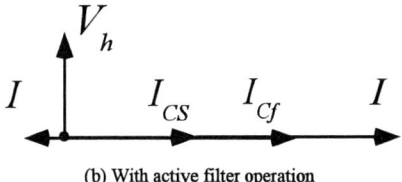

(b) With active filter operation

Fig. 15 Vector diagram of harmonic currents

TABLE I. CHANGE RATIO OF VOLTAGE AT IG

| f [Hz] | Without Active Filter | With Active Filte |
|--------|----------------------|-------------------|
| 300 | 0.681 | 0.543 |
| 250 | 0.825 | 0.774 |
| 200 | 0.863 | 1.135 |
| 150 | 0.745 | 1.719 |

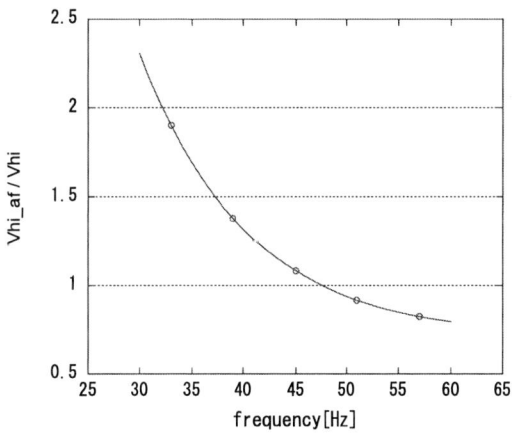

Fig. 16 changing ratio of Voltage at Induction Machine

TABLE II. PARAMETER OF INDUCTION MACHINE

| | |
|---|---|
| stator resistance | 0.35Ω |
| rotor resistance | 0.5Ω |
| stator leakage inductance | 0.54mH |
| rotor leakage inductance | 0.84mH |
| excitation reactance | 28.1mH |
| excitation resistance | 0.027Ω |
| slip | -0.05 |

TABLE III.     CIRCUIT PARAMETER

| | |
|---|---|
| $V_{dc}$ | 1090V |
| $V_{IG}$ | 10× ･ [V] |
| $L_{inv}$ | 0.084mH |
| $C_S$ | 460μF |
| $L_{rec}$ | 5.67μH |
| $C_f$ | 306.7μF (D connected) |
| $L_{dc}$ | 4.55μH |
| Carrier frequency | 20kHz |

Fig. 17 shows the block diagram of the controller using proportional control for the active filter action. The voltage across $L_{inv}$ is selected as an input. Then VSC1 generates the harmonic components additively and makes the harmonic voltage components across $L_{inv}$ to be 0[V].

Fig. 18 shows simulation results with the proportional control method. Fig. 18(a) shows the results without active filter action. The peak current value is 25[A]. Fig. 18(b) shows the suppression effect of the active filter action. The peak current value is less than 15[A]. However, high gain controller becomes unstable as shown in Fig. 18(c).

In steady state operation, $5^{th}$ harmonic and $7^{th}$ harmonic components are reduced to 1/5 by using the active filter action of proportional control.

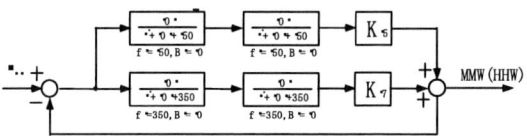

Fig. 17 Block diagram of active filter action of control system proportional control method

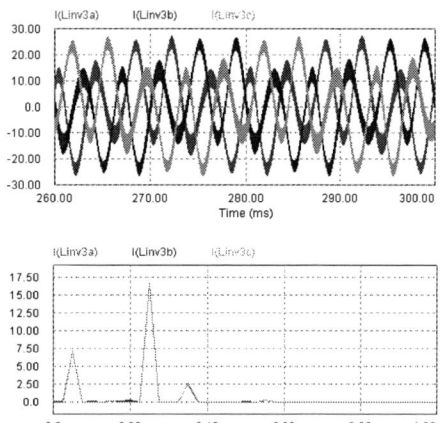

(a) $k_{P5} = 0$   $k_{P7} = 0$

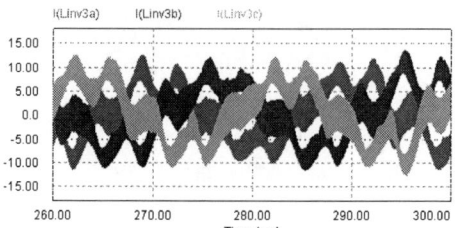

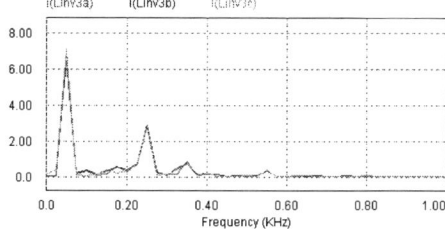

(b) $k_{P5} = 4/500$    $k_{P7} = 2/500$

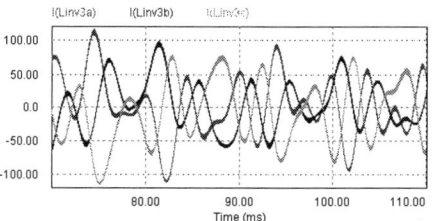

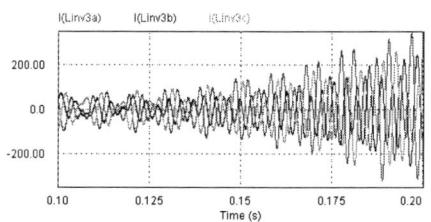

(c) $k_{P5} = 7/500$    $k_{P7} = 4/500$

Fig. 18 simulation results of proportional control method

But, this active filter action cause the second harmonic component at a certain gain. This may be caused by the delay in the band pass filter. Hence the fluctuation such as second harmonic component cannot be suppressed properly.

So, another controller for the active filter action was investigated. In this controller, VSC1 detect $V_{hi}$ in the voltage at the induction generator and output $V_{hi}$ is directly calculated for the modulation. Using DC side voltage of VSC1 $V_{DC}$, modulation factor $M$ is calculated from the output voltage $V_{AC}$ as following equations.

$$V_{AC} \ \square \ 0.5 M V_{DC} \quad ...(8)$$

**1888**

$$M \Box 2 \frac{V_{hi}}{V_{DC}} \qquad ...(9)$$

This controller was named the direct controller.
Figure 19 shows the block diagram of the direct controller. Figure 20 shows simulation results with the direct controller.

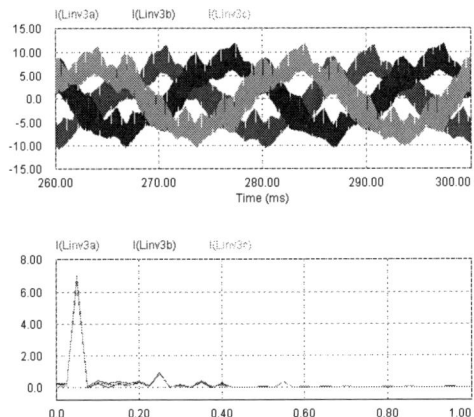

Fig. 19 Block diagram of active filter operation of control system direct control method

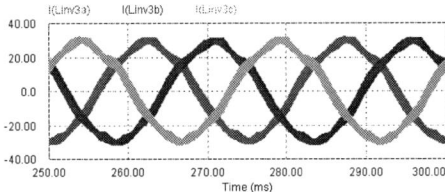

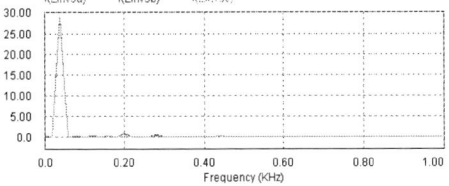

(a) Results of VSC1 current at 40Hz

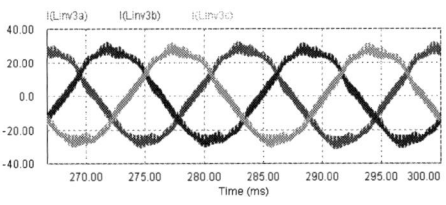

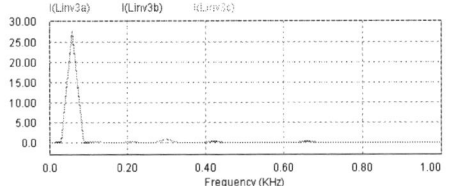

(b) Results of VSC1 current at 60Hz

Fig. 21 Simulation result of VSC1 current

Fig. 20 Simulation results of direct control method

· In steady state operation, 5th harmonic and 7th harmonic components are reduced to 1/10 by using the active filter action of the direct controller. The current of VSC1 becomes less than 20A by active filter action at 50Hz. The results at 40Hz and 60Hz are also shown in Fig. 21. In both cases, the peak current value is suppressed less than 40[A], which is 10[%] of the rated current peak.

## V. CONCLUSION

One of the merits of proposed wind power generation system is reduction of the total cost of the system. Therefore, the reduction of the rated power of the VSC is indispensable matter.

This paper investigated the control method to minimize the current from the VSC1. Fundamental component of the current of VSC1 can be lowered by the duty factor control of the PFC converter. However, the current from the VSC1 contains large 5th and 7th harmonics. Active filter action of the VSC1 is proposed in this paper. Two control methods were investigated and it is noted that the direct control method is more effective for the active filter action than the proportional control method. The peak current of the VSC1 becomes less than 10% of the full rated power. In next step, the filtering of the harmonic components when the frequency were changed.

## References

[1] S. Meier, et.al, "New Voltage Source Converter Topology for HVDC Grid Connection of Offshore Wind Farms", EPE-PEMC-2004 Riga, Latvia. (European Association of Power Electronics and Applications - Power Electronics and Motion Control conference), Paper No.A131501, 2004.9.2-4.

[2] L. Leclercq, et.al, "Grid Connected or Islanded Operation of Variable Speed Wind Generators Associated with Flywheel Energy Storage Systems", EPE-PEMC-2004 Riga, Latvia. (European Association of Power Electronics and Applications - Power Electronics and Motion Control conference).

[3] Noriyuki Kimura, Toshimitsu Morizane, Katsunori Taniguchi, Tomoyuki Hamada, " Wind Power Generation System with Induction Machine and Diode Rectifier ", EPE 2007, 12th European Conference on Power Electronics and Applications, No.0436, (2007-9)

[4] Powersim Inc. Home Page, http://www.powersimtech.com/

[5] Noriyuki Kimura, Toshimitsu Morizane, Katsunori Taniguchi, Tomoyuki Hamada, " Control of PFC Converter with Inverter Excited Induction Generator for Advanced Wind Power Generation System ", Proceedings of PESC'08 "The 39th IEEE Power Electronics Specialists Conference", June 15-19, 2008.

# The Operation of Power Electronic Converters in Photovoltaic Drive Systems

Marek Niechaj

Lublin University of Technology/Department of Electrical Drive Systems, Lublin, Poland, e-mail: *m.niechaj@pollub.pl*

*Abstract*—**The paper presents most important inferences, that have arisen from the research about technical, energetical and economical properties of Photovoltaic Drive Systems (PDS), especially at an angle of conditions of use of different power electronic converters and their operation terms in such systems. The research has been done by availing material models (laboratory stands) and mathematical models (computer simulations). The short description of the most important features of different PDS structures, with presenting their block diagrams, is done. The schematic diagram of simple, and relatively inexpensive, stand-alone PDS without electrical buffer source (e.g. electrochemical accumulator), and with DC commutator motor with permanent magnet excitement, is presented. This system may be applied for irrigation of cultivable lands, or for water circulation and filtration for swimming pools or fountains.**

*Keywords*—**photovoltaic, electrical drive, converter circuit, control of drive.**

## I. BASIC ASPECTS OF PHOTOVOLTAIC DRIVE SYSTEMS OPERATION

Photovoltaic Drive Systems (PDS) always contain a photovoltaic (PV) generator, as the basic energy source. It directly converts sunlight energy into electrical energy. PDS also must contain basic energy receiver – an electrical machine (motor), that converts electrical energy into mechanical energy. The mechanical energy is spent then by motor's mechanical load, for instance: pump, fan, conveyor or vehicle wheels.

For PDS, the best kind of electrical machine is DC commutator motor with permanent magnet excitement, and also DC brushless motor (with electronic commutation) [1]. Also, quite good are: DC commutator motor with series excitement and AC three-phase low-voltage induction motor. The AC one-phase induction motor is not recommended, but its application in PDS is possible. But DC motors with parallely or separately supplied excitement circuit, and also all low-efficiency motors that are not adapted for direct supply from DC voltage sources, are decidedly not recommended for PDS applications.

There are some examples of possible practical applications of PDS presented below. Such systems could be justifiable, of technical and often economical features (now or in the near future), for driving:

*1)* Pumps for: irrigation of cultivable lands and greenhouses; securing water in farms and recreation buildings for drinking and other purposes needs; water filtration and circulation for swimming-pools, fountains and fish nurseries; circulation in solar collector systems.

*2)* Compressors in conditioning units and in refrigerators.

*3)* Fans for agricultural products drying, climatisators condensers cooling, aeration of fish nurseries or buildings.

*4)* Feeders in animal and fish farms.

PV generator of low power (up to about 1kWp – kilowatt-peak) is DC-current energy source, typically low-voltage, with significant and random variations of its output parameters, especially of output electrical power. The variations are usually slow, but sometimes, very rarely, they may reach tens of percents in one second. Such changes extort specific operating conditions of units that match PV generator's energy parameters to motor's demands. In PDS, these adapting units (AU) almost always contain power electronic converters.

The random variations of PV generator's power often extort buffer energy sources applying, especially while motor's load supply reliability demands are high. There are two types of buffer energy sources: electric and non-electric. Electric buffer sources are able to collect electrical energy for a moment while PV generator's output power is greater than the energy demanded by motor, and they can give back this electrical energy while the demands are higher than PV generator's producing possibilities. Non-electric buffer sources collect energy previously converted by motor's load. For instance, for PDS with pumps, they collect water in hydrophores or in tanks placed on certain height, and for cooling systems they accumulate cold in thermal capacity properties of cooled room.

The main factors, influencing PV generators output parameters, are:

*1)* The value of insolation on the surface of PV cells forming PV generator. Generator's short-circuit current almost linearly depends on the value of insolation. The insolation level is influenced by: time of the day, thickness of cloud cover, PV modules arrangement (tilt angle and azimuth angle), shading of PV cells by surrounding objects, dirt and snow covering surfaces of modules. The last two elements cause serious problems in generator's operation, because they cause non-uniform insolation of different PV cells. It may, consequently, significantly reduce generator's conversion efficiency, and even cause the damage of poorly insolated cells (the so-called "hot-spot effect"). To prevent this, PV modules should be placed in proper place, and their surface should be cleaned from time to time.

*2)* The point of operation on present current-voltage output characteristic of PV generator. By reducing the value of resistance loading PV generator, its point of operation is moving from no-load state, by Maximum Power Point (MPP) while the generator is giving maximum output power possible to obtain for actual insolation and temperature values, to short-circuit state. Because of the high cost of PV-generated energy, it is very important to assure the operation of PV generator in MPP at any time. Power electronic converters with special control are often used to attain this purpose. These converters are called MPP Trackers (MPPT) [2].

*3)* The temperature of PV cells. It doesn't significantly affect on the value of short-circuit current, but it affects on generator's idle voltage. The temperature's growth by 2,5°C results in idle voltage, and consequently in output power, drop by about 1%. This is why PV modules should be placed in airy places.

*4)* The aging of PV cells, that results in changing of the shape of output cells characteristics, and consequently in generator's output power reducing.

The PDS structures may be classified as:

*1) Grid-connected PDS.* The operation of these systems depends from the connection to electroenergetic grid, usually to AC one-phase or three-phase grid, rarely to DC grid, for instance in public transport.

*2) Stand-alone PDS.* They can operate without the presence of electroenergetic grid. Generally, stand-alone system may comprise, except PV generator, the second generator – so-called "back-up generator". This is the electric generator driven by fuel engine or wind turbine. Such system is called "hybrid system". But however, in PDS it is not reasonable to convert mechanical energy into electrical energy, and then convert it again into mechanical energy. This is why hybrid PDS are not recommended in practice.

Stand-alone PDS may be classified as:

*2A)* Systems with electric buffer sources.

*2B)* Systems without electric buffer sources. Sometimes they may comprise non-electric buffer sources.

The structures mentioned above have been described in the next chapters.

## II. GRID-CONNECTED PHOTOVOLTAIC DRIVE SYSTEMS

The most typical structure of grid-connected PDS is shown in Fig. 1. It comprises PV generator (PVG) connected by the inverter AU1 to electroenergetic AC-current grid EG. Also the electrical machine EM with its load ML must be found in such struture. Even for DC motors with series or permanent magnet excitement, the structure must contain adapting unit AU2 – DC chopper (almost always step-down one) for commutator DC motors, or independent DC/AC inverter for AC induction motors.

In such PDS structure, the inverter AU1 has to enable bilateral power flow. For low insolation levels of PV generator's modules, and while the motor is operating, AU1 is acting as AC/DC rectifier converting AC energy from grid into DC energy, that is then completing the deficit of energy produced by PVG. For high insolation levels, or while the motor isn't operating, AU1 is acting as DC/AC inverter, transmitting surplus of PV generator's energy into AC grid.

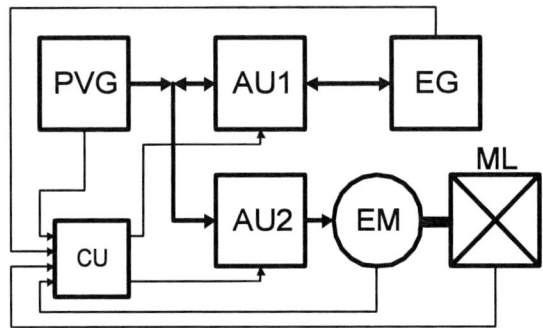

Fig. 1. Block diagram of grid-connected PDS.
CU – Controlling Unit

In simple systems, AU1 unit is a grid-commutated inverter, for instance 4T structure (Graetz bridge with four thiristors) for one-phase grid or for low nominal PVG power (about to 1kWp), or 6T bridge for three-phase grid and higher nominal power of PVG. To make possible bilateral power flow, it must be equipped with a switch changing its polarization connection to PV generator, while passing from rectifier to inverter state of work, and inversely. To make possible proper operation of AU2 converter while changing switch position, there must be a capacitor with adequate capacity connected parallely to PV generator.

The second very important function of AU1 (except of converting energy from DC to AC type and inversely) is MPP tracking. It means keeping PV generator's output voltage equal to the value, that is enabling obtaining the MPP on output PVG characteristic for any terms of insolation and temperature, and also independently of changes of power transmitted to the motor.

The functions of AU2 converter are:

*1)* Limiting motor's angular velocity, to protect the motor and its load from mechanical damage. It is done by limiting average value of supply voltage for DC motor, or limiting value of frequency for AC motor.

*2)* Limiting motor's current, to protect motor's armature windings from overheating while starting-up, or while the torque of motor's load ML is higher than motor's nominal torque for longer period of time.

*3)* The possibility of starting-up and stopping of motor.

*4)* Converting electrical energy type from DC to AC, for induction or DC-brushless motors.

Grid-connected structure presented in Fig. 1 assures extremely low probability of loss of supply for motor, lower than while supplying the motor only from grid, because there are two energy sources in the system. Each of them can independently supply motor, even while the other source is not operating, for instance while the grid is damaged, or while the nighttime for PV generator.

Because of this good point of the structure shown in Fig. 1, the structure shown in Fig. 2 is, to be sure, not recommended. In fact, the PV system structure shown in Fig. 2 is not a PDS structure, because PV-supplied part (PVG, AU1) and drive part (AU2, EM, ML) always operate independently. Furthermore, grid failure in such system makes the operation of grid-commutated AU1 inverter impossible, so PV generator's energy cannot be transferred not only to grid, but also to motor, even during sunny day.

1891

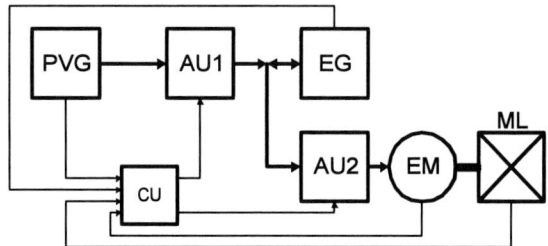

Fig. 2. Block diagram of non-PDS grid-connected system (not recommended structure).

But however, sometimes the structure from Fig. 2 could be applied, because energetic grid owners prefer it. With such structure they can get greater profits, as they can purchase greater amount of PVG energy with "green certificate", than transferred into the grid in system with structure shown in Fig. 1. This is because the transferred energy measurement must be located at output (grid) terminals of AU1 unit, in both structures.

The most important shortcoming of grid-connected systems is decidedly higher cost of energy generated by PV generator, comparing with the cost of energy delivered by grid. This is why such systems are still not profitable for most popular possible applications.

## III. STAND-ALONE PHOTOVOLTAIC DRIVE SYSTEMS WITH ELECTRIC BUFFER SOURCES

The most typical structure of stand-alone PDS with electric buffer source and without back-up generator is shown in Fig. 3. This structure always comprises PV generator (PVG), connected to electric buffer source EBS by AU1 unit. The motor is also connected to EBS, by AU2 unit. According to the type of EBS, stand-alone systems with EBS may be classified as:

*1) SA-A* (Stand-Alone with electrochemical Accumulator, usually of lead-acid type, as EBS). This is the most popular stand-alone structure, as yet, because of the lowest probability of loss of supply for the motor, and possibility of motor's operation with constant velocity for long periods of time (greater than several minutes).

*2) SA-EF* (Stand-Alone with Electrolyser and Fuel Cells Set). In such EBS, surplus of electrical energy from PVG is being converted into hydrogen in electrolyser, the hydrogen is being accumulated in special tank, and finally this gas can be converted into DC electrical energy in the set of fuel cells. Because of very high cost of such EBS, this kind of buffer source is not recommended for PDS, so it is not discussed in the paper.

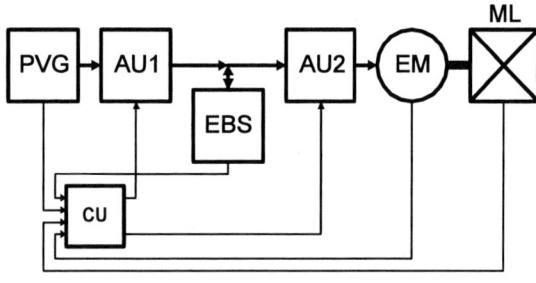

Fig. 3. Block diagram of stand-alone PDS with Electric Buffer Source EBS.

*3) SA-C* (Stand-Alone with Capacitor as EBS). The capacitor can be of electrolytic type, or it can be a super-capacitor. This is quite modern structure [3], and its applications will grow in future certainly, because it is less expensive and less problematic in installation and exploitation than SA-A structure. And on the other hand, it can assure constant velocity of motor up to several minutes, that is impossible to assure in the system without EBS. Finally, this results in the best efficiency of energy conversion in the whole system, comparing with SA-A, SA-EF and all structures without EBS.

In SA-A systems, AU1 unit may operate as usual switch, to disconnect accumulator from PVG while the accumulator is fully loaded and the motor is not working. This protects accumulator from damage caused by overloading. Also at night, AU1 is disconnecting accumulator from PVG, to prevent accumulator from discharging by the cells of PV generator. But however, in SA-A systems with PV generator of nominal power greater than about 100Wp, it is recommended to apply the DC/DC step-down chopper as AU1. This converter would assure, except of the mentioned above features, the operation of PVG in its MPP, and so it would increase generated power up by even 25%, especially in cold days.

Accumulators and capacitors are the devices with some different basic electrical properties. While transferring energy to or from the accumulator, its terminal voltage is changing of little importance limits, but to charge or discharge the capacitor its terminal voltage must be changing in significant limits. This is why the AU1 unit in SA-C systems must be always a DC/DC chopper, not a switch. But it may be not only of step-down type, but also a step-up (if needed) converter.

In SA-A systems with DC commutator motor type, AU2 unit may be an usual switch finally. But however, it is always recommended to apply power electronic converter as AU2: DC/DC chopper for DC motor or DC/AC independent inverter for AC motor. It must be born in mind, that in SA-C systems this converter must be adapted to operate with changes of DC input voltage in significant limits (tens of percents). The functions of AU2 are: limiting motor's angular velocity and current, assuring soft start-up and stopping of motor, and also converting energy from DC to AC – for AC motors.

In SA-A structure, the selection of technical parameters of system elements for short-period operation times is simple, and resolves itself into well-known aspect of supplying electrical machine from constant-voltage, DC energy source. This feature appears from very low internal resistance of accumulator, and from its relatively large capacity. This is why the motor's operation is independent from PV generator's output power fluctuations in short time periods of system's operation (up to several minutes).

However, the analysis of SA-A system operation in long-time periods, and also the selection of PV generator's nominal power and accumulator's nominal capacity, is difficult in calculations. The specialist computer simulation software is recommended for solving such problems.

The basic fault of SA-A systems is relatively high cost of installation and exploitation. This is why the structures described in the next chapter are more recommended for many stand-alone applications.

## IV. STAND-ALONE SYSTEMS WITHOUT ELECTRIC BUFFER SOURCES

The simplest structure of PDS is presented in Fig. 4a. It contains only PV generator, DC commutator motor (with permanent magnet or series excitement), and hand-operated switch AU. In spite of its simplicity (no power electronic converters, no controlling unit) and resulting low failure frequency, the basic fault of this structure is the impossibility of matching PV generator's characteristics to motor's input characteristics for wide range of PVG modules insolation and temperature changes [4]. It results in extremely low efficiency of energy conversion in the whole system, because PV generator's operation close to MPP is possible only in the proximity of one operating point, for instance for high insolation and high temperature of PV modules. The least mismatch of PVG and EM characteristics can be obtained for motor's load ML characterized by fan characteristic (fan, centrifugal pump), where the torque is proportional to square of angular velocity, so for motor with permanent magnet excitement, armature's current is proportional to square of armature's voltage average value.

Because of the fault mentioned above, and because of many other faults of this structure, it is not recommended for practical applications, except for systems with very low PV generator's nominal power (up to several Wp), and for loads with low starting torque, for example for fans driven by 12V DC motors. For all other applications, where it is not necessary to apply electrical buffer source, the structure presented in Fig. 4b is recommended.

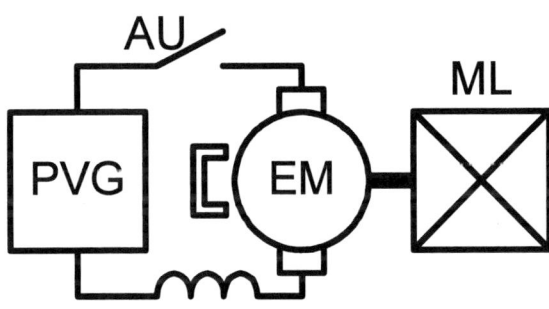

a) Without Buffer Source, and with direct connection of EM to PVG.

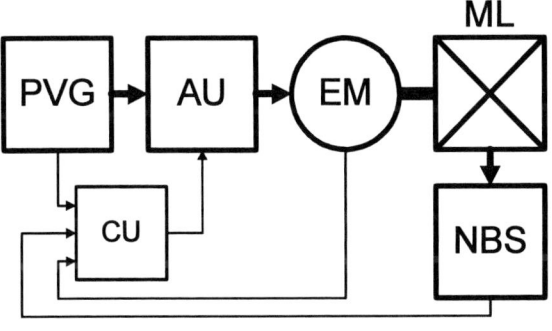

b) With Non-electric Buffer Source NBS, and with converter AU.

Fig. 4. The diagrams of stand-alone PDS without Electric Buffer Sources.

This system contains power electronic converter AU (its functions have been presented in chapter V of the paper). It is a DC chopper for DC commutator motor (may be of step-down or step-up type), or voltage inverter (one or three-phase) for AC induction motor.

Such system may be equipped with non-electric buffer source (NBS), if necessary. For instance, in the system applied to secure water for stock-farm, the water for animals can be gathered in water tank.

Systems with structure presented in Fig. 4b are very good for driving fans or water pumps with low level pumping heights, for instance for irrigation of cultivable lands, where the application of NBS is not necessary. They may be also applied for driving some refrigerating and air-conditioning units, but in such cases the systems with SA-C structure should be rather applied.

## V. AN EXAMPLE OF SIMPLE STAND-ALONE SYSTEM WITHOUT ELECTRIC BUFFER SOURCE

A diagram of example stand-alone PDS without EBS is presented in Fig. 5. This system has been completely projected, modeled, constructed and tested in the Department of Electrical Drive Systems at Lublin University of Technology. The system has been designed to enable different motor loads driving, but especially for driving centrifugal water pumps. This system may be equipped with NBS (e.g. water tank for pump), but it can also operate without any buffer source, e.g. while irrigation of cultivable land. The DC commutator motor with permanent magnet excitement has been applied, but the series motor may also operate in it.

The PV generator consists of parallely connected branches, and each branch is composed of the same number of PV modules connected in series (in the prototype system: six 55Wp modules, three branches, two modules in each branch).

Elements: C1, T1 (p-MOSFET transistor), D1 (Shottky diode) and the inductance of motor's EM armature, form main part of the step-down chopper (AU unit). Elements: K (comparator) and TG (triangle-shape wave generator) enable transforming DC voltage on non-inverting input of comparator, into rectangular wave with constant frequency, but variable pulse duration proportional to this DC voltage.

Controlling Unit of the chopper enables accomplishment of four tasks, mentioned below. Of course, because CU has only one output (this output is placed at the gate of transistor T1, for controlling the chopper), only one task can be done at any time:

*1)* Starting-up of the motor (while opening switch SW) and its stopping (while switching on SW).

*2)* Limiting motor's angular velocity. It is necessary to protect motor and its load ML from mechanical damage during periods of good insolation of PV modules, especially while the temperature of modules is low and load's torque is low (for instance while the operation of pump with no water flow). Velocity limiter is formed with elements: R4, R5, C3, D2, R8, T2. The first three elements form voltage divider with low-pass filter for obtaining average value of motor's armature voltage, that is approximately linearly dependent from motor's angular velocity. While output voltage from divider is exceeding breaking voltage of Zener diode D2,

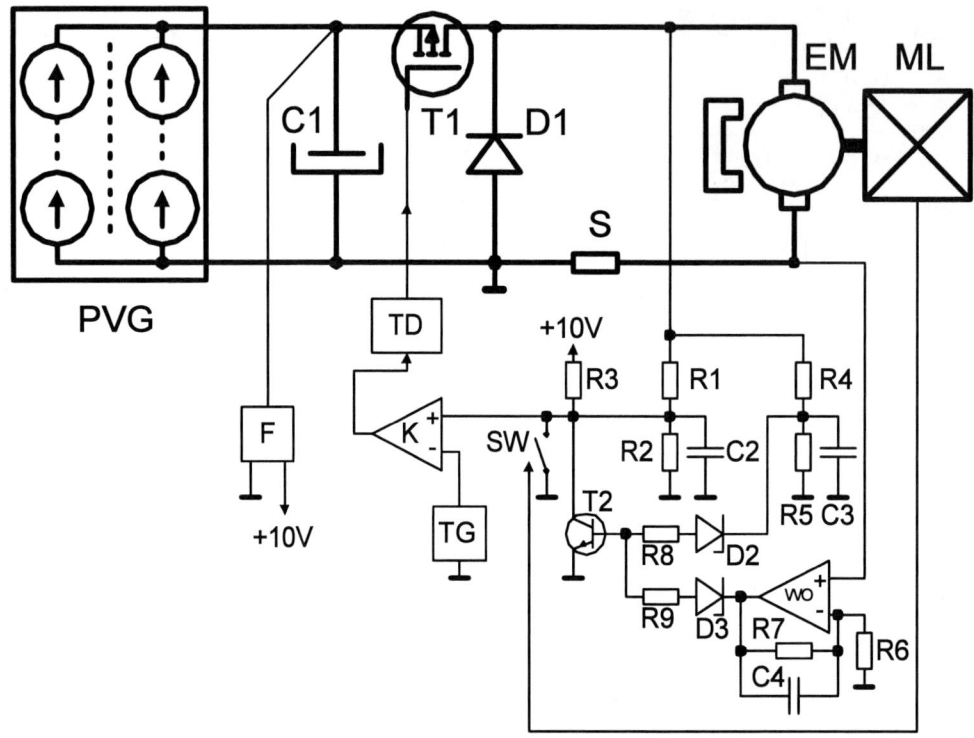

Fig. 5. Schematic diagram of Stand-Alone PDS without Electric Buffer Source, with step-down chopper between PVG and DC commutator motor with permanent magnet excitement.

F – feeder of Control Unit elements, TD – transistor's T1 driver.

transistor T2 is reducing pulse-duty factor of the chopper, that is reducing motor's armature average voltage, and consequently motor's velocity. While the application of series motor, voltage limiter would be more complex, because the value of limited voltage should depend from motor's current. For less current, and resulting less excitation flux, there should be less value of limited voltage, to achieve approximately constant value of limited velocity.

*3) Limiting motor's armature current.* In general, it protects armature windings from overheating, while the anti-torque of motor's load ML is higher than motor's nominal torque for longer period of time. But if motor's load is typified by univocal characteristic (e.g. fan characteristic, where the anti-torque is always proportional to the square of angular velocity), all current limiter elements can be replaced by one element – a fuse in the place of shunt S. Such simplification of the system is possible for permanent magnet excitement motors driving fans or water pumps, but cannot be applied for instance for conveyors and feeders. In such simplificated structure, the voltage (and consequently velocity) limiting, is automatically limiting the value of armature's current during normal operation of the system. Only during the breakdown of some elements in the system (e.g. locking of pump's rotor), the current could exceed nominal value. This is why the presence of fuse is necessary.

In simplified system, to enable limiting of motor's current during start-up of the motor, the value of C2 capacity must be selected properly. Increasing this

capacity, the maximum value of current during start-up is becoming less, but also all dynamic states in the system, including start-up process, are becoming longer. Sometimes it may result in instability of the system.

Current limiter is formed with elements: S, WO, R6, R7, C4, D3, R9 and T2. The first five elements form amplifier of shunt's S voltage (that is proportional to motor's current), with low-pass filter for reducing interferences, principally created by fast current switching between transistor T1 and diode D1. When motor's armature current is exceeding its nominal value, shunt's amplified voltage is exceeding Zener diode D3 breaking voltage, and transistor T2 is reducing pulse-duty factor of the chopper, that is reducing motor's armature average voltage, and consequently motor's velocity. For centrifugal pumps and fans it is consequently reducing their anti-torque, which is proportional to square of velocity. It is finally reducing motor's armature current to nominal value.

But for conveyors and feeders, the velocity doesn't significantly affect their anti-torque, so motor's velocity would be reduced to zero, and its current to nominal armature's value. It is not a problem for motors without forced cooling (such are low-power motors and motors driving submerge pumps), but for motors with self-forced cooling the current limiter should be more complex. In such case, for less velocity (and resulting worse cooling) there should be less value of limited current, than while operation with nominal velocity.

*4)* Securing the operation of PV generator in MPP. Elements R1, R2 and R3 form MPPT circuit, with positive feedback from average value of motor's armature voltage. The principle of operation of this MPPT has been described in details in [2] and [5], and is also shortly presented below.

MPPT circuit enables maximization of PVG output power indirectly, because it can maximize motor's input power. But in fact, it is better than direct maximization of PVG power, because not PVG energy quantity, but motor's energy quantity is more important from user's point of view.

The described MPPT type can be applied in PDS with load ML characterized by strictly monotonic dependence of ML's mechanical input power from ML's angular velocity. Such requirement comply both:

*1)* Centrifugal pumps or fans. For them, the input power is proportional to the third power of angular velocity.

*2)* Displacement pumps, lifts or conveyors. For them, the input power is almost linearly dependent from angular velocity.

In the systems that comply with the above requirement, the average value of armature's voltage is strictly monotonic function of PVG output power, because:

*1)* For DC motor with permanent magnet excitement, operating with load torque (and armature's current) not greater than nominal, the dependence of average armature's voltage from angular velocity is strictly monotonic, increasing function.

*2)* ML's mechanical input power is strictly monotonic, increasing function of angular velocity, as proved above.

*3)* For not very weak values of PVG output power (i.e. while motor's operation), ML's mechanical input power is strictly monotonic, increasing function of PVG output power. This arises from the power balance in the whole system.

Finally, this is why the tracking for maximum of PVG output power can be resolved itself into tracking for maximum of average value of motor's armature voltage.

Referring to Fig. 5, the principle of operation of the MPPT is as follows:

At the beginning, switch SW is shut. Opening it is resulting in rising of C2 capacitor voltage (C2 is being charged by resistor R3). Consequently, the pulse-duty factor of the chopper is rising. It is resulting in increasing of PVG output power, and in increasing of motor's voltage average value, that is strictly monotonic function of PVG output power. Thanks to positive feedback (by resistor R1), increasing motor's voltage is resulting in increasing C2 voltage, pulse-duty factor, motor's average voltage and C2 voltage again. As one can see, the positive feedback enables starting-up of the motor. While this process, PVG's voltage is decreasing from idle voltage value to MPP voltage value.

But if PVG's voltage had decreased below MPP voltage, the output power of PVG also would have decreased. That would have resulted in decreasing of motor's voltage, and consequently in decreasing of C2 voltage and pulse-duty factor. Finally, the PVG voltage would have increased back to MPP voltage. As one can see, for PVG voltage values below MPP voltage value, the same feedback is able to stabilize the operation of PVG close to MPP.

### REFERENCES

[1] M. Niechaj, "Możliwości zastosowań różnego rodzaju silników elektrycznych w fotowoltaicznych systemach napędowych", *IX Ogólnopolskie Forum Odnawialnych Źródeł Energii OZE*, Zakopane-Kościelisko, 2003, pp.230–236.

[2] M. Niechaj, "Maksymalizacja mocy generatora w fotowoltaicznych systemach napędowych", in *Electric Driving Systems Supplied from Unconventional Power Sources*, seria wydawnicza *Postępy Napędu Elektrycznego i Energoelektroniki*, PAN, Lublin, 2000, pp. 24–39.

[3] M. Niechaj, "Analiza pracy silnika magnetoelektrycznego zasilanego w systemie fotowoltaicznym z kondensatorem jako buforowym źródłem zasilania", *Postępy w Elektrotechnice Stosowanej PES-2*, Zakopane-Kościelisko, 1999, pp. 25–32.

[4] M. Niechaj, "Optymalizacja pracy maszyny elektrycznej w fotowoltaicznym systemie napędowym", in *Zeszyty Problemowe BOBRME "Maszyny elektryczne"*, vol. 77/2007, pp. 91–95.

[5] M. Niechaj, J. Kolano, P. Filipek, "Maximum power point tracking for PV generator feeding different DC motor drive systems", *Opto-Electronics Review*, vol. 8, COSiW SEP, 2000, pp. 398–401.

# Experimental results of a hybrid wind/hydro power system connected to isolated loads

*Mehdi NASSER[1], Stefan BREBAN[1,2], Vincent COURTECUISSE [1], Arnaud VERGNOL[1], Benoît ROBYNS[1], Mircea M. RADULESCU[2]*

[1]Laboratoire d'Electrotechnique et d'Electronique de Puissance (L2EP), Lille, France
Ecole des Hautes Etudes d'Ingénieur (HEI), 13, rue de Toul, F-59046 Lille Cedex, France
[2]Technical University of Cluj-Napoca, C. Daicoviciu St., 15, RO-400020 Cluj-Napoca, Romania

*Abstract*—**Variable speed hydro power plants allow to compensate and to smooth the fluctuating wind power when feeding isolated loads. This is shown experimentally in this paper with the help of a 3 kW laboratory test bench able to emulate a hybrid wind/hydro power system. The test bench is equipped of a doubly fed induction generator based small hydro power plant emulator associated with a fixed-speed wind generator emulator. This kind of wind generator is often used in small or medium power isolated networks. Tests show that the frequency and voltage are well controlled in spite of load variations.**

*Keywords*— **Hybrid system, variable-speed hydropower plant, fixed speed-wind generator, doubly-fed induction generator, permanent-magnet synchronous machine.**

## I. INTRODUCTION

Production means based on renewable energies feature a small rated power (compared to classical plants) and often a dependence on primary energy availability, when it comes from natural process. The example of wind generators is significant because wind is fluctuating and not easily predictable from one day to another. The same comments may be put forward for solar energy in temperate climate countries. On the contrary, small-hydro stations can be considered as an exception. Indeed the water flow does not generally vary as quickly as the wind speed or solar radiance except during outstanding meteorological events like violent storms. Flow curves often show seasonal variations. Moreover a small-hydro plant may produce power over the whole year, except if a severe dryness period occurs. Small-hydro demonstrates its ability to complement usefully the wind production **[1, 2]**. It may compensate for wind fluctuations, and then may support wind generators. Consequently, the concept of multi-source system, with integrated and optimized management energy, eventually associated with various storage systems **[3]**, is considered like a strong reply to these problems.

**Fig. 1** shows the micro hydro power station associated to a fixed-speed wind generator considered in this paper. Fixed speed wind generator is often used in small or medium power isolated networks. Variable speed hydro power plant allows to compensate and to smooth the fluctuating wind power when feeding isolated loads. This is shown experimentally in this paper with the help of a 3 kW laboratory test bench able to emulate a hybrid wind/hydro power system. The test bench is equipped of a doubly fed induction generator based small hydro power plant emulator **[4, 5, 6]** associated with a fixed-speed wind generator emulator. After a presentation of both renewable sources and a description of the test bench, simulations and tests under variable loads are shown.

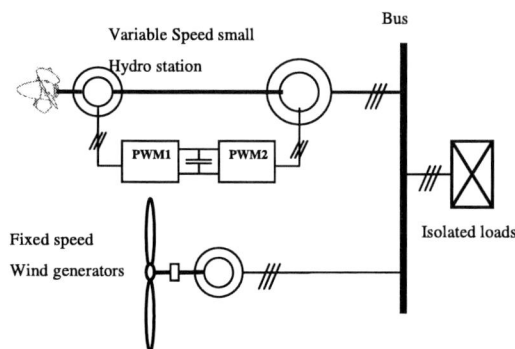

Fig. 1 Wind/hydro power system

## II. VARIABLE-SPEED SMALL HYDRO PLANT ([4], [5], [6])

The scheme of the proposed small hydropower station is represented in Fig.2. Like most of these plants, the studied one is considered as run-of-river leading to the use of a Kaplan hydraulic turbine well suited for low water heads. The turbine is associated with a gear box because of its small rotating speed. It drives a doubly-fed induction generator (DFIG) whose excitation is supplied on its rotor by a permanent-magnet synchronous machine (PMSM) mounted on the same shaft. Two back-to-back PWM power electronic converters, connected by way of a DC bus, carry out the electric link between the machines. Converter 1 controls the DC-link voltage leading to the balance between the DFIG-rotor active power and the PMSM one. Converter 2 is dedicated to the control of the DFIG, in order to achieve operation of this generator on isolated loads or on a power grid.

It may be emphasized that this considered configuration is different from the most common DFIGs whose rotor windings are connected, ever with the help of power electronic converters, to stator ones. This electromechanical set, when compared to classical structures dedicated to small hydroelectric power, feature several interesting characteristics within the context described above : it can operate in an autonomous way due to the PMSM allowing the DC-link capacitor stand-alone charging when the system starts ; power electronic converters increase the plant control possibilities – for instance, additional capacitors coupled with squirrel-cage induction machines are no more necessary – and dynamics, as they replace mechanical controls; converter and PMSM rated powers are only around 30% of the plant rated power, representing the typical average slip power **[7]**.

978-1-4244-1741-4/08/$25.00 ©2008 IEEE

Papers [4], [5] and [6] deal with the structured model of the system connected to isolated loads in which the DFIG stator voltage rms value and frequency are controlled.

## III. MODELLING OF THE HYBRID SOURCE

### A. Hydro power turbine model

As mentioned before, a Kaplan turbine is considered in this paper. It is submitted to a fixed head. It is assumed that water flow variations are very slow compared to the drive dynamics. The turbine model is a basic one, i.e. it includes neither blade pitch control nor upstream guide vane one. According to these assumptions, hydro-power turbine behaviour may be taken into account by means of simplified static mechanical characteristics, as these presented in Fig.3 for a fixed rate of flow.

Turbine torque ($T_t$) vs. speed ($\Omega$) characteristic is assumed to be a straight line. Torque becomes null for a rotating speed value $\Omega_e$ which is the runaway speed, i.e. speed when no-load torque is applied on the shaft. $\Omega_e$ is a turbine parameter, and a value of 1.8 times the turbine rated speed $\Omega_n$ is assumed [8]. Torque vs. speed characteristic equation under rated water flow and head is given below [8]:

$$T_t = T_n \left( 1.8 - \frac{\Omega}{\Omega_n} \right), \tag{1}$$

where subscript "n" is used for rated values.

Mechanical power ($P_{mec}$) simplified characteristic is, consequently, a parabola. Taking into account the water wheel efficiency depending on the rate of flow and on the rotating speed, this power results from the hydro power ($P_{hyd}$) which is expressed as follows:

$$P_{hyd} = \rho g H q , \tag{2}$$

where $\rho$ is the water density, $g$, the gravity acceleration, $H$, the water head and $q$, the water rate of flow.

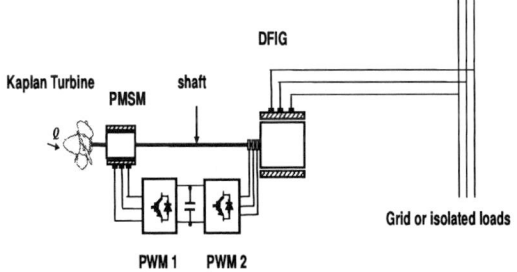

Fig.2. Considered stand-alone variable-speed drive

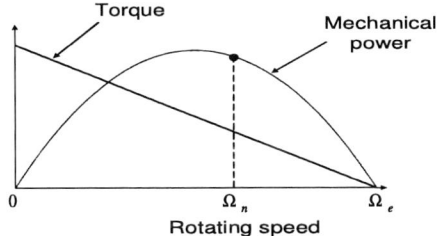

Fig.3. Hydro power turbine torque and mechanical power vs rotating speed simplified characteristics for given water flow

### B. DFIG Park Model

The DFIG is expected to be connected either to isolated loads or to a power grid. In the first situation, the rms value and frequency of the generator stator voltages must be regulated. In the second one, assuming that the grid voltage and frequency are fixed, the DFIG active and reactive powers are controlled. It is therefore convenient to refer to the stator electric equations which may be written as follows in a Park reference frame linked to the rotating field:

$$v_{sd} = R_s i_{sd} + \frac{d\Phi_{sd}}{dt} - \omega_s \Phi_{sq} \tag{3}$$

$$v_{sq} = R_s i_{sq} + \frac{d\Phi_{sq}}{dt} + \omega_s \Phi_{sd} \tag{4}$$

In these equations $R_s$ is the DFIG stator winding resistance and $\omega_s$ the stator electric pulsation; $\Phi_{sd}$ and $\Phi_{sq}$ refer to d,q stator magnetic flux. From the relations between flux and currents, $\Phi_{sd}$ and $\Phi_{sq}$ can be written as follows [8]:

$$\Phi_{sd} = \sigma L_s i_{sd} + \frac{M}{L_r} \Phi_{rd} \tag{5}$$

$$\Phi_{sq} = \sigma L_s i_{sq} + \frac{M}{L_r} \Phi_{rq} \tag{6}$$

$\sigma = 1 - \frac{M^2}{L_s L_r}$ is the DFIG scattering coefficient. $\sigma L_s$ is then the stator leakage inductance.

Therefore (3) and (4) become:

$$v_{sd} = R_s i_{sd} + \sigma L_s \frac{di_{sd}}{dt} + e_{sd} \tag{7}$$

$$v_{sq} = R_s i_{sq} + \sigma L_s \frac{di_{sq}}{dt} + e_{sq} \tag{8}$$

With e.m.f. $e_{sd}$ and $e_{sq}$ defined as:

$$e_{sd} = \frac{M}{L_r} \frac{d\Phi_{rd}}{dt} - \omega_s \Phi_{sq} \tag{9}$$

$$e_{sq} = \frac{M}{L_r} \frac{d\Phi_{rq}}{dt} + \omega_s \Phi_{sd} \tag{10}$$

Assuming that steady state is reached, (7) and (8) become:

$$v_{sd} = R_s i_{sd} + e_{sd} \text{ with } e_{sd} = -\omega_s \Phi_{sq} \tag{11}$$

$$v_{sq} = R_s i_{sq} + e_{sq} \text{ with } e_{sq} = \omega_s \Phi_{sd} \tag{12}$$

(11) and (12) show that, in isolated mode, e.m.f. must be controlled taking into account voltage drop due to winding resistance.

### C. Fixed-speed wind generator

For the fixed-speed turbine, the induction generator (IG) is directly connected to the isolated loads, according to Fig. 4. The squirrel-cage induction generator is a 4-pole machine, and the necessary reactive power for its magnetisation is produced by the DFIG.

The simplified dynamic model of a wind turbine is given by:

$$J \frac{d\Omega}{dt} = T_g - T_{em} - T_{viscous} \qquad (13)$$

Where the inertia (J) represents the total inertia and $\Omega_t$ is the mechanical speed.

$T_{em}$ is the electromagnetic torque and $T_{viscous}$ is the viscous torque. The generator torque ($T_g$) obtained from the aerodynamic torque, which is transmitted by a gearbox of gain G, is related to the aerodynamic torque:

$$T_g = \frac{T_a}{G} \qquad (14)$$

A wind turbine is characterized by its aerodynamic torque, which is given by:

$$T_a = C_p \frac{\rho}{2} A v^3 \frac{1}{\Omega_t} \qquad (15)$$

The power coefficient ($C_p$) represents the aerodynamic efficiency of the wind turbine Fig. 5. It depends on the blade design, the tip speed ratio ($\lambda$) and the pitch angle of the blades ($\beta$). The tip speed ratio is defined as follows:

$$\lambda = \frac{R\Omega_t}{v} \qquad (16)$$

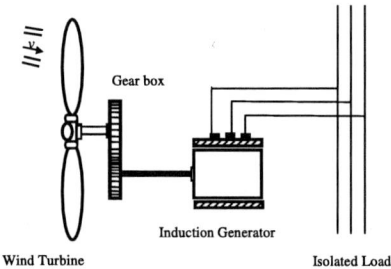

Fig.4. Fixed-speed generator system

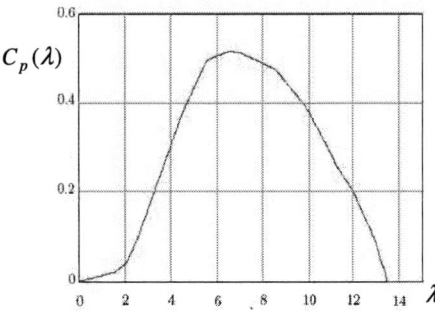

Fig 5 Power coefficient for given pitch angle

## IV. CONTROL OF THE DFIG IN ISOLATED MODE

In this mode the plant must control its voltage rms value and frequency. Due to the fact that the DFIG is controlled by way of its rotor, it is convenient to express the stator e.m.f. versus rotor state variables. Rotor fluxes,

which are naturally linked to e.m.f., are the control variables. Taking (5) and (6) into account, e.m.f. equations (9) and (10) evolve as follows:

$$e_{sd} = \frac{M}{L_r} \frac{d\Phi_{rd}}{dt} - \omega_s \frac{M}{L_r} \left( \Phi_{rq} + \sigma \frac{L_s L_r}{M} i_{sq} \right) \qquad (17)$$

$$e_{sq} = \frac{M}{L_r} \frac{d\Phi_{rq}}{dt} + \omega_s \frac{M}{L_r} \left( \Phi_{rd} + \sigma \frac{L_s L_r}{M} i_{sd} \right) \qquad (18)$$

To regulate voltage frequency at its nominal value, the rotor frequency $f_r$ is adjusted according to the induction machine classical frequency law:

$$f_r = \left| f_s - \frac{p_{dfig}\Omega}{2\pi} \right| \qquad (19)$$

$p_{dfig}$ is the DFIG pole pair number.

Fig.6 depicts the DFIG reversed EMR "Energetic Macroscopic Representation" diagram in which subscript "ref" indicates a reference value. Each storage element, i.e. crossed rectangle, in the direct EMR is normally associated to a control loop in the reversed one in order to regulate its output by means of one of its inputs, the second one being considered as a disturbance associated to a dotted line [4].

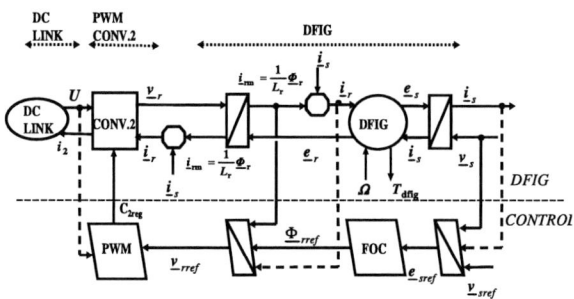

Fig.6. Isolated mode: DFIG stator voltage control strategy [4].

However it may be noted that causal law implies that voltage, which must be regulated, is not an output but an input. Acting on the DFIG stator currents would be a valid theoretical solution but these currents are forced by the load. The only possibility, as showed by the Park model relationships (17) and (18), is therefore to control stator emf, considering stator currents as disturbances. Indeed $\sigma L_s$, which is the origin of the considered storage element, is the stator leakage inductance whose value is frequently neglected. The result of stator voltage control loop is then an emf reference value as shown in Fig.6. (11) and (12) underline that at steady state the difference between stator emf and voltages is only due to stator winding resistance. Consequently an integrator is chosen as the controller of this external loop. It may be noticed that the static voltage drop generated by the leakage inductance is taken into account in the emf expressions as disturbance terms $\omega_s \sigma L_s i_{sq}$ and $\omega_s \sigma L_s i_{sd}$ as shown by (17) and (18). Emf regulation is performed by means of Field Oriented Control

(FOC) method which allows the determination of rotor reference fluxes of internal rotor flux loops whose controller outputs are rotor voltages [4].

More details about the control procedure of the DFIG and of the PMSM, which control the DC voltage, are presented in [4] and [5].

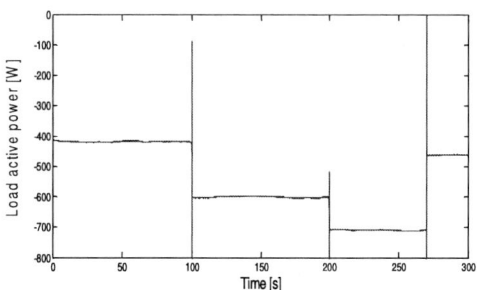

Fig.9. Active power absorbed by loads

## V. SIMULATION OF THE POWER SYSTEM

Simulations are carried out using the Matlab/ Simulink$^{TM}$ software. A 3 kW hydro generator referred to fixed water head and flow is considered. The system is based on a 4-pole DFIG and a Kaplan turbine, whose speed parameters are: $\Omega_e$ = 2700 rpm and $\Omega_n$ =1500 rpm. The nominal wind power system is 3 kW. Fig.7 to Fig.12 show respectively the power produced by the wind power plant for a strong wind, the hydro generator active power, the load active power, the load voltage frequency, the load rms voltage and the DC link voltage. Fig. 9 shows that the hydro plant has very good dynamics, and succeeds in compensating the wind power variations. The reference total active power has been established according to wind power generation and hydropower station's availability. It demonstrates the ability of the hybrid drive to deliver a constant power absorbed by load. It can be noticed from figure 10 and figure 11 that the drive operates well at fixed frequency and rms voltage. The DC link voltage is well maintained at its rated value (Fig. 12).

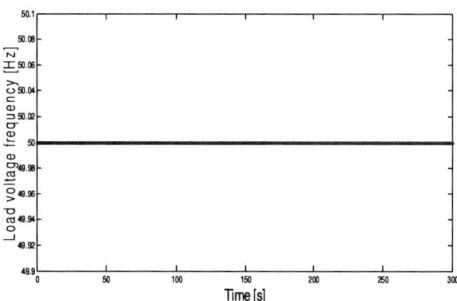

Fig10. Load voltage frequency [Hz]

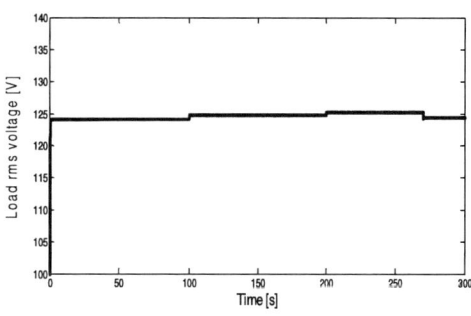

Fig11. Load rms voltage [V]

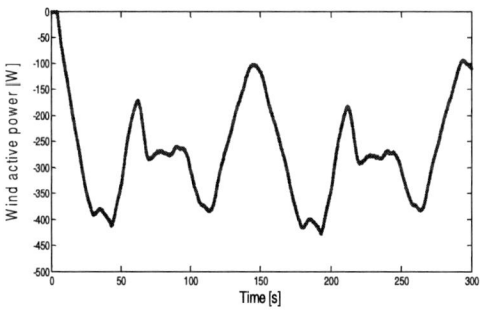

Fig.7. Wind active power

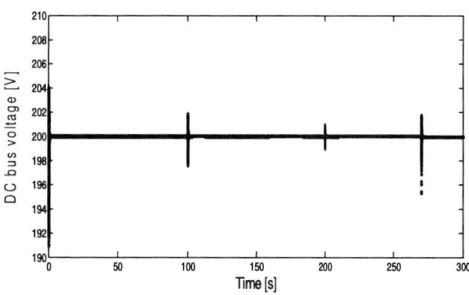

Fig12. DC bus voltage [V]

## VI. EXPERIMENTAL RESULTS

A 3 kW test bench dedicated to power generation is available at the Ecole des Hautes Etudes d'Ingénieur (HEI), Lille, France. This experimental tool has been developed for several years to assess studies on the control of wind generators associated with a kinetic energy storage

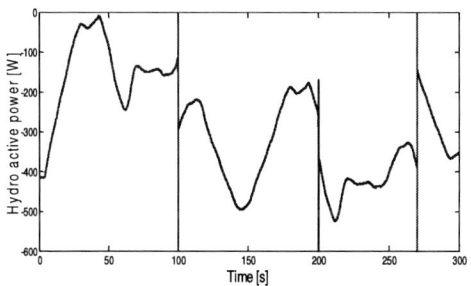

Fig.8. Hydro generator active power

system [9]. For hydro power concerns the bench is composed of a hydraulic turbine emulator, based on a torque controlled DC machine, a PMSM and a 3 kW (4 poles) doubly fed induction machine mechanically coupled to the DC machine and the PMSM. Two converters make the link between the induction machine rotor and the PMSM. Power converters components switches are controlled by DSPACE™ cards. Fig.13 shows the bench structure dedicated to the study discussed in this paper. As DFIG stator and rotor coils are star-connected with neutral point isolated, measuring two stator and two rotor currents is necessary and sufficient. Rotor currents are as well useful to compute rotor fluxes. Stator instantaneous voltage value is also measured to calculate the active and reactive powers transmitted by the generator to isolated loads, to perform RMS value and frequency regulation.

On the other side, a PMSM, that drives the IG, is used to emulate the behavior of the fixed-speed wind generator associated to one power converter. System control is achieved by using DSPACE™ controller boards. The PMSM currents and the shaft rotational speed are measured; its torque is controlled with the help of a classical vector control.

Tests are performed over an interval of 300 seconds

The hybrid system is connected to variable resistive load whose variations can only be processed step by step and not continuously. The load line-to-line reference voltage is 225 V rms. The water rate of flow is considered constant for this interval of time because its variability has a time scale of hours or days.

Results are presented in Fig.14 to Fig.22. They show respectively, the active powers generated by wind and hydro generators (Fig.14 and Fig. 15) and absorbed by loads (Fig. 16), the reactive power absorbed by the IG (Fig.17), the load voltage frequency (Fig.18), the load rms voltage (Fig.19), the wind and hydro generators rotating speeds (Fig.20 and 21), and the DC-link Voltage (Fig.22).

The hydro-power system compensates the fluctuation of the wind power by controlling the voltage and frequency of the loads. Fig. 21 confirms the ability of the hydro power plant to work at variable speed (under and over synchronism). Fig.18 and Fig.19 demonstrate the effectiveness of the hydropower system control strategy even if wind active power (Fig. 14) is very fluctuant and load variations (Fig. 16) are sudden. Another advantage of this hybrid system is the capability of the hydropower station to feed the reactive power needed to magnetize the induction generator (Fig.17) and so to eliminate the necessity to install capacitors banks.

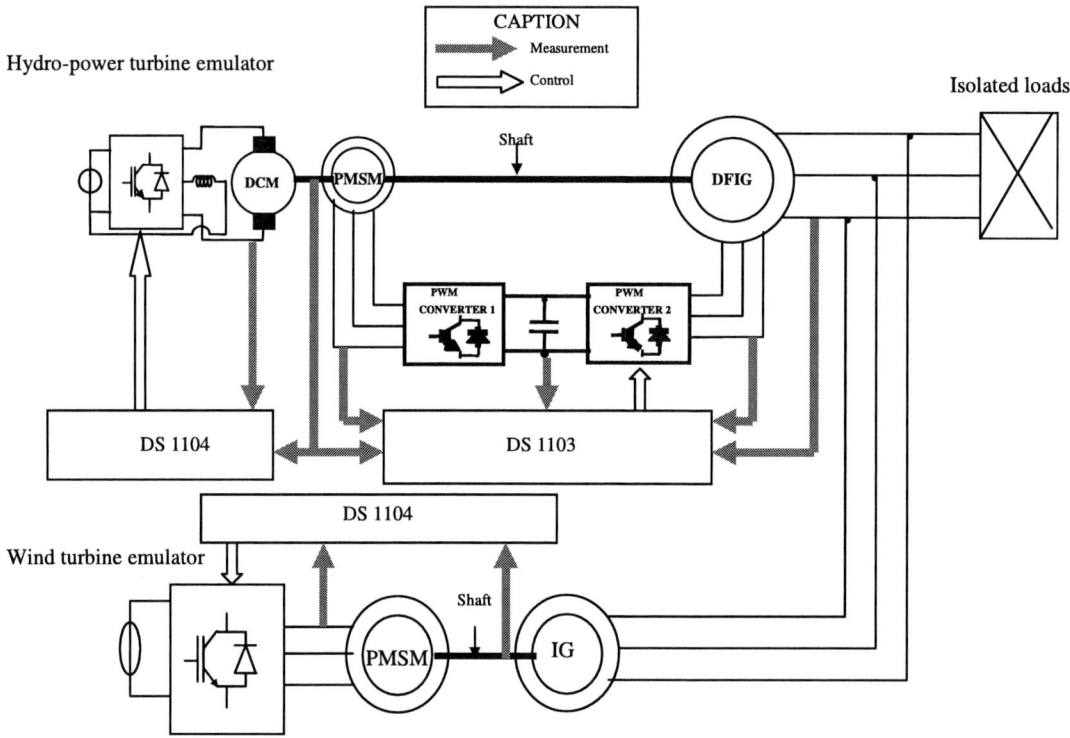

Fig.13 Test bench suited for wind/hydro power generator study.

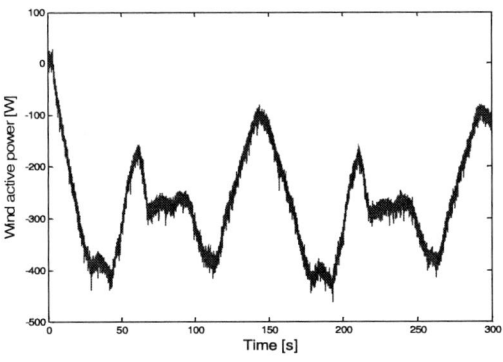

Fig. 14 Wind active power

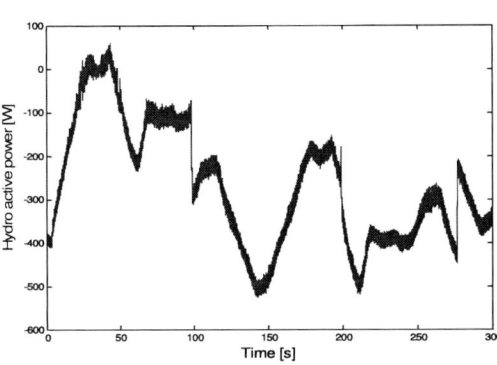

Fig. 15 Small hydro generator active power

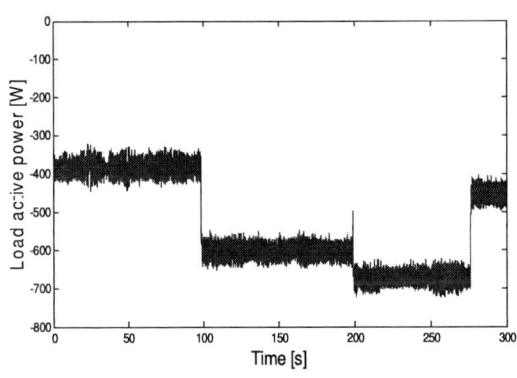

Fig. 16 Active power absorbed by loads

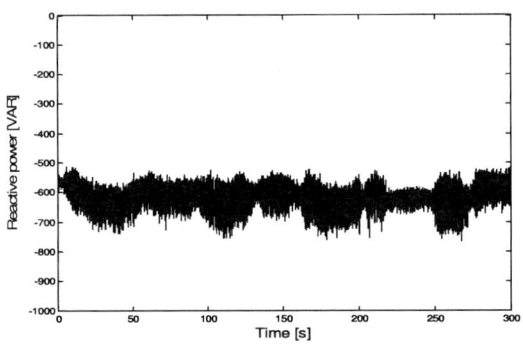

Fig. 17 Reactive power absorbed by IG

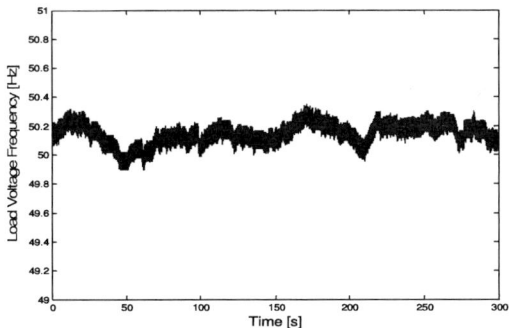

Fig. 18 Load voltage frequency

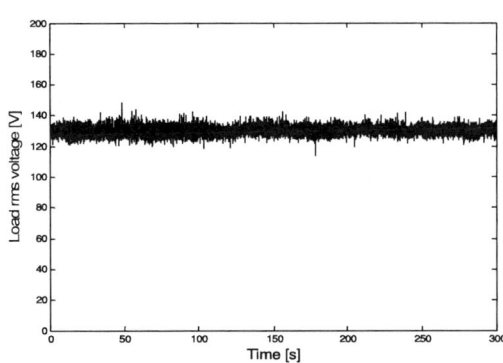

Fig. 19 Load rms voltage

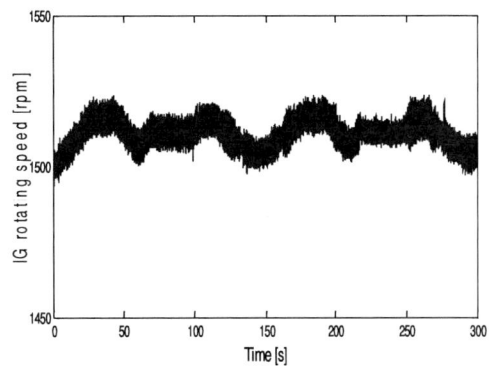

Fig. 20 Wind generator rotating speed

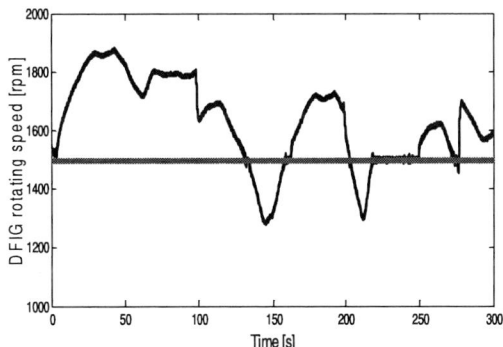

Fig. 21 DFIG rotating speed

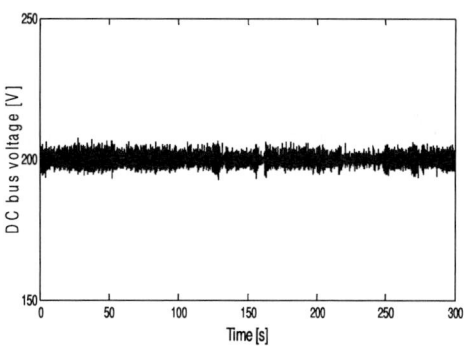

Fig. 22 DC link voltage

## VII. CONCLUSIONS

Variable speed hydro power plants allow to compensate and to smooth the fluctuating wind power when feeding isolated loads. This has been shown experimentally in this paper with the help of a 3kW laboratory test bench able to emulate a hybrid wind/hydro power system. The test bench is equipped of a doubly fed induction generator based small hydro power plant emulator associated with a fixed-speed wind generator emulator. This kind of wind generator is often used in small or medium power isolated networks. Tests under variable loads have been shown.

## VIII. APPENDIX

### a) DFIG Parameters

Rated power: $3kW(220/380V, 50Hz)$;

Number of poles: $2p = 4$;

Stator resistance: $R_s = 1.6\Omega$;

Rotor resistance: $R_R = 0.4\Omega$;

Magnetizing inductance: $M = 55mH$;

Stator inductance: $L_S = 150mH$;

Rotor inductance: $L_R = 23mH$;

Inertia: $J = 0.01kgm^2$.

### b) IG Parameters

Rated power: $3kW(220/380V, 50Hz)$;

Number of poles: $2p = 4$;

Stator resistance: $R_s = 0.75\Omega$;

Magnetizing inductance: $M = 78mH$;

Stator inductance: $L_S = 81mH$;

Rotor inductance: $L_R = 81mH$;

Inertia: $J = 0.0085kgm^2$.

### c) PMSM Parameters

Rated power: $2.87kW(3000rot/min)$;

Number of poles: $2p = 6$;

Stator resistance: $R_s = 0.94\Omega$;

Direct-axis inductance: $L_d = 14.4mH$;

Rotor inductance: $L_q = 25mH$;

Back-emf coefficient: $K_e = 0.78V * s * rad^{-1}$;

Inertia: $J = 0.0014kgm^2$.

## REFERENCES

[1]. E. D. Castronuovo, J. A. Peças, "Bounding active power generation of a wind-hydro power plant", Proc. 8th Conference on Probabilistic Methods Applied to Power Systems, Iowa, USA, 2004, pp. 705-710.
[2]. V. Courtecuisse, S. Breban, M. Nasser, A. Vergnol, B. Robyns, M.M. Radulescu "Supervision d'une centrale multi sources basée sur l'association éolien, micro hydraulique et stockage d'énergie". EF 2007, Toulouse, France

[3] C. Abbey, G. Joos, "Energy storage and management in wind turbine generator systems", EPE Journal. Vol.17; n° 4, p6-p12; January 2008.
[4]. A. Ansel, B. Robyns, "Modelling and simulation of an autonomous variable speed micro hydropower station," *Mathematics and Computers in Simulation*, vol. 71, n° 4-6, 2006, pp. 320-332.
[5] A. Ansel, B. Robyns, "Small hydroelectricity: from fixed to variable speed electromechanical drives," *Electromotion*, vol.13, n°2, 2006.
[6] S. Breban, M. Nasser, A. Ansel, C. Saudemont, B. Robyns, M. Radulescu, "Variable speed small hydro power plant connected to AC grid or isolated loads" EPE Journal; Vol. 17; no 4, p29-36; January 2008
[7] C.R. Kelber, W. Schumacher, "Adjustable speed constant frequency energy generation with doubly-fed induction machines," European Conference Variable Speed in Small Hydro - VSSHy 2000, Grenoble, France.
[8] "Petites centrales hydrauliques – Turbines hydrauliques," *Report of PACER Program*, Switzerland.
[9] G.O. Cimuca, C. Saudemont, B. Robyns, M.M. Radulescu, "Control and performance evaluation of a flywheel energy-storage system associated to a variable-speed wind generator.", IEEE Trans. Ind. Elect., vol.53, n°4, 2006, pp. 1074-1085

## THE AUTHORS

**Mehdi Nasser** was born in Mogadiscio, Somalia. He received the electrical engineering degree from CNAM, Lille, France, in 2001. He is currently working as a teacher in Electrotechnics and Power Electronics at HEI and is pursuing the Ph.D. degree at the Laboratory of Electrotechnics and Power Electronics of Lille, France.

His research interests are Supervision of a hybrid wind/hydro power system for electricity production and decentralized energy supply

**Stefan Breban** was born in 1981 in Baia-Sprie, Romania. He received the M.S. degree in electrical engineering from the Electrical Engineering Faculty, Technical University of Cluj-Napoca, Cluj-Napoca, Romania, in 2005. From October 2005, he is a Ph.D. student at Electrical Engineering Faculty, Technical University of Cluj-Napoca, Romania, and at Ecole Nationale Superieure d'Arts et Metiers (ENSAM) de Lille, France. He is working within the Laboratory of Electrotechnics and Power Electronics of Lille on the site of HEI, France, and he is engaged in theoretical and experimental research on small hydropower plants

**Vincent Courtecuisse** received the « Master de Recherche en Energie Electrique et Développement Durable » from the Université des Sciences et Technologies de Lille (USTL), France. He is currently a Ph.D student at the laboratory of electrotechnics and power electrotechnics of lille, degree at the Electrotechnical Department of the Ecole des Hautes Etudes d'Ingénieur (HEI) of Lille, France. His research interests include wind generator, power system and control

**Arnaud Vergnol** was born in Clermont-Ferrand, France in 1984. He received his M.Sc. degree in Electrical Energy and Sustainable Development from the University of Science and Technology of Lille (USTL), France, in 2007. He is currently a Ph.D. student at the Ecole Centrale de Lille (ECL) in France. His main research topics within the Laboratoire d'Electrotechnique et Electronique (L2EP) de Lille and the Hautes Etudes d'Ingénieur (HEI), are on the large-scale integration of wind power in the electricity network and on the market of the electricity

**Benoît Robyns** was born in Brussels, Belgium, in 1963. He received the Ingénieur Civil Electricien and Docteur en Sciences Appliquées degrees from the Université Catholique de Louvain, Louvain-la-Neuve, Belgium, in 1987 and 1993, respectively, and the Habilitation à Diriger des Recherches degree from the Université des Sciences et Technologies de Lille, Lille, France, in 2000.

From 1988 to 1995, he was with the Laboratory of Electrotechnics and Instrumentation, Faculty of Applied Sciences, Catholic University of Louvain, as an Assistant. Since 1995, he has been with the Ecole des Hautes Etudes d'Ingénieur, Lille, France, where he is currently the Director of Research. Since 1998, he has also been with the Laboratory of Electrotechnics and Power Electronics of Lille, France, as a Researcher, where he is currently the Head of the Electrical Network and Energetic Systems research team. He is the author and coauthor of more than 120 papers and 1 book in the fields of digital control of electrical machines, renewable energies, and distributed generation.

Prof. Robyns is a member of the IEEE society, the Société Française des Electriciens et des Electroniciens, the Société Royale Belge des Electriciens, and the European Power Electronics Association.

**Mircea M. Radulescu** was born in Cluj-Napoca, Romania, on September 4, 1954. He received the Dipl.-Ing. degree (with honors) from the Technical University of Cluj-Napoca, Romania, in 1978 and the Dr.-Ing. degree from the Polytechnic University of Timisoara, Romania, in 1993, both in electrical engineering.

In 1983, he joined the Faculty of Electrical Engineering, Technical University of Cluj-Napoca, where he is currently a Full Professor in the Department of Electric Machines and the Head of the Small Electric Motors and Electric Traction Group. He was an Invited Research Associate with the Laboratoire d'Electromécanique et de Machines Electriques, Ecole Polytechnique Fédérale de Lausanne, Switzerland, during 1990-1991 and with the

Laboratoire d'Electrotechnique de Grenoble, Institut National Polytechnique de Grenoble, France, during 1992-1993. He was an Invited Professor with Helsinki University of Technology, Finland, in 1997; Rheinisch Westfälische Technische Hochschule Aachen, Germany, in 1999; the University of Akron,OH, in 1999 and 2001; the Université 'Pierre et Marie Curie' (Paris VI), France, in 2002; and the Université de Picardie 'Jules Verne' Amiens, France, in 2003. He is the author and coauthor of ten scientific monographs, multiauthor books, and textbooks, and of more than 100 published scientific papers in refereed technical journals and international conferences and symposium proceedings. His teaching and research activities include computer-aided design of electromechanical devices; field analysis of electromagnetic structures; design and control of small electric motors; actuators and mechatronic drives; design, control, and electromagnetic compatibility of electric traction systems; and ferrohydrodynamics.

Prof. Radulescu is a Foundation Member of the Romanian Association of Small Electric Machines Builders, an Associate Editor of the international scientific quarterly ELECTROMOTION, and a Member of the International Steering Committee of several conferences and symposia in the field of

# Grid Connection of Multi-Megawatt Clean Wave Energy Power Plant under Weak Grid Condition

Kai Rothenhagen*, Marek Jasinski[†], and Marian P. Kazmierkowski[+]

* Institute for Power Electronics and Electrical Drives / University of Kiel, Kiel, Germany, *kro@tf.uni-kiel.de*

[†] Institute of Control and Industrial Electronics / Warsaw University of Technology, Warsaw, Poland, *mja@isep.pw.edu.pl*

[+] Institute of Control and Industrial Electronics / Warsaw University of Technology, Warsaw, Poland, *mpk@isep.pw.edu.pl*

*Abstract*— The WaveDragon is a 7 Megawatt Wave Energy Converter, that is currently developed for clean offshore energy production. The system will be floating several kilometres off the Pembrokeshire coast, Wales, UK, and transfers energy using a submarine power cable. Key interest is the control of the power take-off system at disturbed voltages and under weak-grid condition due to long submarine cable and remote, rural location. A multi-reference frame controller is implemented for harmonics compensation. A parameter estimator is implemented to estimate the grid impedance. Simulation and measurements results from a laboratory test setup that illustrate properties of developed method are shown.

*Keywords*— Renewable energy systems, Voltage Source Converter (VSC)

## I. INTRODUCTION

It has been widely discussed that the use of clean renewable energy is one of the important future task our society has to take on. The remarkable progress in the field of Wind Energy may serve as an example for other sources of clean, carbon-dioxide free power sources. Next to bio-fuels, solar and hydropower, wave energy is a very interesting and not yet intensively covered field. Oceans cover roughly 75 percent of the earth surface, and offer a tremendous amount of Renewable Energy in form of waves, which may serve as an extension to the current energy mix. Waves contain an average of 75 kW per meter [1]. Table 1 shows the potential of Renewable Energy according to [1].

TABLE I. RENEWABLE ENERGY RESOURCES [1]

|  | Available [$10^{18}$ J] | Current Use [$10^{18}$ J] |
|---|---|---|
| Biomass | 283 | 50 |
| Hydro | 50 | 10 |
| Solar | 1570 | 0.2 |
| Wind | 580 | 0.2 |
| Goethermal | 1401 | 2 |
| Ocean | 730 | 0 |
| Total | 4614 | 62.4 |

In order to harvest this energy and convert into AC electric power, wave energy plants are needed. Several different types have been researched in the last decades, such as the Pelamis [12] and AquaBuoy [13] which are floating devices, or so called Oscillating Water Column (OWC) devices located on-shore or off-shore such as described in [14] and [15].

One very promising approach is the Wavedragon, which uses water turbines. Converting the rotational energy of these turbines into the public grid is most efficiently done by using a variable speed generator. This generator is connected to the grid using a PWM inverter, a DC-link and a PWM converter. A submarine cable of about 5 km length links the Wave Dragon to the point of common coupling at the coast. A transformer will be used to step up the voltage from 690 V rms converter output to 11 kV submarine cable voltage.

The long submarine cable and the weak grid impose two difficulties to the converter control. Firstly, harmonics in the grid voltages lead to non-sinusoidal currents. Secondly, the parameters of the cable and the grid should be known for high control performance and unity power factor at the point of common coupling.

The goal of this paper is to analyze the problem and present a solution to achieve sinusoidal currents and unity power factor. A multi-reference frame controller and a parameter estimator is used to achieve this.

It is shown that it is easily possible to compensate the current harmonics by implementing a harmonic compensator. It is further shown that parameter estimation shows limited performance. It is analysed what factors are responsible for this and what can be done to improve parameter estimation.

The paper is organised as follows. An introduction was given in I. The Wave Dragon and its control scheme is described in II. Harmonics compensation is presented in theory in III, while simulation and measurements are shown in IV. Parameter estimation is presented in theory in V, while measurements and Simulation results are shown in VI. The paper is summed up in a conclusion in VII. References are given.

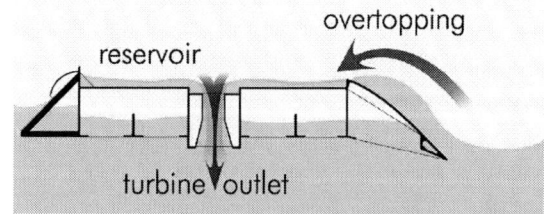

Fig. 1 Wave Dragon Energy Converter [1]

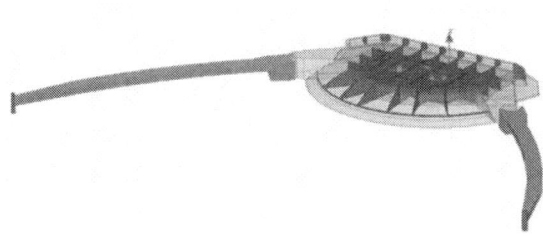

Fig. 2: The WaveDragon with the water reservoir and the wave focussing arms.

## II. WAVE DRAGON AND POWER TAKE OFF SYSTEM

### A. Description of the Wave Dragon

The Wave Dragon is extracting energy from the waves using the overtopping principle to store water in a reservoir above the sea level. The water is released downwards through low head turbines, much like in small hydro power plants, as is sketched in fig. 1.

In total, up to 20 individually controllable turbines will be used, which are switched on or off depending on the amount of incoming water. Two arms extend the structure to both sides, focussing the waves towards the ramp and thereby increase their height. The Wave Dragon, as shown in figure fig. 2, is 300 m wide at the tips of the arms, each arm has a length of 145 m. It is 170 m long and raises up to 6 m over the sea level.

### B. Direct Power Control Space Vector Modulation

Because of the good experience in past work, a Direct Power Control using Space Vector Modulation (DPTC-SVM) is used for the control of the converters, as is presented in fig. 3. The control scheme has been proved reliable and offers unity power factor towards the grid. Since the power generated by the turbines is known, it is fed forward using an Active Power Feedforward (AF) in order to decrease the actual DC-link voltage fluctuation during transients [2].

For the grid side converter control, the grid currents, DC-link voltage and grid voltages are measured. Grid voltage sensorless operation is generally possible with this scheme [3], but for the benefit of grid impedance estimation, which is based on calculating $\Delta U/\Delta I$ in a synchronous reference frame, voltage sensors are used on the capacitor of the LCL-filter. These voltage sensors may also be used to derive the grid angle by means of a Phased Locked Loop, which is needed for coordinate transformation $\alpha$-$\beta$/d-q and d-q/$\alpha$-$\beta$.

The angle derived from the voltage measurement is not only used for power control, but also for transformation into the reference frames synchronous to the 5th and 7th harmonic.

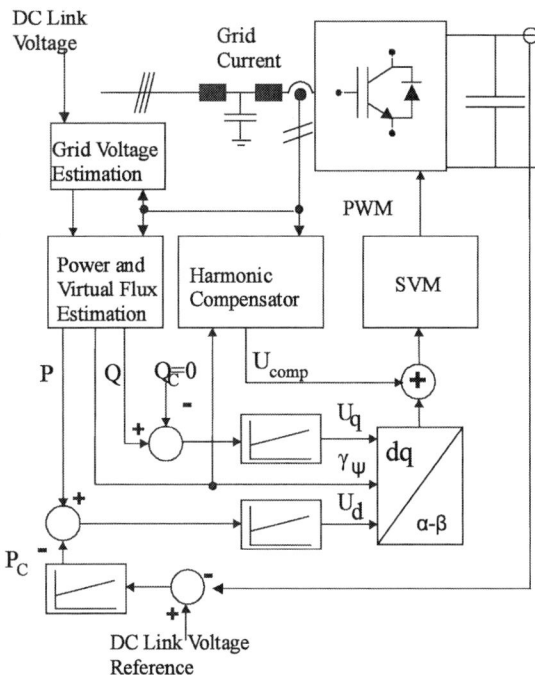

Fig. 3: The grid side control scheme used for Wave Dragon power takeoff system.

## III. PWM RECTIFIER UNDER DISTORTED VOLTAGES

The system is likely to work under distorted grid voltage condition, meaning that 5th, 7th and other harmonics are present in the grid voltages $U_G$ due to nonlinear loads as presented in Eq. (1). These harmonics lead to distortion in the grid currents with the same frequencies, if the control of the active rectifier is not altered to compensate these voltage harmonics. This is explained in fig. 5 and fig. 6.

In order to feed sinusoidal currents into the power grid, a harmonic compensator, also called multi reference frame controller, is used for each of the harmonics that shall be compensated. Note that the commonly present 5th harmonic is always a negative sequence, while the 7th is a positive sequence. The grid rotational frequency is denoted as $\omega_g = 2\pi f_g$. The harmonic compensator consists of a vector rotation into a reference frame that is synchronously rotating with the harmonic to be compensated, as is shown in fig. 4. This way, the harmonic is transformed to a DC component. An integral controller is used to control this component to zero, thus eliminating the respective current harmonic. The controller output generates the necessary compensating voltage $U_{Comp}$ in synchronous reference frame. A second vector rotation transforms it back to the stationary reference frame. It is then added to the converter reference voltage, which is used by the PWM modulator to derive the duty cycles of the IGBT-converter bridges.

Note that the current signal in the rotating reference frame will contain the fundamental now at six times the grid frequency, as is shown in Eq. (2). Therefore, the controller should be sufficiently slow, thus not trying to pick up on these frequencies. The same is true for the reference frame synchronous to the 7th harmonic Eq. (3). A similar scheme may be used for 11th and 13th harmonic. Other authors propose a resonant controller for harmonics compensation, which does not need vector transformation [4], [5].

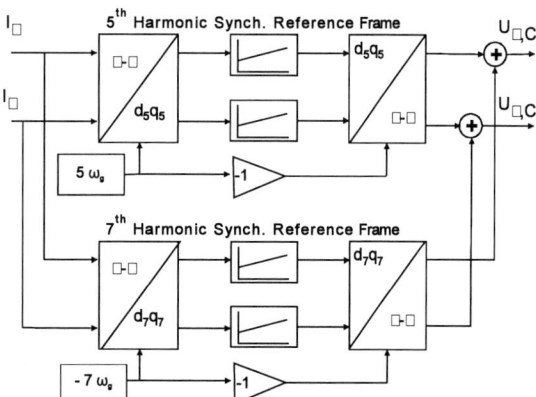

Fig. 4 Harmonic compensator for the 5th and 7th harmonics

$$U_G = U_1 e^{j\omega_g t} + U_5 e^{-j5\omega_g t} + U_7 e^{j7\omega_g t} \quad (1)$$

$$U_G e^{j5\omega_g t} = U_1 e^{j6\omega_g t} + U_5 + U_7 e^{j12\omega_g t} \quad (2)$$

$$U_G e^{-j7\omega_g t} = U_1 e^{-j6\omega_g t} + U_5 e^{-j12\omega_g t} + U_7 \quad (3)$$

TABLE II. HARMONIC CONTENT IN EXPERIMENT AND SIMULATION

| Frequency | Amplitude Simulation | Amplitude Lab |
|---|---|---|
| 1st Fundamental | 400 V rms | 150 V rms |
| 5th Harmonic | 2.9 % | 2.9 % |
| 7th Harmonic | 2.7 % | 2.7 % |
| 11th Harmonic | 0 | 1.1 % |
| 13th Harmonic | 0 | 0.3 % |

## IV. SIMULATION AND MEASUREMENT OF GRID HARMONICS COMPENSATION

Simulations have been carried out to test the harmonic compensator. The results are shown in fig. 7 and fig. 8 for generating mode. The total harmonic distortion is decreased from 11.16% to 6.97%. Good performance is also achieved for motoring mode, as shown in fig. 11 and fig. 12.

The Harmonic Compensator has also been implemented in a laboratory test setup. It is used on the grid side converter of a back to back Squirrel Cage Induction Machine variable speed drive, which is controlled by the DPTC-SVM control scheme [2].

The Harmonic Compensator is implemented as is shown in fig. 4. The transformation angle is derived as a multiple of the fundamental grid angle, which is derived by a Phased Locked Loop. The PLL is implemented on the LCL filter voltage using the dq-PLL-method described in [10].

Measurements are taken via the dSPACE DS1103 Rapid Prototyping System, which is equipped with enough analogue digital converters, encoder inputs and PWM outputs to control a back to back frequency converter, as is needed for a variable speed generation system. A 5 kHz sampling frequency is used.

The drive is a 3 kW squirrel cage induction machine. Since the focus of this work is not on the drive itself but on the grid connection, the drive can be regarded as an active load.

The back to back converter consists of two Danfoss VLT 5005 frequency converters with adapters, which allow external control using the dSPACE system. For the measurements, a California Instruments 5000 iX AC power supply has been used. Using this device, harmonics were implemented into the grid voltage according to table II. The converter DC link voltage was set to 500 V, the load torque in generating mode to –20 Nm at a rotational speed of 1420 rpm. Results are given in fig. 9 and fig. 10. Harmonic distortion is decreased from 10.6 % to 6.0 %, which also fit very well to the simulation.

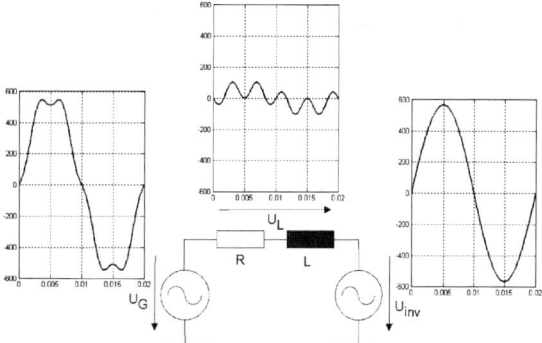

Fig. 5: Distorted Grid Voltage and Sinusoidal Converter Reference lead to Distorted Inductor Voltage, and thereby to distorted current.

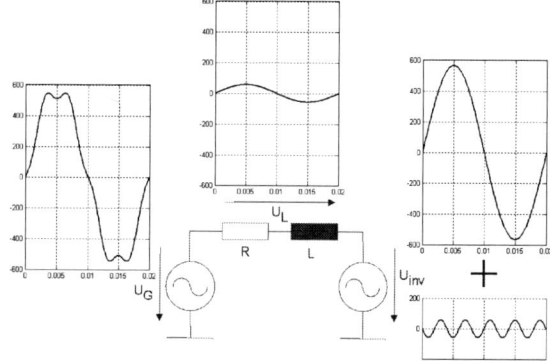

Fig. 6: Compensating the distortion leads to sinusoidal Inductor Voltage

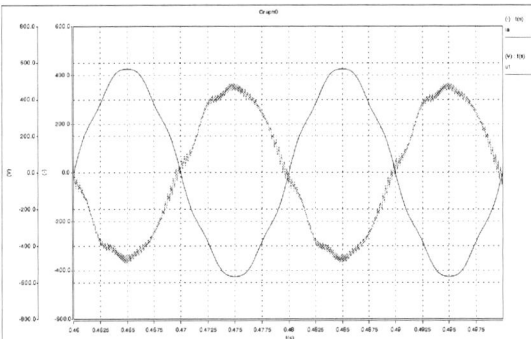

Fig. 7: Disturbed currents due to 2.9% of 5th and 2.7% of 7th harmonic. Simulated THD of line currents is 11.16%.

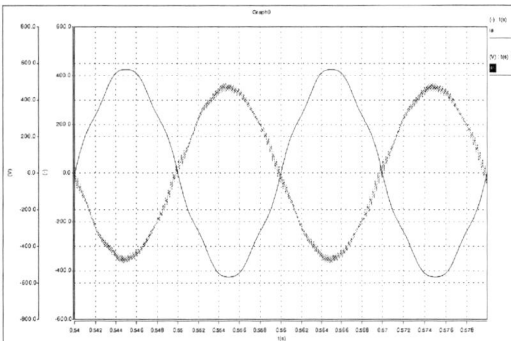

Fig. 8: Sinusoidal Currents with Harmonic Compensator: Simulated THD of line currents is 6.97%.

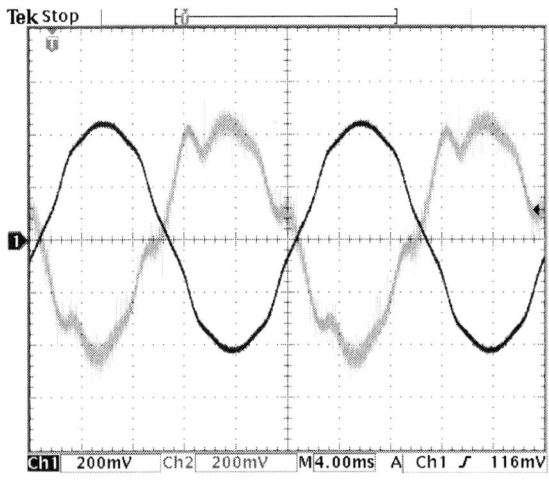

Fig. 9: Disturbed currents due to 2.9% of 5th and 2.7% of 7th harmonics Measured THD of line current is 10.6%.

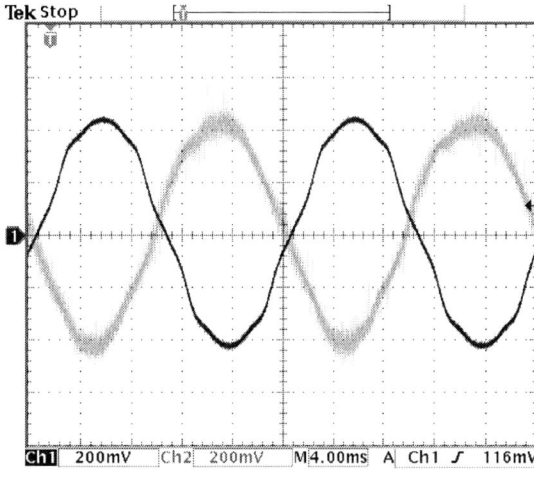

Fig. 10: Currents with Harmonic Compensator. Measured THD of line current is 6.0%.

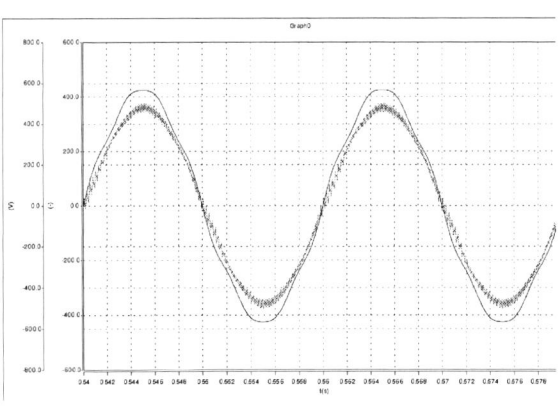

Fig. 11: Motoring mode: Sinusoidal Currents after activating the Harmonic Compensator. THD of simulated current is 7.01%

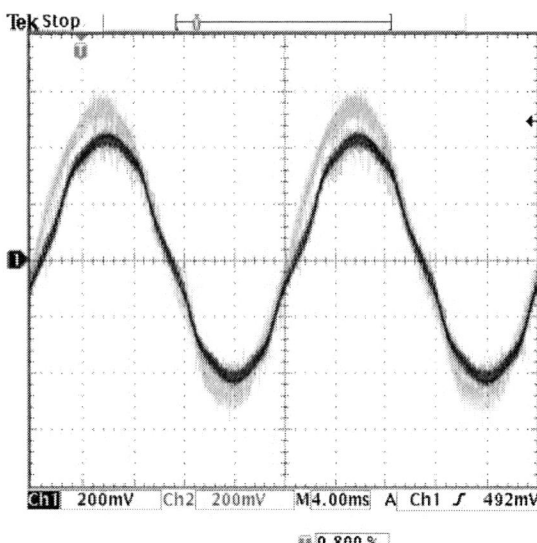

Fig. 12: Motoring mode: Sinusoidal Currents after activating the Harmonic Compensator. THD is 4,6%

## V. GRID PARAMETER ESTIMATION

In order to estimate the grid impedance, the method proposed by [6] is first tested in simulation and later implemented in a laboratory test setup. Unlike as in this method, the LCL Filter voltage is used rather than the grid voltage. Other methods, such as [7] and [8], employ complicated extra devices or extensive algorithms and are therefore not suitable for implementation on a floating device, where ruggedness is required. Most work on Grid Impedance Detection focuses on anti-islanding detection, for example for distributed energy production. One approach [9] excites oscillations in an LCL filter, and estimates the grid impedance using the resonance frequency. This approach is also most likely too complicated for industrial use. Basically, the voltage drop over the grid impedance is used to estimate its value. Therefore, a step in the grid current (power) is required to eliminate the unknown actual grid voltage from the equations. Since the grid side converters real power controls the DC-Link voltage, the reactive power is increased to provide a current step without influencing the active power control. Note that this is not possible whenever the current rating of the converter is already used by active current, e.g. in full power operation.

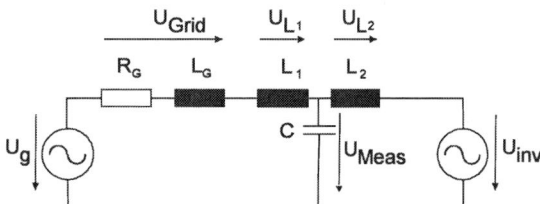

Fig. 13: Voltage drop on grid impedance used for grid impedance estimation.

$$R = \frac{\Delta U_d \Delta I_d + \Delta U_q \Delta I_q}{\Delta I_d^2 + \Delta I_q^2} \quad (4)$$

$$L = \frac{\Delta U_q \Delta I_d - \Delta U_d \Delta I_q}{\left(\Delta I_d^2 + \Delta I_q^2\right)\omega} \quad (5)$$

As it can be seen in fig. 13, there is a voltage drop on the grid impedance denoted $U_{Grid}$. There is also a voltage drop on the known grid side inductor of the used LCL-filter titled $U_{L1}$. The measured Voltage $U_{Meas}$ will therefore show a drop whenever the current is increased.

A three phase PLL is used to transfer the three phase voltages in a two phase synchronous reference frame, where $U_q$ is always zero. The voltage and currents in synchronous reference frame are low pass filtered to remove ripple which will distort and influence measurement accuracy. The values are measured twice at different grid current. The difference that results from this is used to calculate the grid impedance according to equations Eq. (4) and Eq. (5). This scheme is shown in fig. 14.

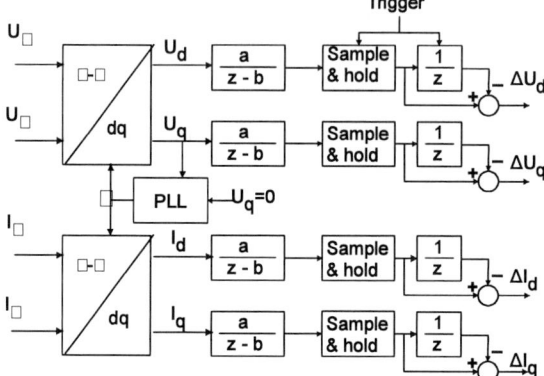

Fig. 14: Calculating the terms needed for equations (4) and (5).

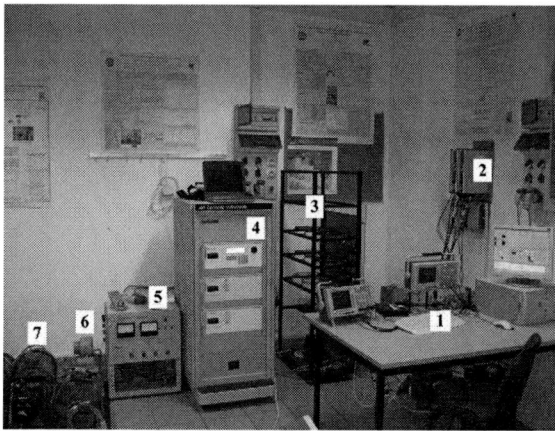

Fig. 15: Pictures of the laboratory setup,
1–isolating interface and control desk,
2– AC/DC/AC converter constructed based on two VSIs (Danfoss VLT 5005 with AAU replaced control boards),
3–resistive load, 4– power supply simulator, 5– reversible rectifier,
6– induction machine set,
7– permanent magnet synchronous machine set

Simulations are carried out with and without anti-aliasing filter. For filtering, a 3$^{rd}$ order chebychev type 1 as shown in fig. 16 filter is chosen, that results in a phase shift of 6.13° for the 50 Hz grid frequency, but suspends the switching frequency by the factor 0.0059. Filter coefficients are given in the appendix.

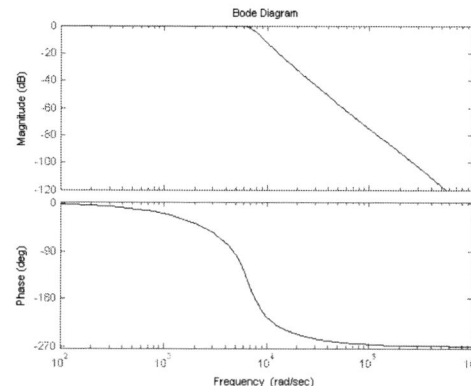

Fig. 16: Chebychev Filter Bode Diagramm

## VI. SIMULATION AND EXPERIMENTAL RESULTS OF GRID PARAMETER ESTIMATION

Simulations were carried out in order to evaluate the performance of the described algorithm. In simulation, the LCL filter parameters were $L_2$=361 µH, $L_1$=288 µH, $R_1$=$R_2$=2 mOhms, and C=84.9 µF. The grid impedance was set to $L_G$=192 µH and $R_G$=2mOhms. The used grid voltage was $U_g$=690 V phase to phase, feeding a current of 100 A to the grid. The reactive power reference was increased to 60 kVAr at time t=0.5 and set back to 0 kVAr at t=0.9. This is shown in fig. 17.

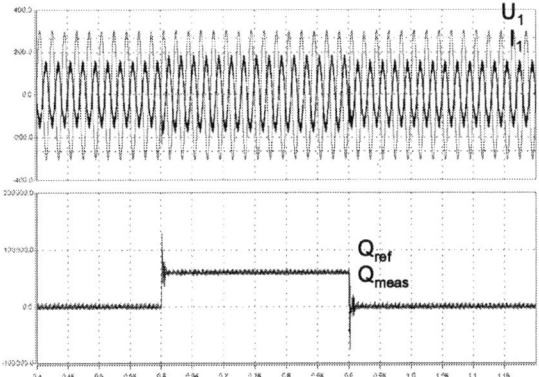

Fig. 17: Step in reactive power used to line parameter estimation

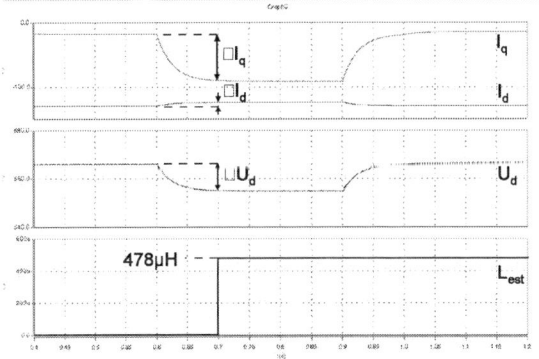

Fig. 18: Step in reactive power used to line parameter estimation

The measured and filtered currents and voltages are shown in fig. 18. During simulation, the quadrature component of the measured voltage stays zero, due to the used PLL. The inductance of the grid is very well estimated to be $L_{est}$=478 µH, compared to $L_G$+$L_1$=480µH. The remaining error results from the ripple found in the measured signals, which could be further reduced using heavier filtering, which in return would also make the system slower. In order to verify that anti-aliasing filtering is necessary for correct estimation, the estimation procedure has also been run without proper filtering, as is shown in table III.

Sampling is synchronized to the PWM of the rectifier.

TABLE III. ESTIMATED INDUCTANCE WITH AND WITHOUT ANTI-ALIASING FILTER

| Implemented Inductance | Calculated Inductance with Anti-Aliasing Filter | Calculated Inductance without Anti-Aliasing Filter |
|---|---|---|
| 480 µH | 478 µH | 771 µH |
| 580 µH | 578.6 µH | 902 µH |
| 680 µH | 679.7 µH | 1020 µH |

Calculation of the ohmic resistance did not deliver usable values on a step of the reactive power.

The same scheme was implemented in a laboratory test setup. The total impedance of grid and grid side inductor is approximately 1.435 mH and 0.745 Ohms. No anti-aliasing filter was used. Using a step in reactive power, an impedance of 1.8mH and 0.022 Ohms was calculated. While the inductance is within a tolerance of 30%, the ohmic resistance is also not well estimated in the laboratory.

## VII. SUMMARY AND CONCLUSION

The Wave Dragon is a 7 MW renewable energy source, which encounters weak grid problems due to the remote nature of an offshore floating platform. Due to the long cable and its inherent phase shift, a parameter estimation of the grid is desirable.

It is shown in this paper that by using a harmonic compensator, it is possible to feed sinusoidal currents into a distorted grid. Estimation of the grid impedance is generally possible using a step response in reactive power, but suffers from inaccuracy in the real world implementation. Further research is necessary to improve this scheme.

### ACKNOWLEDGEMENTS

The authors gratefully acknowledge the financial support of the European Union FP6 (contract no. 019983 Wave Dragon MW).

APPENDIX

TABLE IV. DATA OF EXPERIMENTAL SETUP

| Control | dSPACE DS1103 |
|---|---|
| Sampling Time | 200µs |
| Inverter (Grid) | IGBT 2-level Voltage Source Inverter Danfoss VLT 5005 |
| DC-Link Voltage | 500 V |

TABLE V. DATA OF PROPOSED SUBMARINE CABLE

| Type | 2XS(FL)2YRAA6/10(12) kV |
|---|---|
| Ohmic Resitance | 0.16 Ohm / km |
| Inductance | 0.34 mH / km |
| Rated Current | 363 A |
| Rated Voltage | 11 kV |
| Length | Approx 5 km |

TABLE VI. CHEBYCHEV FILTER COEFFICIENTS

| Numerator | [0 0 0 17755e11] |
|---|---|
| Denominator | [1 7872 60595243 17755e11] |

## REFERENCES

[1] L. Christensen, E. Friis-Madsen, J. Kofoed: *The Wave Energy Challenge The Wave Dragon Case*, PowerGen 2005 Europe conference

[2] M. Jasinski, *Direct Power and Torque Control of AC/DC/AC Converter-Fed Induction Motor Drives*, PhD-thesis Warsaw University of Technology, Warsaw 2005.

[3] M. Malinowski, *Sensorless Control Strategies for Three-Phase PWM Rectifiers*, PhD-thesis Warsaw University of Technology , Warsaw 2001.

[4] H Jae-Wong, M. Winkelnkemper, P. Lehn: Design of an Optimal Stationary Frame Controller for Grid Connected AC-DC Converters, IECON 06, pp 167-172, Paris.

[5] R. Teodorescu, F. Blaabjerg, U. Borup, M. Liserre: *A New Control Structure for Grid-Connected LCL PV Inverters with Zero Steady-State Error and Selective Harmonic Compensation*, 19th annual IEEE Applied Power Electronics Conference and Exposition, pp 580-586, 2004.

[6] M. Ciobotaru, R. Teodorescu, P. Rodriguez, A. Timbus, F. Blaabjerg Online grid impedance estimation for single-phase grid-connected systems using PQ variations, PESC'07, Orlando, Florida.

[7] M. Harris, A. Kelley: Instrumentation for Measurement of Line Impedance, Proc. On APEC'94, Vol. 2, pp 887-893, 1994

[8] B. Palethorpe, M. Sumner, D. Thomas: System Impedance Measurement for use with active filter control, Proc. Of Power Electronics and Variable Speed Drives, pp 24-28, 2000.

[9] M. Liserre, F. Blaabjerg, R. Teodorescu: *Grid Impedance Detection via Excitation of LCL Filter Resonance*, IEEE Transactions on Industry Applications, vol 45, no 5, pp 1401-1407, 2005

[10] A. Timbus, R. Teodorescu and F. Blaabjerg, M. Liserre: *Synchronization Methods for Three Phase Distributed Power Generation Systems An Overview and Evaluation*, 36th PESC 07, pp 2474-2481, 2005.

[11] Cichowlas, PWM Rectifier with Active Filtering, Warsaw University of Technology, Ph.D. Thesis, Warsaw, Poland, 2004.

[12] A. Weinstein, G. Frederikson, M. Parks, K. Nielsen: *AquaBuoy – The Offshore Wave Energy Converter Numerical Modeling and Optimization*, OCEANS'04, MTS/IEEE Techno-Ocean'04, Vol. 4, pp 1854-1859, 2004.

[13] www.oceanpd.com, website of the Pelamis manufacturer.

[14] L. Christensen, E. Friis-Madsen, J. Kofoed: *The Wave Energy Challenge The Wave Dragon Case*, PowerGen 2005 Europe conference

[15] H. Polinder, M. Scuotto: *Wave Energy Converters and Their Impact on Power Systems*, International Conference on Future Power Systems, 2005.

**Improved sizing method of storage units for hybrid wind-diesel powered system**
A.M. Tankari, B. Dakyo, C. Nichita
GREAH University of Le Havre, Le Havre, France, e-mail: *dakyo@univ-lehavre.fr*

*Abstract:* **This paper deals with the segmentation of frequency domain to achieve DC bus stability by efficient design of storage and control principle. The wind energy disturbances measured on DC link are actively filtered by flywheel and battery storages to provide smoothed and secured behaviour of diesel engine. This is obtained by cascaded low pass filters applied to currents control when the dc link voltage is assumed constant. Thus, the templates, the degree of filters selectivity influence the sizing of storage units. This study focuses on the flywheel and battery sizing taking in to account wind energy path dynamics. Simulation results provide first step considerations for energy management tied to energy periodicity and system effectiveness.**

*Keywords:* **Battery management systems (BMS). DC power supply. Flywheel system. Energy system management. Energy storage. Renewable energy systems. Wind generator systems.**

## I. INTRODUCTION

Energy storage device sizing techniques are studied first. These techniques are used to design a hybrid system which includes wind turbine, diesel engine, flywheel, battery, capacitor and emulated load. The maximum power of the wind turbine and the load is respectively 15kW and 76 kW.

According to energy storage issue, the properties of available technologies can be spread on dual energy-power definition related to time horizon of effectiveness.

From 10 to 100kw output power depending on energy amount to be stored, we have proposed the following Storage Units Frequency Allocation Table (SUFAT) [1]

TABLE 1
Storage Units Frequency Allocation Table (SUFAT)

| Frequency range | low | Medium | high |
|---|---|---|---|
| technology | battery | flywheel | capacitor |

The aim of storage elements and diesel engine coupling is based on the frequency that each source can support with long lifetime guarantee. An overview of known models of single path or element of the system is presented. The energy management strategy based on a frequency approach is explained for each section. Filtering principle is presented and Storage elements (battery and flywheel) are then sized according to adopted energy management strategy. Batteries lifetime prediction is included as a sample case.

## II. WIND SPEED MODEL

The wind speed is considered, as usual, as a random process. This process is assumed to have low frequency components and the turbulence components.
The method used for the modelling of the wind speed is the wind spectral characteristic of Van Der Hoven [2]. Wind speed is obtained by means of direct discretization of the power spectral characteristic $S_{vv}$.

The wind speed $v(t)$ is the sum of the harmonics characterised by the magnitudes $A_i$, the pulsation $w_i$ and the phase $\varphi_i$ generated randomly.

$$v(t) = v_l(t) + v_t(t)$$
$$= \frac{2}{\pi} \sum_{i=0}^{N_l} A_i \cos(w_i t + \varphi_i) + \frac{2}{\pi} \sum_{N_t}^{N} A_i \cos(w_i t + \varphi_i) \tag{3}$$

$N_l$ are Samples for the slow component $v_l(t)$ and $N-N_t$ are Samples for the component of turbulence $v_t(t)$.

## III. POWER SOURCES MODELS

### A. Wind turbine model

We have considered that the blades are rigidly attached to the wind turbine; consequently the pitch angle of the blades is constant. The wind turbine characteristics modelling have been made by a six order polynomial regression [2]. The power coefficient characteristic $Cp$ is a function of tip speed ratio $\lambda$ and in this case is given by:

$$C_p(\lambda) = \sum_{i=0}^{n} a_i \lambda^i \tag{4}$$

$$\lambda = R\Omega / v \tag{5}$$

Where $R$ is the radius of the rotor, $\Omega$ is the mechanical angular velocity of the rotor and $v$ is the wind speed.
The $a_i$ parameters *(i=0...6)* are determined by a Matlab computing program [2]. The output power of the wind turbine is calculated from the following equation:

$$P_t = (1/2) \rho A v^3 C_p(\lambda) \tag{6}$$

where $\rho$ is air density in kg/m$^3$ and $A$ is the frontal area of the wind turbine in m$^2$. The torque developed by the wind turbine is expressed by :

$$\Gamma_t = \frac{P_t}{\Omega} = (1/2) \rho A R v^2 C_\Gamma(\lambda) \tag{7}$$

Where $\dfrac{C_p(\lambda)}{\lambda}$ is the torque coefficient.

## B. Wind turbine Control strategy

The generator torque is controlled assuming that the permanent magnet generator torque is proportional to the current (8). Consequently, The current reference frame of the wind generator (9) is calculated. For the steady state points the turbine and the generator torques are equals (8).

$$\Gamma_{t\_opt} = \Gamma_g = \frac{3}{2} p \Phi_{max} I_s \tag{8}$$

$$I_{dc\_ref} = \sqrt{\frac{3}{2}} \left(\frac{1}{3} p \Phi_{max}\right) C_\Gamma \left(\lambda_{opt}\right) \rho \pi R^3 v^2 \tag{9}$$

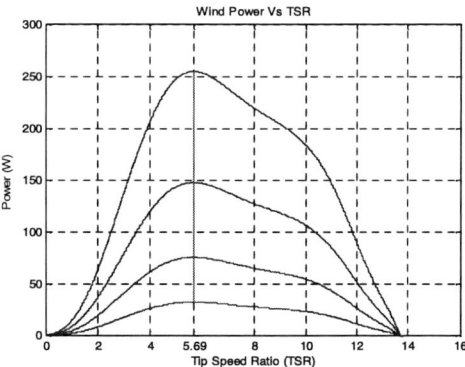

Fig. 1 Power characteristics versus turbine rotational speed

## C. Diesel engine model

The diesel engine includes a gasoline engine and an electric generator. The torque produced by the engine is given by the following equation:

$$T_d = \frac{K}{\tau_1 p + 1} X(p) \tag{10}$$

$\tau_1$ is the time constant of the combustion and K is a gain that adapts the torque and the fuel consumption.
The engine is speed controlled by an outer loop. It drives a synchronous generator (GAP) attached to a rectifier-filter-chopper path.

### IV. ENERGY STORAGE DEVICES

The hybrid system presented in figure 2 is DC bus linked. Flywheel and battery are used as storage devices.
The flywheel energy storage system is based on switched reluctance machine (SRM). In the SRM, phase flux linkage is a function of phase current and rotor position. This is the result of inherent stator and rotor salient poles and iron saturation that impose the flux linkage to be a nonlinear function of both position and current. The phase flux linkage $\psi$ is related to the inductance $L$ and the current $i$ as

$$\psi_k(i_k, \theta) = L(i_k, \theta) i_k \tag{11}$$

Where $\theta$ is the rotor position. The calculation of the inductance is obtained using polynomial interpolation;
The instantaneous voltage across the terminals of each SRM phase (k=1, 2, 3, ...m) is given by [3]:

$$v_k = r_k i_k + \frac{d\psi(i_k, \theta)}{dt} \tag{12}$$

$$\Gamma(i_k, \theta) = \sum_{k=1}^{3} \frac{\partial W_k'(y_k, i_k)}{\partial \theta} \tag{13}$$

The co-energy is defined by

$$W_k'(\psi_k, i_k) = \int_0^{i_k} \psi(i_k, \theta) di_k \tag{14}$$

The converter supplying the SRM is H-hybrid asymmetric bi-directional converter. This converter allows operating the SRM in motor or generator modes.

$$J \frac{d\Omega}{dt} = \Gamma(i_k, \theta) - f \Omega \tag{15}$$

The flywheel associated with static converter is linearized into a first order transfer function $G_{SRM}(p)$ [3] presented to the equation (16).

$$G_{SRM}(p) = \frac{K_{SRM} e^{-\gamma p}}{\tau_{SRM} p + 1} \tag{16}$$

With, $K_{SRM}$ The process gain, $\tau_{SRM}$ the time constant and $\gamma$ the delay time.

The impedance spectroscopy shows that at low frequencies (f <1 hertz), the impedance of the battery can be represented by a resistor and a capacitor in series [4]. Therefore, the battery can be modelled by an internal source $E_{bat}$, in series with a capacity $C_{bat}$ and a resistor $R_{bat}$.

$$G_{bat}(p) = \frac{p C_{bat}}{1 - \tau_{bat} p} \tag{17}$$

With $\varepsilon = Vbat - Ebat$ and the time constant $\tau_{bat} = R_{bat} C_{bat}$.

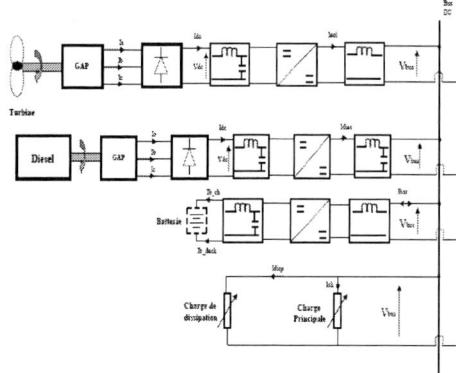

Fig 2. hybrid system WITHOUT FLYWHEEL

### V. HYBRID SYSTEM WITHOUT FLYWHEEL

The hybrid system is shown in Figure 2, the principle of regulating the voltage bus, and energy transfer between the components described in the next paragraph.

Batteries are used to smooth the interactions of frequencies above the dynamics of diesel generator, according to an outline of the figure 2.

Reflected currents exchanged between the turbine, diesel and load are presented in Figure 3.
With this configuration, power demand from the load is satisfied and DC bus voltage is regulated as well (fig.4).

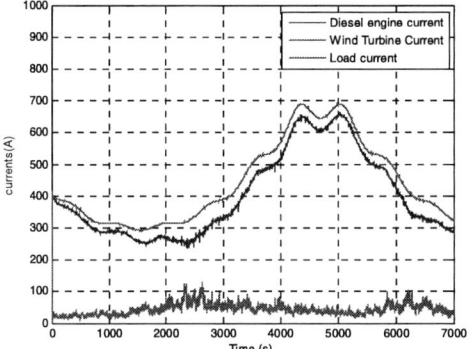

Fig. 3. Currents from the diesel, the wind generators and the load

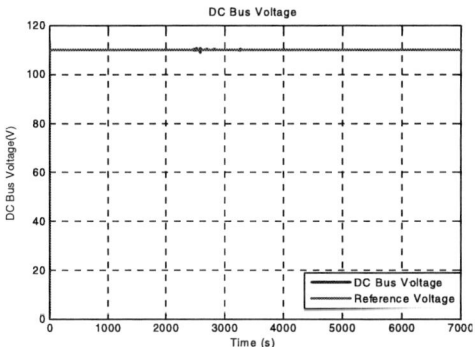

Fig.4 Dc bus voltage

### A. Storage devices sizing

A good design of storage devices requires definition of periodicity of the energy to be stored. To show feasibility, the proposed approach in this paper is to process the signal from storage device (flywheel, battery) and to make the signal periodicity match the window of data collected. The used computer capability must be considered to avoid data overflow.

Pure fluctuating value $\tilde{i}(t)$ (i.e. zero mean value) of instantaneous reference current can be calculated by equation (24).

$$\tilde{i}(t) = i(t) - mean(i(t)) \qquad (18)$$

Equation (24) allows calculating the magnitude of stored energy variations of each storage unit in the assumed period.
Energy exchanged by storage device with DC bus is expressed as:

$$e(t) = \int v(t) * i(t) * dt \qquad (19)$$

With $v(t)$ and $i(t)$ are voltage and current of the storage source.

Equation (26) allows calculating the energy capacity of the storage units.

$$Energy = Max(e(t)) - Min(e(t)) \qquad (20)$$

Equations (19) and (20) are used to calculate battery exchanged energy that is presented in figure 5 for 2h simulation of system operation considered as wind turbine period. Most important consideration is that the energy level at the beginning and at the end of assumed period is the same.
The energy Ah capacity of the battery and capacitor bus continuous tension are reported in Table 2.

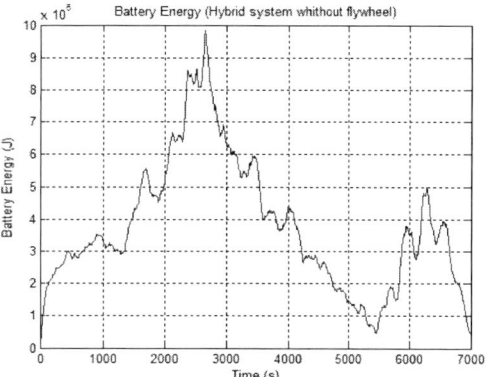

Fig.5. Battery stored energy

TABLE 2.
Storage units capacities.

|  | Battery | Capacitor (35mF) |
|---|---|---|
| Capacity (Ah) | 57 | 0.0283 |

Batteries are one of the key weak links in the long-term operation of fluctuating power systems, such as renewable energy sources or locomotives. They act not only on the performance and operation of the system, but also greatly on the life cycle cost of the system.

Indeed, the performance of the battery is affected by degradation more or less strong as profiles of energy absorbed, resulting in longer or shorter life [4][5][6].

### B. Lifetime estimation

To estimate the battery lifetime the Umass battery lifetime model [4] is used. This model incorporates the main mechanisms of degradation of the battery and uses the method of counting cycles, known as rainflow cycle counting, usually applied in calculating the fatigue of materials. The assumption is that the battery lifetime impact factor is similar to fatigue in material subjected to vibrations.

From battery charges and discharges currents (powers), the lifetime model calculates the time series of state of charge, which will undergo two treatments. A first algorithm is used to identify extreme values (peaks and valleys). This gives a new but short-time series in which a second algorithm is applied to determine the individual cycles. The total discharge range is divided into equal size in 16 to 20 bins, with the final bin corresponding to complete discharge and recharge from a full battery. The cycles determined are classified by bin, the number of cycles per bin calculated and coefficients are used to estimate the lifetime.

So, two independent models are used to estimate the lifetime of the battery: the wind - diesel hybrid system model and the battery lifetime model. The first model (fig.2) allows the calculation of the battery power which will serve as an input to the lifetime model.

Standard hybrid system simulation results show the battery power capacity (fig. 6) that will be used in lifetime model. The battery life is estimated this case to 1.0726 years.

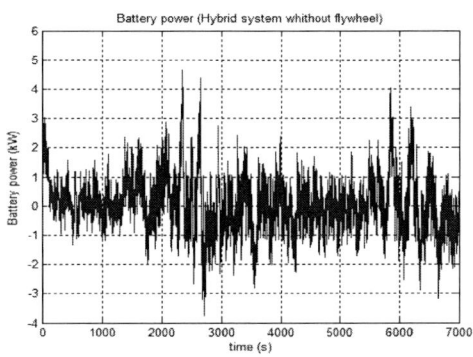

Fig.6. Battery power

Figure 7 shows that the cycles of energy at low amplitudes are the most numerous and that the maximum amplitude is about 10% of the energy capacity of the battery.

The battery is thus subjected to partial charges and discharges cycles that can affect its lifetime.

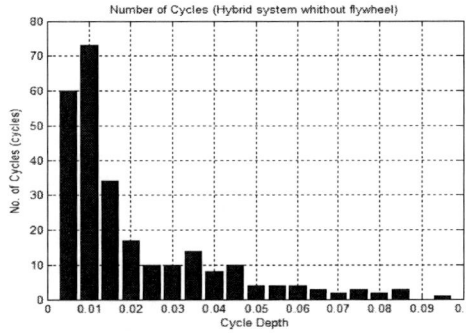

Fig.7 Battery number of cycles

The optimal operation of a wind-diesel hybrid system is characterized by the satisfaction of the load demand, almost constant tension of the DC bus and optimization of lifetime (and therefore cost) of storage devices.

To optimise battery lifetime, a new configuration of the hybrid system is proposed in the following study. It involves inserting a flywheel and makes a frequency distribution.

## V. HYBRID SYSTEM WITH FLYWHEEL

The strategy for managing energy transfer obviously depends on the intrinsic characteristics of each source. As shown in figure 6, energy storage devices are classified according to their energy density, their power density and according to the dynamics they provide [7][8]. Batteries are actual energy sources providing few slow dynamic cycles (1000 cycles). Figure 3 shows the state of the art in the field of relevant storage technologies, gathered in the Ragone plots. This can be use to indicate theoretically and approximately frequency capabilities of the devices.

Figure 7 shows the Nyquist plan of a lead-acid battery impedance and frequency distribution of degradation mechanisms [5]. The main physical effects which affect the lifetime of the battery are the electrical and chemical effects (very fast effects), the effects of operating principles, such as the effects of mass transport and double-layer effects and long - term effects caused by operating systems (ageing, stratification, SOC,...).

The best way for improving their lifetime is to prevent them from fast dynamic currents and high number of cycles [4][5][6]. On the opposite, flywheels and capacitors are capable of absorbing currents with fast dynamics, and to provide a significant number of cycles (50000 cycles) [7][9].

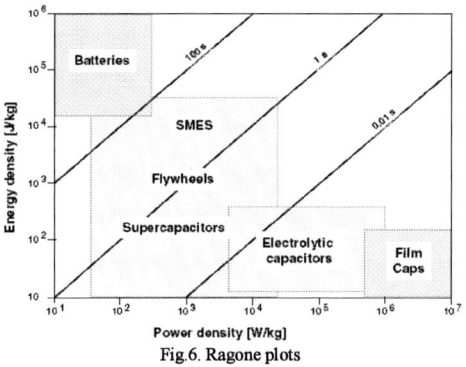

Fig.6. Ragone plots

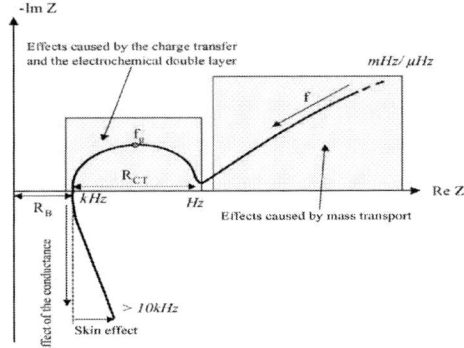

Fig.7 typical Nyquist plot for a lead-acid battery

An optimised configuration of hybrid system including a flywheel is proposed in Figure 8. Development of the strategy of energy transfer presented here follows previous work in [1] [10]. References [8] [11] apply a similar principle in order to make a division between energy sources frequency, storage devices and load. The high frequencies are handled by a SMES in [8] and a super capacitor in [11].

Ref. [11] applies the strategy to a system whose storage capacity is initially defined, which entails the use of fuzzy logic to prevent saturation and depletion of energy. The diesel may be sought by high frequencies Ref [8]. The sizing process of storage devices takes into account batteries and super capacitors efficiency factors.

The design principle presented in this paper is based on the quantification of energy exchanges. It is assumed that the first case is ideal case without any critical damage. A simulation is carried out over a period of wind speed and battery power is used in a second system that integrates the stress factors and calculates the lifetime of the battery.

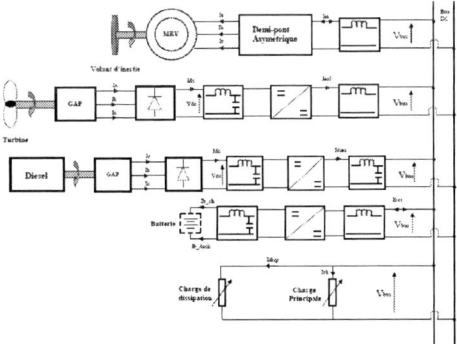

Fig 8. hybrid system with flywheel

The strategy for regulating the DC bus voltage and energy transfer between different components is based on the filtering of the current supplied on DC bus by the wind turbine assuming that it is the main source of disturbances.

## VI. FILTERING PRINCIPLE

Two first order low-pass filters are used to generate the current reference frames of storage devices. The transfer function of a first order low pass filter is expressed by the equation (21).

$$H_{LPF}(p) = \frac{1}{1 + p\tau_c} \tag{21}$$

With, $fc = \frac{1}{2\pi\tau_c}$ the filter cut-off frequency and $\tau_c$ the time constant.

The coupling principle of sources on DC bus is depicted in Figure 9.

The low-pass filters transfer functions $H_{LPF1}(p)$ and $H_{LPF2}(p)$ in the respective cut-off frequencies $f_{c1}$ and $f_{c2}$ can generate the references of flywheel and battery regulator current, with $f_{c1} > f_{c2}$. The cut-off frequency of the filter is equal to the maximum frequency currents that will be covered by the battery and the filter frequency is related to the dynamics of diesel and will correspond to the allocated maximum frequency currents. This principle I based on cascaded filtering of wind turbine measured current $I_{wind\_opt}$.

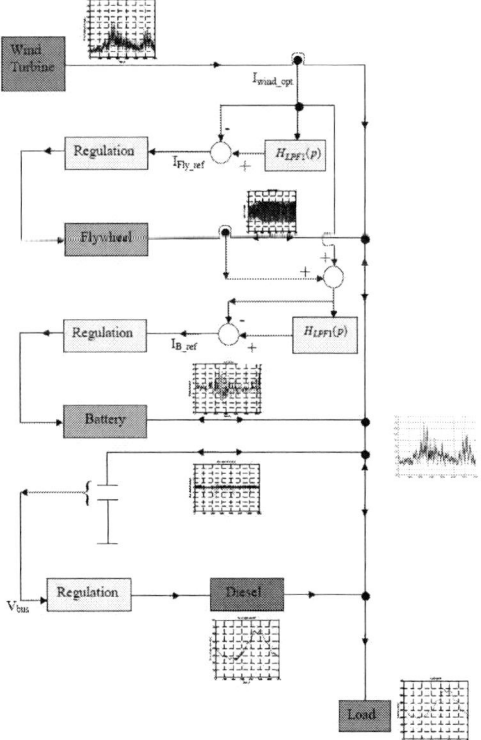

Fig.9. filtering principle

We must keep in mind that very high frequencies are absorbed instantly by the capacitors without a special need of control.

The wind turbine current is forced to maximum $I_{wind\_opt}$ on the DC bus. The current is measured and filtered through the low pass filter $H_{LPF1}(p)$ . This filter output is subtracted from $I_{wind\_opt}$ to get the reference frame current $I_{Fly\_ref}$ of the regulator of flywheel. Thus the current $I_{Fly\_ref}$ contains only high frequency fluctuations devoted to flywheel compensation. Assuming that the final operation is effective, the difference between the actual wind generator current and the $I_{Fly\_ref}$ , provides the reference frame $I_{bat\_ref}$ of the battery regulator.

Currents exchangeable by the flywheel and battery with the DC bus have expressions (22) and (23).

$$I_{Fly}(p) = G_{SRM}(p) * I_{Fly\_ref}(p) \qquad (22)$$

$$I_{bat}(p) = G_{bat}(p) * I_{bat\_ref}(p) \qquad (23)$$

Reference currents are presented by equations (24) and (25).

$$I_{Fly\_ref}(p) = I_{eol\_opt}(p) * \left(1 - H_{LPF1}(p)\right) \qquad (24)$$

$$I_{bat\_ref}(p) = I_{eol\_opt}(p) * \left[1 - G_{SRM}(p) * (1 - H_{LPF1}(p))\right] * (1 - H_{LPF2}(p)) \qquad (25)$$

The resulting current $I_{eol\_BF}$ from the smoothing of wind turbine current fluctuations by the DC bus capacitor and storage elements contains low frequencies components.

$$I_{eol\_BF}(p) = I_{eol\_opt}(p) * \left[1 - G_{bat}(p) * \left[1 - G_{SRM}(p) * (1 - H_{LPF1}(p))\right] * (1 - H_{PLF2}(p))\right] \qquad (26)$$

Diesel compensates the difference between the load demand and the current $I_{eol\_BF}$ supplied by the turbine while maintaining DC bus voltage almost constant (fig. 4). Resulting currents exchanged between the turbine, diesel and load are presented in Figure 3.

After the implementation of an efficient strategy of transfer of energy in a hybrid system across a DC bus, we introduce a method of sizing of the storage units based on simulation results.

*Simulation results*

In our study case, battery allocated band is assumed to be between 0.0013Hz and 0.02 Hz. Lower frequencies are devoted to the diesel engine compensation, high frequencies to the flywheel one, and higher frequencies to the capacitors. Wind turbine is controlled to operate at maximum power point all the time.
Figure 10 shows the spectra of electricity produced by wind and the load divided into four frequency bands supported by diesel, battery, flywheel and capacitors.

There is a wide swath of high-frequency; very high frequencies do not appear because of their very low magnitude. The flywheel sizing requires quantification of energy resulting from such wide fluctuations to ensure stability of the bus voltage and energy availability on load side.

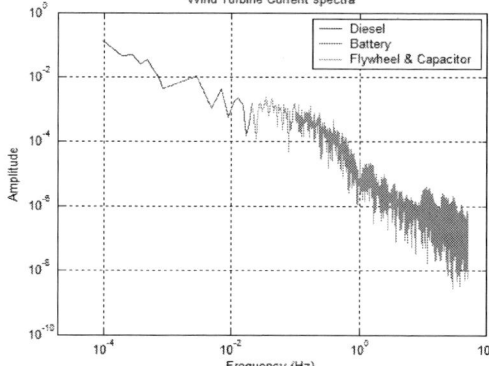

Fig.10. wind turbine current spectrum

Equations (19) and (20) are used to calculate battery and flywheel exchanged energy that are presented respectively in figure 11 and 12 for 2h simulation of system operation considered as wind turbine period. Most important consideration is that the energy level at the beginning and at the end of assumed period is the same.

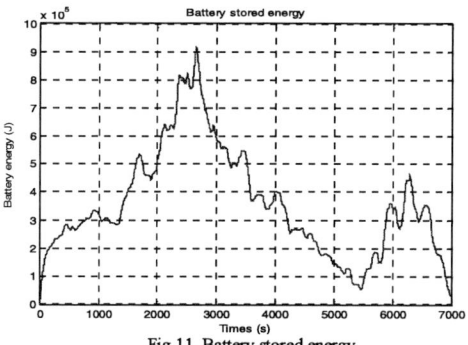

Fig.11. Battery stored energy

Fig.12. Flywheel stored energy

With used wind velocity profile, required storage devices capacities are presented by table 1.

TABLE 3.
Storage units capacities.

|  | Battery | Flywheel (0.5kg m²) | Capacitor (35mF) |
|---|---|---|---|
| Capacity (Ah) | 53 | 10 | 0.0283 |

The allocation of frequency assignments of storage devices aims to optimize their lifetime.
The power resulting from the battery (Fig. 13) is simulated in the Umass model and the lifetime is estimated to 4.3129 years.

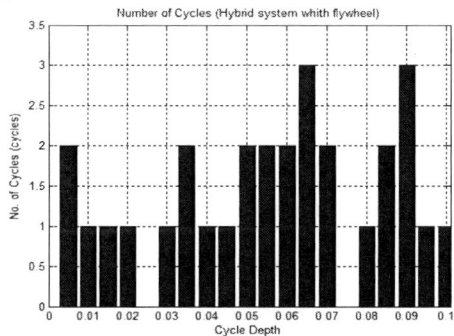

Fig. 14. Battery number of cycles (Hybrid system with flywheel)

Differently from the previous case "without flywheel" the battery is characterized by a smaller number of partial energy cycles with larger amplitudes (Fig. 14), multiplying the battery lifetime by four (4).
Thus, the battery lifetime is affected by the number and magnitude of partial cycles. Large number of cycles and small amplitudes reduce battery lifetime.

## VIII. CONCLUSION

Energy management strategy presented in this study gives the opportunity of discussion on energy storage units sizing procedure. It has been demonstrated the interest of the use of flywheel to increase the lifetime of the battery which is known to be affected by the magnitude and the number of partial cycles of charges and discharges.

We have considered a hybrid system including diesel engine, wind turbine, flywheel, battery, and capacitor with DC point of common coupling. To emphasis the smoothing effect of the wind disturbances, the measurement of the current is a rather original step of management of the transfer of energy and compensation of fluctuations in such a system. Despite its effectiveness, this method includes some limitations which will be analyzed in a next future study. The main subject will be the examination of uncertainties on the evaluation of periodicities and the selectivity of frequency bands of the filter tied to experimental dynamics of devices.

REFERENCES

[1] B. Dakyo, El Mokadem, C. Nichita, W. Koczara "A new method to define power and energy share in a DC link Hybrid wind-diesel powered system by means of storage and dual time-frequency approach" EPE conference Aalborg 2-5 September 2007.

[2] Cristian Nichita, Dragos Luca, Brayima Dakyo and Emil Ceanga, "Large band simulation of the wind speed for real time wind turbine simulators", IEEE Transactions on energy conversion, Vol 17, No. 4, Dec 2002;.

[3] M. El Mokadem, C. Nichita, B. Dakyo, and W. Koczara, "Maximum wind power control using torque characteristic in a wind diesel system with battery storage", SPRINGER MONOGRAPH, Recent Developments of Electrical Drives, August, 2006.

[4] H.Bindner and al., "Lifetime Modelling of Lead Acid Batteries", Risø National Laboratory, Roskilde Denmark, April 2005.

[5] Andreas Jossen, « Fundamentals of battery dynamics», Journal of Power Sources 154 (2006) 530–538.

[6] Heinz Wenzl, and al. "Life prediction of batteries for selecting the technically most suitable and cost effective battery", Journal of Power Sources 144 (2005) 373–384.

[7] T. Christen, Martin W. Carlen, "Theory of Ragone plots", Journal of Power Sources 91, pp210–216, March 2000.

[8] C.R. Akli, X.Roboam, B. Sareni, A. Jeunesse, "Energy management and sizing of hybrid locomotive", EPE conference Aalborg 2-5 September 2007.

[9] Alex Rojas, "Flywheel Energy Matrix Systems – Today's Technology, Tomorrow's Energy Storage Solution", Beacon Power Corp.

[10] A.M. Tankari, B. Dakyo, C. Nichita, "Hybrid wind-diesel powered system DC link stability analysis taking into account load Disturbances", EUROCON2007.

[11] Toshifumi Ise, Masanori Kita, and Akira Taguchi, "A Hybrid Energy Storage With a SMES and Secondary Battery", IEEE TRANSACTIONS ON APPLIED SUPERCONDUCTIVITY, Vol. 15, No. 2, JUNE 2005.

[12] J. F. Manwell et al., "Hybrid2 – A hybrid system simulation model – theory manual", NREL, June 30, 2006.

[13] R. Kaiser et al. , " Development Of Battery Lifetime Models For Energy Storage Systems In Renewable Energy Systems (Res)", Store Conference, Aix en Provence, 20-21 October 2003, France

# A Research Platform for a Smart-Blade Wind Generation System

J. Davey[1], Udaya K. Madawala[1] *(SMIEEE)* and R. Sharma[2]

[1]Department of Electrical & Computer Engineering
[2]Department of Mechanical Engineering
The University of Auckland, Auckland, New Zealand

*Abstract—* **A research platform, which is designed for a novel smart-blade wind turbine system, is presented. The proposed platform, which allows for the integration of different renewable sources, consists of a smart-blade wind turbine, a Permanent Magnet (PM) generator, a rectifier, a hybrid super-capacitor/battery energy buffer, and a boost and bi-directional converter. The PM generator, designed for the power profile of the smart-blade turbine, is driven by an induction motor drive to emulate the wind conditions. The research platform is implemented in a Simulink/MATLAB environment with a dSPACE interface for real time control. It has a controller for peak power point tracking (PPPT) and a user interface for real time metering, and is expected to facilitate rapid prototyping and energy related research.**

*Keywords—* **Wind turbine, bi-directional converter, peak power point tracking, PM generator, research platform.**

## I. Introduction

New global protocols that have been implemented have resulted in the current electricity generation trends. These recent protocols require reduction in green house gas emissions. Electricity generation has a large $CO_2$ footprint with much generation coming from coal and gas fired power stations. In New Zealand a high portion of generation comes from the burning of fossil fuels, which was 32.5% in 2002 [1]. Distributed generation (DG) at residential (micro) level is expected to increase, and would be well suited for uneconomical supply situations that arise due to unique topographical nature of many countries. There has been a dramatic increase in energy related research, both from industry and institutions. This increase is attributed to the awareness of current problems, which arise from traditional methods of electricity generation. The proposed Smart-Blade wind turbine and the research platform are expected to provide a flexible environment for future energy related research amongst various disciplines.

## II. System Overview

The proposed platform is designed, adopting a modular based approach to allow for easy integration of a variety of renewable sources. It is a hybrid energy conversion system, and can be divided into functional blocks, all which play an integral role in the functionality of the overall system. Fig. 1 shows the proposed research platform, consisting of a smart-blade wind turbine, a Permanent Magnet (PM) generator, a rectifier, a hybrid super-capacitor/battery energy buffer, and a boost and bi-directional converter. The smart-blade turbine utilises an extendable blade system, the length of which can be changed according to the wind speed. This functionality produces a linear maximum power trajectory, which allows for simple control strategies to be implemented with peak power extraction with relative ease over a wider wind range. Analysis shows that the annual energy yield of a smart-blade system can be more (as high as 200 %) than that of a similar sized conventional system [2]. The power generated by the PM generator, which is directly coupled to the turbine and designed for the smart-blade power profile, is rectified before being boosted to the DC link voltage. The DC link, shown in Fig. 1, is universal in the sense that any generation source can be connected with a suitable voltage converter such as PV cells and biomass or diesel generators. The DC link will be kept at a voltage sufficient for inversion or alternatively used in a DC load.

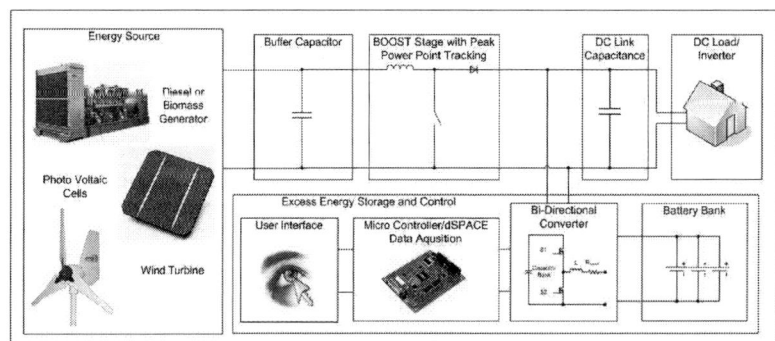

Fig. 1: System Overview

## III. MODULAR DESIGN CONSIDERATIONS

### A. Smartblade System Considerations

The power extracted from the wind is based on the wind velocity v (ms⁻¹), radius of turbine blades r (m), air density ρ (kgm⁻³) and the power coefficient Cp. The power coefficient is based on the blade angle β and the tip speed ratio λ where ω is the rotational speed (rads⁻¹).

$$P = \frac{1}{2}\rho\pi r^2 v^3 Cp(\lambda,\beta) \qquad (1)$$

$$\lambda = \frac{\omega r}{v} \qquad (2)$$

The smart-blade system allows extension and contraction of the turbine blades, enabling the variability of the radius in (1) and (2). Thusly the power produced from the turbines can then be maximized to within the limits of the system. Average wind conditions follow a Weibull distribution, where for Auckland in New Zealand, the majority of the wind speeds fall below 8 ms⁻¹. Fig. 3 shows the power profiles of the smartblade system, the matched generator and a typical fixed blade system correlating to the Weibull distribution. The smart-blade system can thus be optimized to extract as much power as possible at these average wind conditions, resulting in a vast increase in the annual energy output (AEO) shown in Fig. 2.

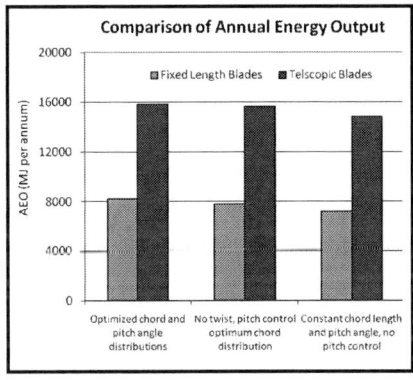

Fig. 2: Smartblade Average Energy Output Comparison

### B. Generator Considerations

For the proposed system to be competitive, the components must be as inexpensive and as reliable as possible. The current trend in wind turbine is to employ PM generators. This is due to their high efficiency, durability, compactness and relatively low cost. The majority of production cost of the PM generator is dictated by the volume and shape of the magnetic material [3]. Thus the generator was designed based on minimising the magnetic material while noting the relationship of the coil emf ε (V) produced with respect to the number of turns N, magnetic flux density B (T), the area of magnetic material A (m²), the rotational frequency ω (rads⁻¹), the number of poles of the machine P and the current in the coils I (A) [4].

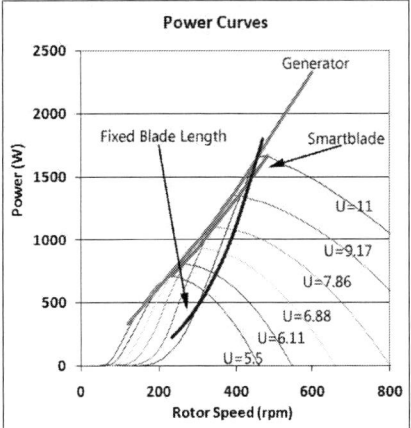

Fig. 3: Power Curve Characteristics

The generator is thus described by,

$$\varepsilon = NBPA\omega \qquad (3)$$

$$\tau = 2NBAIP \qquad (4)$$

JMAG magnetic modelling software was utilised to investigate different designs of the generator. These included inner vs. outer rotor, varying the number of poles, the general thicknesses of the magnets, the number of conductors, thickness of conductors and how the conductors are wound whether this is using a pressed winding for inner rotor designs or bobbins, etc. To obtain a suitable design, all the aspects alluded to must be considered and investigated. The design process was as follows:

1. Investigate the number of poles
2. Investigate inner vs. outer rotor designs
3. Investigate thickness of components
4. Choose winding method

To investigate the number of poles, all other variables were kept constant. These include dimensions, total conductors and rotational frequency. The turbine rotates at a low speed 3-10 Hz, at which the maximum power to be extracted. Thus to have a high electrical frequency either a gear box must be employed or the number of poles of the generator increased. Fig. 4 shows the results of increasing the number of poles with respect to the emf produced for three different magnet thicknesses.

There is a very pronounced peak in Fig. 4 in terms of emf produced at each magnet thickness and number of poles. This peak position also varies depending on the thickness of the magnets. The magnet volume must be kept to a minimum as such consideration of the possible configurations either pressed or bobbin wound resulted in a magnet thickness of 10 mm being chosen. There is also a large voltage increase from 5 mm to 10 mm, but only a slight increase from 10 mm to 15 mm.

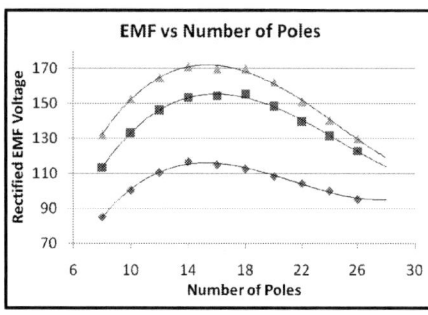

Fig. 4: EMF vs. Number of Poles. Triangle 15mm, square 10mm and diamond 5mm.

From Fig. 4 the optimal position for the 10 mm magnet thickness is approximately 16 poles. This would produce a sufficiently high electrical frequency reducing the ripple after rectification. There is a compromise between the ripple voltage and performance, as such a 10 pole machine was chosen as configuration constraints played a key role. The bobbins and pressed inner winding configurations require plastic spacers, which consume space for windings. This machine produces an electrical frequency of 40 Hz at rated 8 Hz rotational frequency. Any lower number of poles would result in an increased ripple voltage, which would be detrimental to control algorithms and a larger capacitor would be utilised.

The torque produced per unit rotor volume was investigated to reinforce the results from the emf production. Fig. 5 shows the results from the 10 mm thickness magnet section with varying numbers of poles. The rotor volume was fixed at 0.002043 m³. A pronounced peak at 14 poles is seen. This peak is at a lower number of poles than the emf peak of 16 poles. The simulation results were obtained with a load of 8 ohms across a rectifier, while spinning the generator at 8 Hz.

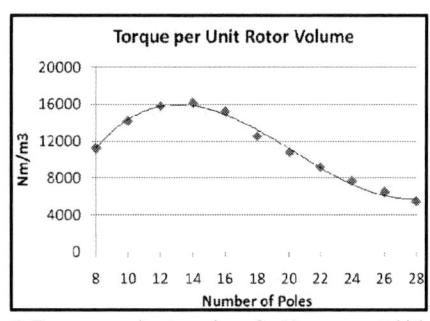

Fig. 5: Torque per unit rotor volume for 10mm magnet thickness

The generator has been designed to extract slightly greater power than that produced by the smart-blade system at rated speed, which is shown in Fig. 3. During the design, consideration has been given to possibility of magnet demagnetization [5], BH curve characteristics, possible saturation of the steel yolks and ease of manufacture. Rare earth Ne-Fe-B magnets are chosen for the generator to minimize the size of the generator [6]. There are two possible configurations for the permanent magnet generator. These are either an inner or outer rotor design. The thicknesses of the stator and rotor back iron are governed by the magnetic flux density in these steel sections. The simulated steel was Nippon 50H600, which has a maximum magnetic flux density of 1.66T [7]. Thus with rare earth magnets, the rotor back iron needed to be sufficiently thick so as not to saturate and introduce added losses. If the total diameter of the generator is kept constant, then the inner rotor design can take full advantage of the plastic section, which can be reduced keeping the relative radial position of the magnets and coil greater than the equivalent outer rotor design.

Fig. 6 shows the typical layout of an inner rotor generator section. The stator steel can encroach upon the plastic section without changing the radial position of the coil and magnets. Thus the emf and torque produced by the generator have a constant radial dimension, which is larger than an outer rotor design with equivalent radius.

Dimensions of two designs, which produce the required 1.7 kW at rated generator current and at maximum wind speed, are shown in Fig. 7 and Table 1.

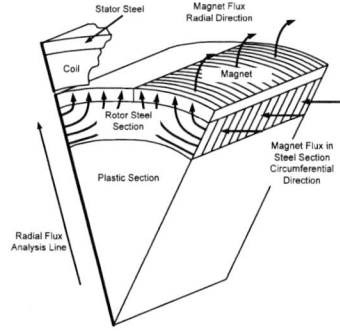

Fig. 6: A rotor sector, containing a magnet and showing the Flux Flow

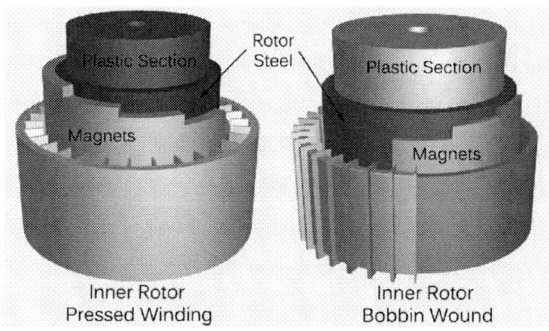

Fig. 7: Inner Rotor Pressed Winding and Inner Rotor Bobbin Wound Permanent Magnet Generator Designs

Table 1: Inner Rotor Flux Flow (dimensions in mm)

| Component | Pressed Winding | Bobbin Wound |
|---|---|---|
| Total Height | 90 | 90 |
| Stator Steel Thickness | 3 | 6 |
| Rotor Steel Thickness | 10 | 10 |
| Air Gap | 2 | 2.5 |
| Outer Radius | 100 | 100 |

These designs have been simulated over the 3-8 Hz range. The output voltage is not critical as this is coupled with the initial boost stage allowing for power to be extracted at any voltage. The generator designs appeared to be efficient and compact.

### C. Matching Smartblades and Generator

For maximum power extraction, it is preferable the generator output power to slightly higher than that produced by the smart-blade turbine. Fig. 3 shows the smart-blade maximum output power, conventional blade output power and the proposed generator (inner rotor pressed winding) output power. It is evident that at certain wind speeds, there is a point of maximum power production from the blades and ultimately the generator. Comparison of the peaks of the smartblade power profile and the generator running at rated output current shows good correlation with the generator producing a greater output at the same rotational speeds.

### D. Converter Electronics

Two converter stages are employed to control the generator and the DC link voltage. These include an initial boost converter, which boosts up the rectified voltage to the desired DC link voltage, and a bi-directional converter which allows for energy supplementation under high loads and storage under light load conditions, while keeping the DC link voltage constant. The prototype power interface containing the two converters, are shown in Fig. 8, which corresponds to the Simulink model in Fig. 9 that was created using the power electronics library. The power electronics library works similar to a SPICE simulation [8]. The Simulink model facilitated the simulation of the control system in conjunction with the bi-directional converter. This was followed by the implementation of the simulated circuit using the dSPACE interface. Simulink provides a dynamic simulation experience compared to that of the static SPICE simulation. The DC Link voltage is monitored to determine the direction of the power flow of the bi-directional converter. When the DL link voltage exceeds a set upper threshold value, the converter is operated in buck mode, storing the excess energy in the battery bank whereas when it falls below a lower threshold voltage then the converter is operated in boost mode, supplementing energy demand to maintain the DC Link voltage. Fig. 10 shows the simulation results of the Simulink model of the converter in Fig. 9. The DC link is restricted to 60 Volts and the inductor current of the bi-directional converter to 30 A. The wind speed is increased at 0.2 seconds to facilitate a change of mode of the bi-directional converter.

The converter in Fig. 8 is operated at a frequency of 100 kHz, and as such two MOSFET drivers are employed to charge the gate capacitance quickly and turn the MOSFET on.

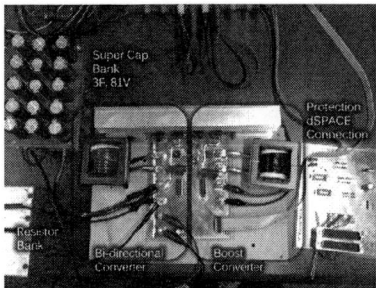

Fig. 8: Power interface

For the boost converter a simple IR4426 dual low side MOSFET driver, which can supply 1.5 A to charge the gate quickly, is used. For the bi-directional converter an IR2183 half bridge MOSFET driver is utilised, and the logic functions are realised through a driver chip. These logic functions allow for a single PWM signal to control both high and low side MOSFETs. This PWM signal is fed into both inputs and operates in the two following configurations:

1. Duty cycle below 50% then the converter is operating in boost mode the low side MOSFET is on for greater than 50% of the time.
2. Duty cycle above 50% operating in buck mode, high side MOSFET is on for more than 50% of the time.

This configuration is turning on the opposite MOSFET while the body diode should be conducting. The chip has an internal dead time compensation of 500 ns so the two MOSFETs are not conducting at the same time. The super capacitor bank amounts to 3F/81V from connecting thirty 90F/2.7V Nesscap capacitors in series. The capacitors each have an internal ESR of 13 mΩ and thus result in an effective ESR of 0.39 Ω, requiring filtering of the resulting ripple voltage. Due to the inherent nature of the current sense resistor of the bi-directional converter an instrument amplifier is utilised to feedback the floating varying voltage. This instrument amplifier is the Burr Brown INA125. The two inductors, used in the initial boost converter and bi-directional converter, are subjected to a high DC bias current, as such the saturation characteristics must be considered so as to decrease the possibility of saturation and reduction of the inductance. The inductors were made using Ferroxcube E65 cores, which have been wound with 150 stranded 0.02 mm diameter litz wire to reduce the skin and proximity effects. The inductors have been calculated based on a 10% current ripple and a maximum duty cycle of 90%, which corresponds to an inductor value of 134 µH at a switching frequency of 100 kHz. Under a DC bias of 20A, the inductance drops in value 98µH, giving a ripple of approximately 14%.

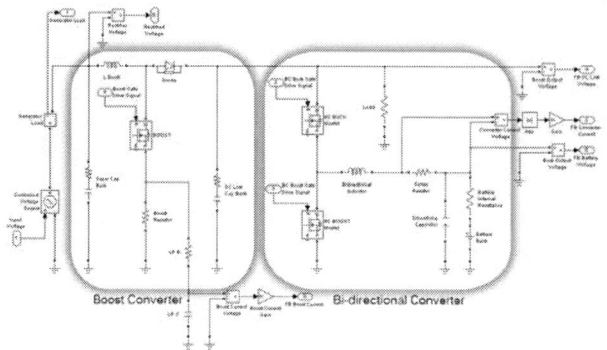

Fig. 9: Simulink Converter Electronics

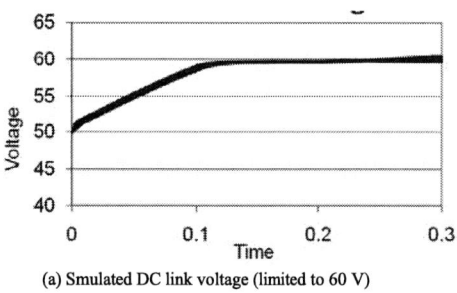

(a) Smulated DC link voltage (limited to 60 V)

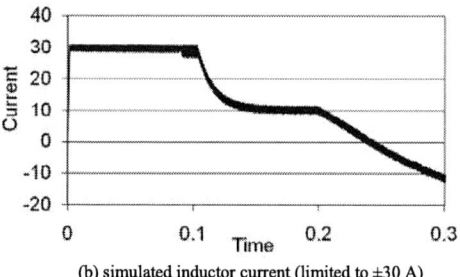

(b) simulated inductor current (limited to ±30 A)

Fig. 10: Simulation Results

### E. Control System

Control system was implemented with the dSPACE DS1104 board to allow for interfacing real time electronics with Simulink. This board in conjunction with Simulink enables rapid prototyping of different control strategies as well as acquisition of performance data for analysis and user interfacing. The DS1104 board is equipped with four 12 bit and 4 16 bit ADC, 8 DAC and 4 single PWM outputs, which is sufficient for feedback of all aspects of the circuit. The control algorithm is based on a current mode controller with voltage feedback of the generator rectified voltage, DC link, and battery bank voltages and current feedback of both the bi-directional converter and initial boost converter. The initial Boost stage is used to monitor and achieve the maximum power extraction from the generator at varying wind speed. The Simulink control block with dSPACE blocks included, is shown in Fig. 12.

This control block uses the perturbation and response method for PPPT, where the rotational speed of the generator is kept constant while the power produced is calculated.

## IV. TEST SYSTEM

A wind emulation system, consisting of a PM generator, which is driven by an induction motor through a variable speed drive (VSD), provides a robust platform for vigorous performance analysis. This emulation system together with the proposed research platform is shown in Fig. 11. The Simulink block diagrams have been set up such that direct wind data can be fed into the wind speed

block and a smartblade torque lookup table will output the required torque at the current rotational speed to the VSD.

Fig. 11: prototype research platform

A 3 kW induction motor is used to emulate the full power range of the PM generator under all conditions and facilitate future upgrades. The fundamental generator equation is described by with $k_{inertia}$ being the inertia of the generator; $T_{Smartblade}$ and $T_{Electrical}$ are the respective torques of the smart-blade system and the electrical load, controlled by the front-end boost converter.

$$\omega_m = \frac{1}{s}\frac{1}{k_{Inertia}}\left(T_{Smartblade} - T_{Electrical}\right) \qquad (5)$$

The Simulink block is shown in Fig. 11 which implements this equation. The motor is controlled via torque methods programmed in the VSD providing the

$T_{Smartblade}$ value. The use of a zero crossing detector on the rectified generator output gives accurate feedback of the rotor speed; this pulsing signal is feed into an interrupt port of the DS1104 board. Thus at a particular wind speed and rotor speed the smart blade output torque can be found from a lookup table reminiscent of Fig. 3 with the power profiles being divided by the rotational speed in rads$^{-1}$ [12].

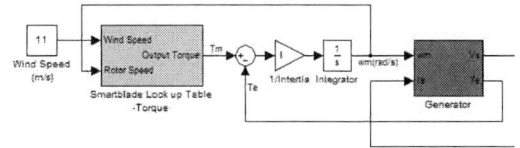

Fig. 12: Simulink generator Block

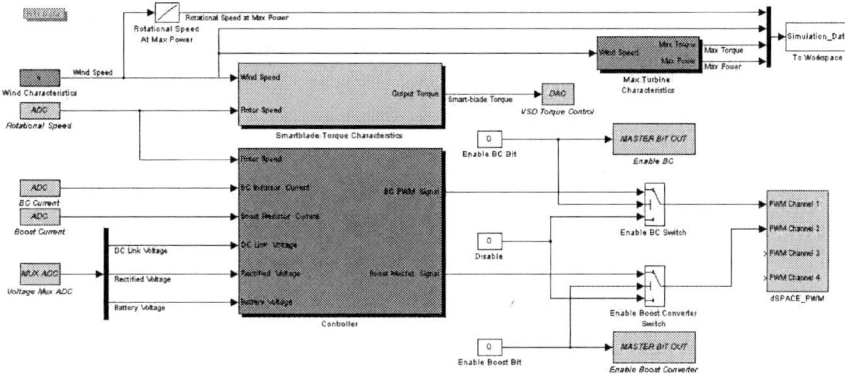

Fig. 13: Simulink Control Block Diagram

## V. CONCLUSIONS

A modular based research platform has been described. The platform has particularly been designed for a novel smart-blade wind turbine system that has extendable blades, but can easily be adopted for other wind turbines. Different PM generator designs that are suitable for the proposed smart-blade wind turbine have also been discussed with detail analysis. A power interface, which has the capability of bi-directional power flow, has been designed, simulated and built to facilitate energy storage or retrieval while maintaining a constant DC link voltage. The front-end boost converter has been designed and constructed to provide PPPT control via manipulation of the duty cycle in a similar manner to the perturbation and response method. The control system has been tested in Simulink, and the user interface has been designed and set up in Controldesk simulation software through a dSPACE interface.

## REFERENCES

[1] "New Zealand Wind Resource." Accessed June 2007 from www.windenergy.org.nz/FAQ/resource.html

[2] Sharma, R. & Madawala, U. K.;'The concept of a Smart Wind Turbine System', *in Proc. of 16th Australasian Fluid Mechanics Conference*, Australia, 2-7 Dec. 2007.

[3] W. Wu, V. S. Ramsden, T. Crawford, and G. Hill, "A low speed, high-torque, direct-drive permanent magnet generator for wind turbines," *Industry Applications Conference*, vol. 1, pp. 147-154, 8-12 October 2000.

[4] T. J. E. Miller, *Brushless permanent-magnet and reluctant motor drives*. Oxford New York: Clarendon Press ; Oxford University Press, 1989.

[5] J. Chen, C. V. Nayar, and L. Xu, "Design and finite-element analysis of an outer-rotor permanent-magnet generator for directly coupled wind turbines," *Magnetics, IEEE Transactions on*, vol. 36, pp. 3802-3809, September 2000.

[6] J. Rizk and M. Nagrial, "Design of permanent-magnet generators for wind turbines," *IPEMC 2000*, vol. 1, pp. 208-212, 15-18 August 2000.

[7] "Grade of Non-Oriented Electrical Systems." vol. 2007: Nippon Steel Corporation, 1999.

[8] J.-H. S. J.-J. C. D.-S. Wu;, "Learning Feedback Controller Design of Switching Converters Via MATLAB/SIMULINK," *Education, IEEE Transactions*, vol. 45, pp. 307-615, 2002.

[9] T. Tafticht, K. Agbossou, and A. Cheriti, "DC Bus Control of Variable Speed Wind Turbine Using a Buck-Boost Converter," *Power Engineering Society General Meeting*, p. 5, 18-22 June 2006.

[10] Neammanee, B. and S. Chatratana, *Maximum Peak Power Tracking Control for the new Small Twisted H-Rotor Wind Turbine*, King Mongkut's Institute of Technology: Thailand.

[11] Esmaili, R., L. Xu, and D.K. Nichols, 'A new control method of permanent magnet generator for maximum power tracking in wind turbine application', *Proc. IEEE Power Engineering Society General Meeting*, 2005, **3**: p. 2090-2095.

[12] M. Chinchilla, S. Arnaltes, and J. L. Rodriguez-Amenedo, "Laboratory set-up for wind turbine emulation," *IEEE ICIT 2004*, vol. 1, pp. 553-557, December 2004.

# Soft Switching Multi-Phase Boost Converter for Photovoltaic System

Joo-Hyuk Lee[*], Jae-Hyung Kim[*], Chung-Yuen Won[*], Su-Jin Jang[**] and Yong-Chae Jung[†]

[*] School of Information and Communication Engineering, Sungkyunkwan University, Suwon, South Korea
e-mail: *leejoohyuk@skku.edu*
[**] Power & Industrial Systems R&D Center, HYOSUNG CORPORATION, South Korea
e-mail: *sjsm@hyosung.com*
[†] Department of Electronic Engineering, Namseoul University, Cheonan, South Korea
e-mail: *ychjung@nsu.ac.kr*

*Abstract*—In this paper, we proposed soft switching multi-phase boost converter. A high efficiency power conversion device is required to obtain constant voltage from solar cells of which output voltage fluctuates very severely by irradiance and temperature. Using the proposed converter, we can control input voltage that fluctuates within operation range to constant output voltage, and it can reduce input current ripple and output voltage ripple. In addition, using ZVS (Zero Voltage Switching) and ZCS (Zero Current Switching), switching losses are reduced. We performed simulation and experiment for the proposed soft switching multi-phase boost converter.

*Keywords*—Interleaved converters, ZCZVS converter, Soft switching, Photovoltaic.

## I. INTRODUCTION

Solar energy comes into the spotlight because of rising oil prices, dwindling energy sources and environmental pollution by increasing consumption of energy. Under active support of Korea Energy Management Corporation, the popularization activity of the photovoltaic system is now progressing. Thus, technical development of the photovoltaic system is urgent.

When we install the photovoltaic system, the expense is very heavy despite of governmental support. Moreover, the photovoltaic system has low efficiency, thus low cost and high efficiency of related technologies is essential to popularization activity.

The technologies related to the photovoltaic system can be divided into two parts, solar cell and power conversion device. With the increase of the power rating, it is often required to associate converters in parallel. In addition, multi-phase boost converter can reduce input current ripple and switching stresses, so the efficiency of the converter is improved [1-4].

In this paper, we propose multi-phase boost converter with soft switching as power conversion device for photovoltaic system. In the photovoltaic system, it requires to boost the output voltage of the solar cells because it is low. Thus, when the multi-phase boost converter applies to the photovoltaic system, low dc voltage from solar cells steps-up. And input current ripples and current stresses of power devices are decreased because the input current is divided into two parallel inductors. Also, the proposed converter can reduce switching losses using ZCS and ZVS.

To verify the operational principle of the proposed converter, we built the prototype circuit with rating power 1.2[kW].

## II. PROPOSED CONVERTER

### A. Configuration

The conventional multi-phase boost converter consists of inductors, switches, diodes and output capacitor. Fig. 1 shows the circuit of the conventional converter.

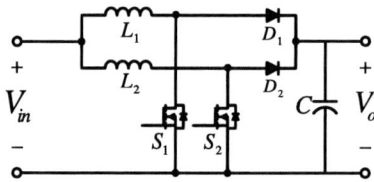

Fig. 1. Conventional multi-phase boost converter.

The multi-phase boost converter is identical with a parallel connection of the single-phase boost converter. The single-phase boost converters are controlled by the interleaved switching signals which have the same switching frequency and the same phase shifting. Each switch turns on with shifting 360°/N (N : the number of phase). As a consequence of the interleaving operation, the multi-phase boost converter exhibits both lower current ripple at the input side and lower voltage ripple at the output side. It has a same output characteristic of the single-phase boost converter, and input current ripple decreases 1/N times of each inductor current ripple because each inductor current is added up.

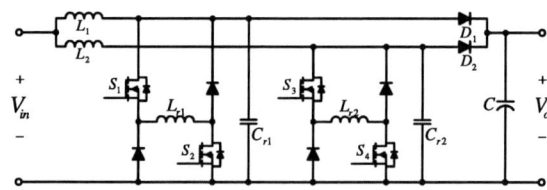

Fig. 2. Proposed soft switching multi-phase boost converter.

Fig. 2 shows the circuit of the proposed soft switching multi-phase boost converter. Switching techniques are identical with conventional multi-phase boost converter. Freewheeling diodes, resonant inductor and resonant capacitor make an addition to the conventional circuit. Resonant inductor and resonant capacitor are used to soft switching (ZVZCS operation) for reduction of switching loss. And two diodes are used to freewheel energy of the resonant inductor, or transmit energy to output side.

### B. Operational Principle

Fig. 3 shows the operational principle of the proposed converter. There are switch states, inductor currents, input current, switch voltage and current.

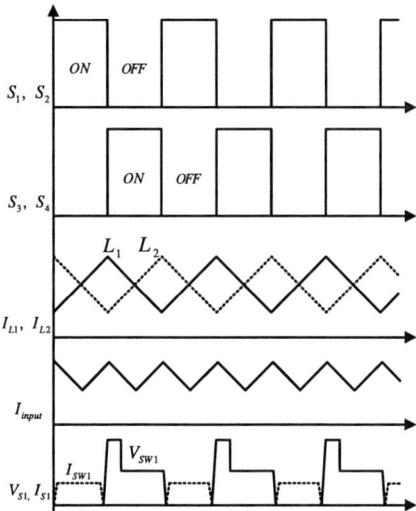

Fig. 3. State, inductor current, input current, switch voltage and current.

Each switch of the proposed converter has switching with a gap of 180° because it consists of 2-phase. Each inductor current repeats linear increment and decrement with a gap of 180° according to switching pattern. Input current ripples are decreased 1/2 times because each inductor current is added up. When switches turn-on, switch current is delayed by resonant inductor $L_r$, thus switches operate with ZCS. Also, when switches turn-off, switch voltage is delayed by resonant capacitor $C_r$, thus switches operate with ZVS.

### C. Operational Modes

The single-phase boost converter which constitutes the proposed converter with parallel is explained for describing soft switching operation using resonant modes. Operation of the single-phase boost converter is divided into 6 modes, as shown in Fig. 4, and their key waveforms are illustrated in Fig. 5.

*1) Mode 1($t_0 \leq t \leq t_1$):* The switches are all turned-off, current doesn't flow through the switches. The main inductor current flows to output side through the output

side diode $D$. And the resonant capacitor $C_r$ is already charged with output voltage $V_o$.

$$v_{C_r}(t) = V_o \qquad (1)$$
$$i_{L_r}(t) = 0 \qquad (2)$$

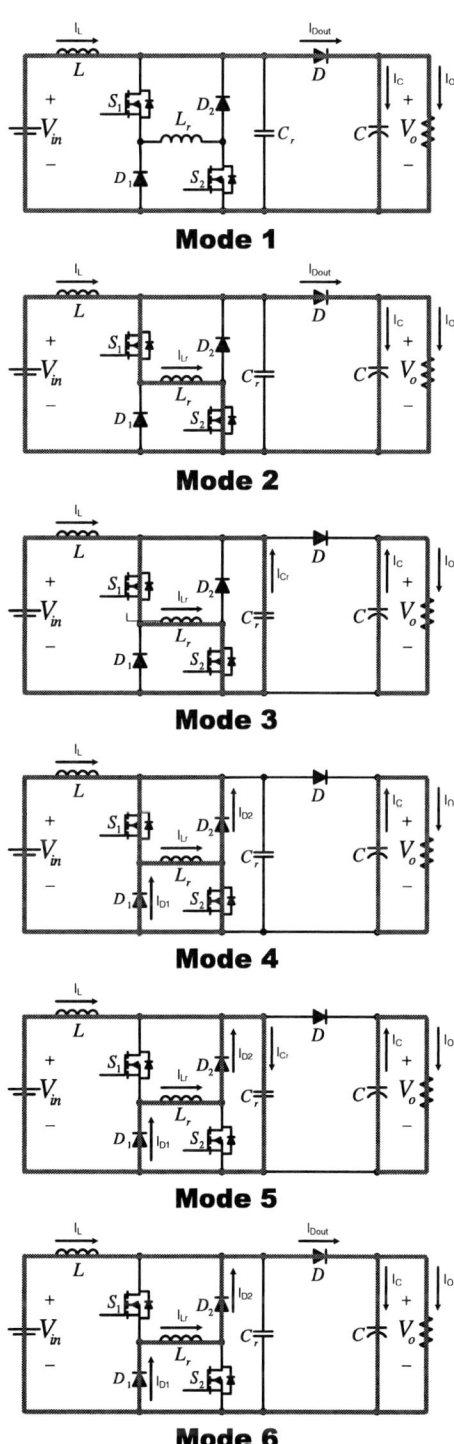

Fig. 4. Operational modes of the single-phase boost converter.

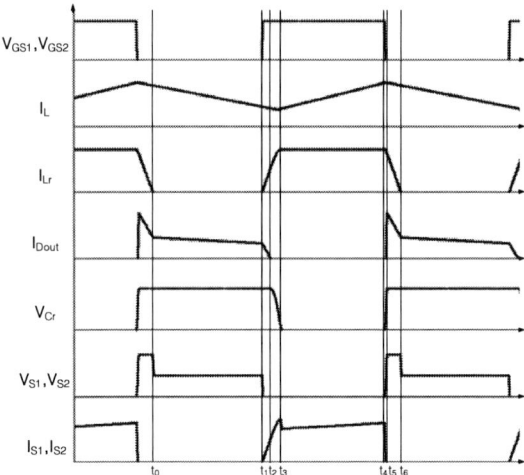

Fig. 5. Key waveforms of the single-phase boost converter.

2) *Mode 2($t_1 \leq t \leq t_2$):* The switches are all turned-on simultaneously, the resonant inductor current $i_{L_r}$ starts flowing. The output diode current decreases linearly, and the resonant inductor current $i_{L_r}$ also increases linearly. The switches are turned-on when the resonant inductor current $i_{L_r}$ is zero, thus the switches operate with ZCS. The resonant inductor current $i_{L_r}$ becomes identical with the main inductor current, and then the output diode current equals zero.

$$v_{C_r}(t) = V_o \tag{3}$$

$$i_{L_r}(t) = \frac{V_o}{L_r}t \tag{4}$$

3) *Mode 3($t_2 \leq t \leq t_3$):* The current which flows to load through the output side diode doesn't flow, and the resonant capacitor $C_r$ starts resonance with resonant inductor $L_r$. The resonant capacitor voltage $v_{C_r}$ falls from $V_o$ to zero, and this mode continues until the resonant capacitor voltage $v_{C_r}$ becomes to zero.

$$v_{C_r}(t) = V_o \cos \omega_r t \tag{5}$$

$$\omega_r = \frac{1}{\sqrt{L_r C_r}} \tag{6}$$

$$Z_r = \sqrt{\frac{L_r}{C_r}} \tag{7}$$

4) *Mode 4($t_3 \leq t \leq t_4$):* Finishing the resonance between $L_r$ and $C_r$, the resonant inductor current $i_{L_r}$ flows through the freewheeling diodes $D_1$ and $D_2$. In this mode, the main inductor current increases linearly.

$$v_{C_r}(t) = 0 \tag{8}$$

5) *Mode 5($t_4 \leq t \leq t_5$):* The switches are all turned-off, the resonant capacitor $C_r$ is charged by two inductor currents. This mode continues until the resonant capacitor voltage $v_{C_r}$ becomes identical with $V_o$. The switches are turned-off with ZVS by the resonant capacitor $C_r$.

$$I_{C_r} = I_L + I_{L_r} \tag{9}$$

6) *Mode 6($t_5 \leq t \leq t_6$):* When the resonant capacitor voltage $v_{C_r}$ identical with $V_o$, this mode is started. The two inductor currents flow to output side through the output side diode $D$. When all of stored energy in resonant inductor $L_r$ transfers to output side, this mode ends.

$$v_{C_r}(t) = V_o \tag{10}$$

$$I_{Dout} = I_L + I_{L_r} \tag{11}$$

## III. SIMULATION RESULTS

The proposed converter has been simulated with PSIM software. Fig. 6 shows simulation schematic, and Table. I shows the design parameters of the soft switching multi-phase boost converter.

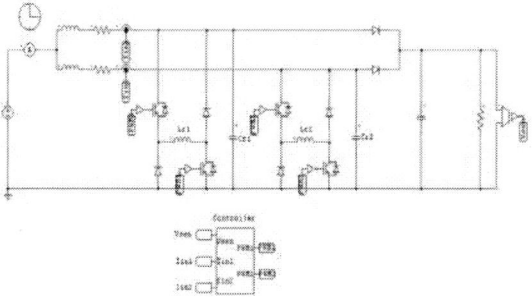

Fig. 6. Simulation schematic of the proposed converter.

TABLE I.
DESIGN PARAMETERS

| Input Voltage($V_i$) | 150~350 [Vdc] |
|---|---|
| Output Voltage($V_o$) | 400 [Vdc] |
| Power($P$) | 1.2 [kW] |
| Inductor($L$) | 500 [μH] |
| Capacitor($C$) | 680 [μF] |
| Resonant Inductor($L_r$) | 40 [μH] |
| Resonant Capacitor($C_r$) | 10 [nF] |
| Switching Frequency($f_s$) | 30 [kHz] |

The PWM signals and current waveforms of the proposed converter are shown in Fig. 7. The input current frequency is two times that of the each inductor current, and input current ripple decreases to half of the each inductor ripple.

Fig. 8 and Fig. 9 show switch voltage and switch current of the proposed converter. When switches turn-on, switch current is delayed by resonant inductor $L_r$, thus switches operate with ZCS, and also when switches turn-off, switch voltage is delayed by resonant capacitor $C_r$, thus switches operate with ZVS.

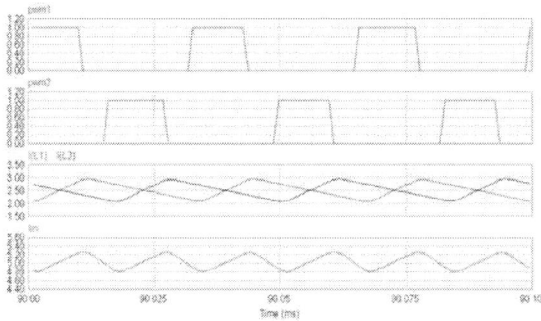

Fig. 7. PWM signals, each inductor current, input current.

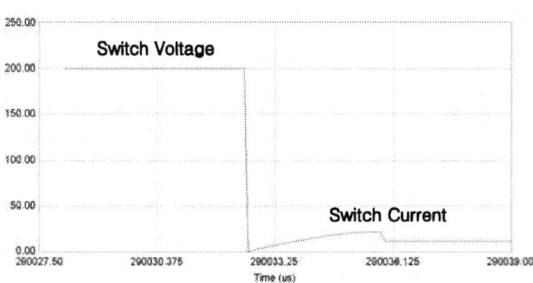

Fig. 8. ZCS turn-on waveform of switch.

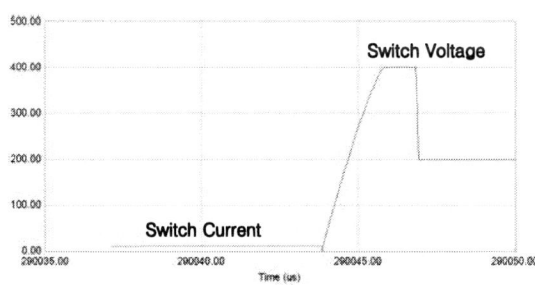

Fig. 9. ZVS turn-off waveform of switch.

## IV. EXPERIMENTAL RESULTS

The experimental prototype circuit, the soft switching multi-phase boost converter, has been built to demonstrate the features and to compare with simulation results. The parameters used are the same as those for the simulation described in Table. I.

The output voltage of solar cell array is used as the input voltage of the proposed converter. Fig. 10 shows solar cell array using the experiment.

Fig. 10. Solar cell array.

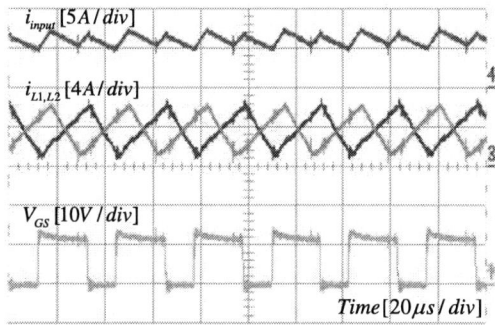

Fig. 11. Input current, inductor current and gate signal.

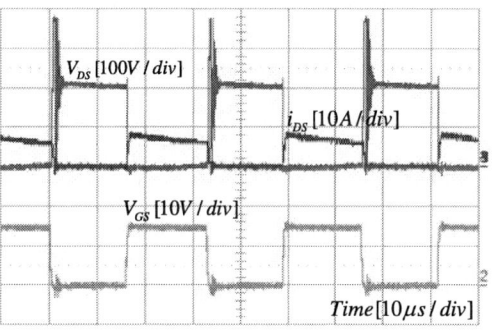

Fig. 12. Switch voltage, switch current and gate signal.

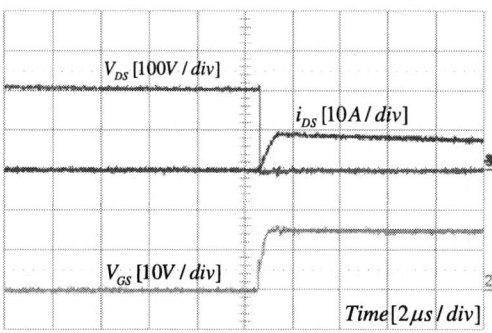

Fig. 13. ZCS operation.

Fig. 11 shows input current and each inductor current, and input current is divided by 2, thus same amount of current flow to each inductor. Fig. 12 shows switch voltage and switch current while switch is turned-on and turned-off. Fig.13 shows the enlarged waveforms of switch voltage and switch current at turn-on of switch. It can be confirmed the switching with ZCS operation.

The efficiency comparison for conventional converter and proposed converter is shown in Fig. 14. It is verified that the proposed converter has higher efficiency more than the conventional converter by decreasing switching loss, and the proposed converter has more than 90% efficiency at above 40% load. Fig. 15 shows the experimental prototype circuit of the soft switching multi-phase boost converter.

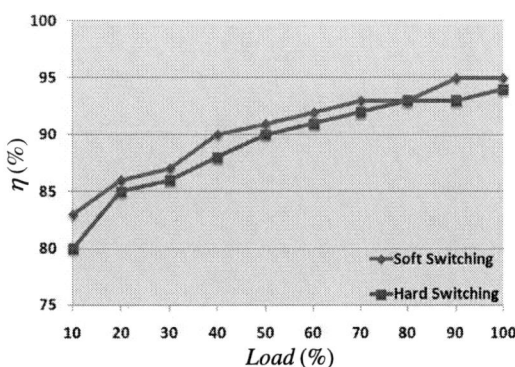

Fig. 14. Efficiency comparison for conventional and proposed converter.

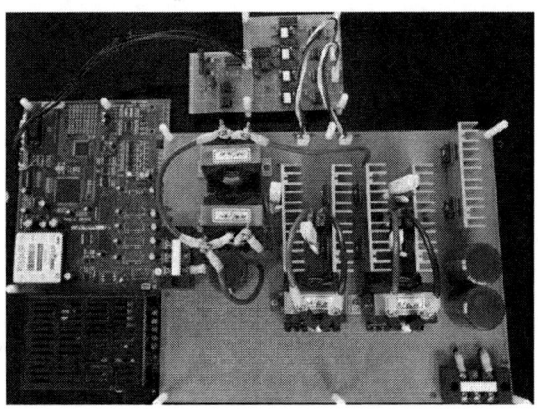

Fig. 15. Experimental prototype circuit.

## V. CONCLUSION

In this paper, the soft switching multi-phase boost converter for photovoltaic system is presented and analyzed. It has smaller ripple of input current and output voltage more than single-phase boost converter. Add to this the proposed converter operates with soft switching using switches, diodes, resonant inductor and resonant capacitor. Thus, the switching loss with hard-switching is decreased. The operational principle is analyzed with operational modes, the simulation and experimental results are given. And the efficiency for the proposed converter compared with the conventional converter.

### ACKNOWLEDGMENT

This work is the outcome of a Manpower Development Program for Energy & Resources supported by the Ministry of Knowledge and Economy (MKE).

### REFERENCES

[1] Gang Yao, Alian Chen and Xianging, "Soft Switching Circuit for Interleaved Boost Converters," IEEE Transactions On Power Electronics, Vol. 22, No. 1, pp.80-86, January 2007.

[2] Po-Wa Lee, Yim-Shu Lee, David K. W. Cheng, Xiu-Cheng Liu, "Steady-State Analysis of an Interleaved Boost Converter with Coupled Inductors," IEEE Transactions On Industrial Electronics, Vol. 47, No. 4, August 2000.

[3] S. Y. Tseng, J. Z. Shiang, H. H. Chang, W. S. Jwo and C. T. Hsieh, "A Novel Turn-On/Off Snubber for Interleaved Boost Converter," IEEE 38th Annual Power Electronics Specialists Conference (PESC '07), pp.2718-2724, 2007.

[4] H. A. C. Braga and I. Barbi, "A 3-kW unity-power-factor rectifier based on a two-cell boost converter using a new parallel-connection technique," IEEE Trans. Power Electron., Vol. 14, No. 1, pp.209-217, January 1999.

# Soft Switching Boost Converter for Photovoltaic Power Generation System

Doo-Yong Jung[*], Young-Hyok Ji[*], Jae-Hyung Kim[*], Chung-Yuen Won[*] and Yong-Chae Jung[†]

[*] School of Information and Communication Engineering, Sungkyunkwan University, Suwon, South Korea
e-mail: *kkc7674@skku.edu*
[†] Department of Electronic Engineering, Namseoul University, Cheonan, South Korea
e-mail: *ychjung@nsu.ac.kr*

*Abstract—* **In this paper, a soft switching boost converter for solar power system is proposed. It is a topology, which raises efficiency for DC-DC converter of photovoltaic PCS (Power Conditioning System). And it minimizes switching losses by adopting soft switching method using resonance. It analyzes the operation characteristic of topology which was suggested through simulation, and proves the validity by analyzing loss and efficiency at real operation with experiment.**

*Keywords—* **Soft switching, Resonant converter, ZCS Converter**

## I. INTRODUCTION

Nowadays, when various uses of photovoltaic system as a distributed generation, in order to provide good quality of electricity to utility system, multifunction, high credibility, high efficiency of Photovoltaic Power Conditioning System, is being needed. Currently, in DC-DC converter of system- applicable type, solar power inverter of PV PCS, IGBT is being used due to the issue of high capacity. There is a limit of switching speed for the traits of IGBT. And it is also hard switching, and could be an obstacle to be high efficiency and miniaturization, along with switching losses [1-2].

This paper proposed soft switching DC-DC boost converter with resonance as an inverter in order to gain stable high-density and high efficiency power.[3] Resonance-type soft switching DC-DC converter can reduce switch loss than hard switching DC-DC converter, and can reduce scale of passive elements (inductor, capacitor) within system by advancing switching frequency, making high efficiency and miniaturization of PV PCS possible.

## II. COMPOSITION OF SYSTEM

### A. Composition of photovoltaic array

In this paper, the module parts are configured on the basis of 900[W] array. The solar cell modules all has serial configurations and are shown in Table 1.

TABLE I

PV module and array parameters

| Module Parameter | | Array Parameter | |
|---|---|---|---|
| Nominal Peck Power(Pm) | 150 [W] | Nominal Peck Power(Pm) | 900 [W] |
| Open-Circuit Voltage (Voc) | 44.2 [V] | Open-Circuit Voltage (Voc) | 265.2 [V] |
| Short-Circuit Current(Isc) | 4.85 [A] | Short-Circuit Current(Isc) | 4.85 [A] |
| Maximum Power Voltage(Vmp) | 34.9 [V] | Maximum Power Voltage(Vmp) | 209.4 [V] |
| Maximum Power Current(Imp) | 4.59 [A] | Maximum Power Current(Imp) | 4.59 [A] |
| Weight | 16kg | Composition | 6series |
| Module eff(%) | 13[%] | Fixing method | Console |

### B. Proposed soft switching boost converter

Fig. 1 shows the proposed soft switching boost converter circuit. One inductor, two capacitors and two diodes are added to the conventional boost converter circuit. On/Off control is done by one switch and the switching loss can be reduced by switching at zero-current and zero-voltage made by resonance of $L_2$ and $C_{r2}$ [4-5].

### C. Mode Analysis of the proposed converter

The soft switching boost converter depicted in Fig. 1 can be analyzed as 7 modes according to the operation conditions.

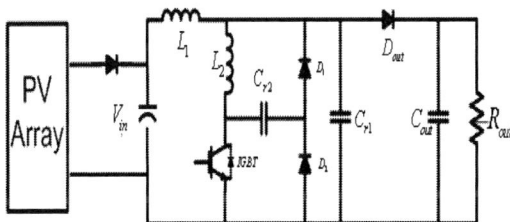

Fig. 1. Proposed soft switching boost converter.

978-1-4244-1741-4/08/$25.00 ©2008 IEEE

## MODE 1 ($T_0 \leq T < T_1$)

The switch is in off state and the DC output of the solar cell array is transmitted directly to the load through $L_1$ and $D_{out}$. During this time, the main inductor voltage becomes $-(V_o - V_{in})$. Thus, the main inductor current decreases linearly.

$$i_{L1}(t) = i_{L1}(t_7) - \frac{V_O - V_{in}}{L} t \qquad (1)$$

$$i_{L2}(t) = 0 \qquad (2)$$

$$v_{Cr1}(t) = V_o \qquad (3)$$

$$v_{Cr2}(t) = 0 \qquad (4)$$

## MODE 2 ($T_1 \leq T < T_2$)

If the switch turns on in condition of ZCS(Zero Current Switching), mode 2 starts. In this case, as the output voltage is supplied to the resonant inductor $L_2$, the current increases linearly. If this current becomes the same as the current of the main inductor $L_1$, the current of the output side diode $D_{out}$ becomes zero.

$$i_{L1}(t) \approx I_{min} \qquad (5)$$

$$i_{L2}(t) = \frac{V_o}{L_r} t \qquad (6)$$

$$v_{Cr1}(t) = V_o \qquad (7)$$

$$v_{Cr2}(t) = 0 \qquad (8)$$

## MODE 3 ($T_2 \leq T < T_3$)

If the current of the output side diode $D_{out}$ becomes zero and turns off, the section of resonant mode starts. While in this section, the auxiliary resonant inductor $L_2$ and the auxiliary capacitor $C_{r1}$ resonate and the voltage of $C_{r1}$ falls to zero at the output voltage. In this case, the current of the main inductor $L_1$ flows through $L_2$ and the switch.

At this time, the load is supplied with power continuously as the voltage charged at $C_{out}$ discharges.

$$i_{L1}(t) \approx I_{min} \qquad (9)$$

$$i_{L2}(t) = I_{min} + \frac{V_o}{Z_{r1}} \sin \omega_{r1} t \;\; , \;\; i_{L2}(t) = i_{L2}(t_3) \quad (10)$$

$$v_{Cr1}(t) = V_o \cos \omega_{r1} t \qquad (11)$$

$$v_{cr2}(t) = 0 \qquad (12)$$

$$\omega_{r1} = \frac{1}{\sqrt{L_2 C_{r1}}} \qquad (13)$$

$$Z_{r1} = \sqrt{\frac{L_2}{C_{r1}}} \qquad (14)$$

## MODE 4 ($T_3 \leq T < T_4$)

If the auxiliary resonant capacitor voltage becomes zero, two auxiliary diodes $D_1$ and $D_2$ turn on, and the mode starts. While in this section, the auxiliary resonant inductor current is divided into two, one is the current of the main inductor $L_1$ and the other is the current turning through two auxiliary diodes. This section is the whole section of the boost converter and the main inductor current increases linearly.

$$i_{L1}(t) = I_{min} + \frac{V_{in}}{L_1} t \;\; , \;\; i_{L1}(t) = i_{L1}(t_4) \qquad (15)$$

$$i_{L2}(t) = i_{L2}(t_3) \qquad (16)$$

$$v_{Cr1}(t) = 0 \;\; , \;\; v_{cr2}(t) = 0 \qquad (17)$$

## MODE 5 ($T_4 \leq T < T_5$)

The switch turns off with zero voltage condition. At this time, there are two current loops. One is the loop of L1-Cr1-Vin and the voltage of the auxiliary resonant capacitor $C_{r1}$ increases linearly from zero to the output voltage Vout. The other is the loop of $L_2$-$C_{r2}$-$D_1$ and the second resonance takes place. The energy stored at $L_2$ moves to $C_{r2}$. If this energy move is finished, the current of $L_2$ becomes zero and the voltage of $C_{r2}$ becomes the maximum value.

$$i_{L1}(t) \approx i_{L1}(t_4) \qquad (18)$$

$$i_{L2}(t) = i_{L2}(t_3) \cos \omega_{r2} t \qquad (19)$$

$$v_{Cr2}(t) = Z_{r2} i_{L2}(t_3) \sin \omega_{r2} t \qquad (20)$$

$$\omega_{r2} = \frac{1}{\sqrt{L_2 C_{r2}}} \qquad (21)$$

$$Z_{r2} = \sqrt{\frac{L_2}{C_{r2}}} \qquad (22)$$

## MODE 6 ($T_5 \leq T < T_6$)

In mode 6, the voltage of $C_{r2}$ decreases, continuously resonates on the loop of $D_2$-$C_{r2}$-$L_2$-$D_{out}$-$C_{out}$ and moves the energy of $C_{r2}$ to $L_2$. If the voltage of $C_{r2}$ becomes zero, the current of $L_2$ flows reversely from the current direction of mode 5. If the voltage of $C_{r2}$ becomes zero, the anti-parallel diode of the switch turns on and it turns over to the next mode.

$$i_{L1}(t) = i_{L1}(t_5) - \frac{V_o - V_{in}}{L}t, \ i_{L1}(t) = i_{L1}(t_6) \quad (23)$$

$$i_{L2}(t) = (\frac{V_o}{Z_{r2}}i_{L2}(t_3))\sin\omega_{r2}t, \ i_{L2}(t) = i_{L2}(t_6) \quad (24)$$

$$v_{Cr1}(t) = V_o \quad (25)$$

$$v_{Cr2}(t) = V_O - (V_O - Z_{r2}I_3)\cos\omega_{r2}t \quad (26)$$

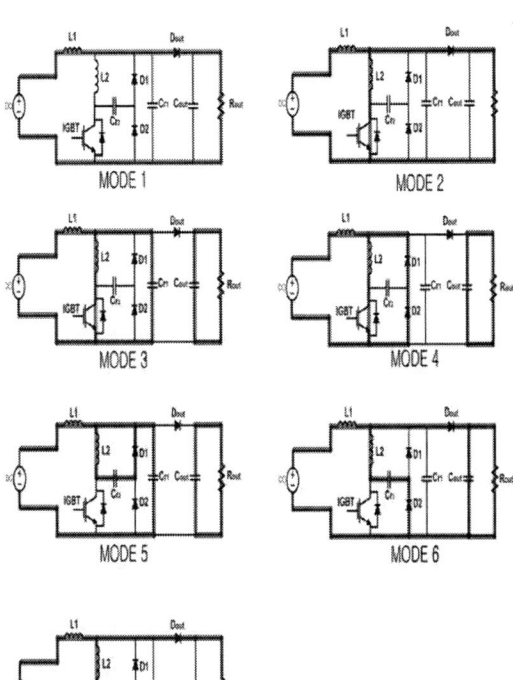

Fig. 2. Operation mode of proposed converter.

## MODE 7 ($T_7 \leq T < T_8$)

In mode 7, there are also two current loops. The current of the main inductor $L_1$ transmits the energy to the output through $D_{out}$ and decreases linearly. The current of the auxiliary resonant inductor $L_2$ transmits the energy to the load through $D_{out}$ and flows through the anti-parallel diode of the switch. If the current of the auxiliary resonant inductor $L_2$ becomes zero, mode 7 ends.

$$i_{L1}(t) = i_{L1}(t_6) - \frac{V_o - V_{in}}{L}t \ , \ i_{L1}(t) = i_{L1}(t_7) \quad (27)$$

$$i_{L2}(t) = i_{L2}(t_6)\frac{V_o}{L_2}t \ , \ i_{L2}(t) = i_{L2}(t_7) = 0 \quad (28)$$

$$v_{Cr1}(t) = V \quad (29)$$

$$v_{Cr2}(t) = 0 \quad (30)$$

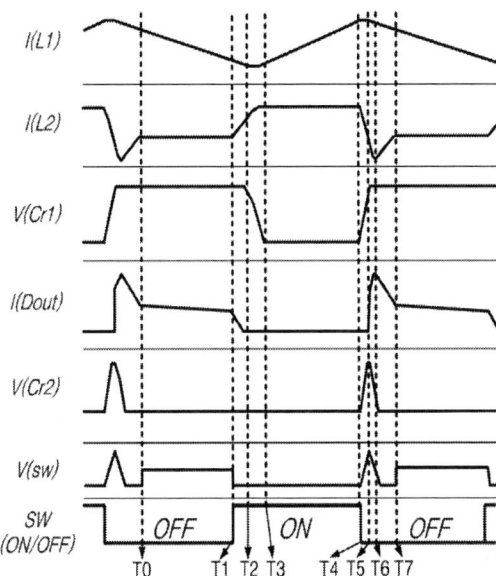

Fig. 3. An each waveforms of proposed circuit.

Fig. 3 shows the waveform of each portion in normal condition of the soft switching boost converter presented in this paper and all the components are supposed to be ideal.

### III. SIMULATION

PSIM is used for the simulation, and the photovoltaic array and MPPT algorithm are composed by DLL in Fig. 4. MPPT of the photovoltaic array output chases the voltage of the maximum power point by detecting the voltage and current of the photovoltaic cells, and the output is compared to the reference values of MPPT, and again compared to the detected voltage of the load passing through PI controller, and then after the output is compared to saw waveform through the second PI controller, the duty ratio of the gate signal of the main switch is decided. The simulation parameter is shown in Table II [6].

In Fig. 5, the inductor current increases and decreases linearly according to On-Off state of the switch. And the auxiliary inductor also accumulates and releases the energy according to the state of the switch.

In Fig. 6, the $V_{sw}$ is referred to voltage of switch, I(IGBT1) for switch current, $V_{gate}$ for PWM signal, and $V_{sw}$ I(IGBT1)*33 is compared with 33 times switch current in order to compare switch voltage and current.

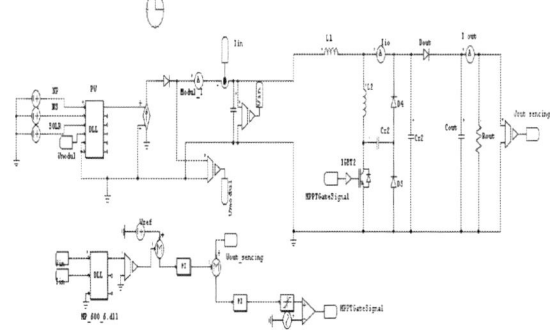

Fig. 4. Simulation circuit.

TABLE II

Simulation parameter

| Photovoltaic array stage | |
|---|---|
| Open voltage ($V_{oc}$) | 265.2 [Vdc] |
| Short current ($I_{sc}$) | 4.85 [A] |
| Maximum power voltage ($V_{mp}$) | 209.4 [Vdc] |
| Maximum power current($I_{mp}$) | 4.59 [A] |
| Soft switching boost converter stage | |
| Input voltage ($V_i$) | 132 - 260 [Vdc] |
| Output voltage ($V_{mp}$) | 380 [Vdc] |
| Capacity (P) | 900 [W] |
| Switching frequency ($f_s$) | 30 [kHz] |

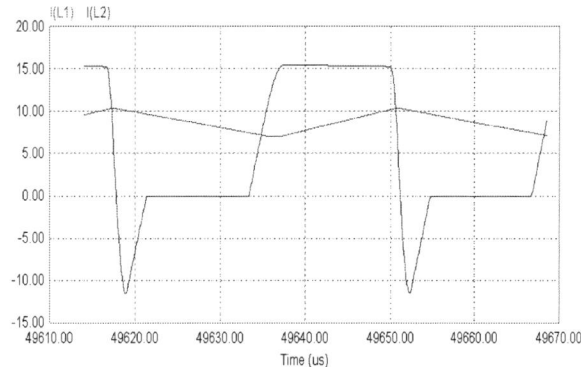

Fig. 5. Current waveforms of L1 and L2.

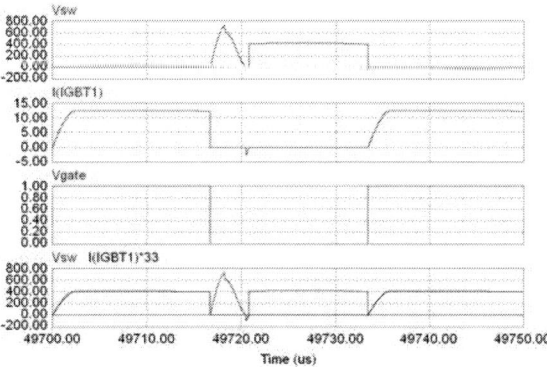

Fig. 6. Each waveform of proposed topology.

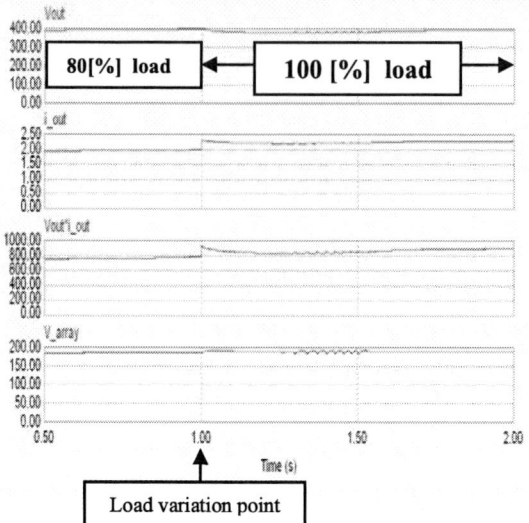

Fig. 7. Output waveforms by load variation.

Fig. 7 shows the output voltage of the presented converter, the output voltage of the modules and the output of MPPT algorithm according to load variation.

If the load increases, the gets larger proportionally according to the size of the load variation

## IV. EXPERIMENTAL RESULTS

To check the utility of suggested topology, a 900W-class trial product was made and tested. IGBT was used as a switching device, and 150W-class KPEM-S150A72 of KPE corp. was formed in the form of 6 series as a solar module.

Photovoltaic array is installed on the roof of present laboratory as Fig. 8 shows, and is connected to the trial product, using 20m cable.

Fig. 8. Photovoltaic array.

Fig.9 shows gate signal(1), switching voltage(2) and main inductor current(3) at full load.

Fig.10 shows enlarged waveform of main switch's current and voltage. It can be seen that ZCS(zero current switching) operation stably through experiment wave form.

Fig. 10 shows current of main inductor L1 and resonance inductor L2. Fig. 12 shows the results of efficiency estimation for the proposed converter according to load variation. It shows over 90% efficiency under the condition of over 40% workload, and sees the 96.85% efficiency at full load. Fig. 13 shows the photograph of a experimental set of soft switching boost converter .

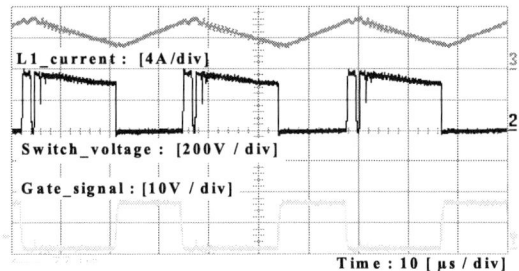

Fig. 9. Experimental waveforms of the proposed topology at full load.

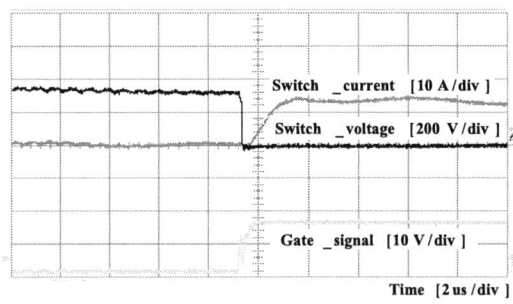

Fig. 10. Zero current switching waveforms.

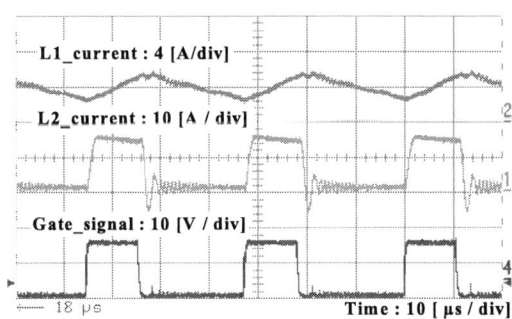

Fig. 11. Experimental result of current waveforms of L1, L2 and gate signal.

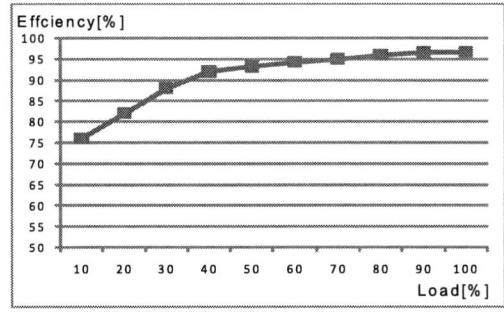

Fig. 12. Measured efficiency.

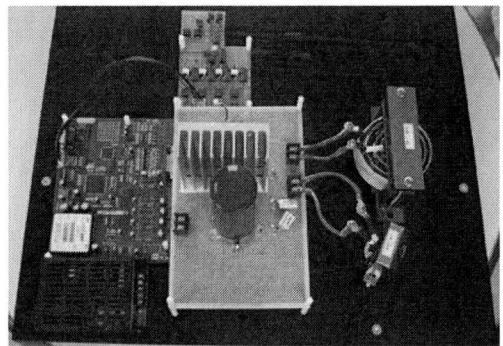

Fig. 13. Experimental set of soft switching boost converter for photovoltaic power generation system.

## V. CONCLUSION

This paper suggests soft switching boost converter using resonance. Circuit composition and operational principle are explained by each mode through mathematical analysis. It verifies the validity of topology suggested through simulation and experiments, and accomplishes the efficiency of over 96% at most.

## VI. ACKNOWLEDGMENT

This work is outcome of a Manpower Development Program for Energy & Resources supported by the Ministry of Knowledge and Economy(MKE)

### REFERENCES

[1] Wakabayashi.F.T, Canesin.C.A "A new family of zero-current-switching PWM converters and a novel HPF-ZCS-PWM Boost converters", IEEJ International Power Electronics conference, pp. 1202~1207, 1995

[2] G. Hua, X. Yang, Y. Jiang, and F.C. Lee, "Novel zero-current-transition PWM converter", IEEE Power Electronics Specialist Conf. Rec., pp. 538~544, 1993.

[3] Doo-yong Jung, Chung-yuen Won, "Soft Switching Boost Converter for Photovoltaic Power Generation System", KIPE Power Electronocs Autumn Conference. Rec., pp.100~102, 2007

[4] David W. Berning and allen R. Hefner, "IGBT model validation for soft-switching applications", IEEE Trans. Industry Applications, Vol. 1, NO. 3-7, 1999.

[5] Yu-Ming Chang, Jia-you Lee "Design and analysis of H-soft-switched converters" IEE Proceeding Electric- .Power Applications, Vol. 142, No. 4, 1995.

[6] Nicola Femia, "Optimization of Perturb and Observe Maximum Power Point Tracking Method", IEEE Trans. on Power Electronics, Vol. 20, NO. 4, 2005.

# OPTIMISATION OF WIND POWER PMSM TO GRID CONVERSION SYSTEM

Ince Kayhan[1], Weiss Helmut[2]

[1] University of Leoben/Institute for Electrical Engineering, Leoben, Austria; e-mail: kayhan.ince@unileoben.ac.at

[2] University of Leoben/Institute for Electrical Engineering, Leoben, Austria; e-mail: helmut.weiss@unileoben.ac.at

*Abstract* — Wind energy conversion systems have become a focal point in the research of renewable energy sources. Due to rapid advances in the size and power rating of wind generators as well as development of power electronics. These electrical utilities generate and provide energy in AC form. As the wind energy to electrical power conversion is rather expensive we concentrate on efficiency increase inside electrical system which can be obtained at moderate costs. The cost per power at grid inversion point shall be as low as possible. This paper describes the implementation, design and performance of a 70 W small wind turbine with two different generator–converter combinations and the analysis of rectifier design with Si-, MOSFET and Schottky diodes for battery charge applications. Three-phase full bridge rectifiers are compared on the basis of topology, cost, efficiency, power consumption and control complexity. The power conversion system includes a pulse-width modulation (PWM) rectifier coupled to a battery load.

*Keywords* — PMSM, MOSFET, converter control, optimization.

## I. INTRODUCTION

Small wind turbines typically employ permanent magnet synchronous machines which need no excitation, exhibit a simple general design and good efficiency. The converter inserted between machine and grid should be optimized for maximum power tracking and for EMC at the grid.

Investigations first lead to an improved rectifier using standard Si-diodes, Schottky rectifier and MOSFET.

We select power electronics components according to minimized losses, because this raises output power at given wind energy input power. Low internal resistance components are used and intended for use in medium voltage operation at high switching frequency circuitries where low losses and low noise are required.

The first step in obtaining knowledge on new ideas about their performance is to design and calculate results.

General objective for this investigation is to identify and explain the control mechanisms of these problems and the interaction of these converters with the network.

A significant efficiency increase of a small wind turbine shall be obtained by using simple and cheap power electronics for generator to load type adaptation.

Measurement of losses in our system is achieved in four steps accompanied by appropriate measurements and has the following features.

However, the experimental verification is obligatory to get reliable results. We use a simple but versatile test bench at small power ratings (Fig. 1).

The reverse recovery current in the silicon diode, like the on-state resistance in the MOSFET, will rise even further with increased temperature.

These increases will cause switching losses and possibly create a thermal runaway situation. The cooling system for this kind of applications can be designed to improve the efficiency of whole system.

Fig. 1. Test bench with generator driven by converter-fed induction (asynchronous) motor

1. Measurement of Schottky rectifier output power.
2. Measurement of MOSFET rectifier output power.
3. Measurement of MOSFET+ Schottky rectifier output power.
4. Measurement of MOSFET+Schottky+MOSFET rectifier output power.

The experiment is related especially with small power applications. The used wind turbine has a power range of 70 W, that feeds 12 V battery (24 V respectively). The power production is only possible if the voltage of generator is greater than battery voltage.

TABLE I.
Maximum power point tracking of Schottky rectifier.

| Hz | Load | $U_{gen}$ | $I_{gen}$ | $U_{rec}$ | $I_{rec}$ | $P_{gen}$ | $P_{rec}$ |
|----|------|-----------|-----------|-----------|-----------|-----------|-----------|
| 10 | 1,07 | 1,57 | 1,29 | 1,77 | 1,65 | 5,81 | 4,59 |
| 20 | 1,62 | 2,88 | 2,65 | 5,54 | 3,41 | 21,86 | 18,94 |

Table I shows the maximum power point tracking of the system with suitable loads. As the frequency increases, the generated power also increases due to high terminal voltage and current, the optimal points of power for each frequency are shown in the table I.

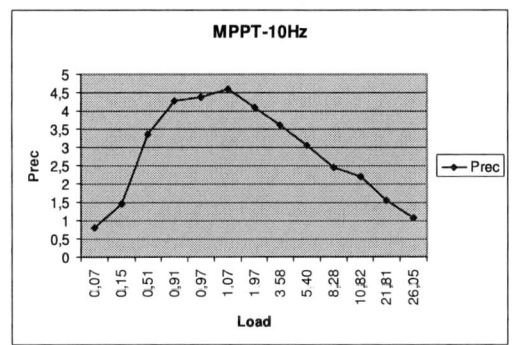

Fig. 1. The optimal power point of system at 10 Hz

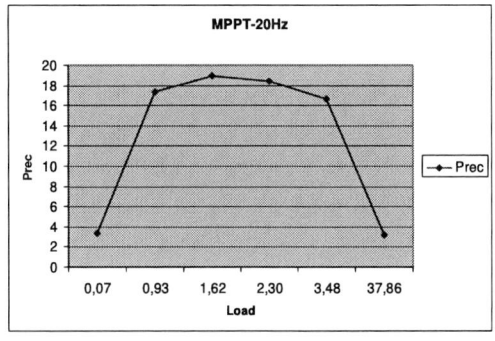

Fig. 2. The optimal power point of system at 20 Hz

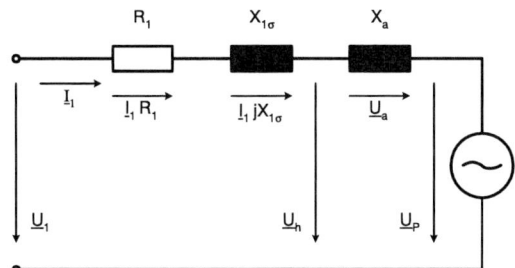

Fig. 3. Equivalent circuit of synchronous machine

with

$$X_d = X_{1\sigma} + X_a$$

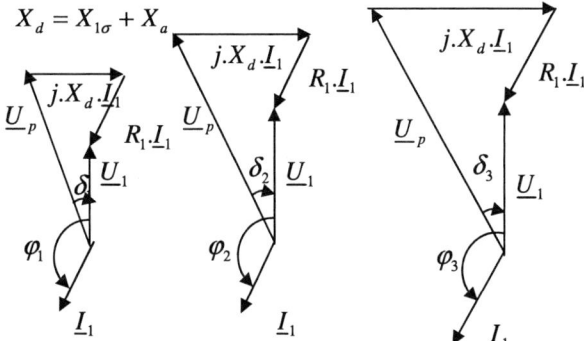

Fig. 4. Phasor diagram of the synchronous machine under load condition at different operation frequencies

Increasing the frequency causes voltage growth at the poles of machine and therefore the stator voltage and stray reactance of stator also increase.

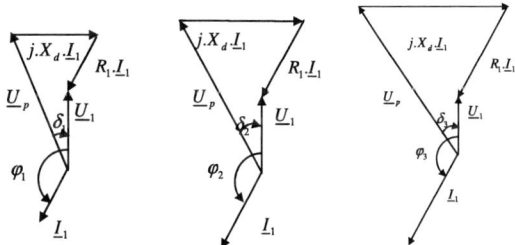

Fig. 5. Phasor diagram of the synchronous machine with battery application at different operation frequencies

Without boost converter (with battery application), the terminal voltage $U_1$ remains constant but current $I_1$ depends on the increasing frequency. The loss power therefore gets bigger and this is the disadvantage of the system and can be overcome finding out the optimal frequency for the MPPT.

The advantage of Schottky diode rectifier with a boost converter against PWM is that there is no need to use controller in case of small power applications. The effort of PWM is unreasonable if generated power is under 100W. The available efficiency gain remains %1-2.

**1935**

It can be implemented a very simple but effective pre-control for MPPT. Instead of buying a new wind turbine, the costs of the system can be reduced with Schottky diode rectifier. For this reason, boost converter for small wind power applications is better than PWM.

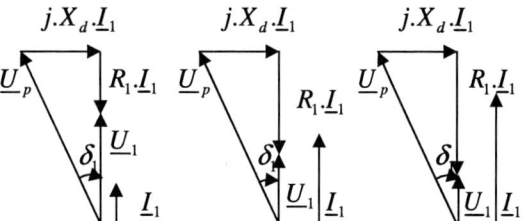

Fig. 6. Phasor diagram of the synchronous machine between idling and short circuit

The MPPT of the system is found out with a suitable load. The optimal operating point is between idling and short circuit.

The phase power:

$$\frac{P_o}{3} = (U_R + U_1).I.\cos\varphi$$

using pole voltage in the formula

$$\frac{P_o}{3} = (U_p.\cos\delta - I.R).I.\cos\varphi$$

$$\cos\delta = \sqrt{1 - \sin^2\delta} = \sqrt{1 - (\frac{X_d.I}{U_p})^2}$$

$$\frac{P_o}{3} = (U_p.(1 - \frac{(X_d.I/U_p)^2}{2}) - I.R).I.\cos\varphi$$

The MPPT can be found out between idling and short circuit states. Actually, load defines the optimal power which is generated from PMSM and flows through the rectifier.

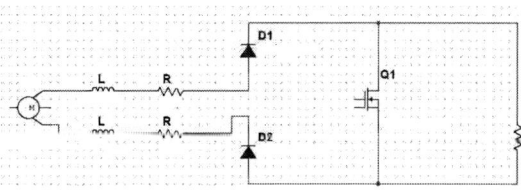

Fig. 7. Current path over diodes and MOSFET

If we assume duty cycle;

D=0,5

$$\eta = \frac{U_{eff}.1,35 - 2.U_{diode} + U_{mosfet}}{U_{eff}.1,35}$$

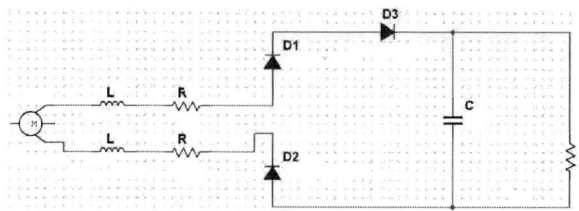

Fig. 8. Current path over diodes

If we assume duty cycle;

D=0,5

$$\eta = \frac{U_{eff}.1,35 - 3.U_{diode}}{U_{eff}.1,35}$$

The total losses of the system is

$$P_{loss} = \frac{U_{condensator}}{\frac{(2.U_{diode} + U_{mosfet}) + (3.U_{diode})}{2} + U_{condensator}}$$

## II. EXPERIMENTAL RESULTS

### A. Motor controller (U/f) operation basics

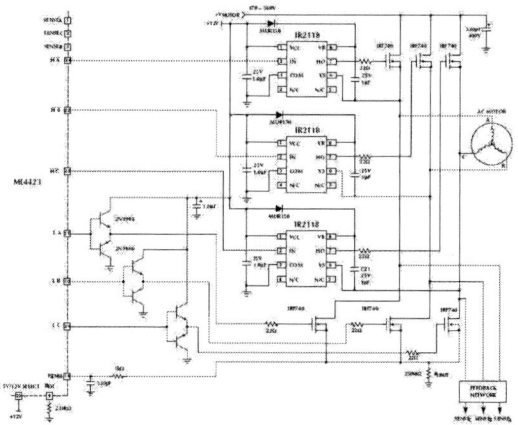

Fig. 9. ML4423 Output Stage Using IR2118 High Side Drivers

The ML4423 provides the PWM sinewave drive signals necessary for controlling three phase AC induction motors. The output variable frequency AC voltages which are produced by PMSM generator are sensed and fed back to the controller to track the sinewave frequency and amplitude set at the speed control input. These sinewaves can be varied in amplitude and frequency by the speed input.

### B. Speed Control

The voltage on $V_{speed}$ controls the sinewave frequency and amplitude. A $160\,kW$ resistor to ground on $R_{speed}$ converts the voltage on $V_{speed}$ to a current which is used to control the frequency of the output PWM sinewaves. The amplitude of the sinewaves increases linearly with $V_{speed}$ until $V_{speed}$ reaches 4.4V. Above this voltage the amplitude remains constant and only the frequency changes as shown in Fig. 10 and 11.

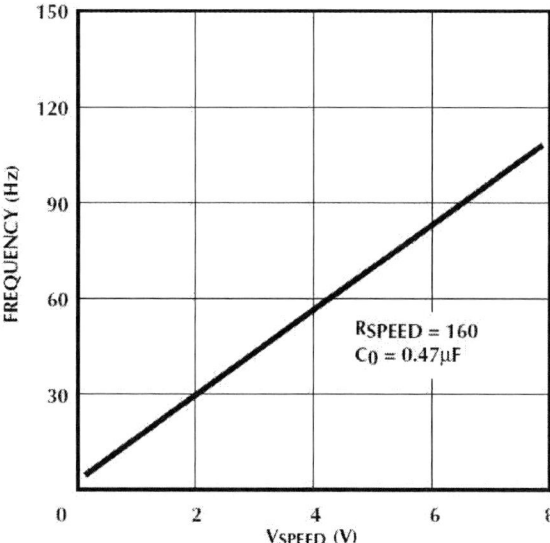

Fig. 10. Frequency vs $V_{speed}$

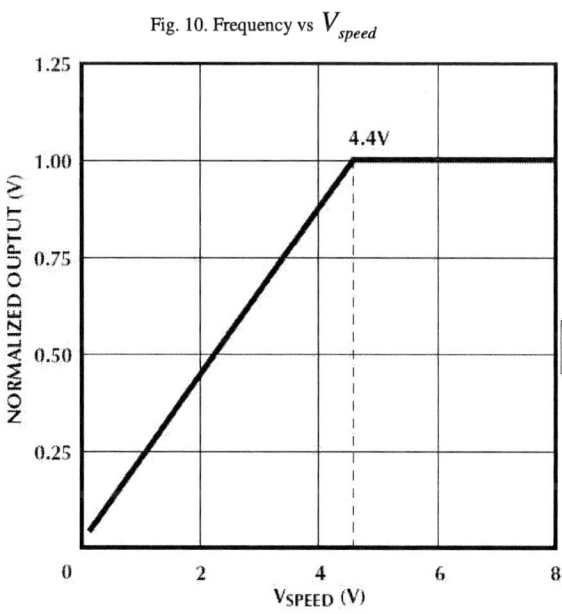

Fig. 11. Normalized Output Voltage vs $V_{speed}$

### C. PWM SINE Controller

This circuit block compares the sinewaves at $Sine_A$ and $Sine_B$ to the sampled inputs $Sense_A - Sense_C$ and $Sense_B - Sense_C$, respectively. The differential signals $Sense_A - Sense_C$ and $Sense_B - Sense_C$ will be approximately 1.7 volts zeropeak maximum at 70 Hz.

The high voltages at the motor terminals are divided down to 1.7V to provide voltage feedback to the controller. In our experiment, the generator voltage is 1.3 V at 20 Hz. The resistors from $Sense_A$ and $Sense_B$ to ground should be $0{,}5k\,\Omega$. $Sense_C$ should have a $250\,\Omega$ to ground because it has 1/2 the input impedance of the other 2 inputs. $V_{peak}$ on the motor is set by the divider ratios.

For the values:
$(V_{speed} = 1.3\text{V})$

$$V_{peak} = 1.7\text{V}\,\frac{49.5k\Omega + 1k\Omega}{1k\Omega}$$
$$= 24 \text{ V (Battery load)}$$

For loss estimation, first we consider the voltage drops on switches in averaged levels. The experimental results will be given at the end of this chapter.

#### 1) Schottky-Diode rectifier:

The losses in the whole system can be described by the path which the current flows through. For %50 MOSFET duty cycle the losses can be described;

$$Current_{path} = 2 * Schottky + \frac{MOSFET + Schottky}{2}$$

As seen below, the current flows for each phase two times over Schottky-diodes rectifier and then is distributed between MOSFET and Schottky diode. Voltage drop estimation of the system gives a rough approach and is proved by the experimental results.

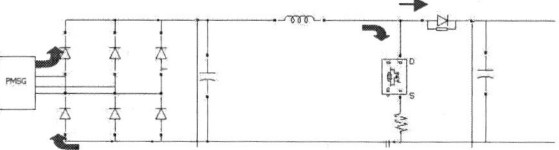

Fig. 12. Rectifier with Schottky-diodes and boost converter

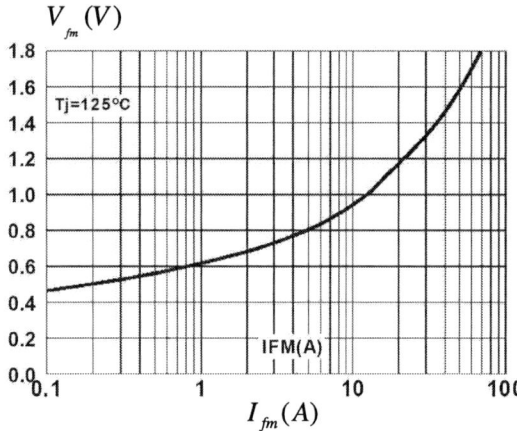

Fig. 13. Forward voltage drop versus forward current
(maximum values)

Typical current value in experimental system is up to 8 A. The worst condition with 10 A causes of the voltage drop at Schottky-diode is 0.96 V.

TABLE II.
Generated power from generator and rectifier.

| f | Generator | | | Schottky-rectifier | | |
|---|---|---|---|---|---|---|
| | P | $U_{RMS}$ | $I_{RMS}$ | P | $U_{RMS}$ | $I_{RMS}$ |
| Hz | W | V | A | W | V | A |
| 20 | 4,3 | 5,2 | 0,3 | 4 | 11,5 | 0,4 |

*2) Schottky Diode + MOSFET:*

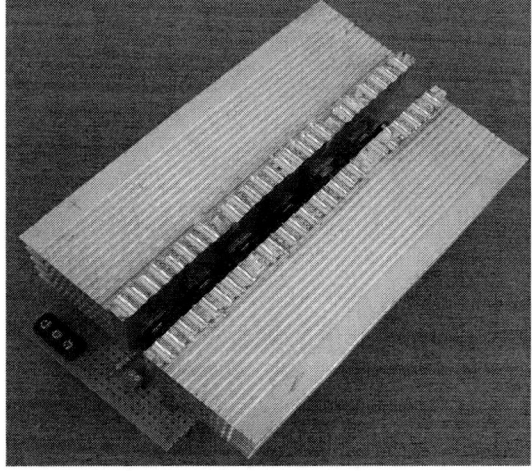

Fig. 14. Construction of rectifier

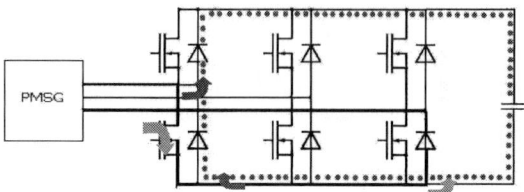

Fig. 15. PWM rectifier with Schottky diodes

Short Circuit: ➡
(On-state of integrated boost converter)

$$Current_{path} = Schottky + \text{MOSFET}$$

In short circuit condition, the current flows over on-state MOSFET due to reverse-blockage of anti-parallel Schottky diode and turns back over Schottky diode of another leg.

Normal Operation: ➡

$$Current_{path} = 2 * Schottky$$

Schottky diodes are coupled to MOSFETS. When the MOSFET is on, it is necessary to prevent the current to flow from the output capacitor or the load through the MOSFET. During the switching transient when the diode is turning off and the N-Bar connected MOSFET is turning on, the reverse recovery current from the diode flows into the MOSFET, in addition to the rectified input current.

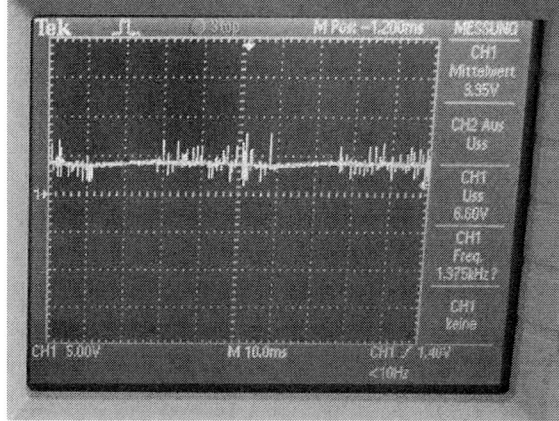

Fig. 16. The output voltage of PWM rectifier

The DC output voltage of rectifier is 3,95 V at 20 Hz. Due to bad generated voltage from the generator, the output dc voltage has ripples.

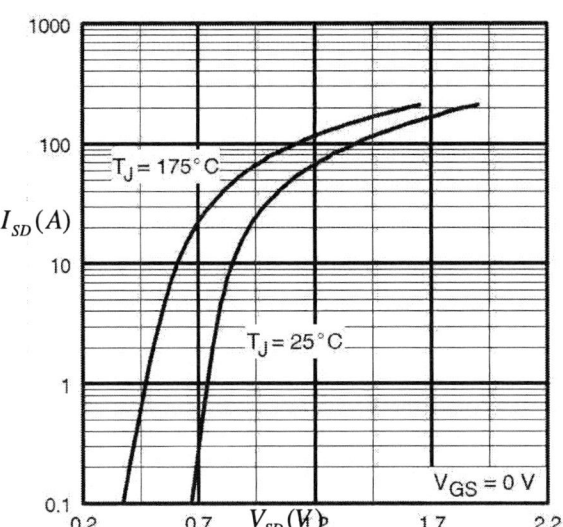

Fig. 17. Typical Source-Drain Diode Forward Voltage

TABLE III.
Generated power from generator and rectifier.

| f | Generator | | | Rectifier | | |
|---|---|---|---|---|---|---|
| | P | $U_{RMS}$ | $I_{RMS}$ | P | $U_{RMS}$ | $I_{RMS}$ |
| Hz | W | V | A | W | V | A |
| 20 | 3,95 | 4,68 | 0,87 | 3,75 | 3,95 | 0,8 |

*3) MOSFET:*

Fig. 18. Construction of rectifier

A controlled MOSFET is used to realize the configuration instead of Schottky-diodes.

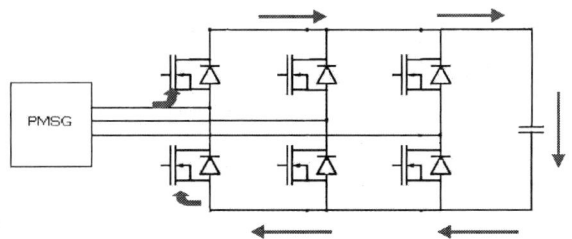

Fig. 19. MOSFET rectifier

$$Current_{path} = 2 * MOSFET$$

TABLE IV.
Generated power from generator and rectifier.

| f | Generator | | | Rectifier | | |
|---|---|---|---|---|---|---|
| | P | $U_{RMS}$ | $I_{RMS}$ | P | $U_{RMS}$ | $I_{RMS}$ |
| Hz | W | V | A | W | V | A |
| 20 | 4,13 | 5,41 | 0,87 | 3,36 | 4,2 | 0,8 |

*4) MOSFET +Schottky+ MOSFET:*

Fig. 20. Construction of rectifier

Short Circuit:

$$Current_{path} = MOSFET + MOSFET$$

Normal Operation:

$$Current_{path} = 2 * MOSFET$$

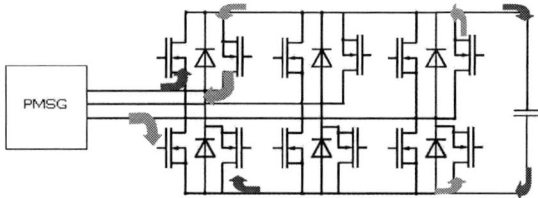

Fig. 21. Schottky + MOSFET rectifier

TABLE V.
Generated power from generator and rectifier.

| f | Generator | | | Rectifier | | |
|---|---|---|---|---|---|---|
| | P | $U_{RMS}$ | $I_{RMS}$ | P | $U_{RMS}$ | $I_{RMS}$ |
| Hz | W | V | A | W | V | A |
| 20 | 4,06 | 5,30 | 0,87 | 3,36 | 4,4 | 0,8 |

These experiments result in a large inrush current into the MOSFET, requiring a substantially larger sized MOSFET than required if the diode had no reverse recovery current. This large MOSFET represents a substantial cost in this circuit but offers lower losses due to over-dimensioning character. These switching losses also limit the frequency of operation in the circuit, and hence its cost, size, weight and volume. A higher frequency would allow the size of the passive components to be correspondingly smaller.

Efficiency gains can be used to allow the delivery of higher power from the same design.

## CONCLUSION

This test circuit employed a 64 A, 55 V International Rectifier MOSFET (IRFZ44N), a 20 A, 200 V high efficiency fast recovery rectifier diode (BYW80F/FP-200) and a 2x15 A, 45 V Schottky diode (STPS3045CT).

Voltage and current measurements were taken on both the MOSFET as well as the diode, in order to estimate the power losses in these components. The input and output power was also measured to calculate the efficiency of the circuit and can be utilized to enhance the performance of the PWM circuit by improving the efficiency, reducing the switching losses in the diode and the MOSFET, reducing the MOSFET case temperature and reducing the number of MOSFETs.

The proposed generation system was developed for laboratory experiments. The performance was evaluated and the effectiveness of the proposed system was confirmed.

TABLE VI.
Comparing of power losses for each configuration.

| | Generator [W] | Power [W] | Efficiency [%] | |
|---|---|---|---|---|
| 1 | 4,3 | 4 | 93 | S |
| 2 | 3,95 | 3,75 | 94 | M+S |
| 3 | 4,13 | 3,36 | 81 | M |
| 4 | 4,06 | 3,36 | 82 | M+S+M |

Table VI shows the efficiency results of rectifier designs. Due to lower power generation of wind turbine, the effect of optimization of the power electronics is also small. The advantage of Schottky diode rectifier with a boost converter against PWM is that there is no need to use a microcontroller.

## REFERENCES

[1] Ince Kayhan, Weiss Helmut, Arslan Seyit, "Practical Efficiency Improvement at Electric Power Conversion System for Small Wind Turbines," 16th Int. Conference on Electrical Drives and Power Electronics Slovakia, 2007.

[2] Alexandru Bitoleanu, Mihaela Popescu, Mircea Dobriceanu, "Power Quality at the Input of DC Motor and Controlled Rectifier Driving Systems," Romania University of Craiova, Electromechanical Faculty, Romania, Portoroz 12th International Conference on Power Electronics and Motion Control, 2006.

[3] Robert Paku, Cristian Popa, Mircea Bojan, Richard Marschalko , "Appropriate Control Methods for PWM ac-to-dc Converters Applied in Active Line-Conditioning," Technical University, Cluj, Romania, 2006, pp 573-579.

[4] http://digchips.com/datasheets/parts/datasheet/456/ BYW80FP-200.php .

[5] http://www.datasheetcatalog.net/de/datasheets_pdf /S/T/P/S/STPS3060CW.shtml.

# Analysis of Wind Farm and Multilevel Converter Interactions in Medium Voltage Networks Under Steady-State and Transient Conditions

J. Sosa-Ruiz[*], E. Moreno-Goytia[*], and O. Anaya-Lara[†]

[*] Instituto Tecnológico de Morelia/Postgraduate Studies in Electrical Engineering, Morelia, México, e-mail:
*julian.sosa.ruiz@ieee.org, elmg@ieee.org*
[†] University of Strathclyde/Institute for Energy and Environment, Glasgow, United Kingdom, e-mail:
*olimpo.anaya-lara@eee.strath.ac.uk*

**Abstract-** **The research work presented in this paper focuses on analyzing the steady-state and transient operation of a wind farm interconnected to the medium voltage network through a Voltage Source Converter with Neutral Point Clamped structure, VSC-NPC. As first part of the analysis, a number of wind speed variations are carried out to observe the steady-state performance of the system. For transient conditions, two particular study cases are performed in the test network: a) the occurrence of voltage sag, b) the occurrence of voltage swell. The simulation results illustrate the favorable impact of the multilevel controller on the distribution network performance.**

**Keywords-: VSC-NPC, multilevel converter, wind generator, distributed generation.**

## I. INTRODUCTION

The massive incorporation of wind generation, along with other renewable sources, to electric systems all around the word is impressively increasing. These new circumstances put pressure to Governmental entities, Industry and Academia to present innovative solutions, new concepts and new rules to play. Although substantial work has already been conducted in planning, controlling and operating distribution networks with distributed generation, it is generally agreed intense research work is required.

The use of power electronics for controlling and operating transmission and distribution networks has significantly increased over the last decade, particularly in applications incorporating distributed renewable generation. Some reasons favouring this tendency are:

a) The continual technological improvements in power electronics devices and PWM schemes,

b) The proven capacities of the VSC topology,

c) The inclusion of sophisticated and intelligent control processes.

However, the escalating integration of distributed generation, mainly renewable such as wind, photovoltaic and biomass, has triggered new challenges in the power electronics field. These include developing better power electronic controller topologies, novel and improved control strategies for enhancing network operation and security, and the power quality delivered to customers.

This paper investigates the steady-state and transient operation of a wind generation system (WGS) interconnected to the distribution network through a five level VSC-NPC. The purpose of this inverter is to support the wind generation system performance and its dynamic interaction with the network. The scenarios considered are: a) WGS start up and normal operation; b) Wind variation conditions, and c) Voltage-sag conditions.

Original contributions by this research work provide a better understanding of the complex interactions arising between the VSC power electronics controllers, embedded wind generation and the distribution network. The results may also provide useful insights for improving protection algorithms and the design of single-purpose/multi-task control platforms.

Figure 1, shows one of the test network with the series connected VSC Controller and the DFIG Wind Turbine to conduct the investigation.

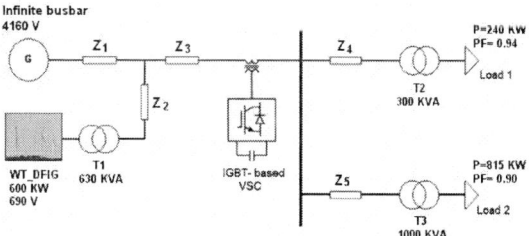

Figure 1. Diagram of the overall systems in the test network

This research work is intended to bring a better understanding of the modern electric networks

## II. SERIES CONNECTED MULTILEVEL VSC-NPC

A series-connected VSC is a powerful controller commonly used for voltage sags mitigation at the point of connection [3, 4]. Figure 2 shows the series-connected VSC structure coupled to the grid.

The VSC topology is a switching power converter behaving as a synchronous voltage source which can be regarded as series reactive and active power compensation scheme producing a controllable voltage in quadrature with the line current. A common type of series connected VSC is the Dynamic Voltage Restorer.

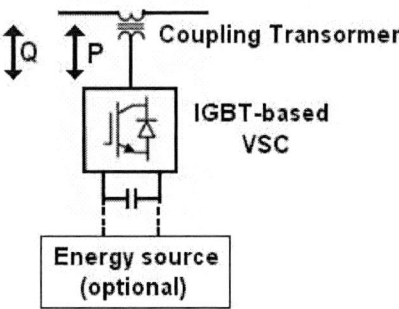

Figure 2. A series connected Voltage Sourced Converter

Recently, the multilevel converter has drawn tremendous interest in the power industry. These converters synthesize a sinusoidal voltage by stepping up and down several voltage levels, typically obtained from the converter DC side capacitor. As the number of levels increases, the synthesised output waveform adds more steps, producing a staircase wave that approaches the sinusoidal wave with minimum harmonic distortion.

The VSC generates a three-phase AC output voltage controllable in phase and magnitude, which is injected into the AC distribution system to maintain a specific parameter under control.

Figure 3 pictures the circuit of a three-level VSC, including the clamping diodes used to transport de DC voltage from the input to the output [5]. The top features associated to this topology are: minimum harmonic content, excellent flexibility and controllability, fast response, and ability to mitigate power quality problems on the grid.

The operation of the IGBT-Based VSC-NPC topology requires a PWM scheme to properly drive the converter switches.

From the three main SPWM schemes suitable for multilevel converters with NPC [6], the Disposition Phase (PD) scheme is used in this paper. SPWM requires all carrier signals to be in phase, as shown in Figure 4c. The other two SPWM techniques, illustrated in Figures 4c and 4b, use different phase correlation for each carrier signal.

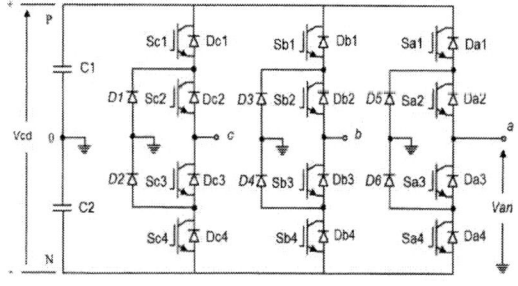

Figure 3. Three phase 3 levels VSC-NPC structure

Figure 5 shows the typical output waveforms and associated harmonic spectra of a three-level VSC-NPC.

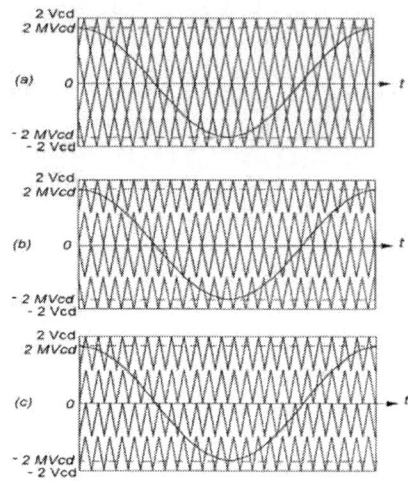

Figure 4. Alternatives for modulation

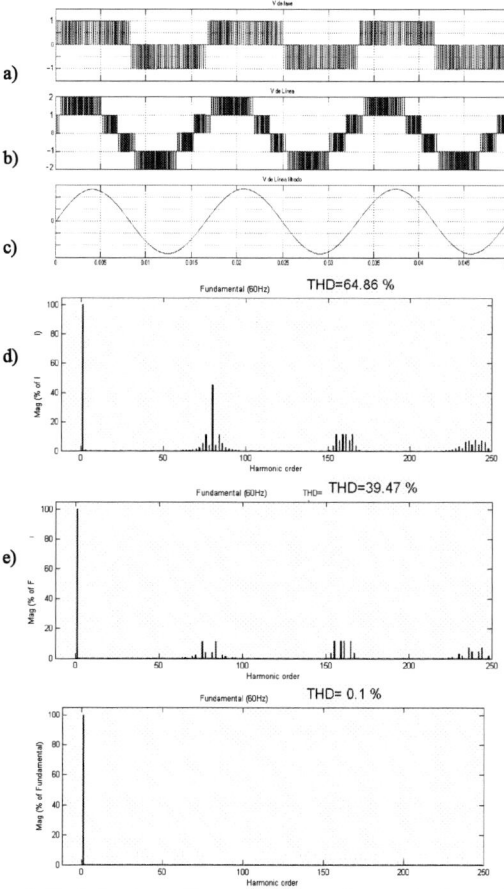

Figure 5. Typical output voltage waveforms and respective harmonic spectrum of three level VSC-NPC: a) phase voltage, b) line voltage, c) line voltage with filter. d) phase voltage harmonics, e) line voltage harmonics, f) harmonic elimination with passive filter.

## III. WIND TURBINE AND DFIG

Wind energy is one of the most important and promising sources of renewable energy, mainly because it is near to zero pollution and potentially can be economically viable in the long term. Wind turbine technology has had an enormous development in the past few years [7]. Today there are wind turbines ranging from a few watts applications up to 5 MW for large scale installations off-shore.

The principal features and the normal steady state operation of the WGS are explored in this section. The WGS is built with a wind turbine with Double Fed Induction Generator (DFIG).

With variable speed operation it is possible to increase the energy captured by the aerodynamic rotor as the optimum power coefficient can be maintain over a wide range of wind speeds. To this to happen, it is necessary to decouple the speed of the rotor from the frequency of the network through the power electronics converters.

The DFIG includes a wound rotor induction generator and an AC/DC/AC IGBT-based PWM converter. The stator winding is connected directly to the grid while the rotor is fed at variable frequency through the AC/DC/AC converter [8]. The DFIG technology allows extracting maximum energy from the wind for low wind speeds by optimizing the turbine speed, while minimizing mechanical stresses on the turbine during gusts of wind. Figure 6 shows the main blocks of the WGS. An advantage of DFIG technology is the ability for power electronic converters to generate or absorb reactive power, thus eliminating the need for installing capacitor banks as in the case of squirrel-cage induction generators.

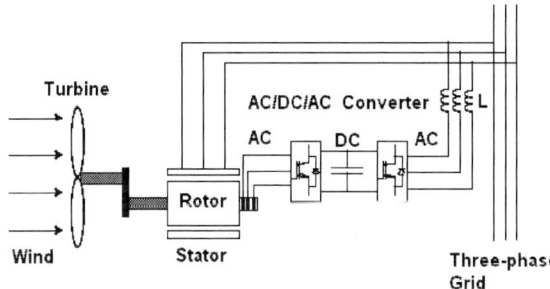

Figure 6. General blocks of the wind turbine and the double fed induction generator

## IV. ANALYSIS OF WIND FARM – VCS INTERACTIONS UNDER STEADY-STATE

The steady state operation of a wind turbine with DFIG is shown in Figure 7. For simulations, wind speed is set at 8 m/s as initial condition, then at $t = 3$ s, wind speed starts increasing up steadily to 14 m/s. At $t = 5$ s, the generated active power starts increasing smoothly (together with the turbine speed) to reach its rated value of 600 kW in approximately $t = 16$ s.

The pitch angle of the turbine blades is zero degree as initial condition and then it is increased from 0 deg to 0.76 deg in order to limit the aerodynamic power. It can be also observed in Figure 7 that the reactive power is controlled to maintain a 1 pu voltage. At nominal power, the wind turbine absorbs 32 kVARs (generated $Q = -32$ kVARs) to control voltage at 1 pu.

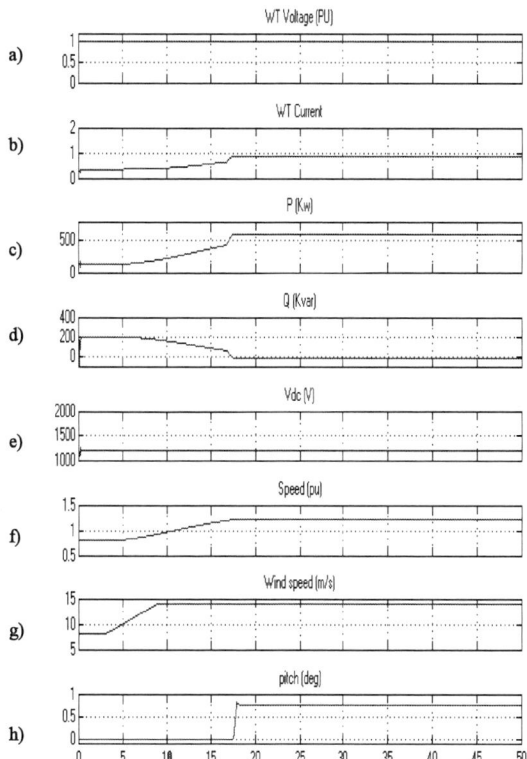

Figure 7. Output waveforms of the wind turbine-DFIG in steady state: a) output voltage, b) output current, c) output active power. d) output reactive power , e) bus of DC, f) rotor speed, g) speed of the wind, h) pitch angle of blades.

This section also presents the steady state operation of the wind farm interconnected to a test network. Figure 8, shows the one-line diagram of the test network. A voltage drop across (Z1+Z3) is included to simulate an electric distance between the infinite bus and N2 node.

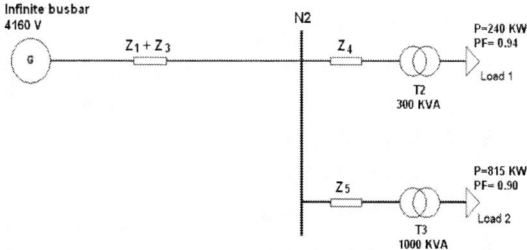

Figure 8. Test network without systems

Simulations show the voltage at node N2 is 0.975 pu as depicted in Figure 9.

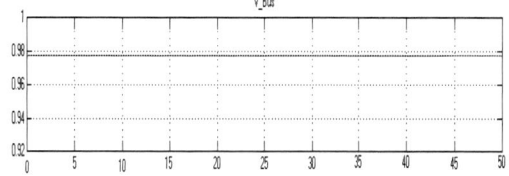

Figure 9. Voltage response at node N2 in test network.

For further clarification, Figure 10 shows the diagram of the test network with wind generation. Simulations show that wind generation supports the voltage drop in node N2, which can be linked to the DFIG self-voltage regulation capability when the power electronics is included. Figure 11 shows the voltage at N2 compensated up to close to 1 pu. This figure illustrates the impact of the 600 kW wind turbine in the first 16 seconds.

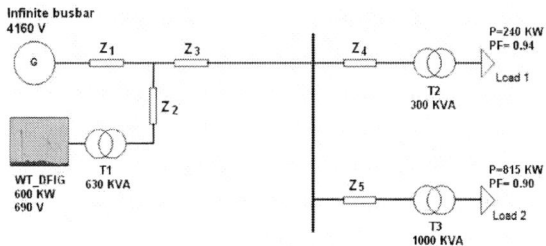

Figure 10. Test network with wind generation

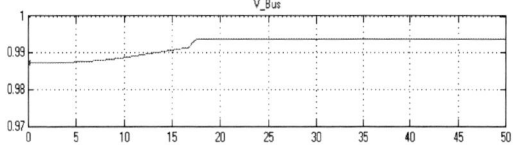

Figure 11. Test network response at N2 with wind generation

## V.  TRANSIENT CONDITIONS

Two test network configurations are used for the study cases. In the first one, the VSC-NPC is in-between the loads and the wind generators. In the second one, the wind generator is in-between loads and the VSC-NPC.

### A.  Study Case for network with VSC-NPC in between

Figure 12 shows the network configuration for the VSC in-between. The first study case is a voltage sag at node N2. The sag propagates from the infinite busbar impacting adversely N2. The sag produces a 10% line drop from its nominal voltage (1 pu). Figure 13 shows the voltage profile and Figure 14 shows the voltage sag waveforms both in the distribution network with duration from $t = 0.085$ to $t = 0.15$ s.

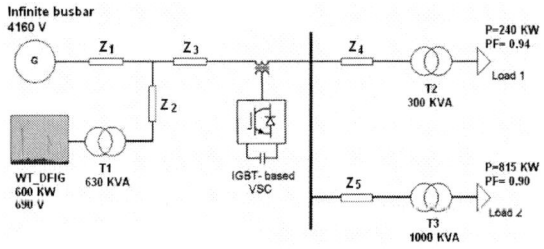

Figure 12.  Test network with VSC in-between

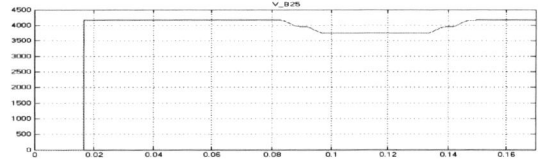

Figure 13. Voltage sag profile in the test network at node N2

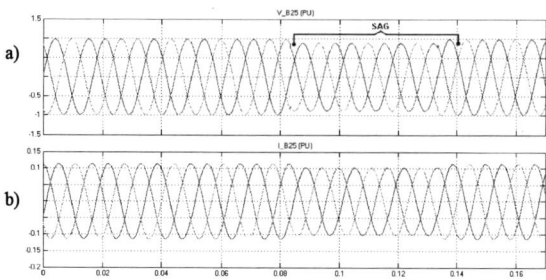

Figure 14. Voltage sag in the test network: a) voltage sage waves, b) line current

Figure 15 shows the voltage sag compensated by the action of the series connected multilevel VSC.

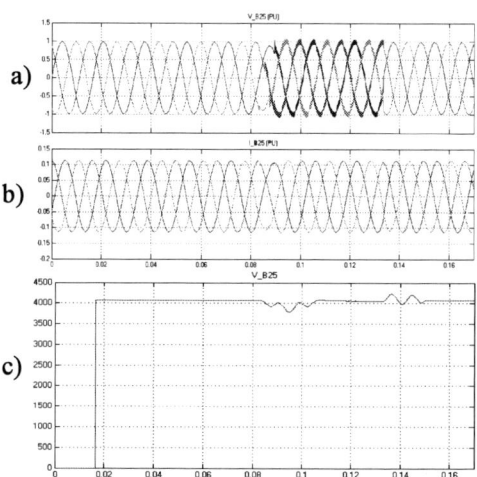

Figure 14. Voltage sag compensated: a) voltage sag compensated waves, b) Line current compensated, c) voltage sag-rms compensated

### B.  Study Case for network with Wind Farm in between

Figure 15 illustrate the network configurations for the wind farm in-between loads and the VSC-NPC

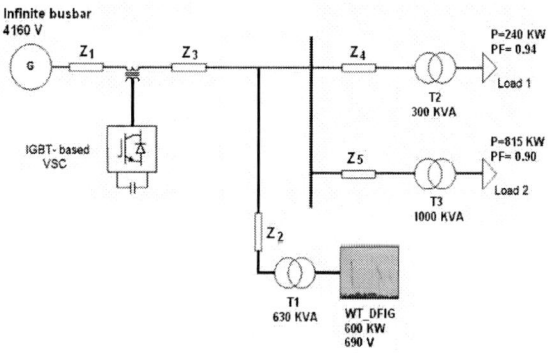

Fig. 15. Test network with wind generator in-between

Figure 16 shows a voltage sag in the distribution network starting at $t = 0.2$ s and ending at $t = 0.25$ s. The voltage sag is of 0.2 pu (0.8pu retained voltage).

1944

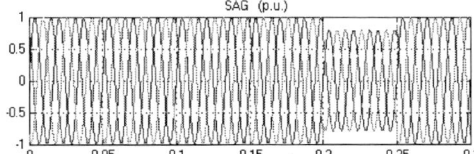

Fig. 16. Voltage sag in the distribution network in t = 0.2-0.25 seg.

This voltage sag is reflected throughout the system, causing a drop in the voltage of the critical loads, and the wind turbine. The impact of the sag on the wind turbine is shown in Figure 17.

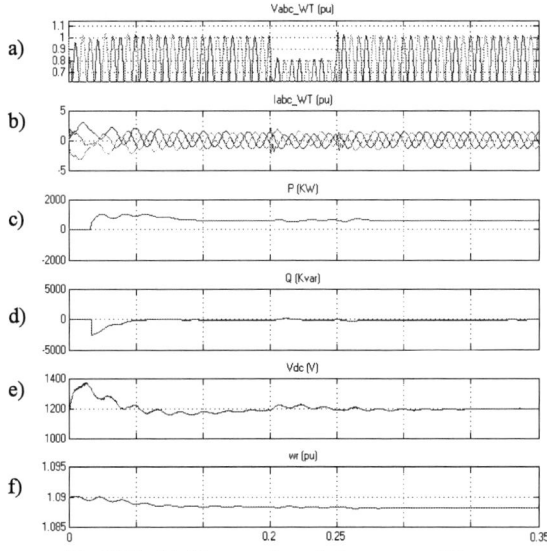

Fig. 17.Wind turbine wave forms in the presence of the sag.

In particular, Figure 17 shows the influence of the voltage sag in the output of the wind turbine, also during the time of the event, the current (b), power (c), Vdc (e) and the rotor speed (f), shows oscillations

The benefits of the operation of the VSC-NPC interacting with wind generator can be seen in Figure 18 where the dynamic compensation of the sag is displayed. For this case no harmonic filter has been installed.

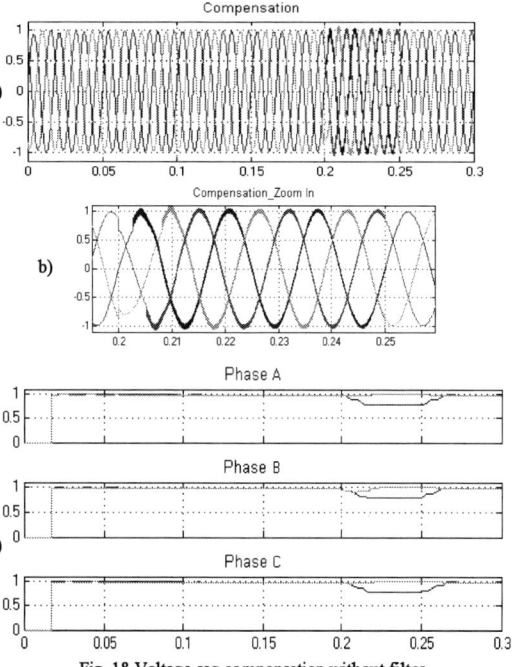

Fig. 18.Voltage sag compensation without filter.

Figure 18a, shows the sag compensation from *t*=0.2 to *t*=0.25. Figure 18b, shows an approach to the time of compensation and it shows a voltage drop for the first moments in the phase voltages. Figure 18c, shows the voltage profile in RMS values before and after the compensation. As additional information Figures 19 and 20 show the Simulink© models for network configurations of Figures 12 and 15.

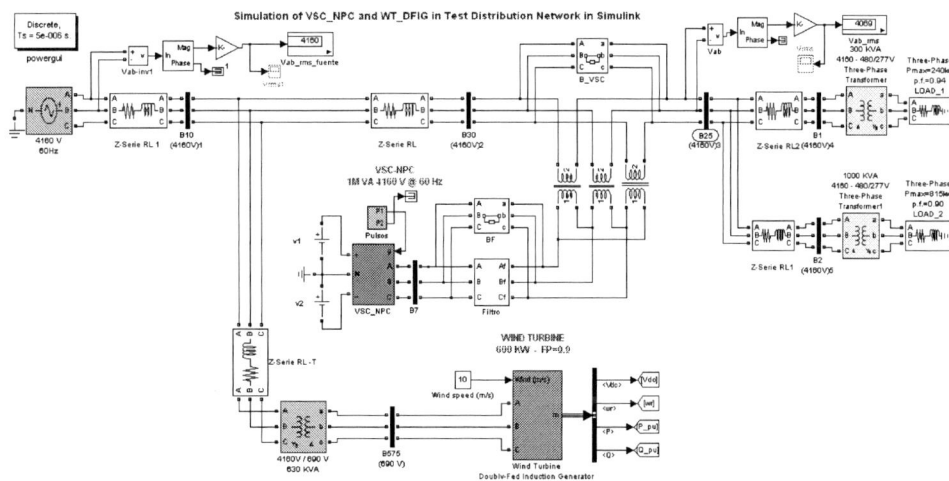

Figure 19. Simulink© model of the network configuration with VSC-NPC in-between

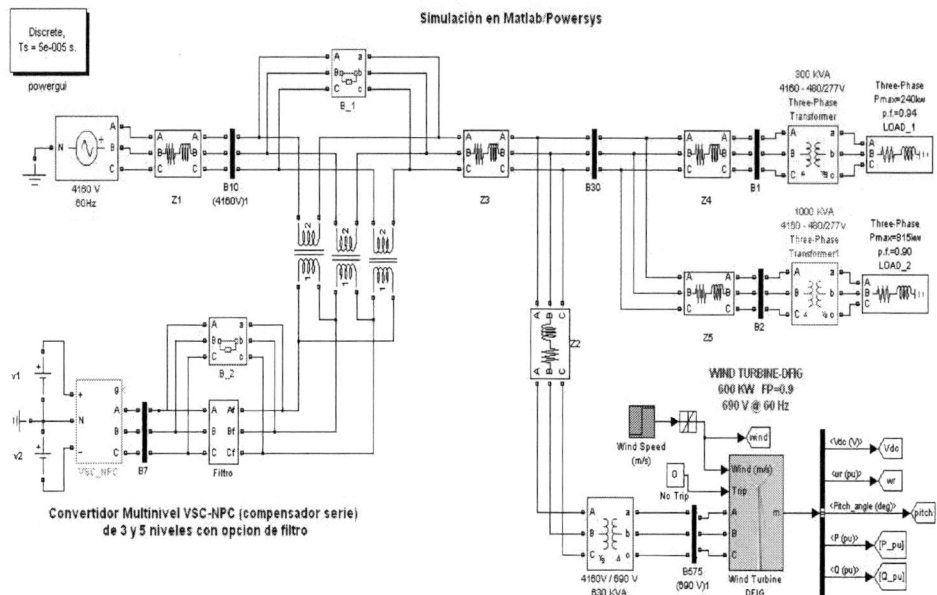

Figure 20. Simulink© model of the network configuration with wind generator in-between

## CONCLUSIONS

The simulation results shows that the interactions between a DFIG wind turbine and the electric network are complex and further research is still needed. Power electronics converters have brought many benefits to the incorporation of wind generation to electrical networks. This paper illustrated some of these benefits through simulation cases in steady and transient conditions, such as the node voltage compensation and timely sag mitigations with no disruption of the wind turbine which open the way to reinforce investigations about the impact of power electronics converters on grid code compliance.

From the point of view of the load, the installation of a VSC in-between with the AC network and the wind is a better option for mitigating voltage sag propagating in the grid or provoked by the wind farm. On the other hand, if the wind farm is allocated in-between the VSC and the grid, but in the load neighbourhood, then both wind farm and load can be alleviated of undesired events from the AC network. However, although the wind farm can provide some voltage support to the load also the load is exposed to undesired event provoked by the wind farm.

## REFERENCES

[1] N. Hingorani, "FACTS—Flexible ac transmission systems," *Proc. IEE 5th Int. Conf. AC DC Transmission*, London, U.K., 1991, Conf. Pub. 345, pp. 1–7.

[2] Petrella, A.J. "Issues, impacts and strategies for distributed generation challenged power systems". Paper no. 300-12, CIGRE Symposium on *Impact of demand side management, integrated resource planning and distributed generation*, Neptun, Romania, 17-19 September 1992.

[3] K. Chan and A. Kara, "Voltage sags mitigation with an integrated gate commutated thyristor based dynamic voltage restorer," *Proc. 8th ICHQP '98*, Athens, Greece, Oc. 1998, pp. 210–215.

[4] Agrawal, B.L., et al., "Advance series compensation (ASC) steady state, transient stability and subsynchronous studies" *Proceedings of Flexible AC Transmission Systems (FACTS) Conference, Boston, MA, May 1992.*

[5] A. Nabae, I. Takahashi, Akagi, " A new neutral point clamped PWM inverter", *IEEE Trans. On Ind. Applic.*, Vol. IA-17, No. 5, sep/oct, 1981, pp. 518 – 523.

[6] B. Peter, D., Grahame H., "Multicarrier PWM strategies for multilevel inverters, "*IEEE Trans. on Ind. Elec.*, Vol. 49, no. 4, August 2002.

[7] Tapia, A., Tapia, G., Ostoloza, J., Sáenz, J. "Modeling and control of a wind turbine drive", *IEEE Trans. On Energy conversion*, Vol. 18, no. 2, June 2003.

[8] MATLAB-SimPowerSystems/Distributed Resources models (Richard Gagnon), Version 7.5.0 (R2007b). The MathWorks.

# A Simple, Low Cost Design Using Current Feedback to Improve the Efficiency of a MPPT-PV System for Isolated Locations

Herman Fernández[*], Abelardo Martínez[†], Víctor Guzmán[~] and María Isabel Gímenez[~]

[*] Unexpo Puerto Ordaz/Departamento Ing. Electrónica, Puerto Ordaz, Venezuela, e-mail: *herman_fernand@yahoo.com*
[†] Universidad de Zaragoza/Centro Politécnico Superior, Zaragoza, España, e-mail: *amiturbe@unizar.es*
[~] Universidad Simón Bolívar/Departamento Electrónica y Circuitos, Caracas, Venezuela, e-mail: *vguzman@usb.ve, mgimenez@usb.ve*

*Abstract*— This work presents an efficiency improvement of a simple maximum power point tracking (MPPT) configuration, achieved by current feedback. The output current provided by the photovoltaic (PV) panel is used as the signal controlling the duty cycle of the PWM converter that regulates PV panel power output. The system performance is analyzed with the help of an accurate PSpice model of the PV cells developed for this application and experimentally improved. The results, both simulation and experimental, show that the proposed improved MPPT circuit is highly efficient. Since this is a simple, rugged and low cost system, it is ideal for applications in isolated locations in the less developed countries.

*Keywords*— MPPT, photovoltaic, DC/DC, current control, PSpice.

## I. INTRODUCTION

Many different maximum power point tracking configurations for photovoltaic applications (MPPT-PV) have been presented in the literature [1]; usually even those proposed for rural applications are relatively complex and require the use of powerful and expensive ICs such as microprocessors or DSPs [2-4]. These designs are not the best for applications in underdeveloped countries, where the cost is a fundamental issue and where highly trained installation and maintenance personnel is not easily available. The more promising simple and low cost MPPT-PV strategies, adequate for use in isolated locations and that do not require the use of complex ICs such as microprocessors or DSPs are: the extremum-seeking method [5], modulating a small-signal sinusoidal perturbation into the duty cycle of the main switch [6], using ripple correlation control [7], one-cycle control [8] and the sensor-less configuration [9].

This work develops the very simple configuration presented in [10]-[11], including current sensing. Experimental tests show that this improved configuration offers a better maximum power point tracking and higher efficiency, while remaining a very simple and low cost circuit.

## II. PROPOSED SYSTEM DESCRIPTION

### A. Simulations results

Figure 1 shows the improved MPPT circuit with current feedback block diagram configuration used for PSPICE testing. The main blocks used in the simulation are: a dedicated SPICE PV array model, the MPPT power stage, represented by the DC/DC average model, the MPPT controller, implemented using a standard PWM regulator, the filter stages, the battery array SPICE model and a resistive load connected in parallel across the battery array. The proposed MPPT configuration works as follows. The load power is given by:

$$P_L = V_{bat} \cdot I_{bat} = V_{bat} \cdot \frac{I_{sal}}{D} \qquad (1)$$

Where $V_{bat}$ and $I_{bat}$ are the battery voltage and current, $I_{sal}$ is the current generated by the photovoltaic array and $D$ is the converter duty cycle. The maximum available power as a function of the cell voltage, $p = f(v)$, is assumed to be a straight line that, taking into account the maximum power, is given by:

$$P_{máx} = P_0 + K \cdot \left(V_{conv} + V_D\right) = P_0 + K \cdot \left(\frac{V_{bat}}{D} + V_D\right) \qquad (2)$$

Where $K$ is the line slope, $P_0$ is the short circuit panel power, $V_{conv}$ is the DC/DC converter input voltage and $V_D$ is the panel's output diode conduction voltage. To ensure that maximum power is extracted from the PV panel, the converter's duty cycle must be dynamically adjusted to the value given by:

$$D = -\frac{K \cdot V_{bat}}{P_0 + K \cdot V_D} + \frac{V_{bat} \cdot I_{sal}}{P_0 + K \cdot V_D} = D_{min} - N \cdot I_{sal} \qquad (3)$$

$$D_{min} > 0, \quad N > 0$$

Where $N$ is a proportional constant and $D_{min}$ is the duty cycle minimum value. This equation gives the required instantaneous duty cycle value as a function of the current generated by the PV panel.

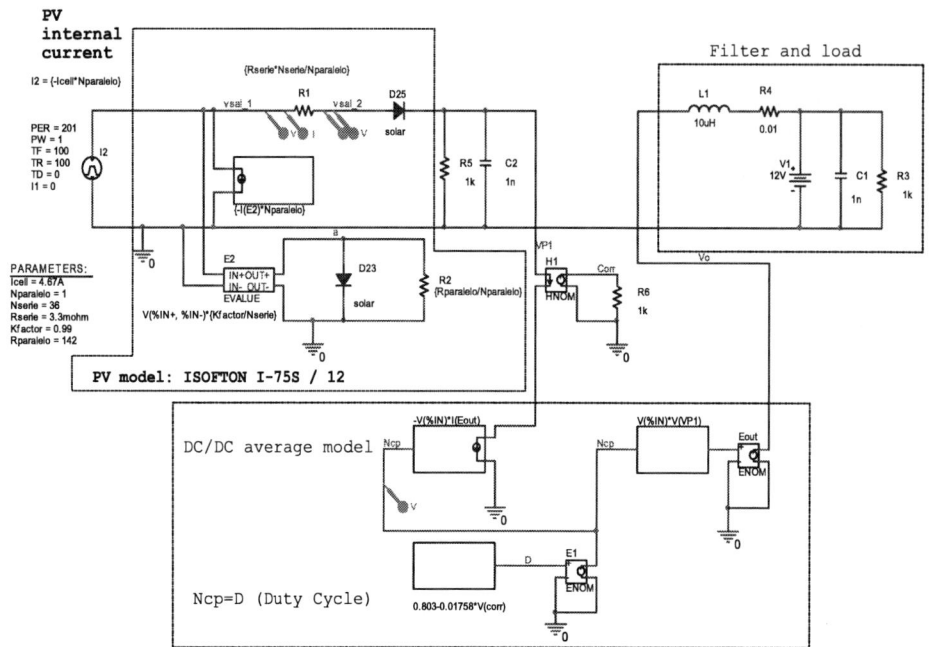

Fig.1. Block diagram of the system used to simulate the proposed MPPT with current feedback strategy in SPICE.

The simulations were run for a 36 cell serial array as in the Isofotón I-75S/12 panel. Solar light intensity variations were simulated using a triangular waveform. The results, shown in Fig.2, are as follows: The theoretical maximum available power is 77.239W, the extracted power is 74.980W, and the MPPT efficiency is 98.546%. In the same conditions, the unimproved version of this MPPT strategy, presented in [10] was 94.35%, hence the improved configuration, taking into account the current signal, shows an increment in efficiency almost equal to 4,2% and is very close to the ideal 100% efficiency. Fig.2 also shows how the duty cycle changes as a function of the changes in solar radiation. As expected, the duty cycle is inversely proportional to the available power. Since the simulations results are very promising, a prototype for field test is under construction.

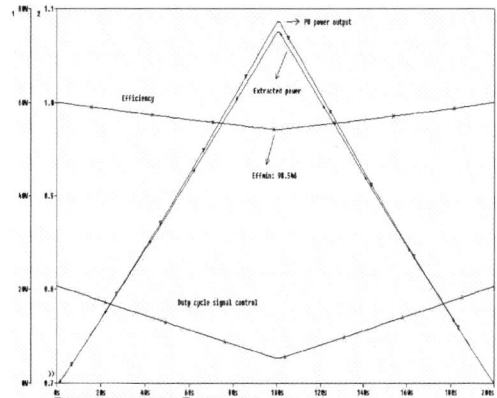

Fig.2 Simulation results showing the calculated PV generated power (PV power output), the power extracted by the MPPT circuit (Extracted power), the extraction efficiency (Effin), and the duty cycle calculated by the MPPT controller as a function of PV power variations.

### B. Experimental MPPT with current feedback circuit

Fig.3 shows the MPPT with current feedback prototype circuit constructed for initial laboratory tests as described below, developed around an integrated PWM IC, type *3524*. This is a low cost, simple IC already in regular use, and available from multiple sources. If necessary, the design can be modified to work with other similar PWM ICs such as the *TL494, LM3525, UA3844*, etc. Current feedback is provided using a Hall sensor monitoring the PV panels output current. This signal, scaled by an amplifier stage, is used as the reference adjusting the PWM controller duty cycle. From the duty cycle equation, and taking into account that the peak amplitude in the internal ramp generated in the LM3524 varies from 1 to 3.5 volts, the control voltage adjusting the duty cycle is given by:

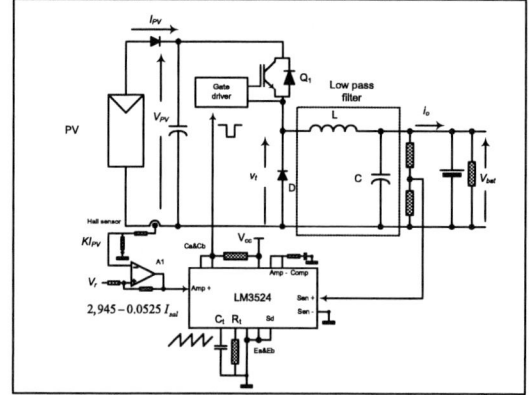

Fig.3 Proposed MPPT with current feedback circuit using the current signal to regulate the PWM stage controlling the PV panel output.

**1948**

$$V_{cont} = 2.945 - 0.0525\, I \qquad (4)$$

Equation (4) is implemented in analog form in the OPAM and its related resistors in Fig. 3. Input "Sen+" in the LM3524 duty cycle limit circuit is used to reduce the duty cycle when over-voltage in the DC/DC converter output is detected. This protection works in two stages. When the voltage at Input "Sen+" reaches 200mV, the duty cycle is reduced to 25% of maximum. If the voltage at Input "Sen+" rises above 200mV, the PWM generator output is blocked.

### C. Experimental MPPT with current feedback circuit test results

To test the proposed MPPT with current feedback circuit in the laboratory, a test rig with a Isofotón I-75S/12 panel, the experimental circuit, a battery bank (50Ah, 24V) and a resistor load were used. Tests were performed with different initial charges in the battery bank, and with different light levels exciting the PV panel. Solar light intensities very close to the maximum were reached in the full light tests, performed at mid-day in the Lab in Ciudad Guayana, about 5° north of the Equator. Results are summed up in Fig.4. The PV output current and output power characteristic curves shown are those calculated by the PSPICE photovoltaic panel model. Output current points measured in the laboratory are shown as small squares in the figure. The line connecting the theoretical maximum power points are drawn, identified as "Ideal". Three different test sets are presented, identified as Exp1, Exp2, Exp3. In Exp1 and Exp2, the current demanded by the load is restricted using a battery bank which is fully loaded before the experiments are performed. These reduced load conditions will be presented early in the afternoon in unclouded days, when overall energy demand by the load is low. Under these reduced current load conditions, the MPPT controller regulates the current required by the load without problems, but the system operates at reduced efficiency since the actual operating points are below the ideal ones. When the resistive load current is increased (Exp2) the MPPT controller reacts increasing the delivered current and moving the operating point towards a higher efficiency one, closer to the ideal operating point; these behavior is observed at all solar light levels. In Exp3 the system is operated at high output demand, with a fully unloaded battery bank and the resistor load connected to the MPPT output. These conditions will be the ones under which the system will usually operate in the mornings, or after a long cloudy interval later in the day. At high load levels the MPPT controller ensures that the PV panel operates with high efficiency and the operating points are close to the theoretical maxima at all illumination levels. Current and power curves shown are those calculated by the PSPICE model for the Isofotón I-75S/12 panel. Small squares mark measured laboratory values. Exp1 and Exp2 show the results obtained when the battery bank is fully loaded, with (Exp2) and without (Exp1) an additional resistive load in parallel with the battery bank. Exp3 shows the results obtained when the battery bank is being charged, raising the current required by the load and allowing for the maximum available system efficiency, approaching the ideal maximum.

### III. CONCLUSIONS

The proposed MPPT based on output current measurements taking into account the theoretical straigh line connecting the maximum power points in the PV panel characteristics has being validated both by numerical simulation and experimental work in a laboratory circuit prototype.

Circuit complexity in the improved MPPT circuit remains low, fulfilling the overall design aim: to produce a simple, efficient and low cost solution, adequate for use in isolated locations in less developed countries.

Maximum power point tracking efficiency is a function of output current demand. Maximum power point operation is not possible at low load levels, but the proposed MPPT based on output current measurements is able to increase the tracking efficiency as the required output current increases, reaching points close to the ideal ones once the required load is close to the maximum.

### REFERENCES

[1] Leyva R., Alonso C., Queinnec I., PastorCid A, Lagarnge D. and Martínez L. "MPPT of photovoltaic systems using extremun-seeking control". IEEE transactions on aerospace and electronics systems, vol 42, No.1. January 2006, pp.249-256.

[2] Shu-Hung H., Tse K.K., Hui R. Mok C. and Ho M. "A novel maximum power point tracking technique for solar panels using a SEPIC or Cuk converter". IEEE transactions on power electronics, vol 18, No.3. May 2003, pp.717-724.

[3] Esram T., Kimball J., Krein P., Chapman P. and Midya P. "Dynamic maximum power point tracking of photovoltaic arrays using ripple correlation control". IEEE transactions on power electronics, vol 21, No.5. September 2006, pp.1282-1291.

[4] Chen Y., and Smedley K. "A cost-effective single-stage inverter with maximum power point tracking" IEEE transactions on power electronics, vol 19, No.5. September 2004, pp.1289-1294.

[5] Rai S., Kumar U., and Naik R. "A novel technique for photovoltaic maximum power tracking system". EPE conference, Dresden, September 2005, pp.1-8.

[6] Koutroulis E., Kalaitzakis K., and Voulgaris N. "Development of a microcontrolled-based photovoltaic maximum power tracking control system. IEEE Transactions on Power Electronics, vol. 16 , No1, January 2001, pp.46-54.

[7] Bose B., Szczesny P, and Steigerwald R. "Microcomputer control of residential power conditioning system". IEEE transactions on industry applications, vol. IA-21, September-October 1985, pp.1182-1191.

[8] Hua C., Lin J. and Shen C. "Implementation of a DSP-controlled photovoltaic system with peak power tracking". IEEE trans. on industrial electronics, vol.45, February, 1998 pp.99-107.

[9] Blaabjerg F., Iov F., Teodorescu R., and Chen Z. "Power electronics in renewable energy systems" EPE-PEMC, Slovenia, September 2006, pp.1-17.

[10] Fernández H., Martínez A., Guzmán V. and Giménez M.I. "Evaluación por simulación de un circuito para la obtención de la máxima potencia de paneles fotovoltaicos por control de corriente" SAAEI 2006, Gijón, Septiembre 2006, pp.1-6.

[11] Fernández H. "Contribución al diseño de células de generación mixta fotovoltaica y eólica para sitios aislados". Tesis de doctorado. Universidad de Zaragoza. Septiembre 2007, pp.83-96.

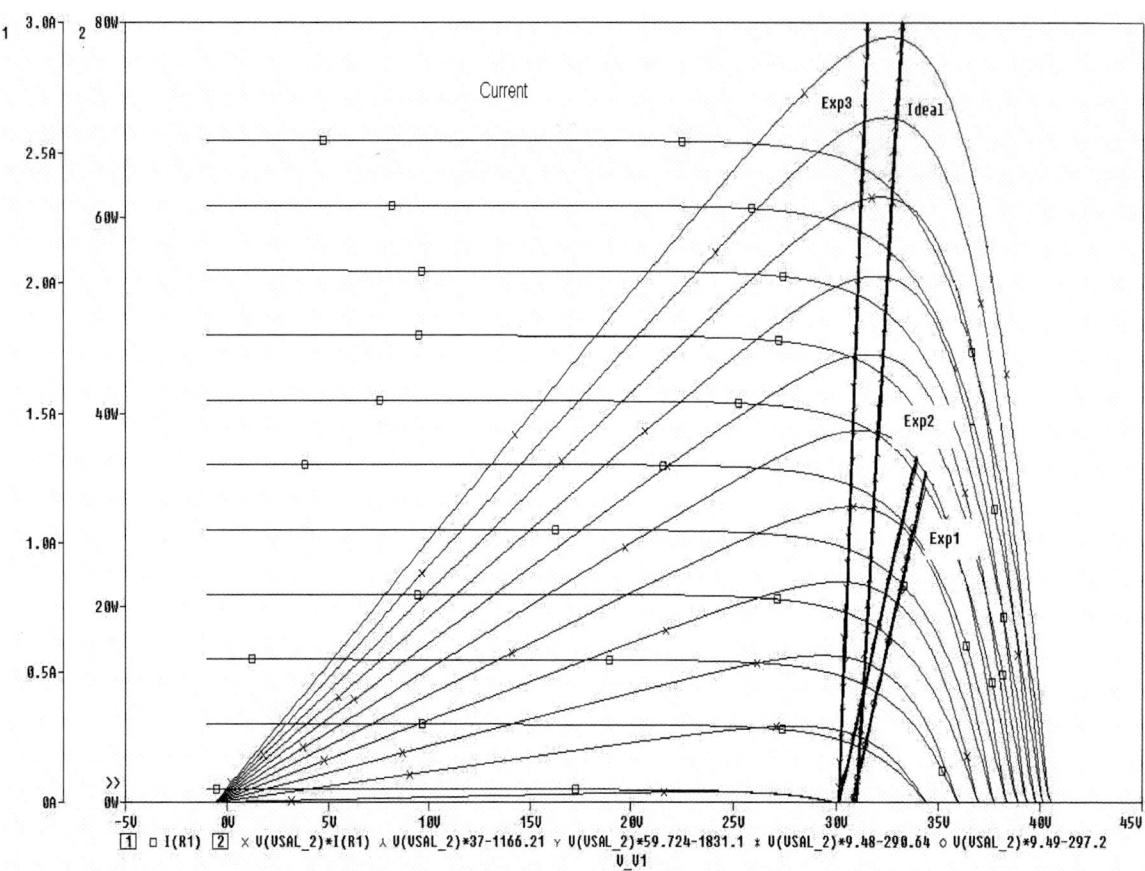

Fig.4. PV panel current and power output showing the ideal MPPT line
and the results obtained with the proposed MPPT configuration under variable load conditions.

# A Single-Phase Active Power Filter Based in a Two Stages Grid-Connected PV System

Kleber C.A. De Souza[*], Denizar C. Martins[*]

[*]Federal University of Santa Catarina, Department of Electric Engineering, Florianópolis, Brazil
email: ksouza@inep.ufsc.br, denizar@inep.ufsc.br

*Abstract* — In this paper a single-phase active power filter based in a two stages grid-connected photovoltaic system with is presented. The proposed system can not only inject PV power into utility but can act always as an active power filter to compensate the load harmonics and reactive power such that the input power factor is unity independently of the solar radiation. In sunny days, the system processes all the reactive and active load power and the excessive power from the PV module can be fed to the utility. On the other hand, on cloudy days for instance, if the PV power is not enough, the system processes all the reactive load power and the shortage of load active power is supplemented by the utility. Besides, just using one current sensor, the control strategy is simpler and of easy practical implementation.

*Keywords*— Single-phase systems, grid-connected PV systems and active power filter.

## I. INTRODUCTION

Photovoltaic (PV) solar energy as an alternative resource has been becoming feasible due to enormous researches and development work being conducted over a wide area [1], [3], [5] and [10].

Some researchers spent efforts in developing PV inverter systems with grid connection and active power filtering features using sensors to measure the load current [2], [6], [8] and [9].

This paper presents a single-phase topology, without load current sensor, composed by a dc-dc converter in cascaded with an inverter, as shown in Fig.1. The system aims transferring the photovoltaic (PV) power to the ac load and paralleled with the utility. The dc-dc converter is used to boost the PV voltage to a level higher than the peak of the voltage utility such that the inverter can provide the ac voltage without requiring the transformer. The dc-dc converter is also responsible for tracking the maximum power point of the PV modules to fully utilize the PV power [7] and [11]. The shortage of load power from the PV module is supplemented by the utility. On the contrary, the excessive power from the PV module to the load can be fed to the utility. The balance of power flow is controlled through the inverter. The inverter is also used to act always as an active power filter to compensate the load harmonics and reactive power such that the input power factor is unity (Fig. 2).

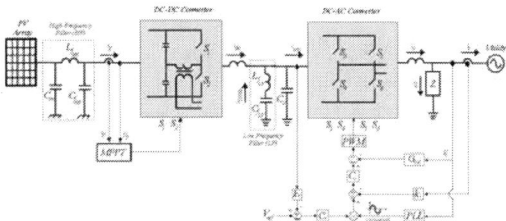

Fig.1. Single-phase two stages Active Power System PV system.

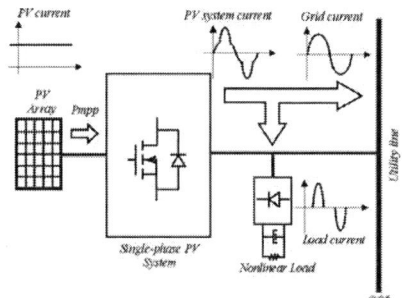

Fig. 2. Power flow of the system with nonlinear load.

## II. SYSTEM CONFIGURATION AND CONTROL

The dc-dc converter power structure used to boost the PV voltage and for tracking the maximum power point of the PV modules to fully utilize the PV power was the Half-Bridge Zero Voltage Switching Pulse Width Modulation (DC-DC HB ZVS-PWM) asymmetrically driven converter [4] shown in Fig. 3.

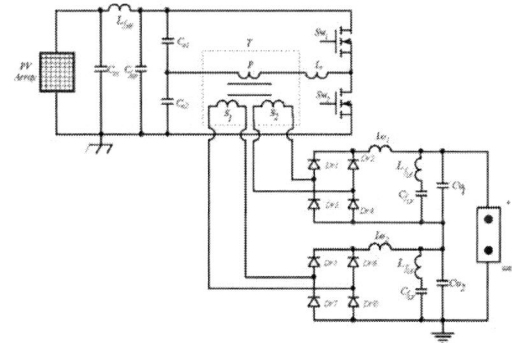

Fig. 3. Half-Bridge Zero Voltage Switching.

The HB ZVS-PWM has been designed considering the reduction of losses and mainly the volume of the magnetic. The right value of the capacitors *Ce1* and *Ce2* can be determined by equations (1) and (2), and the best choice for the value of the resonant inductor $L_r$ and of the transformer ratio (*n*) will be given by the intersection points between the curves (Fig. 4) obtained by the equations (3) (curve b) and (curve a).

$$Ce_1 = \frac{P_o}{f_s \cdot \Delta V_{C_{ieq}} \cdot Vi} \cdot (1-D) \quad (1)$$

$$Ce_2 = \frac{P_o}{f_s \cdot \Delta V_{C_{ieq}} \cdot Vi} \cdot D \quad (2)$$

$$Lr < \left[ \frac{\left( \frac{1}{2} + \frac{1}{2} \cdot \sqrt{1 - \frac{8 \cdot I_{0min} \cdot Lr \cdot f_s}{n \cdot Vi_{máx}} - 2 \cdot \frac{n \cdot V_0}{Vi_{máx}}} \right) Vi_{máx}}{\left( 1 - \sqrt{1 - \frac{8 \cdot I_{0min} \cdot Lr \cdot f_s}{n \cdot Vi_{máx}} - 2 \cdot \frac{n \cdot V_0}{Vi_{máx}}} \right) \cdot I_{0min}} \right]^2 \cdot (C1+C2) \quad (3)$$

$$Lr_{máx\_HB}(n) \leq \left[ 2 \cdot D_{máx} \cdot (1-2 \cdot D_{máx}) - \frac{V_o \cdot n}{Vi_{min}} \right]^2 \cdot \frac{Vi_{min} \cdot n}{4 \cdot fs \cdot I_0} \quad (4)$$

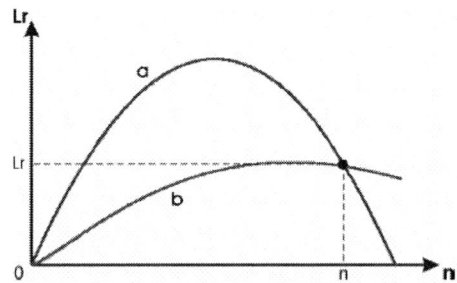

Fig. 4. Optimum adjust for the resonant inductor and the transformer ratio for the Half Bridge converter

Fig. 5 presents the full bridge topology with the inductor *L* connected between the grid ($V_o(t)$) and the inverter, the capacitor $C_i$ in the structure input representing the DC voltage source and a current source ($I_i(t)$), that can be either the output of the DC-DC converter or an array of photovoltaic panels.

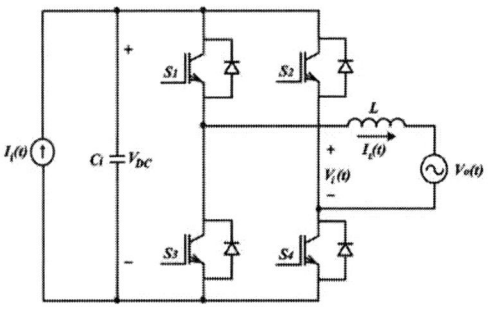

Fig. 5. Full bridge inverter.

Considering the self-commutated inverter switching at high frequency and using three level PWM technique, it is possible to represent the four equivalent circuits of the switching modes (Fig. 6), defined by the combination of

the switches possible states with the two possible direction of the output current.

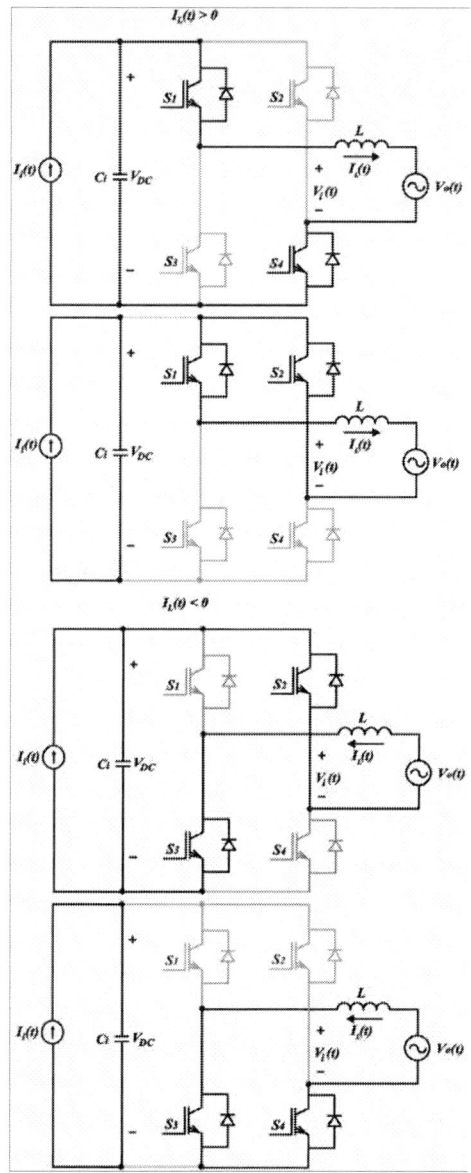

Fig. 6. Four equivalent circuits of the switching modes.

From the operational stages, it can be observed that when $I_L(t)>0$ and the switches $S_1$ e $S_4$ are on, voltage $V_i(t)$ has its polarity defined by the direction of the output current, with its absolute value equal to input voltage $V_{DC}$, whose amplitude should be larger than the peak value of the output voltage, $V_o(t)$. In this manner, the voltage polarity across the inductor causes its current's absolute value to increase. During this stage, energy from the input source, $V_{DC}$, along with part of the energy stored in the inductor, is transferred to the grid. When $I_L(t)>0$ and the switches $S_2$ and $S_3$ are on, voltage $V_i(t)$ is zero. In this manner, the voltage polarity across the inductor is inverted, causing its current's absolute value to decrease. Indeed, the output current is controlled by imposing the derivative of the current through the inductor, or, put differently, by imposing the voltage across the inductor $L$. In this manner, the structure of the converter shown in Fig. 5 can be represented, without loss of generality, as the controlled voltage source $V_i(t)$, presented in Fig. 7 where the link inductors are represented by the inductor $L$, $V_o(t)$ is the utility voltage and $I_L(t)$ is the output PV system current.

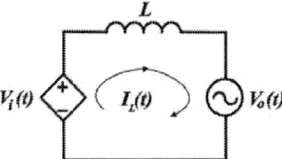

Fig. 7. Simplified equivalent inverter circuit.

In Fig. 7 the energy flow is controlled by the current $I_L(t)$. However, this current is defined by the difference of voltage between the sources $V_i(t)$ and $V_o(t)$, applied across the impedance. In this case, as the impedance is a pure inductance, the current will be equal the integral of the voltage across it. As $V_o(t)$ is known, once it is the utility voltage itself, $V_i(t)$ is imposed and therefore $V_L(t)$, in a convenient form, in a way to obtain the output current desired across the inductor. Thus:

$$V_L(t) = V_i(t) - V_o(t) \qquad (5)$$

PWM defines a modulated signal composed of the reproduction of the modulating signal's spectrum, whose amplitude is defined by the modulation, added to harmonic components of frequencies that are multiples of the switching frequency. Ignoring the effect of the harmonic components of the switching frequency on voltage $V_i(t)$, once the inductor works as a low pass filter for the current, the voltage imposed across the inductor is represented simply by (5). Fig. 8 shows the manner in which the converter allows the voltage to be imposed across the inductor, as shown in the equivalent circuit of Fig. 7.

$$V_i(t) \xrightarrow{+} \sum \xrightarrow{V_L(t)} \boxed{\frac{1}{L}\int V_L(t)} \rightarrow i_L(t)$$

Fig. 8. Block diagram of the simplified equivalent circuit.

Indeed, the output current is desired to be a mirror of $V_o(t)$ as expressed in (6). Nevertheless, according to equation (7), the inductor voltage is the derivative of the current across itself. Therefore, equation (8) describes the voltage $V_i(t)$, which, in effect, is defined by the control loop, should present a sine, in order to annul the effect of $V_o(t)$, and a cosine, which, by composition, will be the resulting voltage imposed across the inductor, therefore, guaranteeing a sinusoidal current. In practice, at the frequency of the grid, the inductor is a very small reactance, causing the voltage drop across the inductor to be smaller than the utility voltage. In other words, the sine of $V_i(t)$ dominates the cosine, demonstrating that the demand on the current loop is much more in favor of annulling the "disturbance" of the utility voltage rather than to effectively control the output current

$$I_L(t) = I \cdot \sqrt{2} \cdot \sin(\omega t) \qquad (6)$$

$$V_L(t) = L \cdot \frac{dI_L(t)}{dt} = L \cdot I \cdot \sqrt{2} \cdot \omega \cdot \cos(\omega t) \qquad (7)$$

$$V_i(t) = L \cdot I_{rms} \cdot \sqrt{2} \cdot \omega \cdot \cos(\omega t) + V_{rms} \cdot \sqrt{2} \cdot \sin(\omega t) \qquad (8)$$

In the classic control strategy, an internal current loop and an external loop to control the input voltage are implemented. The voltage loop defines the amplitude of the reference current by multiplying its control signal by a "waveform", which can be a sample of the output voltage or a digitally generated sinusoid, generating the output current reference. Fig. 9 demonstrates how the classic control strategy is implemented, in which $V_i(t)$ is determined by the current error signal passing through the compensator and the error signal is the difference between a sample of the current and its reference. Fig. 10 shows the same block diagram simplified.

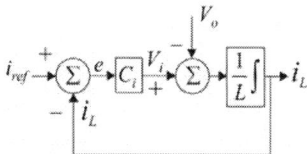

Fig. 9. Block diagram of classical control strategy current loop.

$$i_{ref} \xrightarrow{+} \sum \xrightarrow{e} \boxed{C_i} \xrightarrow{V_i} \sum \xrightarrow{} \boxed{\frac{1}{L}\int} \rightarrow i_L$$

Fig. 10. Simplified block diagram of classical control strategy current loop.

It is observed, however, that the output voltage $V_o(t)$ appears as a disturbance in the simplified traditional model. Therefore, voltage $V_i(t)$, which, in effect, is defined by the control loop should present a sine, in order to annul the effect of $V_o(t)$, and a cosine, which, by composition, will be the resulting voltage imposed across the inductor, therefore, guaranteeing a sinusoidal current. In practice, at the frequency of the mains, the inductor is a very small reactance, causing the voltage drop across the inductor to be much smaller than the voltage of the mains. In other words, the sine of $V_i(t)$ dominates the cosine, demonstrating that the demand on the current loop is much more in favor of annulling the "disturbance" of the output voltage rather than to effectively control the output current.

Rewriting equation (5) as in (9) [8]:

$$L \cdot \frac{di_L(t)}{dt} = k \cdot v_{control}(t) - V_o(t) \qquad (9)$$

$$k = \frac{V_{DC}}{V_{tri}} \qquad (10)$$

Where $V_{tri}$ is the peak of the triangular carrier signal and $v_{control}$ is the control signal witch shapes the sinusoidal

**1953**

current to the utility line. From the block diagram, the current signal error is equal the equation (11).

$$e(t) = i_{Lref}(t) - i_L(t) \qquad (11)$$

Sense a perfect sinusoidal current to the utility line is a designed goal, $e$ must naturally approach zero. Thus, deriving (11) and substituting (9) into (12):

$$\frac{de(t)}{dt} = 0 = \frac{di_{Lref}(t)}{dt} - \frac{di_L(t)}{dt} \qquad (12)$$

$$v_{control}(t) = \frac{L}{k} \cdot \frac{di_{Lref}(t)}{dt} + \frac{1}{k} \cdot V_o(t) \qquad (13)$$

As the disturbance is measurable, the utility voltage disturbance controller $G_{cd}$ is used to reduce de disturbed voltage component. The new block diagram that contains this feed-forward controller is presented in Fig. 11. From Fig. 11, it can be seen that:

$$i_L = \frac{k \cdot C_I}{sL + k \cdot C_I} \cdot I_{Lref} + \frac{k \cdot \left( G_{cd} - \frac{1}{k} \right)}{sL + k \cdot C_I} \cdot V_o \qquad (14)$$

From (14), when $G_{cd} = 1/k$, the disturbance from $V_o$ can be eliminated, and if $kC_i \gg |sL|$, than $I_L = I_{ref}$, identifying accurate current control effect for $I_{ref}$.

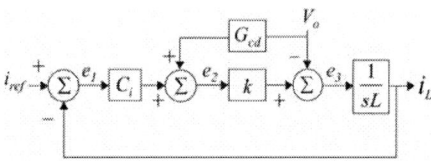

Fig. 11. Block diagram containing the feed-forward controller.

Repeating the same analysis, but now considering the connection of any load between the system and the commercial electric grid, a new configuration, presented in Fig. 12, is obtained. It can be observed that now the inductor current is the load current plus the utility current. Again, as sinusoidal current to the utility line is a designed goal, adding a sample of the load current to the inductor current reference it is possible to control the inductor current and still guarantee a sinusoidal utility current. A new block diagram representing the system is shown in Fig. 12.

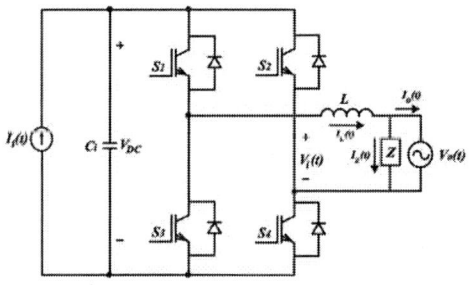

Fig. 12. PV system with any load connected.

However, in this case, besides the current sensor used to sample the current in $L$, it is the necessary another sensor

to sample the current in the load. Another disadvantage in this configuration is that as the control is done monitoring the load current ($I_z$), when the system is operating Active Power Line Conditioner Mode (low sun light), is necessary before to extract the fundamental component of the load current for later to find the reference current. So, it is necessary to observe a period of the grid at least.

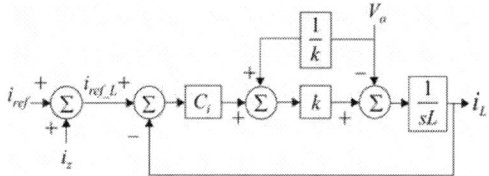

Fig. 13. New block diagram of the inductor current loop.

However, sensing the AC mains current instead of the inductor current, and considering that the difference between the inductor current and the load current is the utility current, the last block diagram can be modified to represent now the utility current loop (Fig. 14). Nevertheless, according to Fig. 9, $i_{o\_ref}(t)$, or, the difference between the reference inductor current ($i_{L\_ref}$) and the load current ($I_z$), is exactly the own $i_{ref}$ defined by the voltage control signal multiplied by a sinusoid in phase with the utility frequency (Fig. 15). Thus, the same system can be controlled observing only the utility current directly ($I_o$), improving the dynamics of the system, once it will be not necessary any previous calculation.

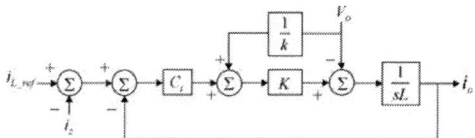

Fig. 14. Block diagram of the utility current loop.

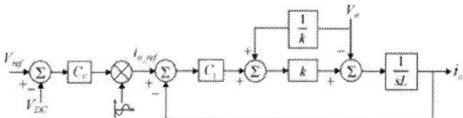

Fig. 15. Utility current control diagram.

Thus, to control the output current in phase with the voltage utility and, to obtain a unity power factor, even with the connection of a load between the grid and the system, it is enough to observe only the AC mains current. In this case, besides doing use of just a sensor one, the control strategy is simpler and of easy practical implementation.

### III. 3 EXPERIMENTAL RESULTS

To demonstrate the feasibility of the discussed PV system, a prototype was designed and implemented. The specifications of the system (Fig. 16 and Fig. 17) are shown below.

Solar array:

- *Number of PV Module: 20;*
- *Rated power: 1002W;*
- *Rated voltage: 83.5V;*

- *Rated current: 12A;*
- *Short-circuit current: 12.4A;*
- *Open-circuit voltage: 107V.*

Dc-dc power converter:

- *Output voltage: 400V;*
- *Switching frequency: 100kHz;*
- *$C_{in}$ and $Cf_{HF}$: 1000μF;*
- *Ce1 and Ce2: 10μF and 5μF;*
- *$Lf_{HF}$ and $L_r$ : 50μH and 680nH;*
- *$C_o$: 1000μF.*

Inverter:

- *Switching frequency: 20kHz;*
- *$Cf_{LF}$: 1000μF;*
- *$Lf_{LF}$: 1.6mH;*
- *$Lo_1$ and $Lo_2$: 1.2mH;*
- *Output voltage: 220V, 60Hz;*
- *Load: 400VA (Capacitivy).*

In the proposed system, a *LC* filter ($Lf_{HF}$ and $Cf_{HF}$) is connected with the PV array output to filter the high frequencies drained by the dc-dc converter, and a series *LC* filter ($Lf_{LF}$ and $Cf_{LF}$) is connected between the converters to support the second-order component (120Hz) presented in the input inverter current.

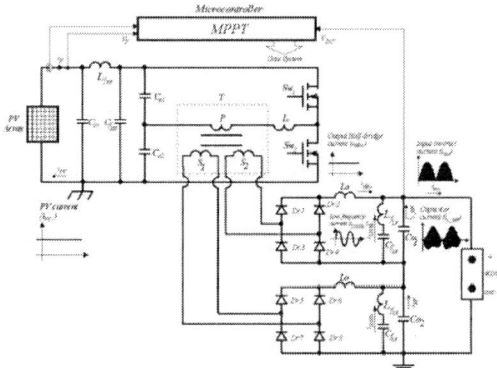

Fig. 16. Half-Bridge Zero Voltage Switching block diagram.

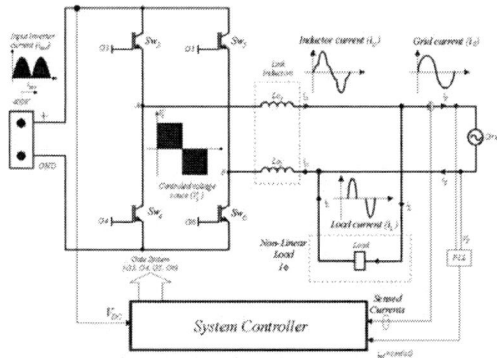

Fig. 17. Simplified block diagram of the Inverter.

Fig. 18 and Fig. 19 depicts, respectively, the load current ($I_L$), the inverter current ($I_o$), the voltage current ($V_{Utility}$) and utility current ($I_S$) with the system operating just as active power line conditioning mode (cloudy day or night). The THD of $I_S$ is 4.0% for a load crest factor of 3.09. Fig. 20 depicts the output inverter current ($I_o$) and the utility current ($I_S$) with the system supplying power to the load and supplying surplus power to the utility line. Fig. 21 shows the utility voltage and the utility current with the system only supplying power to the utility (THD = 2.5%).

## IV. CONCLUSION

This paper presents a single-phase topology for transferring the photovoltaic (PV) power to the ac load and paralleled with the utility. The proposed PV system has the advantage of act always as an active power filter to compensate the load harmonics and reactive power such that the input power factor is unity.

The simplicity in the strategy of the output current control, in other words, current injected in the electric system, is another advantage of the study, because besides doing use of just one current sensor, it is of easy practical implementation

### ACKNOWLEDGMENT

The authors would like to thank the Brazilian agencies CNPq for financial support.

### REFERENCES

[1] S. S. Bahu, S. Palanichamy (1996). "PC based controller for utility interconnected photovoltaic power conversion system"; *Proc. IEEE Power Electronics Specialists Conf.*, vol. 1, Jan., pp. 101–106.

[2] L. Cheng, R. Cheung, K. H. Leung (1997). "Advanced photovoltaic inverter with additional active power line conditioning capability"; Proc. IEEE Power Electronics Specialists Conf.; vol. 1; pp. 279–283.

[3] S. J. Chiang, K. T. Chang, C. Y. Yen (1998). "Residential photovoltaic energy storage system", *IEEE Trans. Ind. Electron.*, vol. 45, no. 3, pp. 385–394.

[4] K. C. A. de Souza, O. H. Gonçalves, D. C. Martins (2006). "Study and optimization of two dc-dc power structures used in a grid-connected photovoltaic system", *Power Electronics Specialists Conference, PESC06*, pp. 1 – 5.

[5] K. Hirachi, T. Mii, T. Nakashiba, K. G. D. Laknath and M. Nakaoka (1996). "Utility-Interactive multi-functional Bi-directional converter for solar photovoltaic power conditioner with energy storage batteries", *Proc. IECON'96 Conf.*, pp. 1693–1698.

[6] S. Kim, G. Yoo, J. Song (1996). "A Bi-functional utility connected photovoltaic system with power factor correction and facility", *Proc. Photovoltaic Specialists Conf.*, pp. 1363–1368.

[7] E. Koutroulis, K. Kalaitzakis, N. C. Voulgaris (2001). "Development of a Microcontroller-Based Photovoltaic Maximum Power Point Tracking Control System", IEEE Transaction On Power Electronics, v. 16, n. 1, pp. 46-54.

[8] Y. C. Kuo, T. J. Liang, J. F. Chen (2001). "Novel maximum-power-point tracking controller for photovoltaic energy conversion system", *IEEE Trans. Ind. Electron.*, vol. 48, n° 3, pp. 594–601.

[9] T. Wu, C. Shen, H. Nein, G. Li (2005). "A 1φ/3W inverter with grid connection and active power filtering based on nonlinear programming and fast-zero-phase detection algorithm", *Power Electronics, IEEE Transactions on*, Volume 20, Issue 1, Jan. 2005 pp. 218 – 226.

[10] M. Yamaguchi, K. Kawarabayashi and T. Takuma (1994). "Development of a new utility-connected photovoltaic inverter line back", *Proc. INTELEC'94 Conf.*, pp. 676–682.

[11] L. Zhang, A. Al-Amoudi, Y. Bai, (2000). "Real-Time Maximum Power Point Tracking for Grid-Connected Photovoltaic Generators", *IEE Power Electronics and Variable Speed Drives Conference – PESC2000*, p. 124-129.

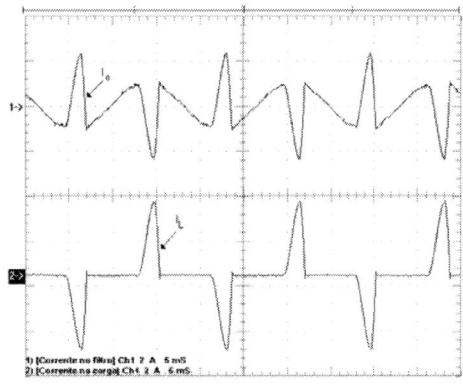

Fig. 18. Waveforms of Load current ($I_L$) and Inverter current ($I_o$) (Ch1 2A/div and Ch2 2A/div).

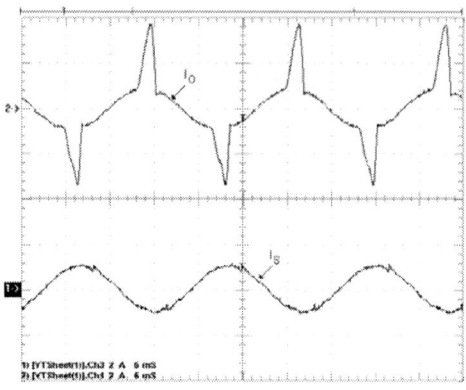

Fig. 20. Waveforms of utility current ($I_S$) and Inverter current ($I_o$) (Ch1 2A/div and Ch2 2A/div).

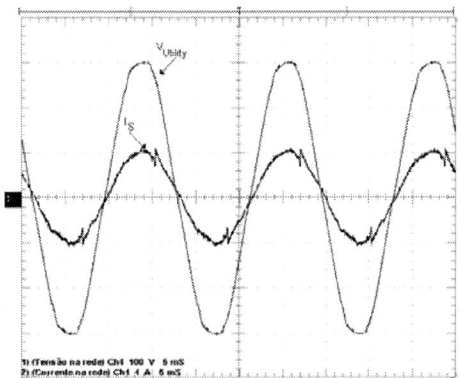

Fig. 19. Utility current ($I_S$) and utility voltage ($V_{Utility}$) with load (Ch1 1A/div and Ch2 100V/div).

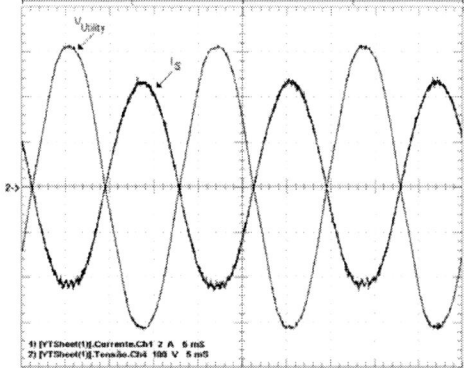

Fig. 21. Waveforms of grid current ($I_S$) and utility voltage ($V_{Utility}$) without load. (Ch1 2A/div and Ch2 100V/div).

# Wide Bandwidth Power Flow Control Algorithm of the Grid Connected VSI under Unbalanced Grid Voltages

Zoran Ivanovic, Marko Vekic, Stevan Grabic, Evgenije Adzic, Vladimir Katic

University of Novi Sad, Faculty of Technical Sciences/Dept. for Power Electronics, Novi Sad, Serbia,
*zorani@uns.ns.ac.yu*

*Abstract*—The main contribution of this paper is the new wide bandwidth power flow control algorithm of the grid connected VSI under unbalanced grid voltage conditions. The algorithm is based on simple control structure and does not demand any filter for obtaining symmetrical components, e.g. notch or anti-resonant filter. Proposed control structure applies dual vector current control, based on regulation of positive and negative sequence components. The main advantages of the suggested control method are wide controller bandwidth, as well as possibility to obtain controller parameters by using standard tuning techniques. Comparing it with the previous solutions, this control strategy eliminates oscillating components in both, active and reactive power. However, as a consequence non-sinusoidal grid current waveforms will appear. The effectiveness of proposed control algorithm is verified by detailed switching simulation model.

*Keywords*—Converter control, fault handling strategy, power quality, voltage source inverter (VSI), wind energy.

## I. INTRODUCTION

The three phase voltage source inverter, shown in Fig.1, is the basic component of most power electronic devices and custom power equipment due to its high controllability and power quality [1]. It is also often used as interface in renewable energy systems, especially in wind turbine applications.

One very important issue deals with operation of such a converter under unbalanced grid voltage conditions, which is quite common case, particularly in week ac system. As a consequence of unbalanced conditions oscillating components at double grid frequency in active and reactive power, as well as in dc link voltage appear [2], [3]. Song and Nam [2] proposed dual vector current controller (DVCC) to achieve robust operation of voltage source inverter under such conditions. It is based on separate regulation of positive and negative-sequence components, allowing the transfer of the active power to the grid at grid frequency, while suppressing the oscillations at double grid frequency and maintaining the desirable average power factor. However, during the severe voltage sags, there is no control over the line currents magnitude, so they could reach values several times higher than nominal [4]. In order to overcome this undesirable effect, modified dual vector current controller was proposed in [4], [5]. The drawback of all these solutions is the lack of ability to eliminate oscillations at double grid frequency in reactive power.

In addition, they require anti-resonant or notch filter in control loops which is proved to be problematic [1]-[5]. Anti-resonant filter introduces non-rational transfer function as it leads to phase delay of input signal [9]. Notch filter for this application usually has the second order transfer function and can cause large phase jump at set frequency. Therefore, application of this filters entail poor dynamic response, and implicate difficult procedure of setting controller gains.

This paper proposes a new control scheme based on instantaneous active and reactive power control, which does not include any filter in control loops. The main advantages of suggested control algorithm are wide controller bandwidth and possibility to obtain controller parameters by using standard tuning techniques. Moreover, using this technique it is possible to achieve decoupled control of active and reactive power with no oscillation in each of them. Active and reactive power is flattened due to new simple current references calculation method and application of the improved control structure.

The drawback of this solution is impossibility to achieve sinusoidal grid currents waveforms during unbalanced voltage sags. Simulation results confirm the proposed control method under severe unbalanced operating conditions.

## II. MATHEMATICAL DESCRIPTION OF THE VSI UNDER UNBALANCED GRID VOLTAGES

An unbalanced system of the three phase voltages, which is general case of grid voltage sags, could be represented with its positive $(u_{dq}^p = u_d^p + j \cdot u_q^p)$ and negative $(u_{dq}^n = u_d^n + j \cdot u_q^n)$ sequence components, as it is given with:

$$u_{\alpha\beta} = e^{j\omega t} \cdot u_{dq}^p + e^{-j\omega t} \cdot u_{dq}^n \qquad (1)$$

where ω is the grid frequency. In such a circuit, unbalanced currents also appear and they could be represented in terms of positive and negative sequence components, in the same way as given in (1):

$$i_{\alpha\beta} = e^{j\omega t} \cdot i_{dq}^p + e^{-j\omega t} \cdot i_{dq}^n . \qquad (2)$$

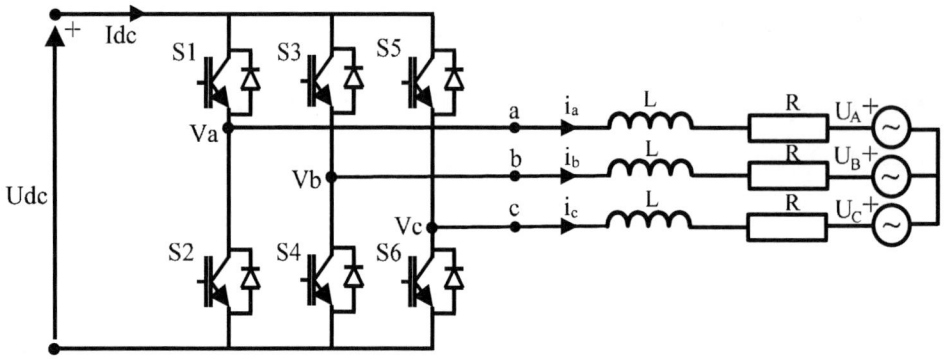

Fig. 1. Structure of VSI PWM converter under unbalanced voltage condition

One case of unbalanced grid voltages in original and dq domain is shown in Fig. 2. It should be noticed that in the positive sequence reference frame positive component appear as a dc, whereas the negative component oscillates at double grid frequency. In negative reference frame it is vice versa what is explained thoroughly in [2].

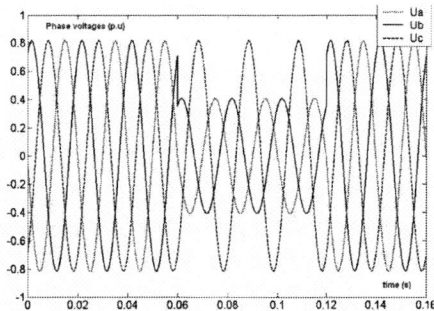

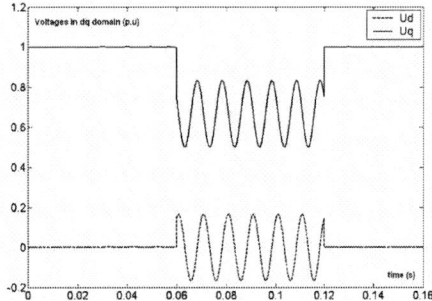

Fig. 2. Phase voltages in original and dq domain

If we want to leave dc reference signals only, it is necessary to use filter in order to extract sequence components. Otherwise, dual current regulator with oscillating reference signals has to be used. The VSI under generalized unbalanced voltages in Fig.1 can be described by the differential equation (3) in the stationary reference frame

$$v_{\alpha\beta} = u_{\alpha\beta} + L \cdot \frac{di_{\alpha\beta}}{dt} + R \cdot i_{\alpha\beta} \qquad (3)$$

where

$$v_{\alpha\beta} = 2/3 \cdot \left( v_a + v_b e^{j2\pi/3} + v_c e^{-j2\pi/3} \right) \qquad (4)$$

$$i_{\alpha\beta} = 2/3 \cdot \left( i_a + i_b e^{j2\pi/3} + i_c e^{-j2\pi/3} \right) \qquad (5)$$

denote converter pole voltages and the line currents, respectively.

Equation (3) can now be transformed and decomposed into two complex equations in positive and negative synchronous rotating reference frames, respectively, as shown in (6) and (7) [2]

$$v_{dq}^p = L \frac{di_{dq}^p}{dt} + Ri_{dq}^p + j\omega Li_{dq}^p + u_{dq}^p \qquad (6)$$

$$v_{dq}^n = L \frac{di_{dq}^n}{dt} + Ri_{dq}^n + j\omega Li_{dq}^n + u_{dq}^n \qquad (7)$$

where $i_{dq} = i_d + ji_q$ and $v_{dq} = v_d + jv_q$ .

Regarding to this, instantaneous apparent power s could be expressed as

$$s = u_{\alpha\beta} \cdot i_{\alpha\beta}^* = p(t) + jq(t) \qquad (8)$$

where active power $P(t)$ and reactive power $Q(t)$ are

$$p(t) = P_0 + P_{C2} \cdot \cos(2\omega t) + P_{S2} \cdot \sin(2\omega t) \qquad (9)$$

$$q(t) = Q_0 + Q_{C2} \cdot \cos(2\omega t) + Q_{S2} \cdot \sin(2\omega t) . \qquad (10)$$

Terms $P_0$ and $Q_0$ designate value of the average power, while $P_{C2}$, $P_{S2}$, $Q_{C2}$ and $Q_{S2}$ are magnitude in power oscillations caused by the unbalance. Detailed expressions for active and reactive power are given in [5].

### III. DVCC WITH CURRENT LIMITATION

Song and Nam recommended dual vector current controller (DVCC) to achieve robust operation of voltage source inverter under unbalanced voltage conditions [2]. Its core is the regulation of positive and negative-sequence components, allowing the transfer of the active power to the grid at grid frequency, while suppressing the

1958

oscillations at double grid frequency and maintaining the desirable average power factor.

Conventional dual vector current controller cannot be applied under the extreme voltage condition [5], [6]. For the severe voltage sag, grid currents could reach unacceptably high values, even several times higher than the nominal one. From the point of reliability and protection issues of the drive, this situation should not be permitted. Therefore, the modified DVCC, with imposed current limitation was recommended [5]. Control structure of that improved system is shown in Fig. 3

Fig. 3. VSI control using modified DVCC

The three-phase grid voltages and currents are measured and transformed into $\alpha\beta$ and dq domain.. As in conventional DVCC, it is necessary to regulate positive and negative sequence components, which are obtained by applying transformation of rotation in both directions. Due to the fact that conventional dual vector current controller operates with dc signals only, anti-resonant filter is involved in order to extract sequence components [5]. Likewise, the filtrated dq components of voltages obtained are needed in current references calculating block.

The current controller (DVCC) used here consists of pair of PI controllers that control the positive- and negative-sequence current separately and are implemented in two different rotating reference frames. Details about controllers and extraction of sequence components can be found in [2] and [5].

In order to generate proper current references we should consider the following equation

$$\begin{bmatrix} I_{GRID}^2 \\ Q_0 \\ P_{C2} \\ P_{S2} \end{bmatrix} = \begin{bmatrix} I_{LIM}^2 \\ 0 \\ 0 \\ 0 \end{bmatrix} = \begin{bmatrix} i_d^{p*} & i_q^{p*} & i_d^{n*} & i_q^{n*} \\ u_q^{pf} & -u_d^{pf} & u_q^{nf} & -u_d^{nf} \\ u_q^{nf} & -u_d^{nf} & -u_q^{pf} & u_d^{pf} \\ u_d^{nf} & u_q^{nf} & u_d^{pf} & u_q^{pf} \end{bmatrix} \cdot \begin{bmatrix} i_d^{p*} \\ i_q^{p*} \\ i_d^{n*} \\ i_d^{n*} \end{bmatrix}$$
(11)

First condition in conventional DVCC, concerning desirable power flow $P_0 = P_0^*$ [2], is now substituted with current limiting condition $I_{GRID}^2 = I_{LIM}^2$, while the actual active power is determined by the grid voltage and set current limit. Using (11), current references are obtained adequately as

$$i_d^{p*} = I_{LIM} \cdot u_d^{pf} / D$$
(12)

$$i_q^{p*} = I_{LIM} \cdot u_q^{pf} / D$$
(13)

$$i_d^{n*} = -I_{LIM} \cdot u_d^{nf} / D$$
(14)

$$i_d^{n*} = -I_{LIM} \cdot u_q^{nf} / D$$
(15)

where $D = \sqrt{(u_d^{pf})^2 + (u_q^{pf})^2 + (u_d^{nf})^2 + (u_q^{nf})^2}$ .

Power which is delivered to the grid is now smaller due to decreased grid voltages and limited line currents.

Using this control structure it is possible to eliminate dc voltage oscillations, as well as oscillations in active power. Unfortunately, with current choice as in (11) it is not possible to eliminate oscillations in reactive power. On the top of everything, this implies that an alternating reactive power exists (oscillating component at double grid frequency), although the average reactive power is equal to zero.

## IV. WIDE BANDWIDTH CONTROL ALGORITHM

Modified DVCC obviously has some drawbacks, listed in previous chapter. Apart from that, design of DVCC with imposed current limitation has proved to be problematic due to anti-resonant, or notch filter used in control loops [5]-[8]. Suh and Lipo [6] established control method based on regulation of positive and negative sequence components, which avoids using filter block for extracting sequence components. Instead of that, dual vector current controller with oscillating reference signals in a hybrid synchronous reference frame has been used. They involved a resonant-gain path to reduce the error caused by oscillating reference signal. Bearing this in mind, it is not possible to eliminate oscillations in reactive power, which could lead not to meet grid requirements, in some cases.

This paper proposes new wide bandwidth power flow control algorithm of the grid connected VSI inverter under unbalanced grid voltage conditions. Current references are set in a way to eliminate active and reactive power oscillating components. Proposed control structure is shown in Fig. 3.

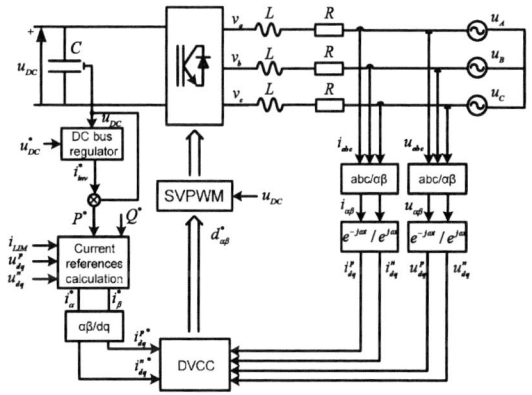

Fig. 4. Wide bandwidth power flow control algorithm

As in previous control structure (Fig.3), the three phase grid voltages and currents are measured and transformed into αβ and dq domain. Here, there is no request for anti-resonant filter, as newly proposed controller operates with sum of dc and ac component. DC bus voltage controller provides proper current reference, applied in current reference calculating block.

In order to achieve proper active and reactive power flow control it is necessary to calculate appropriate current references. Expressions for active and reactive power could be given as:

$$p = u_\alpha \cdot i_\alpha + u_\beta \cdot i_\beta \qquad (16)$$

$$q = u_\beta \cdot i_\alpha - u_\alpha \cdot i_\beta . \qquad (17)$$

If we supply references for active and reactive power together with grid voltages in αβ domain, we can obtain grid current references combining previous two equations as

$$i_\alpha^* = \frac{p^* \cdot u_\alpha + q^* \cdot u_\beta}{u_\alpha^2 + u_\beta^2} \qquad (18)$$

$$i_\beta^* = \frac{p^* \cdot u_\beta - q^* \cdot u_\alpha}{u_\alpha^2 + u_\beta^2} . \qquad (19)$$

If current references are set in this way, it is obviously that oscillations in active and reactive power, caused by unbalanced voltage conditions, can be eliminated in contrast to conventional and modified DVCC [5], [6]. Likewise, references are transformed into dq reference frame and supplied to dual vector current controller, which operate with sum of dc and ac signals.

When voltage disturbance is sensed, current limitation can be achieved by proper setting of active power reference

$$p^* = u_{dq}^p \cdot I_{LIM} . \qquad (20)$$

i.e. using positive sequence voltage component $u_{dq}^p$ and desired current limit $I_{LIM}$. Proposed DVCC control structure is given in Fig. 5.

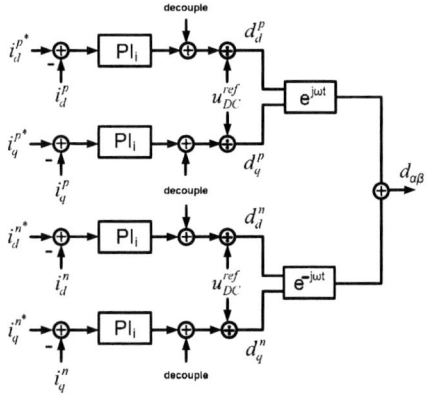

Fig. 5. Wide bandwidth dual vector current controller

It consists of two decoupled controllers, one for positive sequence component (two PI regulators for d and q axis respectively), and one for negative sequence component. According to (1) and Fig. 3 there are dc and ac components at the input of each controller. Fig. 6 shows that dc component from input of PI controller $i_d^p$ (dashed line signals) appears as ac component at the input of lower PI controller (solid line signals). In steady state each of PI controllers will bring corresponding current dc component to its reference value (for example $i_d^p$ to $i_d^{p*}$). As this component appears as ac in other reference frame it will be automatically tuned to desired value.

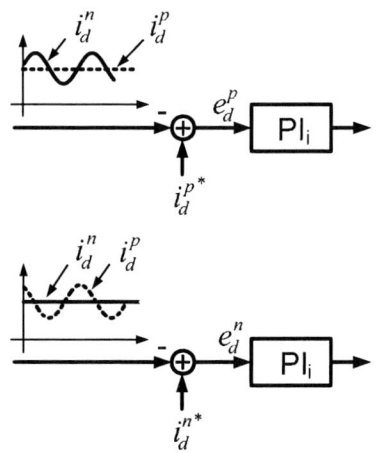

Fig. 6. Operation principle of wide bandwidth DVCC

The main advantage of the proposed solution is avoiding the introduction of filter in obtaining symmetrical components needed in closed control loops. Such approach alleviates procedure for setting controller gains and increases margin of system stability.

Current references do not have to comprise sine or cosine oscillating signal, which is in accordance with (18) and (19). Therefore, in original domain grid current waveforms do not have to be sinusoidal.

## V. CONTROLERS SETTING

### A. Dual Current Controllers

In previous work, design of dual vector current controller has proved to be quite challenging due to anti-resonant filter used in feedback control loops. The consequence of using such a conventional filter is time delay thus making bad transient response. Apart from this, it does not include transfer function which could lead to phase lag of a control signal. As explained in previous chapter, we use classical PI regulator. System dynamic is represented by equations (6) and (7). Applying them, controller gains could be obtained easily. For example, controller for $i_d^p$ can be constructed as shown in Fig. 7.

Term $\omega \cdot L \cdot i_d^p$ is inserted to decouple d-q axes dynamic.

Dual current controller with no filter in feedback control loop provides high bandwidth and improves system transient response.

1960

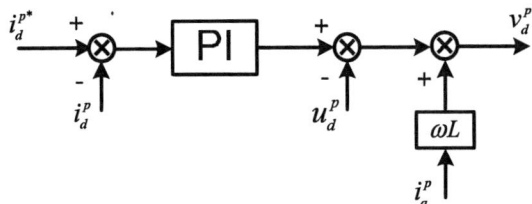

Fig. 7. Current controller setup

Proportional and integral gains could be obtained based on symmetrical criterion as: $K_{pi} = L / 2 \cdot Z_B \cdot T_s$ and $K_{ii} = L / 4 \cdot Z_B \cdot T_s^2$ respectively. For the rest three PI controller parameters are the same.

### B. DC-link Voltage Controller

Control structure of dc voltage regulator is described in Fig. 7.

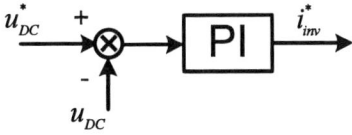

Fig. 8. DC voltage controller

Current loop is now approximated with first order delay block and it is supposed that does not affect dc voltage control loop. Parameters gains are set also based on symmetrical criterion and they are calculated as: $K_{pDC} = C \cdot Z_B / 2 \cdot T_u$ and $K_{iDC} = C \cdot Z_B / 4 \cdot T_u^2$. System parameters are given in the following chapter.

## VI. SIMULATION RESULTS

In order to verify control principle proposed in this paper, detailed model of the system in Matlab/Simulink has been developed, employing space vector modulated inverter. The system data are shown in Tab. 1. Simulations have been carried out for boat control structure, modified DVCC and wide bandwidth power flow controller. Obtained results are compared in details.

TABLE I.
SYSTEM DATA

| Constant | Symbol | Value | Value in pu |
|---|---|---|---|
| Nominal AC voltage | U | 6 KV | 1.0 |
| Nominal grid frequency | $f_n$ | 50 Hz | |
| Nominal dc voltage | $U_{DC}$ | 10.8 kV | 1.8 (dc) |
| DC link capacitance | C | 2 mF | |
| Grid resistance | R | 0.01 Ω | 0.0064 |
| Grid reactance | L | 4 mH | 0.08 |
| Base impedance | $Z_B$ | 15.8Ω | 1 |
| Switching frequency | fs | 4 kHz | |
| Switching period | Ts | 250 µs | |

Simulation results shown in Fig. 9 and 10 illustrate performance of grid connected VSI with proposed modified DVCC under unbalanced voltage conditions. It is supposed that voltage dip starts at about 40 ms and has duration of 100 ms. It is a type D (two face drop) [3] with a magnitude of 50 %.

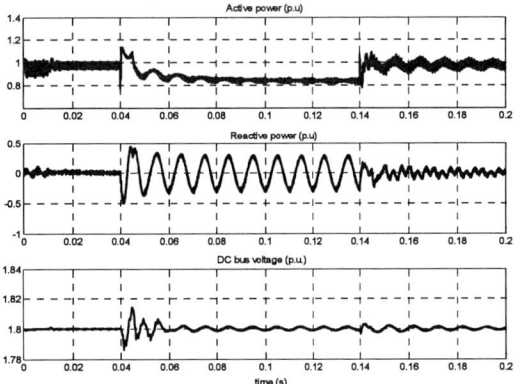

Fig. 9. Active, reactive power and DC bus response – modified DVCC

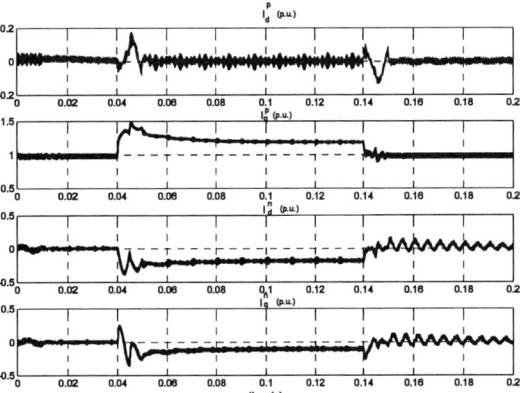

Fig. 10. Grid current components responses – modified DVCC

It could be noticed that modified DVCC can eliminate active power oscillations during unbalanced grid voltage conditions. However, it is not possible to eliminate oscillations at double grid frequency in reactive power, (Fig. 9) which is in accordance with explained control method. During the disturbance, modified DVCC in an effective way limits the magnitude of the grid current to $I_{LIM} = 1.2 \ (p.u.)$. Lower grid voltage and limited grid current imply decrease in active power flow. Stable DC voltage response indicates that the power transfer is correctly managed. Fig. 10 illustrates grid current component responses. It can be seen that they are dc values, which is due to anti-resonant filter employed in feedback control loops. Still, disadvantage of this solution is bad transient response, caused by anti-resonant filter time delay

Simulation results given in Fig. 11 and 12 show wide bandwidth dual vector current controller response to unbalanced voltage sag, when one phase voltage fall to 40 % of nominal value.

It should be noticed that there are no oscillations in active, as well as reactive power during the sag, what is in accordance with predefined aims. Proposed control system also in an effective way limit grid current and keeps DC bus voltage level stable. Active power is lower due to current limitation. Fig. 12 illustrates current components response. As a consequence of this current, waveforms can not be sinusoidal. During the voltage sag it is impossible to achieve both, reactive power equal to zero and sinusoidal current waveforms, what is depicted in figure above. However, this controller has much higher bandwidth which contributes to overall stability of the system.

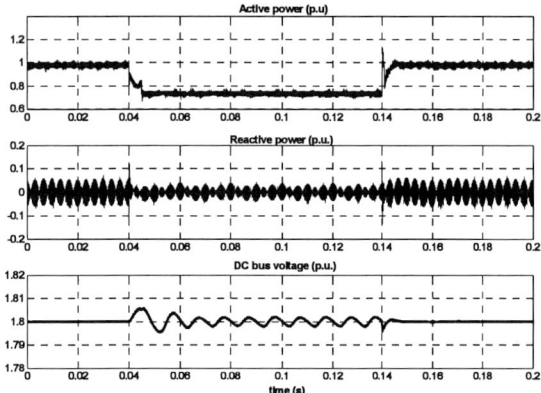

Fig. 11. Active, reactive power and DC bus response – wide bandwidth power flow control algorithm

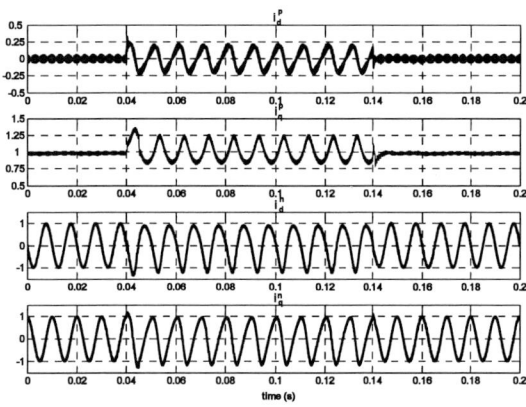

Fig. 12. Grid current component responses – wide bandwidth power flow control algorithm

## VII. CONCUSIONS

This paper proposed new wide bandwidth power flow control algorithm of the grid connected VSI under unbalanced grid voltage conditions.

Using this control method it is possible to eliminate oscillations in active and reactive power at double grid frequency, comparing it with the modified DVCC control method. This is obtained using new instantaneous current reference calculation. Moreover, the advantage of the proposed solution is avoiding the introduction of filter in obtaining symmetrical components needed in closed control loops. Such approach alleviates procedure for setting controller gains and increases margin of system stability. The drawback of this method is impossibility to achieve sinusoidal grid currents waveforms during the sag.

## REFERENCES

[1] F. Magueed, A. Sannino, J. Svensson, "Transient performance of voltage source converter under unbalanced voltage dips", *35th Annual IEEE Power Electronics Specialists Conference, Aachen, Germany*, June 2004, pp. 1163–1168.

[2] H.Song, K.Nam, "Dual current control scheme for PWM converter under unbalanced voltage conditions", *IEEE Trans. on Industrial Electronics*, vol 46, pp.953–959, October 1999.

[3] G.Saccomando, J.Svensson, A.Sannino, "Improving voltage disturbance rejection for variable-speed wind turbines", *IEEE Trans. on Energy Conversion*, vol.17, pp. 422–428, September 2002.

[4] A. Mullane, G. Lightbody, R. Yacamini, "Wind-turbine fault ride-through enhancement", *IEEE Trans. on Power Systems*, vol. 20, no. 5, pp. 1929–1937, Nov. 2005.

[5] Z. Ivanovic, M. Vekic, S. Grabic, V.Katic "Control of multilevel converter driving variable speed wind turbine in case of grid disturbances", *12th International Power Electronics and Motion Control Conference - EPE-PEMC, Portoroz, Slovenia*, Aug./Sep.2006, pp. 1569–1573.

[6] Y. Suh, T. Lipo, "Control scheme in hybrid synchronous stationary frame for PWM AC/DC converter under generalized unbalanced operating conditions", *IEEE Trans. on Industry Applications*, vol. 42, no. 3, pp. 825–835, May/June 2006.

[7] Y. Suh, V. Tijeras, T. Lipo, "A control method in dq synchronous frame PWM boost rectifier under generalized unbalanced operating conditions", *IEEE PESC Conference, Queensland, Australia*, June 23–27, 2002.

[8] Y. Suh, T. Lipo, "A control scheme of improved transient response for PWM AC/DC converter under generalized unbalanced operating conditions", *35th Annual IEEE Power Electronics Specialists Conference, Aachen, Germany*, June 2004, pp. 189–195.

[9] S. Vukosavić, M. Stojić, "Suppression of torsional oscillations in a high-performance speed servo drive", *IEEE Trans. on Industrial Electronics*, vol. 45, no. 1, Feb. 1998, pp. 108–117.

[10] F. Magueed, A. Sannino, J. Svensson, "Design of robust converter interface for wind power applications", *Nordic Wind Power Conference, Chalmers University of Technology*, March 2004. pp. 1–4.

# The use of Switched Reluctance Generator in wind energy applications

Eleonora Darie*, Costin Cepişcă[†] and Emanuel Darie[**]

*Technical University of Civil Engineering/ Electrotechnical Department, Bucharest, Romania, e-mail:
*eleonora_darie@yahoo.com*
[†] University Politechnica of Bucharest/ Electrotechnical Department, Bucharest, Romania, e-mail: *costin@wing.ro*
[**]Police Academy Bucharest/ Engineering Department, Bucharest, Romania, e-mail: *e_darie@yahoo.com*

*Abstract*—Using of the wind energy has become increasingly important as a renewable energy source and therefore is an increasing interest in exploiting it using a Switched Reluctance Machine as a generator and optimize its characteristics in this domain. This work analyzes the generator mode of the Switched Reluctance Machine in the direct coupling to the turbine shaft and coupled to the shaft through a gearbox.

*Keywords*— Control of Drive, Direct torque and flux control, Electrical Drive, Power Converter, Switched reluctance drive, Wind Energy.

## I. INTRODUCTION

In the last decades the Switched Reluctance Machine (SRM) has become an important alternative in various applications in the industrial and domestic markets, namely as a motor showing good mechanical reliability, high torque-volume ratio and high efficiency, plus low cost. Although less evangelized as a generator, there are a few studies of its application in the aeronautical industry and in integrated applications in wind based energy generators.

The development of power electronics and specially the advancements in the field of semiconductors brought improvements in the command and control technology of this machine, thus spearheading a diversified application of Switched Reluctance Machine.

The principles of operation of this machine are simple, well known and based on reluctance torque. The machine has a stator of wound-up salient poles that after energizing synchronized with the position of the rotor develops a torque that tends to align the poles in a way that diminishes the reluctance in the magnetic circuit [2].

Currently the synchronous and induction machines dominate the market of wind energy applications, although, the SRM has been the subject of current investigation and it shows to be a valid alternative for this field [2], [3] and [4].

Comparing with the classical solutions of machines integrated in wind applications, the Switched Reluctance Generator (SRG) indicates the simplified construction associated with the inexistence of permanent magnets or conductors in the rotor, which results in lower manufacturing costs; in addition both the machine and the power converter are robust. The low inertia of the rotor allows the machine to respond to rapid variations in the load.

Associated with these characteristics, these machines have a control system that allows rapid changes in the control strategy such that the performance of the machine is optimized.

The structure of the SRM is not as stiff as the synchronous machines and due to its flexible control system; it is capable of absorbing transient conditions, thus supplying more resilience to the mechanical system [7]. The machine has an inherent fault tolerance, especially when under an open-coil fault (in the windings) and in the power converter (external faults) [4]. Under normal operation, each phase of SRG is electrically and magnetically independent from others.

The SRM is generally felt to be louder than conventional machines. However an adequate mechanical design can do a lot to improve these figures and new control techniques (current control strategy with a torque reference) permits further improvements.

## II. MODE OF OPERATION BY SRG

### A. *SRG – Characteristics*

In electrical drives with variable reluctance (Figure 1), the torque is function of the regular position of the rotor due the double salient poles. The operation of the machine as a generator is obtained by energizing the windings of the stator when the salient poles of the rotor are away from their aligned position due to the rotating motion of the prime mover.

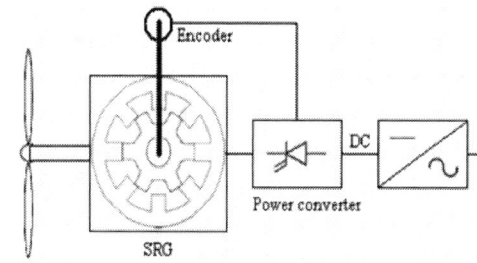

Fig. 1. The Switched reluctance generator in the wind turbine.

The SRM is characterized by the mode of controlling its phase current. For this problem the power electronic converter is used, which functions in a way that the phase currents of the machine are imposed for certain positions of the rotor. In this work is used the standard topology of the converter usually applied in SRM drives, given that it provides a greater flexibility regarding its control and better fault tolerance. The control system of this converter must regulate the magnitude and even the wave shapes of the phase currents to fulfill the requirements of torque and output power available and to ensure safe operation of the generator. This implies that the electronic switches associated with the controller are fully controlled devices.

The topology (Figure 2) used power transistors (IGBT or MOSFET) that work as electronic switches. The capacitor shown in this topology prevents fluctuations in the voltage $V_s$.

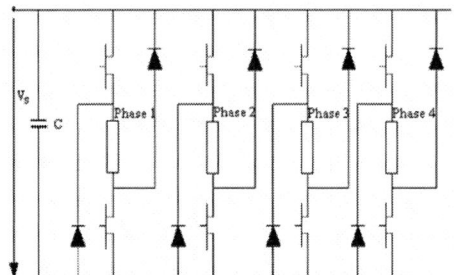

Fig. 2. Circuit diagram of the four phase converter for SRG.

If losses are neglected the output energy over each stroke exceeds the excitation by the mechanical energy supplied [6]. On considers that there is no magnetic saturation and each phase is magnetically independent from others.

In these terms, the expression of the instantaneous power, p, available in the SRG is expressed as follow:

$$p\left(\theta, i_1, i_2, \cdots i_n\right) = \frac{1}{2}\left[\sum_{j=1}^{n} \frac{dL_j(\theta)}{d\theta} i_j^2\right]\omega, \quad (1)$$

where: $n$ - the number of phases; $j$ – the phase number; $\theta$ – rotor position; $\omega$ – rotor speed; $i_j$ – the current phase, $L_j(\theta)$ – the inductance of phase $j$ as the function of $\theta$.

The average of power available $P$, resulting from the operation of the machine as a generator, is (with excluding the losses) equal to the mechanical power. The values can be obtained from the expression of the average value of the torque $T_m$ using (2) and (3):

$$P = T_m \, \omega, \quad (2)$$

$$T_m = \frac{N_r}{2\pi} \int_{0}^{2\pi/N_r}\left(\sum_{j=1}^{n} \frac{1}{2} \frac{dL_j}{d\theta} i_j^2\right)d\theta, \quad (3)$$

where: $N_r$ is the number or rotor poles.

The above equations enable us to infer that the obtained power is approximately constant and it reaches a maximum when the dwell angle is located, in the descending section of the phase inductance profile, which corresponds to the highest average torque [5], [7].

For this type of machines the torque ripple appears mainly in the commutation zones related with the sequential process of establishing and removing the phase currents.

The imposition of phase current waveform using the current control with an adjusted hysteresis band and a sufficient input voltage, allow the torque ripple reduction.

In this way the ripple can be minimized, thus controlling the phase's currents commutation precisely phased relative to the rotor position. For that effect, the current control is done is done using the trapezoidal phase reference torque model [8], two adjacent phases can be supplied at the same time to ensure the continuity in the generated torque.

The SRM is capable of operating continuously as a generator by keeping the dwell angle so that the bulk of the winding conduction period comes after the aligned position, when $\frac{dL_j}{d\theta} < 0$.

The waveforms of the phases reference current $i_j^*$, results from the desired torque $T^*$ and is calculated by the following equation:

$$i_j^* = \sqrt{\frac{2T_j^*}{\frac{dL_j(\theta)}{d\theta}}}, \quad (4)$$

and are themselves the reference signals to be treated using the feedback pulse with modulation (PWM) with adjusted hysteresis band.

### B. SRG – The Current Control

The block diagram from the Figure 3, indicates the current control with the torque reference applied to the 8/6 SRG. The waveforms of the reference currents, $i_1^*, i_2^*, i_3^*$ and $i_4^*$, on calculated using the trapezoidal model torque associated to each phase, $T_1^*, T_2^*, T_3^*$ and $T_4^*$.

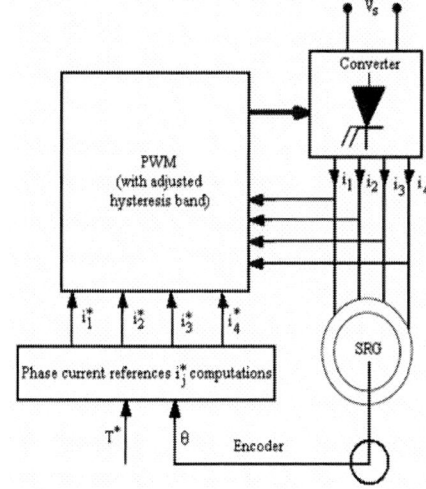

Fig. 3. The current control with the torque reference. applied to the 8/6 SRG.

## C. SRG – Simulations

On used for simulations an 8/6 SRG, with $P_n$=2.4 kW, 4 phase. In these simulation examples of the SRG operation, the converter voltage used was $V_s$=800 V, which allow reduced torque ripple and the rotor speed is 1000 rpm.

The Figure 4 shows the phase current resulting from the trapezoidal phase torque.

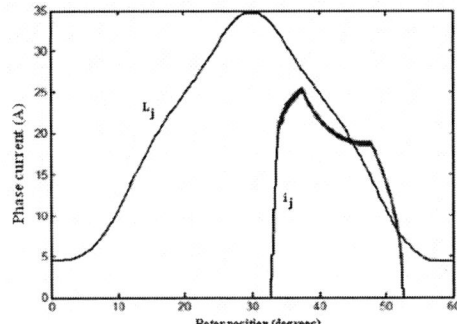

Fig. 4. Phase current.

In figure 5, is indicated the total instantaneous torque for the 8/6 SRG.

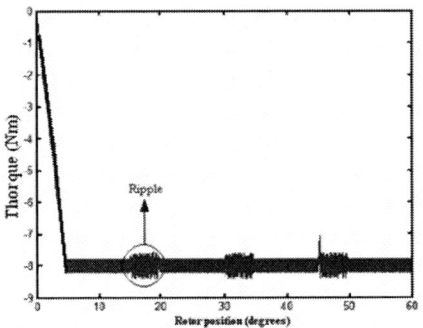

Fig. 5. Total torque

In order to achieve higher performance in SRG operation and higher efficiency in the conversion on includes optimal dwell angle control to further reduce the torque ripple.

## III. ABOUT CONVERSION METHODS OF WIND ENERGY

The capture of the wind energy, in an efficient way, requires the existence of a constant wind flow sufficiently strong [7].

Currently wind turbines are designed to achieve a maximum power at wind speeds above 10 m/s. However, they can be adjusted to the local wind profile.

The maximum theoretical efficiency for the wind to energy conversion is 59.3% (Betz's Limit). The effective efficiency conversion is given by the Power Coefficient ($C_p$), which is expressed by the following, where $P_{mec}$ is the mechanical power of the turbine and $P_w$ is the available wind power.

$$C_p = \frac{P_{mec}}{P_w}. \qquad (5)$$

The power $P_w$ is related with the wind speed $V_w$ calculated by (6),

$$P_w = \frac{1}{2}\rho A V_w^3, \qquad (6)$$

where $\rho$ is the air density ($\rho$ = 1.225 kg/m3) and $A$ is the cross-sectional area of the turbine rotor.

When considering the generator efficiency ($\eta$), the output power is given by (7).

$$P_{out} = \frac{1}{2}\rho A V_w^3 \left(\eta\, C_p\right), \qquad (7)$$

$C_p$ (8) varies with the Speed Ratio ($\lambda$), given in (9):

$$C_p = 0,22\, \rho\, V_w^3 \left(\eta\, C_p\right), \qquad (8)$$

$$\lambda = \frac{r\omega}{V_w}, \qquad (9)$$

where: $r$ is the rotor radius, $\omega$ is the rotor speed.

The low rotor speeds of the turbine bring about small turbulences in the air flow. With high speeds the turbine behaves as a wall for the wind. Therefore the priority is to adapt the wind speed to the rotor speed with the purpose of obtaining a greater conversion efficiency, which results in a maximum $C_p$ [1].

## IV. WIND SYSTEM SIMULATION

This work presents two modes of mechanical coupling of the turbine to the generator: the direct coupling to the turbine shaft, direct - drive wind turbine (Figure 6) and the SRG coupling to the turbine shaft through a gearbox (Figure 8) [1].

### A. Turbine generator direct coupling

The rotor speed $\omega$ of approximately 100 rad/s is too high and not compatible for this type of wind turbines, in normal wind conditions.

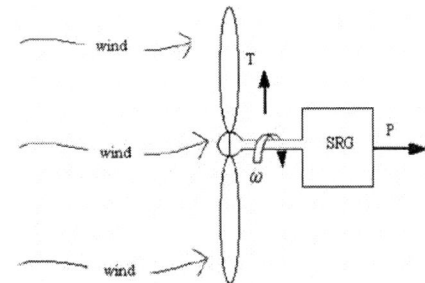

Fig. 6. Direct drive wind turbine with SRG.

The Figure 7 shows the electric power generated by the machine coupled with this turbine, where its average power value corresponds to the power of the system excluding losses in the generator.

case and the power available in the turbine is close to the rated power in the generator.

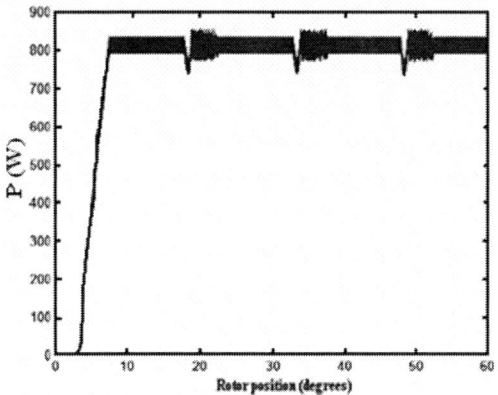

Fig. 7. The 4 –phase SRG instantaneous power versus rotor position.

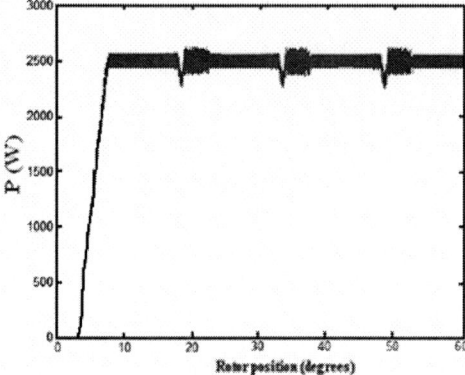

Fig. 8. The 8/6 SRG instantaneous power versus rotor position.

Associated with the required high rotor speed for the good performance of the SRG, the fact that the rotor diameter is small brings about the problem that the wind speed is not sufficient to overcome the combined turbine-generator inertia, namely at the starting stage.

*B. Indirect coupling with gearbox*

The Figure 8 indicates the SRG coupling to the turbine shaft through a gearbox.

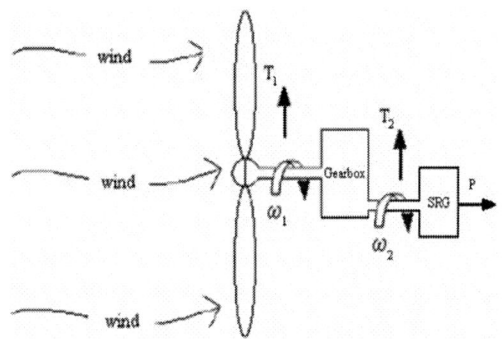

Fig. 8. The Indirect coupling with gearbox.

Assuming that the losses in the gearbox are negligible, and given that the input and output power ($\omega_1 T_1 = \omega_2 T_2$), the transmission ratio $r_t$, varies in the inverse of the torque's ratios:

$$r_t = \frac{\omega_1}{\omega_2} = \frac{T_2}{T_1}. \qquad (10)$$

Figure 8 shows the behavior of the electric power generated by the machine, when coupled with a turbine having a rotor diameter of 5m for a constant wind speed of 8m/s.

With the gearbox the rotor speed of the turbine was reduced to less than half of the value obtained in the first

Considering this SRG is a low power machine and therefore produces low torques, it was expectable that too high rotor speeds would develop, which are hardly compatible with the wind speeds typical of this type of energy conversions and with the wind turbines available for these applications [8].

## V. CONCLUSION

The SRG is a valid alternative in wind energy applications. Therefore it is reasonable to foresee that in the medium power wind systems, the SRG allow good performance in extracting the energy carried by the wind. On the downside we can point out the fact that the SRG is noisier than the other conventional systems.

Nevertheless the current control based on torque reference covered in this paper attenuates this problem; especially via a reduction of the torque ripple.

### REFERENCES

[1] G. Gail, A.D. Hansen, "Controller design and analysis of a variable speed wind turbine with doubly fed induction generator", European Wind Energy Conference, pp. 500-508, 2006.

[2] P. Lobato, A.J.Pires, "Methodology based on energy-conversion diagrams to optimize switched reluctance generators control", *ICEM 2004 Proceedings*, no. 158, pp. 700-705, Sept. 2004.

[3] P. Chancharoensook, M.F. Rahman, "Control of a Four-Phase Switched Reluctance Generator: Experimental Investigations", *IEMDC 03 Proceedings*, vol. 2, pp. 842-848, June 2003.

[4] P. Lobato, A.J.Pires, "A New Control Strategy Based on Optimized Smooth-Torque Current Waveforms for Switched reluctance Motors", *Electromotion'03 Proceedings*, vol. 2, 610-615, Nov. 2003.

[5] V. Akhmatov, A.H. Nielsen, "Variable speed wind turbines with multipole synchronous permanent magnet generators", Proceedings Wind Energineering 2001, vol. 27, no. 6, pp. 531-548, 2003.

[6] A. D. Hansen, "Wind models for predictions of power fluctuations from wind farms", Proceedings APCWEV 2001, no.89, pp. 9-18, 2001, Kyoto, Japan

[7] H. Henao, E. Bassily, "A new control angle strategy for switched reluctance motor, Proceedings EPE '97, vol.3, pp. 613-618, September 1997, Trondheim, Norway.

[8] A. Grauers, "Efficiency of three wind energy generator systems", IEEE Transactions on Energy Conversion, vol. 11, no. 3, pp. 650-657, 1997.

# Active Line Shaping of a Single Phase Rectifier using the Switching Function Technique

Christos Marouchos

Cyprus University of Technology, Department of Electrical Engineering and Information Technologies, Lemesos, Cyprus.
E-mail: christos.marouchos@cut.ac.cy

***Abstract* -** Operation of the Active Line Shaping Single Phase Rectifier with the switching instances pre-calculated is presented. The required input current at unit power factor to feed the dc load is set. The Switching Function Technique is applied to derive the necessary modulation function of the dc side switch. The analysis leads to the calculation of the parameters of the modulation function and the switching instances of the dc side switch. An analytical equation of the line current is also derived. Optimisation is then possible in order to achieve the minimum switching losses by minimising the switching instances for the best quality of input line current. The results are supported by PSPICE simulations.

**Keywords- Active filter, Converter control, Modelling, Power factor correction, DC power supply**

## I. INTRODUCTION

Poor Power factor and high Total Harmonic Distortion (THD) characterises rectifiers with a smoothing capacitor on the dc side. The problem is well defined and solutions are suggested by many authors [1-4] in the recent past. In this paper the application of the switching function [5] for the analysis of the single phase rectifier with boost converter [4,8] is considered. This leads to a new method of operation where the switching instances are pre-calculated and stored.

The application of the Switching Function Technique for the analysis of voltage fed power electronic circuits where the load is separated from the voltage source by the switching arrangement is relatively straight forward [6]. When the switching arrangement lies between impedances as is the case of the single phase rectifier, Fig.1, with active line shaping, a more rigorous procedure is required [5,7]. The suggested procedure is to derive the modes of the circuit and write the voltage-current equations for each mode. The appropriate switching function for each mode is identified and the mode sequence is established. Unified equations are derived by employing the switching functions. A mathematical model can be built for complicated circuits. Combination and expansion of the unified equations leads to the calculation of voltages and

currents at every point of the circuit and the modulating function of the dc side switch. The modulation function is encoded as a PWM signal and the switching function of the switch is derived [5]. The switching instances are then calculated. The line current is also derived by employing the calculated switching function. This method offers the advantage of optimising the number of switching instances for the best line current quality improvement.

## II THE UNIFIED VOLTAGE EQUATION

The current through the inductor in Fig.1 can be discontinuous or continuous. Discontinuous inductor current though causes distortion of the input current. This analysis is limited to continuous operation but in principle it can be expanded to non-continuous operation.

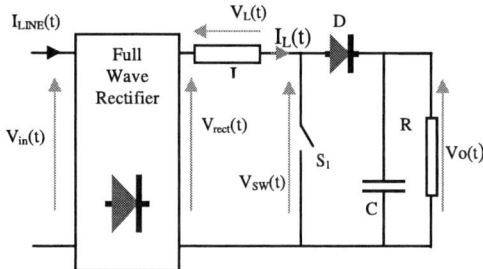

Fig.1 The active line current shaping circuit single phase rectifier

Effectively there are two switches in this circuit. Switch $S_1$ is either a power MOSFET or an IGBT operated by the application of appropriated signals at its gate. The diode D is conducting when it is forward biased. $S_1$ is assumed ideal and the voltage drop across the diode is neglected for the purpose of this investigation.

A switching function is a signal which consists of a series of unit magnitude pulses and of varying width. For every mode of the circuit there is a switching function associated it. When it takes the value of 1, the mode exists, when it takes the value of zero the mode does not exist. There are two modes and two switching functions in the circuit of Fig.1: The switching function $F_1(t)$ takes the value of 1

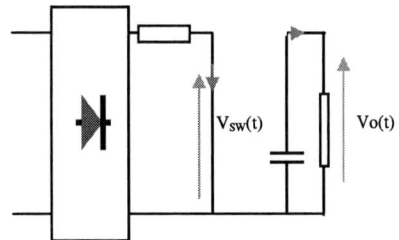

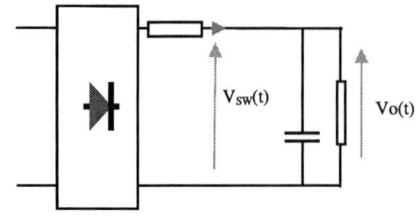

Fig.2a  Mode I of the Active Line Current Shaping Circuit Single Phase Rectifier

Fig.2b  Mode II of the Active Line Current Shaping Circuit Single Phase Rectifier

Table 1
Modes and state of Switching Functions

| MODE | $V_{sw}(t)$ | $F_1(t)$ | $F_2(t)$ |
|------|-------------|----------|----------|
| I | 0 | 1 | 0 |
| II | $V_o(t)$ | 0 | 1 |

during Mode I, Fig.2a  and the switching function $F_2(t)$ takes the value of 1 during Mode II, Fig.2b. During Mode I, Fig.2a, the switch is closed and the voltage across it, $V_{sw}(t)$ is zero. The switching function, $F_1(t)$ associated with this mode represents the switching action of the switch and it takes the value of 1.   The diode is not conducting. During Mode II, Fig.2b, the switch is open ($F_1(t) = 0$) and the voltage across it is Vo(t). The diode is conducting and the switching function, $F_2(t)$ associated with  this mode represents the action of the diode and it takes the value of 1. Table 1 summarises these findings.

The mode sequence is very simple: The two modes are succeeding each other and the voltage across the switch, $V_{sw}(t)$  is made up from the contributions of both modes, hence

$$V_{sw}(t) = 0 \times F_1(t) + Vo(t) \times F_2(t) \qquad (1)$$

Simplified to

$$V_{sw}(t) = Vo(t) \times F_2(t) \qquad (2)$$

The diode is conducting when the switch is off and the switch is conducting when the diode is off: they are working in anti-parallel [5]. Therefore the two switching functions $F_1(t)$ and $F_2(t)$ are non-overlapping and there are no dead periods as the current is continuous. Hence $F_2(t)$ is the inverse of $F_1(t)$ and they are related by

$$F_2(t) = 1 - F_1(t) \qquad (3)$$

Hence equation (2) becomes

$$V_{sw}(t) = Vo(t) \times [1 - F_1(t)] \qquad (4)$$

The output voltage, Vo(t), has a dc component, $V_{dco}$, and an ac component, $Vo_{AC}(t)$. It is desired that the ac component is as small as possible and the smoothing capacitor, C, in Fig.1 is chosen to be of such large value to keep it within Acceptable limits. For many practical purposes the ac component may be ignored. Hence (4) becomes

$$V_{sw}(t) = V_{dco} [1 - F_1(t)] \qquad (5)$$

The inner loop in Fig.1 comprises the voltage at the output of the rectifier $V_{rect}(t)$, the voltage across the inductor $V_L(t)$ and the voltage across the switch, $V_{sw}(t)$.

$$V_{rect}(t) = V_L(t) + V_{sw}(t) \qquad (6)$$

Replacing (5) into (6) the unified equation for the circuit is

$$V_{rect}(t) = V_L(t) + [1 - F_1(t)] V_{dco} \qquad (7)$$

The voltage at the output of the rectifier, $V_{rect}(t)$ is the result of an amplitude modulation of the input voltage, $V_{in}(t)$ and the switching function of the bridge, $F_B(t)$ as

$$V_{rect}(t) = F_B(t) V_{in}(t) \qquad (8)$$

The switching function $F_B(t)$ is a bipolar transparent switching function and it represents the action of the bridge [5]. It is a square wave at the same frequency as the input voltage

$$F_B(t) = 4 \sum_{n=1}^{\infty} \frac{\sin(n\pi/2)}{n\pi} \cos(n\omega t - n\,n\pi/2) \quad (9)$$

*n is an odd integer and $\omega$ is the mains frequency*

## III THE MODULATING FUNCTION

In order to facilitate the calculation of the parameters of the switching function $F_1(t)$, its high frequency components (related to its switching nature) are neglected at the moment. It is therefore replaced by a modulation function $M(t)$. The modulation function contains only the low frequency components of the switching function.

Replace $F_1(t)$ with $M(t)$ into (7)

$$V_{rect}(t) = V_L(t) + [1 - M(t)] V_{dco} \qquad (10)$$

Rearranging (10)

$$M(t) = 1 + [V_L(t) - V_{rect}(t)] / V_{dco} \qquad (11)$$

The wanted input line current $I(t)$, must be at unity power factor i.e. at fundamental frequency $\omega$, free from harmonics and in phase with the supply voltage.

$$I(t) = I_p \sin\omega t \qquad (12)$$

*Where $I_p$ is the peak value of the input line current*

The input line current is a reflection of the output current of the bridge by the bridge switching function $F_B(t)$ [5]. Hence the output current through the inductor $I_L(t)$ is defined by

$$I_L(t) = I_p \sin\omega t \qquad for \quad 0 < \omega t < \pi \qquad (13)$$

Repeated for $\pi < \omega t < 2\pi$, $2\pi < \omega t < 3\pi$ ....

The voltage across the inductor L is a function of the current through it. In the steady state is given by

$$V_L(t) = I_p \omega L \cos\omega t \qquad for \quad 0 < \omega t < \pi \qquad (14)$$

Repeated for $\pi < \omega t < 2\pi$, $2\pi < \omega t < 3\pi$ ....

The rectified voltage at the output of the rectifier, $V_{rect}(t)$ is defined as

$$V_{rect}(t) = V_p \sin\omega t \qquad for \quad 0 < \omega t < \pi \qquad (15)$$

Repeated for the period $\pi$ to $2\pi$, $2\pi$ to $3\pi$ ......

From (14) and (15) the modulation function, $M(t)$ equation (11) is defined in the range $0 < wt < \pi$ as

$$M(t) = 1 + D_2 \cos\omega t - D_1 \sin\omega t \qquad (16)$$

Where

$$D_1 = \frac{V_p}{V_{dco}} \quad and \quad D_2 = \frac{I_p\omega L}{V_{dco}} \qquad (17)$$

The parameters of the modulation function are therefore directly calculated from the values of he input voltage ($V_p$), the output voltage ($V_{dco}$), the input current ($I_p$) and the value of the dc side inductor (L).

Fig.3a displays the modulation function. It has the period of the input supply voltage and it exhibits periodic symmetry: the two half-cycles are exactly the same.

## IV. THE SWITCHING FUNCTION AND THE SWITCHING INSTANCES OF THE DC SIDE SWITCH

The modulating function $M(t)$ is encoded as a PWM signal by dividing its period, Fig.3a into K sectors. Each sector has the width $T_r$ [5]. In Fig.3, the number of sectors is 20, K =20.

$$T_r = \frac{2\pi}{K} \qquad (18)$$

The area, $S_A(k)$, under the curve of $M(t)$ for the period *k-1* to *k*,

$$S_A(k,t) = \int_{(k-1)T_r}^{kT_r} M(t)d\omega t \qquad (19)$$

$$S_A(k,t) = T_r + D_2[\sin(k\,T_r) - \sin\{(k-1)\,T_r\}] $$
$$+ D_1[\cos(k\,T_r)] - \cos(k-1)\,T_r] \qquad (20)$$

The area of the $k^{th}$ pulse, $P_A(k)$ of the switching function,

$$P_A(k) = 2\,\delta(k)\ x\ 1 \qquad (21)$$

*$\delta(k)$] - half the on period of the $k^{th}$ pulse*

For the purpose of deriving the PWM signal, the area under the curve of $M(t)$ is equated to the area of the switching function pulse in the same sector.

$$S_A(k) = P_A(k) \qquad (22)$$

Hence the half-pulse-width, $\delta(k)$, is

$$\delta(k) = \{T_r + D_2[\sin(k\,T_r) - \sin\{(k-1)\,T_r\}] $$
$$+ D_1[\cos(k\,T_r)] - \cos(k-1)\,T_r]]\}/2 \qquad (23)$$

The phase delay, $\beta_1$ of the pulse is

$$\beta_1(k) = T_r k - \frac{T_r}{2} \qquad (24)$$

The switching function is constructed on the basis of the period of one complete cycle of the mains frequency. Since the modulation function exhibits periodic symmetry, for every pulse in the first half-cycle there is a similar one with same pulse width in the second half-cycle ie $\pi$ latter. Hence the component switching function is a switching function with two similar pulses separated by 180° with a switching frequency $\omega$, the mains frequency.

$$F_{comp}(t) = K_o + 2\sum_{n=1}^{\infty} K_n \{\cos(n\omega t - n\beta_1) + \cos[n\omega t - (\pi - n\beta_1)]\} \qquad (25)$$

The switching function, $F_1(t)$ contains **K** pulses per mains cycle. Hence the number of component switching functions required for the construction of $F_1(t)$ is K/2. The PWM switching function of the composite signal, $F_1(t)$ is given by

$$F_1(t) = 2\sum_{k=1}^{K/2}\sum_{n=1}^{\infty} K_n \{\cos(n\omega t - n\beta_1) + \cos[n\omega t - (\pi - n\beta_1)]\} \qquad (26)$$

The switching instances for the switch on the dc side are derived from (26). The centre of the $k^{th}$ pulse is given by

$$T_{centre}(k) = [kT - T/2] / 2\pi f \qquad (27)$$
$$f - \text{mains frequency}$$

The switch-on instant, $T_{ON}(k)$, of the $k^{th}$ pulse is given in seconds, by:

$$T_{ON}(k) = [kT - T/2 - \delta(k)] / 2\pi f \qquad (28)$$

The switch-off instant of the same pulse, $T_{OFF}(k)$, in seconds is given by:

$$T_{OFF}(k) = [kT - T/2 + \delta(k)] / 2\pi f \qquad (29)$$

## IV LINE CURRENT

The line current is derived from the inductor current.
According to the Switching Function Technique of analysis [5], the line input current, $I_{LINE}(t)$ to a bridge is a reflection of the output current, in this case the inductor current, $I_L(t)$.

$$I_{LINE}(t) = I_L(t) F_B(t) \qquad (30)$$

The inductor current can now be derived from (7) by considering first the inductor voltage.

$$V_L(t) = V_{rect}(t) - [1 - F_M(t)] V_{dco} \qquad (31)$$

The ac component of the inductor current in the steady state is found by dividing the inductor voltage, $V_L(t)$ by the harmonic impedance of the inductor

$$I_{Lac}(t) = \frac{V_L(t)}{X(w,n)} \qquad (32)$$

The average value of the inductor current is

$$I_{LDC} = \frac{2I_p}{\pi} \qquad (33)$$

Hence

$$I_L(t) = I_{Lac}(t) + I_{Ldc} \qquad (34)$$

The frequency content of the line current can now be derived from (30).

Fig. 3a displays the modulation function, M(t)  Fig.3b the component switching function, $F_{comp}(t)$  and Fig.3c the PWM switching function $F_1(t)$.

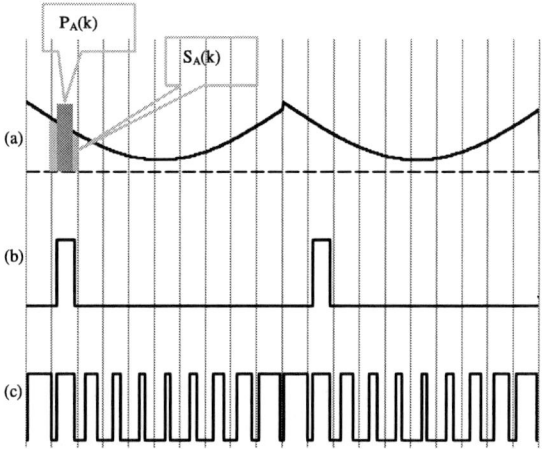

Fig.3. Generation of the PWM switching function. (a) the modulation function, M(t)  (b) the component switching function, $F_{comp}(t)$ (c) the PWM switching function $F_1(t)$.

## V RESULTS

A single phase 1.5KW  bridge rectifier rated at 120V, 50Hz with its output voltage set to 211V and a 30 ohms load is investigated. Both the switching function using Mathcad and PSPICE are employed. The dc side inductor is set to 2mH.  In order to apply the Switching Function Technique,

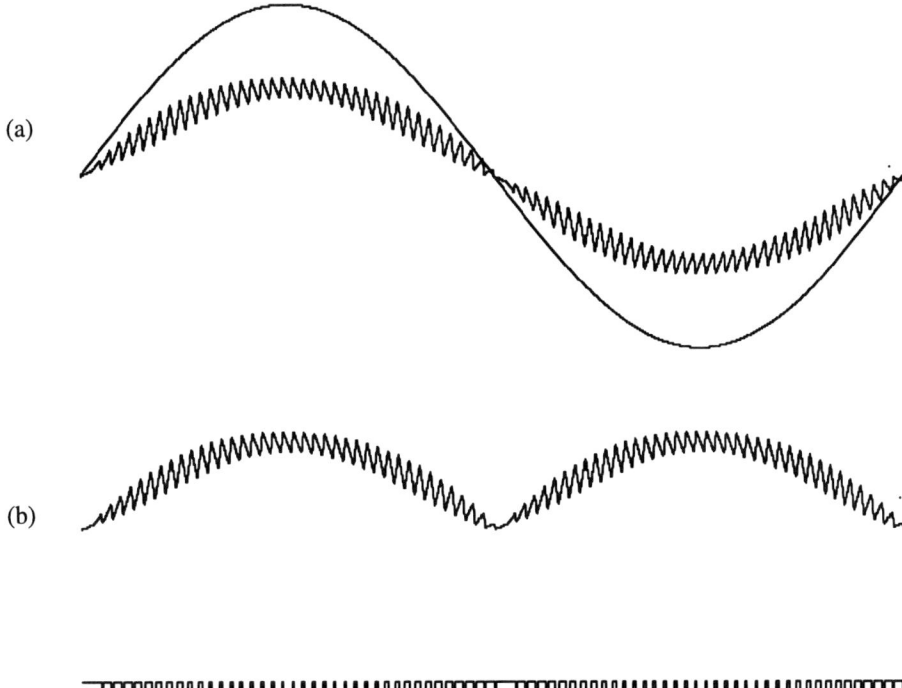

Fig.4 . Mathcad results (a) Supply voltage and input current I(t), (b) Inductor current $I_L(t)$ (c) Switching Function of the switch, $F_1(t)$ (80 pulses)

the coefficients of the modulating function are derived from (17)

$$D_1 = 0.806 \text{ and } D_2 = 0.052$$

By equating input and output power, the peak value of the input current is found as

$$I_p = 17.46A$$

The average value of the inductor current is found from (33) as

$$I_{Ldc} = 11.12A$$

A low switching frequency ($f_{sw}$) at 4KHz is chosen

A switching frequency of 4KHz gives 80 switching pulses per mains cycle. The line current waveform is derived from (30) and displayed in Fig.4a together with the supply voltage using Mathcad. The inductor current waveform is derived from (34) and displayed in Fig.4b and the waveform of the switching function of the switch is derived form (26) and displayed in Fig.4c. The harmonic content of the line current is extracted from (30). The 3$^{rd}$ harmonic is limited to 0.29% and the 5$^{th}$ to 0.22% of the fundamental. There are higher order line current harmonics due to the switching action of the semiconductor switch.

The lowest band of these harmonics is located at ($f_{sw}$) ± 50 and ($f_{sw}$) ±150 . Therefore their order s easily controlled by selecting a higher switching frequency for ease of filtering. For a switching frequency of 4KHz they are at 3.850KHz

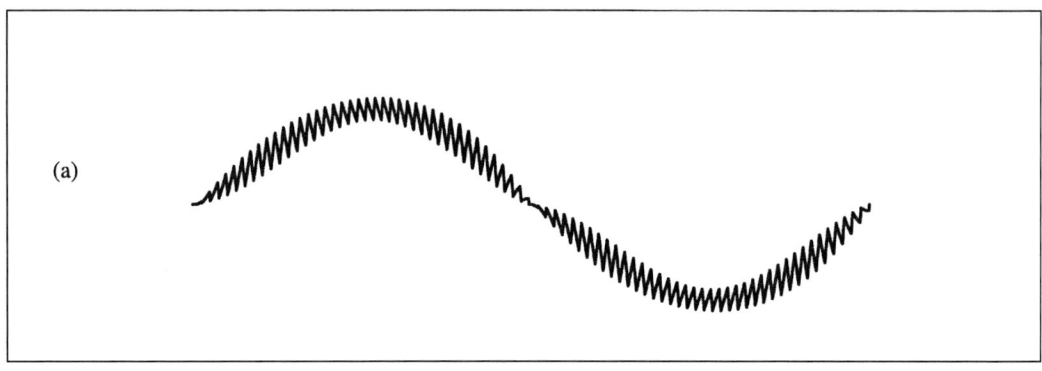

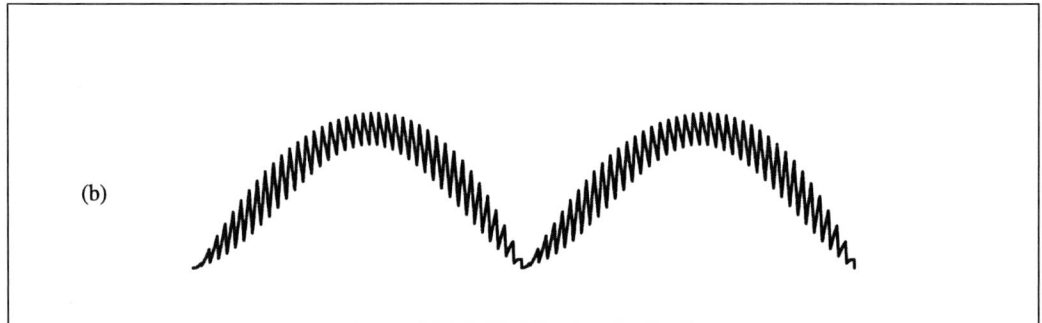

Fig.5. PSPICE results (a) line current (b)) inductor current

3.950 KHz, 4.050KHz and 4.150KHz. Their magnitude is at 6.9%. Filtering them should be easy because of their high order. Their order can be shifted even higher by choosing a higher switching frequency.

The switching instances for the switch $S_1$ are calculated from (28) and (29). Table 2 presents the first 10 switch-on and switch-off instances of the semiconductor switch. The circuit is simulated in PSICE for an 80 pulse system and a switching frequency of 4KHz. The line and inductor current waveforms from the PSPICE simulation are displayed in Fig.5. The harmonic content of the line current is 3.7% for the third harmonic and for the fifth it is 0.99%. The frequency components of the switching frequency are at 6.5%.

Active line current shaping of the single phase bridge rectifier is usually achieved through a feedback circuit. The inductor current is sensed and it is subtracted from a reference to form an error signal. The error signal is used to produce the PWM signal driving the switch on the dc side [4]. This circuit configuration is simulated by PSICE as well. The line and inductor current waveforms are shown in Fig.6. The harmonic content of the line current is 0.43% for the third harmonic and 0.61% for the fifth. There are about 260 switching pulses per mains cycle.

## VI CONCLUSIONS

An alternative method of operation of the Active Line Shaping Single Phase Rectifier was presented. The required input current at unity power factor to feed the dc load is set and the necessary modulation function of the dc side switch is derived by applying the Switching Function Technique. The modulation function is then encoded as a PWM signal to form the switching function of the switch. The switching instances are directly derived from the switching function.

Results from both the Mathematical Modelling using the Switching Function Technique and PSPICE simulations are derived and they are in close agreement. The switching pulses are 80 per mains cycle a dramatic reduction from the 260 of the feedback method of driving the circuit. Using this technique, optimisation is possible in order to achieve the minimum switching losses by minimising the switching instances for the best quality of the input line current.

The displacement power factor is unity and the low order current harmonics ($3^{rd}$, $5^{th}$ etc ) are suppressed to manageable levels. Further tuning of the switching function driving the switch can reduce them further. The high order line current harmonics due to the switching frequency can easily be removed with passive filters.

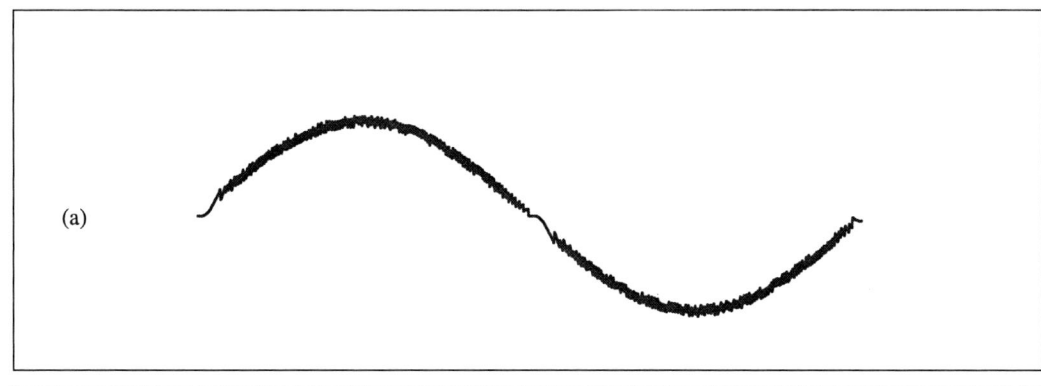

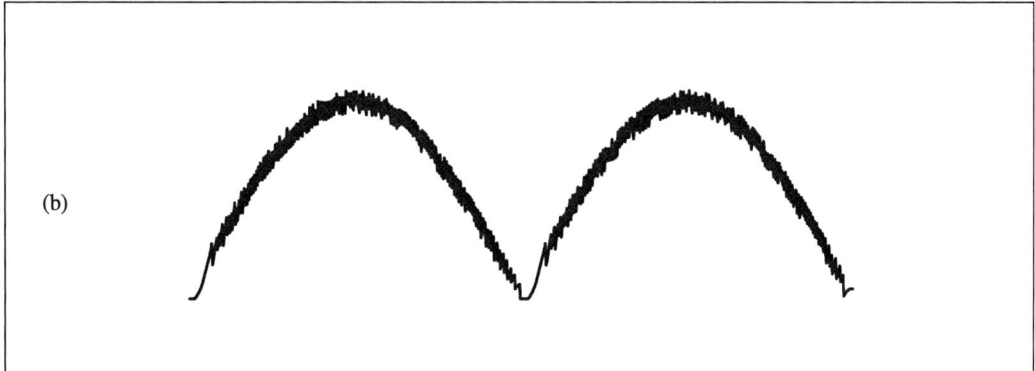

Fig.6 PSPICE results of line current shaping circuit with current feedback for the generation of the PWM signal (a) Line current (b) Inductor current

Table 2
The first 10 switch-on and switch-off instances of the semiconductor switch as calculated by athcad

| Switch-on (µS) | 2.8 | 534.4 | 1065 | 1594 | 2121 | 2645 | 3165 | 3681 | 4193 | 4700 |
|---|---|---|---|---|---|---|---|---|---|---|
| Switch-off (µS) | 497.2 | 965.7 | 1435 | 1906 | 2379 | 2855 | 3335 | 3819 | 4307 | 4800 |

## References

[1] Tae-Sung Kim; Gwan-Bon Koo; Gun-Woo Moon; Myung-Joong Youn "A single-stage power factor correction AC/DC converter based on zero voltage switching full bridge topology with two series-connected transformers"
Power Electronics, IEEE Transactions,
Volume 21, Issue 1, Jan. 2006 Page(s): 89 - 97

[2] Phattanasak, M.; Chunkag, V. "Paralleling of single-phase AC/DC converter with power-factor correction" Power Electronics Specialists Conference, 2004. PESC 04. 2004 IEEE 35th Annual Volume 2, Issue , 20-25 June 2004 Page(s): 1576 - 1580.

[3] Uthen Kamnarn[1] and Viboon Chunkag[1] King "Analysis and Design of a Parallel CUK Power Factor Correction Circuit Based on Power Balance Control Technique" IEEJ Transactions on Industry Applications Vol. 126 (2006), No. 5 pp.533-540

4] Huai Wei, Guangyong Zhu and Peter Kornetsky "Single witch AC-DC Converter with Power Factor correction", IEEE Transactions on Power Electronics May 2000Page(s) 15(3): 421-430

5] Marouchos, CC : "The switching Function: Analysis of Power Electronic Circuits", IEE ,London,2006 , ISBN-10 086341351X, ISBN-13: 978-086341351-3

[6] Wood P.: "Switching power converters", Van Nostrand Reinhold Company, New York, 1981

[7] C. Marouchos [*], M. K. Darwish, M. El-Habrouk "Variable Var Compensator Circuits" IEE Proc Electr Power Appl Vol 153 Vol 5 Sept 2006

[8] MOHAN, N., UNDERLAND, T.M and OBBINS, W.P": "Power Electronics", John Willey and Sons end edition.,1995

[

# Control of Reactive Power in Double-Fed Machine Based Wind Park

Elżbieta Bogalecka[*], Michał Kosmecki[†]

[*]Gdansk University. of Technology/Department. of Electrical and Control Engineering, Gdańsk, Poland,
*e.bogalecka@ely.pg.gda.pl*
[†] Institute of Power Engineering Gdansk Branch, Gdańsk, Poland, *m.kosmecki@gmail.com*

*Abstract*—Different aspects of reactive power regulation problem in wind farms are presented in this paper. In the first part some background of the reactive power control in wind parks is presented, including motivations for its use and methods that can be used. In general, there are active (wind generator, compensator) and passive methods (L, C) of reactive power compensation. A simulation model of the wind park (model of doubly fed induction generator (DFIG) based wind turbines, transformers, cable lines between wind farm and PCC, control system). has been done to verify the efficacy of proposed methods. Conclusions about need and possibilities of reactive power control have been drawn From simulations in steady and transient state. Next features with special regards to limits of the DFIG working as wind generator are presented. In the final part of the paper selected methods of reactive power control are described and simulation results are presented.

*Keywords*—wind generator systems, power factor correction, doubly fed induction motor, simulation.

## I. INTRODUCTION

A wind power system consists of two main parts: the wind turbine and the electrical generator. Wind turbine, mostly horizontal and three-bladed, converts the energy of the blowing air into mechanical energy of the rotating shaft. The generator converts this energy into electrical energy that is sent to the power system. Both energy conversion processes have to be made with maximum efficiency. In most applications, the wind generator works only as a source of the power. It does not take part in the voltage and frequency direct control.

To connect wind generator to the grid a few additional conditions should be fulfilled. They are related to energy quality and machine behavior during short circuit and voltage dips. Some are listed below:
- limited reactive power consumption; limited tgφ,
- limited current harmonics,
- limited flickers level,
- smooth active power production in spite of wind gusts,

As a result requirements for the control system can be formulated as follows: optimal wind power conversion, compensation of torque ripples caused by wind gusts, stability, controlled reactive power, decoupled control of active and reactive power, immunity from faults or short circuits and machine sensorless control. To fulfill all conditions the machine, the converter and the control system should be tailored to meet the requirements. Typical hierarchical structure of the wind turbine control system is presented in Fig.1.

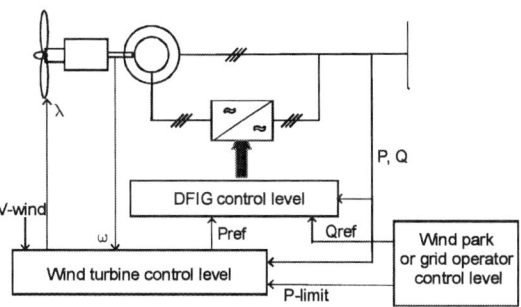

Fig.1.Structure of the wind turbine control system (based DFIG)

The energy captured from the wind reaches the maximal value if the rotor speed is adjusted to wind speed changes. For winds in the range 4-15 m/s the rotor speed should change like 1:2. To make it possible, modern wind generators are switched to the power system through the converter. Nowadays there are two dominant wind power technologies. The main is based on a geared high speed slip ring asynchronous machine with the converter in the machine rotor circuit, called double fed induction generator (DFIG). The second is the gearless low speed synchronous machine (permanent magnet or excited) with the power converter in the main circuit.

Slip ring induction generator DFIG with the stator connected to the grid and rotor fed by a bi-directional AC/AC converter is a good solution for wind power stations. The main advantage of the DFIG is the reduced converter size. The converter size depends on the machines slip and usually does not exceed 25-30% of the machine's nominal power. In this paper, wind power plant with double fed machines is considered into account.

Modern wind farms are large with the power of hundreds of megawatts and consist of many wind turbines of the power 2-3 MW. Each generator is connected to the medium voltage grid through LV/MV transformer. In a typical configuration the wind farm is connected to the grid in the point of common coupling (PCC). The distances between the wind park and the PCC may be long (for example 10-20 km). Internal power transmission in the park is realized by a transformers and medium voltage cables. Transmission elements are the source of reactive power (capacitive or inductive). The value of reactive power depends on wind conditions. Due to generation/absorption of reactive power by cables and transformers, the power factor at PCC changes even if each wind generator works with a unit power factor. As a result, the power factor in PCC is no more equal to one, which might lead to a reduction in profit. Although modern wind turbines (WT) are able to alter their power

978-1-4244-1741-4/08/$25.00 ©2008 IEEE

factor, a common practice of already installed wind farms is to set their WTs' power factor to unit. Unitary power factor is preferred because it is the active power production that is rewarded. The second reason for the unit WT power factor is the reduction of installation costs. The magnitude of the machine, converter, transformer and cables depend on current RMS value (with both active and reactive components).

There are two main motivations for the reactive power control: economical and technical.

If the wind park power factor exceeds the limits the owner of the wind park is obliged to pay for produced/consumed reactive power, especially if the power factor is capacitive. From the owner point of view the wind park reactive power production/absorption that causes additional costs is the main reason for a need of the power factor control. Optimal power transmission between wind park and the grid requires equal inductive and capacitive parts of reactive power.

Large power wind parks which are substantial power source in the local grid, should take part in grid regulation process. The power system operator should handle it like conventional power station and control the state of the wind park, including reactive power. Local grid code defines the wind park power factor range at PCC (for example cos φ from 0.975ind. to 0.975cap.).

Therefore, control of the wind park power factor or reactive power becomes necessary. This task is not easy to perform; there are a few limitations. Wind park power factor and reactive power (at PCC) are wind dependent. Distances between turbines in the wind park are large, which complicates measurements, communication and control system structure. In addition, wind generator limitations (machine stator and rotor current limits, control system possibilities) should be taken into account.

This paper presents various aspects of reactive power regulation problem in wind farms. Reactive power is produced only in special situations. The main goal for the compensation methods is to allow the wind farm to operate with power factor (at PCC) variable in specified range (defined by grid code). Comparison of the features, especially effectiveness of two compensation methods: passive (additional inductance and capacitance) and active (DFIG as a compensator) is presented.

Reactive power compensation through capacitor banks or inductors is simple and widely used today but is ineffective and inaccurate. Therefore to enable more accurate and smoother control of reactive power, active compensation methods should be used. One solution is the use of additional reactive power compensator (e.g. STATCOM). The second solution is the utilization of a wind generator as a source of compensating power: capacitive or inductive.

Active compensation requires measuring reactive power at PCC and sending its value to the wind farm's controller, which regulates it by means of PI action and a reference value. It may also be realized on the basis of a $Q=f(V_w)$ curve obtained beforehand. Then the WT reactive power reference value is calculated according to actual wind conditions. Both methods were tested via simulations and the results are presented at the end of this paper.

The consideration of the transmission elements allows for maintaining an optimal power factor.

## II. WIND PARK MODEL

Wind farm is a complex object that consists of a few wind turbines with LV/MV transformer each, MV cable lines inside the park, one or a few (depending on wind park structure and power) MV/HV transformer and a cable line between transformer and PCC. A simple model has been built to investigate the main features and possibilities of wind park reactive power control.

In figure 2 a model of a single wind turbine connected to the grid is presented. The generator is connected to the HV grid by LV/MV and MV/HV transformers and a cable line. The wind generator is modelled as a controllable source of active and reactive power, which includes static aerodynamic characteristics of wind turbine $P=f(V_w)$ and dynamic characteristics of DFIG (including control system). Transformers are modelled as inductance (equal to transformer leakage inductance that is equal to short circuit voltage in p.u.). A cable line is modelled as a chain of $R, L, C$ elements that parameters depend on cable type and length.

The wind turbine power factor at PCC depends on total value of inductance $L$ and capacitance $C$. The value of reactive power $Q_L$ absorbed by inductance $L$ is equal to:

$$Q_L = \omega L I^2, \tag{1}$$

and the $Q_C$ value:

$$Q_C = -\omega C U^2 \tag{2}$$

where: $U$ denotes voltage, $I$ current, $\omega$ grid pulsation.

From these formulae it follows that $Q_C$ power is (for constant voltage) almost constant, but $Q_L$ power depends on current or WT active power or wind speed. Total reactive power:

$$Q_T = Q_L + Q_C \tag{3}$$

measured at PCC is equal to zero for only one wind speed. The value of this speed depends on system parameters.

In most applications, a wind generator is used only as a source of power. It does not take part in voltage and frequency control directly. The values of active and reactive power are the reference values for the machine control system. The reference value for active power depends on the wind speed, while the reference value for reactive power is often set to zero, but can be imposed by system operator or by wind park control system.

Using the presented model of a wind turbine active and reactive power measured at PCC as a function of wind speed have been calculated (parameters in Appendix).

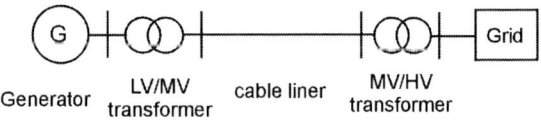

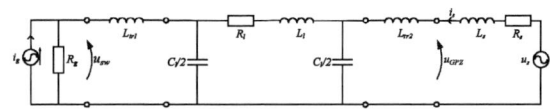

Fig.2. A simplified diagram of single WT

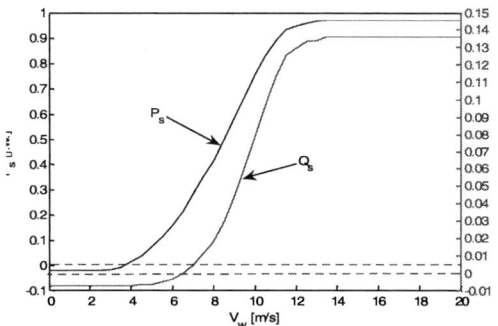

Fig.3. Active (blue line) and reactive power (green line) at PCC as a function of wind speed (for cosφ_WT=1)

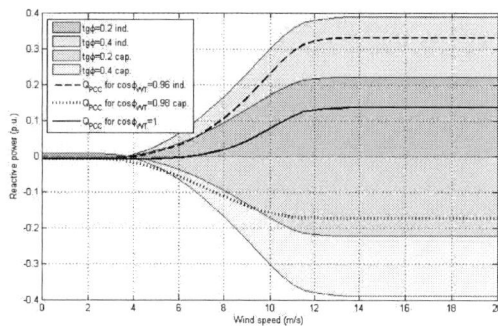

Fig.5. Reactive power at PCC (black solid and dotted lines) for wind turbine different power factor. (colored areas – limits resulting from economical and technical reasons)

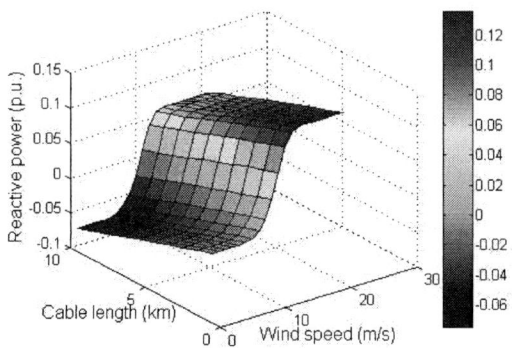

Fig. 4. Reactive power at PCC as a function of wind speed and cable line lenght

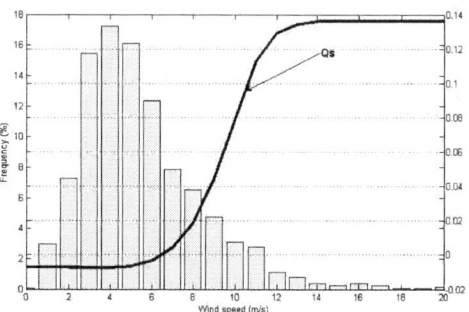

Fig.6. Wind profile and reactive power Q_PCC

Results for unit machine power factor are presented in Fig.3. For strong winds (in this example, stronger than 6.5 m/s) the total reactive power is inductive but for weak winds the total reactive power is capacitive. Only for one wind speed the power factor at PCC is equal to one. The value of reactive power depends on system parameters: transformers leakage inductance, cable line length and type. In Fig. 4 a graph showing the influence of cable line length on reactive power value at PCC is presented.

As it was mentioned above the value of reactive power measured at PCC causes additional costs that may cause reduction in profits. Exceeding of power factor limits (defined in contract, e.g. tgφ=±0.4) causes extra charge). In addition system operator should have possibility to change wind park power factor (measured at PCC) within the defined range (e.g. tgφ=± 0.2). These limits are shown in Fig.5 as colored areas. In Fig.5 the values of reactive power measured at PCC (as a function of wind speed; black solid and dotted lines) for different WT power factors (inductive and capacitive) are presented. In technical specifications producer defines the range of wind turbine power factor changes (e.g. from cosφ=0.98_cap to cos φ=0.96_ind.).

This figure (Fig.5) shows the necessity and possibilities of the reactive power compensation. Wind turbine can produce reactive power only during active power production (this results from generator technical description) and to reduce capacitive power produced by a cable line for wind speeds lower than 3.5 m/s either

additional compensator has to be used (WT starts at about 4-5 m/s) or WT has to be disconnected at PCC. The same situation occurs for the wind speeds stronger than 9 m/s. Inductive reactive power in this example can be compensated by WT. These results relate to data given in Appendix.

Total cost (that results from reactive power consumption/absorption) and the necessity of reactive power compensation depend on wind profile and on frequency of reactive power limits violations. In figure 6 an example of a wind profile and a curve Q_PCC=f(V_w) is presented.

## III. DFM AS A GENERATOR

### A. Double Fed Machine model

Differential equations of a double fed machine (DFM) obtained from the space vector theory developed by Kovacs and Racz are as follows:

$$u_s = R_s i_s + \frac{d\psi_s}{d\tau} + j\omega_a \psi_s, \qquad (4)$$

$$u_r = R_r i_r + \frac{d\psi_r}{d\tau} + j(\omega_a - \omega_r)\psi_r, \qquad (5)$$

$$J\frac{d\omega_r}{d\tau} = \text{Im}\left|\psi_s^* i_s\right| - m_0, \qquad (6)$$

where $\psi_s$, $\psi_r$ are the stator and rotor flux vectors, $i_s$, $i_r$ are the stator and rotor current vectors, $u_s$, $u_r$ are the stator and rotor voltage vectors, $R_s$, $R_r$ are the stator and rotor resistances, $J$ is moment of inertia, $\omega_r$ is the rotor angular velocity, $\omega_a$ is the angular velocity of frame of references and $\tau$ is the relative time. All variables are expressed in p.u. system.

Most of the known up to now, control systems of the DFM are based on the machine vector control theory. The control system presented in this paper is based on multiscalar model (MM) of the induction machine applied in [6] for the DFM. The new variables are defined to obtain the multiscalar model:

$$z_{11} = \omega_r, \qquad z_{12} = \psi_{sx}i_{ry} - \psi_{sy}i_{rx}, \qquad (7)$$

$$z_{21} = \Psi_s^2, \qquad z_{22} = \psi_{sx}i_{rx} + \psi_{sy}i_{ry} \qquad (8)$$

The variables defined by (7,8) do not depend on the frame of references. The variables $z_{12}$ and $z_{22}$ may be interpreted as the rotor current vector components defined in the frame of references tied to the stator flux vector and scaled by $\Psi_s$.

Application of the nonlinear feedback to the machine model based on z variables (4)–(6) transform the DFIG model into two linear subsystems:

- mechanical subsystem:

$$\frac{dz_{11}}{d\tau} = \frac{L_m}{JL_s}z_{12} - \frac{1}{J}m_0, \qquad (9)$$

$$\frac{dz_{12}}{d\tau} = \frac{1}{T_v}(-z_{12} + m_1), \qquad (10)$$

- electromagnetic subsystem

$$\frac{dz_{21}}{d\tau} = -2\frac{R_s}{L_s}z_{21} + 2\frac{R_sL_m}{L_s}z_{22} + 2u_{sf2}, \qquad (11)$$

$$\frac{dz_{22}}{d\tau} = \frac{1}{T_v}(-z_{22} + m_2). \qquad (12)$$

where $m_1$, $m_2$ are new inputs to the plant and $u_{sf2}$ is the disturbance.

The $m_1$ input controls the machine torque but the $m_2$ input controls the $z_{22}$ variable. The name reactive torque is sometimes used in the literature. The system is decoupled and linear, however in the second equation a disturbance $u_{sf2}$ appears:

$$u_{sf2} = u_{sx}\psi_{sx} + u_{sy}\psi_{sy}. \qquad (13)$$

The influence of this disturbance is mainly visible during transients and causes the weak damped oscillations of the machine stator flux.

The DFIG considered in this paper is used as a generator connected to the grid and works as a source of power: active and reactive. The expressions for instantaneous active p and reactive q powers in steady state (for new z variables) take the form:

$$P = -\frac{L_m}{L_s}\omega_U z_{12}, \qquad (14)$$

$$Q = \frac{\omega_U}{L_s}z_{21} - \frac{L_m}{L_s}\omega_U z_{22}. \qquad (15)$$

It means that the active power depends mainly on the $z_{12}$ variable and the reactive power depends on the $z_{22}$ variable. In transient states the expressions (14, 15) are more complicated.

## B. DFIG Control System

The stator of the DFIG is directly connected to the grid and a rotor is connected to the back-to-back converter. A bi-directional power converter consists of two PWM voltage source converters. The size of the converter depends on the generator power and selected speed range. In wind applications, the speed ratio in the generator's operating range is 1:2. Because the DFIG can work as a generator in both sub- and over synchronous area, the power of the converter is usually equal to about 30% of the generator's nominal power. In addition DFIG with converter in the rotor circuit can produce or absorb reactive power. DFIG is controlled from the rotor side by forcing the rotor currents. Many control structures can be constructed and are described in the literature. The structure of the control system should enable decoupled active and reactive power control and rotor position sensorless control. Usually the reference value for the reactive power is set to zero and the reference value for active power is set to the maximal value that can be extracted from the wind.

The full DFIG control system is presented in Fig. 7. The active and reactive powers are controlled and in the system the additional subordinated controllers of the $z_{12}$ and $z_{22}$ variables exist. In a decoupling block a non-linear feedback is realized. Described above MM model is used for the control system synthesis. Nonlinear feedback based on this model enables decoupled control of the $z_{12}$ and $z_{22}$ variables and also, as it results from (14, 15), the active and reactive power decoupled control.

The control system has to be equipped with a rotor position sensor or estimation algorithm. A sensorless system is preferred. Methods of DFIG sensorless control have been developed and successfully solved [2,3]. Laboratory obtained results of the active power control during constant wind speed are presented in fig.8. Settings of PI controllers assure well damped active and reactive power transients, without overshoots. Machine is connected to the power system directly, without any additional elements like e.g. cable line.

Properties of the control processes are quite good, but at the same time the control system influence on the other uncontrolled variables is limited. Control system structure enables the stabilization of only the inner loop variables. The influence on $z_{11}$ is evident, but on the $z_{21}$ variable acts additional signal $u_{sf2}$, that can be treated as a disturbance. The additional coupling exists and during transient state weak damped oscillations of the stator flux vector $z_{21}$ appears. The amplitude is small, but may cause small active and reactive power oscillation in steady state.

There are limitations in the reactive power production by the DFM. Machine stator power and stator current is

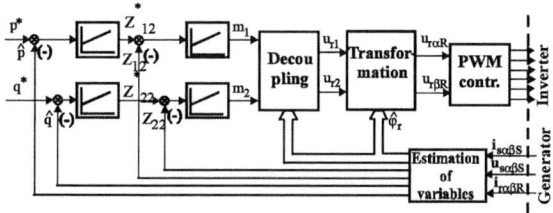

Fig. 7. DFIG control system structure

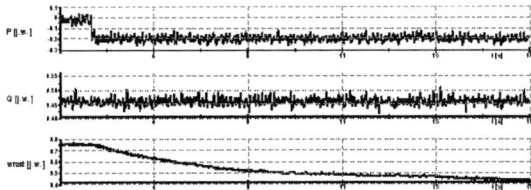

Fig. 8. Transients during power control (experiment)

controlled through the rotor current. In steady state the dependence between stator and rotor current amplitude is as follows (if neglected stator resistance):

$$I_r^2 = I_s^2 + I_m^2 - 2I_s I_m \sin \alpha \qquad (16)$$

where: $I_m = U_s / X_m$ is the magnetizing current and $\alpha$ is the angle between stator voltage and stator current. From this equation (fig.2) results that for limited (to 1 p.u.) value of the rotor current amplitude only bounded range of the $\alpha$ angle is reachable.

The green area denotes DFIG working range (from wind generator data) and is limited by machine power factor (usually 0.98cap. to 0.96 ind.). But it follows from the figure that this area can be enhanced to ABCDE area. Only limits for stator current value, rotor current value and reactive power value limits are the borders of this area. Finally, the possibility of reactive power production depends on the actual active power production. A machine can work in point P where both stator and rotor currents do not exceed limits. Enhancement of the reactive power production value brings about undesirable effects: increase of the machine rotor current and converter power.

## IV. REACTIVE POWER CONTROL

Reactive power compensation through capacitor banks or inductors is ineffective and inaccurate, but still widely used today. Often only inductors located at PCC are used. As a result wind park power factor is always inductive, but charges are smaller. Therefore to enable more accurate way of controlling reactive power, active compensation has been investigated. The main goal of the compensation method is to allow the wind farm to operate with a power factor (at PCC) which is variable within the specified range. Wind farm equipped with such a control system could take part in power system regulation.

Active compensation requires measuring reactive power at PCC and sending its value to the wind farm's central controller which decides: whether and how to control reactive power. Next central controller calculates reactive power reference values for every wind turbine or for selected wind turbines, taking into account WTs actual state. From the control theory point of view the wind park power factor control system is an open loop control system. Distance between wind turbine and PCC and transmission time delay makes closed loop control almost impossible. In such a system not one but a few wind turbines works as compensator and central controller generates only reactive power reference values. The reference value is regulated usually by means of PI action (see fig.4). Closed loop control is possible if only one generator works as a compensator or if additional static compensator located at PCC is used.

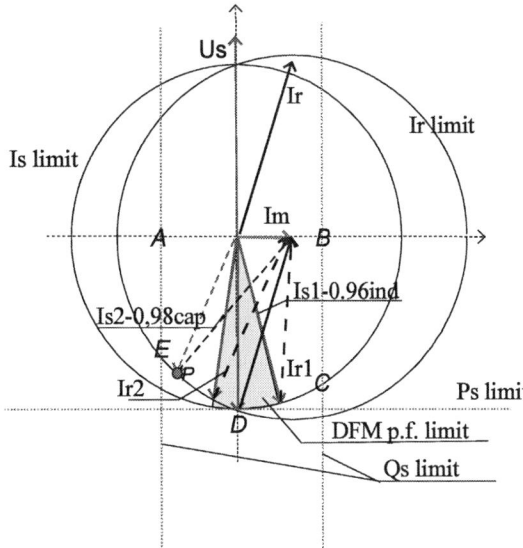

Fig.9. DFIG working area

Both control methods: active and passive were tested via simulations. Structure of the systems is presented in Fig.10 and selected simulation results are presented in Fig. 11.

## V. CONCLUSIONS

Simulation results prove that the reactive power can be compensated effectively with the use of passive methods comprising of both capacitor banks and inductors. However, since the compensation proceeds in a discrete manner, a compensation error will occur when the reactive power resulting from current wind conditions will differ from the reactive power generated by the capacitors. This drawback is not present in active methods. Figures 6a and 6b depict a comparison between active and passive compensation methods. In figure 6b transients during power factor step change are presented. In spite of well damped transients at generator ends, additional oscillatory component that results from transmission line, appears in reactive power at PCC. To reduce oscillations the DFIG controller structure should be changed; nonlinear controller (fuzzy or neural) is justified.

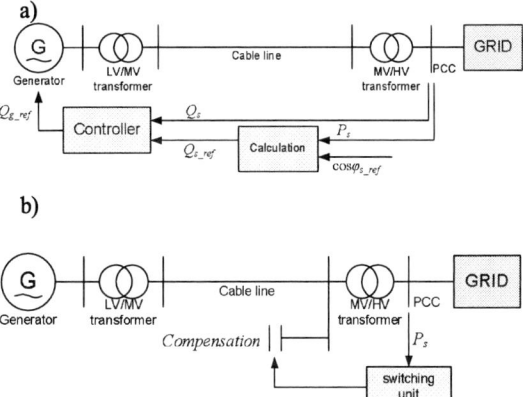

Fig.10. Reactive power compensation methods a) with feedback, b) passive

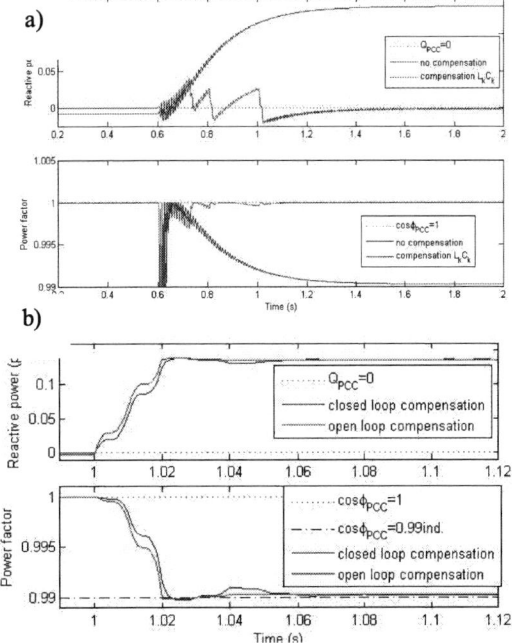

Fig.11. Passive a) and active b) method operation. Transients during wind step change from 0 to 15m/s (a) and during power factor step change from 1 to 0.99

Active methods ensure full and fast compensation. Their effectiveness depends on the type of employed wind turbines and the capability of generating/absorbing reactive power at given active power that they provide.

The methods are constrained by allowable value of wind turbine's power factor and current-carrying capacity.

## VI. REFRRENCES

[1] Kosmecki M., Analiza porównawcza metod kompensacji mocy biernej elektrowni wiatrowej, Diploma thesis, GUT, 2007

[2] Bogalecka E., Krzeminski Z., Sensorless control of DFM for wind power generators, Proc. of 10th Int. Conference EPE-PEMC, 2002

[3] Hansen A., Sørensen P., Iov F., Blaabjerg F., Centralised control of wind farm with doubly-fed induction generators, PCIM 05, Nuremberg, Germany

[4] Schulz D. Wendt O., Hanitsch R., Improved wind park power factor management – a case study, PCIM 05, Nuremberg, Germany

[5] Jermuts S., Rozenkrons J., Influence of independent producer's Power Plants on reactive power balance at load centers in Latvia, EPE'05 Conference

[6] Hansen A., Ion F., Sorensen P., Blaabjerg F.: Overall control strategy of variable speed DFIG wind turbine, Nordic wind power conference, 2004

## VII. APPENDIX

Simulation parameters:
Voltage: $U_{LV}$=690 V; $U_{MV}$=15kV; $U_{HV}$=110kV,
Wind turbine nominal power: $P_{WT}$=2MW,
Transformer leakage reactance: $X_{LV/MV}$=0.06 p.u.,
$X_{MV/HV}$=0.04 p.u.
Cable line:
$R_l$=0.64Ω/km, $L_l$=0.44mH/km, $C_l$=0,21µF/km, l=1km,
Grid:
$S_G$=100MVA; $X_G$=0.04 p.u.
Simulations were made in p.u. system.

# A Novel Hybrid Modulation Method for Cascaded H-bridge Active Power Filter

Yonggang Chen[1,2], Ping Wang[1], Yaohua Li[1], Zixin Li[1,2], Longcheng Tan[1,2]

1, Institute of Electrical Engineering Chinese Academy of Sciences
2, Graduate University of Chinese Academy of Sciences
P.O. Box 2703, 100190, Beijing, P.R.China
E-mail: chenyg@mail.iee.ac.cn

**Abstract:** In this paper, a novel hybrid modulation method is presented based on the analysis of the existing modulation method, the phase-shifted PWM modulation and the level-shifted PWM modulation. The novel modulation is best fitted for the active power filter based on the cascaded H-bridge converter. It's supported by the simulation with Matlab/Simulink. The simulation results show that the efficiency of the APF can be improved greatly with the novel hybrid modulation method.

**Keywords:** cascaded H-bridge, modulation, active power filter

Nowadays, the research on the application of the multilevel converter in active power filter (APF) is done more and more [1~3]. With the application of the multilevel converter in APF, 1) the output transformer is not needed in the high voltage application; 2) the contradiction between the capacity and the switching frequency of the switches is hushed up. The cascaded H-bridge converter is the best choice for the APF for:

1、The dc voltage source is abandoned on the dc side. A high capacity capacitor can substitute for the source for the APF hardly consume any energy, it just exchange the reactive power with the net.

2、Compared with the diode-clamped multilevel converter and flying-capacitor converter, the number of the switch is the least, etc.

The most popular modulation method used by the cascaded H-bridge converter is phase-shifted PWM modulation (PSPWM) and level-shifted PWM modulation (LSPWM) [4].

In this paper, based on the analysis of the above two modulation methods, a novel hybrid modulation for cascaded H-bridge APF is presented. Simulation is done to approve the novel hybrid modulation.

## I、The analysis of the APF based on the cascaded H-bridge converter

Fig. 1 shows the APF topology diagram based on the five-level cascaded H-bridge converter.

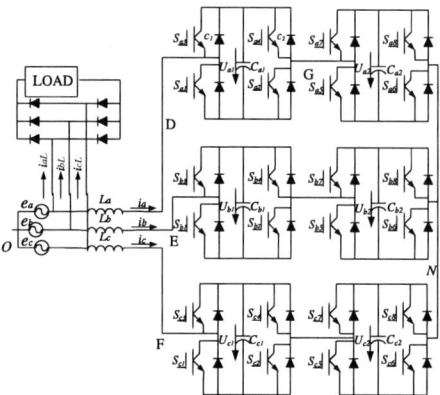

Fig. 1 the APF based on the cascaded H-bridge converter

The APF works as follows: the harmonics of the load current $i_L$ are detected by the harmonic detecting circuit. The compensated currents which has the same value but opposite in phase with the load harmonic currents are generated by the APF and then inject into the net. So the load harmonic currents are compensated.

From the fig.1, it gets:

$$L\frac{di_a}{dt} + R_L i_a = e_a - (V_{DN} + V_{NO})$$

$$L\frac{di_b}{dt} + R_L i_b = e_b - (V_{EN} + V_{NO}) \qquad (1)$$

$$L\frac{di_c}{dt} + R_L i_c = e_c - (V_{FN} + V_{NO})$$

where $R_L$ is the equivalent resistance of switching device; $L$ is the equivalent inductance. $i_m$ is the output current of the converter; $V_{nN}$ is the output voltage; $e_m$ is the net voltage; $V_{NO}$ is the voltage between the neutral point of the converter and the neutral point of the net; $m$=a, b, c; $n$=D, E, F.

From equation 1, the compensation current of APF is decided by the phase voltage of the net and the phase voltage output by the converter. The phase voltage of the converter is different with different modulation methods. The following is the analysis of different modulation

978-1-4244-1741-4/08/$25.00 ©2008 IEEE          1981

method and a novel hybrid modulation method is present.

## II、 A novel hybrid modulation method

### 1、 PSPWM modulation method

In general, a cascaded H-bridge converter with m voltage levels requires (m-1) triangular carriers. In the phase shifted multi-carrier modulation, all the triangular carriers have the same frequency and the same peak-to-peak amplitude, but there is a phase shift between any two adjust carrier waves, given by

$$\theta_{cr} = 360° / (m-1)$$

The principle of this modulation method is shown in fig. 2.

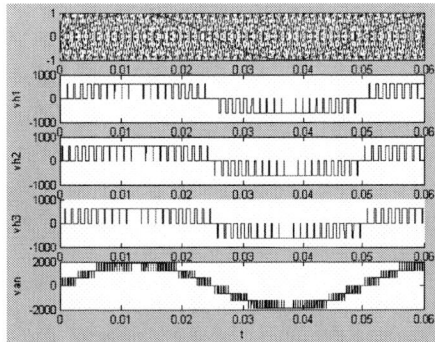

Fig. 2 the principle of PSPWM

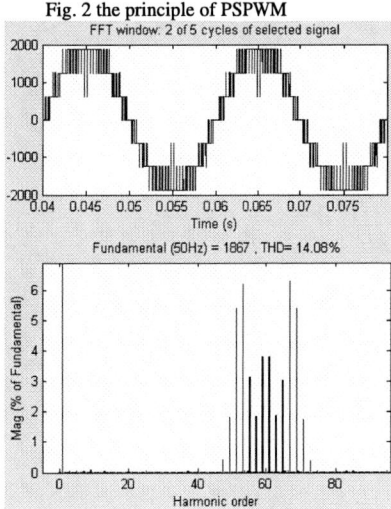

Fig. 3 the output phase voltage with PSPWM

Fig. 3 shows the simulated voltage waveforms and its harmonic content of the seven level converter operating under the condition of $m_a = 1$, $m_f = 10$. $m_a$ is amplitude modulation index and $m_f$ is frequency

modulation index. The waveforms of vh1, vh2, and vh3 are almost identical except for a small phase displacement caused by the phase shifted carriers. The converter phase voltage van does not contain any harmonics of the order lower than $4m_f$, which leads to a significant reduction in THD.

### 2、 LSPWM modulation method

Similar to PSPWM, an m level cascaded H-bridge converter using level shifted multi-carrier modulation scheme requires (m-1) triangular carriers, all having the same frequency and amplitudes. The (m-1) triangular carriers are vertically disposed such that the bands they occupy are contiguous. There are many schemes for the level shifted multi-carrier modulation, such as: in-phase disposition (IPD), where all carriers are in phase; alternative phase opposite disposition (APOD), where all carriers are alternatively in opposite disposition, etc. Fig. 4 shows the principle of the IPD modulation.

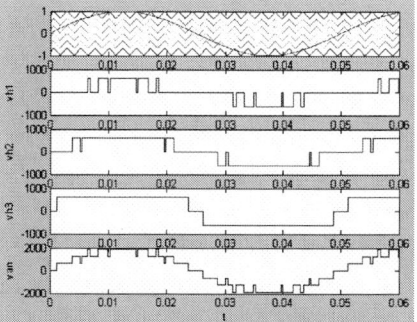

Fig.4 the principle of IPD

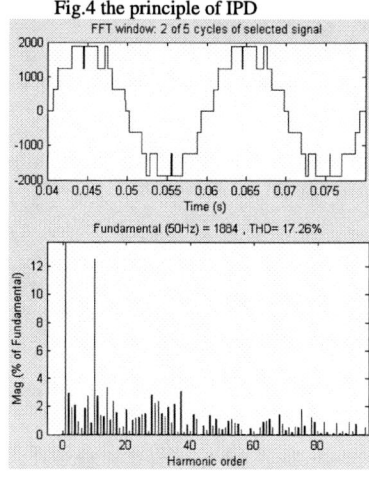

Fig. 5 the output phase voltage with LSPWM

Fig. 5 shows the simulated waveforms for a seven level converter operating under the condition of $m_a = 1$,

$m_f = 10$. Form the fig. 4 and 5, the output voltages of the H-bridge cells, vh1, vh2, and vh3, are all different, signifying that the IGBTs operate at different switching frequencies with various conduction times. Similar to the voltage waveforms produced by the PSPWM modulation, the converter phase voltage is composed of seven voltage levels. The dominant harmonics in van appear as sidebands centered around $m_f$.

3、 A novel hybrid modulation method

The PSPWM and LSPWM are quite popular when the cascaded H-bridge converter's DC side is connected with a dc source. When the dc source is replaced by a capacitor, such as the converter is used as an APF, the LSPWM can not be used for the capacitor voltage can't erect. In fig. 4, it's easy to see that there is a long time the switch hasn't any action. The voltage can be set up with the PSPWM, But as the APF is a system with fast response, a high switching frequency is needed. The switches in each cells turn on or off once in a switching period with PSPWM. The loss of turn on and off is quite large when the work voltage is high.

In this paper, a novel hybrid is presented. Like the LSPWM, two carriers disposed vertically with same frequency and amplitude but opposite in phase for one cell of the cascaded H-bridge. For any two adjust cells, a phase shift is given by

$$\theta_{cr} = 360° / n$$

where n is the cells number. Fig, 6 shows the principle of the novel hybrid modulation.

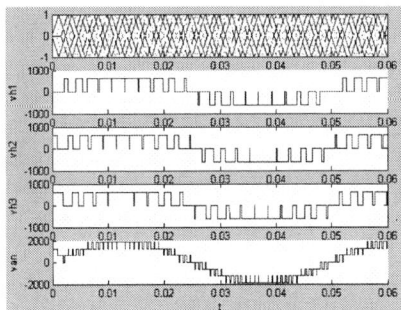

Fig.6 the principle of hybrid PWM modulation

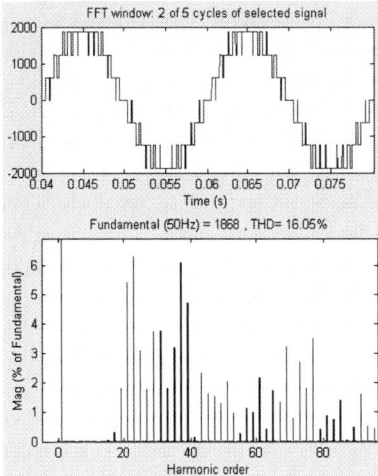

Fig. 7 the output phase voltage with hybrid PWM modulation

Fig. 7 shows the simulated waveforms for a seven level converter operating under the condition of $m_a = 1$, $m_f = 10$. From fig. 6 and 7, the output voltages of the H-bridge cells, vh1, vh2, and vh3, are almost identical except for a small phase displacement caused by the phase shifted carriers. The dominant harmonics in van appear as sidebands centered around $3m_f$. But the switching times of each cell is reduced to a half compared with PSPWM.

Table 1 the comparison of three modulation method

| comparison | PSPWM | LSPWM | Hybrid PWM |
|---|---|---|---|
| Device switching frequency | high | low | medium |
| Device conduction period | same | different | same |
| Equivalent switching frequency | （m-1）times | One times | n times |
| THD of output voltage of converter | low | high | medium |

### III、 The control of APF

The control of APF is a challenging task, due to the non-sinusoidal current that has to be generated. There are two important controlling parts to the shunt APF design. The first is the harmonic extraction and the second is the current control. The compensation principle of APF is shown in fig. 8.

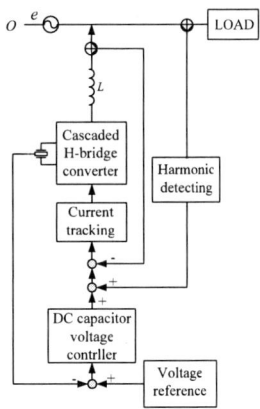

Fig.8 the compensation principle of APF

In this paper, the harmonic detecting circuit is designed an depicted in [5]. The method is based on the instant reactive power theory.

As for the current control of the APF, it can be implemented in various reference frames: stationary, fundamental, or harmonic frame. The stationary reference frame current control includes: hysteresis control [6, 7], dead-beat control [8], and linear equivalent PI control [9]. A comparison of several controllers is done in [10]. In this paper, the linear equivalent PI control is selected for its fastest possible speed of response and simple implementation. The control diagram of APF with linear current control is shown in fig. 9.

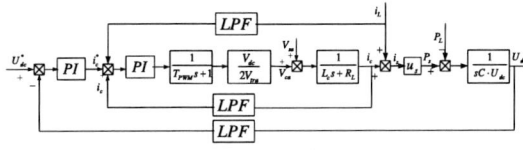

Fig. 9 the control diagram of APF

It can be seen from fig. 9 that the dc capacitor voltage is controlled via a PI controllers. The output of the PI controller is the real energy for the capacitor to maintain its voltage. Added with the harmonic current, which is come from the harmonic detecting circuit, the result is $i_c^*$ which is the reference inputs of the current loop.

The result of $i_c^*$ minus the feedback compensation current $i_c$ is controlled by the current controller which is another PI controller. The output of the controller is the modulation wave which will be modulated by carrier waves with the hybrid modulation method.

## IV、 The simulation results with the hybrid

## modulation method

In this paper, the APF based on the 5 level cascaded H-bridge converter is simulated with the novel hybrid modulation method. The simulation software is Matlab/Simulink. In the simulation, the load is a rectifier with resistance load. The line-to-line voltage of the net is 600V, the frequency is 50Hz. The inductor in the AC side of the converter is 4mH. The capacitor in the DC side of the converter is 1000uF. The reference voltage is 500V. The switching frequency is 5000。The simulation results are as follows.

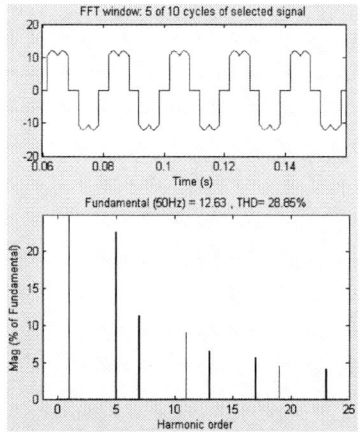

Fig.10 load current and its THD

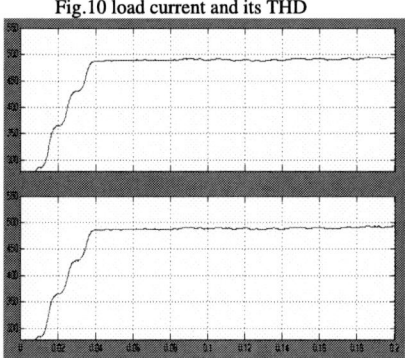

Fig. 11 the voltage waveform of DC side capacitors

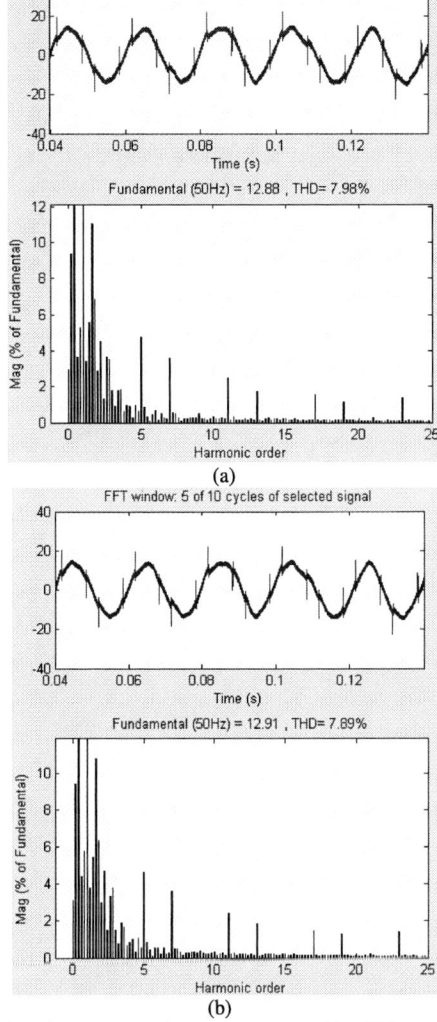

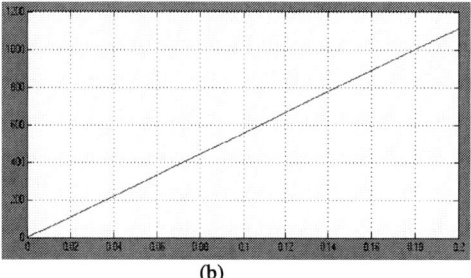

(b)

Fig.13 the comparison of switching times
(a) with hybrid modulation
(b) with PSPWM

Fig.10 shows the load current and its THD is 28.86%. fig. 11 shows the voltage waveforms of the DC side capacitors of the 5 level cascaded H-bridge converter. It can be seen that the voltage is maintained at the voltage reference. Fig. 12 shows the compensated source current and its THD. Fig. 12 (a) shows the current with hybrid modulation method, THD is 7.98%; (b) shows the current with PSPWM modulation, its THD is 7.89%. Fig. 13 shows the comparison of the switching times. Fig. 13 (a) shows the switching times with hybrid modulation method, the switching times is about 550 at 0.2 seconds; (b) shows the switching times with PSPWM modulation method, and it's about 1100. What deserves special mention is that the diagrams and parameters of the two simulations is the same besides the different modulation methods.

It can be seen from the simulation results given above that the compensation ability of the APF with the two modulation methods is quite close. However the switching times with hybrid method is half of the times with PSPWM modulation method. That means the losses of turn-on and turn-off of the switching reduce to almost a half and the efficiency of the APF is approved.

## V、 Conclusion

In this paper, a novel hybrid modulation method is presented for the APF based on the cascaded H-bridge converter. The hybrid modulation method has both the advantages of PSPWM and LSPWM modulation method. The simulation results show that the switching times of APF reduce to a half with the novel modulation method without any loss on the compensation ability. The efficiency of the APF is approved.

## References:

[1] Alan J. Watson, Patrick W. Wheeler, Jon C. Clare. A Complete Harmonic Elimination Approach to DC Link Voltage Balancing for a Cascaded Multilevel Rectifier. IEEE TRANSACTIONS ON INDUSTRIAL ELECTRONICS, VOL. 54, NO. 6, DECEMBER 2007. Page(s): 2946-2953

[2] G. Escobar, M. F. Mart´ınez-Montejano, M. Hern´andez-G´omez. A model-based controller for the cascade h-bridge multilevel converter used as a shunt

Fig. 12 compensated source current and its THD
(a) with hybrid modulation
(b) with PSPWM

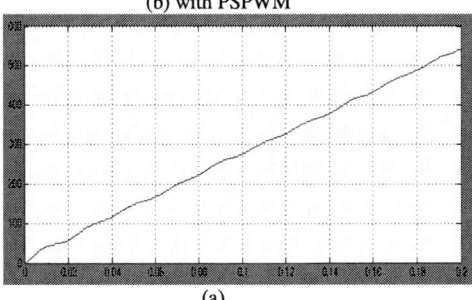

(a)

active filter. Power Electronics Specialists Conference, 2006. PESC ' 06. 37th IEEE. 18-22 June 2006. Page(s):1 – 5

[3] Miguel F. Escalante, Juan Jose Arellano. Harmonics and Reactive Power Compensation Using a Cascaded H-Bridge Multilevel Inverter. IEEE ISIE 2006, July 9-12, 2006, Montreal, Quebec, Canada. Page(s): 1966~1971

[4] Bin Wu. "High-power converters and AC drives". IEEE Press. Page(s): 127-131

[5 ]Gokhale K P,Kawamura A. Dead beat microprocessor control of PWM inverter for sinusoidal output waveform synthesis. IEEE Trans Ind Appl, 1987, 23(5). Page(s): 901-910

[6] S. Buso, L. Malesani, and P. Mattavelli, "Comparison of current control techniques for active filters applications," IEEE Trans. Ind. Electron., vol. 45, no. 5, Oct. 1998. pp. 722–729

[7] S. Buso, S. Fasolo, L. Malesani, and P. Mattavelli, "A dead-beat adaptive hysteresis current control," IEEE Trans. Ind. Appl., vol. 36, no. 4, pp. 1174–1180, Jul./Aug. 2000.

[8] L. Malesani, P. Mattavelli, and S. Buso, "Robust dead-beat current control for PWM rectifiers and active filters," IEEE Trans. Ind. Appl., vol.35, no. 3, pp. 613–620, May/Jun. 1999.

[9] X. Yuan, W. Merk, H. Stemmler, and J. Allmeling, "Stationary-frame generalized integrators for current control of active power filters with zero steady-state error for current harmonics of concern under unbalanced and distorted operating conditions," IEEE Trans. Ind. Appl., vol. 38, no. 2, pp. 523–532, Mar./Apr. 2002.

[10] S. Buso, L. Malesani, and P. Mattavelli, "Comparison of current control techniques for active filters applications," *IEEE rans. Ind. Electron.*, vol. 45, no. 5. Page(s). 722-729, Oct. 1998.

# Apparent Power Ratio of the Shunt Active Power Filter

A. Kouzou[*†], B.S Khaldi[*†], S. Saadi[**], M.O. Mahmoudi[†], M.S. Boucherit[†]

[*] Electrical Department, University Centre of Djelfa, Ain Chih  BP 3117,  Djelfa,  Algeria
E-mail: kouzouabdellah@yahoo.fr
[**] C.R.NB Birine Ain oussera, Djelfa, Algeria
[†] LCP, National Polytechnic School, Hassen Badi BP 182 El Harrach, Algiers, Algeria

*Abstract*— The main purpose of this paper is the evaluation of the Shunt Active Power Filter apparent power ratio between the apparent power generated by the source power system to the load and the compensation apparent power produced by the shunt APF to force the currents circulating toward the source to be sinusoidal and balanced. This ratio allows defining the  compensations capability for different perturbations in AC power system such as current unbalance, phase shift current and undesired harmonics generated by non linear load. This capability is determined by the maximum rate of the apparent power that can be delivered. In this paper a method of evaluation of this ratio is proposed, it is based on the definition of the effective apparent power as defined in IEEE 1459-2000 which was proved to be a suitable amount to be concerned in the design process of different devices.

*Keywords*—Active Power filter, harmonics, design, three phase system; power quality.

## I. Introduction

The evolution of industrial power electrical equipments, due to the large demands and requirements of different consumers has contributed intensively into the degradation of the power quality in AC power system in a drastically manner. The proliferation of nonlinear load, unbalanced load, large single phase loads, and voltage unbalance of one or more phases can frequently occur [1]. The current distortion and/or unbalance may cause undesirable effects on the power system operation, especially when the sensitive loads are used. Furthermore, an unbalanced power system voltage can worsen drastically the power quality, practically with power electronics converters, Ac machines and other equipments. Thus the power ratings of filters and switches are increased due to the power supplied by the source. In the other side an unbalanced voltage system supplied to an AC machine generates large negative-sequence which can increase the machine losses and reduces the machine use qualification [2,3]. A solution for the elimination of the undesired current disturbances followed toward the source or power supply is the use of The Shunt APF. A perfect power quantity and quality supply would be one that is always available, always within voltage and frequency tolerances, and has a pure sinusoidal wave shape; the deviation value from perfection which can be accepted depends on the user's application and their requirements. Users are faced extremely with the exact need for making design investment decisions about the quantity of the compensation power of the Shunt APF required to achieve the quality of power delivered from the power system source. This study will give an approach for the evaluation of the power compensation in a way to allow for the manufacturer to dimension the Shunt APF devices and to the users to get an optimum technical economical choice. Here the new definition of the apparent power is used to avoid the errors which were made in the last years when the apparent power was evaluated using classical definitions [4,5]. In this study differently from [15], it is supposed that the power system voltage is unbalanced.

## II. Shunt Active Power Filter [3-8]

Active power filter (APF) is a power electronics device based on the use of power electronics inverters ( Fig. 1 and 2). The Shunt Active Power Filter is connected in a common point connection between the source of power system and the load system which presents the source of the polluting currents circulating in the power system lines. This insertion is realized via low pass filter such as, L, LC or LCL filters.

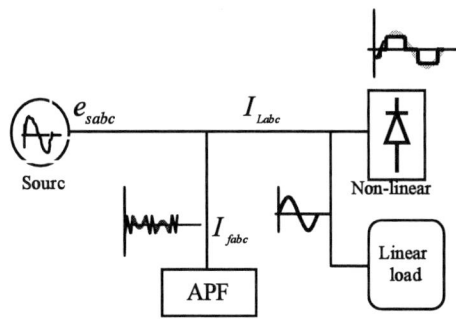

Fig. 1.  Shunt Active Power Filter principle schematics.

The Shunt APF injects current components in the power system in a small amount of power by ratio of the power delivered from source to load. The compensating power can dynamically suppress the distorted current component, eliminate the components contributing in the current unbalance and make the currents circulating toward the power source to be in phase with the direct voltage sequence of the power system voltage. The result of this is that the utility currents after the compensation become sinusoidal, balanced and with the desired amplitudes and shift phase. The fundamental equation representing the principle of the shunt APF compensation is given by:

$$I_{Labc} = I_{fabc} + I_{Sabc} \tag{1}$$

Where:

$$I_{Labc} = \begin{bmatrix} I_{La} \\ I_{Lb} \\ I_{Lc} \end{bmatrix}, I_{fabc} = \begin{bmatrix} I_{fa} \\ I_{fb} \\ I_{fc} \end{bmatrix}, I_{Sabc} = \begin{bmatrix} I_{Sa} \\ I_{Sb} \\ I_{Sc} \end{bmatrix} \tag{2}$$

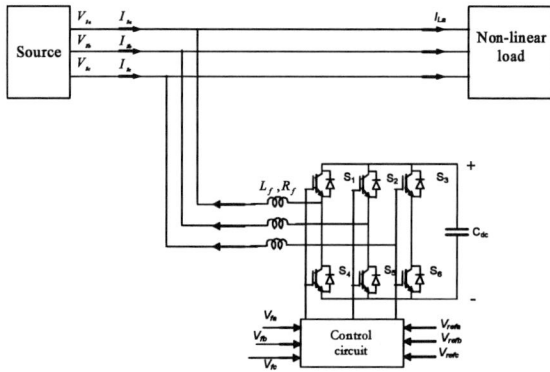

Fig. 2. Three wire schematics of the Shunt APF.

### III. SHUNT ACTIVE POWER FILTER APPARENT POWER

To clarify this study a general case was studied theoretically for three phase three wire systems, and then special cases which can be occur in industrial loads were derived, such as current harmonics, unbalance and/or distorted currents. Practically each case has its calculation to achieve exactly the compensation needed to improve the power quality from the source power system. The power system voltages are supposed to be unbalanced and presented by:

$$
\begin{aligned}
v_a &= \sqrt{2}k_a'V_1 \sin\left(\omega t + \varphi_a\right) \\
v_b &= \sqrt{2}k_b'V \sin\left(\omega t + \varphi_b - \frac{2\pi}{3}\right) \\
v_c &= \sqrt{2}k_c'V \sin\left(\omega t + \varphi_c + \frac{2\pi}{3}\right)
\end{aligned} \tag{3}
$$

$k_a', k_a', k_a'$ are the magnitude voltage unbalance factors;

$\varphi_a, \varphi_a, \varphi_a$ are the shift phase unbalance for the phases a,b and c.

The $h$ component of the load currents are then defined as follow:

$$
\begin{aligned}
i_{ah} &= \sqrt{2}k_a k_a' I_{Mh} \sin\left(h\omega t + \gamma_{ah}\right) \\
i_{bh} &= \sqrt{2}k_b k_a' I_{Mh} \sin\left(h\left(\omega t - \frac{2\pi}{3}\right) + \gamma_{bh}\right) \\
i_{ch} &= \sqrt{2}k_c k_a' I_{Mh} \sin\left(h\left(\omega t + \frac{2\pi}{3}\right) + \gamma_{ch}\right)
\end{aligned} \tag{4}
$$

The necessary apparent power which responds to the load requirement is [10-18]:

$$S_e = 3V_e I_e \tag{5}$$

Where $I_e$ is the effective current defined as follow:

$$I_e = \sqrt{\frac{I_a^2 + I_b^2 + I_c^2}{3}} = \sqrt{I_{e1}^2 + I_{eh}^2} \tag{6}$$

$$I_{e1} = \sqrt{\frac{I_{a1}^2 + I_{b1}^2 + I_{c1}^2}{3}} \tag{7}$$

$$I_{eh} = \sqrt{\frac{I_{ah}^2 + I_{bh}^2 + I_{ch}^2}{3}} \tag{8}$$

So the effective current of the fundamental component is:

$$I_{e1} = \frac{I_M}{\sqrt{3}}\sqrt{k_a^2 k_a'^2 + k_b^2 k_b'^2 + k_c^2 k_c'^2} \tag{9}$$

The effective voltage of the power system is:

$$V_e = \frac{V}{3}\sqrt{2\left(k_a^2 + k_b^2 + k_c^2\right) - \Delta V} \tag{10}$$

$$\Delta V = \sum_{\substack{i,j=a,b,c \\ i \neq j}}^{a,b,c} k_i k_j \cos\left(|\varphi_i - \varphi_j| + \frac{2\pi}{3}\right) \tag{11}$$

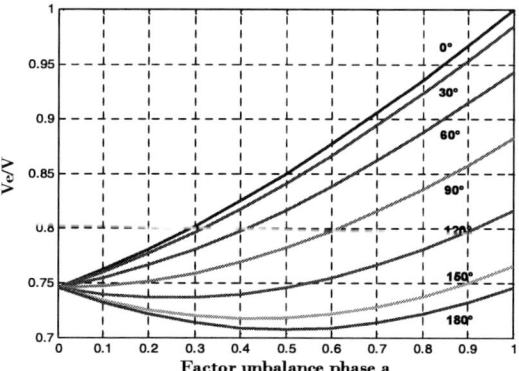

Fig.3. The effective voltage when one phase unbalance is occurred.

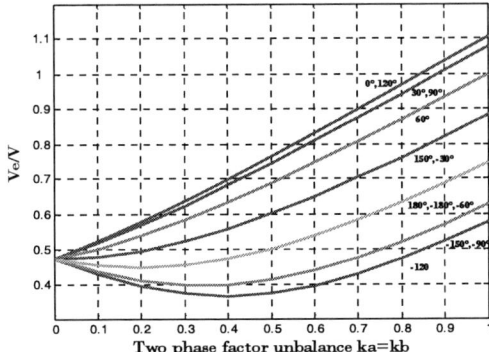

Fig.4. The effective voltage when similar Two phase unbalances are occurred.

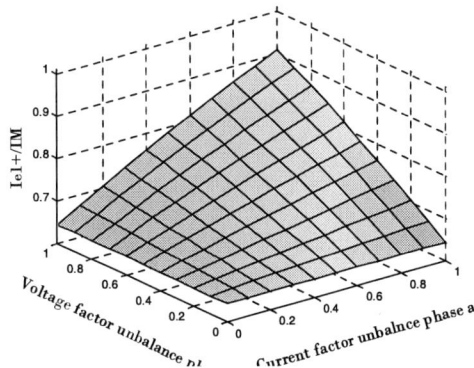

Fig.5. The fundamental positive sequence effective current for the same one phase voltage unbalance and current unbalance.

The apparent power is then:

$$S_e^2 = 9V_e^2 I_e^2 = 9V_e^2 I_{e1}^2 + 9V_e^2 I_{eh}^2 \qquad (12)$$

$$S_e^2 = S_{e1}^2 + S_{eh}^2 \qquad (13)$$

$S_{eh}$ is the apparent power responsible of different harmonics contained in the load current.

The apparent power due to the fundamental component of the current is calculated as follow:

$$S_{e1}^2 = 9V_e^2 I_{e1}^2 \qquad (14)$$

This power contains two parts:

- A component due to the fundamental positive component of current, it is the one generated by the power system to the load. This power is given by:

$$S_{e1}^+ = 3V_e \, I_{e1}^+ \qquad (15)$$

- A component due to the negative and zero components of the current, it is the one responsible of the unbalance and distortion in the load side. The Shunt APF must produce and inject this power to eliminate the unbalance and distortion of the current absorbed from the source of the power system. This power is given by:

$$S_{unb}^2 = S_{e1}^2 - S_{e1}^{+2} \qquad (16)$$

The fundamental positive component of the current is given by:

$$I_{e1}^+ = \frac{I_M}{3} \sqrt{k_a^2 k_a'^2 + k_b^2 k_b'^2 + k_c^2 k_c'^2 + \Delta I} \qquad (17)$$

$$\Delta I = \sum_{\substack{i,j=a,b,c \\ i \neq j}}^{a,b,c} k_i k_i' k_j k_j' \cdot \cos\left(\varphi_i + \gamma_i - \varphi_j - \gamma_j\right) \qquad (18)$$

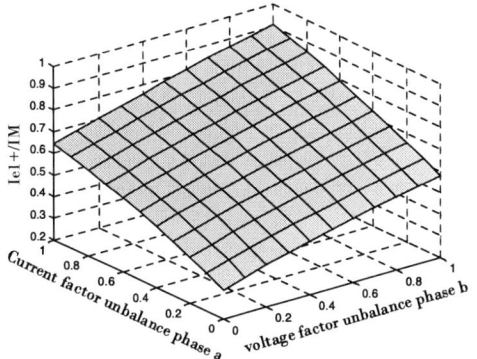

Fig.6. The fundamental positive sequence effective current for the same one phase voltage unbalance and current unbalance of two different phases.

This leads to the apparent power responsible of the unbalance in the load currents:

$$S_{unb} = 3V_e \cdot \sqrt{I_{e1}^2 - I_{e1}^{+2}} \qquad (19)$$

It can be written as:

$$S_{unb} = V_e \cdot I_M \cdot \sqrt{2 \cdot \Delta k - \Delta I} \qquad (20)$$

Where:

$$\Delta k = k_a^2 k_a'^2 + k_b^2 k_b'^2 + k_c^2 k_c'^2 \qquad (21)$$

The power responsible of different harmonics contained in the load current is given by:

$$S_{eh} = 3V_e \cdot I_{eh} \qquad (22)$$

Where the effective harmonic current is:

$$I_{eh} = I_h \sqrt{\frac{k_a^2 k_a'^2 + k_b^2 k_b'^2 + k_c^2 k_c'^2}{3}} \qquad (23)$$

$$I_{eh} = I_h \sqrt{\frac{\Delta k}{3}} \qquad (24)$$

From equations (09) and (21) equation (24) can be written as:

$$I_{eh} = I_h \cdot \frac{I_{e1}}{I_M} = THD_e^I \cdot I_{e1} \qquad (25)$$

$THD_e^I$ is the total harmonic distortion of the load current, it is presented by $\sigma$ so :

$$S_{eh} = 3 V_e \cdot \sigma \cdot I_{e1} \qquad (26)$$

$$S_{eh} = \sqrt{3} V_e I_M \cdot \sigma \cdot \sqrt{\Delta k} \qquad (27)$$

Finally in order to achieve a unite power factor in the source, the reactive power needed by the load have to be canceled from the fundamental components of voltage and current. Thus the Shunt APF has to generate the apparent power needed so that the voltages in the three phases have the same shift phase angles as the currents absorbed from the source by the load in the corresponding three phases. The following schematics shows the case of phase a.

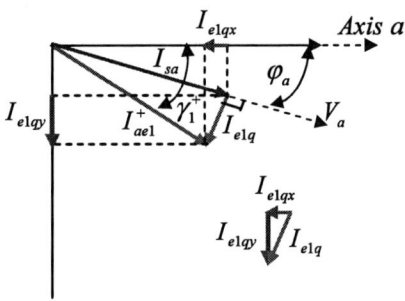

Fig. 7. Principle of canceling the shift phase between voltage and current of phase a.

The magnitude of the positive sequence of the current is the same as the magnitude of the effective positive sequence:

$$\begin{bmatrix} I_{a1}^+ \\ I_{b1}^+ \\ I_{c1}^+ \end{bmatrix} = \begin{bmatrix} I_{e1}^+ \\ I_{e1}^+ \\ I_{e1}^+ \end{bmatrix} \qquad (28)$$

The currents needed to achieve the elimination of the reactive power to be absorbed from the power system are $I_{a1q}^+, I_{b1q}^+$ and $I_{c1q}^+$. To obtain the minimum magnitude

of these components they must be perpendicular on the source currents of the corresponding phases as it is shown on Fig. 3, the magnitude of these currents are then:

$$\begin{bmatrix} I_{a1q}^+ \\ I_{b1q}^+ \\ I_{c1q}^+ \end{bmatrix} = I_{e1}^+ \cdot \begin{bmatrix} \sqrt{1 - \cos(\varphi_a - \gamma_1^+)} \\ \sqrt{1 - \cos(\varphi_b - \gamma_1^+)} \\ \sqrt{1 - \cos(\varphi_c - \gamma_1^+)} \end{bmatrix} \qquad (29)$$

The effective current of these components can be evaluated as:

$$I_{e1q}^+ = I_{e1}^+ \cdot \sqrt{1 - \frac{\sum\limits_{i=a,b,c} \cos(\varphi_i - \gamma_1^+)}{3}} \qquad (30)$$

Or:

$$I_{e1q}^+ = I_{e1}^+ \cdot \Delta q \qquad (31)$$

Where:

$$\Delta q = \sqrt{1 - \frac{\sum\limits_{i=a,b,c} \cos(\varphi_i - \gamma_1^+)}{3}} \qquad (32)$$

The corresponding effective apparent power is:

$$S_{e1q}^+ = 3 \cdot V_e . I_{e1q}^+ \qquad (33)$$

This leads to the following expression:

$$S_{e1q}^+ = S_{e1}^+ \cdot \sqrt{1 - \frac{\sum\limits_{i=a,b,c} \cos(\varphi_i - \gamma_1^+)}{3}} \qquad (34)$$

Finally:

$$S_{e1q}^+ = V_e I_M \cdot \sqrt{\Delta k + \Delta I} \cdot \Delta q \qquad (35)$$

The total apparent power necessary to achieve a good compensation is then calculated by:

$$S_{comp} = \sqrt{S_{unb}^2 + S_{eh}^2 + S_{e1q}^{+\,2}} \qquad (36)$$

So:

$$S_{comp} = V_e I_M \sqrt{S_{comp1} + S_{comp2}} \qquad (37)$$

Where:

$$S_{comp1} = \Delta k \cdot \left(2 + 3 \cdot THD_e^{I\,2} + \Delta q^2\right) \qquad (38)$$

$$S_{comp2} = (\Delta q^2 - 1) \cdot \Delta I \qquad (39)$$

The positive apparent power ratio is supposed as:

$$R_p = \frac{S_{e1}^+}{S_{e1}} \qquad (40)$$

This can be written as:

$$R_p = \frac{I_{e1}^+}{I_{e1}} \qquad (41)$$

It leads to:

$$R_p = \frac{1}{\sqrt{3}} \cdot \sqrt{1 + \frac{\Delta I}{\Delta k}} \qquad (42)$$

Where:

$$0 \prec R_p \leq 1 \qquad (43)$$

But practically values of $R^+$ are not far from 1. The Shunt active power filter is then characterized by the ratio:

$$R = \frac{S_{comp}}{S_s} \qquad (44)$$

Where:

$$S_s = 3 \cdot V_e \cdot I_{se} \qquad (45)$$

Presents the apparent power delivered by the power system (source) to the load with an optimized cost. $I_{se}$ is the effective current circulating from the source to the PCC, it can be calculated by:

$$I_{se} = \sqrt{\frac{I_{sa}^2 + I_{sb}^2 + I_{sc}^2}{3}} \qquad (46)$$

Where:

$$\begin{bmatrix} I_{sa} \\ I_{sb} \\ I_{sc} \end{bmatrix} = I_{e1}^+ \cdot \begin{bmatrix} \cos(\varphi_a - \gamma_1^+) \\ \cos(\varphi_b - \gamma_1^+) \\ \cos(\varphi_c - \gamma_1^+) \end{bmatrix} \qquad (47)$$

The effective source current is then:

$$I_{se} = I_{e1}^+ \cdot \Delta\beta \qquad (48)$$

Where:

$$\Delta\beta = \sqrt{\frac{\sum_{i=a,b,c} \cos^2(\varphi_i - \gamma_1^+)}{3}} \qquad (49)$$

The apparent power of the source is:

$$S_s = 3 \cdot V_e \cdot I_{e1}^+ \cdot \Delta\beta = S_{e1}^+ \cdot \Delta\beta \qquad (50)$$

The compensation apparent power products by the active power filter is:

$$S_{comp} = \frac{S_{e1}^+}{R_p} \sqrt{1 + \sigma^2 + R_p^2 \cdot (\Delta q^2 - 1)} \quad (51)$$

The apparent power ratio of the Shunt APF can then be written by the following expression:

$$R = \frac{1}{R_p \cdot \Delta\beta} \cdot R_0 \qquad (52)$$

Where:

$$R_0 = \sqrt{1 + \sigma^2 + R_p^2 \cdot (\Delta q^2 - 1)} \qquad (53)$$

$R$ gives a clear idea of the Shunt Active Power filter dimension to fulfill the desired compensations, it can also be used in the process design of the devices used in this compensators.

In this study the loses due to the devices operations such as the switching lose of static switches weren't taken into account, as it is neglected beyond the apparent power needed for the compensation.

IV. EVALUATION OF THE APPARENT POWER RATIO OF THE SHUNT ACTIVE POWER FILTER WITH DISTURBANCES OF PARTICULAR CASES

The following cases give examples of the apparent power ratio of the Shunt APF in special cases of disturbances in three phase three wire AC power system. It is supposed in these cases that the power system voltages are balanced.

**Case I:** one phase unbalanced load. The following values are taken to calculate the different parameters used in the evaluation of the power ratio:

$$k_a = k_b = k_c = 1, \ k_b' = k_c' = 1, \ k_a' = k$$
$$\varphi_a = \varphi_b = \varphi_c = 0, \ \gamma_b = \gamma_c = 0, \ \gamma_a = \gamma, \sigma = 0.$$

The following figures give the values of the power ratio $R$ of the Shunt APF, and the positive sequence apparent power ratio $R_p$. It is clear from Fig. 8 that the power ratio equals 0, when there is no unbalance in phase a, this means that the load is linear and the load currents are balanced. The compensation needed from the Shunt APF is nil. The positive sequence power ratio can be written as:

$$R_p = \frac{1}{\sqrt{3}} \cdot \sqrt{1 + \frac{4k_a' \cdot \cos(\gamma_a) + 2}{k_a'^2 + 2}} \qquad (53)$$

It give a clear image about the effect of the unbalance in the load current, when this ratio is near to unit, the compensation apparent power of the Shunt APF is near to zero and the compensation needed is minimal.

$R$ is depending on $R_p$, $\Delta\beta$, $R_0$ where:

$$\begin{cases} \Delta\beta = \cos\left(\gamma_1^+\right) \\ \gamma_1^+ = a\tan\left(\dfrac{k_a' \cdot \sin(\gamma_a)}{k_a' \cdot \cos(\gamma_a)+2}\right) \\ R_0 = \sqrt{1+R_p^{\,2} \cdot \left(\Delta q^2 - 1\right)} \\ \Delta q = \sqrt{1-\cos\left(\gamma_1^+\right)} \end{cases} \quad (54)$$

$\Delta\beta$ presents the effect of the shift phases of voltages power system and the fundamental positive sequence of the current. Therefore, in the present case if $\gamma_a = 0$ then:

$$\begin{cases} \Delta\beta = 0 \\ \Delta q = 0 \\ R_0 = \sqrt{1-R_p^2} \end{cases} \quad (55)$$

Finally:

$$R = \sqrt{\frac{1}{R_p^2}-1} \quad (56)$$

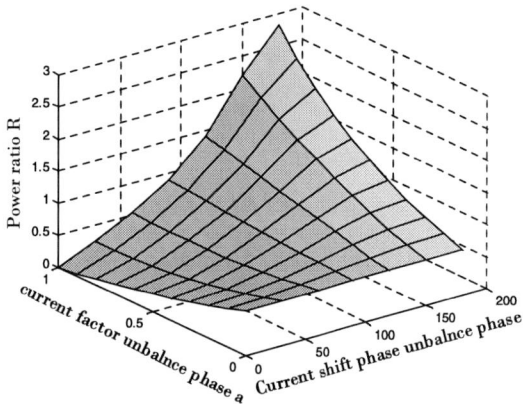

Fig.8. Power ratio when one phase unbalance current is occurred.

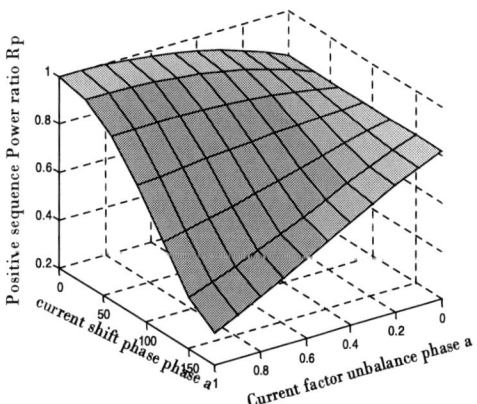

Fig.9. Positive sequence power ratio when one phase unbalance current is occurred.

It is clear that $R$ depends only on $R_p$. In Fig. 8 and 9 it is shown clearly for $k = 1$ (balanced load currents) that:

$$\begin{cases} R_p = 1 \\ R = 0 \end{cases} \quad \text{Hence no compensation is needed.}$$

**Case II:** Two phase unbalanced load. The following parameters are used in the evaluation of the power ratio:

$$k_a = k_b = k_c = 1,\ k_b' = k,\ k_a' = k, k_c' = 1$$
$$\varphi_a = \varphi_b = \varphi_c = 0, \gamma_a = \gamma_b = \gamma_c = 0, \sigma = 0.$$

It is clear in Fig. 10 that the power ratio equals 0 for $k_b' = k_a' = 1$, this means no compensation is needed for balanced power system voltages and balanced linear load currents. This value is maximal when the currents of phases b and c are nil, in this case the power compensation needed from the Shunt APF is nearly twice the power produced from the power system to make the power system currents to be balanced, these results are given with linear loads, but such constraints are so far from practical cases and leads the shunt APF to be useless. The positive sequence power ratio in this case is given by:

$$R_p = \frac{1}{\sqrt{3}} \cdot \sqrt{1+\frac{2k^2+4k}{2k^2+1}} \quad (57)$$

$R$ is depending in this case only on $R_p$ and $R_0$ :

$$R = \frac{R_0}{R_p} \quad (58)$$

Where:

$$\begin{cases} \Delta\beta = 1 \\ \gamma_1^+ = 0 \\ R_0 = \sqrt{1-\dfrac{1}{3}\left(1+\dfrac{2k^2+4k}{2k^2+1}\right)} \\ \Delta q = 0 \end{cases} \quad (59)$$

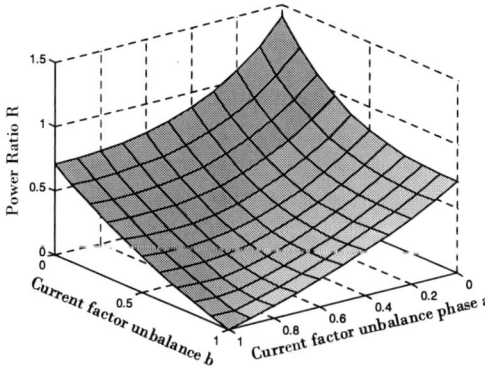

Fig.10. Power ratio when two phase unbalance current are occurred.

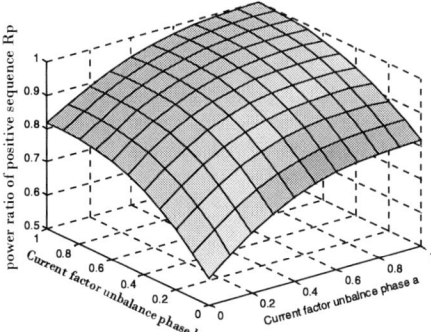

Fig.11. . Power ratio of the positive sequence when two phase unbalance current are occurred.

From (57) and (59) for $k = 1$ the following values are obtained :

$$\begin{cases} R_0 = 0 \\ R_p = 1 \qquad \text{No compensation is needed.} \\ R = 0 \end{cases}$$

These results can be shown clearly in Fig. 10 and 11.

**Case III:** One phase unbalanced with non linear load. The following values are taken to calculate the different parameters used in the evaluation of the power ratio:

$$k_a = k_b = k_c = 1, \ k_a^{'} = k , k_c^{'} = k_b^{'} = 1$$

$$\varphi_a = \varphi_b = \varphi_c = 0, \gamma_a = \gamma_b = \gamma_c = 0, \ \sigma \succ 0 .$$

The following figures give the values of the power ratio $R$ of the Shunt APF, and the positive sequence apparent power ratio $R_p$. It is clear in Fig. 12 that the power ratio equals $0$ for $k = 1$ and $\sigma = 0$, this means no compensation is needed from the Shunt APF for balanced power system voltages and balanced linear load currents. This value is maximal when the current of phases is nil and $\sigma$ equals to unit, in this case the power compensation needed from the Shunt APF is greater than the power produced from the power system to improve the quality of the currents circulating toward the power system to be balanced. It is clear that these results are given with non linear loads with the same high level harmonics distortion in the three phases, but such constraints are so far from practical cases ( where $\sigma \prec 1$ ) and leads the shunt APF to be useless. The same remark can be noticed for the positive sequence power ratio as in case I which is presented by:

$$R_p = \frac{1}{\sqrt{3}} \cdot \sqrt{1 + \frac{4k + 2}{k^2 + 2}} \qquad (60)$$

$R$ is depending in this case only on $R_p$ and $R_0$ :

$$R = \frac{R_0}{R_p} \qquad (61)$$

Where:

$$\begin{cases} \Delta\beta = 1 \\ \gamma_1^+ = 0 \\ R_0 = \frac{1}{\sqrt{3}} \sqrt{2 + 3\sigma^2 - \frac{4k + 2}{k^2 + 2}} \\ \Delta q = 0 \end{cases} \qquad (62)$$

From (62) if $k = 1$ (balanced load currents) then:

$$\begin{cases} R_p = 1 \\ R = R_0 = \sigma \end{cases} \qquad (63)$$

Hence, the compensating power needed is:

$$S_{comp} = \sigma \cdot S_s \qquad (64)$$

It is depending on the quality of the non linear load. This presents the curve where the factor unbalance is equal to unit.

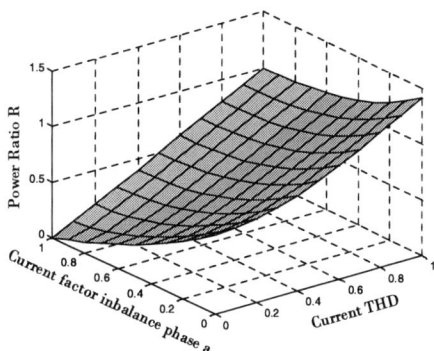

Fig.12. . Power ratio of the positive sequence when one phase unbalance current is occurred for non linear load.

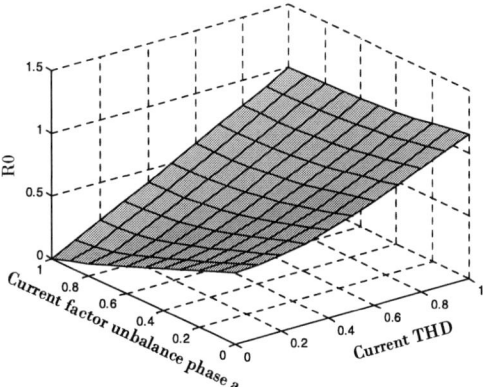

Fig.13. R0 when one phase unbalance current is occurred for non linear load.

## IV. AN OPTIMAL EVALUATION OF THE SHUNT ACTIVE POWER FILTER APPARENT POWER RATIO

For a given characteristics of the load, the apparent power ratio can be evaluated using its constraints, furthermore, an optimal values can be obtained where $R$ is the function to be minimized. Indeed $R$ is depending on several variables presenting the parameters of the load and power system voltages. For the power system voltages the variables are:

$$\begin{cases} k_a, k_b, k_c \\ \varphi_a, \varphi_b, \varphi_c \end{cases} \qquad (61)$$

For the load the variables are:

$$\begin{cases} k_a', k_b', k_c' \\ \gamma_a, \gamma_b, \gamma_c \\ \sigma \end{cases} \qquad (62)$$

To simplify the calculation the following variables can be used, these variables are resulting from the previous variables where:

The variables vector $= \begin{bmatrix} \Delta k & \Delta I & \Delta q & \sigma & \Delta \beta \end{bmatrix}$

The objective function to be minimized is then:

$$R = f(\Delta k, \Delta I, \Delta q, \sigma, \Delta \beta) \qquad (63)$$

Under the constraints:

$$\begin{cases} \Delta k_{min} \leq \Delta k \leq \Delta k_{max} \\ \Delta I_{min} \leq \Delta I \leq \Delta I_{max} \\ \Delta q_{min} \leq \Delta q \leq \Delta q_{max} \\ \sigma_{min} \leq \sigma \leq \sigma_{max} \\ \Delta \beta_{min} \leq \Delta \beta \leq \Delta \beta_{max} \end{cases} \qquad (64)$$

The min and max values of these variables are calculated from the basic variables. Finding $R_{min}$ has an important use in the process design, therefore to minimize the cost of the necessary Shunt APF.

## VI. CONCLUSION

This paper deals with the evaluation of the Shunt Active Power Filter apparent power ratio which contributes in the determination of the used devices dimensions. This obtained value defining the needed compensating apparent power for special loads or special consumer needs, the objective is to avoid over or/and under dimensions. This is important for the manufacturers and users to minimize economically the burdens of production and the use of such equipments. This study shows the errors which were made for the evaluation of the apparent power to dimension these equipments using the classical definitions those are correct for sinusoidal balanced systems of voltages and currents. The definition of the equivalent or effective apparent power is used to avoid the errors mentioned. The main aim is to evaluate the optimal apparent power ratio following to the approach presented in this paper. It is important to clarify that the evaluation of the equipments needed by the users is determined by the constraints of the loads to be fed and also by the constraints of the power system source constraints. The approach given in this paper can achieve this aim.

## REFERENCES

[1] Lee C Y, Chen B K, Lee W J, Hsu YF."Effets of Various unbalanced voltages on the operation performane of an induction motor under the same voltage unbalnce factor ondition". Annual Meeting, IEEE 1997: 51- 59.

[2] Souto CN, De Oliveira JC, Ribeiro Pf, Neto LM."Power quality impact on performance and associated osts of three-phase inction motors",Proeeding Harmonis and Quality of power 1998, Vol. 2,791-797.

[3] Jan Svensson, a. Sannino, "Ative Filtering of Supply Voltage with Series-Connected Voltage Soure Inverter," EPE 2001 Graz.

[4] H. Akagi,Y. Kanazawa, A. Nabea,"Generalized theory of the instantaneous reactive power in three-phase circuits,"IPEC'83-Int. Power Electronics Conf., Tokyo, Japan, 1983, pp. 1375-1386.

[5] H.akagi,Y. Kanazawa, A. Nabea, "Istanteneous reactive power compensators comprising switching devices without energy storage components," IEEE Trans Ind. Appl, Vol .I4-20, p.625, 1984.

[6] M.A.E. Alali, S. Saadate, Y. A. Chapuis, "Energetic Study of a Series active power conditioner compensating voltage dips, unbalanced voltage and voltage harmonis,"IEEE-IEP-2000 Aapulo, Mexio, p. 80-86, October 2000.

[7] M.A.E. Alali, S. Saadate, Y. A. hapuis, "Control and analysis of Series and shunt active power filters with SABER,"IPEC , Tokyo, Japan,pp. 1467-1472, April 2000.

[8] H. Watanabe, M. Aredes, "New concepts of instantaneous active and reactive powers in electrical systems with generic loads," IEEE, Trans on Power Delivry, Vol. 8, N°, 2, April 1993.

[9] Moleykutty George and 2Kartik Prasad Basu, "Modeling and Control of Three-Phase Shunt Active Power Filter", American Journal of Applied Sciences 5 (8): 1064-1070, 2008

[10] J. L.Willems, J. A. Ghijselen, and A. E. Emanuel, "The apparent power concept and the IEEE standard 1459-2000," IEEE Trans. Power Del., vol. 20, no. 2, pp. 876–884, Apr. 2005.

[11] Definitions for the Measurement of Electric Quantities Under Sinusoidal, Nonsinusoidal, Balanced or Unbalanced Conditions, IEEE Std. 1459-2000, Jun. 21, 2000.

[12] A. E. Emanuel, "Reflections on the effective voltage concept," in Proc. 6th Int. Workshop on Power Definitions, Milano, Italy, Oct. 2003, pp.1–7.

[13] J. L. Willems, Jozef A. Ghijselen, A. E. Emanuel, " Addendum to the Apparent Power Concept and the IEEE Standard 1459-2000", IEEE Transactions on Power Delevery, vol. 20, no. 2, April 2005.

[14] Pajic, S. Emanuel, A.E. ,"A comparison among apparent power definitions", IEEE Power Engineering Society General Meeting, 18-22 June 2006.

[15] Pajic, S. Emanuel, A.E. ,"Modern Apparent Power definitions: Theoretical Versus Practical Approach-The General case" IEEE transactions on power Delivery, Vol.21,NO.4 ,October 2006.

[16] A. Kouzou. B.S. Khaldi. M.O. Mahmoudi. M.S. Boucherit " Shunt Active Power filter Apparent Power For Design Process" SPEEDAM 2008. Ischia , Italy. pp. 1402-1408.

[17] A. Kouzou. B.S. Khaldi. M.O. Mahmoudi. M.S. Boucherit "The Effect of the Zero Sequence Component on the Evaluation of the Series APF Apparent Power " ICEEE 2008. Okinawa , Japan. 6-7 July.

[18] A. Kouzou. B.S. Khaldi. M.O. Mahmoudi. M.S. Boucherit "Apparent Power Evaluation of Series Active Power Filter with Recent Definitions" IEEE SSD'08 2008. Amman , Jordan. 20-23 July.

# Shunt Active Power Filter with Improved Dynamic Performance

## Krzysztof Piotr Sozanski*

*University of Zielona Gora, Institute of Electrical Engineering, Zielona Gora, Poland,
e-mail: *K.Sozanski@iee.uz.zgora.pl*

*Abstract*—**This paper describes a shunt active power filter (APF) with improved dynamic performance. When the value of load current changes rapidly, the typical APF transient response is too slow, and the line current suffers from dynamic distortion. This distortion causes an increase in harmonic content in the line current, which is dependent on a time constant. The APF control current dynamics is dependent on the inverter output time constant consisting of APF output inductance and resultant impedance of load and mains. In the proposed circuit transient performance of APF is improved using a modified output inverter. According to this modification the APF dynamics are improved. The Matlab simulation results of a modified APF are also presented in the paper.**

*Keywords*—**Power quality, DSP, pulse width modulation (PWM).**

## I. INTRODUCTION

Shunt active power filters (APF) are one of the best devices for compensating the harmonics and asymmetries of the mains currents caused by nonlinear loads [1]. A harmonic compensation circuit with current-fed active power filter with open loop (with unity gain) is depicted in Fig. 1, where $Z_s$ represents the mains power line impedance, $Z_l$ represents nonlinear load, $u_s$ represents mains power line voltage. The shunt active power filter injects AC power current $i_c$ to cancel the main AC harmonic content. The line current $i_s$ is the result of summing the load current $i_l$ and the compensating current $i_c$

$$i_S = i_L + i_C . \qquad (1)$$

A simplified block diagram of the active power compensation circuit with the parallel APF for a power of 75 kVA is depicted in Fig 2. This APF was built in the Institute of Electrical Engineering at the University of Zielona Gora [7]. The APF is used for checking the control algorithm. The APF's circuit consists of a power part with a three-phase IGBT power transistor bridge IPM (intelligent power module) PM300DSA120 from Mitsubishi [5], connected to the AC mains through an inductive filtering system composed of inductors $L_{C1}$, $L_{C2}$, $L_{C3}$.

The APF circuit contains DC energy storage, ensured by two capacitors, $C_1$ and $C_2$. The control circuit is realized using the digital signal processor ADSP-21364 (EZ-KIT Lite). The active power filter injects the harmonic currents $i_{C1}$, $i_{C2}$, $i_{C3}$ into the power network and

offers a notable compensation for harmonics, reactive power and unbalance.

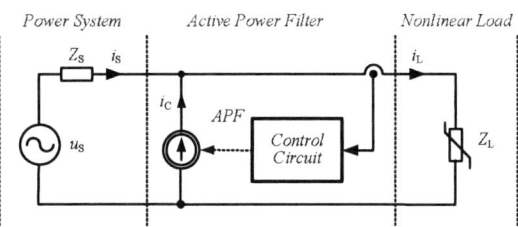

Fig. 1. Harmonic compensation circuit with current-fed active power filter with open-loop (with unity gain).

The APF control current dynamic is dependent on the inverter output time constant, itself resulting from APF output inductance and resultant impedance of load and mains power line (Fig. 2). When the value of load current changes rapidly, as in current $i_l$ in Fig. 3, the APF transient response is too slow [9] and the line current $i_s$ suffers from dynamic distortion. This distortion causes an increase in harmonic content in the line current, which is dependent on a time constant. In the APF shown in Fig. 2 the THD ratio is increased by about 10%.

The loads can be divided into two main categories: predictable loads and noise-like (unpredictable) loads. Most loads belong to the first category. For this reason it is possible to predict current values in subsequent periods, after a few periods of observation [9], [4].

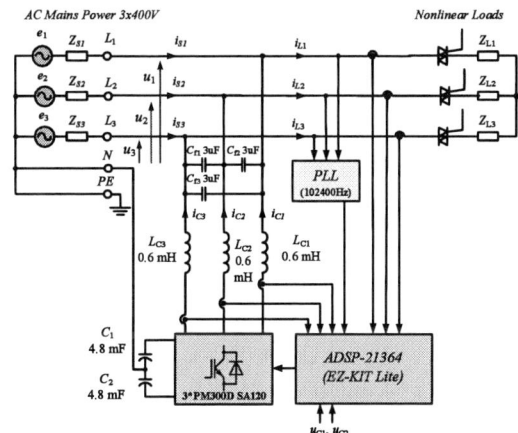

Fig. 2. Test circuit of classical three-phase shunt active power filter.

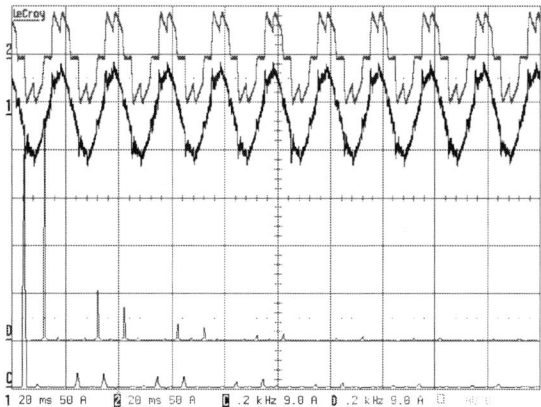

Fig. 3. Experimental waveforms of active power filter in steady-state with the resistive load: load current $i_l$ (red), line current $i_s$ (blue).

TABLE I.
DISCUSSION OF POSSIBLE SOLUTIONS

| For predictable loads | For unpredictable loads |
|---|---|
| 1. Typical APF with non-causal control algorithm.<br>2. Typical APF with repetitive control algorithm. | 1. High speed APF.<br>2. Set of two APFs:<br>- high power low speed APF,<br>- low power high speed APF.<br>3. APF with modified output inverter. |

Possible solutions for improvement of APF dynamic are shown in Table I. This paper describes an APF with modified output inverter [8]. Presented solution is especially good suitable for unpredictable loads.

## II. APF OUTPUT INVERTER

A diagram of a simple power inverter model connected to the mains power is shown in Fig. 4a, where $C_c$ is an output filter capacitance, for dumping modulation components. In this circuit the time constant is mainly dependent on inductor $L_c$ value. Therefore when the transistor switching period $T_s=1/f_s$ is much less than the circuit main time constant $\tau_c$ it is possible to simplify the circuit (Fig. 4a) to the circuit shown in Fig. 4b. The circuit resultant resistance $R_c$ mainly depends of the resistance of inductor $L_{C1}$,

$$R_C \cong R_{LC} \quad . \quad (2)$$

For switching state $S_1=1$, $S_2=0$ compensating current $i_C$ can be calculated by formula

$$i_C(t) = \frac{u_{DCp}(t_0) - u_S(t_0)}{R_c}\left(1 - e^{-\left(\frac{R_c t}{L_c}\right)}\right) + i_C(t_0)e^{-\left(\frac{R_c t}{L_c}\right)} \quad , (3)$$

and for state $S_1=0$, $S_2=1$ $i_C$ can be calculated by formula

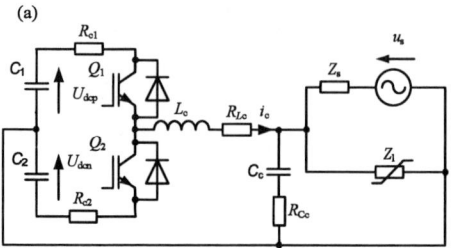

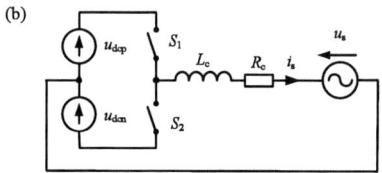

Fig. 4. Diagrams of APF output inverter connected to the mains power: (a) simplified circuit, (b) simplified inverter model connected to the mains power, used for current ripple calculation.

$$i_C(t) = \frac{-u_{DCn}(t_0) - u_S(t_0)}{R_c}\left(1 - e^{-\left(\frac{R_c t}{L_c}\right)}\right) + i_C(t_1)e^{-\left(\frac{R_c t}{L_c}\right)} \quad , (4)$$

where:

$$\tau_C = \frac{L_C}{R_C} \quad . \quad (5)$$

If it is assumed that

$$f_s \gg \frac{1}{\tau_C} \quad \text{and} \quad f_s \gg \frac{1}{T_M} \quad , \quad (6)$$

where: $T_M$ – mains period,
during the switching period $T_s$. and voltages $u_S$, $u_{DCp}$, $u_{DCn}$ are constant, and that average current $i_C$ is constant too, then the output current can be calculated by simplified equations: for state $S_1=1$, $S_2=0$

$$i_C(t_1) = \frac{u_{DCp} - u_S(t_0)}{L_C}t_1 + i_{Cn} \quad , \quad (7)$$

and for state $S_1=0$, $S_2=1$

$$i_C(t_2) = i_{Cp} + \frac{-u_{DCn} - u_S(t_0)}{L_C}t_2 \quad , \quad (8)$$

where: $t_1$–$t_0$ switch-on time for $S_1$, $t_2$-$t_1$ switch-on time for $S_2$.

A time diagram of idealized compensating current $i_C$ is shown in Fig. 5. Output ripple can be calculated by the equation

$$\Delta i_C(t) = \left|\frac{u_{DCp} - u_S(t_0)}{L_C}t_1\right| = \left|\frac{-u_{DCn} - u_S(t_0)}{L_C}t_2\right| \quad . \quad (9)$$

The voltage value at capacitors $C_1$ and $C_2$ is stabilized by a voltage controller and is equal to $u_{DC}$; this is why it is possible to describe $u_{DC} = u_{DCp} = u_{DCn}$. To achieve high dynamic performance the output current $i_C$ slew rate must be high. The slew rate can be calculated by the formula

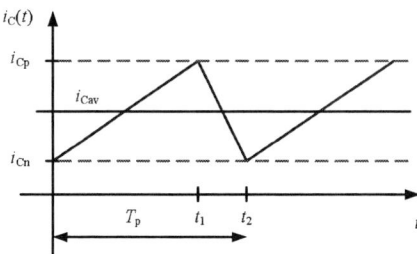

Fig. 5. Time diagram of idealized compensating current $i_c$

$$\left|\frac{\mathrm{d}\,i_{\mathrm{C}}(t)}{\mathrm{d}t}\right| = \left|\frac{\pm u_{\mathrm{DC}} - u_{\mathrm{S}}(t)}{L_{\mathrm{C}}}\right| \quad . \qquad (10)$$

Currently IGBT transistors are mostly used as switching elements in the inverters. For the ordinary IGBT the maximum switching frequency is equal to 20 kHz [5] and around 60 kHz [6] for fast IGBT. The transistor switching power losses can be approximately described by the formula

$$P_{\mathrm{tot}} \approx f_{\mathrm{k}} E_{\mathrm{k}} + P_{on} \quad , \qquad (11)$$

where: $E_{\mathrm{k}}$ - energy lost in a single switching cycle, $P_{on}$ - power losses in switched-on state.

So it is possible to assume that transistor power losses are proportionally dependent on the switching frequency.

According to the above-mentioned problem, choosing the right value of inductor $L_c$ is very difficult. Discussion about selecting the right value of APF output inverter inductor is shown in Table II. For a higher $L_c$ value, the time constant is higher and dynamic distortion is bigger, while for a lower $L_c$ value, circuit dynamic distortions are smaller, but the value of compensating current ripple $i_c$ is higher. One of the ways to decrease the dynamic distortion and keep current ripple at a reasonable value is to increase transistor switching frequency, but in this case there are increased switching losses and an influence from the switching transition.

### III. APF WITH MODIFIED OUTPUT INVERTER

Given that high dynamic performance is necessary only for approximately 10% of the time in the mains power period, increasing the switching time to 60 kHz seems to be ineffective. Therefore, the author is proposing an inverter output stage (Fig. 6) with two sets of transistors (fast and slow) and inductors. The circuit has common DC bank ($C_1$, $C_2$) for both parts of inverter. The simplified version of this proposition is shown in Fig. 7.

The circuit consists of two output stages [11], [8] : one with switches $S_{s1}$, $S_{s2}$ and inductor $L_{cs}$, and a second with switches $S_{f1}$, $S_{f2}$ and inductor $L_{cf}$. The first works continuously with the slowest switching frequency $f_{p1}$. The value of inductor $L_{cs}$ is designed to achieve a low $i_c$ current ripple. In the second, switches $S_{f1}$ and $S_{f2}$ work with a several-times higher frequency only in the case when output current changes very quickly (typically 10% of mains power period). The value of inductor $L_{cf}$ was designed to achieve a fast response in the output current.

TABLE II.

DISCUSSION OF INDUCTOR VALUE

| Bigger value of inductor | | Lower value of inductor | |
|---|---|---|---|
| Positives | Negatives | Positives | Negatives |
| - lower current ripple,  - lower transistor switching frequency. | - slower transitions response,  - bigger cost and weight. | - faster transitions response,  - lower cost and lower weight. | - higher value of current ripple,  - higher switching frequency,  - bigger influence from the switching transition. |

At the beginning a hysteresis digital modulator was designed for controlling the modified inverter. Taken into consideration during the simulation analysis were the modified inverter and classical inverter. The simulation parameters are: $L_{Cf} = 0.5$ mH, $L_{Cs} = 2.5$ mH, $u_{DC} = 390$ V, $f_{p2} = 51200$ Hz, $f_{p1} = 25600$ Hz. Currently being built at our Institute is a three-phase modified inverter for a 75 kVA shunt active filter (Fig. 8). This APF will be used for testing APF control algorithms.

A simplified diagram of the modified output inverter simulation circuit is shown in Fig. 9. The control algorithm of two hysteresis digital modulators with additional conditional control logic implemented in Matlab is shown in Listing I. Step responses for modified inverter and classic inverter are shown in Fig. 10. The classic inverter response time is about 420 μs, and is near 70 μs for the modified inverter.

The hysteresis digital modulator is one of the simplest and safest, especially at the early experimental stage. It has a lot of disadvantages, especially for digital implementation [3], therefore during future investigations other modulator control algorithms will be designed and implemented.

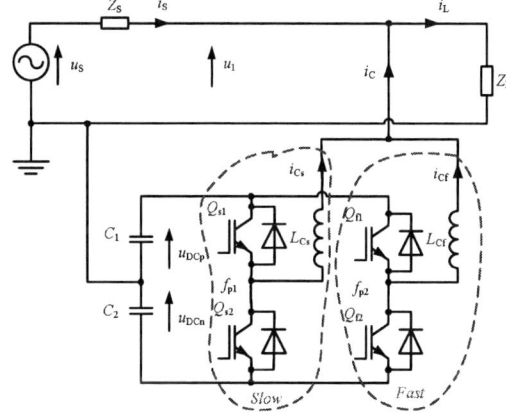

Fig. 6. Single-phase active power filter with modified inverter, test circuit

**1997**

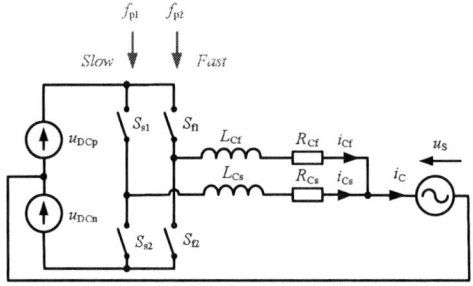

Fig. 7. Simplified diagram of modified inverter model connected to the mains power

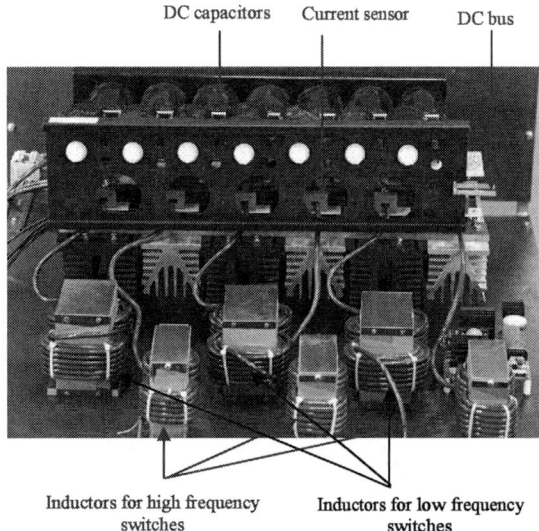

Fig. 8. Three-phase active power filter modified inverter.

LISTING I.

MATLAB PROGRAM OF TWO HYSTERESIS MODULATORS

```
e=i_cref(n)-i_c(n)*kr;
 e_s=i_ref(n)-i_cs(n)*kr;
 if u_cs>=0
  if e_s>kh*h
   u_cs=u_DCp-u_s(n);
   if e<-h&&u_cf>=0
    u_cf=u_DCn-u_s(n);
   elseif e>h&&uc_f<=0
    u_cf=u_DCp-u_s(n);
   end
  else
   u_cf=0;
   if blad<-h
    u_cs=u_DCn-u_s(n);
   else
    u_cs=u_DCp-u_s(n);
   end
  end
 else %u_cs <0
  if e_s<-kh*h
   uc_s=u_DCn-u_s(n);
   if e<-h&&u_cf>=0
    u_cf=u_DCn-u_s(n);
   elseif e>h&&u_f<=0
    u_cf=u_DCp-u_s(n);
   end
  else
   u_cf=0;
```

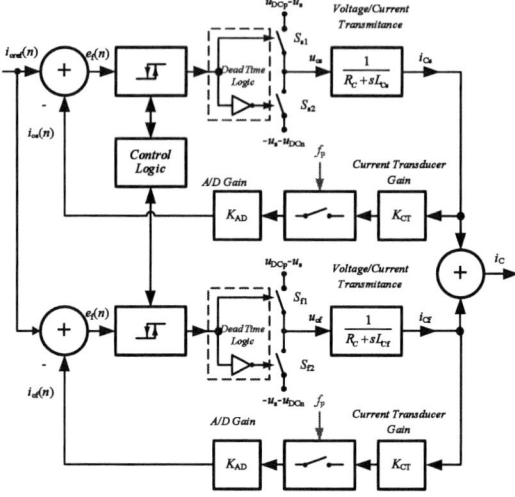

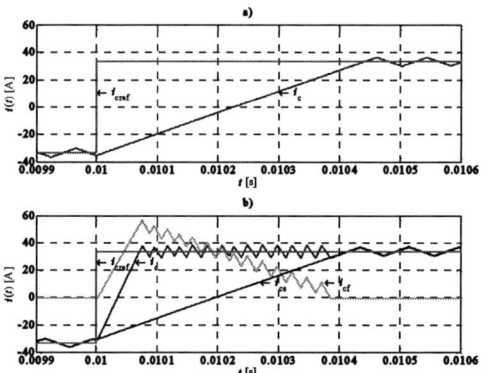

Fig. 9. Simplified block diagram of the output inverter simulation circuit

Fig. 10. Step responses of two types of inverters: a) classical inverter $i_{cref}(t)$ – red, $i_c(t)$ – blue, b) modified inverter $i_{cref}(t)$ – red, $i_c(t)$ – blue, $i_{cf}(t)$ – green, $i_{cs}(t)$ – black

```
    if e>h
     u_cs=u_DCp-u_s(n);
    else
     u_cs=u_DCn-u_s(n);
    end
   end
  end
```

## A. APF Simulation Results

A block diagram of the control circuit for the considered APF is presented in Fig. 11. The control algorithm uses sliding DFT [2], [9], [10], for the load current first harmonic detection. In respect to sliding DFT characteristics control circuits have to be synchronized to line voltages $u_1(t)$, $u_2(t)$, $u_3(t)$. This is why the synchronization unit is one of the most important parts of the control circuit. The digital control circuit is synchronized with the mains voltage by a synchronization unit. It consists of a low-pass filter and phase-locked loop circuit (PLL).

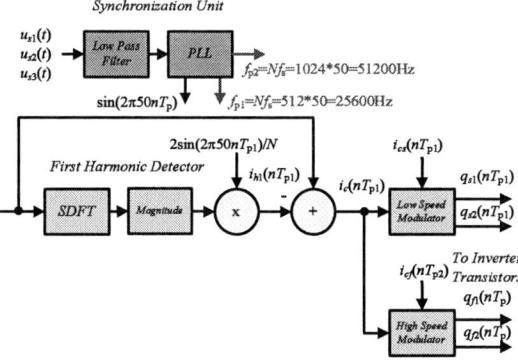

Fig. 11. Block diagram of APF control algorithm.

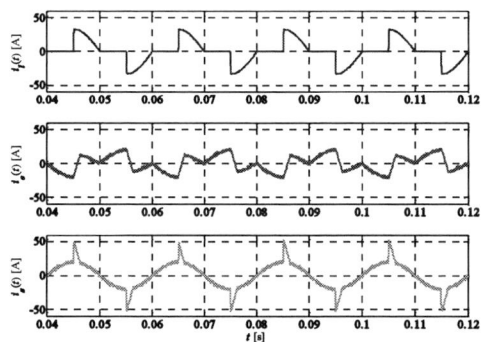

Fig. 12. Simulation waveforms of single-phase active power filter in steady-state with the resistive load, classical inverter: load current $i_l$, compensating current $i_c$, line current $i_s$.}

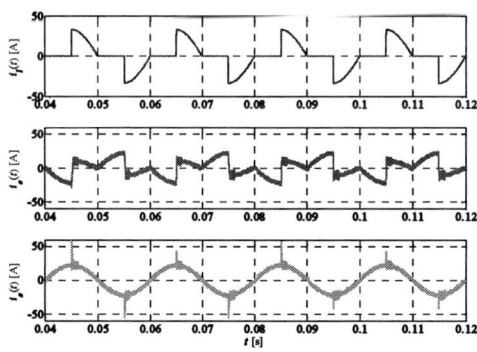

Fig. 13. Simulation waveforms of single-phase active power filter in steady-state with the resistive load, with modified output inverter: load current $i_l$, compensating current $i_c$, line current $i_s$.}

Figures 12 and 13 show the simulation waveforms in the same steady-state conditions, for classical circuit (Fig. 12) and for circuit with modified inverter (Fig. 13). Depicted are the following waveforms: load currents $i_l$, compensating currents $i_c$, line currents $i_s$. Using the modified inverter it is possible to decrease the harmonic

contents in power line currents from about THD=15% to about THD=5%.

The results of the simulation analysis confirm good dynamic performance of the modified inverter used in a shunt active power filter. The presented solution will be employed together with the non-causal algorithm described, while for predictable loads there will be used only a working non-causal algorithm, but for unpredictable rapid change of load current the fastest part of the modified inverter will be working.

For assumed simulation parameters current ripples are higher when the fastest part of the inverter is switched-on, but the resultant value of THD ratio is less if compared to the classical inverter.

## IV. CONCLUSION

The presented solution is good for noise type nonlinear loads (such as in an arc furnace), where the load currents are non periodic and stochastic. Using the proposed APF with improved dynamic performance it is possible to decrease harmonics contents of line current.

## REFERENCES

[1] H. Akagi, E., H. Watanabe, M. Aredes, ``Instantaneous power theory and applications to power conditioning,'' *Wiley-Interscience a John Wiley & Sons, Inc., Publication*, 2007.

[2] E. Jacobsen, R. Lyons, "The sliding DFT", *Signal Processing Magazine*, IEEE, Vol. 20, No. 2, March 2003.

[3] M. Kazimierkowski, L. Malesani, "Current Control Techniques for Three-Phase Voltage-Source Converters: A Survey," *IEEE Transactions on Industrial Electronics*, vol.45 N0 5, October, 1998.

[4] S. Mariethoz, A. Rufer, "Open Loop and Closed Loop Spectral Frequency Active Filtering," *IEEE Transactions on Power Electronics*, vol.17, N0 4, July, 2002.

[5] Mitsubishi, ``Mitsubishi Intelligent Power Modules, PM300DSA120,'' *Data Sheet*, September 2000.

[6] Mitsubishi, ``Mitsubishi Intelligent Power Modules, CM200DU-24NFH,'' *Data Sheet*, February 2004.

[7] K. Sozanski, R. Strzelecki, A. Kempski, ``Digital Control Circuit for Active Power Filter with Modified Instantaneous Reactive Power Control Algorithm,'' *IEEE 33rd Annual IEEE Power Electronics Specialists Conference - PESC 2002*, Conference proceedings. Cairns, Australia, 2002.

[8] K. Sozanski, "The Shunt Active Power Filter with Better Dynamic Performance," *Power Tech 2007 Conference*, Lausanne, Switzerland, 2007.

[9] K. Sozanski, "Harmonic Compensation Using the Sliding DFT Algorithm," *35rd Annual IEEE Power Electronics Specialists Conference - PESC '04*, Aachen, Germany, 2004.

[10] K. Sozanski, "Sliding DFT Control Algorithm for Three-Phase Active Power Filter," *21rd Annual IEEE Applied Power Electronics Conference - APEC '06*, Dallas, Texas, USA, 2006.

[11] S. Watanabe, P. Boyagoda, H. Iwamoto, M. Nakaoka, H. Takano, "Power conversion PWM amplifier with two paralleled four quadrant chopper for MRI gradient coil magnetic field current tracking implementation", *Power Electronics Specialists Conference*, 1999. PESC 99, 30th Annual IEEE Volume 2, Issue , 1999 Page(s):909 - 913 vol.2.

# The Research on the Active Power Filter Based on the Cascaded H-bridge Converter

Yonggang Chen[1,2], Junling Chen[1], Ping Wang[1], Yaohua Li[1], Longcheng Tan[1,2], Zixin Li[1,2], Wei Xu[1,2]

1, Institute of Electrical Engineering Chinese Academy of Sciences
2, Graduate University of Chinese Academy of Sciences
P.O. Box 2703, 100190, Beijing, P.R.China
E-mail: chenyg@mail.iee.ac.cn

*Abstract*- The model and the normal control method of the cascaded H-bridge active power filter is described firstly in this paper. However, problems exist in the normal control. One of the main problems is the unbalance of dc capacitor voltages of the H-bridge units. In this paper, an improved method for dc capacitor voltages control of the cascaded H-bridge active power filter is presented. With slightly shifting the modulation waves, it makes the dc capacitor voltages balancing in the accepted tolerance without any additional units. The method is validated by simulation and experiment.

*Keywords*- Active filter, cascaded H-bridge converter, dc voltage balance control

## I. INTRODUCTION

Harmonic mitigation in power system is a goal which has occupied a great deal of research since the early 60s. Passive filters are used for harmonic mitigation due to their advantages of simplicity, low cost, and easy maintenance. But the disadvantage that these filters introduced are numerous [1], [2]. Nowadays, the attention of researchers has been drawn to the active power filter (APF). The shunt active filter can be used for harmonic and reactive power compensation. It acts as a harmonic current source which injects an anti-phase but equal magnitude to the harmonic and reactive load current to eliminate the harmonic and reactive components of the supply current. There are two ways to apply the shunt APF in the medium voltage range, either by using a transformer or by extending the semiconductor switches rating (via multilevel converters or series connection of semiconductor devices). For the transformer case, the problem lies in the high cost for medium voltage transformers. Also, as the voltage decreases at the secondary side, the current increases which mean parallel operated converters may be needed.

For the second solution there are two aspects. For the first part, series operation of semiconductor switches faces problems of unbalances static and dynamic voltage sharing due to the spread of device characteristic and/or mismatch of drive circuits. For the second part, multilevel converters achieve high voltage switching by means of a series of accumulated voltage steps, each of which lies within the rating of the individual power devices. Also lower total harmonic distortion can be achieved. There are three basic types of multilevel converters: cascaded type multilevel,

neutral point clamped, and flying capacitor converters. It has been recognized that the cascaded H-bridge inverter is the promising topology for the APF application due to its attractive features of the modular structure and high level output voltage without switches in series. In this paper, as shown in Fig. 1, the cascaded type, seven-level inverter is used as a shunt APF. [3]

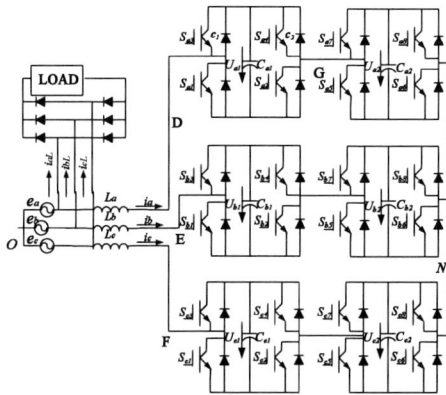

Fig. 1 seven-level inverter used as a shunt APF

Unfortunately, problems exist in the cascaded H-bridge APF. One of the main problems is the unbalance of dc voltages in H-bridge units. Many reasons may contribute towards unbalanced dc capacitor voltages: unequal delays in switching, shunt loss, and hybrid loss, etc. The dc capacitor-voltage unbalance may result in uneven voltage stress on switches and the distortion of inverter output voltage due to the degradation of total harmonic distortion (THD) factor. [4]

Many methods is proposed to maintain the dc voltages balance: to exchange the digital output signals to the semiconductor devices between each unit of the three H-bridge units in each phase[3]; to add an additional unit to exchange the energy of each dc capacitor of the H-bridge units in each phase. In this paper, a new method based on the control system of the APF is proposed. There are no additional units needed.

## II. THE MODEL OF CASCADED H-BRIDGE APF

The main circuit of the cascaded H-bridge APF is shown in

Fig. 1. The circuit will be analyzed under the following assumptions:

1) The object of the dc capacitor voltage control is to balance the capacitor voltages, here the capacitor voltages of each H-bridge is equivalent to $V_{dc}$ at first.

2) The utility is a three phase balanced, sinusoidal voltage source

3) The filter inductors are linear; saturation is not considered.

As shown in Fig. 1:

$$L\frac{di_a}{dt} + R_L i_a = e_a - (V_{DN} + V_{NO})$$

$$L\frac{di_b}{dt} + R_L i_b = e_b - (V_{EN} + V_{NO}) \qquad (1)$$

$$L\frac{di_c}{dt} + R_L i_c = e_c - (V_{FN} + V_{NO})$$

where $R_L$ is the equivalent resistance of switching device and inductor $La$.

Firstly, the output voltage of the first H-bridge of phase A is analyzed. Phase-shifted multi-carrier modulation is used to generate the PWM signals [5]. So the output voltage of the H-bridge is the subtraction of the output of the two arms of the bridge. As for the upper arm, when switch $S_{a3}$ is on and switch $S_{a1}$ is off, the switching function is

$$c_1 = 1, \text{ and } V_{DN} = v_{dc}.$$

Otherwise

$$c_1 = 0, \text{ and } V_{DN} = 0.$$

As for the lower arm of the bridge, when switch $S_{a4}$ is on and switch $S_{a2}$ is off, the switching function is

$$c_2 = 1, \text{ and } V_{DN} = v_{dc}$$

Otherwise

$$c_2 = 0, \text{ and } V_{DN} = 0$$

So the output voltage of the H-bridge is

$$V_{DG} = (c_1 - c_2)V_{dc}. \qquad (2)$$

The output voltage of the phase A is the addition of the output voltage of the two H-bridge. It is

$$V_{DN} = (c_1 + c_3 - c_2 - c_4)V_{dc} \qquad (3)$$

where $c_3$, $c_4$ are the switching functions of the corresponding

switches of the second H-bridge of phase A.

Correspondingly,

$$V_{EN} = (c_5 + c_7 - c_6 - c_8)V_{dc}$$
$$V_{FN} = (c_9 + c_{11} - c_{10} - c_{12})V_{dc} \qquad (4)$$

where $c_5$ to $c_{12}$ are the switching functions of the corresponding switches of the upper arms of the other two phase. So (1) becomes

$$L\frac{di_a}{dt} + R_L i_a = e_a - \left[(c_1 + c_3 - c_2 - c_4)V_{dc} + V_{NO}\right]$$

$$L\frac{di_b}{dt} + R_L i_b = e_b - \left[(c_5 + c_7 - c_6 - c_8)V_{dc} + V_{NO}\right]$$

$$L\frac{di_c}{dt} + R_L i_c = e_c - \left[(c_9 + c_{11} - c_{10} - c_{12})V_{dc} + V_{NO}\right]$$

(5)

The voltage $V_{NO}$ can be obtained by adding the three equations of (5) together:

$$V_{NO} = -\frac{1}{3}\sum_{m=0}^{5}(c_{2m+1} - c_{2m+2}) \cdot V_{dc} \qquad (6)$$

Each capacitor voltage of the cascaded H-bridge APF is

$$C\frac{dv_{dc}}{dt} = (c_{2m+1} - c_{2m+2})i_n \qquad (7)$$

where $m=0,1,\ldots,5$, $n=a,b,c$. The formula above is based on the 5 level converter. It can be used to the n level converter as well. To introduce the control more popularly, the control of this type converter discussed below is based on the 7 level converter.

### III. CONTROL OF CASCADED H-BRIDGE APF

There are two important controlling parts to the shunt APF design. The first is the harmonic extraction and the second is the current control. The harmonic extraction method employs the method discussed in [8] which is based on the instant reactive power theory.

As for the current control, there are three methods which are popular [9]. In this paper, linear current control is used. Phase shifted multi-carrier modulation is used as a pulse width modulation (PWM) technique [6, 7]. Using a seven-level inverter reduces the ratings of the semiconductor switches and improves total harmonic distortion. Fig. 3 shows the control diagram of the cascaded H-bridge APF. It contains two control loops: the current loop and the voltage loop. When the dynamic response of the current loop is fast enough, the current loop can be substituted by a gain.

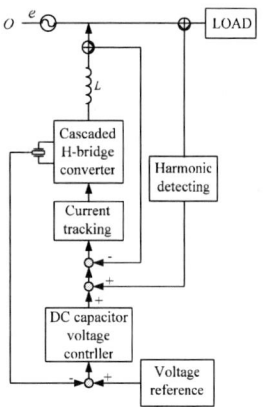

Fig.2 the control topology of APF

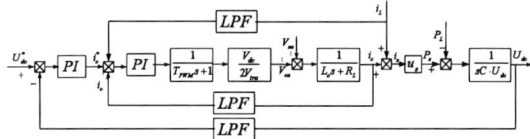

Fig.3 the control diagram of APF

As shown in fig.4, the dc capacitor voltages are controlled via PI controllers. To simplify the diagram, only one phase is offered. The outputs of the controllers are the real energy for the capacitor to maintain its voltage. With harmonic current, which is come from the harmonic detecting circuit, minus the output of the voltage loop controller, the value is the reference inputs of the current loop.

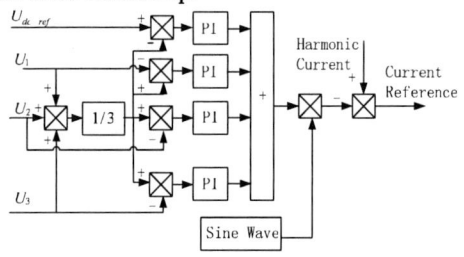

Fig.4 the control diagram of voltage loop

## IV. BALANCE CONTROL OF THE INDIVIDUAL H-BRIDGE DC VOLTAGES

In order to implement the individual dc capacitor voltage balance control, every dc capacitor voltage has to be measured by a voltage sensor and controlled separately. In this paper, a new method is presented as follows.

As for a capacitor voltage of a H-bridge unit in the seven-level APF, The capacitor voltage is:

$$U_{dc}(t) = \frac{1}{C} \int_{-\infty}^{t} i_{dc}(t)dt \qquad (8)$$

where $U_{dc}$ is the capacitor voltage, C is the capacitance of the capacitor, and $i_{dc}$ is the current flowing through the capacitor and equals:

$$i_{dc} = SW \cdot i_c \qquad (9)$$

where $SW$ is the switching function of the H-bridge, and

$$SW = \begin{cases} -1 & i_c > 0 \\ 1 & i_c < 0, \\ 0 & i_c = 0 \end{cases}$$

where $i_c$ is the compensation current. Substituting (9) into (8) gives:

$$dU_{dc} = \frac{1}{C} SW \cdot i_c \cdot dt \qquad (10)$$

The average voltage of per phase in the proposed APF is:

$$U_{av} = \frac{U_1 + U_2 + U_3}{3} \qquad (11)$$

where $U_{av}$ is the average voltage, $U_1$, $U_2$, $U_3$ is the capacitor voltage of each H-bridge respectively. To balance the dc capacitor voltages of a phase, each capacitor voltage should reach the average voltage. The differences of each capacitor are:

$$U_{dff1} = U_1 - U_{av}$$
$$U_{dff2} = U_2 - U_{av} \qquad (12)$$
$$U_{dff3} = U_3 - U_{av}$$

So:

$$U_{dff1} + U_{dff2} + U_{dff3} = 0 \qquad (13)$$

As the period time is too short, $i_c$ can be seen as a constant. Equation (10) shows that the voltage across each capacitor can be maintained by prolonging or shortening the current flowing through it. This can be achieved by slightly shifting the modulating wave of phase-shifted PWM. Equation (13) means the total energy flowing in per phase is equal in a period. The energy used to maintain the capacitor voltage balance flows inside the phase.

Substituting (12) into (8), the delay time for each capacitor are:

$$|\Delta t_1| = \left| \frac{C \cdot U_{dff1}}{i_c} \right|$$

$$|\Delta t_2| = \left| \frac{C \cdot U_{dff2}}{i_c} \right| \qquad (14)$$

$$|\Delta t_3| = \left| \frac{C \cdot U_{dff3}}{i_c} \right|$$

where $\Delta t1$, $\Delta t2$, $\Delta t3$ is the delay time for each capacitor respectively. The shifted distances are proportional to the delay time. It is

$$\Delta U_m = k \cdot \Delta t_m \qquad (15)$$

where $\Delta U_m$ is the shifted distance, $k$ is proportional gain between $\Delta U_m$ and $\Delta t_m$, $m$=1, 2, 3.

The direction of phase shift follows:

$$sign(\Delta U_m) \cdot sign(U_{dffm} \cdot i_c) = -1 \qquad (16)$$

That means the direction of phase shift is the opposition of the production of $U_{dff}$ and $i_c$. With the method, the modulation wave for one phase becomes three waves. Fig. 5 shows the modified modulation waves.

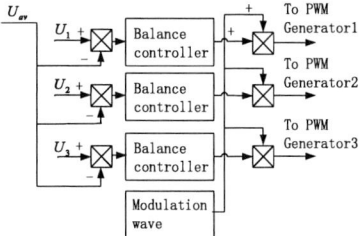

Fig.5 the modified modulation waves

## V. SIMULATION AND EXPERIMENT RESULTS

Simulations on a cascaded H-bridge active power filter are studied. MATLAB Simulink is adopted to do the simulation. The parameters are the same when the simulations are done with the dc capacitor voltage control and without it.

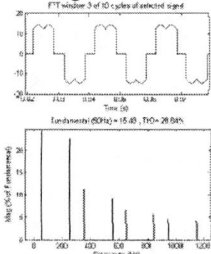

Fig.6 the load current and its THD

Fig.6 shows the load current and its THD. It shows that the THD is more than 28% without the compensation current generated by the APF.

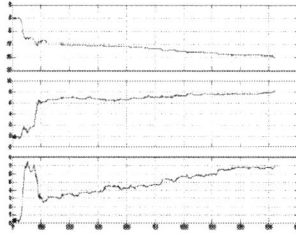

Fig. 7 the differences between per dc capacitor voltages and their average without the balance control

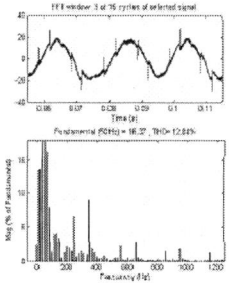

Fig.8 the compensated source current without balance control and its THD

Fig.7 shows the differences between the three capacitor voltages and their average value and Fig. 8 shows the compensated source current and its THD without dc capacitor voltages balance control. It can be seen that the differences are not in the accepted tolerance. The source current is compensated to 12.84%.

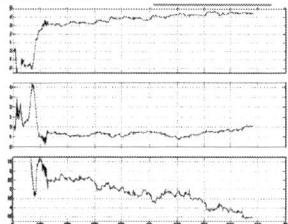

Fig.9 the differences between per dc capacitor voltages and their average with the balance control

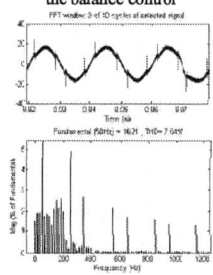

Fig.10 the compensated source current with balance control and its THD

Fig. 9 shows the differences between the three capacitor voltages and their average value and Fig. 10 shows the compensated source current and its THD with dc capacitor voltages balance control. Compared to fig. 7 and 8, it's obvious to see that the differences are reduced and the source current is compensated more accurately to 7.84%.

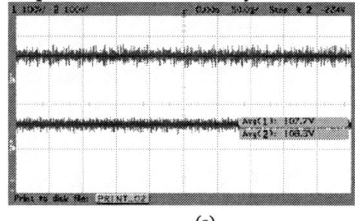

(a)

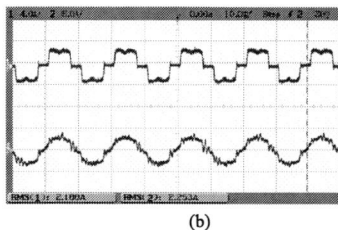

(b)

Fig.11 the experiment result

(a) the capacitor voltage of the two H-bridge converter of phase a

(b) the load current and the compensated source current of phase a

The experiments on an APF based on five-level H-bridge converter are done. The capacitor voltage reference of the APF DC side is set to 100 Volts. In fig.11, the capacitor voltage of the two H-bridge converter, the load current, and the compensated source current of phase a is shown. It can be seen that the voltages of the two H-bridge converter cells is balanced very well, and the compensated source current becomes sinusoidal.

## VI. CONCLUSIONS

In this paper, an improved method to control the dc capacitor voltages balance is presented. The simulation and experiment results validate the method and show that the dc capacitor voltages are controlled in the accepted tolerance and the harmonic currents is well compensated than without the balance control. Though it's validated with seven-level/five-level converter, the method can be applied to n-level converter.

## REFERENCES

[1] Moran. L.; Diaz M.; Higucra. V.; Wallacc. R.; Dixon. J.. "A three-phase active power filter operating with fixed switching frequency for reactive power and current harmonic compensation". IEEE Ind. Electron., Control, Instrumentation, and Automation, 1992. Proceedings of the 1992 International Conference on Vol.1, pp: 362-367

[2] Pittorino L.A.; Horn A.; Ensin J.H.R.. "Power theory evaluation for the control of an active power filter". AFRICON, 1996. IEEE AFRICON 4th, Vol. 2, 24-27 Sep. 1996. pp: 678-681

[3] A.M. Massoud, S.J. Finney, and B.W. Williams. "Seven-level shunt active power filter". 2004 11th International Conference on Harmonics and Quality of Power. pp:136-141

[4] George Zhou, Bin Wu, Donglai Xu. "Direct power control of a multilevel inverter based active power filter". 2004 IEEE International Conference on Industrial Technology (ICIT). Pp: 498-503

[5] Bin Wu. "High-power converters and AC drives". IEEE Press. Pp: 127-131

[6] Hill W A, Harbourt C D. Performance of medium voltage multilevel inverters. in Proc. IEEE Industrial Application. Soc. Conf., Phoenix, AZ, 1999, 1186-1192

[7] Zhang Zhongchao, BoonTeck Ooi. Multi-modular current source SPWM converter for SMES, IEEE Trans. Power Electron., 1993, 8(3):250-256

[8] Gokhale K P,Kawamura A. Dead beat microprocessor control of PWM inverter for sinusoidal output waveform synthesis. IEEE Trans Ind Appl, 1987, 23(5): 901-910

[9] S. Buso, L. Malesani, and P. Mattavelli, "Comparison of current control techniques for active filters applications," IEEE rans. Ind. Electron., vol. 45, no. 5, pp. 722-729, Oct. 1998.

# E-laboratory in the Field of Electrical Drives

H.Hõimoja* A.Rosin* T.Möller* M.Müür*
* Tallinn University of Technology/Electrical Drives and Power Electronics, Tallinn, Estonia

*Abstract*—Modern information and communication technologies have changed the concept of conducting laboratory experiments and learning the theories behind them. New terms like "distant learning", "remote laboratories", "virtual learning environments" etc have emerged. To follow the mainstream and improve practical aspects of given education, the TUT has started designing its own remotely controlled electrical drives laboratories where students can make experiments on real objects. Designing a remote laboratory presumes finding solutions to problems regarding technical, didactic, security and financial requirements as well as the integration into a wider distant learning environment. These requirements give a basis for laboratory structures and methodology.

*Keywords*—*Education methodology, electrical drive, test bench, virtual instrument.*

## I. INTRODUCTION

Energy technologies remain essential regardless of currently prioritized sciences, therefore academic education and practice training in this field to produce qualified specialists is a must for modern society. A traditional university training course consists not only of lectures, but also of exercises, laboratory and industrial practice. The laboratories' primary task has been promoting specific skills and knowledge, at universities also scientific dimension is added, while further development of technology is unthinkable without experimental part.

Local laboratories have been the most common way of experimenting and relating theoretical knowledge to real objects. Unfortunately they possess some limitations related to fixed time and place as well as the number of experimentation sets and participating students. Empowered by information and communication technologies (ICT), the new approach is to conduct remote experiments at a distance from the actual experimental setup over the World Wide Web (WWW). Existing software and hardware needs to be adapted to suit tutors' and students' needs, the schedule should be more flexible and allow repetitions [1]. Remote learning features should be applicable both in academic and industrial context like in the form of complementary courses. Remote laboratories have been already utilized in several universities all over the world; the feedback has been mainly positive [1], [4]. Encouraged by different success stories, the TUT has decided to introduce its own electrical drives remote laboratory; the background and considerations are discussed in the next chapters of current paper.

## II. PRESENT SITUATION

Existing electrical drives laboratory equipment at the TUT has been partly procured in 1950's - 1960's and therefore become obsolete. Laboratories carried out so far are in most cases outdated and do not correspond to the requirements of modern labor market. For example, the current experiments in general course of electrical drives comprise:

Taring the load machine

Here the DC motor no-load losses and relationship between armature current and torque are estimated.

DC motor characteristics

This experiment includes DC motor speed control with armature voltage regulation, series resistance in the armature circuit and field weakening. Electrical and mechanical values are measured or calculated and compared with theoretical ones.

Induction motor characteristics

Here the motor speed, voltages, currents and electrical power of a squirrel-cage induction motor are measured at different loads. Based on the results, speed, efficiency and power factor as a function of the load are represented graphically and compared with theoretical values.

Induction motor transients

Using an oscilloscope, induction motor speed and current transients starting from standstill are recorded and compared with theoretical graphs.

Getting acquainted with a frequency converter

Here the students learn to parameterize a frequency converter, test different acceleration-deceleration profiles and braking modes.

As seen, four of five laboratories described above are solely motor-based, though the motor is only a part of a drive system. Obviously, there should be more experiments regarding design and control, including sensors, power converters and controllers. These considerations were the main starting point when planning the renewal and upgrading laboratory experimentation sets.

## III. CONSIDERATIONS WHILE UPGRADING ELECTRICAL DRIVES LABORATORIES

### A. Objectives

First of all, main objectives while selecting new equipment must be defined:
1. Developing a remote laboratory to support learning and R&D activities in the electrical drives and power electronics disciplines.

2. Increasing quality of teaching at the TUT and its colleges.
3. Complementary training for industry specialists.

Based upon those objectives, new ideas can be generated.

### B. *Questions to be answered*

Different electrical drives courses are included in the curricula of two TUT departments:

 department of electrical drives and power electronics;
 department of mechatronics.

Remarkably, students having electric drives and power electronics as their main subject have previously undergone electrical machines' courses with corresponding laboratories, thus an unnecessary duplication might occur while laying more stress on a driven motor than the whole drive system, whereas mechatronics students have only basic knowledge of electrical circuits and electromechanics. Consequently, before undertaking next steps, one must find answers to following questions:

1. To whom is this project targeted?
2. What types of experimentation sets are needed?
3. What can be taught on these sets?
4. What experimentation sets are commercially available?
5. Is there any existing equipment that can be integrated into new installations?

As defined in objectives, the main target group consists of TUT's, its colleges' and institutes' students both in B.Sc., M.Sc. and PhD levels (0). Industry specialists can also participate in the courses; localized Web user interfaces enable involvement of foreign partner universities in the framework of student and knowledge exchange.

Following new demands, a state procurement to furnish the new drives laboratory was announced in June 2007. The aim was to provide courses with up-to-date laboratory benches with modern power conversion and control systems, at least three of the newly procured benches are planned to be connected to the Internet.

### IV. AN OVERVIEW OF WEB-BASED LABORATORIES

### A. *Virtual and remote laboratories*

One can distinguish between two basic types of Web-based laboratories:

 virtual laboratories;
 remote laboratories.

TABLE 1. REMOTE LABORATORY TARGET GROUPS

| Target group | Trainees per year |
| --- | --- |
| TUT department of electrical drives and power electronics | 120 |
| TUT department of mechatronics | 50 |
| TUT Virumaa College | 175 |
| TUT Kuressaare College | 30 |
| TUT Institute of Sustainable Technology | 20 |
| Estonian industry | 25 |
| Total | 420 |

A good example of a virtual laboratory is *MatLab*'s embedded Web server, developed by *MathWorks* which enables simulations to be carried out remotely using a Web browser. Distant Web-based *MatLab* experiments have already been implemented at some universities [2]; unfortunately the software producer has discontinued this product. Wolfram Research company has its own *webMathematica* software package, which enables online calculations and visualizations of previously defined processes, where the user can check the outcome by different input values. Well-known *MathCAD* software has also its own remote Web interface [3]. Simulations are undoubtedly the cheapest and safest way to model the processes in existing systems, including electrical drives. There is no physical threat to laboratory equipment like overloads, short circuits, etc.

As the students must have real world experiences, the virtual labs based solely on simulations are less than a half of a solution. While arranging multiple experiments is expensive and even impossible in full scale, a more advanced way is to conduct real experiments in laboratory, which are remotely controlled and monitored by students using standard Web tools. The provision of remotely controlled experiments accessible over the internet or university intranet can potentially address the issue of access to practical exercises in a number of ways [4]:

 By giving access to experiments over a longer time frame and at times preferred by students.
 By sharing expensive resources between institutions.
 In giving access to safety critical and expensive equipment with reduced risk.
 By offering improved access for disabled students.
 By facilitating greater access to experimental work in distance education.

Usually the Web-based laboratories are a combination of both basic types, where the outcomes from real experiments can be compared to simulation results.

### B. *Application areas*

The increasing popularization of distance learning as well as the availability to as many students as possible is the main reason for adoption of remote laboratories. Four application areas have been pointed out [1]:

1. Shared remote laboratories can be established to access a single expensive or rare experimentation set that is only available at a distant university.
2. Localized remote laboratories allow to carry out experiments according a more flexible schedule and place or to repeat a missed laboratory session.
3. Distant remote laboratories are mainly used in distant education to replace difficult to attend classroom experiments.
4. Technical review laboratories allow industry specialists to test new products without attending traditional workshops and seminars.

### C. *Requirements*

The design of a properly functioning remote laboratory assumes fulfilling a set of requirements. These demands are mostly derived from practical experience and divided into four main categories [5]:

1. Technical requirements state the local equipment to be maintenance-free during the remote experiment. The installation should support full real-time feedback regarding control commands.
2. Security requirements pose demands mainly on the software. The installation must be protected against wrong commands, exceeding allowable limit values and network attacks.
3. Didactical requirements define student-tutor interactions. Every student's results and solutions must be recorded for evaluation purposes, the files and folders must be private, their access controlled by means of individual username and password.
4. Financial requirements above all state the end user software to be as common and license-free as possible, in this respect the usual Web browser is a perfect solution. Locally applied software and control apparatus should also be commercially available and cost-efficient.

All these demands cannot be satisfied simultaneously, so some compromises must be made excepting security. For example, communication over HTTP is not as fast as over TCP/IP, but the latter has some access problems regarding software and ports disabled by firewall.

## V. REMOTE LABORATORY STRUCTURES

There are several approaches to the definition of a remote laboratory; one of the most generalized architectures can be described as consisting of core components [5], [6]: HMI, physical process, applications, didactical content and client software with user interface (Fig. 1).

HMI is the central element communicating with all other components, driving the signals and allowing students to perform various experiments. Usually it is an interface program running on a server computer.

Physical process can be any remotely controlled physical device or process, including also data acquisition and control functionalities.

Applications mean the tools and services for human communication, collaboration, production, etc. They deal

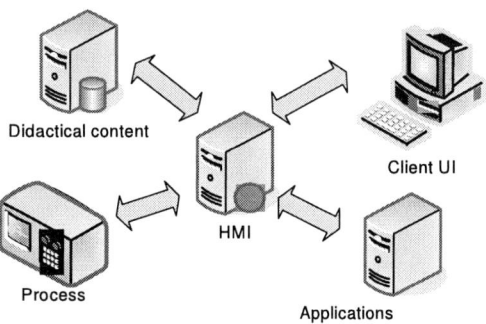

Fig. 1. Core components of a remote laboratory

with the communication between the laboratory and students. The didactical content consists of the textual, audio, video or animated instructions and exercises.

### A. The four dimensions

Whatever an approach to a remote laboratory structure might be, one can describe it four-dimensionally [7]:

1. Remote manipulation dimension includes necessary functionalities between controlled object and HMI to remotely manipulate the system under research.
2. Didactic dimension corresponds to the educational viewpoint, regarding teaching methodology.
3. Communication dimension deals with the interaction between laboratory counterparts, including students, tutors and administrators.
4. Administration dimension features schedule management, access control and site maintenance.

### B. Methodical structure

Carrying out a set of remote laboratory experiments, each experiment can be roughly divided into eight parts with approximate durations (Fig. 2):

1. Methodical preparation - up to 1 hour
2. Registration and authentication - up to 15 minutes
3. Theoretical preparation - 2 hours
4. Local activities and experiments - 2 hours
5. Preparation assessment - up to 30 minutes
6. Remote experiments - up to 3 hours
7. Final report - up to 1 hour
8. Final assessment - up to 30 minutes

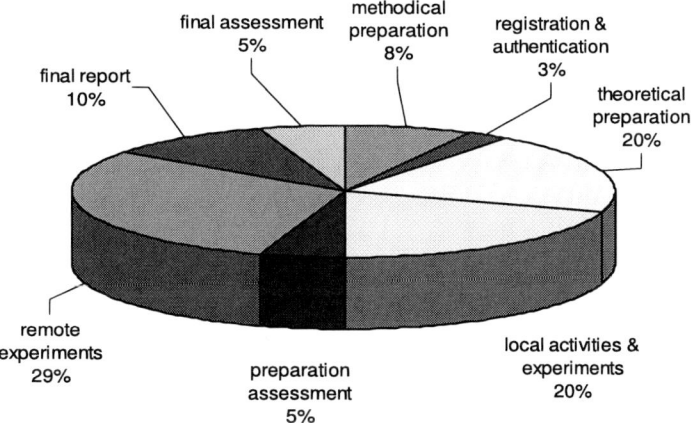

Fig. 2. Methodical structure

So the overall maximum duration of an experiment will be about 10 hours, 80% of which shall be done over the Internet.

### C. Hardware and software structures

To guarantee fast and effective maintenance, the hardware components should be as simple and available as possible [8]. The Web-enabled PLC used in designed remote laboratory environment has data acquisition interface, allowing it to be connected to an experimentation set's (Fig. 3):

1) voltage, current and movement sensors;
2) actuators in the form of converters;
3) measuring instruments.

The PLC can communicate with control objects over Ethernet or using digital and analogue I/Os. A visual image of the installation is obtained via a webcam set.

Some data monitoring and plotting can also be realized over frequency converter's or soft starter's remote HMI, where IP addresses and other network parameters can be specified, the HMI is also used for local control, monitoring and parameterization, like defining IP addresses. Simulation and other auxiliary programs are stored in an application server, the SQL server houses various databases and processes queries. Firewall router protects the installation from outer intrusions. To conduct a remote experiment, the user needs to perform following steps [9]:

1. Connecting the webpage of the remote laboratory.
2. Authorization and registration.
3. Selecting the desired and available experiment.
4. Displaying corresponding Web pages.
5. Conducting a remote experiment.
6. Terminating the session with logout.

Local configurations at the laboratory site are carried out using manufacturer's own application software, like HMI parameterization and PLC programming. All laboratories are planned to have Web-based instructions, which can be linked to interactive calculation and simulation programs hosted inside an application server and assembled in Java or other programming environment.

## VI. PLANNED EXPERIMENTS

The e-lab project includes seven experimentation sets, all these enable local experimentation and remote preparatory learning. Sets with remote control and monitoring are marked in the list below with italics:

1. Self-commutated converter.
2. Frequency converter drive with asynchronous motor.
3. Servo drive.
4. Stepper motor drive.
5. *Frequency converter with vector control.*
6. *Soft starter.*
7. *Torque control bench.*

Covered topics are explained in the next subchapters.

### A. *Self-commutated converter*

IGBT fundamentals.
Control principles: PWM, DC chopper controller in 1-, 2- and 4-quadrant operation.
Low-frequency AC voltage PWM.
Circuits: step-down controller, H-bridge, inverter.
Resistive, capacitive and inductive loads.
Suppressor, link and free-wheeling circuits.
Control characteristics and operating graphs.
Computer-assisted measurements.
Fourier analysis of harmonics.

### B. *Frequency converter drive with asynchronous motor*

Fundamentals of inverters with voltage and frequency control.
Analysis of U/f ratios.
Stator resistance compensation.
Characteristics of the inverter-fed drive.
Computer-assisted parameter setting and animation.

### C. *Servo drive*

Computer-aided commissioning and parameter setting of a linear axis servo drive.
Positioning and sequence control.
Setting parameters of the position and speed controller using industrial software.
Investigating the effects of various controller settings.

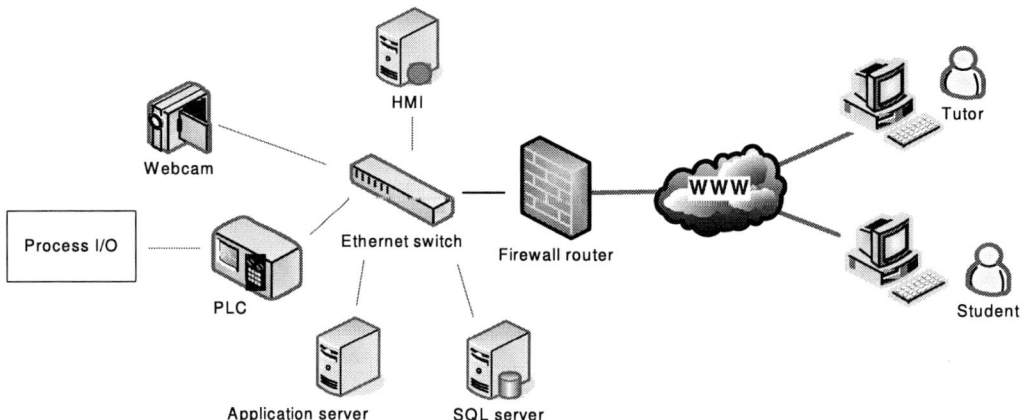

Fig. 3. Hardware structure

### D. Stepper motor drive

Stepper motor fundamentals.
Introduction to CNC programming.
Speed and position control.
Influence of different acceleration-deceleration ramps.

### E. Soft starter

1) Investigation of acceleration-deceleration ramps.
2) Soft start with current limitation.
3) DC dynamic braking features.
4) Voltage, current and speed transients.

The *Altistart 48* soft starter is controlled via PLC's digital outputs and monitored by analogue and digital inputs by means of voltage and current measuring transducers. The settings are viewed and changed over Modbus RTU protocol. For speed acquisition, a tachogenerator has been installed. In Fig. 5 starting curves of a 4 kW asynchronous drive are depicted with following settings:

1) acceleration time $t_{acc} = 2.0$ s,
2) applied torque $T_L = 0.25\ T_N$.
3) no torque boost.

### F. Frequency converter with vector control

Comparison of acceleration and deceleration profiles.
Speed, current and torque transients.
Differences between scalar frequency and flux vector control.
Behavior in constant torque and weakened field regions.
Different braking modes: freewheel, ramp stop and dynamic DC braking.

The frequency converter test bench comprises two similar converter-fed 0.75 kW asynchronous drives with mechanically linked shafts, whilst one drive acts as the working machine and the other as load in generating quadrant. The load is controlled with resistive torque reference, which can be made dependent on drive's speed using a PLC subroutine. This way several load characteristics can be simulated. The DC links are connected together to enable reuse of generated energy by the drive in motoring quadrant.

In Fig. 6, the remotely obtained results of an acceleration-deceleration experiment with following parameters are shown:

1) acceleration time $t_{acc} = 2.0$ s,
2) deceleration time $t_{dec} = 2.0$ s,
3) ramp type: S-spline,
4) applied torque $T_L = 0.7\ T_N$.

### G. Torque control

Load sharing.
Frequency converter's behavior in master and slave modes.
DC bus sharing.
Driving multiple motors in parallel.

The laboratory stand consists of three *Altivar 71* frequency converters, which are connected to common DC bus and three asynchronous motors with their shafts mechanically linked by pulleys' transmission belt. Remote control and monitoring is similar to vector controlled frequency converter's experimentation set.

The three last test benches can be switched between local and remote control modes. The start of a remote experiment is signaled at the site audibly and visually by a buzzer and strobe driven by PLC's digital outputs.

## VII. DATA PROCESSING

The SQL server saves time-based values in real time on its storage media. These values can be later downloaded by the user in HTML (Hypertext Markup Language), CSV (Comma Separated Values) or XLS formats to be later handled, especially in Excel worksheets, like shown in Fig. 5 and Fig. 6. After composing the graphs demanded in the task, the trainee sends them to tutor for assessment. During the experiment, a nearly real-time graph is running in the user window for feedback purposes only.

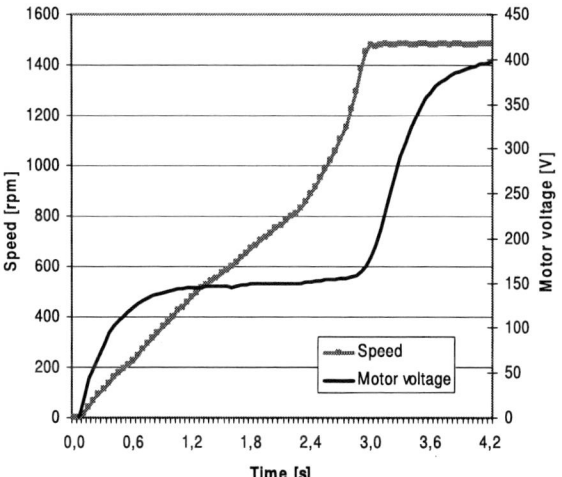

Fig 5. Remotely obtained soft starter curves

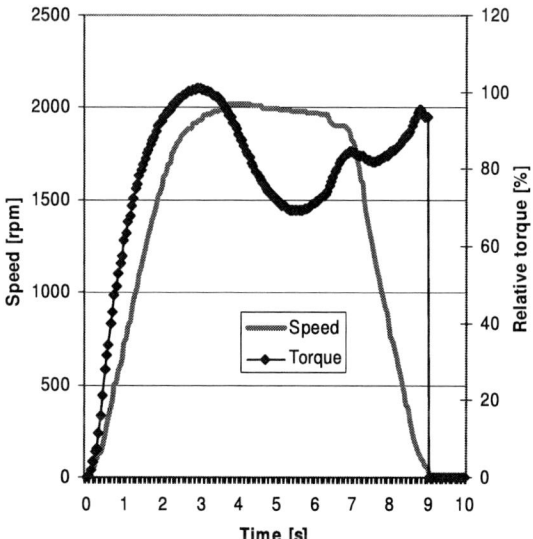

Fig. 6. Remotely obtained frequency converter's speed diagram

## VIII. EXPERIENCED PROBLEMS AND OBSTACLES

Obviously, one could expect a laboratory test bench to cover a possibly wider amount of experiments. These experiments must be safely conducted without permanent local supervision and eliminate any serious human error. As test benches' PLCs are freely programmable, new experiments may be flexibly added to the exercise lists. As for every ambitious project, there have emerged certain major and minor obstacles, described below:

1. Hardware problems are mainly related to I/O scanning capabilities. The data refresh rate is important while recording transients and other short duration processes, however the Modbus protocol used for communication between the master PLC and frequency converter allows the fastest rate of 50 ms. Besides control hardware, the controlled load still remains the issue to be solved. There are thoughts of variable pitch or choked fans.

2. Software obstacles, confusing the end user, are caused by the behavior of Java applets to run correctly only on the Internet Explorer browser. A really user-friendly user environment must be OS and browser-independent.

3. Network constraints mainly consist of bandwidth and security issues. The user expects the system to react and respond without a remarkable time lag, caused by bottlenecks between the installation and the student. Security solutions must protect the installation against unauthorized access and hacking.

## IX. CONCLUSION

Present experiences have shown that the best way to explain theory is not by difficult formulas, but hands-on experiments. Modern tendencies are towards reducing the number of lectures in favor of exercise and laboratory classes, where the students can fix existing and obtain new knowledge through practical results of their personal work. Empowered by increasing broadband Internet connections and software platforms development, many laboratory and exercise classes can be carried out remotely, where theoretical background can be delivered and assessment given by automated program applets.

Present tendency at the TUT implementing e-studies in laboratories has shown that expenditure of time in Web-based courses has paid off. Animated diagrams and exercises help to understand functional principles, so the students are better prepared for real experiments. Thanks to more flexible schedule the working students do not interrupt their courses so easily as before. So it might be summarized that implementing e-labs as a part of e-studies has become inevitable due to increasing amount of extramural and complementary courses as well as the fact that 60% of students must share their time between studies and daily work.

To enhance efficiency, remote laboratories must be integrated into bigger managed learning environments. Laboratory control applets must have links to methodical and theoretical materials, facilitating understanding of performed experiments. In longer terms, the remote laboratories can contribute to [10]:

1) easier understanding the performance of electric, electronic, electro-mechanical and control circuits;
2) by understanding physical phenomena in circuits developing analytical thinking without learning a set of difficult formulas;
3) developing engineering skills through circuit synthesis.

As described, the TUT plans only to conduct remote experiments on AC induction drives. Although these constitute the majority of all drives, remote research of other motor drives, like DC, synchronous or SRM must also not be neglected. In this case, co-operation with our foreign partner universities to mutually utilize the remote laboratory resources would be a solution.

The designed remote laboratory will be fully launched in autumn 2008. Another important part besides machinery, hardware and software installations would be final choice of remote laboratory exercises, preparation of dedicated methodology, composing interactive learning materials and sharing gained experiences with other interested academic and industrial counterparts.

## REFERENCES

[1] Deniz, D.Z., Bulancak, A., Özcan, G.: A Novel Approach to Remote Laboratories. *FIE 2003*, Boulder, November 5-8, 2003, pp T3E-8 - T3E-12.

[2] Abdel-Hamid, A. M., Zein El-Din A.S., Tibken, B.: Distance Teaching of Electrical Power Engineering via Matlab Web Server. *EPE-PEMC 2004*, Riga, September 2-4, 2004.

[3] Humar, I., Sinigoj, A., Hagler, M.: Mathematical Tools for Supporting Web-Based Education of Electromagnetics. *ITHET 2004*, Istanbul, May 31 - June 2, 2004, pp 111-116.

[4] Cooper, M.: Remote laboratories in Teaching and Learning – Issues Impinging on Widespread Adoption in Science and Engineering Education. *iJOE*, Vol. 1, No. 1, 2005, http://www.ijoe.org/

[5] Chiculita, C., Frangu, L.: A Web Based Remote Control Laboratory. *SCI 2002*, Orlando; July 14-17, 2002.

[6] Benmohamed, H., Leleve, A., Prevot, P.: Remote Laboratories: New Technology and Standard Based Architecture. *ICTTA 2004*, Damascus, April 19-23, 2004, pp. 101-102.

[7] Leleve, A.; Prevot, P. et al: Generic e-Lab Platforms and eLearning Standards. *CALIE 2004*, Grenoble, February 16-18, 2004.

[8] Schmid, C.: Web-based Remote Experimentation. *IFAC TA 2001*, Weingarten, July 24-26, 2001, pp 443-447.

[9] Chen, S.H., Chen, R. et al: Development of Remote Laboratory Experimentation through Internet. *SRC 1999*, Hong Kong, July 2-3, 1999, pp 756-760.

[10] Fedàk, V., Bauer, P. et al: Interactive e-Learning in Electrical Engineering. *EDPE 03*, High Tatras, September 24-26, 2003, pp. 368-373.

# Laboratory Setup for Studying Ultracapacitors in Industrial Applications

I. Roasto, *D. Vinnikov, and **T. Lehtla

Tallinn University of Technology, Ehitajate tee 5, 19086, Tallinn, Estonia
E-Mail: indrek.roasto@ttu.ee, *dm.vin@mail.ee, **tlehtla@staff.ttu.ee

*Abstract*—**An innovative teaching topic of ultracapacitors was worked out in Tallinn University of Technology. For that purpose, a special test bench was built and tested. The test bench has been designed in view of student's safety and includes numerous protections. This paper describes the structure of the test bench and explains the working principles of different components. To demonstrate and test the functionality of the test bench, three tests were made: charging, discharging, and UPS (uninterruptible power supply) mode.**

*Keywords*— **DC power supply, energy storage, teaching, test bench, uninterruptible power supply (UPS)**

## I. INTRODUCTION

Ultracapacitors are modern energy storage devices that fill the gap between batteries and conventional capacitors. They suit best in the situations where short term energy bursts are needed. Ultracapacitors have many applications in modern electrical systems, the widest field being transportation. Bombardier Transportation has equipped one bogie of a light rail vehicle for public transportation in Mannheim with an ultracapacitor energy storage module. The vehicle has been in daily operation since 2003 [1][2]. The Nissan Diesel Motor Company developed an ultracapacitor-based hybrid medium-duty delivery truck, which combines a diesel engine with an electric motor and an ultracapacitor-based energy storage system. At the Frankfurt Motor show 2005, BMW presented its new hybrid car X3 with ultracapacitors [3]. Honda FCX is another good example of ultracapacitors use in vehicles [3][4]. Based on the examples above, it is clear that the industry is increasingly interested in the ultracapacitors. The ultracapacitors have already become an inseparable part of the modern energy storage technology. Thus it is also important to teach topics of ultracapacitors in the universities. All the graduates of electrical engineering should have some knowledge of modern energy storage technologies, including that of ultracapacitors.

It was the first effort at the Department of Electrical Drives and Power Electronics of Tallinn University of Technology to introduce ultracapacitors as a topic for teaching. Ultracapacitors will compose a module of the power electronics course. The ultracapacitor module includes both theory and practical tests. For that purpose, a completely new ultracapacitor test bench was built, where students can carry out tests and measurements. The ultracapacitor lecture module consists of three parts: theory, practical work and evaluation of the results. In the theoretical part, the ultracapacitor will be introduced and its working principles will be explained. The theoretical part is followed by the practical part. That can be divided into two sections: computer simulation and measurements on the test bench. Simulation is always the first step in any engineering project. Therefore it is important to teach some widely used simulation software to the students. One purpose of simulation is also to demonstrate the differences that may occur in a real-life system. For our simulation, a Matlab toolbox SimPowerSystems will be used [6][7][8].

## II. ULTRACAPACITOR TEST BENCH

The ultracapacitor test bench is based on a 19-inch rack system, as shown in Fig. 1. The technical parameters are given in Table I. The electrical principle of the test bench is quite simple, as shown in Fig. 3. It consists of three blocks: a power supply, a buck-boost converter of the ultracapacitors and a buck-boost converter of the load. The power supply consists of a transformer, a bridge rectifier and an LC filter. The maximal input power is 3000 VA. The transformer provides the needed isolation between the input and the output. For the interface between the ultracapacitor tank (UCT) and the DC link, different solutions can be considered [5]. In the current project one of the simplest was chosen, a topology that uses a single leg buck-boost converter. The main benefits of this solution are its low price, simple construction and a small number of IGBTs. The drawback is the uncontrollable discharge current through the diode (D1).

Fig. 1. Ultracapacitor test bench

978-1-4244-1741-4/08/$25.00 ©2008 IEEE

Fig. 2. Ultracapacitormodule PCM14014

The ultracapacitor tank consists of eight in series connected ultracapacitor modules. The parameters of one module are given in Table I. The ultracapacitor tank has a capacitance of 18.9 F and an internal resistance of 0.098 Ω, the maximal charging voltage being 112 V [9][10].

The switching frequency of the IGBTs is 1 kHz. The frequency was chosen considering the simulation speed and time step. Higher frequency requires a smaller time step, which reduces the simulation speed drastically, e.g. a simulation of 5 s by 1 kHz lasts about 18 s in real life. The same simulation by 10 kHz takes already 9 min.

Two kinds of load can be used: a resistive load or a DC motor.

The test bench is equipped with two voltage and three current sensors. Voltage is measured on the terminals of the ultracapacitor and on the load. Current is measured with three LEM current sensors LT 100-S/SP30 in the points In, W, U, as shown in Fig. 3.

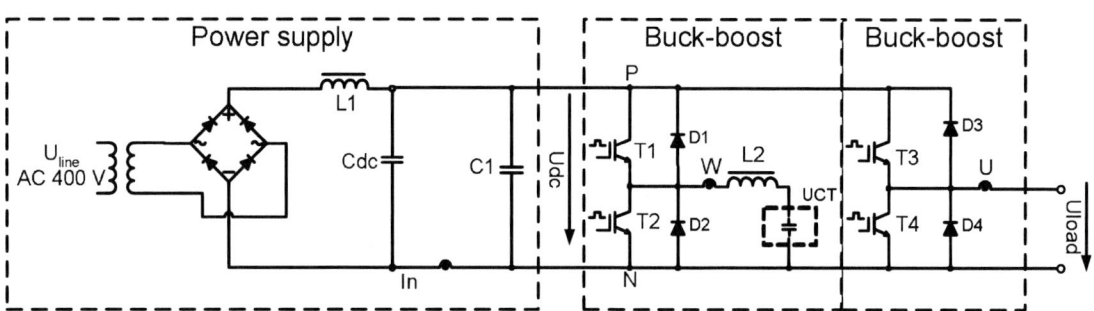

Fig. 3. Electrical circuit diagram of the ultracapacitor test bench

TABLE I. TECHNICAL PARAMETERS OF THE TEST BENCH

|  | Maximal | Typical |
|---|---|---|
| Input |  |  |
| Input 2 Phase AC, 50 Hz, $U_{line}$ [V] | 400 | 400 |
| Input power S [VA] | 3000 |  |
| DC link |  |  |
| DC link voltage, mean value $U_{dc}$ [V] | 170 | 100 |
| DC link current $I_{dc}$ [A] | 30 |  |
| Ultracapacitormodule |  |  |
| Type | PCM14014 |  |
| Ultracapacitor voltage $U_{uc}$ [V] | 14 |  |
| Capacitance $C_{uc}$ [F] |  | 151.62 |
| Internal resistance $r_{uc}$ [mΩ] |  | 12.19 |
| IGBT module |  |  |
| Type | PM100CSA120 |  |
| Collector current $I_c$ [A] | 200 | 100 |

## III. USER INTERFACE

As a user interface, a PC (personal computer) will be used. Data exchange between the PC and the test bench is realized via serial communication interface (RS232). There are several possibilities to read and write information from the serial port, e.g. using Matlab Real-Time Workshop, a GUI (graphical user interface) or the communication software HyperTerminal. Matlab Real-Time Workshop has an advantage that it can draw voltage and current diagrams in real time, which then can be compared to simulation results. Moreover, the simulation and the testing would be executed in the same environment. However, using Real Time Workshop assumes programming of according subfunctions. Neither is it easy to program a GUI that takes time. The simplest solution is to use the communication software HyperTerminal. It requires no additional programming and does not cost extra, since it is a part of MS Windows. The user interface will be displayed in the HyperTerminal window, as shown in Fig. 4. It is based on a menu system. The sub menus can be entered just by pushing the corresponding number on the keyboard. There are four choices in the menu: charging, discharging, UMD (uninterruptible motor drive), and error code. The charging and discharging modes allow manual pulse width adjustment. In the UMD mode it is done by a regulator. The error code in the fourth menu helps to identify errors occurring. As it was mentioned before, the test bench is equipped with three current and two voltage sensors. The sensor information will not be displayed. It is entirely used by the regulator and protection algorithms. To measure current or voltage, students will have to connect some measurement instruments to the test bench [11][12].

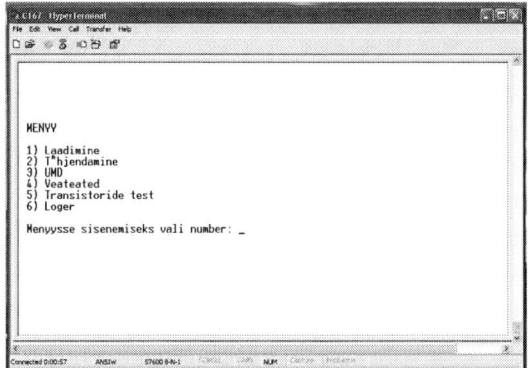

Fig. 4. The user interface will be displayed in the MS Windows program HyperTerminal

## IV. PROTECTIONS AND SAFETY

A test bench for teaching purposes should be safe for the students to use and also protected against all kinds of failures, which may be accidentally caused by the students. The key for safety is a completely isolated and closed system. That was the principle followed also in the current test bench. The front side of the test bench is completely covered, as shown in Fig. 5. All needed measurement points are brought to the front panel using special safety type lab sockets. The students can safely connect different measurement instruments and do the measurements without getting into contact with dangerous voltages. This kind of closed and isolated system has the drawback that the students cannot see the real components. In this test bench the problem is partly solved by placing a transparent cover in front of the ultracapacitors, as shown in Fig. 5.

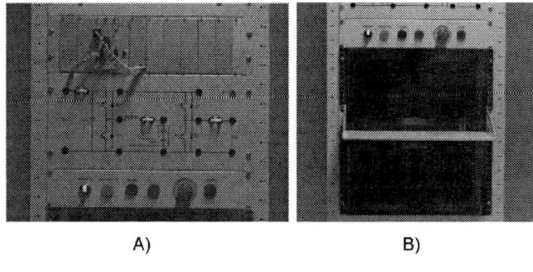

Fig. 5. The front side of the test bench is completely covered: measurement points (A), transparent cover of ultracapacitor tank (B)

The protection system has mainly two tasks: short circuit prevention and overvoltage protection of ultracapacitors. The protection system can be divided into two parts: software and hardware protections. The software includes overvoltage protection and cross conduction prevention.

The hardware serves as a second level protection against software errors. The hardware part consists of cross conduction protection and dead time generation. A simple logical circuit was used for cross conduction protection (Fig. 6 (a)). By replacing four logic ICs with one "XOR" (exclusive OR) element, the presented circuit can be further simplified. The "pull down" resistors R3 and R4 make sure that the PWM outputs are "pulled down" during a microcontroller reset or a failure.

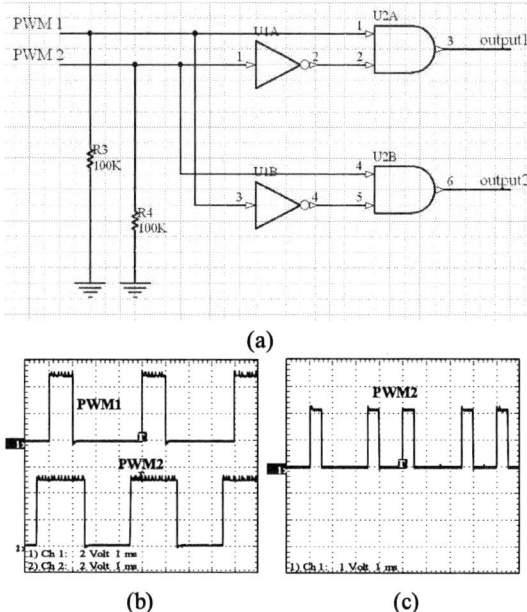

Fig. 6. Hardware-based cross conduction protection circuit (a) and the demonstration of a hardware-based cross conduction protection: input signals (b), output signal (c)

Simultaneous switching of IGBTs is demonstrated in Fig. 6 (b). Microcontroller generates two PWM signals. In the output only the PWM2 will occur, since the PWM1 is blocked completely. Also, the cross conduction part will be blocked (Fig. 6, c).

## V. EXPERIMENTS

The following tests have been made on the test bench: charging, discharging and UPS (uninterruptible power supply) mode. In the first test ultracapacitors will be charged with constant pulse width for 20 s. The voltage and current shapes of the ultracapacitor tank were taken up with an oscilloscope, as shown in Fig. 7. At the beginning the current rapidly reaches 30 A and then slowly shrinks to 1.5 A. The voltage grows from 1 V to 10 V. At the beginning of charging the voltage grows almost instantaneously from 1 V to 3.8 V. This is typical for ultracapacitors and is caused by the voltage drop on the internal resistance

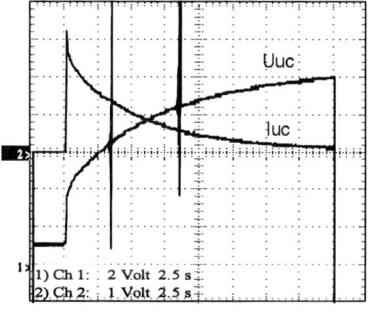

Fig. 7. Voltage of the ultracapacitor (Ch1) and the charging current (Ch2)

In the second test, ultracapacitors will be discharged over a resistor for 20 s. The ultracapacitor voltage and current shapes were taken up with an oscilloscope, as shown in Fig. 8. At the beginning, the current rapidly reaches 14 A and then slowly shrinks to 6.5 A. The voltage shrinks from 28.5 V to 16.0 V. Also, here the typical voltage drops at the beginning and the end of the discharging process can be seen. It is due to the internal resistance of the ultracapacitor tank.

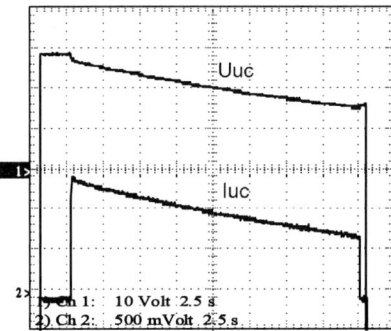

Fig. 8. Voltage of the ultracapacitor (Ch1) and the discharging current (Ch2)

In the last test, ultracapacitors work in the UPS mode as shown in Fig. 9. A resistor was used as a load. Ultracapacitors start discharging if the input current of the test bench drops to zero, i.e. power break down. The line voltage interruption takes about 12.5 s. During this time the load is supplied with ultracapacitors energy. If the input current is recovered, the discharging is stopped and charging follows.

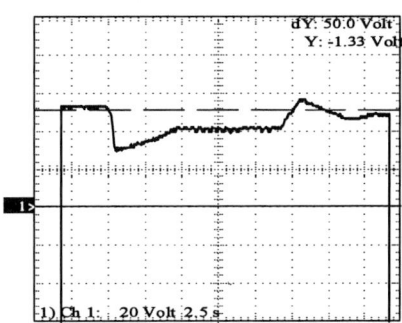

Fig. 9. DC link voltage in a power interruption situation

## VI. CONCLUSION

Ultracapacitors are very attractive energy storage devices in today's modern technology. Ten years ago they were mostly an object of research but today ultracapacitors have found their way into the industry. This emphasizes the significance of teaching ultracapacitors also at the universities. However an internet search reveals that that ultracapacitors are still used only for research purposes and no serious equipment for study purposes has been built. The Department of Electrical Drives and Power Electronics of Tallinn University of Technology has overcome this shortcoming by building a special ultracapacitor test bench for teaching purposes. As it was explained in this paper, the test bench is designed with regard to student's safety. All current conducting parts are covered. Special safety type lab sockets are used as measurement points. The test bench is

provided with all necessary protection functions like short circuit, over voltage, cross conduction prevention. The tests gave positive results. All sensors and control algorithms were working properly.

## REFERENCES

[1] Steiner, M., Klohr M., Pagiela, S.: Energy Storage System with UltraCaps on Board of Railway Vehicles, EPE 2007

[2] Steiner, M., Scholten, J.: Energy storage on board of DC fed railway vehicles, PESC 2004, Vol 1, pp 666 – 67.

[3] Marei, M.I., Samborsky, S.J., Lambert, S.B., Salama, M.M.A.: On the Characterization of Ultracapacitor Banks Used for HEVs, VPPC 2006, pp 1-6.

[4] Honda Fuel Cell Power – FCX, Press Release, http://world.honda.com/FuelCell/FCX/FCXPK.pdf

[5] Rufer, A., Hotellier, D., Barrade, P.: Power-Electronic Interface for a Supercapacitor Based Energy Storage Substation in DC Transportation Networks, IEEE Transactions on power delivery 2004, Vol 19, pp 629 – 636.

[6] Ozatay, E., Zile, B., Anstrom, J., Brennan, S.: Power Distribution Control Coordinating Ultracapacitors and Batteries for Electric Vehicles, American Control Conference 2004, Vol 5, pp 4716-4721.

[7] Youngho, K.: Ultracapacitor Technology Power Electronics Circuits, Power Electronics Technology, Oktober 2003, www.powerelectronics.com.

[8] Corley, M., Locker, J., Dutton, S., Spee, R.: Ultracapacitor-Based Ride-Through System for Adjustable Speed Drives, PESC 1999, Vol 1, pp 26-31.

[9] Becker, K.-P., Späth, H.: Short-Time-Storage-System with Double-Layer-Capacitors,connected to the DC-Link of Voltage Source Converters, PCIM 2001.

[10] Dougal, R. A., Gao, L., Liu, S.: Ultracapacitor model with automatic order selection and capacity scaling for dynamic system simulation, Journal of Power Sources, February 2004, Vol 126, pp 250-257.

[11] Yoon-Ho Kim; Soo-Hong Kim; Sung-Chan Rho; Hyun-Wook Moon; Kee-Hwan Kim; Instantaneous Voltage Drop Compensation for UPS System Connected in Parallel with Batteries and Ultracapacitors, PESC '06. 37th IEEE, June 2006, Page(s): 1 – 5.

[12] Krishnan, R., Lee, S.: Uninterruptible motor drives: a case study with switched reluctance motor drives, IECON 1994, Vol 1, pp 220-225.

# Synchronous machine direct axis parameters estimation module from an iterative strategy

Emile Mouni*, Slim Tnani*, Gérard Champenois*
*University of Poitiers, Laboratoire d'Automatique et d'Informatique Industrielle,
40, avenue du recteur Pineau, 86000 Poitiers, France,
e-mail: *emile.mouni@ieee.org*.

*Abstract*—In this paper, an original method of facilitating the synchronous generator parameters estimation, is proposed. This new method, based on the IEEE standards, makes the parameters calculation as easy as the data filling. It consists in an interface, the function of which is to allows the user to be free of the design algorithm complexity. The experimental results are very satisfactory and the usefulness of such an interface is presented at the end of the paper.

keywords: Synchronous machine, modelling, parameters estimation and Graphical User Interface.

## I. INTRODUCTION

This paper provides an academic tool usable in engineering school by last year students for synchronous machine study. Indeed, this tool uses a simple interface to implement complex identification algorithms in order to facilitate the students' work. The identification procedure consists in finding a set of parameters and then a numerical model which fits the most with the involved process. Several algorithms have been developed in this domain such as least square algorithm, output error algorithm, Prony algorithm, ...[1], [2]. The approach, involved in this paper, combines the IEEE standards [3] and an iterative method based on least square strategy to determine the synchronous machine parameters. Because of the complexity of the design procedure, one develops an interface thanks to Matlab$^{TM}$ which allows to identify the machine parameters in an easy way.

This paper is organized as follows. In the section II, the synchronous generator modelling will be presented. This model is based on the classical approach using the electrical circuits. Then, the IEEE standards on the machine parameters estimation will be given. The proposed algorithm and the academic interface will appear in section IV and the experimental test bench presentation, the experimental results and the discussion on the usefulness of the approach will end the paper.

## II. SYNCHRONOUS GENERATOR MODELLING

Considering a synchronous generator with dampers at the rotor, the simplified scheme of the machine is given in Fig. 1:

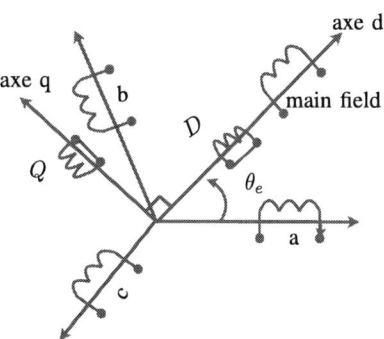

Fig. 1. Synchronous generator winding with dampers

$D$ and $Q$ represent respectively d-axis and q-axis dampers. $a$, $b$ and $c$ are the three phases of the synchronous generator and $\theta_e$ is its electrical angle depending on the poles number. As we can see on Fig.1, dampers are synthesized by short circuited elements. From this figure and adopting generator convention, we can write machine equations in three axis frame as follows:

$$
\begin{aligned}
v_{abc} &= -r_s.i_{abc} + \tfrac{d}{dt}\Psi_{abc} \\
v_f &= r_f.i_f + \tfrac{d}{dt}\Psi_f \\
0 &= r_D.i_D + \tfrac{d}{dt}\Psi_D \\
0 &= r_Q.i_Q + \tfrac{d}{dt}\Psi_Q
\end{aligned}
\tag{1}
$$

where:
$i_D$ and $i_D$ are direct and transverse dampers' currents, $\Psi_D$ and $\Psi_Q$ are the direct and transverse dampers' total flux, $\Psi_{abc}$ is the stator total flux, $\Psi_f$ is the main field total flux.

The study will be done in Park's frame thanks to Park's matrix defined below with the electrical angle of the machine $\theta_e$:

$$
P(\theta_e) = \begin{pmatrix} \cos(\theta_e) & \cos(\theta_e - \tfrac{2\pi}{3}) & \cos(\theta_e + \tfrac{2\pi}{3}) \\ -\sin(\theta_e) & -\sin(\theta_e - \tfrac{2\pi}{3}) & -\sin(\theta_e + \tfrac{2\pi}{3}) \end{pmatrix}
$$

such as:

$$
P(\theta_e).v_{abc} = v_{dq}
\tag{2}
$$

Then, the equation (1) becomes

$$
\begin{aligned}
v_d &= -r_s.i_d + \tfrac{d}{dt}\Psi_d - \omega_e.\Psi_q \\
v_q &= -r_s.i_d + \tfrac{d}{dt}\Psi_q + \omega_e.\Psi_d \\
v_f &= r_f.i_f + \tfrac{d}{dt}\Psi_f \\
0 &= r_D.i_D + \tfrac{d}{dt}\Psi_D \\
0 &= r_Q.i_Q + \tfrac{d}{dt}\Psi_Q
\end{aligned}
\tag{3}
$$

To determine synchronous generator equivalent circuit, a two salient poles machine will be considered. The scheme of this one is given in Fig. 2:

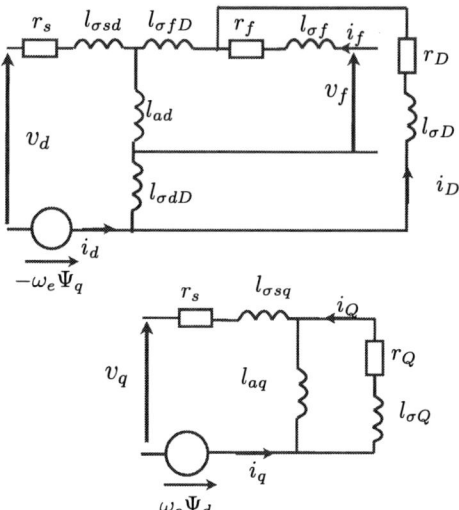

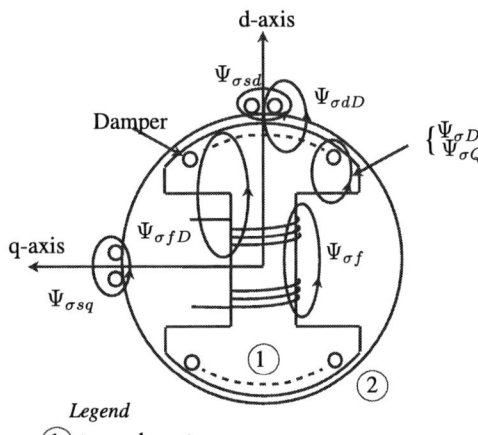

Fig. 2. Synchronous generator leakage and linkage flux

where:
$\Psi_{\sigma D}$ and $\Psi_{\sigma Q}$ are direct and transverse dampers leakage flux, $\Psi_{\sigma sd}$ and $\Psi_{\sigma sq}$ are direct and transverse stator leakage flux, $\Psi_{\sigma dD}$ is linkage flux between direct axis and direct dampers, $\Psi_{\sigma fD}$ linkage flux between main field and direct dampers, $\Psi_{ad}$ and $\Psi_{aq}$ are direct and transverse main flux (implicitly omitted in Fig.2).
The main field influence on the stator ($\Psi_{\sigma fs}$) is neglected, then the following relationships can be written

$$
\begin{cases}
\Psi_d = \Psi_{ad} + \Psi_{\sigma sd} + \Psi_{\sigma dD} \\
\quad = l_{ad}.(-i_d + i_D + i_f) - l_{\sigma sd}.i_d + l_{\sigma dD}.(i_D - i_d) \\
\Psi_q = \Psi_{aq} + \Psi_{\sigma sq} \\
\quad = l_{aq}.(-i_q + i_Q) - l_{\sigma sq}.i_q \\
\Psi_f = \Psi_{ad} + \Psi_{\sigma f} + \Psi_{\sigma fD} \\
\quad = l_{ad}.(-i_d + i_D + i_f) + l_{\sigma fD}.(i_f + i_D) + l_{\sigma f}.i_f \\
\Psi_D = \Psi_{ad} + \Psi_{\sigma fD} + \Psi_{\sigma dD} + \Psi_{\sigma D} \\
\quad = l_{ad}.(-i_d + i_D + i_f) + l_{\sigma fD}.(i_f + i_D) + l_{\sigma D}.i_D \\
\quad + l_{\sigma dD}.(-i_d + i_D) \\
\Psi_Q = \Psi_{aq} + \Psi_{\sigma Q} \\
\quad = l_{aq}.(-i_q + i_Q) + l_{\sigma Q}.i_Q
\end{cases} \quad (4)
$$

From these equations, we can deduce the synchronous generator electrical scheme as follows [4], [5]:

Fig. 3. d-axis and q-axis electrical equivalent circuits

where:
$l_{\sigma sd}$ and $l_{\sigma sq}$ are direct and transverse stator leakage inductances,
$l_{\sigma f}$ is main field leakage inductance,
$l_{ad}$ and $l_{aq}$ are direct and transverse stator main inductances,
$l_{\sigma fD}$ is linkage inductance between rotor and the direct damper,
$l_{\sigma D}$ and $l_{\sigma Q}$ are dampers leakage inductances.
As reactance ($x$) and inductance ($l$)are linked by $x = l\omega$ and thanks to the equation (4) we can deduce the following relationships between main and mutual inductances on one hand and reactances on the other hand:

$$
\begin{cases}
l_{ad} = \frac{x_{ad}}{\omega_e} \\
l_{aq} = \frac{x_{aq}}{\omega_e} \\
l_d = \frac{x_d}{\omega_e} = \frac{x_{ad} + x_{\sigma sd} + x_{\sigma dD}}{\omega_e} \\
l_q = \frac{x_q}{\omega_e} = \frac{x_{aq} + x_{\sigma sq}}{\omega_e} \\
l_D = \frac{x_{ad} + x_{\sigma D} + x_{\sigma fD} + x_{\sigma dD}}{\omega_e} \\
l_Q = \frac{x_{aq} + x_{\sigma Q}}{\omega_e} \\
l_f = \frac{x_{ad} + x_{\sigma f} + x_{\sigma fD}}{\omega_e} \\
m_{sf} = \frac{x_{ad}}{\omega_e} \\
m_{sQ} = \frac{x_{aq}}{\omega_e}
\end{cases} \quad (5)
$$

The relations between main reactances and machines parameters are:

$$
x_{ad} = \sqrt{T'_{d0}.r_f.\omega_e.(x_d - x'_d)} \quad (6)
$$

$$
x_{aq} = \sqrt{x_q.r_Q.\omega_e.(T''_{q0} - T''_q)} \quad (7)
$$

where:
$x_d$ is steady state reactance,
$x'_d$ is direct transient reactance,
$x_d''$ is direct sub transient reactance,
$x_q''$ is transverse sub transient q-reactance,
$T'_d$ is direct transient time constant,

2016

$T_d''$ is direct sub transient time constant,
$T_{d0}'$ is open direct transient time constant,
$T_{q0}''$ is open transverse sub transient time constant,
$\omega_e$ is machine electrical speed corresponding to the time derivative of $\theta_e$.
Thus, from equations (6) and (7), the machine can be represented by reactances and time constants.

## III. IEEE STANDARDS ON SYNCHRONOUS GENERATOR PARAMETERS DETERMINING

The determining of synchronous generator parameters is based on the sudden short-circuit test which can be described as follows:
The machine is involved at the rated speed without load until it reaches the steady state in terms of output voltage. During this steady state, a short circuit is performed on its three phases and currents and voltages are measured during that test. Then theories can be applied to determine the generator parameters. After performing of sudden short circuit, the current on each phase can be described as:

$$
\begin{aligned}
i = V_m \cdot [\frac{1}{x_d} + (\frac{1}{x_d'} - \frac{1}{x_d}). \exp(-\frac{t}{T_d'}) + \\
(\frac{1}{x_d''} - \frac{1}{x_d'}). \exp(-\frac{t}{T_d''})].cos(\omega.t + \theta_0) + \\
V_m \cdot [(\frac{1}{x_d''} + \frac{1}{x_q''}). \exp(-\frac{t}{T_a}).cos(\theta_0) + \\
(\frac{1}{x_d''} - \frac{1}{x_q''}). \exp(-\frac{t}{T_a}).cos(2\omega.t + \theta_0)]
\end{aligned} \quad (8)
$$

where:
$V_m$ is the voltage maximum value prior the short circuit applying.
Notice that the above expression can be simplified in considering the current which aperiodic component is null. Among the parameters we used in the simulation, transverse sub transient q-reactance $x_q''$ and direct sub transient d-reactance $x_d''$ are the same. Definitely the equation (8) becomes:

$$
\begin{aligned}
i = V_m \cdot [\frac{1}{x_d} + (\frac{1}{x_d'} - \frac{1}{x_d}). \exp(-\frac{t}{T_d'}) + \\
(\frac{1}{x_d''} - \frac{1}{x_d'}). \exp(-\frac{t}{T_d''})].cos(\omega.t + \theta_0)
\end{aligned} \quad (9)
$$

The IEEE standards, [3], [6], recommend to draw an envelope which fits the best with output current. Then calculations are made with this last one. The current corresponding to the phase which aperiodic component is null has been chosen to achieve calculations. Thus the equation (9) becomes:

$$
\begin{aligned}
i_{env} = V_m \cdot [\frac{1}{x_d} + (\frac{1}{x_d'} - \frac{1}{x_d}). \exp(-\frac{t}{T_d'}) + \\
(\frac{1}{x_d''} - \frac{1}{x_d'}). \exp(-\frac{t}{T_d''})]
\end{aligned} \quad (10)
$$

Notice that in these standards only the upper envelope is used. But The lower one is also valid. In the algorithm presented further, the reader can chose the type

of envelope that he wants for the parameters estimation. This allows to get very good results even if the aperiodic component of the current is not null.

### A. Direct steady and transient parameters estimation

The steady reactance is easy to calculate. It corresponds to the reactance of the machine when it works at steady state. In the equation (10), as the exponential functions are decreasing, direct steady state reactance can be deduced as following:

$$
x_d = \frac{V_m}{i_{steady}} \quad (11)
$$

Once the steady current value is found, it is substract to $i_{env}$ as presented below:

$$
\begin{aligned}
i_{env} - i_s = V_m \cdot [(\frac{1}{x_d'} - \frac{1}{x_d}). \exp(-\frac{t}{T_d'}) + \\
(\frac{1}{x_d''} - \frac{1}{x_d'}). \exp(-\frac{t}{T_d''})]
\end{aligned} \quad (12)
$$

One can then use a semi logarithmic frame to determine reactances and time constants. This allows to go from an exponential curve to a sum of real straight curves leading to easier calculations. Nevertheless an assumption is made on the time constants:
*assumption*: Transient time constant is very high besides sub transient time constant $(T_d' >> Td'')$
*Consequently* The sub transient component decreases quickly in relation to the transient one. So its effects can be neglected from a certain time. This assumption leads to make an approximation of the above current difference $i_{env} - i_s$:

$$
i_{env} - i_s \approx V_m \cdot (\frac{1}{x_d'} - \frac{1}{x_d}). \exp(\frac{t}{T_d'}) \quad (13)
$$

Using semi logarithmic method, we can say that it exists two quantities $A$ and $B$ such as

$$
\ln(i_{env} - i_s) \approx A.t + B \quad (14)
$$

$A$ is the slope and $B$ the value at the frame origin.
The transient parameters can be then obtained by solving the following equations system:

$$
\begin{cases}
T_d' = -\frac{1}{A} \\
\ln(V_m \cdot (\frac{1}{x_d'} - \frac{1}{x_d})) = B
\end{cases} \quad (15)
$$

### B. Direct sub-transient parameters estimation

The calculation method of sub-transient parameters is the same as described above. Therefore the approximation is made on the quantity bellow:

$$
\ln(i_{env} - i_s - i_{trans}) \approx A'.t + B' \quad (16)
$$

where
$i_{trans}$ is the transient current calculated above with steady

and transient parameters.
This leads to the following new system of equations:

$$\begin{cases} T_d^{''} = -\frac{1}{A'} \\ \ln(V_m \cdot (\frac{1}{x_d^{''}} - \frac{1}{x_d'})) = B' \end{cases} \quad (17)$$

Note that, several other methods are used to determine
direct axis parameters, see [7], [8] and [2]. The one
presented in this paper is easy to implement and the
results that we got are satisfactory.

### C. Time constants in open circuit

Some direct axis parameters cannot be directly calculated but they can be deduced from the above calculations.
According to the IEEE standards, to calculate open circuit
time constants, the following relationships are used:

- Open circuit direct transient $T_{d0}'$

$$\frac{x_d}{x_d'} = \frac{T_{d0}'}{T_d'} \quad (18)$$

- Open circuit direct sub transient $T_{d0}^{''}$

$$\frac{x_d'}{x_d^{''}} = \frac{T_{d0}^{''}}{T_d^{''}} \quad (19)$$

Thanks to all these relationships, one achieves an algorithm in order to resolve the problem. In the following
section details will be given on the involved iterative
strategy.

## IV. PARAMETERS ESTIMATION USING ITERATIVE STRATEGY

In order to perform the synchronous generator parameters estimation, an algorithm is elaborated according to
the figure (4). After the initialization of the algorithm,
an interface is created to allow the user to enter data.
Then calculations can be started. The program begins by
checking the validity of these data, so in the case of a
non matching value error messages are displayed. Then,
the user is asked to correct the wrong values and to try
again. Once this step is verified, the algorithm calculates
three important values: the signal frequency ($f$), the short-
circuit instant($t_{sc}$) and the sampling time ($t_s$). At this
stage, the main part of the algorithm concerning the direct
axis parameters computation can be performed using an
iterative strategy and results are displayed via numerical
tables and figures.

### A. Interface design

The interface generated by the program is shown by
Fig.5.

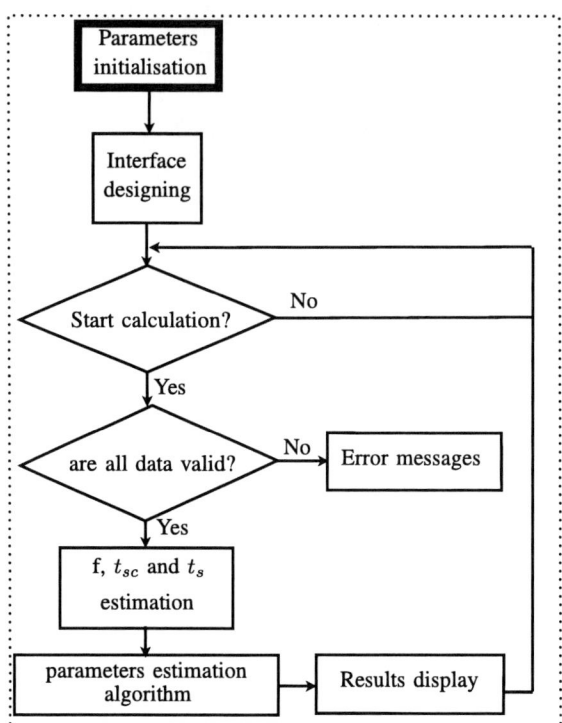

Fig. 4.   General algorithm flow chart

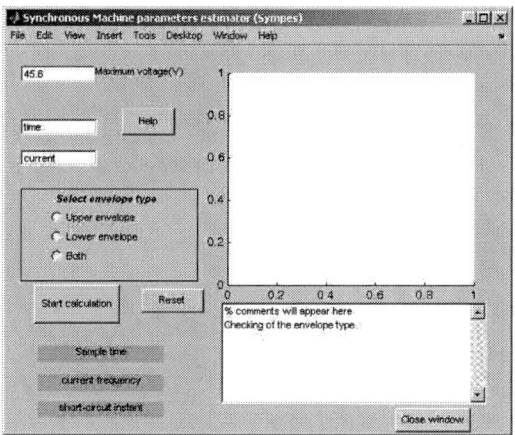

Fig. 5.   Graphical interface for parameters estimation

The three first cases allows the user to enter data: the
maximum voltage value, the current and time vectors.
These last vectors must be **.mat** format to be accepted
by the algorithm. The button **Help** gives the general
information to the user. A reserved area is made for
figures. On this area the current and its different envelopes
will be plotted. The interface allows to chose the type of
used envelope. Indeed, one can decide to use the upper
envelope, the lower or both upper and lower envelopes.

**2018**

The **Reset** button brings back the interface to its default settings. Once the calculation is started by the push-button **Start Calculation**, the algorithm state is given in the comments' area. The closing of the interface can be done by the button **Close window**.

### B. Parameters estimation algorithm

Its rule is to take into account the data provided by the user and to determine the different parameters of the machine. The figure (6) shows the strategy used to implement the algorithm.

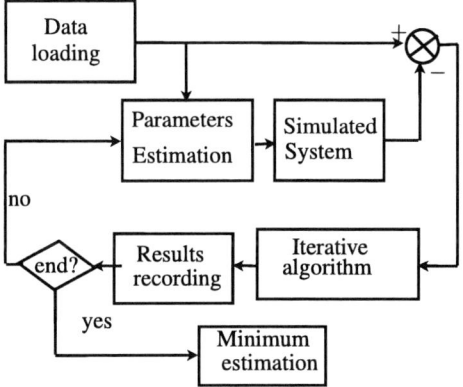

Fig. 6.   Parameters estimation diagram

Initially, a sudden short circuit test is performed on the synchronous machine and the useful data are obtained. The estimation method is based on the IEEE standards [3]. Each set of parameters is used to simulate a new system. From this last one, an error between the actual data and the simulated model is calculated such as:

$$J(k) = \sqrt{\sum_{i=1}^{N}(i_{actual}(i) - i_{sim}(i,k))^2} \qquad (20)$$

As we can see this cost is about the $k^{th}$ set of parameters. where:
$k$ the iteration order or set of parameters order, $i_{actual}(i)$ is the actual current value for $i^{th}$ sampling point, $i_{sim}(i,k)$ is the simulated current value for $i^{th}$ sampling point with the $k^{th}$ set of parameters, $N$ is the number of sampling points, $J(k)$ is the criterion value with the $k^{th}$ set of parameters.
For each iteration the criterion value, extended to the whole range of variation is stored in a vector. The operation is done again until the instruction *end* is reached. From there, the vector $J$ allows to get the optimal set of parameters. The flow chart Fig.7 explains how the algorithm is performed:

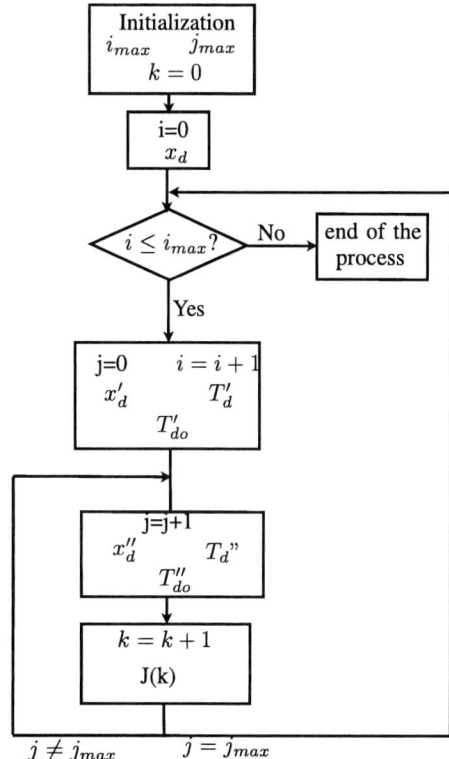

Fig. 7.   Flow chart for parameters estimation

First of all, initialization is done: $i_{max}$, $j_{max}$, $i = 0$, ... Then steady reactance is calculated and some ranges are defined to determine transient set of parameters. These ranges are chosen far enough from short circuit start point and their width depends on the machine output current frequency. As regards to sub transient ranges they are chosen near the start point. The figure (8) shows the subdivisions made on the current envelope.

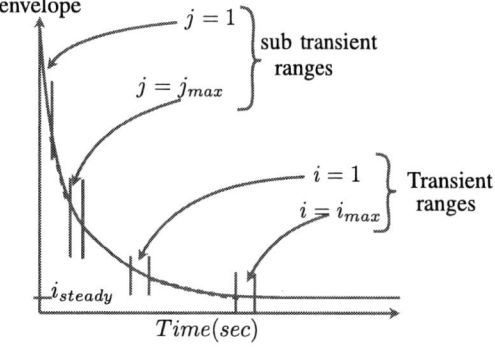

Fig. 8.   Transient and sub transient ranges subdivision

From Fig.7, we can notice that for each transient range, corresponding to one transient set of parameters, $j_{max}$ sets of sub-transient parameters are calculated. At the end of the process, the criterion $J$ is a vector which length is about $i_{max} \times j_{max}$. Each value of this vector corresponds to a particular set of parameters. Minimizing this criterion allows to get the parameters which fit the best with accurate values. The developed algorithm, works in order to find a particular value $k_{opt}$ corresponding to the general minimum such as:

$$J(k_{opt}) = Min \left( \sqrt{\sum_{i=1}^{N} (i_{actual}(i) - i_{sim}(i,k))^2} \right)_{k \in K} \quad (21)$$

where: $K$ is the iterations group which length is defined above $(i_{max} \times j_{max})$, Min denotes the minimum operator. When this optimum iteration $(k_{opt})$ is found, the different validation curves can be plotted to check whether actual quantities and estimated ones match each other.

## V. EXPERIMENTAL RESULTS

### A. Test bench presentation

The two machines coupling and the final experimental test bench, which are used to validate algorithms, are presented by figures (9) and (10)

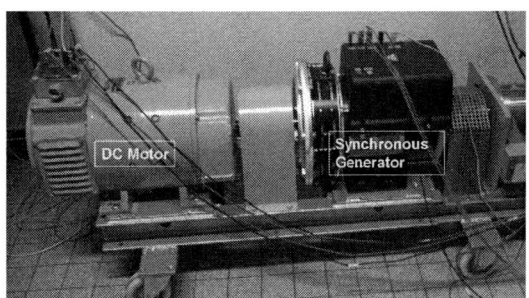

Fig. 9.   The machines coupling

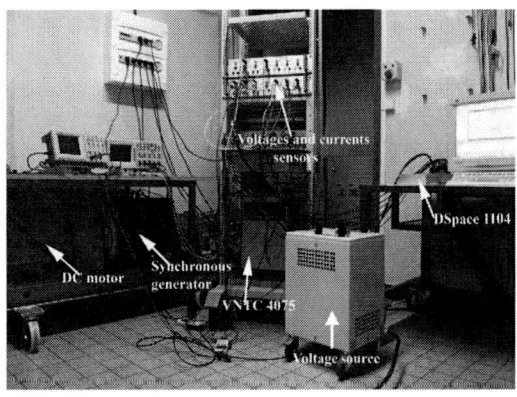

Fig. 10.   Experimental test bench

A DC motor is used to involve the synchronous generator. This motor is speed-controlled thanks to a motor drive: the VNTC4075 from Alstom company. Thanks to a right parametrization and controller'settings the synchronous generator's speed is 1500rpm by feeding its exciter and armature. This allows to specify the synchronous

generator's output voltage frequency to 50Hz. A voltage source is used to feed the excitation of the synchronous generator. One can see on Fig.10, the voltages and currents sensors boxes which allow to measure and exploit data thanks to the Dspace 1104.

### B. Results and discussion

Hereafter is the developed interface overview

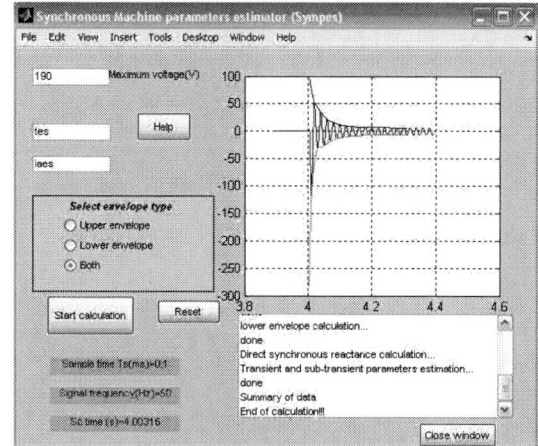

Fig. 11.   Interface global overview

As we can see on the figure (11), the signal frequency, the sampling time and the short-circuit instant are well estimated and comments on the calculation state are displayed below the curves. The synchronous generator is involved at rated speed thanks to the DC motor and excited by the voltage source described above, until its output voltage (open circuit) reaches a steady state, then a sudden short circuit test is performed and data are recorded. Thank to a series of sudden circuits tests and the developed algorithms, the results are given in TABLE.I.

TABLE I
EXPERIMENTAL RESULTS INVOLVING 7 TESTS

| Labels | $T_1$ | $T_2$ | $T_3$ | $T_4$ | $T_5$ | $T_6$ | $T_7$ |
|---|---|---|---|---|---|---|---|
| $v_{max}$(V) | 108.5 | 130.5 | 162.6 | 217 | 250.5 | 279.5 | 310.5 |
| $x_d(\Omega)$ | 30.36 | 29.47 | 28.93 | 28.04 | 26.53 | 24.49 | 21.75 |
| $x'_d(\Omega)$ | 4.25 | 4.67 | 4.6 | 4.47 | 4.56 | 4.16 | 3.58 |
| $x''_d(\Omega)$ | 1.21 | 1.07 | 0.98 | 0.72 | 0.67 | 0.63 | 0.61 |
| $T'_d$(ms) | 42.3 | 43.3 | 40.9 | 40.8 | 38.6 | 34.5 | 35.2 |
| $T''_d$(ms) | 8.79 | 8.97 | 9.03 | 8.39 | 9.01 | 8.87 | 8.23 |
| $T'_{do}$(ms) | 301 | 273 | 258 | 256 | 225 | 203 | 214 |

Where: $x_d$, $x'_d$ and $x''_d$ are respectively the direct, transient and the sub transient synchronous reactances . $T'_d$, $T''_d$ and $T'_{do}$ are respectively transient, sub transient and open circuit direct transient time constants.
This table shows results for seven tests ($T_1$ through $T_7$) performed on the real machine (the lsa371 from Leroy Somer). In this table, $v_{max}$ denotes the output voltage maximal value prior the sudden circuit test, it depends on the excitation voltage magnitude. As one can notice, some parameters vary according the excitation voltage. This is

**2020**

due to the saturation of the magnetic circuit [9]. Thus, thanks to the interface, the students can easily identify and study the synchronous machine behavior according to the magnetic circuit state.

## VI. CONCLUSION

In this paper an academic tool of electrical engineering is proposed. The originality of this paper is mainly based on the approach. Indeed some problems are faced by students, teachers and even manufacturers, when the machine is to be identified. Some methods exists but are sometimes unknown or difficult to use. Our approach is to make easy-use these methods by using an interface. All the algorithms are made using the Matlab$^{TM}$ M-file. Then the only action to do by the user is to execute the file and fill the required boxes: signal vector, time vector, prior voltage value. Another interesting property of this interface is the possibility given to the user to choose the type of envelope involved in the calculation. According to the short-circuit instant, one can find benefit to use upper, lower or both envelope.

The experimental results are very satisfactory and the interface friendliness allows us to confirm its application to teaching domain. Thanks to such an interface, students can easily identify synchronous machine and study its parameters variation according to the excitation voltage.

## REFERENCES

[1] C. Xingang, J. Luming, W. Xusheng, Y. Kaisheng, and W. Zhifei, "A new approach to determine parameters of synchronous machine using wavelet transform and prony algorithm," *2004 International Conference on Power System Technology - POWERCON 2004 Singapore*, 2004.

[2] A.Tumageanian and A. Keyhani, "Identification of synchronous machine linear parameters from standstill step voltage input data," *IEEE Transactions on Energy Conversion, Vol. 10, No. 2, June 1995*, vol. 10, 1995.

[3] S. IEEE, "Test procedures for synchronous machines ieee standards 1158," *IEEE guide*, 1995.

[4] J. Verbeeck, *Standstill frequency response measurement and identifictaion methods for synchronous machines*. Thesis , Vrije Unversiteit Brussel, 2000.

[5] H. Guesbaoui and C. Iung, "Reduced models and characteristic parameters of the synchonous machine obtained by a multi-time scale simplification," *IEEE Industry Applications Society Annual Meeting, 1994., Conference Record of the 1994 IEEE 2-6 Oct. 1994 Page(s):262 - 271 vol.1*, 1994.

[6] J. D. GHALI, *Détermination des paramètres de la machine synchrone*. Ecole Polytechnique Fédérale de Lausanne, 1994.

[7] S. Moreau, R. Kahoul, and J. LOUIS, "Parameters estimation of permanent magnet synchronous machine without adding extra signal as input excitation," *ISIE'04 Ajaccio*, vol. 1, pp. 371 – 376, 2004.

[8] F. L. Alvarado and C. cañizares, "Synchronous machine parameters from sudden short tests by back solving," *IEEE transactions on energy conversion, Vol. 4, No. 2, June 1989*, vol. 4, 1989.

[9] R. Escarela-Perez, T. Niewierowicz, and E. Campero-Littlewood, "A study of the variation of synchronous machine parameters due to saturation: a numerical approach," *Electric Power Systems Research*, vol. 72, pp. 1–11, 2004.

# Determination of the Characteristic Life Time of Paper-insulated MV-Cables based on a Partial Discharge and tan(δ) Diagnosis

I. Mladenovic, Ch. Weindl

Institute of Electrical Power Systems, University of Erlangen-Nuremberg, Erlangen, Germany, e-mail:
*mladenovic@eev.eei.uni-erlangen.de*
*weindl@eev.eei.uni-erlangen.de*

*Abstract* – **Monitoring the condition of electrical equipment insulation by means of partial discharge measurement and analysis is an effective diagnostic tool for the prediction of service failures. In this paper an accelerated aging system specialized for paper insulated lead covered (PILC) cables, with a highly sensitive and selective PD-detection/localization and tan(δ) measurement is presented. It facilitates the monitoring of the entire aging process and the later development of sophisticated diagnostic criteria based on the physical aging process.**

*Keywords* – **Partial Discharge, Diagnostics, Insulation, Prognosis, Software for Measurements, Measurement, Maintenance, Estimation Technique**

## I. INTRODUCTION

In most developed countries, cable systems dominate the electrical distribution networks in the medium voltage range. Therefore diagnostic analyses based on partial discharge (PD) measurement are a widely used tool for the prediction of failures, its localization and selective replacement of defective cable sections.

Cables and their equipment represent the biggest capital in Germany's distribution networks [3]. Moreover, cables and especially joints are the prime origins of failure and, so the main sources of supply discontinuity. Therefore the condition of this equipment and the sophisticated prediction of failures are of the highest importance.

Most of the test systems established on the market are very suitable for cross-linked polyethylene (XLPE), thermoplastic polyethylene (PE) and ethylene-propylene rubber (EPR) insulated cable systems. In many regions worldwide more than 50 percent of the medium voltage systems are still based on paper insulated lead covered (PILC) cables. Also, an age, which is commonly referred to as the end of electrical equipment life, is reached by about 30% of plants in distribution networks of Germany [3]. In case of lead shielded cable types, the above mentioned test and prediction systems do not deliver reliable prognosis results, because the physical and chemical backgrounds are not comparable to those of XLPE cable systems. Nevertheless, trusted studies and the appropriate outcomes are necessary for a reliable maintenance strategy and investment planning.

According to the study of an international cable diagnostic and service company, it has been computed that the replacement of a 2,5 km long 20 kV paper insulated cable results in average costs of 175,000 EUR (example taken from Germany). If this cable is PD-tested, the sections to be replaced are reduced to small portions with a total length of e.g. 20 m. In this case the total costs represent the fees for the cable diagnostics, the replacement of the cable portions and the necessary joints. Therefore the savings are expected to be about 80%, [2].

Based on these facts, a new accelerated aging project was started in 2006. Its aim is to find suitable parameters describing the aging process of PILC cables by a constant monitoring of the relevant aging parameters. On this data basis, a new accelerated aging test should be developed which facilitates a better evaluation of the time to failure of cable samples.

In order to obtain a realistic but accelerated aging process, the artificially made conditions should be comparable to normal service conditions at 50 or 60 Hz. In this paper an aging and monitoring system is described, starting with the voltage and current generation, the PD detection, its location, the tan(δ)-measurement, the control system and finally the data acquisition, correlation and storage.

## II. ACCELERATED AGING METHOD

Materials, which are used for power cable insulation, gradually deteriorate with time due to electrical stress activity, moisture and temperature stress. There are a lot of other factors that could also have a strong influence on the rate of aging. For example load cycles, transient over-voltages, ground-temperature, ground characteristics, seasonal conditions, parallel traces and cross lines to other power cables have the significant influence on cable status. In addition the cable handling and the ambient temperature during the laying of the cables could result in a decline of the breakdown voltage or an increase of the failure rate during normal operation.

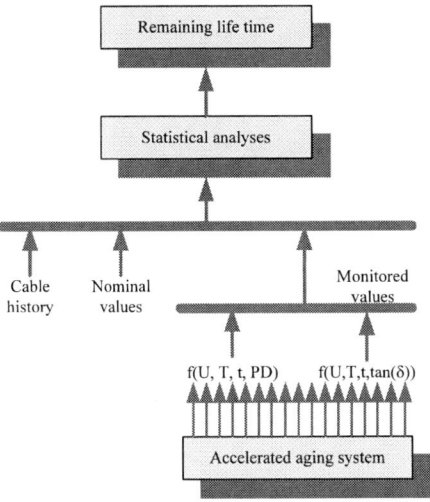

Fig. 1 Accelerated aging method

Using a statistical approach based on at least two different pre-aged cable sets, the aging process has to be qualified. Different cable sets and their history must also assure that at the very end of the aging process the statistical time to cable failure can be evaluated most exactly, as it is shown in figure 1.

Basic data, like manufacturing year, average use and load configuration are also necessary. Databases should be elaborated as much as possible, so the most exact approximation to the statistical time to failure can be made.

## III. PRINCIPLE OF THE DESIGN OF AN ACCELERATED AGING AND MONITORING SYSTEM FOR PILC CABLES

One important part of the accelerated aging system is shown in figure 2. The schematic diagram represents the main parts of the controllable voltage source. It consists of a series resonant circuit with a variable inductance, its resistive components and the cable capacitance. In this way a variable and efficient voltage generation is supplied with a software controllable time profile.

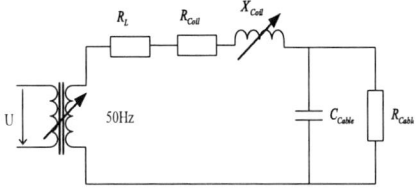

Fig. 2 Resonant circuit

The cables lay in a water filled tank, whose size and materials are carefully chosen according to reference [5]. It has been shown that in the tests, which use water-filled tanks, the most important parameters are electrical stress, temperature, water characteristics, and time.

In order to reach the demanded cable temperature and its allocation an additional current transformer (CT) is used. The conductor temperature is controlled by a regulation of the CT's secondary current (I_sec, see fig. 3).

In this way load variations can be simulated, according to the normal or average field operation. Furthermore, the cable temperature and the water temperature are controlled and kept constant to pre-defined values.

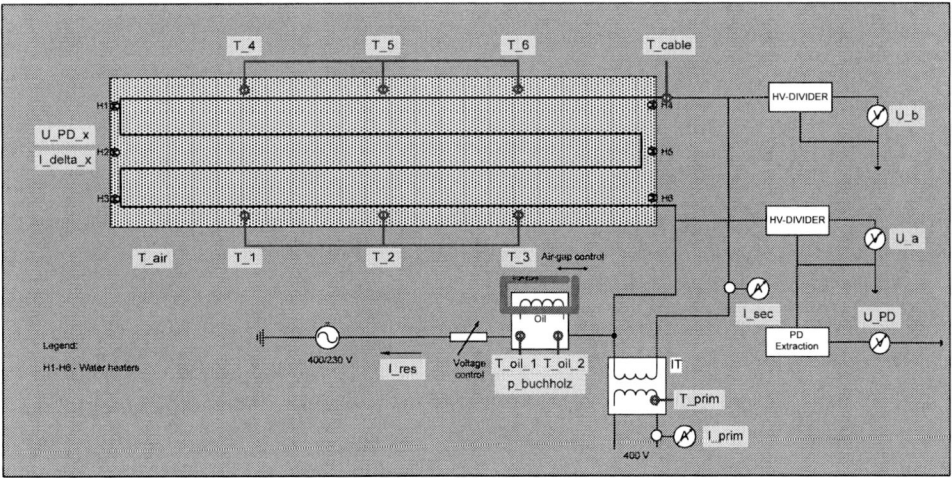

Fig. 3 Simplified structure of the measurement and control system

Thus defined load variations can be simulated under well defined conditions using a specially designed measurement and software system.

The water temperature is measured at 6 different locations (T1-T6) to ensure a minimal aberration of the environmental conditions, as shown in fig. 3. Additionally the conductor temperature is measured directly and the terminator temperature is acquired using a suitable IR sensor. All temperature differences and their deviations are monitored by the acquisition system and controlled according to user defined values. In this way, a defined temperature gradient and a controllable temperature profile in the cable insulation can be obtained. The water can be cooled or heated using a computer controlled conditioning system (H1-H6). By a programmable circulation of the water flow using a high power pump, a relatively constant water temperature is achieved.

Fig. 4 Construction of the coil

The other important aging parameter, the applied voltage, is obtained using an oil-insulated and actively cooled resonant coil that makes up a resonant circuit of 50 Hz with the cable capacitance. One of the main design criteria of the developed voltage source was to achieve a duty cycle of 100% which exceeds the requirements of most of the standard measurement systems. Additionally a wide range of inductance accommodation for different cable setups and numbers has to be provided. This could be reached by an external, motor controlled air gap in the iron core. The possible inductance ranges from 47 H (necessary for 45 cable samples) up to 580 H (4 cable samples). The aging voltage of up to 50 kV is applied between the cable conductor and ground; it is transmitted to the monitoring system by using high-precision voltage dividers. Also the mainly capacitive leakage current through the cable insulation is measured to facilitate both, PD and tan $\delta$ monitoring. In fig. 4 one steep in the realization of the coil is shown. Fig. 5 gives an overview of some of the main components of the accelerated aging equipment.

Fig. 5 Parts of the accelerated aging equipment in the laboratory (under construction)

All measurement and control tasks are accomplished by a specially equipped Personal Computer. The data acquisition is done by two sophisticated PCI DAQ extension boards. The first one is a high precision analogue-digital computer interface that provides 32 analogue input channels, 4 analogue output channels and 32 digital I/O-channels. The second one is a digital computer interface providing 64 additional digital I/O-channels. Most of the measurement and control software is written in a graphical programming system for data acquisition purposes (LabView, National Instruments). Some of the security and time relevant tasks, like a watchdog function for increased reliability and security are developed using Microsoft Visual Studio in C# as a system service.

The LabView program carries out the acquisition and monitoring of 9 AC-values and 13 DC-values every second. Each value is written to memory with a time stamp and is then periodically saved to hard disk in separate files for each day. Moreover, 11 digital inputs and more than 50 internal system state variables have to be watched. The whole system is controlled by more than 40 outputs. Every transaction and incident is written in a log file so critical system states and faults can be reconstructed and analyzed.

The system is freely configurable in many parts to accommodate to the aging processes. For example the upper and lower limits for each measured value can be defined. If the limits are exceeded, alarm messages are generated and transferred as a SMS (short message service) to a cellular-phone emergency number and/or the entire system is automatically shut down. Furthermore, different levels for the aging voltage and the corresponding time intervals can be selected, as well as the sequence intervals for partial discharge and tan $\delta$ measurement, the time intervals and temperature limits for water control and the limit for the correction of the aging voltage etc. Also the PD and tan($\delta$) measurements are done independently for each cable sample by the use of the DAQ board. To achieve a high accuracy, especially for the tan(δ) measurement, a further developed filter algorithm has been implemented.

## IV. FUNDAMENTAL CORRELATIONS TO PREDICT THE AGING STATUS AND TIME TO FAILURE OF PILC CABLES

The main goal of the entire project is to find reliable correlations between field measured values, such as tan(δ) or PD-levels, and the time to failure ( $t_{RL}$ ) of the analyzed cable tracks. Therefore, the following general equation holds:

$$t_{RL} = f(\tan \delta, PD) \qquad (1)$$

All other physical and operational parameters are assumed to be included in the basic parameters tan(δ) and PD-level.

A complete monitoring of all relevant cable and aging parameters provides a substantial database for the following statistical and mathematical approach. Finally, a reliable postulation has to be determined for the time to failure as a basis for the investment and maintenance planning of the distribution companies.

The presentation of the principal mathematical approach follows. It can be subdivided into three major steps:

### A. Determination of accelerating factor (AF)

If $t_f$ is the time-to-failure under the test conditions and $t_{RL}$ is the time to failure for the same cable in real use in the field (see fig. 6) then the correlation between them is:

$$t_{RL} = AF \cdot t_f . \qquad (2)$$

The simplest and ideal case to mark out the value of $AF$ would be with two cables: one brand new with the corresponding parameters $c_N$ and a second cable, which was $t_j$ years in use (parameters $c_j$). After time $\tau$ under the test conditions the $c_N$ parameters will be equal to the parameters that $c_j$ had at the beginning. In that ideal case $AF$ is real number:

$$AF = \frac{t_j}{\tau} . \qquad (3)$$

In reality, the usage of a cable system is influenced by many time depending parameters such as load curves or patterns, thermal effects, environmental influences, etc. Therefore, especially the parameters of the pre-aged cables are based on stochastic processes and, as a result, it cannot be expressed exactly with single deterministic values. On the other hand, also brand new cables slightly differ with respect to many parameters. Thus, even new cable samples may have a different expected life-time since small environmental, production and laying dependant effects accumulate to differing initial conditions prior to the first operational use. As a consequence, the mathematical approach cannot be based on only one idealized cable sample but on the statistical results of a greater number of cable systems.

In order to specify $AF$, in this project, a set of cables with different usage history are tested under the same accelerated aging conditions. Parameters of old-unused, brand new and used cable samples are recorded and graphically analyzed, so the AF is determined as accurately as possible.

During the accelerated aging process, the value of AF is mainly influenced by the applied temperature and voltage profiles. It is to be chosen in a way that faults in at least some of the test samples happen within the expected project duration.

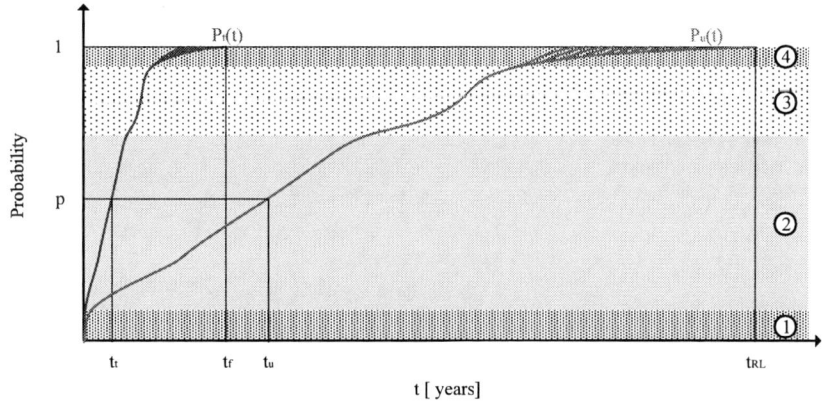

Fig. 6 *Conceptional representation of P(t)*
$P_f(t)$ - probability of fault after time t in test condition,
$P_u(t)$ - probability of fault after time t in real use in field

Otherwise, the time schedule of the accelerated aging process must enable a good extrapolation to the aging scenario under real load conditions.

Additionally, the influence of the applied test voltages on the cable status and the rest lifetime must be limited to prevent an instant or uncoordinated damaging of the cable samples.

*B. Defining the function for the calculation of the time-to-failure (t$_f$) under accelerated aging conditions*

In a basic assumption the time $t_f$ is a function of more variables that have an independent influence on the time-to-failure. That's why the principal equation for defined voltage ($U$) and temperature ($T$) stress levels can be presented as:

$$t_f = f_1(U) \cdot f_2(T) \qquad (4)$$

where:

$f_1(U)$ is the ageing function only under the voltage stress,
$f_2(T)$ is the ageing function only under the load stress.

In this first theoretical approach, time dependant effects resulting e.g. from varying load or voltage stress conditions are neglected. So the shape of the probability functions shown in fig. 6 are defined by the physical and chemical processes caused by the given constant stress conditions.

The operating conditions of cable tracks in distribution networks are statistically defined by the load demands of the supplied customers.

Therefore during the accelerated aging and also in real field tests the aging conditions depend on many external factors like load patterns, over-voltages etc.

In cable diagnostic field tests only some of the physical and chemical parameters resulting from a certain cable status can be approximated analyzing certain discrete electrical parameters. For that reason these measurements are mostly based on the tan(δ) and *PD* intensity at different test-voltage levels. The principal dependency is shown in equation (5):

$$t_f = f_3(\tan \delta, PD, U) \qquad (5)$$

where $f_3(\tan \delta, PD, U)$ is the ageing dependency on the tan(δ) and PD levels at different test-voltages *U*.

It is obvious that $f_3$ is a highly non-linear function, which can be determined only by a statistical approach using a great number of different aged cable samples. Hence, the electrical, physical and chemical status of the cable, the time-to-failure ($t_f$) and therefore the rest lifetime under average load conditions can be determined only by certain accuracy.

*C. Determination of the probability of failure for a aged cable sample P(t)*

Because of the statistical background of the fault events described above, a distribution function must be applied to estimate the probability of failure for given cable samples.

In many technical fields the normal or Gaussian distribution is used to estimate the behavior of different processes. The described aging process and the resulting cable faults must be described in a different way, thus the life-time of the cable systems is not unlimited.

The Weibull distribution, which is in fact a special case of the extreme value distribution, delivers a more suitable solution. Depending on the quality and the amount of the collected data, the estimation of the Weibull distribution parameters could be very complex.

Additionally the probability functions shown e.g. in fig. 6 must be separated into different areas:

1- Fault caused during the manufacturing process, the installation, by mechanical damages, etc.
2- Quasi-linear part of probability function.
3- Stabilization area.
4- Fault could be caused by unpredictable smaller or bigger changes of the operational conditions (over-voltages, flicker etc.).

The results delivered by the Weibull distribution are limited to the intervals 2 and 3, because the events causing the faults in areas 1 and 4 are not predictable. The interval borders and the functional relations are dependent on the given stress levels of the aging process.

Once, when the Weibull distribution is known and also AF has been determined, it is possible to extrapolate the results from the accelerated aging to the rest life-time respective the failure probability of field samples as follows:

$$P_u(t) = P_t(t/AF) \qquad (6)$$

where index "u" signify "in use" and index "t" means "test".

## V. Project Overview

In principle, the entire project can be separated into three major phases:

1. Development, implementation and setup of the accelerated aging, the control and measurement systems.
2. Accelerated aging of different cable sets (unused, brand new and used cable samples, see figures 7 and 8) over a period of at least one year. In parallel the data acquisition, monitoring and formation has to be implemented. Cable faults have to be recognized, located and analyzed. The

corresponding samples must be replaced according to a database.

3. Estimation of the acceleration factor and evaluation of the important degradation processes by the identification of the relevant electrical, physical and chemical parameters. Identification of the cable conditions. Diagnostics and knowledge rules have to be implemented for different aging conditions.

Afterwards, the results of diagnostic cable measurements in a field test should be used to verify the predicted time to failure of the analyzed cables.

Fig. 7 Exemplary set of cable samples for the accelerated aging

Fig. 8 Sample of an old-unused cable from 1990

## V. CONCLUSION

In this paper the principal structure and functionality of a specially designed accelerated aging system for PILC medium voltage cables is described. Based on a continuous sample-wise monitoring of the partial discharges and the $\tan(\delta)$ values and their development during the ageing process, an "aging" database is obtained. This is used as a foundation for the determination of the

relevant electrical, physical and chemical parameters.

The realized multi-channel acquisition system for PD-detection & localization and $\tan(\delta)$-measurement, as well as the software control environment and the necessary test field equipment (test transformers, voltage dividers etc.) are presented. On behalf of the created database, criteria for the development of a calculation and diagnosis system shall be derived. In this paper the principal correlations are described.

Therefore, a condition-oriented maintenance strategy and asset management can be made, and also investments can be rightly planned. Moreover, this is one way to increase the power quality and reliability in today's MV distribution networks.

## ACKNOWLEDGMENT

The authors would like to thank for the financial and organizational support of the entire project to the following cooperating companies:

- N-ERGIE AG (Germany),
- IMCORP Europe B.V.L.A. (Belgium),
- N-ERGIE Service GmbH (Germany),
- Bayerische Kabelwerke AG, (Germany).

## REFERENCES

[1] Freitag, C.: „Entwicklung und Implementierung eines Steuerungs-, Regelungs- und Messsystems zur Realisierung einer automatisierten Versuchsanlage für die beschleunigte Alterung von Mittelspannungskabeln", Diploma Thesis, University of Erlangen-Nürnberg, 2008

[2] Imcorp - Instrument manufacturing company, „IMCORP Teilentladungsmessverfahren als Zustandsdiagnose für Mittel- und Hochspannungskabelanlagen", Feb. 2003

[3] „Zustandsorientierte Instandhaltung von Mittelspannungsnetzen", Forschungsreport, FGH – Forschungsgemeinschaft für elektrische Anlagen und Stromwirtschaft, 2002

[4] "IEEE Guide for Field Testing and Evaluation of the Insulation of Shielded Power Cable Systems", IEEE Std 400-2001

[5] „IEEE Trial-Use Guide for Accelerated Aging Tests for Medium-Voltage Extruded Electric Power Cables Using Water-Filled Tanks", IEEE Std 1407-1998

# Elimination of Increased Excitation of Common-Mode Oscillations in Electrical Drive Systems with Active Front End and Long Motor Cables

Thomas Weidinger

Chair of Electrical Drives, University of Erlangen-Nuremberg, Erlangen, Germany,
e-mail: *weidinger@eas.eei.uni-erlangen.de*

*Abstract*— In this paper the analysis of common-mode oscillations in electrical drive systems is presented. The analysed oscillations occur especially in electrical drive systems with active front end, common DC link and long motor cables. Because of resonant frequencies at 10 to 40 kHz low damping occurs. Some motors with concentrated winding show resonant behaviour in this frequency range and increase the overvoltages from the terminals to the star point. For example a torquemotor is analysed and modelled. In the concerned frequency range switching operations generate a significant excitation. As the focus of this paper a special coupling mechanism is described. This mechanism enforces the excitation by the switching of the active front end. Afterwards remedies are analysed. Simulations and measurements on sample systems with and without remedies show apparent voltage stressing of motors, common-mode losses and currents that disturb the drive system.

*Keywords*— common-mode noise, EMC, EMI, modelling, torquemotor, overvoltages

## I. INTRODUCTION

Electrical drive systems as shown in Fig.1 with active front end and common DC-link are often used for industrial applications with several axes, owing to the high potential for energy-saving. In widely-ramified facilities the length of the motor supply cables can thereby achieve several hundred metres. To reduce electromagnetic interference (EMI) usually shielded cables are used. They represent parasitic capacitances to ground. Other parasitic capacitances to ground also exist in power semiconductor devices and motor winding. Together with the sum of these capacitances $C_{par\Sigma}$ the choke used in the active front end builds an inductance for a common-mode LC resonant circuit.

The damping of this circuit is almost only represented through the damping appearing in the choke. Excitation is generated through switching in the rectifier of the active front end and the inverter. Owing to resonant frequencies at 10 to 40 kHz low damping occurs. As a consequence of these oscillations high voltages appear on the motor terminals.

At this point it is important to have a closer look at the high-frequency behaviour of the motor in the common-mode region. Owing to parasitic capacitances between winding and grounded core and the inductances in the winding a common-mode resonant behaviour also exists. Most types of AC-Motors with distributed winding

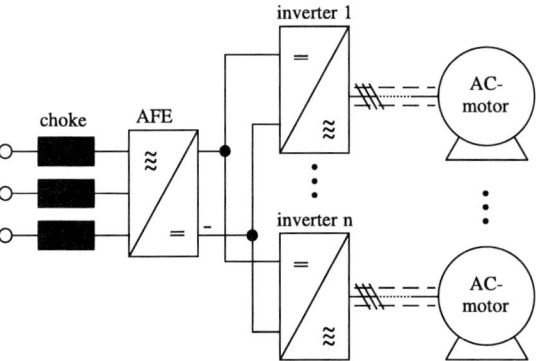

Fig. 1. Electrical drive system with active front end (AFE), common DC-link and long shielded motor cables

have a resonant frequency much higher than the resonant frequency of the drive system. For this reason the motors can be modelled as capacitances to ground. Hence the voltage at the star point is equal to that on the terminals.

Nowadays especially, permanent magnet synchronous motors are often produced with concentrated stator coils owing to the lower winding losses and easy mounting possibilities [1]. This type of winding is also used in in the so-called 'direct drives' like torque- or linearmotors. These motors typically have low speed values and high values of torque respectively force and therefore high numbers of poles. Because of the high parasitic capacitances in the coils the resonant frequencies may lie in the range of the system resonant frequency. As consequence the voltage is strongly elevated from the motor terminals to the star point.

Most analyses of common-mode perturbations, such as those by [2]–[4], describe oscillations from 100 kHz to several MHz. These oscillations are characterised by relatively low amplitudes of the common-mode current and relatively high damping compared with the analysed problem.

As regards overvoltages at the motor terminals reflections are a well-known problem often described, e.g. in [5]–[7]. This phenomenon occurs because of high $du/dt$ at the inverter and the high frequency characteristic of the shielded motor cables. A maximum voltage of the double

978-1-4244-1741-4/08/$25.00 ©2008 IEEE

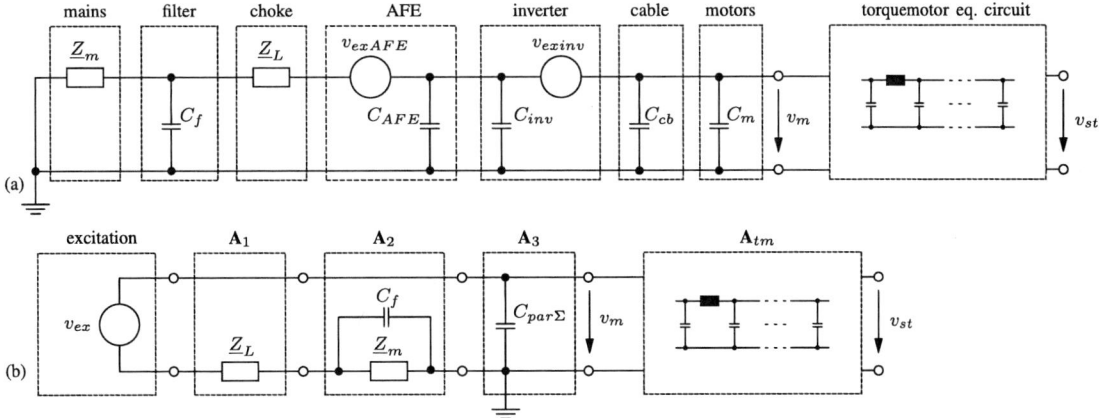

Fig. 2. Common-mode system model: (a) detailed model with separate modelling of every element of the electrical drive system (b) simplified model for further investigations, arranged as a daisy-chain of two-ports with the input signal $v_{ex}$ and the output $v_m$

of the height at the inverter may therefore appear on the motor terminals. The overvoltages at the motor terminals described in this paper are independent of the reflection phenomenon. The overvoltages owing to reflections may appear additionally and increase the maximum voltages.

As regards the problem described, only a few publications, such as [8]–[11], propose remedies. Thereby analysis of the oscillations is also made. Since a detailed analysis is lacking, complex investigations were attempted to achieve exact and detailed information about the system behaviour and important dependences on parameters. The outcomes concerning drive systems without AC motors with low frequency resonances are presented in [12].

In this paper the modelling of components of a typical drive system is shown first. Especially detailed analysis and modelling of an AC motor with comparatively low resonant frequency is presented. On the basis of the system model, which consists of component models, the system behaviour in the frequency domain is analysed. Subsequently a closer look is taken at the excitation of the system in the active front end and the inverter. Thereby a special coupling mechanism leads to a strongly enforced excitation by the switching of the active front end. Afterwards remedies are analysed. Finally results of simulations and measurements are shown for verification of the theoretical analysis and the proposed remedies. Evident overvoltages, common-mode currents and losses are presented.

## II. SYSTEM ANALYSIS

### A. Modelling

To obtain a detailed description of the system behaviour, all components of an electrical drive system were modelled as common-mode models in the frequency domain. The derived system model is shown in Fig. 2 (a). Important aspects of the component models are explained below.

*1) Mains:* concerning the mains a common-mode impedance $\underline{Z}_m$ is derived from the parameters of the

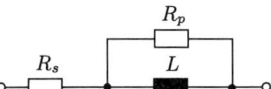

Fig. 3. Simple common-mode model of the choke

leakage impedances of the in-feed transformer and the impedances of the supply cables. The type of grounding must therefore be considered carefully. For modelling the EMC-filter in the common-mode domain only the capacitors to ground are important. The common-mode choke gets saturated very fast because of high common-mode currents appearing owing to the described oscillations.

*2) Choke:* the analysis and modelling of the choke are important elements of this investigation. Damping of the system is almost only produced in the choke, so the parameters of the choke have an important influence on the system's behaviour. The damping of the choke results largely in iron losses in the core, which cannot be calculated accurately. For this reason extensive measurements were attempted on different chokes to determine the common-mode impedance at the considered frequencies depending on current and temperature. An example of the common-mode impedance of an iron powder core choke is shown in [12]. Two different models were derived. One extensive model including current dependency of the inductance and frequency dependency of the resistance was derived for time-domain simulation. One simple common-mode model, as shown in Fig. 3, was created for the frequency domain, including only the frequency dependency of the resistance.

*3) Active Front End:* the model of the active front end (AFE) consists of a capacitance $C_{AFE}$ to ground that shows the parasitic capacitance to the heat sink in the semiconductor devices. The voltage source $v_{exAFE}$ represents the exciting voltage that is generated through the switching operations. An important point for the excitation is the generation of the switching commands in the control unit. This item is analysed in section III.

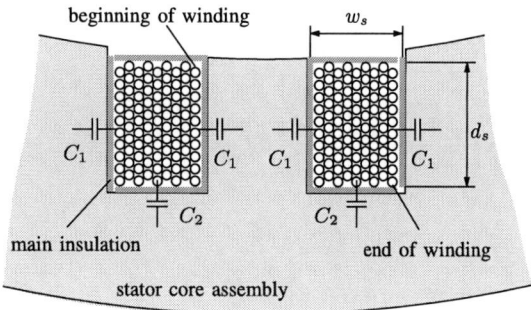

Fig. 4. Cross section of a stator coil of a concentrated stator winding

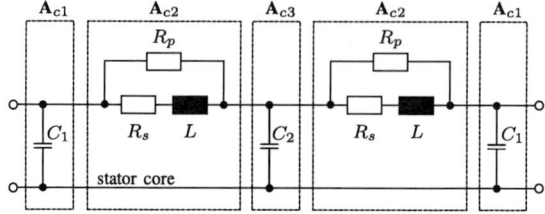

Fig. 5. Equivalent circuit of a coil, consisting of twoports $\mathbf{A}_{c1}, \mathbf{A}_{c2}$. and $\mathbf{A}_{c3}$

*4) Inverter:* owing to its similarity to the active front end the model of the inverter shows almost the same structure. The exciting voltage is called $v_{exinv}$ and the parasitic capacitance has the value $C_{inv}$.

*5) Cable:* as regards the motor supply cables, only the parasitic capacitances to the grounded shield are modelled as $C_{cb}$. High-frequency phenomena like reflections are neglected owing to the comparatively high frequency range. The capacitance $C_{cb}$ depends especially on the structure and the materials used for the cables. The most important factor, however, is the length of the cables.

*6) Motors:* concerning the motor it is important to have a closer look at the high-frequency behaviour in the common-mode region. Owing to parasitic capacitances between winding and grounded core and the inductances in the winding a common-mode resonant behaviour also exists. Most types of AC-motors with distributed winding have a resonant frequency much higher than the resonant frequency of the drive system. For this reason these motors are modelled as a simple capacitance to ground $C_m$ that shows the parasitic capacitances to the grounded stator core.

In contrast, 'direct drives' like torquemotors or linear motors may especially show a resonant behaviour with resonant frequencies below 40 kHz [8]–[11]. This leads to an voltage increase of the ground voltage from the terminals to the star point.

The resonant behaviour originates from the stator winding that consists of concentrated coils. This type of winding is often preferred today owing to its low copper losses and easy mounting possibilities [1]. In Fig. 4 a cross-section of a typical stator coil of a single layer winding is shown. From it a physically-motivated equivalent circuit of the coil can be derived. Unlike the frequently-used distributed winding schemes, where the single turns are accidentally distributed, in this coil the different turns are equally wound in different layers. All turns of the first layer have parasitic capacitances to the grounded stator core. These capacitances are summarised in $C_1$. Owing to the approximitely same area to the slot the capacitance of the last layer to the grounded core equals the same value $C_1$. Other parasitic capacitances can be found on the bottom of the slot. They are summarised as $C_2$. Together with the inductance of the coil the

equivalent circuit is shown in Fig. 5. Two resistances $R_p$ and $R_s$ are used to model the damping of the inductance. Parasitic capacitances between the different windings can be neglected, owing to the concerned frequency range. The proposed model of a coil has similiar structure to commonly-used motor models [13]–[15]. It can be clearly seen that the circuit consists of a series connection of two LC resonant circuits.

To derive the complete motor model the four-terminal networks of the coils can be connected the same way as the coils are connected in the motor. To combine parallel branches the parameters have to be adjusted. The model of the complete torquemotor is shown in Fig. 6 (a). Owing to the fact that the three phases are lying in parallel, they can also be combined into a common-mode model of the torquemotor shown in Fig. 6 (b). This model represents a series connection of several resonant circuits with the same parameters and therefore the same resonant frequency.

The matrix $\mathbf{A}_c$ describing the system behaviour of a single coil, can be calculated as product of the chain matrices:

$$\mathbf{A}_c = \mathbf{A}_{c1} \cdot \mathbf{A}_{c2} \cdot \mathbf{A}_{c3} \cdot \mathbf{A}_{c2} \cdot \mathbf{A}_{c1} \qquad (1)$$

The product of the combined elements $\mathbf{A}_{ccm}$ hence is describing the system behaviour of the complete motor

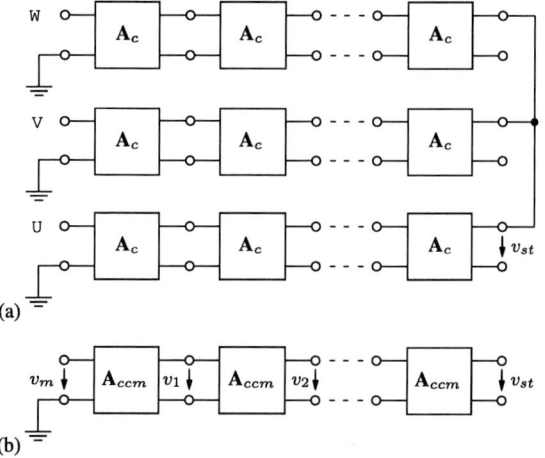

Fig. 6. Models of the torquemotor: (a) full motor model (b) simplified common-mode model

TABLE I
DATA OF INVESTIGATED TORQUEMOTOR

| nominal power | 24 kW |
|---|---|
| nominal speed | 76 1/min |
| branches per phase m | 2 |
| coils per branche n | 6 |

in the common-mode region:

$$\mathbf{A}_{tm} = \mathbf{A}_{ccm}^{n} \qquad (2)$$

whereof the voltage transfer function $\underline{H}_{tm}(2\pi f)$ can be derived:

$$\underline{H}_{tm}(2\pi f) = \frac{\underline{V}_{st}(2\pi f)}{\underline{V}_m(2\pi f)} = \frac{1}{\mathbf{A}_{tm(1,1)}(2\pi f)} \qquad (3)$$

$|\underline{H}_{tm}(2\pi f)|$ shows the voltage increase from the terminals to the star point where $V_m$ and $V_{st}$ represent the Fourier transform of $v_m$ and $v_{st}$.

To obtain the parameters of the model, measurements of several torquemotors were investigated. For example the results of one motor are shown. In Table I the data of this torquemotor are given.

From the common-mode impedance $\underline{Z}_{cm}$ measured with the set-up shown in Fig. 7 (a) the sum of all parasitic capacitances to ground $C_{tm}$ can be derived. The sum of the inductances of two phases $2\,L_{ph}$ is investigated with the measurement set-up shown in Fig. 7 (b). The parameters $C_1$, $C_2$ and $L$ of the matrix $\mathbf{A}_{ccm}$ can therefore be calculated as:

$$C_1 = \frac{C_{tm}}{n} \frac{d_s}{(2d_s + w_s)} \qquad (4)$$

$$C_2 = \frac{C_{tm}}{n} \frac{w_s}{(2d_s + w_s)} \qquad (5)$$

$$L = \frac{L_{ph}}{2n} \qquad (6)$$

where $n$ is the number of coils per branche (see Table I), $d_s$ is the depth of a slot and $w_s$ is the width of a slot.

The value of the resistance $R_s$ is optimised to represent the damping in the low-frequency region, whereas the

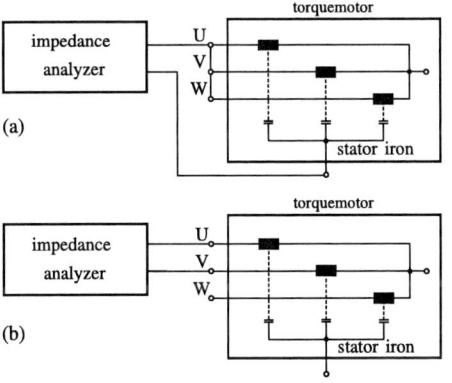

Fig. 7.   Measurement setup: (a) common-mode impedance, (b) impedance between two phases.

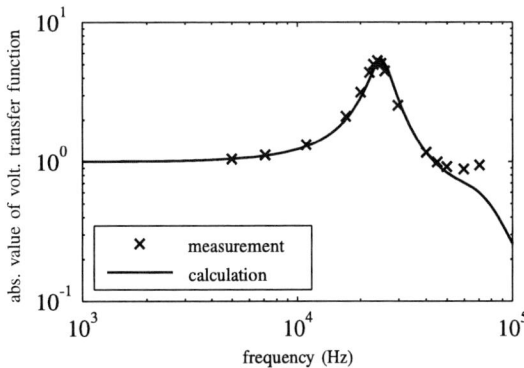

Fig. 8.   Absolute value of voltage transfer function of torquemotor: measurement and calculation results

resistance $R_p$ is optimised to represent the damping behaviour for the resonant frequency.

As a result the absolut value of voltage transfer function $\underline{H}_{tm}(2\pi f)$ is shown in Fig. 8 and compared with the measurement results. It shows the voltage increase from the terminals to the star point of the torquemotor. In the concerned frequency range up to 50 kHz good accordance can be derived. It can be seen that as consequence of the serial connection of the resonant circuits in every coil the voltage is increased from the terminals to the star point for up to 5,3 times at the resonant frequency of about 27 kHz.

B. System behaviour in the frequency range

As regards the system behaviour, several simplifications like the combination of the two excitation sources and the combination of all parasitic capacitances to ground lead to the equivalent circuit shown in Fig. 2 b), where $C_{par\Sigma}$ is:

$$C_{par\Sigma} = C_{AFE} + C_{inv} + C_{cb} + C_m \qquad (7)$$

The complete system model is formed as an iterative network. Chain-matrix descriptions of the different elements $\mathbf{A}_1$, $\mathbf{A}_2$, $\mathbf{A}_3$, $\mathbf{A}_{tm}$ lead to a system Matrix $\mathbf{A}_{sys}$:

$$\mathbf{A}_{sys} = \mathbf{A}_1 \cdot \mathbf{A}_2 \cdot \mathbf{A}_3 \cdot \mathbf{A}_{tm} \qquad (8)$$

whereof the voltage transfer function $\underline{H}(2\pi f)$ can be derived:

$$\underline{H}(2\pi f) = \frac{\underline{V}_{st}(2\pi f)}{\underline{V}_{ex}(2\pi f)} = \frac{1}{\mathbf{A}_{sys(1,1)}(2\pi f)} \qquad (9)$$

$|\underline{H}(2\pi f)|$ shows the voltage increase from the excitation to the star point of the motor where $V_{ex}$ and $V_{st}$ represent the Fourier transform of $v_{ex}$ and $v_{st}$.

Given a system with a certain configuration of a choke and inverter, the parasitic capacitances of the motors and the length of the motor cables hence $C_{par\Sigma}$ is the only parameter to be varied. Thereby systems with and without motors with low resonant frequency have to be distinguished.

In Fig. 9 the system behaviour of sample systems with iron powder core choke are shown as absolute value of

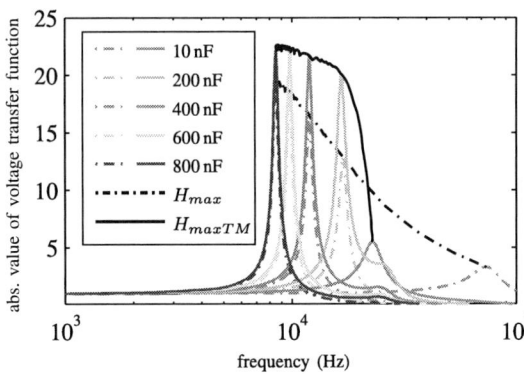

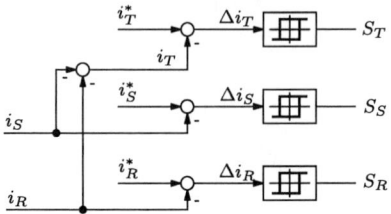

Fig. 10. Hystheresis current control of the active front end

**Fig. 9.** Description of the system behaviour: absolute value of the voltage transfer function $|\underline{H}(2\pi f)|$ of the system at different sums of parasitic capacitances $C_{par\Sigma}$. Dashed lines describe systems without, and normal lines systems with, the analysed torquemotor.

the transfer function $|\underline{H}(2\pi f)|$ with parameter $C_{par\Sigma}$. Systems with the analysed torquemotor are shown as normal lines and systems with only motors with resonant frequencies high above the concerned frequency range are marked with dashed lines. It can be seen that an increment of the parasitic capacitance leads to both an increment of the maximum voltage increase at the resonant frequency and to a decrease of the resonant frequency. At the same value of the parasitic capacitance $C_{par\Sigma}$ systems with torquemotor show higher voltage increase. The maximum values of the voltage increase vary from two at $C_{par\Sigma} = 10\,\mathrm{nF}$ and twenty at $C_{par\Sigma} = 800\,\mathrm{nF}$ in systems without torquemotor and between five and twenty-three in systems with the torquemotor. Especially for parasitic capacitances between $200\,\mathrm{nF}$ and $800\,\mathrm{nF}$ high values of voltage increase are reached.

## III. Excitation Mechanism

As regards excitation, as well as switching in the active front end, as switching in the inverter generates significant excitation. As shown in [12] the modulation index has strong influence on the excitation. The highest excitation is generated with modulation indices near zero, because the three phases often switch at the same time and voltage steps with the height of the full DC voltage may appear. Whereas at high modulation indices always only one phase switches at the same time and so excitation steps of one third of the DC voltage appear.

### A. Active Front End

The aim of the control scheme in the active front end is on the one hand to generate a DC-link voltage that is above the naturally-rectified DC-voltage and on the other hand to reach an approximately sinusoidal line current. Possible control structures are on the one hand a superior control scheme for DC-voltage and line current coupled with sine-triangle, space-vector or other modulation generators. On the other hand a superior control scheme for the DC-voltage can be coupled with a hystheresis control for the line current. The first structure generates periodic

switching commands which create a periodic excitation voltage $v_{exAFE}$ with different harmonics that excite the common-mode system. The second structure generates commands that appear non-periodically. Although this seems to be a lower excitation, a special coupling mechanism may generate an enforced excitation. Thus in this paper the second structure is investigated.

For control purposes the line currents have to be measured. To reduce hardware costs the current is often only measured in two phases and calculated in the third phase on the assumption that the common-mode current can be neglected and the sum of all phases equals zero. In many cases this assumption is justified because of a common-mode voltage with low amplitude and a frequency high above the cut-off frequency of the current measurement. As regards the analysed problem, this supposition leads to a significant error, owing to the fact that common-mode currents with high amplitude and comparatively low frequencies appear. Supposing that the common-mode current $i_{cm}$ is equally distributed to the three phases, the currents of the three phases can be written as

$$i_R = i_{Rdm} + \frac{i_{cm}}{3} \tag{10}$$

$$i_S = i_{Sdm} + \frac{i_{cm}}{3} \tag{11}$$

$$i_T = -i_R - i_S = i_{Tdm} - \frac{2}{3}i_{cm} \tag{12}$$

whereas $i_{xdm}$ builds the differential mode part of the current in phase x. Comparison of the calculated current $i_T$ with the currents $i_R$ and $i_R$ shows it has a common-mode fraction of double amplitude and a negative algebraic sign.

In Fig.10 the structure of the current control is shown, where the values $i_R^*$, $i_S^*$ and $i_T^*$ represent the set point values from the superior control of the DC-voltage. In normal operation this control scheme generates a current with approximately linear sections always lying in-between an allowed current band. The width of the current band thereby influences directly the number of the switching commands which appear. If high common-mode currents appear, the amplitude of the common-mode current fraction in the calculated current $i_T$ may be higher than half of the hysteresis width. In this case, in phase T switching commands appearing with the common-mode frequency are generated.

As shown in [12], every switching command produces a voltage step of a third of the DC-link voltage. Depending on the algebraic sign this step either leads to a damping or

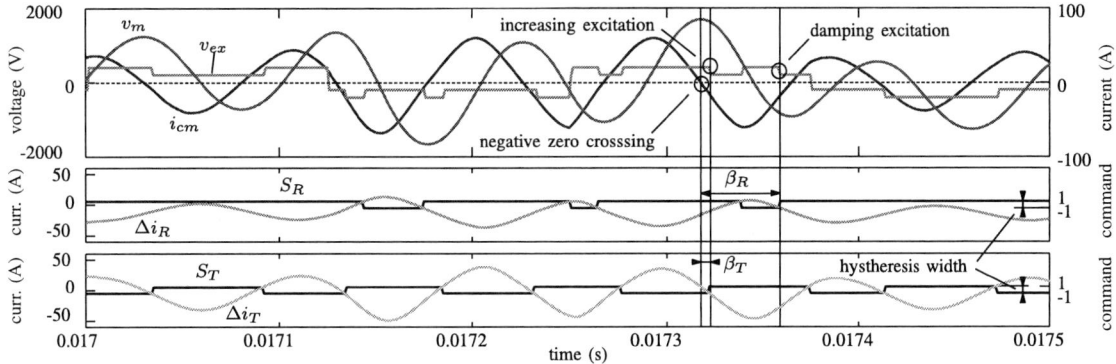

Fig. 11. Simulation for analysis of switching commands in the active front end: depending on the angle $\beta_{cmx}$ a switching command may work as increasing or damping excitation. Typical waveforms of $v_m$, $v_{ex}$ and $i_{cm}$ are also shown.

an enforced excitation of the system. In Fig. 11 examples are given for such switching cases. Because of the different common-mode fraction in the calculated current $i_T$ in some cases a special phase relation between $i_{cm}$ and $S_T$ may appear that causes almost only switching commands that excite the system. This phase relation is not always apparent and depends on the resonant frequency, the amplitude of the common-mode current and the phase of the line voltage. The indicator is the angle $\beta_x$ between the positive zero crossing of the common-mode current and a switching command with a positive voltage step, between the negative zero crossing and a switching command with a negative voltage step respectively. The nearer this angle is to zero, the more exciting the switching command is. If it is about 180° it has a damping characteristic. Investigations show that only in phase T does the angle get near zero. In the other phases the angle is near 180°, owing to the fact that phase T has inverted common-mode fraction. As shown in Fig. 11 the switching commands of phase T are predominantly increasing excitation whereas the switching commands of phase R are predominantly damping excitation.

*B. Inverter*

As regards the inverter the control scheme generates the switching commands. Here different control schemes are typically used [16]. Most of them consist of a superior control that calculates a voltage phasor and passes it to a pulse width modulator with a definite switching frequency. Different types of modulators produce switching commands. As shown in [12] the highest excitation at most of the frequently-used modulator schemes like space vector modulation or sinus triangle modulation appears with modulation index zero. Accordingly, the switching commands of the inverter are simplified to those of a modulator with modulation index zero.

The exciting voltage $v_{exinv}$ then has a rectangular sequence with the switching frequency $f_s$ and an amplitude of $V_{DC}/2$. Fourier analysis of this these voltage characteristics leads to the harmonics that excite the common-mode system. In Table II the most significant ones are shown. Assuming e.g. a switching frequency of

TABLE II
HARMONICS OF THE EXCITING VOLTAGE AT INVERTER $v_{exinv}$

| | Frequency | Amplitude |
|---|---|---|
| 1 | $f_s$ | $2/\pi \, V_{DC} = 0,64 \, V_{DC}$ |
| 3 | $3\,f_s$ | $2/(3\,\pi) \, V_{DC} = 0,21 \, V_{DC}$ |
| 5 | $5\,f_s$ | $2/(5\,\pi) \, V_{DC} = 0,13 \, V_{DC}$ |
| 7 | $7\,f_s$ | $2/(7\,\pi) \, V_{DC} = 0,09 \, V_{DC}$ |
| 9 | $9\,f_s$ | $2/(9\,\pi) \, V_{DC} = 0,07 \, V_{DC}$ |

$f_s = 4\,\text{kHz}$ the third and the fifth Harmonic are lying in the frequency range of very high voltage increase of the systems. Hence they might produce overvoltages at the motor.

## IV. REMEDIES

As shown above, oscillations in the common-mode system of electrical drive systems are originated on the one hand by the low damped resonant behaviour of the system and the motor and on the other hand by the excitation through the active front end and the inverters. Remedies can therefore either improve the system behaviour or reduce the excitation. Two examples are investigated. One reduces the enforced excitation in the hysteresis control and the other increases the damping of the system.

*A. Optimisation of Hysteresis Control*

As regards the described enforced excitation in the hysteresis current control of the active front end with two phase current measurement, the measurement of the third phase would be the easiest solution. To avoid increasing costs a cheaper solution has to be found. Two ways are proposed to avoid the enforced excitation. If a low pass filter as shown in Fig. 12 (a) is inserted into the path of the calculated current $i_T$ the common-mode part of the current that is high frequency with respect to the main differential mode part can be reduced. The cut-off frequency is chosen at 5 kHz with the aim on the one hand to guarantee a stable control and on the other hand to yield a sufficient reduction of the high frequency common-mode parts. A second way is to invert the high frequency part of the calculated current $i_T$ as shown in Fig. 12 (b). A further gain is used to optimise the

2033

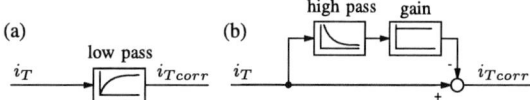

Fig. 12. Optimisation of hysteresis control: (a) low pass (b) inversion of high frequency parts

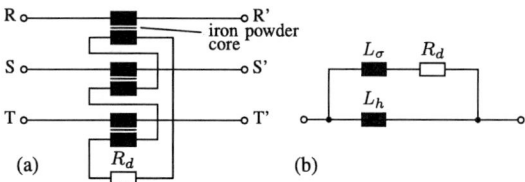

Fig. 13. Choke with transformational damping in the common-mode system: (a) circuit (b) common-mode equivalent circuit

**TABLE III**

DATA OF SIMULATED AND MEASURED ELECTRICAL DRIVE SYSTEMS

| Active Front End | |
|---|---|
| nominal power | 34 kW |
| current control | hystheresis control |
| **Choke** | |
| type | iron powder core |
| nominal current | 62 A |
| **Inverter** | |
| DC - voltage | 600 V |
| modulation | sine triangle |
| modulation index | 0 |
| switching frequency | 4 kHz |
| **Axes** | |
| number of axes | $1 \cdots 21$ |
| sum of cable lengths | $20\,\mathrm{m} \cdots 1300\,\mathrm{m}$ |
| sum of parasitic capacitances | $10\,\mathrm{nF} \cdots 800\,\mathrm{nF}$ |

amplitude of the inverted part of the current. With the aid of this manipulation the common-mode part of the current $i_{Tcorr}$ equals almost the common-mode part of the measured currents and in consequence should lead to a damping excitation. The optimal value of the gain was found as two.

### B. Choke with Transformational Damping

As second remedy a choke with transformatorial damping in the common-mode region is analysed. This choke was proposed by [8] and substitutes the normal choke before the active front end. As shown in Fig. 13 (a) the choke consists of three iron powder cores similar to the choke proposed in section II-A.2. Every core has two windings and thereby works as a transformer. The second windings of the three phases are connected in series and also connected to a damping resistor $R_d$. As a consequence the resistor acts only in the common-mode system. In the differential mode system the choke acts only as the main inductance $L_h$.

The equivalent circuit for the impedance $\underline{Z}_L$ in the common-mode system is therefore shown in Fig. 13 (b). It consists of a transformer model with the main inductance $L_h$, the leakage inductance $L_\sigma$ and the damping resistor $R_d$. This parallel circuit acts below the corner frequency as inductance $L_h$ and above as as serial connection of $R_d$ and $L_\sigma$. The corner frequency is given through the equality of both branches. It has to be chosen low enough to guarantee good damping in the concerned frequency range. The value of the damping resistor on the other side decides the damping behaviour of the system. With the investigated choke a value of $R_d = 16\,\Omega$ can be realised.

### V. SIMULATION AND MEASUREMENTS

In consequence of the oscillations high voltage appears at the motor terminals: in the case of the investigated torquemotor it is also on the star point and stresses the winding isolation. Other consequences of the oscillations are common-mode losses in the choke that decrease the efficiency of the inverter system and raise the temperature

of the choke. Furthermore, common-mode currents elevate the current loading of the semiconductor switches. To determine the resulting voltage stressing, losses and common-mode currents simulations were investigated.

The main parameters concerning the system behaviour are the sum of the parasitic capacitances $C_{par\Sigma}$, the inductance and the damping behaviour of the choke. So for a given choke numerous simulations were processed over the full range of possible $C_{par\Sigma}$. The simulated system was modelled in the time domain and simulated for a sufficient amount of time. To get the results of the worst case operating point the modulation index of all inverters was set to zero. As a result the absolute maximum value of the voltage $V_{mmax}$ at the motor terminals, in the case of systems with the torquemotor at the star point $V_{stmax}$, the average value of the common-mode losses $P_{cm}$ and the root-mean-square value $I_{cmrms}$ of the common-mode current were saved. The simulation parameters are generated from measurements on parts of a given drive system, that was also used for measurement of the system behaviour. In Table III the data of the simulated and measured systems are given.

### A. Simulation Results

In Fig. 14 the results of simulations are shown. Concerning the maximum values of the voltage on the motor isolation, it is compared with the recommended maximum voltage $V_{maxIEC} = 1,56\,\mathrm{kV}$ for designing winding isolation in AC-motors in the international standard IEC/TS60034-25:2004. If exceeded, partial discharges may appear, degrading the winding isolation and abbreviating the lifetime [7]. The higher the voltage, the worse the damage. In the simulations and measurements overvoltages owing to reflections on motor cables are not regarded. So, if they appear, they would additionally elevate the overvoltages, but only on the terminals of the motors. The overvoltages at the star point of the torquemotor are not elevated owing to the fact that the overvoltages of reflections only reach the first winding turns.In Fig. 14 (a) and (b) the overvoltages of systems with and without the investigated torquemotor are shown. As regards systems without torquemotor, in the case of systems without remedy in a wide range the proposed maximum value

**2034**

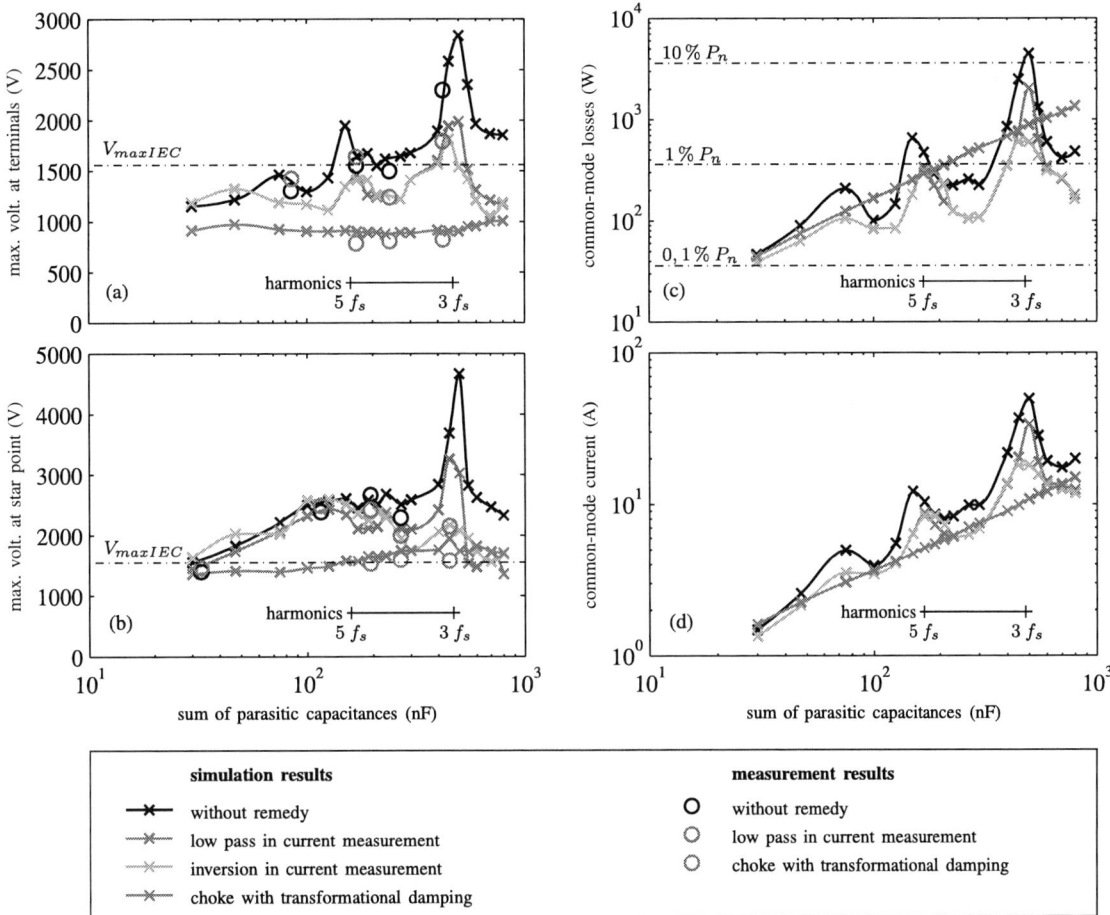

Fig. 14. Results of simulations and measurements with and without remedies: (a) maximum voltage appearing at the motor terminals $V_{mmax}$ in systems without torquemotor, (b) maximum voltage appearing at the star point of the torquemotor $V_{stmax}$ in systems with torquemotor, (c) common-mode losses $P_{cm}$ and (d) trms-value of common-mode current $I_{cmrms}$ in systems without torquemotor

$V_{maxIEC}$ is exceeded especially where harmonics of the inverter excitation are lying near the resonant frequency of the system. The proposed remedies of the low pass and the inversion in the current measurement ameliorate the system behaviour and reduce the overvoltages. Only in the region of the third harmonic do the overvoltages exceed the proposed maximum value. Comparing the inversion and the low pass, we see that both reach almost the same results, except near the third harmonic where the inversion shows better results. The use of the choke on the other hand reduces the overvoltages in the full range of possible $C_{par\Sigma}$ to 1000 V. Concerning systems with the investigated torquemotor in Fig. 14 (b), in general the voltages appearing at the star point of the torquemotor are much higher than in systems without torquemotor. Without remedy the proposed maximum value $V_{maxIEC}$ is exceeded in the full range of possible $C_{par\Sigma}$. In a wide range the overvoltages are lying between 2400 V and 4700 V which is a multiple of the proposed voltage $V_{maxIEC}$. Owing to the high voltage increase in the torquemotor the remedies of low pass and inversion in

the current measurement do not reach sufficient reduction of the overvoltages. The choke with the transformational damping greatly reduces the overvoltages at the star point. For high values of $C_{par\Sigma}$ however, the voltage $V_{maxIEC}$ is exceeded in a low value.

As regards common-mode losses shown in Fig. 14 (c), they are compared with the nominal power $P_n = 34$ kW of the active front, owing to the fact that they are extracted from the DC-capacitor and therefore from the mains. Here systems without torquemotor are shown, because both cases show similar results. Concerning systems without remedy, high values appear. The losses reach a maximum of about 12 % of the nominal value $P_n$ and a strong dependency on the parasitc capacitance $C_{par\Sigma}$. Especially near the harmonics they are additionally elevated. The remedies that treat the current measurement reduce the losses over the full range in allmost the same quantity whereas the inversion shows slightly better results near the third harmonic. Systems with the choke with transformatorial damping show only dependency on the parasitic capacitance. Apart from cases with increased excitation through the harmonics the values are lying above the values without remedy.

Concerning the common-mode currents, the results of the simulations show similar dependences to the losses, except for the fact that the current of systems with the choke with transformational damping has almost the minimum values over the full range of $C_{par\Sigma}$.

*B. Measurement Results*

To prove the simulation results, measurements on an electrical drive system were made. The modelled system corresponds to the measured one except for the exact values of $C_{par\Sigma}$. Several configurations with different $C_{par\Sigma}$ were investigated. As a result of the measurements the values of the maximum voltages appearing at the terminals $V_{mmax}$ and at the star point of the torquemotor $V_{stmax}$ were determined and compared with the simulation results in Fig. 14 (a) and (b). As regards systems without remedies, the measurements do match the simulations. Even very high values of 2300 V without torquemotor and 2700 V at the starpoint of the torquemotor were measured. At the maximum point of the simulation results no measurement was done owing to possible destruction of the torquemotor. Concerning the remedy with the current measurement only the low pass was realised owing to the easy way of implementation. The measurement results are lying a little bit higher for small values of $C_{par\Sigma}$ and exactly match the simulation results for high values of $C_{par\Sigma}$ in systems without torquemotor. In systems with torquemotor the measurement results are lying almost equal or below the simulation results. Measurements on systems with the choke with transformational damping show voltages lying always a little under the simulations. In the case of systems with torquemotor particulary the values are lying equal to or under the proposed maximum voltage $V_{maxIEC}$.

As regards the effectiveness, the remedies concerning the current measurement eliminate the increased excitation. In systems without torquemotor this is sufficient to reach maximum voltages below the proposed maximum voltage $V_{maxIEC}$ in a wide range of $C_{par\Sigma}$. At the third harmonic, however, this voltage is exceeded. In systems with torquemotor these remedies reduce the maximum voltages, but not enough. Only the choke with transformational damping shows a good reduction of overvoltages in both cases. At high values of $C_{par\Sigma}$ the simulation results exceed the proposed maximum voltage somewhat, although measured values are lying below. Even though the remedies in the current measurement are not sufficient in all possible cases they are very easy to realise and thereby reduce the currents and losses effectively. On the other hand the use of the choke with transformational damping needs much greater amounts, but guarantees sufficient reduction in all possible cases, as can be seen in the measurement results.

## VI. CONCLUSION

By means of system analysis reasons for the appearance of respectively low-frequency common-mode oscillations in electrical drive systems were derived. In addition a common-mode resonant behaviour in motors with concentrated windings was investigated and modelled. The enforced excitation of switching commands of an active front end with hysteresis current control and two-phase current measurement and the harmonics of the inverter switchings particulary excite the common-mode system and lead to high voltages on the motor terminals. Different remedies were analysed. Low pass or inversion in the current measurement of the active front end reduces the enforced excitation. A choke with transformational damping elevates the damping of the system and therefore reduces the overvoltages. Simulations and measurements are presented that show the resulting voltages stressings at the motor winding and leading to a reduction of the isolation lifetime. Additionally, as shown, the common-mode losses and currents are not neglectable and mark significant disturbances to the electrical drive system. Furthermore simulations and measurements prove the effectiveness of the analysed remedies.

## REFERENCES

[1] P. Salminen, *Fractial Slot Permanent Magnet Synchronous Motors for low Speed Application.* Phd thesis, University of Technology, Lappeenranta, 2004.

[2] M. Cacciato, A. Consoli, G. Scarcella, and A. Testa, "Reduction of common- mode currents in pwm inverter motor drives," *IEEE Transactions on Industrial Applications*, vol. 35, no. 2, pp. 469–476, 1999.

[3] A. Muetze, *Bearing currents in Inverter-Fed AC-Motors.* Phd thesis, University of Darmstadt, Darmstadt, 2004.

[4] F. Costa, C. Vollaire, and R. Meuret, "Modeling of conducted common mode perturbations in variable-speed drive systems," *IEEE Transactions on Elektromagnetic Compability*, vol. 47, no. 4, pp. 1012–1021, 2005.

[5] P. Finlayson, "Output filter considerations for pwm drives with induction motors," *IEEE Techical Conference on Textile, Fiber and Film Industry*, Atlanta, USA, 1996.

[6] R. Essig, *Reduction of voltage stressings of inverter driven motor winding (in german).* Phd thesis, Universiy of Erlangen-Nuremberg, Erlangen, 1994.

[7] M. Kaufhold and G. Börner, "Longtime behaviour of the isolation of inverter driven asynchronous machines (in german)," *Elektrie*, vol. 47, no. 3, pp. 90–95, 1993.

[8] R. Dilling, B. Segger, and R. Steinmueller, "German patent application publication: De 10064213 a1," 2000.

[9] J. Heidenhain GmbH, "German registered design: De 023 11 104," 2003.

[10] N. Huber, "United states patent application publication: Us 2005/0013145 a1," 2005.

[11] S. Raith and B. Segger, "German patent application publication: De 100 59 334 a 1," 2002.

[12] T. Weidinger and B. Piepenbreier, "Analysis and modelling of common-mode oscillations of electrical drive systems with active front end and long motor cables," *Speedam International Symposium on Power Electronics, Electrical Drives, Automation and Motion*, Ischia, Italy, 2008.

[13] B. Mirafzal, S. G.L., R. Tallam, D. Schlegel, and R. Lukaszewski, "Universal induction motor model with low-to-high frequency-response characteristics," *IEEE Transactions on Industry Applications*, vol. 43, no. 5, pp. 1233–1246, 2007.

[14] R. Nuscheler and D. Potoradi, "Transient overvoltages at the terminals of inverter fed induction motors," *EPE*, Lausanne, Switzerland, 1999.

[15] A. Boglietti, A. Cavagnino, and M. Lazzari, "Experimental high-frequency parameter identification of ac electrical motors," *IEEE Transactions on Industry Applications*, vol. 43, no. 1, pp. 23–29, 2007.

[16] M. Kazmierkowski, R. Krishan, and F. Blaaberg, *Control in Power Electronics.* San Diego: Academic Press, 2002.

# Internal Short Circuit in a Tooth Wound PMSM with Stranded Conductors

Damien Birolleau*[†], Christian Chillet[†] and Laurent Albert*

* Renault SAS, Guyancourt, France, e-mail: *damien.birolleau@renault.com ; laurent.albert@renault.com*
[†] Grenoble Electrical Engineering Laboratory, Grenoble, France, e-mail: *christian.chillet@g2elab.inpg.fr*

*Abstract*— This paper deals with the modelization of an internal short circuit in a coil of a permanent magnet synchronous motor with a tooth wound stator. It focuses on the simulation of windings made of stranded conductors. The method of modelization is described, validated and applied to a case study. Different configurations of short-circuits and their consequences are investigated.

*Keywords*— Permanent magnet motor, Stranded conductors, Short circuit.

## I. INTRODUCTION

Car manufacturers try to develop technologies in order to make a "dry car". Making a dry car means to have no liquid necessary to ensure main functionality in the car. Currently, it is common to find hydraulic power steering or hydraulic braking in a car.

As in the aero-spatial industry with the "more-electric aircraft", which consists in replacing hydraulic mechanism by electric actuators and cables, car manufacturers investigate the concept of "more electric car". This started with the electric power steering (EPS), which is already in production on many cars, and can go as far as steer or brake by-wire.

However, an important issue is the dependability of such systems. In current EPS, the strategy is to make a fail silent actuator. The aim is to limit the influence of a faulty actuator on the system, but this leads to the loss of the steering assist function.

Therefore, even if the reliability is an important aspect, the availability of these systems has to be developed. Availability is here defined has the capacity of a system to ensure its function even in case of a fault.

The main difference, considering the "more electric" concept, between a car and an aircraft is the place available for the system and its cost. If in an aircraft it is possible to place multiple actuators to ensure a function even in case of failures, this redundancy is more difficult in a car.

In order to limit the number of actuators needed for the system dependability, fault-tolerant actuators are necessary. This paper deals with the electric motor.

Permanent magnet synchronous machines (PMSM) have been chosen because of their torque to size ratio. A tooth wound stator is interesting because electromagnetic mutuals between phases are lower than in a full pitch winding.

A fault tolerant electrical motor needs to sustain different kinds of failures [1] [2]. Among possible faults, an internal short-circuit in the winding has been identified as one of the most critical to detect and sustain [3] [4] [5]. Indeed, this fault leads to high currents and braking torque. However, before finding solutions to tolerate this fault, its consequences must be understood and quantified.

This paper focuses on the particular case of a short circuit in a coil of a tooth wound stator with stranded conductors. The motor is star connected and all teeth of one phase are connected in series. Fig.1 shows the example of a tooth coil with a conductor made of two strands. The three different short-circuit configurations studied are represented.

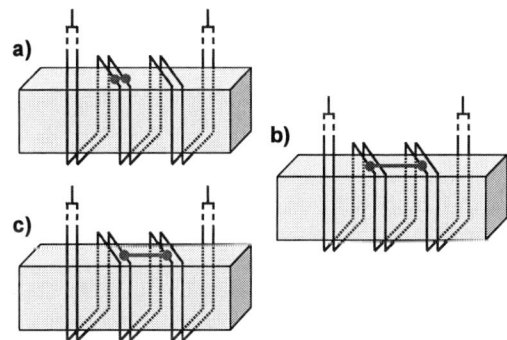

Fig. 1. Different configurations of short circuit on a two strands conductor ; a) between two strands on the same turn ; b) on the same strand on different turns ; c) between two strands on different turns.

The short circuit is studied without command currents in phases. This allows understanding consequences when the rotor is rotating due to an external load.

Firstly, a simplified model of the winding with a short-circuit is suggested. This analytical model gives currents in strands and the mean value of the generated braking torque.

Secondly, the model is validated with finite Element Software "Flux" and applied. Different configurations of short-circuit are investigated, including variation in the number of short-circuited turns and in the number of strands per conductor.

978-1-4244-1741-4/08/$25.00 ©2008 IEEE

## II. MODELIZATION

### A. General Model

The first step of modelization, shown in Fig.2, consists in considering each strand turn as a resistance $R_{turn}$, an inductance $L_{turn}$ and a no load electromotive force $E_{nl}$ in series. The inductance includes electromagnetic mutuals with all other turns of the motor.

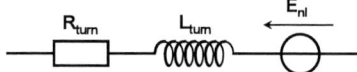

Fig. 2. Equivalent model of a strand turn

A phase winding, constituted of $N_{tot}$ turns and $N_{strands}$ strands is represented in Fig.3. In this figure, a short circuit between two different strands is shown. The short circuit is considered with no resistance.

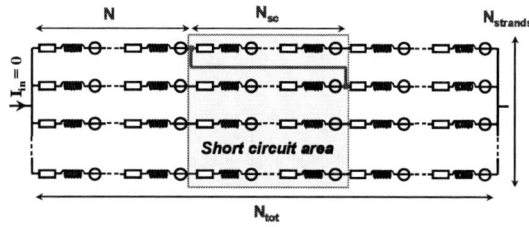

Fig. 3. Phase winding with a conductor made of $N_{strands}$ strands and a short circuit of $N_{sc}$ turns between two strands. N represents the short circuit position in the winding.

### B. Model simplification

A phase turn has as many currents flowing in as the number of strand turns that composes it. In a phase turn outside the short-circuit area (defined in Fig.3), sum of these currents is null (strands are connected at the winding edge). So, flux generated by inductances in this area are canceling themselves. Therefore, they do not influence short circuit currents. Moreover, strands of a phase turn are on the same tooth, so electromotive forces induced in these strands outside the area are balanced. As a result, outside turns are equivalent to their resistances.

The Fig.4 presents the simplified model. Strands that are not touched by the short circuit are gathered in one with an equivalent resistance $R_5$. No load electromotive force by strand turn $E_{nl}$ and a turn inductance have been gathered in the "in load" electromotive force by strand turn E. Because strand turns in the short circuit area are on the same tooth, they see the same flux variation, so their electromotive forces are equals.

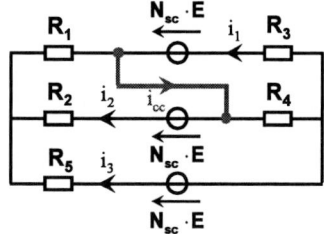

Fig. 4. Simplified model used

### C. Currents in the strands

To define all currents in the model, only $i_1$, $i_2$ and $i_3$ are needed. They are obtained by using the superposition principle on the three voltage sources. Equation (1) gives an important relation between currents in strands and the short circuit current $i_{cc}$.

$$i_1 + i_2 + i_3 = i_{cc} \tag{1}$$

Because the electrical circuit of Fig.4 is composed only of resistances, currents are proportional to E (2).

$$\begin{cases} i_1 = K_1 \cdot N_{cc} \cdot E \\ i_2 = K_2 \cdot N_{cc} \cdot E \\ i_3 = K_3 \cdot N_{cc} \cdot E \end{cases} \tag{2}$$

Equation (3) gives the formula of the "in load" electromotive force E.

$$E = -\frac{N_{cc}}{\Re} \frac{di_1 + i_2 + i_3}{dt} - \frac{d\varphi_{aim}}{dt} = -\frac{N_{cc}}{\Re} \frac{di_{cc}}{dt} - \frac{d\varphi_{aim}}{dt} \tag{3}$$

$\Re$ : Reluctance of the magnetic circuit

$\varphi_{aim}$ : Flux generated by rotor magnets in the faulty tooth

It is now possible to write the differential equation (4) to calculate $i_{cc}$.

$$i_{cc} = i_1 + i_2 + i_3 = (K_1 + K_2 + K_3) \cdot N_{cc} \cdot E$$
$$i_{cc} = (K_1 + K_2 + K_3) \cdot N_{cc} \cdot \left( -\frac{N_{cc}}{\Re} \cdot \frac{di_{cc}}{dt} - \frac{d\Phi_{aim}}{dt} \right) \tag{4}$$

This differential equation can be formally or numerically solved.

### D. Mean value of braking torque

If friction and iron losses are neglected, when the rotor turns inside the stator the only losses are copper losses $L_{copper}$ in the winding. Therefore, the mean value of the braking torque $\langle T_{brake} \rangle$ is obtained with (5).

$$\langle T_{brake} \rangle = \frac{\langle L_{Copper} \rangle}{\Omega} \tag{5}$$

$\Omega$ : rotational speed of the rotor in rad/s

### E. Model possibilities and limits

This simple analytical model allows calculating from one strand per conductor.

Limits of this model are the followings:
- The model does not take into account the iron saturation
- This model gives values at constant speed
- Simulation of parallel winding is not considered
- Resistances are constant (influence of temperature is not taken into account)

## III. APPLICATION

### A. Case study

Characteristics of the three phase tooth wound PMSM used are given in Table I.

TABLE I
CASE STUDY CARACTERISTICS

| | |
|---|---|
| Number of pole pair | 3 |
| Number of teeth | 9 |
| Number of turns per coil | 8 |
| Wire diameter for 1 strand per conductor | 2 mm |
| Phase resistance | 14 mΩ |
| Magnets remanent induction at 20°C | 1 T |
| Maximum output torque | 3.5 Nm |
| Maximum RMS phase current in nominal | 60 A |

### B. Finite Elements Validation of the model

The analytical model is validated for one strand and two strands per conductor. Finite Elements simulation software "Flux" [6] has been used to simulate short circuits on different numbers of turns. Fig.5 shows the modelized motor for a one strand or a two strands conductor with areas in slots assigned to short-circuited turns.

*1) One strand conductor:* Fig.6 shows the short circuit current comparison between both methods for 500 rpm and 1000 rpm (see Fig.3 for variables definition). For the mean torque values, Fig.7 refers to 500 rpm while Fig.8 refers to 1000 rpm.

Even if short-circuit current values are good, an important difference appears for the mean torque at 1000 rpm. This difference is mainly due to the tooth shoe saturation of the faulty tooth. This saturation is not taken into account in the linear analytic model.

Still, results are enough precise to have a good order of magnitude and analyze tendencies, even if results have to be used with caution at high speed.

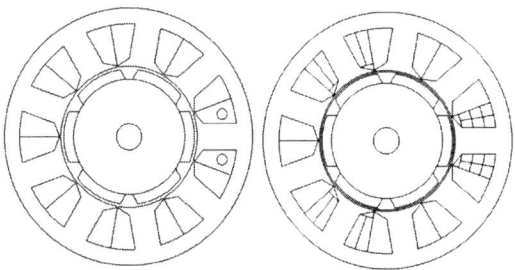

Fig. 5. Modelizations of the motor in "Flux" for a one-strand conductor on the left, and a two strands conductor on the right.

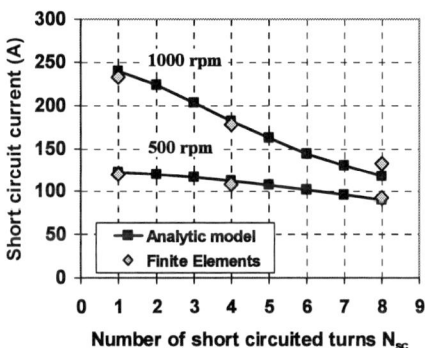

Fig. 6. Results comparison for the short circuit current.

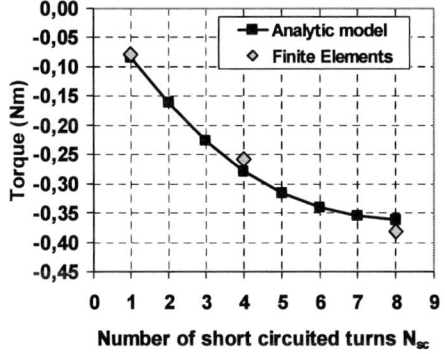

Fig. 7. Results comparison for the mean torque at 500 rpm.

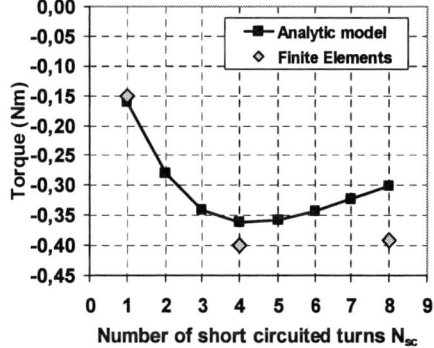

Fig. 8. Results comparison for the mean torque at 1000 rpm.

*2) Two strands conductor:* Simulations for a short circuit on the same strand or between two different strands have been made. Current in a strand and mean torque are compared. For a same strand fault (Fig.9 and Fig.10) as for a two strand one (Fig.11 and Fig.12), comparisons show very good results.

For a fault between two different strands, results are only given for 1000 rpm, which is the most critical simulation made. On Fig.11 and Fig.12, curves with four or eight short circuited turns have fewer points because defaults between two different teeth are not considered in this study. Moreover, values for N equal to 0, 23 or 24 are not represented. Indeed, these cases are equivalent to short circuits on the same strand. On these figures, faults on the three teeth of a phase wound in series are represented.

These results validate the proposed no load modelization of the two different kinds of short circuit in a stranded conductor investigated. The model is now used to study the influence of different winding parameters.

## IV. RESULTS

### A. Results conditions

For comparison purpose between different numbers of strands per conductor, turn resistances are calculated in order to keep a constant global resistance of the winding. The motor design is never modified. The configuration 'a' of Fig.1 is not studied because it does not induce currents in strands.

Two cases of windings connections are investigated. The first one considers that strands are connected between each tooth. In this case, only the faulty tooth is concerned. The second one considers that all teeth of one phase are wound in series without strands connection between each tooth. In this case, other teeth resistances are taken into account.

The study is carried out for a conductor made of 1, 2 or 3 strands. The rotor speed is 1000 rpm. Three values are observed: the amplitude of the short-circuit current $i_{cc}$, the maximum RMS current density in the strands and the generated braking torque. Variables definition of following figures is given in Fig.3.

*Short-circuit on the same strand*

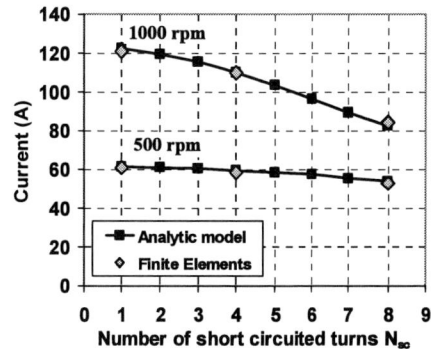

Fig. 9. Results comparison for a strand current

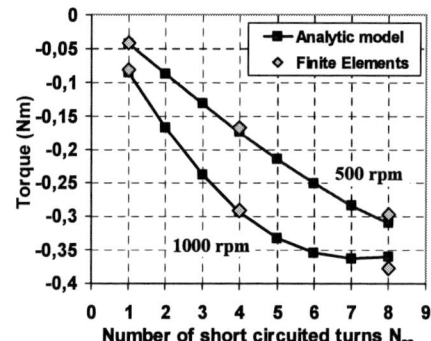

Fig. 10. Results comparison for the mean torque

*Short-circuit between two different strands*

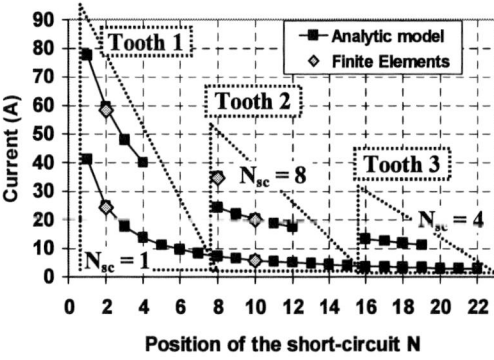

Fig. 11. Results comparison for a strand current at 1000 rpm

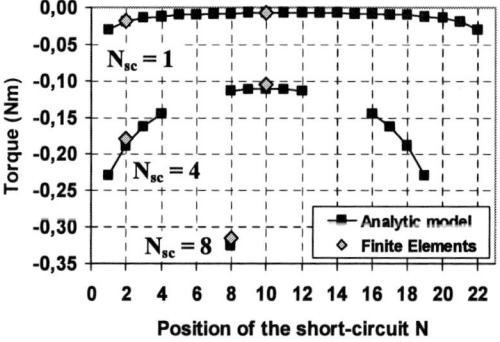

Fig. 12. Results comparison for the mean torque at 1000 rpm

## B. Short circuit on the same strand

Fig.13 to Fig.18 show the amplitude of the short circuit current $i_{cc}$, the mean torque and the maximum RMS current density in strands for a fault on the same strand.

With a stranded conductor, currents flowing through the short circuit are equal or smaller than for a one strand conductor. However, the maximum mean torque amplitude is the same. The difference is the number of shorted turns corresponding to this maximum.

For one strand conductor, when the number of short-circuited turns grows, the fault current $i_{cc}$ decreases. Otherwise, the mean torque amplitude increases until the circuit inductances become preponderant in front of resistances (Here for $N_{sc}$ equal to 4).

For a stranded conductor, the short circuit current $i_{cc}$ decreases as the number of strands increases. Moreover, the effect of the inductances is delayed to a greater number of short-circuited turns. These effects are even more important in case of not connected strands. This is due to added resistances of others teeth coils.

Maximum RMS current density in strands is around 50 A/mm² for a short circuit on one turn. At the opposite of the short circuit current $i_{cc}$, for an equivalent $N_{sc}$ the current density grows as the number of strands increases. However, the maximum value remains unchanged.

So, using a stranded conductor does not introduce a worst case than with a one strand conductor.

### Strands connected between each tooth

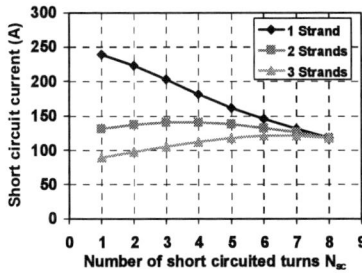

Fig. 13. Short circuit current amplitude for a fault on the same strand

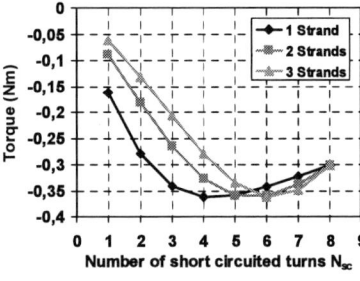

Fig. 14. Mean torque for a short-circuit on the same strand

Fig. 15. Maximum RMS Current density for a short-circuit on the same strand

### Strands not connected between each tooth

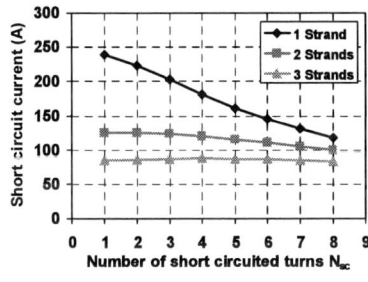

Fig. 16. Short circuit current amplitude for a fault on the same strand

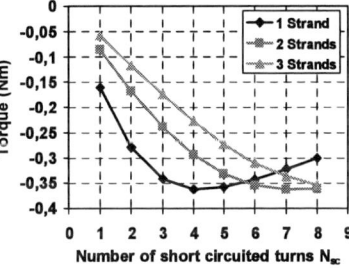

Fig. 17. Mean torque for a short-circuit on the same strand

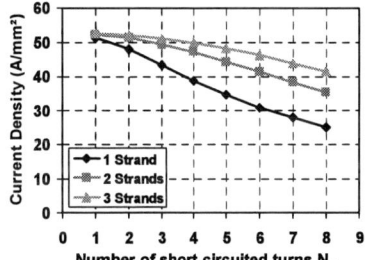

Fig. 18. Maximum RMS Current density for a short-circuit on the same strand

## C. Short circuit between two strands

Short circuit current $i_{cc}$ and braking torque for a conductor made of 2 or 3 strands are compared in the case of a two strands short circuit (Case 'c' of Fig.1). Results for both type of winding connection are given in Fig.19 to Fig.22.

These figures show that the position of the short circuit in the winding, which is represented by the variable N, has an important impact. It is particularly true for multiple teeth wound in series without strands connection between each tooth. Indeed, it can be seen in Fig.21 and Fig.22 that teeth on the extremities of the winding (teeth 1 and 3) are more critical for current and torque than tooth in the middle.

The comparison between both cases shows that current is always smaller for not connected strands. The influence of added resistances of other teeth is one more time the reason.

Contrary to a fault on the same strand, more short-circuited turns $N_{sc}$ means higher short-circuit currents. Another aspect is the influence of the number of strands. Like in the case of a short circuit on the same strands (Part 4-B), Amplitude of $i_{cc}$ becomes smaller as the number of strands grows. Nevertheless, maximum current densities in strands (not represented here) stay equivalent.

In comparison with a short circuit on the same strand, maximum short circuit current amplitude is smaller, but braking torque reaches equivalent values.

*Strands connected between each tooth*

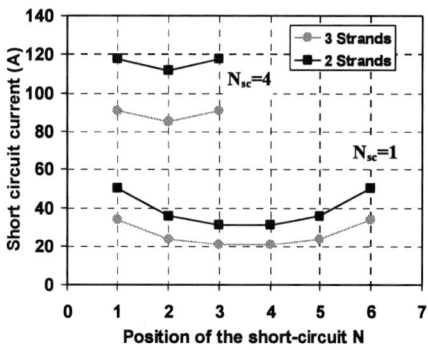

Fig. 19. Short circuit current amplitude for a fault between two strands

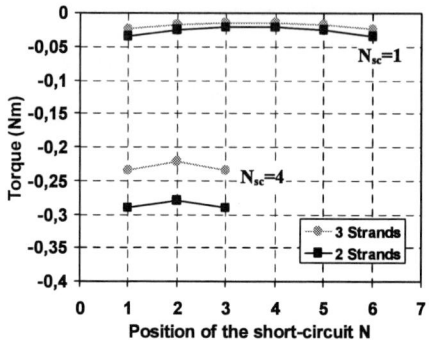

Fig. 20. Mean torque for a short-circuit between two strands

*Strands not connected between each tooth*

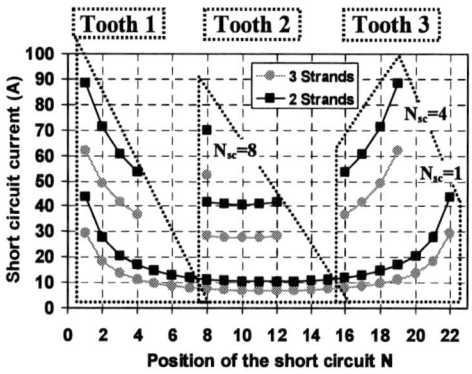

Fig. 21. Short circuit current amplitude for a fault between two strands

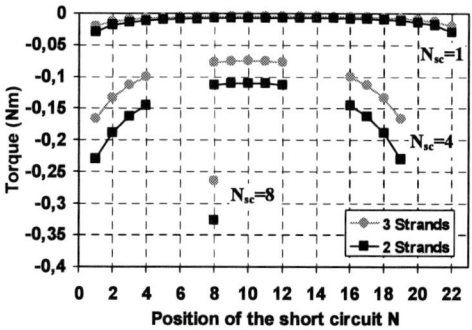

Fig. 22. Mean torque for a short-circuit between two strands

## V. CONCLUSION

An analytical model of short circuit in PMSM winding with stranded conductors has been made and validated.

The results obtained with this model shows lower short circuit current and braking torque with a multiple strands conductor than with a one strand conductor. Even if having a stranded conductor implies an equal or higher current density, it is important to note that it does not introduce a worst case than with a one strand conductor.

Consequences of a short circuit on the same strand or between two have been compared. It revealed that for a short circuit between two strands, the place in the winding has an important influence, contrary to a short circuit on the same strand.

Differences between two winding connections have been studied. The interest to wind without strands connection between teeth has been highlighted. Finally, results in the case of a winding without strands connection between teeth underlined that tooth on the extremities of the winding are more critical for current and braking torque. This is even more important for the first turns of the phase winding.

As a conclusion, this model is a good tool to have orders of magnitude of the fault currents and of the generated braking torque. It also gives evolution tendencies of these values according to multiple parameters as the number of strands or the place of the short circuit in the winding.

## REFERENCES

[1] B.C. Mecrow, A.G. Jack, J.A. Haylock, J. Coles, "Fault-tolerant permanent magnet machine drives", IEE Proceedings, Electric Power applications, 1996, Vol. 143, n°6, pp. 437-442

[2] B.C. Mecrow, A.G. Jack, D.J. Atkinson, , J.A. Haylock, "Fault tolerant drives or safety critical applications", IEE Colloquium on New Topologies for Permanent Magnet Machines, Juin 1997, pp. 511-517

[3] A. Haylock, B.C. Mecrow, A.G. Jack, D.J. Atkinson, "Operation of fault tolerant machines with winding failures", IEEE Transactions on Energy Conversion, 1999, Vol. 14, n°4, pp. 1490-1495

[4] B.A. Welchko, T.M. Jahns, W.L. Soong, J.M. Nagashima, "IPM Synchronous Machine Drive Response to Symmetrical and Asymmetrical Short Circuit Faults", IEEE Transactions on Energy Conversion, Juin 2003, Vol. 18, n°2, pp. 291-298

[5] C. Gerada, K. Bradley, M. Sumner, "Winding Turn-to-Turn Faults in Permanent Magnet Synchronous Machine Drives", IAS 2005, Hong Kong, Octobre 2005

[6] "Flux", Cedrat, http://www.cedrat-groupe.com/en/software-solutions/flux.html

# Implementation of a Virtual Laboratory for Low Power Electrical Drives

Gh. BALUTA*, V. HORGA* and C. LAZAR†

* "Gh. Asachi" Technical University/Faculty of Electrical Engineering, Iasi, Romania, *gbaluta@tuiasi.ro*
† "Gh. Asachi" Technical University/Faculty of Automatic Control and Computer Engineering, Iasi, Romania, *clazar@ac.tuiasi.ro*

*Abstract*—E-learning has introduced new opportunities in teaching and learning electrical engineering subjects. This paper presents the architecture of a virtual engineering laboratory, designed to allow remote training in the control of DC, brushless DC, and stepper servomotors. An example of DC servomotors control on virtual laboratory is also presented.

*Keywords*—education tool, DC machine, control of drive.

## I. INTRODUCTION

Traditionally, the main objective of education consisted in acquiring knowledge by the students. The assessment was based on testing whether students could reproduce the acquired knowledge. By the end of last century engineering educators began to realize that the demands from industry changed. There should be more emphasis in skill and deep understanding rather than knowledge [1]. A general problem in electrical engineering is the fact that it deals with rather abstract notions such as current, voltage, resistance, capacitance, etc. These electrical quantities and phenomena are not directly observable and can only be made observable by means of measurements. Using new media and IT-technologies in classroom enables not only make studying more attractive to the student, but also might make teaching much easier. Complex technical problems have to be presented in a way that is easy to follow and understand. However, even if computer animations are used, student cannot grasp the details in a short time, since the teacher only once or twice shows examples or animations. There remains a need for repetition and exercises.

E-learning courses offer user-controlled elements that just aren't feasible in regular training classes. This self-paced element is one of the things that make e-learning so effective. It can lead to increased retention and a stronger grasp on the subject. This is because there is the ability to revisit or replay sections of the training that might not have been clear the first time around. Another element that an e-learning offers is the fact that it can work from any location and any time.

E-learning has introduced a new access to engineering subject learning; by using interactive animation and simulation it enable to create interactive training environment helping partly to replace laboratories [2]. Software simulations exercises are certainly helpful. But it is an absolute necessity that hardware-based laboratory experiments be performed concurrently by the students. It is a dangerous trend in many universities to move away from hardware to purely software-based laboratories; such

an approach fails to excite students and also does not prepare them for the "real word" where they will design, build, test or use real hardware [4].

It is important to give to the students a real world experience. However, to build an experiment is expensive and it is impossible for an educational institute to have the complete scale of experiments. The hardware experiments should therefore be redesigned in such a way that they also can be accessed in the Web and possibly integrated in e-learning. It must be a real electrotechnical experiment conducted in the laboratory but remotely controlled and monitored by e-learning Web-based tools.

A Virtual Laboratory (VL) is an electronic workspace for distance collaboration and experimentation in research or other creative activity, to generate and deliver results using distributed information and communication technologies. A VL is not viewed as a replacement for, or a competitor with, a Real Laboratory (RL). Instead, VLs are possible extensions to RLs and open new opportunities not realizable entirely within a RL at an affordable cost. They will probably have an important role in the future because they integrate the technical, financial and human resources by sharing data, information, documents, multimedia means, etc., that is, the knowledge base.

This paper presents the architecture of a virtual engineering laboratory, designed to allow remote training in the control of DC, brushless DC, and stepper servomotors. An example of DC servomotor control on VL is also presented.

## II. THE ARCHITECTURE OF THE VL FOR TRAINING IN THE DC, BRUSHLESS DC, AND STEPPER SERVOMOTOR

Using mechanical actuators based on electric motors is one of the most widely used options in industrial and consumer applications. This growing demand, together with the call for better performance in systems using electric motors, explain why more and more electronic devices (controllers, drivers, etc.) are marketed, designed to facilitate and improve the control of electric servomotors [3], [7].

### A. The Architecture of the VL

The VL for Training in the DC, Brushless DC and Stepper Servomotors is an Internet-based Virtual Learning and Training Environment where learners can interact with lab equipment, regardless of geographical constraints. Fig. 1 illustrates the concept of a virtual engineering laboratory on a network that allows for geographically separated users to have access to real devices at different sites [9].

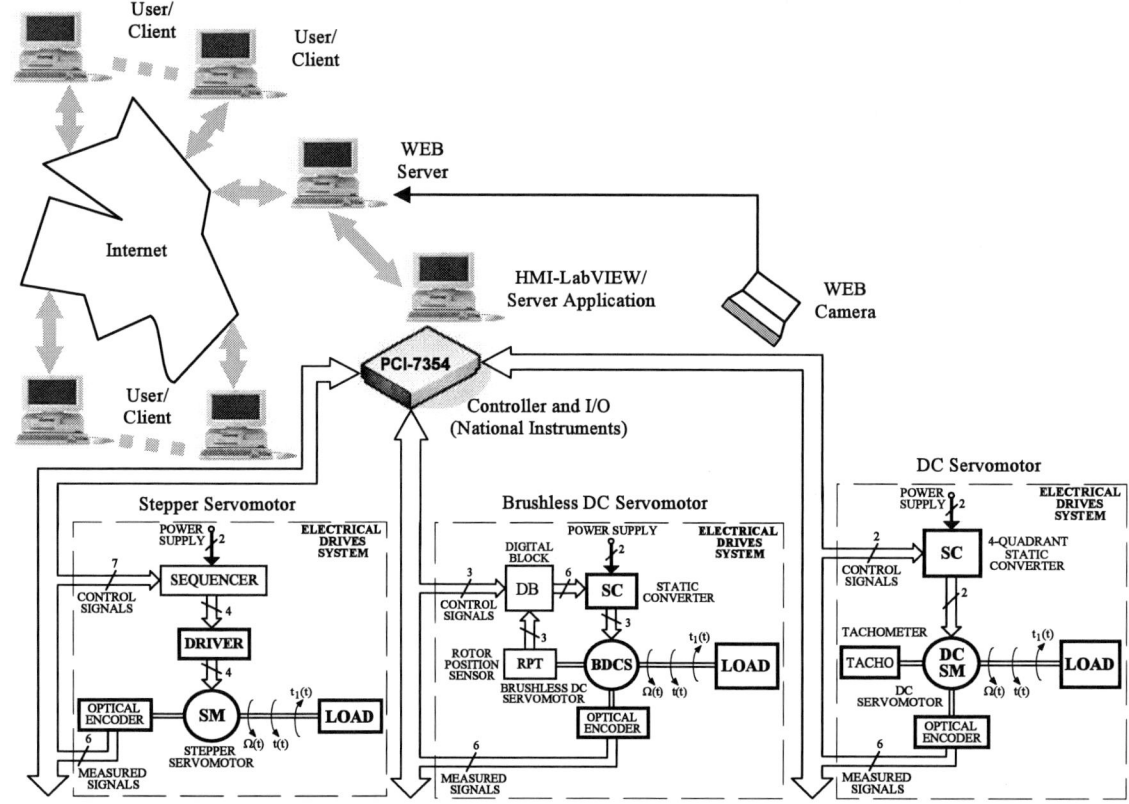

Fig. 1. The virtual electrical engineering laboratory concept.

The training equipment aims to make easier the task of instruction on methods and devices for the control and driving of servomotor, offering in a single environment, the possibility of testing different types of servomotor and different means of controlling each one.

### B. Hardware and Software Resources of Server Application

In the first stage the VL include three representative types of electrical drives (stepper/DC/brushless DC servomotors). Each one of these VL components has own specific actuator, controlled loading device as well as transducers for electrical and mechanical quantities and protection devices. In this way all of three types of servomotors can be controlled in open loop or closed loop.

DSP controllers available today are able to perform the computation for high performance digital motion control structures for different motor technologies and motion control configuration. The level of integration is continuously increasing, and the clear trend is towards completely integrated intelligent motion control [5]. Highly flexible solutions, easy parameterized and "ready-to-run", are needed in the existent "time-to-market" pressing environment, and must be available at non-specialist level.

Basically, the digital system component implements through specific hardware interfaces and corresponding software modules, the complete or partial hierarchical motion control structure, i.e., the digital motor control functionality at a low level and the digital motion control functionality at the higher level (Fig. 2).

The National Instruments PCI-7354 controller is a high-performance 4-axis-stepper/DC/brushless DC controller. This controller can be used for a wide variety of both simple and complex motion applications. It also includes a built-in data acquisition system with eight 16-bit analog inputs as well as a host of advanced motion trajectory generator and triggering features. Through four axes, individually programmable, the board can control independently or in a coordinated mode the motion. The board architecture, which is build around of a dual-processors core, has own real-time operating system (Fig. 3). These board resources assure a high computational power, needed for such real-time control.

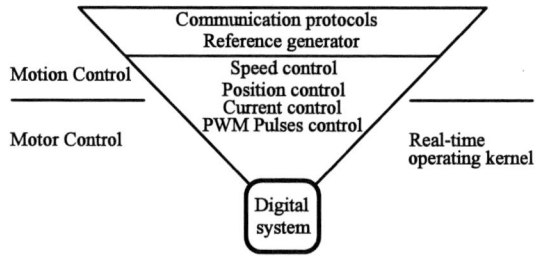

Fig. 2. Motion system structure hierarchy.

Functionally, the architecture of the NI PCI-7354 controller is generally divided into four components (Fig. 4): supervisory control, trajectory generator, control loop, and motion I/O. Supervisory control performs all the command sequencing and coordination required to carry

2044

out the specified operation. Trajectory generator provides path planning based on the profile specified by the user, and control loop block performs fast, closed-loop control with simultaneous position, velocity, and trajectory maintenance on one or more axes, based on feedback signals.

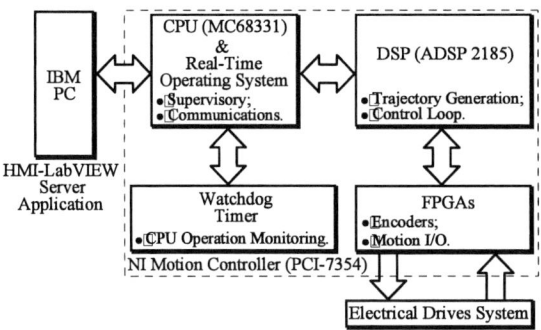

Fig. 3. Motion controller board structure.

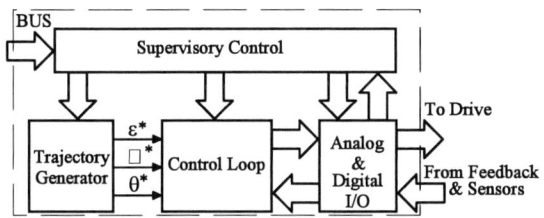

Fig. 4. Functional architecture of the NI PCI-7354.

SCADA software enables programmers to create distributed control applications having supervisory facilities and a Human-Machine Interface (HMI). As SCADA software, LabVIEW is used. The development environment used to complete the applications is LabVIEW 7.0, which beside the graphic implementation that gives easy use and understanding takes full advantage of the networking resources.

In the following an example is presented to illustrate the manipulation of instruments and real devices for low power electrical drives education on the Web. It illustrates the laboratory works dedicated to control an electrical drives system with DC servomotor.

### III. SET-UP DC SERVOMOTOR ELECTRICAL DRIVES SYSTEM

The detailed layout is presented in Fig. 5 and the set-up for DC servomotor control can be inspected in Fig. 6.

The rated parameters of the used separately-excited DC servomotor are presented in Tab. I. It is driven by a reversible chopper with the L292 specialized integrated circuit (made by SGS-THOMSON, Microelectronics Company). The servomotor's shaft is connected with an incremental optical encoder that gives 1000 pulses/rev. (this means 4000 pulses/rev. at the output of the NI PCI-7354 controller) [8].

TABLE I.
THE RATED PARAMETERS OF THE DC SERVOMOTOR

| Rated parameters | Symbol | Value | Unit |
|---|---|---|---|
| Armature voltage | $U_n$ | 20 | V |
| Armature current | $I_n$ | 2 | A |
| Speed | $n_n$ | 2400 | rpm |
| Armature winding resistance | $R_a$ | 1.31 | $\Omega$ |
| Armature winding inductance | $L_a$ | 7.58 | mH |

The DC servomotor is loaded using a controlled loading device. Among the features of the loading system, the authors emphasize [8]:

-the possibility to load the electric drive motor with a reactive load torque in a wide range of speed, including very low speeds;

-the possibility to impose a constant load torque operation mode or an overload operation mode.

The driven axis with DC servomotor can be controlled in open loop or closed loop.

The implemented control system uses the algorithm from Fig. 7, which allows to generate in real-time the move trajectory and to change the motion parameters.

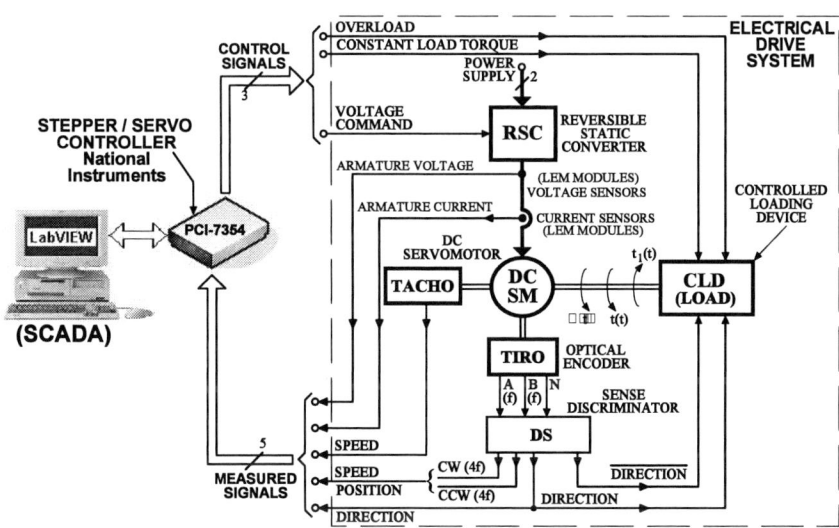

Fig. 5. The block diagram of the DC electrical drives system.

Fig. 6. General view of the DC electrical drives system.

In order to program the motion control system, the developer employs LabVIEW environment with specific virtual instruments for motion control from the FlexMotion library. For closed-loop operating mode, the associated graphic program is shown in Fig. 8.

```
┌─────────────────────────────┐
│   Read Maximum Velocity      │
└─────────────────────────────┘
              ↓
┌─────────────────────────────┐
│  Read Maximum Acceleration   │
│  Read Maximum Deceleration   │
└─────────────────────────────┘
              ↓
┌─────────────────────────────┐
│      Set Operation Mode      │
└─────────────────────────────┘
              ↓
┌─────────────────────────────┐
│      Read Final Position     │
└─────────────────────────────┘
              ↓
┌─────────────────────────────┐
│         Start Motion         │
└─────────────────────────────┘
              ↓
┌─────────────────────────────────┐
│  Loop Waiting for Move Complete │
│  ┌───────────────────────────┐  │
│  │  Read (Update) Position and│  │
│  │  and Move Constraints      │  │
│  └───────────────────────────┘  │
│  ┌───────────────────────────┐  │
│  │  Start Motion (Optional)   │  │
│  └───────────────────────────┘  │
│  ┌───────────────────────────┐  │
│  │  Perform Non-Motion        │  │
│  │  Measurement               │  │
│  └───────────────────────────┘  │
└─────────────────────────────────┘
```

Fig. 7. Position-based straight-line move algorithm.

The vast majority of motion control algorithms employed in industrial applications are of two forms [6]:

-the well-known PID position loop (Fig. 9);

-an average velocity loop cascaded with a position loop (Fig. 10).

On the other hand, in order to achieve near zero following or tracking error, feedforward control is often employed. A requirement for feedforward control is the availability of both the velocity, $\Omega^*(t)$, and acceleration, $\varepsilon^*(t)$, commands synchronized with the position commands, $\theta^*(t)$. An example of how feedforward control

is used in addition to the second servocontrol topology is shown in Fig. 11. The NI PCI-7354 controller can be configured for any of the above servocontrol topologies.

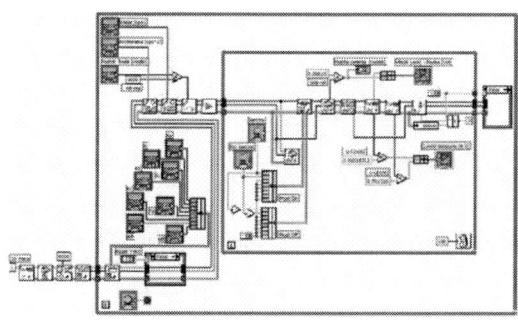

Fig. 8. LabVIEW diagram for single axes positioning system with DC servomotor (closed-loop operating mode).

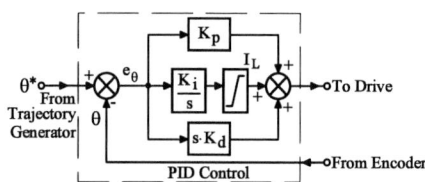

Fig. 9. PID servocontrol topology.

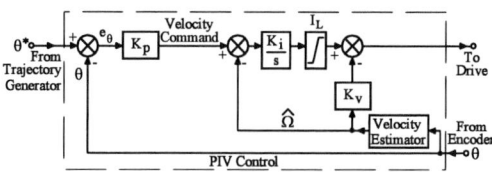

Fig. 10. PIV servocontrol topology.

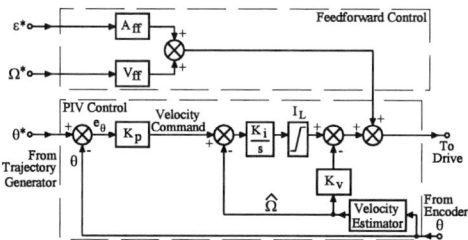

Fig. 11. Basic feedforward and PIV control topology.

In order to build a positioning application, the developer has to follow the stages shown in Fig. 12, which represent the steps required to design a motion application.

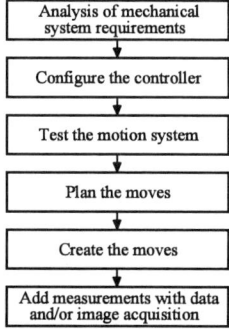

Fig. 12. Generic steps for designing a motion application.

To introduce a student to motor control, the DC servomotor set-up provides the opportunity to learn about this motor without ever attending a laboratory session at their institution. Since most engineering students cannot procure expensive electronic instruments (oscilloscope, voltmeters, etc.) and since the majority of students do not have prior technical experience, this remote set-up is very economical and offers students a great tool for learning about the servomechanism and data acquisition.

The learner can also perform the second to fifth stages of motion application design (Fig. 12). To ensure proper performance for servomechanism the motion control system must be tuned and tested. Then the type of move profile is planed. Motion constraints are the maximum velocity, acceleration, deceleration and jerk that the system can handle. Trajectory parameters are expressed as a function of motor shaft revolutions. The trajectory generator takes into account the type and the constraints of motion and generates values of instantaneous trajectory parameters in real-time.

Basically, the student can learn about the parameters of the DC servomechanism by tuning the control loop parameters and setting the motion constrains to make execute the operations, i.e. change the rotation direction, position and speed. Also during the experiment, the student can visualise critical input and output points in the DC servomotor electronic interface using virtual oscilloscopes. Since this system is in real-time mode, the learner observes changes in position, direction, and velocity as the motor rotates after being commanded.

The graphical user interfaces specific to this application are shown in Fig. 13 (open-loop operating mode) and Fig 14 (closed-loop operating mode).

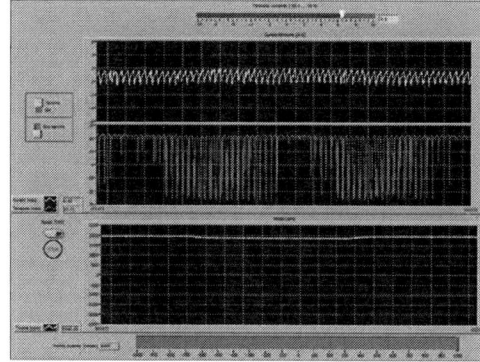

Fig. 13. Main panel for DC servomotor electrical drives system (open-loop operating mode).

The first one reveals the experimental results being obtained for the open-loop control of the electrical drives system when the user applies a load torque disturbance to the controlled servomotor shaft. The variations of the armature current and modulated voltage applied to its terminals (for a command signal of 6.5 V) are presented by means of the first virtual oscilloscope. The speed drop caused by the load torque disturbance can be noticed on the second virtual oscilloscope. Thus the user can deduce the causal link between the disturbance acting over the system and its effects over the output measure.

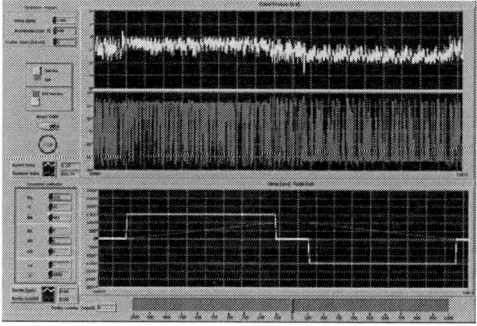

Fig. 14. Main panel for DC servomotor electrical drives system (closed-loop operating mode).

The user graphical interface presented in Fig. 14 shows the experimental results obtained for a classical position control structured by using a PID controller (the motion control algorithms from Fig. 9, Fig. 10 and Fig. 11). The tuning parameters of the control law are presented in the bottom-left area reserved for the controller settings and those related to the motion in the top-left part. The review of the results certifies the fact that the positioning within the range [0÷1000] rev. is completed in full accordance with the requirements imposed by the motion constraints and with optimal dynamic behaviour performances that is little and well harmonized transitory regimes.

2047

## IV. THE WEB PAGE

The WEB Server shown in Fig. 1 and accessed to the address http://www.lvaemp.tuiasi.ro, hosts the WEB page created for this VL. The WEB page (Fig. 15) allows the users to remote control the applications of RL. But in order to access the current configured application the user must to register an account and to be authorized by site administrator. If the access is granted the user has the possibility to read the breviary related to the applications of the RL and/or to test. On the other hand, if the experimental set-up is accessing by another user then the last in user can only to assist at the remote control of the application and to the experimental data analyzing, waiting in a queue the release of the application in order to have full access.

Fig. 15. The WEB page of VL.

The design of WEB pages is based on PHP and AJAX programming languages and their publishing is made using the CGI scripts. In this way, the potential user can access the VL from any location without other supplementary software resources.

Even the used browser can be Internet Explorer we strongly recommend Netscape Navigator because this support the *.monitor* mode, which permits the images update periodically irrespective of users.

## V. CONCLUSIONS

This paper focuses on the use of a VL concept to show that it is now possible to bring remote access instrumentation and control techniques within an educational framework.

Through the user interfaces, the students acquire the following abilities:

-enabling the DC servomotor and resetting the control system;

-calibrating a DC servomotor system;

-controlling the rotation direction of the servomotor (clockwise or counterclockwise);

-moving the servomotor shaft from one arbitrary position to another;

-accelerating the servomotor and maintaining a constant velocity as well as decelerating and bringing it to a complete stop;

-visualizing intermediate servomotor control and encoder signals;

-understanding the relationship between the external loads and command effort.

This type of laboratory has some advantages compared to RL, such as:

-it is not limited in time, as students can exercise at any hour;

-if a university has just one expensive equipment it could share it with other universities and so students can practice on different equipment;

-students can review the laboratory session that has been made earlier as many times they want and in a relaxed environment;

-VL can be a good alternative in distance learning system because it can fill the absence of practical session.

In the case of electrical drives system real-time control is assured by the hardware and software resources of the NI PCI-7354 board. On the other hand, the remote monitoring and parameterization is depended on the speed of telecommunication network.

Work is in progress to upgrade the laboratory with new electrical drives system [10]: brushless DC servomotor electrical drives system.

## REFERENCES

[1] V. Fedok and P. Bauer, "E-learning in Education of Electrical Drives and Power Electronics: Opportunities and Challenges", *Proceedings of the 15-th International Conference on Electrical Drives and Power Electronics*, Dubrovnik, Croatia, pp. 1-9, September 2005.

[2] P. Bauer and V. Fedak, "E-learning for Power Electronics and Electrical Drives", *Proceedings of the 13-th International Conference on Electrical Drives and Power Electronics*, High Tatras, Slovak Republic, pp. 567-572, September 2003.

[3] M. Mazo, J. Urena, F. J. Rodriguez, J. J. Garcia, J. L. Lazaro, E. Santiso et. al., "Teaching Equipment for Training in the Control of DC, Brushless and Stepper Servomotors", *IEEE Transaction on Education*, vol. 41, No. 2, pp. 146-158, May 1998.

[4] V. Trifa, C. Marginean, and C. Rusu, "Aspects Concerning the Implementation of a Virtual Laboratory for Reluctance Motors Using the Internet", *Proceedings of the 6-th Symposium on Advanced Electromechanical Motion Systems*, Lausanne, Switzerland, pp. 1-6, September 2005.

[5] L. Kreidler, "DSP Solution for Digital Motion Control", *Journal of Electrical Engineering*, Vol. 2, pp. 1-9, June 2002.

[6] G. Ellis and R. D. Lorentz, "Comparison of Motion Control Loops for Industrial Applications", *Proceedings of IEEE IAS Annual Technical Conference*, Phoenix, AZ, USA, Vol. 4, pp 2599-2605, October 1999.

[7] H. H. Saliah, E. Nurse, and A. Abecassis, "Design of a Generic, Interactive, Virtual and Remote Electrical Engineering Laboratory", *Proceedings of the 29-th ASEE/IEEE Frontiers in Education Conference*, San Juan, Puerto Rico, Session 12c6, pp. 18-23, November 1999.

[8] Gh. Baluta, *Low Power Electrical Drives. Applications* (in Romanian), Iasi: Politehnium, 2004, pp. 40-90.

[9] Gh. Baluta and C. Lazar, "Remote Control Laboratory Development for Electrical Drive Systems", *Proceedings of the 13-th IEEE/IFAC International Conference on Methods and Models in Automation and Robotics*, Szczecin, Poland, pp. 393–398, August 2007.

[10] Gh. Baluta and V. Horga, "Control of Stepper Motors on Virtual Laboratory", *Proceedings of the 15-th IMEKO TC4 International Symposium on Novelties in Electrical Measurement and Instrumentations*, Iasi, Romania, vol. II, pp. 393–398, September 2007.

# DQ-Transformation Approach for Modelling and Stability Analysis of AC-DC Power System with Controlled PWM Rectifier and Constant Power Loads

K-N Areerak, S.V. Bozhko, G.M. Asher, and D.W.P. Thomas

School of Electrical and Electronic Engineering, The University of Nottingham, Nottingham, UK.
*Corresponding Author E-mail:* eexka1@nottingham.ac.uk

*Abstract-* **In this paper a technique for analysing aircraft frequency wild power systems with constant power loads is developed and demonstrated. Power electronic based loads often behave as constant power loads, especially when feeding machine or actuator drives under current and speed control. The constant power (CP) loads can affect the stability of the power system. The problem is a particular issue in aircraft power systems as the proportion of CP loads increases with the advent of the more-electric aircraft. This paper deals with stability analysis of a three-phase frequency-wild AC power system with a CP load fed through a vector-controlled front-end PWM converter. The mathematical model suitable for stability analysis is derived using the DQ-transformation method. It is shown how the proposed approach can be applied to study the power system behaviour under different loads and parameters variations, as well as to assess the power system stability margins. The study is supported by intensive simulations that verify the reported theoretical results.**

## I. INTRODUCTION

It is well known that power electronic loads often behave as constant power (CP) loads especially when feeding machine or actuator drives under current and speed servo control. These CP loads can affect the stability of the power system [1], [2]. This is a particular issue for aircraft power systems since the more-electric aircraft concept is leading to a large proportion of its loads being of constant power.

Two approaches are commonly used for analysing the performance of power converter systems. The first is the generalized state-space averaging (SSA) method, which has been used to analyze many power converters in DC distribution systems [3], [4], [5], uncontrolled and controlled rectifiers in AC distribution systems [6], and 6- and 12-pulse diode rectifiers in three phase systems [7]. Another technique, for AC systems, is through DQ transformation [8], [9], [10] in which power converters can be treated as time-varying transformers. Application of this approach to analyze an uncontrolled 6-pulse rectifier in three-phase AC systems has first been reported in our previous papers [11]. This paper logically continues the previous study and deals with modelling and stability analysis of a three-phase AC frequency-wild power system feeding a vector controlled front-end PWM converter with a CP load installed at the DC-side, together with a filter and DC-link capacitor. The DQ-transformation method is applied for analysis; that is more applicable to a converter controlled in terms of rotation *dq*-frame aligned with the grid voltage vector. The stability study also takes into account a phase shift between the AC supply and the converter AC terminals that is normally ignored in many previous works [6], [7]. The paper also addresses the effect of a mixed load (i.e. a variable proportion of CP load to ordinary load) on the power system stability. As has been found, the bandwidth of DC-link voltage control loop has a substantial effect on system stability, therefore this is considered in detail in this paper.

The paper is organized as follows. In section II, the DQ method is illustrated. An eigenvalue theorem and linearization using the first order terms of a Taylor expansion are used for stability analysis and this is addressed in section III. The results of the stability studies are presented in section IV, including small signal simulation to consolidate the analytical results. Finally, section V concludes and discusses the power of the DQ approach to predict stability boundaries.

## II. DQ METHOD

### A. The Power System Considered

The power system studied in this paper is shown in Fig.1. It consists of a three-phase source, transmission line, filters and a front-end PWM converter under standard close-loop control [12]. The CP load is fed through a DC bus. It is assumed that the three-phase source is balanced. $R_{eq}$, $L_{eq}$, and $C_{eq}$ are the equivalent parameters of a transmission line, $r_F$ and $L_F$ are AC filter parameters, and $C_F$ is a DC-link capacitor. The PWM converter considered uses six ideal switches as IGBTs and assuming no loss and no system harmonics.

The PWM converter considered consists of vector-controllers to keep the voltage across the DC-link constant, here equal to $E^*_{dc}$ voltage command, and to keep a unity power factor at the AC bus by setting $I^*_{in,q}$ equal to zero.

978-1-4244-1741-4/08/$25.00 ©2008 IEEE

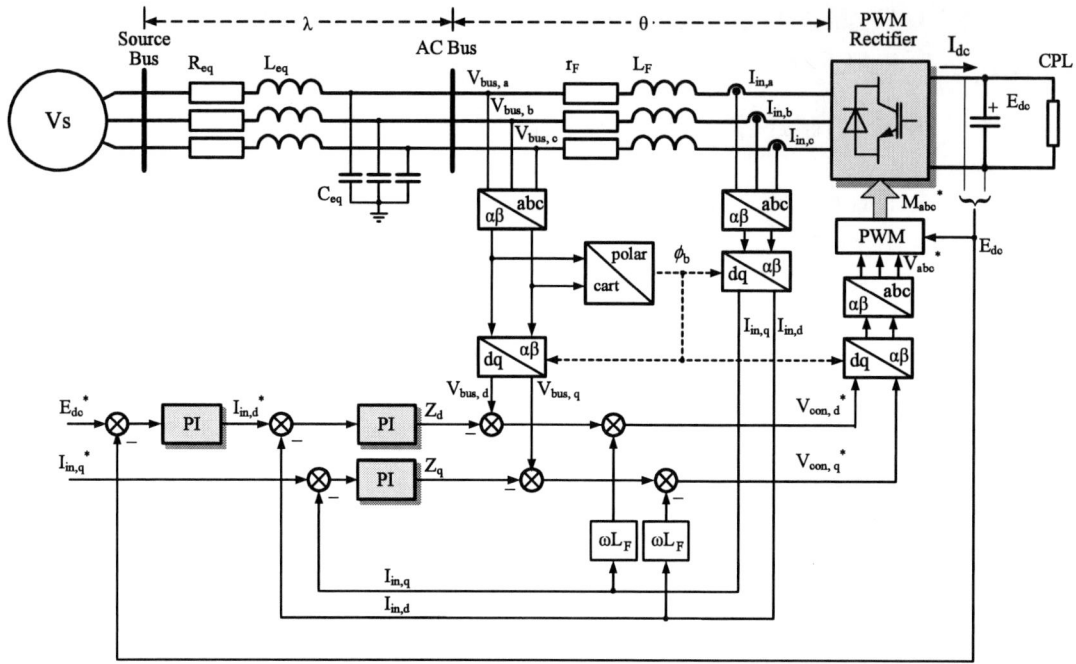

**Fig.1.** The power system studied

The vector-control is in terms of the reference frame aligned to the AC bus voltage vector. Detailed description of the control strategy including controller design can be found in [12]. Note here that the controllers on the direct axis regulate the DC-link voltage and the control along quadrature axis regulates the PWM converter's reactive power as shown in Fig. 2.

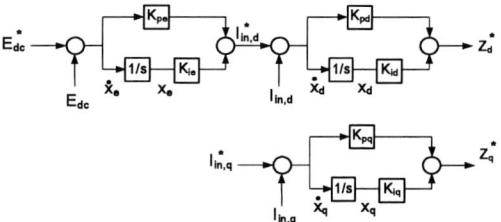

**Fig. 2.** The schematic of controllers

### B. The equivalent circuit of the system on the DQ frame

In this paper, the DQ-transformation approach is applied to derive a mathematical model for the power system in Fig. 1 including vector-controllers where the PWM converter can be modelled on the DQ frame as a transformer. The vector diagram for DQ-transformation is depicted in Fig. 3. Note that phase $\varphi_{con}$ of the input terminal voltage $V_{con}$ lags phase $\varphi_b$ of the bus voltage $V_{bus}$ by an angle $\theta$, while phase $\varphi_b$ of the bus voltage $V_{bus}$ lags phase $\varphi_s$ of the source voltage $V_s$ by an angle $\lambda$ as shown in Fig.1.

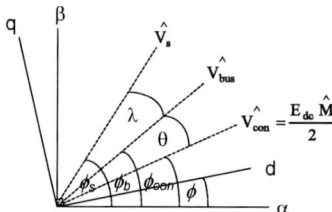

**Fig. 3**. The vector diagram for DQ-transformation

For the PWM converter, the voltage source equation, bus voltage equation, terminal voltage of converter equation and switching signal equation in the DQ frame at rotating frequency $(\omega t + \phi_b)$ using peak convention fixed to the ac bus voltage vector $(\phi = \phi_b)$ are given in (1)-(4) respectively:

$$\mathbf{V_{s,dq}} = \hat{V_s}\, e^{j(\phi_s - \phi_b)} = \hat{V_s}\, e^{j\lambda} \qquad (1)$$

$$\mathbf{V_{bus,dq}} = \hat{V_{bus}}\, e^{j(\phi_b - \phi_b)} = \hat{V_{bus}}\, e^{j0} \qquad (2)$$

$$\mathbf{V_{con,dq}} = \hat{V_{con}} e^{j(\phi_{con} - \phi_b)} = \hat{V_{con}} e^{-j\theta} \quad (3)$$

$$\mathbf{M_{dq}} = \hat{M} e^{j(\phi_{con} - \phi_b)} = \hat{M} e^{-j\theta} \quad (4)$$

From (3) and (4), the relationship between the input and output terminals of the PWM converter on the DQ frame is

$$\mathbf{V_{con,dq}} = \frac{E_{dc} \mathbf{M_{dq}}}{2} \quad (5)$$

According to (1)-(5), the equivalent circuit of the system in the DQ frame is shown in Fig.4, where the PWM converter can be modelled as a transformer. It can be seen in Fig. 4 that $Z^*_{dq}$ can be derived by using the diagram of the controllers in Fig. 2. In addition, the CP load can be considered as a voltage-dependent current source given by

$$I_{CPL} = \frac{P_{CPL}}{E_{dc}} \quad (6)$$

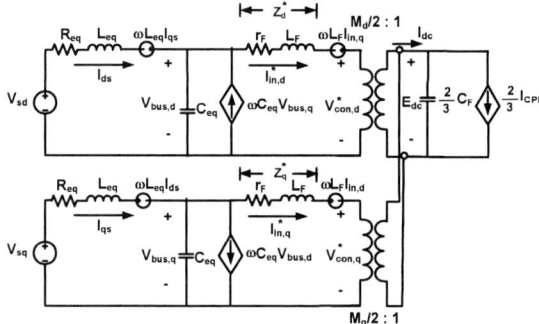

**Fig. 4.** The equivalent circuit of the power system in the DQ frame

### C. Dynamic Model of the System

From Fig. 2, $Z^*_{dq}$ can be derived and given in (7) and (8), respectively.

$$Z^*_d = -K_{pd} I_{in,d} - K_{pe} K_{pd} E_{dc} + K_{ie} K_{pd} X_e \\ + K_{id} X_d + K_{pe} K_{pd} E^*_{dc} \quad (7)$$

$$Z^*_q = -K_{pq} I_{in,q} + K_{iq} X_q + K_{pq} I^*_{in,q} \quad (8)$$

Applying KVL and KCL to the circuit in Fig. 4 with (7) and (8) determines the set of nonlinear differential equations to describe the power system dynamics:

$$\dot{I}_{ds} = -\frac{R_{eq}}{L_{eq}} I_{ds} + \omega I_{qs} - \frac{1}{L_{eq}} V_{bus,d} + \frac{1}{L_{eq}} V_{sd}$$

$$\dot{I}_{qs} = -\omega I_{ds} - \frac{R_{eq}}{L_{eq}} I_{qs} - \frac{1}{L_{eq}} V_{bus,q} + \frac{1}{L_{eq}} V_{sq}$$

$$\dot{V}_{bus,d} = \frac{1}{C_{eq}} I_{ds} + \omega V_{bus,q} - \frac{1}{C_{eq}} I_{in,d}$$

$$\dot{V}_{bus,q} = \frac{1}{C_{eq}} I_{qs} - \omega V_{bus,d} - \frac{1}{C_{eq}} I_{in,q}$$

$$\dot{I}_{in,d} = -\frac{(r_F + K_{pd})}{L_F} I_{in,d} - \frac{K_{pe} K_{pd}}{L_F} E_{dc} + \frac{K_{ie} K_{pd}}{L_F} X_e \\ + \frac{K_{id}}{L_F} X_d + \frac{K_{pe} K_{pd}}{L_F} E^*_{dc}$$

$$\dot{I}_{in,q} = -\frac{(r_F + K_{pq})}{L_F} I_{in,q} + \frac{K_{iq}}{L_F} X_q + \frac{K_{pq}}{L_F} I^*_{in,q}$$

$$\dot{E}_{dc} = \frac{3}{2 C_F E_{dc}} (V_{bus,d} I_{in,d} - K_{pe} K_{pd} I_{in,d} E^*_{dc} + K_{pe} K_{pd} I_{in,d} E_{dc} \\ - K_{ie} K_{pd} I_{in,d} X_e + K_{pd} I^2_{in,d} - K_{id} I_{in,d} X_d + V_{bus,q} I_{in,q} \\ - K_{pq} I_{in,q} I^*_{in,q} + K_{pq} I^2_{in,q} - K_{iq} I_{in,q} X_q) - \frac{P_{CPL}}{C_F E_{dc}}$$

$$\dot{X}_e = -E_{dc} + E^*_{dc}$$

$$\dot{X}_d = -I_{in,d} - K_{pe} E_{dc} + K_{ie} X_e + K_{pe} E^*_{dc} \quad (9)$$

$$\dot{X}_q = -I_{in,q} + I^*_{in,q}$$

This set of equations can not be solved analytically. Therefore the stability analysis is performed using small-signal linearization. The linear model can be derived using the first-order terms of the Taylor expansion of (9), so as to achieve a set of linear differential equations around an equilibrium point. Setting the state variables $x$, input $u$ and output vector $y$ as

$$x = < I_{ds}, I_{qs}, V_{bus,d}, V_{bus,q}, I_{in,d}, I_{in,q}, E_{dc}, X_e, X_d, X_q >$$

$$u = < \hat{V_s}, E^*_{dc}, I^*_{in,q}, P_{CPL} > \quad , \quad (10)$$

$$y = < E_{dc} >$$

the DQ linearized model of (9) can be written in the following form:

$$\dot{\delta \mathbf{x}} = \mathbf{A}(\mathbf{x_o}, \mathbf{u_o}) \delta \mathbf{x} + \mathbf{B}(\mathbf{x_o}, \mathbf{u_o}) \delta \mathbf{u}$$
$$\delta \mathbf{y} = \mathbf{C}(\mathbf{x_o}, \mathbf{u_o}) \delta \mathbf{x} + \mathbf{D}(\mathbf{x_o}, \mathbf{u_o}) \delta \mathbf{u} \quad (11)$$

where

$$\delta \mathbf{x} = [\delta I_{ds} \ \delta I_{qs} \ \delta V_{bus,d} \ \delta V_{bus,q} \ \delta I_{in,d} \ \delta I_{in,q} \ \delta E_{dc} \ \delta X_e \ \delta X_d \ \delta X_q]^T$$

$$\delta \mathbf{u} = [\delta V_s \ \delta E^*_{dc} \ \delta I^*_{in,q} \ \delta P_{CPL}]^T, \quad \delta \mathbf{y} = [\delta E_{dc}]$$

$$\mathbf{A}(\mathbf{x_o}, \mathbf{u_o}) = [\ ]_{10 \times 10}, \quad \mathbf{B}(\mathbf{x_o}, \mathbf{u_o}) = [\ ]_{10 \times 4}$$

$$\mathbf{C}(\mathbf{x_o}, \mathbf{u_o}) = [0 \ 0 \ 0 \ 0 \ 0 \ 0 \ 1 \ 0 \ 0 \ 0]$$

$$\mathbf{D}(\mathbf{x_o}, \mathbf{u_o}) = [0 \ 0 \ 0 \ 0]$$

The **A** and **B** matrices are too awkward to put in this paper.

## III. The Stability Analysis

The eigenvalue theorem is used for stability analysis. The eigenvalue can be calculated from the Jacobian matrix, $A(x_o, u_o)$ in (11), by using (12):

$$\det[\lambda I - A] = 0 \qquad (12)$$

and the system is stable if

$$real\ \lambda_i < 0 \qquad (13)$$

where $i = 1, 2, \ldots, n$ ($n$ - the number of state variables) and (11) is derived for a particular operating point, or a steady state regime. Therefore, the model (11) needs a steady state to be defined in order to implement the eigenvalue analysis. In this study we use the power flow equation in order to determine the steady state values. A phase shift between the AC source and the AC bus is taken into account and this has not been considered in previous publications.

## IV. The Example System

The set of parameters for the example power system according to Fig. 1 is given as follows: $V_s$=230$V_{rms/phase}$, $\omega$ = $2\pi\cdot400$rad/s, $R_{eq}$=0.01$\Omega$, $L_{eq}$=30$\mu$H, $C_{eq}$=2nF, $r_F$=0.1$\Omega$, $L_F$=100$\mu$H, $C_F$=1000$\mu$F, $\zeta_e$=0.7 and $\omega_{h\_e}$ = $2\pi\cdot10$ rad/s for voltage control loop ($K_{pe}$=0.0541, $K_{ie}$=2.4279), $\zeta_i$=0.8 and $\omega_{h\_i}$= $2\pi\cdot500$rad/s for current control ($K_{pd}$=$K_{pq}$=0.403, $K_{id}$=$K_{iq}$=986.96), $M_{max}$=1.15, $E^*_{dc}$=600V, and $I^*_{in,q}$=0A.

### A. Small Signal Simulation

The small signal simulation of the power system with the parameters given above is used to verify the accuracy of the mathematical model. Fig. 5 shows the step response of the system output, $E_{dc}$, delivered by the three-phase benchmark nonlinear model (using SimPowerSystem$^{TM}$ in MATLAB), and those of the DQ linearized model. They are compared together when $P_{CPL}$ changes step-wise from 5kW to 5.5 kW. In addition, Fig. 6 also shows the compared response when the bandwidth for voltage loop control is changed from 10 Hz to 20 Hz and $P_{CPL}$ changes step-wise from 5kW to 5.5 kW. It can be seen from the results that excellent agreement between both models is achieved.

### B. Stability Analysis

For stability analysis, the eigenvalues of the system are calculated from the matrix $A(x_0, u_0)$ in (11). The eigenvalue plot for the power system with the above parameters when $P_{CPL}$ varies from 0 kW to 322 kW is shown in the Fig.7. Note that the maximum power for eigenvalue plot depends on the value of $M_{max}$ (modulation index limitation), here equal to 322 kW for $M_{max}$ =1.15. In this case, Fig.7 shows that the system is always stable. However, if the voltage control loop bandwidth is increased up to 100Hz, one can

get to an unstable point at certain a CP load level. The eigenvalue plot for this case is shown in Fig.8 where some of the eigenvalues go to the right-hand semiplane. Fig. 9 shows the zoomed area of interest: as one can see, the system becomes unstable when the CP load exceeds 320kW. Fig. 10 shows the step response of the system to verify there is good agreement between the analytical results and the simulation results for unstable conditions.

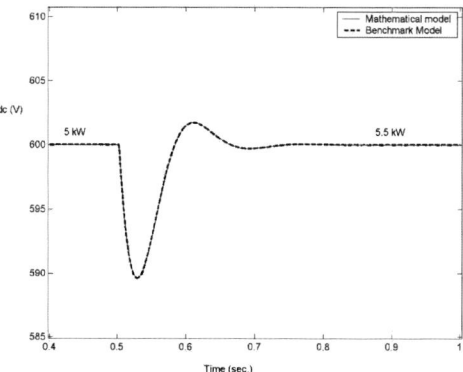

**Fig. 5.** Compared step response for $\omega_{h\_e} = 2\pi\cdot10$ rad/s

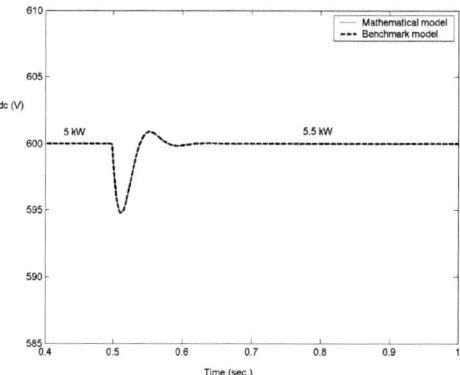

**Fig. 6.** Compared step response for $\omega_{h\_e} = 2\pi\cdot20$ rad/s ($K_{pe}$= 0.1082 , $K_{ie}$=9.712)

### C. Variations in System Parameters

In this section, the dynamic model is used to predict instability for variations in the system parameters. It can be seen from the results in section B that the $\omega_{h\_e}$ is very crucial in terms of stability. Therefore, Fig. 11 shows the eigenvalue plot when the $\omega_{h\_e}$ varies from 0.5Hz to 100Hz with all other parameters fixed. As can be seen from Fig. 11, the higher $\omega_{h\_e}$ can make the system unstable. The other parameters do not affect the stability results.

2052

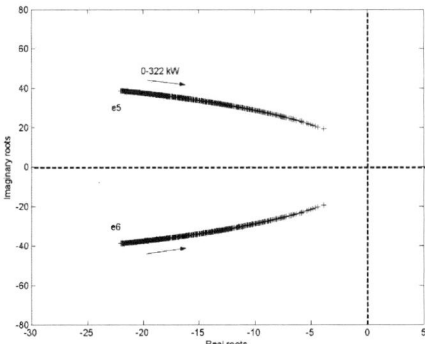

**Fig. 7.** Eigenvalue plot for $\omega_{n\_e} = 2\pi \cdot 10$ rad/s

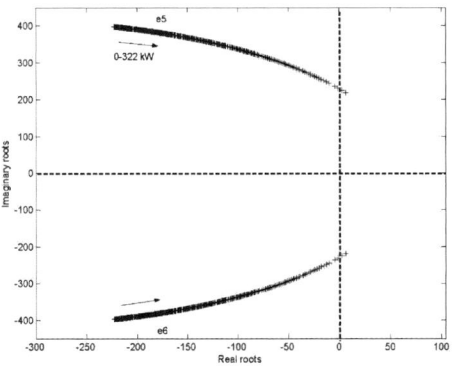

**Fig. 8.** Eigenvalue plot for $\omega_{n\_e} = 2\pi \cdot 100$ rad/s ($K_{pe}$= 0.541 , $K_{ie}$=242.79)

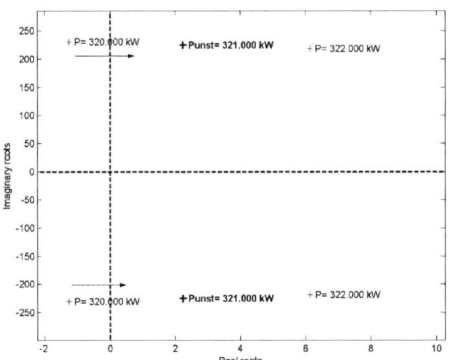

**Fig. 9.** Zoomed area of interest from Fig.8

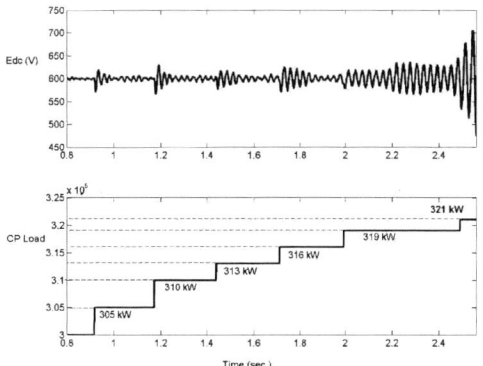

**Fig. 10.** Step response for unstable condition ($\omega_{n\_e} = 2\pi \cdot 100$ rad/s)

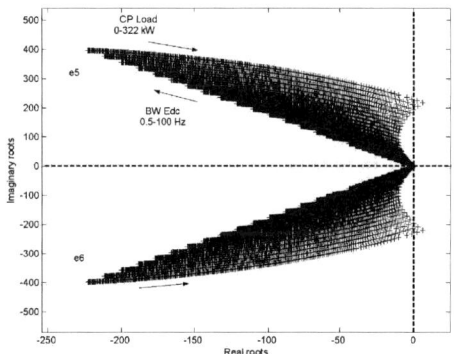

**Fig.11.** Eigenvalue plot for varying $\omega_{n\_e}$ from 0.5 to 100 Hz

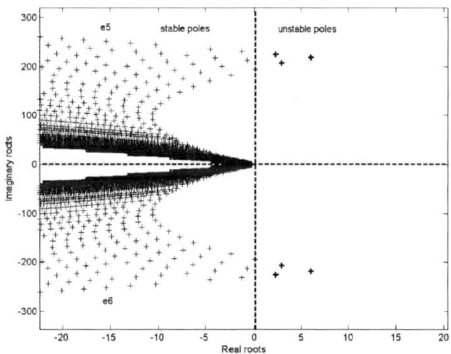

**Fig. 12.** Zoomed area of interest from Fig. 11

## V. CONCLUSIONS

In this paper, the DQ-transformation method is applied for modelling and analysis of a three-phase AC distribution power system with CP load fed through the front-end PWM power converter under conventional vector control. The mathematical model is derived, validated and used to predict the unstable operation of the power system under certain CP load levels and parameters variations. The reported model can be used for thorough investigations of aircraft power system stability for representative architectures and worst-case operational modes.

## REFERENCES

[1] C. Rivetta, G.A. Williamson, and A. Emadi, "Constant Power Loads and Negative Impedance Instability in Sea and Undersea Vehicles: Statement of the Problem and Comprehensive Large-Signal Solution," in *Proc. IEEE Electric Ship Tech. Symposium.*, Philadelphia, PA USA, July 2005, pp.313-320.

[2] A. Emadi, A. Khaligh, C.H. Rivetta, and G.A. Williamson, "Constant Power Loads and Negative Impedance Instability in Automotive Systems: Definition, Modeling, Stability, and Control of Power Electronic Converters and Motor Drives," *IEEE Trans. on Vehicular Tech.*, vol. 55, no. 4, pp.1112-1125, July 2006.

[3] J. Mahdavi, A. Emadi, M.D. Bellar, and M. Ehsani, "Analysis of Power Electronic Converters Using the Generalized State-Space Averaging Approach," *IEEE Trans. on Circuit and Systems.*, vol. 44, pp.767-770, August 1997.

[4] A. Emadi, "Modeling and Analysis of Multiconverter DC Power Electronic Systems Using the Generalized State-Space Averaging Method," *IEEE Trans. on Indus. Elect.*, vol. 51, no. 3, pp. 661-668, June 2004.

[5] M.M. Jalla, A. Emadi, G.A. Williamson, and B. Fahimi, "Modeling of Multi-Converter More Electric Ship Power Systems Using the Genearalized State Space Averaging Method," in *Proc. The 30$^{th}$ Annual Conf. of the IEEE Indus. Elect. Soc.*, Busan, Korea, Nov. 2004, pp.508-513.

[6] A. Emadi, "Modeling of Power Electronic Loads in AC Distribution Systems Using the Genearlized State-Space Averaging Method," *IEEE Trans. on Indus. Elect.*, vol. 51, no. 5, pp. 992-1000, October 2004.

[7] L. Han, J. Wang, and D. Howe, "State-space average modelling of 6- and 12-pulse diode rectifiers," in *Proc. The 12$^{th}$ European Conf. on Power Elect. and Appl.*, Aalborg, Denmark, Sep. 2007.

[8] C.T. Rim, D.Y. Hu, and G.H. Cho, "Transformers as Equivalent Circuits for Switches: General Proffs and D-Q Transformation-Based Analyses," *IEEE Trans. on Indus. Appl.*, vol. 26, no. 4, pp. 777-785., July/August 1990.

[9] C.T. Rim, N.S. Choi, G.C. Cho, and G.H. Cho, "A Complete DC and AC Analysis of Three-Phase Controlled-Current PWM Rectifier Using Circuit D-Q Transformation," *IEEE Trans. on Power Electronics*, vol. 9, no. 4, pp. 390-396., July 1994.

[10] S.B. Han, N.S. Choi, C.T. Rim, and G.H. Cho, "Modeling and Analysis of Static and Dynamic Characteristics for Buck-Type Three-Phase PWM Rectifier by Circuit DQ Transformation," *IEEE Trans. on Power Electronics*, vol. 13, no. 2, pp.323-336., March 1998.

[11] K-N Areerak, S.V. Bozhko, G.M. Asher, and D.W.P. Thomas, "Stability Analysis and Modelling of AC-DC System with Mixed Load Using DQ-Transformation Method," in *Proc. The IEEE International Symposium on Industrial Electronics*, Cambridge, UK, 30 June-2 July, 2008.

[12] R. Pena, J.C. Clare, and G.M. Asher, "Doubly fed induction generator using back-to-back PWM converters and its application to variable-speed wind-energy generation," *IEE Proc.-Electr. Power Appl.*, vol. 143, no. 3, pp.231-241., May 1996.

# Genetic Identification of Parameters the Sandwich Piezoelectric Ceramic Transducers for Ultrasonic Systems

Paweł Fabijański[*], Ryszard Łagoda[†]

[*]Institute of Control and Industrial Electronics/Warsaw University of Technology, Warsaw, Poland
e-mail: *pawel@isep.pw.edu.pl*
[†] Institute of Control and Industrial Electronics/Warsaw University of Technology, Warsaw, Poland
e-mail: *lagoda@isep.pw.edu.pl*

*Abstract*— In this paper the genetic algorithm is used to optimization of parameters the equivalent electrical circuit of piezoelectric ceramic. Sandwich type piezoelectric ceramic transducer are the source of power ultrasounds when the vibration is activated by a voltage inverter. The output frequency of inverter most by equals the mechanical resonance frequency of the transducer and most by tuning with very high precise. In real circuits the mechanical resonant frequency is function of many parameters of piezoelectric material. The equivalent electrical circuit of transducer transforms this parameters in electrical value. This electrical circuit consist of connection in parallel: Co end RLC elements and this units are part of the oscillating circuit of inverter. The values of equivalent electrical circuit varies during the operation of transducer in function extending parameters, among others, the most important are temperature and time. In this situation to obtain the optimal value of the output voltage frequency of inverter most by know electrical parameters equivalent circuit of transducer. In the resolution of difficult problems to by effective the genetic algorithm.

*Keywords* — **AC/AC converter, generator excitation system, piezo actuators, transducer**

## I. INTRODUCTION

The Sandwich-type piezoelectric ceramic transducers are the most frequently applied sources of ultrasound.

They have the ability to radiate in an ultrasonic medium with maximum acoustic power when the vibration is activated by a current whose frequency equals the mechanical resonance frequency of the transducer. Typical units of the ultrasonic generators feeding these transducers operate at frequencies between 20 kHz and 100 kHz, with output power in the range of 20 W to 5 kW.

In resonant converters, the SANDWICH-type transducer units are a part of the oscillating circuit. This transducer made of piezoelectric ceramics PZT and is a combination of steel-ceramics-aluminum blocks connected by one or several screws.

## II. POWER CIRCUIT CONFIGURATION

Block diagram and the main circuit of the converter with piezoelectric ceramic transducer (Fig. 1) consists of: converter AC/DC, full-bridge inverter FBI, isolating transformer T, where $z_2/z_1=n_1$, $z_1/z_3=n_2$, special filter F, transducer PT, sensor of vibrations S, control and identification system.

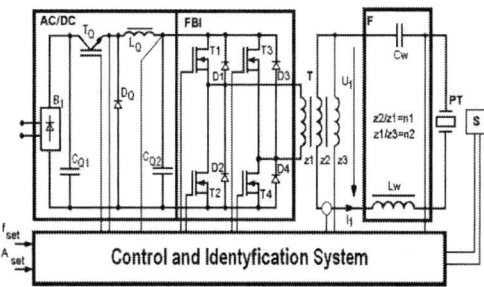

Fig. 1. The main circuit of the ultrasonic generator.

The control unit consists of two independently parts: in FBI inverter, is the frequency feedback control loop and in AC/DC converter, is the amplitude feed-back control loop. Signal $f_{set}$ make possible to set up manually frequency switching inverter FBI and signal $A_{set}$ establish amplitude ultrasonic oscillation.

## III. DIGITAL MODEL OF INVERTER-SPECIAL FILTER-TRANSDUCER GROUP SYSTEM CIRCUIT

To the identification of structure and parameters of piezoelectric ceramic transducer is used Genetic Algorithm. Simple genetic algorithm was applied and shown on Fig. 2.

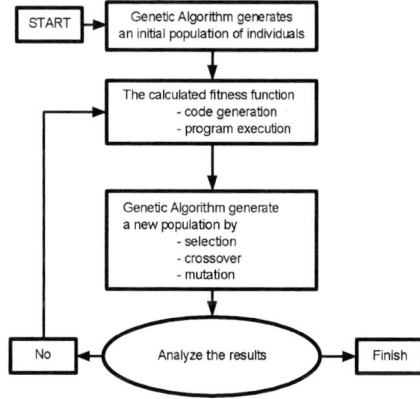

Fig.2. Simply Genetic Algorithm.

In resonant state the transducer are the oscillating element, for which equivalent electrical circuit consist of connection in parallel: $C_0$ and RLC, this simple non-linear circuit is numerically oppressive. To modification the basic Euler's interpolation algorithm is proposed classical genetic algorithm. Our genetic algorithm was implemented in DSP simulation programming language.

*//Program piezoelectric ceramic transducer*
*input { Circuit parameters [ R,L,C,u(t)];*

*Simulation parameters( $t_p$, $t_k$, Δ,υ(0));*

*Evolution parameters(Num,Size, NumM, Initial,NumG,stop);)*

*output{ t, u; $t_p$; $u_o$ =u(o);};*

*for (t, $t_k$, t++){ New function form;*

*start initial population;}*
*for(integer i=1,Num,i++){New population,*

*Select the best}*
where:
Genotype dimension: Num,
Size of the generation created by mutation: Size,
Number of the best selected genotypes for mutation: NumM,
Number of the best selected genotypes for crossing: NumC,
initial parameter mutation range: initial.

The linear equivalent circuit of the piezoelectric ceramic transducers with frequency close to resonant frequency is shown in Figure 3, where: $C_0$ - static capacity of the transducer, C - equivalent mechanical capacity, L - equivalent mechanical inductance, R - equivalent resistance,

$R = R_m + R_a$,
where: $R_m$ - equivalent mechanical loss resistance,
$R_a$ - equivalent acoustic resistance.

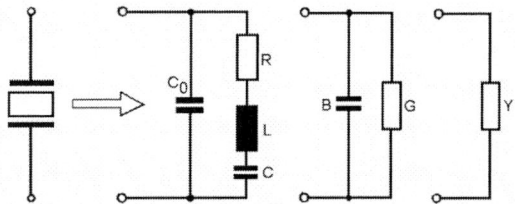

Fig. 3. The equivalent circuit of piezoelectric transducer.
Exemplary values Co = 4 nF, L = 246 mH, C = 182 pF, R = 392 •.

The susceptance B, conductance G, admittance Y of the transducers in frequency f function may be expressed by the following equations:

$$B(\omega) = \mathrm{Im}Y(\omega) = \omega C_0 + \frac{\omega C(1 - \omega^2 LC)}{\omega^2 C^2 R^2 + (1 - \omega^2 LC)^2} \quad (1)$$

$$G(\omega) = \mathrm{Re}\,Y(\omega) = \frac{\omega^2 C^2 R}{\omega^2 C^2 R^2 + (1 - \omega^2 LC)^2} \quad (2)$$

$$Y(\omega) = \sqrt{G^2 + B^2} \quad (3)$$

where $\omega = 2\pi f$.

The functions B(f), G(f) and Y(f) are given in Fig. 4. Maximum value of function G(f) is for the frequency $f_m$.

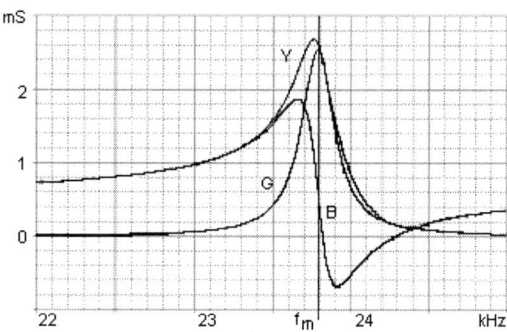

Fig. 4. The functions B(f), G(f) i Y(f).

The frequency $f_m$ of mechanical vibration may be calculated by equation:

$$f_m = \frac{1}{2\pi}\sqrt{LC} = 23{,}786\,kHz \quad (4)$$

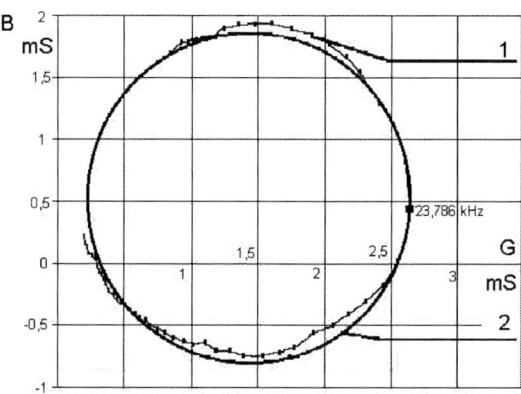

Fig. 5. Admittance characteristics of transducer:
1-real characteristic (in air)
2-teoretical characteristic

Amplitude of mechanical vibration is directly proportional to the RMS value of current $I_2$ in RLC elements (Fig. 6).

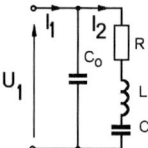

Fig. 6. Voltage and currents of the transducer.

Very important transmittances to the analysis of the fluency of the parameters transducers for the waveform current and voltage in inverter are:

$$G_1(s) = \frac{I_1(s)}{U_1(s)} = \frac{s^2 L + sR + \dfrac{1}{C}}{s^3 LC_0 + s^2 C_0 R + s\dfrac{C_0}{C} + 1} \quad (5)$$

and

$$G_2(s) = \frac{I_2(s)}{U_1(s)} = \frac{1}{s^2 L + sR + \dfrac{1}{C}} \quad (6)$$

Exemplary Special Filter-Transducer Group system circuit are presented in Fig. 7.

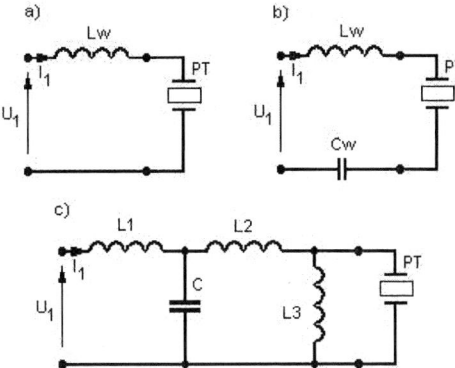

Fig. 7. Special Filter-Transducer Group system circuit.

This special filter make possible:
- stabilization amplitude of mechanical vibration independent the stage a load of the transducer.
- to extend the pass band Special Filter-Transducer Group system. In the case untuning the output frequency of inverter according to mechanical frequency of piezoelectric ceramic transducer, the range of frequency in area safety work the inverter is greater.
- to forming quasi sinusoidal of current and voltage waveform and elimination unfavorable charging impulse current in capacity $C_0$ which appear in the commutation time of the output voltage $V_1$.

For the circuits, which are presented In Fig.7 the best property's have the system of series connected elements $L_W$ and $C_W$ (Fig. 7b).

A optimum value inductance of choke Lw of special filter may by expressed by the equation:

$$Lw = \frac{4\pi^2 f_m^2}{C_0} \qquad (7)$$

and value of capacity $C_w$ is equals $Cw = C_0 \cdot n_2$.

The full block diagram of the ultrasonic generator are presented in Fig. 8.

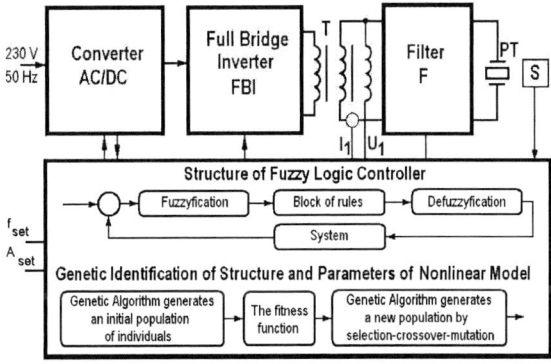

Fig. 8. The block diagram of the ultrasonic generator.

The presented system of filter-transducers make possible to supply the piezoelectric ceramic transducer

with the quasi-sinusoidal current and voltage and to determine a extreme values of parameters the elements of the converters FBI in case of tuning and untuning resonance frequency of converter and piezoelectric ceramic transducer ($f \approx f_m$).

The digital model of the inverter-special filter-transducer group system circuit, work in PSpice language. Exemplary results for output voltage frequency of inverter equal $f_m$ are presented in Fig. 9, where:
$U_1$ - output voltage waveform of inverter FBI,
$U_{PT}$, $I_{PT}$ – voltage and current waveform of the transducer.

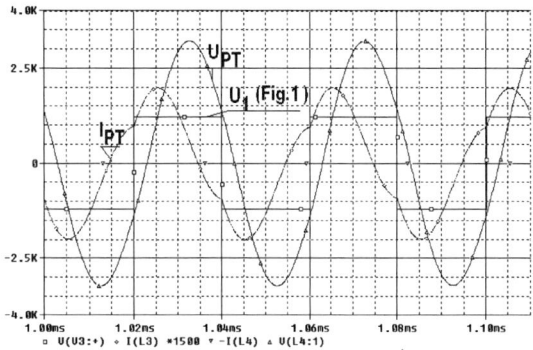

Fig. 9. Current and voltage waveform for $f \approx f_m$.

## IV. FREQUENCY FEED-BACK CONTROL LOOP

To obtain the maximum value of converter efficiency is necessary to assure the optimal frequencies of inverter. Changing the vibration state results in the variations of output power of inverter and in output resonant frequency we have four case (Fig. 10).

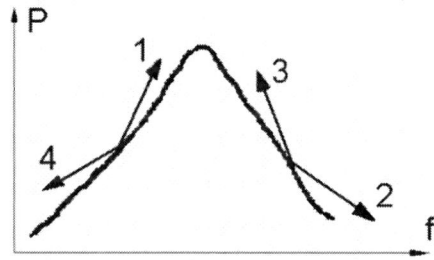

Fig. 10. Four case variations of change output power of inverter and resonant frequency.

To obtain the maximum value of converter efficiency, its important role of fuzzy logic control system. The logic control system define derived mark of signal amplitude proportional to output power and output resonant frequency and change adequate the output frequency of inverter. The relationship between the input and output variable contain Tab. 1 and fuzzy controller is shown in Fig. 11.

Table 1. Control rules

|   | derived of power | derived of frequency | rules of logic control |
|---|---|---|---|
| 1 | dP/dt > 0 | df/dt > 0 | frequency increase |
| 2 | dP/dt < 0 | df/dt > 0 | frequency decrease |
| 3 | dP/dt > 0 | df/dt < 0 | frequency decrease |
| 4 | dP/dt < 0 | df/dt < 0 | frequency increase |

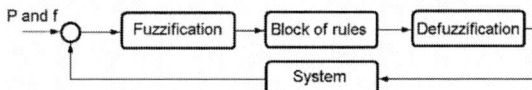

Fig. 11. Structure of Fuzzy Controller.

## V. EXPERIMENTAL RESERCH

The main circuit of the series-resonant converter with piezoelectric ceramic transducer and special filter system (Fig. 1) was modeling, build and tested.

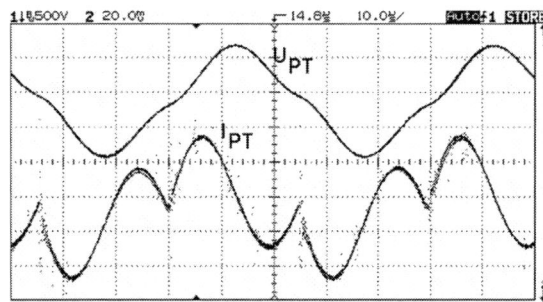

Fig. 12. Current and voltage waveform in case when $f < f_m$.

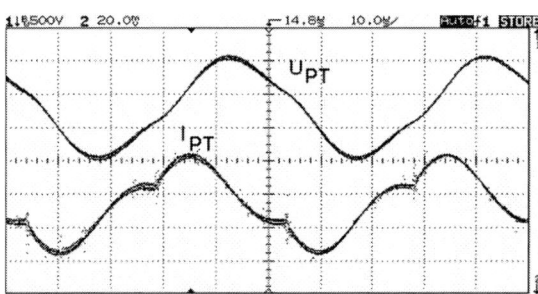

Fig. 13. Current and voltage waveform in the case when $f = f_m$.

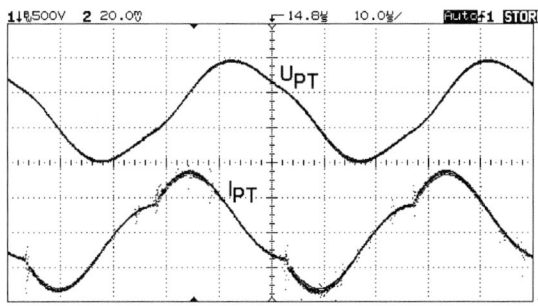

Fig. 14. Current and voltage waveform in the case when $f > f_m$.

The experimental result was obtain in the case of tuning and untuning of output frequency of inverter and the mechanical frequency of the transducer. Exemplary experimental current and voltage waveform of transducer in the case when $f = f_m$ are shown in the Fig. 13.

## VI. CONCLUSION

The presented control method make possible to supply the piezoelectric ceramic transducers group with the quasi-sinusoidal current and voltage waveform with self-tuning frequency to mechanical resonance. To the identification of structure and parameters of piezoelectric ceramic transducer was used Genetic Algorithm. The frequency feedback control system was to test in the case of tuning the generator frequency to the mechanical resonance frequency of transducer.

In the case of untuning the output frequency of inverter according to piezoelectric ceramic transducer mechanical frequency the current and voltage waveforms of piezoelectric transducer are non-sinusoidal. Especially current waveform is deformation.

The results of analysis of piezoelectric transducer and of the system of the resonance converter with control loop of frequency have been compared with experimental results in real piezoelectric transducer system and satisfactory results have been obtained.

### REFERENCES

[1] P. Fabija□ski: Characteristics of Self-tuning Output Frequency Series Resonant Converter, 16 – th IASTED International Conference MODELLING, IDENTYFICATION AND CONTROL, Insbruck - 1997.

[2] P. Fabija□ski, R. □agoda: Design of Series Resonant Converter with Sandwich – Type Piezoelectric Ceramic Transducer. 9 – th International Conference on Power Electronics and Motion Control EPE – PEMC Kosice 2000.

[3 L. Faa-Jeng, W. Rong-Jong, S. Kuo-Kai, L. Tsih-Ming: Recurrent Fuzzy Neutral Network Control for Piezoelectric Ceramic Linear Ultrasonic Motor Drive. IEEE Transactions on Ultrasonic, Ferroelectrics and Frequency Control, Vol. 48, No. 4/ 2001.

[4] P. Fabija□ski, R. □agoda: Control and Application of Series Resonant Converter in Technical Cleaning System. Proceedings of the IASTED, International Conference Control and Application, Cancun, Mexico, (2002)

[5] P. Wnuk: Genetic optimization of structure and parameters of TSK fuzzy models. Elektronika 8-9/2004

[6] P. Fabija□ski, R. □agoda: Theoretical & Experimental Analysis of Series Resonant Converter in Technical Cleaning System. Przegląd Elektrotechniczny 4/2004.

# The Impact of Higher-Order Harmonics on Tripping of Residual Current Devices

## Stanislaw Czapp

Gdansk University of Technology, Faculty of Electrical and Control Engineering, Gdansk, Poland, s.czapp@ely.pg.gda.pl

*Abstract*—Distorted line to earth voltage on the output terminals of the inverter of variable speed drives results in an earth fault current comprising harmonics. Higher-order harmonics in the earth fault current have a negative effect on tripping of residual current devices. In some cases sensitivity of residual current devices rapidly decreases and protection against electric shock is not effective. In the paper results of the laboratory testing of residual current devices sensitivity are described.

*Keywords*—Converter circuit, harmonics, protection device, safety.

## I. INTRODUCTION

### A. Residual Current Devices Construction and Operation

The most frequently used means of protection against indirect contact in low voltage systems is automatic disconnection of supply [1], since it is a simple and economically attractive way to achieve safety in an electrical installation. As a protective device, residual current devices (RCDs) are commonly used. In certain conditions using the RCDs is necessary [2].

A residual current device usually comprises coils on a magnetic circuit to carry the phase (three-phase) and neutral current in opposing directions (Fig. 1).

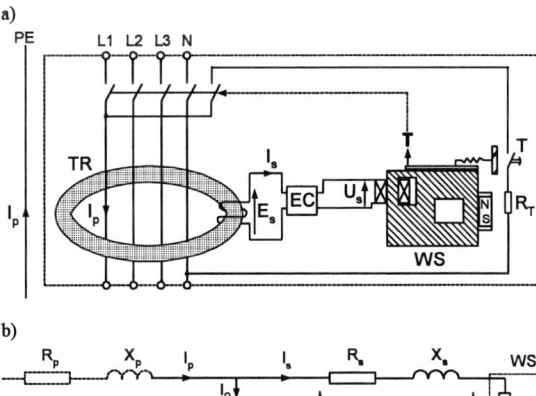

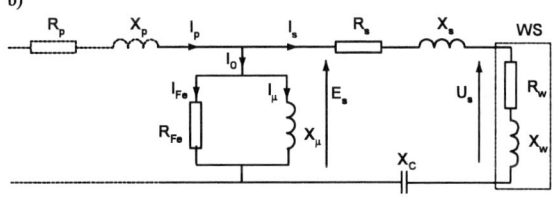

Fig. 1. Residual current device: a) simplified diagram, b) equivalent circuit. Main symbols: *TR* – current transformer, *WS* – electromechanical relay, *N-S* – permanent magnet, *T* – test button, *EC* – additional electronic components, $I_p$ – primary (earth fault) current, $I_s$ – secondary current, $E_s$ – induced secondary voltage, $U_s$ – electromechanical relay voltage

In the balanced conditions no magnetic flux is generated, but if an earth fault occurs in the system, the phase and neutral current imbalance induces an electromagnetic force in the secondary circuit, openning the main circuit.

In the nowadays developed construction of residual current devices, the secondary circuit contains electromechanical relay with a permanent magnet. The permanent magnet keeps the moving armature on the yoke and the spring pulls the moving armature in the opposite directions. In a balanced condition the magnetic circuit of the electromechanical relay is closed. The earth fault current (residual current), transformed by the current transformer, in the one half-wave amplifies the magnetic flux of the permanent magnet but in the second half-wave that flux in reduced. If the residual current reaches a predetermined level, the magnetic flux derived from that current is high enough to reduce the magnetic flux of the permanent magnet to the level in which the spring is able to pull out the moving armature of the electromechanical relay and RCD opens the main circuit. In a graphical way, the relationships among elements described above are presented in Fig. 2.

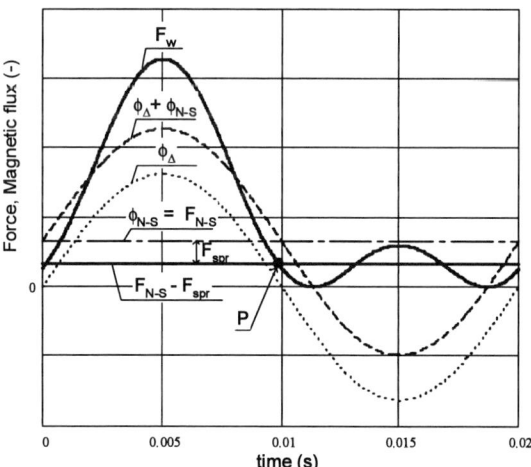

Fig. 2. Magnetic flux and magnetic forces in the electromechanical relay of residual current devices for 50 Hz alternating earth fault current
Symbols:

$F_w$ – resultant force,
$\phi_\Delta$ – magnetic flux from the earth fault current,
$\phi_{N-S}$ – magnetic flux from the permanent magnet,
$F_{N-S}$ – magnetic force from the permanent magnet,
$F_{spr}$ – magnetic force from the spring,
$P$ – point, at which the moving armature of the electromechanical relay moves and RCD opens

---

**978-1-4244-1741-4/08/$25.00 ©2008 IEEE**

Basically residual current devices do not operate if the residual current is equal to or less than $0.5 \cdot I_{\Delta n}$, while their operation is required if the residual current equal to or more than $I_{\Delta n}$ occurs. There are three different types of RCDs in terms of the sensivity to the earth fault current shape:
– AC– for alternating earth fault current (50/60 Hz),
– A – for alternating and pulsating direct earth fault current,
– B – for alternating earth fault current, pulsating direct earth fault current and smooth direct earth fault current.

Manufacturers of the residual current devices are obliged to test their product according to the standards [3, 4]. However, program of the test does not cover all residual currents which can occur in practice. RCDs are not tested under a variable frequency residual current and current containing harmonics.

### B. Earth Fault Current in Circuits with Frequency Converters

Nowadays frequency converters are commonly used to control the speed of squirrel cage motors. The first part, starting from the supply side, is a rectifier (Fig. 3). In the second part (intermediate circuit), pulsating DC voltage produced by the rectifier is filtered. The last main part of the frequency converter is inverter which uses DC current or voltage from the intermediate circuit to produce AC current or voltage of a desired frequency.

Taking the above into account, the earth fault current may be from low to high frequency [5]. The output earth fault current is also distorted and comprises harmonics, whose order mainly depends on the PWM (Pulse Width Modulation) frequency. The earth fault current comprises a low frequency component which depends on the desired motor speed, a constant 150 Hz component as well as a component at the PWM frequency and its multiple and also its interharmonics. The amplitude of the current of motor frequency and the PWM current changes as a function of the motor speed and frequency. For the 50 Hz operating frequency the amplitude of 50 Hz component exceeds the amplitude of PWM component but for the low motor frequency the participation of the above mentioned components varies in the opposite ways [6, 7]. For very low motor frequency the amplitude of PWM component significantly exeeds the amplitude of current of motor frequency and the 150 Hz component.

Fig. 4 presents earth fault current simulation and Fig. 5 this current harmonics analysis. Both were performed using TCad software [8]. Simulation was performed for motor freqency equal to 50 Hz and PWM frequency equal to 1 kHz or 3 kHz. In the earth fault current following frequencies dominate:

– 50 Hz, 150 Hz, 1 kHz and its harmonic and interharmonics – for the PWM frequency equal to 1 kHz,
– 50 Hz, 150 Hz, 3 kHz and its harmonic and interharmonics – for the PWM frequency equal to 3 kHz.

Fig. 6 presents earth fault current shape and current harmonics analysis in the case of earth fault (under laboratory condition) in the motor teminals. The earth fault was performed through the 1000 Ohm resistor. This resistor reflected typical human body resistance. The test was carried out for motor frequency 50 Hz, 25 Hz, 1 Hz and the inverter PWM frequency 3 kHz.

Oscillographs in Fig. 6 show that the lower motor frequency the higher is participation of the higher-order harmonics. For the very low motor frequency only higher-order harmonics dominate (PWM frequency, its multiple and interharmonics).

The earth fault current waveform shape influences the residual current device tripping [9, 10, 11, 12]. RCDs may operate improperly if the earth fault current is significantly distorted and higher-order harmonics dominate. The following paragraphs present results of the theoretical analysis and laboratory tests of the residual current devices operation. They prove that in some cases for many residual current devices the real tripping current increases to an unacceptable level and protection against electric shock is not effective.

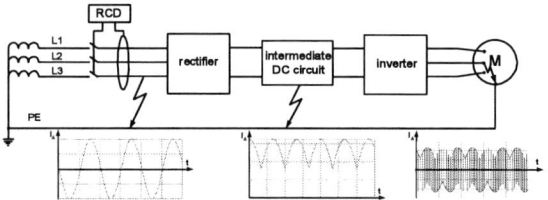

Fig. 3. Earth fault current waveform shape in the circuit with frequency converter [5]

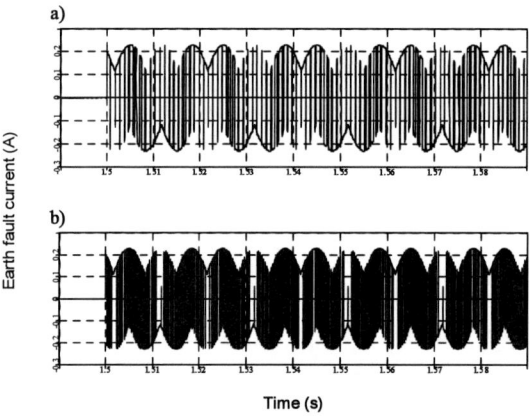

Fig. 4. Earth fault current in case of fault (through the 1000 Ohm resistor) in the output terminals of the frequency converter (TCad simulation). Motor frequency 50 Hz, PWM frequency a) 1 kHz, b) 3 kHz

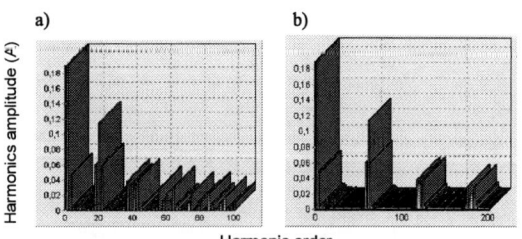

Fig. 5. Harmonic analysis of the earth fault currents presented in Fig. 4. Motor frequency 50 Hz, PWM frequency: a) 1 kHz, b) 3 kHz

a)

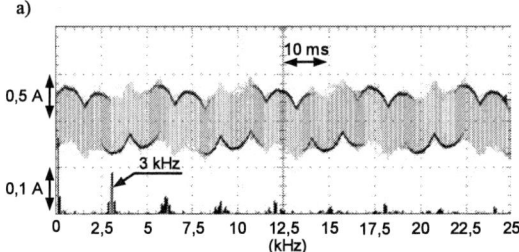

b)

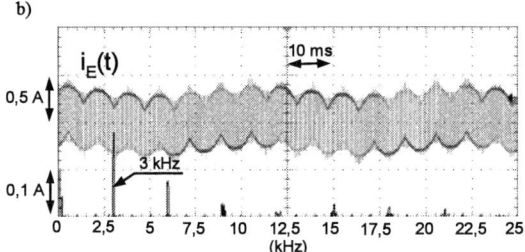

c)

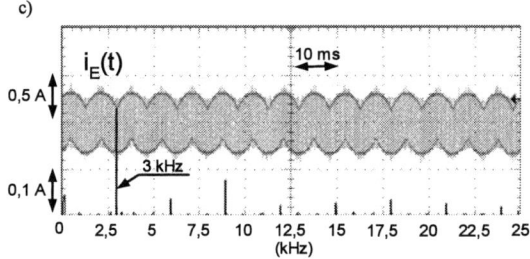

Fig. 6. Earth fault current in case of fault in the output terminals of the frequency converter and current harmonics analysis (laboratory test), PWM frequency 3 kHz, motor frequency: a) 50 Hz, b) 25 Hz, c) 1 Hz

## II. THEORETICAL ANALYSIS

### A. General Information

In the theoretical analysis it is assumed that earth fault current comprises the fundamental (50 Hz) and one of the odd harmonics. It is also assumed that earth fault current and secondary current of the current transformer have the same waveform shape. Analysis is performed according to the criterion that the "rms" value of the sinusoidal current and distorted current is constant. Current which flows in the coil of the electromechnical relay is described by the following equation:

$$i_{s\Delta}(t) = \sqrt{2} \cdot I_{s\Delta} \cdot [\sin(\omega t + \alpha_1) + A_3 \sin(3\omega t + \alpha_3) + \qquad (1)$$
$$+ A_5 \sin(5\omega t + \alpha_5) + ... + A_n \sin(n\omega t + \alpha_n)]$$

where:

$A_3, A_5, ... A_n$ – amplitude-to-fundamental of 3rd, 5th, ..., $n$-th harmonic,

$\alpha_3, \alpha_5, ... \alpha_n$ – phase angle of 3rd, 5th, ..., $n$-th harmonic.

The following theoretical analysis does not take into account the fact that the secondary circuit winding and electromechanical relay impedance vary with the frequency and harmonics. Only impact of the phase angle of particular harmonic on the electromechanical relay operation is considered.

### B. Comparision of the Impact of the 3rd or 23rd Harmonic

The impact of 3rd or 23rd harmonic on the electromechanical relay operation is considered for following amplitude-to-fundamental of harmonic: $A_3 = 0.1$ and $A_{23} = 0.1$. The 3rd harmonic (or 23rd harmonic) is in phase with the fundamental $\alpha_3 = 0°$ or the phase angle is 180° ($\alpha_3 = 180°$).

Symbols in Fig. 7 and 8 indicate:

$F_{w1}$ – resultant force for sinusoidal wave (dashed line),

$F_{w1+n}$ – resultant force for fundamental and $n$-th harmonic (solid line),

$F_{N-S}$ – magnetic force from the permanent magnet,

$F_{spr}$ – magnetic force from the spring,

$P$ – point, at which the moving armature of the electromechanical relay moves and RCD opens.

Horizontal axis in Fig. 7 and 8 is scaled in relative units.

The analysis results presented in Fig. 7 show that the phase angle of 3rd harmonic influences electromechanical relay and in consequence the RCDs operation. The 3rd harmonic may decrease (Fig. 7a) or increase (Fig. 7b) the sensitivity of RCDs. In Fig. 7a there is no point $P$ because the $F_{w1+3}$ line does not cross the $(F_{N-S} - F_{spr})$ line. In order to cross that line, the earth fault current must have a higher value. Waveform shape presented in Fig. 7b increases the sensitivity of RCDs. In such a case RCDs trip out for the "rms" earth fault current value which is lower than for the pure sinusoidal current.

a)

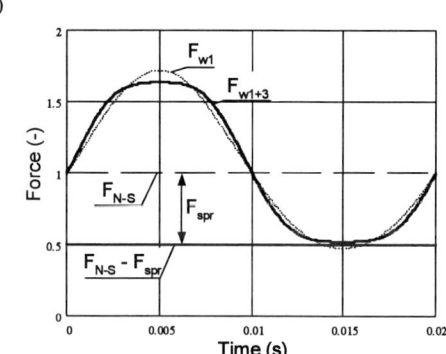

b)

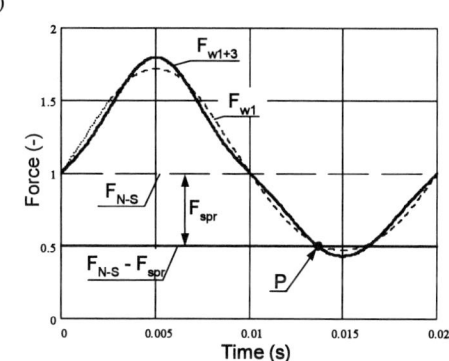

Fig. 7. Magnetic forces in the electromechanical relay of residual current devices for earth fault current comprising fundamental frequency and 3rd harmonic: $A_3 = 0.1$; a) $\alpha_3 = 0°$, b) $\alpha_3 = 180°$

a)

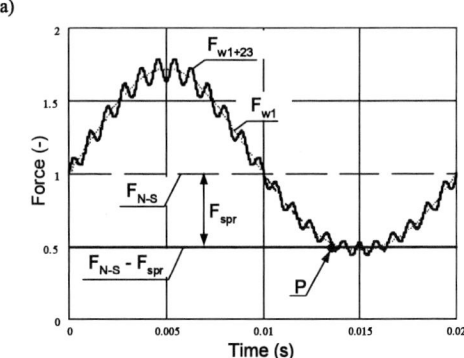

b)

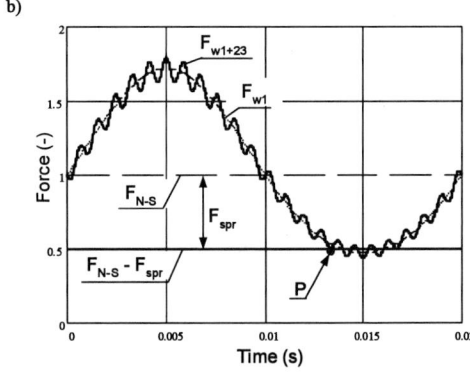

Fig. 8. Magnetic forces in the electromechanical relay of residual current devices for earth fault current comprising fundamental frequency and 23rd harmonic: $A_{23} = 0.1$: a) $\alpha_{23} = 0°$, b) $\alpha_{23} = 180°$

If the earth fault current contains harmonic of a considerably higher-order, for example 23rd, the influence of phase angle of that harmonic on RCDs operation is negligible. Fig. 8 presents the analysis results. In both cases ($\alpha_{23} = 0°$, $\alpha_{23} = 180°$) the electromechanical relay trips out at the same value of earth fault current. Unfortunately, if a high-order harmonic occurs and its participation is also high, the RCDs sensitivity may rapidly decrease. A high-order harmonic – similar to the high frequency earth fault current – impacts the current transformer behaviour and impedance of the secondary circuit winding and electromechanical relay. Laboratory tests which were carried out by the author and results presented in [9] showed that histeresis loop of some types of current transformers is significantly wider for higher frequency. It is a negative effect, because to achieve the same level of induction in the current transformer core a higher value of earth fault current is necessary. Also the higher earth fault current frequency the higher is the secondary circuit winding and electromechanical impedance. Higher impedance of these elements makes the secondary current decrease.

## III. RESULTS OF THE LABORATORY TESTS

In order to check the influence of harmonics on tripping of the residual current devices, over twenty RCDs with rated operating residual current $I_{\Delta n} = 30$, 100 and 300 mA were tested. They were two-pole, four-pole, type AC, type A and type B, without intentional time

delay, short-time-delayed and time-delayed (selective) devices. Fig. 9, 10 and 11 show chosen results of the test which was carried out under laboratory conditions.

During the test the programmable current source produced earth current comprising the fundamental (50 Hz) and one harmonic. The test was carried out for a few selected harmonics within the range of from 3rd to 50th. The "rms" value of earth fault current rose and real operating current was recorded. The phase angle of harmonic of earth fault current varied with 45° step. Participation of the particular harmonics was the following: 0.10; 0.50; 1.00, 2.00; (in Fig. 9, 10 and 11 it is marked, respectively, 10%, 50%, 100%, 200%; "sin 50Hz" describes pure sinusoidal earth fault current).

The laboratory test proves results of the theoretical analysis. If the earth fault current comprises a low-order harmonic the real tripping current may be higher or lower than for the pure sinusiodal current. It depends on the harmonic phase angle. For a higher-order harmonic and its high participation, because of the magnetic core and electromechanical relay properities, the real tripping current of RCDs may change many times.

a)

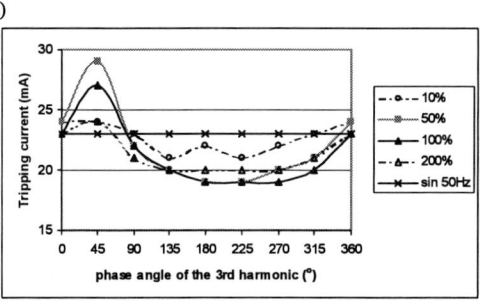

b)

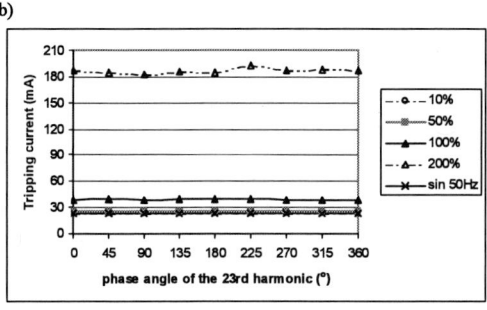

c)

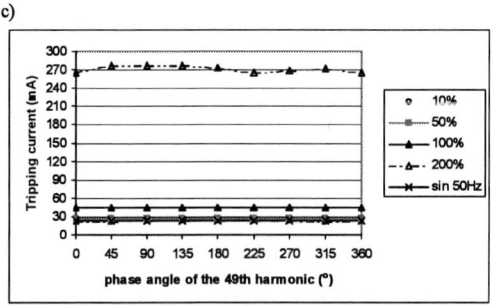

Fig. 9. Real tripping current of 30 mA type AC RCD. Earth fault current comprises harmonics: a) 1st+3rd, b) 1st+23rd, c) 1st+49th. The percentage of harmonic-to-fundamental: 10%, 50%, 100%, 200%

a)

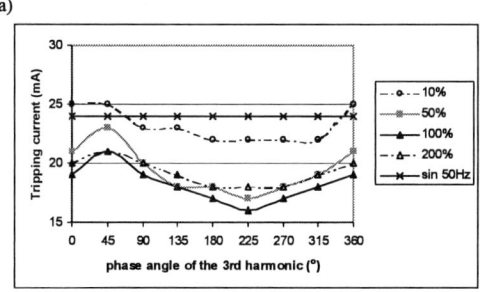

b)

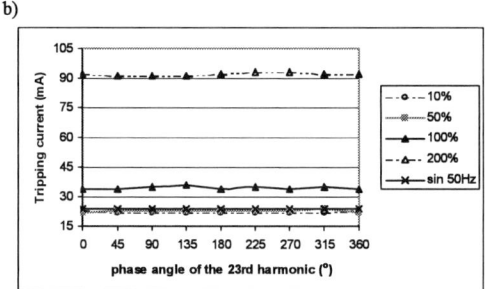

c)

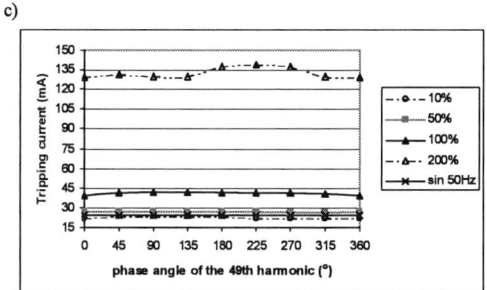

Fig. 10. Real tripping current of 30 mA type A RCD. Earth fault current comprises harmonics: a) 1st+3rd, b) 1st+23rd, c) 1st+49th. The percentage of harmonic-to-fundamental: 10%, 50%, 100%, 200%

a)

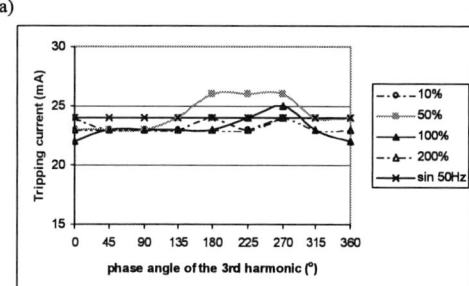

b)

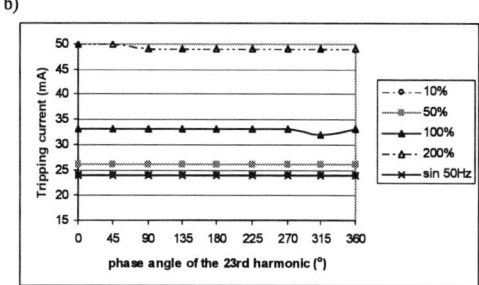

c)

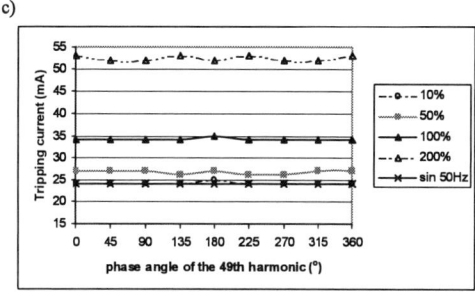

Fig. 11. Real tripping current of 30 mA type B RCD. Earth fault current comprises harmonics: a) 1st+3rd, b) 1st+23rd, c) 1st+49th. The percentage of harmonic-to-fundamental: 10%, 50%, 100%, 200%

The presented example (Fig. 9c) shows that tripping current of the RCD increases twelve times (from 23 mA to 276 mA) if the earth fault current contains 49th harmonic with 200% participation. That decreasing sensitivity of the residual current devices makes protection against electric shock not effective. In many tested residual current devices the real tripping current significantly exceeded the rated value for high participation of the 23rd or 49th harmonics, even for the most technologically advanced type B residual current devices.

Tripping current of the residual current devices was also checked for the earth fault current comprising several harmonics. Using laboratory generator this part of the test was performed for three types of the currents. The first type of the current comprises harmonics which dominate in the earth fault current in case of fault in the output terminals of frequency converters for motor frequency equal to 50 Hz. This type of test current is presented in Fig. 12 and comprises following frequencies:

– 50 Hz – motor frequency,

– 150 Hz – 3rd harmonic in respect to the motor frequency and frequency of voltage of the rectifier neutral point,

– 1000 Hz – inverter PWM frequency,

– 900 and 1100 Hz – main interharmonics of PWM frequency.

The second type of the test current reflects the earth fault current in case of earth fault in the output terminals of frequency converters for motor frequency equal to 25 Hz and is presented in Fig. 13. This current comprises following frequencies:

– 50 Hz – motor frequency,

– 75 Hz – 3rd harmonic in respect to the motor frequency,

– 150 Hz – frequency of voltage of the rectifier neutral point,

– 1000 Hz – inverter PWM frequency.

The third type of the test current reflects the earth fault current for motor frequency equal to 1 Hz (Fig. 14). This current comprises following frequencies:

– 150 Hz – voltage of the rectifier neutral point,

– 1000 Hz – inverter PWM frequency.

The frequency equal to 1 Hz is avoided since its participation with respect to the other components is negligible.

Captions of the Fig. 12, 13 and 14 contain information about the participation of all the current components. Reffered participation is adopted for frequency 50 Hz (100 %) in the waveform marked "50+PWM". Every frequency participation is compared to this value.

The aim of this test is to find the tendency in residual current devices tripping in case of earth fault in the output circuit of the frequency converter for various motor speed.

Results of the laboratory test of 30 mA RCDs are presented in Fig 15, 16 and 17. Fig. 15 shows real tripping current of chosen residual current devices type AC. Real tripping current of the RCD1 and RCD3 exceed $I_{\Delta n}$ even for respectively low distorted earth fault current ("50Hz+PWM"). For waveform "25+PWM" real tripping current increased over two times in respect to pure sinusoidal waveform. Very dangerous situation occurred for waveform "1Hz+PWM". The RCD2 RCD3 RCD4 do not tripped out even for current equal to 5 A "rms".

Similar effect was observed for the tested type A RCD5, RCD6, RCD7 (Fig. 16). For waveform "1Hz+PWM" which mainly contains PWM component equal to 1 kHz the tested type A residual current devices do not tripped out for 5 A current.

The tested type B residual current devices tripped out for all the test currents but the RCD9 real tripping current significantly increased if the test current mainly contained high frequency (PWM) component.

In this part of the test respectively low PWM frequency (1 kHz) was simulated. If the PWM frequency changes to a higher value it is expected that the negative effect of PWM component on tripping of residual current devices increases.

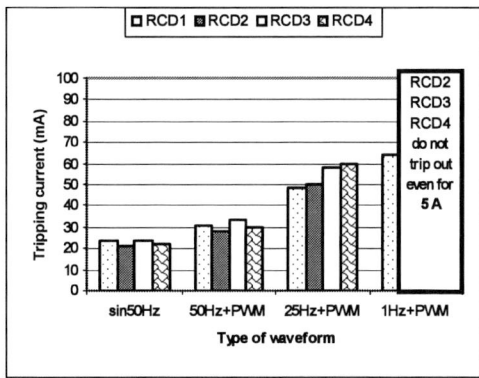

Fig. 15. Real tripping current of 30 mA type AC RCDs

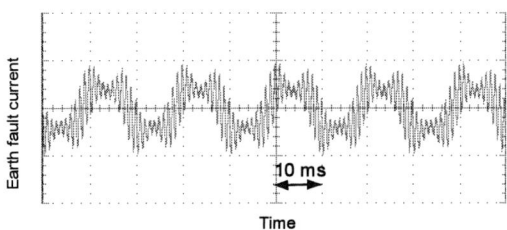

Fig. 12. The test earth fault current (marked "50Hz+PWM") comprising frequencies: 50 Hz – amplitude 100 %, 150 Hz – amplitude 25 %, 1000 Hz – amplitude 70 %, 900 and 1100 Hz – amplitude 25 %

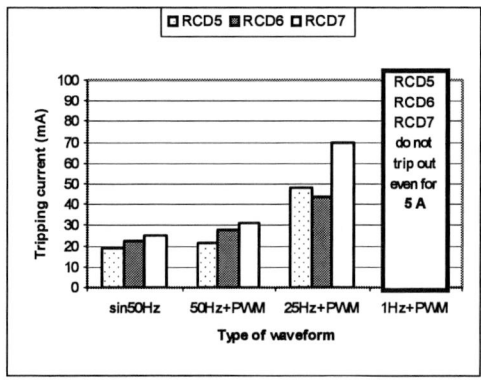

Fig. 16. Real tripping current of 30 mA type A RCDs

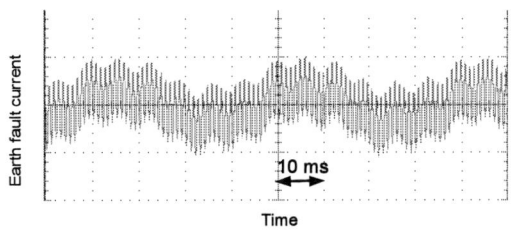

Fig. 13. The test earth fault current (marked "25Hz+PWM") comprising frequencies: 25 Hz – amplitude 60 %, 75 Hz – amplitude 10 %, 150 Hz – amplitude 25 %, 1000 Hz – amplitude 110 %

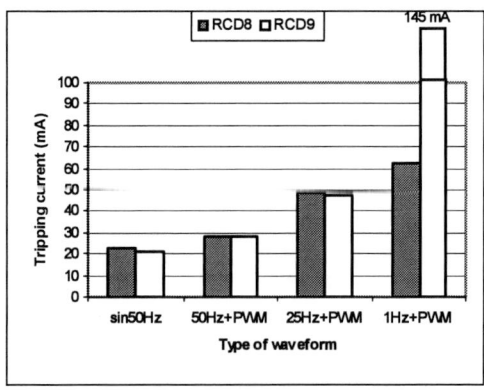

Fig. 17. Real tripping current of 30 mA type B RCDs

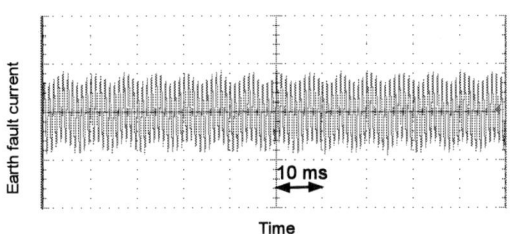

Fig. 14. The test earth fault current (marked "1Hz+PWM") comprising frequencies: 150 Hz – amplitude 25 %, 1000 Hz – amplitude 150 %

## IV. CONCLUSION

In order to achieve effectiveness of protection against electric shock by automatic disconnection of supply in low voltage systems proper coordination between tripping current of the protective device and earth path impedance shall be fulfilled. The above presented results of the laboratory tests indicate that harmonics may change the residual current device tripping current. A higher-order harmonic basically increases tripping current in respect to sinusoidal earth fault current. If the participation of that harmonic is very high, tripping currrent may increase to an unacceptable level and protection against electric shock is not effective since automatic disconnection of supply does not occur in a specified time. Under the extreme condition residual current devices may not operate at all.

Residual current devices are also applied as additional protection. Then residual current devices shall operate in case of failure of other protective measures or carelessness of users. In such a situation all fault current is a body current. If the residual current device tripping current moves towards a higher value or residual current device does not trip at all, as a result of earth fault current distortion, ventricular fibrillation and even negative thermal effects may occur.

For proper selection of the residual current devices in circuits with variable frequency and distorted earth fault current, real operational characteristics of a particular RCD should be known. Manufacturers are not obliged to deliver such characteristics but it is necessary to obtain them in a specialized laboratory. Thus, modification of the international standards concerning the operation and tests of residual current devices is necessary. Obligatory test of the selected types of residual current devices should cover wide frequency and distorted residual currents.

Actually the author develops his work towards the construction of the voltage independent residual current device which enables to eliminate negative effect of higher-order harmonics.

## ACKNOWLEDGMENT

This paper is a result of research project financed by the Ministry of Science within the 2005-2008 programme.

## REFERENCES

[1] IEC 60364-4-41 *Low-voltage electrical installations – Part 4-41: Protection for safety – Protection against electric shock.*

[2] IEC 60364-7-7xx *Electrical installations of buildings. Requirements for special installation or locations.*

[3] IEC60755 *General requirements for residual current operated protective devices.*

[4] EN 61008-1 *Residual current operated circuit breakers without integral overcurrent protection for household and similar uses (RCCB's). Part 1. General rules.*

[5] L. Nowak, "Modelling of leakage and earth fault current in circuits with frequency converters". Master thesis 2007 (in polish). Gdansk University of Technology. Faculty of Electrical and Control Engineering, (in Polish).

[6] J. Schoneck, Y. Nebon, "LV protection devices and variable speed drives", *Cahier technique no. 204.* Schneider Electric 2002.

[7] G. Grünebast, "Allstromsensitive Fehlerstromschutzeinrichtungen. Teil 2: Vorschriftsmässiger Einsatz", *Elektropraktiker* vol. 2 (62), 2008.

[8] L. Beldycki, "Anaysis of the effectiveness of protection against electric shock in circuits with frequency converters". (Master thesis in progress). Gdansk University of Technology. Faculty of Electrical and Control Engineering, (in Polish).

[9] S. Czapp, "Analysis of the residual current devices independent trip for the residual current frequency higher than rated value". XIII International Scientific Conference „Present-Day Problems of Power Engineering APE'07", Gdansk–Jurata, vol. 4, p. 283-290,. 13-15 June 2007 (in Polish).

[10] S. Czapp, The Impact of DC Earth Fault Current Shape on Tripping of Residual Current Devices. *Elektronika ir Elektrotechnika,* vol. 4 (84), pp. 9-12, 2008.

[11] V. Cohen, "Why electronic and not electromechanical ELCBs?", IEEE AFRICON 4th vol. 2, pp 715 – 719, 24-27 Sept. 1996.

[12] T.M. Lee, T.W. Chan, "The effects of harmonics on the operational characteristics of residual current circuit breakers". Int. Conf. on Energy Management and Power Delivery, Proc. of EMPD'9, vol. 2, pp. 715-719, Nov. 1995.

# Estimation of the Untapped Regenerative Braking Energy in Urban Electric Transportation Network

Leonards Latkovskis, Linards Grigans

Institute of Physical Energetics, Aizkraukles 21, Riga, Latvia,
e-mail: *leonsl@edi.lv; linardsgrigans@gmail.com*

*Abstract* — The paper presents an estimation of the reserves of untapped regenerative braking energy in the power supply network of urban electric transport. The estimation is made by a method based on probabilistic principles, transport timetable and measured power diagrams. The amount of unused regenerative energy is calculated for two substations in the electric transport network in Riga, with power measurement technique described for *T3A* tramcars. The results of the work allow calculation of economical aspects of the regenerative braking energy utilization not only for the current situation but also for future, taking into account growth in the transport fleet and changes in its organization.

*Keywords*–regenerative power, estimation technique, rail vehicle.

## I. INTRODUCTION

The public transport company is the major consumer of electric energy in the Riga city. With renovation of *Tatra T3* trams and purchase of new modern trolleybuses a new type of electric transport has been introduced in Riga. These trams and trolleybuses are capable of recuperating the obtained kinetic energy back to the overhead power supply line thus allowing up to 40% of the consumed energy to be saved [1]. However, it is impossible to return this energy back to the grid because none of the substations is equipped with a reversible rectifier and the regenerative braking energy is dissipated in braking rheostats. This does not happen only when there is another electric vehicle consuming energy in the same section of a power supply line – otherwise the braking energy is converted into heat and is lost. This wasted energy is the untapped energy reserve which could be utilized.

Estimation of this energy is not as simple as it seems at first sight. Direct measurement of the energy dissipated in the braking rheostats is not the best choice due to the following:

- vehicles with a slightly lower threshold voltage setting can dissipate in their braking rheostats the energy from other vehicles with a higher threshold setting. To avoid obtaining false results it is necessary to install energy loggers on every vehicle. Extraction of the measurement data obtained during a long time for a particular section of the power supply network would be a complicated technical task;

- braking rheostats are also used for switching rails. This could be a source of errors at estimation of the braking energy;

- direct measurements of the dissipated energy would give suitable estimation of untapped energy for the existing situation, however they could be inapplicable in new situations, e.g., if a new type of electric vehicle is introduced or the traffic organization is changed.

The latter consideration is especially important, since the transport fleet in Riga is persistently being renewed by replacing obsolete vehicles with new ones that are capable of recuperating the braking energy back to the overhead line with the maximum efficiency. Fig. 1 shows the rate of increasing the number of vehicles capable of recuperating braking energy in the Riga city for the last 3 years (as percentage of the total number of vehicles).

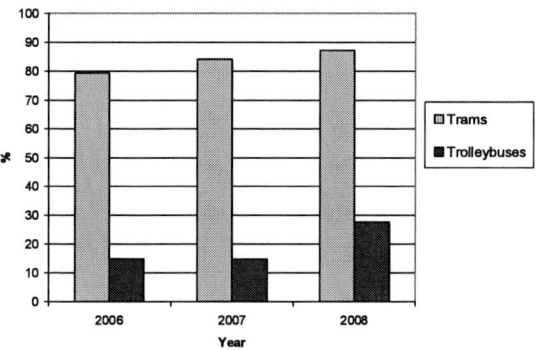

Fig. 1. Vehicles capable of recuperating braking energy in Riga over last 3 years (as percentage of the total number of vehicles)

Taking into account the development plans of Riga's public transport, it can be estimated that within the next 2 years all trams will be recuperating. For trolleybuses, such a progress is expected in the 10-12 years' time. As the number of recuperating vehicles increases every year, the untapped energy will be also increasing.

The purpose of this work is to estimate the amount of untapped energy, applying the method described in [2, 3]. The concept of the method is to measure the power consumption of an electric vehicle and then, according to the electric transport timetable, calculate the unused energy for a particular section of power supply.

## II. ASPECTS CONSIDERING UNTAPPED ENERGY SAVING SOLUTIONS

There are three basic solutions for saving the untapped braking energy:

- modification of substations by replacing old rectifiers with reversible ones;

- installation of onboard energy storage devices on the electric vehicles;
- installation of stationary energy storages at substations or near the optimal connection points of the overhead power supply line [4].

For energy storage, supercapacitors are considered the most promising means due to their large power capability [5,6]; however this solution has to be evaluated in comparison with the flywheel technology [6,7].

Equipment of substations with reversible rectifiers has several drawbacks:

- the necessity to modify substations, including replacement of power transformers or installation of additional ones;
- simple reversible thyristor rectifiers have a low power factor and distorted line current, while transistor rectifiers with a sinusoidal current waveform are rather complicated and expensive;
- reversible rectifiers are not capable of leveling out the power consumed from the grid; on the contrary, with the opposite power flow they even increase power fluctuations in the grid;
- failure of a reversible rectifier causes interruption of supply in the overhead power network.

The most effective way of using the recuperated energy with the least losses is the installation of onboard energy storage devices in electric vehicles. Several studies have been conducted regarding this issue, e.g. [8,9]. This solution, being most efficient from the viewpoint of energy use, is however the most expensive; this could be considered when designing new electric vehicles, but to install them in already running vehicles is problematic. For example, a 300kW/900kJ energy storage device needed to store braking energy for a *T3A* tramcar weighs up to 1.2 tons and occupies ~ 1m$^3$ of space. This tramcar has no space under the floor for this device, while its installation in the passenger compartment or on the roof of a tramcar would entail its overhaul and, consequently, large expenses. Such a rise in the cost of electrical equipment would not pay back in the remaining post-renovation service time of the vehicle.

In the Riga city, the most optimal solution is to install stationary energy storage devices which would be connected to the overhead power supply network. Such devices for an overhead section fed from one substation can be installed at several optimally chosen connection points so as to provide the least energy losses, though the minimum capital investments would be required for installation at substations. An additional energy storage device could be installed (for instance, at the end-of-line for a longer section) in the second stage if its usefulness is economically justified. At a substation such energy storage is able to store braking energy of several vehicles operating within the section of overhead line fed from the same substation in the meantime returning this energy back to the overhead line. During peak hours it can perform the opposite function – reduction of power fluctuations in the grid by returning energy at the maximum power consumption time but in the meanwhile accumulating it from the power supply grid, thus leveling out the power consumption of the substation.

The proposed solution has the following advantages in comparison with that using reversible rectifiers:

- no need to modify substations,
- installed storage system does not affect reliability of power supply – in the case of its failure the overhead line power supply is not cut off,
- a higher power factor for the substation and smaller losses in transformers,
- possibility to use the energy storage system for reducing the peak power demand charges,
- whereas in comparison with the onboard energy storage installation:
- considerably better operational conditions and therefore longer operation time,
- the energy storage equipment built for stationary applications is cheaper as compared with a transport version of the equal power capability,
- the energy storage is more loaded and has shorter idle time,
- the total costs are considerably lower.

Further research focuses on the scenario where an energy storage device is installed at a substation. Taking into account that there are considerably more trams than trolleybuses capable of recuperating energy, two substations – № 8 and № 10 supplying overhead sections of tram lines № 11 and № 6, respectively, were selected for the research.

## III. POWER CONSUMPTION MEASUREMENT FOR *T3A* TRAMCAR

The simplest way to record power consumption of the tramcar is by utilizing already onboard installed voltage and current sensors. A simplified power circuit diagram of the tramcar T3A in the running mode a) and the braking mode b) is depicted in Fig. 2. The figure shows also the location of installed measuring sensors. The following voltages and currents are measured for the tramcar's control:

$V_{line}$ – line voltage,

$V_C$ – filter capacitor voltage,

$I_{aux}$ – current of auxiliary consumers,

$I_{line}$ – line current,

$I_{M1}, I_{M2}$ – currents of the traction motors.

Besides, measurements have been performed of currents IQ1, IQ2 of the traction converter's main IGBT Q1 and braking IGBT Q2, respectively. The currents are measured for protection purposes. IGBT Q2 operates only in the braking mode when capacitor voltage VC reaches 780V. This happens when other consumers cannot use all braking energy – its excess is dissipated in rheostat Rbr1.

When a tram is braking at a high speed, the counter EMF of motors exceeds the supply voltage. To make the braking process controllable, an additional braking rheostat (Rbr2) is included into the power circuit, which is short-circuited by contactor Kbr when the speed falls below 30km/h. The portion of braking energy dissipated in Rbr2 is technological, and for this type of drive may not be considered as untapped.

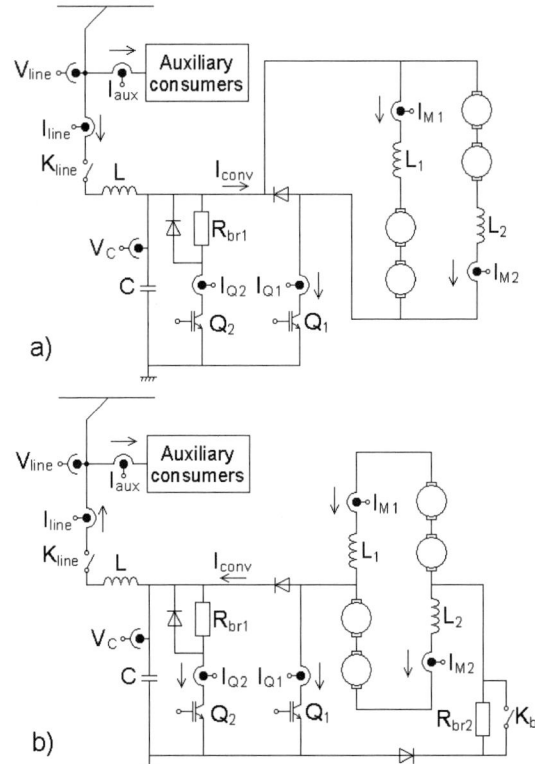

**a)**

**b)**

Fig. 2. A simplified power circuit diagram of the *T3A* tramcar, a) in the running mode, b) in the braking mode of operation.

Power $P_{tr}$ of the tramcar is the sum of traction power $P_{conv}$ of the converter and auxiliary (heating, lighting, ventilation, control system) power $P_{aux}$:

$$P_{tr} = P_{conv} + P_{aux} = V_C I_{conv,av} + V_{line} I_{aux}, \qquad (1)$$

where $I_{conv,av}$ is the average value of the current $I_{conv}$ over a switching period.

Unfortunately, current $I_{conv}$ is not measured, but currents $I_{M1}$, $I_{M2}$ and $I_{Q1}$ cannot be used for its calculation, because the relation between these currents depends on the operation mode, i.e. in the running mode

$$I_{conv,av} = (I_{M1} + I_{M2})D = I_{Q1,av}, \qquad (2)$$

and in the braking mode

$$I_{conv,av} = -(I_{M1} + I_{M2})(1-D) = -(I_{M1} + I_{M2} - I_{Q1,av}), \quad (3)$$

where D is a duty factor.

However, as can be seen from Fig. 2,

$$P_{conv} = V_{line} I_{line} - V_C I_{Q2,av} \qquad (4)$$

and the tramcar power can be expressed as

$$P_{tr} = V_{line}(I_{line} + I_{aux}) - V_C I_{Q2,av}. \qquad (5)$$

Evidently, to estimate the tramcar power it is necessary to record the measurements of two voltage and three current sensors with a succeeding data processing in accordance with (5). It should be noted that the calculated $P_{tr}$ is the tram's own consumed or returned power. In the cases when a tram in its rheostat $R_{br1}$ dissipates the braking energy of another tram, this power is not measured, since in (5) it appears twice, with opposite signs. The recording of the measured data is done with the

help of a 16-bit multifunction USB data acquisition module (USB-4716) using five input channels with the differential inputs connected to the sensors' load resistors located on the tram control cards VMT-1 and RT-1 (see Fig. 3). Signal of the current "$I_{line}$" is bipolar, being positive in the running mode; the other signals are unipolar.

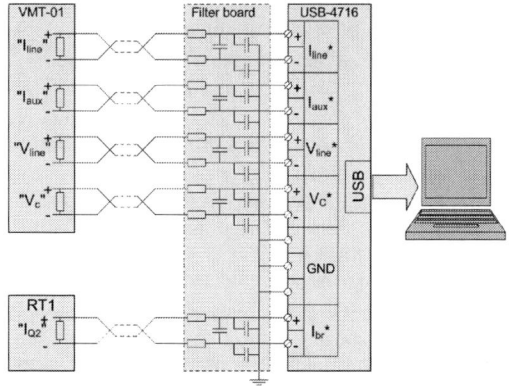

Fig. 3. Connection of data logger to tramcar's sensors.

The RC filter board is used for reducing noise and averaging current $I_{Q2}$ pulsed with a frequency of 1000Hz. This allows reducing the sample rate without losing accuracy. Taking into account the gain of sensors, the tramcar power is calculated as

$$P_{tr} = 200V_{line}^*(200I_{line}^* + 20I_{aux}^*) - 12000V_C^* I_{Q2}^*, \quad (6)$$

where $V_{line}^*, I_{line}^*, I_{aux}^*, V_C^*, I_{Q2}^*$ are the measured signals (in volts).

## IV. BRIEF DESCRIPTION OF THE METHOD FOR ESTIMATION THE UNTAPPED ENERGY

The proposed in [2,3] method for estimation of untapped energy is based on the assumption that at any freely chosen time moment the power consumed by a vehicle has a random nature. Like any random variable, it can be characterized with the probability density function (pdf), which allows determination of the probability for the power to be within some interval.

Applying the method to the recorded data of the power consumed by an electric vehicle one can calculate the discrete power probability density function $p[k]$ where $k \in [-N; N]$ is the number of the power interval $P_k \in [(k-0.5)\Delta P; (k+0.5)\Delta P]$, $N = |P_{max}|/\Delta P$, $P_{max}$ is the maximum consumed/returned power, $\Delta P$ is the width of power interval.

If in one section of the overhead power supply network there are simultaneously two vehicles of the same type, e.g. type "A", assuming that their power consumptions are random functions and these processes are independent, the total probability density $p_{2A}$ can be calculated using the convolution [10,11]:

$$p_{2A}[k] = \sum_{m=\max(-N;k-N)}^{\min(N;k+N)} p_A[m] p_A[k-m], \qquad (7)$$

$$k \in [-2N; 2N]$$

If there are simultaneously several vehicles of different type in one section of the overhead power supply network, then the total power density function is obtained step-by-step. For example, if there are $l$ vehicles of type "A" and $i$ vehicles of type "B", consecutively $p_{2A}$, $p_{3A}$,... $p_{lA}$, $p_{2B}$, $p_{3B}$... $p_{iB}$ probability density functions are calculated, and the total $p_{lAiB}$ pdf will be:

$$p_{lAiB}[k] = \sum_{m=\max(-M;k-M)}^{\min(M;k+M)} p_{lA}[m]p_{iB}[k-m],$$
$$k \in [-M;M], \qquad (8)$$
$$M = (l+i)N$$

The average power consumed by a vehicle is calculated as

$$P_{av} \approx \Delta P \sum_{k=-N}^{N} k \cdot p[k], \qquad (9)$$

and average regenerated power $P_{r,av}$ can be calculated applying (9) for the negative power region:

$$P_{r,av} = -\Delta P \sum_{k=-N}^{0} k \cdot p[k]. \qquad (10)$$

The average untapped power for $q$ vehicles is

$$P_{un,q,av} = -\Delta P \sum_{k<0} k \cdot p_q[k], \qquad (11)$$

and untapped energy $E$ is obtained as

$$E = \sum_q P_{un,q,av} T_{\Sigma q} = \sum_q E_q , \qquad (12)$$

where $p_q[k]$ – the power pdf for $q$ vehicles,
$T_{\Sigma q}$ – the total time when $q$ vehicles are within the limits of the power supply line's section.

## V. EXPERIMENTAL RESULTS AND CALCULATION OF THE UNTAPPED REGENERATIVE ENERGY

The measuring equipment shown in Fig. 3 was installed in $T3A$ trams running on routes № 6 and №11; one-day measurement data were recorded. In order to relate the recorded data to the topography of the routes, also GPS receivers were installed in the trams. The feeding section of substation № 8 on route №11 is 3.5 km long, and that of substation №10 on route № 6 – 3.8 km long. In this section there are also routes №1 and № 3.

At calculation of the average tram's power pdf for the relevant feeding section, the data for tram's location outside the feeding section were excluded. The graphical representation of average power pdf of tram № 6 for the feeding section of substation №10 is shown in Fig. 4.

Based on the tram schedule for working days and weekends, the total time $T_{\Sigma 1}$, $T_{\Sigma 2}$ ...$T_{\Sigma q}$ was calculated for situations when $1,2...q$ trams are running in the substation feeding section. For substations № 8 and № 10 they are shown in Table I and Table II, respectively. It could be seen that in the feeding sections of both substations the number of simultaneously running trams does not exceed 6 and most often there are $3 - 4$ trams.

Assuming that only recuperating $T3A$ trams are running on both routes and applying (7), (11) and (12), the average untapped power $P_{un,q,av}$ and the daily untapped energy for a summer season is calculated. These data for substations № 8 and № 10 are summarized in Tables I and II.

TABLE I.
DAILY UNTAPPED ENERGY FOR SUBSTATION № 8

| q | $P_{un,q,av}$ (kW) | Monday-Friday | | Saturday | | Sunday | |
|---|---|---|---|---|---|---|---|
| | | $T_{\Sigma q}$ (h) | $E_q$ (kWh) | $T_{\Sigma q}$ (h) | $E_q$ (kWh) | $T_{\Sigma q}$ (h) | $E_q$ (kWh) |
| 1 | 17.9 | 0.90 | 16.1 | 1.52 | 27.2 | 1.52 | 27.2 |
| 2 | 23.6 | 3.97 | 93.8 | 4.47 | 105.6 | 4.47 | 105.6 |
| 3 | 24.4 | 5.28 | 128.9 | 5.10 | 124.4 | 5.10 | 124.4 |
| 4 | 23.2 | 6.95 | 161.5 | 5.12 | 118.9 | 5.12 | 118.9 |
| 5 | 21.4 | 1.62 | 34.7 | 1.85 | 39.7 | 1.85 | 39.7 |
| 6 | 19.5 | 0.00 | 0.0 | 0.03 | 0.6 | 0.03 | 0.6 |
| 7 | 17.6 | 0.00 | 0.0 | 0.00 | 0.0 | 0.00 | 0.0 |
| | | Total: | 434.9 | Total: | 416.4 | Total: | 416.4 |

TABLE II.
DAILY UNTAPPED ENERGY FOR SUBSTATION № 10

| q | $P_{un,q,av}$ (kW) | Monday-Friday | | Saturday | | Sunday | |
|---|---|---|---|---|---|---|---|
| | | $T_{\Sigma q}$ (h) | $E_q$ (kWh) | $T_{\Sigma q}$ (h) | $E_q$ (kWh) | $T_{\Sigma q}$ (h) | $E_q$ (kWh) |
| 1 | 19.7 | 1.32 | 26.0 | 1.33 | 26.3 | 1.72 | 33.9 |
| 2 | 28.5 | 4.13 | 117.7 | 6.65 | 189.4 | 5.65 | 160.9 |
| 3 | 31.7 | 5.42 | 171.7 | 5.40 | 171.2 | 6.00 | 1902 |
| 4 | 32.2 | 5.77 | 185.5 | 3.30 | 106.2 | 3.42 | 109.9 |
| 5 | 31.3 | 2.10 | 65.7 | 1.15 | 36.0 | 0.80 | 25.0 |
| 6 | 29.8 | 0.13 | 4.0 | 0.10 | 3.0 | 0.00 | 0 |
| 7 | 28.0 | 0.00 | 0 | 0.00 | 0 | 0.00 | 0 |
| | | Total: | 570.7 | Total: | 532.1 | Total: | 520.0 |

The energy consumption in spring/autumn and winter seasons differs mainly due to additional energy used for heating. Tram's power pdf for these seasons can be approximately calculated adding the average daily power for heating $\Delta P_{aux}$ to the experimentally obtained auxiliary power for summer seasons $P_{aux}$. We assume that this power for heating in a spring/autumn season is 9kW, whereas in winter months it is 20kW.

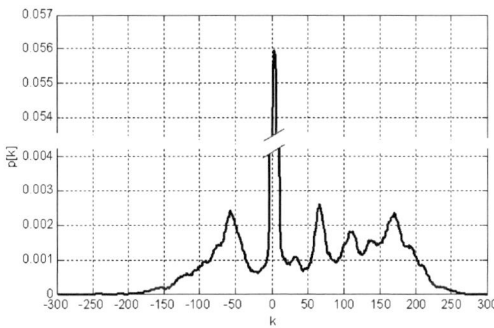

Fig. 4. Tramcar's power pdf $p[k]$ at $\Delta P=1$kW for feeding section of substation № 10.

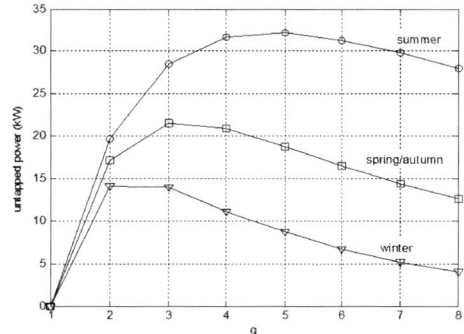

Fig. 5. Untapped average power vs number of trams $q$ for different seasons.

2069

The calculated average untapped power in different seasons versus the number of trams $q$ running in the feeding section of substation № 10 is shown in Fig. 5. In summer the power has maximum at $q=5$ and decays with $q$ growing. In winter the untapped power is much less, with a maximum at $q=2-3$. The calculated daily untapped energy for different seasons is presented in Table III, and the annual total untapped energy – in Table IV.

TABLE III.
DAILY UNTAPPED ENERGY FOR DIFFERENT SEASONS

| Subst. | Season | Daily untapped energy (kWh) | | |
|---|---|---|---|---|
| | | Mon.–Fri. | Sat. | Sun. |
| №8 | summer | 434.9 | 416.4 | 416.4 |
| | spring/autumn | 259.1 | 254.2 | 254.2 |
| | winter | 129.2 | 131.7 | 131.7 |
| №10 | summer | 570.7 | 532.1 | 520.0 |
| | spring/autumn | 369.7 | 361.2 | 353.8 |
| | winter | 201.5 | 208.6 | 204.9 |

TABLE IV.
ANNUAL ESTIMATED TOTAL UNTAPPED ENERGY FOR SUBSTATIONS
№.8 AND №.10

| Season | Months | Total estimated untapped energy (kWh) | |
|---|---|---|---|
| | | Substation №8 | Substation №10 |
| summer | 5 | 81194 | 84851 |
| spring/autumn | 4 | 38960 | 44556 |
| winter | 3 | 15655 | 18525 |
| **Total** | **12** | **135810** | **147932** |

Assuming the efficiency of the energy storage system to be 0.85, the annual amount of saved energy can reach 115 MWh for substation № 8 and 125.7 MWh for substation № 10.

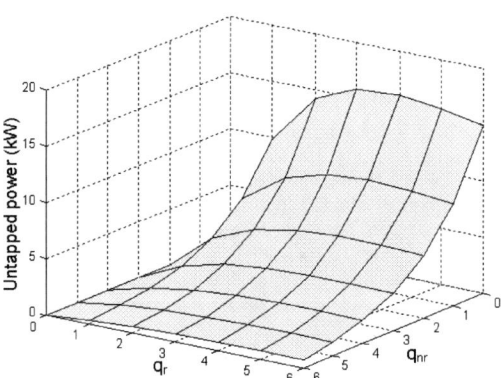

Fig. 6. Untapped average power vs. number of recuperating trams $q_r$ and non-recuperating trams $q_{nr}$.

Fig. 6 shows the impact of a non-recuperating vehicle running in the feeding section on the untapped average power. The power pdf of such a non-recuperating vehicle was obtained from the pdf shown in Fig. 4 taking only positive $k$ values. One can see that even small number of non-recuperating vehicles considerably reduces the untapped energy. If non-recuperating vehicles dominate, the untapped energy becomes negligible.

## VI. CONCLUSIONS

1. The amount of untapped braking energy within the limits of one feeding substation depends on the season and the number and type of the vehicles in a given feeding section.
2. The maximum untapped energy is observed when there are 4-6 recuperating vehicles simultaneously present in the feeding section in summer, 3-4 – in spring or autumn, and 2-3 – in winter. As this number grows, the amount of untapped energy tends asymptotically to zero.
3. The presence of even a small amount of non-recuperating vehicles reduces considerably the untapped energy.
4. It is advisable, first, to install the energy storage equipment at the substations where the number of vehicles simultaneously running in the feeding section does not exceed 10 and the portion of recuperating vehicles is close to 100%.

## REFERENCES

[1] I. Rankis, V. Brazis, "Simulation of Tramcar's Energy Balance", 2nd Intern. Conf. "Simulation, Gaming, Training and Business Process Reengineering in Operations" Riga, 2000, pp. 160–163.

[2] L. Latkovskis, L. Grigans, "A Method for Estimation of the Unutilized Regenerative Braking Energy in the Urban Public Electric Transport", (in Latvian) Latvian Journal of Physics and Technical Sciences, 2007, № 4, pp. 21–28.

[3] L. Latkovskis, L. Grigans, "A Method for Estimation of the Untapped Regenerative Braking Energy in Urban Electric Transport", CD-ROM of Conference of Young Scientists on Energy Issues CYSENI 2008, Kaunas, May 2008, pp. IV-41 – IV-48.

[4] A. Rufer et al, "A Supercapacitor-based Energy-storage Substation for Voltage-compensation in Weak Transportation Networks", Power Tech Conference Proceedings, 2003 IEEE Bologna.

[5] A. Rufer, "Power-Electronic Interface for a Supercapacitor-Based Energy-Storage Substation in DC-Transportation Networks", EPE Conf. proceedings Toulouse, 2003, pp. D1–D8

[6] E. R. Furlong, M. Piemontesi, P. Prasad, De Sukumar, "Advances in Energy Storage Techniques for Critical Power Systems", Proceedings of Battcon 2002 Conference, Ft. Lauderdale, FL pp. 10-1 – 10-8.

[7] P. Barrade, A. Rufer, "The use of Supercapacitors for Energy Storage in Traction Systems", In Proceedings of IEEE Vehicular Power and Propulsion Symposium, (VPP), Paris, France, 6-8 October,2004.

[8] L. Latkovskis, V. Bražis, "Application of Supercapacitors for Storage of Regenerative Energy in T3A Tramcars", Latvian Journal of Physics and Technical Sciences, 2007, №. 5, pp.13-23.

[9] L. Latkovskis, V. Bražis, "Simulation of the Regenerative Energy Storage with Supercapacitors in Tatra T3A Type Trams", Proceedings of the Tenth International Conference on Computer Modeling and Simulation (UKSIM 2008), Cambridge, UK, 1-3 April 2008, pp. 398-403.

[10] E. J. Borowski, J. M. Borwein, "Collins Dictionary of Mathematics", Caledonial International Book Manufacturing Ltd, Glasgow, 1989, p.659.

[11] T. T. Soong, "Fundamentals of Probability and Statistics for Engineers", John Wiley & Sons, Ltd, Chichester, 2004, p. 408.

# Author Index

## A

Abbatelli, L. .................................................61
Abbey, Chad ...........................................2178
Abdelhamid, Tamer H. .............................606
Abdellatif, Meriem .................................938
Abe, Seiya ...............................................393
Abourida, Simon ...................................1077
Abroshan, Mohammad ...........................1117
Abroushan, Mohammad...........................365
Abuishmais, Ibrahim..............................867
Abu-Rub, Haithem ....................1084, 1382
Adabi, Jafar...............................718, 903
Adamidis, Georgios ..............................1840
Adamowicz, Marek................................1729
Adzic, Evgenije......................................1957
Ahmadi, Muhammad ..............................1847
Ahmed, M.M.R..........................1866, 2472
Ahn, Jonng-Bo .......................................2524
Ahn, ong-Bo ..........................................2492
Ait-Ahmed, Mourad .............................1740
Akhondi, Hamidreza...............................2071
Alarcón, E. .............................................2108
Albert, Laurent.......................................2037
Al-Diab, Ahmad ...................................1710
Alexandrov, Alexandar...........................787
Al-Khayat, Nazar ...................................2150
Allard, Bruno .......................................2457
Al-Othman, A. K. ....................................606
Amelon, Nicolas .....................................1740
Anaya-Lara, O. ........................1784, 1941
Andersen, Michael A.E.............................127
Ando, Kenji .............................................614
Andrzejewski, Andrzej ...........................1090
Areerak, K-N .........................................2049
Arellano-Padilla, J. ...............................1173
Arellano-Padilla, Jesus...........................769
Armstrong, S. .......................................1688
Aroudi, A. El .........................................2108
Aroudi, Abdelali El..................2115, 2120
Arshad, Waqas M. ...................................867
Asher, G. M. .........................................2261
Asher, G.M. ..........................................2049
Asher, Greg...........................................2300
Asiltürk, Ilhan .........................................967
Aurel, Campeanu ....................................893
Averberg, Andreas ..................................213

## B

Baalbergen, Freek J.F. ..........................2170
Baghaee, H.R. .............313, 629, 750
Bahri, I. ................................................1365
Bailey, Chris ............................................76
Bakas, Panagiotis .................................1840
Balazovic, Peter .....................................1402

Balouktsis, Anastasios ..........................1840
Baluta, Gh..............................................2043
Ban, Drago............................................818
Barai, Mukti ...........................................674
Barakat, Georges .....................1834, 810
Baranowski, Jerzy ......................1432, 1446
Barbosa, Fabián H. ..................................637
Barlik, Roman...........................................84
Barrero, R. ...........................................1512
Bartelt, R. ...............................................521
Baskys, Algirdas ....................................1140
Bastiani, A. ...........................................1293
Baszynski, Marcin .................................1779
Bauer, Pavel...........................................422
Bauer, Pavol ...........2170, 2354, 2368, 2371
Beck, Hans-Peter ...................................1243
Bekbudov, Radiy .....................................337
Bekishev, Anatoly ...................................663
Bélanger, Jean............................1077, 1475
Belfkira, Rachid......................................1834
Belkhodja, I. Slama ...............................1149
Bellini, Armando .....................................490
Bellmunt, Oriol ........................................731
Belter, D. .............................................1044
Benadero, Luis ......................................2115
Bendkowski, Lukas .................................250
Benecke, Marcel ....................................1280
Benkhoris, Mohamed-Fouad ..................1740
Bennani, A.Ben Abdelghani ..................1149
Beran, Leos ............................................782
Bergas-Jane, Joan ..................................731
Bergogne, Dominique.............................2457
Berthon, A. ...........................................1542
Bertoluzzo, Manuele ..............................1491
Bertram, Torsten ....................................1215
Betz, R.E. .............................................1293
Bevilacqua, Pascal .................................2457
Bifaretti, Stefano..................1771, 490, 561
Binder, Andreas ...........................1625, 2385
Binkowski, T. ..........................................714
Birolleau, Damien...................................2037
Biswas, Jayanta.......................................674
Bizon, Nicu ............................................621
Blahník, Vojtech ....................................1535
Blanco, M. ............................................2481
Blazic, B. .............................................2510
Böcker, J. .............................................1598
Böcker, Joachim .....................................159
Bodora, A. .............................................326
Bogalecka, Elzbieta ...................1975, 804
Bojda, Petr .............................................422
Bolgov, Viktor ........................................154
Bolognani, Silverio ...............................1097
Boora, Arash A .......................468, 723
Bossche, A. Van den..............................1326
Botan, Corneliu .....................................1111

# Author Index

Botsali, FatihM. .................................................. 949
Bouafia, Abdelouahab .......................................... 703
Boucherit, M.S. ................................................... 1987
Bouhalli, Nadia ................................................... 281
Bozhko, S.V. ....................................................... 2049
Brand1tetter, Pavel ............................................. 1375
Braslavsky, I.Ya. ................................................ 1050
Breban, Stefan .................................................... 1896
Brown, Neil L. .................................................... 2150
Bruno, Francois ................................................... 2205
Bucher, Alexander ........................................ 244, 250
Buja, Giuseppe .................................................... 1491
Bukatov, Alexander ............................................. 1872
Bulic, Neven ....................................................... 556
Buonomo, S. ....................................................... 61
Buss, Martin ....................................................... 2312

## C

C., Ilioudis Vasilios ............................................. 1105
Caballero, M. ...................................................... 1555
Cabrita, Carlos M. P. .......................................... 1646
Calado, Maria R. A. ............................................ 1646
Camara, M.B. ..................................................... 1542
Cambál, Marek .................................................... 982
Candusso, Denis .................................................. 734
Cartes, D.A. ........................................................ 793
Case, Michael James ........................................... 1798
Castaing, Ambroise .............................................. 2464
Catalão, J. P. S. .................................................. 1682
Cédl, Marek ................................................ 1593, 372
Ceglia, Gerardo ................................................... 268
Cepisca, Costin ............................................ 1963, 908
Cernohorský, Josef .............................................. 1009
Cerovský, Zdenk ................................................. 982
Cha, Gil-Ro ......................................................... 383
Champenois, Gérard ............................................ 2015
Chan, Paul K.W. ................................................. 1688
Chang, Hao-Chi ................................................... 1652
Chang, Lon-Kou ................................................... 320
Chang, Yuan-Chih ............................................... 456
Chante, Jean-Pierre ............................................. 2457
Charaabi, L. ........................................................ 1365
Chekhet, Eduard .................................................. 307
Chen, Anyuan ..................................................... 799
Chen, Junling ...................................................... 2000
Chen, Yonggang .................... 1981, 2000, 405, 515
Chen, Zhe ........................................................... 2325
Chen, Zong-Jie .................................................... 1704
Cheng, K.W.E. .................................................... 576
Cherif, M. Ghodbane ........................................... 1149
Cheung, N. C. ..................................................... 1221
Chien, Sywe-Bin .................................................. 1652
Chillet, Christian ................................................. 2037
Chimento, F. ....................................................... 61

Chlodnicki, Zdzislaw ........................................... 2150
Choi, Heung-Kwan ............................................... 2524
Choi, Jaeho ......................................................... 2498
Choi, Uk-Don ...................................................... 1421
Chou, Ming-Chang ............................................... 1652
Chrenko, Daniela ................................................. 2156
Chrzan, Piotr J. ................................................... 144
Chudzik, Piotr ..................................................... 1568
Chun, Tae-Won .................................................... 1421
Clare, J. .............................................................. 1326
Clare, Jon C ........................................................ 207
Clare, Jon C. ................................................ 229, 561
Clare, Jon ................................................... 1771, 307
Comnac, Vasile ................................................... 1748
Cook, B.J. ........................................................... 1293
Cook, D. ............................................................. 1326
Coquery, Gérard .................................................. 2192
Correa, Pablo ............................................... 451, 699
Courtecuisse, Vincent ................................... 1896, 2184
Cousineau, Marc .................................................. 281
Cuk, Vladimir ..................................................... 1426
Cychowski, Marcin ............................................... 2241
Czapp, Stanislaw ................................................. 2059

## D

Dabroom, A.M. .................................................... 1337
Dakyo, B. ............................................................ 1911
Dannehl, J. .......................................................... 444
Darie, Eleonora ............................................ 1963, 908
Darie, Emanuel ............................................. 1963, 908
Davey, J. ............................................................ 1918
De Bernardinis, Alexandre .................................... 2192
De Castro, M.R. ................................................... 2126
De Gersem, Herbert ............................................. 2385
de Kock, H.W. ..................................................... 859
De Souza, Kleber C.A. .......................................... 1951
Deaconu, Sorin .................................................... 1409
Debowski, Andrzej ....................................... 1568, 2289
Degeratu, Sonia ................................................... 893
Delaney, Kieran ................................................... 2241
Demenko, Andrzej ............................................... 2412
Denny, Ernest Edward .......................................... 1798
Depernet, Daniel .................................................. 734
Derbel, Nabil ...................................................... 2120
Deskur, Jan ................................................. 1204, 2227
Deuse, Jacques .................................................... 2184
Dheilly, Nicolas ................................................... 2457
Di, Lu ................................................................ 2205
Dianov, Anton ..................................................... 1002
Díaz, Nelson L. .................................................... 637
Diblík, Martin ..................................................... 1676
Diguet, Marc ....................................................... 1382
Dilevs, Guntis ..................................................... 1811
Dimitrakakis, Georgios S ....................................... 1301
Dinkhauser, Vincenz ............................................ 1819

# Author Index

Dobrucky, Branislav ................................................ 1402
Dockhorn, Matthias .................................................. 1734
Dodds, Stephen J. .......................................... 2551, 2559
Dodds, Stephen James ............................................. 2543
Doebbelin, Reinhard ................................................. 1280
Doi, Nobuaki ............................................................. 744
Dong, P. ..................................................................... 576
Dong, Wei ............................................................... 1716
Dontchev, Dimitar ...................................................... 787
Drábek, Pavel .................................................... 1593, 372
Draganov, Denis ..................................................... 1610
Dubowski, Marian Roch .......................................... 1090
Dudak, Juraj ............................................................ 2368
Dudrik, Jaroslav ........................................................ 295
Duerbaum, Thomas ............................................ 244, 250
Dufour, Christian ............................................ 1077, 1475
Duke, Richard ........................................................... 528
Dumur, Guillaume ................................................... 1475
Durovsky, Frantisek .................................................. 961
Dybkowski, Mateusz ......................................... 2211, 2306
Dzieniakowski, Maciej A. ........................................ 2082

## E

Eberhard, Andreas ................................................... 1371
Eckel, Hans-Guenter ................................................... 48
Edrington, C.S. ......................................................... 793
Egan, Michael G. ..................................................... 1249
Egorov, Mikhail ...................................................... 1257
Ehsan, Mehdi .......................................................... 1847
Eilenberger, Andreas ................................................ 945
Elmoctar, Mohamed Y. Ould ...................................... 810
Empringham, Lee ...................................... 207, 229, 388
Endo, Tsunehiro ....................................................... 924
Eno, Otu A. .............................................................. 114
Erceg, Gorislav ......................................................... 556
Etxeberria-Otadui, I. .............................................. 1555

## F

Fabianowski, Jan ..................................................... 2082
Fabijanski, Pawel ........................... 1040, 2055, 2087
Fahrni, C. ................................................................. 256
Fakham, Hicham ...................................................... 2142
Fan, Yue .................................................................. 1771
Farhangi, Sh. ........................................................... 173
Farshad, Siamak ...................................................... 1575
Fedák, Viliam .......................................................... 2354
Fedak, Viliam ............................................................ 961
Fedyczak, Zbigniew ........................................... 165, 236
Feki, Moez .............................................................. 2120
Fernández, Herman ................................................. 1947
Fernandez-Mola, Josep-Maria ................................... 731
Ferreira, Jan Abraham .............................................. 187
Ferreira, Luís António Fialho Marcelino ................... 2076
Fetyko, Jan ............................................................... 961
Filchev, T. ............................................................... 1326

Filho, Braz Jesus Cardoso ....................................... 1345
Filka, Roman ........................................................... 1402
Fisher, R. ................................................................ 1293
Fleisch, Karl .............................................................. 48
Fodor, D. ................................................................ 2096
Foft, Jiří ................................................................. 1593
Foo, Gilbert ............................................................ 2269
Forster, Stefan ......................................................... 2420
Fotouhi, Reza .......................................................... 1575
Francois, Bruno ............................................... 2142, 2184
Franke, W. Toke ........................................................ 69
Franko, Marek ......................................................... 2538
Friedli, T. .................................................................. 27
Fröhleke, Norbert ..................................................... 159
Fuchs, F.W. ............................................................. 444
Fuchs, Friedrich W. .............................. 1390, 1819, 69
Fujita, Y. ................................................................. 275
Fukushima, Kentaro ................................................. 148
Funabashi, Toshihisa ....................................... 2478, 2487
Funato, Hirohito ...................................................... 479
Funayama, Koichi .................................................... 1020
Futami, Motoo ........................................................ 2337

## G

Gabriela, Petropol Serb ............................................ 893
Gan, W. C. .............................................................. 1221
Gao, Fanqiang ......................................................... 515
Gao, Q. ................................................................... 2261
Gao, Qiang .............................................................. 1058
García-Tabarés, L. .................................................. 2481
Gardecki, Arkadiusz ................................................ 1193
Gasiewski, Marcin ................................................... 1562
Gaubert, Jean-Paul ................................................... 703
Gavranic, Ivica ........................................................ 818
Gaztañaga, H. ......................................................... 1555
Gelezevicius, Vilius Antanas .................................... 1144
Gennadevich, Kiselev Michail ................................... 428
Gennadevich, Lepanov Michail .................................. 428
Gerada, C. .............................................................. 1173
Gerada, Chris .............................. 1058, 388, 769, 887
Ghaedi, Azam .......................................................... 1054
Gharehpetian, G.B. ............................ 313, 629, 750
Ghosh, Arindam .............................. 468, 723, 903
Gímenez, María Isabel ............................................ 1947
Giménez, María ....................................................... 268
Giral, Roberto ........................................................ 2115
Gizinski, Zygmunt ................................................... 1562
Glasberger, Tomá1 .................................................. 1268
Glavin, M.E. ........................................................... 1688
Glushkin, Evgeny .................................................... 1872
Gnacinski, Piotr ....................................................... 826
Gobis, Vitoldas ....................................................... 1140
Goeldel, C. ............................................................. 2126
Gomis-Bellmunt, Oriol ............................................ 1670
González-Hernández, S. ........................................... 1784

# Author Index

Gorbounov, Yassen................................787
Goto, Hiroki........................... 1163, 1168
Grabic, Stevan....................................1957
Grad, M.............................................714
Grecki, Filip.....................................1440
Grigaitis, Arunas................................1144
Grigans, Linards ................................2066
Grossi, Federica..................................874
Grzesiak, Lech M.................................1071
Grzesik, B..........................................956
Gualous, H........................................1542
Guo, Hai-Jiao ......................... 1163, 1168
Gustin, F..........................................1542
Gustin, Frederic ..................................734
Guy, Owen J......................................2464
Guzinski, Jaroslaw......................... 1382, 994
Guzmán, Víctor........................... 1947, 268
Gwózdz, Michal....................................728

## H

Haan, Sjoerd de...................................187
Habetler, Thomas G................................21
Hadas, Zdenek ...................................1665
Hadjov, Kliment...................................787
Hájek, Vítezslav..................................2371
Halasz, S...........................................682
Halgos, Jan.......................................2368
Hamada, Tomoyuki ...............................1884
Hamar, J..........................................1755
Hameyer, K........................................2393
Hameyer, Kay......................................2412
Harada, Yosuke ...................................148
Hartansky, Rene ..................................2368
Hartnett, Kevin J.................................1249
Hasegawa, Masaru .................................614
Hashimoto, Seiji ..................................932
Hayashi, Kenta....................................589
Hayashi, Yusuke...................................2445
Hayes, John G.....................................1249
Heising, C.........................................521
Helmut, Weiss ............................ 1722, 1934
Henrotte, F.......................................2393
Ilcnzc, Olaf......................................2385
Hercog, Darko.....................................2349
Hicham, Fakham....................................2205
Himmelstoss, Felix. A..............................331
Hiraki, Eiji.............................. 119, 1877
Hirokawa, Masahiko................................393
Hissel, Daniel....................................2156
HISSEL, Daniel....................................734
Hmasic, N.........................................2134
Ho, S.L...........................................576
Hoffmann, Frank...................................1215
Hõimoja, H........................................2005
Hõimoja, Hardi....................................1581

Hojo, Masahide....................................2487
Holtz, Joachim....................................1084
Holub, Marcin.....................................195
Horen, Yoram......................................776
Horga, V.........................................2043
Horga, Vasile....................................1111
Hrasko, Martin...................................2538
Hu, Weihao.......................................2325
Hubik, Vladimir..................................1620
Huiqing, Wen ............................. 1518, 417
Hurley, W.G......................................1688
Huttin, N........................................1523

## I

I., Margaris Nikolaos.............................1105
Iannuzzi, Diego..................................1469
Ibach, Robert ...................................2082
Ibáñez, Fernando .................................268
Ichinokura, Osamu.................... 1163, 1168, 758
Ichinose, Masaya .................................2337
Ide, Kazumasa ...................................2337
Igic, P. M.......................................2464
Iida, Takahiko....................................595
Ikeda, Yoshiko....................................498
Ikhouane, Faycal.................................1670
Iman-Eini, H.....................................173
Inoue, Yukinori..................................1859
Ion, Petropol Serb................................893
Iov, Florin.............................. 1771, 561
Ishikawa, Kazumi.................................1020
Iskhakov, Albert..................................663
Ito, Fumio.......................................1309
Itoh, Jun-ichi....................................581
Itoi, M..........................................275
Ivanovic, Zoran..................................1957
Iwaji, Yoshitaka..................................924
Iwanski, Grzegorz......................... 1440, 2164
Iwase, Yuta......................................2487
Izadbakhsh, Alireza..............................2102

## J

Jalakas, T.......................................1263
Jalakas, Tanel ...................................1257
Jan, Mucko.......................................1316
Ján, Vittek......................................2219
Jang, Gil-Soo....................................2498
Jang, Su-Jin.....................................1924
Jansen, Uwe.......................................88
Janson, Kuno......................................154
Járdán, Rafael K..................................916
Jardan, Rafael Kalman............................2360
Jasim, O.........................................1173
Jasim, Omar......................................887
Jasinski, Marek..................................1904
Javurek, Jiri....................................1465

# Author Index

Jedryczka, Cezary ....................................2406
Jennings, Michael R. ..............................2464
Jeon, Jin-Hong ...........................2492, 2524
Jezernik, Karel ................2283, 2349, 432
Ji, Young-Hyok .......................................1929
Jian, Xiao ...............................................1722
Jin, Zhao ................................................1128
Johnson, C Mark .........................................76
Joós, Géza ..............................................2178
Joost, M. ................................................1064
Judek, Slawomir ....................................1497
Jufer, Marcel ...............................................1
Jun, Liu .......................................1518, 417
Jung, Doo-Yong .....................................1929
Jung, Yong-Chae ...........181, 1924, 1929, 383

## K

Kalatchikov, P. ........................................837
Kalisiak, Stanislaw .................................195
Kallaste, Ants..........................................154
Kallenbach, E. .......................................1598
Kalyoncu, Mete....................1132, 949, 974
Kamata, Yuki...........................................498
Kamiski, Bartlomiej...............................2378
Kamper, M.J. ...........................................859
Kampisios, Konstantinos .........................887
Kaneko, Daigo .........................................924
Kanerva, Sami..........................................867
Kaplon, Andrzej .......................................377
Karaffy, Z. .............................................2096
Karsli, Vedat M. .......................................850
Karwowski, Krzysztof .............................1497
Kasa, Nobuyuki .......................................595
Kasinski, A. ...........................................1044
Kasprowicz, Andrzej ..............................1332
Katic, Vladimir ......................................1957
Kato, Koji ...............................................581
Katsura, Seiichiro ..............1187, 1604, 1614
Kawamura, Atsuo .............................7, 924
Kayhan, Ince .........................................1934
Kazimierz, Jaracz.....................................912
Kazmierkowski, Marian P. ...........1548, 1904
Kelemen, Franjo .......................................855
Kennel, R.M. ...........................................859
Kennel, Ralph ........................................1239
Khaldi, B.S ............................................1987
Kim, Eel-Hwan ......................................2498
Kim, Heung-Gun .....................................1421
Kim, Jae-Hong .......................................2498
Kim, Jae-Hyung .............................1924, 1929
Kim, Jong-Yul ........................................2492
Kim, Se-Ho .............................................2498
Kim, Seul-Ki................................2492, 2524
Kimura, Kensuke .....................................1168
Kimura, Noriyuki .....................................1884

Kinoshita, Hirotaka................................2337
Kireev, V. ...............................................1598
Klimczak, Pawel .......................................108
Klug, O. .................................................2096
Klyachko, Leonid .....................................663
Klytta, Marius .........................................165
Knop, André ............................................69
Kobayashi, Yukinori .................................479
Kobougias, Ioannis C. ............................1274
Koczara, Wlodzimierz ..........1440, 2150, 2164, 2254
Koda, Noriaki .........................................1877
Kolar, J. W. ..............................................27
Kolesnikov, Artem..................................1872
Kolomeitsev, L. ......................................1598
Kompa, K. ...............................................695
Komura, Akiyoshi ...................................2337
Kondo, Masaki .......................................1614
Koneke, Thies.........................................1458
Kong, S.T. ...............................................43
Konstantinovich, Rozanov Yurie..............428
Korondi, Peter........................................2360
Korotyeyev, Igor .....................................236
Koskin, Y. ...............................................837
Kosmecki, Michal ..................................1975
Kostylev, A.V. ........................................1050
Kotodziejek, Piotr ....................................804
Kouzou, A. .............................................1987
Kowalski, Czeslaw T. ..............................1359
Kraeftner, Wilhelm ..................................331
Kraynov, D. ...........................................1598
Krettek, Johannes...................................1215
Krim, Fateh ............................................703
Krismer, F. ...............................................27
Krykowski, K. ..........................................326
Krystkowiak, Michal ...............................728
Krzeminski, Zbigniew .................1382, 2294
Kubiak, Andrzej......................................2452
Kubin, Jiri ..............................................1815
Kubota, Sachio .......................................1309
Kuchta, Jozef .........................................2538
Kudarauskas, Sigitas...............................2200
Kuebrich, Daniel .....................................244
Kuhn, Harald .........................................1458
Kuisma, M. ............................................1233
Kulka, Arkadiusz .....................................657
Kumar, Dinesh ........................................207
Kuperman, Alon ......................................776
Kurokawa, Fujio ...........................2434, 2504
Kürschner, Daniel ..........................1696, 1734
Kuß, H. ..................................................695
Kusserow, Wolf ......................................1239
Kütt, Lauri .............................................154
kuwata, M. .............................................275
Kyritsis, A.C. .........................................1287

A-5

# Author Index

## L

Laczynski, Tomasz ................................ 569, 649
Lafoz, M. ................................................... 2481
Lagoda, Ryszard ...................... 1040, 2055, 2087
Laloya, Eduardo ........................................ 845
Lange, E. ................................................... 2393
Lapointe, Vincent ...................................... 1077
Lastowiecki, Jozef .................................... 1440
Latka, M. .................................................. 714
Latkovskis, Leonards ................................ 2066
Laugis, J. ................................................. 1263
Laugis, Juhan ........................................... 1017
Laur, R. .................................................... 1064
Lazar, C. .................................................. 2043
Lazar, Mihai ............................................. 2457
Ledwich, Gerard ..................... 468, 723, 903
Lee, Joo-Hyuk ........................................... 1924
Lee, Tzung-Lin ......................................... 1704
Lehtla, Madis ............................................ 1581
Lehtla, T. ................................................. 2011
Leidhold, Roberto ..................................... 1353
Leszek, Szychta ........................................ 2091
Leuchter, Jan ............................................ 422
Levins, Nikolajs ....................................... 1811
Lewandowski, Daniel ...................... 2289, 669
Lewicki, Arka diusz .................................. 1382
Lewis, A.W. ............................................. 1790
Leyva, R. ................................................. 2108
Li, Kaihang .............................................. 97
Li, Rongyuan ............................................ 159
Li, Yaohua .................... 1981, 2000, 405, 515
Li, Zixin ..................... 1981, 2000, 405, 515
Liaw, Chang-Ming ...................... 1652, 456
Lie, Xu .................................................... 229
Liffran, Florent ........................................ 409
Lillo, Liliana de ....................................... 388
Lindemann, Andreas ................... 1280, 2420
Lingemann, M. .......................................... 2134
Lis, Jacek D. ............................................ 1359
Lisik, Zbigniew ......................................... 2452
Lisowski, Grzegorz .................................... 669
Liu, Congwei ............................................ 405
Liu, Li ..................................................... 793
Lladó, Juan .............................................. 845
Lodzinski, Michal ..................................... 2464
Lopez-de-Heredia, A. ................................ 1555
Lorenz, Robert D. ..................................... 903
LU, Di ..................................................... 2142
Lu, Hua ................................................... 76
Lu, Y. ...................................................... 1221
Luft, Miroslaw .......................................... 463
Luiz, Alex-Sander Amavel ......................... 1345
Luniewski, Piotr ....................................... 88
Lyons, Brendan J. ..................................... 1249
Lyskawinski, Wieslaw ............................... 2406

## M

Macek-Kaminska, Krystyna ........................ 1193
Madawala, U. K. ....................................... 139
Madawala, Udaya K. ................................. 1918
Maga, Dusan ............................................ 2368
Mahmoudi, M.O. ...................................... 1987
Mahyob, Amin .......................................... 810
Mailat, Adrian .......................................... 1748
Majidi, Behrooz ........................................ 763
Maksimovic, Dragan .................................. 498
MAKYS, Pavol .......................................... 2538
Malekian, Kaveh ............. 1117, 1123, 2071, 365, 763
Malska, W. .............................................. 714
Man, T.K. ........................................... 400, 475
Mandache, Lucian ..................................... 1585
Mandra, Slawomir ..................................... 1071
Mandrek, Slawomir .................................... 144
Marek, Stulrajter ...................................... 2219
Margaliot, M. ........................................... 260
Mariano, Sílvio José Pinto Simões .............. 2076
Marouchos, Christos ................................. 1967
Martín-del-Brío, Bonifacio ........................ 845
Martínez, Abelardo ...................... 1947, 845
Martinez, Itziar ........................................ 437
Martins, Denizar C. ................................... 1951
Masada, E. ............................................... 1755
Mascibrodzki, Ireneusz ............................. 1562
Mathis, W. ............................................... 132
Matsui, Keiju ............................................ 614
Matsui, Nobumasa ..................................... 2504
Mawby, P.A. ............................................. 2472
Mawby, Philip A. ...................................... 2464
McEachern, Alex ....................................... 1371
Mecke, Rudolf .......................................... 1734
Melício, R. ............................................... 1682
Mendes, V. M. F. ...................................... 1682
Mertens, A. ............................................... 132
Mertens, Axel ................. 1458, 213, 569, 649
Meuret, R. ................................................ 1523
Meynard, Thierry ...................................... 281
Michalík, Jan ...................................... 1535, 550
Michalke, N. ............................................ 695
Mierlo, J. Van .......................................... 1512
Milanovic, Miro ........................................ 301
Milimonfared, Jafar .............. 1117, 2071, 365, 763
Mimura, Yasuhiro ..................................... 2434
Mirsalim, M. ...................... 313, 629, 750
Mirsalim, Mojtaba .................................... 1123
Mirzaeva, G. ............................................ 1155
Mishima, Tomokazu ................................... 119
Mitani, Tetsuya ........................................ 2428
Mladenovic, I. .......................................... 2022
Mohd, A. ................................................. 2134
Mokrovica, Josipa ..................................... 855
Mõlder, Heigo .......................................... 154

# Author Index

Molinas, Marta ................................................. 2318
Möller, T. ........................................................... 2005
Mollov, Stefan V. ............................................... 350
Molnár, Jan ......................................... 1535, 550
Mondzik, Andrzej ............................................. 345
Monmasson, E. ................................................ 1365
Montesinos-Miracle, Daniel .............. 1670, 731
Morel, Herve ..................................................... 2457
Moreno-Font, Vanessa ..................................... 2115
Moreno-Goytia, E. ............................... 1784, 1941
Morimoto, Shigeo ............................................ 1859
Morino, Kimio .................................................. 2478
Morizane, Toshimitsu ...................................... 1884
Morton, D. ......................................................... 2134
Mouni, Emile .................................................... 2015
Mukhopadhyay, Siddhartha ............................ 485
Munk-Nielsen, Stig .......................................... 108
Murata, Toshiaki .............................................. 2337
Musallam, Mahera ............................................. 76
Mustonen, P. .................................................... 1233
Musumeci, S. ...................................................... 61
Muszynski, Roman ........................................... 2227
Mutschler, Peter .............................................. 1353
Müür, M. ............................................................ 2005
Mysinski, Wojciech ......................................... 1321

## N

Nagy, I. .............................................................. 1755
Nagy, Istvan ..................................................... 2360
Nagy, István ....................................................... 916
Naka, Toshiyuki ............................................... 498
Nakagawa, Akio ............................................... 498
Nakamura, Kazutoshi ...................................... 498
Nakamura, Kenji .............................................. 758
Nakaoka, M. ..................................................... 275
Nakaoka, Mutsuo ............................................. 119
Nakayama, Hiroaki .......................................... 1877
Nanakos, Anastasios Ch. ................................. 1827
Naouar, M-W. ................................................... 1365
Narayanan, E.M. Sankara ................................. 43
Narjiss, Abdellah ............................................. 734
Nasser, Mehdi .................................................. 1896
Navarro, Daniel ................................................ 437
Nawaz, Muhammed .......................................... 2472
Nekoui, Mohammad Ali ................................... 1054
Ngwendson, L. ................................................... 43
Ni, Bingchang .................................................. 2331
Nichita, C. ........................................................ 1911
Nichita, Cristian .............................................. 1834
Nicolae, Ileana-Diana ...................................... 1585
Nicolae, Petre Marian ...................................... 1585
Nicolae, Petre-Marian ...................................... 1181
Niechaj, Marek ................................................ 1890
Niemelä, Markku .............................................. 1763
Nikolic, Aleksandar ......................................... 1426

Nilssen, Robert .................................................. 799
Ninomiya, Tamotsu ............................... 148, 393
Nishida, Yasuyuki ............................................ 2530
Nishikata, Shoji ............................................... 2343
Nishimiya, Ayumu ........................................... 1163
Nishioka, Kunihiro .......................................... 1309
Nitta, Mayumi .................................................. 932
Noda, Shuji ...................................................... 1877
Norigoe, Isami ................................................. 148
Novák, Jaroslav ................................................ 982
Novák, Martin .................................................. 982
Nowak, Lech .................................................... 2400
Nowak, Mietek .................................................. 84
Numata, Shigeo ................................................ 2478
Nuutinen, Pasi ................................................. 1763
Nyczkowski, Lukasz ......................................... 740
Nymand, Morten ............................................... 127
Nysveen, Arne .................................................. 799

## O

O'Sullivan, D.L. ............................................... 1790
Ogiwara, H. ...................................................... 275
Ohashi, Hiromichi ....................... 2428, 2445, 54
Ohishi, Kiyoshi ..................... 1187, 1604, 1614
Ohsaki, H. ........................................................ 1755
Ohyama, Kazuhiro ........................................... 2300
Okamatsu, Masashi .......................................... 2434
Oleschuk, Valentin .......................................... 1548
Omari, O. ......................................................... 2134
OMORI, Hideki ................................................ 2530
Ondrusek, Cestmir .......................................... 1665
ONEN, Umit ..................................................... 949
OPROESCU, Mihai ........................................... 621
Orlik, B. ................................................ 1064, 830
Orlowska-Kowalska, Teresa ............... 2211, 2306
Ortjohann, E. ................................................... 2134
Oyarbide, Estanislao ....................................... 845

## P

Pacas, Mario .................................................... 2248
Pajchrowski, Tomasz .......................... 1198, 1204
Pakhomin, S. .................................................... 1598
Palis, Frank ..................................................... 1610
Palis, Stefan .................................................... 1660
Panoiu, Caius .................................................. 1409
Panoiu, Manuela .............................................. 1409
Papanikolaou, N.P. ........................................... 1287
Papic, I. ........................................................... 2510
Paquin, Jean-Nicolas ....................................... 1475
Park, JuneHo .................................................... 2492
Park, Sang-Hoon ................................... 181, 383
Park, So-Ri ...................................................... 181
Parkatti, P. ....................................................... 201
Parker-Allotey, N-A. ....................................... 2472
Patel, N. D. ...................................................... 139

# Author Index

Patra, Pradipta.................................485
Patra", Amit....................................485
Pavelka, Jiri...................................221
Pavelka, Jirí...................................988
Pavlitov, Constantin............................787
Pavlovsky, Martin.................................7
Pavol, Makys..................................2219
Pavoni, Alessandro.............................1491
Peftitsis, Dimosthenis.........................1840
Peltoniemi, Pasi...............................1763
Peplinski, Marcin...............................826
Pera, Marie-Cecile.............................2156
Perez, Francisco................................845
Perez-Tomas, Amador............................2464
Peric, Nedjeljko...............................2235
Peroutka, Zdenek........1268, 1529, 1535, 550
Peter, Bris...................................2219
Peter, Zaucher.................................1722
Petit, Marc...................................2184
Petrella, Roberto..............................1097
Petrisor, Anca..................................893
Piatek, Pawel..................................1446
Pietrzak-David, Maria...........................938
Piróg, Stanislaw...............................1779
Pittermann, Martin........................1593, 372
Planson, Dominique.............................2457
Poljugan, Alen.................................1058
Pollán, Tomás...................................845
Popa, Anca Sorana..............................1225
Popa, Mircea...................................1225
Porada, Ryszard.................................740
Pospelov, Vladimir..............................663
Pronin, M......................................837
Pugachevs, Vladislavs..........................1811
Pyrhönen, Juha.................................1763

## Q

Quiroga, J.....................................793

## R

Rabkowski, Jacek................................84
Raciti, A.......................................61
Radomski, Grzegorz..............................504
Raducu, Marian..................................621
Radulescu, Mircea M............................1896
Rafecas-Sabate, Josep...........................731
Rafiei, S.M.R..................................2102
Rahman, M.F....................................2269
Rahnamaee, Arash...........................1117, 365
Rao, Sachit...................................2312
Rathge, Christian..............................1696
Ratoi, Marcel..................................1111
Rawicki, Stanislaw.............................1481
Raynaud, Christophe............................2457
Rednov, F......................................1598

Reghem, Pascal............................1834, 810
Rerucha, Vladimir..............................422
Rezaei, Mohammad Mehdi.........................1123
Reznikov, B....................................260
Ribickis, Leonids..............................1811
Richter, F.....................................1398
Riipinen, T....................................1233
Risteiu, Mircea................................1243
Riz, A........................................2096
Roasto, I.....................................2011
Robert, B.G.M..................................2126
Robert, Bruno Gerard Michel....................2120
Robinson, Jonathan.............................2178
Robyns, B......................................1523
Robyns, Benoît.................................1896
Robyns, Benoit.................................2184
Rodic, Miran..................................2283
Rodriguez, E..................................2108
Rodriguez, Jose...........................451, 699
Rojas, A......................................1155
Rojko, Andreja................................2349
Rolek, Jaroslaw.................................377
Rompelman, Otto...............................2354
Ronkowski, Mieczyslaw..........................880
Rosin, A......................................2005
Rothenhagen, Kai..........................1390, 1904
Round, S. D.....................................27
Ru1scin, Vladimír..............................295
Ruderman, A....................................260
Ruderman, Michael..............................1215
Rufer, A.......................................256
Ruger, N. E....................................132
Rusinov, Radoslav..............................787
Rylko, Marek S................................1249
Ryvkin, Sergey................................1505
Rzasa, Janina..................................357

## S

Saadi, S......................................1987
Sabirin, Chip Rinaldi..........................1625
Saito, Makoto.................................2439
Saito, Tsuyoshi................................744
Sajkowski, M...................................956
Sakamoto, Kiyoshi..............................924
Sakamoto, Yosei................................288
Salo, M........................................201
Salonen, Pasi..................................1763
Samanta, Susovon...............................485
Samuelsen, Dag................................1416
Sanada, Masayuki...............................1859
Sánchez, Beatriz...............................845
Sánchez, Carlos................................268
Sang-Joon, Lee................................1002
Sang-Taek, Lee................................1002
Sanjari, M. J.........................313, 629, 750

# Author Index

San-Sebastian, J. .................................. 1555
Santo, António Espírito ....................... 1646
Sarraute, Emmanuel ............................ 281
Sasaki, Masahiro .................................. 2434
Sato, Muneo ........................................ 1309
Saudemont, C. .................................... 1523
Sayed, Mahmoud A. ............................ 542
Sayeef, S. ............................................ 2269
Schallschmidt, Thomas .............. 1610, 1660
Schanen, JL. ...................................... 173
Schmelter, A. ...................................... 2134
Schmid, Markus .......................... 244, 250
Schmidt, Istvan .................................. 1803
Schmidt-Obermoeller, Richard ........... 1505
Schmitt, Günter .................................. 1239
Schneider, T. ...................................... 1598
Schnick, O. ........................................ 132
Schrödl, Manfred ................................ 2275
Schroedl, Manfred .............................. 945
Schuffenhauer, U. .............................. 695
Scollo, R. ............................................ 61
Sengupta, Sabyasachi ......................... 674
Seppä, L. ............................................ 1233
Shao, S. .............................................. 1293
Shapoval, Ivan .................................... 307
Sharma, R. .......................................... 1918
She, X. ................................................ 710
She, Yun ............................................. 710
Shieh, Fa-Hwa .................................... 1652
Shimaoka, Yoshihiro ......................... 1309
Shimizu, Takaaki ............................... 498
Shimizu, Toshihisa .......... 2428, 2445, 288, 600
Shimoda, Eisuke ................................. 2478
Shiraishi, Keiichi ............................... 2504
Shonin, O. .......................................... 837
Shoyama, Masahito ..................... 148, 393
Shyu, Juei Lung .................................. 643
Siatkowski, M. ................................... 830
Silea, Ioan .......................................... 1225
Silventoinen, P. .................................. 1233
Simetzberger, Christian ...................... 2275
Simon, Miklós G. ............................... 916
Singule, Vladislav ...................... 1620, 1665
Sinsukthavorn, W. .............................. 2134
Siostrzonek, Tomasz .......................... 1779
Sîrbu, Ioana-Gabriela ................. 1181, 1585
Siroky, Peter ...................................... 2368
Sitar, Jan ............................................ 2368
Sivkov, Oleg ...................................... 221
Skovpen, Sergey ................................ 663
Skuta, Ondřej ..................................... 1375
Slama-Belkhodja, I. ........................... 1365
Slama-Belkhodja, Ilhem ..................... 938
Smet, Bart .......................................... 102
Sobczuk, Dariusz ............................... 2378
Sobczynski, D. ................................... 714

Sochacki, Mariusz .............................. 2452
Soltani, Hamid .................................... 718
Song, Sang-Hoon ............................... 383
Song, Seung-Ho ................................. 2498
Soroudi, Alireza ................................. 1847
Sosa-Ruiz, J. ...................................... 1941
Souad, Rafa ........................................ 1209
Sourkounis, C. ................................... 1398
Sourkounis, Constantinos ..... 1633, 1710, 2331
Sozanski, Krzysztof Piotr ................... 1995
Stadler, Paul Andreas ........................ 2543
Stala, Robert ............................. 1852, 345
Stamann, Mario .................................. 1660
Stanescu, Dan-Gabriel ....................... 1181
Staudt, V. ........................................... 521
Staudt, Volker .................................... 2371
Stefanutti, Fabio ................................ 1097
Steimel, A. ......................................... 521
Steimel, Andreas ...................... 1505, 2371
Stenzel, T. .......................................... 956
Stepanyuk, D.P. ................................. 1050
Stepien, P. .......................................... 1293
Stocco, Piero ...................................... 1097
Strac, Leonardo .................................. 855
Strzelecki, Ryszard Michal ................. 1332
Strzelecki, Ryszard ............................ 1729
Stumpf, P. ........................................... 1755
Stumpf, Péter ..................................... 916
Sugai, T. ............................................. 275
Sugimasa, Junji Tamura Masatoshi ..... 2337
Suissa, Uri ......................................... 776
Sulkowski, Waldemar ......................... 1416
Sumida, Yuichi ................................... 2434
Sumina, Damir .................................... 556
Sumiyoshi, Shinichiro ........................ 2530
Summers, T.J. .................................... 1293
Sumner, M. ................................. 1173, 2261
Sumner, Mark .................... 1058, 2300, 769
Sun, Z. G. .......................................... 1221
Susluoglu, Berrin ............................... 850
Suul, Jon Are ..................................... 2318
Sveda, Martin ..................................... 1620
Sweet, M. ........................................... 43
Sykulski, Jan K. ................................. 2383
Szabat, Krzysztof ..................... 2211, 2241
Szamel, Laszlo ................................... 1033
Szczeniak, Pawel ............................... 165
Szczepankowski, Pawel ...................... 1332
Szczesniak, Pawel .............................. 236
Szelag, Wojciech ............................... 2406
Sziebig, Gabor .................................... 2360
Szmidt, Jan ........................................ 2452
Szubert, Krzysztof ............................. 536
Szweda, Mariusz ................................ 826
Szychta, Elzbieta ............................... 463
Szychta, Leszek .................................. 463

# Author Index

Szymanski, B. J. .................................... 695

## T

Tackoen, X. ....................................... 1512
Tae-Ho, Yoon ...................................... 1002
Taguchi, Toyoki .................................... 498
Takahashi, Nobuo .................................. 1877
Takahashi, Rion ................................... 2337
Takao, Kazuto ................................ 2445, 54
Takeshita, Takaharu ................................ 542
Takeuchi, Nobuhito ................................ 614
Takeuchi, Toshihiro ................................ 924
Tan, Longcheng ................... 1981, 2000, 405, 515
Tanabe, Takayuki .................................. 2478
Tanaka, Toshihiko ................................. 1877
Taniguchi, Katsunori .............................. 1884
Taniguchi, Satoshi ................................. 600
Tankari, A.M. ..................................... 1911
Tao, Zhou ........................................ 2205
Tapuchi, Saad ..................................... 776
Tarczewski, Tomasz ................................ 1071
Tatakis, E.C. ..................................... 1287
Tatakis, Emmanuel C. .............. 1274, 1301, 1827
Tatsuta, Fujio .................................... 2343
Theodoridis, Michael P. ............................ 350
Thomas, D.W.P. ................................... 2049
Thomas, David W.P. ................................ 1716
Thompson, David S. ................................ 114
Tinkir, Mustafa ...................... 1132, 949, 974
Tnani, Slim ...................................... 2015
Tournier, Dominique ............................... 2457
Tran, Quang-Vinh ................................. 1421
Trentin, Andrew ................................... 887
Trujillo, Cesar L. ................................. 637
Tsai, Jih-Run ..................................... 1652
Tseng, K.J. ...................................... 2516
Tsukakoshi, Kenta ................................. 148
Tsuruta, Yukinori ................................... 7
Tulbure, Adrian ................................... 1243
Turner, Robert W. ................................. 528
Tutaj, Andrzej .................................... 1432
Tuusa, H. ........................................ 201

## U

Ueda, Yoshinobu ............................ 2478, 2487
Ummaneni, Ravindra. B. ............................ 799
Undeland, Tore ............................... 2318, 657
Ünüvar, Ali ...................................... 967
Urabe, R. ........................................ 275
Urbanski, Konrad ................................. 1454
Utkin, Vadim ............................... 2312, 512

## V

Väisänen, V. ..................................... 1233

Valchev, V. ...................................... 1326
van Duivenbode, Jeroen ............................ 102
Vasak, Mario ..................................... 2235
Vedrana, Jerkovic ................................. 690
Vekic, Marko ..................................... 1957
Vergnol, Arnaud .................................. 1896
Veszpremi, Karoly ................................. 1803
Vicuña, Javier .................................... 845
Villanueva, Elena ................................. 451
Villwock, Sebastian ............................... 2248
Vinnikov, D. ............................... 1263, 2011
Vinnikov, Dmitri .................................. 1257
Viscarret, U. .................................... 1555
Vittek, Jan ...................................... 2551
Vladimír, Vavrus ................................. 2219
Vodovozo, Valery ................................. 1017
Vorontsov, A. .................................... 837
Vrana, Petr ...................................... 1465

## W

Wada, Keiji ................................ 2428, 288, 600
Walas, K. ........................................ 1044
Walter, Julio .................................... 268
Walton, Simon .................................... 528
Wang, Ping ..................... 1981, 2000, 405, 515
Wang, Yi ......................................... 187
Wang, Yue ....................................... 2325
Wang, Zhaoan .................................... 2325
Weidinger, Thomas ................................ 2028
Weiland, Thomas .................................. 2385
Weindl, Ch. ..................................... 2022
Werner, Timur .................................... 649
Wheeler, P. ...................................... 1326
Wheeler, Patrick W. ............................... 207
Wheeler, Patrick W. ............................... 229
Wheeler, Patrick .................................. 388
Wiktor, Hudy ..................................... 912
Willis, K. ....................................... 1293
Winternheimer, Stefan ............................. 1872
Wisniewski, Janusz ................................ 2254
Wlas, Miroslaw ................................... 1084
Won, Chung-Yuen ............... 181, 1924, 1929, 383
Wong, L.K. .................................. 400, 475
Wu, Dongming ...................................... 97

## X

XiaoyanHuang, .................................... 388
Xu, Wei ..................................... 2000, 405
Xuhui, Wen .................................. 1518, 417
Xuhui, Zhang ................................. 1518, 417

## Y

Yaguchi, Hiroyuki ................................ 1020
Yamanouchi, Wataru ............................... 1187

# Author Index

Yang, Lingling ...................................................97
Yang, Liu ........................................................1128
Yang, Ru-Shiuan ...........................................320
Yin, Chunyan ...................................................76
Yokokura, Yuki ...............................1187, 1604
Yokoyama, Tomoki ..........................589, 744
Young-Kwan, Kim ........................................1002
Yousefi, Ashkan ..........................................1847

## Z

Zakrzewski, Zbigniew ...............................1332
Zamma, Toshihiro ......................................1020
Zanasi, Roberto .............................................874
Zanchetta, Pericle ...........1716, 1771, 561, 887
Zare, Firuz ......................468, 718, 723, 903
Zaring, Carina ............................................2472
Zarko, Damir ..................................................855
Zarko, Damirarko ...........................................818
Zaskalicka, Maria ..........................................899
Zaskalicky, Pavel ...........................................899
Zatocil, Heiko ..............................................1024
Zawirski, Krzysztof ..............1198, 1204, 1454
Zdenek, Jiri ..................................................1638
Zdravko, Valter ..............................................690
Zeljko, Spoljaric ............................................690
Zeman, Karel ...............................................1529
Zeroug, Houcine ..........................................1209
Zhang, H. .....................................................1523
Zhang, S. ......................................................2516
Zhao, S. W. ...................................................1221
Zhou, Tao ....................................................2142
Zhu, Haibin ....................................................515
Zielinski, K. ................................................1064
Zigic, Aleksandar ........................................1426
Zinoviev, Genady Stepanovic ....................1332
Zlosnikas, Valerijus ...................................1140
Zouhar, Jan .................................................1665
Zulawnik, Marcin .......................................1562
Zych, Michal ................................................1562
Zymmer, Krzysztof .........................1332, 1562

9781424417414